LABORATORY MANUAL

Seeley's Anatomy AND Physiology

ELEVENTH EDITION

Eric Wise
SANTA BARBARA CITY COLLEGE

LABORATORY MANUAL, SEELEY'S ANATOMY & PHYSIOLOGY, ELEVENTH EDITION

This book is printed on acid-free paper.

4 5 6 LMN 19 18 17

ISBN 978-1-259-67129-6
MHID 1-259-67129-1

Senior Vice President, Products & Markets: *Kurt L. Strand*
Vice President, General Manager, Products & Markets: *Marty Lange*
Vice President, Content Design & Delivery: *Kimberly Meriwether David*
Managing Director: *Michael S. Hackett*
Brand Manager: *Amy Reed*
Director, Product Development: *Rose Koos*
Product Developer: *Mandy C. Clark*
Marketing Manager: *Jessica Cannavo*
Digital Product Developer: *John J. Theobald*
Director, Content Design & Delivery: *Linda Avenarius*
Program Manager: *Angela R. FitzPatrick*
Content Project Manager: *Jayne Klein*
Buyer: *Laura M. Fuller*
Design: *David W. Hash*
Content Licensing Specialists: *Lori Hancock/Lorraine Buczek*
Cover Image: *Stand up paddle yoga @Jack Affleck/Getty Images/RF; colored scanning electron micrograph (SEM) of Escherricha*
Compositor: *SPi Global*
Printer: *LSC Communications*

Some of the laboratory experiments included in this text may be hazardous if materials are handled improperly or if procedures are conducted incorrectly. Safety precautions are necessary when you are working with chemicals, glass test tubes, hot water baths, sharp instruments, and the like, or for any procedures that generally require caution. Your school may have set regulations regarding safety procedures that your instructor will explain to you. Should you have any problems with materials or procedures, please ask your instructor for help.

mheducation.com/highered

Contents

Laboratory Exercises

Instructor Preface

Anatomy and physiology can be the crown jewel in our students' education or the bane of their college career. As an instructor of anatomy and physiology for many years, I decided to write a lab manual that was student-friendly and with a singular focus on the lab portion of the course. This lab manual was written for the undergraduate student of anatomy and physiology, and it consists of 43 exercises designed to help students learn basic human anatomy and the practical lab applications in physiology.

The diversity of interests in today's anatomy and physiology students is due, in part, to the number of majors that either require or recommend the subject. This lab manual provides a framework for understanding anatomy and physiology for students interested in nursing, radiology, physical or occupational therapy, physical education, dental hygiene, and other allied health majors.

This manual was written to be used with *Seeley's Anatomy & Physiology,* eleventh edition, by VanPutte, Regan, and Russo. The illustrations are labeled; therefore, students do not need to bring their lecture text to the lab. The lab manual accompanies the lecture text and lecture portion of the course and can be used in either a one-term or a full-year course. The illustrations are outstanding, and the balanced combination of line art and photographs provides effective coverage of material. The amount of lecture material in the manual is limited, so there is little material included that is not part of the lab experience.

Practical lab experience is an invaluable opportunity to reinforce lecture concepts, enrich students' understanding of anatomy and physiology, and allow them to explore new dimensions in the subject area. The educational benefit of reinforcing lecture material with hands-on experiments and acquiring knowledge with a learn-by-doing philosophy makes the anatomy and physiology lab a very special educational environment. Many of us use lab experiences to present conceptually difficult material in physiology and to provide students with different learning styles another avenue for learning.

The exercises in this lab manual provide a comprehensive overview of the human body. Each exercise presents the core elements of the subject matter. You can tailor this manual to match your own vision of the course or use it in its entirety. There are significant differences among anatomy and physiology laboratories, and the advances in physiology equipment, especially computer modules, are numerous and continually evolving. The materials section in each lab is designed for a lab of 24 students and includes the amounts and types of reagents to be used. The labs generally take between 2 and 3 hours to complete.

This lab manual was written for three types of anatomy and physiology courses. For courses that use the cat as the primary dissection animal, cat dissections or mammalian organ dissections follow the material on human anatomy. For courses that use models or charts, numerous cadaver photographs are included so that students can see the representative structures as they exist in the cadaver material. Finally, for courses that use cadavers, this lab manual can be used by studying the human material and omitting the cat dissection sections.

(A)[1] This edition has activity symbols throughout the lab exercises for easier guidance by the lab instructor through the lab. Review sections have all labeling terms listed in alphabetical order for easier use.

Exercise 1—Condensed and streamlined the charts. Revised the section on hydrogen bonding.
Exercise 2—Rewrote using the Microscope section. Rearranged the text for better flow.
Exercise 3—The prophase section of mitosis was rewritten. New pictures were added for the osmosis and diffusion section.
Exercise 6—The section on Bone Structure was rewritten for clarity. The material on Bone Cells was moved ahead of Decalcified Bone for better flow.
Exercise 7—Rearrangement of the Objectives section. Some of the directions of what to do in lab were moved to the beginning of the section instead of at the end.
Exercise 9—Updated the skull artwork to correspond to the lecture text new edition. Moved the fetal fontanel section from the articulations exercise to this exercise.
Exercise 10 Moved the fontanel section to Exercise 9. Added new illustration for tendon sheaths and bursae.
Exercise 11—Rewrote all of the simulated muscle physiology was rewritten. The Biopac exercise was updated.
Exercise 12—Reorganized introduction for better flow.
Exercise 13—Updated figure 13.5 to match new edition of text.
Exercise 14—Updated terms and figures to match new edition of lecture text.
Exercise 16—Changes in nomenclature and figures to match new edition of text.
Exercise 17—Changes in nomenclature and rearrangement of exercise for clarity.
Exercise 18—Reorganized introduction figures for clarity.
Exercise 19—Replaced some figures to update illustrations.
Exercise 21—Updated the figure on taste buds.
Exercise 22—Updated 2 figures to match new edition of text. Minor rewrite for clarity.
Exercise 23—Updated 2 figures to match new edition of text.
Exercise 24—Rewrote the gonadotropin and colloid sections.
Exercise 26—Revised blood clinical values.

Exercise 27—Updated several illustrations to match lecture text.
Exercise 28—Updated Biopac section and illustrations.
Exercise 29—Updated virtual lab for frog heart. Updated illustration.
Exercise 32—Replaced some of the artwork to match new edition of text.
Exercise 36—Minor changes in nomenclature made. Updated figures to match text.
Exercise 37—Added Waist/Hip ratio Section as indicator of health.
Exercise 40—Updated the coronal section of kidney illustration to match text.
Exercise 43—Updated the Anatomy of the Female Breast illustration to match text.

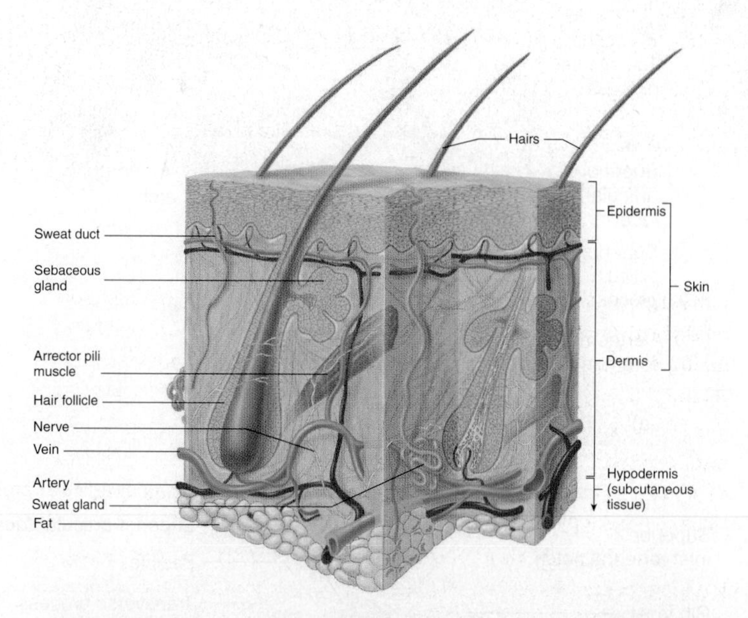

KEY FEATURES

1. **Dynamic art program.** The figures in this laboratory manual have been carefully rendered to convey realistic, three-dimensional detail.
2. **Instructional photographs.** Numerous full-color photomicrographs and dissection images prepare students for what they will encounter in the lab or can supplement discussions when hands-on labs are not available. These labeled photographic references also preserve a record of the lab experience long after it has passed.
3. **Labels.** Illustrations are labeled for students to learn the names and terminology by looking at real-life examples or models and by referring to the illustrations in the manual.
4. **Focus on the laboratory.** This manual focuses primarily on the material necessary for the laboratory and does not repeat the material presented in the lecture text, with the expectation that students can look up material in the lecture text when necessary.
5. **Integrated use of the cat for dissection specimen.** The cat is used as the dissection animal; however, it is integrated with material on human anatomy, so that animals do not have to be relied upon as dissection specimens if so desired.
6. **Safety.** Safety guidelines appear in the inside front cover for reference. The international symbol for caution (⚠) is used throughout the manual to identify material that the reader should pay close and special attention to when preparing for or performing the laboratory exercise.

7. **Cleanup.** At the end of many laboratory exercises, an icon for cleanup (🖐) reminds the student to clean up the laboratory. Special instructions are given where appropriate.
8. **Data collection.** Collection of data is embedded within each exercise, as opposed to in a separate table at the back of the manual.
9. **User-friendly format.** Each exercise begins on a right-hand page, and the pages are perforated to allow students to more easily remove the exercises to turn them in and later store them.
10. **Key terms.** Current anatomical terminology is used throughout the laboratory manual. Key terms are boldfaced.

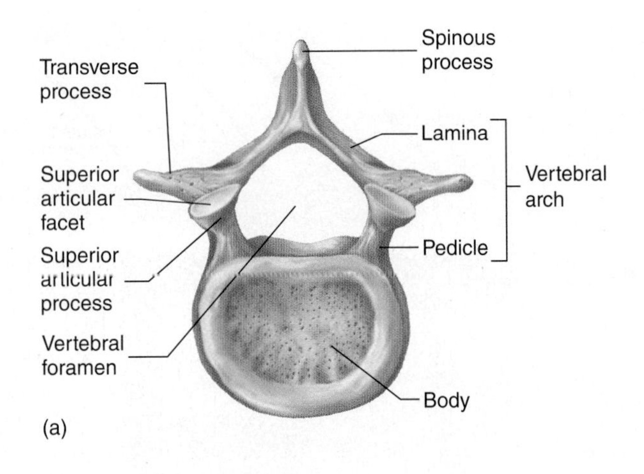

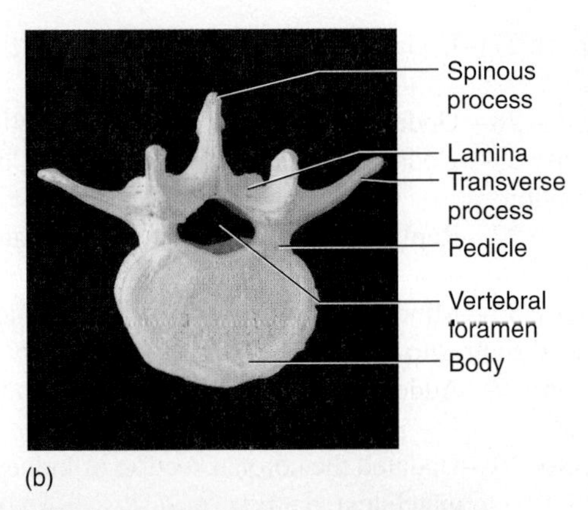

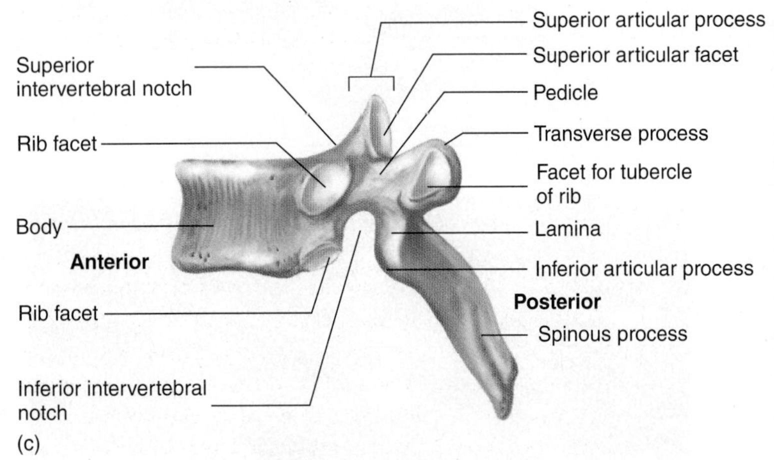

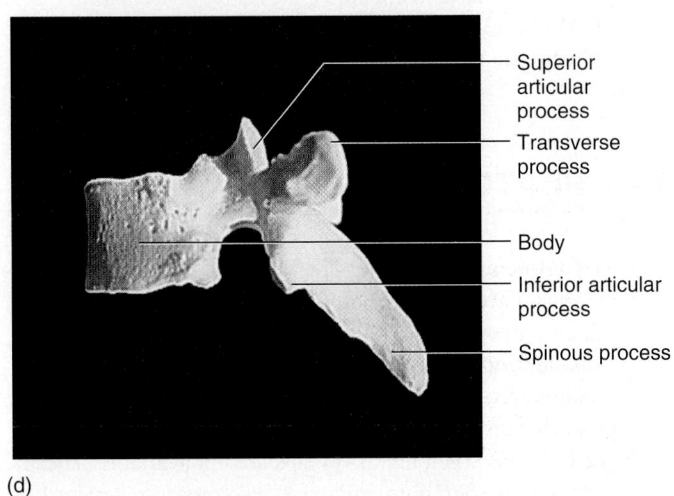

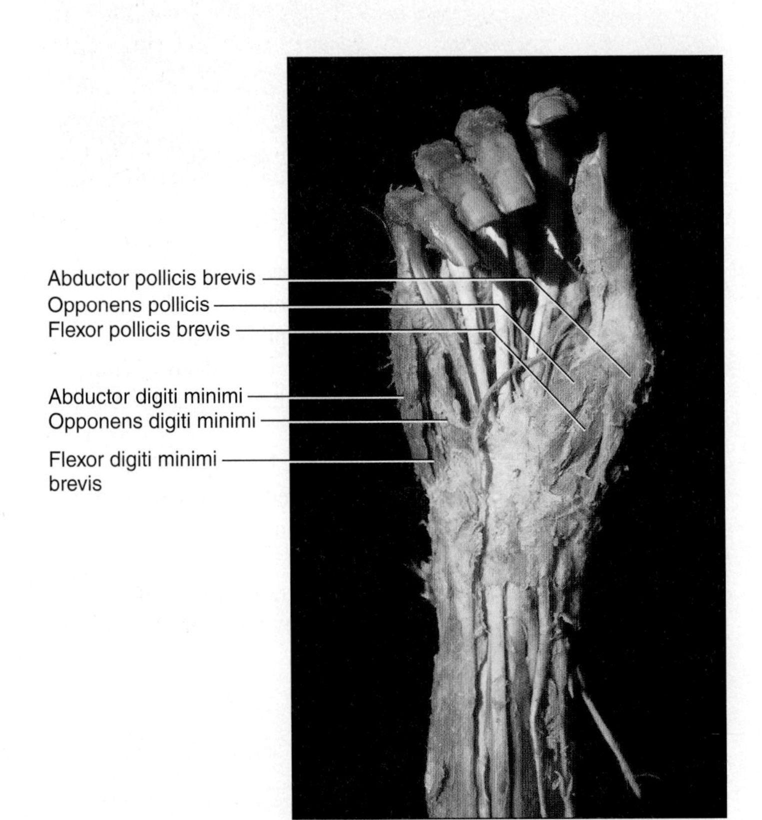

MCGRAW-HILL TEACHING AND LEARNING TOOLS

Instructor's Manual for the Laboratory Manual

To view the instructor's manual for the eleventh edition, please go to Instructor Resources. This helpful preparation guide includes suggestions for coordinating lab exercises with the textbook, set-up instructions and materials lists, and answers to the laboratory review questions at the end of each exercise.

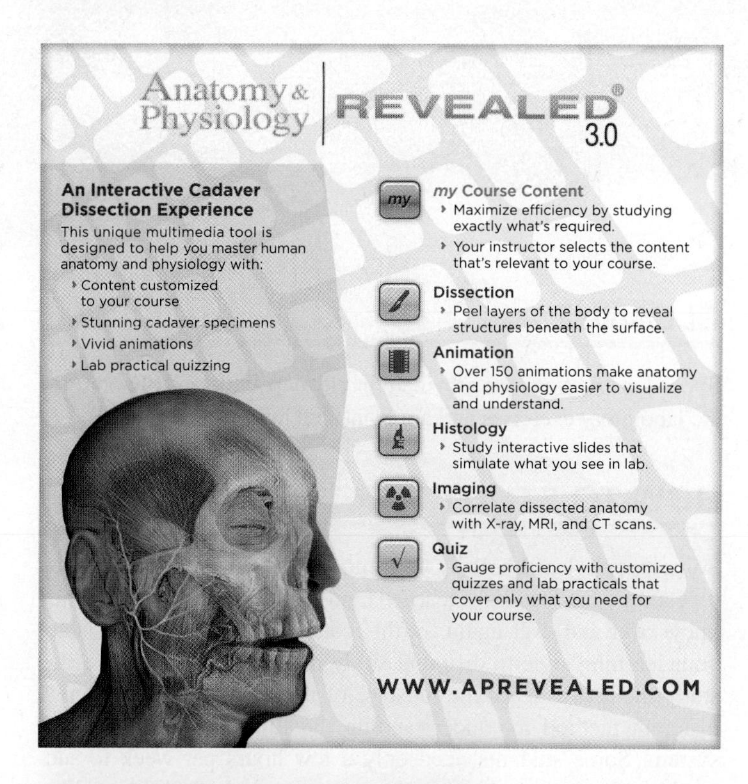

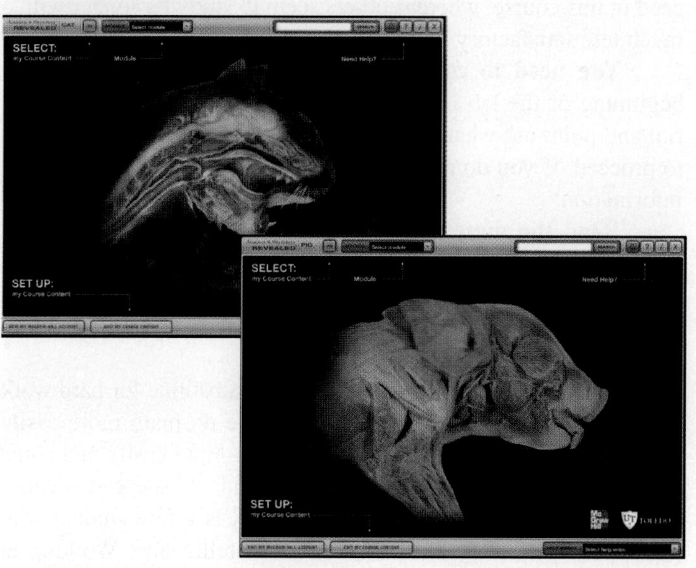

Physiology Interactive Lab Simulations (Ph.I.L.S.) 4.0

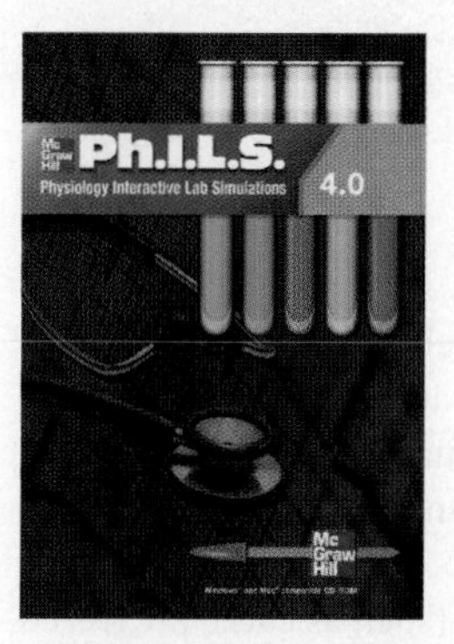

Ph.I.L.S. 4.0 is the perfect way to reinforce key physiology concepts with powerful lab experiments. Created by Dr. Phil Stephens at Villanova University, this program offers 42 laboratory simulations that may be used to supplement or substitute for wet labs. All 42 labs are self-contained experiments—no lengthy instruction manual required. Users can adjust variables, view outcomes, make predictions, draw conclusions, and print lab reports. This easy-to-use software offers the flexibility to change the parameters of the lab experiment. There are no limits!

ACKNOWLEDGMENTS

Many people have been involved in the publication of this lab manual. I would like to thank the editorial and marketing teams at McGraw-Hill, including Amy Reed, James Connely, Mandy Clark, Jessica Cannavo and Jean Schmieder. Thanks also go to the McGraw-Hill production team, including Jayne Klein, Lori Hancock and David Hash, for their input and encouragement. I would like to dedicate this edition to my wife, Ashley Wise.

Please feel free to write me or email me with your comments, suggestions, and criticisms. I value your input and hope that your comments will lead to an even better revision of this laboratory manual.

Eric Wise
Santa Barbara City College
721 Cliff Drive
Santa Barbara, CA 93109
wise@sbcc.edu

Student Preface

This laboratory manual was written to help you gain experience in the lab as you learn human anatomy and physiology. The 43 exercises explore and explain the structure and function of the human body. You will be asked to study the structure of the body using the materials available in your lab, which may consist of models, charts, mammal study specimens (such as cats), preserved or fresh internal organs of sheep or cows, and cadaver specimens. You may also examine microscopic sections from various organs of the body. You should familiarize yourself with the microscopes in your lab very early so that you can take the best advantage of the information they can provide.

The physiology portion of the course involves experiments including those that you will perform on yourself or your lab partner. They may also involve the mixing of various chemicals and the study of the functions of live specimens. Because the use of animals in experiments is of concern to many students, a significant attempt has been made to reduce (but, unfortunately, not eliminate) the number of live experiments in this manual. Until there is an effective replacement for live animals, their use will continue to be part of the college physiology lab. Your instructor may have alternatives to live animal experimentation exercises, such as computer simulations. It is important to get the most out of what live specimen experimentation there is. Coming to the lab unprepared and then sacrificing a lab animal while gaining little or no information is an unacceptable waste of life. Use the animals with care. Needless use or inhumane treatment of lab animals is not acceptable or tolerated.

As a student of anatomy and physiology, you will be exposed to new and detailed information. The time it takes to learn the information will involve *more* than just time spent in the lab. You should maximize your time in the lab by reading the assigned lab exercises before you come to class. You will be doing complex experiments, and, if you are not familiar with the procedure, equipment, and time involved, you could ruin the experiment for yourself and/or your lab group. The exercises in physiology are written so that you can fill in the data as you proceed with the experiment. At the end of each exercise are review sheets that your instructor may wish to collect.

The illustrations are labeled, except on the review pages. All review materials can be used as study guides for lab exams, or they may be handed in to the instructor.

The anatomy exercises are written for cat and human study, although these exercises can be used with or without cats or cadavers. Get *involved* in your lab experience. Don't let your lab partner do all of the dissections or all of the experimentation; likewise, don't insist on doing everything yourself. Share the responsibility and you will learn more.

CLEANUP

Special instructions are provided for cleanup at the end of appropriate laboratory exercises and are identified by a unique icon ().

HOW TO STUDY FOR THIS COURSE

Some people learn best by concentrating on the visual, some by repeating what they have learned, and others by writing what they know over and over again. In this course, you will have to adapt your learning style to different study methods. You may use one study method to learn the muscles of the body and a completely different method for understanding the function of the nervous system. Some students need only a few hours per week to succeed in this course, whereas others seem to study far longer with a much less satisfactory performance.

You need to come to class. Come to lab on time. The beginning of the lab is when most instructors go over the material and point out what material to omit, what to change, and how to proceed. If you do not attend lab, you do not get the necessary information.

Read the material ahead of time. The subject matter is very visual, and you will find an abundance of illustrations in this manual. Record on your calendar all of the lab quizzes and exams listed on the syllabus provided by your instructor. Budget your time so that you study accordingly.

Work hard! There is absolutely no substitute for hard work to achieve success in a class. Some people do math more easily than others, some people remember things more easily and some people express themselves better. Most students succeed because they work hard at learning the material. It is a rare student who gets a bad grade because of a lack of intelligence. Working at your studies will get you much further than worrying about your studies.

Be actively involved with the material and you will learn it better. Outline the material after you study it for a while. Read your notes, go over the material in your mind, and then make the information your own. There are several ways that you can get actively involved.

Draw and doodle a lot. Anatomy is a visual science, and drawing helps. You do not have to be a great illustrator. Visualize the material in the same way you would draw a map to your house for a friend. You do not draw every bush and tree but, rather, create a schematic illustration that your friend could use to get the pertinent information. As you know, there are differences in maps. Some people need more practice than others, but anyone can do it. The head can be a circle, which can be divided into pieces representing the bones of the skull. Draw and label the illustration after you have studied the material and without the use of your text! Check yourself against the text to see if you really know the material. Correct the illustration with a colored pen so that you highlight the areas you need to work on. Go back and do it again until you get it perfect. This does take some time, but not as much as you might think.

Write an outline of the material. Take the mass of information to be learned and go from the general to the specific. Let's use the skeletal system as an example. You may wish to use these categories:

1. Bone composition and general structure
2. Bone formation
3. Parts of the skeleton
   a. Appendicular skeleton
      (1) Pectoral girdle
      (2) Upper extremity
      (3) Pelvic girdle
      (4) Lower extremity
   b. Axial skeleton
      (1) Skull
      (2) Hyoid
      (3) Ribs
      (4) Vertebral column
      (5) Sternum

An outline helps you organize the material in your mind and lets you sort the information into areas of focus. If you do not have an organizational system, then this course is a jumble of terms with no interrelationships. The outline can get more detailed as you progress, so you eventually know that the specific nasal bone is one of the facial bones and the facial bones are skull bones, which are part of the axial skeleton, which belongs to the skeletal system.

Test yourself before the exam or quiz. If you have practiced answering questions about the material you have studied, then you should do better on the real exam. As you go over the material, jot down possible questions to be answered later, after you study. If you compile a list of questions as you review your notes, then you can answer them later to see if you have learned the material well. You can also enlist the help of friends, study partners, or family (if they are willing to do this for you). You can also study alone. Some people make flash cards for the anatomy portion of the course. It is a good idea to do this for the muscle section of the class, but you may be able to get most of the information down by using the preceding technique. Flash cards take time to fill out, so use them carefully.

Use memory devices for complex material. A mnemonic device is a memory phrase that has a relationship to the study material. For example, there are two bones in the wrist right next to one another, the trapezium and the trapezoid. The mnemonic device used by one student was that trape*zium* rhymes with th*umb* and it is the one under the thumb.

Use your study group as a support group. A good study group is very effective in helping you do your best in class. Hang out with people who will push you to do your best. If you get discouraged, your study partners can be invaluable support people. A good group can help you improve your test scores, develop study hints, encourage you to do your best, and let you know that you are not the *only* one living, eating, and breathing anatomy and physiology.

Just as a good study group can really help, a bad group can drag you down further than you might go on your own. If you are in a group that constantly complains about the instructor, that the class is too hard, that there is too much work, that the tests are not fair, that you don't really need to know this much anatomy and physiology for your own field of study, and that this isn't medical school, then get yourself out of that group and into one that is excited by the information. Don't listen to people who complain constantly and make up excuses instead of studying. There is a tendency to start believing the complaints, and that begins a cycle of failure. Get out of a bad situation early and get with a group that will move forward.

Do well in the class and you will feel good about the experience. If you set up a study time with a group of people and they spend most of the time talking about parties, sports, or personal problems, then you aren't studying. There is nothing wrong with parties, sports, or discussions about personal problems, but you need to address the task at hand, which is learning anatomy and physiology. Don't feel bad if you must get out of your study group. It is *your* education and, if your partners don't want to study, then they don't really care about your academic well-being. A good study partner is one who pays attention in class, who is prepared ahead of the study time session, and who can explain information that you may have gotten wrong in your notes. You may want to get the phone number of two or three such classmates.

LABORATORY SAFETY GUIDELINES

- The following is a partial list of safety guidelines for you to follow in the anatomy and physiology lab. More complete descriptions of safety procedures are found throughout the manual. The international symbol for caution (!) is used throughout the manual to identify material that you should pay close and special attention to when preparing for or performing laboratory exercises.
- Read all of the lab material prior to coming to class. This is a safety issue. Failure to read or understand the lab can result in hazards. Unauthorized experiments are not allowed in the lab.
- Locate the first aid-kit, eyewash station, shower station, fire blanket, fire extinguisher, and other safety areas in the lab prior to beginning the first lab. Be familiar with how to use the equipment in the event of an emergency.
- Clean up spills. Inform your instructor of any spill in the lab. Be careful if the material is toxic or caustic. If you are not sure if the material is hazardous, ask your instructor for the proper procedure for the cleanup.
- Assume all bodily fluids in the lab are infectious. Follow precautions when handling bodily fluids, such as wearing protective gloves, lab coats, and protective eye wear. Never use any instrument twice that comes into contact with bodily fluid! Once the instrument is used, dispose of it in either a biohazard bag or in a container of 10% bleach or other disinfecting solution. Clean all

lab surfaces with a bleach or disinfecting solution at the end of a lab involving bodily fluids, even if you think no fluid has come in contact with the table surface.
- Keep the lab clean and free of clutter. Place all backpacks, purses, and umbrellas in safe areas and not on the lab tables.
- Do not eat, smoke, or chew gum in the lab. Many reagents in the lab are toxic, so do not drink them. Never pipette anything by mouth. Use a pipette bulb or pipette pump when pipetting.
- Keep your hair secured so it does not catch fire or dip into beakers containing solutions. Never heat volatile material over an open flame. An explosion might occur.
- Do not wear contact lenses in the lab. Notify your instructor if you wear contact lenses.
- Do not throw sharp material such as glass or cutting blades in the normal trash containers in the lab. They are to be disposed of in an appropriate container such as a "sharps" container. Report any glassware breakage to your instructor, and dispose of it in the appropriate container.
- Never point a test tube that is heating over a Bunsen burner in the direction of someone else. Never walk away from anything that is being heated. Pay attention to material on hot plates and remove material with appropriate mitts or tongs. Heat material only in appropriate heat-resistant containers. Turn off and unplug hot plates immediately after use. Dissect with the blade cutting away from you and your lab partners. If you do cut yourself, make sure you wash the wound well with soap and water and notify your instructor.
- If you have an allergic reaction to the preserving fluid (usually restricted breathing, a flushed feeling, or a skin rash), notify your instructor immediately. Notify your instructor if you are pregnant or have any medical condition.
- Do not apply cosmetics in lab.
- Wear closed-toed shoes in lab, not sandals.
- Wash your hands thoroughly after lab, especially before eating or going to the restroom.

WORKING IN THE LAB

The lab is a busy place, and your first priority in lab is to have a safe laboratory experience. You should read the Laboratory Safety Guidelines in the Student Preface and follow all of the safety directions that your instructor provides you. Know where the closest phone is in the event of an emergency, and make sure that you understand any specific emergency procedures for your lab.

Working in the science lab requires you to focus on the procedures and materials at hand. You may work as part of a group in some labs, and it is important that you read your lab material before coming into the lab. Some of the materials you work with may be dangerous, and a thorough prior knowledge of the lab exercise will ensure a safer lab.

Pay attention to the experiment and what is to be done and when. Casual observation and carelessness may lead to incorrect results. Establish a procedure for conducting experiments. If you are working with one or more lab partners, divide up responsibilities before the experiment begins. If you are responsible for a particular portion of the experiment, make sure your lab partners see the results. Make a careful record of the results of your experiment.

Be honest. Fudging data is not tolerated in the scientific community. Record your data as you measure them. If your results do not seem to be what they should, then discuss this with your instructor. Never record data that you think you should get; instead, record the observed data.

SUPPLEMENTAL MATERIALS

A variety of materials can be purchased separately to supplement this laboratory manual. Please see the **instructor preface** for a list and description of these items.

TEST-TAKING

Finally, you need to take quizzes and exams in a successful manner. By doing practice tests, you can develop confidence. Do well early in the semester. Study extra hard early (there is no such thing as overstudying). If you fail the first test or quiz, then you must work yourself out of an emotional ditch. Study early and consistently, and then spend the evening before the exam going over the material in a general way and solving those last few problems. Some people do succeed under pressure and cram before exams; however, the information is stored in short-term memory and does not serve you well in your major field. If you study on a routine basis, then you can get up on the morning of a test, have a good breakfast, listen to some encouraging music, maybe review a bit, and be ready for the exam. The morning is a great time to study.

Your instructor is there to help you learn anatomy and physiology, and this laboratory manual was written with you in mind. Relate as much of the material as you can to your own body and keep an optimistic attitude.

Please feel free to write me or email me with your comments, suggestions, and criticisms. I value your input and hope that your comments will lead to an even better revision of this laboratory manual.

Eric Wise
Santa Barbara City College
721 Cliff Drive
Santa Barbara, CA 93109
wise@sbcc.edu

Exercise 1

Introduction to Lab Science, Chemistry, Organs, Systems, and Organization of the Body

Introduction

Science is the study of physical phenomena and follows specific guidelines that make it unique from other disciplines. The human body's structure and function fall within the realm of scientific investigation; for example, **human anatomy** is the study of the structure of the human body, while **human physiology** involves the function of the body. There are many ways that people learn about anatomy and physiology. One of these involves the use of the **scientific method.**

Scientific understanding frequently begins with a question. An example of a question might be how the body digests food. The next part in this understanding frequently involves the development of a **hypothesis,** which is a testable proposal that seeks to explain a scientific question. The testing done to prove or disprove the hypothesis is usually carried out in the form of an **experiment.** A more lengthy discussion of the scientific method is provided in the Seeley text in chapter 1, "The Human Organism."

Much of science involves measurement and the collection of **data.** Experimental data are the pieces of information, or "facts," obtained and later examined to support or reject the proposed hypothesis. In this exercise, you collect data and examine how to graph data, so that the information is easier to comprehend. You will also look at what type of measurement is done in science.

A basic knowledge of chemistry is essential for physiology. The very brief introduction to chemistry in this exercise starts the learning process.

The study of the human body requires you to understand both how the body or organ is oriented and how it is presented in terms of body regions. In this exercise, you examine the major organ systems of the body, the directional terms, and the levels of organization of the body, from the subatomic level to the whole organism. You also describe the major regions of the body.

Objectives

At the end of this exercise, you should be able to

1. describe the advantages of the metric system over the U.S. customary system;
2. define *independent variable* and *dependent variable;*
3. list four base units of the metric system;
4. convert fractions into decimal equivalents and convert large or small numbers to scientific notation;
5. collect and graph data taken in class;
6. discuss pH, acid, base, ionic bond, and covalent bond;
7. list the levels of anatomical organization from smallest to largest;
8. list the 11 organ systems of the body;
9. place major organs, such as the heart, lungs, and stomach, in the proper organ system;
10. describe anatomical position;
11. give directional terms that are equivalent to up, down, front, back, toward the midline, and toward the surface of the body;
12. determine from an illustration whether a section is in the frontal, transverse, or sagittal plane;
13. identify the major body cavities;
14. identify the quadrants and nine regions of the abdomen.

Materials

Acid/Base

Safety goggles and gloves
Five 10 mL test tubes
Test tube rack
10 mL graduated cylinder
Permanent marker
Distilled water in dropper bottle
0.1 M HCl in dropper bottle
0.1 M NaOH in dropper bottle
Baking soda (sodium bicarbonate)
Sodium chloride (table salt)
Wide-ranging pH paper (pH 1–14)
Parafilm®
Small metal spatula
Balance and weigh paper

Ionic and Covalent Molecules

18-gauge wire
Alligator clips
9-volt battery
6-volt flashlight bulb
Miniature screw lamp receptacle
(Carolina # 756481 or Sargent Welch # CP 33008-00)
Two 50 mL beakers
15% sucrose solution in dropper bottle
15% sodium chloride solution in dropper bottle

Hydrogen Bonds

Graduated cylinder
Two 50 mL beakers
Small bottle of distilled water
Small bottle of ethanol (70% or greater)
Hot plate (do not use open flame)

Heart Rate and Exercise

Clock or watch with accuracy in seconds
Calculator

Organ Systems Section

Models of human torso
Charts of human torso

TABLE 1.1	Metric System and Equivalents	
Quantity	**Base Unit**	**U.S. Equivalent**
Length	Meter (m)	1.09 yards (39.4 inches)
Volume	Liter (L)	1.06 quarts
Mass	Gram (g)	.036 ounce (1/454 of a pound)
Time	Second (s)	Second

Procedure

Measurements in Science

Members of the scientific community and people of many nations of the world use the **metric system** to record quantities such as length, volume, mass (weight), and time. This is because the metric system is based on units of 10, and conversion to higher or lower values is relatively easy when compared to using the U.S. customary system. For example, assume you are working on a bicycle and are using a ½-inch wrench. If you need a larger wrench you move to a 9/16-inch, then a 5/8-inch, then an 11/16-inch, or perhaps as large as a ¾-inch wrench. This requires a bit of computation as you move from one size to the next. On the other hand, if you are using the metric system and a 12-millimeter (mm) wrench is too small, you progressively move to a 13 mm, 14 mm, or 15 mm wrench.

The same idea can be applied to volume, temperature, or weight. In the case of volume, there are 8 ounces per cup, 128 ounces per gallon. The calculation for the number of ounces in 7 gallons is a little cumbersome (7 gallons × 128 ounces). In the metric system, there are 1,000 milliliters in 1 liter, so there are 7,000 milliliters in 7 liters. The conversions are much easier. Medical dosages are given frequently in milliliters or cubic centimeters (cc). One milliliter occupies 1 cubic centimeter, so these values are interchangeable. Examine table 1.1 and compare the quantity, base unit, and U.S. equivalent. Additional conversions can be found in Appendix A. If the quantity measured is much larger or smaller than the base unit, then the base unit can be expressed in multiples of 10. For example, if you had 1,000 grams, you would have a **kilo**gram. If you had one-thousandth of a gram (1/1,000 gram), you would have a **milli**gram. Examine table 1.2 as you answer the following questions.

What is 1/100 gram? ______________________

What is 1,000 seconds? ______________________

What is 10 meters? ______________________

What is 1/1,000,000,000 liter? ______________________

As you can see from table 1.2, some measurements in science are very small. For example, the amounts of hormones circulating in the blood are very minute. To provide a shortened notation of very large or small numbers we use **scientific notation.** A number such as 60,000 is written as 6×10^4. You move the decimal point four places to the left and thus the superscript above the 10 is a 4. Write 6,000 in scientific notation. ______________ For numbers less than 1, the superscript is written as a negative number. The number 0.00006 is written as 6×10^{-5}, as you move the decimal point five places to the right. (A)[1] Convert the following numbers into scientific notation:

4,300,000 4.3×10^6

0.000034 3.4×10^{-6}

2,200 2.2×10^3

0.0019 1.9×10^{-4}

Working in Lab

Working in the science lab requires you to focus on the procedures and materials at hand. You may work as part of a group in some labs, and it is important that you read your lab manual *before* coming into the lab. Some of the materials you work with may be dangerous, and a thorough prior knowledge of the lab exercise

TABLE 1.2	Decimals of the Metric System			
Name	**Description**	**Fraction**	**Multiple**	**Scientific Notation**
Kilo	One thousand times greater		1,000	1.0×10^3
Deca	Ten times greater		10	1.0×10^1
Base Unit				
Deci	One-tenth as much	1/10	0.1	1.0×10^{-1}
Centi	One-hundredth as much	1/100	.01	1.0×10^{-2}
Milli	One-thousandth as much	1/1,000	.001	1.0×10^{-3}
Micro	One-millionth as much	1/1,000,000	.000001	1.0×10^{-6}
Nano	One-billionth as much	1/1,000,000,000	.000000001	1.0×10^{-9}

will ensure a safer lab. Pay attention to the experiment, what is to be done, and when. Casual observation and carelessness may lead to incorrect results. Establish a procedure for conducting experiments. If you are working with one or more lab partners, divide responsibilities before the experiment begins. If you are responsible for a particular portion of the experiment, make sure your lab partners see the results. Make a careful record of the results of your experiment. Be honest. Fudging data is not tolerated in the scientific community. Record your data as you measure them. If your results do not seem to be what they should, then discuss this with your instructor. Never record data that you think you *should* get but, rather, *record the observed data.*

Data Collection

Scientists experiment by altering one variable and seeing what effect occurs. The variable that scientists change in an experiment is called the **independent variable.** The effect or result caused by this change is known as the **dependent variable.** For example, if we look at the amount of food a person eats versus weight gain, the amount of food is the independent variable. Weight gain occurs due to the amount of food eaten; therefore, weight gain is the dependent variable.

(A)[2] Measuring Heart Rate

In this part of the exercise you will look for a correlation between heart rate and exercise. Although you may be familiar with this response, make a hypothesis about the relationship between heart rate and exercise.

Record your hypothesis in the following space.

____

____

By graphing data you can more easily see potential correlations in a sample size. One problem in sampling is that you need to have a large enough number to have a valid sample. Let's suppose that one-half of the basketball team is enrolled in your lab section. This might have a rather unusual effect on your graph (figure 1.1). On the other hand, if you were able to sample your entire school, the effects of the size of the basketball players in the sample would be minimized (figure 1.2).

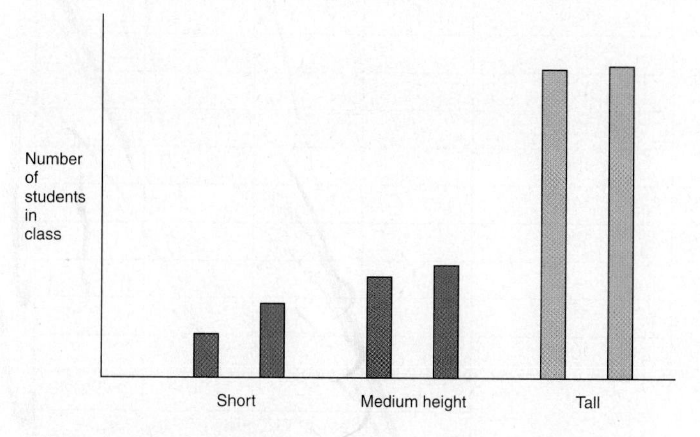

Figure 1.1 Distribution of Students in a Class

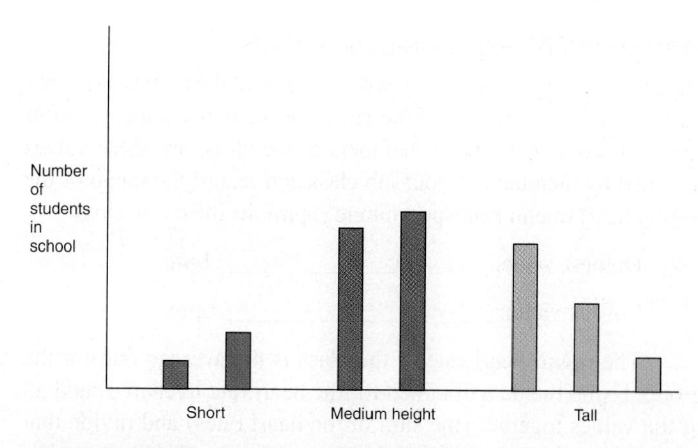

Figure 1.2 Distribution of Students in Entire School

While you are sitting quietly in lab, have your lab partner measure your heart rate by placing his or her fingers on the thumb side of your wrist. The heart rate should be counted for a full minute (beats per minute or bpm). You should not have done any strenuous exercise for at least 10 minutes prior to taking your heart rate. Write your results on the board or a designated piece of paper in lab. Record the distribution of the entire class in terms of resting heart rate in chart 1.

CHART 1.1 Heart Rate

	Subject Resting Heart Rate	Number of Seconds of Exercise				Percent Change in Heart Rate
0		30	60	90	120	
1						
2						
3						
4						
5						
6						
7						
8						
9						
10						
11						
12						
13						
14						
15						
16						
17						
18						
19						
20						
21						
22						
23						
24						

Range and Mean in Measurements

The extremes of measurement, which represent the highest and lowest measurements, represent the **range** of the measurements. After chart 1 is completed by all members of the class, use these values obtained by members of your lab class and record the range of the resting heart rate in beats per minute (bpm) for the entire class.

Highest value: ______________________ bpm

Lowest value: ______________________ bpm

The **mean** heart rate of the class is the average rate for the group. Using the data obtained for the heart rate in chart 1, add all of the values together (the sum of the heart rates) and divide that number by the number of individuals in the class. This is the mean heart rate. Record the number in beats per minute.

Mean heart rate: ______________________ bpm

(A)[3] *Heart Rate and Exercise*

Caution!

If you have a heart condition or any other reason for which you should not do exercise, please let your instructor know.

Your instructor should divide the class into four groups for the next part of the exercise. These groups should be balanced, so that there is an even distribution of students who are athletic and students who do not exercise in each group. Have the four groups do the following activity.

Group one does jumping jacks or other similar exercise for 30 seconds.

Group two does jumping jacks or other similar exercise for 60 seconds.

Group three does jumping jacks or other similar exercise for 90 seconds.

Group four does jumping jacks or other similar exercise for 120 seconds.

Immediately after finishing your prescribed exercise, record your heart rate for 1 minute.

Length of time of exercise: ______________________

Heart rate after exercise: ______________________

Determine the percent increase or decrease in your heart rate after exercise by dividing the rate after exercise by the resting rate and multiplying it by 100, as illustrated in the following equation.

$$\frac{\text{Heart rate after exercise}}{\text{Resting heart rate}} \times 100 = \text{Percent change in heart rate}$$

Write your results on the board or a designated piece of paper in lab. Record the distribution of all members of the entire class in terms of percent change in heart rate after exercise in chart 1. Make sure that you enter the data in the column that correlates to the amount of exercise done (30, 60, 90, or 120 seconds of exercise).

Take the results from each group (1–4) and determine the mean percent change in heart rate for each group. This is done by adding all of the percent change in heart rate values for individuals who did the exercise and dividing by the total number of individuals who did that exercise.

Make a bar graph by drawing a line representing the mean in chart 2 and shading in the area under the line.

Lab Reports

Your instructor may ask you to write up your results for specific experiments. This process is valuable in that you evaluate your experimental data and form an understanding of the process that comes from the results of your experiment. Writing up scientific experiments generally follows a very specific process, and this is described in appendix C at the back of the lab manual. You should refer to it before you begin your lab write-ups.

Scientific Words

In science, many terms are derived from Greek or Latin words. Remembering a word is easier if its meaning is understood. A word may consist only of a **root word.** For example, a *gastric* ulcer refers to an ulcer of the stomach. The term *gastric* comes from *gastro,* meaning stomach. In addition to the root word, **prefixes** or **suffixes** may be added. A prefix is added to the front of the root word. The word *epigastric* means on top of the stomach (*epi* = upon). *Hypogastric* means under the stomach (*hypo* = below). A suffix is added to the end of a root word. *Gastritis* is an inflammation of the stomach (*itis* = inflammation).

Sometimes you might feel like you are learning a new language and, in many respects, this is true. There is an explanation of many of these word elements inside the back cover. Refer to this section if you need to look up a word for its meaning.

Chemistry

In order to fully understand the functions of the body, a fundamental knowledge of chemistry is essential. For those of you who have studied chemistry the following pages are a simplified review.

CHART 1.2 Mean Percent Change in Heart Rate

% Change				
200				
190				
180				
170				
160				
150				
140				
130				
120				
110				
100				
	30	60	90	120
	Number of Seconds of Exercise			

For those of you who have never had chemistry a small chemistry book or online resources may be invaluable to you for the rest of the course. In the following section you will experiment with acids, bases, and chemical bonds.

Acid/Base Relationships

Human cells exist within a range of temperature, salinity, and pH. Significant changes in any of these environmental conditions can lead to the death of the cell. In terms of pH, solutions can be **acidic, neutral,** or **basic (alkaline).** A solution with a pH of 7 has a neutral pH because, in pure water, the number of hydrogen ions equals the number of hydroxyl ions (see figure 1.3). As the solution becomes more acidic, the hydrogen ion concentration increases and the pH decreases. Adding more hydrogen ions to the solution, such as when the stomach wall adds hydrochloric acid (HCl) to the stomach cavity, causes the pH to drop to about 2 (in the case of the stomach, the pH drops to about 2). If you increase the alkalinity of the solution (such as adding ammonia to water), then the pH increases. Household ammonia has a pH of about 11.

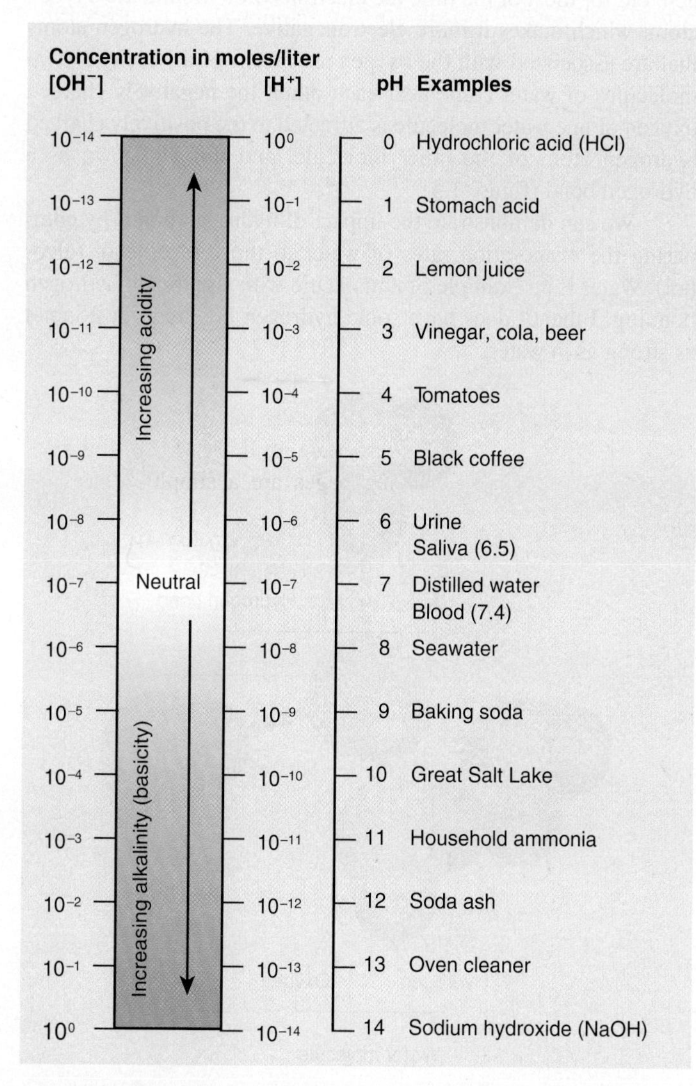

Figure 1.3 The pH Scale

The pH of 7 is neutral. The lower the number the more acidic. The higher the number the more basic. Examples are listed on the right.

Solutions such as hydrochloric acid and carbonic acid are common. Basic solutions, such as bicarbonate and ammonia, also play important roles in the body. The acidic or basic condition of a solution is measured in **pH** units. The abbreviation pH reflects the hydrogen ion concentration. In simple terms this concentration differs by a factor of 10 between whole pH values. Thus, a solution of pH 5 is 10 times more acidic than a solution of pH 6.

Buffers are materials that resist changes in pH. Cells often secrete buffers so that when conditions around them change, the pH is "buffered," and great fluctuations in the acid or base conditions do not occur. Introductory chemistry books will give you a more detailed description of pH, and there are many online tutorials that you can find if you search using the words *acid base.*

Caution!

When using acids and bases, make sure to wear safety goggles and gloves. This material is potentially caustic and can cause skin burns. Do not drink any of the solutions. If you spill acids or bases, notify your instructor immediately for the proper way to clean up the spill.

(A)[4] For this experiment select five test tubes and label them 1–5 with a permanent marker. Place them in a test tube rack. Use pH paper to measure pH.

1. Use a graduated cylinder and put 6 mL of distilled water into each of the five test tubes.
2. Measure the pH of test tube 1 (distilled water). Record this pH value ______.
3. To test tube 2, add 10 drops of HCl. Place a small square of Parafilm® over the top and mix the contents by inverting the test tube. Do not shake the tube! Measure the pH and record the value ______. Is this solution acidic, basic, or neither (circle one)?
4. To test tube 3, add 0.1 g sodium bicarbonate. Place a small square of Parafilm® over the top and mix the contents. Record the pH ______.
5. To test tube 3, add 10 drops of HCl. Place a small square of Parafilm® over the top and mix the contents as in step 3. Record the pH of the solution. Does the pH drop to the same level as test tube 2? ______.
6. Add 10 more drops of HCl to test tube 3 and record the pH value ______. How might you describe the action of sodium bicarbonate with respect to HCl? ______
7. To test tube 4 add 10 drops of sodium hydroxide. Place a small square of Parafilm® over the top and mix the contents. Record the pH value ______.
   Is this solution acidic, basic, or neither (circle one)?
8. To test tube 5 add 0.2 g of sodium chloride (table salt). Place a small square of Parafilm® over the top and mix the contents. Record the pH value ______. Is this solution acidic, basic, or neither (circle one)?
   Tissue fluid has a pH between 7.35 and 7.45. What was the range of pH in your experiment? ______.

Chemical Bonds

Most of the material around us is held together by chemical bonds. These bonds can be strong, intermediate, or weak. There are several types of bonds, but the three that we will examine are covalent bonds, ionic bonds, and hydrogen bonds. **Covalent bonds** are generally thought of as those where bonded atoms share electrons. These are strong bonds. The nucleus of each atom is positively charged due to the presence of protons. The positive charges of two separate atoms repel one another, but when an electron is shared between these atoms the electrons are attracted to both of the positive nuclei, and this holds them together.

In **ionic bonds** electrons from one atom are transferred to another atom and the result is an atom with a positive charge and an atom with a negative charge. These oppositely charged particles, called **ions** or **electrolytes,** attract one another and form ionic bonds. The bond is due to the electrostatic attraction between the two ions. These are also considered strong bonds.

You can examine the differences between covalent and ionic bonds with a simple experiment. Ionic bonds are pulled apart (dissociate) by the action of water. Electrolytes (ions) in solution conduct electricity. Molecules that are held together with covalent bonds do not conduct electricity so easily. This can be demonstrated with the use of a 6-volt flashlight bulb, appropriate cables and switches, and a DC generator, or a 9-volt battery. Examine figure 1.4 for the proper setup.

Caution!

Be careful when working with electricity. You will be using low-voltage direct current. Do NOT use any voltage higher than 10 volts for this experiment, and do not put any part of your body into the solutions being tested.

(A)[5]

1. Connect a 9-volt battery terminal, using one alligator clip and wire, to the 6-volt flashlight bulb apparatus.
2. Connect the other terminal to an insulated wire that is placed in a glass beaker with 25 mL of 15% sucrose solution. The wire should have about 2 cm of insulation removed.
3. Connect the vacant terminal of the flashlight bulb setup to an insulated wire.
4. Do not touch the ends of the wires together but insert the ends of the insulated wire into the water.

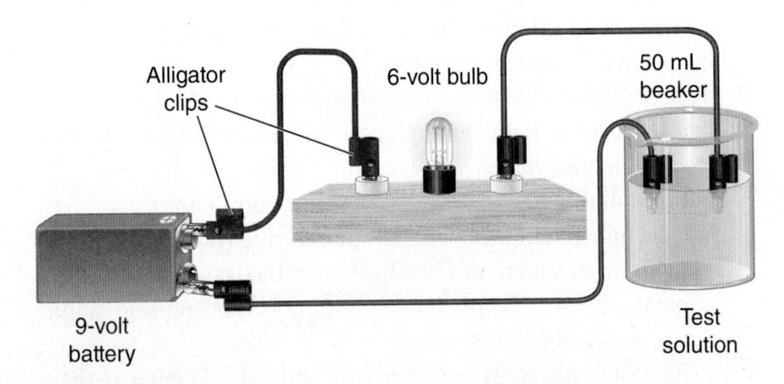

Figure 1.4 Electrolyte Conductivity Setup

5. Record the results (light on or light off).
6. Repeat the experiment but insert the wires into a beaker with 25 mL of 15% sodium chloride solution.
7. Record the results (light on or light off).
   Which one of the solutions has solute particles composed of ions? Which solution has solute particles with covalent bonds?

Note

Despite the fact that pure water, or water with covalent molecules, does not conduct electricity as well as water with many electrolytes it still can conduct electricity and, when the voltage is high enough, can kill you. This is why you do not swim outside during lightning storms or place electrical appliances on the edge of a bathtub when you are taking a bath.

Hydrogen Bonds

(A)[6] A prime example of a molecule that has hydrogen bonds is water. Two hydrogen atoms share their electrons with oxygen, but for most of the time the electrons orbit around the oxygen atom, which makes it more electronegative. The hydrogen atoms that are associated with the oxygen are electropositive. When two molecules of water come near each other, the negatively charged oxygen of one water molecule is attracted to the positively charged hydrogen atom of the other molecule, and this is known as a hydrogen bond (figure 1.5).

We can demonstrate the impact of hydrogen bonds by comparing the evaporation rates of water to those of ethanol (alcohol). Water is an example of a molecule with significant hydrogen bonding. Ethanol does have some hydrogen bonding but it is not as strong as in water.

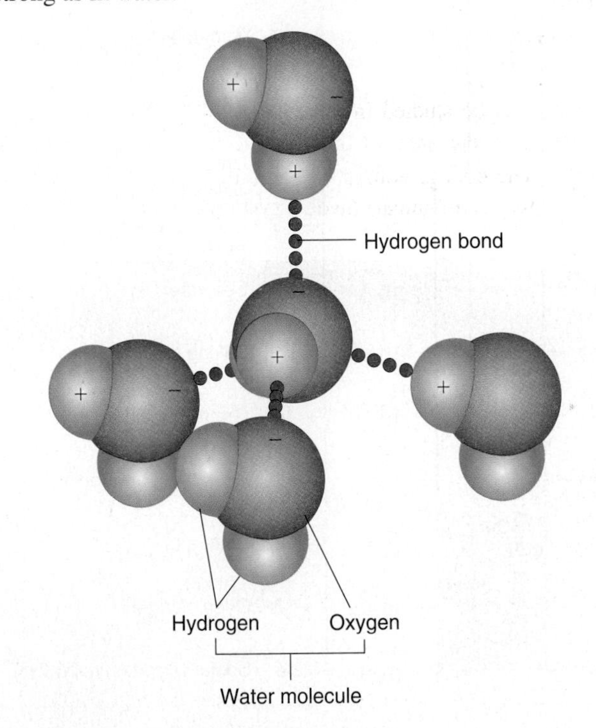

Figure 1.5 Hydrogen Bonding

Positive ends of hydrogen atoms are attracted to negative ends of oxygen atoms.

1. Pour 5 mL of water into a 50 mL beaker labeled "water."
2. Pour 5 mL of 100% ethanol into a 50 mL beaker labeled "alcohol."
3. Make sure that both liquids are at room temperature before you begin the experiment.
4. Warm the two beakers on a hot plate (NOT an open flame).

NOTE: You can use 70% lab alcohol but the remaining 30% is water; therefore, the results will not be as dramatic. Which fluid evaporates more slowly, alcohol or water (circle one)?

Slow evaporation is due to the greater number of hydrogen bonds "holding" the molecules together.

Levels of Organization

The human body can be studied from a number of perspectives. The earliest study involved **gross anatomy,** or cutting up part or all of the body and examining its details. As more sophisticated equipment was developed, other levels of organization became apparent. Today, the manipulation of atomic nuclei under magnetic fields has led to magnetic resonance imaging (MRI) studies that do not depend on dissection of the body (figure 1.6). The following list shows the levels of organization, along with examples of each:

Level	Examples
Chemical	Oxygen, carbohydrates
Organelle	Mitochondrion, Golgi apparatus
Cellular	Neuron, red blood cell
Tissue	Nervous, muscular
Organ	Stomach, kidney
Organ system	Digestive system, urinary system
Organism	*Homo sapiens*

Organ Systems

Anatomy can be studied in many ways. **Regional anatomy** is the study of particular areas of the body, such as the head or leg. Most undergraduate college courses in anatomy and physiology (and the format of this lab manual) involve **systemic anatomy,** which is the study of **organ systems,** such as the skeletal system and the nervous system. Although organ systems are studied separately, it is important to realize the intimate connections among the systems. If the heart fails to pump blood as part of the cardiovascular system, then the lungs do not receive blood for oxygenation and the intestines do not transfer nutrients to the blood as fuel. The brain is no longer capable of functioning, and the result is death. From a clinical standpoint, the failure of one system has impacts on many other organ systems.

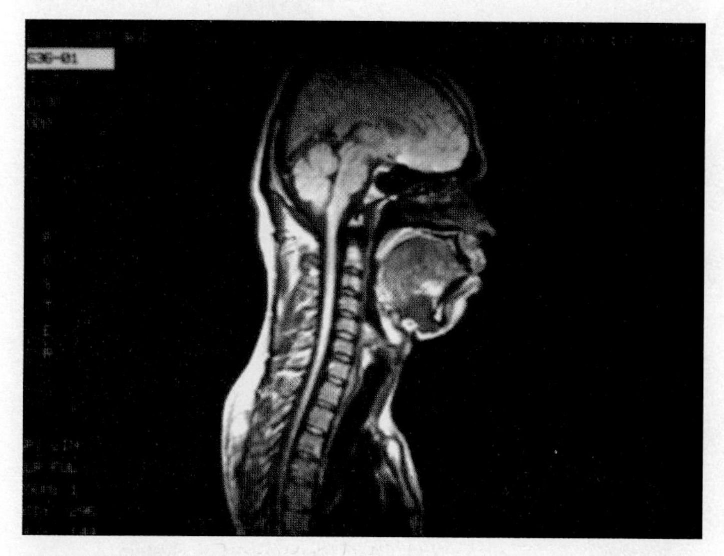

Figure 1.6 MRI of the Neck

(A)[7] Examine the torso models and charts in the lab and locate various organs. Using figure 1.7 find the following organ systems:

Reproductive	Lymphatic
Urinary	Integumentary
Nervous	Digestive
Muscular	Endocrine
Respiratory	Cardiovascular
Skeletal	

Nose
Nasal cavity
Pharynx (throat)
Larynx
Trachea
Bronchi
Lungs

1. respiratory System

Kidney
Ureter
Urinary bladder
Urethra

2. urinary System

Figure 1.7 Organ Systems of the Human Body
Write the name of the system underneath the figure representing it.

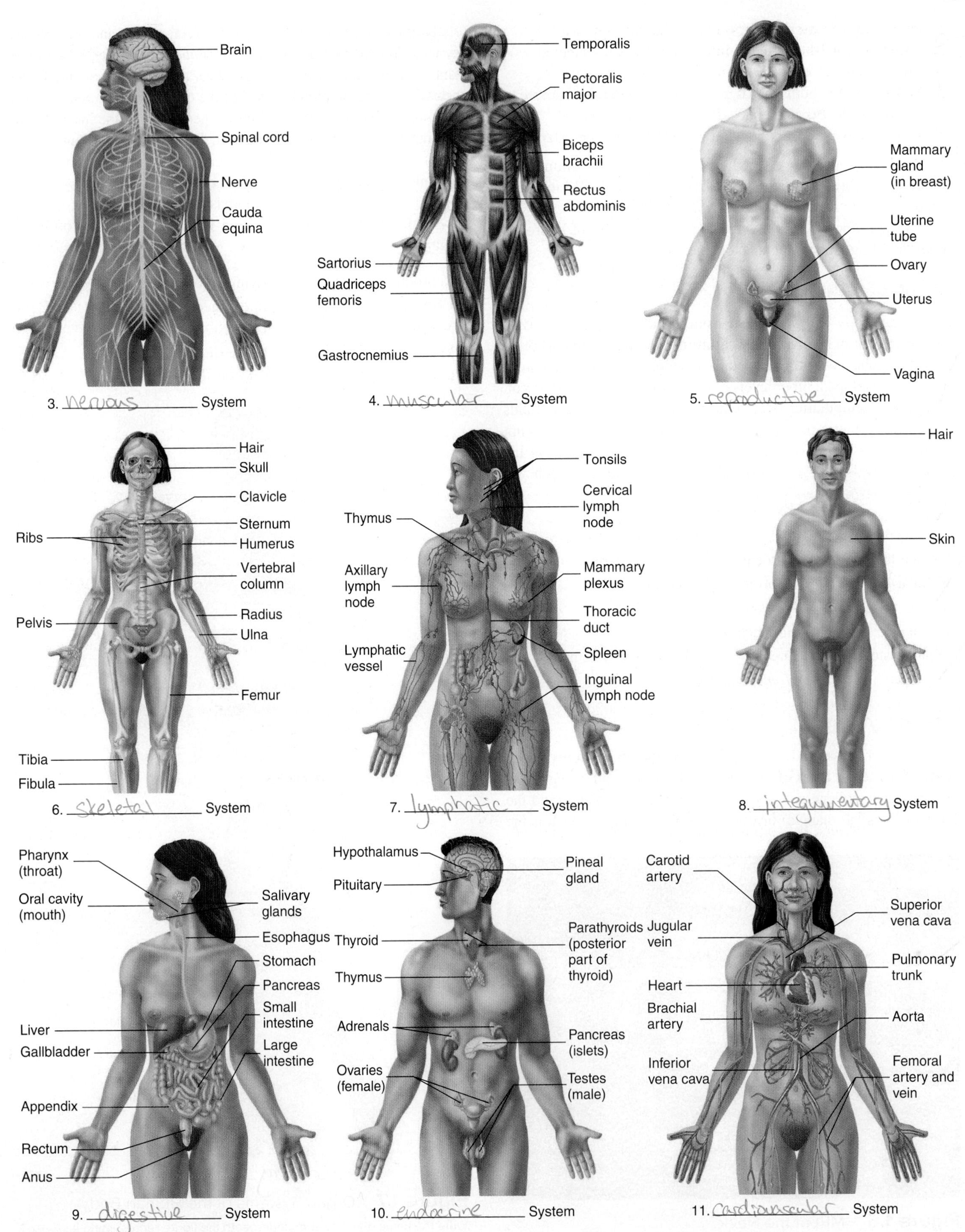

Figure 1.7 *Continued.*

The **reproductive system** is responsible for the maintenance of the species. The sex cells from the male join with the sex cells of the female and produce offspring. The main organs in the system are the ovaries, uterine tubes, uterus, and vagina in females and the testes, ductus deferens, glands producing seminal fluid, and penis in males. The **lymphatic system** cleanses and returns tissue fluid to the cardiovascular system and assists the body in protecting itself from foreign organisms. The lymphatic system (immune system) consists of lymphatic vessels, along with such organs as the thymus, spleen, and tonsils. The **urinary system** rids the body of waste products; it consists of the kidneys, ureters, urinary bladder, and urethra. The **integumentary system** provides the protective covering of the body and is mostly formed by the skin. The **nervous system** is well developed in humans, allowing us to interact with and interpret our environment. The brain, spinal cord, and peripheral nerves make up the system. The **digestive system** is responsible for providing nutrition to the tissues. The mouth and salivary glands, along with the esophagus, stomach, intestines, and associated organs, such as the liver, are part of the digestive system. The **muscular system** moves the body and consists of the individual muscles, such as the biceps brachii and the gluteus maximus muscles. In the **endocrine system,** the individual organs produce hormones. Organs such as the hypothalamus, pituitary, thyroid, pancreas, and gonads are endocrine organs. The **respiratory system** takes oxygen to the body and releases carbon dioxide. The nose, pharynx, larynx, trachea, and lungs are the organs of the respiratory system. The **cardiovascular system** is one primarily of transport. The heart and blood vessels are the organs of this system. The **skeletal system** provides a framework for movement and a mechanism for protecting the body. The individual bones of the body, such as the humerus and femur, are the organs of the system.

A quick way to remember all 11 systems is to remember the phrase "Run Mrs. Lidec." Each letter of the phrase represents the first letter in the name of one of the organ systems.

Anatomical Position

In clinical settings, it is vital to have a proper orientation when dealing with patients. If two physicians are operating on a patient and one tells the other to make an incision to the left, the physician making the cut does not have to ask, "My left or your left?" because the cut is always to the *patient's* left side. When referring to the human body, you will orient the body in the anatomical position. In this position, the body is upright, facing forward, arms and legs straight, palms facing forward, feet flat on the ground, and eyes open (figure 1.8a).

Directional Terms

With the body in the anatomical position, there are specific terms to describe the location of one part with respect to another. Table 1.3 lists the directional terms used for humans.

There are other terms for location that have unique meanings. For the digestive system, **proximal** refers to regions closer to the mouth, while **distal** is in reference to regions closer to the anus. **Parietal** is in reference to the body wall when compared to **visceral,** which refers to areas closer to the internal organs. The heart, for example, has a visceral layer closer to the heart proper called the **visceral pericardium,** while it also has a parietal layer farther from the heart called the **parietal pericardium.** Likewise, the lungs have a **visceral pleura** and a **parietal pleura.**

Ipsilateral refers to being on the same side of the body, and **contralateral** refers to being on the other half (left side/right side). The right hand and right arm are ipsilateral, while the ears are contralateral. Directional terms do not change if the body position changes. If you stand on your head, it is still superior to your feet because you reference the body as if it were in anatomical position.

In quadrupeds (four-footed animals), the directional terms are somewhat different. Note in figure 1.8c that quadrupeds do not have a superior/inferior designation. *Dorsal* refers to the back, and ventral is on the belly side. Anterior (or cephalic) is toward the front, or head, end of the animal and posterior (or caudal) is toward the rear, or tail, end of the animal.

Planes of Sectioning

When you view a picture of an organ that has been cut, it is important to understand how the cut was made. Just as an apple looks very different when cut crosswise as opposed to lengthwise, so

TABLE 1.3 Directional Terms Used for Humans

Term	Meaning	Example
Superior	Above	The nose is superior to the chin.
Inferior	Below	The stomach is inferior to the head.
Medial	Toward the midline	The sternum is medial to the shoulders.
Lateral	Toward the side	The ears are lateral to the nose.
Superficial	Toward the surface	The skin is superficial to the heart.
Deep	Toward the core	The lungs are deep to the ribs.
Ventral (or anterior)	To the front	The toes are ventral/anterior to the heel.
Dorsal (or posterior)	To the back	The spine is dorsal/posterior to the sternum.
Proximal	For extremities, meaning near the trunk	The elbow is proximal to the wrist.
Distal	For extremities, meaning away from the trunk	The toes are distal to the knee.

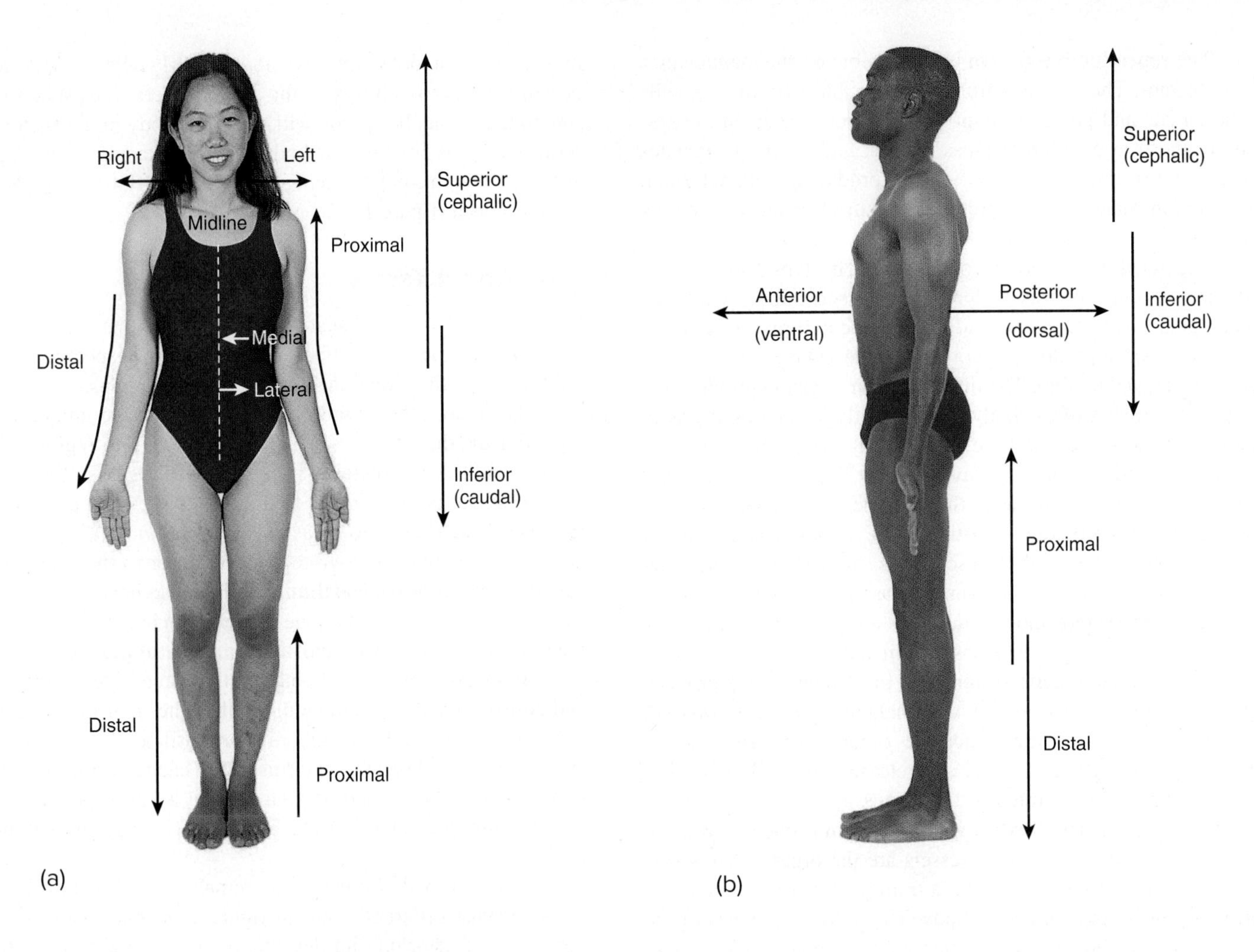

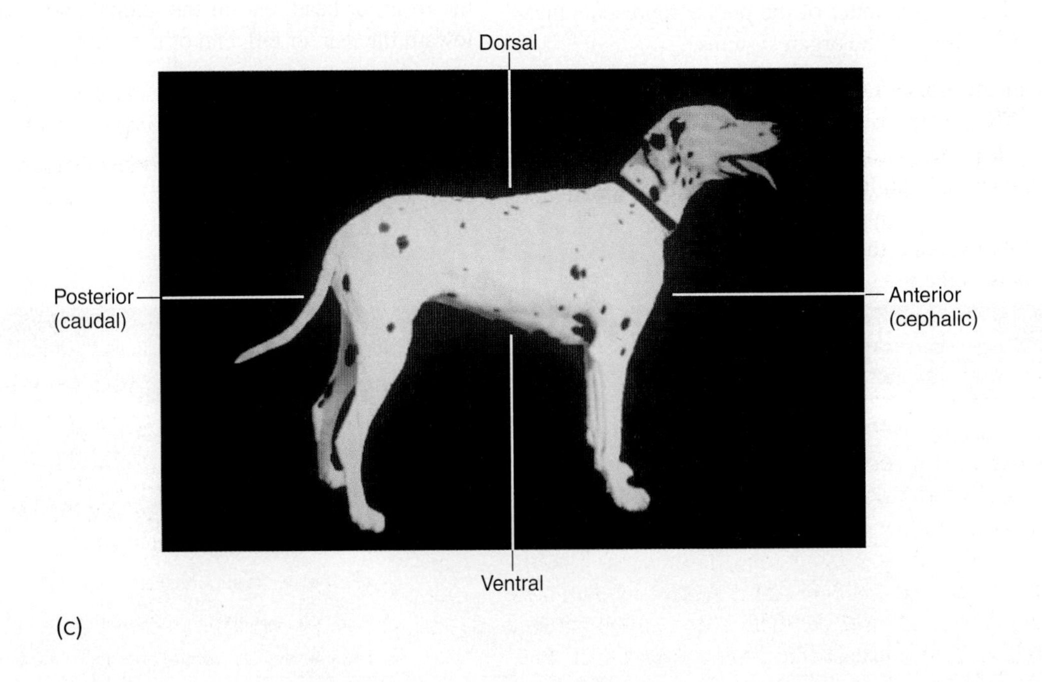

Figure 1.8 Anatomical Position and Directional Terms
(a) For humans, anterior view; (b) for humans, lateral view; (c) for quadrupeds, lateral view.

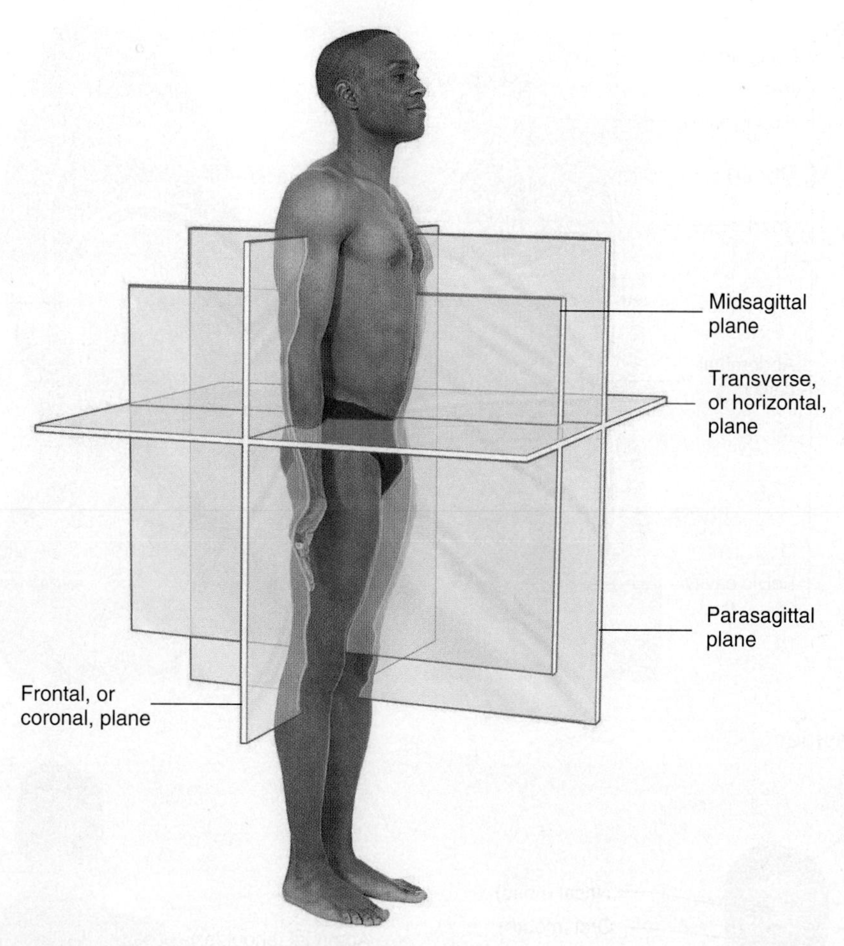

Figure 1.9 Sectioning Planes

do some organs. Examine figure 1.9 for the following **sectioning planes.** A cut that divides the body or organ into superior and inferior parts is in the **horizontal,** or **transverse, plane.** A cut that divides the body into anterior and posterior portions is in the **frontal,** or **coronal, plane.** A cut that divides the body into left and right portions is in the **sagittal plane.** A cut that divides the body equally into left and right halves is in the **midsagittal** (or **median plane),** whereas one that divides the body into unequal left and right parts is in the **parasagittal plane.**

Major Body Cavities

A cavity is an enclosed space inside the body. The brain is located in the cranial cavity, the tongue is located in the oral cavity, and the stomach is found in the abdominal cavity. The three largest cavities that do not open to the exterior environment are the **thoracic cavity,** the **abdominal cavity,** and the **pelvic cavity** (figure 1.10). Compare the models and charts in the lab with this figure. The thoracic cavity is located directly above the diaphragm and is further divided into the **mediastinum** and the **pleural cavities.** The mediastinum is the region between the lungs; it contains, among other things, the heart, esophagus, and trachea. The pleural cavities are on each side of the mediastinum. Below the diaphragm is the **abdominopelvic cavity,** which can be subdivided into the abdominal cavity and the pelvic cavity. The abdominal cavity contains the stomach, the small intestine, most of the large intestine, and various digestive organs, such as the liver and the pancreas. The pelvic cavity begins at the region of the hips and contains the lower part of the large intestine and some of the reproductive organs (such as the uterus and ovaries) of the female reproductive system. All of these cavities are lined with **serous membranes.**

Regions of the Body

Examine figure 1.11 for specific areas of the body. You will refer to these areas throughout this lab manual, so a complete study of these regions here is essential. Anatomical regions are listed first, with the commonly used nouns (if appropriate) listed in parentheses. In anatomy, some regions of the body are described differently than you might expect. In anatomical usage, the **arm** is the region between the shoulder and the elbow, and the **leg** is the region between the knee and the ankle.

Cephalic (head)
- Frontal (forehead)
- Orbital (eye)
- Nasal (nose)
- Buccal (cheek)
- Oral (mouth)
- Mental (chin)

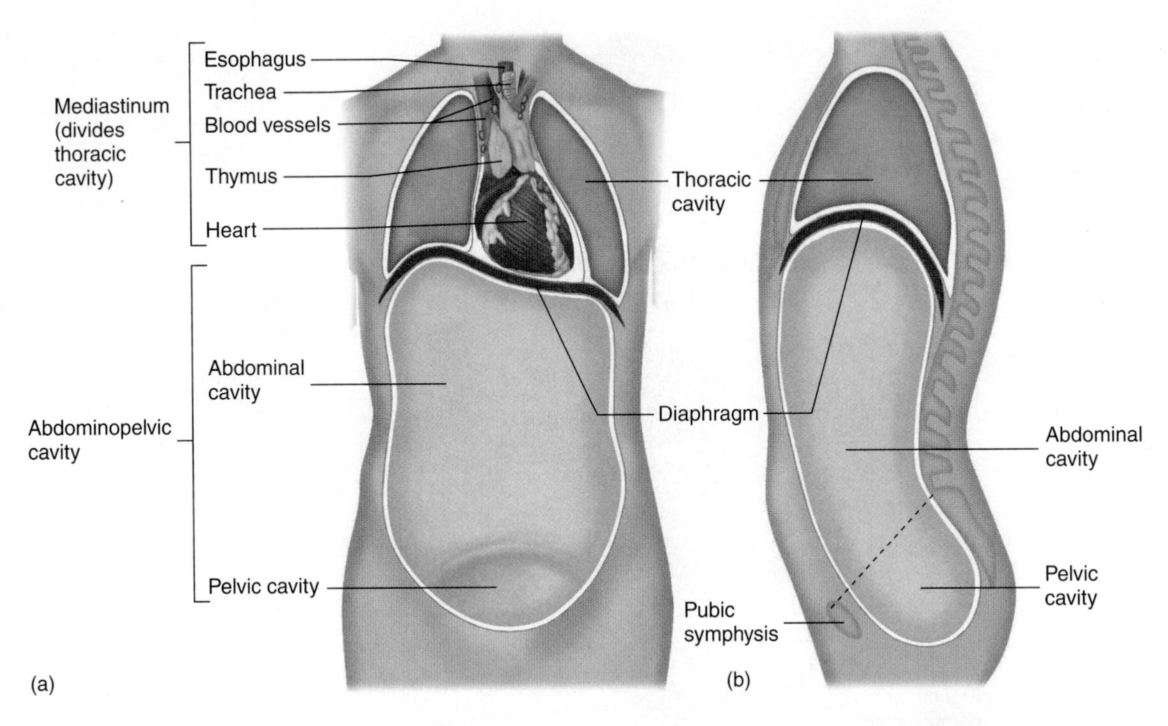

Figure 1.10 Trunk Cavities

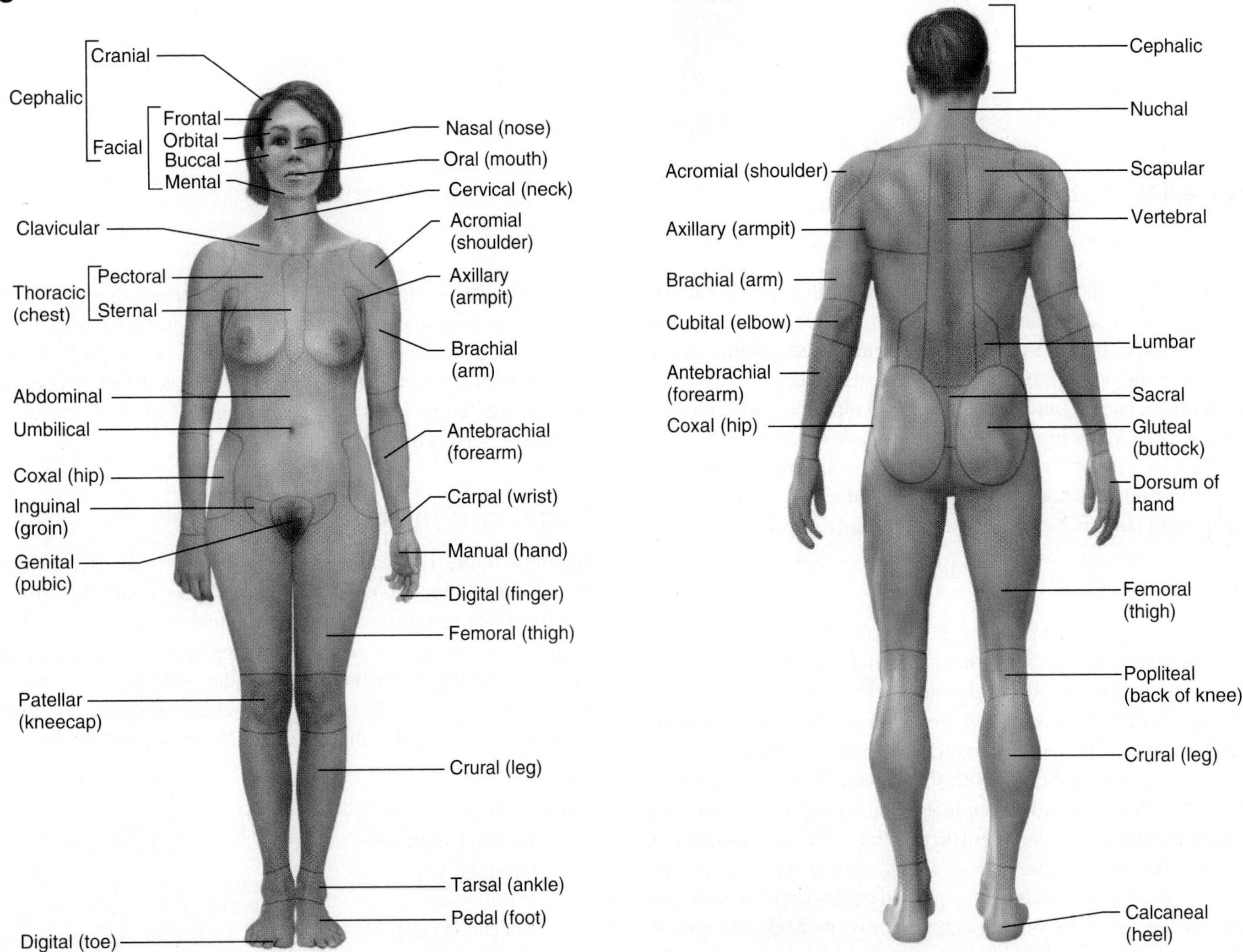

Figure 1.11 Regions of the Body

Cervical (neck)
Trunk
Thoracic (chest)
Pectoral
Sternal
Clavicular
Acromial (shoulder)
Abdominal (belly)
Inguinal (groin)
Genital (pubic)
Coxal (hip)
Upper extremity
Axillary (armpit)
Brachial (arm)
Cubital (elbow)
Antebrachial (forearm)
Carpal (wrist)
Manual (hand)
Digital (finger)
Lower extremity
Femoral (thigh)
Patellar (kneecap)
Crural (leg)
Tarsal (ankle)
Pedal (foot)
Digital (toe)

(A) 8 Find the following locations on your body and provide the appropriate anatomical description for these regions.

Shin ______________________________

Elbow ______________________________

Neck ______________________________

Toes ______________________________

Shoulder ______________________________

Thigh ______________________________

Kneecap ______________________________

Abdominal Regions

The abdomen can be further divided into either four quadrants or nine regions (figure 1.12). In clinical practice, the abdomen is divided into quadrants. In anatomical studies, the nine-region approach is often used. Examine figure 1.12 and locate the following regions:

Four Quadrants Approach
Right-upper quadrant
Left-upper quadrant
Right-lower quadrant
Left-lower quadrant

Nine Regions Approach
Right hypochondriac
Left hypochondriac
Epigastric
Right lumbar
Left lumbar
Umbilical
Hypogastric
Right iliac
Left iliac

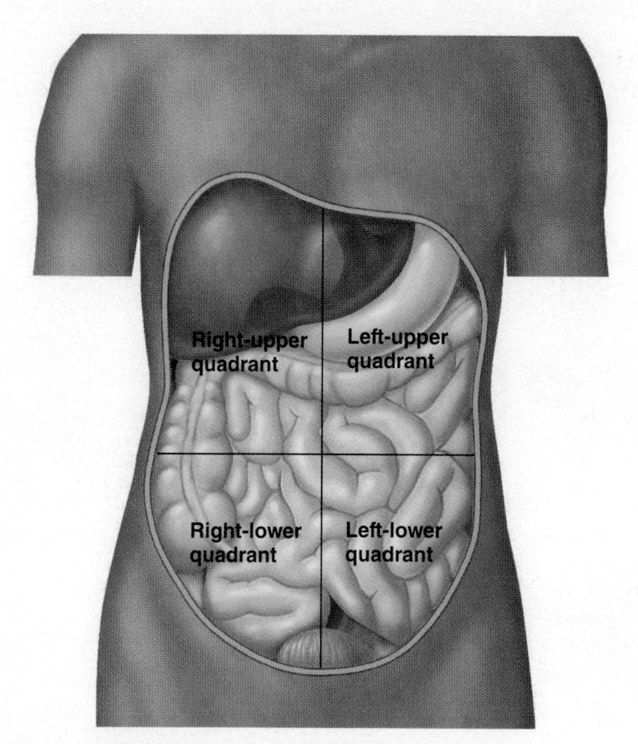

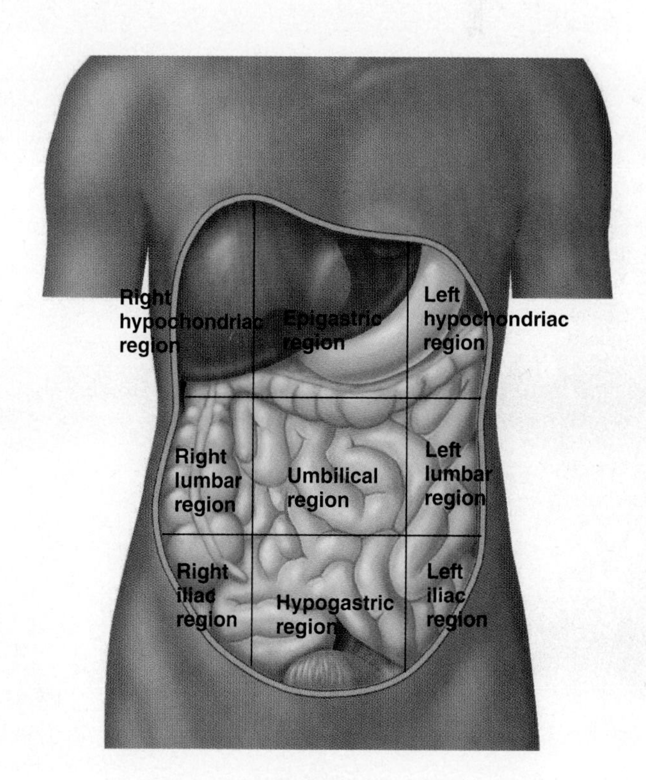

Figure 1.12 Abdominal Regions

NOTES

Exercise 1 Review

Name: ______________________

Lab time/section: ______________________

Date: ______________________

Introduction to Lab Science, Chemistry, Organs, Systems, and Organization of the Body

Review Questions

1. In terms of base units
   a. What is the base unit of length in the metric system?
   b. What is the base unit of volume in the metric system?

2. How many cubic centimeters are there in 200 milliliters?

3. Assume a pill contains 350 mg of medication. How much medication is this in grams?

4. How would you write 0.000345 liter in scientific notation?

5. How many milligrams are there in 4.5 kilograms?

6. How much of a meter is 250 millimeters?

7. If given a length of 1/10,000 of a meter
   a. Convert this number into a decimal:
   b. Convert it into scientific notation:

8. Use a word to describe
   a. One-thousandth of a second:
   b. One thousand liters:
   c. One-hundredth of a meter:

9. Did you see a trend in your results with heart rate and exercise? If so, what do you predict for additional exercise? Would the trend continue indefinitely? Why or why not?

10. In terms of heart rate and exercise, which one is the dependent variable and which one is the independent variable?

11. According to your bar chart (chart 2) what would be the mean heart rate at 150 seconds of exercise?

12. Define the term *buffer.*

13. What is the neutral pH?

14. Is a pH of 8 more basic or more acidic than a pH of 6?

15. As a solution becomes more acidic, what happens to the concentration of hydrogen ions?

16. The term *electrolyte* is derived from the words *electro* ("electricity") and *lyte* ("to separate"). How does this term correlate with your experiment?

17. In which bond are electrons significantly shared between atoms?

18. Which bond—covalent, ionic, or hydrogen—is a weak bond?

19. The scientific discipline that studies the function of the human body is known as ____________________.

20. Organs are associated into functionally related groups called ____________________.

21. In terms of reference, the body is placed in what position? ____________________

22. What body cavity lies directly inferior to the diaphragm? ____________________

23. The specific body cavity that is enclosed by the rib cage is known as the ____________________.

24. The body cavity surrounded by the hip bones is called the ____________________.

25. The term *arm* in anatomy refers to the region between the
    a. shoulder and elbow.
    b. elbow and wrist.
    c. shoulder and wrist.
    d. shoulder and hand.

26. The term *leg* in anatomy refers to the region between the
    a. hip and knee.
    b. knee and ankle.
    c. hip and ankle.
    d. ankle and foot.

27. A mitochondrion belongs to which level of organization?
    a. cellular
    b. tissue
    c. organelle
    d. organ system

28. The liver occupies what two regions of the abdomen? ____________________

Use correct anatomical terminology to describe the following relationships.

29. In terms of up and down, the head is ____________________ to the toes.

30. In terms of nearness to the trunk, the fingers are ____________________ to the arm.

31. In terms of nearness to the surface, the brain is ____________________ to the scalp.

32. In terms of front to back, the nipples are ____________________ to the shoulder blades.

33. The lungs belong to the ____________________ system.

34. The liver belongs to the ____________________ system.

35. The pectoralis major belongs to the ____________________ system.

36. If you were to sit on a horse's back, you would be on the ____________________ aspect of the horse.
    a. anterior
    b. ventral
    c. posterior
    d. dorsal

37. What is the difference between the abdomen and the abdominal cavity? ____________________

38. Complete the illustration by correctly placing the following terms:

abdominal	acromial	antebrachial	axillary	brachial	carpal	cephalic	cervical
coxal	crural	femoral	frontal	genital	pectoral	pedal	sternal

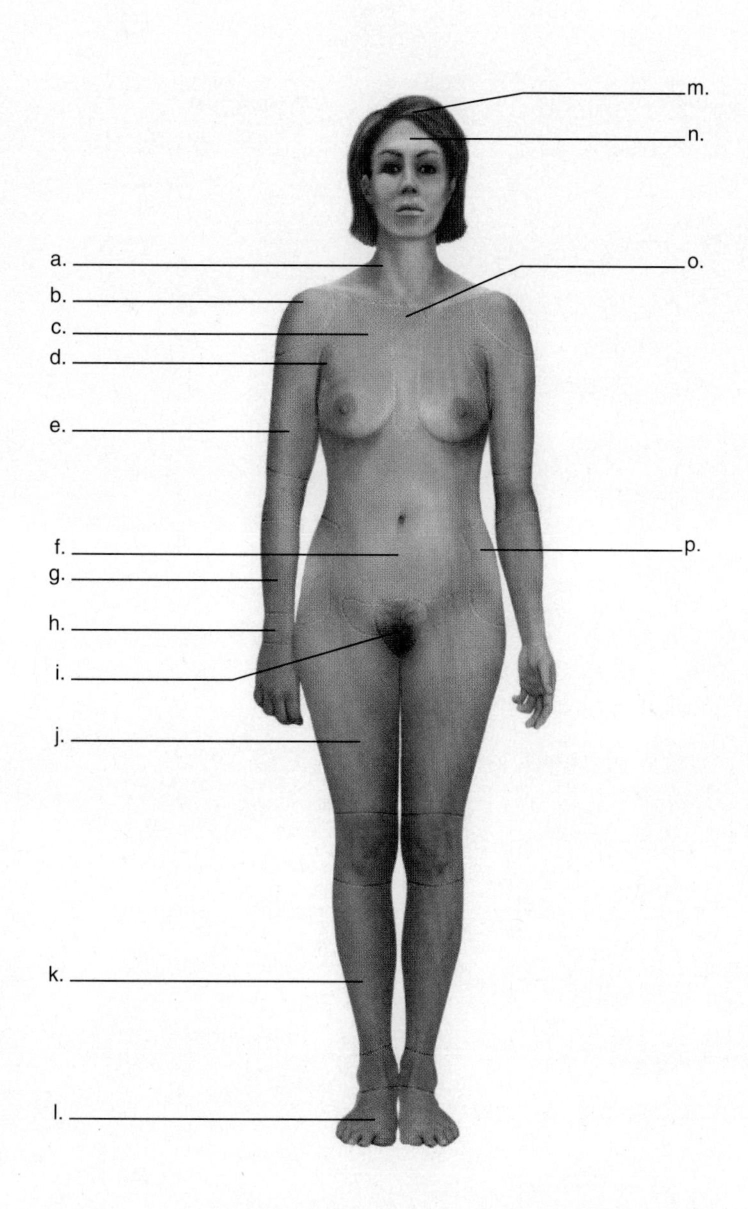

39. In the illustration of the brain, place the appropriate terms next to the planes that represent them.

frontal midsagittal (median) transverse

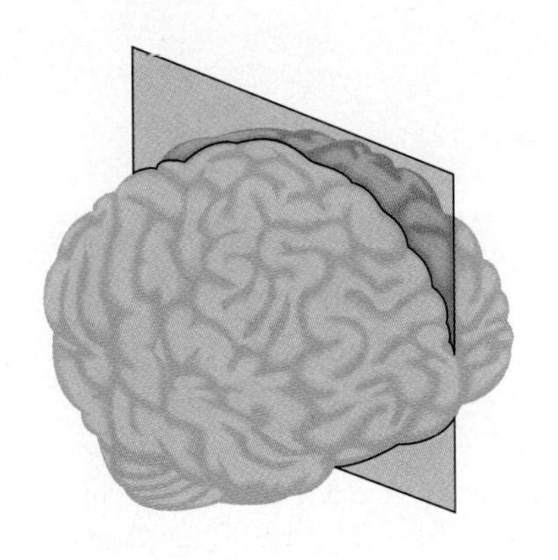

(a) ______________________

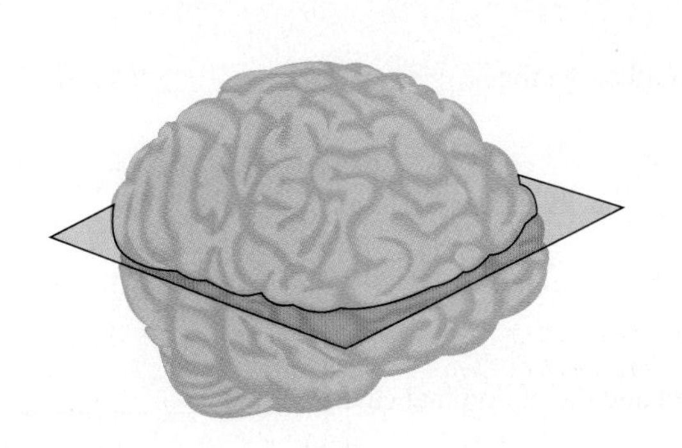

(b) ______________________

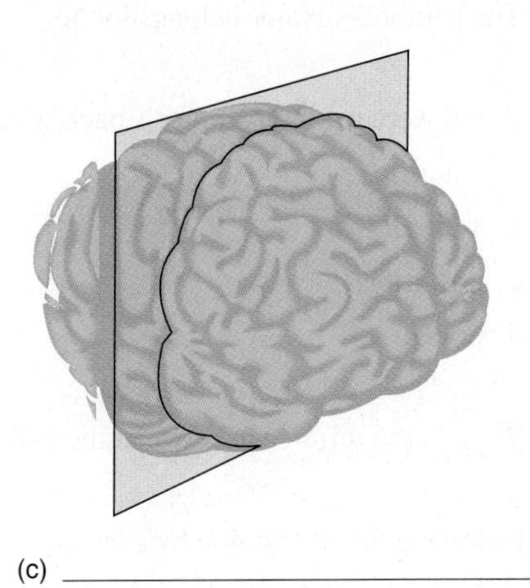

(c) ______________________

Exercise 2

Microscopy

Introduction

Originally, the study of anatomy and physiology was based on macroscopic, or gross, observation. This study was limited by the **resolution** of the human eye, which is the ability to distinguish two objects as separate. With the invention and use of the compound light microscope (microscopes increase resolution), much greater detail was seen, and thus began the study of cells and tissues. **Light microscopy** involves the use of visible light and glass lenses to magnify and observe a specimen.

The various types of microscopes are discussed in the text in chapter 3, "Cell Biology." Chapter 3 also has many electron microscope photographs that show details of cellular structure. This exercise covers how to use the compound light microscope, how to examine prepared slides under the microscope, and how to make simple slides for study.

Objectives

At the end of this exercise, you should be able to

1. list the rules for proper microscope use;
2. name the parts of the microscope and their functions presented in this exercise;
3. demonstrate the proper use of the compound light microscope;
4. place a slide on the microscope and observe the material, in focus, under all magnifications of the microscope;
5. calculate the total magnification of a microscope based on the lenses used;
6. prepare a wet mount for observation.

Materials

Compound light microscope
Prepared slide with the letter *e* (or newsprint and razor blades)
Transparent metric ruler
Glass microscope slides
Coverslips
Lens paper
Kimwipes or other cleaning paper
Lens cleaner
Small dropper bottle of water
1% methylene blue solution
Toothpicks
Histological slides of kidney, stomach, or liver
Prepared slide of silk threads

Procedure

Care of the Microscope

Microscopes are very expensive pieces of equipment, and you should always take great care handling them. There are a few rules concerning microscopes that you should always observe:

1. When carrying the microscope, hold it securely with two hands—one hand under the base and one on the arm.
2. Keep the microscope upright at all times.
3. Keep microscope lenses clean with lens cleaner and lens paper. Do *not* use paper towels or your clothing.
4. Use only the fine-focus knob when using the high-power objective lens.
5. Remove slides from the microscope before putting it away.
6. Secure the cord with a rubber band or wrap the cord gently around the microscope.
7. Store the microscope with the low-power objective lens in place.
8. Lower the stage.
9. Put the microscope away in its proper location.

Using the Microscope

(A)[1] 1. Examine figure 2.1 and familiarize yourself with the parts of the microscope.

2. Remove the microscope from the storage area. Never tilt the microscope from an upright position, as lenses or filters may fall and break.
3. Take the microscope to your desk. Unwrap the electrical cord from around the microscope and compare the one that you have in lab to the one illustrated in figure 2.1. There may be differences between your microscope and the one in figure 2.1, but you should be able to locate the parts listed in the microscope parts checklist. Use the checklist to make sure that you find all of the listed parts. Place a check mark next to the appropriate space when you locate each part of the microscope.

Microscope Parts Checklist

_____ Base	_____ Arm	_____ Objective lens
_____ Condenser	_____ Iris diaphragm lever	_____ Body tube
_____ Nosepiece	_____ Coarse-focus knob	_____ Fine-focus knob
_____ Stage	_____ Ocular (eyepiece) lens	_____ Light source

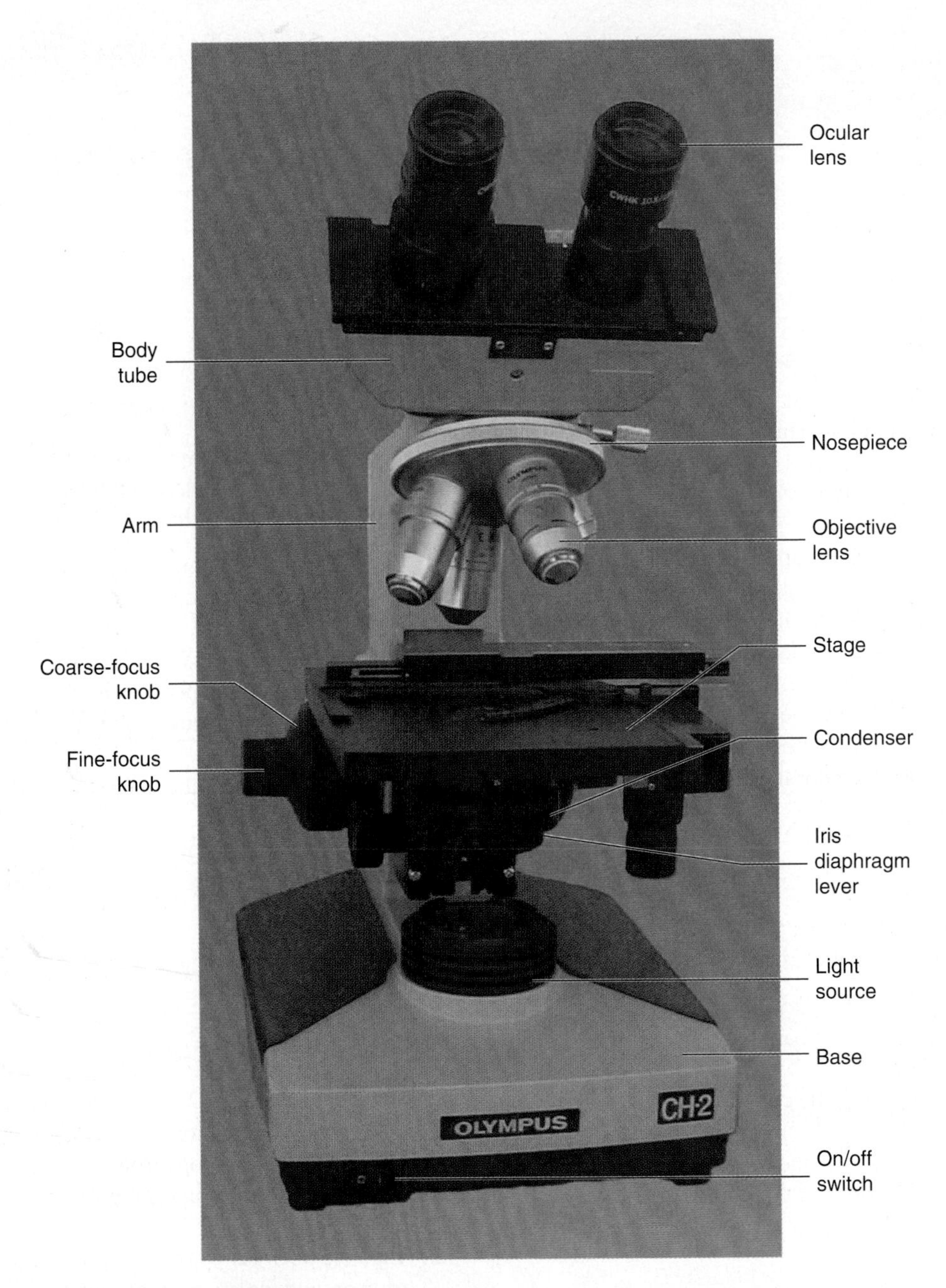

Figure 2.1 Compound Light Microscope
(Courtesy Olympus Microscopes)

4. With the microscope in front of you, plug it in, making sure the ocular lens or lenses are facing you. Make sure the cord is not hanging over the counter or in the aisle where someone might trip on the cord or pull the microscope off the counter. If the microscope has an illuminator (light source) dial, make sure that the setting is on the lowest level.
5. The low-power objective lens (typically 4 magnifications or 4×) should be facing down. This lens is often the shortest one and, in many microscopes, has a red line around the barrel of the lens. If it is not in place, turn the nosepiece until the low-power objective clicks into place. Always use the low-power objective lens when you first look at a microscope slide.
6. Use the knob on the microscope to raise the condenser lens so that it is at its highest level.
7. Obtain a prepared slide with the letter *e* on it and go to step 10 or take a piece of newsprint and, using a razor blade or scissors, cut out a single letter from the paper. Place the newsprint letter on a glass microscope slide and add a drop of water to the piece of paper on the slide.
8. Place a thin coverslip on the slide by touching one edge of the coverslip to the water and lowering it slowly over the newsprint specimen as seen in figure 2.2. If you drop the coverslip on top of the slide, you will probably trap several air bubbles, which may obscure some of your specimen.
9. Locate the light switch on your microscope and turn it on.
10. Place the slide on the microscope stage with the coverslip on the top and the letter centered in the circle on the stage of the microscope. Make sure that the letter is situated so you

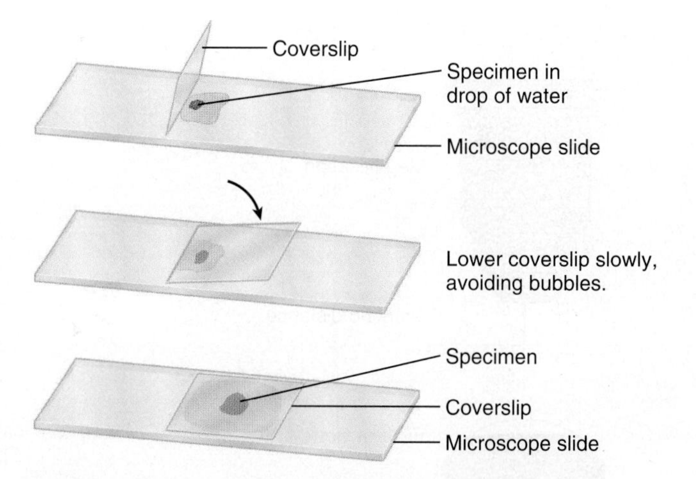

Figure 2.2 Preparation of a Wet Mount

can read it from where you are sitting. Most microscopes have a slide clip that holds the microscope slide in place. Use the stage control knobs to move the specimen so that light is coming through the specimen. This may require that you look at the specimen from the side to see if it is centered in the middle of the microscope stage.

11. Examine the specimen under low power. Use the coarse-focus knob to bring the specimen into focus. This is done by raising the mechanical stage of the microscope completely and then slowly lowering the stage as you look into the ocular lenses. On some microscopes it is the objective lens that is raised or lowered with the coarse focus knob.

    Focusing the microscope requires a little bit of patience. When first looking at microscope slides, always use the low-power objective lens. Make sure the specimen is on the stage and centered in the open circle on the stage. There should be light coming through the specimen. The coverslip should be close to the objective lens. The distance between the objective lens and the coverslip is known as the **working distance** as seen in figure 2.3. Look through the ocular lens and rotate the coarse-focus knob slowly so the objective lens and the slide begin to move away from each other. This should bring the object into focus in the **field of view**. The field of view is the circle that you see as you look into the microscope.
12. Adjust the ocular lenses. Binocular microscopes have one fixed ocular and another that you can adjust. Focus the microscope so that the image in the nonadjustable ocular lens comes into focus. This is usually the right one. Once it is in focus, use the knurled ring on the adjustable ocular (typically the left one) to bring that lens into focus for the other eye.
13. Adjust the light. Too much or too little light makes a specimen difficult to see. You can change the light level by either adjusting the light from the rheostat knob on the base or adjusting the iris diaphragm lever in the front of the microscope. The condenser lens will also help with lighting. For most observations put the condenser lens close to the specimen.
14. Draw what the specimen looks like in the space provided. Does the letter appear right side up, or is the image inverted?

    ____

    Is the letter oriented correctly, or is the image flipped horizontally? ______________________________

    How much of the letter occupies the field of view?

    ____

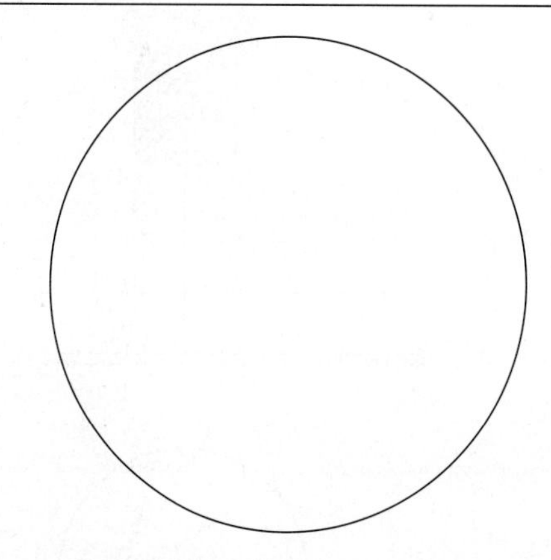

15. After you get the specimen in focus under low power, you should examine the slide under the next higher power. This is done by centering the image that you observe in the field of view and then switching the objective lens to the next higher power. The image should be pretty well in focus if your microscope is *parfocal*. Do not adjust the height of the mechanical stage. The next-higher power objective lens should clear the slide. Once you rotate the lens, adjust the focus by using the fine-focus knob. You can look at the subject under high power by the same procedure by turning the objective lens to the high-dry lens. If you cannot focus on the high power, or have lost the image that you were looking for, you should return to low power and try the process again. If you still cannot find the object under high power, you should ask your instructor to help you.

Microscope Troubleshooting

If you are having a difficult time seeing anything or seeing things in focus, there could be several reasons. Use the troubleshooting tips listed in table 2.1 to correct the problem.

Cleaning the Microscope

Smudges on the images you view through the microscope may be due to several things. There may be makeup or dirt on the ocular

TABLE 2.1 Troubleshooting Tips

Problem	Solution
Nothing is visible in the lens.	Plug in the microscope.
	Turn the power supply on.
	Rotate the objective lens until you hear it click into place.
	The bulb is burned out; replace the bulb.
You see a dark crescent.	The objective lens is not in proper position; click the lens into place.
All you see is a light circle.	The microscope is out of focus; adjust the coarse-focus knob.
	The light is up too high; turn down the light.
	The iris diaphragm is open too much; close it down.

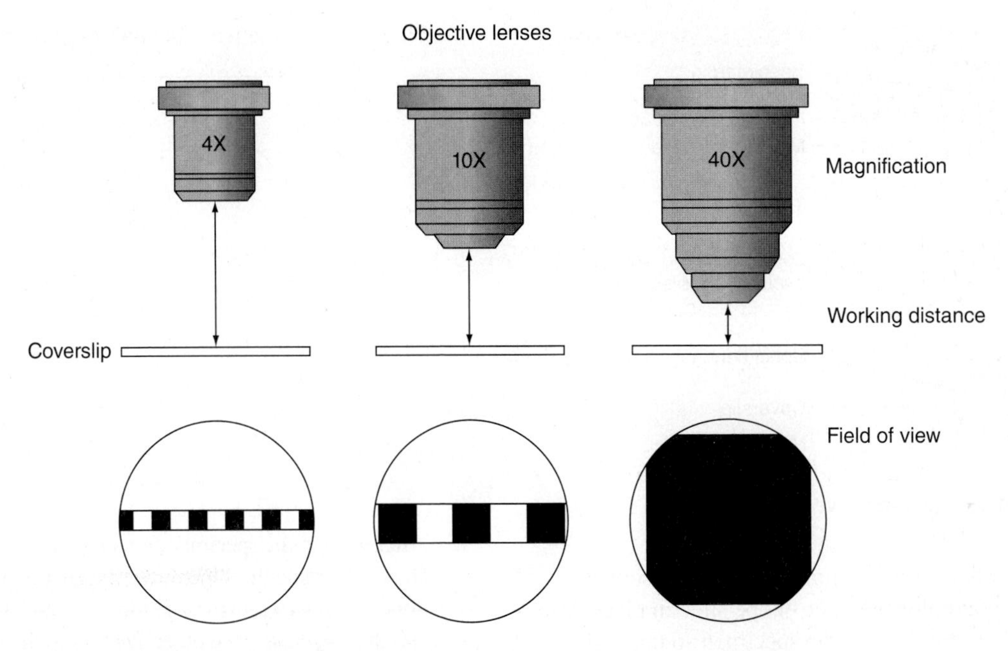

Figure 2.3 Increasing Magnification and Decreasing Field of View
As magnification increases (seen by increasing numbers on the objective lenses), both the working distance (distance between the objective lens and microscope slide) and the field of view decrease.

lens (or lenses). There may be dirt, oil, salt, stains, or other material on the objective lenses. To clean a lens, place a small amount of lens-cleaning fluid on a clean sheet of lens paper. Make one circular pass on the lens and throw the paper away. If you continue to clean the lens with the same lens paper, you can grind dirt or dust into it. Use a fresh piece of lens paper and repeat the procedure if further cleaning is needed.

You may want to clean the microscope slide before you examine it. Use a cleaning paper, such as a Kimwipe, to clean oil or dust from the slide.

Finally, dust may have collected inside the microscope over the years or the lenses may be scratched. There is nothing you can do about this, though you may want to bring this to your instructor's attention.

Magnification and Field of View

Ⓐ[2] You can determine the size of the object under view if you know the diameter of the field of view. The field of view can be measured directly under low magnification by using a clear ruler. If higher magnifications are used, rulers won't work, and you have to calculate the field of view. There is a relationship between the diameter of the field of view and the magnification used. First calculate the total magnification using the following procedure. Look at the barrels of your lenses and determine the magnification of each lens.

Eyepiece (ocular) magnification: ____________________

Low-power objective magnification: ____________________

Total magnification (= ocular magnification × objective magnification): ____________________

Place a transparent ruler on the stage of the microscope. The space between the dark vertical lines is 1 millimeter (mm). Count the number of millimeters at the broadest part of the field of view and enter this number as the diameter of the field of view in the following space.

Diameter of the field of view (mm): ____________________

You can calculate the length of an object by determining how much of the diameter of the field of view it occupies. Let's say that the diameter of the field of view is 10 mm. If an object takes up one-half of the field of view, then you can estimate its size at 5 mm. If the object takes up only one-third of the field of view, how large is it? Record your answer in the following space.

Object size (mm): ____________________

As the magnification increases, the field of view decreases proportionally. Thus, if the diameter of the field of view is 10 mm at one magnification and you double the magnification by changing lenses, the field of view is reduced to a diameter of 5 mm. If you switch to a new lens and increase the magnification by 10 times, then the field of view is reduced to one-tenth of the original field of view. Look at figure 2.3 for a representation of this.

Keep the transparent ruler under the microscope and increase the magnification to the next higher power by moving the next larger objective lens into place. Record the total magnification of your microscope with this objective lens.

Total magnification: ____________________

Examine the ruler under the microscope and record the diameter of the field of view in millimeters.

Diameter of the field of view in millimeters: ____________________

Has the increase in magnification produced a decrease in the field of view proportional to the magnification?

____

Now calculate the total magnification of the microscope using the high-power objective lens.

Magnification with high-power objective lens: __________

You will not be able to measure the field of view accurately under high power with a ruler; however, you should be able to calculate the diameter of the field of view. For example, if the diameter of the field of view is 2.5 mm at 40 power (40×), then it is 0.25 mm at 400 × (10 times more magnified yet one-tenth the field of view). Calculate the diameter of the field of view under high power.

Diameter of the field of view under high power: _________

Proper Lighting

Too much or too little light makes a specimen difficult to see. You can adjust the light by adjusting either the illuminator or the iris diaphragm lever. On some faintly stained specimens, the material may be difficult to see on low power. One trick is to locate the edge of the coverslip, center it in the field of view, and turn the coarse-focus knob up and down until the edge is in sharp focus. This lets you know that you are in the approximate focal plane for examining the material on the slide. Move the slide to where the specimen should be, adjust the light, and reexamine it.

(A)[3] Examination of a Prepared Slide

Examine three crossed threads under low power. As you focus on the threads, which color thread is on top? _______

Which color thread is in the middle? ______________

Which color thread is on the bottom? ______________

Under low power, how many threads or how much of one thread appears to be in focus? ______________

As you switch to the next higher power, how many threads or how much of one thread appears in focus?

____

The **depth of field** is the thickness of the visible plane of the material that is in focus under a particular magnification. If you are looking at tissue that has multiple layers of cells, you can examine only one layer of cells at a time. What happens to the depth of field when you increase the magnification in the microscope? ______________

____

Caution!

Methylene blue is a stain. Be careful to not get it on your skin or in your eyes

Preparation of a Wet Mount

You can make relatively quick and easy observations under the microscope as long as the material is thin enough and small enough. One technique for cell examination is to examine cells from the inside of the oral cavity. Do the following procedure.

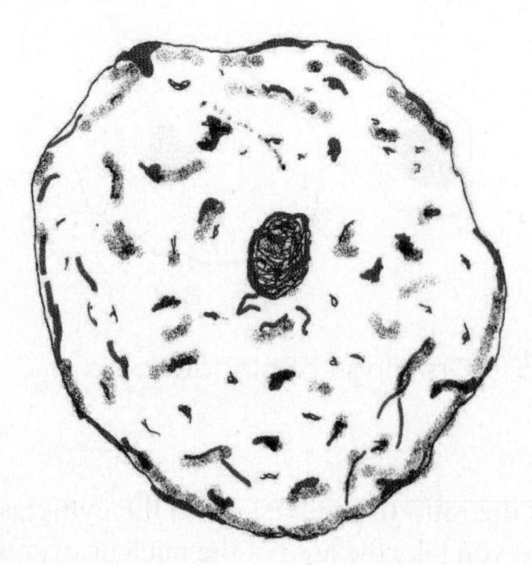

Figure 2.4

(A)[4] **1.** With a toothpick, gently scrape the inside of your cheek.

2. Smear the cheek material from the toothpick on a clean microscope slide.

3. Place a drop of methylene blue (a stain) on the smear.

4. Place a coverslip on the slide by touching one edge of the coverslip to the slide and slowly lowering it on the slide and examine it under the microscope. The small, oval structures inside the cells are the nuclei. The angular lines around the nuclei are the plasma membranes.

5. Draw a single cell in the space provided and label the plasma membrane, nucleus, and cytoplasm.

Your illustration of a cheek cell:

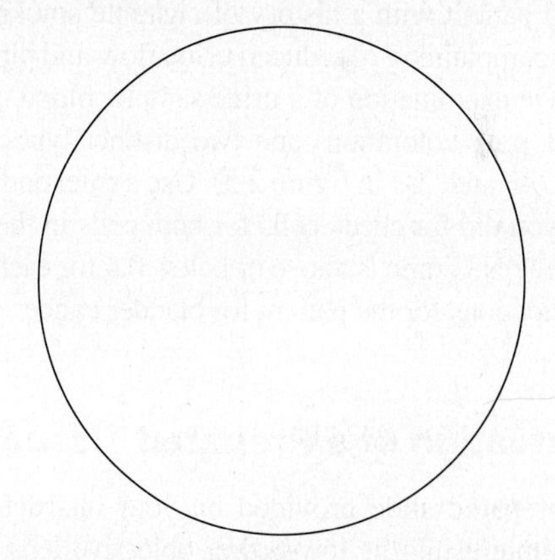

6. Determine the area of a typical cheek cell under the microscope. If you have no way to measure the cell in the microscope use a ruler and measure figure 2.4, which was drawn from an image of an actual cheek cell.

You can get a rough approximation of the area by multiplying the height of the cell by the width of the cell. Record the area in the following space:

Area of the cheek cell: ______________________

Now determine the area of the nucleus of the cheek cell under the microscope by multiplying the height by the width. Record the area in the following space:

Area of the nucleus: ______________________

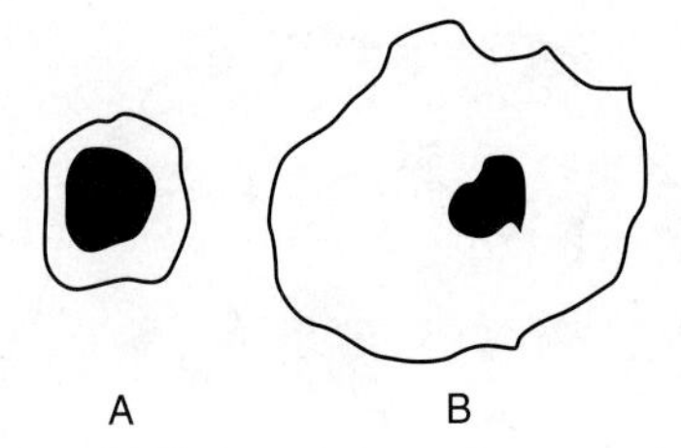

Figure 2.5 Drawings of bladder cells

To determine the ratio of the nucleus to the cytoplasm of the cell (the N:C ratio) you take the area of the nucleus over the area of the cell subtracted from the nucleus as described by the formula:

$$N/C = \frac{\text{Area nucleus}}{\text{Area cell} - \text{Area nucleus}}$$

If it is well under 40% (0.4) then you are probably looking at a normal cheek cell. There is a condition known as squamous cell carcinoma (a kind of cancer) and cells in this condition often show a high nuclear to cytoplasmic ratio due to the rapid cell division and increased genetic material in the cell.

This test is done for illustrative purposes only. Please do not self-diagnose your epithelial slide, and recognize that professionals examine many cells (because there are normally unusual cells in your cheek) and perform other tests to determine whether an abnormal condition is present.

Case Study

A 56-year-old patient with a history of cigarette smoking comes to the physician complaining of reduced urine flow and pink coloration of the urine. On examination of a urine sample, blood is detected in the urine (the pink coloration) and two distinct types of cells are detected (see "A" and "B" in figure 2.5). Use a ruler and measure the N:C ratio as you did for cheek cells for both cells in the illustration. Determine if the N:C ratio is above or below 0.4 for each cell and the possible implications for the patient for bladder cancer.

(A)[5] Observation of a Prepared Tissue Slide

Examine a prepared slide provided by your instructor. Examine the entire sample using the low-power objective lens. You should scan the entire area, looking for areas you want to observe more closely. Move to the next higher power and adjust the focus using the fine-focus knob only. Finally, examine the material with the high-power objective lens and draw what you see in the space provided.

Your illustration of material from a prepared slide:

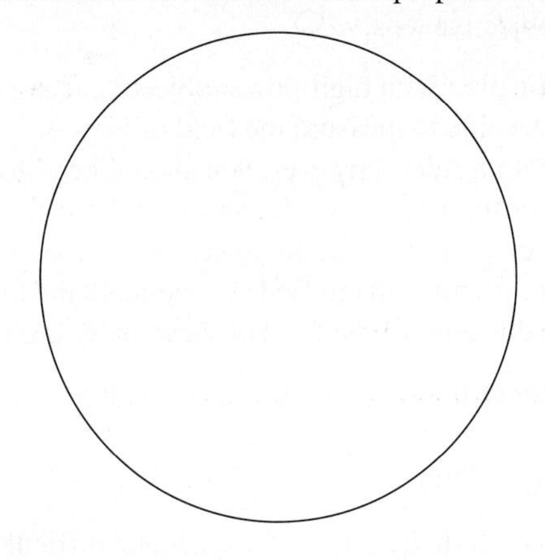

Name of the sample drawn: ______________________________

You should use the iris diaphragm lever to adjust the amount of light that strikes the specimen. Too much or too little light produces significant changes in the observed image.

Oil Immersion Lens

The objective lenses you have used so far are called "dry" lenses. Your lab may be equipped with microscopes that have oil immersion lenses. Oil causes the light to bend (refract) differently. The techniques for using these lenses are somewhat different from those for dry lenses. Once you have examined the specimen using the high-power dry lens, find the spot you want to examine and center it in the field of view. Add a drop of immersion oil on top of the coverslip and carefully swing the oil immersion lens into place. Use the fine focus only; otherwise you might drive the oil immersion lens through the slide and break it. Once you have examined the slide, swing the lens away and remove the slide, carefully wiping away the immersion oil with a clean piece of lens paper. Use only lens paper to clean the oil from the oil immersion lens. Use another piece of lens paper to remove any remaining oil, if needed.

Cleanup

Make sure to read the "Care of the Microscope" section at the beginning of the exercise. When you put away the microscope, follow these directions and any supplemental directions your instructor gives you. Remember to remove any slides from your microscope, wrap the cord loosely around the scope or secure it with a rubber band, carry the scope with two hands, and put it away in the appropriate place.

Exercise 2 Review

Name: ____________________

Lab time/section: ____________________

Microscopy

Date: ____________________

1. If the ocular lens is 10×, what is the total magnification for the following objective lenses?
   a. 7 × ____________________
   b. 15 × ____________________
   c. 20 × ____________________

2. What is the name of the thin glass plate that is placed on top of a specimen? ____________________

3. The microscope that you use in this lab is a(n)
   a. compound microscope.
   b. dissecting microscope.
   c. electron microscope.

4. What is the name of the circle you see when you look through the ocular lens of the microscope? ____________________

5. What is the function of the iris diaphragm of the microscope? ____________________

   ____

   ____

6. If the diameter of the field of view is 5.6 mm at 40×, what is the diameter at 80×? ____________________

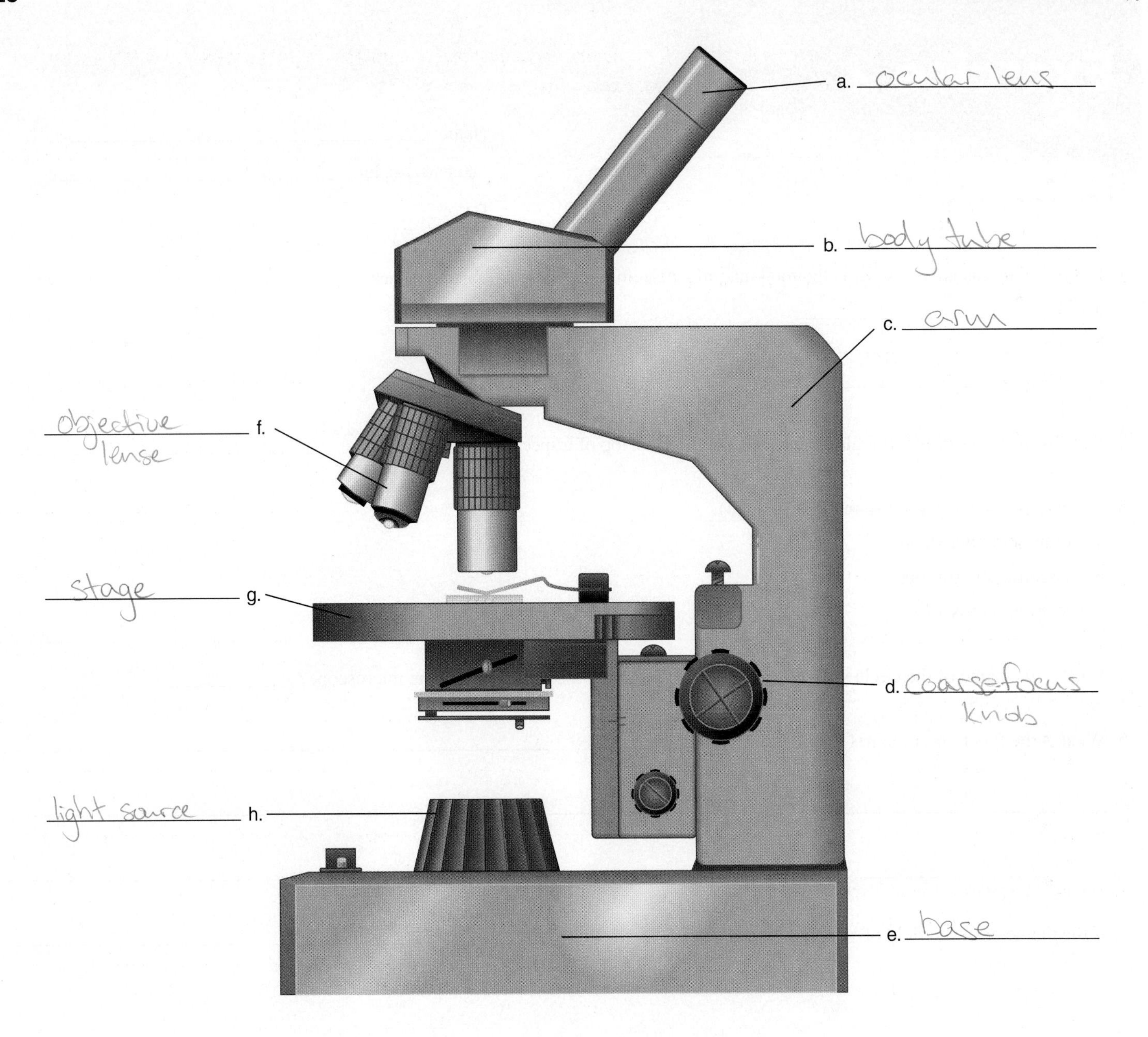

7. Place the following labels correctly in the microscope illustration.

   arm
   base
   body tube
   coarse-focus knob
   light source
   objective lens
   ocular lens
   stage

8. When you switch from a low-power objective lens (for example, 4×) to a higher-power objective lens (for example, 10×), what happens to the working distance between the lens and the coverslip?

____

____

9. When you change from a low-power objective lens to a higher-power one, what happens to the field of view? Does it increase or decrease?

____

10. When should you use the low-power objective lens on the microscope?

____

____

____

11. How should you clean the lenses of a microscope?

____

____

12. What is the proper way to carry the microscope in lab?

____

____

13. Examine the following "field of view" and determine what the size of the object is.

____

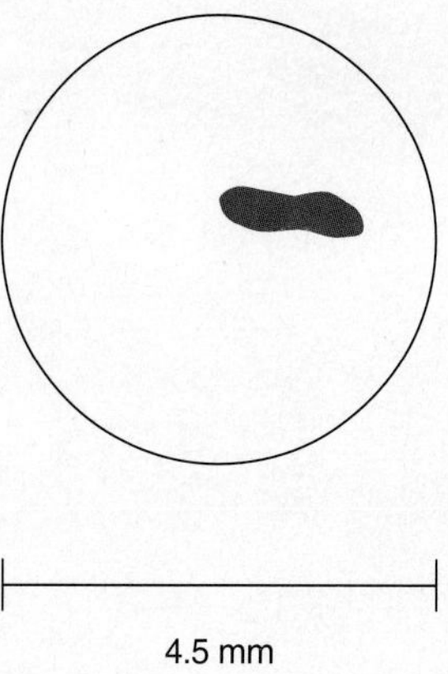

NOTES

Exercise 3

Cell Structure and Function

Introduction

The cell is the structural and functional unit of the human body and is the fundamental unit of living organisms. Cells perform many functions, some of which are unique to the particular organ where they are found. Generally speaking, however, cells grow, divide, acquire nutrients, release wastes, and respond to local stimuli. Many diseases can be traced to some type of cellular change. **Cytology** is the scientific study of cells.

The membrane of the cell (plasma membrane) is the dynamic interface between the internal environment of the cell and the external environment. In humans, most cells are bathed in a liquid medium called extracellular fluid (ECF), which provides nutrients, oxygen, hormones, water, and other materials to the cell. From the interior of the cell, the cell releases ammonia, carbon dioxide, and other metabolic products into this liquid. The plasma membrane is important in the exchange of materials between the cell's interior and the environment surrounding it. The exchange of materials between the cell and the ECF maintains the homeostatic balance the cell must have to survive. Even small changes in the concentration of certain materials in the cell might lead to cellular death, so the constant adjustment of water, ions, and other metabolic products is extremely important. In large part, the plasma membrane actively regulates what enters and what leaves the cell. This is done passively in some cases and, in others, ATP is used to actively transport material. The membrane also has embedded enzymes that play a significant role in cellular function.

In this exercise, you examine the structure of animal cells, learn how cells of the body divide to make new cells, look at the physical processes that influence plasma membrane dynamics, and study the nature of membrane transport. These topics are discussed in chapter 3, "Cell Biology," in the Seeley text.

Objectives

At the end of this exercise, you should be able to

1. describe the importance of cells to the body;
2. describe the plasma membrane in terms of location, composition, and function
3. list the function of the nucleus;
4. describe the structure and function of the organelles;
5. describe the two stages of the cell cycle;
6. name the four phases of mitosis and what occurs during each phase;
7. describe the processes by which substances move across membranes;
8. describe the differences between *hypertonic, hypotonic,* and *isotonic;*
9. explain *diffusion, osmosis,* and *filtration;*
10. describe the movement of water across a selectively permeable membrane;
11. compare and contrast diffusion and osmosis.

Materials

Cell Structure

Models or charts of animal cells
8.5 × 11 blank notebook paper
Electron micrographs of cells or a textbook with electron micrographs
Prepared slides of whitefish blastula
Microscopes
Modeling clay (Plasticine™)—two colors
Marbles

Brownian Motion

India ink in dropper bottles
Dropper bottle of water
Microscopes
Microscope slides
Coverslips
Hot plate

Diffusion Demonstration

Potassium permanganate crystals
100 mL beaker (one per table)
Water
Small tweezers or small spatula

Diffusion

Agar plates (three per table)
0.01 Molar (M) potassium permanganate solution in dropper bottles
0.01 M methylene blue solution in dropper bottles
Dishpan filled with crushed ice
Plastic drinking straws
Millimeter ruler
Fine probe, spatula, or small forceps
Warming tray (35°–40°C) or incubator

Osmosis Demonstration

Two thistle tubes
Rubber bands
Dialysis tubing

Dark corn syrup or concentrated, colored sucrose solution (20%)
1% starch solution
Ring stand and clamp
250 mL beaker
Distilled water
Permanent marker

Osmosis Experiment

Four strips of 20 cm-long dialysis tubing (one set of four per table)
Four 200 mL beakers
String
Scissors
Four solutions (2 L each) of 0%, 5%, 15%, and 30% sucrose (color each one with a different color of food coloring)
One liter of 15% sucrose solution
Balances
Towels
Pipettes (10 mL)
Pipette pumps

Osmosis and Living Cells

Clean glass microscope slides
Coverslips
5 mL of mammal blood (check with local veterinarian's office)
Distilled water in dropper bottle (one per table)
0.9% saline solution in dropper bottle (one per table)
5% sodium chloride solution in dropper bottles (one per table)
Protective gloves

Virtual Lab

Ph.I.L.S. 4.0 software and computer

Filtration

Filter paper
Funnel
Ring stand with ring clamp
10 mL graduated cylinder
500 mL beaker
Iodine solution in dropper bottles (one per table)
Filtration solution (500 mL of 1% starch, 1% charcoal, and 1% copper sulfate), consists of 5 g each of starch, charcoal, and copper sulfate in one bottle or flask
Stopwatch or clock with second hand

Procedure

Overview of the Cell

There are many types of cells in the body. Some are long and thin, some are spherical, and some are flat. You will examine a representative cell as an example, but realize that there is tremendous diversity in cell shapes and functions.

Cells consist of two main parts, the **plasma** (or **cell**) **membrane** and the **cytoplasm.** The plasma membrane is the outer boundary of the cell, and, although it cannot be seen using the light microscope, its location is determined by the difference in color between the cytoplasm and the surrounding liquid on the microscope slide. The cytoplasm is the portion of the cell in which water, dissolved materials, and small cellular **organelles** are found. The fluid in which the organelles are suspended is called the **cytosol.** Small filaments and tubules make up the **cytoskeleton** (also considered part of the cytoplasm). The nucleus directs the cell's activities and stores its genetic information. It is visible with the light microscope and frequently appears as a spherical or an oblong structure.

(A)[1] Figure 3.1 shows the plasma membrane and cytoplasm. Locate these on the model or charts in the lab. Take a single sheet of notebook paper and make a sketch of a typical cell with all of the membranes and organelles. Use the entire sheet for your drawing.

Plasma Membrane

The plasma membrane is the "gatekeeper" of the cell. It is selective in what it allows into the cell. The plasma membrane is composed of a **phospholipid bilayer,** proteins, cholesterol, and other molecules. Phospholipids consist of a hydrophilic phosphate group attached to hydrophobic lipid groups (figure 3.2). Interspersed among the phospholipid molecules are **cholesterol molecules,** which provide stability to the membrane or, in larger concentrations, can make the membrane more fluid. Two major types of proteins are also found in the membrane. **Peripheral proteins** are found on the inner or outer surface of the membrane, whereas those passing into the membrane are known as **integral proteins.** Integral proteins may have carbohydrates or other molecules associated with them and frequently serve as cell markers. Some integral proteins function as channels by which specific materials can pass through the membrane. Proteins may anchor one cell to another, provide a place for metabolic reactions to take place, act as cell markers that identify a particular cell, or act as membrane receptors or channels. The plasma membrane is important in establishing electrochemical charge differentials, which allows for signals to be passed along the membrane. It is also a selectively permeable membrane that provides an entrance to or exit from the cell for some materials while excluding other material from entering or exiting the cell's interior.

Cytoplasm

Most of the inside of the cell is cytoplasm, which consists of the inner fluid portion of the cell, known as the cytosol; the inner framework of the cell, called the cytoskeleton; and the small, specialized units of the cell, called organelles. The cytosol is composed of water with dissolved materials, such as sugars, ions, proteins, and amino acids.

Figure 3.3 illustrates the cytoskeleton (not visible under the light microscope), which consists of microtubules, microfilaments, and intermediate filaments, all of which provide shape to the cell, a place to anchor organelles, and resistance to gravitational and other forces acting on the cell. **Microtubules** are made of the protein tubulin and are approximately 25 nanometers (nm) in diameter. **Intermediate filaments** are composed of fibrous proteins and are approximately 10 nm in diameter. **Microfilaments** are made of **actin** and are approximately 8 nm in diameter.

Organelles

The term *organelle* literally means "small organ." Look at illustrations of organelles in figure 3.1. Each organelle has a

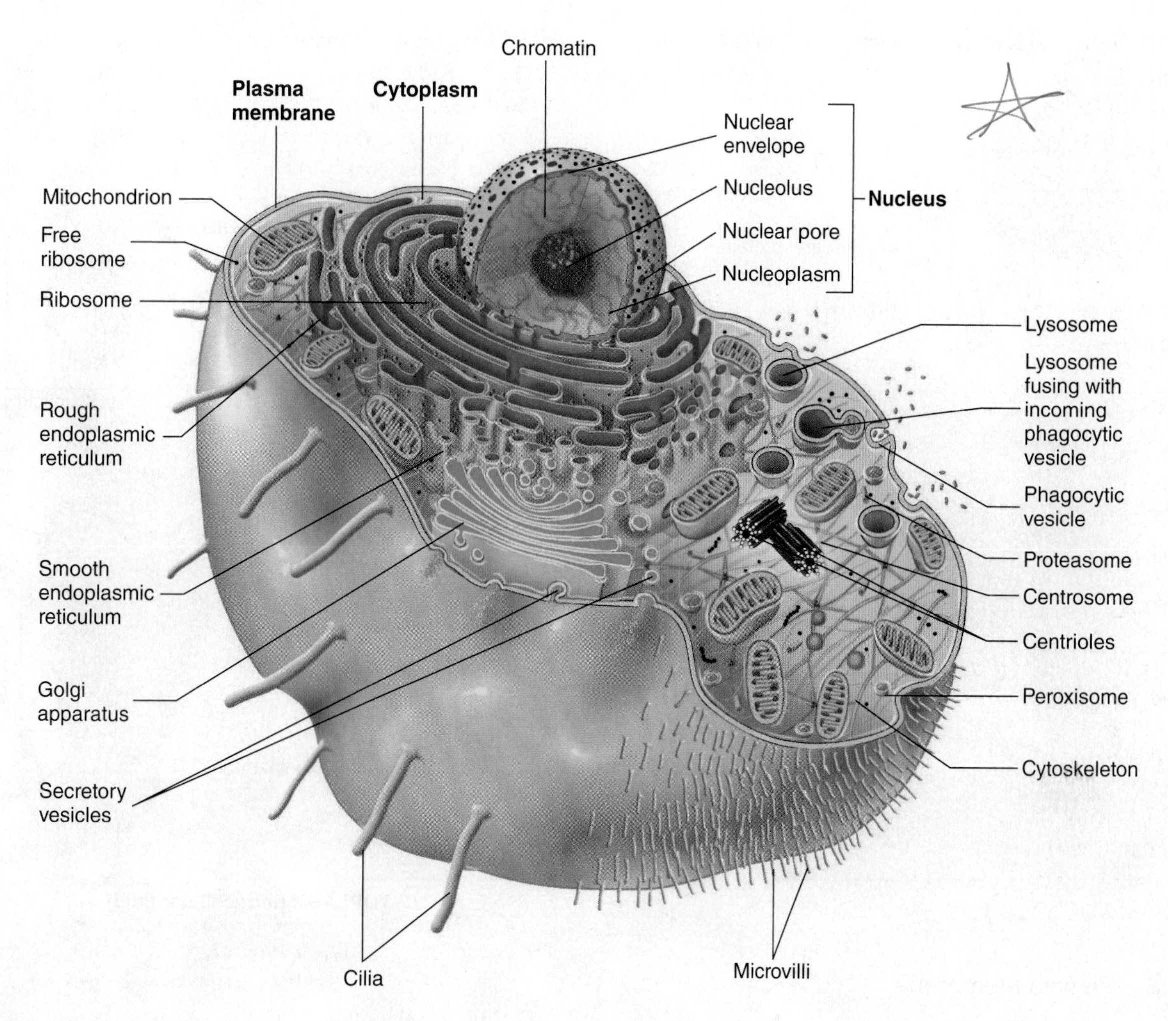

Figure 3.1 Overview of a Cell

particular function in the cell. There are two types of organelles—membranous and nonmembranous. Organelles represent a wonderful example of **specialization** on a microscopic scale—the function of the organelle is determined by its structure. **Mitochondria** are rod-shaped organelles with a double membrane. The major function of the mitochondrion (plural *mitochondria*) is to convert the stored chemical energy in food molecules to stored chemical energy in molecules of adenosine triphosphate (ATP). **Ribosomes** are the smallest of the organelles (about 25 nm in diameter) and are nonmembranous. Their function is to produce proteins. The **endoplasmic reticulum** is an organelle composed of a network of enclosed channels. There are two types of endoplasmic reticulum: rough endoplasmic reticulum, which has attached ribosomes, and smooth endoplasmic reticulum, which does not have ribosomes on its surface. The **rough endoplasmic reticulum (RER)** is associated with the nucleus and produces proteins for transport and use outside the cell. The **smooth endoplasmic reticulum (SER)** is a distal extension of the rough endoplasmic reticulum. It produces lipid compounds (including phospholipids and steroids) and detoxifies material. The **Golgi apparatus** receives material from the endoplasmic reticulum and other parts of the cytoplasm and serves as an assembly and packaging organelle. It has **cisternae,** which are flattened, membranous sacs, and it forms vesicles to transport the molecules it assembles.

Nucleus

The nucleus has two major functions: One is to house the genetic information of the cell, and the other is to direct many cellular functions. These two functions are carried out by DNA (deoxyribonucleic acid), which combines with proteins to form a material called chromatin in the nucleus. The nucleus is bounded by a **nuclear envelope,** which is a double membrane. The nuclear envelope contains nuclear pores, which allow the movement of materials into or out of the nucleus.

One or more structures known as the nucleoli (sing. *nucleolus*) are inside the nucleus. The nucleoli consist of portions of chromosomes and thus contain DNA and protein. They make rRNA, which forms ribosomes, the protein-producing organelles in the cytoplasm of the cell. Examine a model or chart in lab and find the nucleus, the nuclear envelope, and the nucleolus. Look at figure 3.1 and table 3.1 for a brief description of cellular structures and their function. Your text has more detailed descriptions of these structures and you should refer to it for more information.

Vesicles

Vesicles are membrane-bound sacs inside the cell; they digest subcellular material, transport material out of the cell, and carry on enzymatic activities. Vesicles can fuse with other organelles or the plasma membrane; thus they protect the integrity of the plasma

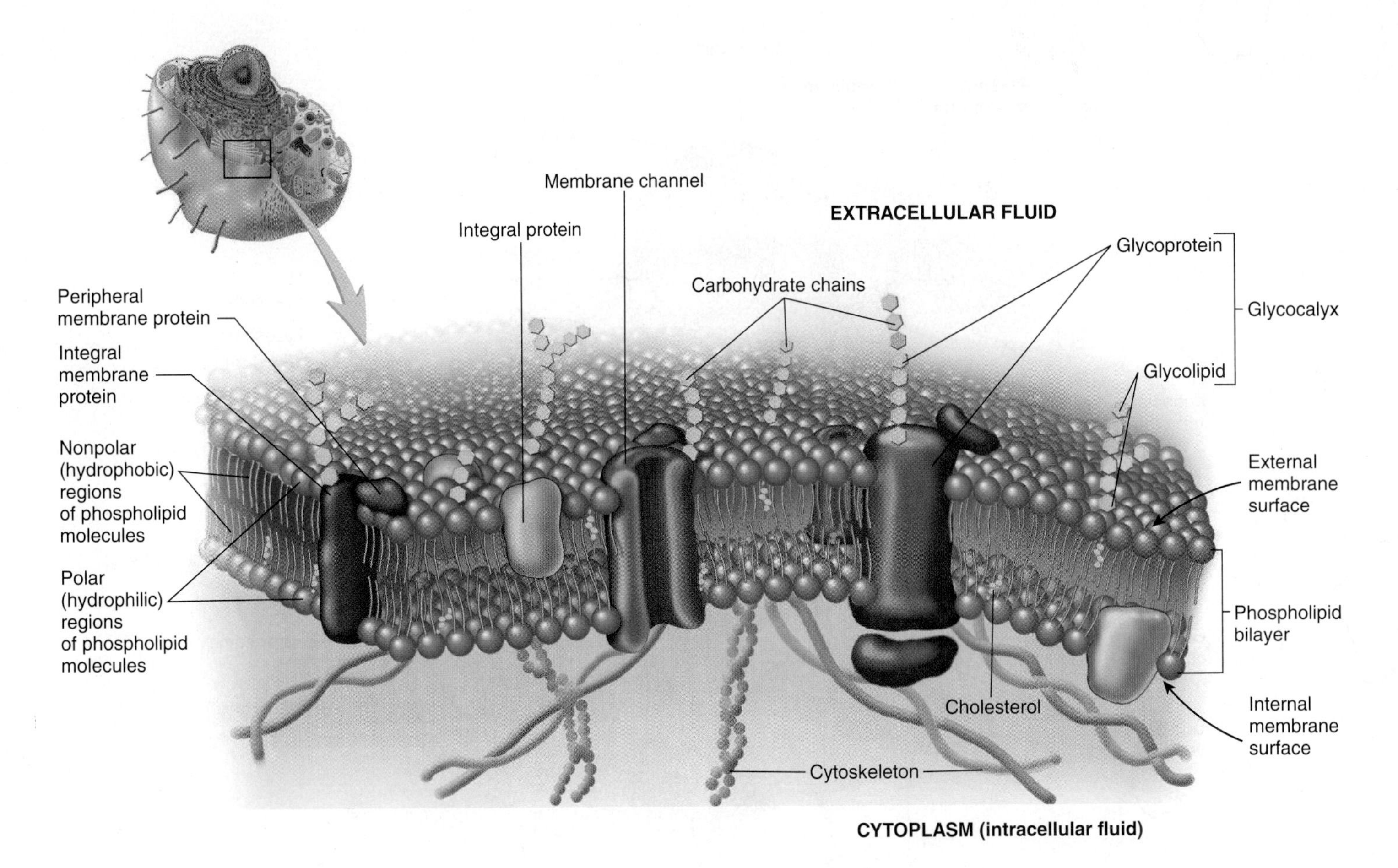

Figure 3.2 Plasma Membrane

membrane. If a substance were to be removed from the cell by simply opening a hole in the plasma membrane, the cell would probably burst. The vesicle fuses with the plasma membrane, ejecting the larger molecules without disrupting the plasma membrane. Two specialized vesicles in the cytoplasm are **lysosomes** and **peroxisomes.** Lysosomes digest material with enzymes in a process known as phagocytosis. Therefore, they are sometimes known as phagocytic vesicles. Peroxisomes use enzymes to convert potentially toxic hydrogen peroxide to water and oxygen. Hydrogen peroxide is formed in cells by the metabolism of fatty acids and amino acids, as well as by the interaction of water with an unstable form of oxygen known as oxygen free radicals.

Ⓐ2 Examine figure 3.1 and find the cellular structures listed in this exercise. Describe their functions in your own words and, afterward, compare these descriptions to table 3.1.

Other Cellular Components

Additional structures occur in the cell that do not fall neatly into the category of plasma membrane, nucleus, or cytoplasm. Cilia and flagella are two such structures. These structures extend from the body of the cell and consist of microtubules covered by the plasma membrane. Most cells have cilia, but flagella are found only in the sperm cells of humans. Cilia are found on cells that are involved in movement, such as the movement of mucus along the free edge of the respiratory passage or on the inner wall of uterine tubes. Cilia and flagella have a similar structure, but cilia are shorter than flagella.

Centrosomes are also unique cellular structures. They are typically found close to the nucleus and contain centrioles, which play a part in microtubule formation. They also are involved in the formation of a structure known as the spindle apparatus, which is associated with cellular division.

Finally, microvilli are small extensions of the surface of some cells that are involved in the absorption of material (such as the microvilli of the digestive or urinary system).

The Cell Cycle

One of the great wonders of science is the mechanism by which a single cell, the result of the fusion of egg and sperm, develops into a complex, multicellular organism, such as a human. Various estimates put the total number of cells in the human body in the trillions. All of these cells come from the first cell, or zygote. In this part of the lab exercise, you examine the mechanism by which this occurs.

Most cells produce more cells by a process known as the cell cycle. The cell cycle can be divided into two stages: **interphase** and **cell division.** These events are illustrated in figure 3.4. Note that, for most of the time in the average life of a cell, it is in interphase.

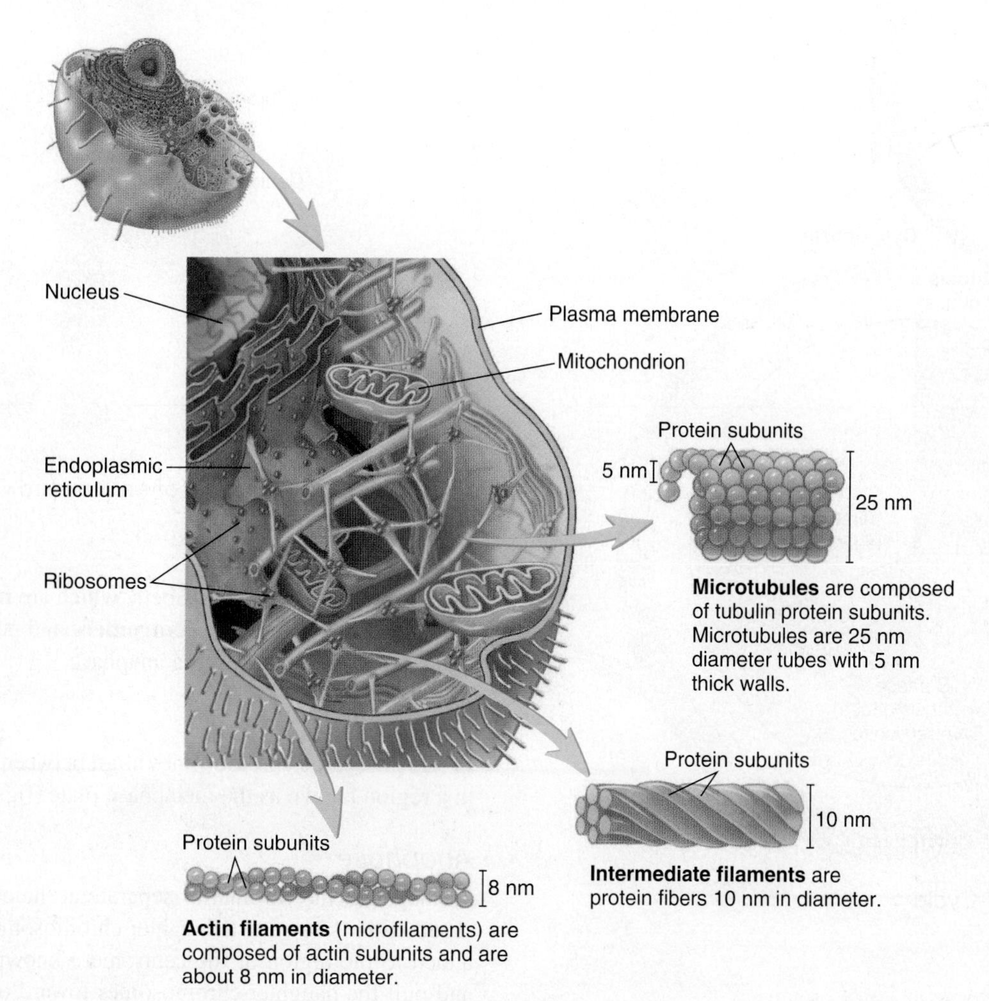

Figure 3.3 Cytoskeleton

Interphase

Interphase is the time when a cell undergoes growth and duplication of DNA in preparation for the next cell division. If a cell is not going to divide any further (as with brain cells and some muscle cells), then interphase is regarded as the time when a cell carries out normal cellular function.

Interphase has three separate phases known as the G_1 phase, S phase, and G_2 phase. In the G_1 phase (*G* stands for *gap*), cells are growing in size and producing organelles. In the S phase (*S* stands for *synthesis*), the DNA of the cell is duplicated. The double helix of the DNA molecule unzips and two new, identical DNA molecules are produced. In the final phase of interphase, the G_2 phase, the cell continues to grow and prepares for the process of mitosis. Some cells, such as muscle cells and brain cells, do not undergo further division and are said to be in the G_0 (G zero) phase. Cells in interphase have a distinct nuclear envelope, and the genetic information is dispersed in the nucleus as chromatin.

TABLE 3.1 Cellular Structures and Their Functions in Cells

Structure	Organization	Function
Mitochondrion	Double membrane, rod-shaped with cristae	ATP production, fatty acid oxidation
Ribosome	No membrane, rRNA subunits	Protein production
Rough endoplasmic reticulum	Single membrane, enclosed channels with ribosomes	Protein production for export
Smooth endoplasmic reticulum	Single membrane, enclosed channels without ribosomes	Lipid and steroid synthesis, detoxification
Golgi apparatus	Single membrane, cisternae as flattened sacs forming vesicles	Assembly of macromolecules, transport from cell for secretion
Lysosome	Single membrane, contains hydrolytic enzymes	Digestion of material
Peroxisome	Single membrane, contains peroxidase	Conversion of H_2O_2 to $H_2O + O_2$
Nucleus	Double membrane with nuclear pores, chromatin and nucleoli	Storage of genetic information, regulation of cellular activity

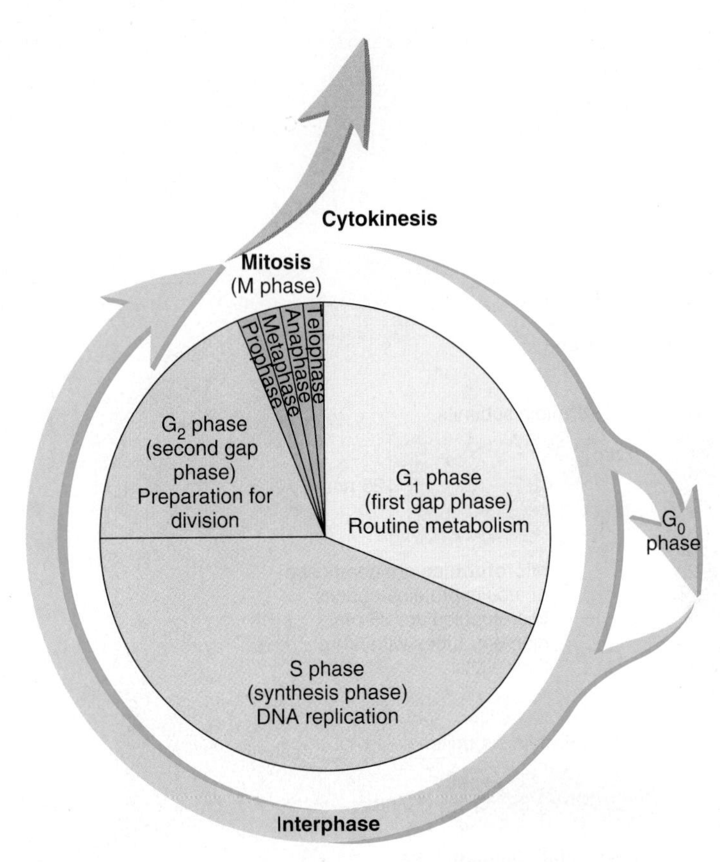

Figure 3.4 Cell Cycle

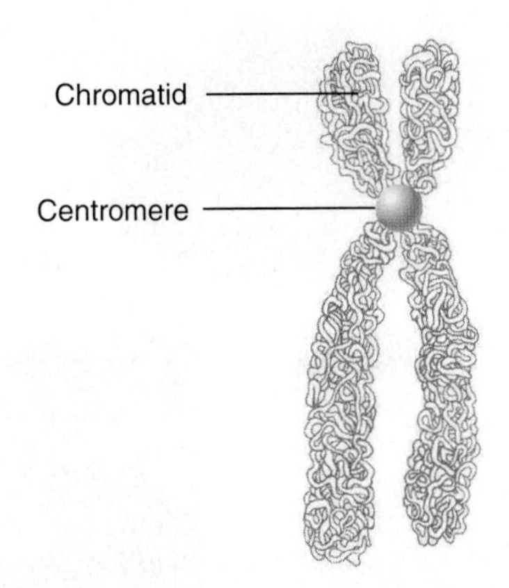

Figure 3.5 Structure of an Isolated Chromosome

Mitosis

Mitosis is a continuous event divided into four distinct phases. Mitosis is nuclear division, and it involves the division of genetic information to produce two identical nuclei. For mitosis to occur, the chromatin in the nucleus of the cell must condense into compact units called chromosomes. A chromosome consists of two chromatids held at the center by a centromere. The two chromatids are identical and result from the duplication of DNA. Examine figure 3.5 for the structure of a chromosome. You should also note the structure of chromosomes as you study the cells undergoing mitosis.

The four phases of mitosis—**prophase, metaphase, anaphase,** and **telophase**—are described next. Refer to figure 3.6 as you read the descriptions.

Prophase

Examine a cell in interphase and note that there are no distinct chromosomes visible (figure 3.6a). The first indication that a cell is undergoing mitosis is the condensation of chromatin into chromosomes (figure 3.6b).

In addition to the thickening of the chromosomes, the nucleolus disappears and the nuclear envelope begins to disassemble. In order for the chromosomes to separate and move away from each other the nuclear envelope, which normally forms a barrier, must not be present. During prophase, the mitotic apparatus becomes apparent. The **mitotic apparatus** consists of two **asters,** which are points of radiating fibers at each end (pole) of the cell, **centrioles** in the middle of the asters, and **spindle fibers,** which attach to chromosomes. The spindle fibers, which are **microtubules,** attach to the chromosomes at the **centromere** and pull the chromosomes to the poles of the cell during anaphase.

Metaphase

In this phase, the chromosomes align between the poles of the cell in a region known as the metaphase plate (figure 3.6c).

Anaphase

In anaphase, the chromatids separate at the centromere, and each chromatid is known as a daughter chromosome. The spindle fibers attach to the region of the centromere known as the kinetochore and pull the daughter chromosomes toward opposite poles of the cell. The centromere region moves first, and the arms of the chromosomes follow (figure 3.6d).

Telophase

Once the daughter chromosomes reach the poles, telophase begins (figure 3.6e). The daughter chromosomes begin to unwind into chromatin, the nucleolus reappears, and the nuclear envelope begins to re-form. The mitotic apparatus disassembles, thus terminating mitosis.

Cytokinesis

The splitting of the cell's cytoplasm into two parts is known as cytokinesis. Although cytokinesis is a distinct process, it frequently begins during late anaphase or early telophase. In late anaphase, as the chromosomes are moving to the poles, the plasma membrane begins to constrict at a region known as the cleavage furrow. This begins the process of dividing the cytoplasm (figure 3.7) as the cell splits into two separate daughter cells. The cytoplasm and the organelles are effectively divided into two parts.

(A) 3 Examine a slide of whitefish blastula and look for the various phases of the cell cycle in those cells. Most of the cells that you see are in a particular part of the cell cycle. What is this phase, and why are most of the cells in this phase?

____

____

____

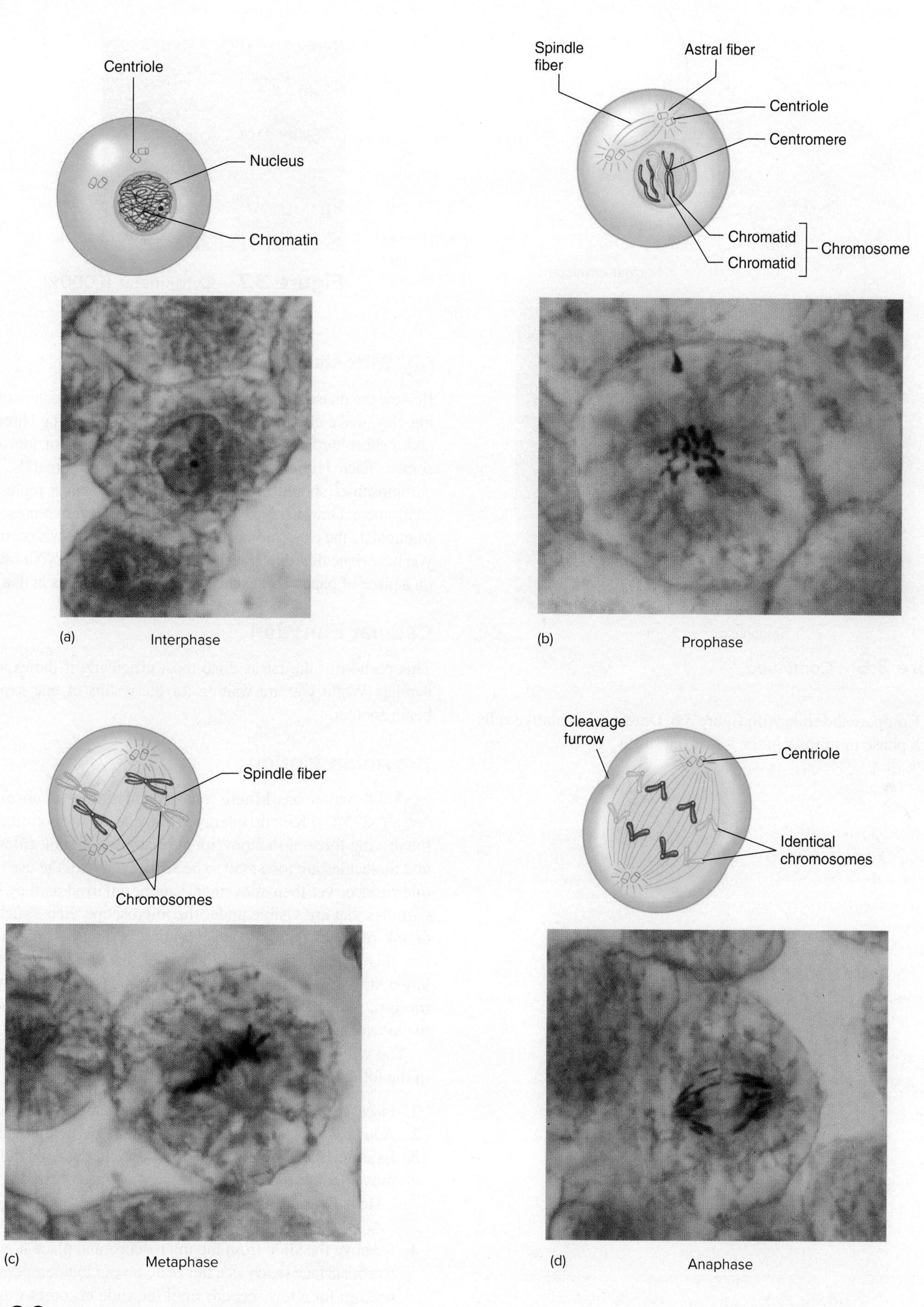

Figure 3.6 The Cell Cycle (1,000×)

Compare the stages of each part of the cell cycle.

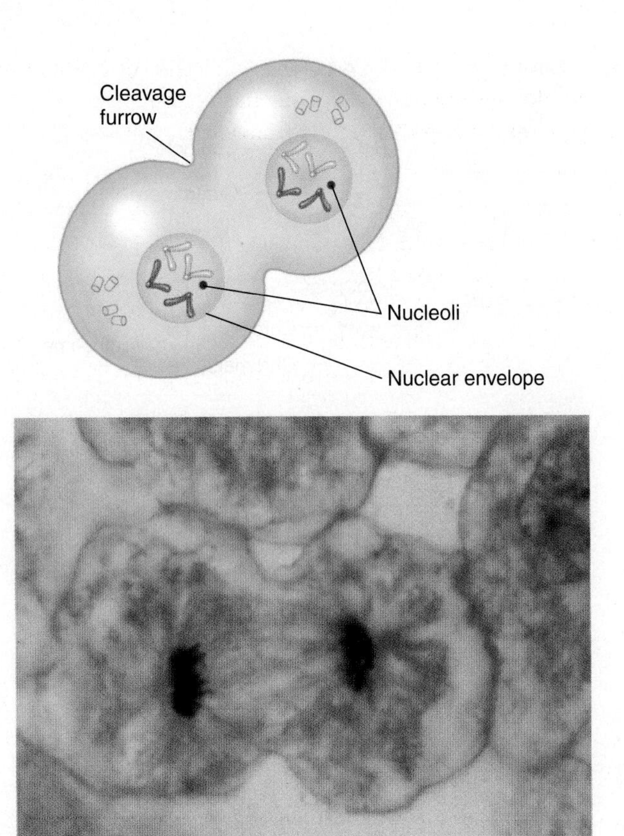

Figure 3.6 *Continued.*

Compare the slide with figure 3.6. Draw representative cells in each phase of mitosis in the following space.

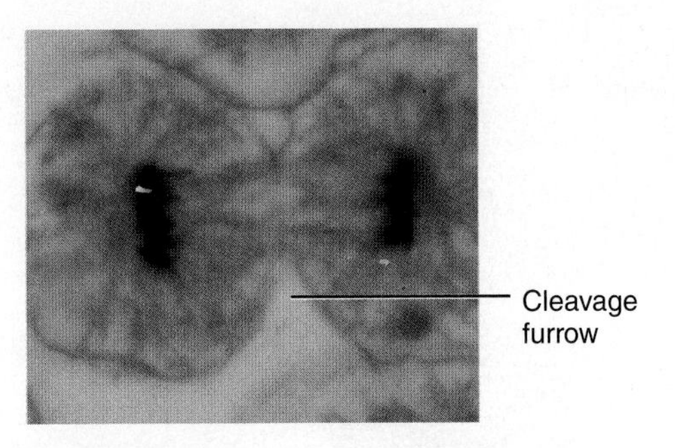

Figure 3.7 Cytokinesis (1,000×)

(A)[4] Mitosis Activity

Review the phases of mitosis in table 3.2. With two colors of modeling clay, make chromosomes. You should have a long chromosome and a short chromosome of each color, for a total of four chromosomes. Each chromosome should have two chromatids, and the chromosomes should be joined by a marble, which represents the centromere. Draw a large circle on a sheet of paper to represent a cell. Manipulate the clay chromosomes to show how mitosis occurs. After you have done this, describe the process of mitosis in your own words on a piece of paper. List each stage and what happens in that stage.

Cellular Function

This portion of the lab is done most efficiently if the experiments overlap. While you are waiting for the results of one experiment, begin another.

Brownian Motion

(A)[5] All matter has **kinetic energy** unless it is at absolute zero (–273°C). Kinetic energy is the energy of motion, and it is the driving force of the movement of atoms and molecules. Atoms and molecules are too small to be seen, even with the use of a light microscope, yet their movement can be inferred as they hit large particles that are visible under the microscope. Items such as dust or ink particles can be seen vibrating or jiggling, and we interpret this as the collective collisions of many molecules striking a larger structure and causing it to move. This is named **Brownian motion,** after botanist Robert Brown, who first described this in the nineteenth century.

You will observe Brownian motion by examining ink particles in the following activity:

1. Place a drop of India ink on a clean glass microscope slide.
2. Add a drop of water and carefully set a coverslip on the slide.
3. Examine the slide under high power and describe the movement of the ink particles in the space provided.
   Movement of ink particles: ______________________________
   ____
4. Remove the slide from the microscope and place it on a warm surface (such as a hot plate on the low-temperature setting) for a few seconds until the slide becomes warm. If you leave it on the hot plate for too long, the water will evaporate and you will see no motion at all.

TABLE 3.2	**Major Events of Mitosis**
Prophase	Chromatin condenses to form visible chromosomes.
	Nuclear envelope disappears.
	Spindle apparatus forms.
	Nucleolus disappears.
Metaphase	Chromosomes align on the metaphase plate.
Anaphase	Chromosomes split and daughter chromosomes migrate to poles; cytokinesis often begins.
Telophase	Chromosomes reach poles; nuclear envelope re-forms.
	Chromosomes unwind to chromatin; cytokinesis divides the cytoplasm.
	Nucleolus reappears.

5. Quickly return the slide to the microscope and note the movement of the particles, compared with the initial observation.
6. Describe any difference in the space provided. ______

______
7. How might kinetic energy play a part in the difference between the first observation and the second observation?

______

______

Diffusion

Kinetic energy moves particles in solution or in a gas, and you can see that a concentration of particles would be struck by chance collisions with some of those particles beginning to disperse. This process is known as **diffusion** and can be defined as the movement of particles from regions of high concentration to regions of low concentration. The difference between the two concentrations is known as the **concentration gradient.** Molecules move down the concentration gradient. If the particles become uniformly dispersed, then the system has reached **equilibrium.** An example of the essential nature of diffusion is the movement of oxygen into the blood vessels of the lungs. Oxygen in the air is at a higher concentration than in the blood of the lung capillaries; consequently, oxygen moves from the air to the blood.

Many solutions consist of a liquid portion, the **solvent,** and the dissolved portion, or **solutes.** If the solutes are concentrated in one area, they will diffuse in the solvent. There are many solvents, but water is a vital solvent for the body, as it is the most abundant solvent and frequently contains sugars, ions, and amino acids.

(A) 6
1. You can demonstrate diffusion by placing a crystal of potassium permanganate in a 100 mL beaker of water. Do this as a group at each table.
2. Fill a 100 mL beaker almost to the top with tap water.
3. Place the beaker on your table and drop a small crystal of potassium permanganate into it.
4. Leave the beaker *undisturbed,* but note the changes that occur after an hour.
5. Record your observations in the space provided. While you are waiting, continue with the other experiments.
Description of potassium permanganate diffusion:

______

______

Many factors can affect the rate of diffusion, including changes in temperature, changes in concentration of the solute, size or weight of the solute particles, and interactions between the solute and the solvent. In the following two experiments, you will examine the effects of the weight of the particle and of temperature on the diffusion rate.

Caution!

Dyes such as potassium permanganate and methylene blue stain clothes, skin, and lab notebooks!

Diffusion Rates and Particle Weight

Unless your instructor states otherwise, do this experiment as a group of three or four students. Agar, a gel, can be used as a medium in which to measure diffusion rates of materials. Agar consists of water and algal polysaccharides. The liquid nature of agar is such that diffusion occurs in the gel at a slow rate.

1. Using a plastic drinking straw and a petri dish filled with agar, gently make two stabs into the dish opposite one another (figure 3.8). Do not twist the straw, or you risk breaking the agar and leaving a crack into which fluid will run.
2. Remove the small plug of agar with a fine probe if needed so that a well is left in the agar.
3. Into one of the wells, place 3 drops of 0.01 M potassium permanganate solution (molecular weight 158).
4. Into the other well, place an equal amount of 0.01 M methylene blue solution (molecular weight 320).
5. Potassium permanganate is a purple solution; methylene blue is blue.

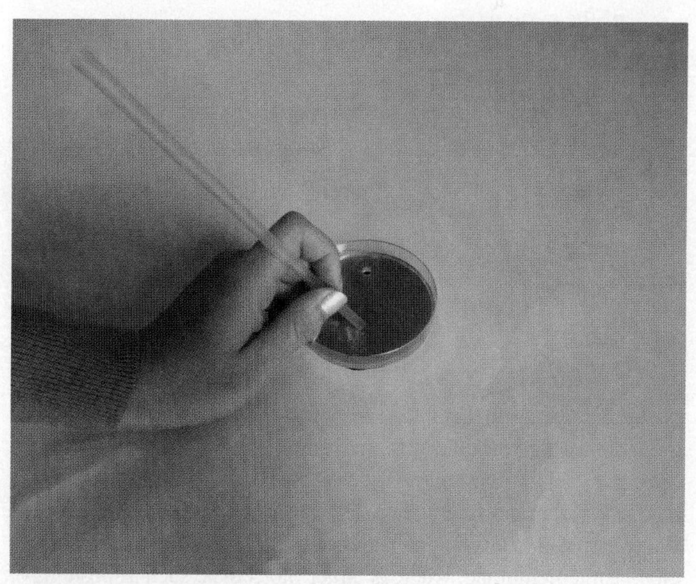

Figure 3.8 Placement of the Wells in Agar

Leave the petri dish on your desk and make observations of the diffusion rate every 20 minutes for 80 minutes. Record the diameter of the diffusion in chart 3.1 and continue with the following experiments.

CHART 3.1	Diffusion Rate and Particle Weight Diameter of Diffusion (in mm)			
	20 min	40 min	60 min	80 min
Potassium permanganate	_____	_____	_____	_____
Methylene blue	_____	_____	_____	_____

Effects of Temperature on Diffusion Rates

A[8] 1. Prepare two more petri dishes in the same way, except this time take one of the petri dishes from a refrigerator and another from a warming tray (such as an electric warming tray on the lowest setting or in an incubator set at 37°C).

2. Make two wells in the cold petri dish and place three drops of the respective dyes, as you did previously.
3. This time, however, place the petri dish in a dish pan filled with crushed ice and examine after 80 minutes. You only need to record the diameter of diffusion at the end of the 80 minutes.
4. Remove a petri dish from the warming tray or incubator, make two wells in the dish, fill them with appropriate solutions, and return the dish to the warming tray or incubator. Examine after 80 minutes.
5. Record your results in chart 3.2.

Compare the diffusion rates of the cold and warm temperatures with the dish left at room temperature for 80 minutes. How does an increase or a decrease in temperature affect the diffusion rate? ____________________

____

What can you say about temperature and the kinetic energy of the system? ____________________

____

CHART 3.2	Diameter of Diffusion (in mm)	
	80 min (Cold)	80 min (Warm)
Potassium permanganate	_____	_____
Methylene blue	_____	_____

Osmosis

In the preceding diffusion experiments, there was no barrier to the movement of the particles. Plasma membranes are barriers to certain molecules, while they allow other molecules to pass through. This type of membrane is called a **selectively permeable membrane.** Water, some alcohols, oxygen, and carbon dioxide move easily across the plasma membrane, whereas larger molecules, such as proteins or charged particles (ions), are prevented from crossing the membrane.

The movement of water across a selectively permeable membrane from regions of higher water concentration (for example, more pure water) to lower water concentration (for example, less pure water) is known as **osmosis.** Osmosis is a type of diffusion, and the process can be viewed from the perspective of the solvent or the perspective of the solute.

The Solvent Perspective of Osmosis

In diffusion, material moves from higher concentrations to lower concentrations. The same can be seen with osmosis except that you measure the movement of water instead of the movement of solutes. If you have two solutions separated by a selectively permeable membrane, a 10% sugar solution (or 90% water) and a 5% sugar solution (or 95% water), then the water will move from higher water concentration (95% water) to lower water concentration (90% water). The greater the difference between the two solutions, the greater the concentration gradient, which, in this case, is known as the **osmotic potential.**

The Solute Perspective of Osmosis

Even though the movement of water is measured in osmosis, it is the concentration of solute particles that influences that movement. A 10% sugar solution has more solutes than a solution of 5% sugar. The 10% sugar solution is said to be **hypertonic** to the 5% sugar solution. The 5% sugar solution is said to be **hypotonic** to the 10% sugar solution. If these two solutions are separated by a selectively permeable membrane, then water flows *from* the hypotonic solution *to* the hypertonic solution. If enough water flows across the membrane and the two solutions reach the same concentration of sugar, an *equilibrium* is established and the net movement of water stops. If solutions have the same concentration of solutes, they are said to be **isotonic** to one another.

Demonstration of Osmosis

A[9] Observe osmosis with an osmometer. This device consists of a dialysis bag attached to a long tube. The tube is filled by using a funnel with dark corn syrup or sugar solution. Sugar does not cross dialysis membranes, but water does, so the dialysis membrane is a selectively permeable membrane. The osmometer is placed in a beaker of water and clamped to a ring stand (figure 3.9). The level of the corn syrup or sugar solution is indicated with a mark from a permanent marker.

Examine the setup throughout the lab period. You may find that the liquid in the tube eventually stops rising. This occurs when the gravitational pressure equals the force exerted by the process of osmosis. The amount of force required to balance, or equilibrate, osmosis is the **osmotic pressure.**

Examine an osmometer (or prepare one if your lab instructor indicates for you to do so) that contains a starch solution.

Does the liquid move up the tube? ____________________

Why or why not? ____________________

____

What does this say about the osmotic activity of starch?

____

Figure 3.9 Osmosis Demonstration

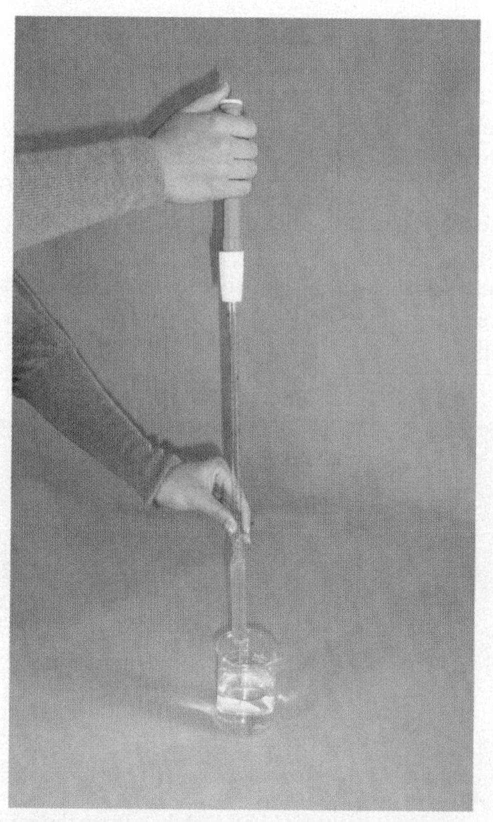

(a)

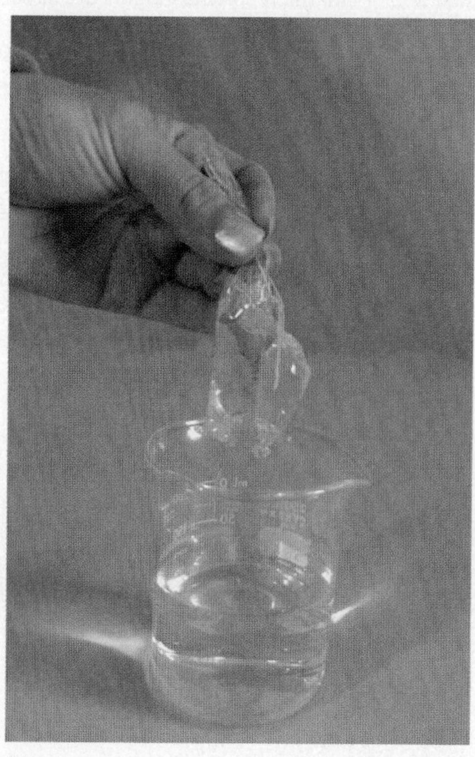

(b)

Figure 3.10 Dialysis Bags

(a) Filling; (b) tying.

Osmosis and the Concentration Gradient

In this experiment, you determine the effects of various concentrations of a sucrose solution on the rate of osmosis by studying the change of weight of bags made from dialysis tubing. This tubing is normally used to help kidney patients as it allows waste products to be removed from the body while retaining red blood cells.

(A) 10

1. Take four dialysis bags that have been soaking in water back to your desk. Secure the bottoms of the bags by tying them with a string or closing them with clamps.
2. Put a pipette pump or pipette bulb on the end of a 10 mL pipette. Draw up 10 mL of a 15% sucrose solution and fill the dialysis bags. Remove as much air as you can from the bags. Tie or clamp the open ends of each bag (figure 3.10).
3. Rinse each bag in distilled water and blot with a towel.
4. Weigh each bag to the nearest tenth of a gram and record the weights in chart 3.3. This weight is the **initial weight** of the bag.
5. Label the beaker with a wax pencil or permanent marker.

CHART 3.3	Initial Weight of Dialysis Bags
Bag 1	______________ grams
Bag 2	______________ grams
Bag 3	______________ grams
Bag 4	______________ grams

6. Place bag 1 in a beaker filled about two-thirds to the top with a 0% sugar solution. Make sure that all the bags are covered with the solution. Leave it there for 20 minutes.
7. Place bag 2 in a 5% sugar solution and leave it for 20 minutes.
8. Place bag 3 in a 15% sugar solution and leave it for 20 minutes.
9. Place bag 4 in a 30% sugar solution and leave it for 20 minutes.
10. Remove the bags from the beakers after 20 minutes, and blot and weigh each bag.
11. Remember to place each bag back in its proper solution! Record the weight of the bags each 20 minutes for a total of 80 minutes in chart 3.4.

CHART 3.4 Weight of Bags (in g) for Each Time Period

Bag	Time			
	20 min	40 min	60 min	80 min
1	______	______	______	______
2	______	______	______	______
3	______	______	______	______
4	______	______	______	______

Calculate the change of weight of each bag from the initial weight for each of the time periods. For example, let's assume a bag initially weighed 20.5 g and the recorded weights are as follows:

20 min	40 min	60 min	80 min
23.5 g	24.2 g	25.0 g	25.6 g

The change in weight is as follows:

3.0 g	3.7 g	4.5 g	5.1 g

Graph the change in weight for each bag of your experiment in chart 3.5 and connect the points with a line. Indicate which line represents which solution.

Which of the bags (if any) gained weight? ______________

Which of the bags (if any) lost weight? ______________

Determine the osmotic relationship (hypertonic, hypotonic, isotonic) of the bag to the solution in the beaker. Record your answer in chart 3.6.

Does the change in weight correlate to what you know about osmosis? If so, how does it correlate? ______________

____

____

Caution!

Make sure you wear protective gloves while conducting the next experiment to avoid any potential transmission of disease.

CHART 3.6 Osmotic Relationship

Bag 1: The bag solution is ______________ to the beaker solution.

Bag 2: The bag solution is ______________ to the beaker solution.

Bag 3: The bag solution is ______________ to the beaker solution.

Bag 4: The bag solution is ______________ to the beaker solution.

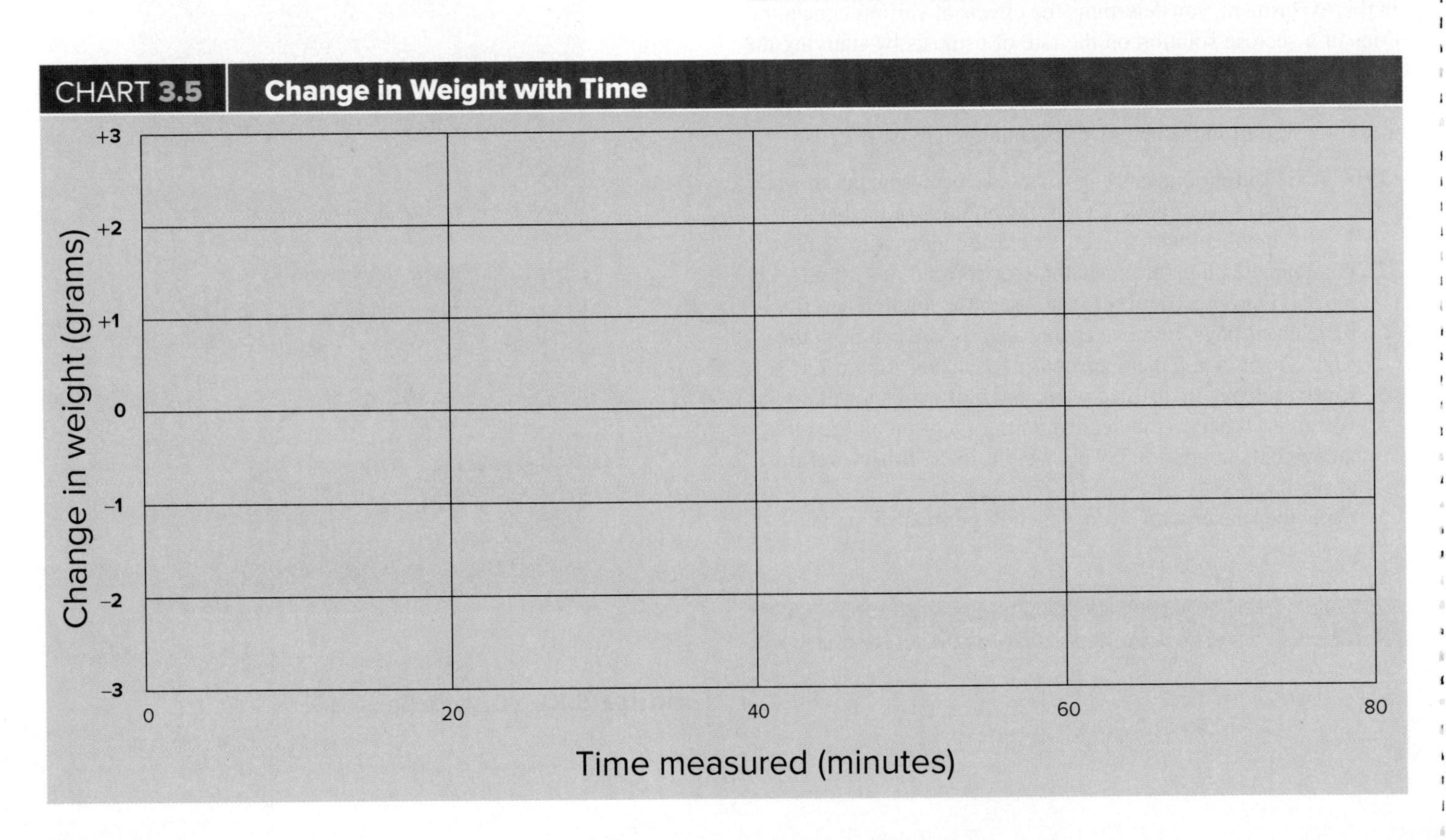

Osmosis and Living Cells

The importance of isotonic solutions can be demonstrated by the following procedure:

(A)[11] 1. Mark each slide with a permanent marker.
2. Place a drop of fresh mammal blood on a slide.
3. To this slide, add a drop of physiological saline (0.9% sodium chloride) and place a coverslip on the slide.
4. Observe the cells on high power under the microscope and note their shape.
5. Describe this in the space provided. ____________
6. Place another drop of blood on slide #2.
7. To this slide, add a drop of 5% sodium chloride (NaCl) solution.
8. Place a coverslip on the slide and record your observations.

____

9. Examine the slide for at least a few minutes or until a change of shape becomes obvious.

Description of the shape of the cells in 5% NaCl: ________

10. For slide #3, add a drop of blood and a drop of distilled water. Immediately observe this slide and then continue to look at it for a few minutes. Record your observations.

Description of the shape of cells in distilled water: ________

When red blood cells lose water, they undergo a process known as **crenation.** The loss of water causes the membrane to look wrinkled.

Did any of the cells show crenation? ____________

Which solution might produce this effect? ____________

When water moves into a cell at a rapid rate, the cell becomes inflated and sometimes bursts in a process known as **hemolysis.**

Did any of the cells undergo hemolysis? ____________

Which solution might produce this effect? ____________

Examine figure 3.11 for the various effects of solutions on red blood cells.

Case Study—Water Balance

Electrolyte imbalance can lead to headache, confusion, and even death. In some cases, participating in a water drinking contest or drinking large amounts of pure water after intensive exercise causes such an imbalance. Water absorbed in the blood increases the flow of water to the brain cells, putting pressure on vital centers and causing them to malfunction.

A 28-year-old woman who is in the final stages of running a marathon on a hot day has regularly consumed water along the path. She begins to have a severe headache and shows symptoms of being confused. She has been sweating profusely, losing electrolytes and, to compensate, has been drinking pure water. Explain the osmotic nature of what is happening in relationship to the movement of water in her blood and brain cells and what should be done to correct the problem. ____________

____

____

____

____

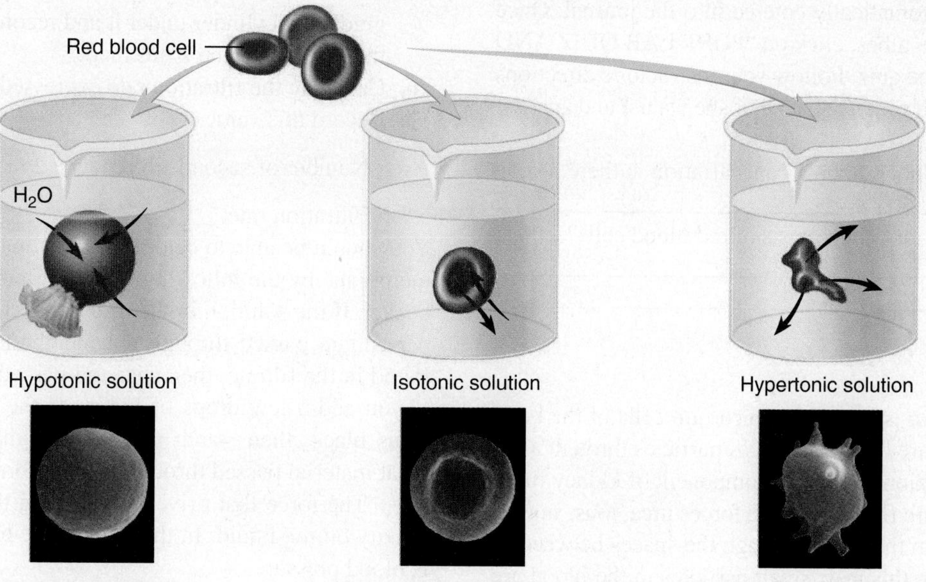

(a) When a red blood cell is placed in a hypotonic solution (one having a low solute concentration), water enters the cell by osmosis (*black arrows*), causing the cell to swell or even burst (lyse; *puff of red in lower part of cell*).

(b) When a red blood cell is placed in an isotonic solution (one having a concentration of solutes equal to that inside the cell), water moves into and out of the cell at the same rate (*black arrows*). No net water movement occurs, and the cell shape remains normal.

(c) When a red blood cell is placed in a hypertonic solution (one having a high solute concentration), water moves by osmosis out of the cell and into the solution (*black arrows*), resulting in shrinkage (crenation).

Figure 3.11 Stages of Red Blood Cells

(A)[12] Virtual Lab—

Open the Ph.I.L.S. program (Ph.I.L.S. 4.0) and select the first simulation, "Osmosis and Diffusion 1. Varying Extracellular Concentration." Read the Objectives and Introduction and take the pre-lab quiz to make sure that you understand the principles of the exercise. Use the scroll bar on the right-hand side of the screen to control the text. Once you have finished reading the lab, either click "Continue" or click on the "PRE-LAB QUIZ" at the top of the screen. Once you have finished the pre-lab portion you can select the "WET LAB" to run the experiment. In the wet lab experiment the use of transmitted light in a spectrophotometer is the way that osmotic relationships are determined between the red blood cell and the solution surrounding the red blood cell. The more light that is blocked, the less transmission of light. If red blood cells burst, then the transmission of light increases as there are fewer cells absorbing the light.

To begin the wet lab, read and follow the instructions. When you click the power switch on the spectrophotometer it will turn green. You can then set the spectrophotometer setting by clicking the up or down arrows on the spectrophotometer. In order to load the pipette you must click and hold the arrow on the pipette. Once it is filled you can click and drag the pipette to the appropriate test tubes. Once you do this three times, the rest of the tubes will automatically fill.

You must set the spectrophotometer to zero. Click on the "zero" button of the spectrophotometer (to the left of the calibrate term) to open the chamber. Click and drag the zero test tube to the holder and use the up or down arrows to set the calibration to 100.

You must double-click the test tube holder to remove the tube. The values are automatically entered into the journal. Once you have sampled all the tubes, click on "POST-LAB QUIZ AND LAB REPORT." Take the quiz. Follow your instructor's directions for printing the results (if that is what he or she wants) and analyze your results.

Between what values of NaCl concentration is there a drop in transmittance? ______________________________

How does this relate to hemolysis of red blood cells? _____

____

____

Filtration

The process of **filtration** is important in certain cells of the body and results as the pressure of a fluid forces particles through a filtering membrane. Filtration is a major component of kidney function. Hydrostatic pressure from the blood forces urea, ions, sugars, and other materials from the blood through the spaces between or pores in kidney cells. In this way, small particles in the blood are forced into kidney tubules, while larger molecules, such as proteins, remain in the blood. In this experiment, you learn the basic principles of filtration, such as the **selectivity** of the filtration membrane and the **filtration rate** of a system.

(A)[13]
1. Fold a piece of filter paper in half and then fold it in half again, so that it forms a cone (figure 3.12).
2. Place the cone in a funnel mounted on a ring stand over a beaker.

Figure 3.12 Filtration Apparatus

3. Into this cone filter, add the filtration solution, which is a mixture of copper sulfate, powdered charcoal, and starch in water.
4. Fill the funnel to near the top of the brim and let the filtrate (the material passing through the filter) collect in the beaker.
5. When the funnel is approximately half full, place a 10 mL graduated cylinder under it and record the time it takes to fill the cylinder to the 2 mL mark.
6. Calculate the filtration rate expressed as mL/minute.
7. Record this value.

Number of seconds to produce 2 mL: ______________

Filtration rate: ______________________ mL/min

You should be able to determine what material passes through the membrane by the following method. Remove the funnel from the beaker. If the solution in the beaker has a blue cast to it, then copper sulfate passed through the membrane. If black particles are found in the filtrate, then charcoal passed through the membrane. If you add a few drops of iodine to the beaker and the solution turns black, then starch passed through the membrane. Record what material passed through the membrane in chart 3.7.

The force that drives filtration in the funnel is the force of gravity on the liquid. In the kidney, the force that drives filtration is blood pressure.

CHART 3.7 Filtration

Substance	Yes	No
Copper sulfate	_____	_____
Activated charcoal	_____	_____
Uncooked starch	_____	_____

Exercise 3 Review

Name: ______________________

Lab time/section: ______________________

Date: ______________________

Cell Structure and Function

1. Cells in the body have a fluid surrounding them. What is the name of this fluid? ______________________

2. The cytoplasm has a liquid portion. What is it called? ______________________

3. Name the major parts of the cell. ______________________

____

4. What structure on the outer surface of a cell is composed mostly of a phospholipid bilayer? ______________________

5. Which organelle is responsible for ATP production? ______________________

6. Which organelle makes protein for use outside the cell? ______________________

7. Which organelle in the cell produces lipids? ______________________

8. The DNA that controls most of the function of the cell is found in what structure? ______________________

9. What cellular structure is responsible for ribosome production? ______________________

10. What are the two stages of the cell cycle? ______________________

11. Of the two stages of the cell cycle, in which one is DNA duplicated? ______________________

12. In what part of the cell cycle do chromosomes first split apart? ______________________

13. What part of the cell cycle involves the division of the cytoplasm? ______________________

14. Describe the four phases of nuclear (mitotic) division and what occurs during those phases. ______________________

____

____

____

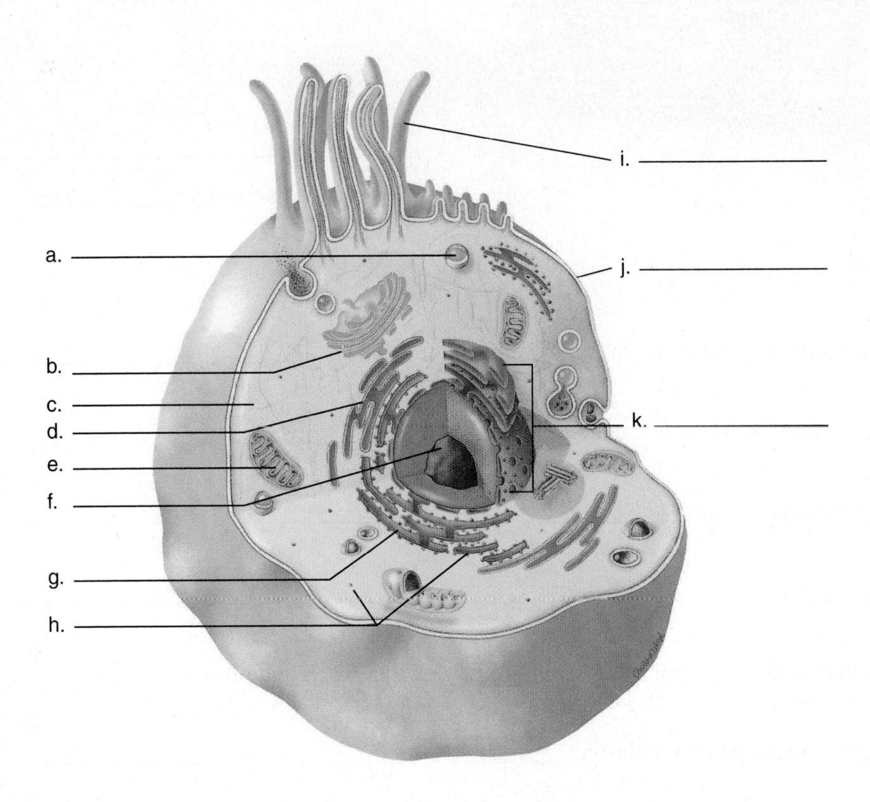

15. Name the cellular structures seen in the illustration using the terms provided.

cilia	cytoplasm	Golgi apparatus	mitochondrion
nucleolus	nucleus	plasma membrane	ribosome
rough endoplasmic reticulum	smooth endoplasmic reticulum	vesicle	

16. Name the phases of interphase and describe what happens during those phases.

____

____

____

17. Particles that are visible in the light microscope and move erratically by molecular collision represent what process? ___________

____

18. Particles in a gas or liquid moving from a region of higher concentration to a region of lower concentration represent what process?

____

19. Name the events of the cell cycle as illustrated. Use the terms provided.

anaphase interphase metaphase prophase telophase

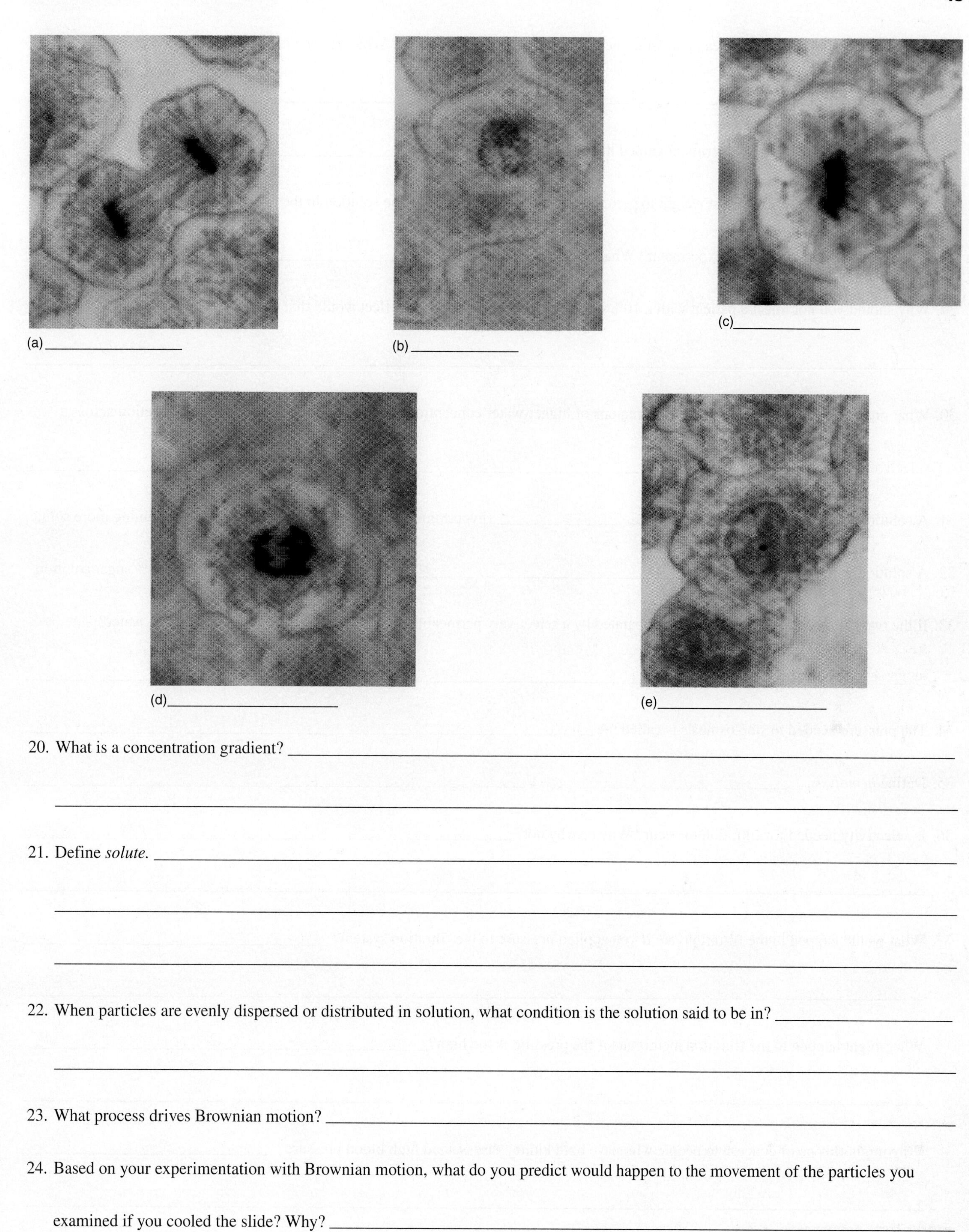

20. What is a concentration gradient? ______________________________

____

21. Define *solute.* ______________________________

____

____

22. When particles are evenly dispersed or distributed in solution, what condition is the solution said to be in? ____________

____

23. What process drives Brownian motion? ______________________________

24. Based on your experimentation with Brownian motion, what do you predict would happen to the movement of the particles you examined if you cooled the slide? Why? ______________________________

25. What particle diffused the farthest, potassium permanganate or methylene blue? What was the factor in the difference in diffusion rates? ______________________

26. Which osmosis bag in your experiment gained the most weight? ______________________

27. Was the bag that gained the most weight hypertonic, hypotonic, or isotonic to the solution in the beaker? ______________________

28. Which bag lost weight in your experiment? What caused the weight loss? ______________________

29. Why should you not inject a patient with a 10% saline solution? What osmotic effect would that cause? ______________________

____

30. What process occurs as water moves from regions of higher water concentration to regions of lower water concentration across a selectively permeable membrane? ______________________

31. A solution containing less solute is ______________________ (hypertonic/hypotonic/isotonic) to a solution containing more solute.

32. A solution with 5% sugar is ______________________ (isotonic/hypertonic/hypotonic) to a 3% sugar solution.

33. If the two solutions in question 32 were separated by a selectively permeable membrane, which solution would lose water?

____

34. The pressure needed to stop osmosis is called the ______________________.

35. Define *hemolysis.* ______________________

36. Is selectivity needed for filtration to occur? Why or why not? ______________________

____

37. What would happen to the filtration rate if you applied pressure to the filtration system? ______________________

____

What might happen to the filtration membrane if the pressure is too high? ______________________

____

Why might this be of concern to people who have both kidney disease and high blood pressure? ______________________

____

Exercise 4

Tissues

Introduction

The study of tissues, called **histology,** is microscopic anatomy. Individual tissues consist of cells and extracellular material that have a particular function. Organs of the body are formed by two or more tissues. The study of histology is very important because many organic dysfunctions of the human body are diagnosed at the tissue level. Surgical specimens are routinely sent to pathology labs so that accurate assessment of the health of the tissue, and consequently the health of the person, can be made.

Stem cells are unspecialized, or **undifferentiated, cells**. Stem cells may divide to produce more stem cells, or they may **differentiate** to become specialized cells of the body, such as those in neural tissue or blood. Developing embryos contain many stem cells **(human embryonic stem cells),** which differentiate into embryonic tissues. Adult stem cells repair old or damaged tissue or generate new cells, such as blood cells. Therapeutic uses of stem cells may hold cures for diseases (such as Parkinson disease), neural injury, or tissue repair in burn patients. Stem cells are currently used to regenerate white blood cells in patients who have been treated for leukemia. As you study the tissues in this exercise remember that they all arose from stem cells.

There are four primary tissue types in the human body: **epithelial tissue, connective tissue, muscular tissue,** and **nervous tissue.** These are discussed in chapter 4, "Tissues," in the Seeley text.

In this exercise, you examine numerous slides of organ tissue and begin an introduction to histology. In later exercises, you revisit histology as you examine various organ systems.

Objectives

At the end of this exercise, you should be able to

1. recognize the various types of epithelial, muscle, nervous, and connective tissue;
2. associate each tissue with an organ, such as kidney or bone;
3. examine a slide under the microscope or a picture of a tissue and name the tissue represented;
4. draw the three parts of a typical neuron;
5. describe the main muscle cell types according to location and structure.

Materials

Microscope
Colored pencils

Epithelial Tissue Slides

Simple squamous epithelium
Simple cuboidal epithelium
Simple columnar epithelium
Stratified squamous epithelium
Transitional epithelium
Pseudostratified columnar epithelium

Muscular Tissue Slides

Skeletal muscle
Cardiac muscle
Smooth muscle
All three muscle types

Nervous Tissue Slides

Spinal cord smear

Connective Tissue Slides

Loose (areolar) connective tissue
Adipose tissue
Reticular connective tissue
Dense regular connective tissue
Dense irregular connective tissue
Elastic connective tissue
Hyaline cartilage
Fibrocartilage
Elastic cartilage
Ground bone
Cancellous bone
Bone marrow
Blood

Procedure

Before you begin this exercise, you should be thoroughly familiar with the microscopes in your lab. If you are not familiar with the use of the microscope, review Laboratory Exercise 2.

Study Hints

As you examine various tissues look for distinguishing features that will identify the tissue. It is a good idea to examine more than one slide of a particular tissue so that you see a range of samples of that tissue. Frequently, the material you see in the lab is from a slice of an organ and as such includes more than one tissue type. For example, a slide labeled *cartilage* taken from the trachea contains epithelial tissue, adipose tissue, and other connective tissues in addition to the cartilage you want to study. If you are looking for smooth muscle from the digestive tract, you may also find both epithelial tissue and connective tissue in the slide. Use the figures in this exercise to help you locate tissues on the prepared slides. Examine each slide by holding the slide up to the light and visually locating the sample. Then put the slide on the microscope and examine it on low power, scanning around the slide. Move to progressively higher powers after you have identified the tissue. If you cannot identify the tissue after some searching, then ask your lab partner or your instructor for help.

Tissues are usually stained so that the details become visible. The most common stain is hematoxylin and eosin (H&E) stain. Hematoxylin stains the nucleus purple, and eosin colors the cytoplasm pink.

Epithelial Tissue

Epithelial tissue, or epithelium, is a highly cellular tissue, meaning that it is composed mostly of cells, with little matrix (an extracellular material). It covers or lines parts of the body (such as the skin on the outside or the lining of the digestive tract on the inside) or makes up glandular tissue, such as the sweat glands or the pancreas. In most cases, epithelial tissue adheres to the underlying layers by way of a **basement membrane,** which is a noncellular adhesive layer. In an organ, such as the skin, the superficial epithelial tissue is connected by the basement membrane to the underlying dermis. Epithelial tissue is classified according to the shape of the cells at its free surface and the number of layers present. The cell shapes are **squamous** (flattened), **cuboidal,** and **columnar.** The arrangement of layers is **simple** (cells in a single layer), **stratified** (cells stacked in more than one layer), or **pseudostratified** (cells in one layer, though appearing to be in many layers). When you look at epithelial tissue under the microscope, examine the edge of the sample because epithelial tissue is frequently found on free surfaces. Epithelial tissue is listed by type in the following discussion. Examine table 4.1 for an overview of epithelium.

Simple Epithelium

Simple epithelium is only one cell layer thick. The cells are located on the basement membrane, which may adhere to connective tissue, muscle, or even other epithelial tissue. Each epithelium is further classified according to shape.

TABLE 4.1 Epithelial Classification

Layers	Cell Shape
Simple	Simple squamous epithelium
	Simple cuboidal epithelium
	Simple columnar epithelium
Stratified	Stratified squamous epithelium
	Transitional epithelium
Pseudostratified	Pseudostratified columnar epithelium

Simple Squamous Epithelium

This epithelial type is composed of thin, flat cells that lie on the basement membrane like floor tiles. From a side view, these cells look flat, and their resemblance to floor tiles is apparent.

(A)[1] Examine a prepared slide of simple squamous epithelium, which is seen in surface view or side view. Simple squamous epithelium lines the air sacs of lungs; it also lines blood vessels, where it is called **endothelium.** It can be found as the surface layer of many membranes in various body cavities, where it is called **mesothelium.** It provides a slippery surface (as found on the inside of blood vessels), allows filtration, and permits diffusion (as in the lungs). Compare your slide with the photograph in figure 4.1. Draw what you see under the microscope in the space provided. Note whether your slide shows the cell as a surface or side view.

Illustration of simple squamous epithelium:

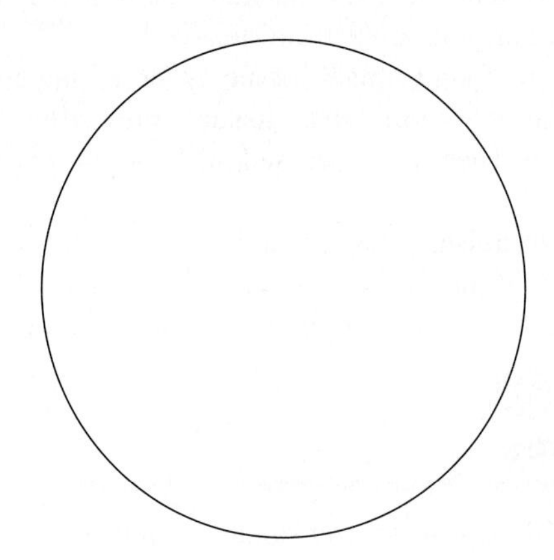

Simple Cuboidal Epithelium

Simple cuboidal epithelium consists of cubelike or wedge-shaped cells that are mostly uniform in dimension. These cells form many of the major glands and glandular organs of the body and are the major tissue in the kidneys. Simple cuboidal epithelium lines tubules. It is often involved in the secretion of fluids (sweat, oil) or in reabsorption (as in the kidneys).

(A)[2] Examine a prepared slide of simple cuboidal epithelium and compare it with figure 4.2. Draw what you see under the microscope in the space provided and label the nucleus of the cell and the basement membrane.

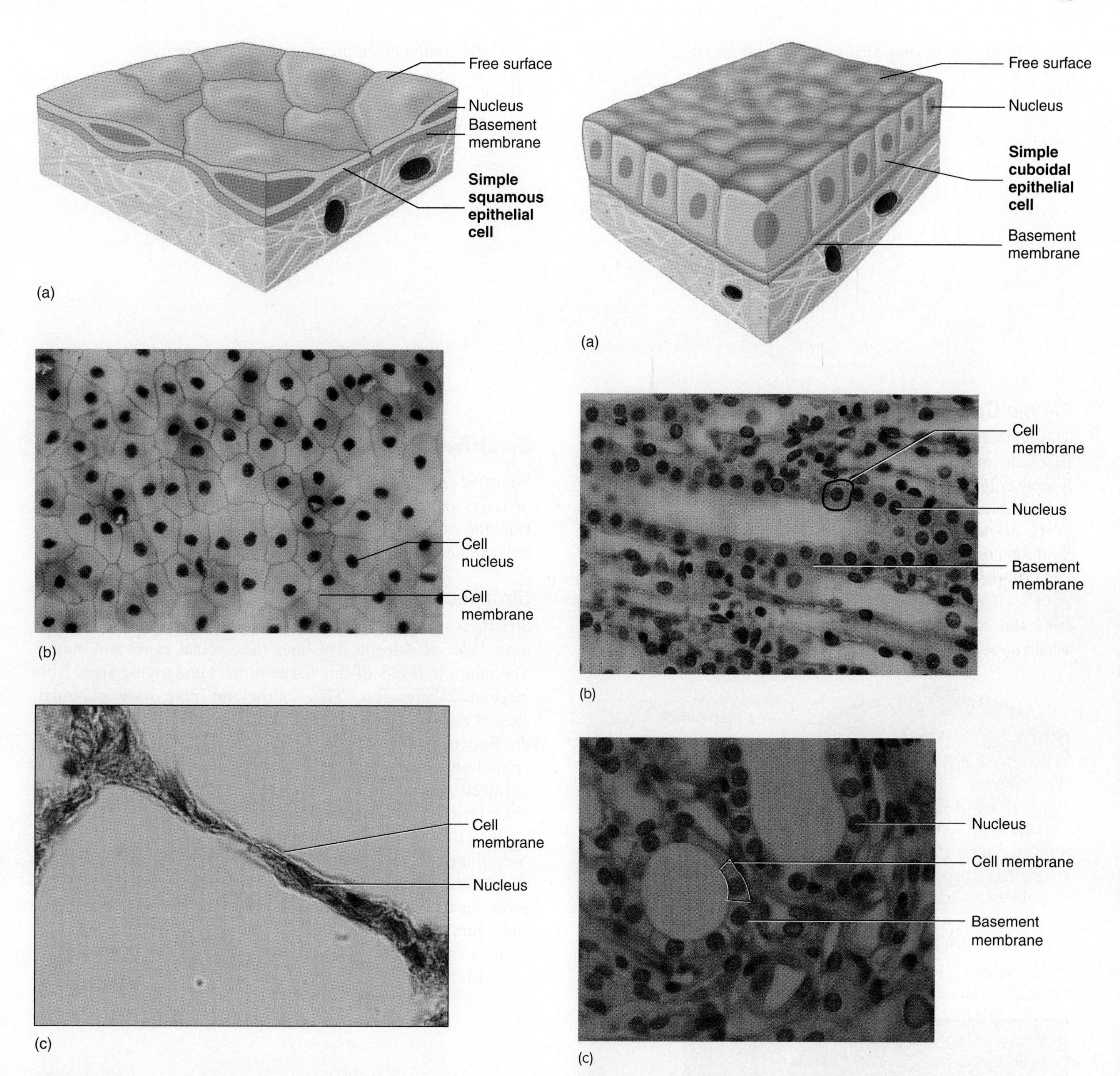

Figure 4.1 Simple Squamous Epithelium

(a) Diagram; (b) photomicrograph of mesothelium, surface view (400×); (c) photomicrograph of lung tissue, side view (1,000×).

Figure 4.2 Simple Cuboidal Epithelium

(a) Diagram; (b) photomicrograph of kidney, long section (400×); (c) photomicrograph of kidney, cross section (400×).

Illustration of simple cuboidal epithelium:

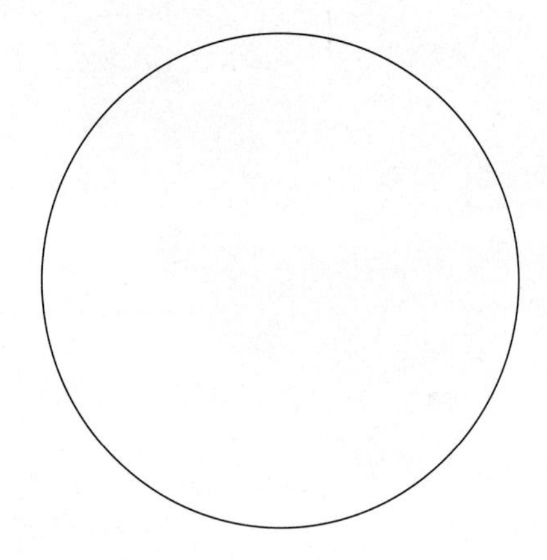

Illustration of simple columnar epithelium:

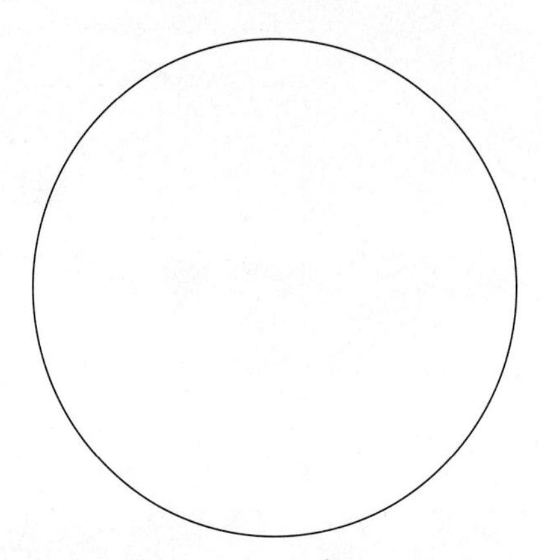

Simple Columnar Epithelium

This epithelium resembles tall columns anchored at one end to the basement membrane. The nuclei are frequently in a row. Simple columnar epithelium can be ciliated on the apical surface or have microvilli. The cells with microvilli line the inner portion of the digestive tract and provide an absorptive area for digested food. Mucus-secreting **goblet cells** are often found with them. Ciliated simple columnar epithelium lines the uterine tubes.

(A) 3 Examine a prepared slide of simple columnar epithelium and compare it with figure 4.3. Draw a representation of what you see in the space provided.

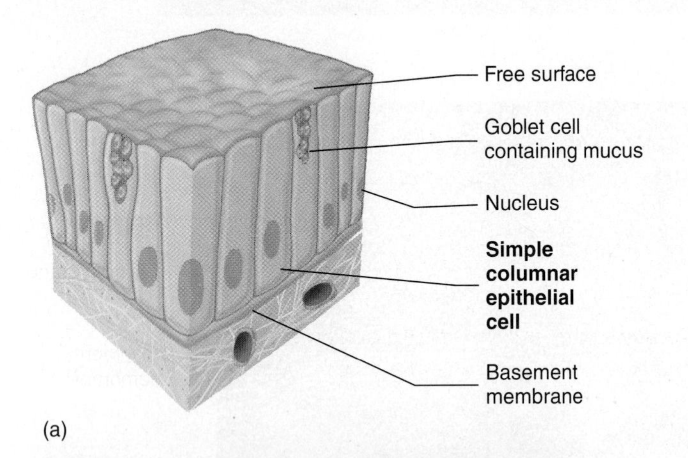

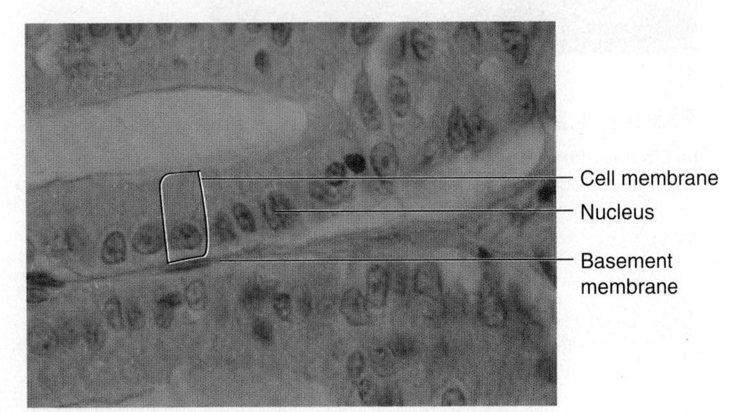

Figure 4.3 Simple Columnar Epithelium
(a) Diagram; (b) photomicrograph of stomach (400×).

Stratified Epithelium

Stratified epithelium is so named because the epithelial cells occur in layers on a basement membrane. The term *stratified* comes from the word *strata* and refers to the layering of these cells. You will examine two common tissues belonging to this group.

Stratified Squamous Epithelium

Stratified squamous epithelium is a tissue that covers the outermost layer of skin. It also lines the vaginal canal and mouth. The multiple layers of this tissue protect underlying areas from mechanical abrasion. This epithelium may have cuboidal-shaped cells at the basement layer, but it derives its name from the flattened cell shape at the free surface. Stratified squamous epithelium comes in two distinct types, keratinized and nonkeratinized (keratin is a tough protein that hardens cells in the outer layer of the skin).

(A) 4 Examine a prepared slide of stratified squamous epithelium and locate the basement membrane. Compare your slide with figure 4.4 and draw what you see under the microscope in the space provided. Locate the basement membrane and count the layers of cells between it and the surface of this tissue.

Illustration of stratified squamous epithelium:

Record the numbers of layers of cells here: ____________________

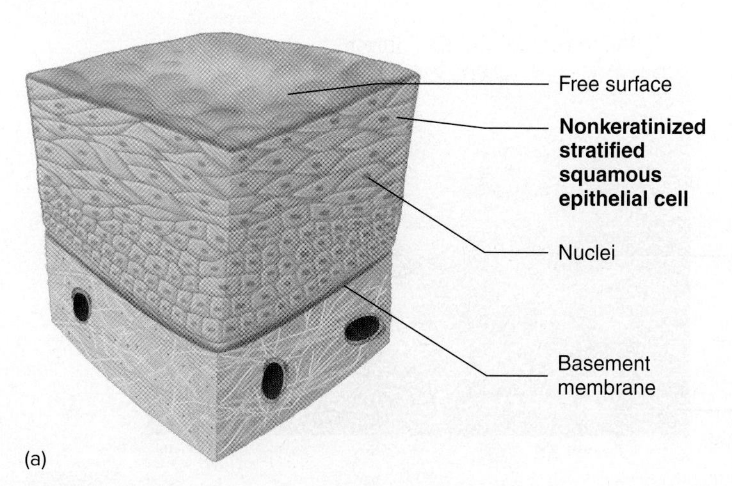

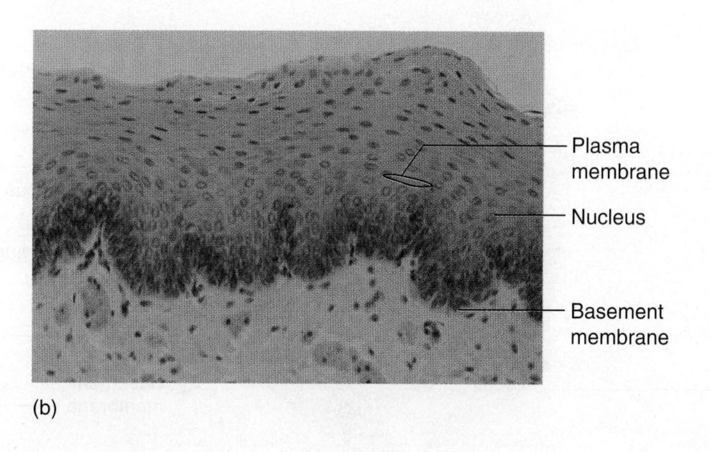

Figure 4.4 Stratified Squamous Epithelium
(a) Diagram; (b) photomicrograph of esophagus (400×).

Transitional Epithelium

Transitional epithelium is unusual in that it has some remarkable stretching capabilities. Transitional epithelium lines the ureter, urinary bladder, and proximal urethra and allows these organs to expand as urine collects within or passes through them.

Ⓐ[5] Examine a prepared slide of transitional epithelium. Look at figure 4.5 and draw the cell type in the space provided.
Illustration of transitional epithelium:

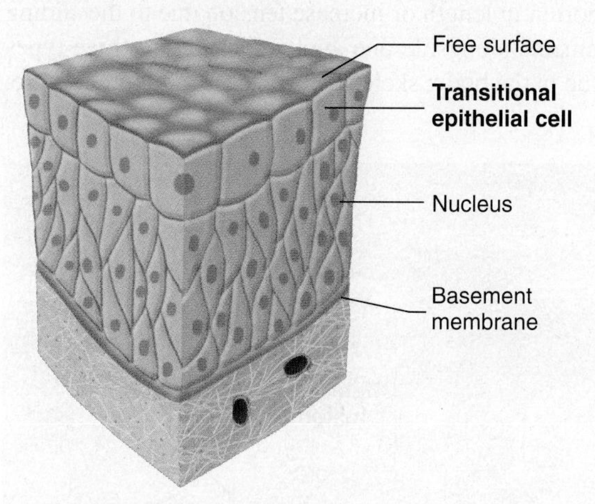

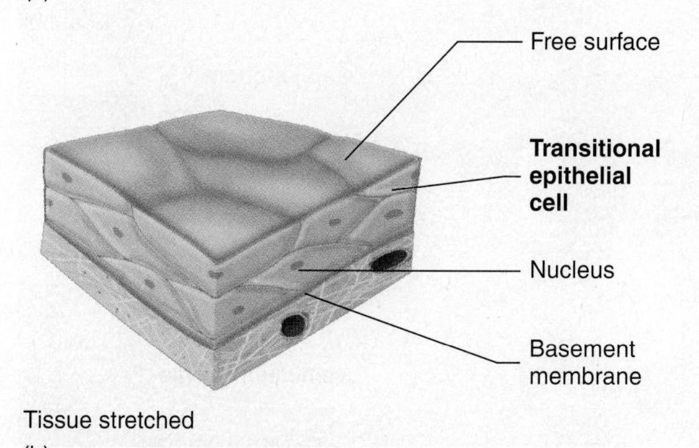

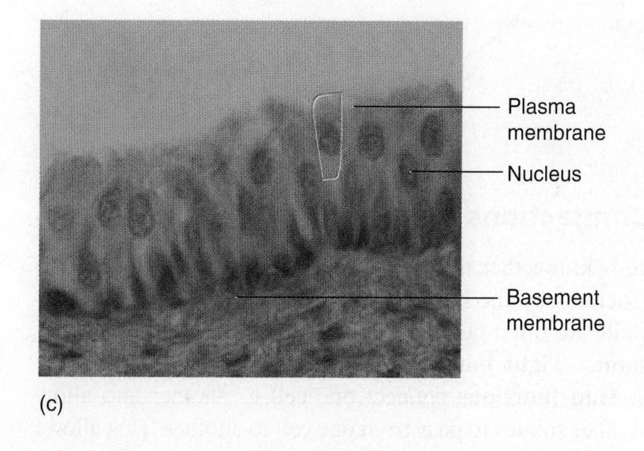

Figure 4.5 Transitional Epithelium
(a) Diagram, relaxed; (b) diagram, stretched; (c) photomicrograph of bladder (400×).

Pseudostratified Columnar Epithelium

This tissue may first look as if it occurs in a few layers, but all of the cells rest on the basement membrane. The nuclei are at different levels in the tissue. Pseudostratified columnar epithelium lines

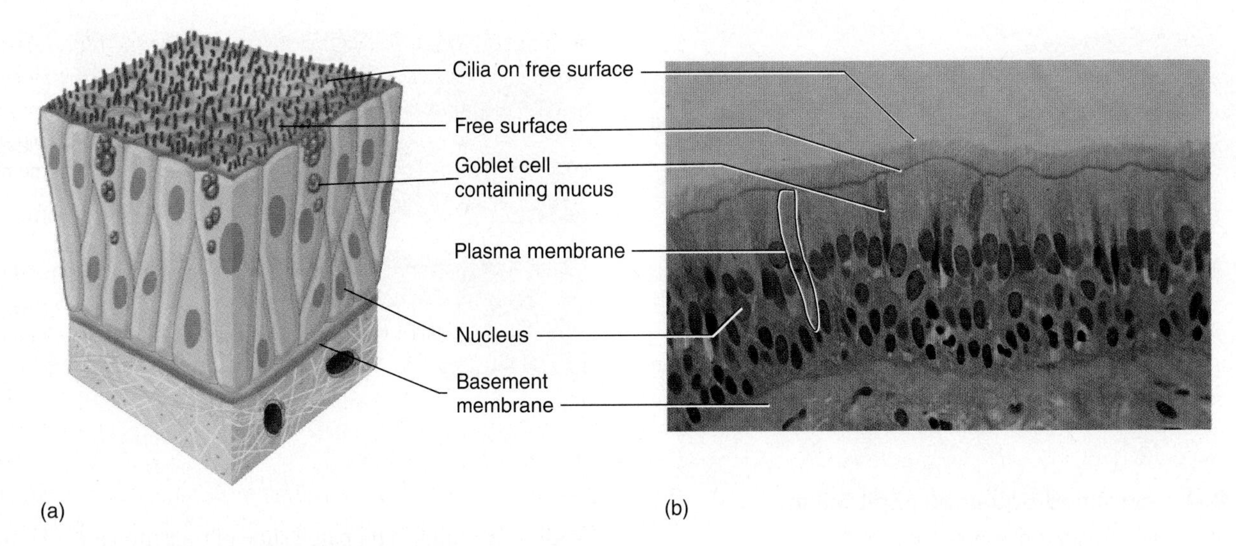

Figure 4.6 Pseudostratified Ciliated Columnar Epithelium
(a) Diagram; (b) photomicrograph of trachea (400×).

some portions of the respiratory passages, where it protects the lungs by trapping dust particles in a mucous sheet and moving the particles away from them.

(A) 6 Examine a prepared slide of pseudostratified columnar epithelium and compare it with figure 4.6. Draw a representative sample of what you see in the space provided.

Illustration of pseudostratified columnar epithelium:

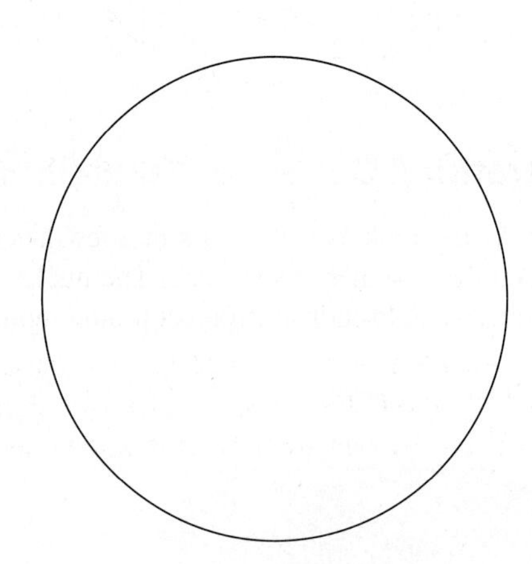

Cell Connections

Cells are held together in many ways. Epithelial cells are held to other structures by the basement membrane, as discussed previously. Cells are often held together at specific regions known as **desmosomes. Tight junctions** hold the cells even more closely together. **Gap junctions** connect one cell to another and allow ions and other solutes to pass from one cell to another. This allows communication between cells. These are illustrated, along with other types of connections, in figure 4.7.

Muscular Tissue

Like epithelial tissue, muscular tissue is a cellular tissue with the tissue having mostly cells and little matrix. These cells are contractile and shorten in length or increase tension due to the sliding of protein filaments alongside one another. There are three types of muscle tissue in the body: skeletal, cardiac, and smooth muscle.

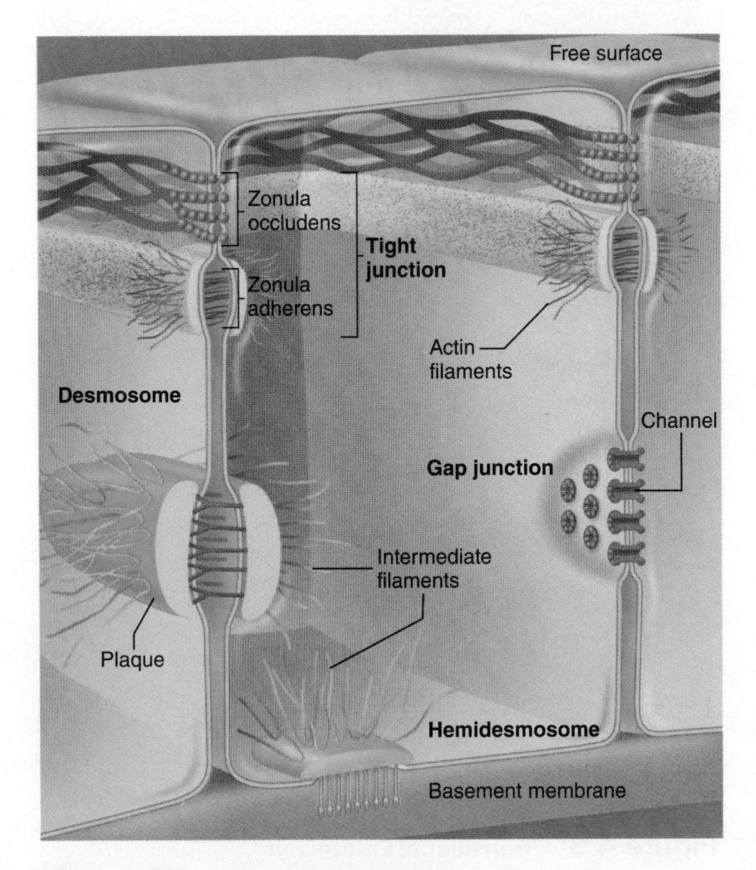

Figure 4.7 Cell Connections

Muscle Types

Skeletal Muscle

The cells of skeletal muscle are called **fibers.** When you refer to the muscles of your body, you are referring to organs made mostly of skeletal muscle. Skeletal muscle is sometimes known as striated muscle because it has obvious **striations** (they look like stripes) in the fiber. Skeletal muscle is **voluntary** because you have conscious control over this type of muscle in your body. The individual muscle cells have many nuclei and are thus called **multinucleate,** or **syncytial.** The nuclei are elongated and occur on the periphery of the cell.

(A) 7 Examine a prepared slide of skeletal muscle under high power. Compare this slide with figure 4.8. Draw a representation of the muscle in the space provided.

Illustration of skeletal muscle:

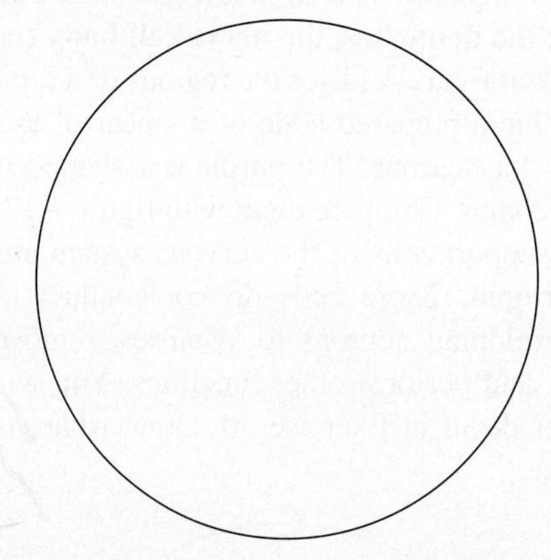

How many widths of muscle cells fit across the field of view of your microscope when viewed at high power?______________

How many muscle fiber widths do you count? __________

By counting the cell widths, you can further distinguish among skeletal, cardiac, and smooth muscle.

Cardiac Muscle

Cardiac muscle is found only in the heart, and it is the main tissue making up that organ. Cardiac muscle is similar to skeletal muscle in that it is **striated,** but the striations are much less obvious, and the individual cell diameter is less than those of skeletal muscle cells. Cardiac muscle cells are called **myocytes.** Another difference between cardiac muscle and skeletal muscle is that cardiac muscle is branched.

(A) 8 Place a prepared slide of cardiac muscle under the microscope and count the number of widths of cardiac muscle fibers across the diameter of the microscope under high power. Cardiac muscle contracts without conscious thought and is therefore called involuntary muscle.

The cells are mostly uninucleate, with the nucleus centrally located and oval in shape. Cardiac muscle cells are joined together by **intercalated disks,** which facilitate the transmission of the electrical impulses in the heart. These intercalated disks have gap junctions. As one muscle cell receives an impulse, it sends it on to the next cell. Look at a prepared slide of cardiac muscle and compare it with figure 4.9, noting the striations, intercalated disks, and nuclei of the cells. Draw the cells in the space provided.

Illustration of cardiac muscle:

Record the number of cell widths that occupy the diameter of your field of view at high power: ________________

Smooth Muscle

Smooth muscle is **nonstriated** in that the fibers do not have stripes perpendicular to the length of the fiber. Smooth muscle, like cardiac muscle, is **involuntary** and is found in the digestive tract, where it propels food along by a process known as peristalsis.

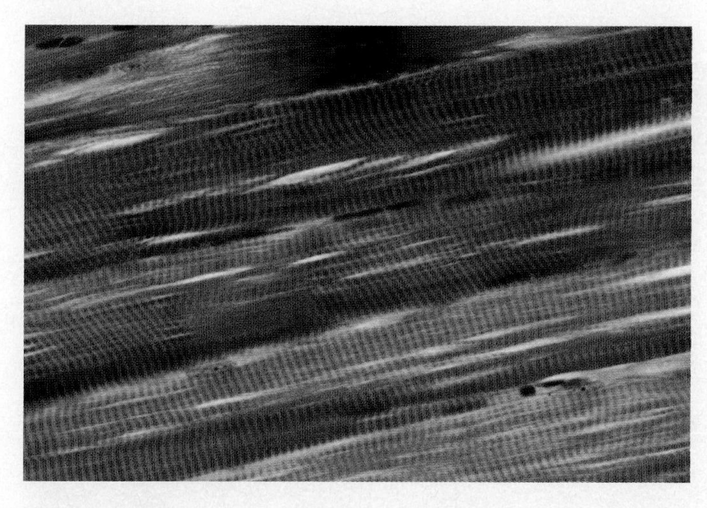

Figure 4.8 Skeletal Muscle (400×)

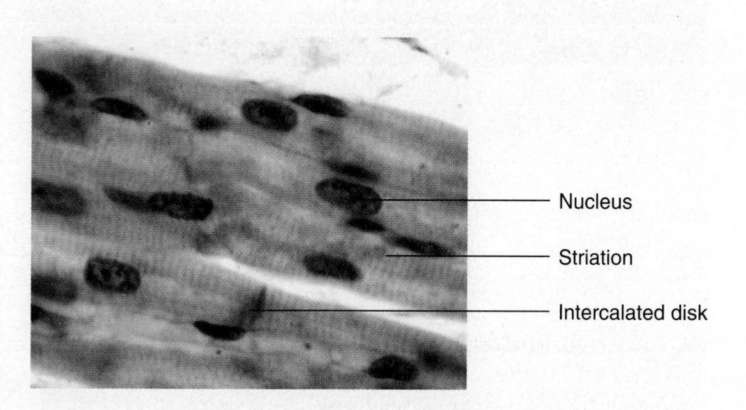

Figure 4.9 Cardiac Muscle (1,000×)

Smooth muscle is also found in abundance in the uterus and is the driving force behind uterine contractions during labor. The cells of smooth muscle are spindle-shaped and uninucleate, and the nucleus is centrally located and elongated in shape when the muscle is relaxed. When the muscle is contracted, the nucleus appears corkscrew-shaped.

(A) 9 Examine a prepared slide of smooth muscle and note the narrow diameter of the fibers. Compare your slide with figure 4.10 and draw an illustration of smooth muscle in the space provided.

Illustration of smooth muscle:

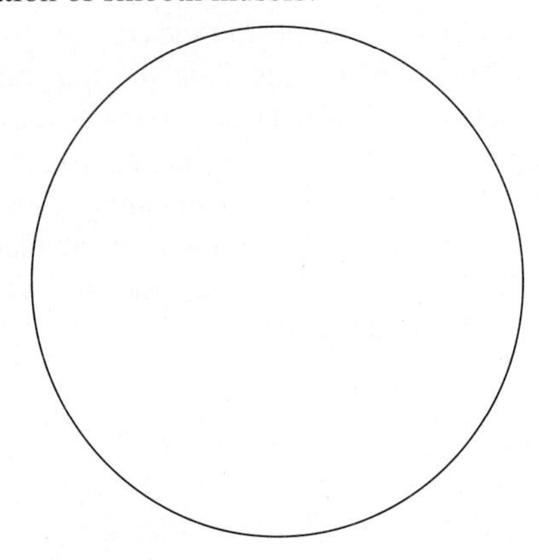

How many fiber widths do you count across the diameter of the field of view of your microscope under high power?________

Number of cell widths under high power: ______________

(A) 10 Examine a prepared slide of all three muscle types, if available. Work with your lab partner and identify each muscle cell. Fill in chart 4.1, comparing the muscle types and their characteristics.

Nervous Tissue

Nervous tissue, like epithelial and muscular tissues, is also mostly composed of cells. Nervous tissue is found in the brain, spinal cord, ganglia, and peripheral nerves of the body. The conductive cell of this tissue is the **neuron,** which receives and transmits

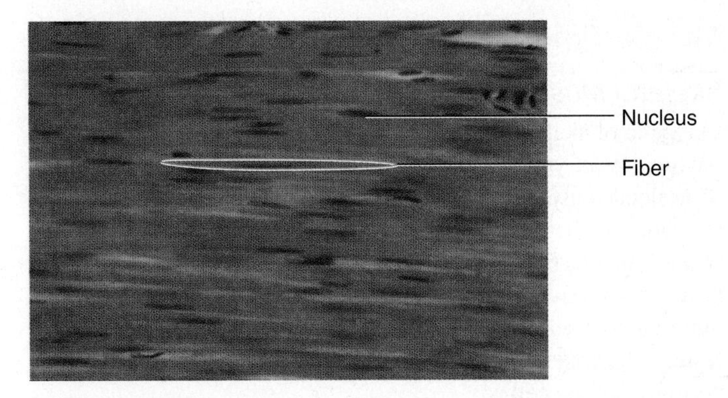

Figure 4.10 Smooth Muscle

Photomicrograph of intestine (400×).

electrochemical impulses. A neuron is a specialized cell with three major regions: the **dendrites,** the **nerve cell body** (or **soma**), and the **axon.** Look at figure 4.11 for the regions of a typical neuron.

(A) 11 Examine a prepared slide of a smear of the spinal cord and look for neurons. The purple star-shaped structures are the nerve cell bodies. Compare them with figure 4.12.

Special support cells of the nervous system are called **glial cells,** or **neuroglia.** These cells do not conduct impulses, but they guide developing neurons to synapses, remove some neurotransmitters, and perform other functions. You will study glial cells in greater detail in Exercise 20. Draw what you see in the following space.

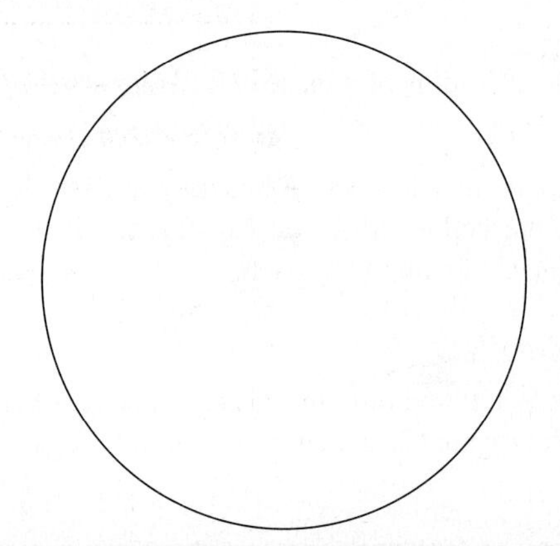

CHART 4.1 Muscle Characteristics

Muscle Type	Number of Nuclei	Location	Voluntary/Involuntary
	Many		
	One		
	Mostly one		

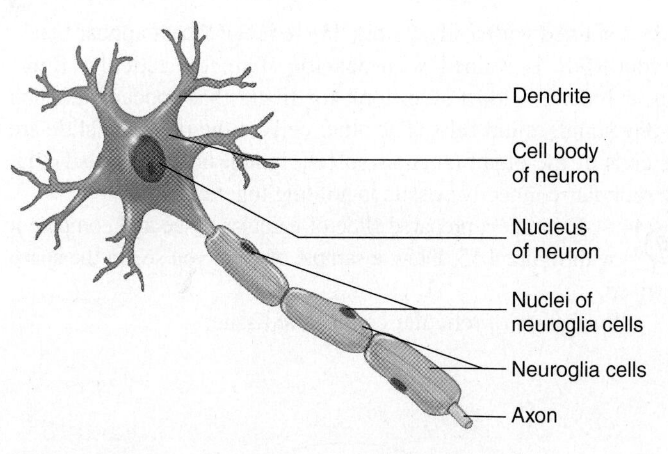

Figure 4.11 Regions of a Typical Neuron

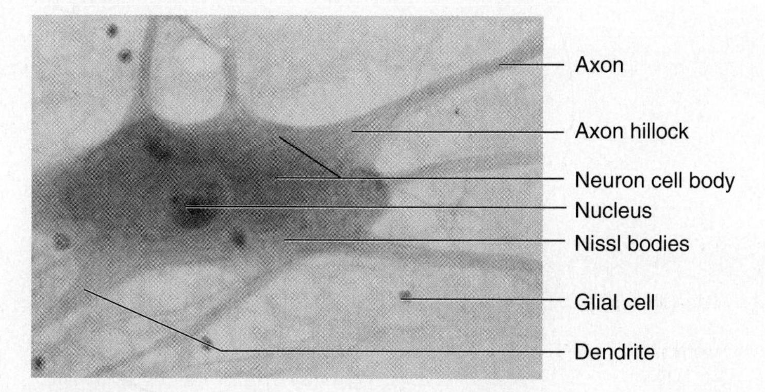

Figure 4.12 Spinal Cord Smear (400×)

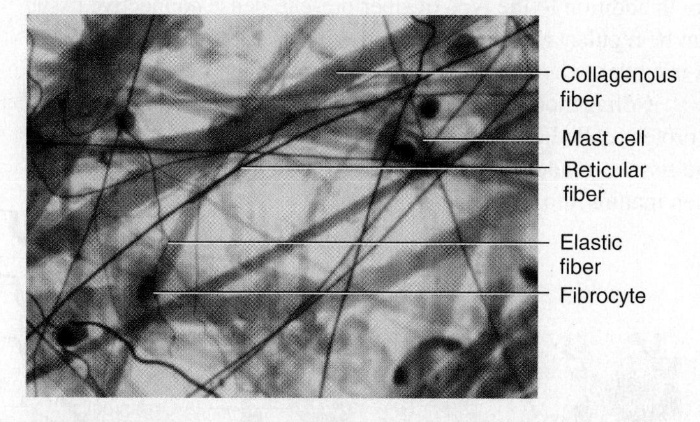

Figure 4.13 Areolar Connective Tissue (400×)

Connective Tissues

The tissues that you have seen so far, epithelial, muscular, and nervous, are composed mostly of cells with very little matrix. This is not the case with connective tissue. There is usually more matrix than cells in connective tissue. **Matrix** is extracellular material, and it usually consists of fibers and fluid, gel, or solid **ground substance.** Because of the abundance of matrix, connective tissue is classified by *specific tissue.*

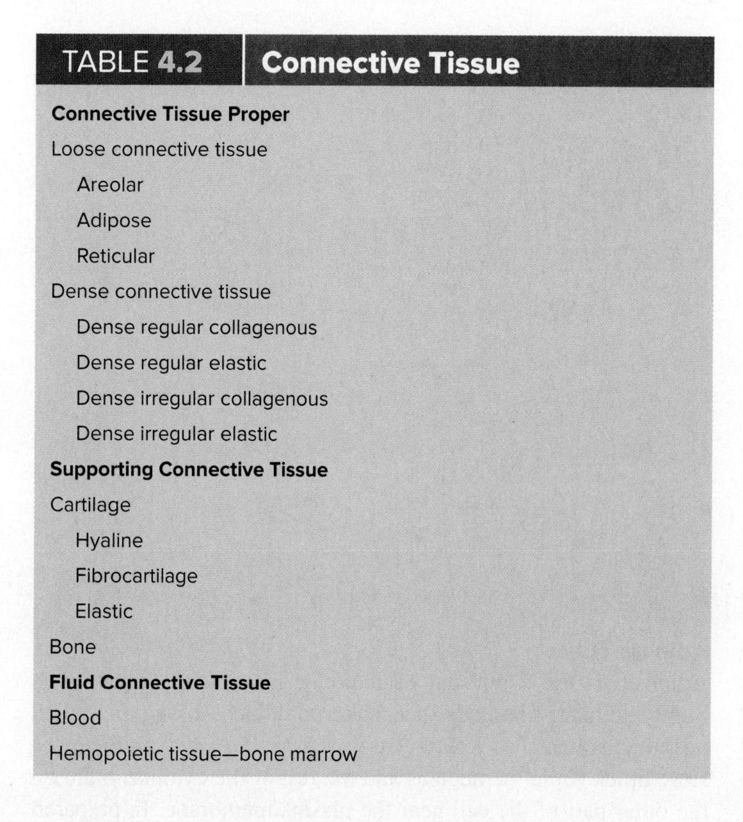

TABLE 4.2 Connective Tissue
Connective Tissue Proper
Loose connective tissue
Areolar
Adipose
Reticular
Dense connective tissue
Dense regular collagenous
Dense regular elastic
Dense irregular collagenous
Dense irregular elastic
Supporting Connective Tissue
Cartilage
Hyaline
Fibrocartilage
Elastic
Bone
Fluid Connective Tissue
Blood
Hemopoietic tissue—bone marrow

Connective tissues have diverse appearances and functions, and they may not seem to have much in common with one another; however, all arise from an embryonic tissue known as **mesenchyme.** Connective tissue is described in more detail next and is outlined in table 4.2.

There are three main categories of adult connective tissue, and there are specific tissues in each. These three categories are **connective tissue proper** (loose and dense), **supporting connective tissue** (cartilage and bone), and **fluid connective tissue** (blood).

Connective Tissue Proper

Two subdivisions of connective tissue proper are **loose connective tissue** and **dense connective tissue.**

Loose Connective Tissue

Areolar Connective Tissue

Areolar connective tissue is a complex collection of fibers and cells. Areolar connective tissue functions as packing material or meshwork providing an internal framework or an external wrapping around organs or as layers, such as the connective tissue between the dermis and muscle. Areolar connective tissue consists of large, pink-stained collagenous fibers; smaller, dark elastic and reticular fibers (fine collagenous fibers); and a collection of cells, including fibroblasts, fibrocytes, mast cells, and macrophages.

(A) 12 Examine a prepared slide of areolar connective tissue and compare it with figure 4.13. Make a drawing of it in the space provided.

Illustration of areolar connective tissue:

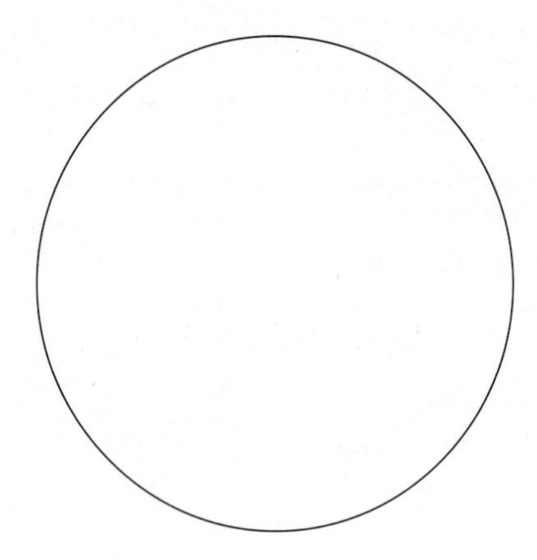

Adipose Tissue

Adipose tissue is unusual as a connective tissue in that it is highly cellular. The cells that make up adipose tissue are called **adipocytes** (which is a technical way to say *fat cells*). Adipocytes store lipids while the nucleus and the rest of the cytoplasm are on the outer part of the cell near the plasma membrane. In prepared slides of adipose tissue, the fat typically has been dissolved in the preparation process.

(A) 13 Locate the large, empty cells with the nuclei on the edges of some of the cells. Compare what you see in the microscope with figure 4.14 and make a drawing in the space provided. You may need to close down the iris diaphragm or reduce the amount of light shining on the specimen to see this tissue best.

Reticular Connective Tissue

Reticular connective tissue has fibroblasts and fibrocytes, as well as **reticular fibers.** These fibers are made of fine strands of collagen. Reticular connective tissue is found in soft internal organs, such as the liver, spleen, and lymph nodes. Reticular connective tissue provides an internal framework for these organs. If your slide is stained with a silver stain, the reticular fibers appear black. If your slide is stained with Masson stain, the reticular fibers appear blue. In either case, look for fibers that appear branched among small, round cells. The other cells in the prepared slide are the cells of the organ (such as spleen, lymph node, or tonsil) that the reticular connective tissue is holding together.

(A) 14 Examine a prepared slide of reticular tissue and compare it with figure 4.15. Draw a sample of what you see in the space provided.

Illustration of reticular connective tissue:

Dense Connective Tissue

Dense connective tissue can either be primarily collagenous or elastic. In addition to the type of fiber present, dense connective tissue can be regular, where the fibers are mostly parallel to one another, or irregular, where the fibers are at various angles to one another.

Collagenous tissue is named for the fibers that are made of a protein called collagen. Fibrocytes and fibroblasts also occur in the tissue, in addition to the fibers. Fibroblasts secrete fibers and then mature into fibrocytes.

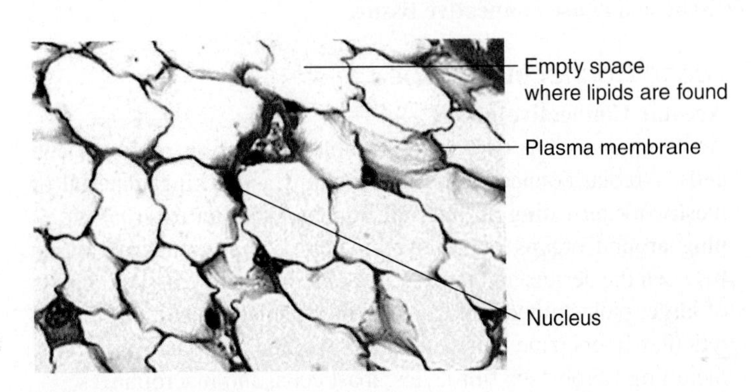

Figure 4.14 Adipose Tissue (400×)

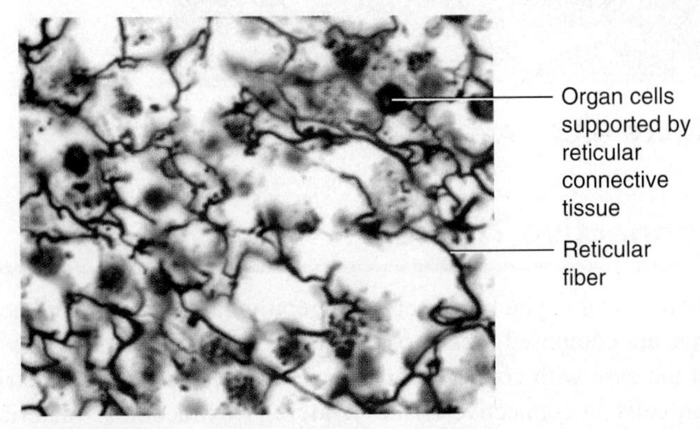

Figure 4.15 Reticular Connective Tissue (400×)

Dense Regular Collagenous Connective Tissue

Dense regular collagenous connective tissue is found in tendons and ligaments and the parallel fibers have great tensile strength. These fibers bind bone to muscle (tendons) or bone to bone (ligaments).

Dense Irregular Collagenous Connective Tissue

This tissue is found in the deep layers of the skin and the white of the eye.

(A) 15 Examine a slide of dense collagenous connective tissue and compare it with figure 4.16. Draw a section of dense regular or dense irregular collagenous connective tissue in the space provided.

Illustration of dense connective tissue:

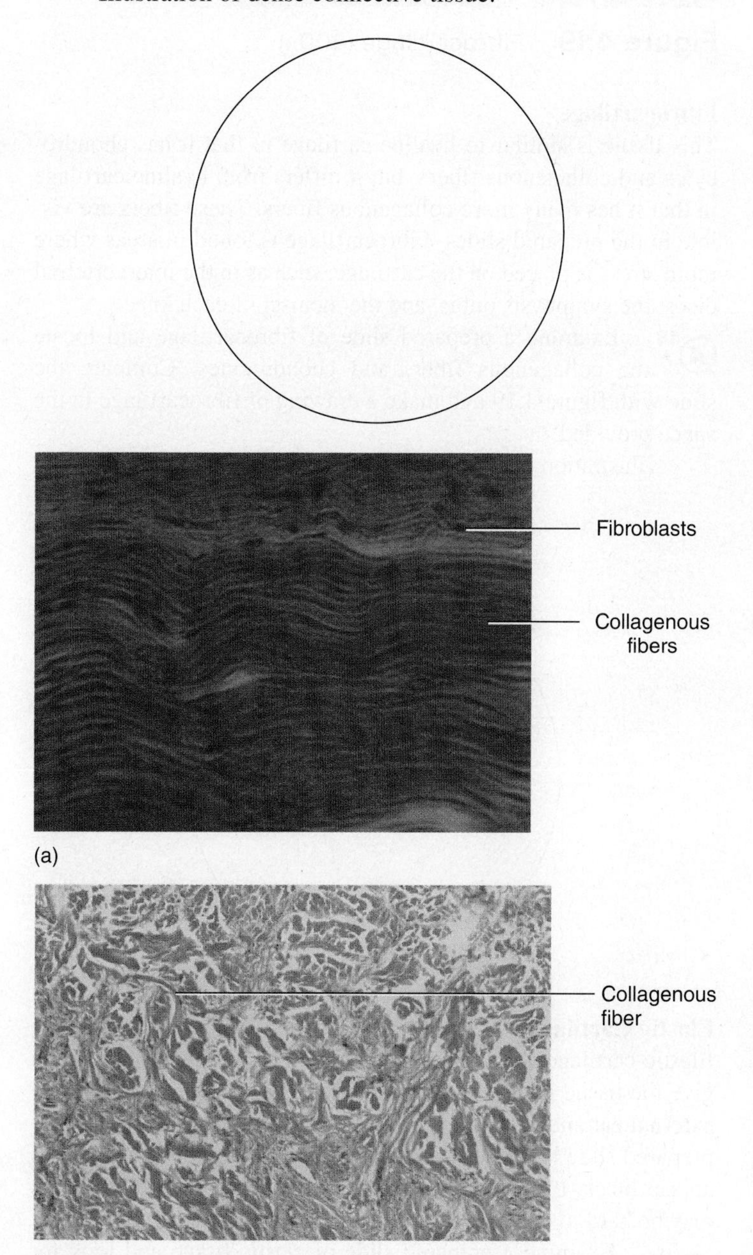

Figure 4.16 Dense Connective Tissue

(a) Dense regular collagenous connective tissue (400×); (b) dense irregular collagenous connective tissue (100×).

Dense Regular Elastic Connective Tissue

This tissue has cellular components of fibroblasts and fibrocytes. The fibroblasts actively secrete elastic fibers and subsequently become fibrocytes. The fibers in this tissue are made of the protein elastin.

Dense Irregular Elastic Connective Tissue

Dense irregular elastic connective tissue occurs in the walls of arteries. It can be identified as dark, squiggly lines in the artery walls. Elastic tissue that forms the vocal cords is known as dense regular elastic connective tissue.

(A) 16 Examine figure 4.17 as you look at a prepared slide of elastic tissue. Then draw the specific tissue in the space provided.

Illustration of elastic connective tissue:

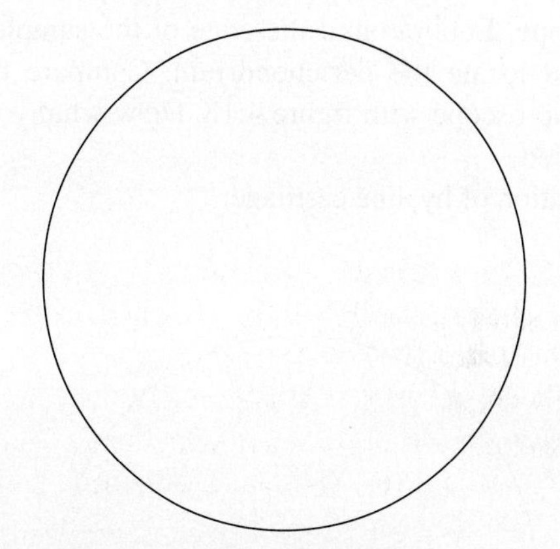

Supporting Connective Tissue

Cartilage (Cartilage and Perichondrium)

Cartilage is a type of connective tissue in which the matrix is composed of a pliable material that allows for some degree of movement. The matrix of cartilage is made of glucosamine and chrondroitin sulfate. These compounds form a semisolid gel in which fibers and cells are found. In prepared slides, the cells sometimes come out of the matrix and leave a cavity. This cavity is known as a **lacuna** and is diagnostic of cartilage tissue. The **perichondrium** is dense irregular connective tissue on the surface of cartilage. There are three types of cartilage in the human body, and they vary by the number and type of protein fibers they contain. These three types are **hyaline cartilage, fibrocartilage,** and **elastic cartilage.**

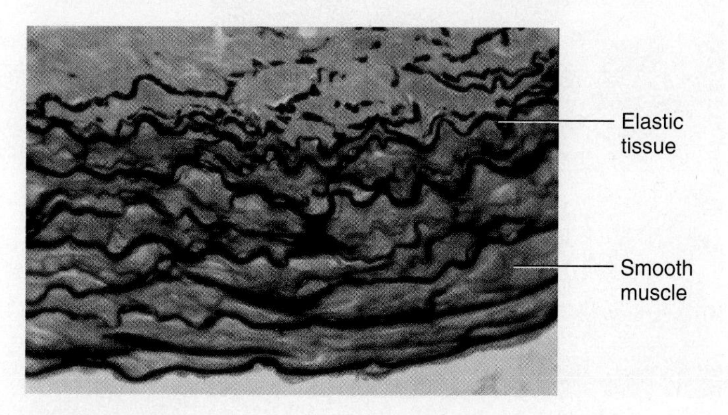

Figure 4.17 Dense Irregular Elastic Connective Tissue (400×)

Hyaline Cartilage

The most common cartilage in the body is **hyaline cartilage,** which is found at the apex of the nose, covering the ends of most long bones at joints, between the ribs and the sternum, in the respiratory passages, and in many other locations. Hyaline cartilage is clear and glassy in fresh tissue but will probably appear light pink in prepared, stained slides. The cells that occur in hyaline cartilage are called **chondrocytes,** and the fibers are **collagenous fibers.**

(A) 17 Find the chondrocytes in a prepared slide of hyaline cartilage. The fibers will not appear distinct because they blend into the color of the matrix. The chondrocytes of hyaline cartilage frequently occur in pairs. Some people say that they look like pairs of eyes staring back at you as you look at them in the microscope. Look around the edge of the sample of hyaline cartilage and locate the perichondrium. Compare the material under the microscope with figure 4.18. Draw what you see in the space provided.

Illustration of hyaline cartilage:

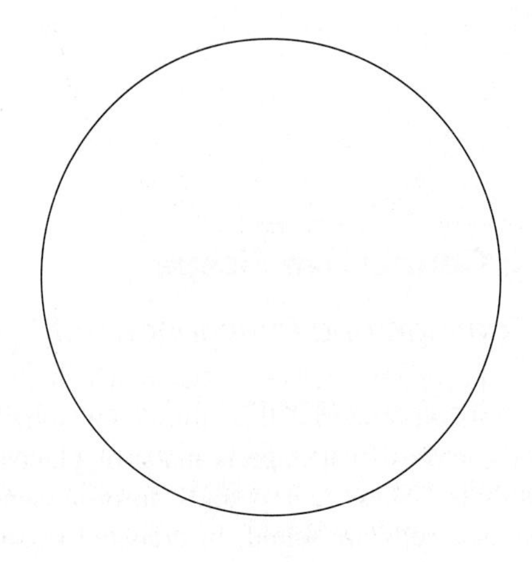

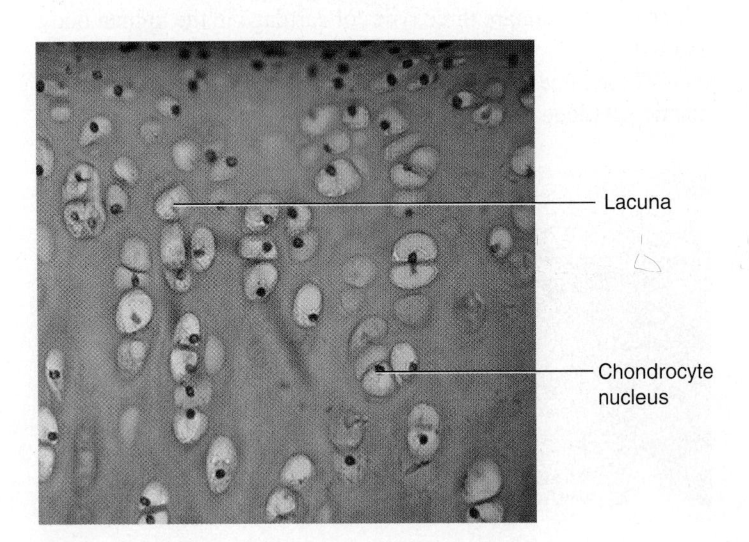

Figure 4.18 Hyaline Cartilage (400×)

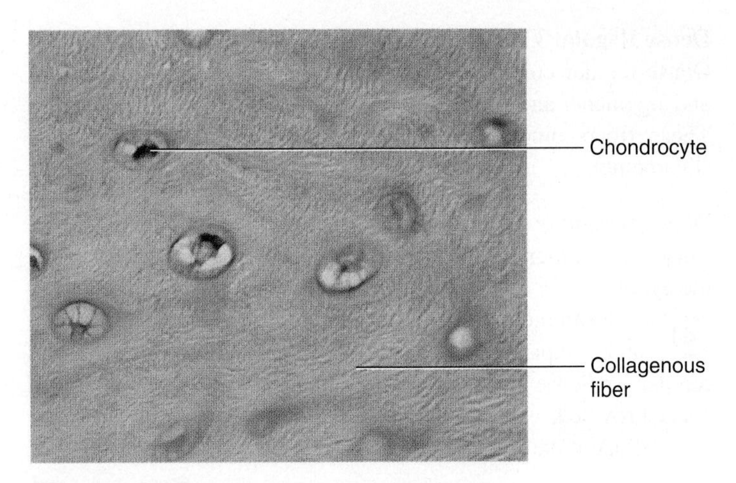

Figure 4.19 Fibrocartilage (400×)

Fibrocartilage

This tissue is similar to hyaline cartilage in that it has chondrocytes and collagenous fibers, but it differs from hyaline cartilage in that it has many more collagenous fibers. These fibers are visible in the prepared slides. Fibrocartilage is found in areas where more stress is placed on the cartilage, such as in the intervertebral discs, the symphysis pubis, and the menisci of each knee.

(A) 18 Examine a prepared slide of fibrocartilage and locate the collagenous fibers and chondrocytes. Compare the slide with figure 4.19 and make a drawing of fibrocartilage in the space provided.

Illustration of fibrocartilage:

Elastic Cartilage

Elastic cartilage is unique in that it contains elastic fibers, which give the tissue its flexible nature. Elastic cartilage is found in the external ear and in a part of the larynx called the epiglottis. If the prepared slide is stained with a silver stain, then the elastic fibers appear black. If the slide is not stained in this way, then the fibers may be hard to distinguish.

(A) 19 Examine a prepared slide of elastic tissue and look for the chondrocytes and elastic fibers. Compare your slide with figure 4.20 and make a drawing of elastic cartilage in the space provided.

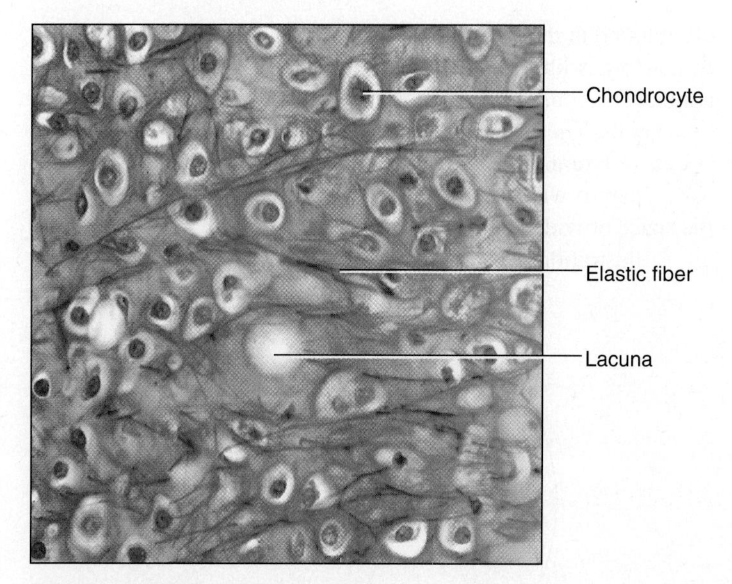

Figure 4.20 Elastic Cartilage (400×)

Illustration of elastic cartilage:

Bone

Bone, or osseous tissue, is a type of connective tissue in which the matrix is made of a mineral component called **hydroxyapatite crystal,** or **bone salt.** Collagenous fibers occur in the tissue along with osteocytes, or bone cells. Two types of bone are common in the body. **Compact bone** is the more common type; it makes up the dense material in a long section of a bone. **Cancellous (spongy) bone** is found in the end regions of long bones and has plates of bone interspersed with bone marrow.

(A) 20 Look at a prepared slide of ground bone and compare it with figure 4.21. Locate the large, circular regions, which are cross sections of **osteons.** They resemble cross sections of tree trunks with the growth rings. The space in the middle is the **central canal** and the dark spots that occur in rings around the central canal are the **lacunae,** spaces where osteocytes are located. A more detailed examination of bone appears in Exercise 6. After you study the slide, draw a sample of it in the space provided.

Illustration of ground bone:

Fluid Connective Tissue

Blood

Blood is a connective tissue in which the matrix is fluid and no fibers are visible. The matrix of the blood is called **plasma,** and the cells of the blood are either **red blood cells (erythrocytes)** or **white blood cells (leukocytes).** Some cellular fragments called **platelets** occur in blood.

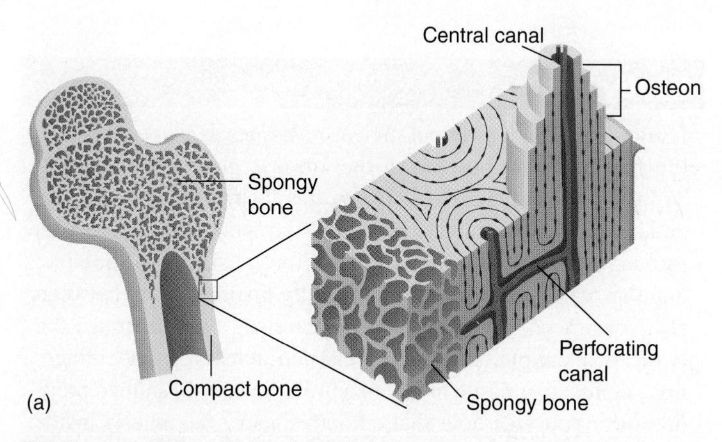

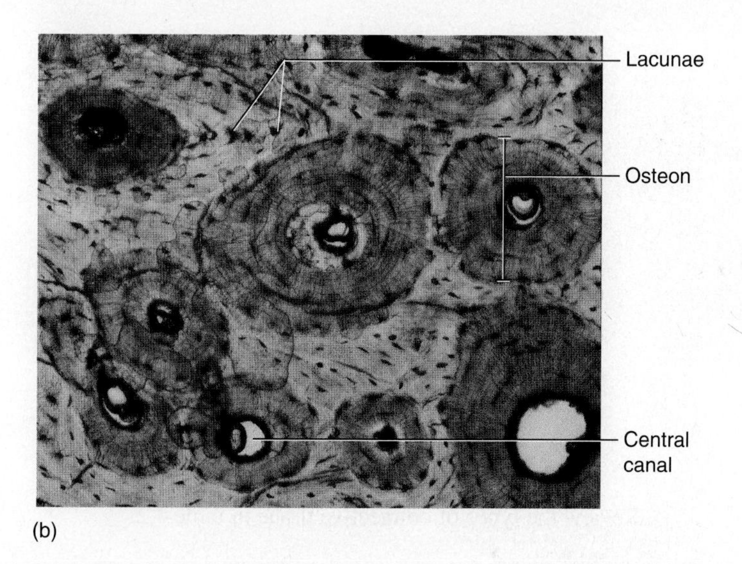

Figure 4.21 Ground Bone
(a) Diagram of ground bone; (b) photomicrograph (100×).

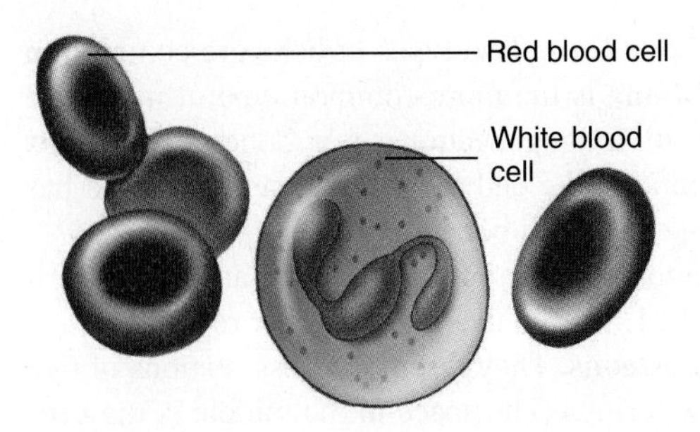

(a)

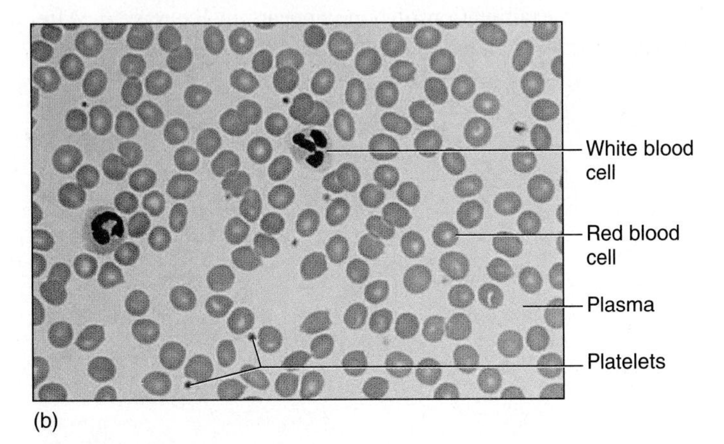

(b)

Figure 4.22 Blood
(a) Diagram; (b) photomicrograph (400×).

Ⓐ[21] Examine a prepared slide of blood and look for the numerous pink disks in the sample. These are the erythrocytes. You may find larger cells with lobed, purple nuclei. These are the leukocytes. The details of blood are studied in Exercise 25. Examine the blood slide under high power and compare it with figure 4.22. Make a drawing of blood in the space provided.

Illustration of blood:

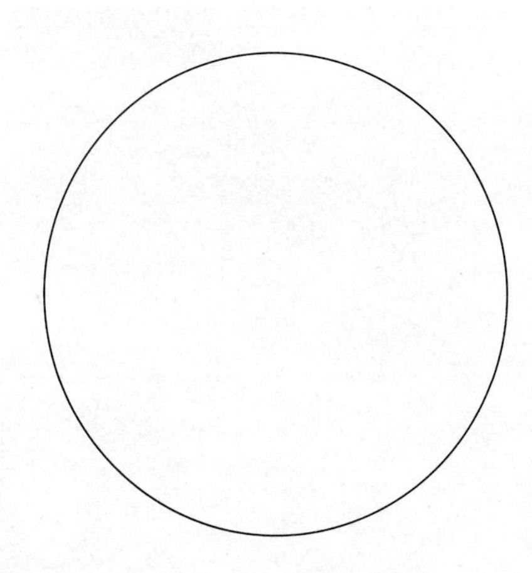

Review the types of connective tissue in table 4.2.

Hemopoietic Tissue—Bone Marrow

Bone marrow is a special connective tissue that contains many cells. **Yellow bone marrow** contains mostly adipose tissue and is often found in the shafts of the long bones of adults. Another type of marrow is known as **red bone marrow,** which, in addition to adipocytes, contains precursor cells of red blood cells and white blood cells. These are the stem cells.

Ⓐ[22] Examine a prepared slide of red bone marrow and compare it with figure 4.23. Draw a sample of what you see in the space provided.

Illustration of red bone marrow:

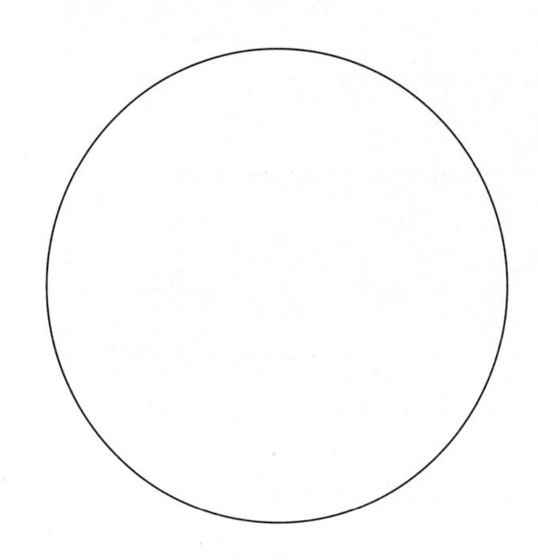

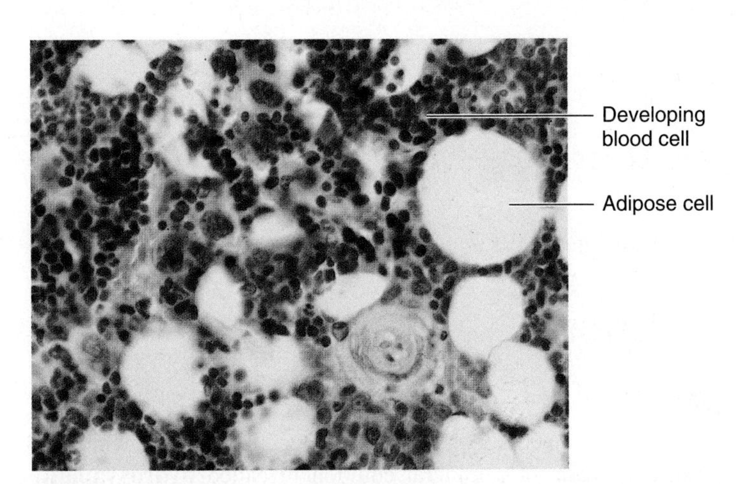

Figure 4.23 Bone Marrow
Photomicrograph of red bone marrow (400×).

Study Hint

Examine as many different images of tissues as you can. Drawing the material observed in the lab is a good memory tool. Try to make associations between the abstract image of tissue and something common. For example, you can think of bone as looking like a series of tree rings. Adipose tissue may look like angular balloons. If you have difficulty distinguishing between two specimens, try to find a characteristic that separates the two. For example, smooth muscle and dense regular collagenous connective tissue may look similar in certain stained preparations. You will note that smooth muscle has nuclei inside the fibers, whereas dense connective tissue has nuclei of the fibrocytes between adjacent fibers.

Exercise 4 Review

Tissues

Name: ______________________

Lab time/section: ______________________

Date: ______________________

1. List the four primary tissue types of the body. ______________________

2. What is the name of the noncellular layer that attaches epithelial tissue to deeper tissues? ______________________

3. How many layers are in a simple epithelium? ______________________

4. Name the three general cell shapes of epithelial tissue. ______________________

5. What is the specific name of a tissue made of a single, flattened layer of epithelial cells? ______________________

6. What functional problem could occur if the digestive tract were lined with stratified squamous epithelium instead of simple columnar epithelium? ______________________

7. What is the specific name of a tissue made of multiple layers of flattened epithelial cells? ______________________

8. Can you make the hairs on your arm "stand on end"? Based on your results, what type of muscle is responsible for doing this? ______________________

9. What tissue lines the inside of the urinary bladder? ______________________

10. What muscle tissues have the following features?
    a. striated and voluntary ______________________
    b. striated and involuntary ______________________
    c. nonstriated and involuntary ______________________

11. What muscle tissue has intercalated disks? ______________________

12. The intestine is lined with what kind of muscle? ______

13. The heart is composed primarily of what kind of muscle? ______

14. The muscles of the arm are composed primarily of what tissue? ______

15. What cell type is responsible for the transmission of electrochemical impulses? ______

16. What is the matrix in blood called? ______

17. Name the three types of fibers in areolar connective tissue. ______

______

18. What type of connective tissue is springlike and found in the middle walls of arteries? ______

19. What is the cell type in adipose tissue? ______

20. Name the outer connective tissue layer around cartilage. ______

21. What kinds of fibers are in fibrocartilage? ______

22. Describe the functional difference between fibrocartilage and hyaline cartilage. ______

______

______

23. What is another name for calcium salts in bone? ______

24. In a cross section of a bone, you can usually see two types of bone tissue. What are these called? ______

______

25. How do you distinguish generally between epithelial tissue and connective tissue? ______

______

______

26. Label the following photomicrographs by tissue type and by using the terms provided.

   a. collagenous fibers, elastic fibers

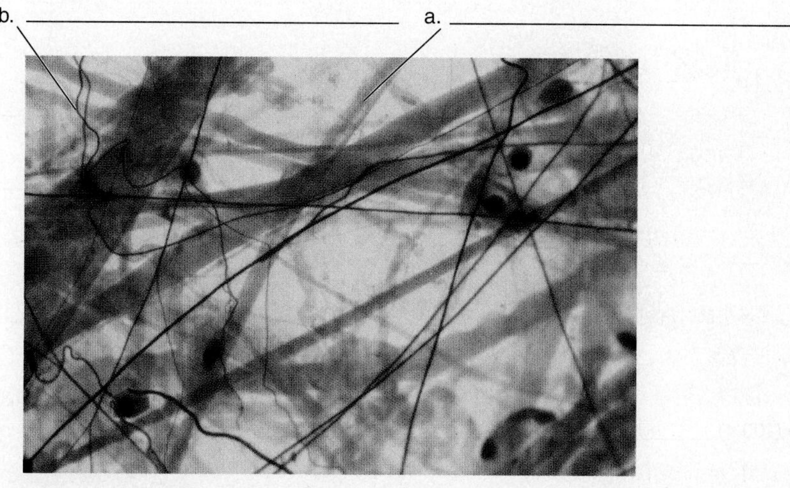

Tissue (400×) ______________________

   b. cilia, basement membrane, nucleus

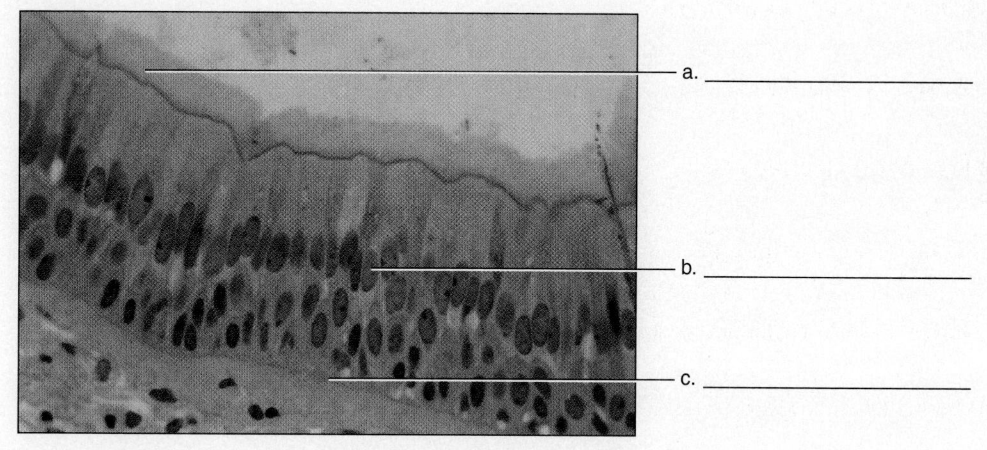

Tissue (400×) ______________________

   c. intercalated disks, striations, nucleus

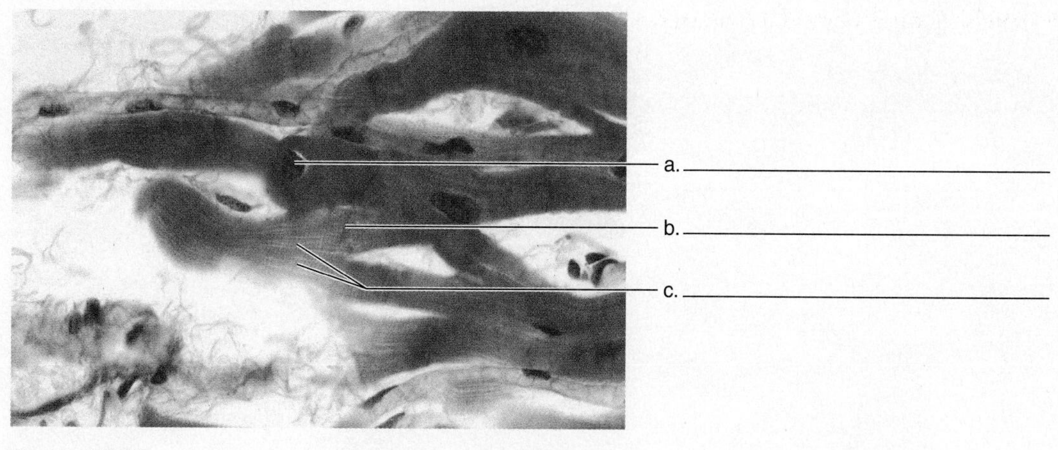

Tissue (400×) ______________________

d. nerve cell body, glial cells

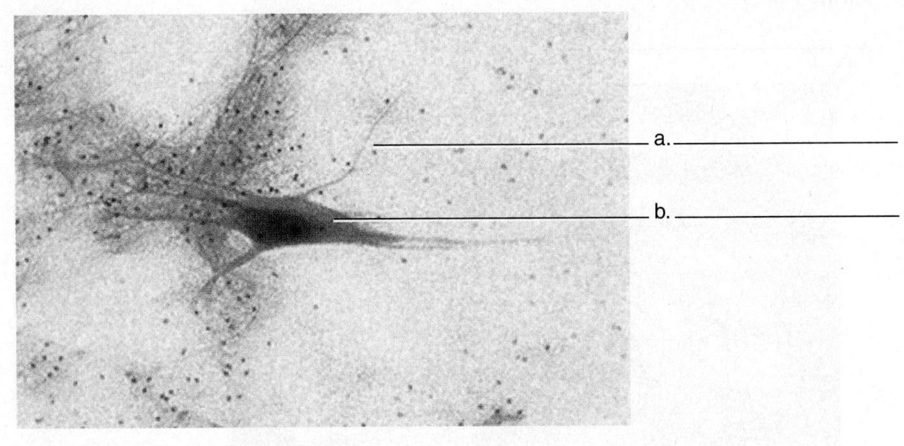

Tissue (100×) ________________

Exercise 5

Integumentary System

Introduction

The integumentary system consists of the skin and associated structures, such as hair, nails, and various glands. The **integument** consists primarily of a cutaneous membrane composed of an **epidermis** and a **dermis.** Below these two layers is an underlying layer called the **hypodermis,** which anchors the integument to deeper structures. The skin is the largest organ of the body and is vital in terms of water and temperature regulation and protection against microorganisms. These topics are discussed in chapter 5, "Integumentary System," in the Seeley text. In this exercise, you examine the structure of skin, hair, and nails and name the layers of the epidermis and dermis.

Objectives

At the end of this exercise, you should be able to

1. list the two layers of the integument;
2. list all of the layers of the epidermis from superficial to deep;
3. describe the structure and function of sweat glands and sebaceous glands;
4. draw a hair in a follicle in longitudinal section;
5. discuss the protective nature of melanin;
6. describe the subcutaneous tissue and its relationship to the integument.

Materials

Models and charts of the integumentary system
Microscopes
Prepared microscope slides of
- Thick skin (with Pacinian corpuscles)
- Hair follicles
- Pigmented and nonpigmented skin

Procedure

Overview of the Integument

(A) 1 Examine models or charts of the integumentary system and locate the two major regions of the integument, the epidermis, and the dermis. The epidermis is a cellular layer and can be distinguished as the most superficial layer of the integument. The dermis is the deeper layer. Also locate the subcutaneous tissue (hypodermis), not a layer of the integument but a structure anchoring the integument to underlying bone or muscle. Compare the models in the lab with figures 5.1 and 5.2.

Locate the **hair follicles** in the dermis, **sebaceous (oil) glands,** and **sudoriferous (sweat) glands.** The sweat glands are connected to the surface by **sweat ducts.**

Microscopic Examination of the Integument

(A) 2 Examine a prepared slide of thick skin under low power. Locate the hypodermis, dermis, and epidermis. The hypodermis can be distinguished by the presence of significant amounts of adipose tissue found there. The dermis consists primarily of collagenous fibers. The epidermis is the most superficial layer and can be determined by the epithelial cells of that layer. You can look for cells that have obvious nuclei. Draw and label the overview of the integument in the following space, locating the epidermis, dermis, and subcutaneous tissue.

Illustration of integument:

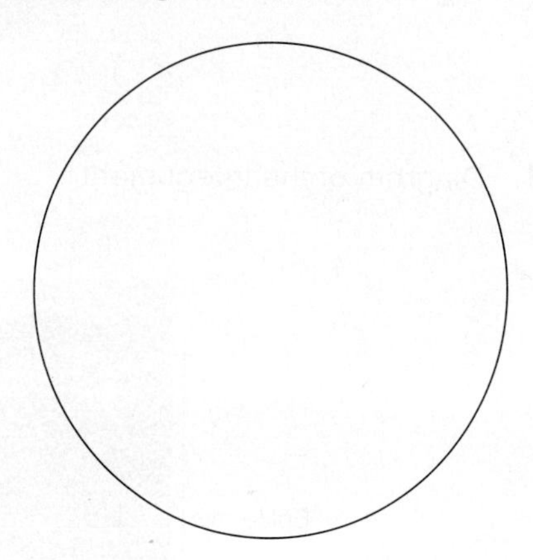

Dermis

The dermis is responsible for the structural integrity of the integumentary system. It is composed primarily of collagenous and other fibers, along with blood vessels, nerves, sensory receptors, hair follicles, and glands. **Blood vessels** are located in the dermis, taking nutrients to both the dermis and the epidermis. The blood vessels do not enter the epidermis, but the nutrients and oxygen diffuse from the dermis into the cells of the epidermis. Blood vessels are also important because they release heat to the external environment. As the internal temperature of the body rises, blood flows into the capillaries of the dermis, allowing heat to radiate from the skin. In cold weather, precapillary sphincters close, decreasing the amount of blood flowing to the dermis, keeping heat in the body core.

There are many **nerves** present in the dermis. Sensory information, such as light touch, changes in temperature, and the feeling of pain, is transmitted from receptors in the dermis to the central nervous system. Receptors that respond to light touch are found closer to the epidermis, while those that respond to pressure

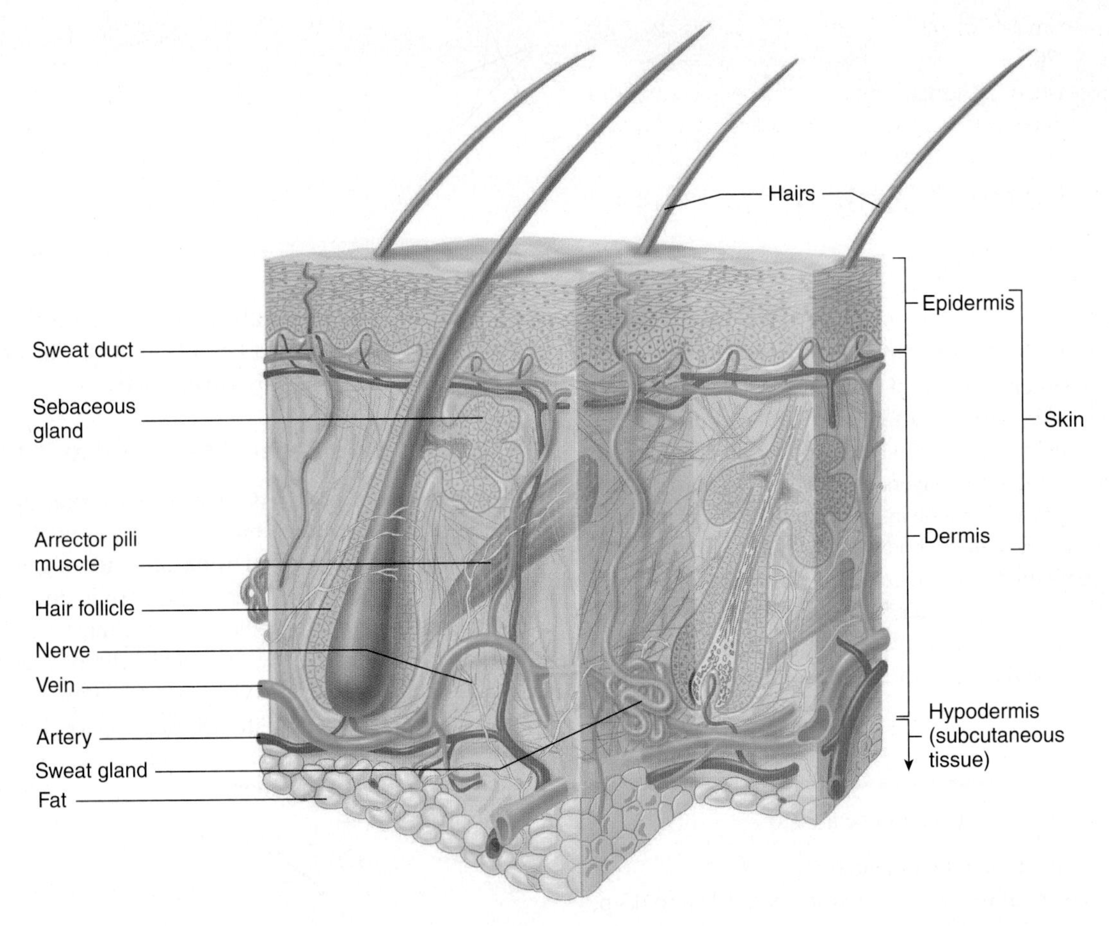

Figure 5.1 Diagram of the Integument

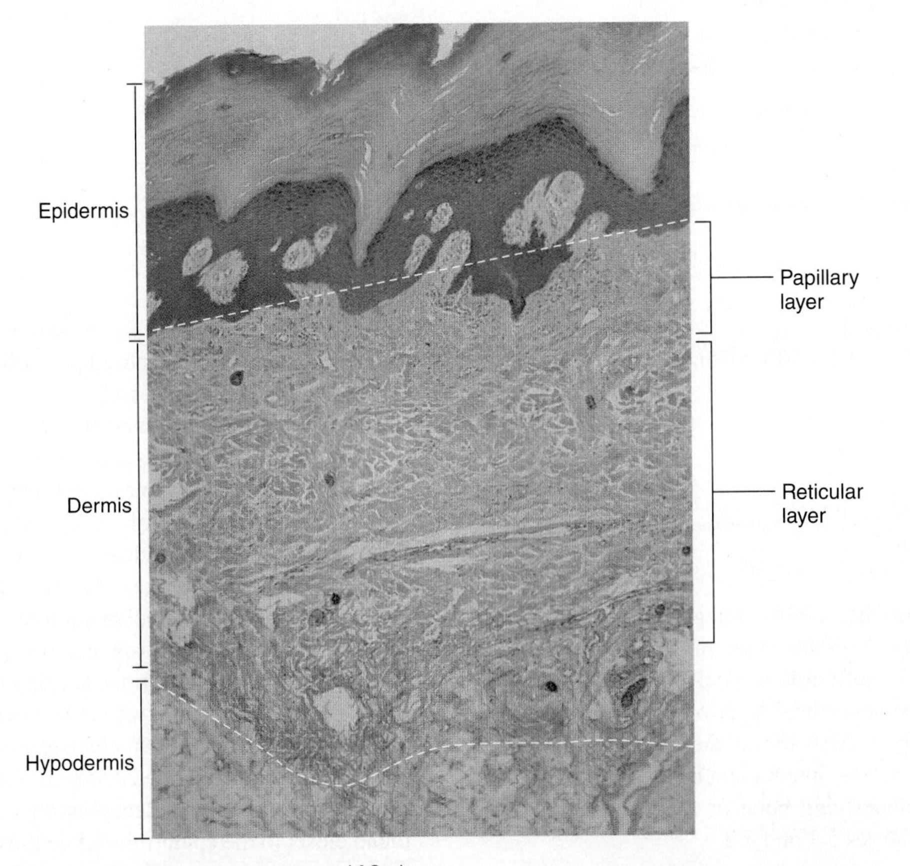

Figure 5.2 Photomicrograph of the Integument (40×)

are found in the dermis. The sensory structures of the skin are covered in Exercise 20.

The dermis consists of two major regions, the deeper **reticular layer** and the superficial **papillary layer.** The reticular layer is the thickest layer of the dermis. It is primarily composed of irregularly arranged collagenous fibers, along with some elastic and reticular fibers. The collagenous fibers provide strength and flexibility. The elastic fibers, as their name implies, provide elasticity to the skin. The papillary layer is found between the deeper reticular layer and the superficial epidermis. It is named for the **papillae** (rete pegs), which are bumps that interdigitate with the epidermis, providing good adhesion between the two layers. The papillae prevent side slippage of the epidermis over the dermis. The papillary layer contains more elastic fibers than the reticular layer.

(A)[3] Examine a slide of thick skin and locate the two layers of the dermis, as seen in figure 5.2. Draw a representation of the dermis in the following space.

Illustration of dermis:

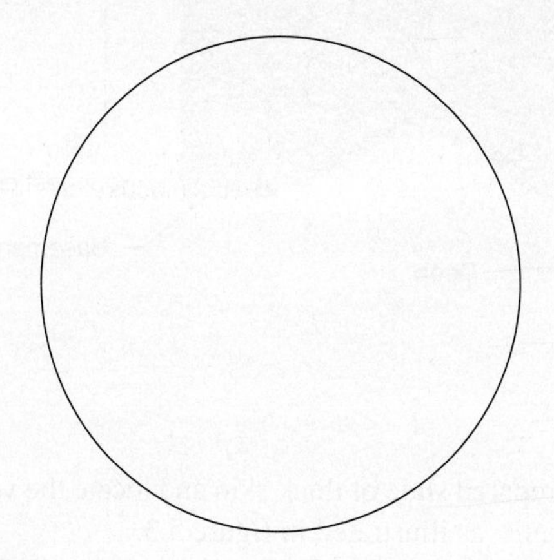

Epidermis

The epidermis is composed of stratified squamous epithelium with the superficial layer infused with the protein keratin. Keratin is a material that toughens the skin and protects the lower layers of the epidermis from abrasion. Keratin also protects the skin from ultraviolet radiation.

In slides stained with hematoxylin and eosin (H & E) stain, the dermis is often colored pink, and the superficial region with multicolored layers is the epidermis. The colors may vary in the slides used in your lab.

(A)[4] Compare these layers in the prepared slide with figure 5.3.

Strata of the Epidermis

In thick skin, the deepest layer of the epidermis is known as the **stratum basale.** This single layer of cells is attached to the dermis by the **basement membrane,** a noncellular layer. Cells in the stratum basale divide repeatedly and push up toward the superficial layers as the cells increase in age. This process is illustrated in figure 5.4. In the stratum basale are specialized cells known as **melanocytes.** These cells produce the brown pigment **melanin.** Melanin protects the stratum basale from the damaging effects of ultraviolet radiation. The number of melanocytes is relatively constant among people of all skin colors, but the amount of melanin produced by the melanocytes is variable. People of darker

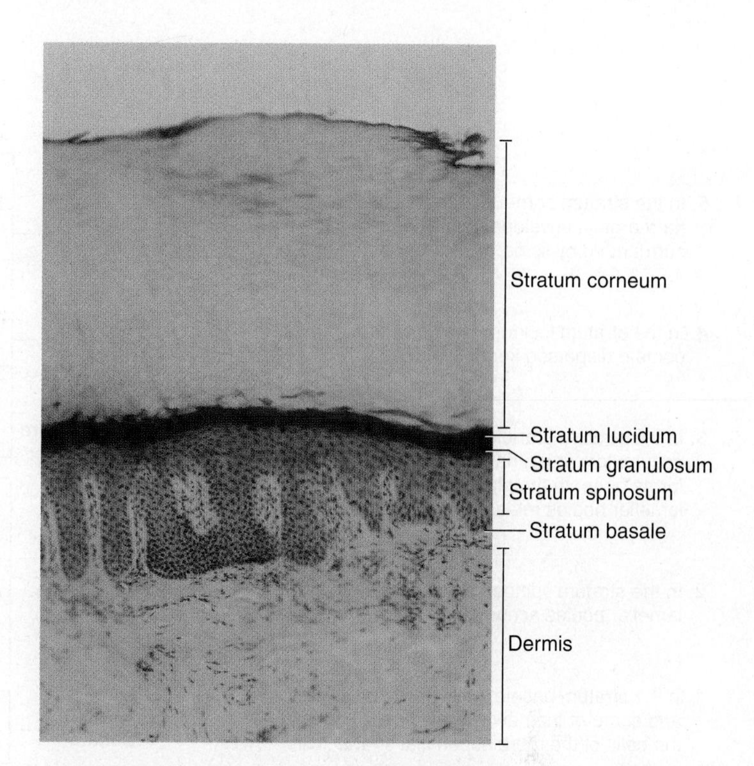

Figure 5.3 Strata of the Epidermis
Photomicrograph (100×).

pigmentation produce more melanin, whereas people of lighter pigmentation have melanocytes that produce less melanin. People of lighter pigmentation are more susceptible to the effects of ultraviolet radiation and skin cancer.

(A)[5] Locate the pigment melanin and the stratum basale and draw them in the following space.

Illustration of lower epidermis:

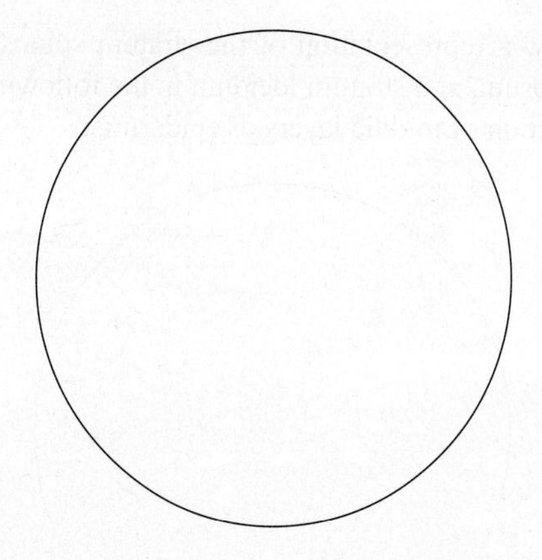

The layer containing many cells superficial to the stratum basale is the **stratum spinosum.** In prepared slides, the cells appear star-shaped because the plasma membrane pulls away from the other cells, except where the cells are joined at microscopic junctions called desmosomes. The stratum basale and stratum spinosum are often classified as the **stratum germinativum.**

A layer of granular cells is present in the next layer, the **stratum granulosum.** The cells have purple-staining keratohyalin granules in their cytoplasm. The granules are precursors of keratin found in the outermost layer of the epidermis.

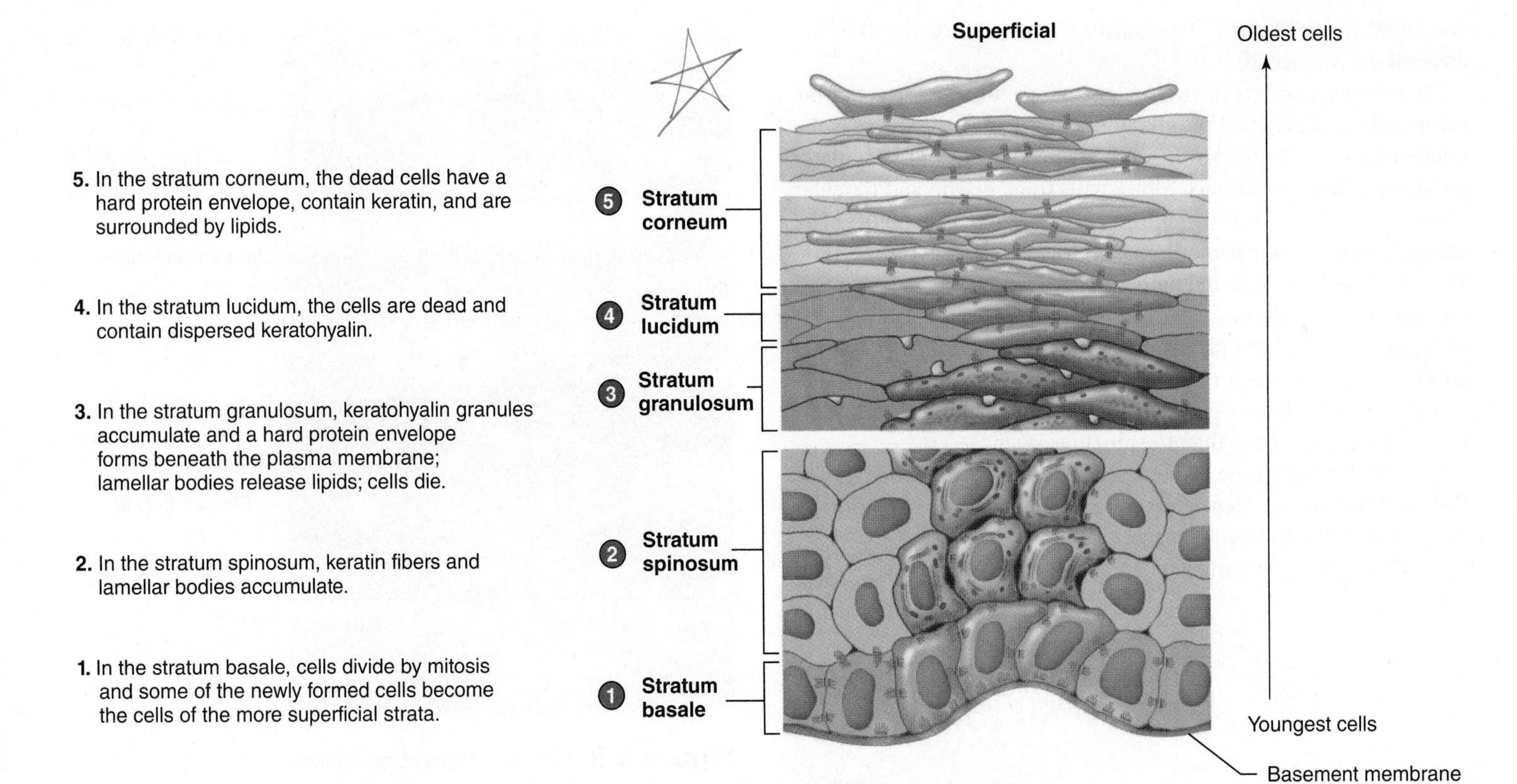

Figure 5.4 Layers of the Epidermis and the Process of Maturation

Superficial to this layer is the **stratum lucidum,** so named (*lucid,* light) because this layer appears translucent in fresh specimens. This layer generally appears clear or pink in prepared slides stained with H & E. It is found only in the palms of the hands and soles of the feet.

(A) 6 Draw a representation of the stratum spinosum, stratum granulosum, and stratum lucidum in the following space.

Illustration of middle layers of epidermis:

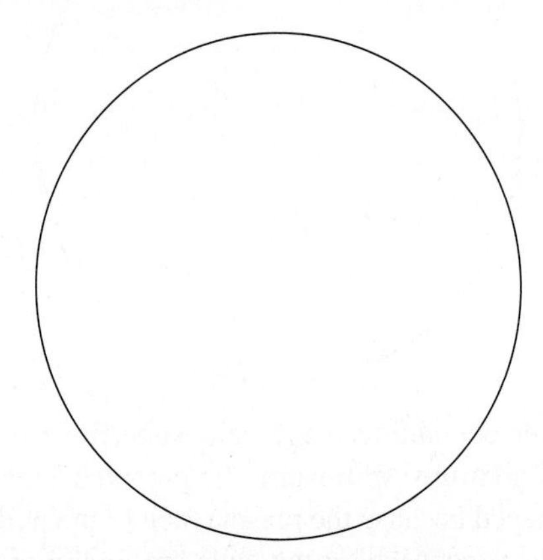

The most superficial layer of the integument is the **stratum corneum,** which is a tough layer consisting of dead, flattened cells that form the surface. Cells of the stratum corneum have been toughened by complete **keratinization;** thus, the epidermis is said to be made of **keratinized stratified squamous epithelium.**

Examine a prepared slide of thick skin and locate the various strata of the epidermis, as illustrated in figure 5.3.

(A) 7 Draw an example of the stratum corneum in the following space.

Illustration of stratum corneum:

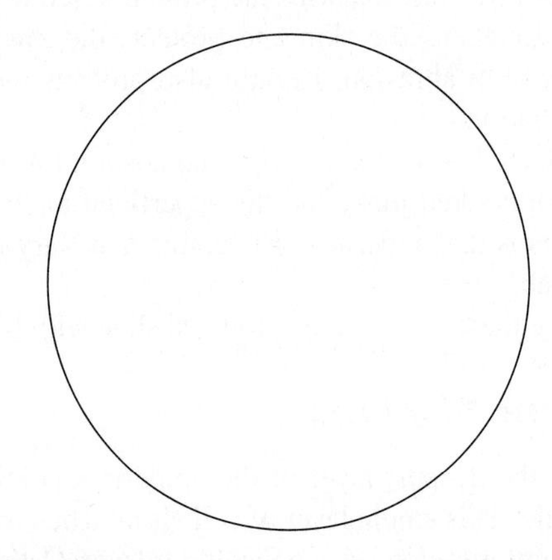

Cells in the epidermis go through three major phases. In the stratum basale, the cells are involved in rapid cell division. Superficial to this layer, cells undergo a process of producing precursor molecules to keratin. The final phase is the completion of keratinization, which leads to the death of the epithelial cells. The most superficial cells of the epidermis are tough. They protect the skin from microorganisms and desiccation and eventually slough off. The epidermis renews itself about every 6 to 8 weeks.

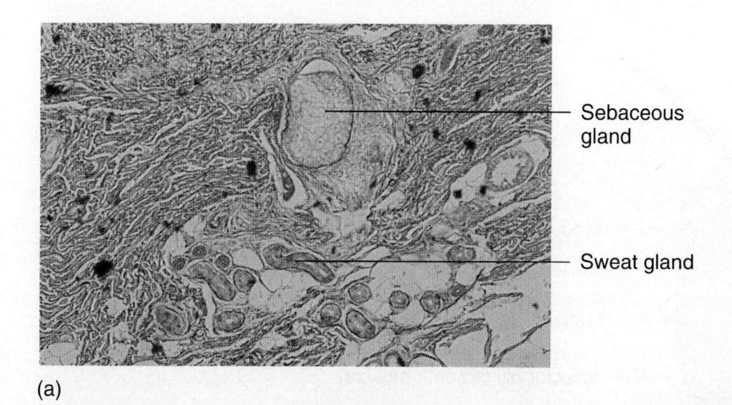

(a)

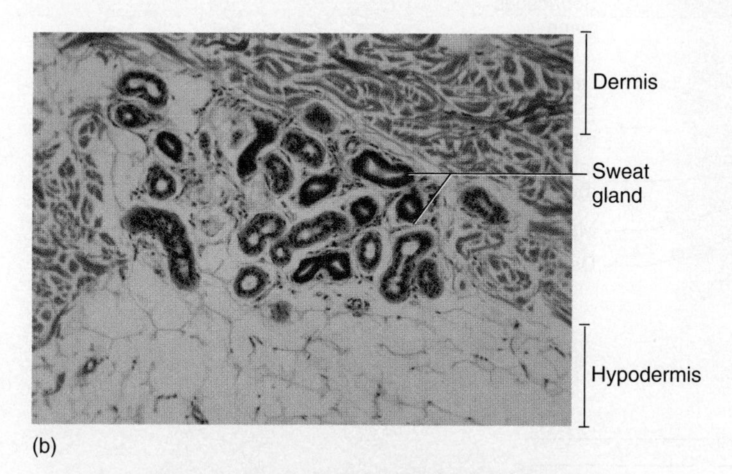

(b)

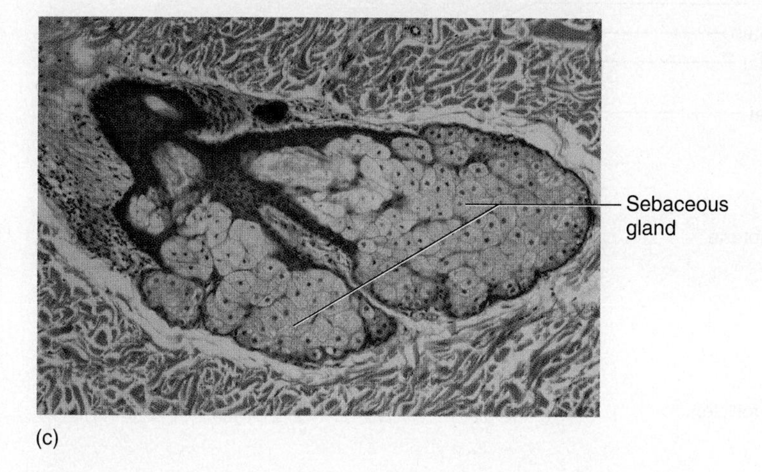

(c)

Figure 5.5 Integumentary Glands
(a) Overview of glands; (b) sweat glands (100×); (c) sebaceous glands (100×).

Integumentary Glands

A significant number of glands occur in the dermis and hypodermis (figure 5.5a). These include **sweat glands, lactiferous (milk) glands, sebaceous (oil) glands,** and **ceruminous (earwax) glands.** The sweat glands are composed of simple cuboidal epithelium and come in two different types. **Eccrine sweat glands,** the most common type, are sensitive to temperature and produce normal body perspiration. Perspiration reduces the temperature of the body by **evaporative cooling. Apocrine sweat glands** secrete water and a higher concentration of organic acids than eccrine glands. The organic acids, along with bacterial action, produce the characteristic pungent body odor. These glands are concentrated in the region of the axilla and groin and are associated with hair follicles.

(A) 8 Examine a prepared slide of skin and look for clusters of cuboidal epithelial cells with purple nuclei in the dermis or hypodermis. These are the sweat glands illustrated in figure 5.5b. Note that the glands are tubules, and when these are cut in section on your slide you are mostly seeing the tubules cut in cross section. Draw a sweat gland in the following space.

Illustration of sweat gland:

Sebaceous glands are typically associated with hair follicles and are larger than sweat glands. You may not see a hair follicle with the sebaceous gland if the section was made through the gland and not through the gland and follicle. These glands secrete an oily material called **sebum,** which lubricates the hair and causes the skin to repel water.

(A) 9 Examine a slide of hair follicles and compare the slide with figure 5.5c. Draw a sebaceous gland in the space provided.

Illustration of sebaceous gland:

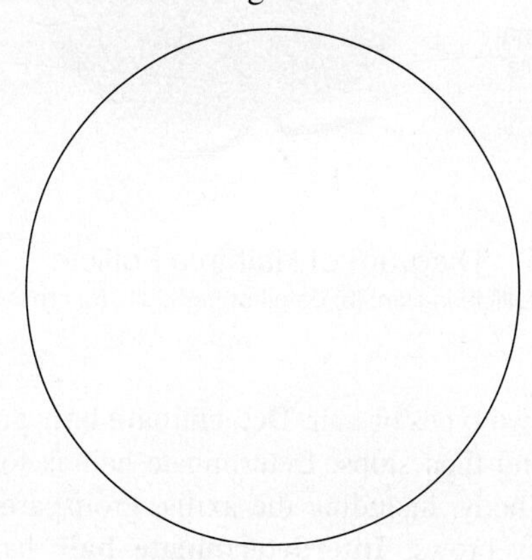

Hair

Hair is considered an accessory structure of the integumentary system; it consists of the **shaft,** a portion that erupts from the skin surface; the **root,** a deeper portion that is enclosed by the **follicle** (a pocket of epidermis that folds into the dermis); and the **hair bulb,** which is the actively growing portion of the hair. At the center of the bulb is a small structure known as the **papilla,** a region with blood vessels and nerves that reach the hair bulb.

(A) 10 Look at the prepared slide of hair, locate these structures, and compare them with figures 5.6 and 5.7.

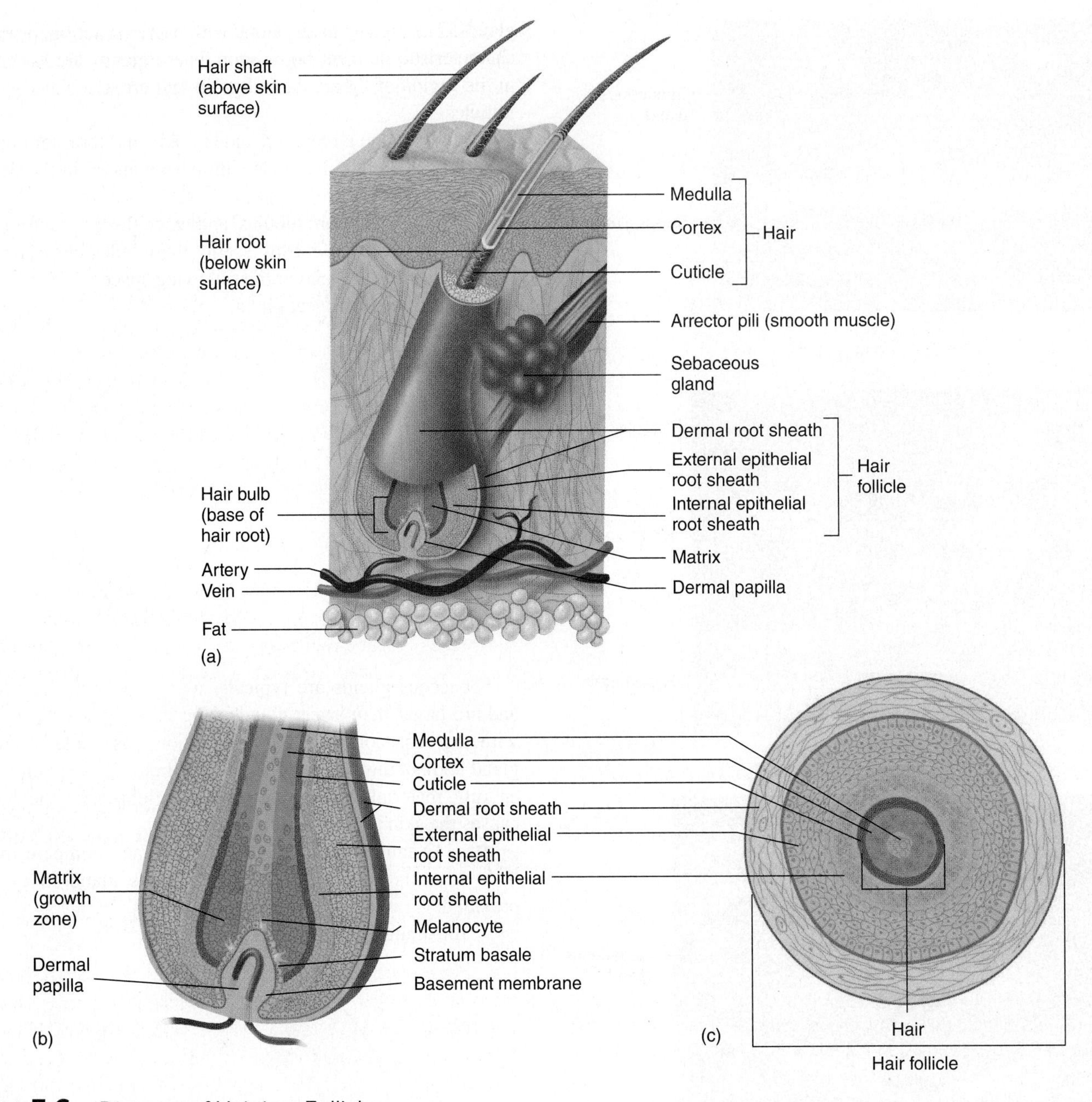

Figure 5.6 Diagram of Hair in a Follicle
(a) Overview of follicle in skin; (b) detail of hair bulb; (c) cross section of hair in follicle.

There are two types of hair. **Determinate hair** grows to a specific length and then stops. Determinate hair is found in many places in the body, including the axilla, groin, arms, legs, eyelashes, and eyebrows. **Interdeterminate hair** has continuous growth. It is found on the scalp and in facial hair in males.

(A) 11 Examine models and charts in the lab and compare them with figure 5.6.

The hair root is enclosed by a follicle, which is a pocket of epidermis in the dermis; thus, hair is considered an epidermal derivative. The follicle is composed of an outer dermal layer of connective tissue and an inner epithelial layer. These form the **root sheath** of the hair. The **follicle** encloses the hair in the same way a bud vase encloses the stem of a rose.

(A) 12 Draw a longitudinal section of hair.

Illustration of hair:

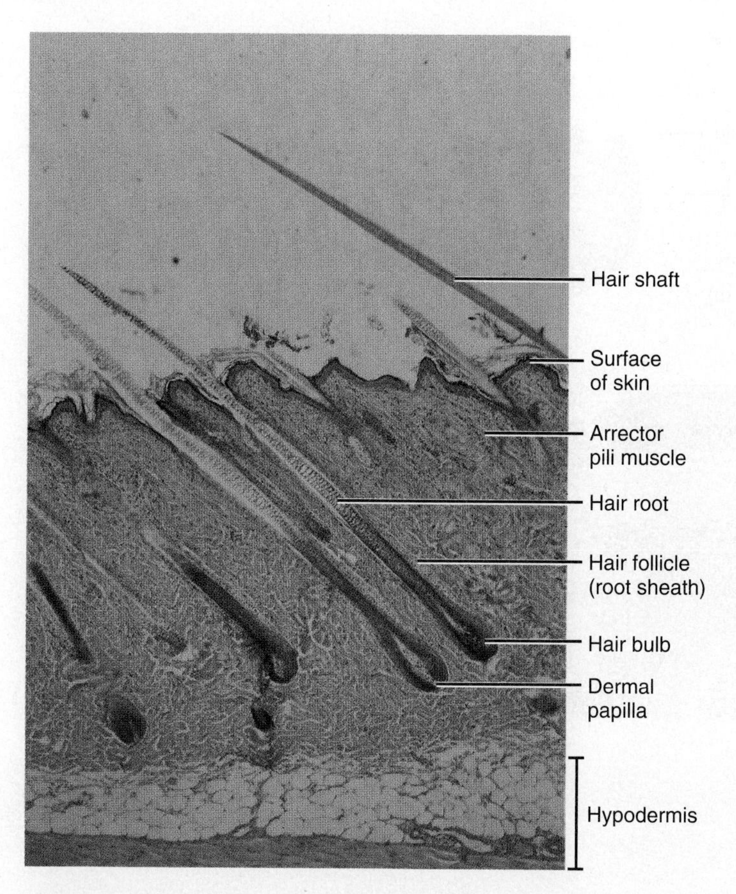

Figure 5.7 Photomicrograph of Hair in a Follicle (40×)

Arrector Pili

When the hair "stands on end," it does so by the contraction of **arrector pili muscles.** An arrector pili muscle is a cluster of parallel smooth muscle fibers that connects the hair follicle to the upper regions of the dermis. In response to cold conditions, the arrector pili muscles contract, producing goose bumps. The arrector pili also contract when a person is suddenly frightened. These muscles are controlled by the autonomic nervous system.

(A) 13 Find an arrector pili muscle in a prepared slide and compare it with figures 5.6 and 5.7. If you have trouble finding an arrector pili muscle, look for slender slips of smooth muscle that angle from the follicle to the superficial regions of the dermis. You may have to look at more than one slide to find them.

Cross Section of Hair

The central portion of the hair is known as the **medulla,** which is enclosed by an outer **cortex.** The cortex may contain a number of pigments, which give the hair its particular color. The brown and black pigments are due to varying types and amounts of melanin. Iron-containing melanin pigments produce red hair. Gray hair is the result of a lack of pigment or the presence of air in the hair root and shaft. The layer superficial to the cortex is the **cuticle** of the hair; it looks rough in microscopic sections. Figure 5.6 shows a cross section of hair.

Nails

Nails, the keratinized cells of the stratum corneum of the skin, are located on the distal ends of the fingers and toes. Most of the nail consists of the **nail body;** as it grows, the **free edge** of the nail is the part you clip. The **cuticle** of the nail is the **eponychium,** and the **nail root** is underneath the eponychium and is the area that generates the nail body. The **nail bed** is the layer deep to the **nail body,** whereas the **lunula** is the small, white crescent at the base of the nail. The **hyponychium** is the region under the free edge of the nail.

(A) 14 Examine figure 5.8 and compare your fingernails with the diagram. On each side of the nail is the **nail groove,** a depression between the body of the nail and the skin of the fingers. The ridge above the groove is the **nail fold.**

Skin Cancer

Skin cancer is a general term for different integumentary carcinomas. The three most common are squamous cell carcinoma, basal cell carcinoma, and melanoma. Although all three types of cancer can occur anywhere on the body, basal cell and squamous cell carcinomas are usually found in areas exposed to sunlight. Melanomas can occur anywhere and can either develop on their own or come from moles.

Basal cell carcinoma occurs as a waxy lump or a flat brown, red, or flesh-colored patch. They are called basal cell carcinomas because they arise in the stratum basale and, if detected early, they can be removed with a high success rate.

Squamous cell carcinoma usually is a red or scaly nodule. They occur in the stratum spinosum and have a low mortality rate if detected early.

Melanomas are the most lethal of the skin cancers, and if the tumor cells migrate out of the epidermis they are difficult to treat. Regular skin examinations are beneficial, especially if there is a family history of melanoma. Large moles, those with an irregular border, asymmetrical moles, and those of variable color should be evaluated as soon as possible by a dermatologist.

Case Study

The patient is a 60-year-old male whose grandmother died of skin cancer. He is active outdoors and has been a bicyclist for years. He presented to the hospital with a mole on the back of his neck, which tends to bleed. On further examination the physician notices lumps in his right axillary lymph node and some in the nodes of the groin. Excision of the mole and analysis have led to the diagnosis of melanoma. What steps would you take to determine whether the cancer is confined to the original site or whether it has metastasized? What physical evidence would you look for to consider whether the mole was possibly malignant or not?

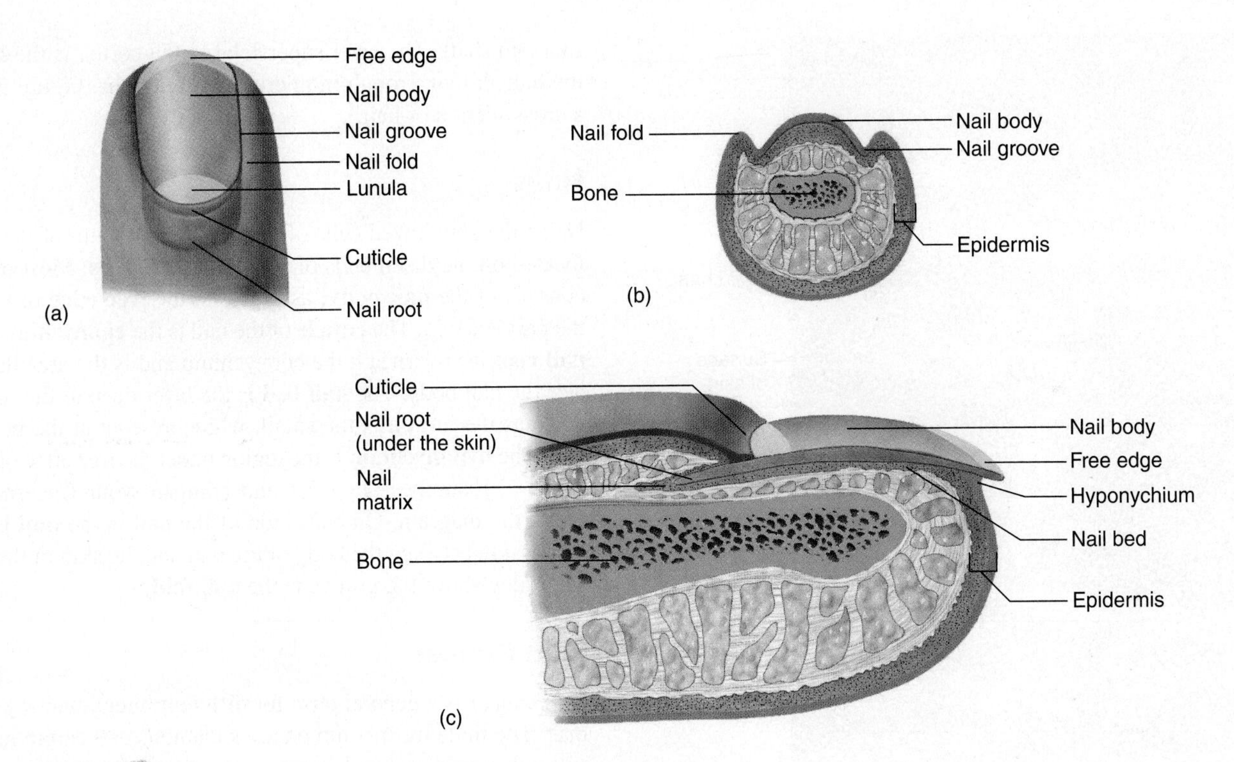

Figure 5.8 Fingernail

(a) Dorsal view; (b) cross section; (c) longitudinal section.

Exercise 5 Review

Name: ______________________

Lab time/section: ______________________

Date: ______________________

Integumentary System

1. The main organ of the integumentary system is the skin. Name three structures that are associated structures in the integumentary system. ______________________

2. Which one of the three layers (epidermis, dermis, subcutaneous tissue) is not part of the integument? ______________________

3. The dermis has two main layers. Which one of these is the most superficial? ______________________

4. What is the most common connective tissue fiber in the dermis? ______________________

5. The release of heat from the body by blood vessels occurs in what main layer of the integument? ______________________

6. Which stratum is the deepest layer of the epidermis? ______________________

7. Which stratum of the epidermis is the most superficial? ______________________

8. Approximately how long does it take for the epidermis to renew itself? ______________________

9. What cell type produces a pigment that darkens the skin? ______________________

10. What material makes the epidermis tough? ______________________

11. What integumentary gland secretes oil? ______________________

12. Which kind of sweat gland (eccrine, apocrine) is involved in evaporative cooling? ______________________

13. What part of the hair is found on the outside of the skin? ______________________

14. What does an arrector pili muscle do? ______________________

15. The outermost portion of the hair is known as the
    a. cuticle. b. lunula. c. root. d. medulla.

16. The hair of the axilla is considered determinate/indeterminate hair (circle the correct answer).

17. Electrolysis is the process of hair removal by using electric current. Explain how this might destroy the process of hair growth in relation to the hair bulb. ______________________________

____

____

18. Since hair color is determined by pigment in the cortex, and the hair shaft is dead, explain the fallacy of a person's hair turning white overnight. ______________________________

19. Since the hair shaft is dead, explain the fallacy of "feeding" the hair, as claimed by some cosmetics firms. ______________________________

20. Label the following illustration using the terms provided.

arrector pili	dermis	epidermis	hair follicle	hair root
hair shaft	sebaceous gland	stratum corneum	subcutaneous tissue	sweat gland

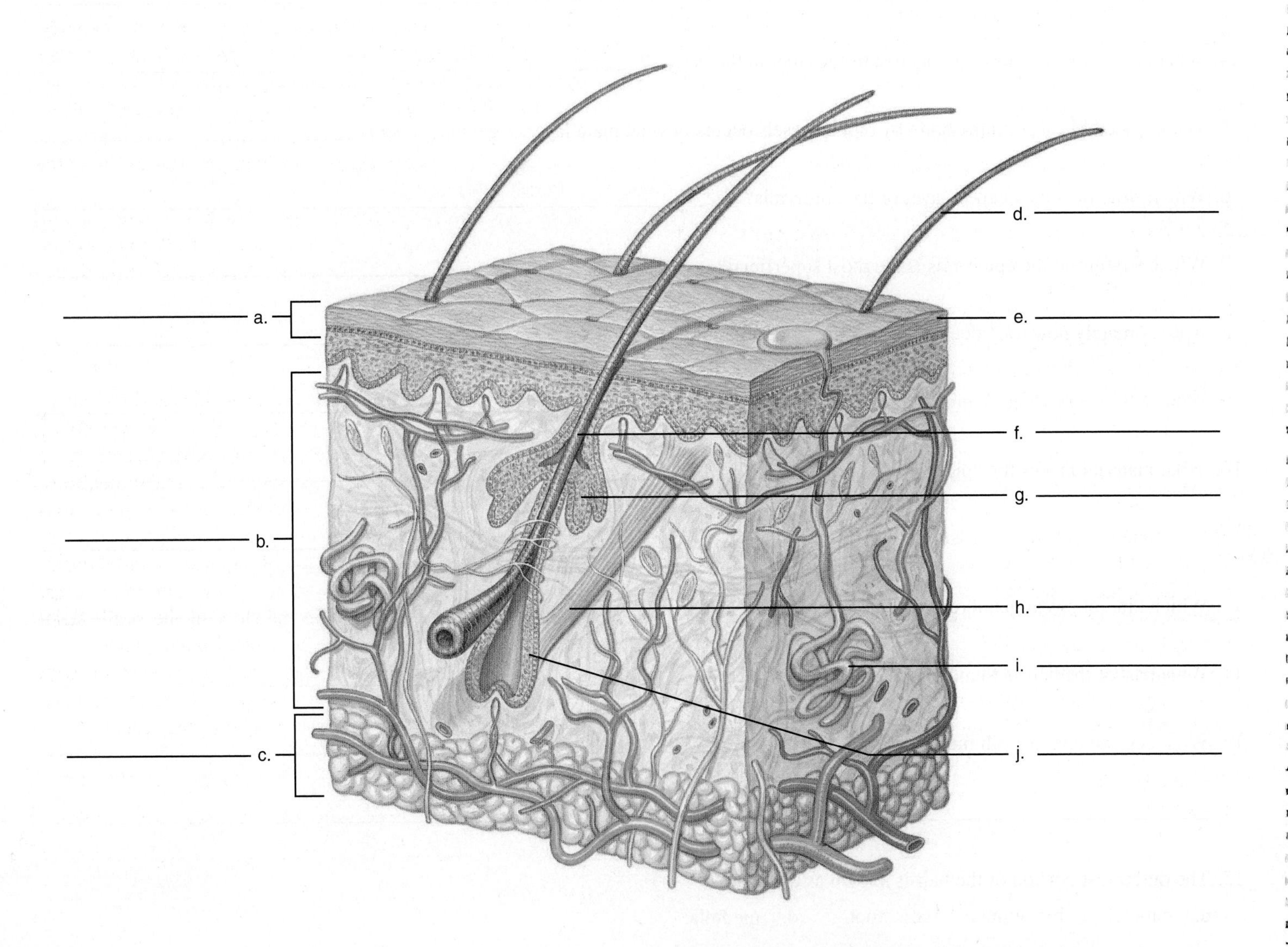

Exercise 6

Introduction to the Skeletal System

Introduction

The skeletal and muscular systems are intimately related. The skeletal system has many functions, including supporting the body; protecting soft tissues, such as the brain and lungs; providing a structure for movement; storing minerals; and forming blood cells. These topics are covered further in the Seeley text in chapter 6, "Skeletal System: Bones and Bone Tissue." The interaction between the skeletal and muscular systems is one where the skeletal system provides the lever system on which the muscles can work, resulting in movement. In this exercise, you examine the structure and composition of the skeletal system.

Objectives

At the end of this exercise, you should be able to

1. describe the composition of bone tissue;
2. discuss how the skeletal system provides for efficient movement;
3. name specific projections and depressions on a bone, such as a tubercle or fossa;
4. list the two major types of bone material found in long bones;
5. draw or describe the microscopic structure of compact bone;
6. relate the anatomy of the bone to its growth;
7. classify a bone according to its shape.

Materials

Chart of the skeletal system
Cut sections of bone showing compact and spongy bone
Articulated human skeleton
Articulated cat skeleton
Disarticulated human skeleton
Prepared slide of ground bone
Prepared slide of decalcified bone
Bone heated in oven at 350°F for 3 hours or more
Bone placed in vinegar for several weeks or 1 N nitric acid for a few days
Microscopes

Procedure

Divisions of the Skeletal System

(A) 1 Examine the skeletons or charts in the lab and locate the division of the skeletal system called the axial skeleton. The **axial skeleton** is so named because it is in the vertical axis of the body. It consists of the skull, hyoid bone, vertebral column, ribs, and sternum. All the other parts of the skeletal system, including the **pectoral girdle, upper limb bones, pelvic girdle,** and **lower limb bones,** belong to the **appendicular skeleton.** These bones are "appended" (attached) to the axial skeleton. Locate the major bones in the lab and compare them with figure 6.1 and table 6.1.

Major Bones of the Body

Figure 6.1 shows the major bones and bone regions of the body. You should be familiar with them before you study the specifics of the individual bones in later exercises. It helps to take a bone from a disarticulated skeleton and compare it with an articulated skeleton.

(A) 2 Locate the major bones in the figure as you examine the bones in the lab.

The typical young adult has about 206 bones, with more in younger individuals and fewer in older individuals. Many developing bones fuse as a person ages, forming larger bones, thus reducing the overall number.

Bone Shape

Bones are classified according to four main shapes: long, short, flat, and irregular. **Long bones** are longer than they are broad. Bones such as those of the arm, forearm, fingers, thigh, and leg are long bones. **Short bones** are more or less equal in all dimensions. Bones of the wrist and ankle are examples of short bones. **Flat bones,** such as bones of the cranium and sternum, appear compressed in one dimension. The ribs and scapulae are also considered flat bones. **Irregular bones** do not fit into any of the other categories. Bones such as those in the floor of the skull, facial bones, vertebrae, and pelvic girdle bones are irregular bones.

(A) 3 Examine bones in the lab and compare them with figure 6.2. Find a bone that represents each of the bone shapes. Write the name of the bone in the following spaces:

Long bone ____________________

Short bone ____________________

Flat bone ____________________

Irregular bone ____________________

Bone Features

Bones have many surface features that have specific names. These features are grouped into **projections** that arise from the surface of

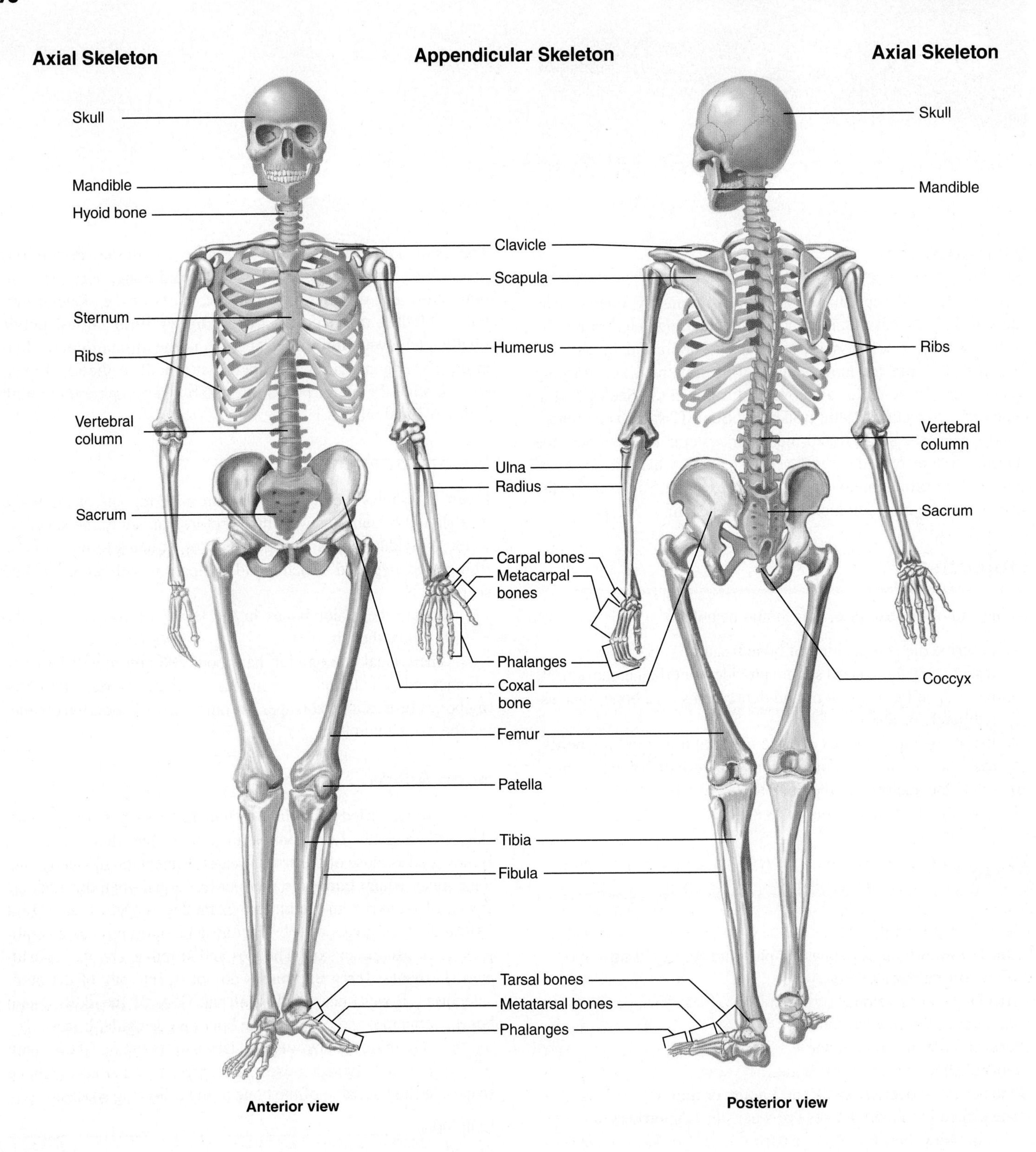

Figure 6.1 Major Bones and Bone Regions of the Body

TABLE 6.1 Axial and Appendicular Skeleton

	Number of Bones		Number of Bones
Axial Skeleton			
Skull	22	Ribs	24
Hyoid bone	1	Sternum	1
Vertebral column	26	Auditory ossicles	6
Appendicular Skeleton			
Pectoral girdle		Pelvic girdle	
Clavicle	2	Coxal bone	2
Scapula	2	Lower limb	
Upper limb		Femur	2
Humerus	2	Patella	2
Radius	2	Tibia	2
Ulna	2	Fibula	2
Carpals	16	Tarsals	14
Metacarpals	10	Metatarsals	10
Phalanges	28	Phalanges	28

the bone and **depressions** and **cavities** in the bone. These are presented in table 6.2.

(A) 4 Study these terms and be able to find them on bones in the lab.

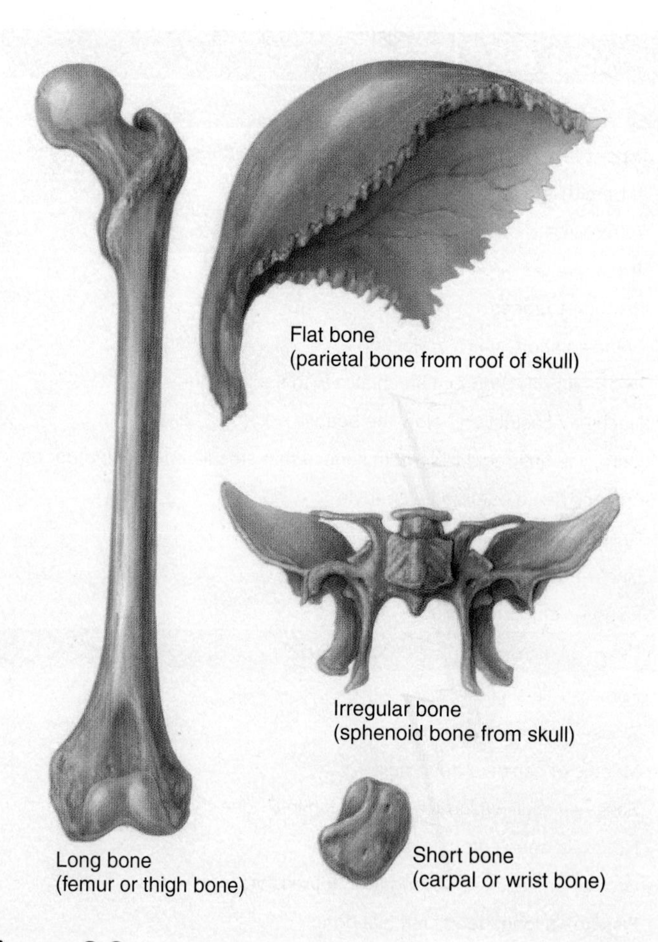

Figure 6.2 Bone Shapes

Composition of Bone Tissue

Bone consists of both organic and inorganic material. The organic matter is composed mostly of cells and collagenous fibers and makes up about 35% of the bone by weight. The inorganic matter makes up the remaining 65% of the bone and mostly consists of **hydroxyapatite,** a complex salt consisting of calcium phosphate and calcium carbonate.

(A) 5 If available in the lab, examine bones that have been soaked in acid. The acid has dissolved the mineral salts, leaving the collagenous fibers as a flexible framework. Examine the bones that were baked in an oven for a few hours. In this preparation, the structural characteristics of proteins have been destroyed, and although the bones remain hard because of the mineral salts, they are brittle due to the heat's destruction of the protein fibers. Your instructor may wish to demonstrate the fragility of specially prepared bones.

Bone Structure

The general anatomy of a long bone consists of the proximal and distal ends of the bone, called the **epiphyses,** and the shaft of the bone, called the **diaphysis** (figure 6.3). The epiphyses of bones have **articular cartilage** at the ends of the bone. The articular cartilage is composed of hyaline cartilage; it helps reduce friction as the joint moves.

The general structure of a long bone consists of the proximal and distal ends of the bone, called the **epiphyses,** and the shaft of the bone, called the **diaphysis.** The epiphyses of bones have **articular cartilage** at the ends of the bone. The articular cartilage is composed of hyaline cartilage and helps reduce friction as the joint moves.

In individuals who are growing in height, there is a plate of hyaline cartilage between the epiphysis and the diaphysis. This is the **epiphyseal plate,** which increases in thickness by division of

TABLE **6.2**

Projections from Bones	Example
Process—a general term for a projection from the surface of the bone	Styloid process of ulna
Tubercle—a relatively small bump on a bone	Greater tubercle of humerus
Tuberosity—a relatively larger bump on a bone	Deltoid tuberosity of humerus
Trochanter—a large bump (on femur)	Greater and lesser trochanter of femur
Ramus—a branch	Ramus of mandible
Spine—a short, sharp projection	Vertebral spine
Head—a projection that articulates with another bone	Head of femur
Neck—a constriction below the head	Neck of rib
Condyle—an irregular, smooth surface that articulates with another bone	Lateral condyle of femur
Epicondyle—a bump on a condyle	Medial epicondyle of humerus
Crest—an elevated ridge of bone	Crest of ilium
Line—an elevation smaller than a crest	Gluteal line of ilium
Facet—a smooth, flat face	Articular facets of vertebrae
Depressions or Cavities in Bone	**Example**
Foramen—a shallow hole	Foramen magnum of occipital bone
Sinus—a cavity in a bone	Maxillary sinus
Meatus or canal—a deep hole	External auditory meatus of temporal bone
Fossa—a shallow surface depression in a bone	Iliac fossa
Notch—a deep cutout in a bone	Greater sciatic notch of ilium
Groove or sulcus—an elongated depression	Intertubercular groove of humerus
Fissure—a long, deep cleft in a bone	Inferior orbital fissure

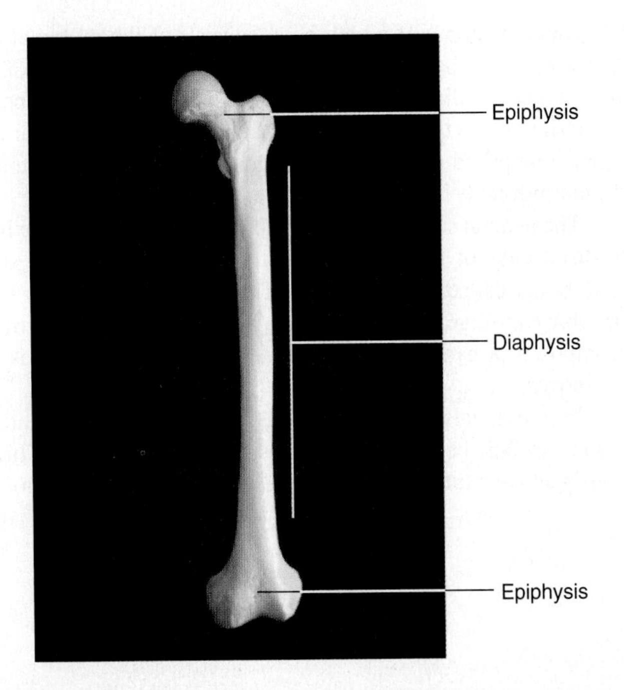

Figure 6.3 Regions of a Long Bone, Anterior View

the chondrocytes. As the cartilage grows, the person increases in height. During growth, the cartilage away from the epiphyseal plate is replaced by bone. When a person reaches adult height, all the cartilage is replaced by bone and the remnants of the plate are seen as the **epiphyseal line** (figure 6.4a).

(A) 6 Examine long bones cut lengthwise for the epiphyseal line. In the cut sections of bone you can also see the hard, **compact (dense) bone** on the outside of the bone and the inner **spongy (cancellous) bone** (fig. 6.4b). The spongy bone is made of **trabeculae,** thin rods or plates of bone that run in the same direction as the stress applied to the bone. Stress on the bone may be in the form of gravity, or it may occur due to a common force applied to the limb. Trabeculae make an internal framework that strengthens the bone.

The innermost section of bone is hollow and is called the **marrow** (or **medullary**) **cavity.** Marrow is the material that occupies this cavity and can be of two types, a hemopoietic **red marrow** or an adipose-containing **yellow marrow.** In flat bones there is an outer and an inner layer of compact bone with spongy bone sandwiched between them. In cranial bones the spongy bone is called **diploe**.

Compare the cut bones in the lab with figure 6.4a and b and note the features listed in the discussion. Note how the compact bone in the middle of the long section of bone is thicker than the

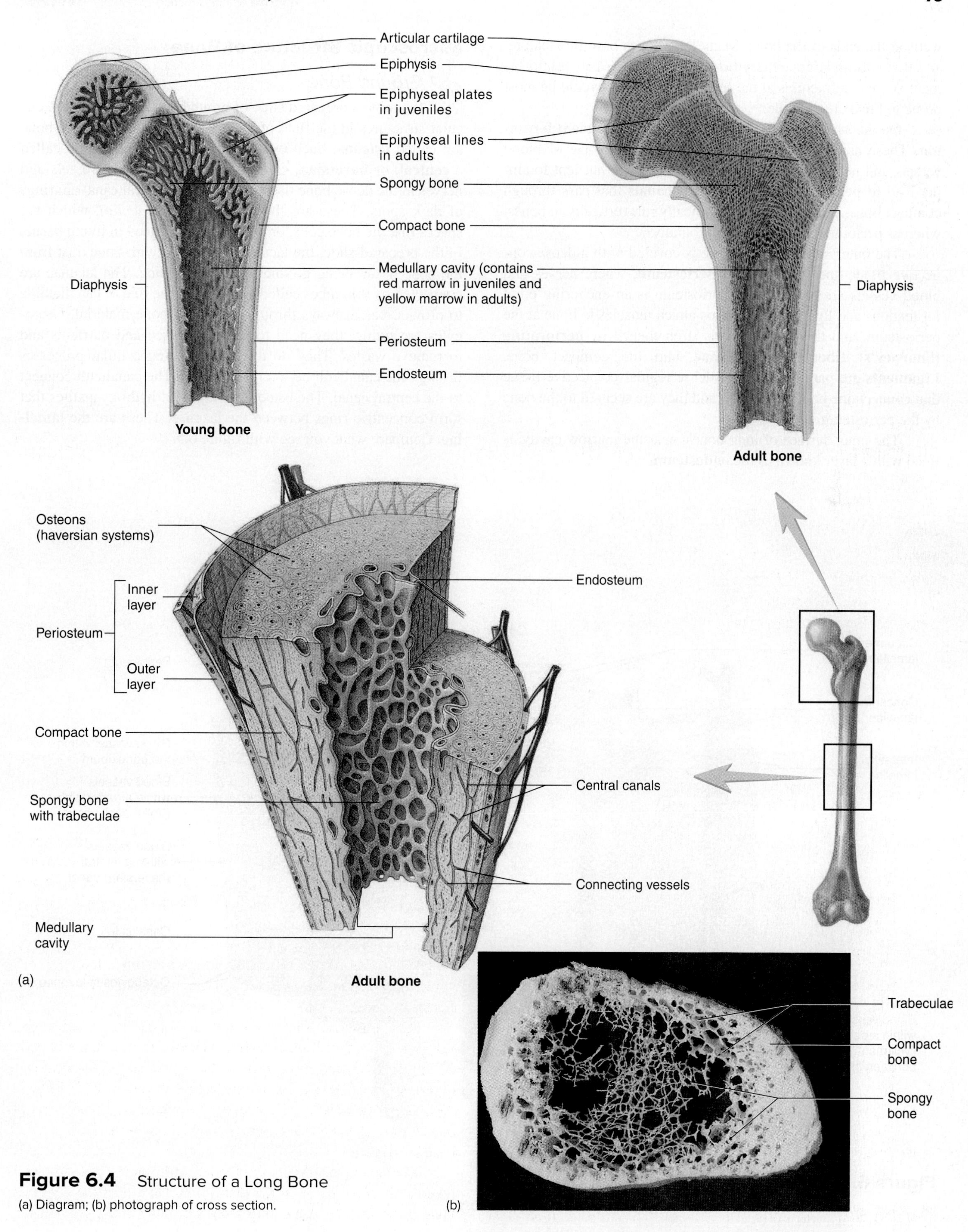

Figure 6.4 Structure of a Long Bone
(a) Diagram; (b) photograph of cross section.

walls at the ends of the bone. Many long bones have this feature, and they resemble an archery bow in this regard. This reinforcement decreases fractures at the region of bone that would be most prone to break, the middle.

On the surface of the diaphysis, locate the **nutrient foramina.** These are small holes that allow for the passage of blood vessels and nerves into and out of the bone. The nutrient foramina lead to **perforating (Volkmann's) canals** that pass through compact bone. The central canals typically run vertically in bones, whereas perforating canals run horizontally.

The outer surface of the bone is covered with a dense connective tissue sheath called the **periosteum,** where nerves and blood vessels are located. The periosteum is an anchoring point for tendons and ligaments. **Tendons** attach muscles to bone at the periosteum, and this attachment is strengthened by **perforating (Sharpey's) fibers** that penetrate into the compact bone. **Ligaments** are parallel straps of dense regular connective tissue that connect one bone to another, and they are secured to the bone by the periosteum.

The inner surface of long bones, near the marrow cavity, is lined with a layer known as the **endosteum.**

Microscopic Structure of Bone

(A)[7] *Ground Bone*

Examine a prepared slide of ground bone and locate the circular structures in the field of view. These modular units of bone are called **osteons.** Each osteon has a hole in the middle called a **central,** or **haversian, canal,** which houses blood vessels and nerves in the dense bone tissue. Around the central canal are rings of dark spots. These are the **lacunae** (sing. *lacuna*), which are spaces that the bone cells, or **osteocytes,** occupied in living tissue. In the prepared slide, the lacunae have filled with bone dust from the bone tissue being ground to make the slide. The lacunae are connected by thin tubes called **canaliculi.** The role of canaliculi is to provide passageways through the dense bone material. Osteocytes are living; they need to receive oxygen and nutrients and to remove wastes. They do this by extending cellular processes through the canaliculi between osteocytes. The canaliculi connect to the central canal. The osteon has layers of hydroxyapatites that form concentric rings between the lacunae. These are the **lamellae.** Compare what you see with figure 6.5.

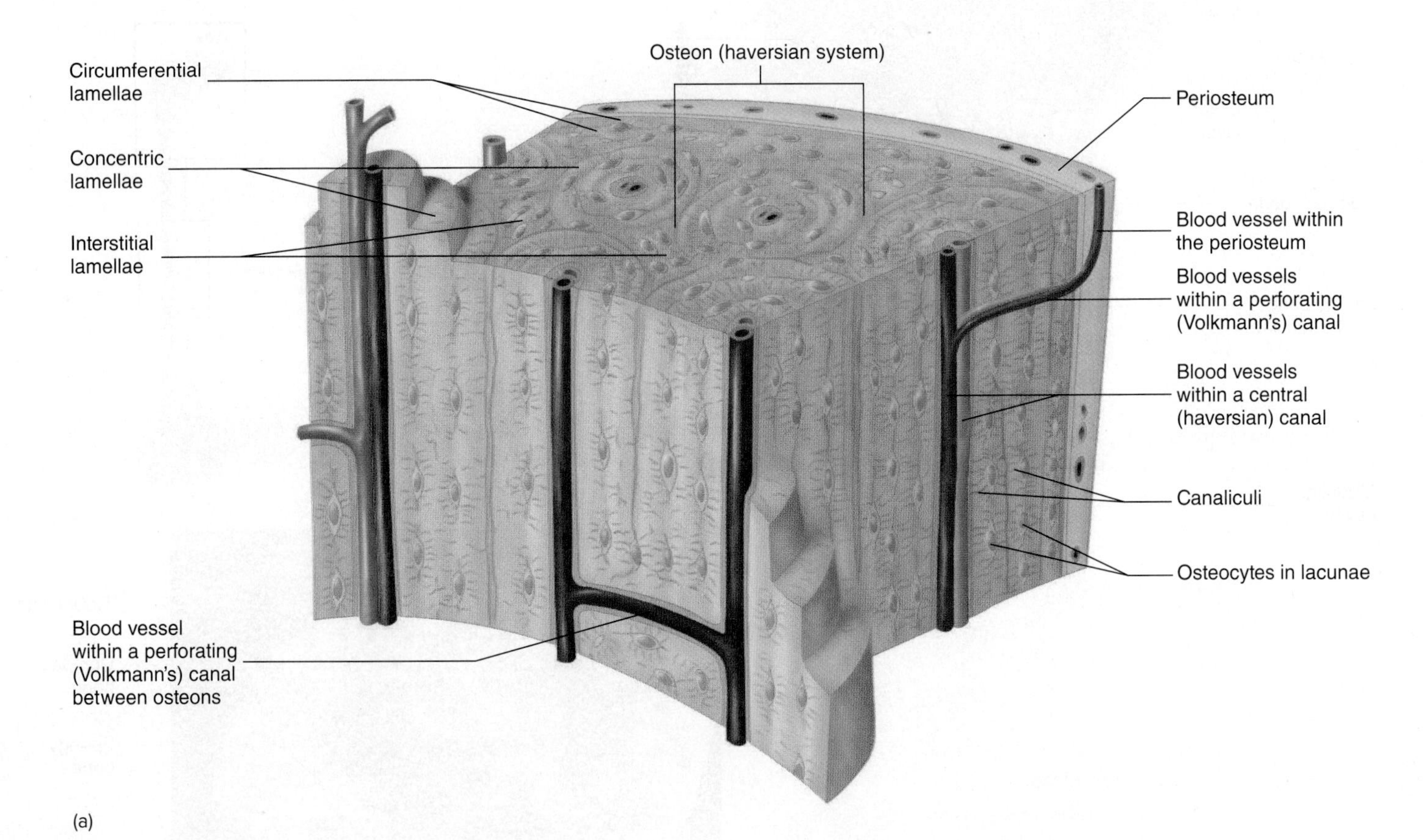

Figure 6.5 Histology of Compact Bone

(a) Diagram; (b) photomicrograph, cross section (100×); (c) photomicrograph (400×).

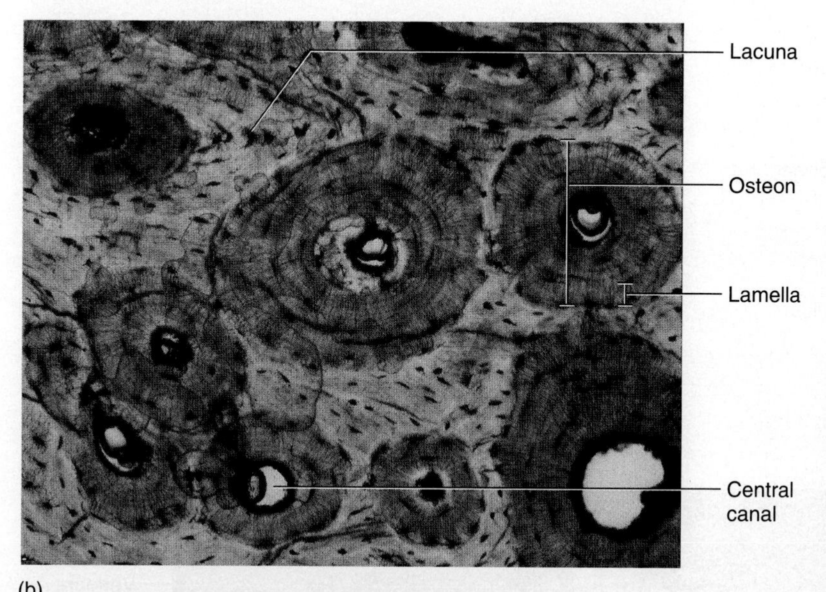

(b)

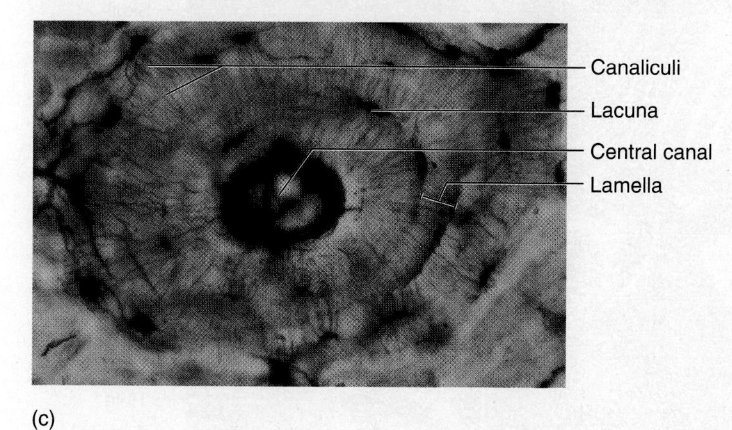

(c)

Figure 6.5 *Continued.*

Bone Cells

There are three main types of bone cells: **osteoblasts, osteocytes,** and **osteoclasts.** Osteoblasts originate from stem cells called **osteogenic progenitor cells.** Osteoblasts produce new bone and become osteocytes, which are mature bone cells. Osteocytes sense the stresses placed on bone and add more bone material if needed. They also become activated in fractures.

Osteoclasts are involved in bone reabsorption. They increase the size of the marrow cavity as a person grows from a child to an adult. Osteoclasts are formed from the fusion of cells, such as macrophages, and each cell has multiple nuclei.

Decalcified Bone

(A) 8 Examine a slide of decalcified bone. In this preparation, the bone salts have been dissolved away and the remaining thin section of bone has been sliced, mounted, and stained. The periosteum is visible on the surface of the bone, with the compact bone in the middle and the endosteum on the inside of the bone (figure 6.6). Locate the periosteum, compact bone, and endosteum.

The cells that are the most common in compact bone are the **osteocytes.** In ground bone, only the spaces where the osteocytes occur are visible (the lacunae), but, in decalcified bone, the actual osteocytes are visible. Look for osteocytes and compare them with figure 6.7.

Osteoblasts are found typically on the outer or inner surface of bone and have one nucleus. Osteoclasts are also found on the surfaces of bone, but each cell has many nuclei. Osteocytes are seen in the lacuna of bone tissue.

Skeletal Anatomy of the Cat

The cat skeleton differs from the human skeleton in several ways. Humans are bipedal, and this is reflected in the vertebral column. In humans, the bodies of the vertebrae are smaller at the superior end and get progressively larger in the inferior regions of the vertebral column. Cats are quadrupeds, so the weight of the vertebral column is distributed more evenly and the vertebrae are more evenly matched in size. Cats also have fully functional tail vertebrae.

Another difference is that humans walk on the entire inferior surface of the foot, whereas cats walk on the distal parts of their feet. Examine a cat skeleton in the lab and note the general features of the skeleton (figure 6.8).

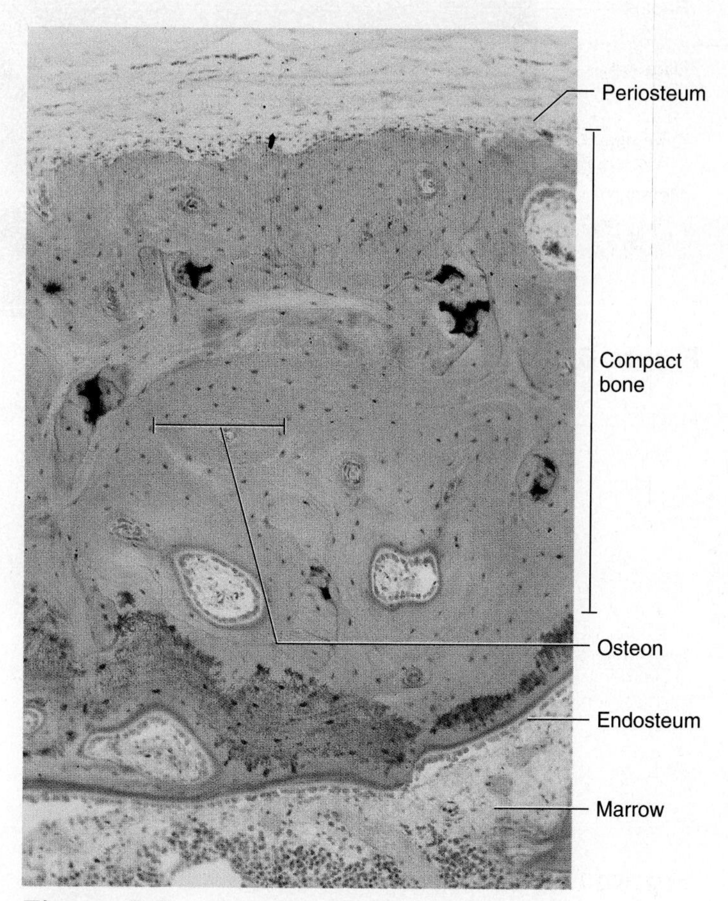

Figure 6.6 Decalcified Bone (100×)

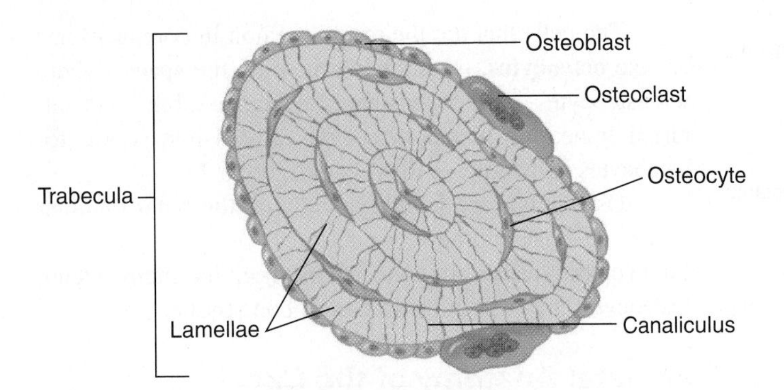

Figure 6.7 Three Types of Bone Cells

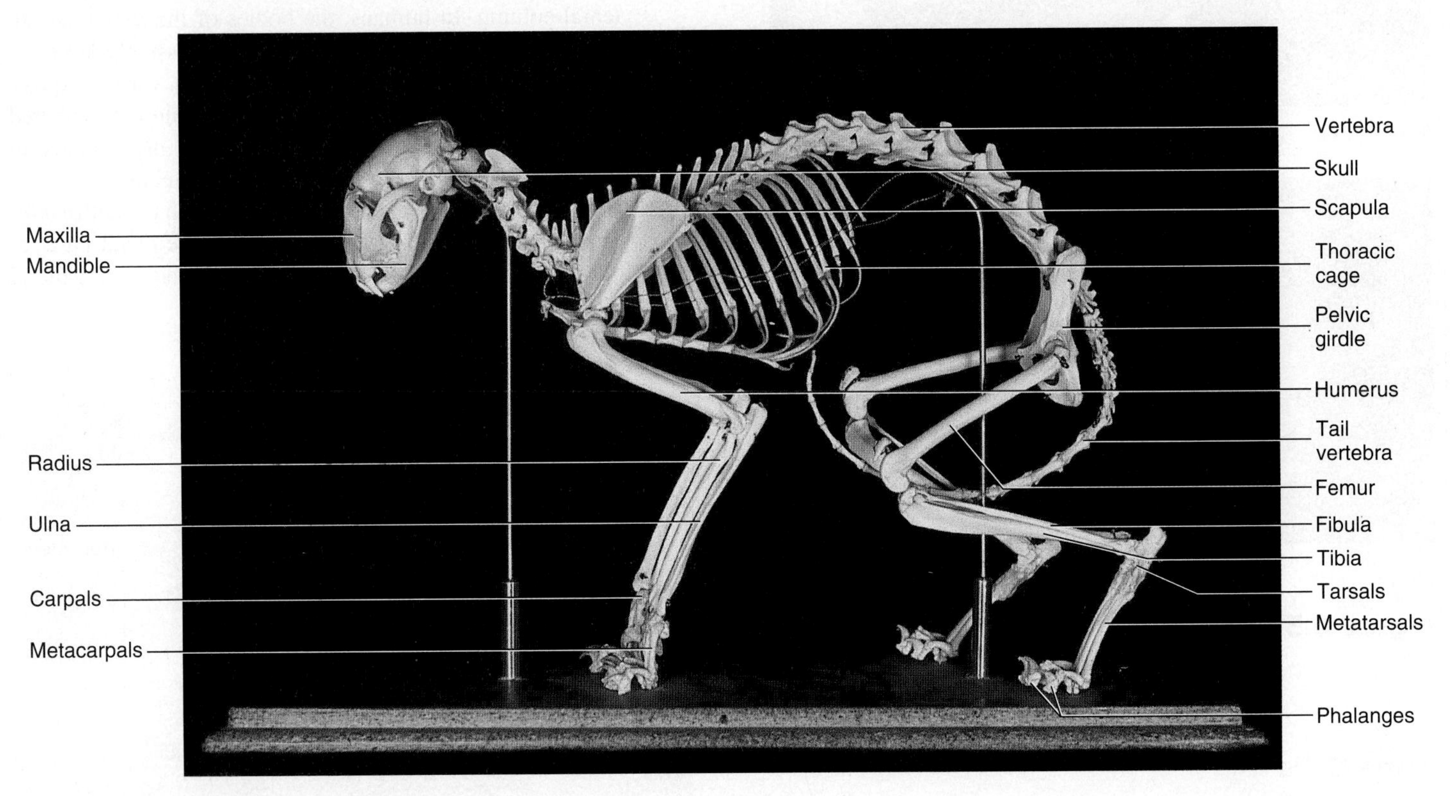

Figure 6.8 Cat Skeleton, Lateral View

Exercise 6 Review

Introduction to the Skeletal System

Name: ______________________

Lab time/section: ______________________

Date: ______________________

1. Label the illustration of the skeletal system using the terms provided.

carpals	clavicle	coxal bone	femur	fibula	humerus	mandible
metacarpals	metatarsals	patella	phalanges (feet)	phalanges (hand)	radius	rib
sternum	tarsals	tibia	ulna	vertebra		

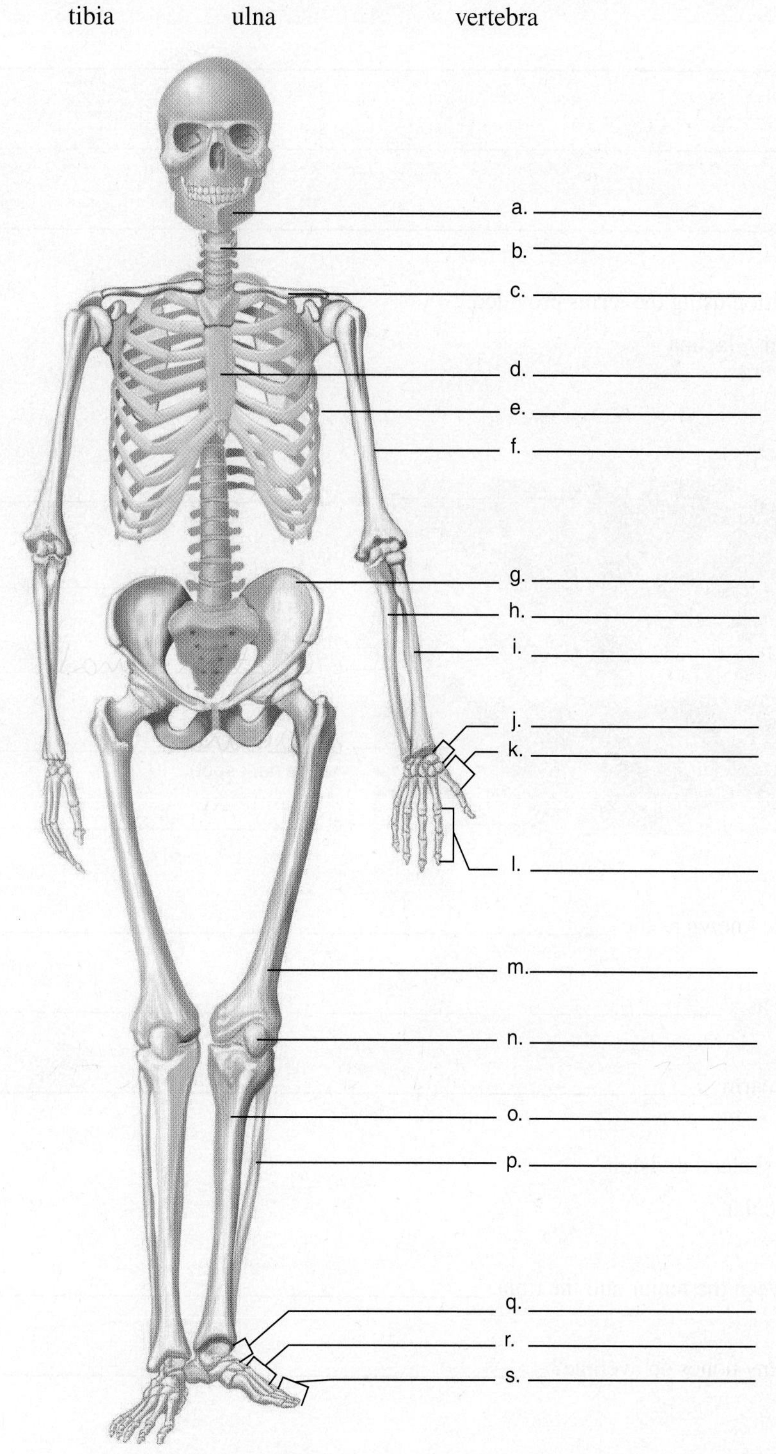

2. The hyoid bone belongs to the
   a. appendicular skeleton. b. axial skeleton. c. upper limb. d. skull.

3. The clavicle belongs to the
   a. axial skeleton. b. pectoral girdle. c. pelvic girdle. d. upper limb.

4. In the disease osteoporosis, there is a significant loss of spongy bone. Explain how the loss of this specific bone material can weaken a bone. ____________________

5. Label the following illustration using the terms provided.

   canaliculi central canal lacuna
   lamellae osteon

a. Osteon
b. Canaliculi
(the thin lines)
c. Central canal
d. lacuna
(the dark spot)
e. lamellae

6. The ends of a long bone are known as the ____________________.

7. A carpal bone is classified as a ____________________ bone in terms of shape.

8. Name two bones of the forearm. ____________________

9. The ribs are part of which skeletal division?
   a. axial b. appendicular

10. Name the bone found between the femur and the tibia. ____________________

11. A young adult has how many bones on average? ____________________

12. The inorganic portion of bone tissue is made of what complex mineral salt? ______

13. In terms of shape, what are the bones of the cranium?
    a. long b. short c. flat d. irregular

14. In terms of shape, what is the sternum?
    a. long b. short c. flat d. irregular

15. In terms of shape, what are the metacarpals?
    a. long b. short c. flat d. irregular

16. In terms of shape, what are the tarsals?
    a. long b. short c. flat d. irregular

17. What is an osteon? ______

    ______

    ______

18. Synthetic bone material known as hydroxyapatite is often molded into the shape of bone. This procedure is done to replace bone that has been diseased, damaged, or surgically removed. Cells in the body remodel the synthetic material and produce new bone. What bone cells do you predict will be involved in this remodeling, and what roles do these cells have?

    ______

    ______

    ______

    ______

19. What role do osteocytes have in bone tissue?

20. How does the shape of a long bone resist breaking when put under stress?

21. Describe the nature of decalcified bone. What was removed in the process of decalcification, and what impact did this have on the bone structure?

22. How does the central canal differ from a lacuna in terms of size, location, and the material found in each respective space?

23. What is another name for the shoulder blade, and what two bones attach to it?

NOTES

Exercise 7

Appendicular Skeleton

Introduction

In this exercise, you examine the appendicular skeleton, which consists of approximately 126 bones in four major groups. The first group is the **pectoral girdle,** which is composed of the scapula and the clavicle on each side. The next is the **upper limb,** which consists of the humerus, radius, ulna, carpals of the wrist, metacarpals of the palm, and phalanges of the fingers. The lower portion of the appendicular skeleton is composed of the **pelvic girdle** (the hipbones). Each hipbone consists of three fused bones, the ilium, the ischium, and the pubis. The **lower limb** is the final group of the appendicular skeleton and consists of the femur, patella, tibia, fibula, tarsals of the ankle, metatarsals of the foot, and phalanges of the toes. These topics are discussed further in the Seeley text in chapter 7, "Skeletal System: Gross Anatomy."

Objectives

At the end of this exercise, you should be able to

1. locate and name all the bones of the appendicular skeleton;
2. name the significant surface features of the major bones;
3. place bones into one of the four major groups;
4. orient individual bones to their proper position in the body;
5. determine whether bones of the scapula, pelvic girdle, and long bones of the upper and lower limbs are from the left or right side of the body;
6. determine the gender of isolated or articulated hipbones;
7. name the individual carpal bones and tarsal bones in articulated hands and feet, respectively.

Materials

Articulated skeleton or plastic cast of articulated skeleton
Disarticulated skeleton or plastic casts of bones
Charts of the skeletal system
Wooden applicator sticks or plastic drinking straws cut on a bias for pointer tips
Foam pads of various sizes (to protect real bones from hard countertops)

Procedure

(A)[1] Locate the bones of the appendicular skeleton on the material in the lab and in Exercise 6 (figure 6.1). Compare an isolated (or disarticulated) bone with the articulated skeleton (one that is joined together). Hold the bone up to the skeleton to see how it is positioned in relation to the other bones of the body. When examining bones in the lab, do not use your pen or pencil to locate a structure. They leave marks on bones. Use a wooden applicator stick or a cut plastic straw to point out structures. Your instructor may want you to place real bone material on foam pads to cushion the bone from the tabletop. Use your time in lab to find the material at hand. Once you are at home, draw or trace material from your lab manual and be able to name all of the parts.

Pectoral Girdle

The pectoral, or shoulder, girdle consists of the right and left **scapulae** (sing. **scapula**) and the right and left **clavicles.** The pectoral girdle provides a movable yet stable support for the upper limb.

Scapula

(A)[2] Locate the features of the scapula in figure 7.1. The scapula, commonly known as the shoulder blade, is found on the posterior, superior portion of the thorax. The scapula is roughly triangular in shape and has three borders, a **superior border;** a **medial,** or **vertebral, border;** and a **lateral,** or **axillary, border.** On the superior border of the scapula is an indentation known as the **scapular notch,** which contains a nerve that innervates shoulder muscles. On the anterior surface of the scapula is a smooth, hollowed depression known as the **subscapular fossa.** The posterior surface of the scapula is divided by the **scapular spine** into a superior depression known as the **supraspinous fossa** and an inferior depression known as the **infraspinous fossa.** The spine of the scapula is actually a crest that runs from the vertebral border to the lateral edge of the scapula, where it expands to form the **acromion** (ă-krō´mē-on) **process.** Another projection from the scapula is the **coracoid process,** which projects anteriorly. The scapula also has two sharp angles known as the **inferior angle** and the **superior angle.** The humerus moves in the shallow depression of the scapula, known as the **glenoid cavity** or **glenoid fossa.** Above the cavity is the **supraglenoid tubercle,** a process to which the biceps brachii muscle attaches. Just below the glenoid cavity is a bump known as the **infraglenoid tubercle,** which is an attachment point for the triceps brachii muscle. The features of the scapula are important to know because many muscles attach to it. You should also be able to distinguish a right scapula from a left one. The spine of the scapula is posterior, the inferior angle is less than 90°, and the glenoid cavity is lateral for the articulation with the humerus.

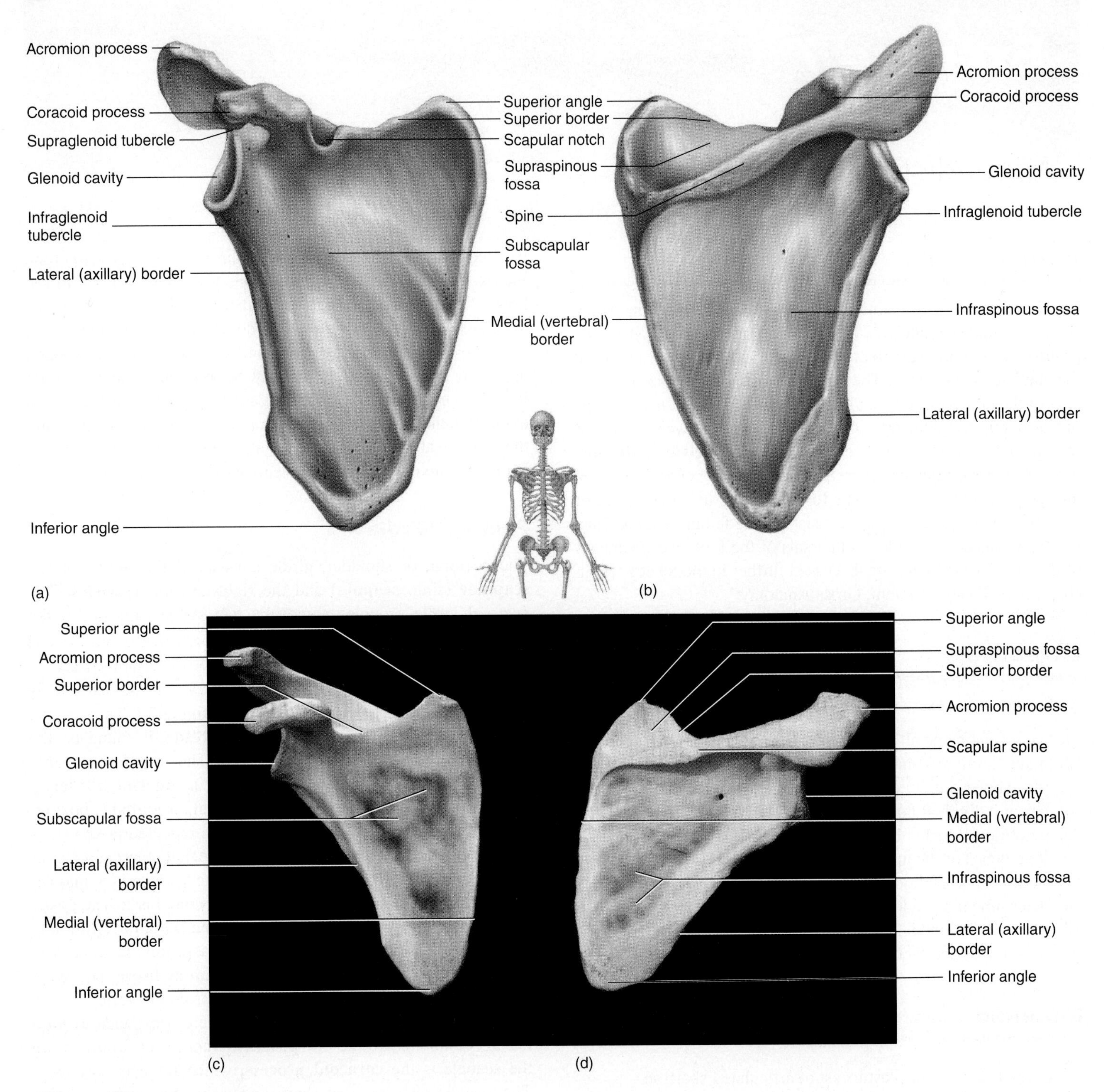

Figure 7.1 Right Scapula

Diagram showing (a) anterior and (b) posterior views; photograph showing (c) anterior and (d) posterior views.

Clavicle

(A) 3 Examine the clavicles in the lab and compare them with figure 7.2. The **clavicle** is commonly known as the collarbone, and it is the most commonly fractured bone in the body. It is a small bone that is a strut between the scapula and the sternum. The blunt, vertically truncated end of the clavicle is the **sternal end,** and the horizontally flattened end is the **acromial end.** Note the **conoid** (kaw´-noyd) **tubercle** on the inferior surface of the clavicle.

Upper Limb

(A) 4 Examine the bones of the upper limb as you read the following material.

Humerus

The **humerus** is a long bone that articulates with the scapula at its proximal end and with the radius and ulna at its distal end. The **head** of the humerus is a hemispheric structure that ends in

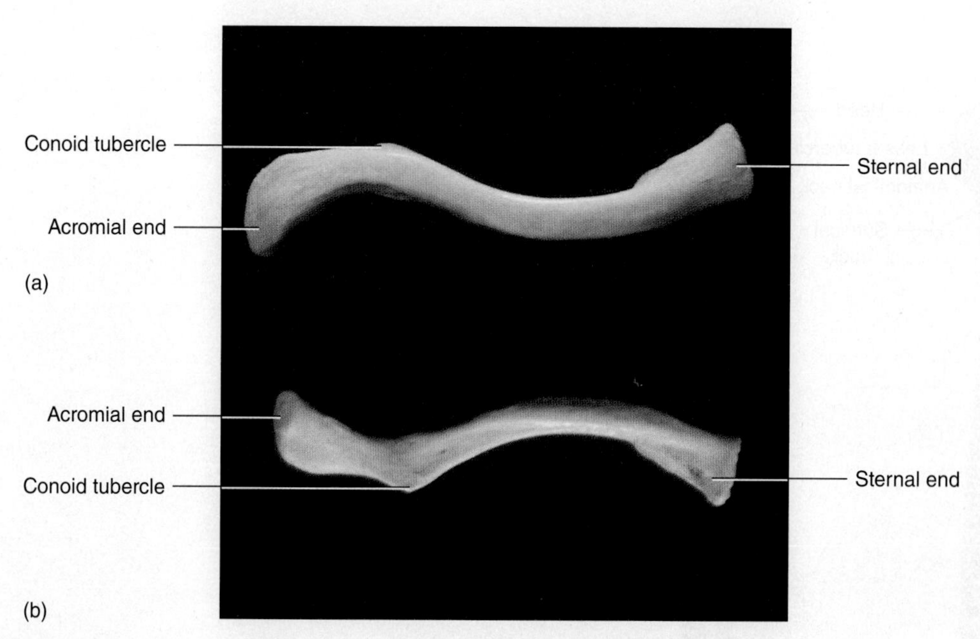

Figure 7.2 Right Clavicle
(a) Superior view; (b) inferior view.

a rim known as the **anatomical neck.** The anatomical neck is the site of the epiphyseal line. The humerus has two large processes: the **greater tubercle,** which is the largest and most lateral process, and the **lesser tubercle,** which is medial and smaller. These processes are attachment points for muscles originating from the scapula. Between the tubercles is the **intertubercular,** or **bicipital** (bī-sip´i-tăl), **groove,** an elongated depression through which one of the biceps brachii tendons passes. The intertubercular groove is anterior and the head of the humerus is medial. These two features orient you to the right or left humerus. Examine figure 7.3 and locate these features. Below the tubercles of the humerus is a constricted region known as the **surgical neck.** The surgical neck is so named because it is a frequent site of fracture. Locate the surgical neck on your arm. A roughened area on the lateral surface of the humerus is the **deltoid tuberosity,** named for the attachment of the deltoid muscle. The majority of the length of the humerus is the **diaphysis,** or **shaft,** and a small hole penetrating the shaft is the **nutrient foramen.** Nutrient foramina occur in many bones; they conduct blood vessels into these bones.

At the distal end of the humerus are the regions of articulation with the bones of the forearm. On the lateral side is a round hemisphere known as the **capitulum** (kă-pit´ū-lum). This is where the head of the radius fits into the humerus. On the medial side is an hourglass-shaped structure called the **trochlea** (trok´lē-ă). The trochlea is where the ulna attaches to the humerus. The capitulum and trochlea represent the **condyles** of the humerus. To the sides of these condyles are the **epicondyles.** The **medial epicondyle** is the bump you can palpate (feel) medial to the elbow, and the **medial supracondylar ridge** forms a small wing of bone extending from the epicondyle to the distal shaft of the humerus.

The **lateral epicondyle** also has a **lateral supracondylar ridge.** The epicondyles and supracondylar ridges are points of attachment for muscles that manipulate the forearm and hand. At the distal end are two depressions in the humerus. The one on the anterior surface is the **coronoid** (kōr´ŏ-noyd) **fossa,** and the one on the posterior surface is the **olecranon** (ō-lek´ră-non) **fossa.**

Locate these structures on specimens in the lab and in figure 7.3.

Forearm

Two bones make up the skeletal portion of the forearm. These are the **radius,** which is lateral (on the thumb side), and the **ulna,** a medial bone. The radius has a proximal **head** that looks like a wheel. Distal to this is the **radial tuberosity,** where the biceps brachii muscle inserts, and at the most distal end is the **styloid** (stī´loyd) **process** of the radius. On the distal end of the radius, there is a medial depression called the **ulnar notch.** This is where the distal part of the ulna joins with the radius. Look at figure 7.4 for these structures and compare them with the bones in the lab.

The ulna has a U-shaped depression (some students remember this as "U for ulna") on the proximal portion, which is the **trochlear,** or **semilunar, notch.** The most proximal portion of the ulna is the **olecranon process,** commonly known as the elbow. The process distal to the trochlear notch is the **coronoid process,** which ensures a relatively tight fit with the humerus and provides an area for muscle attachment. The ulna has a *distal* **head,** unlike the radius (the head of the radius is proximal), and a **styloid process** distal to the head. On the proximal part of the ulna, where the head of the radius fits into the ulna, is a depression known as the **radial notch.** If you know that the trochlear notch is anterior and the radial notch is lateral, then you can determine if you are looking at a left or right ulna. In the following space, list what side of the body the bones of the forearm you are studying come from.

Side of the body: ______________________________

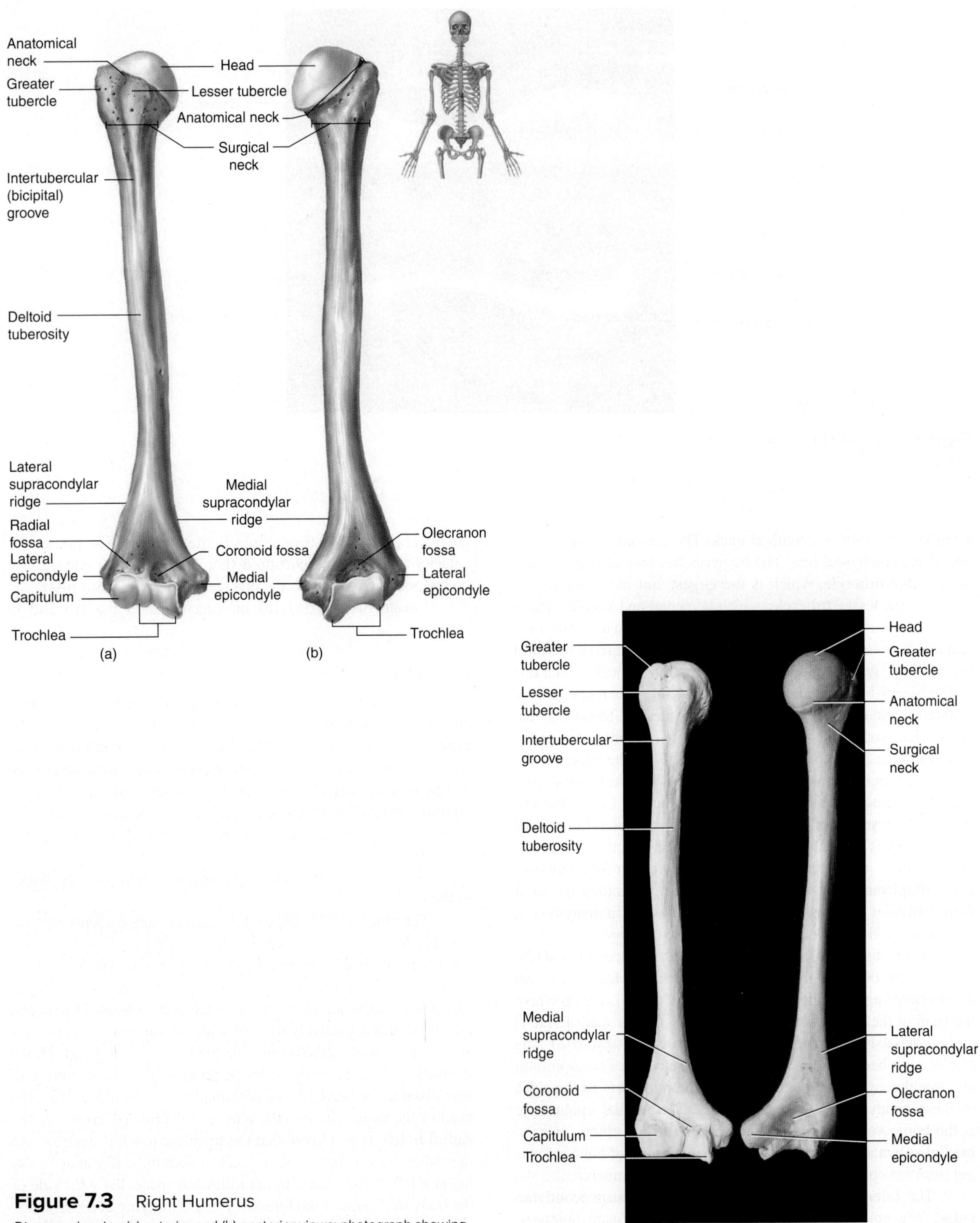

Figure 7.3 Right Humerus

Diagram showing (a) anterior and (b) posterior views; photograph showing (c) anterior and (d) posterior views.

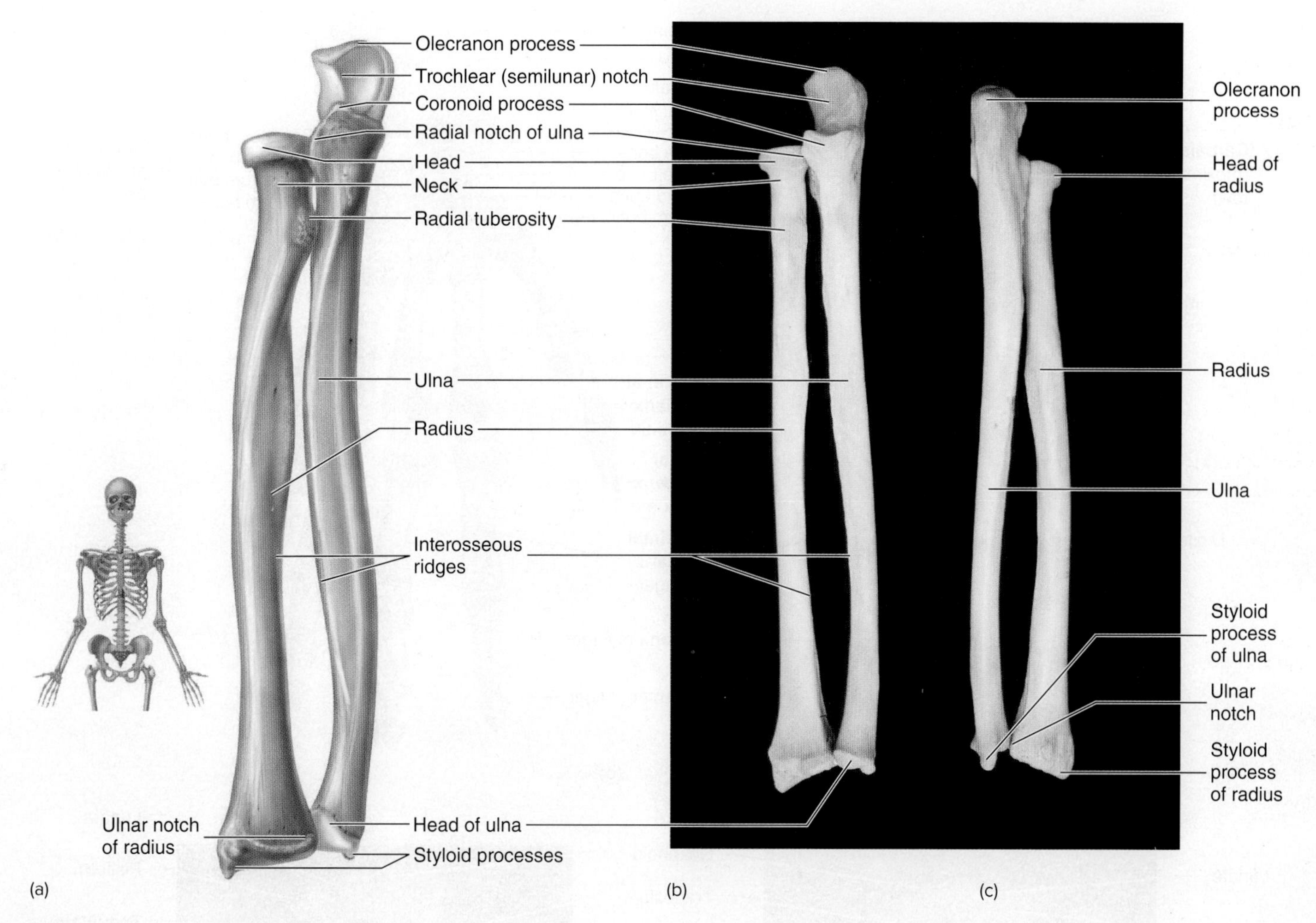

Figure 7.4 Right Radius and Ulna
(a) Diagram, anterior view; photograph (b) anterior view; (c) posterior view.

Hand

Ⓐ 5 Place an articulated hand in front of you and look for the following features. Each hand consists of three groups of bones: the **carpals,** the **metacarpals,** and the **phalanges** (fălan´jēz; sing. *phalanx*). The human hand is composed of eight carpal bones (figure 7.5). These bones can be roughly aligned in two rows. The proximal row from lateral to medial includes the **scaphoid** (navicular), **lunate, triquetrum** (trī-kwē´-trum; triangular), and **pisiform** (pī´sih-form). The distal row includes the **trapezium** (greater multangular), the **trapezoid** (lesser multangular), the **capitate,** and the **hamate.** On the hamate is a hooklike projection called the **hamulus of the hamate.** The carpal bones are named for their shapes, with the scaphoid resembling a boat (*scaphos* = boat), the lunate named for the crescent moon, the triquetrum (a triangle), the pisiform being pea-shaped, the trapezium rhyming with *thumb* and named for a four-sided shape, the trapezoid named for another four-sided shape, the capitate being "headlike," and the hamate resembling a hook (*hamus* = hook). Another way to remember the bones in their proper order is with a mnemonic device that uses the first letter of each carpal bone (*S* for scaphoid, *L* for lunate, etc.) in a sentence such as "Say Loudly To Pam, Time To Come Home." The carpal bones are regions of attachment for forearm muscles and for intrinsic muscles of the hand.

The **metacarpals** are the bones of the palm. Each metacarpal consists of a proximal **base,** a **shaft,** and a distal **head.** The metacarpals are labeled by either Arabic (1–5) or Roman (I–V) numerals. Metacarpal 1 is proximal to the thumb, and 5 is proximal to the little finger. Metacarpals 2 through 5 have little movement, but metacarpal 1 allows for significant movement of the thumb. Locate metacarpal 1 on the skeletal material and on your own hand, and note the substantial range of movement. This type of movement is covered in greater detail in Exercise 10.

Distal to the metacarpals are the **phalanges.** There are 14 bones that make up this group. Each finger of the hand has three phalanges—a **proximal,** a **middle,** and a **distal phalanx**—except for the thumb. The thumb has only a proximal and a distal phalanx. The thumb is also known as the **pollex.** You should identify each bone by noting the digit it comes from and whether it is proximal, middle, or distal. Therefore, the tip of the little finger is the distal phalanx of digit 5. Like the metacarpals, each phalanx also has a **base, shaft,** and **head** and, like the metacarpals, phalanges are classified as long bones.

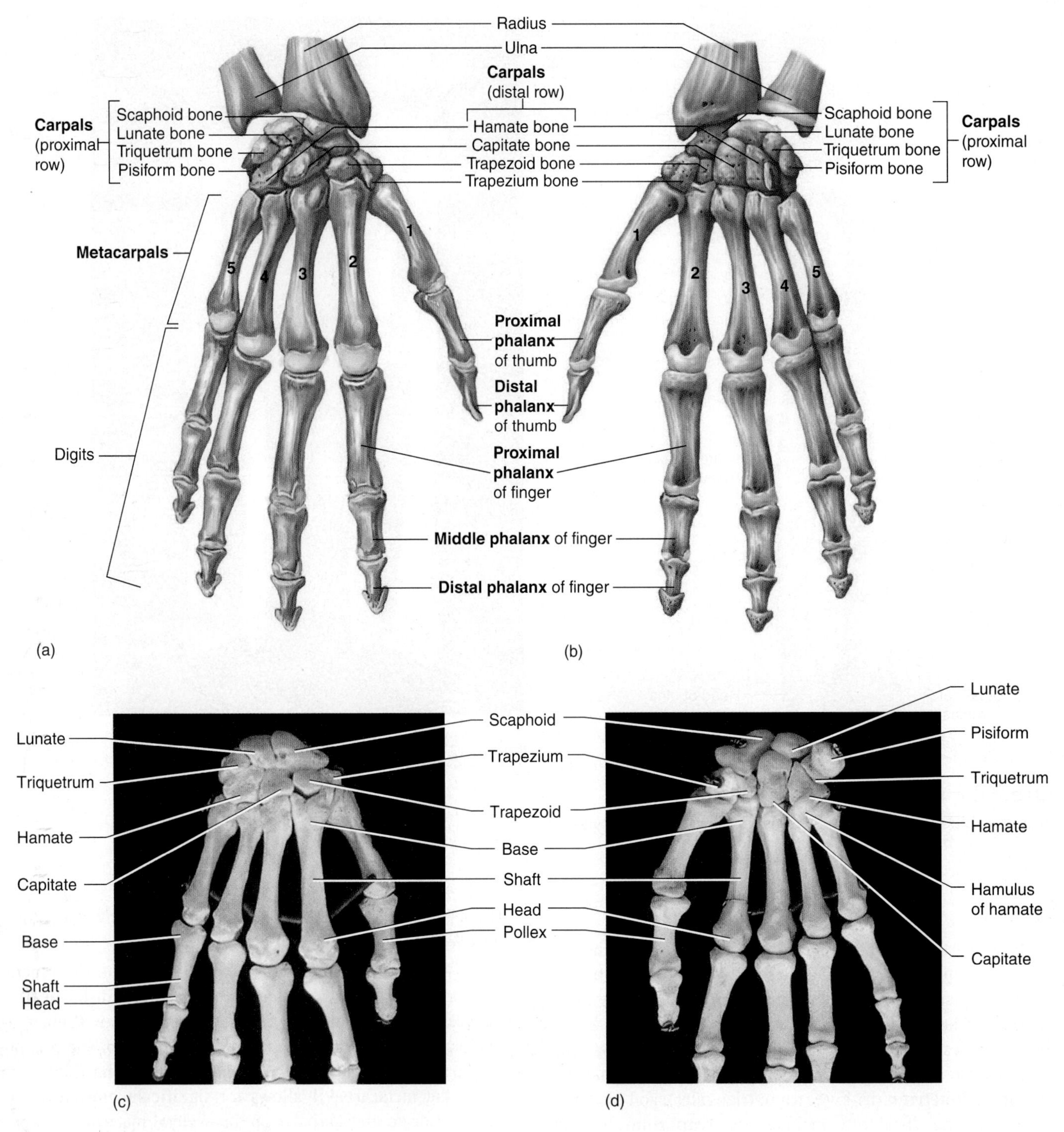

Figure 7.5 Bones of the Right Hand

Diagram (a) posterior view; (b) anterior view; photograph (c) posterior view; (d) anterior view.

Pelvic Girdle

Ⓐ6 Examine the hip bones available in lab and compare them to figure 7.6 as you read the following material. The pelvic girdle consists of the two **coxal,** or **innominate, bones.** Each coxal bone results from the fusion of three bones: the ilium, the ischium, and the pubis. The coxal bones are joined in the front by the **symphysis pubis,** or **pubic symphysis,** which is a fibrocartilage pad. Below the symphysis pubis is the **subpubic angle.** The **sacrum,** a wedge-shaped section of the vertebral column, attaches to the coxal bones in the posterior and is part of the pelvis, but it is not considered a bone of the pelvic girdle.

Ilium

The most superior coxal bone is the **ilium** (il´ē-um). If you feel the top part of your hip, you are feeling the **iliac crest,** which is a long, crescent-shaped ridge of the ilium. The two processes that jut from the ilium in the front are the **anterior superior iliac spine** and the

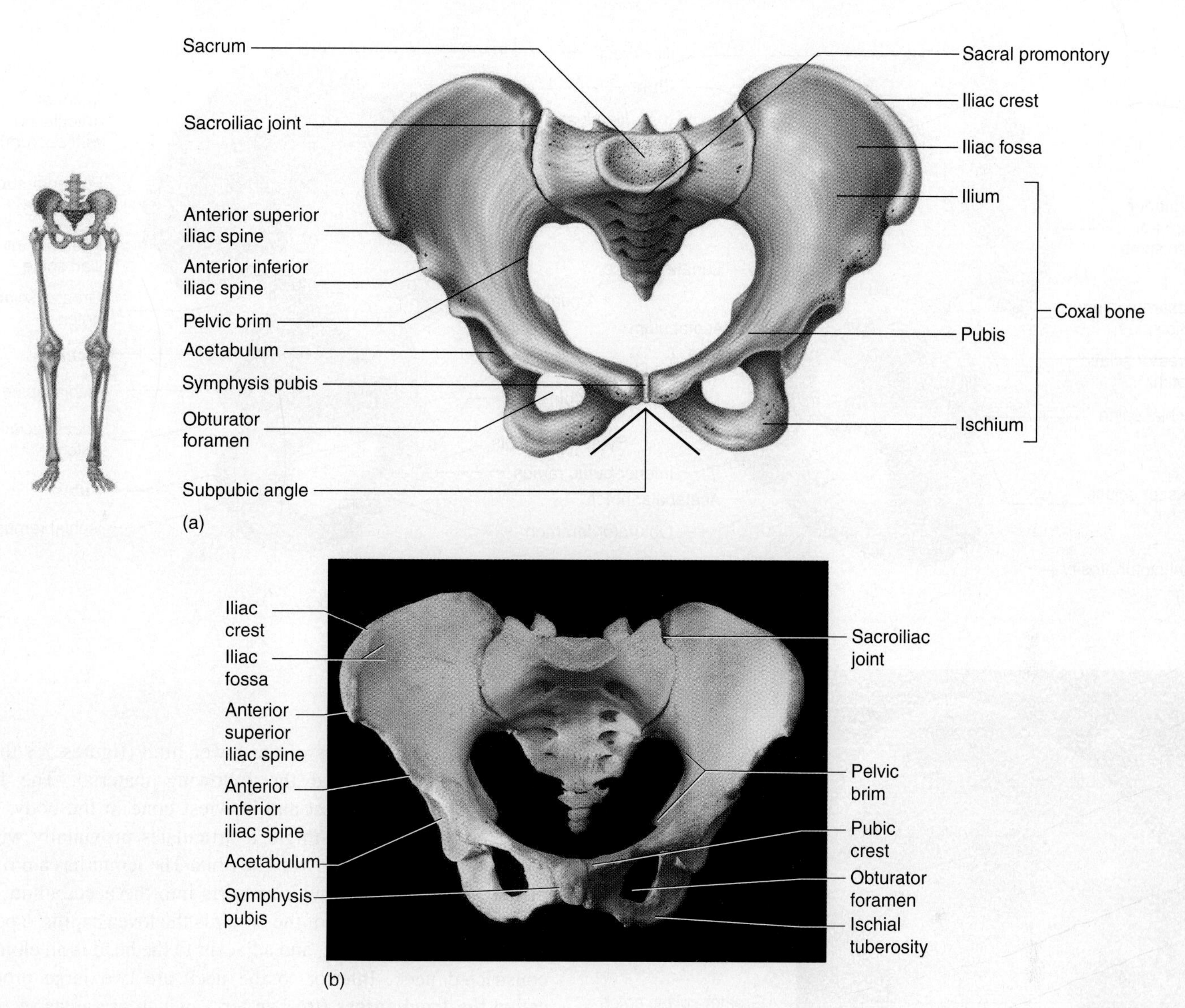

Figure 7.6 Bones of the Pelvic Girdle
Anterior view (a) diagram, (b) photograph.

anterior inferior iliac spine. These are attachment points for some of the thigh flexor muscles. The posterior ilium has the **posterior superior iliac spine** and the **posterior inferior iliac spine.** Inferior to the posterior inferior iliac spine is a large depression known as the **greater sciatic** (sī-at´ĭk) **notch.** The outer surface of the ilium is an attachment point for the gluteal muscles. The interior, shallow depression of the ilium is known as the **iliac fossa.**

Ischium and Pubis

The **ischium** (is´kē-ŭm) is inferior to the ilium and is the part of the coxal bone on which you sit. The **ischial spine** is a sharp projection on the ischium, and the **ischial tuberosity** is an attachment site for the hamstring muscles. Anterior to the ischial tuberosity is the **obturator** (ob´too-rā-tŏr) **foramen,** a large hole underneath a cuplike depression known as the **acetabulum** (as-ĕ-tab´ū-lŭm; a region where all three coxal bones are joined). The acetabulum is the socket into which the femur fits. The ischium connects to the pubis by an elongated portion of bone known as the **ischial ramus.** This ramus connects to the **inferior pubic ramus,** which is posterior to the symphysis pubis.

If the two coxal bones are articulated, you can easily see the upper basin of the pelvis, which is known as the **false,** or **greater, pelvis.** The false pelvis is medial to the iliac fossa. A rim of bone separates the upper basin from a deeper, smaller basin known as the **true pelvis.** This rim is known as the **pelvic brim** in an intact pelvis or the **arcuate line** on an isolated pelvic bone. The true pelvis is medial to the obturator foramen. You can determine if you have a left or right bone, and you can determine the person's gender. In a female pelvis, the greater sciatic notch has a fairly broad angle, which approximates the arc inscribed by your outstretched thumb and index finger. In a male, the greater sciatic notch is less than 90° and approximates the angle if you were to separate the index finger and the middle finger. The subpubic angle is greater than 90° in females. Features relating to the gender of the individual are seen in figure 7.7.

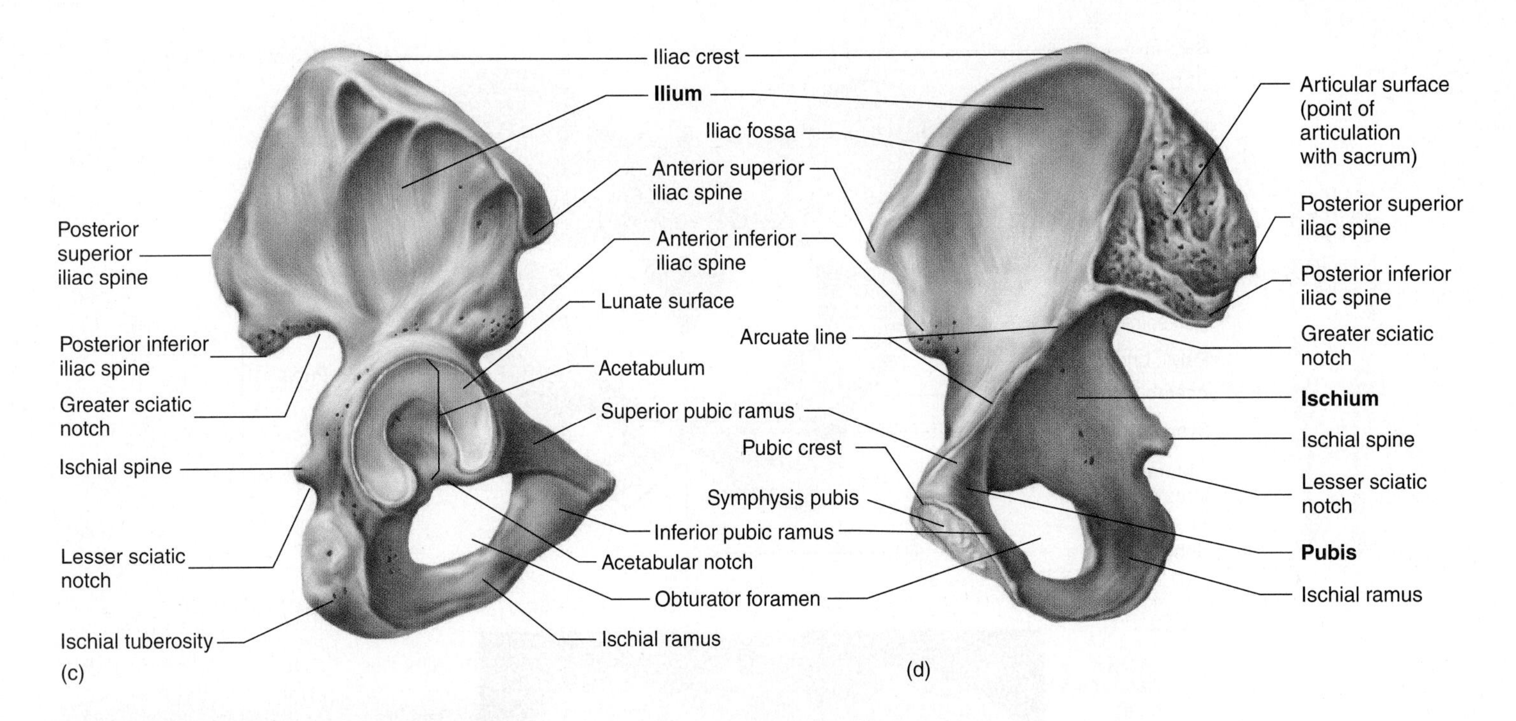

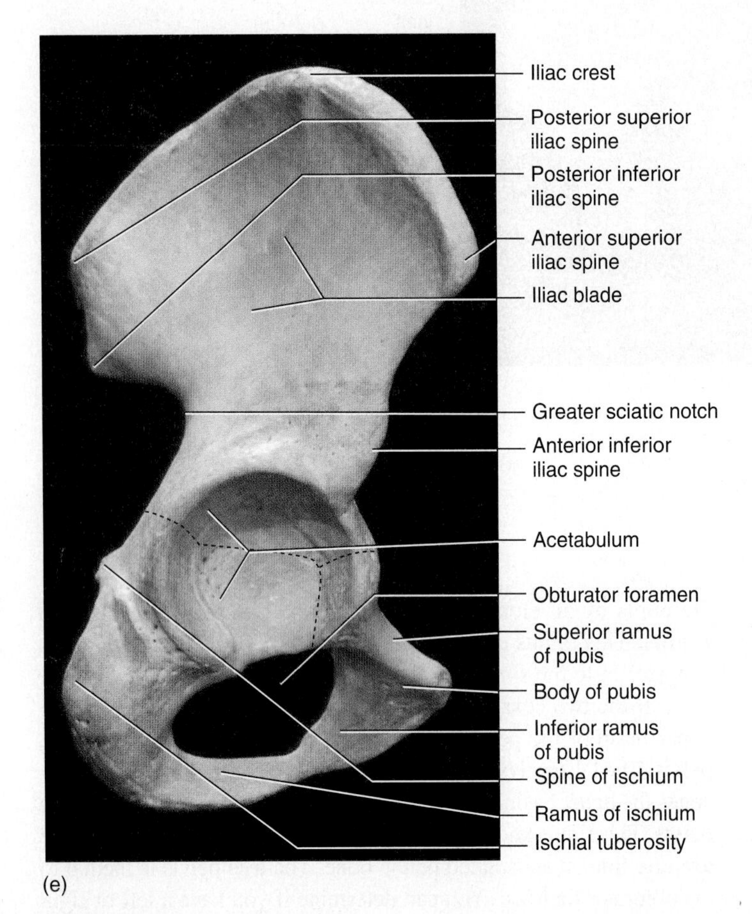

Figure 7.6 *Continued.*

(c) diagram of lateral view of right hip; (d) diagram of medial view, right hip; (e) photograph lateral view (dotted lines show fusion of the ilium, ischium, and pubis).

Lower Limb

Femur

Ⓐ7 Examine the bones of the lower limb (figures 7.8 through 7.11) as you read the following material. The **femur** (figure 7.8) is the longest and heaviest bone in the body, and it is the only bone of the thigh. It articulates proximally with the coxal bone and inferiorly with the tibia. The femur has a proximal, medial spherical **head,** which inserts into the acetabulum of the coxal bone. On the head of the femur is the fovea capitis, a point of attachment of a ligament, and adjacent to the head is an elongated, constricted neck. Inferior to the neck are two large processes called the **trochanters** (trō-kan´terz), which are areas of muscle attachment. The larger, anterior process is the **greater trochanter,** and the smaller, posterior process is the **lesser trochanter.** Look for the **intertrochanteric crest,** which is a posterior ridge between the two trochanters and the anterior **intertrochanteric line.** In the following space, indicate which is more broad, the intertrochanteric crest or the intertrochanteric line.

Broadest surface feature: ______________________

The shaft of the femur is curved and bows anteriorly with the posterior shaft marked by the **linea aspera** (lin´ē-ah as´per-ah; rough line). At the proximal portion of the linea aspera is a roughened area called the **gluteal tuberosity,** an attachment site for the gluteus maximus muscle. The distal portion of the femur consists of **medial** and **lateral condyles.** These two smooth processes come into contact with the condyles of the tibia. To the sides of the condyles are bulges called the **epicondyles.** Locate the **lateral epicondyle** and the **medial epicondyle** on the femur. Also note the location of the **adductor tubercle,** a small, triangular process proximal to the medial epicondyle. This is one of the attachment points for the adductor magnus muscle. The head of the femur is medial and the linea aspera is posterior. These features help you distinguish whether you have a left or right femur.

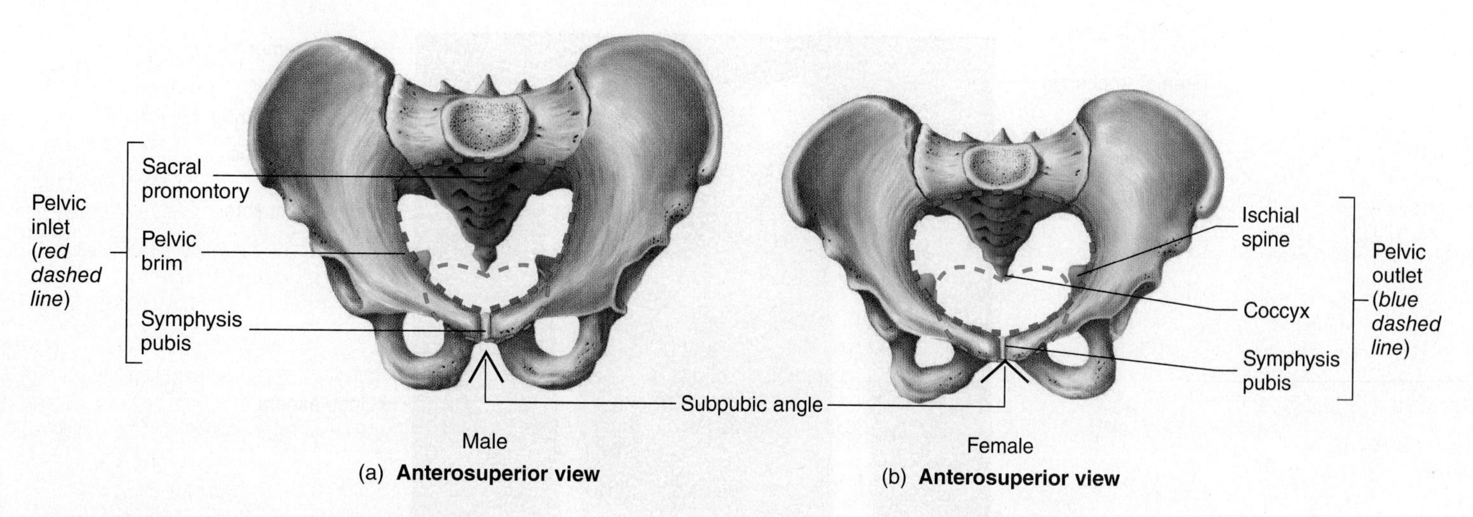

Figure 7.7 Male and Female Pelves

(a) Male pelvis, with narrow subpubic angle and small pelvic outlet; (b) female pelvis, with greater subpubic angle and larger pelvic outlet.

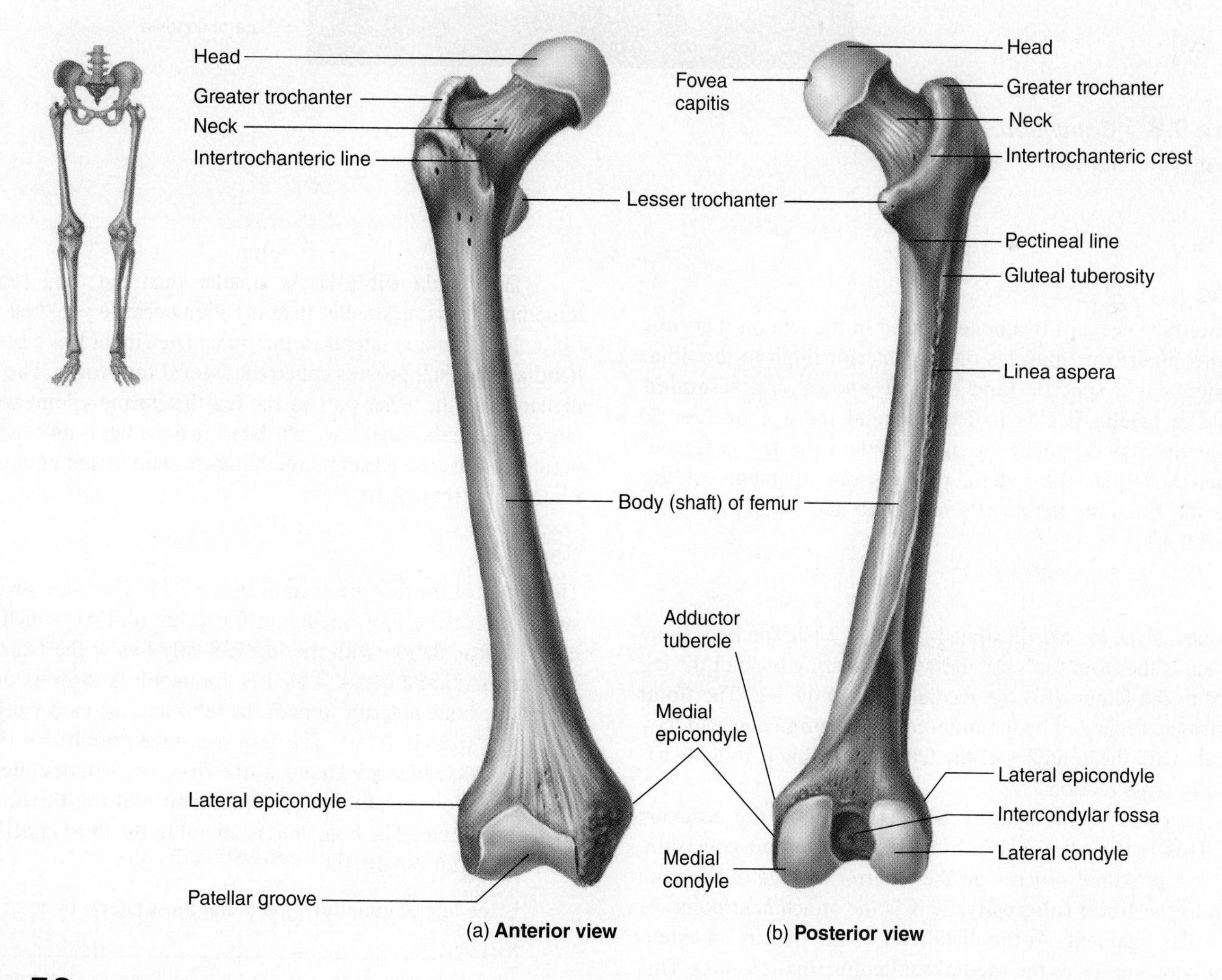

Figure 7.8 Right Femur

Diagram (a) anterior view; (b) posterior view.

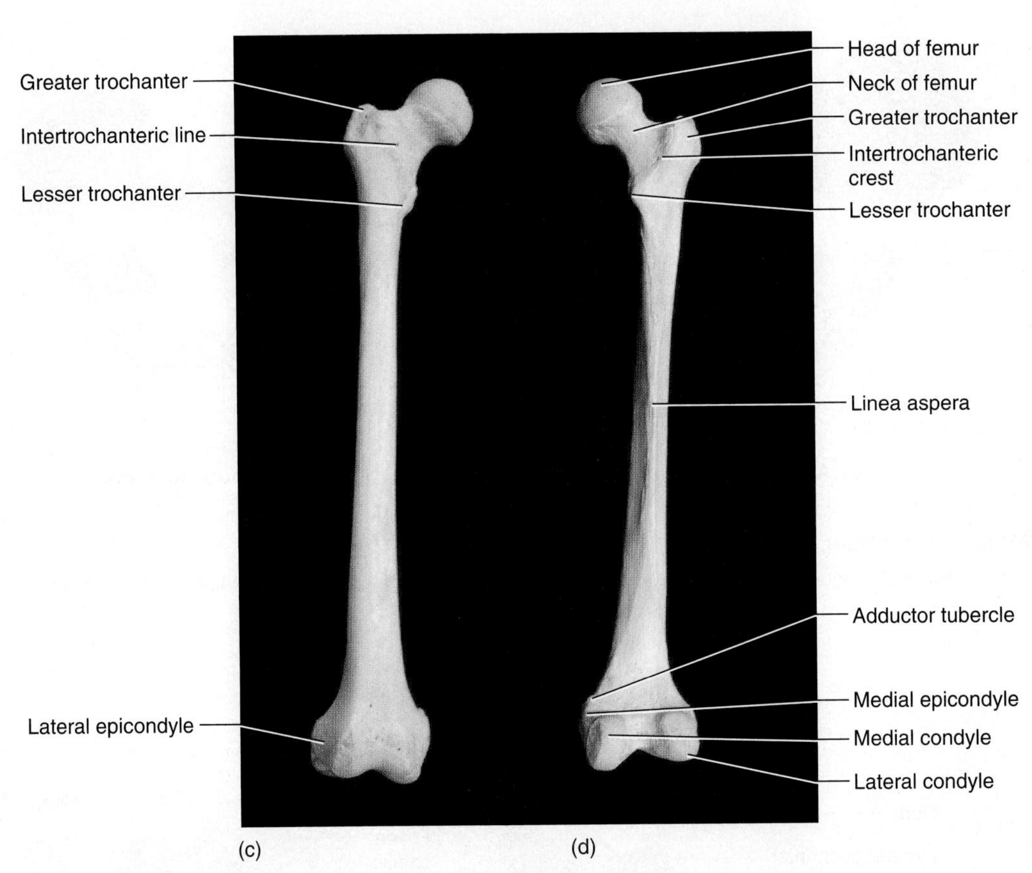

Figure 7.8 *Continued.*
Photograph (c) anterior view; (d) posterior view.

Patella

The **patella** (kneecap) is a bone formed in the tendon that runs from the quadriceps muscles on the anterior thigh to the tibia. The patella is a specific kind of bone known as a **sesamoid bone.** The patella begins to form around the age of 2 or 3, and usually it is complete by age 6. When the leg is flexed, as when kneeling, the patella protects the ligaments of the knee joint. Examine the patella in the lab and compare it with figure 7.9.

Leg

The bones of the leg are illustrated in figure 7.10. The large bone of the leg is the **tibia** (tib´ē-ă), the weight-bearing bone of the leg inferior to the femur. It is the medial bone of the leg. The **tibial condyles** are separated by the **indercondylar eminence,** and they articulate with the condyles of the femur. The tibia is roughly triangular in cross section.

The ridge on the anterior tibia is known as the **anterior crest.** This is the crest that is bruised when you hit your shin. There is a proximal process on the anterior surface of the tibia known as the **tibial tuberosity.** This is the attachment point for the patellar ligament. At the distal end of the tibia is an extension of bone known as the **medial malleolus** (ma-lē´ō-lŭs). This bump is one part of the ankle joint as it articulates with the talus of the foot.

The **fibula** (fib´ū-lă) is smaller than the tibia (you can remember that it is smaller than the tibia because you "tell a little fib"). The fibula is lateral to the tibia. The fibula has a proximal **head** and a distal process called the **lateral malleolus.** The lateral malleolus is the other part of the leg that forms a joint with the talus. The fibula is not a weight-bearing bone but is an attachment point for muscles. Examine the bones in the lab and compare the fibula with figure 7.10.

Foot

The bones of the foot are seen in figure 7.11. There are seven **tarsal bones** of the foot, including the **talus** (tă´lus), which is the bone of articulation with the leg. Directly below the talus is the **calcaneus** (kal-kā´neus), which is commonly known as the heel bone. The bone anterior to both the talus and the calcaneus is the **navicular** (nă-vik´ū-lar). The foot has three **cuneiform** (kūnē´i-fōrm) **bones,** which are known as the **first,** or **medial, cuneiform;** the **second,** or **intermediate, cuneiform;** and the **third,** or **lateral, cuneiform.** The bone that is lateral to the third cuneiform is the **cuboid** (kū´boyd).

Is the lateral cuneiform bone the most lateral bone of the foot? ______________________________

One way to remember the tarsal bones is with the mnemonic "children that never march in line cry" where *C* stands for

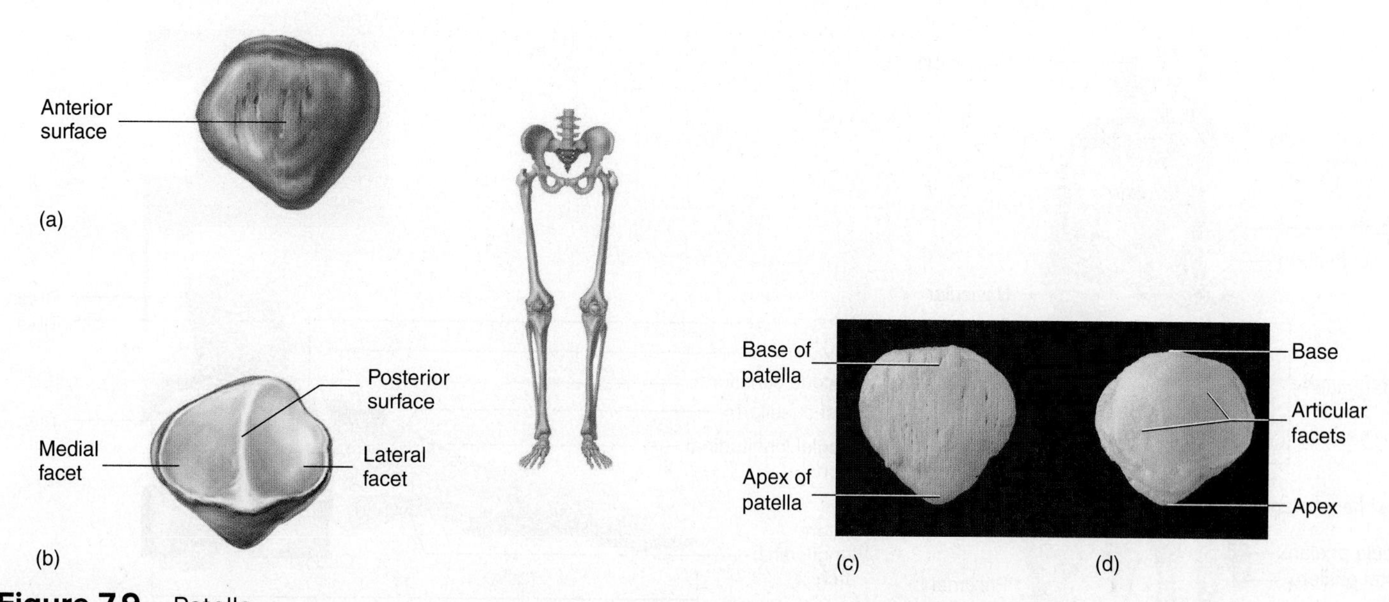

Figure 7.9 Patella

Diagram (a) anterior view; (b) posterior view; photograph (c) anterior view; (d) posterior view.

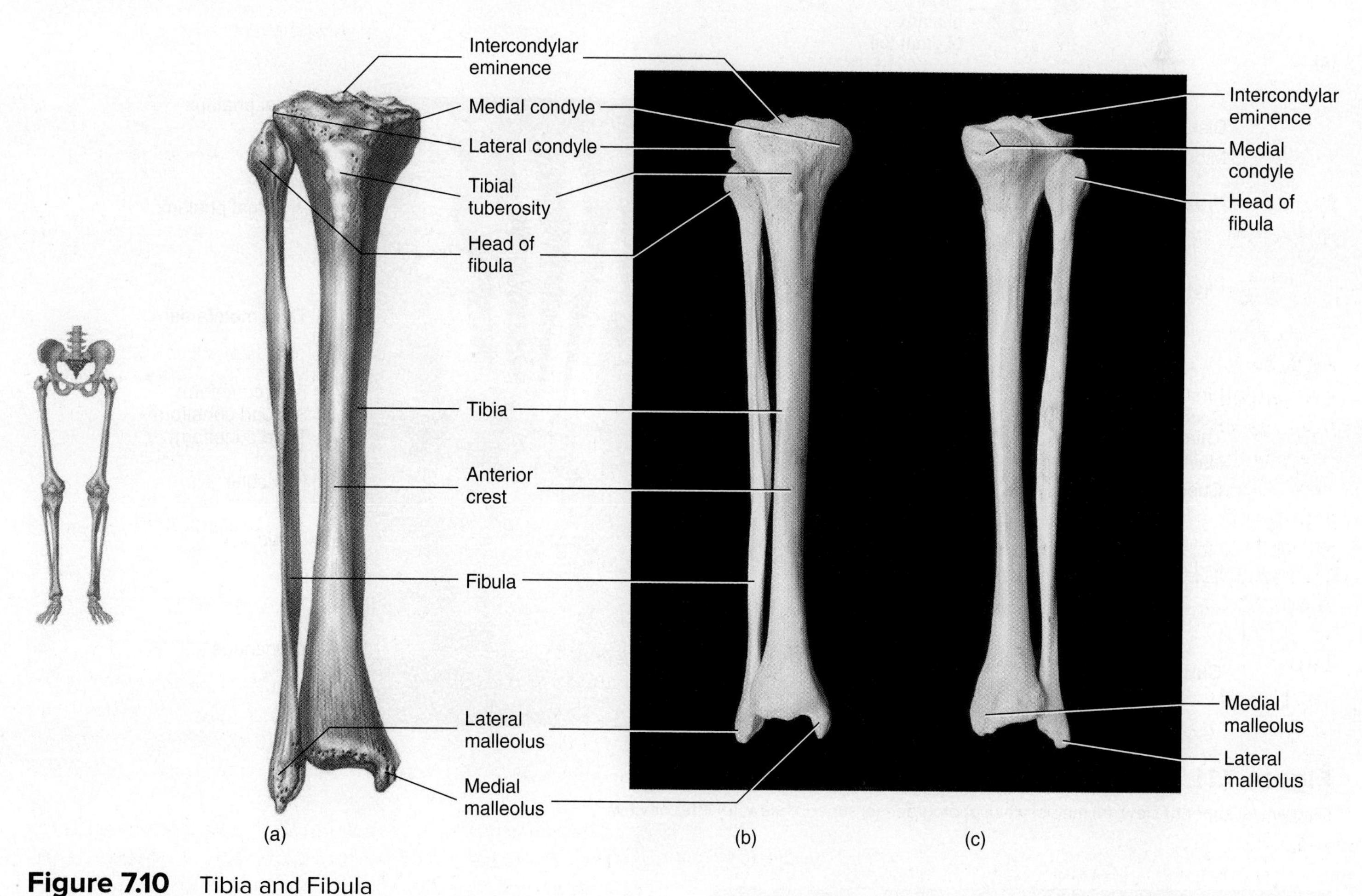

Figure 7.10 Tibia and Fibula

Diagram (a) anterior view; photograph (b) anterior view; (c) posterior view.

calcaneus, *N* stands for Navicular, etc. The beginning letter of each word is the same as a tarsal bone. The **metatarsals** are similar to the metacarpals in that the first metatarsal is under the largest digit, the big toe, and the fifth metatarsal is under the smallest digit. The pattern of the phalanges of the foot is the same as that of the hand, with toes 2 through 5 having a **proximal, middle,** and **distal phalanx** each and the big toe, or **hallux,** having a proximal and a distal phalanx.

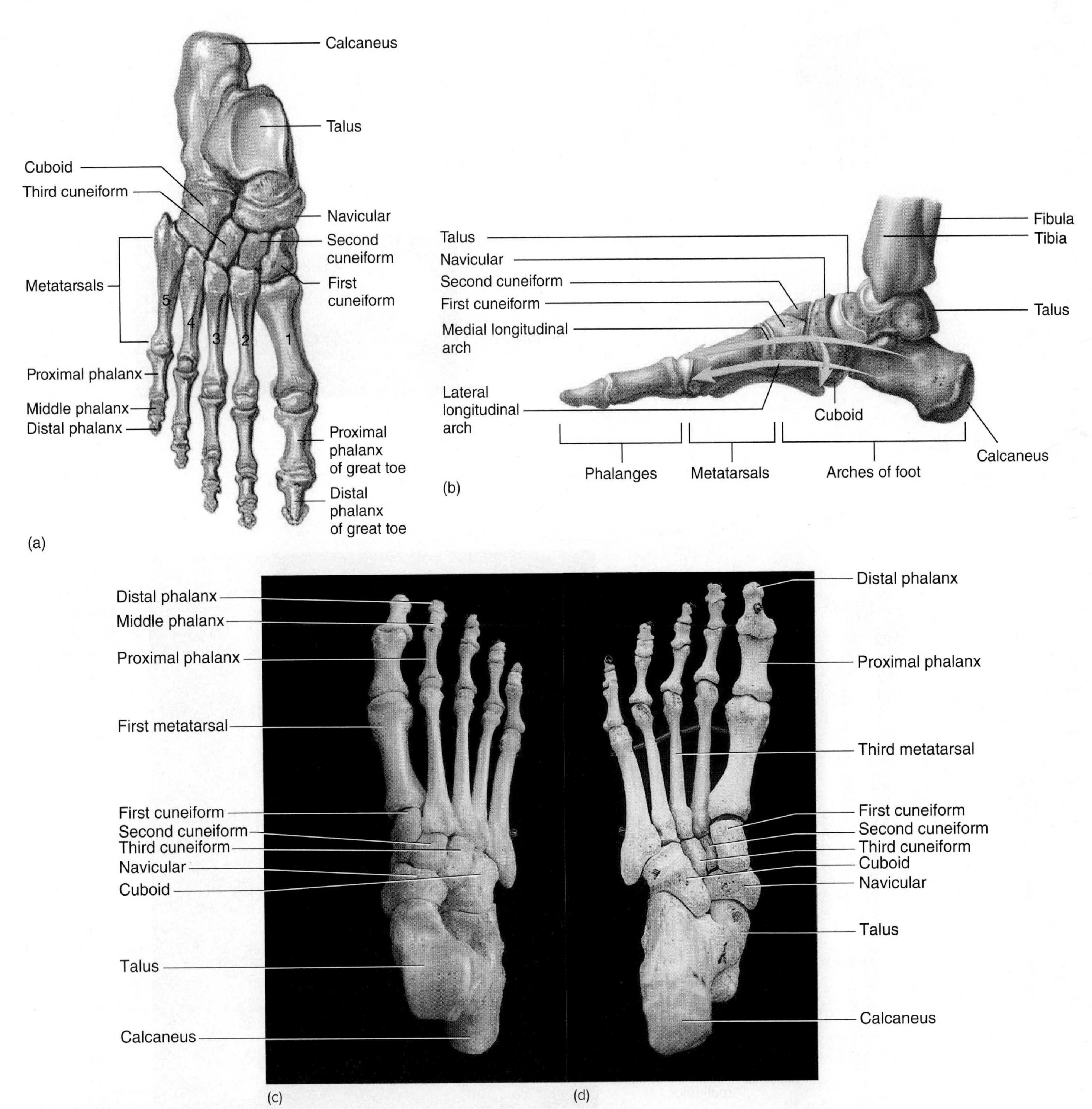

Figure 7.11 Bones of the Foot

Diagram (a) superior view; (b) medial view; photograph (c) superior view; (d) inferior view.

Study Hint

When you have finished locating the major features of the bones of the appendicular skeleton, test yourself and your lab partner by pointing to various structures and seeing how accurate you are at identifying the structures.

Exercise 7 Review

Appendicular Skeleton

Name: ______________________________

Lab time/section: ______________________________

Date: ______________________________

1. Name the anterior depression on the scapula. ______________________________

2. The humerus fits into what depression of the scapula? ______________________________

3. What specific part of the clavicle attaches to the scapula? ______________________________

4. Frequently, the clavicle is broken when the arms are extended to brace a fall. Explain how hitting the ground with your hands can fracture the clavicle. ______________________________

5. The epiphyseal line on the humerus has what other name? ______________________________

6. The condyles of the humerus have specific names. What are they? ______________________________

7. Name the ridge of bony tissue proximal to the lateral condyle of the humerus. ______________________________

8. What is the name of the lateral bone of the forearm? ______________________________

9. Name the depression on the ulna into which the humerus inserts. ______________________________

10. Name the bony process that extends distally from the head of the ulna. ______________________________

11. How does the head of the ulna differ in position from the head of the radius? ______________________________

12. Each metacarpal bone consists of three major regions. What are these? ______________________________

____

13. Name the carpal bone at the base of the thumb. ______________________________

14. The ______________________________ joins the two coxal bones at the anterior junction of these bones.

15. What curved area occurs on the coxal bones directly inferior to the posterior inferior iliac spine? ______________________________

16. What is the cuplike depression of the coxal bones into which the head of the femur fits? ______________________________

17. Explain the difference between the false pelvis and the true pelvis. ______________________________

18. Name the roughened line that runs along the length of the posterior femur. ______________________________

19. Name the sesamoid bone that forms in the quadriceps tendon. ______________________

20. What is the name of the process on the distal portion of the tibia? ______________________

21. What is the function of the fibula? ______________________

22. What is another name for the heelbone? ______________________

23. What is the name of the bone of the foot that joins with the tibia and fibula? ______________________

24. Match the bone with the region it comes from.

a. hallux	coxal
b. ilium	hand
c. clavicle	upper limb
d. scaphoid	pectoral girdle
e. radius	foot

25. How many bones are in the ankle versus the wrist? ______________________

26. Label the parts of the scapula in the following illustration using the terms provided.

acromion process	coracoid process	inferior angle
infraspinous fossa	lateral border	medial border
scapular spine	supraspinous fossa	

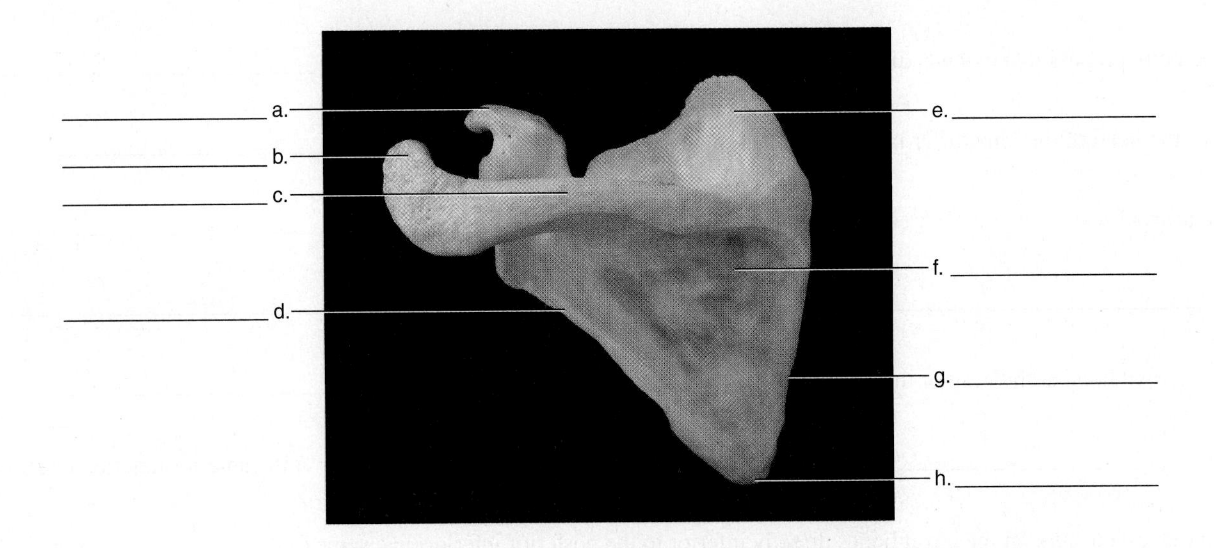

27. Label the parts of the ulna as illustrated, and determine if the bone is from the left or right side of the body.

coronoid process	head
olecranon process	radial notch
styloid process	trochlear notch

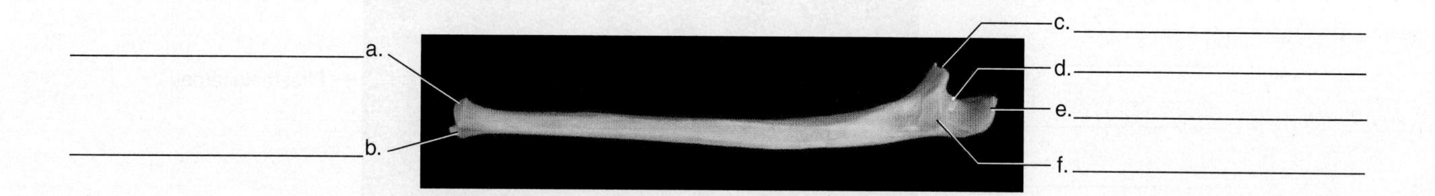

28. Name the bone at the tip of the middle finger. ____________________

29. Label the following illustration of the foot using the terms provided.

calcaneus	cuboid	first cuneiform	navicular
second cuneiform	talus	third cuneiform	

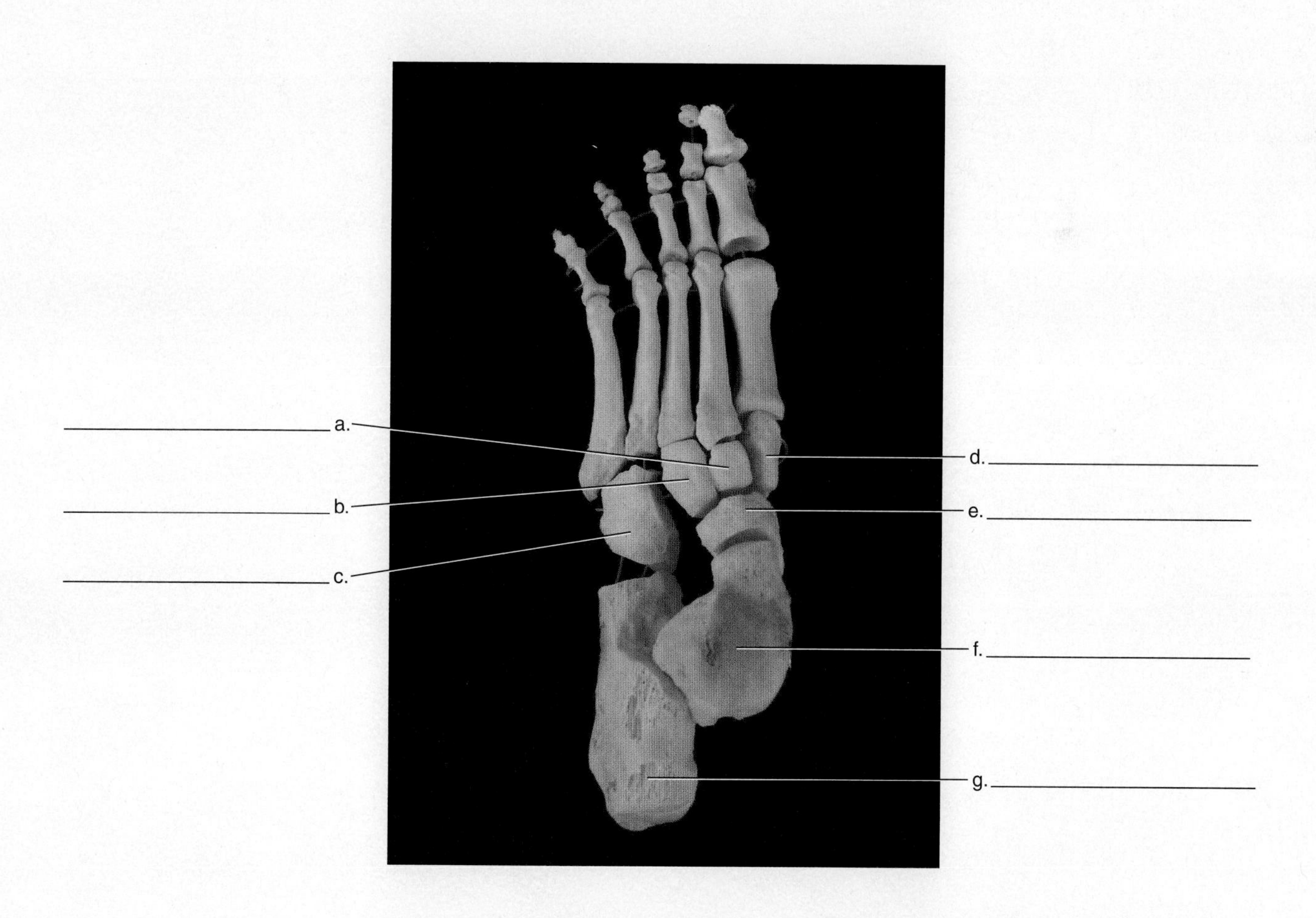

30. Label the following illustration of the hand using the terms provided.

capitate hamate lunate pisiform trapezoid

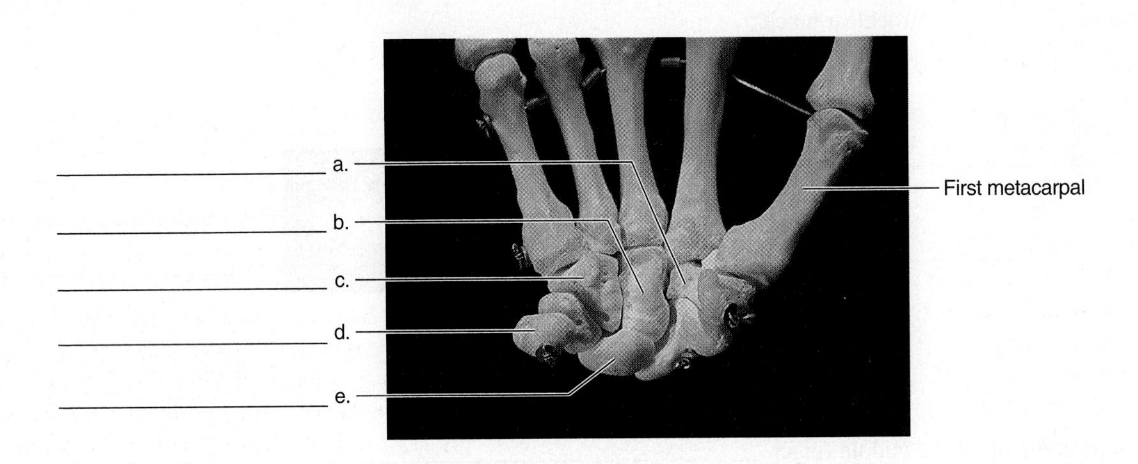

Exercise 8

Axial Skeleton: Vertebrae, Ribs, Sternum, Hyoid

Introduction

The axial skeleton consists of 80 bones, including the ribs, hyoid, skull, vertebral column, and sternum. In this exercise, you examine all parts of the axial skeleton except for the skull, which will be covered separately in Exercise 9. The **vertebral column** consists of the 33 **vertebrae,** some of which are fused into larger structures, such as the sacrum. The regions of the vertebral column include the cervical, thoracic, lumbar, sacral, and coccygeal.

The hyoid is a small, floating bone between the floor of the mouth and the superior anterior neck. It assists the tongue in swallowing and the larynx in speech.

The thoracic cage consists of the 12 pairs of ribs and the sternum, both of which protect the lungs and heart yet provide for flexibility during breathing. These topics are discussed in the Seeley text in chapter 7, "Skeletal System: Gross Anatomy."

Objectives

At the end of this exercise, you should be able to

1. find a specific vertebra (such as T6) on an articulated vertebral column;
2. name a selected vertebra on an articulated vertebral column;
3. name the bony features of an individual vertebra and determine the region of the vertebral column to which it belongs;
4. describe the features that determine cervical, thoracic, lumbar, sacral, and coccygeal vertebrae and identify the atlas and the axis;
5. describe the normal and abnormal spinal curvatures;
6. distinguish between the different types of ribs, the markings of the individual ribs, and whether a rib comes from the left or right side of the body;
7. identify the bony features of the sternum.

Materials

Articulated skeleton or plastic casts of a skeleton
Disarticulated skeleton or plastic casts of bones
Charts of the skeletal system
Wooden applicator sticks or plastic straws cut on a bias for pointer tips
Foam pads of various sizes (to protect bone from hard countertops)
Cardboard box or piece of wood about 1 foot square and 3 inches deep

Procedure

(A) 1 Review the bones of the axial skeleton in the lab and in Exercise 6 (figure 6.1). In this exercise, you study the bones of the axial skeleton by examining both the disarticulated bones and the articulated skeleton. Hold each bone up to the skeleton to see how it is positioned in relation to the other bones of the body. When examining bones in the lab, do not use your pen or pencil to locate a structure. These leave marks on the bones. Carefully use a wooden applicator stick or plastic straw to point out structures. Your instructor may want you to place real bone material on foam pads to cushion the bone from the tabletop.

Vertebrae

Overview of the Vertebrae

The **vertebral column** in humans is much different than in other mammals, because we are the only habitually bipedal mammal. Between the vertebrae are fibrocartilaginous pads known as **intervertebral discs.**

(A) 2 Examine a vertebra for the features represented in figure 8.1. Place the vertebra in front of you so that you can see through the large hole known as the **vertebral foramen,** where the spinal cord is located. The diameter of the vertebral foramen increases from the sacral region to the cervical region. This is due to the greater number of neural fibers in the spinal cord as sensory information is transmitted to the brain and motor information is transmitted from the brain. Note the **body** of the vertebra, which supports the weight of the vertebral column and is in contact with the intervertebral discs. Find the **vertebral,** or **neural, arch,** which consists of two **pedicles** (ped´i-kls) and two **laminae** (lam´i-nay). The pedicles are the parts of the arch that extend from the body of the vertebra to the two lateral projections called the **transverse processes.** Each lamina is a broad, flat structure between the transverse process and the dorsal **spinous process.** If you rotate the vertebra as illustrated in figure 8.1, you should see the vertebral body in lateral view with the **superior articular process** and the **superior articular facet** (fas´et) visible. The superior articular facet is a smooth face that articulates with the vertebra above. There is also an **inferior articular process** and an **inferior articular facet** that articulates with the vertebra below. The inferior articular facet of a superior vertebra joins with the superior articular facet of the next inferior vertebra. If you put two adjacent vertebrae together and look at the lateral aspect, you can see the **intervertebral foramen,** a hole that allows the spinal nerve to exit from the vertebral column.

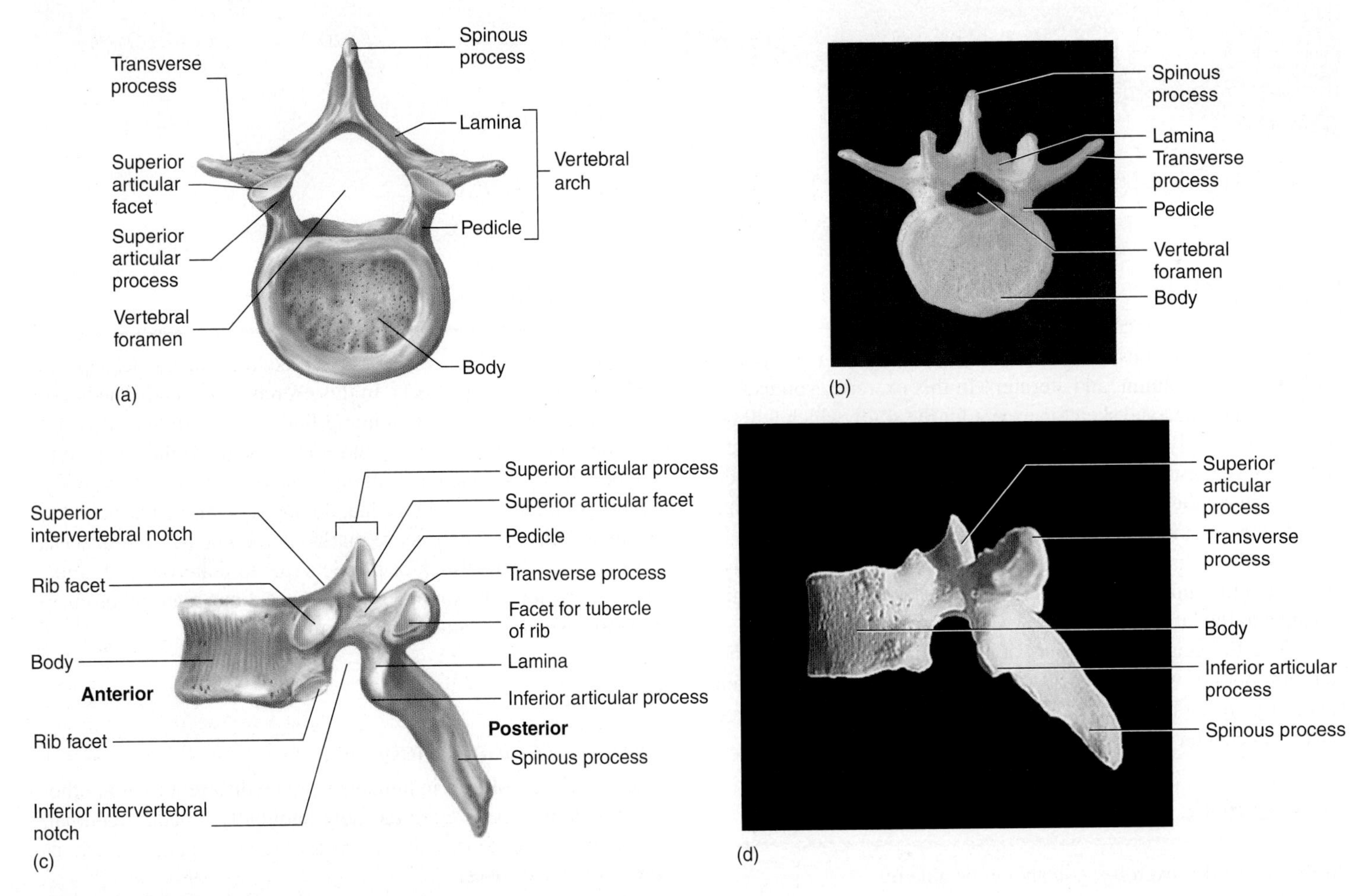

Figure 8.1 Features of a Typical Vertebra
(a) Diagram, superior view; (b) photograph, superior view; (c) diagram, lateral view; (d) photograph, lateral view.

Spinal Curvatures

Ⓐ 3 Locate the spinal curvatures in figure 8.2. The spine has four curvatures, which alternate from superior to inferior. The **cervical curvature** is convex (bowed forward) when seen from anatomical position. The **thoracic curvature** is concave, whereas the **lumbar curvature** is convex and the **sacral (pelvic) curvature** is concave. These allow for balance when standing upright. In addition to the spinal curvatures just described, there are abnormal spinal curvatures. **Scoliosis** (skō´lē ō´sis) is a lateral curvature. **Kyphosis** (kī fō´sis) is an exaggerated thoracic curvature, and **lordosis** is an exaggerated lumber curvature.

Cervical Vertebrae

Ⓐ 4 Examine the isolated cervical vertebrae in the lab and compare them with figure 8.3. There are seven **cervical** (sĕr´vĭ-kal) **vertebrae,** and these can be distinguished from all other vertebrae in that each vertebra has three foramina. The **vertebral foramen** is the largest opening, and the two **transverse foramina** are specific to the cervical vertebrae. The transverse foramina house the vertebral arteries and vertebral veins. Cervical vertebrae 2–6 have **bifid** (bī´fid) **spinous processes** (*bifid* = split in two), and the bodies of the cervical vertebrae are less massive than inferior vertebrae.

The first cervical vertebra (C1) is known as the **atlas** and is the only cervical vertebra without a body. The atlas carries the weight of the head and was named after the Greek mythological figure Atlas, a giant who carried the weight of the heavens on his shoulders. The atlas joins with the head and provides for range of motion, as when you move your head to indicate "yes." The second cervical vertebra (C2) is the **axis,** and it has a unique process called the **dens,** or **odontoid** (ō-don´toyd) **process,** which runs superiorly through the atlas. The odontoid process allows the atlas to rotate on the axis and provides for the range of motion in your head when you rotate your head to indicate "no." The seventh cervical vertebra (C7) is known as the **vertebra prominens.** It has a spinous process that projects sharply in a posterior direction and can be palpated (felt) as a significant bump at the posterior base of the neck. Compare the atlas, an axis, and a vertebra prominens in the lab with figure 8.4.

Thoracic Vertebrae

Ⓐ 5 Examine figure 8.5 and note the characteristics of the vertebrae as seen in the lab. There are 12 thoracic vertebrae. The **thoracic** (thōr ăsĭk) **vertebrae** are distinguished from all other vertebrae by their markings on the posteroloateral surface of the body of the vertebrae, which are attachment points for the ribs.

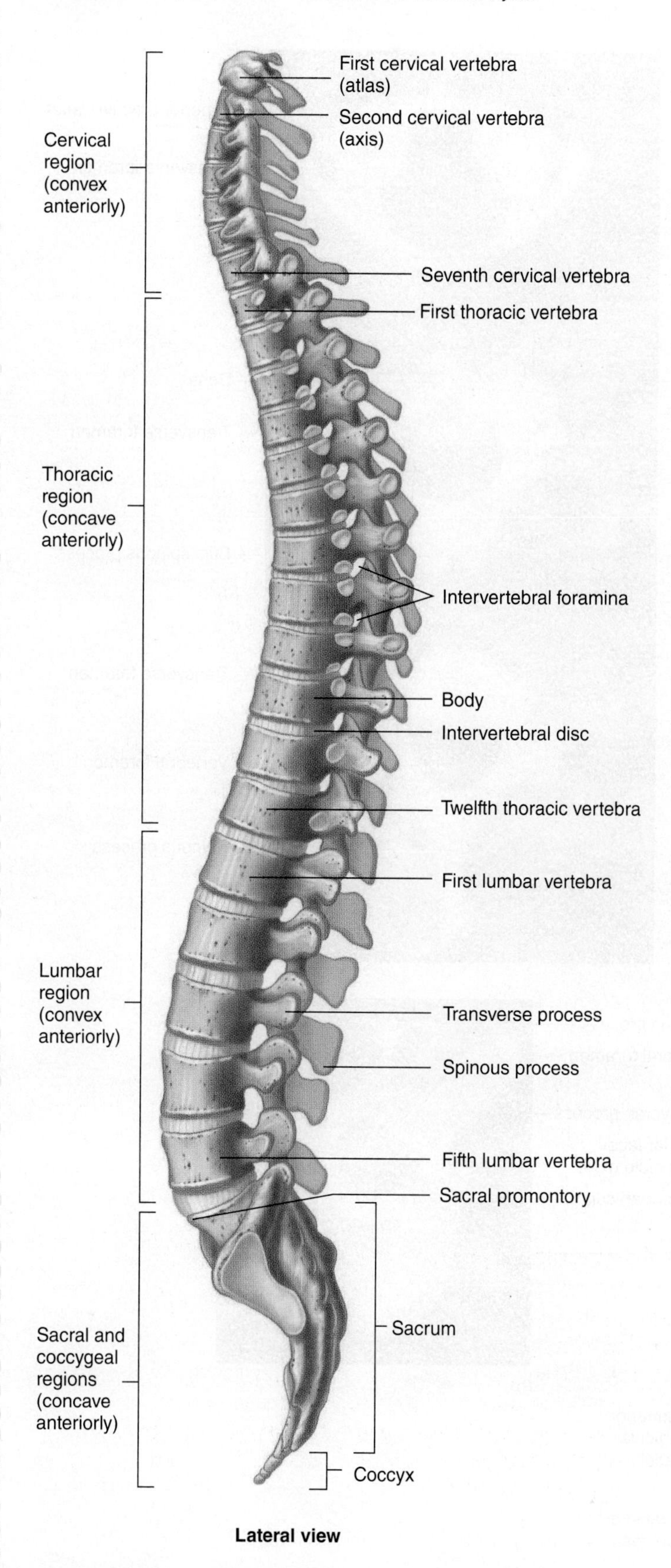

Figure 8.2 Vertebral Column, Lateral View

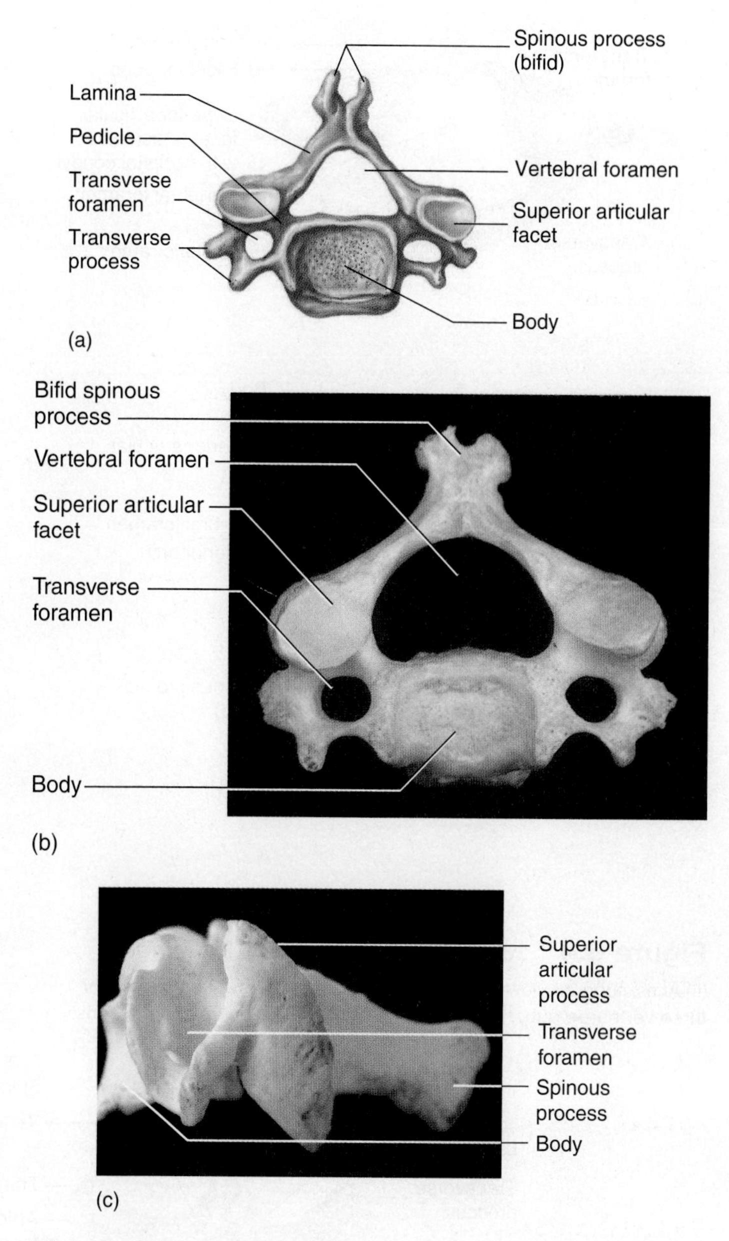

Figure 8.3 Cervical Vertebra, General Features
(a) Diagram, superior view; (b) photograph, superior view; (c) photograph, lateral view.

Sometimes a rib attaches to just one vertebra, in which case the marking on the vertebral body is known as a **rib facet.** In some sections of the thoracic region, the head of a rib is attached to two vertebrae. In this case, each vertebra has a smooth attachment point known as a **demifacet.** The head of the rib spans both of the vertebrae and articulates with the demifacet of the superior and inferior vertebrae.

The thoracic vertebrae also have longer spinous processes than the cervical vertebrae, and the spinous processes of the thoracic vertebrae tend to point inferiorly.

Lumbar Vertebrae

(A)[6] Look at the lumbar vertebrae in the lab and compare them with figure 8.6. There are five lumbar vertebrae. The **lumbar** (lŭm´băr) **vertebrae** are distinguished from the other vertebrae by

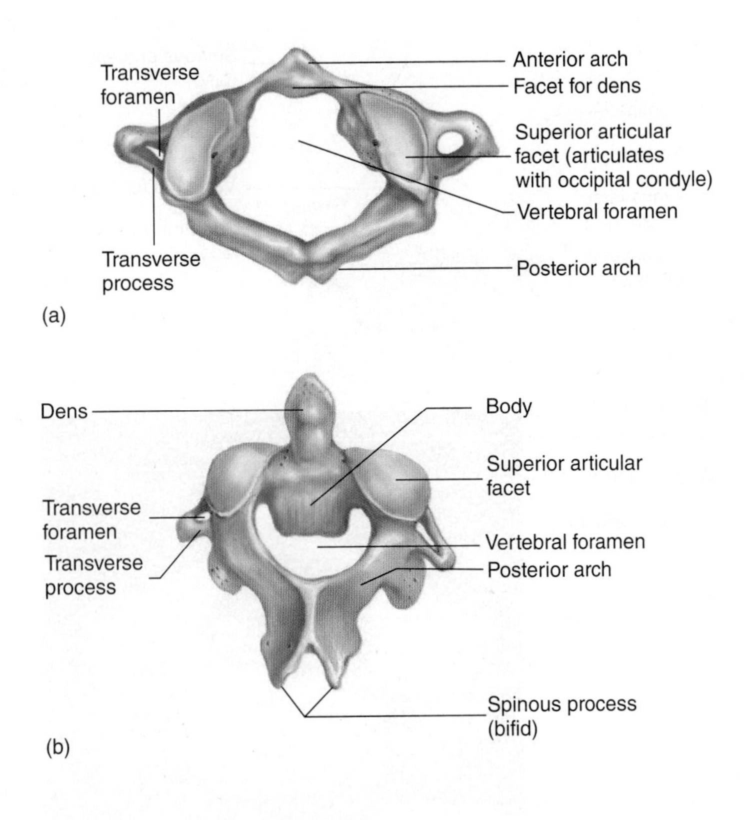

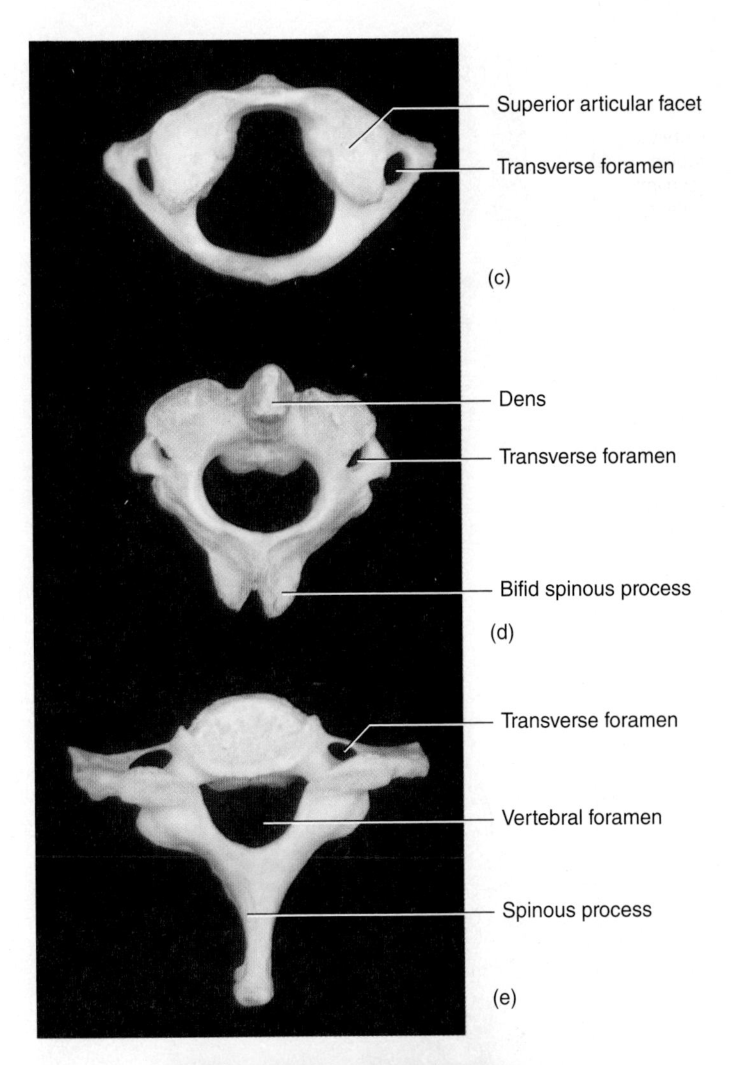

Figure 8.4 Atlas, Axis, and Vertebra Prominens

(a) Atlas, superior view; (b) axis, supero-posterior view. Photographs of three vertebrae, superior view; (c) atlas; (d) axis; (e) vertebra prominens.

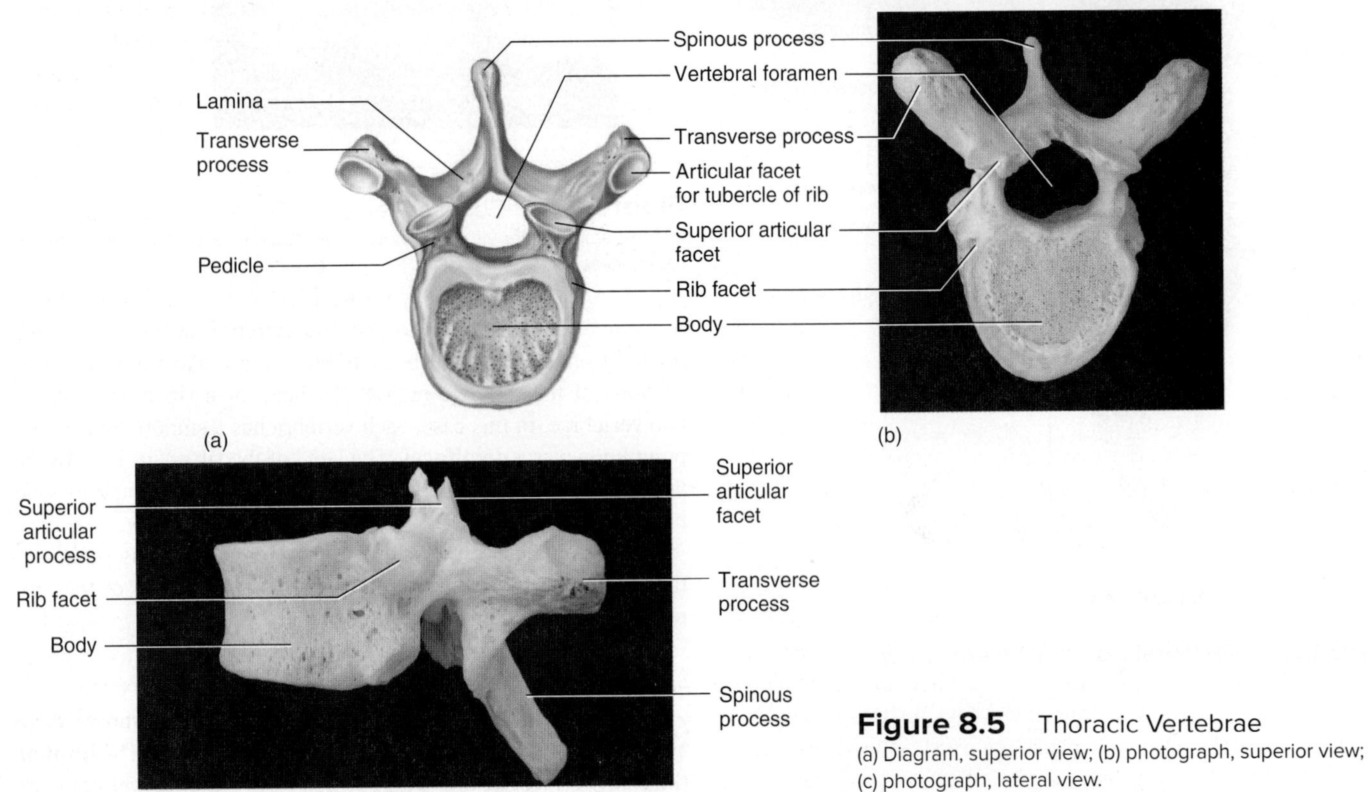

Figure 8.5 Thoracic Vertebrae

(a) Diagram, superior view; (b) photograph, superior view; (c) photograph, lateral view.

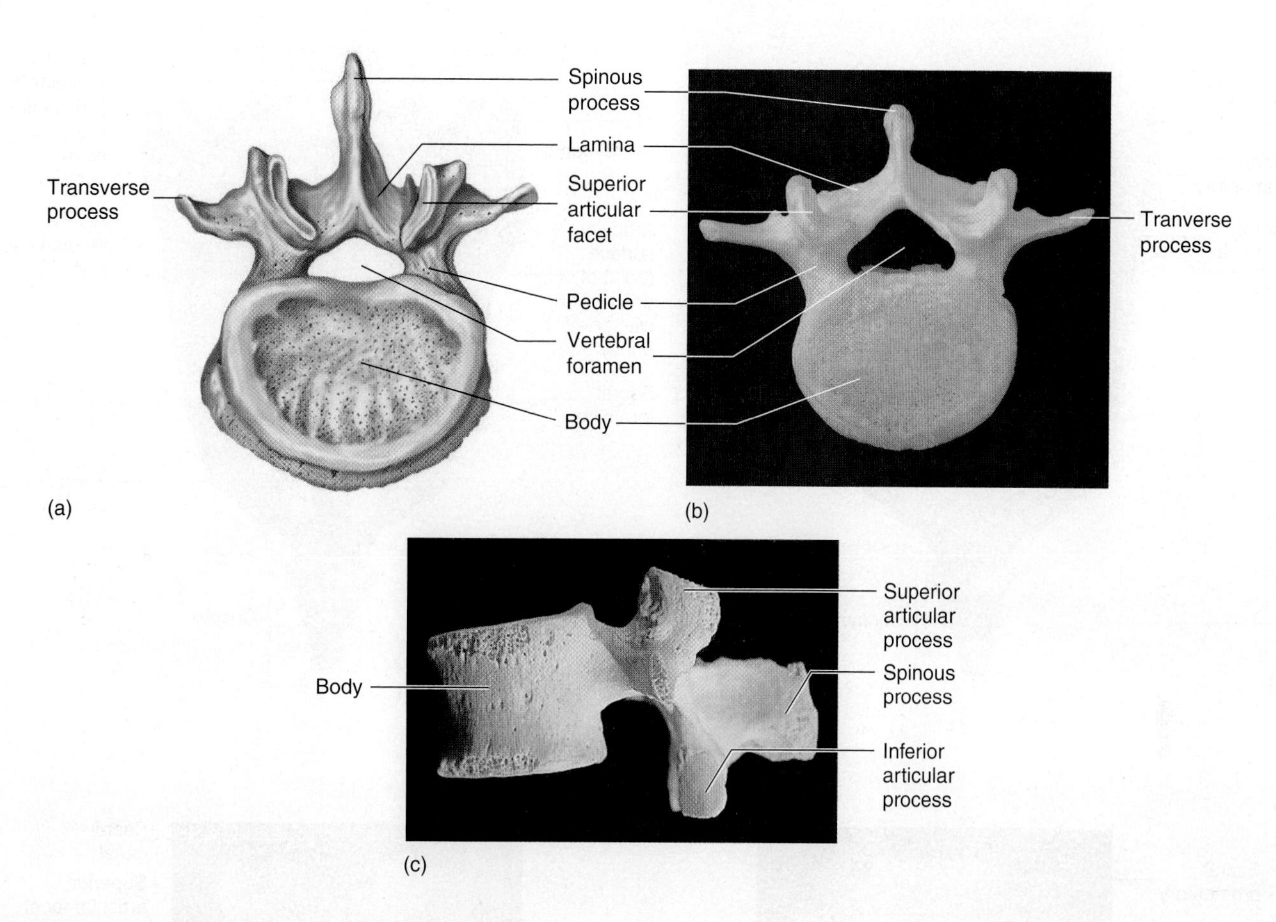

Figure 8.6 Lumbar Vertebrae
(a) Diagram, superior view; (b) photograph, superior view; (c) photograph, lateral view.

having neither transverse foramina nor rib facets. The spinous processes of the lumbar vertebrae tend to be more horizontal than the thoracic vertebrae, and the bodies of the lumbar vertebrae are larger, since they carry more weight than the thoracic vertebrae, yet the twelfth thoracic and the first lumbar vertebrae are remarkably similar. The last thoracic vertebra has rib facets on it, and the first lumbar does not.

Sacrum

(A) 7 Compare a sacrum in the lab with figure 8.7. The **sacrum** (sā´krum) is a large, wedge-shaped bone composed of five fused vertebrae. The lines of fusion are called **transverse lines,** and these can be seen on both the anterior and posterior sides. Notice how the sacrum is shaped like a shallow, triangular bowl. If you place the sacrum in front of you, as you would a cereal bowl, the shallow depression is the **anterior surface.** The two rows of holes you see are the **anterior sacral foramina.** On the posterior surface are the **posterior sacral foramina.**

The **sacral promontory** is a rim on the anterior superior part of the sacrum, and the **ala** (ah´lah) are two expanded regions of the sacrum lateral to the promontory. The roughened areas lateral to the ala are the **articular,** or **auricular, surfaces** of the sacrum; each joins with the ilium to form the **sacroiliac** (sāk´rō-ĭl´ē-ăk) **joint.** As additional force is applied to the sacrum, it wedges itself tighter into the ilium. If you examine the posterior surface of the sacrum, you will notice the **posterior sacral foramina,** the **median** and **lateral sacral crests,** and the superior and inferior openings of the **sacral canal.** The **sacral hiatus** (hī-ā´tŭs) is the inferior opening of the sacrum. As with the other vertebrae, the **superior articular processes** and the **superior articular facets** join with the next most superior vertebra (the fifth lumbar vertebra).

The sacrum in humans is wedge-shaped; this allows the upper body to carry more weight than if the sacrum were box-shaped, as it is in many quadrupeds. Hold a block of wood or a box between your hands, as illustrated in figure 8.8a. Have your lab partner push down on the box and see how much force is required for the box to slip through your hands. This would be similar to the forces acting on a rectangular sacrum. Turn the box or block of wood so that it forms a wedge in your hands, as illustrated in figure 8.8b. Have your lab partner push down on the box. Is it easier or harder to dislodge the "sacrum" in this way?

Coccyx

(A) 8 Examine the coccyx in the lab and compare it with figure 8.7. The **coccyx** (kok´siks) is the terminal portion of the vertebral column, and it usually consists of four fused vertebrae. The coccyx may be fused with the sacrum in some women who have had vaginal deliveries or in individuals who have fallen backward and landed on a hard surface.

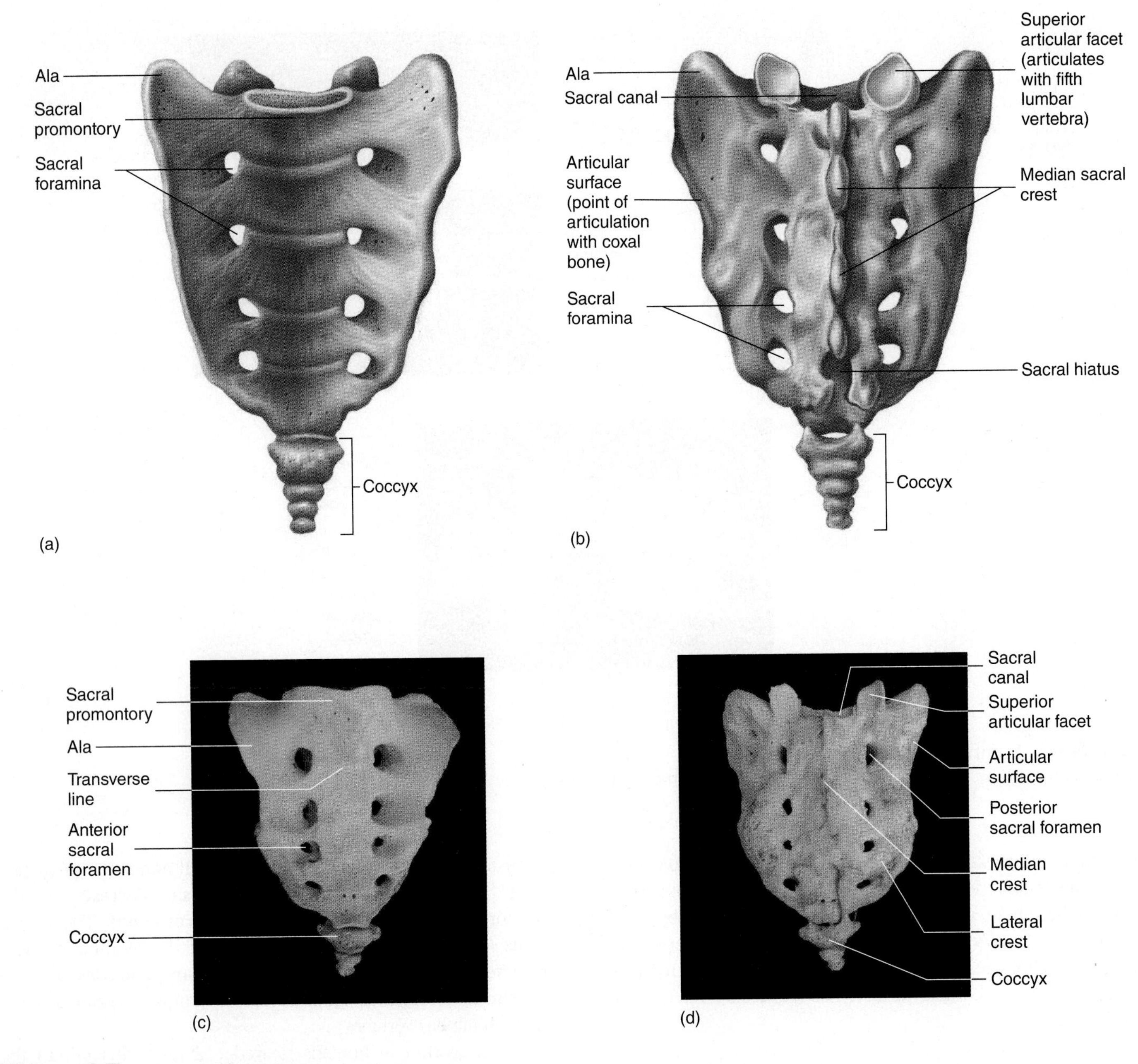

Figure 8.7 Sacrum and Coccyx
Diagram (a) anterior view; (b) posterior view; photograph (c) anterior view; (d) posterior view.

Hyoid

Ⓐ 9 Examine a hyoid bone in lab (figure 8.9). The **hyoid** is a floating bone (it has no direct bony attachments) at the junction of the floor of the mouth and the neck. The hyoid is anchored by muscles from the anterior, posterior, and inferior directions and aids tongue movement and swallowing. Locate the central **body** of the hyoid, the **greater cornua** (kor´nū-ah; sing. **cornu**), and the **lesser cornua** on the material in the lab.

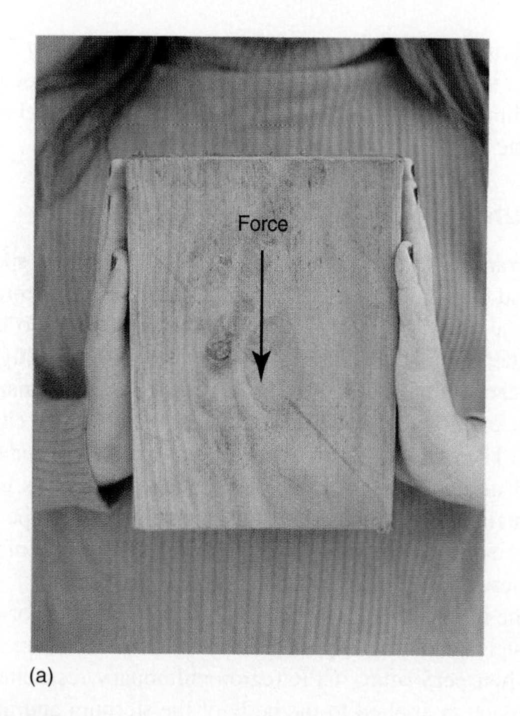

Figure 8.8 Forces Acting on the Sacrum

(a) Rectangular sacrum; (b) wedge-shaped sacrum.

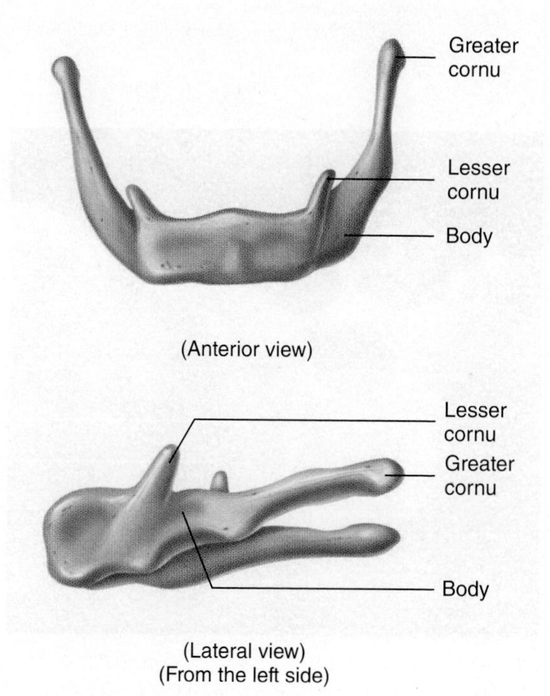

Landmarks seen on this figure:

Body: major portion of the bone

Greater cornu: attachment point for muscles and ligaments

Lesser cornu: attachment point for muscles and ligaments

Special features: one of the few bones of the body that does not articulate with another bone; it is attached to the skull by muscles and ligaments

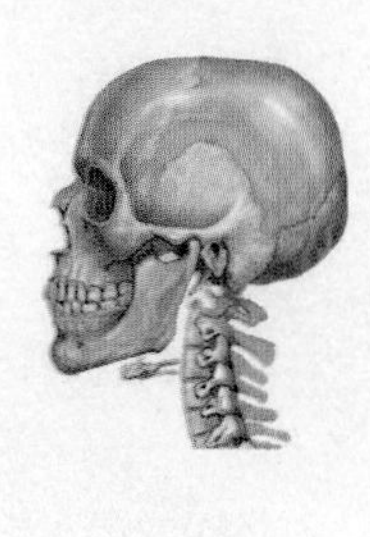

Figure 8.9 Hyoid Bone

Ribs

Ⓐ **10** Take an isolated rib and compare it to figure 8.10. There are 12 pairs of **ribs** in the human; along with the sternum, they make up the **thoracic cage.** Each rib has a **head** that articulates with the body of one or more vertebrae. On the head is a **facet,** which is the site of articulation with the vertebral body. The **neck** is a constricted region near the head of the rib. The process near the neck on ribs 1–10 is the **tubercle** of the rib. This tubercle articulates with the transverse process of the vertebra. The **angle** of the rib **(costal angle)** is the part of a rib that makes a sharp bend at about the same distance from the midline as the inferior angle of the scapula. Some ribs have a truncated **sternal end** that attaches to a costal cartilage prior to joining the sternum. The superior edge of the rib is more blunt than the inferior edge. The depression that runs along the inferior side of each rib is known as the **costal groove.** With the costal groove in an inferior position and the blunt sternal end toward the midline, determine whether you are looking at a left rib or a right rib.

Examine the ribs in an articulated skeleton and compare them with figure 8.11. The first 7 pairs of ribs are **true,** or **vertebrosternal, ribs.** A true rib is one that attaches to the sternum by its own cartilage. Ribs 8 through 12 are **false ribs** because they do not attach to the sternum by their own cartilage. Ribs 8 to 10 attach to the sternum by way of the cartilage of rib 7 and are known as **vertebrochondral ribs.** Ribs 11 and 12 do not attach to the sternum at all and are known as **floating,** or **vertebral, ribs** (a specific type of false rib).

Sternum

Ⓐ **11** Examine the structures of the sternum on lab specimens and in figure 8.11. The **sternum** is composed of three fused bones. The superior segment is the **manubrium** (mă-noo´brē-ŭm), and the depression at the top of the manubrium is the **jugular,** or **suprasternal, notch.** The lateral indentations on the manubrium are sites of articulation with the clavicles known as **clavicular notches.** The main portion of the sternum is the **body,** and between the body and the manubrium is the **sternal angle.** This is a landmark for finding the second rib when using a stethoscope to listen to heart sounds. Locate the **costal notches** on the body of the sternum. These are where the cartilages of the ribs attach.

The narrow, bladelike part that is the most inferior segment of the sternum is the **xiphoid** (zī´foyd) **process.** Care must be taken when performing CPR (cardiopulmonary resuscitation) so that pressure is applied to the body of the sternum and not to the xiphoid process. If the force were applied to the xiphoid process, it could be fractured and driven into the liver.

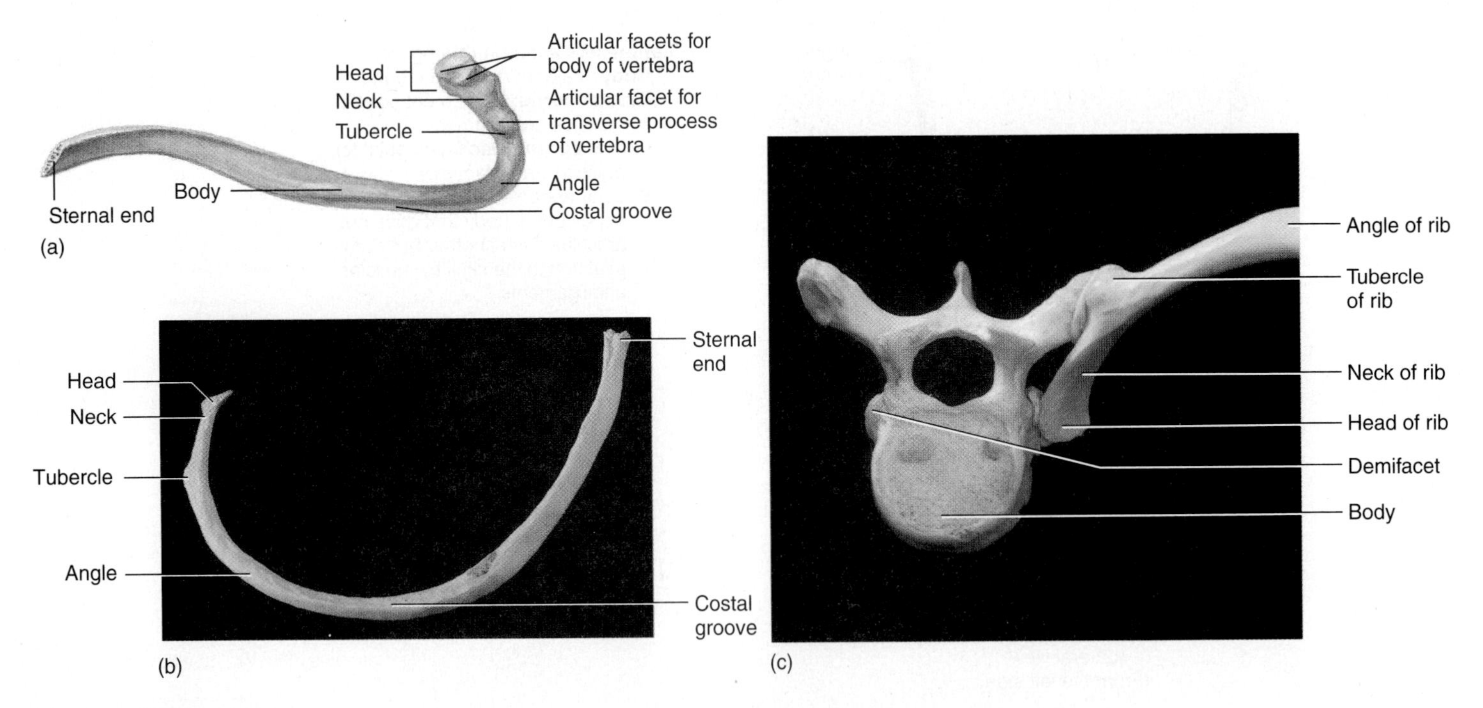

Figure 8.10 Rib
(a) Diagram, anterior view; (b) photograph, inferior view; (c) photograph of rib and vertebra, superior view.

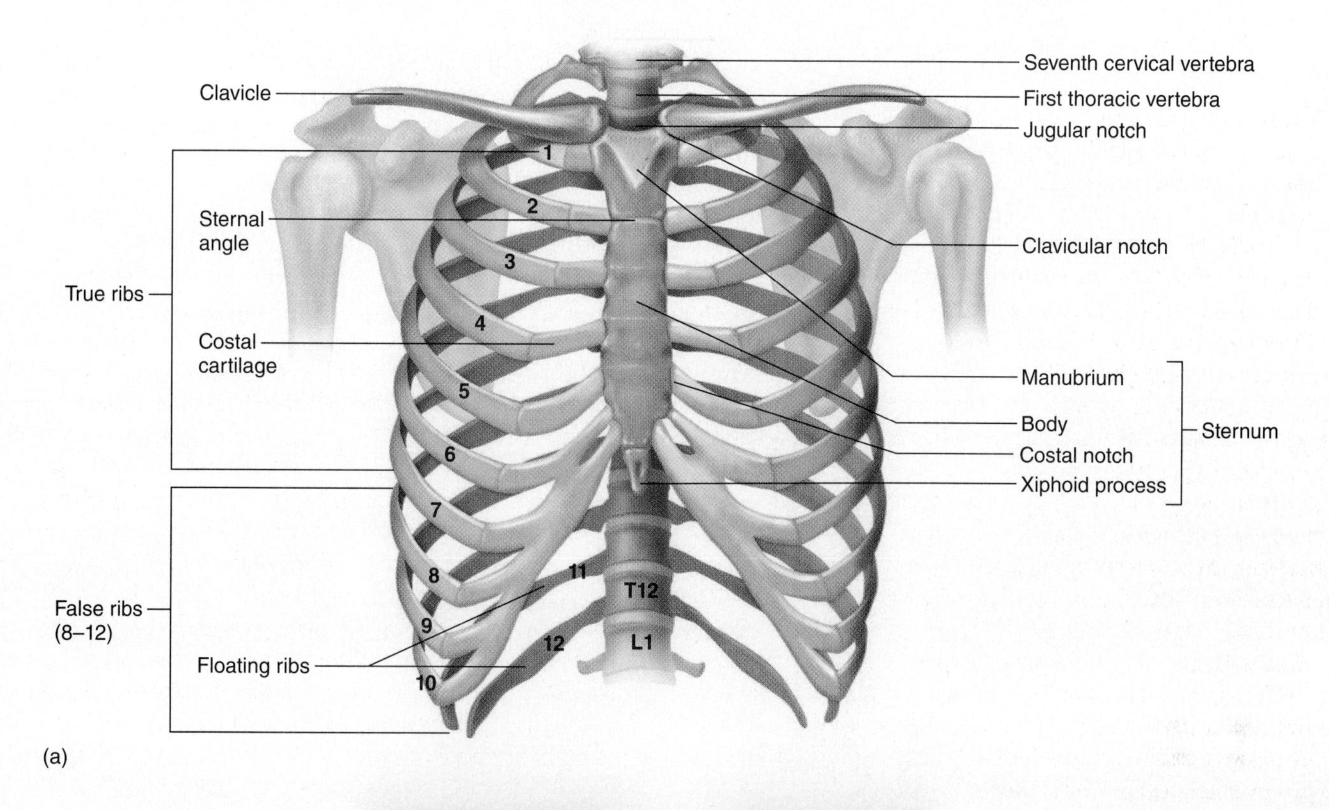

(a)

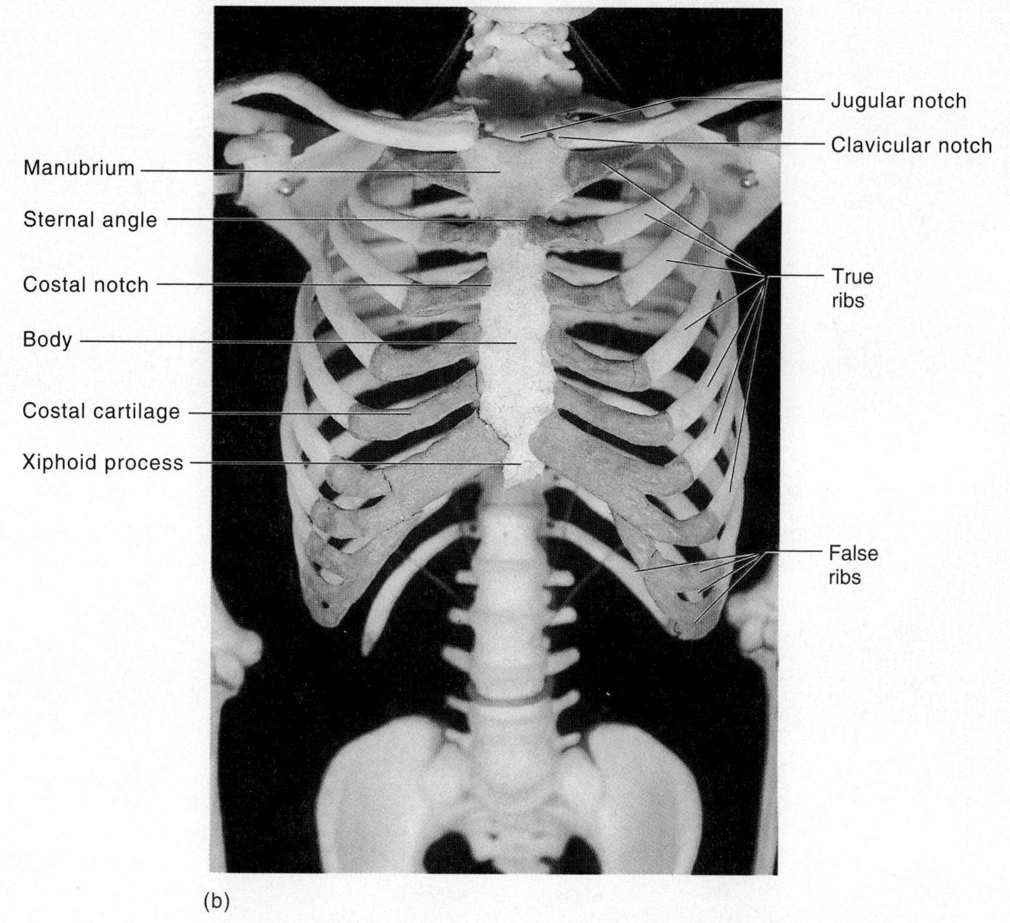

(b)

Figure 8.11 Thoracic Cage
(a) Diagram; (b) photograph.

NOTES

Exercise 8 Review

Name: ______________________

Lab time/section: ______________________

Date: ______________________

Axial Skeleton: Vertebrae, Ribs, Sternum, Hyoid

1. Label the following illustration using the terms provided.

lamina	pedicle
spinous process	transverse process
vertebral arch	vertebral body

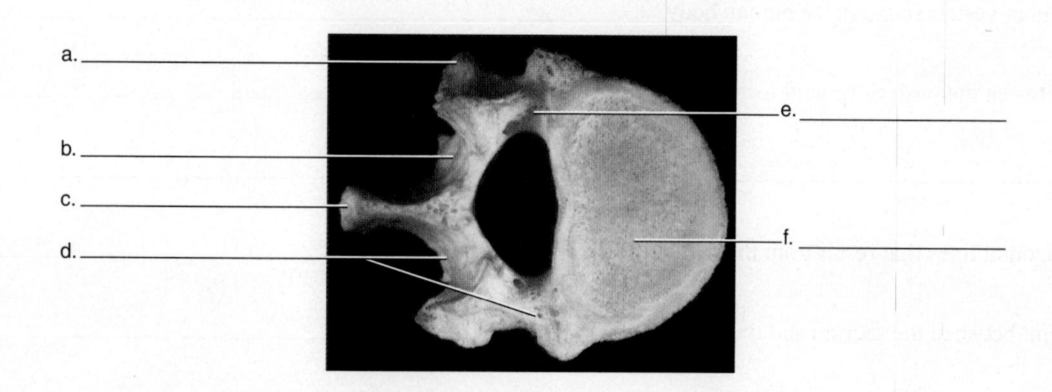

2. What are the names of the fibrocartilage pads between adjacent bodies of the vertebrae? ______________________

3. What structures make up the vertebral arch? ______________________

____

4. The superior articular process of a vertebra articulates (joins) with what specific structure? ______________________

5. Which spinal curvature is the most superior one? ______________________

6. On what vertebra would you find the odontoid process? ______________________

7. Which vertebra has no body? ______________________

8. What two features do cervical vertebrae have that no other vertebrae have? ______________________

____

9. Match the cervical vertebrae with their numeric descriptions.

axis	C1
vertebra prominens	C2
atlas	C7

10. The vertebra in the following illustration comes from what part of the spinal column? ______

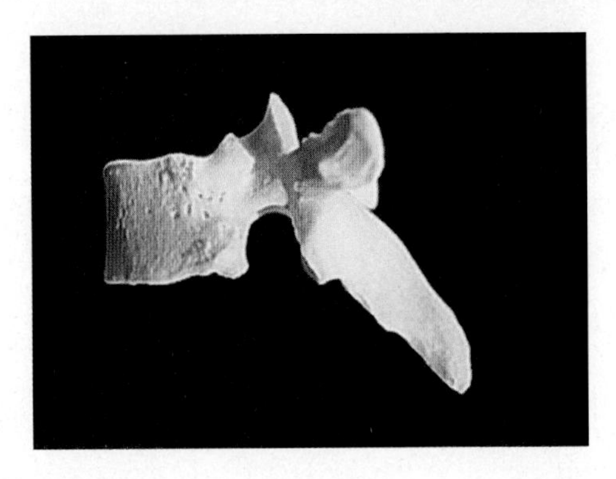

11. How many lumbar vertebrae are in the human body? ______

12. Distinguish between the posterior sacral foramina and the sacral canal. ______

______

13. Name the horizontal lines that result from the fusion of the sacral vertebrae. ______

14. What is the joint between the sacrum and the hipbones called? ______

15. Match the region, or vertebrae from the region, with the features or structures found there.

Region/Vertebrae	Features/Structures
_____ cervical	a. ala
_____ thoracic	b. vertebrae with the largest vertebral bodies
_____ lumbar	c. vertebrae with rib facets
_____ sacral	d. typically, four fused vertebrae
_____ coccyx	e. transverse foramina

16. Name the major features of the hyoid bone. ______

______

17. Does the hyoid bone have any solid bony attachments? ______

18. A rib that attaches to the sternum by the cartilage of rib 7 has what name? ______

19. Is the angle of the rib on the anterior or posterior side of the body? ______

20. Which ribs (by number) are the floating ribs? ______

21. What part of a rib articulates with the transverse process of a vertebra? ______

22. Label the following illustration using the terms provided.

    angle costal groove head neck tubercle

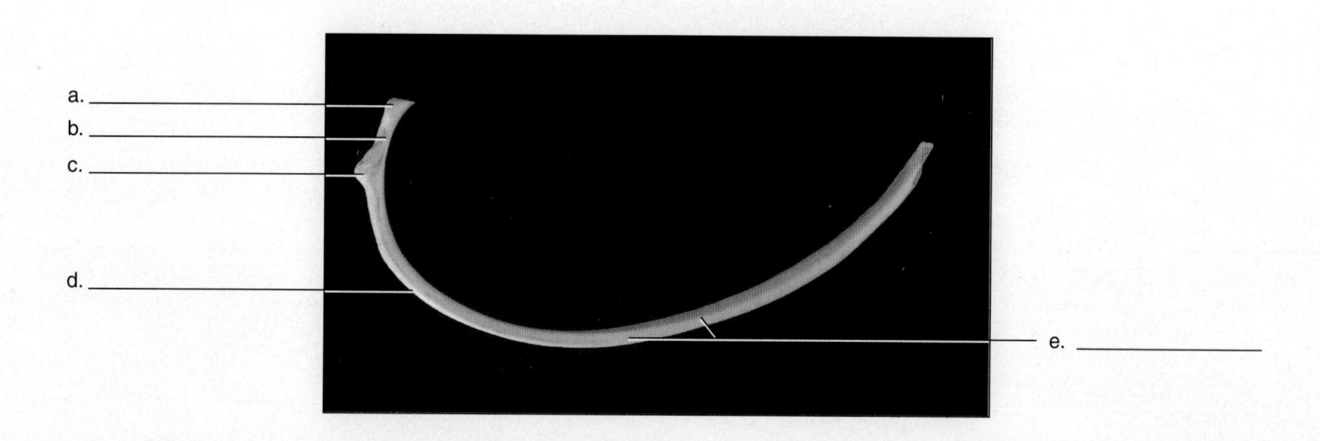

23. The most inferior portion of the sternum is the

    a. body. b. manubrium. c. angle. d. xiphoid.

24. The superior portion of the sternum is called the ______________________________.

NOTES

Exercise 9

Axial Skeleton: Skull

Introduction

The bones of the skull can be divided into three groups: bones of the cranial vault, bones of the face, and bones of the middle ear. The skull is the most complex region of the skeletal system and not only houses the brain but also contains a significant number of sense organs. In this exercise, you learn the details of the skull, particularly those of the cranial vault and the face. These topics are discussed in the Seeley text in chapter 7, "Skeletal System: Gross Anatomy." You will study the bones of the middle ear in this lab manual in Exercise 23. Be careful if you handle real bone material, as bone is fragile and can easily be broken.

Objectives

At the end of this exercise, you should be able to

1. list all the bones of the cranium and the major features of those bones;
2. list all the bones of the face and the major features of those bones;
3. name all the bones that occur singly or in pairs in the skull;
4. locate the major foramina of the skull;
5. find the specific bony markings of representative bones;
6. name the major sutures of the skull;
7. point out all the fontanels of the fetal skull.

Materials

Disarticulated skull, if available
Articulated skulls with the calvaria cut
Fetal skulls
Foam pads to cushion skulls from the desktop
Pipe cleaners, wooden applicator sticks, or plastic straws

Procedure

Overview of Cranial Vault

(A) 1 Examine a skull and locate the bones listed in figure 9.1. The skull consists of the **cranium** and the **face.** The skull bones in the cranium are listed next with a number, in parentheses, after the name of the bone, indicating whether the bone occurs singly or in pairs.

Frontal (1) Parietal (2)
Occipital (1) Temporal (2)
Sphenoid (1) Ethmoid (1)

Overview of Face

(A) 2 Locate bones of the face on a skull as shown in figures 9.1 and 9.2. The numbers in parentheses represent whether these facial bones are paired or occur singly in the skull.

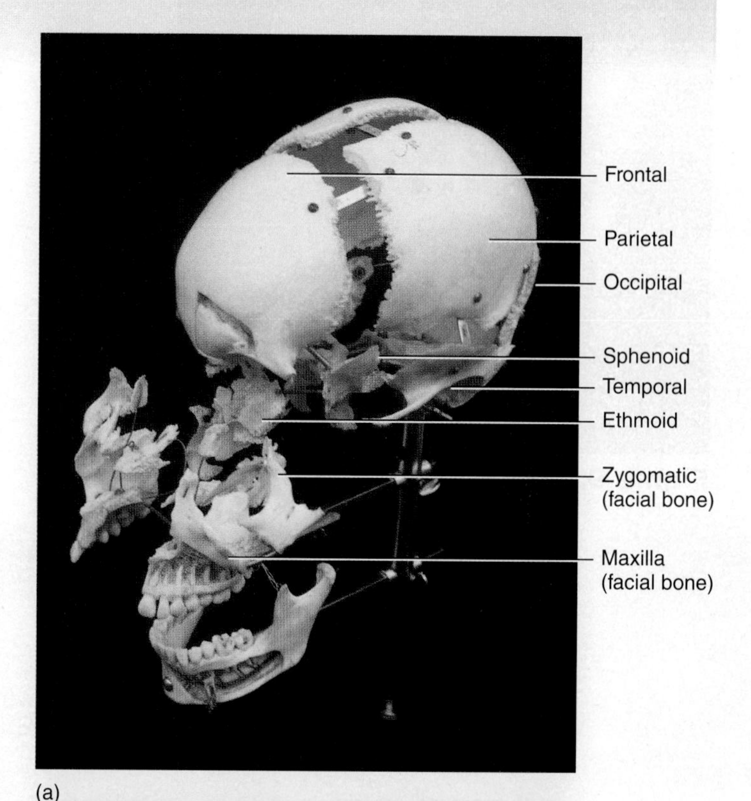

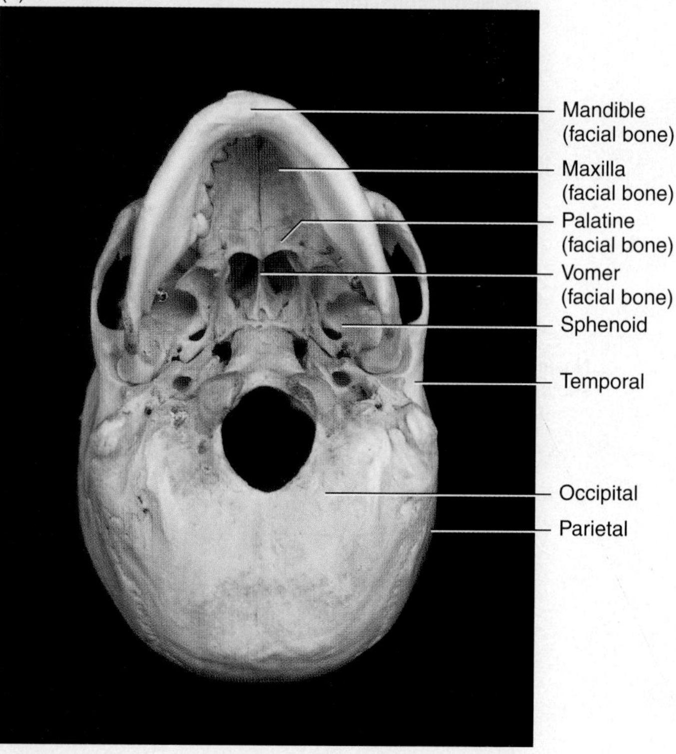

Figure 9.1 Bones of the Cranium
(a) Disarticulated skull, 3/4 view; (b) inferior view.

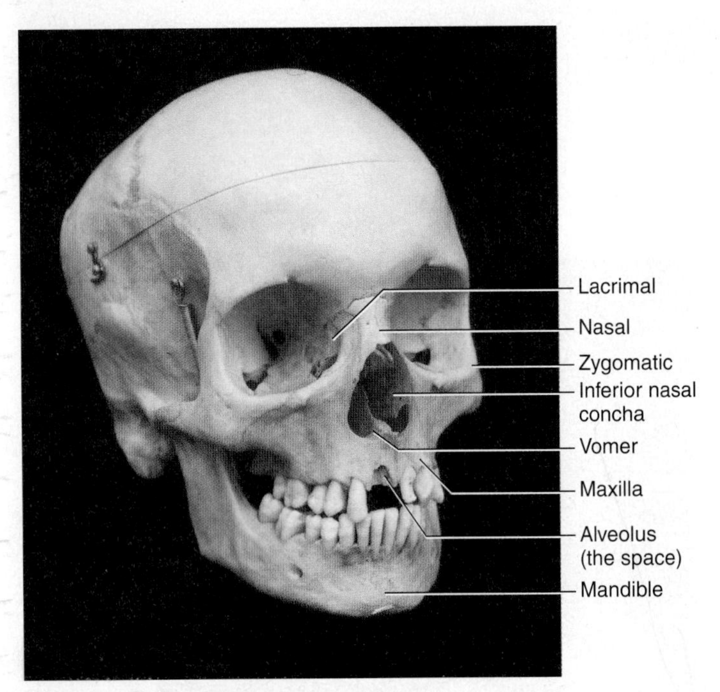

Figure 9.2 Bones of the Face

Maxilla (2)	Nasal (2)
Mandible (1)	Palatine (2)
Vomer (1)	Zygomatic (2)
Lacrimal (2)	Inferior nasal concha (plural *conchae*) (2)

As you locate the various bones on a skull in the lab, make sure you do not poke pencils, pens, or fingers into the delicate regions of the eyes or nasal cavity. Be very careful handling skulls. Do not break the delicate structures in these regions. Once you have found the major bones of the skull, use the following descriptions and illustrations to find the specific structures. The skull is described from a series of views, and you should locate the anatomical features listed for those views.

Anterior View

(A)[3] Examine a skull from the front and compare it to figure 9.3. The large bone that makes up the forehead is the **frontal bone.** It is a single bone that makes up the superior portion of the **orbits** (eye sockets) and has two ridges above the eyes called **supraorbital margins** (eyebrows). There are holes in each of these ridges, called **supraorbital foramina,** where nerves and arteries

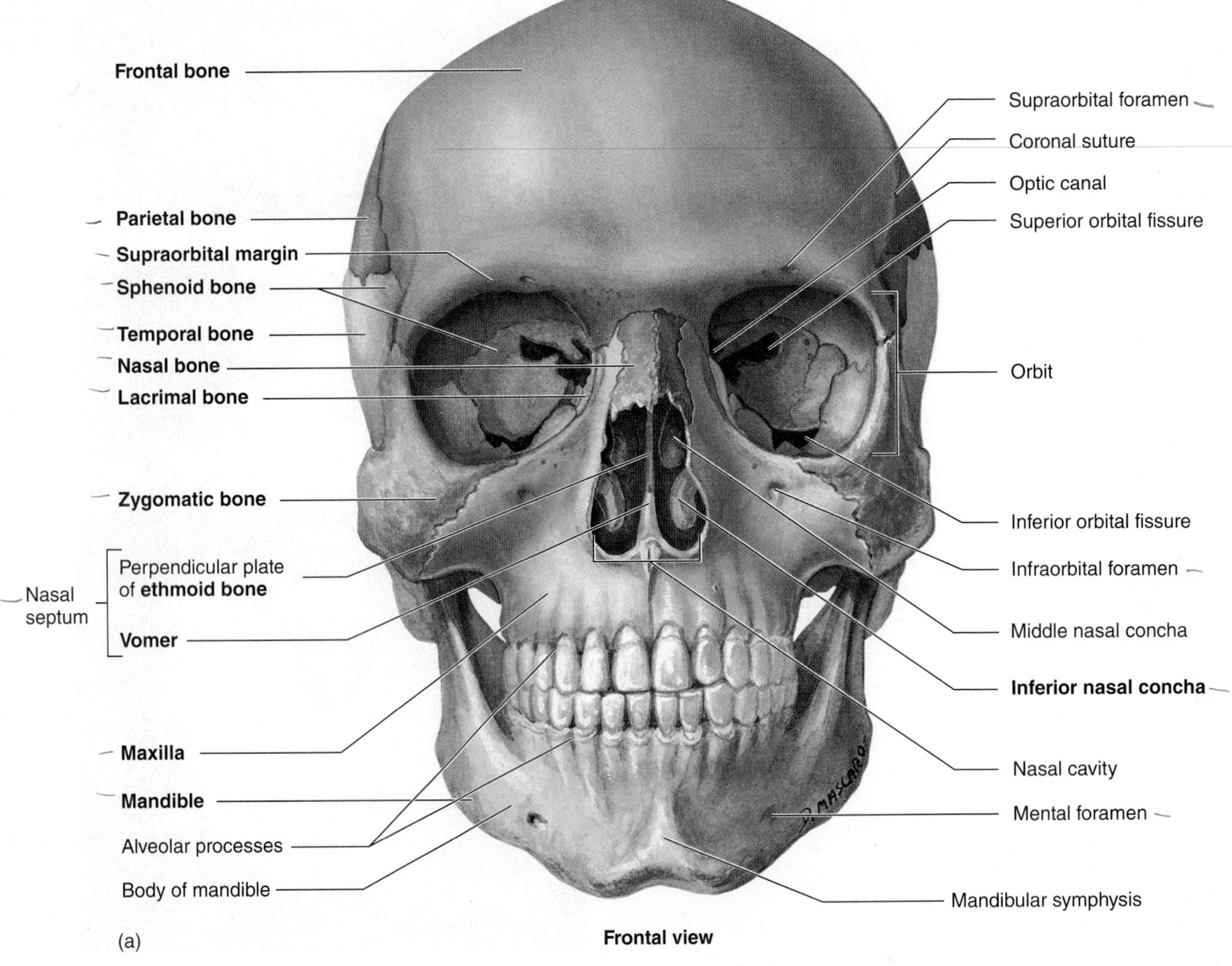

Figure 9.3 Skull, Anterior View
(a) Diagram; (b) photograph.

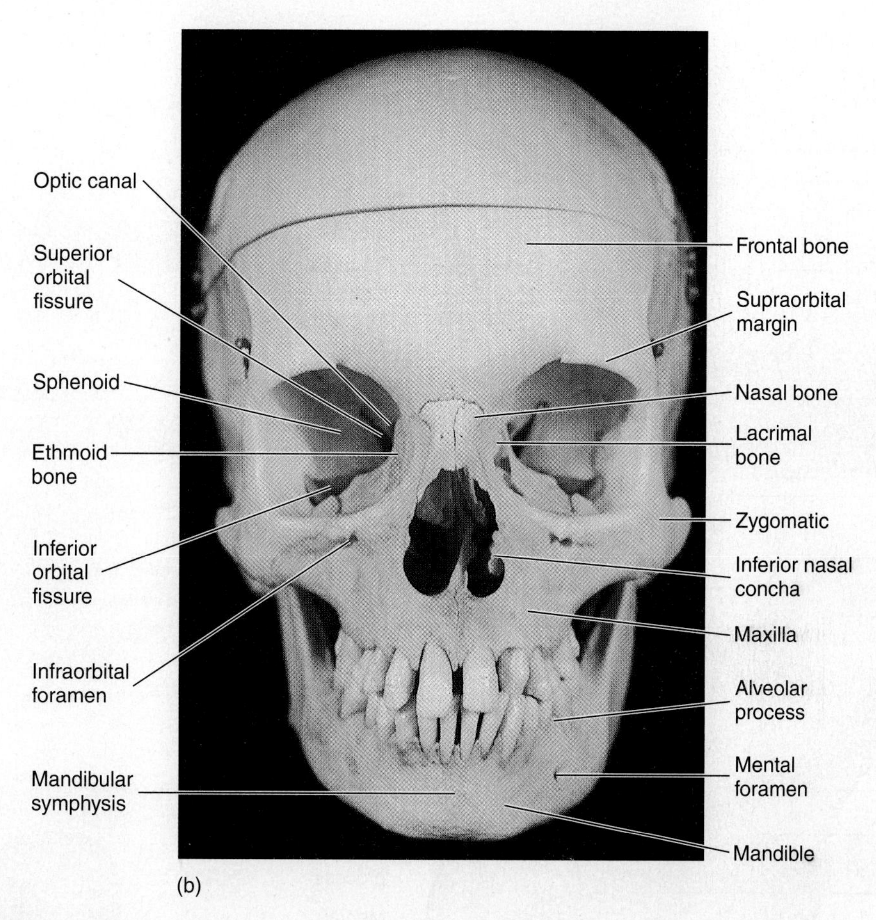

Figure 9.3 *Continued.*

reach the face. Most of what you can see inferior to the frontal bone are facial bones. The bony part of the nose is composed of paired **nasal bones,** which join with the cartilage (not present in bony skulls) that forms the tip of the nose. Posterior to the nasal bones are thin strips of the upper maxillae, bones that hold the upper teeth. Posterior to the maxilla are the thin **lacrimal** (lăk´rĭh-mul) **bones,** which contain the **naso-lacrimal canal,** a tube that drains tears from the eye to the nose. Posterior to the lacrimal bones is the **ethmoid** (ĕth´moyd) **bone,** which is very delicate and frequently broken on skulls that have been mishandled. Posterior to the ethmoid is the **sphenoid** (sfē´noyd) **bone,** which forms the posterior wall of the orbit and contains not only the **optic canal** (a passageway for the optic nerve) but also the **superior orbital fissure** and the **inferior orbital fissure.** Note the bones on the lateral side of the orbit. These are the **zygomatic bones.**

The major bone below the orbit is the maxilla, which also makes up the floor of the orbit. The two maxillae have sockets called **alveoli** (ăl´vē-oh-lī) (sing. **alveolus**), which contain the teeth and extensions of bone between each pair of sockets called **alveolar processes.** The **infraorbital foramen,** a small hole below the eye in the maxilla, is a passageway for nerves and blood vessels. The most inferior bone of the face is the **mandible,** commonly known as the lower jaw. The mandible also has alveoli and alveolar processes (figure 9.4). The mandible begins as two bones in utero and fuses at the midline of the chin. This fusion results in the **mandibular symphysis** (sĭhm´fah-sĭs). Lateral to the mandibular symphysis are the **mental foramina,** which conduct nerves and blood vessels to the tissue anterior to the jaw.

Superior View

(A) 4 Locate the major suture lines of the skull from the superior view, as seen in figure 9.5. The frontal bone is separated from the pair of parietal bones by the **coronal suture.** The parietal bones are separated from one another by the **sagittal suture.** The parietal bones are separated from the occipital bone by the **lambdoid** (lamb´doyd) **suture.** The lambdoid suture is named after the Greek letter lambda, which is Y-shaped. There may be small bones between the occipital bone and the parietal bones (or between other skull bones), and these are known as **sutural,** or **Wormian, bones.**

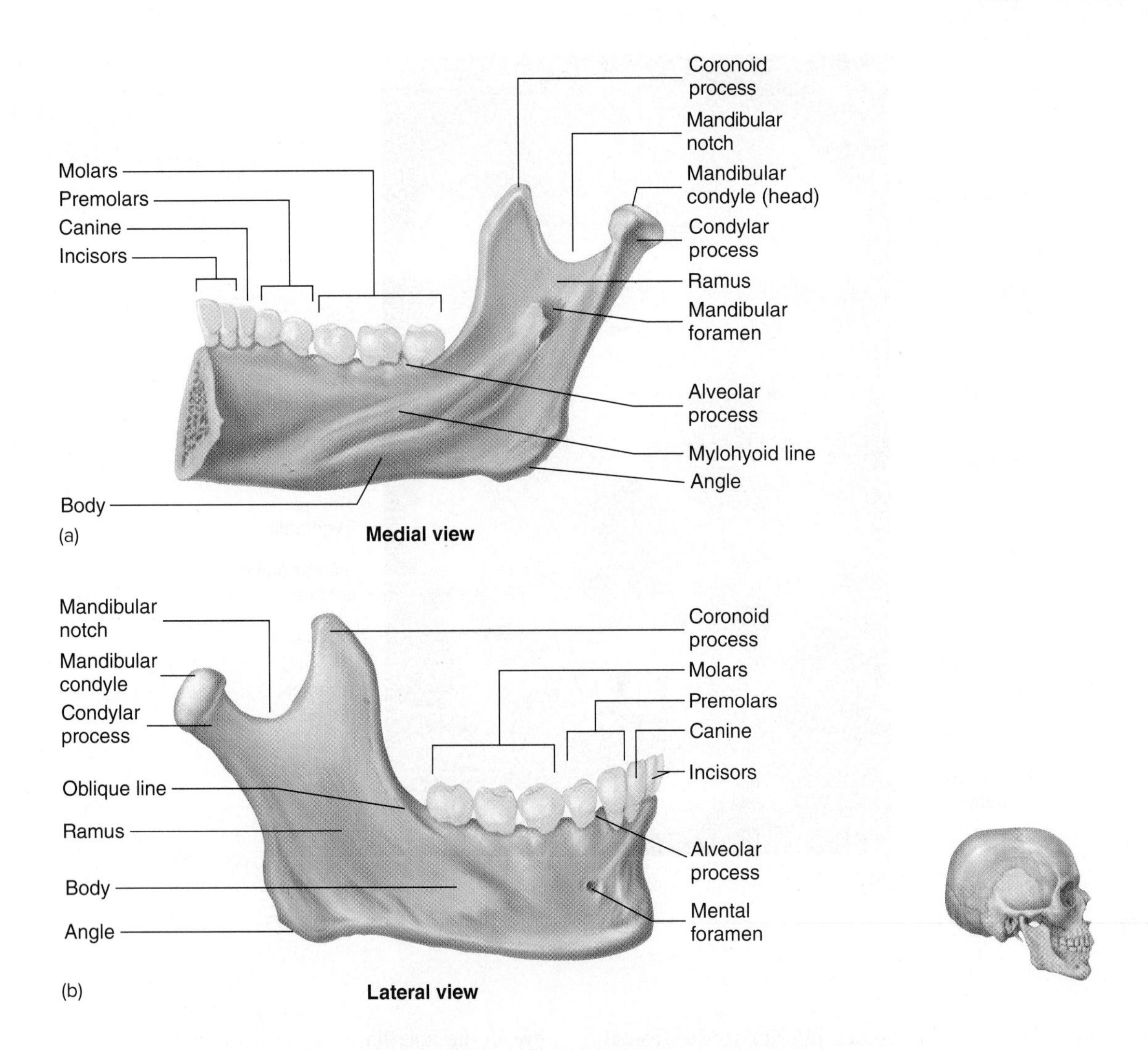

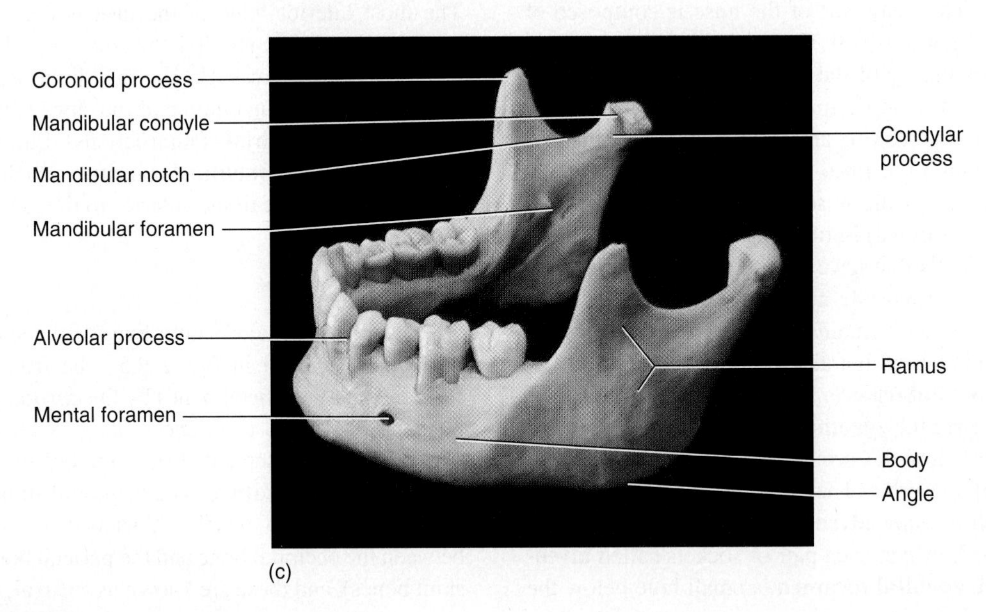

Figure 9.4 Mandible
Diagram (a) medial view; (b) lateral view. (c) Photograph, lateral view.

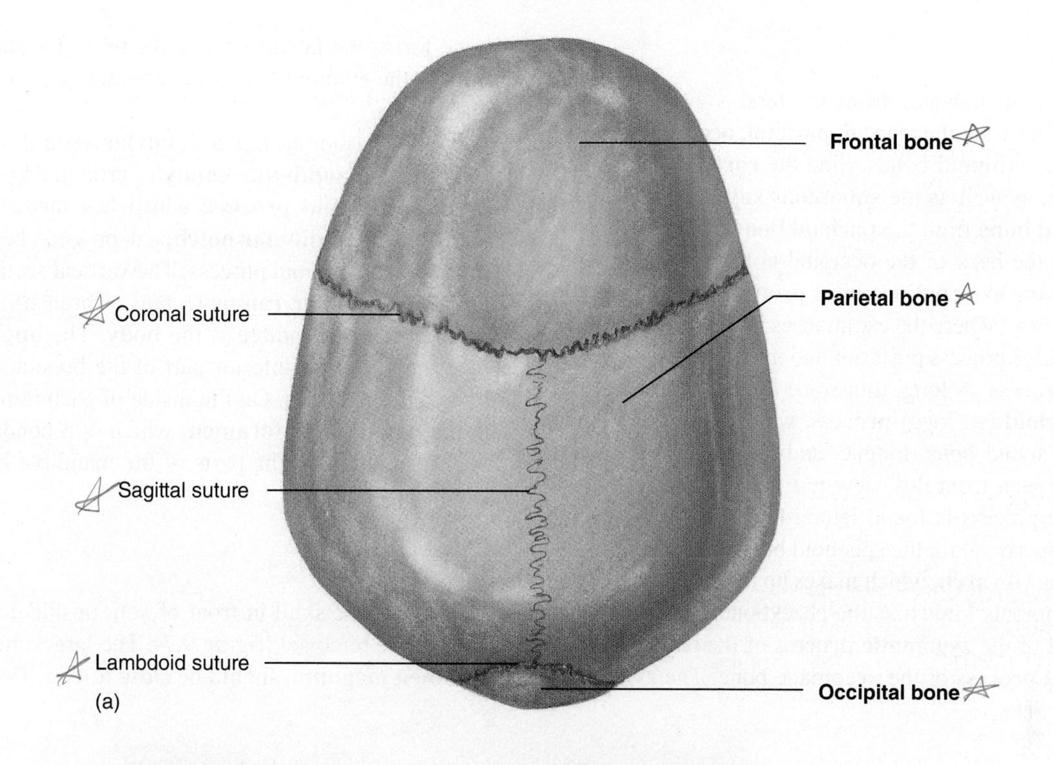

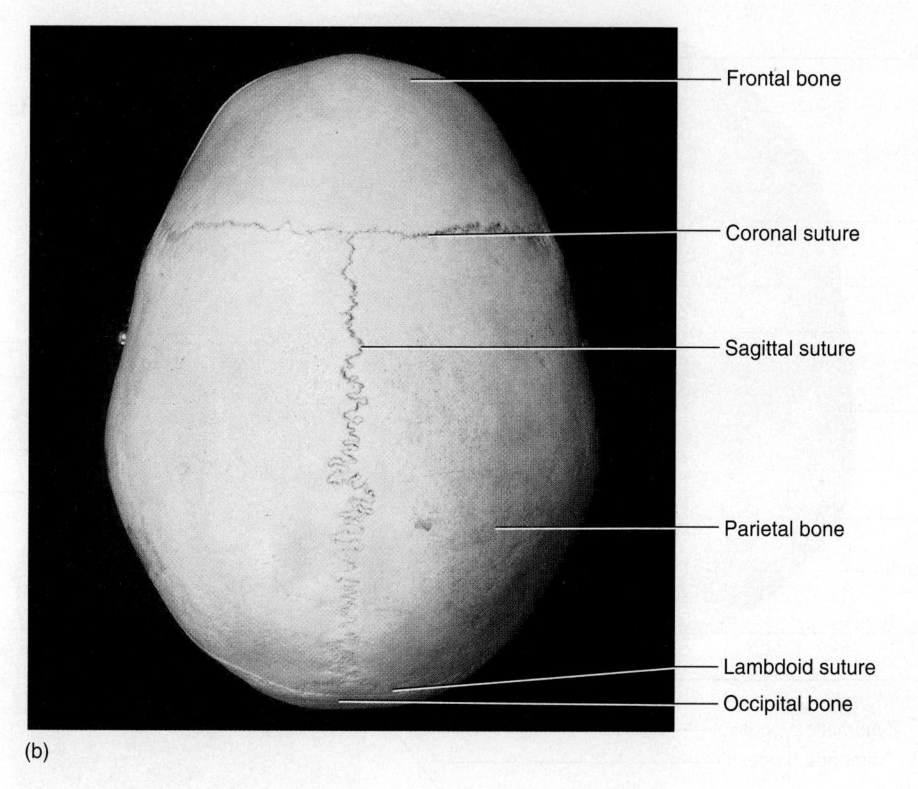

Figure 9.5 Skull, Superior View
(a) Diagram; (b) photograph.

Lateral View

Ⓐ 5 Locate the cranial bones from a lateral view as seen in figure 9.6. These are the **frontal, parietal, occipital, temporal, sphenoid,** and **ethmoid** bones. Find the **coronal suture** and **lambdoid suture,** as well as the **squamous suture,** which separates the **temporal bone** from the **parietal bone.** You may be able to see a bump at the back of the occipital bone from this angle. This is known as the **external occipital protuberance.** The hole on the side of the head where the ear attaches is the **external auditory canal.** The large process posterior and inferior to this opening is the **mastoid process.** A long, thin spine medial to the mastoid process is the **styloid** (stī´loyd) **process,** which has muscles that connect it to the hyoid bone, tongue, and larynx. The sphenoid bone can also be seen from this view just anterior to the temporal bone. A bony process is found lateral to, and forming a thin bridge of bone superficial to, the sphenoid bone. This is the **zygomatic** (zī´-goh-mă´tik) **arch,** which makes up the upper part of the cheek and is commonly known as the cheekbone. The zygomatic arch is composed of the **zygomatic process** of the temporal bone and the **temporal process** of the zygomatic bone. The zygomatic bone forms the lateral wall of the orbit. On the inner wall of the orbit is the ethmoid bone, the lacrimal bone, the maxilla, and the nasal bone.

The mandible has a **condylar** (kon´dih-lar) **process** with a terminal **mandibular condyle,** articulating with the temporal bone; a **coronoid process,** which lies medial to the zygomatic arch; and a **mandibular notch,** a depression between the condylar process and coronoid process. The vertical section of the mandible is the **mandibular ramus** (*ramus* = branch), and the horizontal portion of the mandible is the **body.** The **angle** of the mandible is at the posterior, inferior part of the bone at the junction of the body and the ramus. On the inside of each ramus of the mandible is the **mandibular foramen,** which is a conduit for an artery, a vein, and a nerve. The parts of the mandible can be identified in figures 9.4 and 9.6.

Inferior View

Ⓐ 6 Place the skull in front of you, upside down, with the mandible removed (figure 9.7). The largest hole in the skull, the **foramen magnum,** should be close to you. The foramen magnum

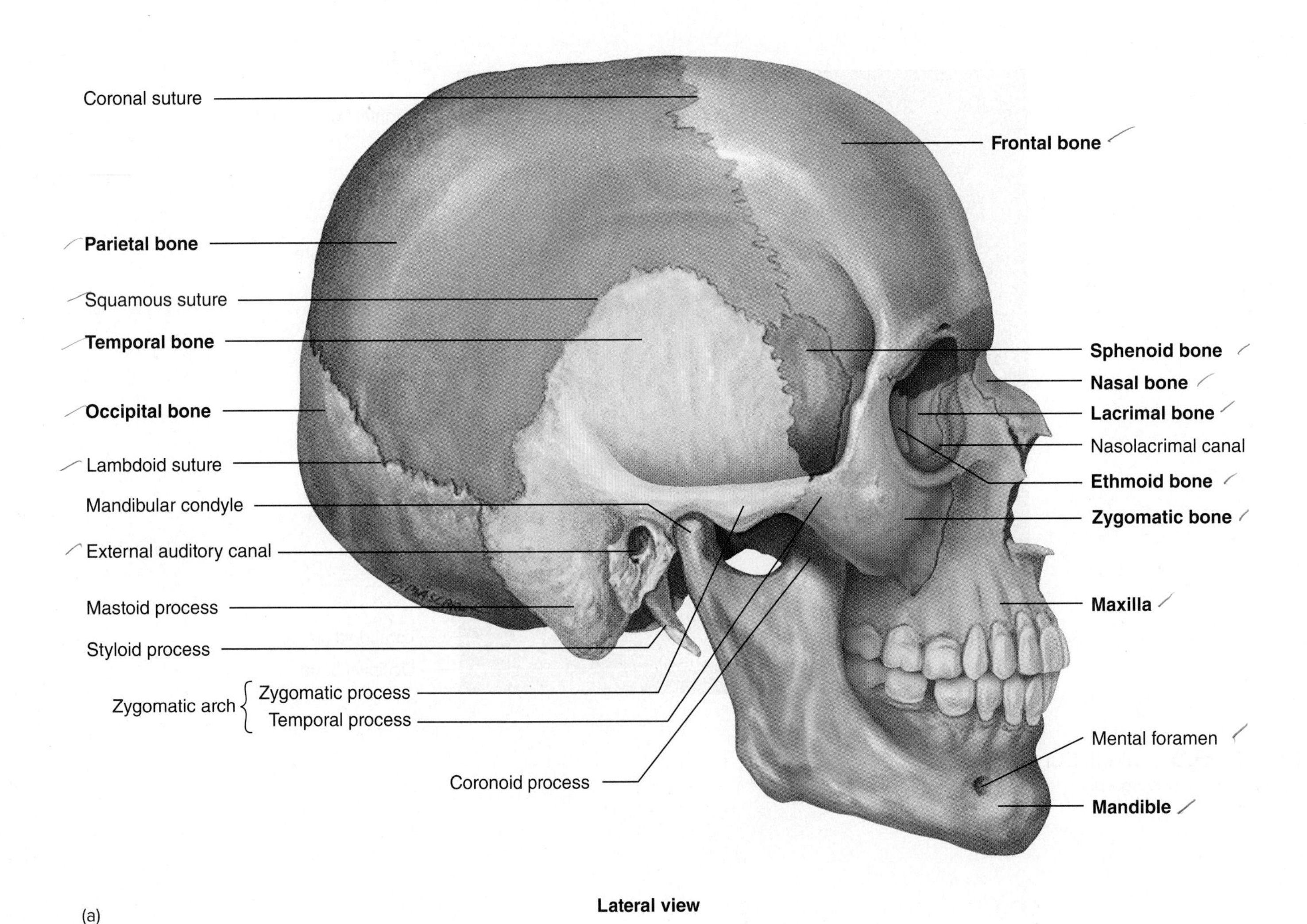

Figure 9.6 Skull, Lateral View
(a) Diagram; (b) photograph.

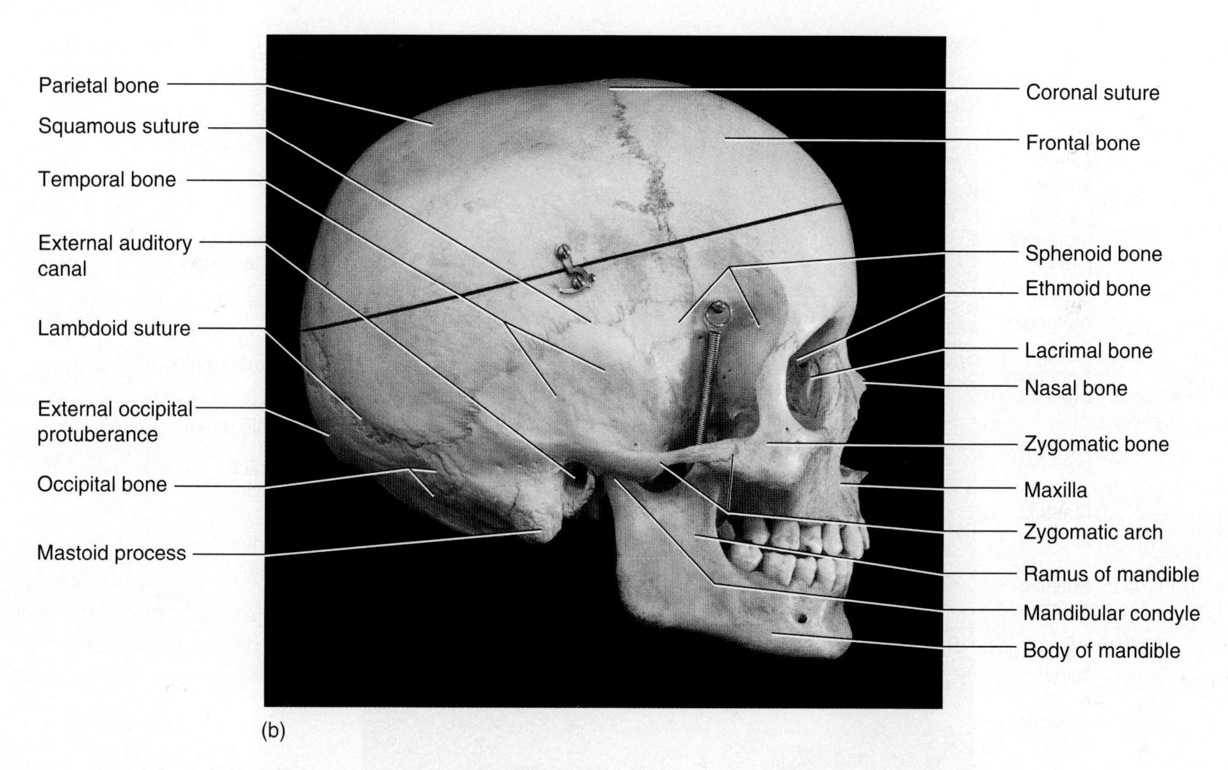

Figure 9.6 *Continued.*

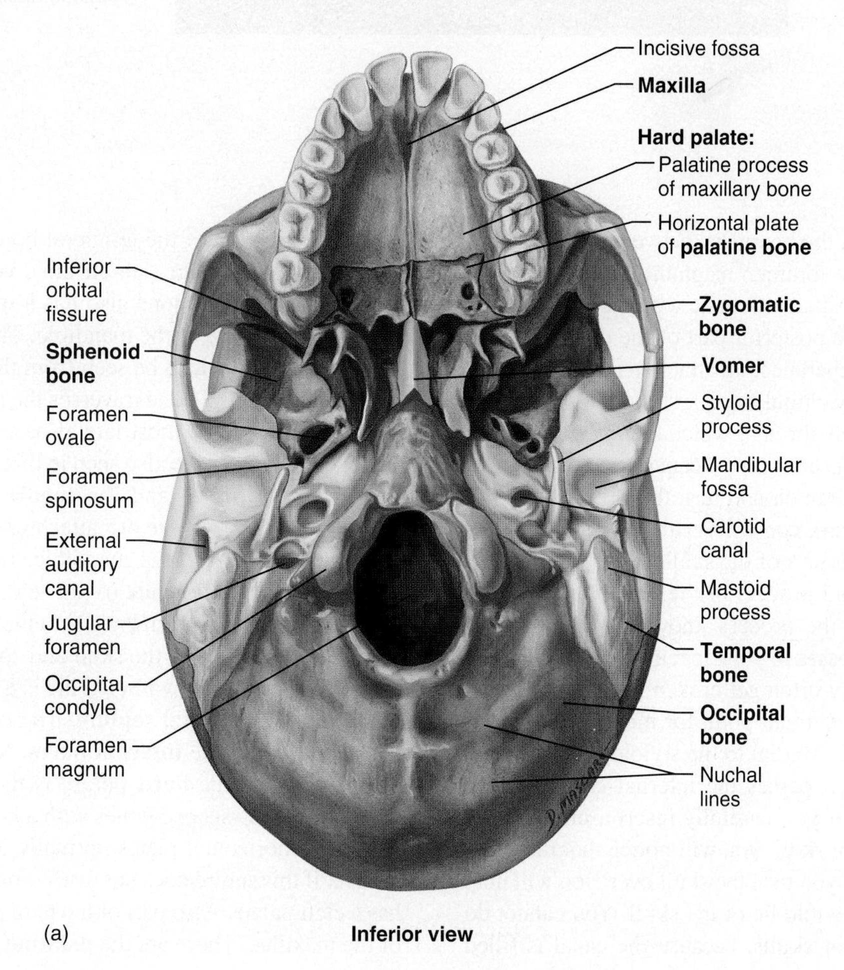

Figure 9.7 Skull, Inferior View
(a) Diagram; (b) photograph.

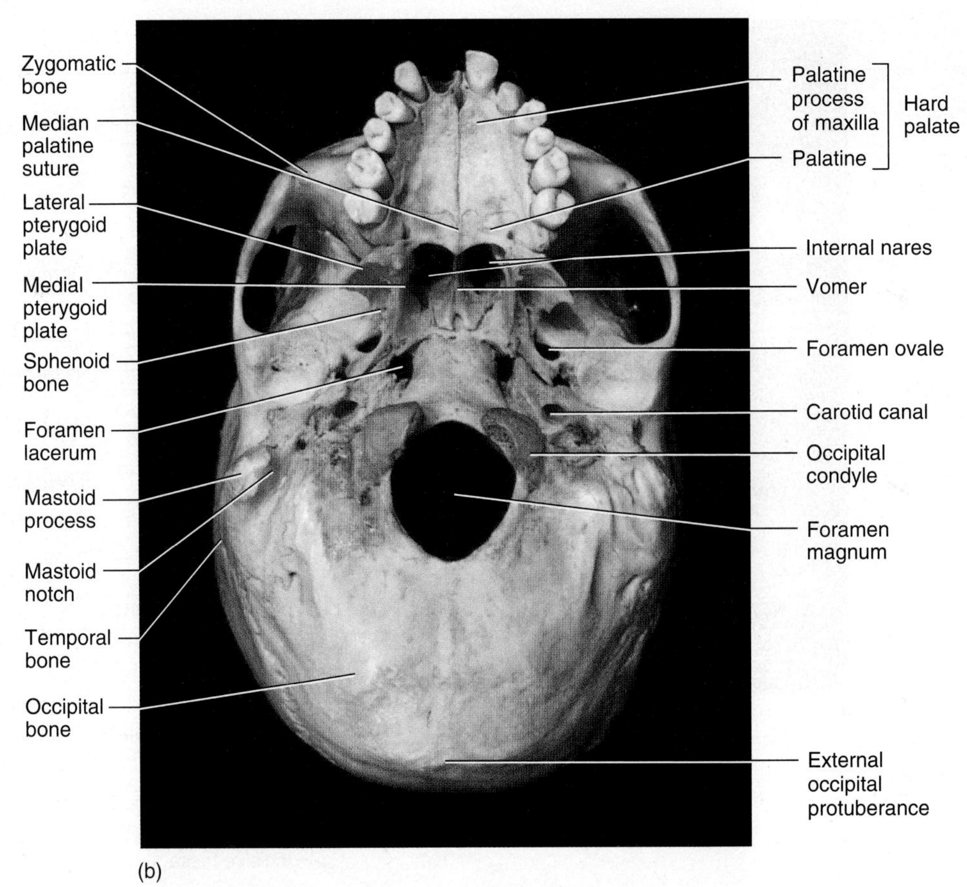

Figure 9.7 *Continued.*

is in the occipital bone and is the boundary between the brain and the spinal cord. Lateral to the foramen magnum are the **occipital condyles,** which are processes that articulate with the first cervical vertebra. A small bump at the posterior part of the occipital bone is the external occipital protuberance, an attachment site for muscles. At the junction of the occipital bone and the temporal bone is the **jugular foramen,** a hole through which the internal jugular vein passes. If you *carefully* insert a pipe cleaner or wooden applicator stick into the jugular foramen and turn the skull over (with the top of the cranium removed), you will see that the jugular foramen comes from the posterior part of the skull.

You can see the mastoid process of the temporal bone and a depression just medial to the process known as the **mastoid notch.** Find the **styloid processes** in your specimen, although they may be hard to locate, as they often get broken in lab specimens. The styloid process is an attachment point for muscles that move the tongue, larynx, and hyoid. Medial to the styloid process is the **carotid canal,** through which passes the internal carotid artery carrying blood to the brain. If you carefully insert a pipe cleaner into the carotid canal of a real skull, you will notice that the canal bends at about a 90° angle; if you turn the skull over, you will note that the opening occurs in the middle of the skull. You cannot do this with most plastic casts of skulls, because the canal is filled in. At the junction of the temporal bone and the sphenoid bone is the **foramen lacerum** (lah-cĭr´um), which is next to the carotid canal. The temporal bone also has a **mandibular fossa,** which is the articulation site of the mandible. The zygomatic process of the temporal bone can also be seen from this view.

The sphenoid bone traverses the skull. The **greater wings** of the sphenoid are the most lateral parts of the bone. Two pairs of flattened processes are also seen in this view, the **lateral pterygoid** (tear´ih-goyd) **plates** and the **medial pterygoid plates** (*pterygoid* = winglike). These are attachments for muscles that extend from the sphenoid to the mandible. Just posterior to a pterygoid plate is the **foramen ovale** (ō-val´-eh), which conducts one of the branches of the trigeminal nerve to the mandible.

In the midline of the skull and sometimes looking like a part of the sphenoid is the **vomer.** This is a single bone of the face that forms part of the **nasal septum.** The two large holes on each side of the vomer are the **internal nares.** Connected to the vomer and forming part of the **hard palate** is the **palatine bone.** The palatine bones are L-shaped bones with a horizontal plate and a vertical plate. The horizontal plates normally join at the **median palatine suture.** If this suture does not fuse completely at birth, an individual has a cleft palate. Also part of the hard palate are horizontal shelves of the maxillae. These are the **palatine processes of the maxillae.**

TABLE 9.1 Openings of the Skull

Opening	Function or Structure in Opening
Carotid canal	Internal carotid artery
External acoustic canal	Opening for sound transmission
Foramen lacerum	Cartilage
Foramen magnum	Spinal cord, vertebral arteries
Foramen ovale	Mandibular branch of trigeminal nerve
Foramen rotundum	Maxillary branch of trigeminal nerve
Foramen spinosum	Meningeal blood vessels
Inferior orbital fissure	Maxillary branch of trigeminal nerve
Infraorbital foramen	Infraorbital nerve and artery for the face
Internal auditory canal	Vestibulocochlear nerve and facial nerve
Jugular foramen	Internal jugular vein, vagus, and other nerves
Mandibular foramen	Mandibular branch of trigeminal nerve
Mental foramen	Mental nerve and blood vessels
Optic canal	Optic nerve
Stylomastoid	Facial nerve exits skull
Superior orbital fissure	Nerves to the eye and face
Supraorbital foramen	Supraorbital nerve and artery for the face

The major openings of the skull are presented in table 9.1. Locate the openings and note the structures that pass through these holes.

Interior of the Cranial Vault

(A)[7] Examine the inside of the skull and compare it to figure 9.8. With the top of the skull removed, you can see that the **cranial cavity** is divided into three major regions. These are the **anterior cranial fossa,** a depression anterior to the lesser wings of the sphenoid; the **middle cranial fossae,** which lie between the **lesser wings of the sphenoid** and the petrous portion of the temporal bone; and the **posterior cranial fossa,** which is posterior to the petrous portion of the temporal bone.

Beginning with the anterior cranial fossa, find the centrally located **ethmoid bone.** A sharp ridge known as the **crista galli** projects from the main portion of this bone. The small, horizontal plate of bone with numerous holes lateral to the crista galli is the **cribriform** (krĭ´bri-form) **plate.** The holes in this plate house nerves that carry the sense of smell from the nose to the brain. If the skull was cut close to the orbit, you can see the **frontal sinus,** a hollow space in the anterior portion of the frontal bone.

The dividing line between the anterior and middle cranial fossae is the sphenoid bone. Locate the lesser wings of the sphenoid and the **sella turcica** (sell´ah-tur´sik-ah) just posterior to it.

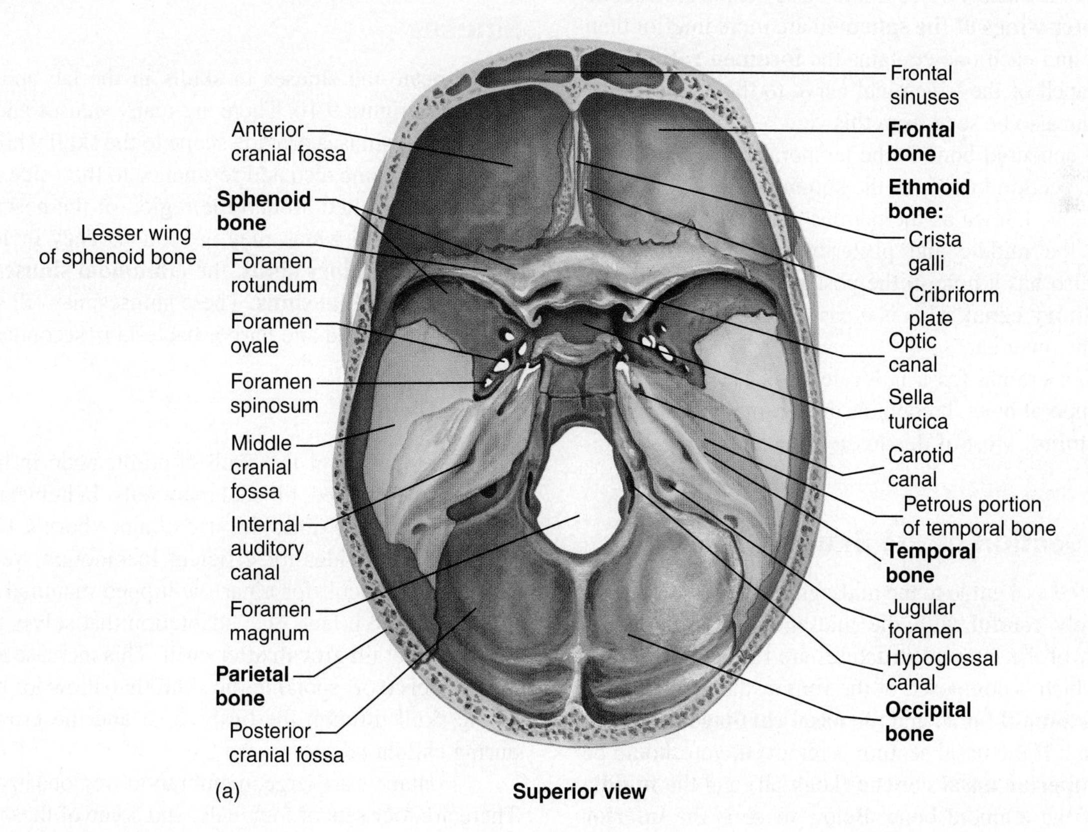

Figure 9.8 Interior of the Cranium
(a) Diagram; (b) photograph.

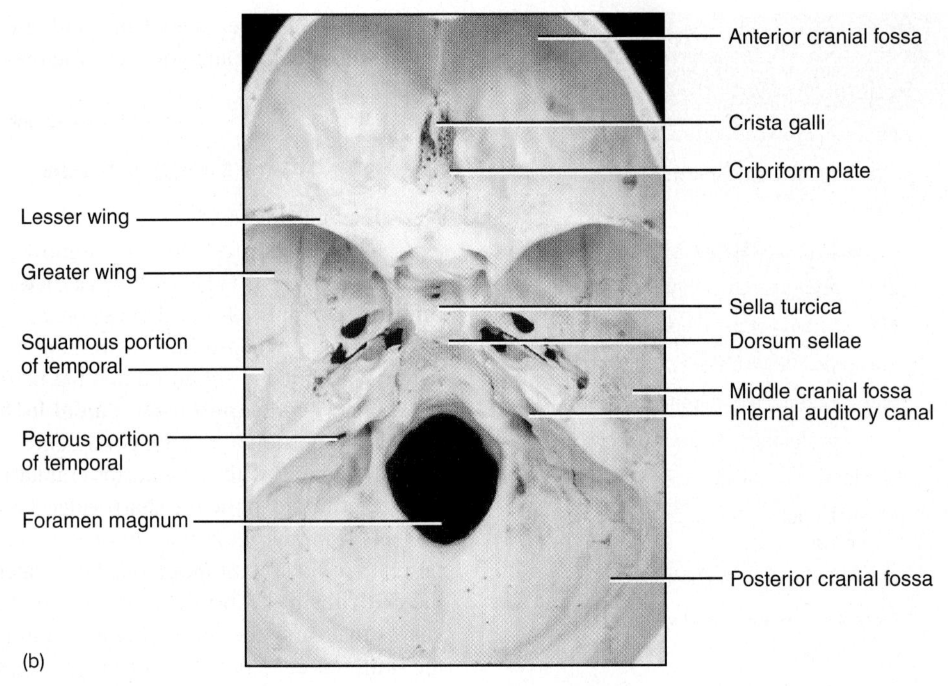

Figure 9.8 *Continued.*

The sella turcica has a small depression called the **hypophyseal** (hī-poh´fah-seal) **fossa,** in which the pituitary gland sits. The posterior, raised part of the sella turcica is known as the **dorsum sellae** (sell´ā). The **greater wings of the sphenoid** are more inferior than the lesser wings, and each one contains the **foramen rotundum,** which takes a branch of the trigeminal nerve to the maxilla. The **foramen ovale** can also be seen from this view.

Behind the sphenoid bone is the temporal bone, which has a flattened lateral section known as the **squamous portion** and a heavier mass of bone known as the petrous portion. The **petrous portion** divides the middle and posterior cranial fossae. The petrous portion also has a hole in the posterior surface, which is the **internal auditory canal.** This is a passageway for the nerves that come from the inner ear.

The posterior cranial fossa is located dorsal to the petrous portion of the temporal bone. It contains the foramen magnum and the **jugular foramina.** Most of this fossa is formed by the occipital bone.

Midsagittal Section of the Skull

(A) 8 Use figure 9.9 as a guide to the midsagittal sections of skulls. Be extremely careful with the midsagittal section of the skull, since many of the internal structures are fragile. Locate the **nasal septum,** which is composed of the **vomer,** the **perpendicular plate of the ethmoid bone,** and the **nasal cartilage** (absent in skull preparations). If the nasal septum is removed, you should be able to see the **superior nasal concha** (konk´ah) and the **middle nasal concha** of the ethmoid bone. Below these is the **inferior nasal concha,** which is a separate bone. Look for the junction between the palatine bone and the palatine process of the maxilla. These two bony plates make up the hard palate. If the mandible is present, locate the **mandibular foramen,** which is on the inner aspect of the mandible and transmits branches of the trigeminal nerve and blood vessels to the mandible.

Sinuses

(A) 9 Locate the sinuses in skulls in the lab and compare them with figure 9.10. There are many sinuses and air cells in the skull. These sinuses provide shape to the skull while decreasing its weight, and some also add resonance to the voice. The **paranasal sinuses** are located around the region of the nose and are named for the bones in which they are found. They include the **frontal sinus,** the **maxillary sinus,** the **ethmoidal sinuses** (or labyrinth), and the **sphenoidal sinus.** These sinuses may fill with fluid when a person has a cold and harbor bacteria in secondary infections.

Fetal Skull

The development of the skull is problematic in humans because we are large-brained, bipedal mammals. In humans the adult brain is more than three times the size of a newborn's. Our hips are narrow, which provides for efficient locomotion, yet our skulls are large. It is difficult for a narrow-hipped mammal to give birth to a large-headed infant. One adaptation that solves this problems is significant brain growth after birth. This increase after birth is due to fontanels (soft spots) in the skull that allow for both the passage of the skull through the birth canal and the growth of the skull during childhood.

Fontanels are large, membranous regions in an infant's skull. There are four sets of fontanels, and some of these permit the passage of the skull through the birth canal by enabling the bones of the cranium to slide over one another. The frontal (anterior) fontanel is between the frontal bone and the parietal bones. The occipital (posterior) fontanel is between the occipital bone and the

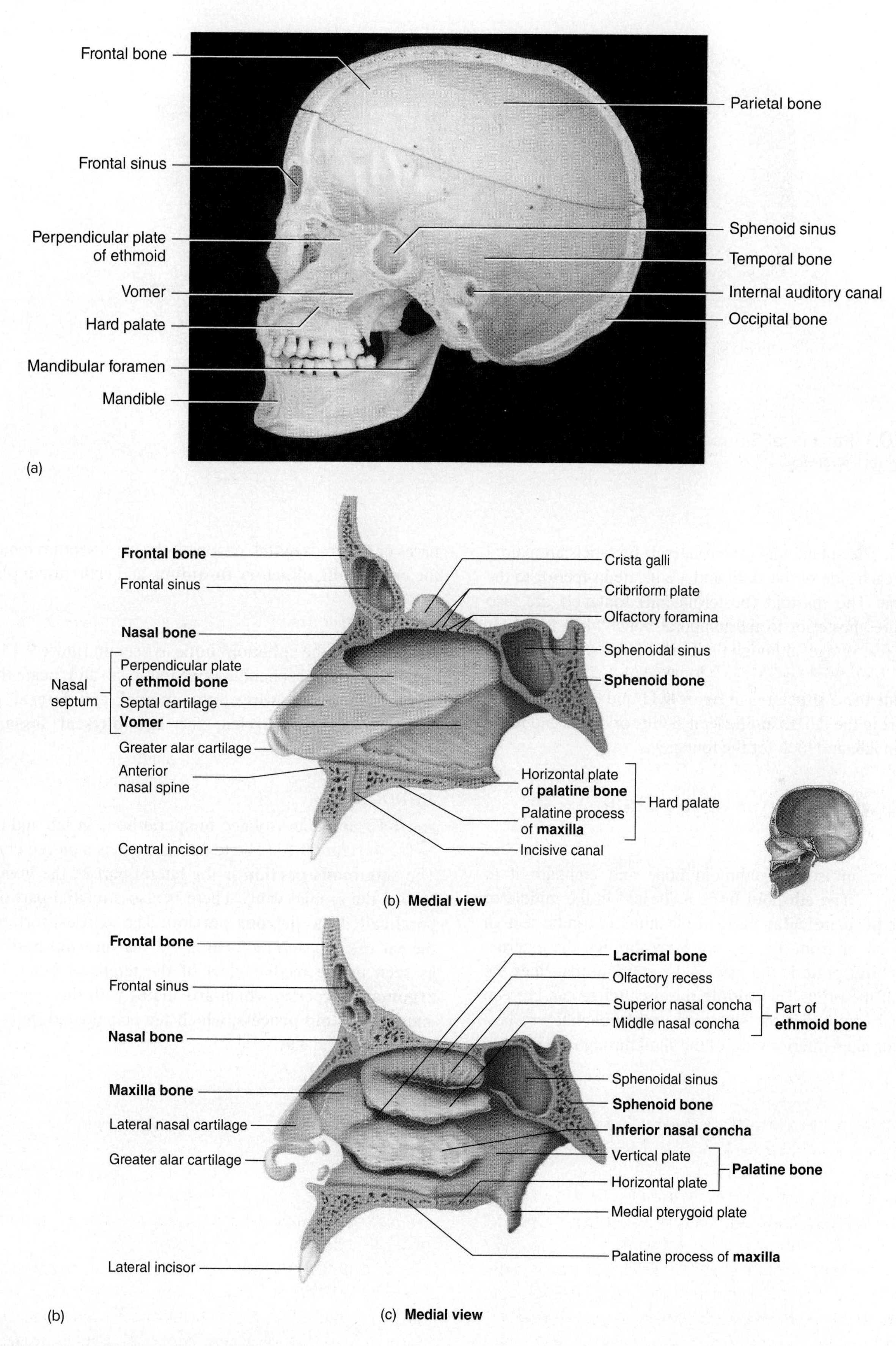

Figure 9.9 Skull, Midsagittal Section
(a) Photograph; (b) nasal septum; (c) wall of nasal cavity.

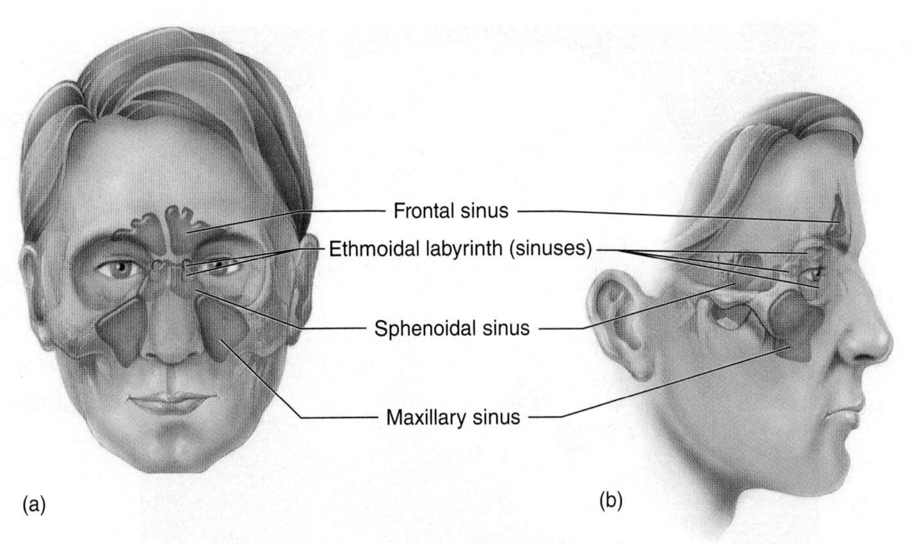

Figure 9.10 Paranasal Sinuses of the Skull
(a) Anterior view; (b) lateral view.

parietal bones. The sphenoidal (anterolateral) fontanels are paired structures on each side of the skull and are located superior to the sphenoid bone. The mastoid (posterolateral) fontanels are also paired structures posterior to the temporal bone. Most fontanels fuse before 1 year of age, although the frontal fontanel may fuse as late as age 2.

(A) 10 Locate these structures in figure 9.11 and on the material available in the lab. Examine fetal skulls, or charts and models of skulls in lab, and look for the fontanels.

Select Individual Bones of the Skull

Ethmoid

(A) 11 Examine an isolated ethmoid bone and compare it to figure 9.12. The **ethmoid bone** is located in the middle of the skull. The **perpendicular plate** of the ethmoid can be seen in midsagittal view or from the anterior view through the external nares. The **orbital plate** is the part of the ethmoid that lines the medial wall of the orbit. The **middle nasal conchae** can be seen from the nasal cavity, but the **superior nasal conchae** are best seen by looking at an inferior view of the skull through the internal nares or at a midsagittal view with the nasal septum removed. Find the **crista galli, olfactory foramina,** and **cribriform plate.**

Sphenoid

(A) 12 Locate the **sphenoid bone** as seen in figure 9.13. Examine an isolated sphenoid bone in the lab and locate the **greater wings,** the **lesser wings,** the **medial** and **lateral pterygoid plates,** the **sella turcica,** the **hypophyseal fossa,** and the **dorsum sellae.**

Temporal

(A) 13 Examine an isolated temporal bone in lab and compare it to figure 9.14. The temporal bone is a paired cranial bone. The **squamous portion** is the lateral part of the bone; it forms part of the cranial vault. There is also a medial part of the temporal called the **petrous portion.** The petrous portion contains the ear ossicles and the opening of the **internal auditory canal** as seen in the medial view of the temporal bone. Locate the **zygomatic process,** which articulates with the zygomatic bone, and the **mastoid process,** which can be palpated (felt) as a bump posterior to the ear.

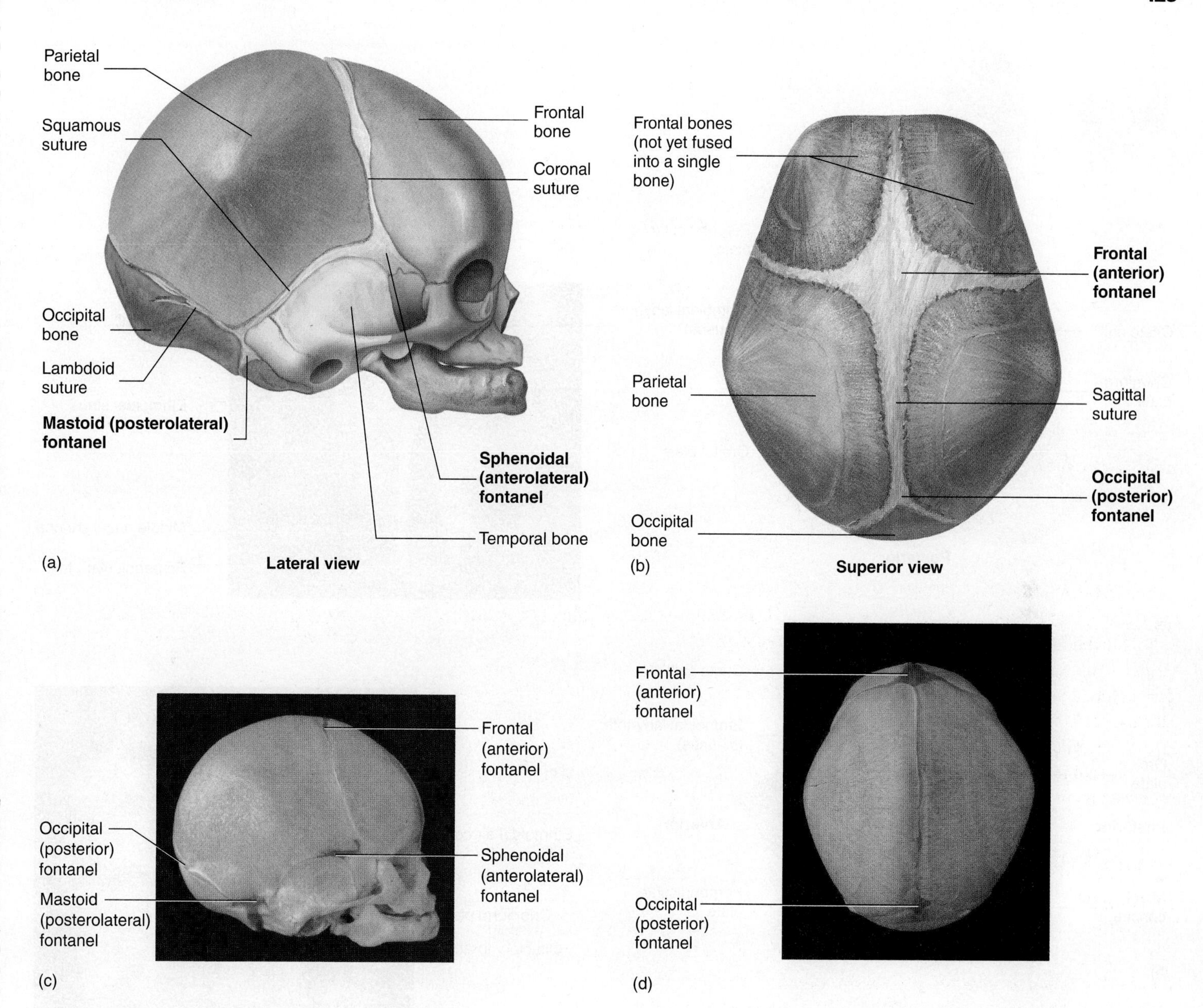

Figure 9.11 Fontanels of the Fetal Skull
Diagram (a) lateral view; (b) superior view. Photograph (c) lateral view; (d) superior view.

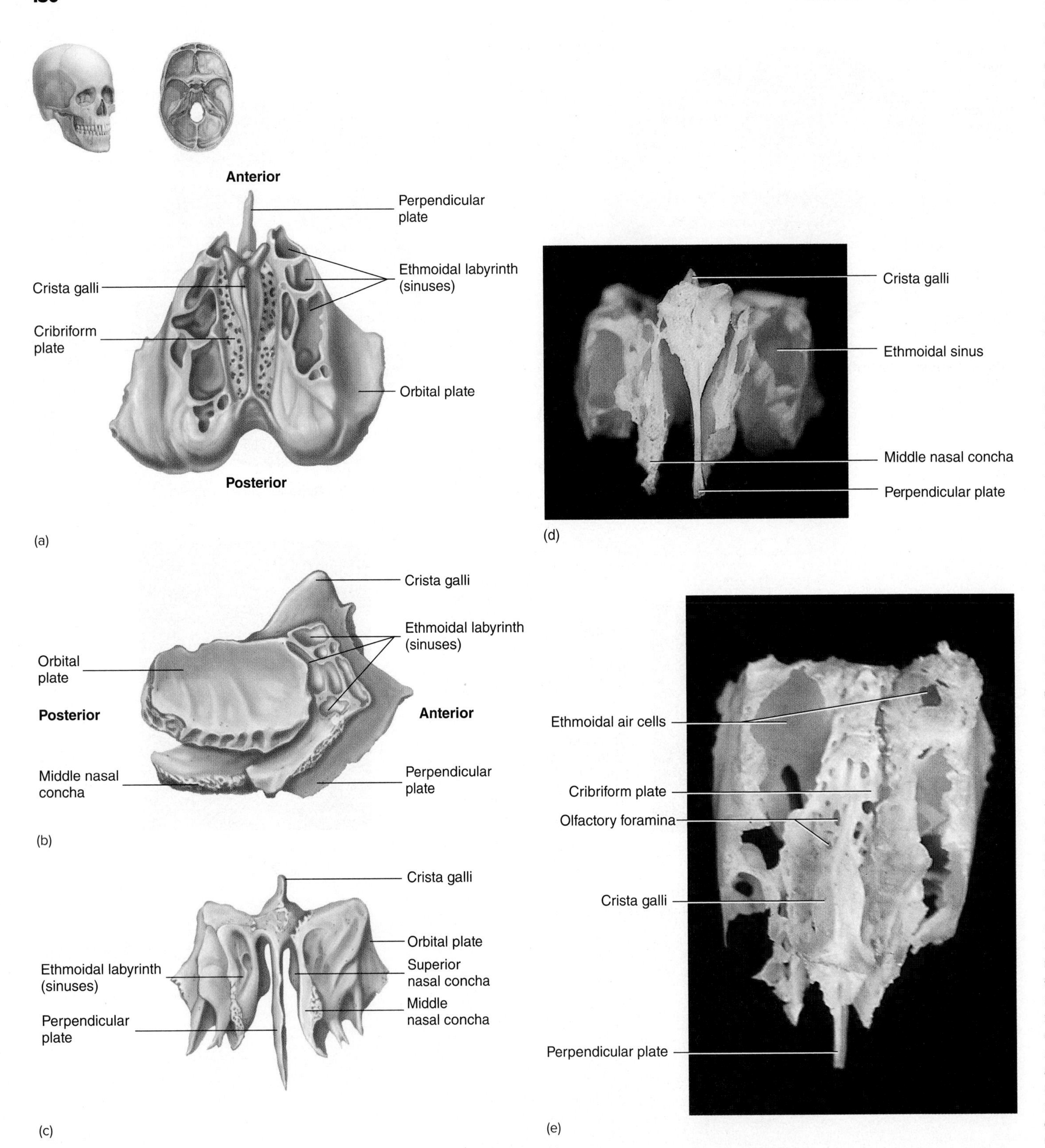

Figure 9.12 Ethmoid Bone
Diagram (a) superior view; (b) lateral view; (c) anterior view. Photograph (d) anterior view; (e) superior view.

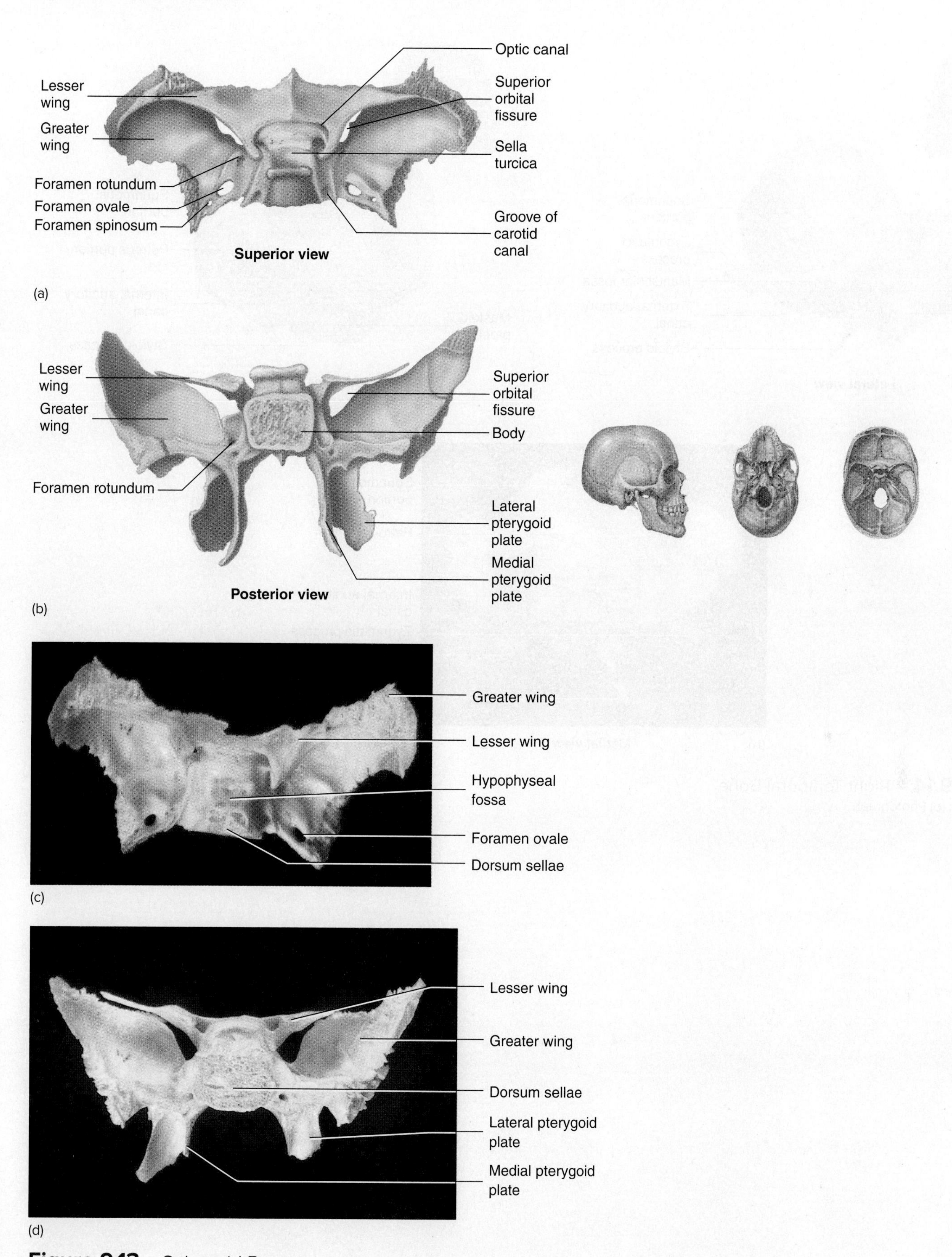

Figure 9.13 Sphenoid Bone

Diagram (a) superior view; (b) posterior view. Photograph (c) superior view; (d) posterior view.

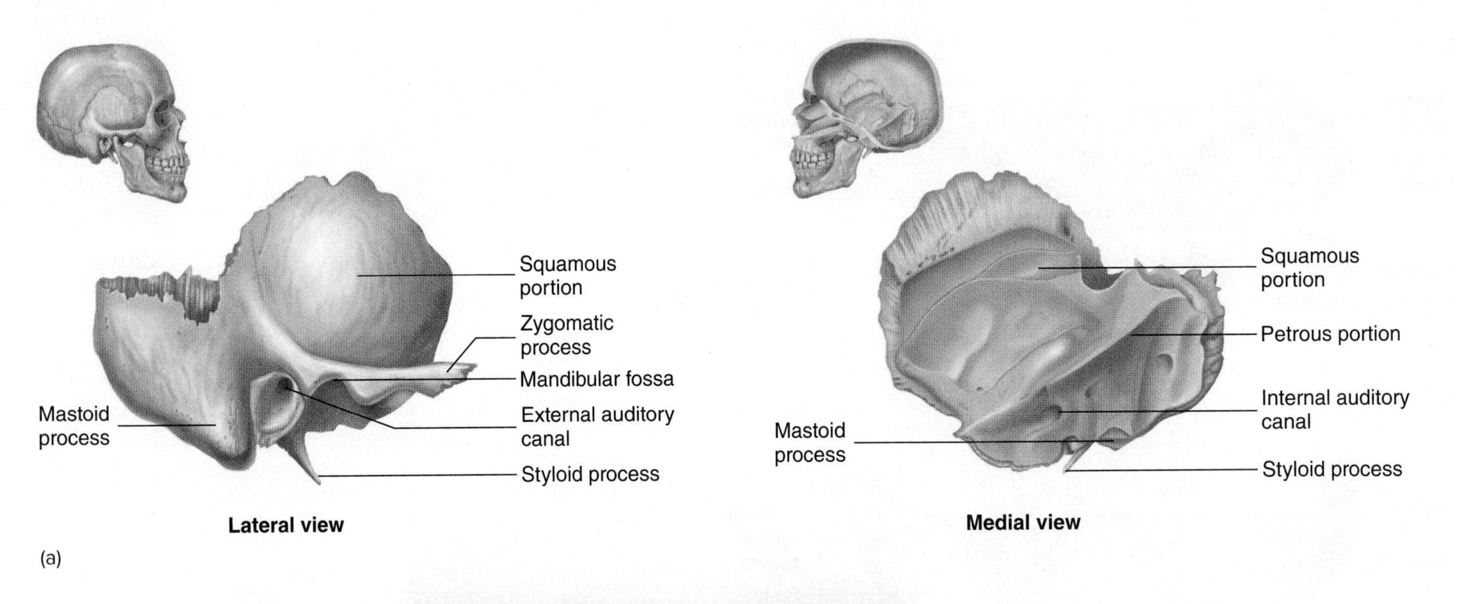

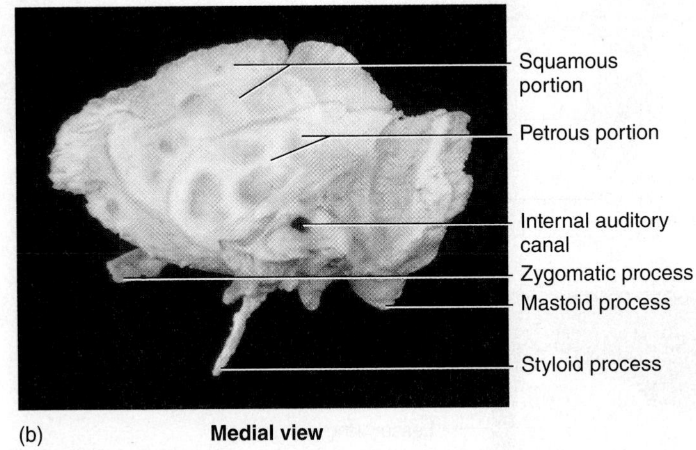

Figure 9.14 Right Temporal Bone
(a) Diagram; (b) Photograph.

Exercise 9 Review

Axial Skeleton: Skull

Name: ____________________

Lab time/section: ____________________

Date: ____________________

1. The eyebrows are superficial to what bone? ____________________

2. What is the common name for the zygomatic bone? ____________________

3. What is the name of the bony process posterior to the earlobe? ____________________

4. The hard palate is made up of what bones? ____________________

5. What are the names of the major paranasal sinuses? ____________________

____

6. The mandible fits into what part of the temporal bone to form the jaw joint? ____________________

7. What bone is found just posterior to the ethmoid bone in the orbit? ____________________

8. The sella turcica is found in what bone? ____________________

9. What is the name of the bone that makes up most of the temple? ____________________

____

10. What are the names of the bones that surround the opening of the nose? ____________________

11. The upper teeth are held by what bones? ____________________

12. In what bone would you find the foramen magnum? ____________________

13. What are the two bony structures that make up the nasal septum? ____________________

14. The sagittal suture separates the ____________________ from the ____________________.
    a. sphenoid, ethmoid
    b. left parietal, right parietal
    c. frontal, parietal
    d. parietals, occipital

15. Which bone is *not* located in the orbit?
    a. maxilla b. zygomatic c. ethmoid d. sphenoid e. temporal

16. Which bone is *not* a paired bone of the skull?
    a. zygomatic b. temporal c. lacrimal d. vomer

17. Label the following illustration using the terms provided.

carotid canal	foramen magnum
foramen ovale	internal nares
lateral pterygoid plate	mastoid process
maxilla	median palatine suture
occipital bone	occipital condyle
palatine	palatine process of maxilla
sphenoid	temporal
vomer	zygomatic bone
zygomatic arch	

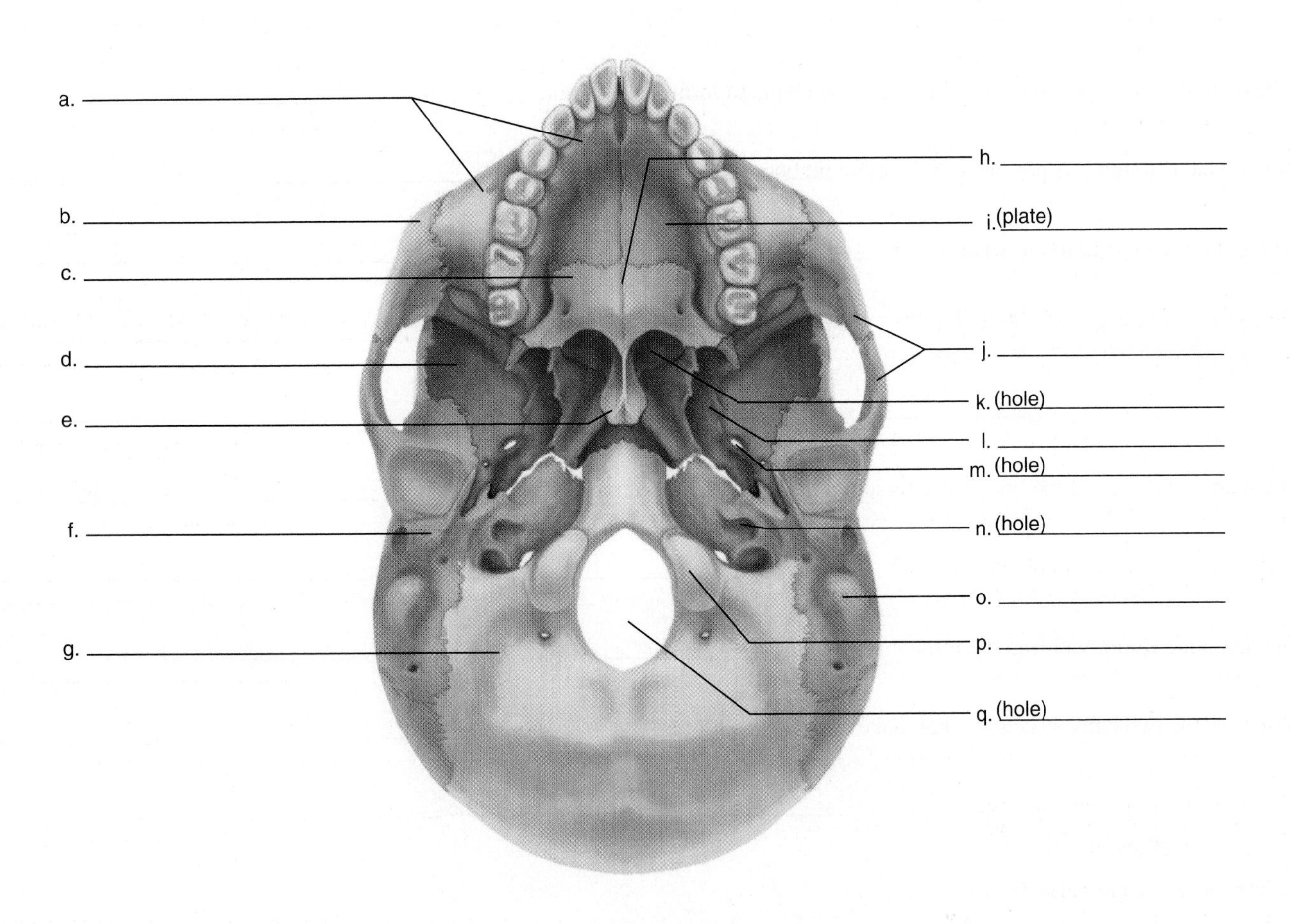

18. Label the following illustration using the terms provided.

alveolar process	angle
body	coronoid process
mandibular condyle	mandibular foramen
mandibular notch	mental foramen
ramus	

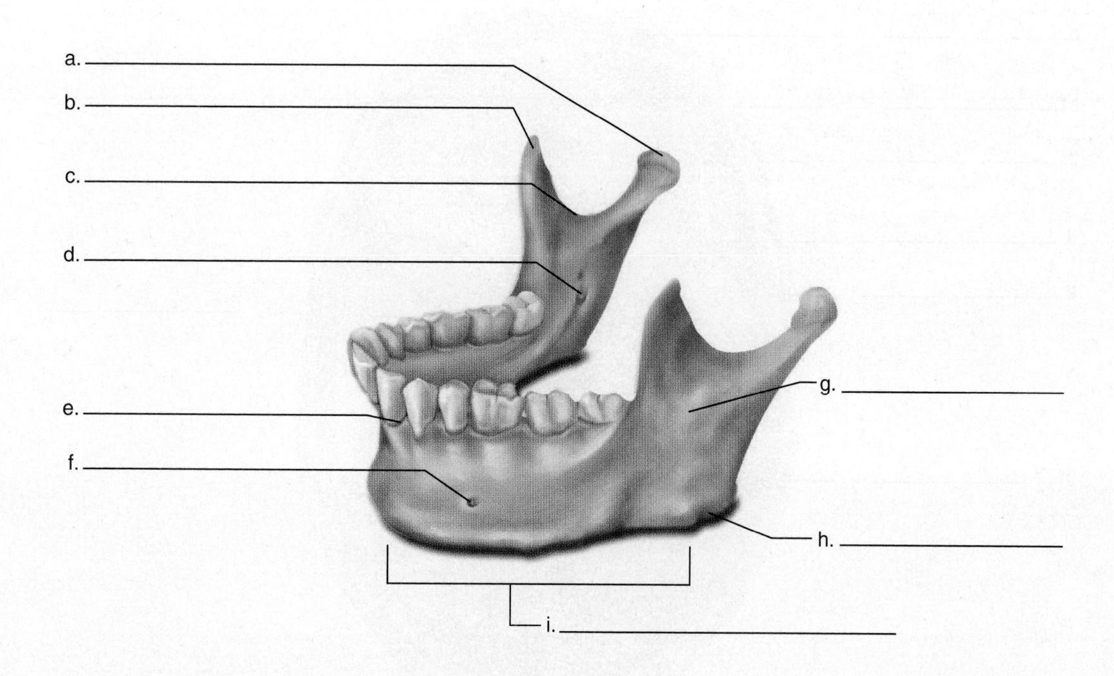

19. Based on what you know about the maxillary sinus, why would a significant impact to the maxilla create a more difficult situation for healing than would the fracture of a long bone? ______________________________

20. Name all the fontanels in the fetal skull. ______________________________

21. Which one of the fontanels is the most dorsal? ______________________________

22. Label the following illustration using the terms provided.

anterior cranial fossa	cribriform plate	crista galli
foramen magnum	foramen ovale	frontal sinus
greater wing of sphenoid	lesser wing of sphenoid	petrous portion of temporal
sella turcica		

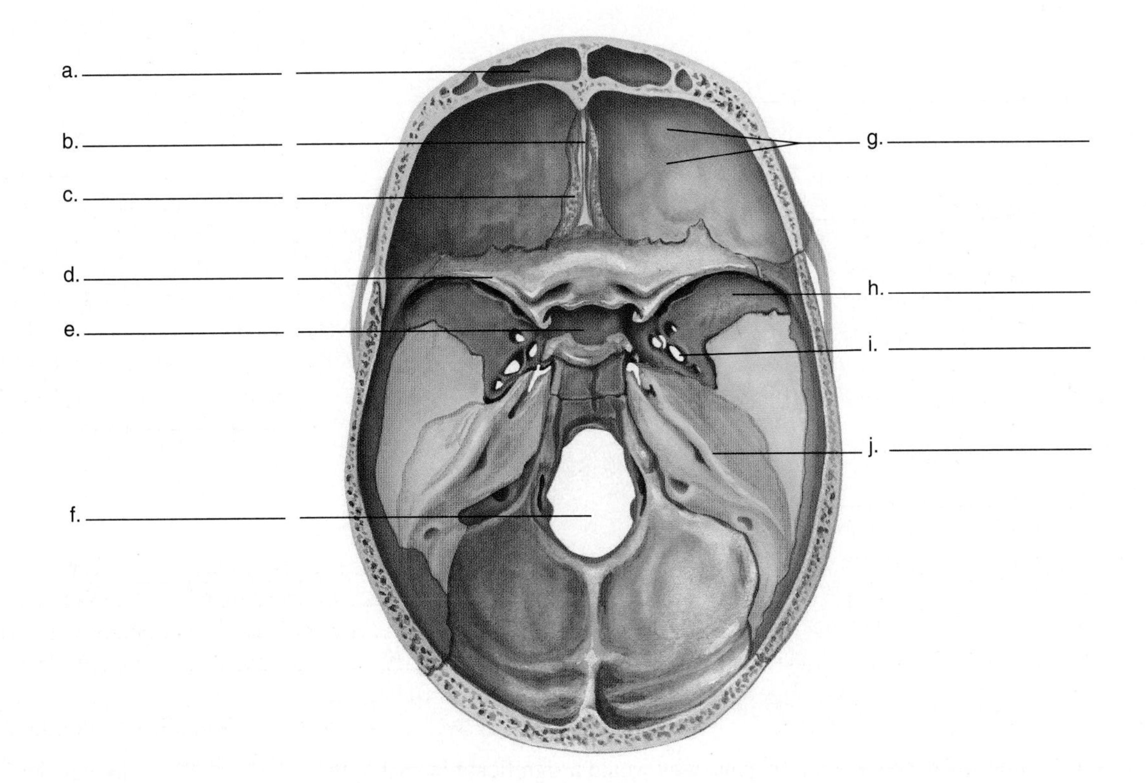

Exercise 10

Articulations

Introduction

The study of the joints between bones is called **arthrology.** The joints themselves are known as **articulations.** Joints are important because they are the points of movement of the body and they can suffer significant trauma or disease. Diseases such as arthritis affect millions of people worldwide. The development of artificial joints is important to replace joints that cause pain, no longer provide a comfortable range of movement, or provide the stability required for support.

In this exercise, you study the various types of joints in the body in terms of their structure. Joints can be classified according to their physical composition, or they can be classified according to their range of movement. In terms of composition, joints are classified as fibrous, cartilaginous, or synovial. In terms of range of movement, joints are immovable, semimovable, or freely movable. As you study the joints in the lab, use those in your body as reference and mimic the action of a specific articulation. These topics are discussed in the Seeley text in chapter 8, "Joints and Movement."

Objectives

At the end of this exercise, you should be able to

1. distinguish among fibrous, cartilaginous, and synovial joints;
2. discuss the nature of a synovial joint;
3. locate the fibrous capsule, synovial membrane, synovial fluid, and articular cartilage in a dissected synovial joint;
4. list five types of synovial joints;
5. explain the structure of the jaw, shoulder, hip, knee, and ankle;
6. point out different articulations on your body and indicate what kinds of joints these are.

Materials

Mammal joint with intact synovial capsule
Dissection tray with scalpel, blunt probe, and protective gloves
Waste container
Model or chart of joints, including those of the shoulder, elbow, knee, ankle, hip, and jaw
Articulated skeleton

Procedure

Types of Joints

The three major groups of joints classified according to composition are fibrous, cartilaginous, and synovial joints. **Fibrous joints** are composed of connective tissue fibers between two bones. They permit little movement. **Cartilaginous joints** consist of cartilage between two bones and are generally more movable than fibrous joints, although some cartilaginous joints may have no movement at all. **Synovial joints** have the most complex structure, including a joint capsule, an inner membrane, and synovial fluid, and they are the most movable of the joints (table 10.1).

Fibrous Joints

Fibrous joints connect one bone to another with collagenous fibers. If the fibers are close together, the joint does not allow for much, if any, movement between the bones. One example of this type of joint is called a **suture,** and it occurs between adjacent bones in the cranium.

(A) 1 Examine an intact skull in the lab and locate the sutures. In these joints, the bones of the cranium are tightly bound together by dense, fibrous connective tissue. As a person approaches the age of 35 or so, some of the sutures of the skull begin to fuse from the region closest to the brain toward the superficial surface of the skull. This fusion leads to the complete union of two bones and is then called a **synostosis.** The two frontal bones in the fetal skeleton later fuse together as a synostosis and form a single frontal bone.

Another type of fibrous joint is a **gomphosis,** which is represented by teeth in the sockets of the maxilla and the mandible. This type of joint is a peg in a socket. A gomphosis connects the bone of the jaw to the tooth by fibrous connective tissue called **periodontal ligaments.**

(A) 2 Examine a jaw or complete skull in the lab and locate a gomphosis. Compare your specimen with figure 10.1.

Another type of fibrous joint is called a **syndesmosis.** The fibrous connective tissue is longer in these joints than in sutures or gomphoses. An example of a syndesmosis is the connection between the distal radius and ulna or the distal tibia and fibula.

(A) 3 Examine an articulated skeleton in the lab and compare it with figure 10.2.

Cartilaginous Joints

If bones are held together by cartilage, the articulation is known as a cartilaginous joint. As in fibrous joints, if the cartilage is thin

TABLE 10.1 Classification of Joints

Fibrous Joints—Joints Held Together by Collagenous Fibers	Examples
Suture	Frontal and parietal bones
Gomphosis	Teeth and mandible
Syndesmosis	Distal tibia and distal fibula
Cartilaginous Joints—Joints Held Together by Cartilage	
Synchondrosis	Epiphyseal plate, humeral head and shaft
	Costal cartilages, first ribs, and sternum
Symphysis	Joint between two vertebral bodies
Synovial Joints—Joints Enclosed by Synovial Capsule	
Gliding	Between carpal bones
Hinge	Humerus and ulna
Pivot	Atlas and axis
Ellipsoid (condyloid)	Radius and scaphoid
Saddle	Trapezium and first metacarpal
Ball-and-socket	Acetabulum and femur

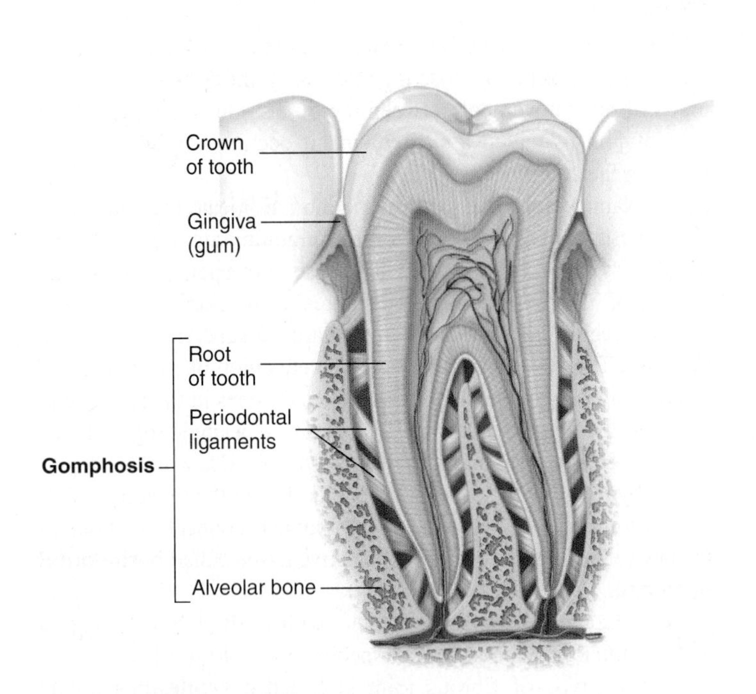

Figure 10.1 Gomphosis

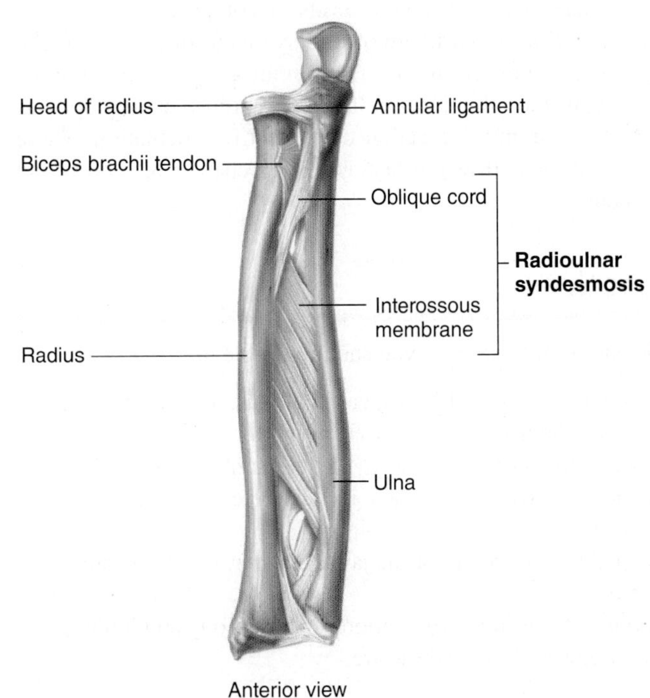

Figure 10.2 Syndesmosis

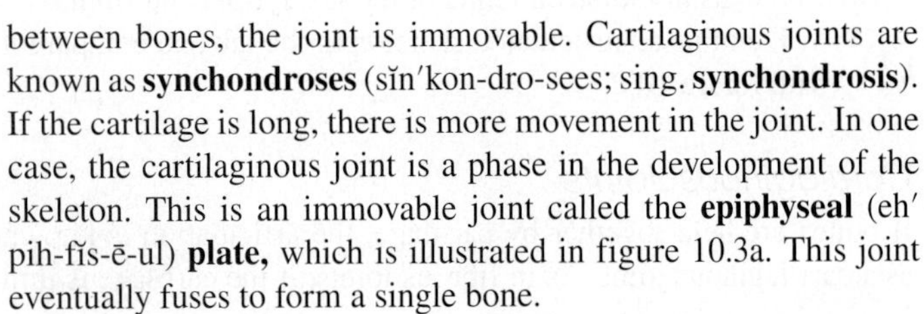

between bones, the joint is immovable. Cartilaginous joints are known as **synchondroses** (sĭn′kon-dro-sees; sing. **synchondrosis**). If the cartilage is long, there is more movement in the joint. In one case, the cartilaginous joint is a phase in the development of the skeleton. This is an immovable joint called the **epiphyseal** (eh′ pih-fĭs-ē-ul) **plate,** which is illustrated in figure 10.3a. This joint eventually fuses to form a single bone.

Another example of a synchondrosis is the articulation between the first ribs and the sternum by the **costal cartilages.** In this joint, hyaline cartilage binds the first rib to the sternum yet lets it move somewhat (figure 10.3b).

(A)[4] Locate a symphysis in lab and compare it with figure 10.4.

A **symphysis** is a **fibrocartilaginous** pad between the pubic bones (for example, the pubic symphysis) or in the intervertebral discs. These joints also allow for some movement. The physical stress on a symphysis is greater than that on the costal cartilages, and the joint reflects this in its fibrocartilage composition. Fibrocartilage endures much greater stress than hyaline cartilage.

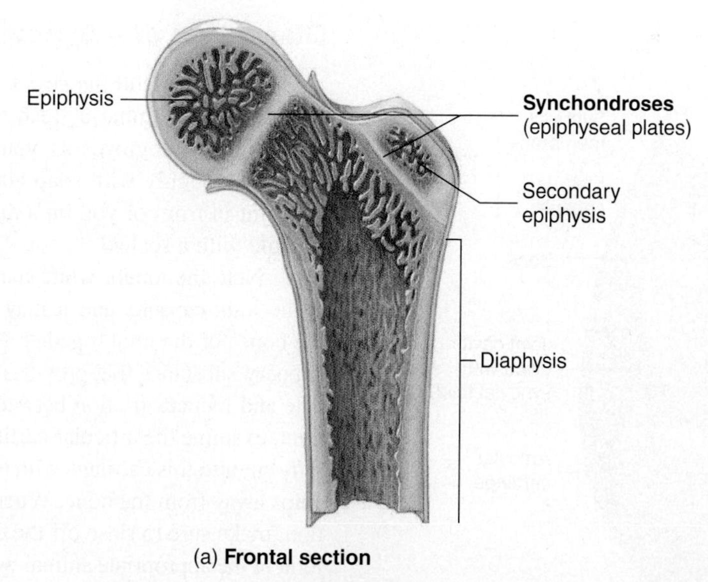

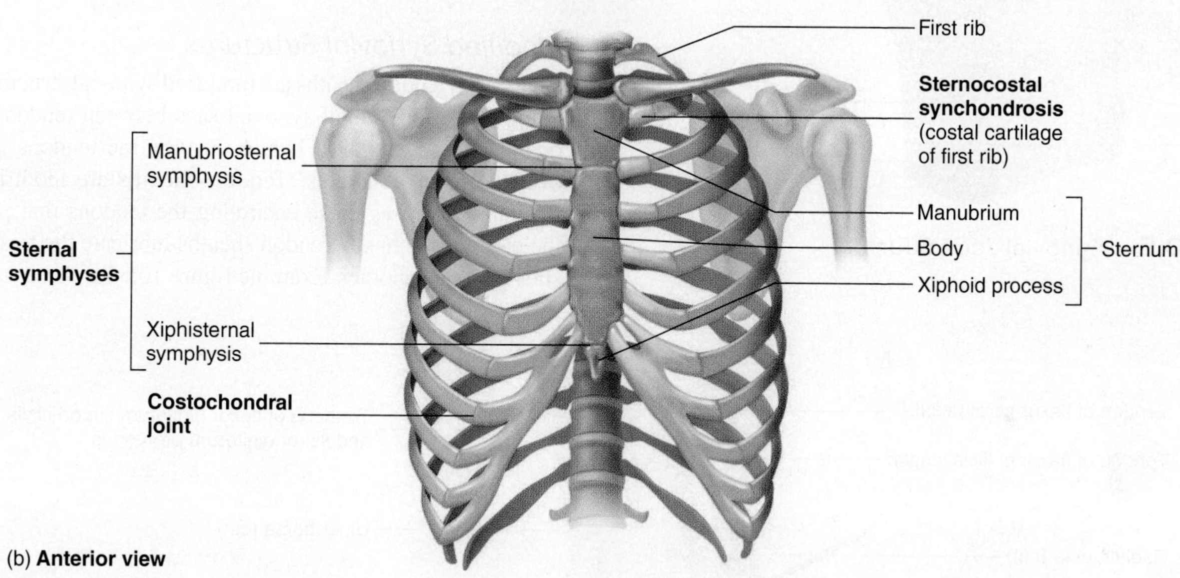

Figure 10.3 Synchondroses
(a) Femur with immovable synchondrosis; (b) thorax with movable synchondroses.

Synovial Joints

(A) **5** Compare the features of the joint in figure 10.5 with models or charts in the lab. Joints that allow for extensive movement are called synovial joints. The outer part of the synovial joint is the **joint capsule,** which is made of an outer **fibrous capsule** and an inner **synovial membrane.** The synovial membrane secretes **synovial fluid,** a lubricating liquid that reduces friction inside the joint. The space inside the joint is called the **synovial cavity,** and each bone of the joint ends in a hyaline cartilage cap called the **articular cartilage.** There are other structures in synovial joints that are characteristic of specific joints, and these are discussed later in this exercise. The shape of the bones determines the type of movement at the articulation. As a general rule, the more movable a joint is, the less stable it is. The stability of a joint depends on the number and types of ligaments, tendons, and muscles and on the way the bones fit together.

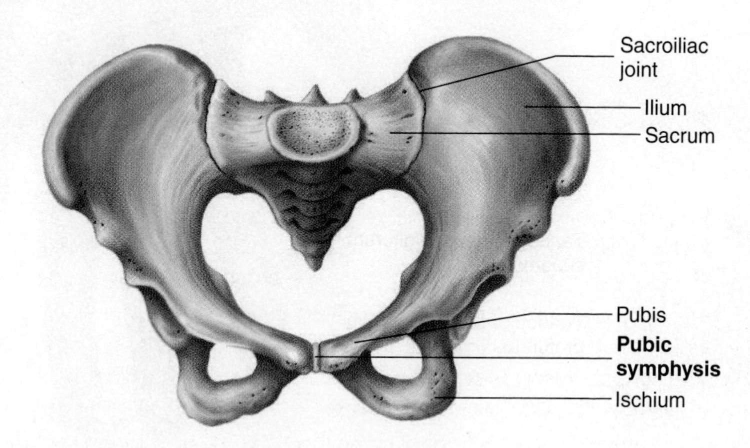

Figure 10.4 Symphysis

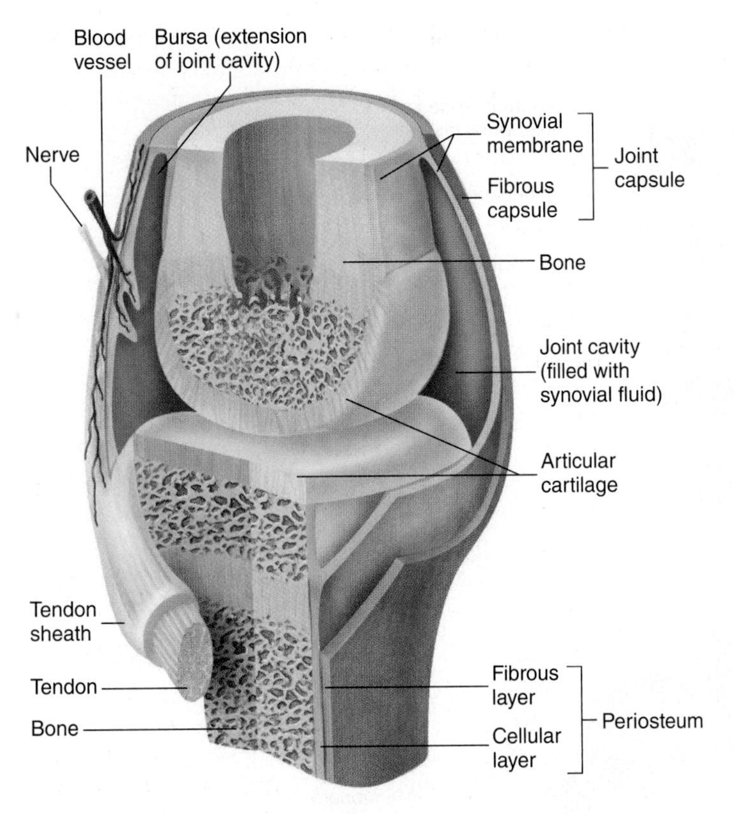

Figure 10.5 Synovial Joint Structure

Dissection of a Synovial Joint

(A) 6 To understand the structure of a synovial joint, it is beneficial to examine a fresh or recently thawed mammal joint. Wear protective gloves as you dissect the joint, and wash your hands thoroughly with soap and water after the dissection. Place the joint in front of you on a dissecting tray and cut into the joint capsule with a scalpel.

Note the tough, white material that surrounds the joint. This is the joint capsule, and it may be fused with **ligaments** that bind the bones of the joint together. Notice the synovial fluid, which is a slippery substance that provides a slick feel to the inside of the capsule and reduces friction between the bones. Once you cut into the joint, examine the articular cartilage on the ends of the bones. *Carefully* cut into this cartilage with a scalpel and notice how the material chips away from the bone. When you have finished with the dissection, make sure to rinse off the dissection equipment, dispose of the joint in the appropriate animal waste container, and wash your hands.

Modified Synovial Structures

Bursae and tendon sheaths are modified synovial structures. **Bursae** (sing. **bursa**) are small synovial sacs between tendons and bones or other structures. The bursae cushion the tendons as they pass over the other structures. **Tendon sheaths** are modified synovial structures, such as those encircling the tendons that pass through the palm of the hand. Tendon sheaths lubricate the tendons as they slide past one another. Examine figure 10.6 for these structures.

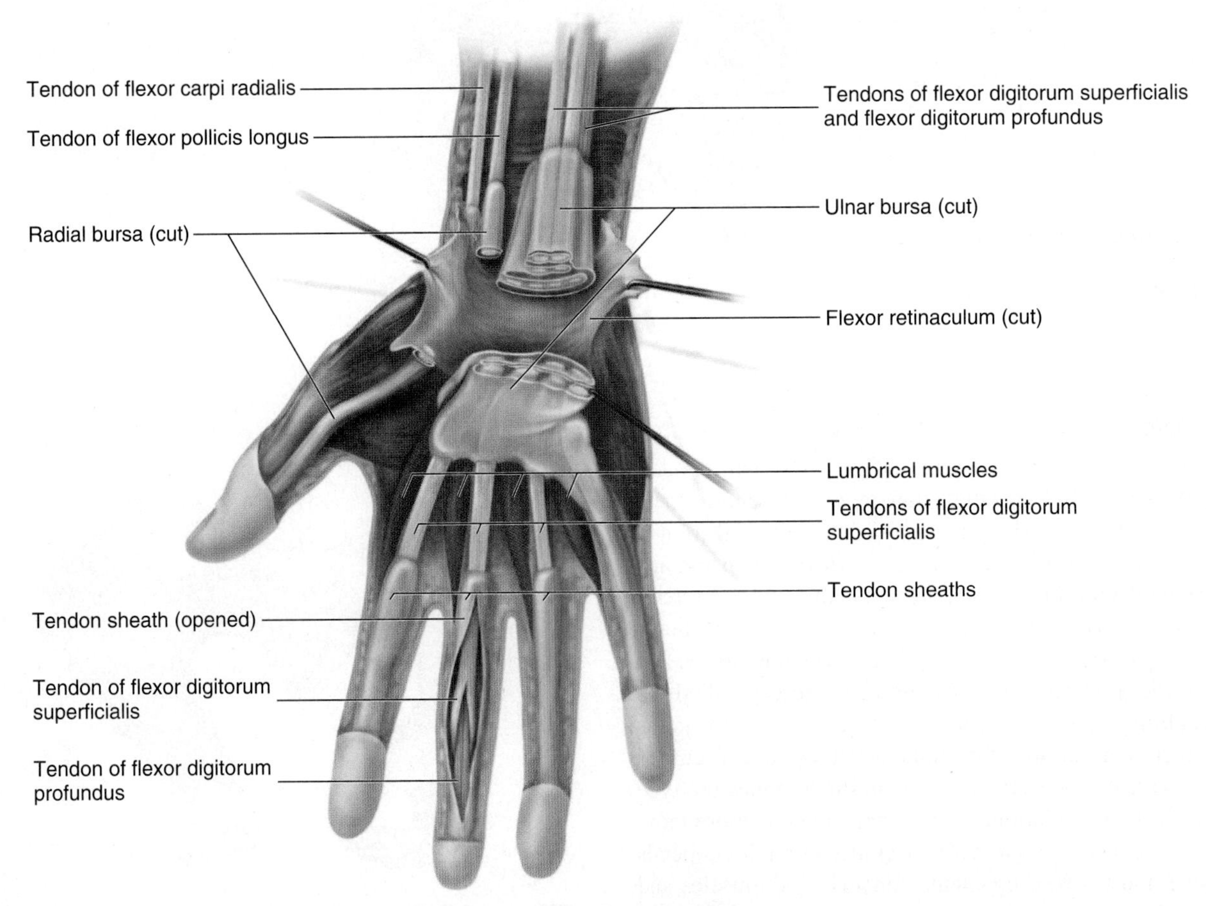

Figure 10.6 Modified Synovial Structures—Tendon Sheaths and Bursae

Joints Classified by Movement

Articulations are immovable, semimovable, or freely movable. Immovable joints are known as **synarthrotic joints.** The bones are tightly bound by fibers or hyaline cartilage. Semimovable joints are known as **amphiarthrotic joints,** and they may be fibrous or cartilaginous. Freely movable joints are known as **diarthrotic joints,** and they are always synovial joints. Specific types of synovial joints are discussed next.

Synovial Joints Classified by Movement

The synovial joints are classified according to the type of movement they allow between the articulating bones. The joints are listed here in the general order of least movable to most movable.

Ⓐ7 Examine the joints on an articulated skeleton in lab and compare them with figure 10.7.

1. **Plane,** or **gliding, joints** allow for movement between two flat surfaces, such as between the superior and inferior facets of adjacent vertebrae or intertarsal or intercarpal joints.
2. **Hinge joints** allow for angular movement, such as in the elbow, in the knee, or between the phalanges of the fingers. You can increase or decrease the angle of the two bones with this joint.
3. **Pivot joints** allow for rotational movement between two bones, such as in the movement of the atlas and the axis when moving the head to indicate "no." They also occur at the proximal radius and ulna.
4. **Ellipsoid,** or **condyloid, joints** allow significant movement in two planes, such as at the base of the phalanges. Condyloid joints consist of a convex surface paired with a concave surface. The junction between the radius and scaphoid bone is a good example of a condyloid joint. Others are located between the atlas and occipital bone or the metacarpals and phalanges of digits 2 through 5.
5. **Saddle joints** have two concave surfaces that articulate with one another. An example of a saddle joint is that between the trapezium and the first metacarpal of the thumb. This provides for more movement in the thumb than the condyloid joint of the wrist.
6. **Ball-and-socket joints** consist of a spherical head in a round concavity, such as in the shoulder and the hip. There is extensive movement in these joints, yet they are inherently less stable due to the freedom of movement that they afford.

Uniaxial joints move in only one plane. Hinge, pivot, and gliding joints are uniaxial joints. Biaxial joints move in two planes. Condyloid and saddle joints are examples of biaxial joints. Multiaxial joints move in many planes. Ball-and-socket joints are multiaxial joints.

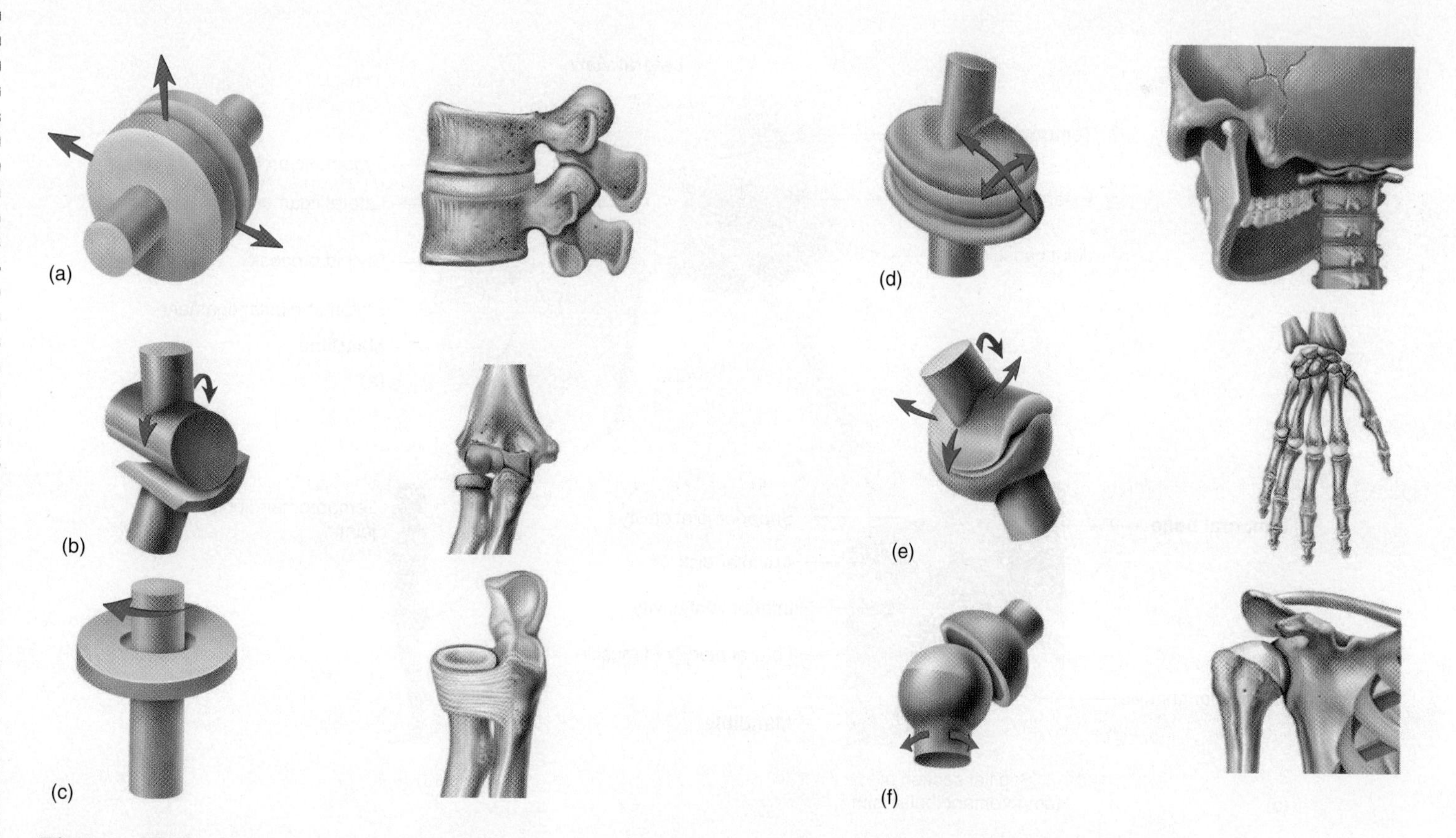

Figure 10.7 Classes of Synovial Joints
(a) Plane, or gliding, joint; (b) hinge joint; (c) pivot joint; (d) ellipsoid, or condyloid, joint; (e) saddle joint; (f) ball-and-socket joint.

Specific Joints of the Body

Temporomandibular

Ⓐ 8 Examine a skull with the mandible attached for the nature of the temporomandibular joint. The **temporomandibular joint (TMJ)** is the only diarthrotic joint of the skull. The **articular disk** is a pad of fibrocartilage that provides a cushion between the mandibular condyle and the temporal bone. This joint is both a hinge and a gliding joint. Numerous ligaments strengthen this joint (figure 10.8).

Shoulder Joint

Ⓐ 9 Compare figure 10.9 with a model or an actual joint in the lab. The **shoulder (humeral) joint** is known as the **glenohumeral (humeroscapular) joint** because the glenoid cavity articulates with the head of the humerus. The shallow glenoid cavity is deepened by the **glenoid labrum,** a cartilaginous ring that surrounds the cavity. Numerous bursae, ligaments, a tough joint capsule, and the **rotator cuff muscles** also stabilize the joint.

Elbow Joint

The **elbow (humeroulnar) joint** is a complex joint having both hinge and pivot characteristics. The ulna locks into the humerus tightly but the **annular ligament** that wraps around the radius is commonly torn from the radial head when the forearm is abruptly pulled (figure 10.10).

Hip Joint

Ⓐ 10 Compare the models, charts, or specimens of a hip joint in the lab with figure 10.11. The **hip joint** is known as the **acetabulofemoral joint.** As with the shoulder joint, the **acetabular labrum** deepens the hip socket. Numerous ligaments, including the iliofemoral, pubofemoral, and ischiofemoral, bind the femur to the hip bone. The **ligamentum teres** is a band of dense connective tissue that attaches the acetabulum to the **fovea capitis** of the femur.

Knee Joint

Ⓐ 11 Compare specimens of the knee joint in the lab with figure 10.12. The **tibiofemoral joint,** also known as the **knee joint,** is the largest, most complex joint of the body. It consists of several major ligaments, including the **tibial (medial) collateral ligament,** the **fibular (lateral) collateral ligament,** the **anterior cruciate ligament,** and the **posterior cruciate ligament.** Another important structure is the **patellar (quadriceps femoris) tendon,** which runs from the quadriceps femoris muscle to the patella. The **patellar ligament** connects the patella and the tibial tuberosity. The last major structures of the tibiofemoral joint are the **medial** and **lateral menisci,** wedge-shaped pads of fibrocartilage that provide a cushion between the femur and tibia. The knee is primarily a hinge joint with a little lateral movement allowed.

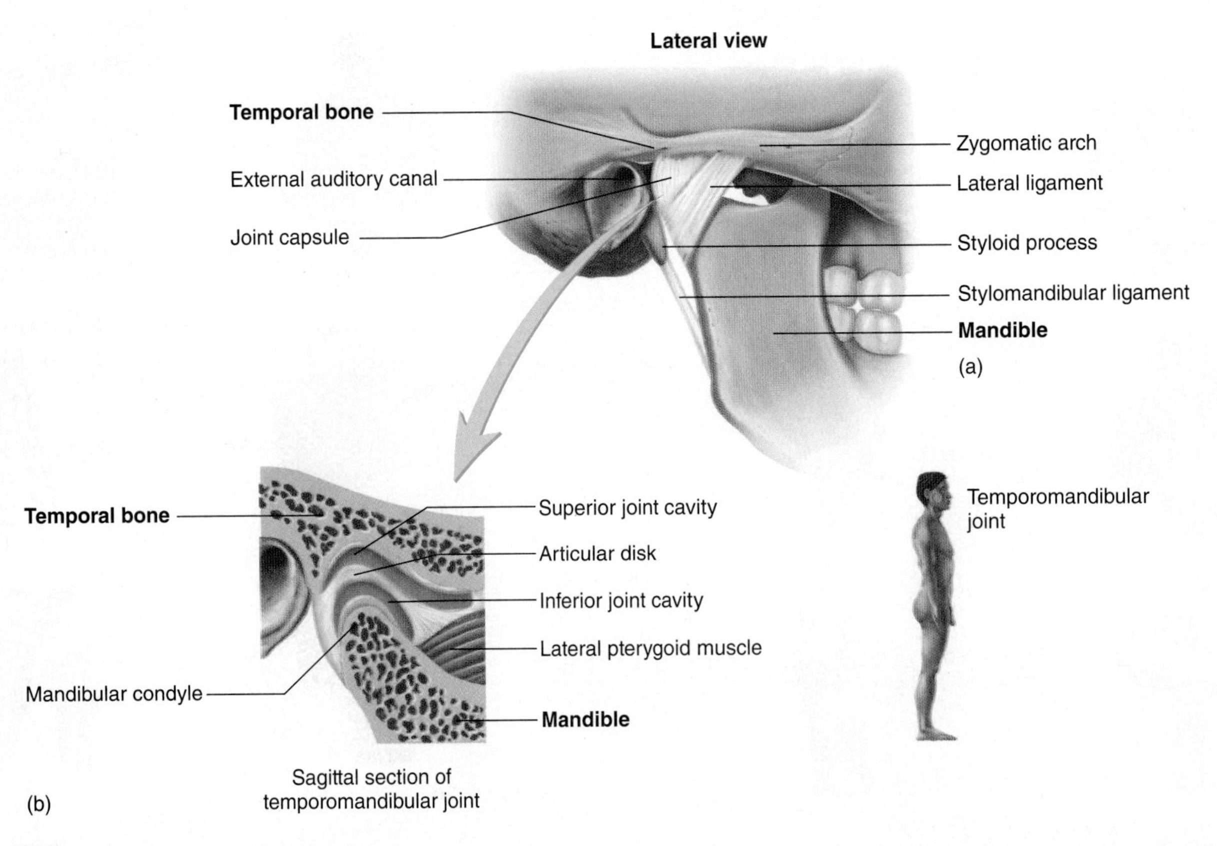

Figure 10.8 Temporomandibular Joint
(a) Lateral view; (b) sagittal section.

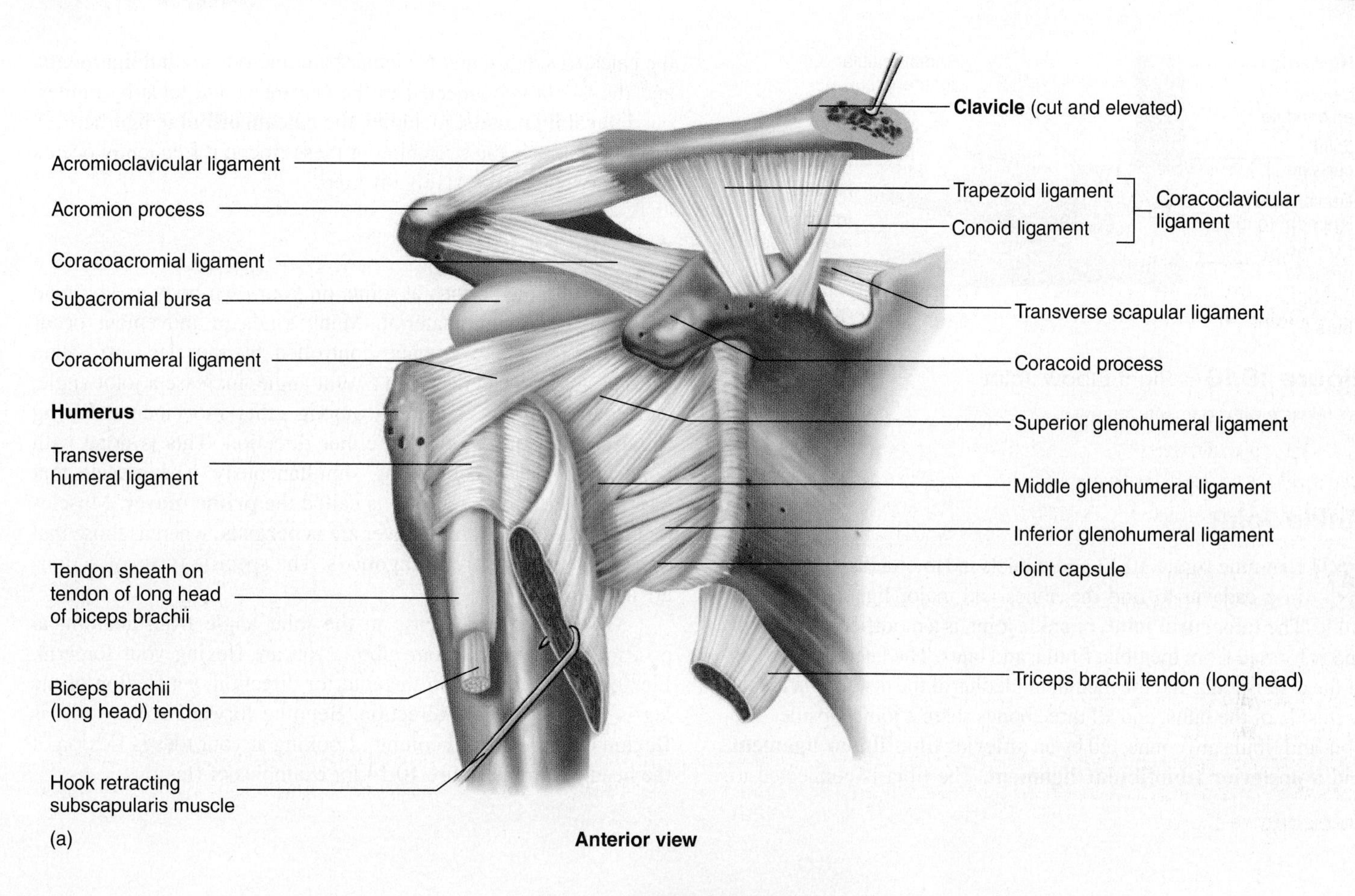

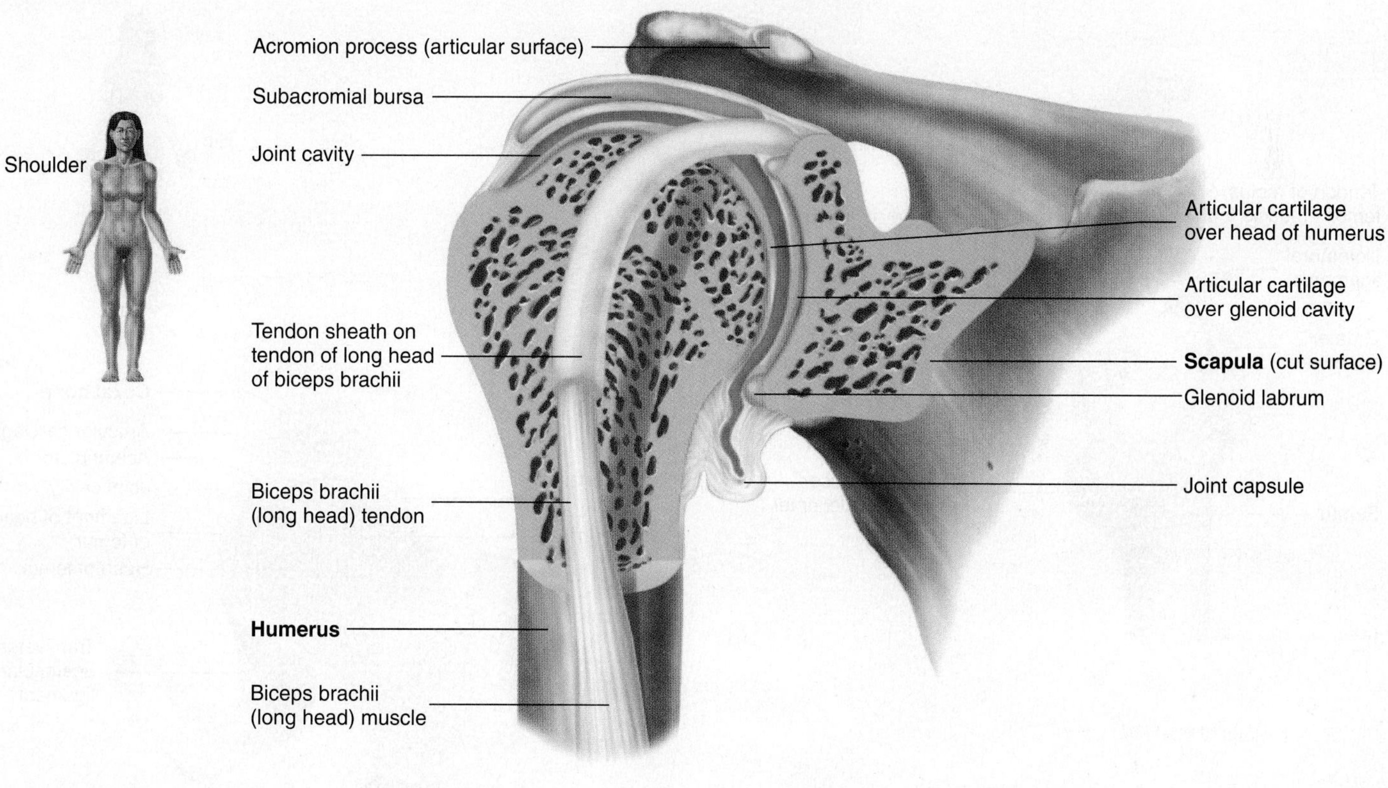

Figure 10.9 Glenohumeral Joint

(a) Anterior view; (b) frontal section.

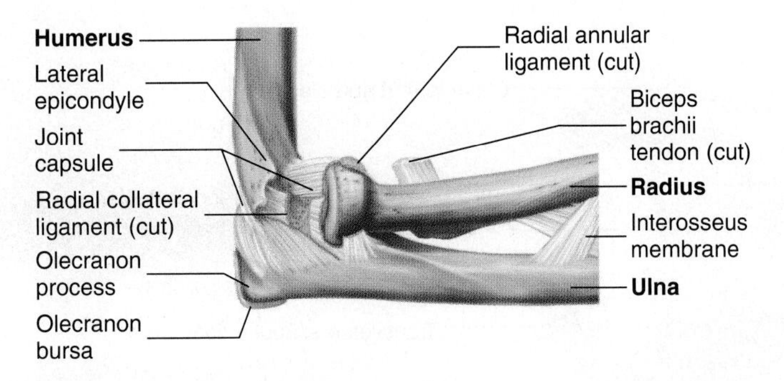

Figure 10.10 Right Elbow Joint
Lateral view with ligaments removed.

Ankle Joint

(A) 12 Examine figure 10.13 and models in lab or a dissected ankle in a cadaver to find the bones and major ligaments of the ankle. The **talocrural joint,** or ankle joint, is a modified hinge joint and is formed from the tibia, fibula, and talus. The lateral malleolus of the distal fibula and the medial malleolus of the distal tibia are on each side of the talus, and all three bones share a joint capsule. The tibia and fibula are connected by an **anterior tibiofibular ligament** and a **posterior tibiofibular ligament.** The tibia is connected to the calcaneus, talus, and navicular by numerous **medial ligaments,** and the fibula is connected to the calcaneus and talus by numerous **lateral ligaments,** including the **calcaneofibular ligament.** A sprained ankle is the stretching of these fibers; it frequently occurs when the foot is excessively inverted.

Actions at Joints

(A) 13 Mimic the actions at joints on your own body as you read the following material. Many kinds of movement occur at joints. These movements, controlled by muscles, are called actions. Actions can decrease a joint angle, increase a joint angle, and cause rotation at a joint, among other movements. **Fixing** a muscle prevents motion in either direction. This is done with opposing muscles contracting simultaneously. The muscle that has the main force on a joint is called the **prime mover.** Muscles that assist with the prime mover are **synergists,** whereas those that oppose the muscle are **antagonists.** The specific types of actions are as follows.

Flexion is a decrease in the joint angle from anatomical position. If you bend your elbow, you are flexing your forearm. Flexion of the thigh is in the anterior direction, yet flexion of the leg is in the *posterior* direction. Bending forward at the waist is flexion of the vertebral column. Looking at your toes is flexion of the head. Examine figure 10.14 for examples of flexion.

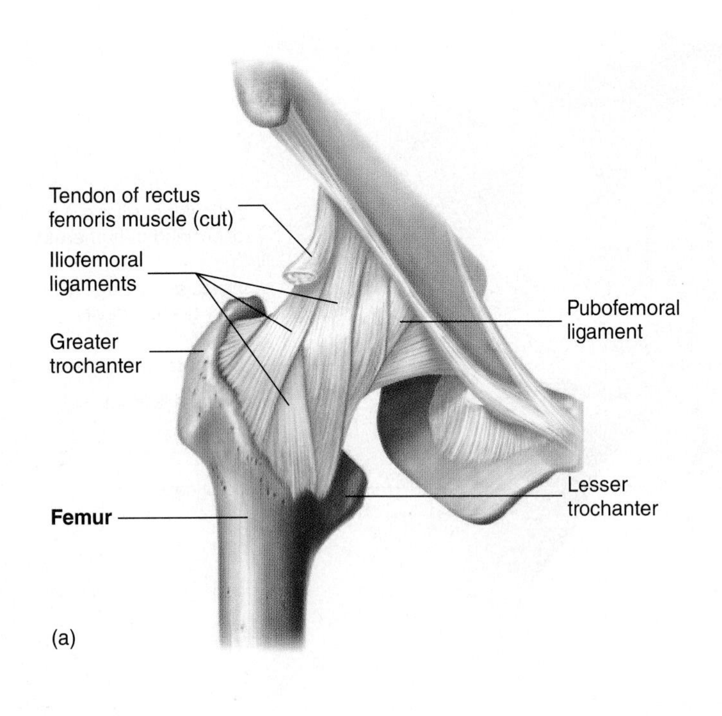

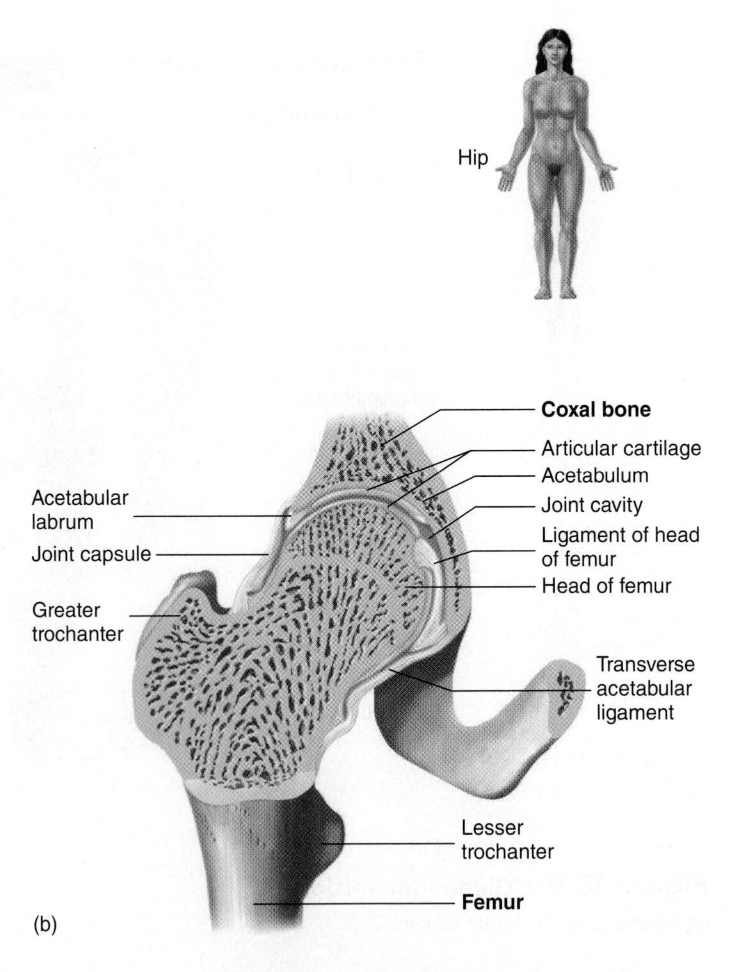

Figure 10.11 Acetabulofemoral Joint
(a) Anterior view; (b) frontal section.

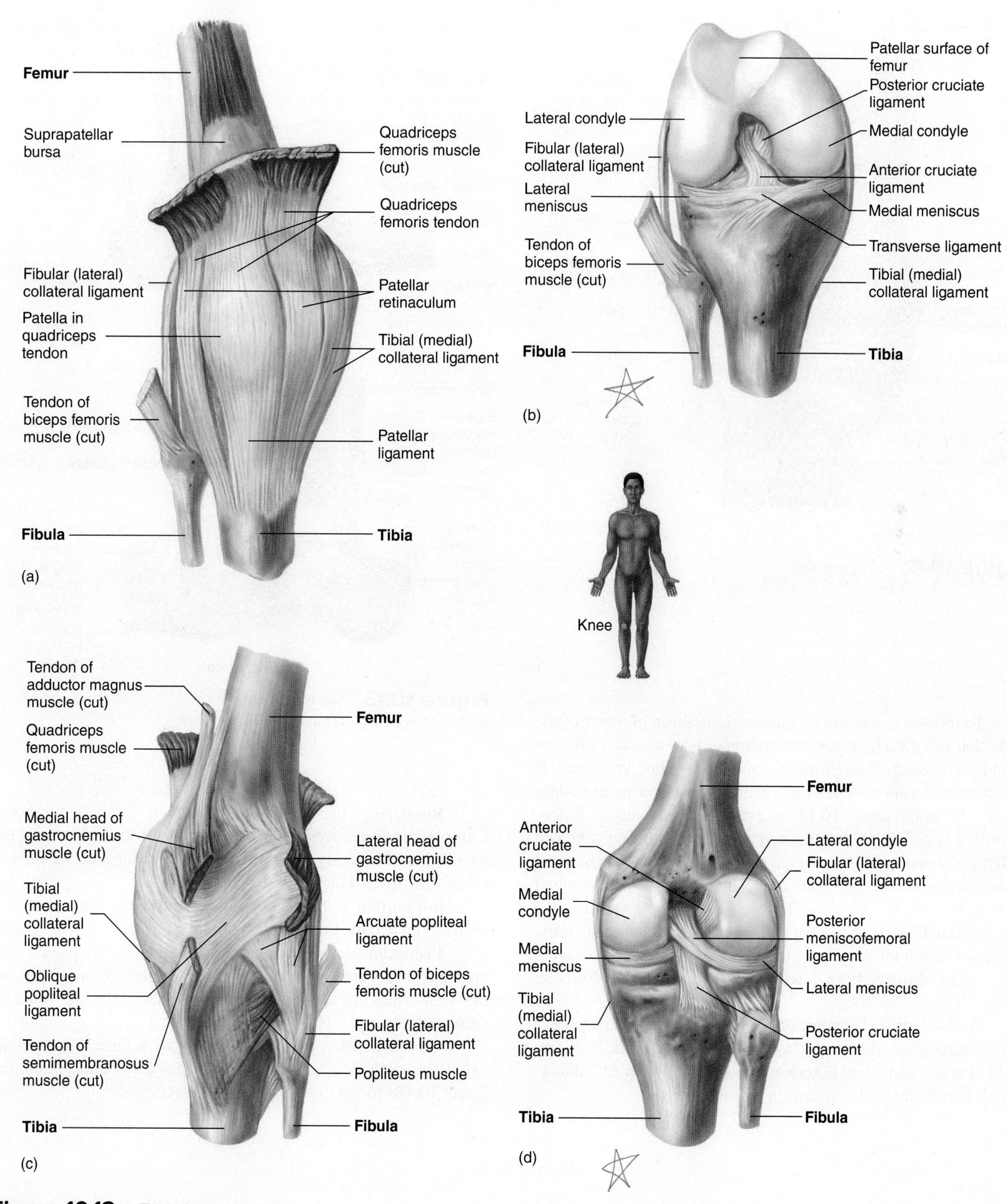

Figure 10.12 Tibiofemoral Joint
(a) Anterior surface view; (b) anterior deep view; (c) posterior surface view; (d) posterior deep view; (e) lateral section.

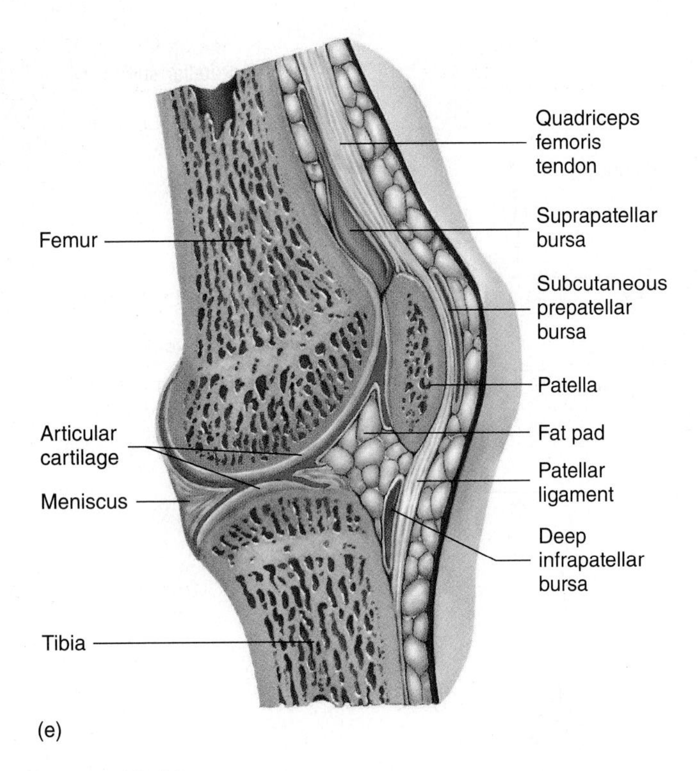

Figure 10.12 *Continued.*

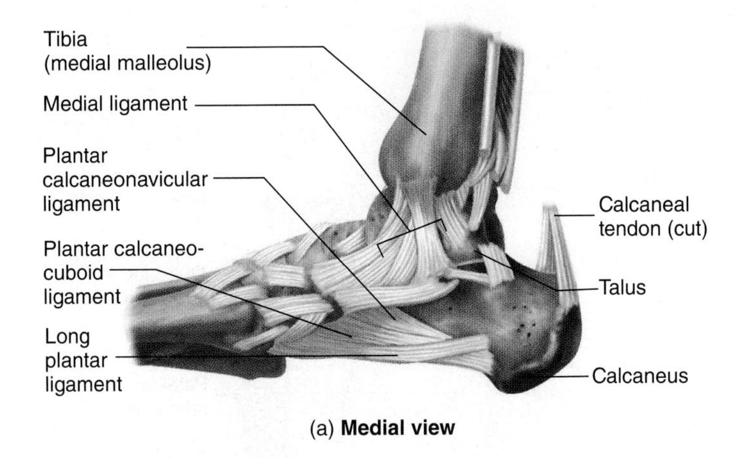

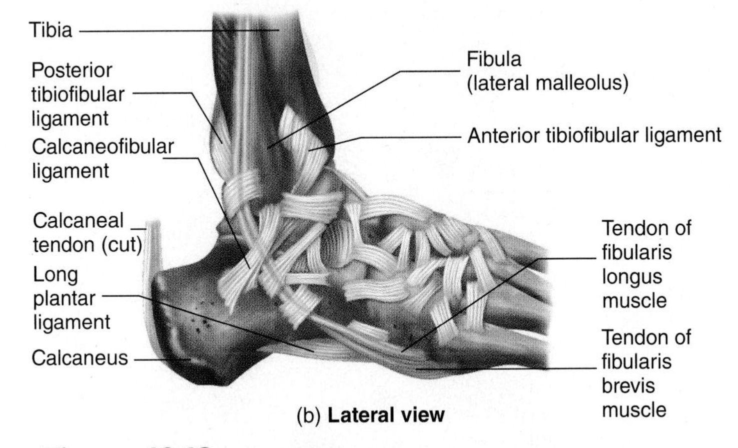

Figure 10.13 Ankle Joint.
Note the many ligaments that stabilize the joint.

Extension is a return to anatomical position of a part of the body that was flexed. If you are looking at your toes and lift your head back to anatomical position, you are extending your head. If you straighten your knee after it is bent (flexed), you are extending the leg. Examine figure 10.14 for examples of extension. Extension of the part of the body beyond anatomical position is known as **hyperextension.** When you are about to roll a bowling ball and your arm reaches the very back of the arc, you are hyperextending your arm.

Abduction is movement of the limbs in the coronal plane away from the body (*abduct* = to take away). Abduction is taking away a part of the body in a lateral direction. This can be seen in figure 10.15.

Adduction is the return of the part of the body to anatomical position after abduction. In doing "jumping jacks," you are abducting and adducting in series. Think of adduction as "adding" a limb back to the body, as seen in figure 10.15.

Rotation is the circular movement of a part of the body. **Lateral rotation** moves the anterior surface of the limb toward the lateral side of the body. **Medial rotation** turns the anterior surface of the limb toward the midline.

Supination is lateral rotation of the hand. The hands are supinated when the body is in anatomical position.

Pronation is medial rotation of the hands. When you turn your palms posteriorly from anatomical position, you are pronating your hands. Rotate various joints of the body and compare them with figure 10.16.

Circumduction is the movement of a muscle in a conical shape, with the point of the cone being proximal. Examine figure 10.16c for an example of circumduction.

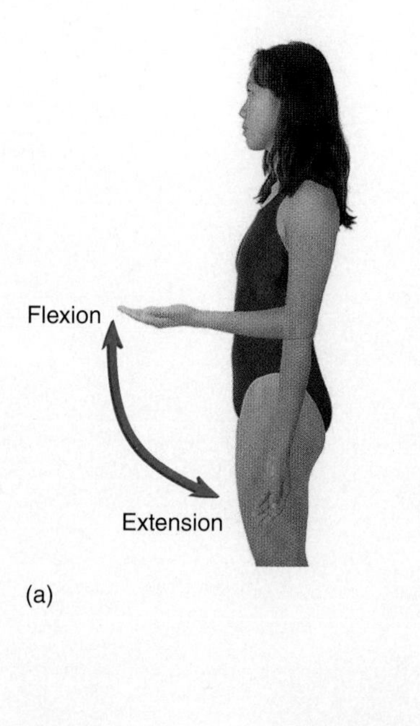

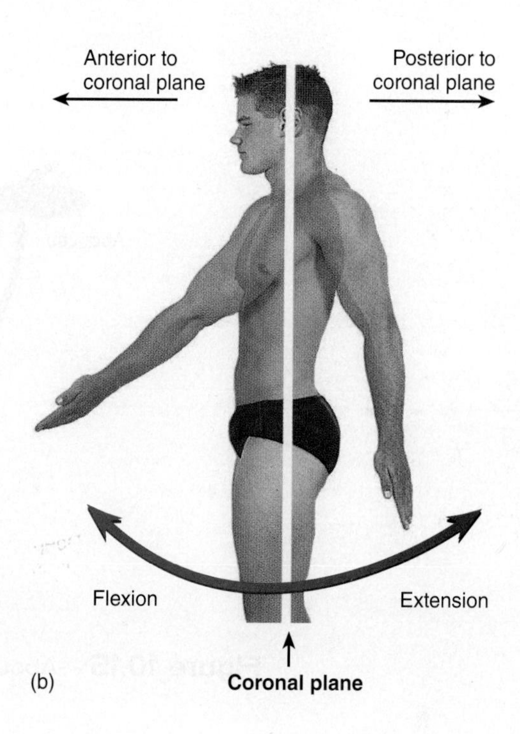

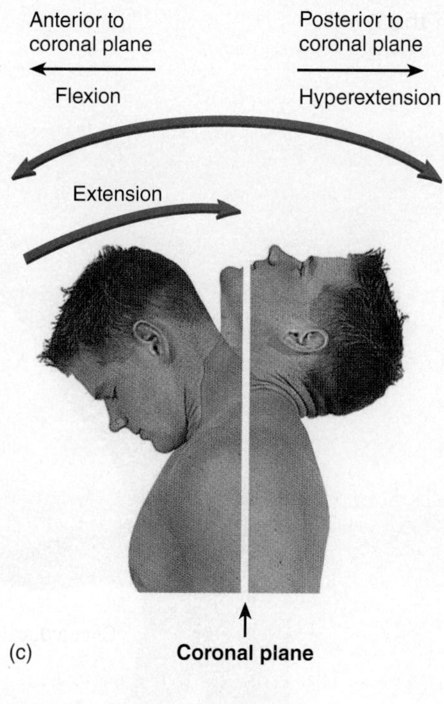

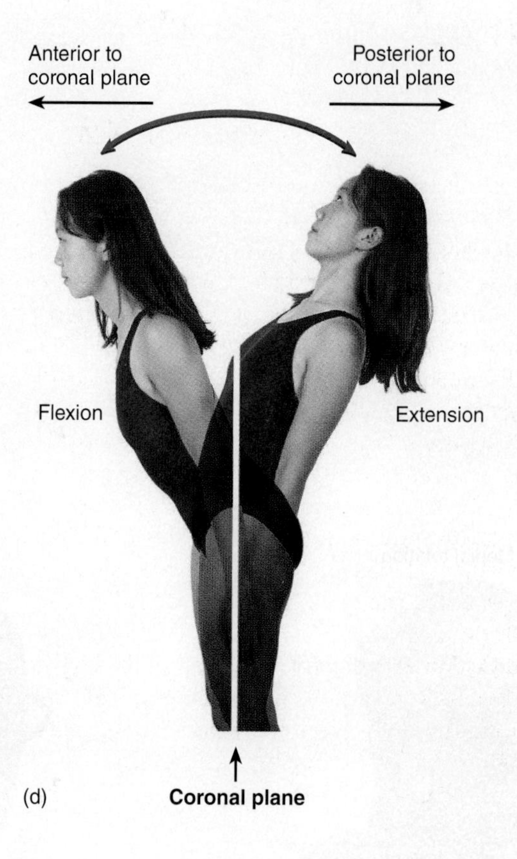

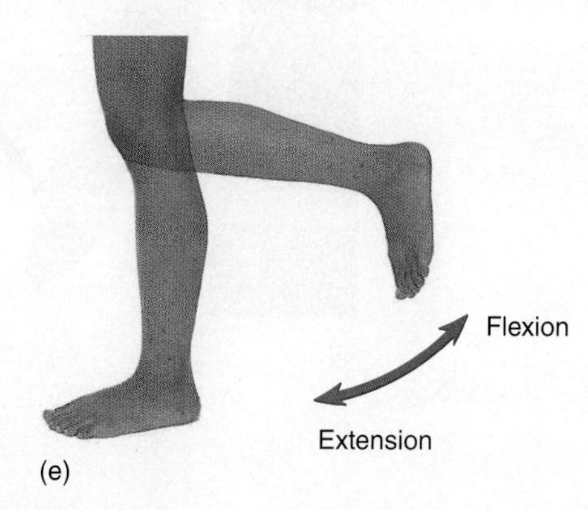

Figure 10.14 Flexion and Extension

Flexion and extension of the (a) forearm; (b) arm; (c) head; (d) trunk; (e) leg.

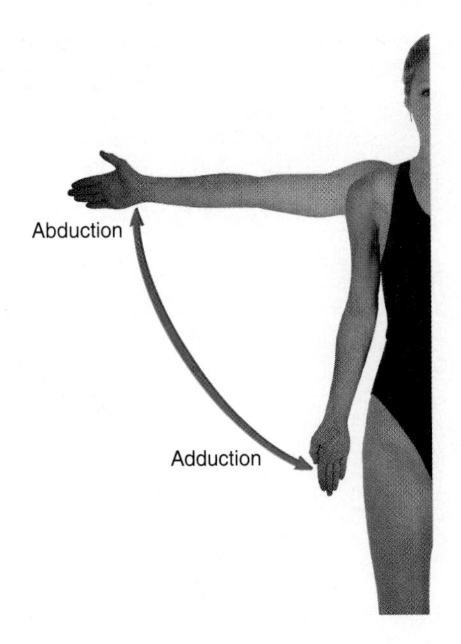

Figure 10.15 Abduction and Adduction of the Arm

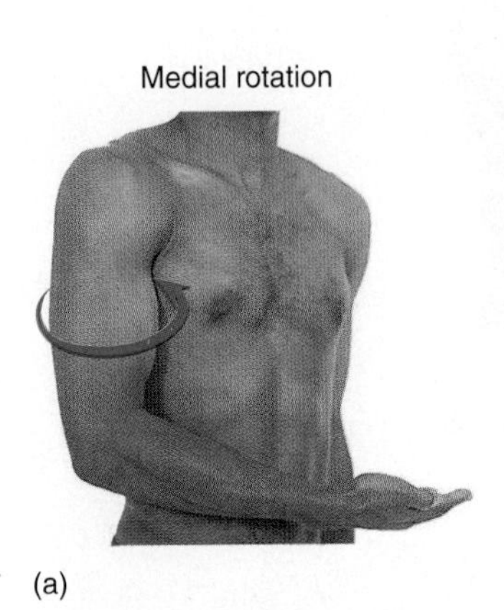

(a)

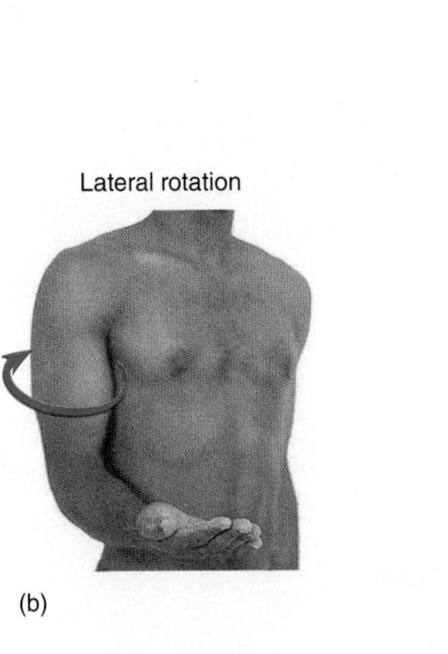

(b)

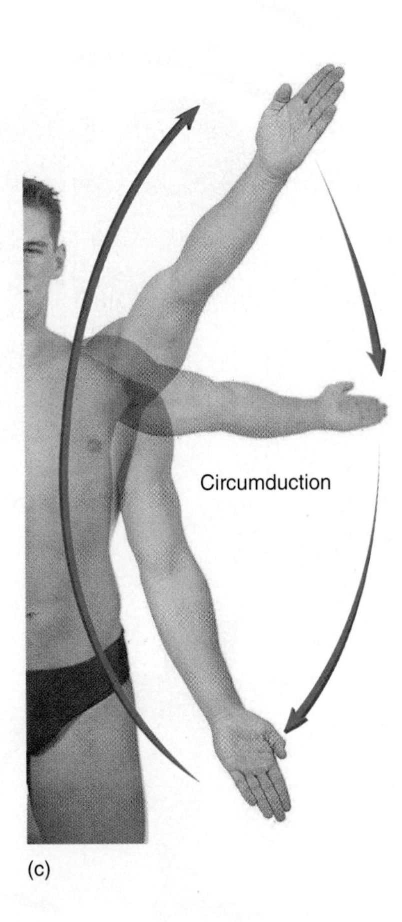

(c)

Figure 10.16 Rotation of Joints

Protraction is a horizontal movement in the anterior direction, as in jutting the chin forward.

Retraction is the reverse of protraction. The jaw that moves from anterior to posterior is retracted. These two actions are illustrated in figure 10.17.

Elevation means to move in a superior direction. Elevation of the shoulders occurs when you shrug your shoulders.

Depression is the opposite of elevation; it is movement in the inferior direction. Elevation and depression are seen in figure 10.18.

Inversion describes movement of the feet—turning the soles of the feet medially so they face each other.

Eversion means turning the soles of the feet laterally. Inversion and eversion can be seen in figure 10.18.

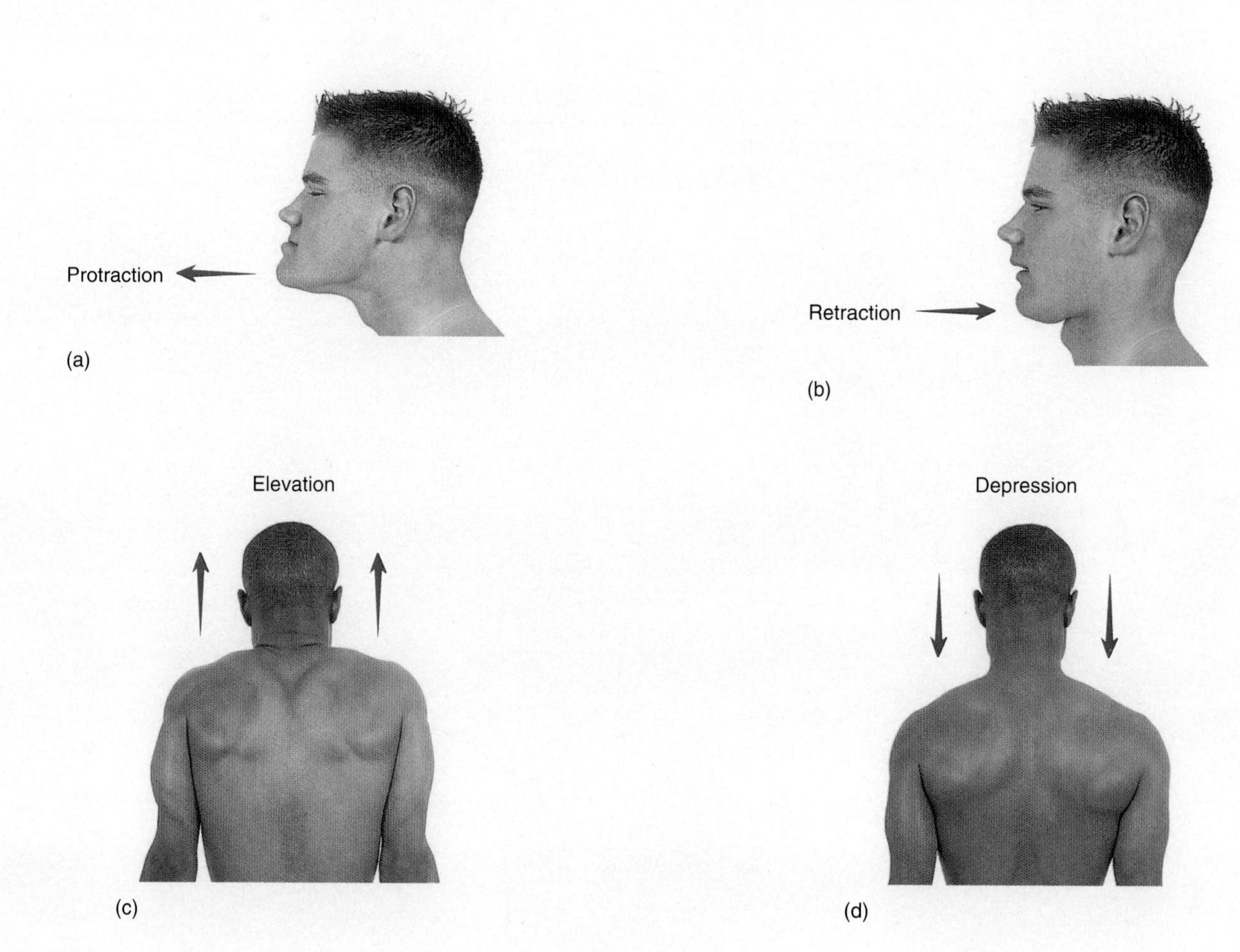

Figure 10.17 Protraction, Retraction, Elevation, and Depression

(a) Protraction of the mandible; (b) retraction of the mandible; (c) elevation of the scapulae; (d) depression of the scapulae.

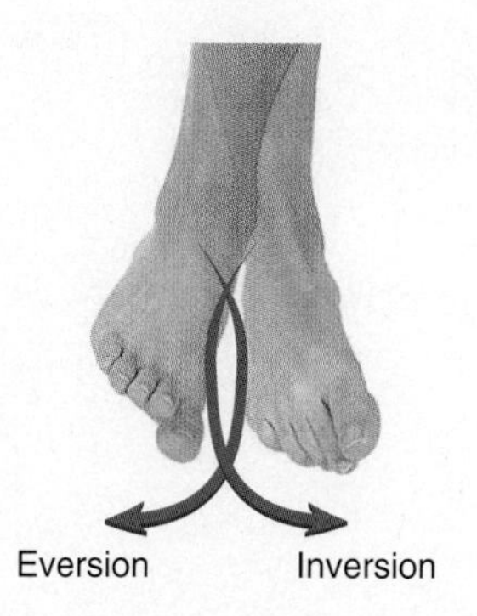

Figure 10.18 Inversion and Eversion of the Foot

NOTES

Exercise 10 Review

Name: ______________________

Lab time/section: ______________________

Date: ______________________

Articulations

1. The study of articulations, or joints, is called ______________________.

2. What kind of joint (based on joint composition or structure) is one in which the bones are held together by collagenous fibers?

____

3. When two bones fuse into a single bone, this union is called a(n) ______________________.

4. The teeth are held in the jaw by what specific kind of joint? ______________________

5. What kind of joint is slightly movable and is held together by fibrous connective tissue? ______________________

6. In terms of their structural composition, bones that are held together by cartilage (cartilaginous joints) are also known as ______________________ joints.

7. The epiphyseal plate is a cartilaginous joint. It is also called a(n) ______________________.

8. Label the following illustration using the terms provided.

articular cartilage

fibrous capsule

joint cavity

periosteum

synovial membrane

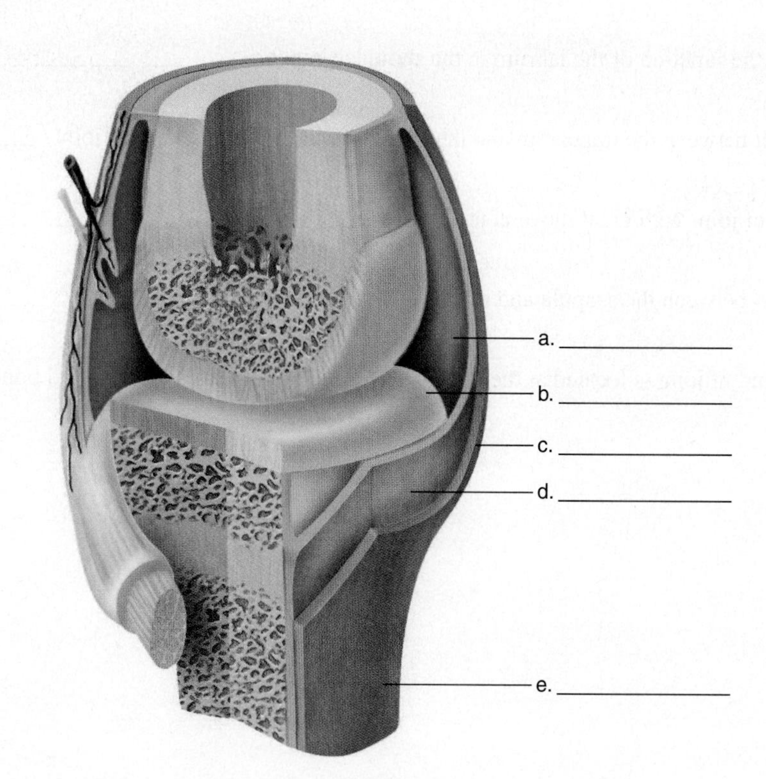

9. What is the name of a joint that is held together by a joint capsule? ______________________

10. Synovial fluid is secreted by what structure? ______________________

11. A skull suture is what kind of joint, in terms of movement? ______________________

12. Rank the following joints in terms of least movable to most movable, with 1 being the least movable and 5 being the most movable.

_______ gliding

_______ saddle

_______ suture

_______ syndesmosis

_______ ball-and-socket

13. Which one of the following joints has the greatest range of movement?

a. gomphosis b. suture c. synchondrosis d. hinge

14. In which of these joints would you find a meniscus?

a. cartilaginous b. fibrous c. synovial

15. What is the function of the meniscus in the knee? ______________________

16. Match the joint in the left column with the type of joint in the right column.

_______ hip	a. hinge
_______ radiocarpal	b. ball-and-socket
_______ tibiofemoral	c. plane
_______ intercarpal	d. ellipsoid

17. What is the function of the labrum in the shoulder joint? ______________________

18. The joint between the trapezium and the first metacarpal is what kind of joint? ______________________

19. A class of joint with great movement is a(n) ______________________.

20. The joint between the scapula and the humerus is what type of joint? ______________________

21. What kind of joint is located at the wrist (between the radius and the scaphoid bones)? ______________________

22. Label the following illustration using the terms provided.
    anterior cruciate ligament
    fibular collateral ligament
    meniscus
    tibial collateral ligament

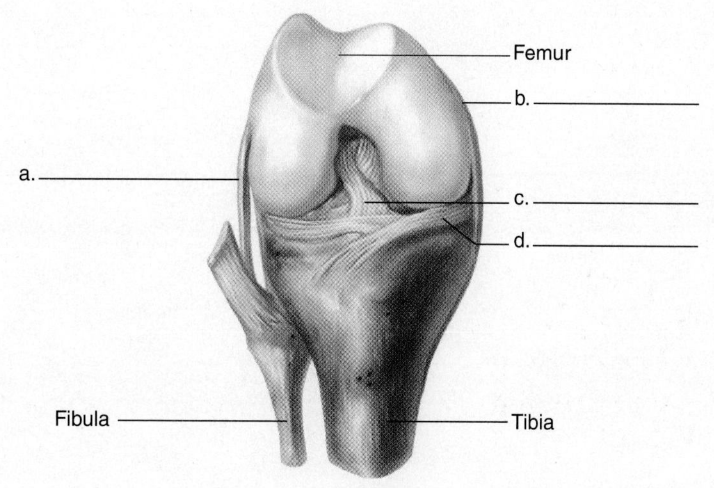

NOTES

Exercise 11

Muscle Physiology

Introduction

Muscles have several functional characteristics. Muscles show **contractility** when they are stimulated. The contraction of muscles can bring about various functions, such as walking, eating, talking, and breathing. Muscles also have **extensibility.** They can be stretched when opposing muscles contract. Muscles have **elasticity** in that they recoil when stretched, and muscles have **excitability** because they respond to stimuli. These topics are covered in the Seeley text in chapter 9, "Muscular System: Histology and Physiology."

Skeletal muscles contract due to stimulation by nerves controlling them. Normally, a series of nerve impulses begins to stimulate a muscle and continues with increasing strength, producing the uniform muscle contraction. When these multiple nerve impulses diminish, the muscle relaxes. In this exercise, you explore the nature of skeletal muscle excitation and contraction as initiated by external electrical stimulation and apply this information to the functions of the skeletal muscle in your body.

Two events are fundamental to an understanding of muscle physiology. One is an electrochemical event that occurs in muscle membranes, and the other is the physical contraction of the muscle itself. The normal contraction of skeletal muscle occurs when an electrochemical **nerve impulse** travels down an axon and reaches the **synapse** between the nerve and the muscle. This synapse between the corresponding neuron and muscle fiber is known as the **neuromuscular junction.** This junction is seen in figure 11.1. **Acetylcholine (ACh)** is released by the terminal regions of the neuron and stimulates an electrochemical impulse, which travels along the length of the muscle fiber.

When the impulse reaches the **transverse,** or **T, tubules** of the muscle, **calcium** is released from the sarcoplasmic reticulum, and the **actin** and **myosin** myofilaments join together, producing a power stroke in the muscle cell. As a muscle fiber is **depolarized,** the entire fiber contracts maximally. This characteristic represents the **all-or-none principle** of muscle fibers. A single skeletal muscle cell is composed of many myofibrils, each of which consists of discrete units called **sarcomeres,** which are connected together at a region known as the **Z disk.** Skeletal muscle is striated because of the regular arrangement of myofilaments. When a muscle contracts, the actin myofilaments and myosin myofilaments slide over one another, causing the muscle to shorten. This is the **sliding filament model,** and it is described in more detail in your textbook.

One neuron stimulates many muscle fibers. This complex of a neuron and the associated muscle cells is known as a **motor unit,** which is illustrated in figure 11.2. When slight contractions occur, few muscle fibers contract, because only a few motor units are activated. When a muscle contracts more forcefully, more motor units are stimulated, and they subsequently contract.

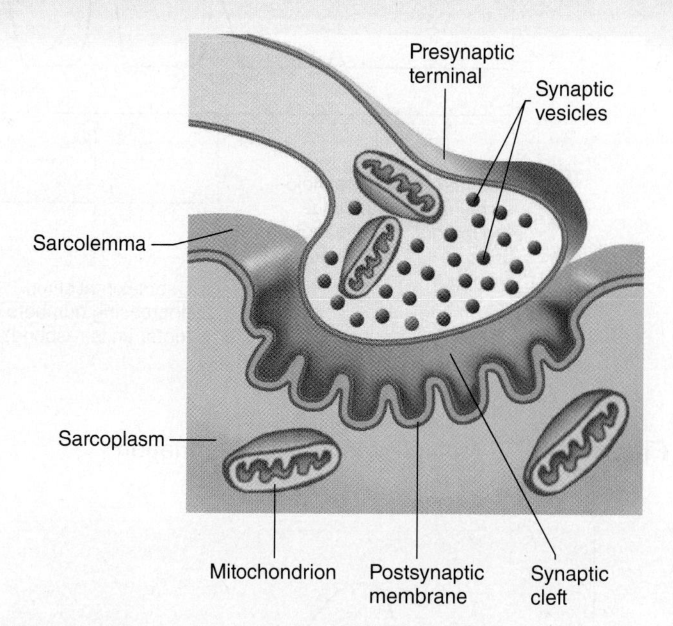

Figure 11.1 Neuromuscular Junction

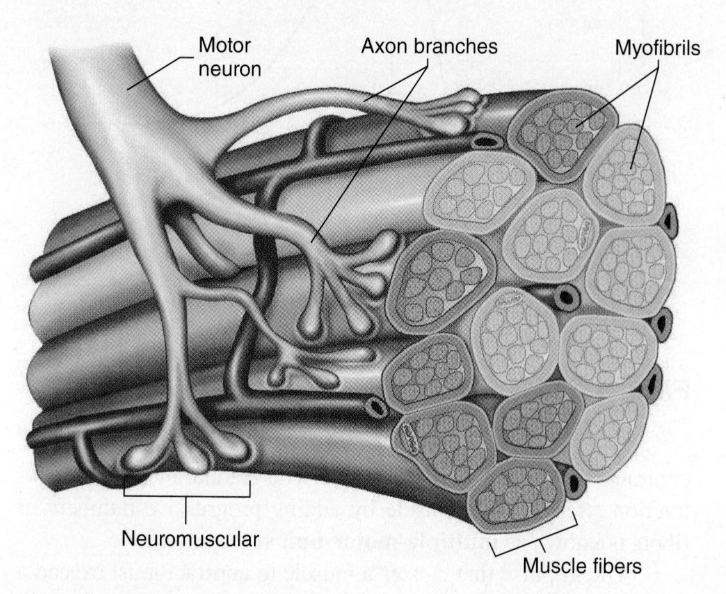

Figure 11.2 Motor Unit

Entire muscles in the body do not contract with an all-or-none response but, rather, exhibit a **graded response,** in which muscles gradually increase from slight to more forceful contractions. An increase in the force of contraction is known as **summation.** Summation can occur by increasing the frequency of impulses to an individual fiber or by increasing the stimuli to more and more muscle fibers.

One way to increase the overall contractile strength of the entire muscle is to stimulate more muscle fibers (each of which

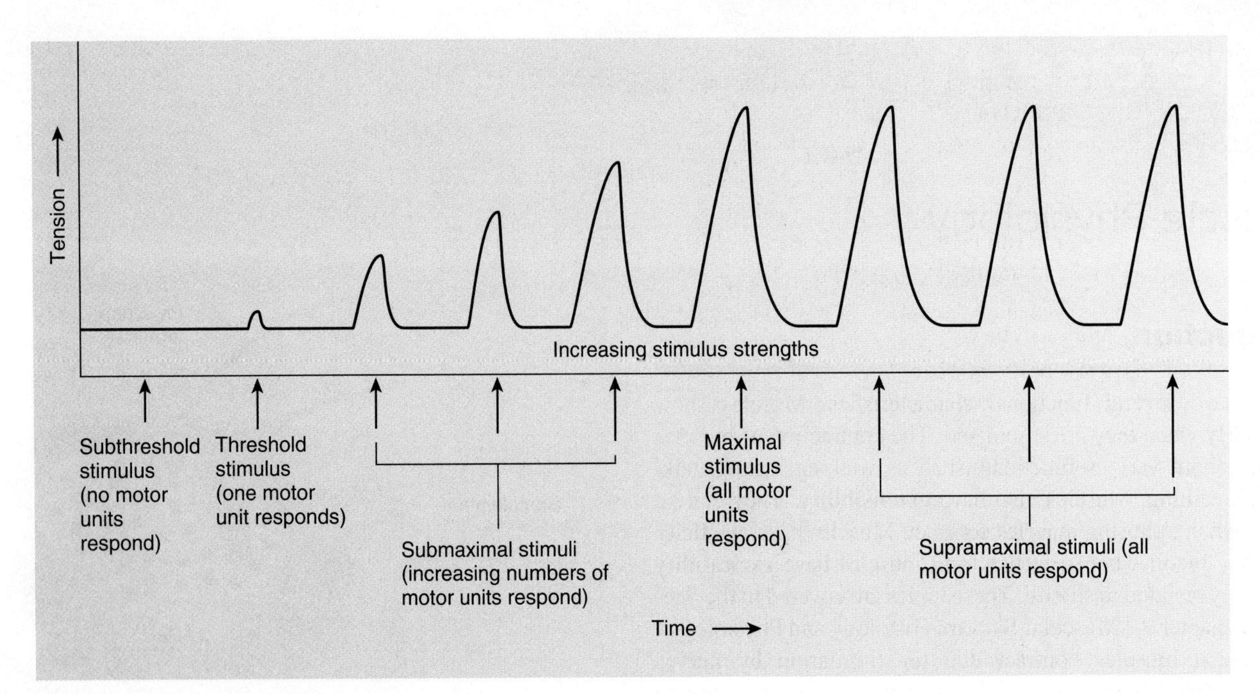

Figure 11.3 Multiple-Motor-Unit Summation

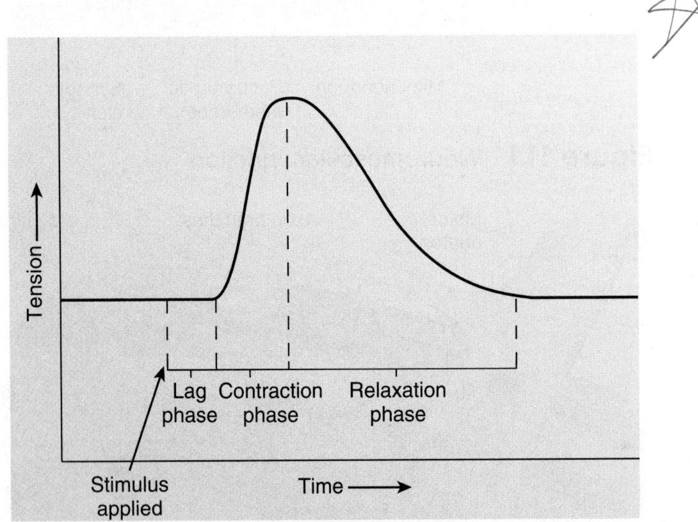

Figure 11.4 Three Phases of a Muscle Twitch

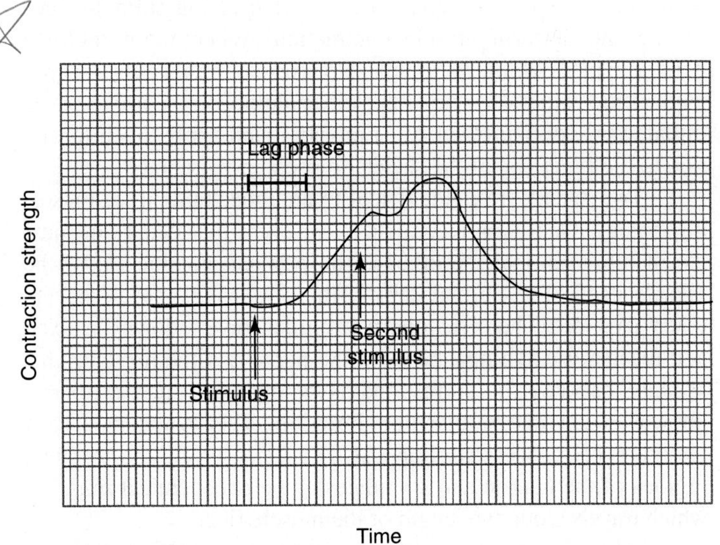

Figure 11.5 Wave Summation

contracts completely) in that muscle. The gradual increase of contraction strength of a muscle by adding progressive numbers of fibers is known as **multiple-motor-unit summation.**

The impulse that causes a muscle to contract must exceed a **threshold** value before any contraction can occur. A **subthreshold stimulus** is one that will not elicit a response in the muscle. Once the muscle contracts, the addition of progressively stronger stimuli results in stronger muscle contractions. The point where all of the muscle fibers are forcefully contracting is known as the **maximal stimulus.** Examine figure 11.3 for the subthreshold, threshold, and maximal stimulus of a muscle in multiple-motor-unit summation.

When a muscle fiber is stimulated with a single, quick electrical impulse, the fiber undergoes a contraction known as a **twitch.** In a twitch, the fiber has three phases: a lag phase, a contraction phase, and a relaxation phase. After the initial impulse, known as a **stimulus,** the muscle fiber undergoes a **lag phase** (figure 11.4). This is the time when the electrical impulse travels across the muscle cell membrane and calcium ions are released from the sarcoplasmic reticulum. No contraction occurs at this time. After this short period, the fiber goes through the **contraction phase** as the myofilaments slide across one another and the muscle fiber shortens. Finally, there is a **relaxation phase,** in which the fiber returns to a resting state.

A stimulus above threshold level that occurs too soon after a preliminary stimulus also does not cause the muscle fiber to contract. This is known as the **refractory period.** It is the time when a stimulus, delivered just after a previous stimulus, produces no contraction. If a stimulus is applied shortly after the refractory period, a muscle contracts, and the contraction strength is more pronounced. This effect is called **wave summation** (figure 11.5).

If rapid, repeated stimuli are sent to a muscle, then the muscle produces a series of contractions called **incomplete tetanus** (figure 11.6). In incomplete tetanus, the muscle does not return to a baseline of relaxation but exhibits a rapid series of contractions. If the frequency (number of pulses per second) of the stimuli increases, the contractions fuse in a smooth contraction of the muscle known as **complete tetanus.** Numerous, sequential stimulations of a muscle produce **temporal summation** in the muscle. Temporal summation results in the smooth, continuous muscle contractions that normally occur in the body. Complete tetanus occurs even in fast muscular movements, such as the flicking of a finger. Examine figure 11.6 for recordings of incomplete tetanus and complete tetanus.

In this exercise, you examine the muscular contraction in a simulated frog or a real frog and discover what occurs as voltage is increased or the frequency is changed. Due to the decline in some frog populations (such as leopard and grass frogs in North America), simulated activities may be preferable to the use of live frogs. Bullfrogs are not only plentiful but are a pest species in many areas and may be used today without significant impact on their population levels.

Objectives

At the end of this exercise, you should be able to

1. demonstrate the procedure to determine threshold stimulus;
2. differentiate between tetanus and treppe;
3. describe incomplete tetanus and complete tetanus;
4. demonstrate maximum recruitment with lab equipment and a frog;
5. compare the lab experiments on frogs with muscular contractions in humans;
6. describe the process of muscle recruitment and fatigue in humans.

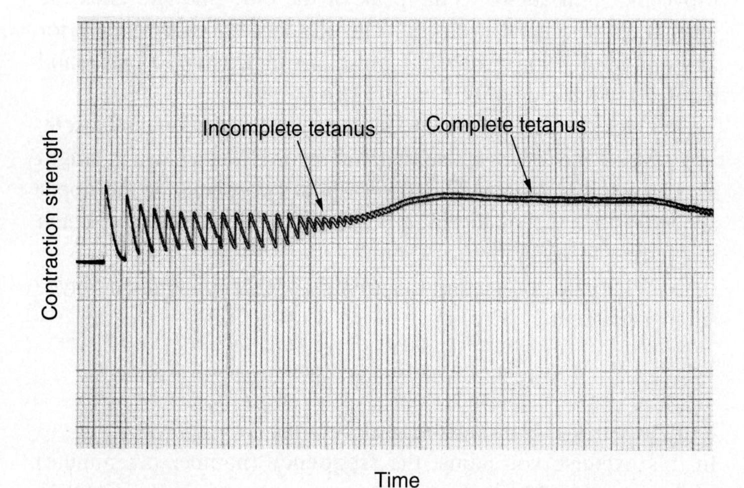

Figure 11.6 Incomplete and Complete Tetanus

Materials

Virtual Experiment Lab

Ph.I.L.S. CD (available through McGraw-Hill). Ph.I.L.S. is also available within Connect.
Compatible computer

Live Frog Lab

Frog (one per experiment)
Frog Ringer's solution in dropper bottles (150 mL per experiment)
Nylon thread
Duograph, physiograph
Myograph transducer
Stimulator and cables
Scissors
Clean, live animal dissection pan
Sharp pithing probes
Glass hooks
Scalpel
BSLSTM stimulator
Force transducer assembly—SS12LA
Pin electrodes ELSTM2
HDW100A tension adjuster
Data acquisition unit MP30
BSL *PRO* template file: FrogMuscle.gtl
Ring stand
Anchoring board
Goggles
Latex gloves
Forceps
Dissection pins
Animal disposal container
Multiple weights (5, 10, 15 lb)

Procedure

Simulated Frog Muscle Experiment—Physiology Interactive Lab Simulations

(A)[1] Open Physiology Interactive Lab Simulations (Ph.I.L.S.) Version 4.0. For all of the virtual muscle experiments you should follow the standard procedure outlined next.

1. Click on the number of the exercise that you want to complete. These are under the **Skeletal Muscle Function** section and are listed as:
   "5. Stimulus-Dependent Force Generation
   6. Weight and Contraction
   7. Length-Tension Relationship
   8. Principles of Summation and Tetanus
   9. EMG and Twitch Amplitude"

   After you have selected your choice you should:

2. Read the objectives and introduction to the lab simulation. You can click on highlighted terms to see pictures of the material in question.
3. Take the pre-lab quiz. If you get a question wrong, the program will let you know and provide you with the correct answer.
4. Click on the Wet Lab tab and read the material.
5. Click on Continue, which opens up the Laboratory Exercise.
6. Turn the power on the computer screen and the Data Acquisition Unit. The green light lets you know it is on.
7. Connect all the plugs to their respective locations. You must drag the tip of the plug to the middle of the adapter in order for it to connect.
8. Adjusting the voltage, frequency, or tension can be done by clicking on the up or down arrow repeatedly or clicking on the arrow and holding it until the desired value is obtained.
9. Take the post-lab quiz to determine your understanding of the lab.

The information for the specific lab exercises follows.

5. Stimulus-Dependent Force Generation

In this laboratory exercise you examine the impact that an increase in voltage stimulus has on the strength of contraction. You should see a picture that shows a virtual muscle, electrodes to stimulate the muscle, and an input that records the "pull" of the muscle after it has been stimulated. Click on the "power switch" to turn on the computer screen. The switch will appear on the lower right of the simulated screen. The red and blue wires deliver electric impulses (shocks) to the frog muscle and the black wire records how much force the muscle exerts. Follow the text directions in the program and attach the plugs to their proper connections. They are all color coded.

The directions ask you to select at least 1 volt to get a response. After you select 1 volt, click on the Shock button. You should see an upward curve.

To record the amplitude you need to select a crosshair cursor on the blue line to the right of the peak and drag it to the highest point of the curve on the graph. Click on the Journal entry icon. You should see the data points automatically plotted on the graph. Click on the X in the upper right corner of the graph table in the journal and you will return to the experiment.

Reduce the voltage to 0. Hit the Start button again. Enter the material in the journal as described above. Do this repeatedly until you fill in the entire journal from 0 to 1.6 volts. Notice that at low voltages the force line stays flat. This is because muscles have a threshold of contraction below which no contraction occurs. Record the threshold voltage (when you first see a contraction) for the muscle in the following space.

Threshold voltage:

What happens to the muscle contraction strength as you increase the voltage?

Response:

Although it was not demonstrated in this exercise, there is a point at which the increase in voltage does not lead to an increase in contraction strength. This is known as **maximum recruitment,** and it is the voltage where all the muscle fibers are contracting.

6. Weight and Contraction

In this part of the exercise you will measure the latent period, the contraction phase, and the relaxation phase. Notice the flat line that precedes the actual contraction of the muscle. This is the **latent (latency) period,** the time between when the stimulus was delivered and when the muscle started to contract. The latent period is the time after the muscle is stimulated when calcium diffuses from the transverse tubules to the myofibrils. Follow the directions in the virtual lab, turn on the power switch on the computer monitor, connect the colored plugs, and set the voltage to 1.0 volt. Click on the crosshair cursor and move it as far to the left as you can on the graph (at the beginning of when the muscle was stimulated). Move the other crosshair cursor to the point where the blue line begins to rise up. This is the latent period. Enter the value in the following space.

Latent period:

The left side of the curve is steeper than the right side of the curve. The left side of the curve is known as the **contraction phase.** How long does this phase take? It is measured by moving the left crosshair cursor to the position on the graph where the line moves sharply up from the baseline and then moving the other crosshair cursor to the peak of the blue line at the top of the curve. If the program will not allow you to enter the data, click Erase and redo the experiment. Select the appropriate areas and click on them.

Enter the time in the following space.

Contraction phase:

Move the cursors so one is at the peak of the graph and the other is at the region of the blue line where it reaches the baseline and record the **relaxation phase.** Enter the time of the relaxation phase in the following space.

Relaxation phase:

Follow the directions and complete the experiment by adding successively more weight to the muscle.

7. Length-Tension Relationship

In this laboratory exercise you examine the strength of contraction based on the initial length of the contracting muscle. Follow the directions as outlined above and in the directions in the program. You must click on the zoom button before you can set the voltage. Move the right cursor to the peak of the blue line and click the journal entry to get the tension. The muscle can be stretched prior to stimulation. Notice how the muscle is attached to a horizontal bar. On the left side of this bar are two arrows.

Click on the upper one to increase the tension on the muscle.

Move the arrow up, so that you record the increase in length in 0.5-mm increments. This takes about two clicks of the upper arrow. What happens to the strength of contraction as you stretch the muscle from 26 mm to 30 mm?

Where do you predict the greatest overlap of actin and myosin myofilaments? Why?

8. Principles of Summation and Tetanus

In this exercise you adjust the **frequency** (number per minute) of the stimulus. Most muscles do not contract by a single twitch. Neurons typically send many impulses that repeatedly stimulate the muscle, producing a smooth contraction. As in the previous experiment, you will need to read the material presented and answer the questions before doing the experiment. Once you get to the experimental stage, turn on the power button to the computer monitor and the data acquisition unit, drag the appropriate cables to their inputs, and set the voltage between 1.6 and 2 volts. If you click and hold the up arrow button it will continue to increase the voltage. The voltage remains the same in all of the trials. Under low frequency (few stimulations per unit time) you will see that the muscle contractions all have the same height and return to the baseline. When you increase the frequency, by using the cursor to decrease the interval by holding the down arrow, the muscle contraction tracing leaves the baseline and you have **summation.** You can see the beginning of summation at an arrow on your graph by clicking on the highlighted term *summation.*

You will have to pay attention to when the blue line does not return to the baseline. Record this value in milliseconds (ms) in the journal and in the following space.

Summation:

As you decrease the interval between stimuli, notice how the peaks begin to merge. If you increase the frequency more, the contractions on the graph occur above the normal contraction peaks, but you still see individual contractions. This is **incomplete tetanus.** Record the time interval in milliseconds for incomplete tetanus in the following space.

Incomplete tetanus:

When the two peaks merge into a straight line, it represents continuous muscle contraction or **complete tetanus.** Record the time interval in milliseconds for complete tetanus in the following space.

Complete tetanus:

9. EMG and Twitch Amplitude

The EMG records the electrical activity of contracting muscles. You will record two events. One is the electrical activity of contracting muscle and the other is the force generated by the muscles of the arm and forearm squeezing on a rubber bulb. As opposed to the earlier exercises where you shock the muscle, in this exercise you are picking up recording information only. In this exercise, in addition to placing the plugs in their respective areas, you must also attach the recording electrodes to the proper location on the subject's right arm. After you press the "start" button, drag the crosshair cursor to the peak of the muscle tension graph, which is the orange line. Then find the peak of the blue line (EMG) and move the crosshair cursor there.

Close the journal box by clicking on the X in the right upper corner before you record the EMG data. Do this for several contractions by adjusting the arrows below the squeeze bulb in the subject's right hand. What is the relationship between the force of contraction and the amount of electrical activity in the muscle? Record your answer in the following space.

Live Frog Experimentation

You may do this experiment in small groups, or your instructor may elect to do a demonstration for the class. If you are doing this experiment as a group, read the entire exercise first and then follow the directions. The directions are first for physiograph recordings. A section on using **Biopac** follows for labs equipped with this hardware. Preparing the frog is the same whether you use a physiograph or the Biopac setup and it is described next.

Frog Preparation

(A)[2] If the frog has not been pithed, follow the directions under number 1. If the frog has been pithed, begin at number 2.

1. The most humane way to conduct frog muscle experiments is to pith the frog quickly by inserting a sharp probe into the braincase.
   a. To do this, firmly grasp the frog and bend the head over with your index finger, as illustrated in figure 11.7.
   b. Insert the probe into the braincase and twirl it around in a conical manner, destroying the brain. This is known as a **single pith.** The frog will be killed at this point yet still have reflexes in the lower limbs.
   c. To stop the reflexes, insert the sharp probe into the vertebral canal and run it toward the caudal end of the frog. This is known as a **double pith.** Be careful not to thrust the probe into your hand in your attempt to locate the vertebral canal.

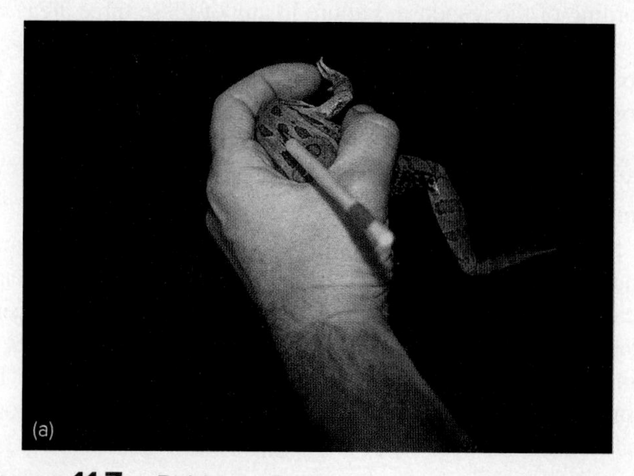

Figure 11.7 Pithing a Frog

(a) Single pith—probe inserted into braincase; (b) double pith—probe inserted into vertebral canal.

Figure 11.8 Preparation of the Frog, Skin Removal

Cut skin at this point and remove skin from thigh and leg.

2. Once the frog is pithed, carefully snip its skin above the hip joint and peel it back to the foot (figure 11.8).
   a. Separate the posterior muscles of the thigh and locate the **sciatic nerve,** which appears as a thin, white, glossy thread that runs along the lateral aspect of the femur. You may have to use a scalpel and tease the muscles away from the sciatic nerve.
   b. Using a glass hook, carefully lift the sciatic nerve away from the thigh muscles, keeping it moist with frog Ringer's solution (figure 11.9).
   c. **Ligate** the nerve by tying a thread around the proximal end of the nerve, near the sacrum, and cut the nerve above the ligature.
   d. Separate the thigh muscles from the femur and remove them, leaving the femur exposed. Be careful not to cut or damage the sciatic nerve as you do this.
   e. Locate the **gastrocnemius muscle** of the frog and tie the tendon of the muscle with thread.
   f. Cut the calcaneal tendon distal to the ligature and lift the gastrocnemius away from the other muscles and from the tibiofibula (a fused bone in frogs).
   g. Cut the muscles and the tibiofibula just distal to the knee, so that you have the femur, the sciatic nerve, the gastrocnemius, and the knee joint intact (figure 11.9).

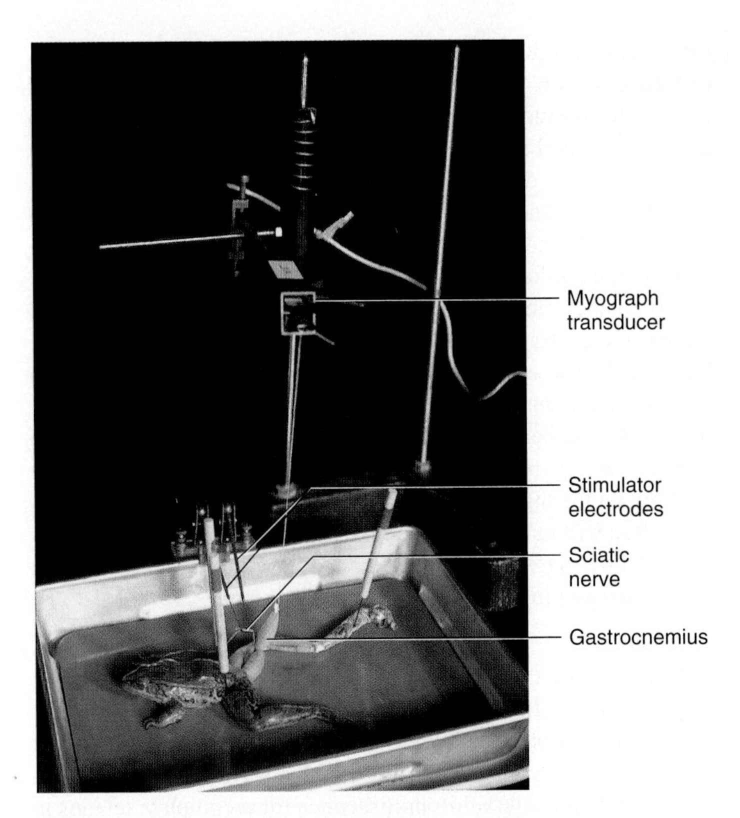

Figure 11.9 Frog Hookup to Recording Apparatus

   h. Anchor the femur to a board or mount it on a tray and lay the sciatic nerve on the gastrocnemius muscle.
   i. Keep the muscle and the nerve moist during the entire experiment. Do not tug on the nerve but gently lay it on the gastrocnemius muscle.
   j. Attach the thread tied to the calcaneal tendon to the end of a myograph transducer leaf (figure 11.9). This should be connected to a recording device, such as a physiograph, duograph, kymograph, or physiology computer. If you lightly tap on the leaf of the muscle transducer, you should see a response in your recording apparatus.

There are three critical areas of concern in this lab: the preparation of the muscle, the stimulation of the muscle, and the recording of the response. Failure in any of these three areas will cause poor results or no results at all. The first of these areas is the preparation of the muscle, which has already been outlined. The second and third areas are discussed next.

Physiograph Setup

Stimulator Setup

The preparation of the stimulator first involves determining how many stimuli you deliver in a particular time. Most stimulators can deliver repeating stimuli or single pulses. These are measured as the frequency of the stimulus, and a good frequency to start with is two pulses per second. Another important factor is the duration of the stimulus. This is how long the stimulus is delivered to the sciatic nerve. A duration of 2 milliseconds (ms) usually produces good results.

Determination of Threshold Stimulus

1. Turn the voltage to zero on the stimulator.

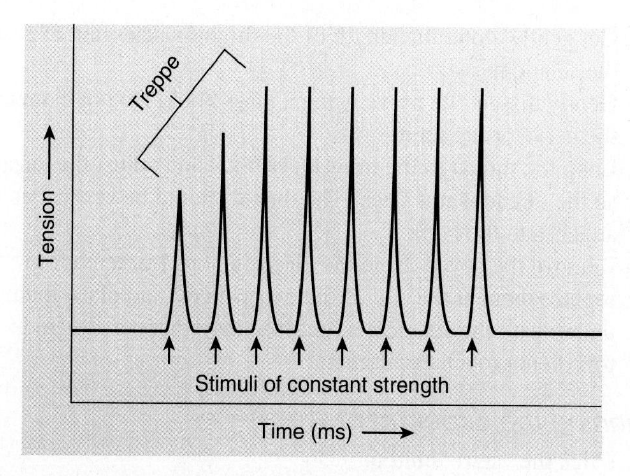

Figure 11.10 Treppe

2. Set the stimulator frequency to two pulses per second.
3. Adjust the impulse duration to 2 milliseconds.
4. Set the switch to *repeat* on the stimulator.
5. Lay the sciatic nerve on the stimulator electrodes and slowly increase the voltage until you see the contraction of the muscle. If you have reached 5 to 8 volts and you still have no response, turn the voltage down, shut the stimulator off, and recheck your connections and settings. Once you see the muscle contract, then the lowest voltage that produces a response is the threshold stimulus.
6. Record this value in the space provided.

Threshold stimulus: ______________________

Recording Apparatus Setup

Your lab may be equipped with one or more different physiologic recorders. Follow your instructor's directions to set up the apparatus if you are to do the experiment in groups, or pay close attention if your instructor demonstrates the experiment. Pay particular attention to the settings of the apparatus.

Spatial Summation (Maximum Recruitment)

(A)[3] The **maximum recruitment** is the lowest voltage stimulus at which *all* of the muscle fibers are stimulated.

1. To demonstrate this, set the duration of the pulse to 2 milliseconds, keep the frequency at two pulses per second, and set the stimulus on repeat.
2. Begin the recording and increase the voltage until you see a response **(threshold)** in the muscle. Continue to increase the voltage slowly, and you should see the contraction force increase as indicated by an increase in the height of the tracing.
3. As you continue to increase the voltage slowly, the contraction tracings will not get any higher. The minimum voltage it takes to produce the maximum height is known as the maximum recruitment voltage.
4. Record the maximum recruitment voltage in the space provided.

Maximum recruitment voltage: ______________________

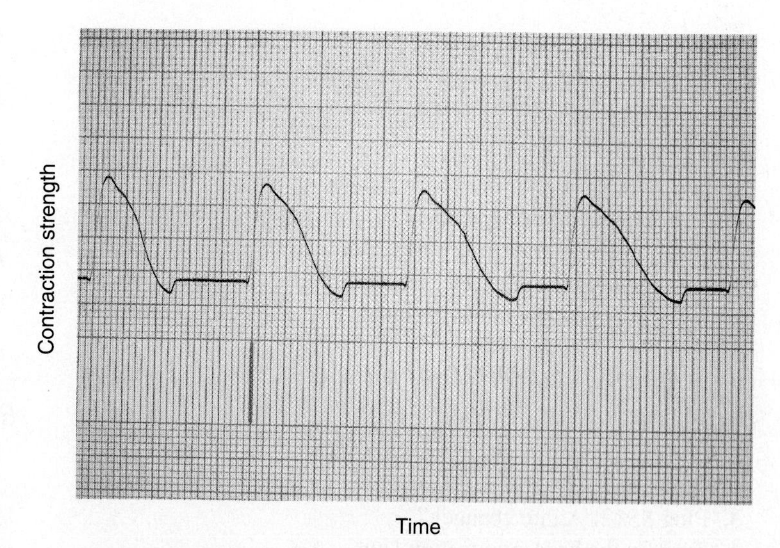

Figure 11.11 Phases of Muscle Contractions in a Single Twitch Contraction

(A)[4] Treppe

1. Set the voltage reading at the maximum recruitment voltage and the frequency at one pulse per second.
2. As you stimulate the muscle, notice that there is a stepwise increase in the peaks of the contractions as the muscle contracts over a period of time. This increase in contraction strength is known as **treppe.** Treppe is thought to occur due to the increased availability of calcium in the muscle fibers and is a possible rationale for "warming up" before exercising. Compare your recording with figure 11.10.

(A)[5] Phases of Muscle Contraction

1. If you increase the chart speed to 50 mm per second, you can obtain tracings of the muscle where the **lag phase, contraction phase,** and **relaxation phase** can be seen. If you can record the time when you stimulate the muscle, then you can calculate the lag phase of muscle contraction.
2. Make a tracing of the contraction and compare it with figure 11.11. If the chart speed is 50 mm per second in your tracing, what is the length of the lag phase? This assumes that you have a mark indicating the time of stimulation.

Lag phase: ______________________

How long is the contraction phase? Record the length of time.

Contraction phase: ______________________

How long is the relaxation phase? Record the length of time.

Relaxation phase: ______________________

Fatigue in Frog Muscle

Fatigue in muscles can be due to a decrease in oxygen to the muscle, a decrease in ATP, an increase in metabolic by-products (such as lactic acid and carbon dioxide), or a combination of all three.

(A)[6] You should set up the stimulator to the maximum recruitment voltage and increase the tension on the frog muscle.

Stimulate the muscle at 20 pulses per second. The muscle should contract and reach a peak and then the tracing should decrease over time. How long does it take for the muscle to fatigue? Record the time in the space provided.

Length of time for muscle to fatigue: ________________

Make sure that all members of your lab group see the frog demonstration and the recordings made from the experiment. Clean your dissection equipment and place the remains of the frog into the appropriate container designated by your instructor.

Biopac Setup for MP30 or MP150 Data Acquisition Unit

1. Plug in Data Acquisition Unit to wall outlet.
2. Connect Data Acquisition Unit to computer.
3. Plug SS12LA into channel 2.
4. Turn on the Data Acquisition Unit.

Preparation of Stimulator

1. Plug stimulator into wall outlet.
2. Attach Trigger to Analog Out of Data Acquisition Unit.
3. Attach ELSTM2 Needle electrodes.
4. Set the voltage range to 10 volts but turn the voltage knob to zero.

Computer Setup

1. Start the BSL *PRO* software. You should see an untitled window.
2. Select the menu and select "Show Stimulator."
3. Go to the File menu and choose **"Graph Template (*GTL) > File Name: FrogMuscle.gtl."**
4. Do NOT close the stimulator window.
5. Select 0–50 grams force range.
6. Select the Setup Channels from the menu and choose the wrench icon from the Channel menu and locate the scaling button. Make sure the stimulator is set to 0 volts (with the knob turned counterclockwise) and the pulse light NOT blinking. Click on the Cal 1 button.
7. Enter the Cal 2 value, which equals Cal 1 Input Value + 50.
8. Click "OK" on the scaling window.
9. Go to the Channel 2 window and locate the wrench icon and the scaling window.
10. With only the S hook on the tension adjuster, click Cal 1.
11. Attach a 50-gram weight to the adjuster and click Cal 2.
12. Click "OK" and exit the scaling window.
13. Attach the tension adjuster to the ring stand with the holes facing down. The adjuster should be set so that, when the frog is attached, the string will be vertical. Attach a small S hook to the adjuster.

Frog Preparation

1. Loop and tie the thread to the calcaneal tendon. Cut the tendon from the calcaneus.
2. Pin the frog down to a board. Make sure that the thigh and leg are anchored. Moisten the muscle with frog Ringer's solution.
3. Cut gently along the length of the thigh muscles and expose the sciatic nerve.
4. Gently dissect the nerve using a glass hook. Do not damage the nerve or tug on it.
5. Loop the thread to the transducer hook and adjust the tension so the thread is not slack. The thread should be vertical as it attaches to the hook.
6. Remove the covers from the electrode tips (remember to replace them at the end of the experiment) and place them underneath the sciatic nerve. Make sure that the electrode tips do not touch each other.

Running the Experiment

1. Click the "start" button.
2. Adjust the level to 0.1 volt.
3. Click the on/off button in the stimulator window.
4. Increase the voltage by 0.2 volt until you see a response in the CH 2 window.
5. When you get a response, this is the threshold voltage.
   Record the value here: ________________________.
6. Continue to increase the voltage until you get no further increase in contraction. This is the maximum recruitment.
   Record the value here: ________________________.
7. Reduce the voltage and stop the recording.

(A)[7] Summation and Tetanus

1. Go to the stimulator window and select Continuous Pulses.
2. Adjust the pulses to 1 Hz with the scroll bar.
3. Adjust the voltage to what you obtained for maximum recruitment.
4. Start the recording and increase the frequency slowly by 1–2 Hz every second until the muscle shows a continuous contraction.
5. Turn off the stimulator and stop the recording.
6. Save your data by selecting the File **menu.** Choose Save As and select the file type:

   BSL Pro files (*.ACQ) File name: (your name). Choose Save.

(A)[8] Temperature Effects on Muscle Contraction

1. Open up a new file in FrogMuscle.gtl.
2. Use the maximum recruitment voltage and stimulate the frog muscle at room temperature. Record the tracing.
3. Add 75 mL of warm frog Ringer's solution to the muscle and stimulate the muscle again. Record the tracing. Describe any change in the contraction strength below.
   Change in contraction strength:
4. Add 75 mL of iced frog Ringer's solution to the muscle and stimulate the muscle again. Record the tracing. Describe any change in the contraction strength below.

Change in Contraction Strength

(A)[9] Fatigue

1. Go to the stimulator window and select Continuous Pulses.

2. Adjust the pulses to 1 Hz with the scroll bar.
3. Adjust the voltage to what you obtained for maximum recruitment.
4. Start the recording and increase the frequency slowly by 1–2 Hz every second until the muscle fatigues.

Human Muscle Physiology

In this part of the exercise, you study the effects of muscle recruitment and fatigue. These experiments are best done with the use of a device that can measure electrical activity, such as a physiograph or computer and hardware unit, such as the Biopac MP30 or MP150 and the EMG1 student module.

(A) 10 *Multiple Motor Unit Summation*

1. Attach the appropriate recording electrodes to the forearm (a positive, negative, and ground) and connect the electrodes to the physiograph or hardware unit.
2. You will first need to establish a baseline recording by adjusting the sensitivity of the equipment to register the maximum grip strength that you can produce. To do this, you should grip your hand tightly while recording the activity and make sure that you have adequate displacement displayed. Ask your instructor for specific directions for the equipment in your lab.
3. Once you have produced a reasonable recording for baseline data, begin the recording by slightly gripping your hand. Pause for a second and then produce a slightly stronger grip. Pause once more for a second and then close your hand as tightly as possible. Stop the recording.
4. Examine the height of the displacement of the physiograph recording or the electronic recording produced. As more and more muscle fibers are recruited, the increase in the electrical activity produces a stronger signal.

What do you predict to be the difference between your favored arm (for example, the right arm in a right-handed person) and your weak arm?________________________

Produce a recording of your *maximum* grip strength in your other arm and compare the two recordings. ____________

Muscle Fatigue

(A) 11 You can examine the effects of muscle fatigue by using the same setup as described previously; instead of using grip strength, examine the nature of muscles when they contract maximally for long periods of time. Hold a 5- or 10-pound weight for a moment to establish your baseline recording. Once you have produced an adequate recording, begin the experiment as follows.

1. Hold a weight (5, 10, or 15 lb) with your forearm held at a 90° angle to your body.
2. Begin the recording and wait for the muscle to begin to fatigue. This is seen when the forearm begins to drop somewhat. Stronger individuals should hold heavier weights.
3. When the weight can no longer be held and the forearm begins to fall significantly, stop the recording.

How does the strength of the signal compare at the beginning of the recording with that at the end of the recording? ________________________________

____

____

Does the stability of the recording stay the same, or is there variation in the electrical activity? ____________________

____

____

How long does it take for the muscle to fatigue? __________

____

____

If you measured fatigue in the frog, compare your results with the frog in terms of time for fatigue to begin.

NOTES

Exercise 11 Review

Muscle Physiology

Name: ______________________

Lab time/section: ______________________

Date: ______________________

1. A skeletal muscle is stimulated to contract by what structure? ______________________

2. What chemical crosses the synapse, causing a muscle to contract? ______________________

3. Where is calcium released inside the muscle fiber to cause muscle contraction? ______________________

4. What are the two types of myofilaments found in muscle cells that cause muscle contraction? ______________________

____

5. Does a single muscle fiber or an entire muscle contract by an all-or-none response? ______________________

6. Name the first phase after a stimulus in a muscle contraction. ______________________

7. Define *subthreshold stimulus.* ______________________

8. What happens to the strength of contraction during wave summation? ______________________

____

9. Describe *tetanus.* ______________________

____

10. What is maximum recruitment? ______________________

____

____

11. How do tetanus and twitch, as demonstrated in the lab, correlate to human muscle contraction? Which one is more reflective of human muscle response? Why? ______________________

____

12. In the following illustration, which one is the contraction phase, the lag period, and the relaxation phase?

a. ______________________

b. ______________________

c. ______________________

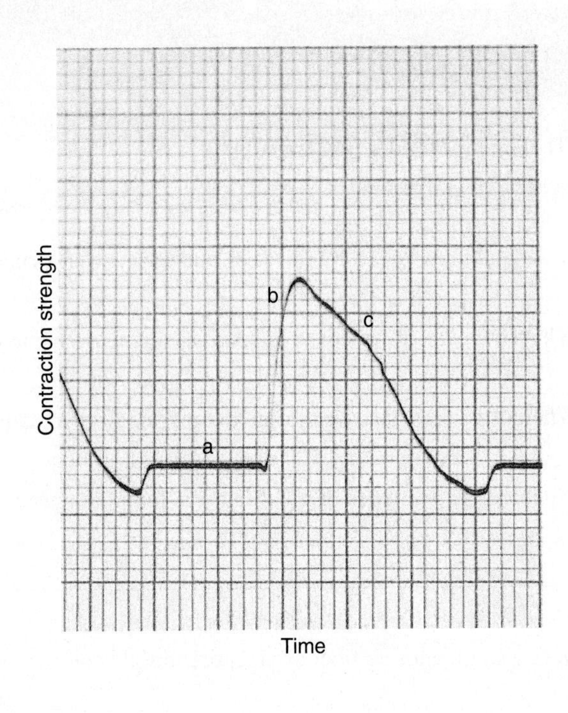

13. What was the threshold stimulus you obtained in lab? ______________________

14. What was the voltage at which you first got maximum recruitment? ______________________

____

____

15. Explain a muscle spasm in terms of recruitment of muscle fibers. ______________________

16. How does your favored arm compare with your weak arm in terms of electrical activity? ______________________

____

17. How does the muscle respond to weight when it is first contracting versus when it is fatigued? ______________________

____

____

18. As an athlete "warms up" before exercising, the muscles increase in temperature. What effect does this have on various phases of muscle contraction? ______________________

Exercise 12

Overview of Muscles and Muscles of the Shoulder and Upper Extremity

Introduction

The next four exercises focus on skeletal muscles. There are over 600 skeletal muscles in the human body. The muscles are grouped according to their locations, with most muscles occurring as pairs. In the following exercises, you learn about some of the major muscles of the body. The muscles presented in the following labs are covered in the Seeley text in chapter 10, "Muscular System: Gross Anatomy." The origin, insertion, action, and innervation are listed for about 100 muscles in this lab manual. Your instructor may wish to customize the list, so that you learn specific muscles or specific things about each muscle. The study of muscles can be both fun and challenging. Begin your study of muscles early and do not wait to learn them until the night before the lab exam.

Some students find the study of human musculature to be a daunting task, whereas others recall the muscle section of the course as their favorite part of the study of human anatomy and physiology. The satisfaction of studying muscles can be increased if you set attainable goals. This is best accomplished by choosing small numbers of muscles to study at a sitting and using flash cards as study aids. It is vital that you know the bones and bony markings before studying muscles. You may also want to make a quick sketch of the individual muscles to help you visualize them.

A thorough study of muscles involves knowing not only the name of the muscle but also its origin, insertion, action, and innervation. The muscles are listed in this exercise by name, followed by their origin. The **origin** is the attachment point of the muscle that typically does not move during muscular contraction. The origin of the muscle is the stable point and it is usually either proximal to the torso or medial in its position. The muscles also are listed with their **insertion,** which is the attachment that usually moves during contraction.

Each muscle has an **action,** which represents the effects the muscle has on a part of the body (such as *flexion* of the arm). To fully understand muscles, you must know their actions at joints and associate the actions with the body in anatomical position. When you are learning the action of a muscle, imagine a piece of string tied between the origin and the insertion of the muscle. Think of how the bones move as you pull the insertion closer to the origin. Mimic the action of the muscle as you learn it. When you perform the action of a particular muscle, you should be able to feel that muscle tighten. Finally, the **innervation** of a muscle is the name of the nerve that controls it. If the motor portion of a nerve is severed, the muscle innervated by that nerve cannot function. Examine an articulated skeleton as you study the muscles to find the origin, insertion, and action. Visualize the muscle as it attaches to the origin and the insertion. As you study each muscle, locate it on your body and try to determine which part is the stable origin and which part is the moving insertion.

Another aid to learning muscles is to understand how they are named. Muscles are named by a number of criteria:

Location: Some muscles are named by their location, such as the **tibialis** anterior, the **frontalis,** or the **temporalis.**

Size: Muscles such as the adductor **magnus** are named for their size. The gluteus **minimus** is the smallest of the gluteal muscles.

Shape: The **deltoid** muscle is shaped like a delta, or triangle.

Orientation: The **transversus** abdominis muscle has fibers that are oriented horizontally, or in a transverse direction.

Origin and *insertion:* The **infraspinatus** muscle originates on the infraspinous fossa of the scapula, whereas the flexor **hallucis** longus muscle inserts on the hallux, or big toe.

Number of heads: The **triceps** brachii muscle has three heads.

Action: The **extensor** digitorum is a muscle that extends the fingers.

Length: Muscles with the term **longus** or **brevis** are named for the overall length of the muscle.

Skeletal muscle is voluntary muscle in that it contracts when consciously stimulated by specific nerves. In this exercise, you are introduced to concepts that apply to all of the muscles, and you then study a specific region, learning the major muscles of the shoulder and upper limb. Some of these muscles originate on the scapula and move or stabilize the humerus. Others originate on the humerus and insert on the forearm.

The muscles of the forearm and hand are very important, particularly in terms of rehabilitating limbs after surgery to relieve carpal tunnel syndrome or after trauma to the limb. You should learn the origins and insertions as you learn the muscles of the forearm, since many of these muscles look quite similar. By knowing the origins and insertions of a muscle, you will not easily mistake it for another muscle. You can compare the musculature of the human with that of the cat.

Objectives

At the end of this exercise, you should be able to

1. locate and name the muscles of the shoulder and upper extremity on a torso model, chart, cadaver, or cat (if available);
2. list the origin, insertion, and action of each muscle presented;
3. describe what nerve controls each muscle;
4. list what muscles function as synergists or antagonists to the prime mover;
5. name all the muscles that have an action on a joint, such as all of the muscles that laterally rotate the arm or muscles that flex the hand;
6. distinguish between the cat musculature and the human musculature and compare them for similarities and differences;
7. reproduce the actions of a muscle by moving the appropriate limbs to demonstrate the actions;
8. name two muscles (prime mover and antagonist) that fix a joint.

Materials

Human torso model and upper extremity model
Human muscle charts
Articulated skeleton
Cadaver (if available)
Cat
Materials for cat dissection:
- Dissection trays
- Scalpel and two or three extra blades
- Gloves (household latex gloves work well for repeated use)
- Blunt (Mall) probe
- String and tags
- Pins
- Forceps and sharp scissors

First aid kit in lab or prep area
Sharps container
Animal waste disposal container
Cat wetting solution

Procedure

Overview of the Muscles

Most people are familiar with the larger skeletal muscles, such as the pectoralis major, gluteus maximus, deltoid, and trapezius. You should examine the charts and models in lab and locate as many of the most common muscles of the body as you can on the charts or models and in figure 12.1. You should refer to this figure in the subsequent exercises if you need an overview of the muscles. You should also review the muscle nomenclature and actions at joints as described in Exercise 10. There are two views as to muscle action. One is that muscles act on an appendage, such as flexion of the hand. Another perspective is that muscles act on a joint, such as flexion of the wrist. These can be seen as equivalent. Muscles can have many actions at a joint. In addition to moving a joint, muscles can also fix a joint.

Human Muscles

Examination of Muscles

(A)1 Look at the charts, models, and cadaver (if available) in the lab and locate the muscles presented in this section. Review the bones of the hand as well as the bones of the arm and forearm in Exercise 7 to locate precisely the bony origins and insertions. Refer to the following descriptions as you examine the muscles. The specific details of the muscles are provided in table 12.1.

Muscles of the Shoulder

Some muscles of the shoulder act on the scapula. Each **trapezius** is a triangle-shaped muscle that originates in the midline of the vertebral column and head and inserts laterally. If the scapula is fixed, the head moves, but, if the vertebral column and head are fixed, the trapezius moves the scapula. Figure 12.2 illustrates the trapezius.

The **rhomboideus** muscles are deep to the trapezius, so reflection (cutting and folding back) of the trapezius is necessary to see them. These muscles originate on the vertebral column and insert on the medial border of the scapula. As they contract, they adduct the scapulae (pull them to the midline). The **rhomboideus major** is larger than and more inferior to the **rhomboideus minor.** Compare the material in lab with figure 12.2.

The **levator scapulae** is named for what it does, elevate the scapula. The levator scapulae originates on the lateral side of the neck and inserts on the upper scapula. If the neck is fixed, the levator scapulae raises the scapulae, as in shrugging. If the scapula is fixed, the muscle rotates or abducts the neck.

The **deltoid** is located on top of the shoulder and abducts the arm, among other actions. Locate the deltoid on material in the lab and compare it with figure 12.3. It is a fan-shaped muscle whose distal attachment partially covers the insertion of the **pectoralis major.** The pectoralis major has fibers that run horizontally across the chest region, and it is a superficial muscle of the chest. Deep to the pectoralis major muscle is the **pectoralis minor** muscle, whose fibers run in a more vertical direction, as seen in figure 12.4.

Another muscle of the posterior surface is the **latissimus dorsi,** which arises from the vertebrae by way of a broad, flat **lumbodorsal fascia.** The latissimus dorsi originates on the back, but the insertion is on the anterior aspect of the humerus. It is a powerful extensor of the arm and is known commonly as the swimmer's muscle (figures 12.2 and 12.5).

Deep to the trapezius and the deltoid are muscles that originate on the scapula proper. The **supraspinatus** is named for its origin on the supraspinous fossa, and it is a synergist to the deltoid. The **infraspinatus** is a muscle originating on the infraspinous fossa. It runs laterally and posteriorly to the humerus, and it laterally rotates the arm. The **subscapularis** is named for its origin on the subscapular fossa. The insertion of the subscapularis is on the anterior surface of the humerus. When it contracts, it medially rotates the arm. The subscapularis originates on the anterior surface of the scapula, between the scapula and the ribs, and thus cannot be seen in a posterior view. Locate these muscles in the lab and compare them with figure 12.5.

The scapula gives rise to two other muscles, the teres minor and the teres major. The **teres minor** appears like a slip of the infraspinatus, whereas the **teres major** crosses to the medial,

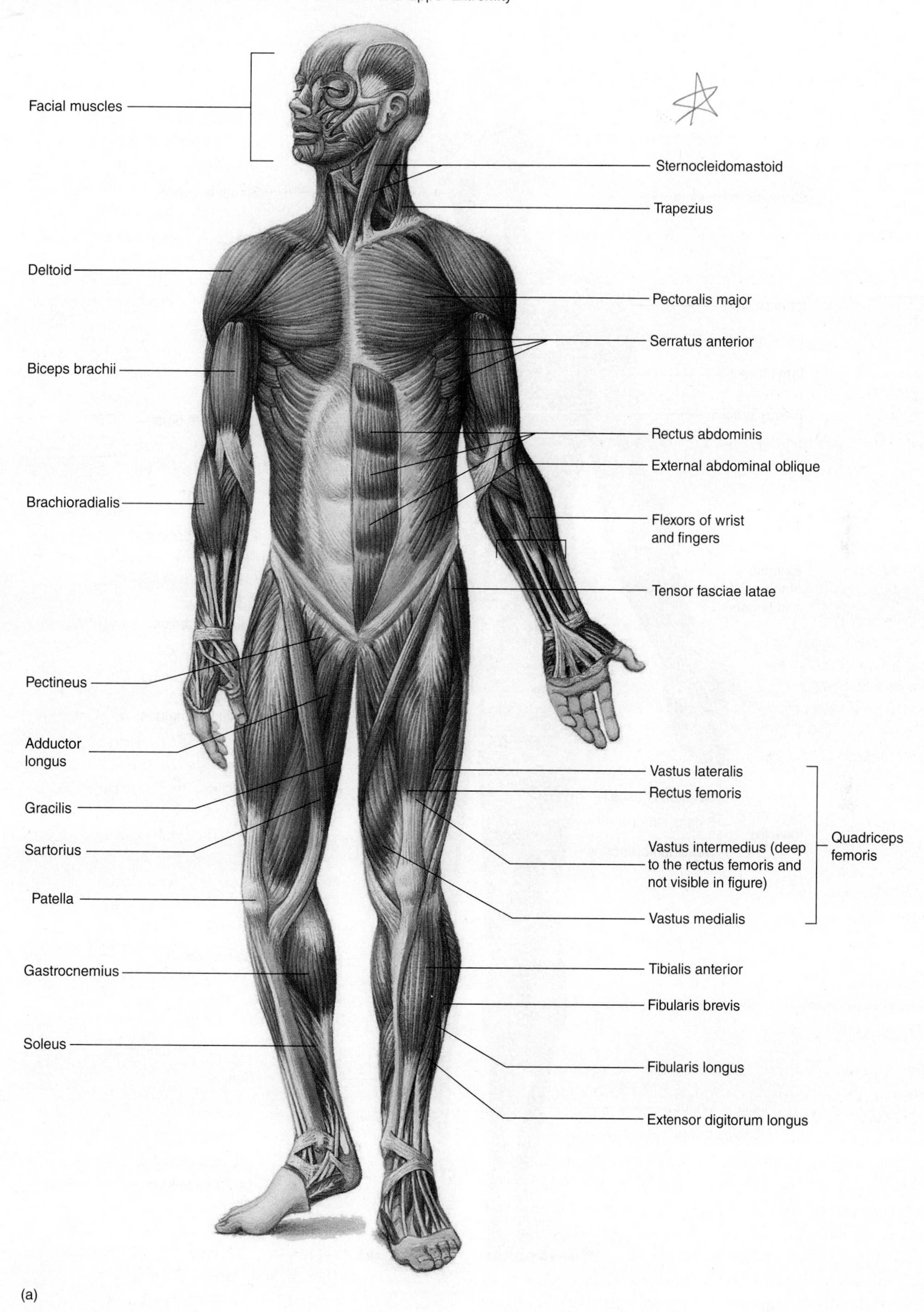

Figure 12.1 Overview of the Muscles of the Body
(a) Anterior view; (b) posterior view.

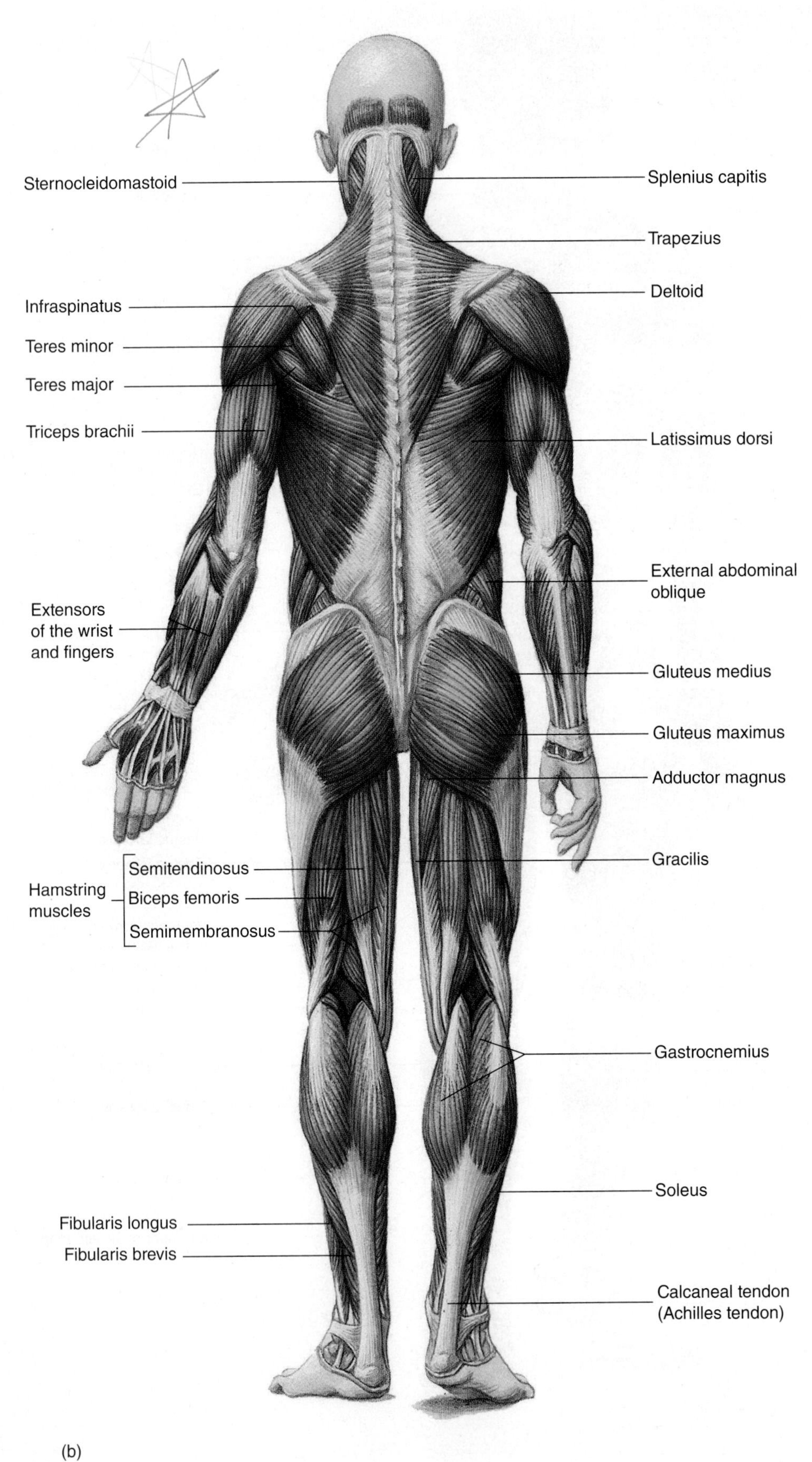

Figure 12.1 *Continued.*

TABLE 12.1 Muscles of the Shoulder and Upper Extremity

Name	Origin	Insertion	Action	Innervation
Posterior Muscles Acting on the Shoulder and Arm				
Trapezius	Occipital protuberance, nuchal ligament, spinous processes of C7–T12	Clavicle, acromion process, spine of scapula	Extends and abducts head, rotates and adducts scapula, fixes scapula	Accessory (XI) nerve
Latissimus dorsi	T7–12, L1–5, sacrum, iliac crest, ribs 10–12	Intertubercular groove of humerus	Extends, adducts, and medially rotates arm; depresses shoulder	Thoracodorsal nerve
Deltoid	Clavicle, acromion process, spine of scapula	Deltoid tuberosity of humerus	Abducts, flexes, and extends arm; medially and laterally rotates arm	Axillary nerve
Rhomboideus major	Spines of T1–4	Inferior, medial border of scapula	Adducts scapula	Dorsal scapular nerve
Rhomboideus minor	Nuchal ligament, spines of C6–7	Medial border of scapula at scapular spine	Adducts scapula	Dorsal scapular nerve
Levator scapulae	C1–4	Superior angle of scapula	Elevates scapula, abducts and rotates neck	Spinal nerves C3–5, dorsal scapular nerve
Supraspinatus*	Supraspinous fossa	Greater tubercle of humerus	Abducts arm (stabilizes shoulder joint)	Suprascapular nerve
Infraspinatus*	Infraspinous fossa	Greater tubercle of humerus	Extends and laterally rotates arm (stabilizes shoulder joint)	Suprascapular nerve
Subscapularis*	Subscapular fossa	Lesser tubercle of humerus	Extends and medially rotates arm (stabilizes shoulder joint)	Subscapular nerve
Teres minor*	Lateral border of scapula	Greater tubercle of humerus	Extends, laterally rotates, and adducts arm (stabilizes joint shoulder)	Axillary nerve
Teres major	Lateral border of scapula	Crest of lesser tubercle of humerus	Extends, adducts, and medially rotates arm	Subscapular nerve
Triceps brachii	Infraglenoid tuberosity of scapula, lateral and posterior surface of humerus	Olecranon process of ulna	Extends and adducts arm, extends forearms	Radial nerve
Anterior Muscles Acting on the Shoulder and Arm				
Pectoralis major	Clavicle, sternum, cartilages of ribs 1–7	Crest of greater tubercle of humerus	Flexes, adducts, and medially rotates arm	Anterior thoracic nerves
Pectoralis minor	Ribs 3–5	Coracoid process of scapula	Depresses glenoid cavity, raises ribs 3–5	Anterior thoracic nerves
Biceps brachii	Supraglenoid tubercle, coracoid process of scapula	Radial tuberosity	Flexes arm, flexes forearm, supinates hand	Musculocutaneous nerve
Coracobrachialis	Coracoid process of scapula	Midmedial shaft of humerus	Flexes and adducts arm	Musculocutaneous nerve
Anterior Muscles Acting on the Forearm				
Brachialis	Anterior, distal surface of humerus	Coronoid process of ulna	Flexes forearm	Musculocutaneous and radial nerves
Brachioradialis	Lateral supracondylar ridge of humerus	Styloid process of radius	Flexes forearm	Radial nerve
Supinator	Lateral epicondyle of humerus, anterior ulna	Proximal radius	Supinates hand	Radial nerve
Pronator teres	Medial epicondyle of humerus, coronoid process of ulna	Lateral, middle shaft of radius	Pronates hand, flexes forearm	Median nerve
Pronator quadratus	Distal part of anterior ulna	Distal radius	Pronates hand	Median nerve (anterior interosseus branch)

TABLE 12.1 Muscles of the Shoulder and Upper Extremity *Continued.*

Name	Origin	Insertion	Action	Innervation
Posterior Muscles Acting on the Shoulder and Arm				
Palmaris longus	Medial epicondyle of humerus	Palmar fascia	Flexes hand	Median nerve
Flexor carpi radialis	Medial epicondyle of humerus	Metacarpals 2 and 3	Flexes and abducts hand	Median nerve
Flexor carpi ulnaris	Medial epicondyle of humerus and ulna	Pisiform, hamate, metacarpal 5	Flexes and adducts hand	Ulnar nerve
Flexor digitorum superficialis	Medial epicondyle of humerus, coronoid process of ulna, proximal radius	Middle phalanges of digits 2–5	Flexes proximal and middle phalanges 2–5, flexes hand	Median nerve
Flexor digitorum profundus	Ulna, interosseus membrane	Distal phalanges of digits 2–5	Flexes phalanges, flexes hand	Median and ulnar nerves
Flexor pollicis longus	Anterior portion of radius and interosseus membrane	Distal phalanx of pollex (thumb)	Flexes thumb and hand	Median nerve
Flexor pollicis brevis	Trapezium	Proximal phalanx of digit 1	Flexes thumb	Median and ulnar nerves
Abductor pollicis brevis	Scaphoid, trapezium	Proximal phalanx of digit 1	Abducts thumb	Median nerve
Opponens pollicis	Trapezium	Metacarpal 1	Opposes thumb	Median nerve
Abductor digiti minimi	Pisiform	Proximal phalanx of digit 5	Abducts digit 5	Ulnar nerve
Flexor digiti minimi brevis	Hamulus of hamate	Proximal phalanx of digit 5	Flexes digit 5	Ulnar nerve
Opponens digiti minimi	Hamulus of hamate	Metacarpal 5	Opposes digit 5	Ulnar nerve
Posterior Muscles Acting on the Forearm and Hand				
Extensor carpi radialis longus	Lateral supracondylar ridge of humerus	Metacarpal 2	Extends and abducts hand	Radial nerve
Extensor carpi radialis brevis	Lateral epicondyle of humerus	Metacarpal 3	Extends and abducts hand	Radial nerve
Extensor carpi ulnaris	Lateral epicondyle of humerus, proximal ulna	Metacarpal 5	Extends and adducts hand	Radial nerve
Extensor digitorum	Lateral epicondyle of humerus	Middle and distal phalanges of digits 2–5	Extends phalanges 2–5, extends hand	Radial nerve
Abductor pollicis longus	Posterior radius and ulna, interosseus membrane	Metacarpal 1	Abducts and extends thumb, abducts hand	Radial nerve
Extensor pollicis longus and brevis	Posterior radius and ulna, interosseus membrane	Proximal and distal phalanges of pollex (thumb)	Extends thumb	Radial nerve

*Rotator cuff muscles

anterior side of the humerus in a way similar to the latissimus dorsi (figure 12.5). The **rotator (musculotendinous) cuff** muscles stabilize the shoulder joint by holding the head of the humerus in the glenoid fossa. Muscles composing the cuff are the supraspinatus, infraspinatus, subscapularis, and teres minor. A rotator cuff injury occurs if there is damage to any of these muscles.

The **biceps brachii** muscle is a two-headed muscle of the arm. It neither originates nor inserts on the humerus, yet it has an action on the arm. Flexion of the forearm occurs primarily from the contraction of this muscle. The biceps brachii has a long tendon that originates on the scapula and runs between the intertubercular groove of the humerus and a short tendon that originates on the coracoid process. The insertion of the biceps brachii is on the radius.

Alongside the short head of the biceps brachii is a small slip of muscle known as the **coracobrachialis.** This muscle is a short, diagonal muscle of the arm. Deep to the biceps brachii on the anterior surface of the arm is the **brachialis** muscle. The brachialis crosses the anterior surface of the elbow joint and flexes the forearm. The **brachioradialis** is a distal muscle of the arm and is the most lateral muscle of the forearm.

The **triceps brachii** is the only major muscle on the posterior surface of the humerus. The triceps brachii extends the arm and forearm and thus is an antagonist to the biceps brachii (figure 12.6).

Examine figures 12.6 and 12.7 for these muscles and compare them with the descriptions in table 12.1.

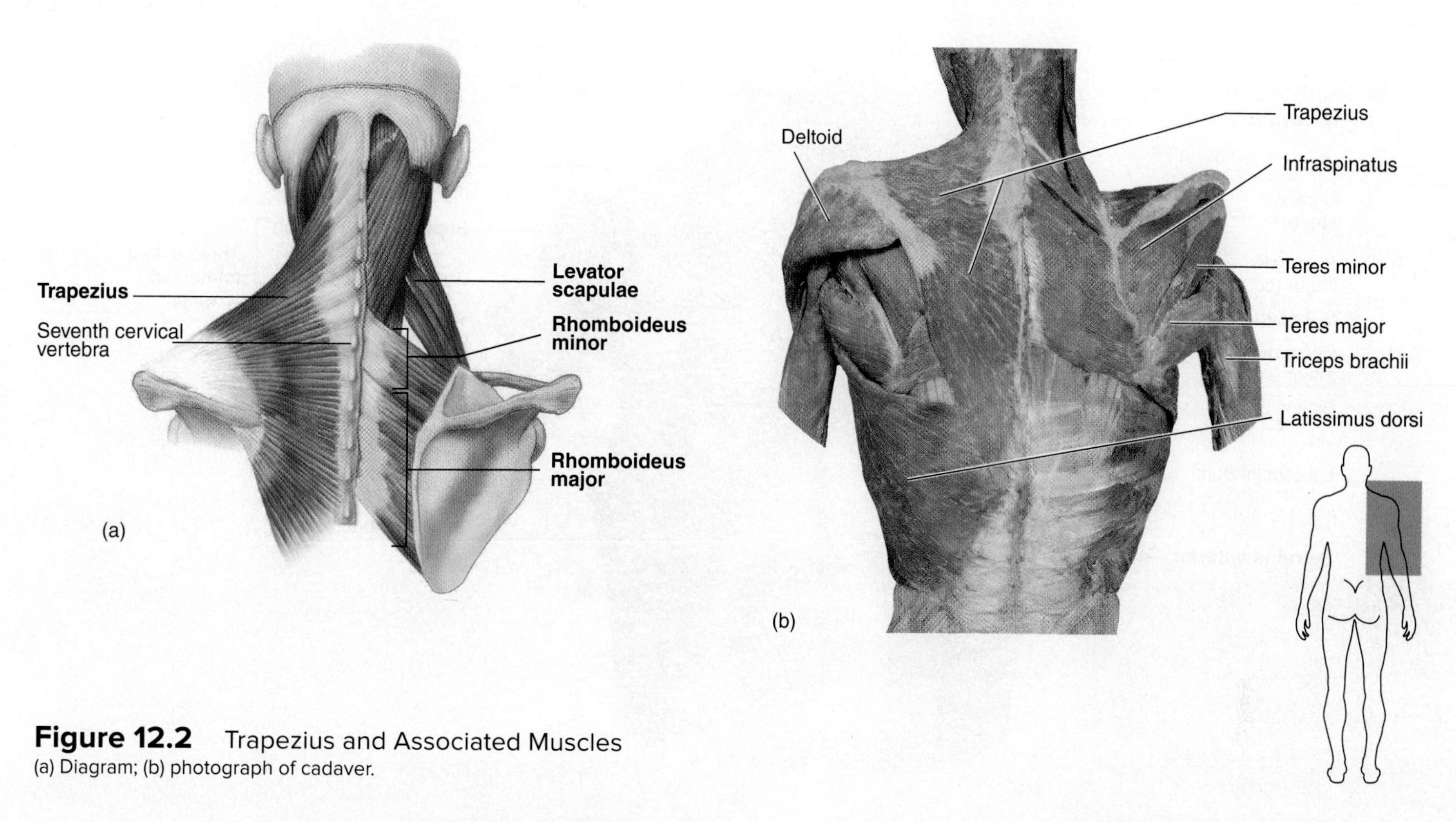

Figure 12.2 Trapezius and Associated Muscles
(a) Diagram; (b) photograph of cadaver.

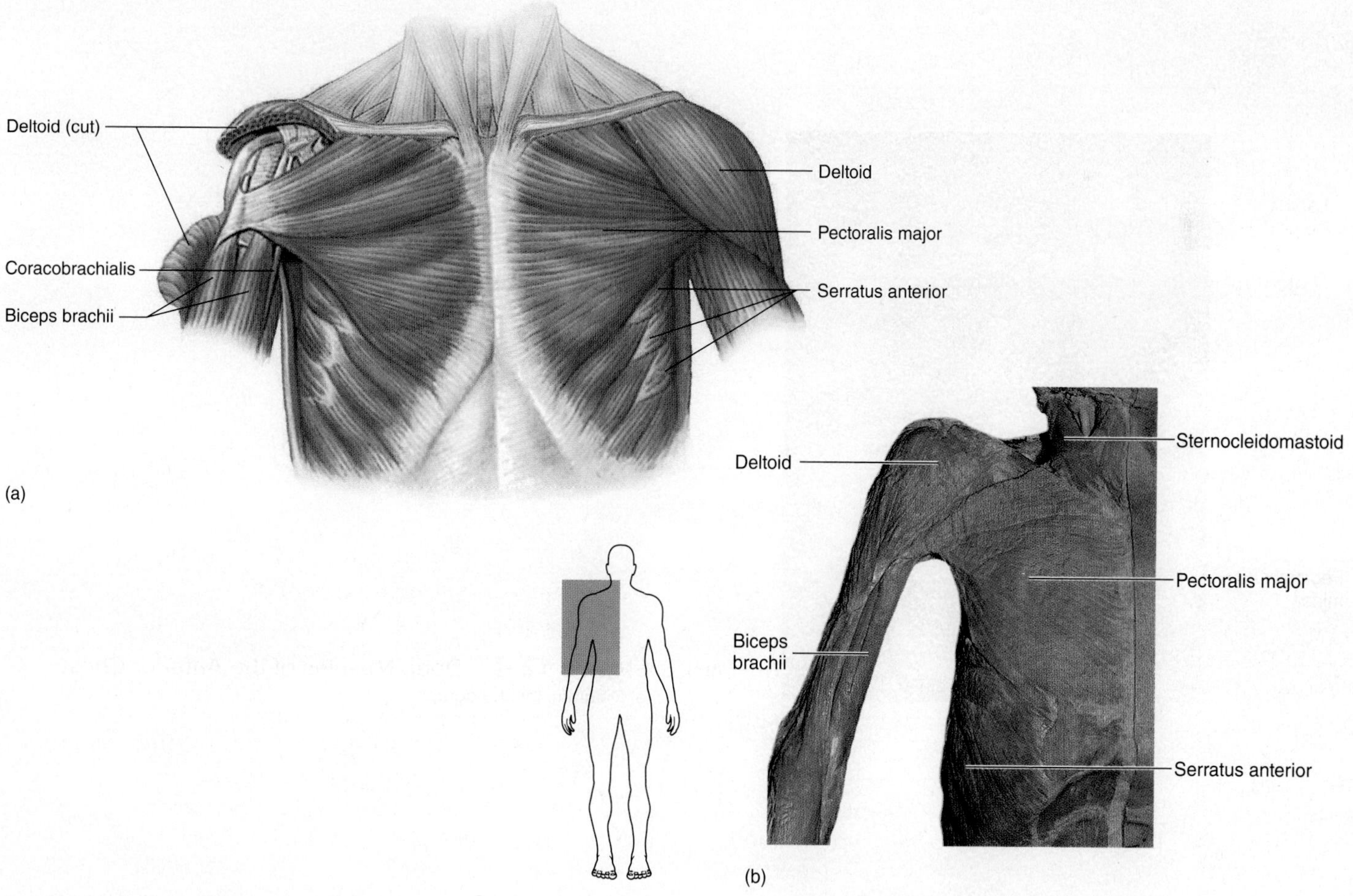

Figure 12.3 Superficial Muscles of the Anterior Chest
(a) Diagram; (b) photograph of cadaver.

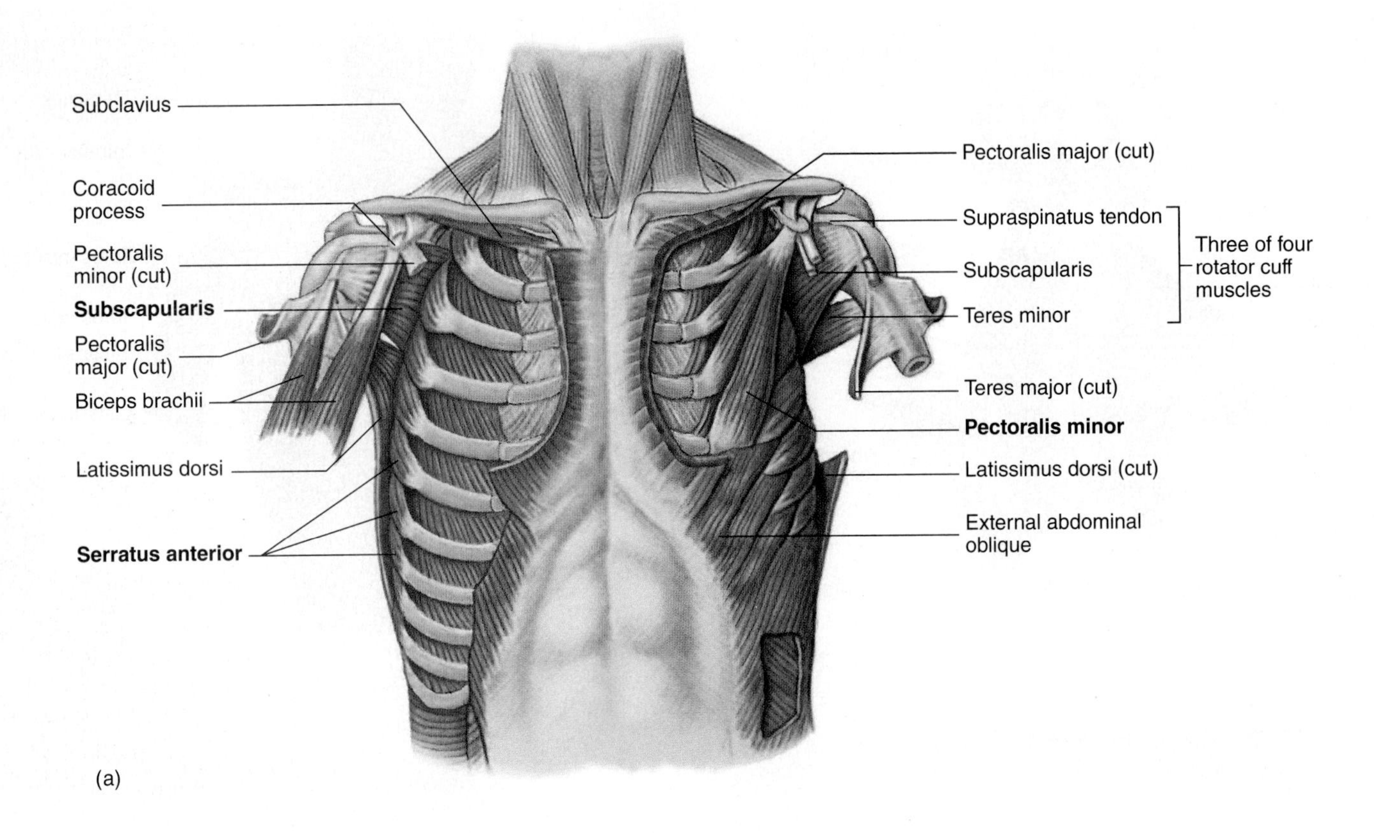

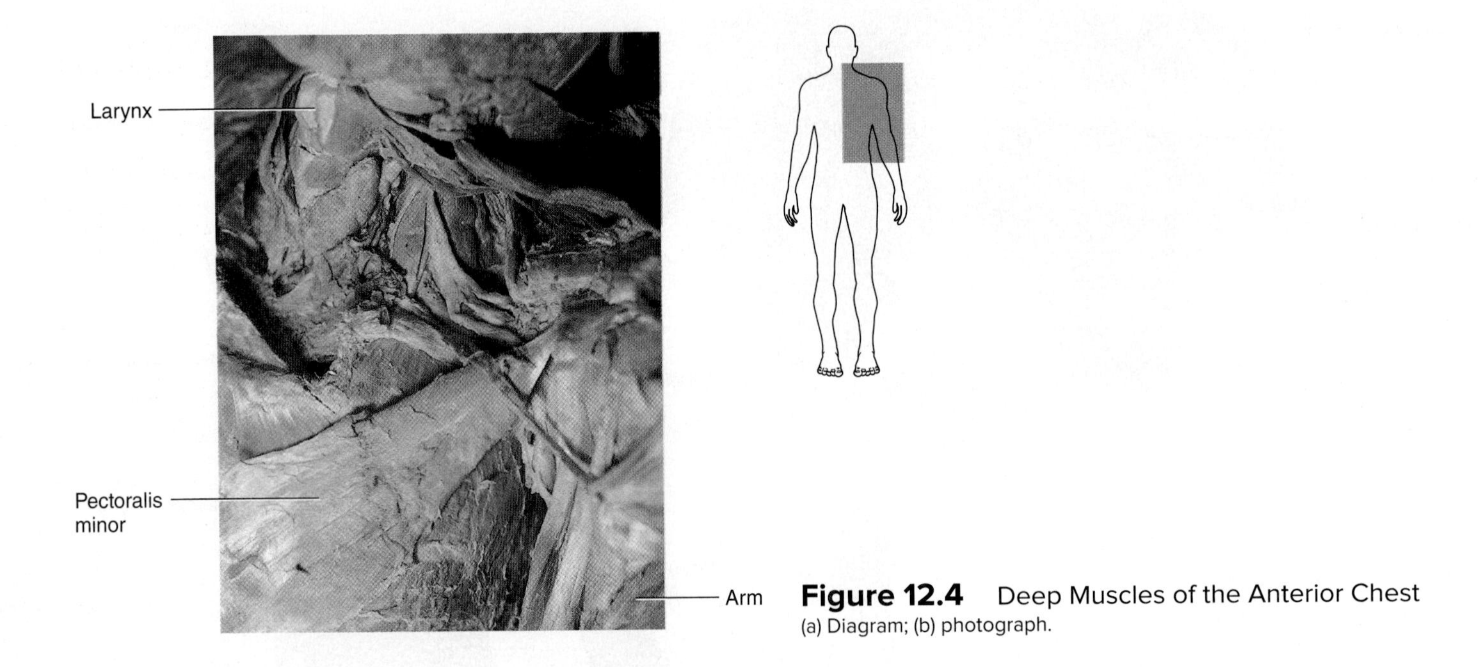

Figure 12.4 Deep Muscles of the Anterior Chest
(a) Diagram; (b) photograph.

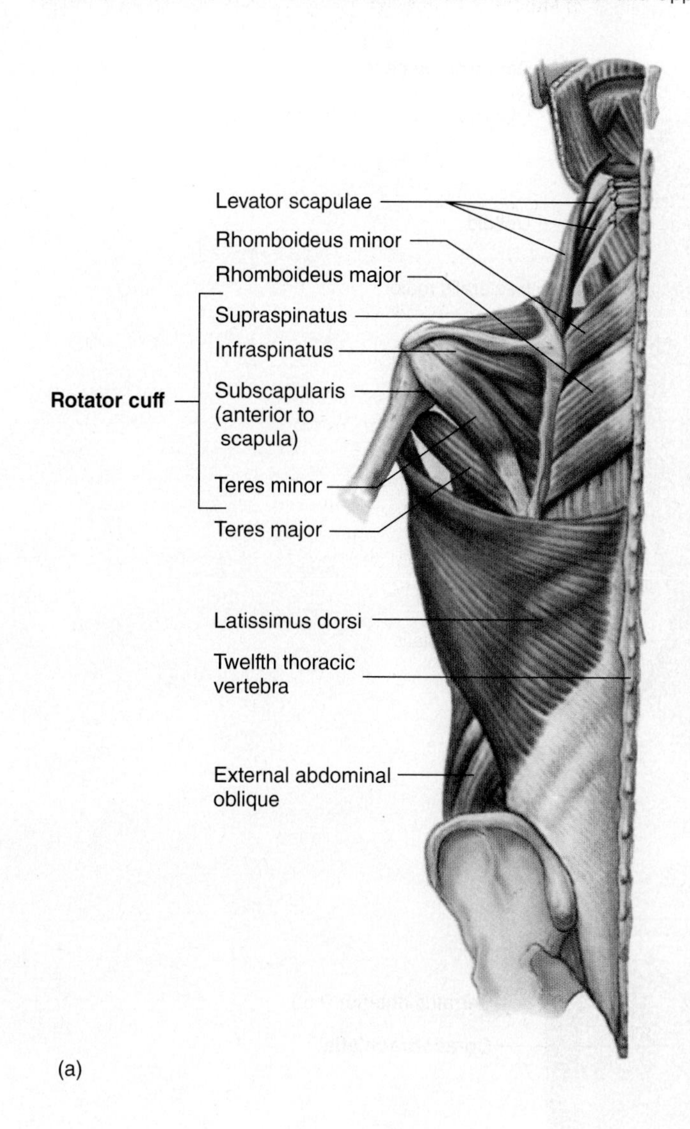

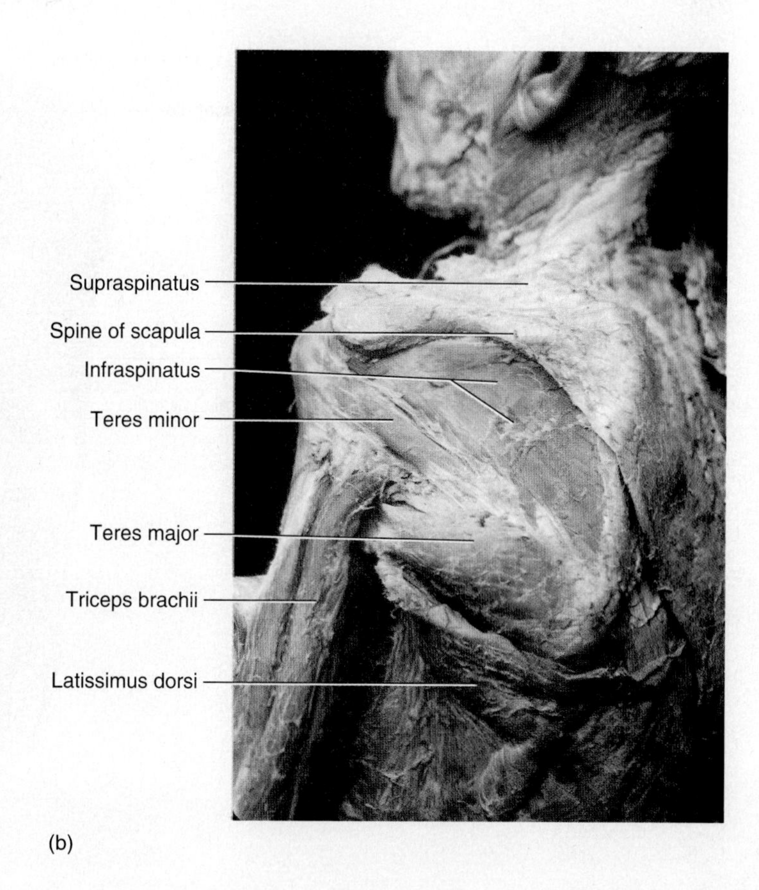

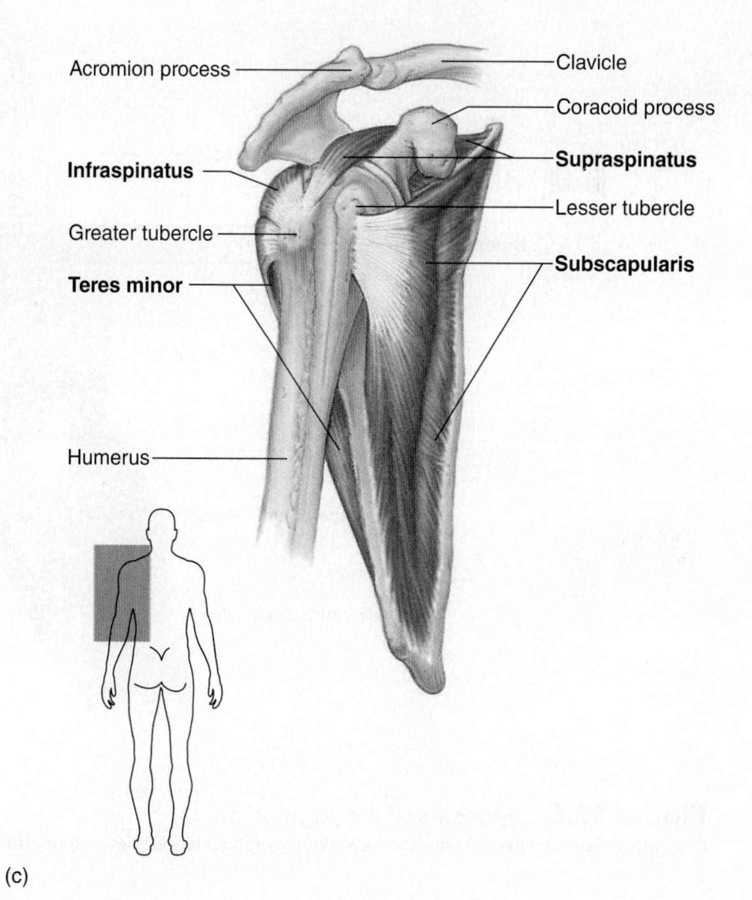

Figure 12.5 Scapular Muscles
(a) Diagram of left posterior view; (b) photograph of left posterior view; (c) diagram of right lateral view.

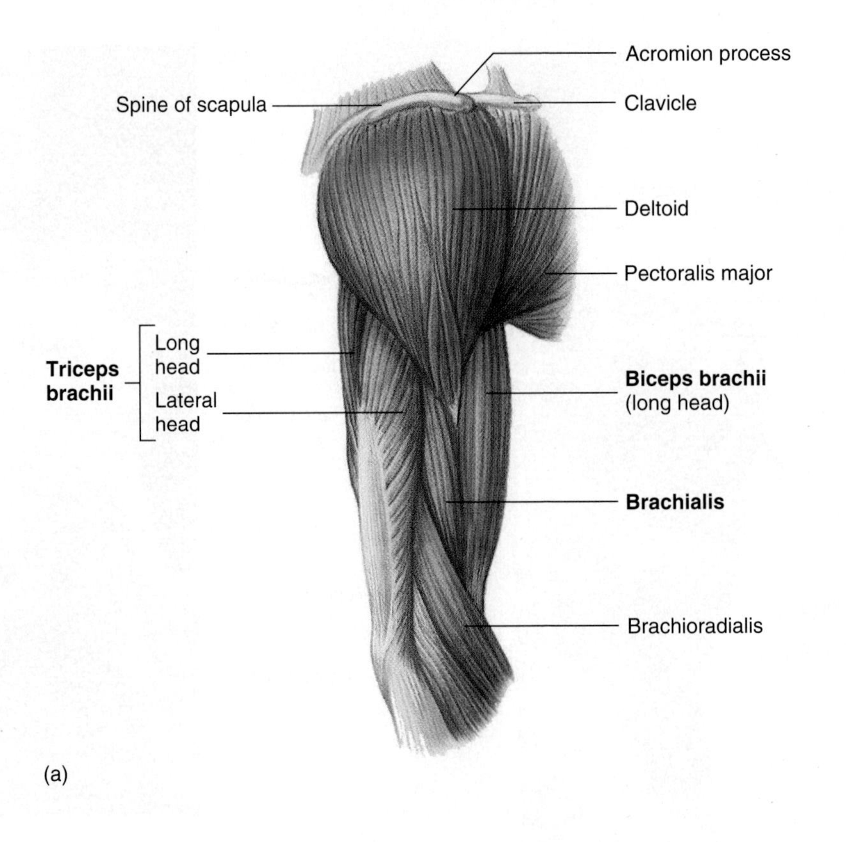

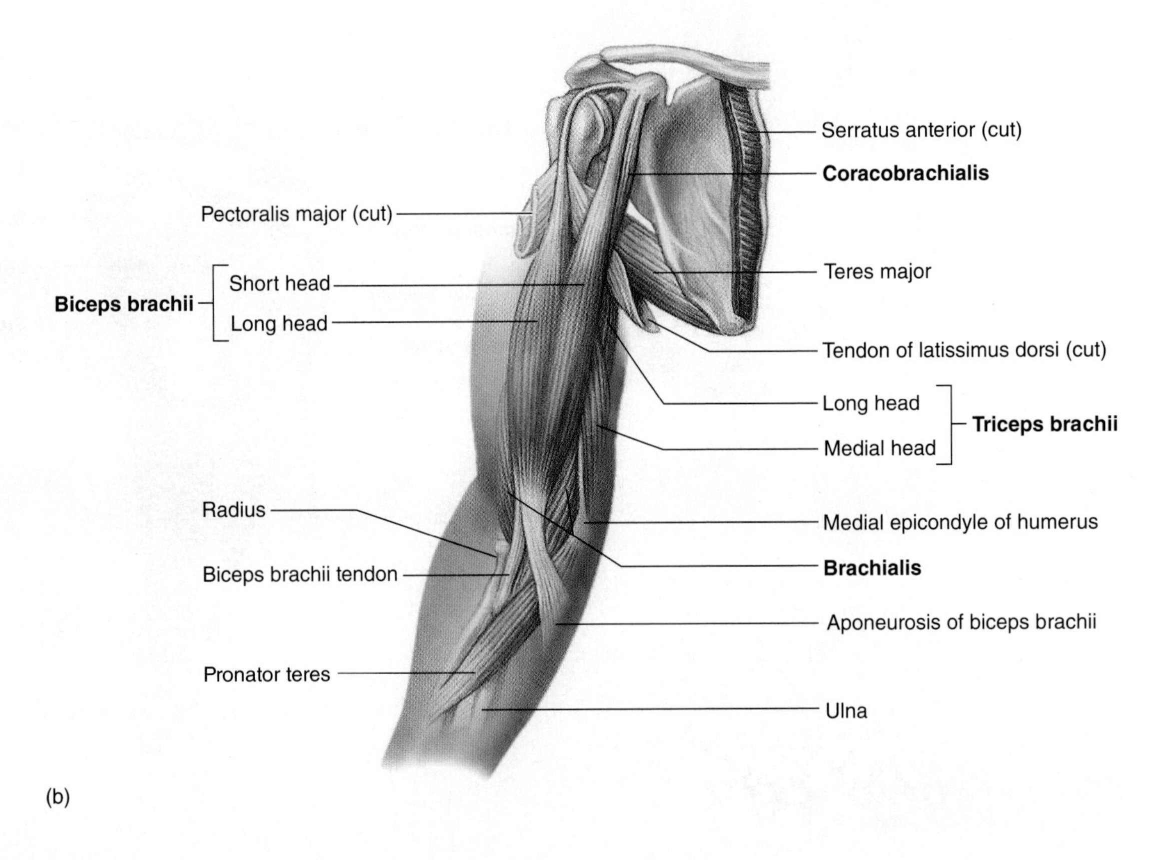

Figure 12.6 Muscles of the Right Arm
Diagram (a) lateral view; (b) anterior view. Photograph (c) lateral view; (d) posterior view.

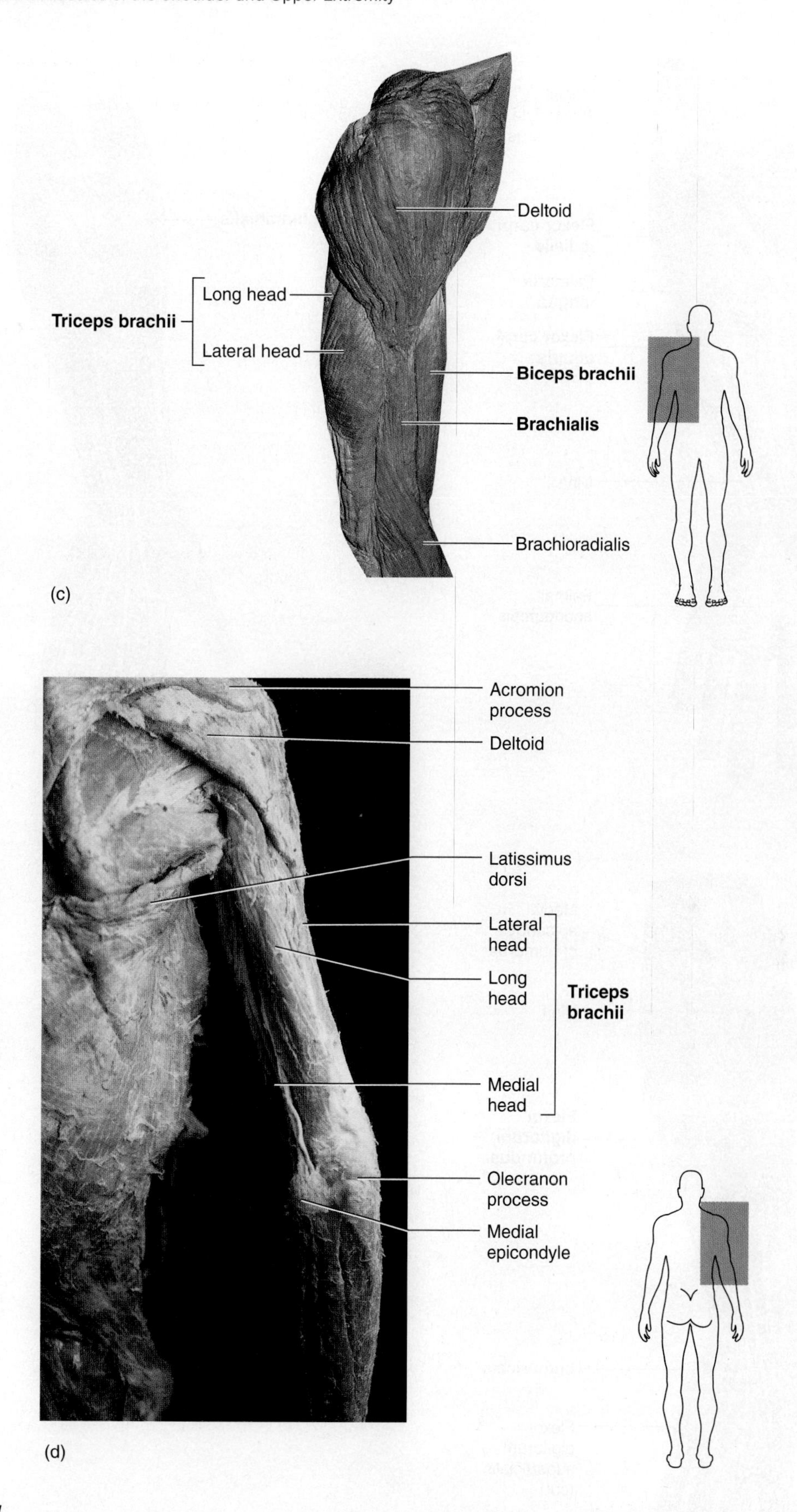

Figure 12.6 *Continued.*

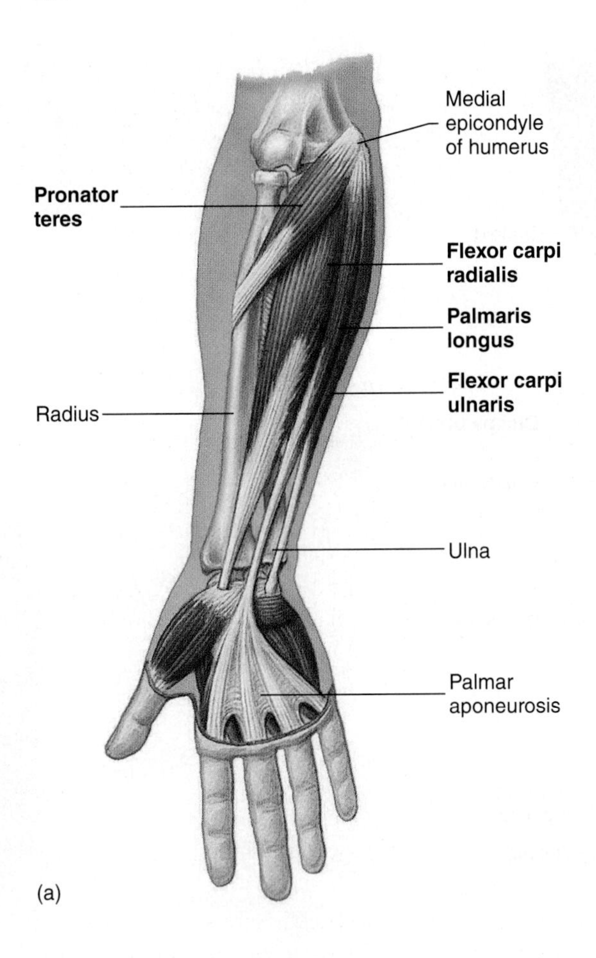

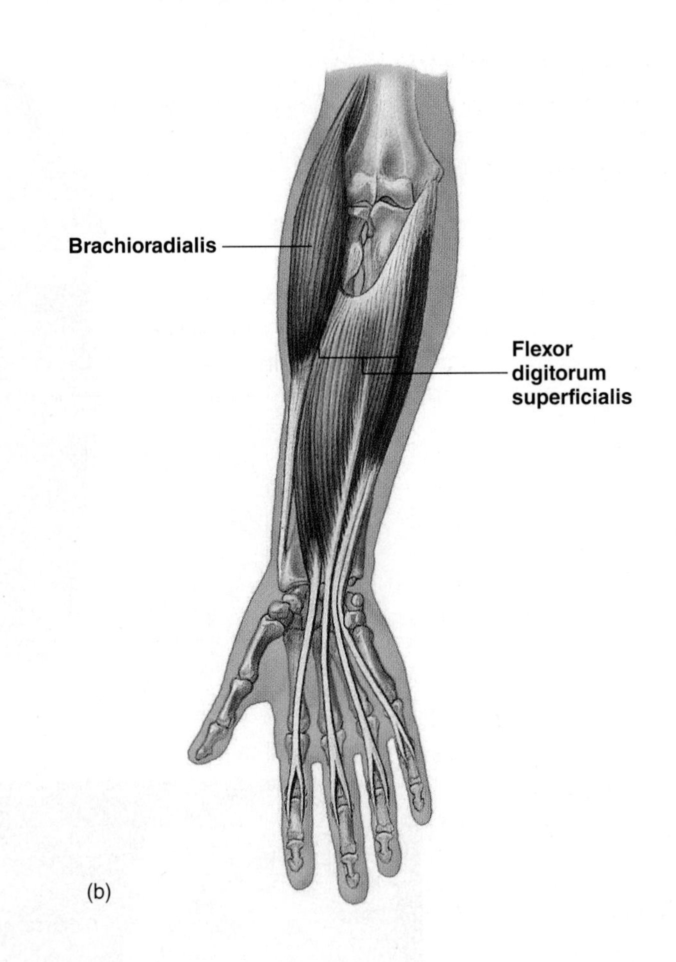

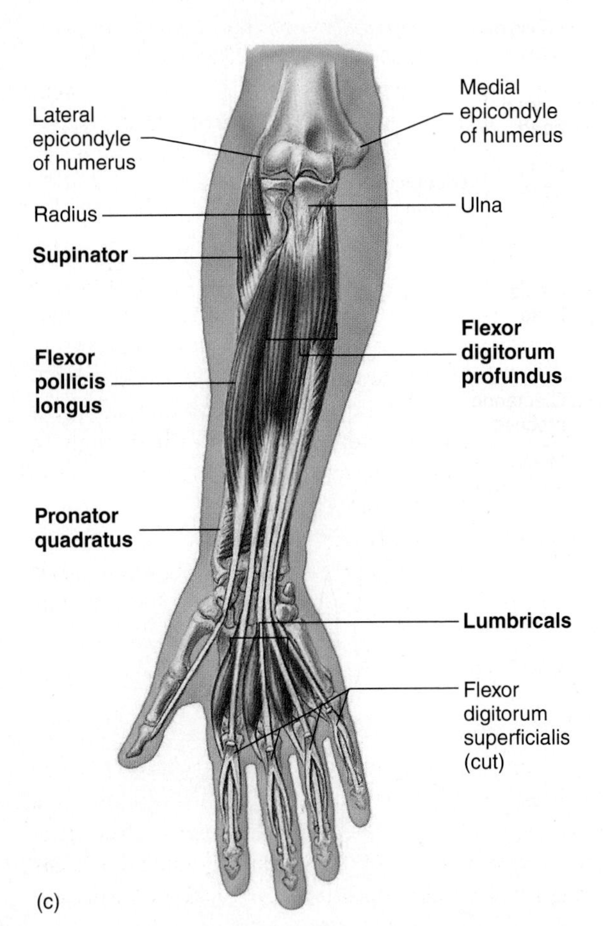

Figure 12.7 Muscles of the Forearm, Anterior View

(a) Superficial group; (b) muscles of middle depth; (c) deep muscles.

Muscles of the Forearm and Hand

The muscles of the forearm, due to their similar appearance, provide greater challenges than do the muscles of the shoulder and arm. (A)2 As you study these muscles, it is important that you locate them and determine their origins and insertions. They will help you determine if you are looking at the correct muscle. The muscles of the forearm and hand are generally named for their action (**pronate** the hand or **flex** the digits) or for their insertion (**carpi** for inserting on carpals or metacarpals, **digitorum** for fingers, and **pollicis** for thumb). In a few instances, they are named for the shape of the muscle (**teres** for round or **quadratus** for square).

Notice that much of the muscle mass for the fingers is located on the forearm. The fingers move by a force applied from a distance, analogous to a puppet controlled by puppet strings. If the muscle mass for the hand muscles were located on the hands proper, then the hands would look like softballs. By having the muscle mass in the forearm, an efficiency of form is achieved that allows for a powerful grip yet precise movements of the fingers.

Muscles That Supinate and Pronate the Hand

The first group of muscles in this study includes those muscles that insert on the radius. The **supinator** is a muscle that originates on the arm and forearm and supinates the hand. It wraps around the radius and is the deepest, proximal muscle of the forearm. Examine this muscle in the lab and compare it with figure 12.7.

The **pronator teres** is a round muscle that pronates the hand. It originates on the medial side of the arm and forearm and inserts on the lateral side of the radius. The pronator teres is a bit different from the other superficial forearm muscles in that it runs at an oblique angle on the forearm, whereas the other muscles run parallel to the bones of the forearm.

The **pronator quadratus** is a square muscle deep to the other forearm muscles on the distal part of the radius and the ulna. It pronates the hand. Compare the two pronator muscles in the lab with figures 12.7 and 12.8.

Flexor Muscles

The next muscles to be studied are grouped by their insertion on the hand. The superficial **palmaris longus** muscle is absent in about 10% of the population. It is centrally located in the middle, anterior forearm and inserts into a broad, flat tendon known as the **palmar aponeurosis.** This aponeurosis has no bony attachment but, rather, attaches to the fascia of the underlying muscles.

The **flexor carpi radialis** muscle inserts on the metacarpals on the radial side of the hand. The pulse of the radial artery is taken at the wrist just lateral to the tendon of the flexor carpi radialis muscle. Most of the flexor muscles of the hand originate from the medial epicondyle of the humerus, including the flexor carpi radialis. The tendons of this muscle run underneath a connective tissue band known as the **flexor retinaculum,** which anchors the tendons to the wrist and prevents the tendons from pulling away from the wrist when the hand is flexed.

The **flexor carpi ulnaris** muscle also has an origin on the medial epicondyle of the humerus and inserts on the carpals and a metacarpal of the ulnar side of the hand. The flexor carpi ulnaris is

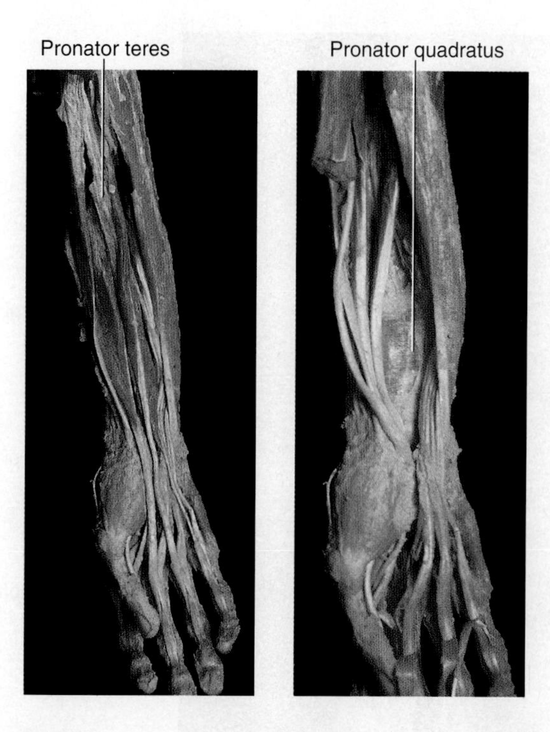

Figure 12.8 Muscles of Pronation, Anterior View

a medial muscle of the forearm. Examine these muscles in the lab and in figures 12.7 and 12.9.

The **flexor digitorum superficialis** is a superficial flexor muscle of the digits. It is not the most superficial muscle of the forearm but is deep to the palmaris longus, the flexor carpi radialis, and the flexor carpi ulnaris. The flexor digitorum superficialis is superficial to the deep digit flexor, which is discussed next. As with the other flexor muscles, the flexor digitorum superficialis has an origin on the medial epicondyle of the humerus. The insertion tendons form a V on the middle phalanges of digits 2 through 5.

The **flexor digitorum profundus** is a deep muscle of the digits. It is an exception to the other hand flexors in that it does not originate on the medial epicondyle of the humerus but on the ulna and the membrane between the radius and the ulna (interosseus membrane). It is a deep muscle of the forearm that runs underneath the flexor digitorum superficialis. At the insertion point, the tendons of the flexor digitorum profundus actually run through the split tendons of the flexor digitorum superficialis and extend to the distal phalanges of digits 2 through 5.

The **flexor pollicis longus** originates on the radius and the interosseus membrane and runs along the lateral forearm to insert on the anterior of the thumb. Flexion of the thumb is the curling of the thumb as if you were ready to flip a coin. Examine these muscles and compare them with figures 12.7 and 12.10.

Intrinsic Muscles of the Hand

There are many muscles that originate on the hand and have an action on the hand. Some of these arise on the pad of muscles at the base of the thumb known as the **thenar eminence.** These muscles are named for their action on the thumb, including the **flexor pollicis brevis,** the **abductor pollicis brevis,** and the **opponens**

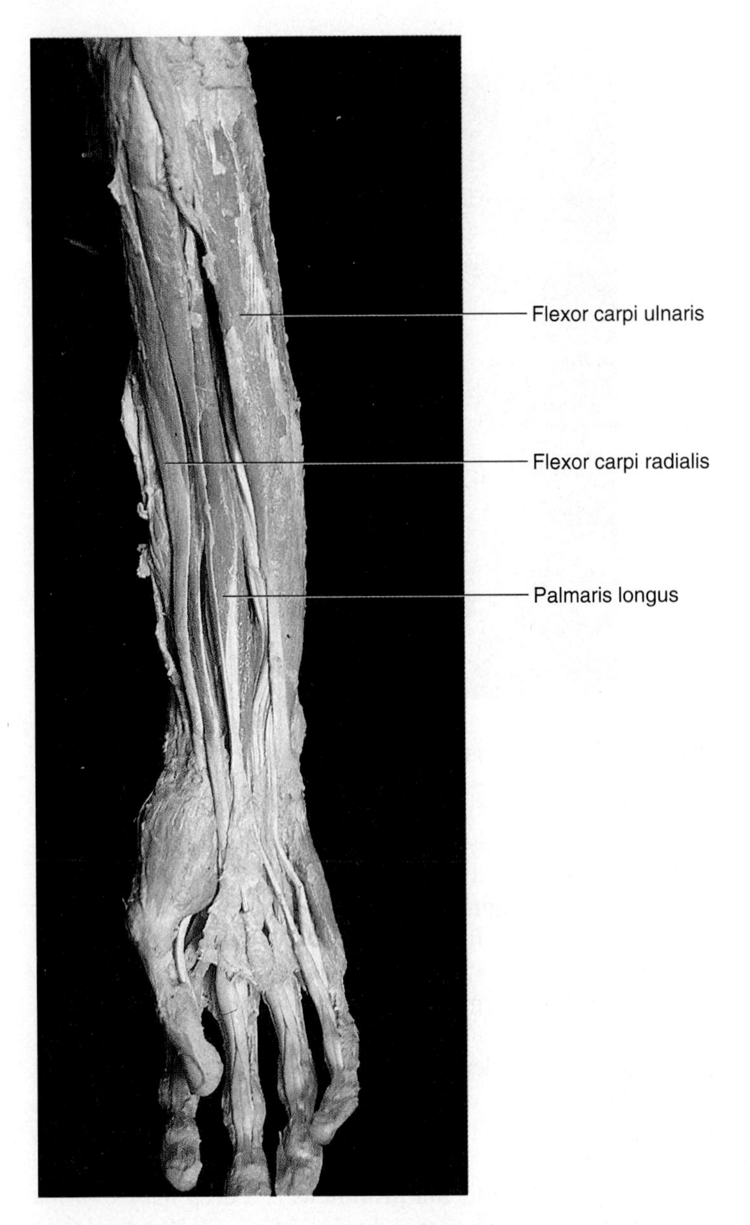

Figure 12.9 Superficial Flexor Muscles of the Right Forearm and Hand, Anterior View

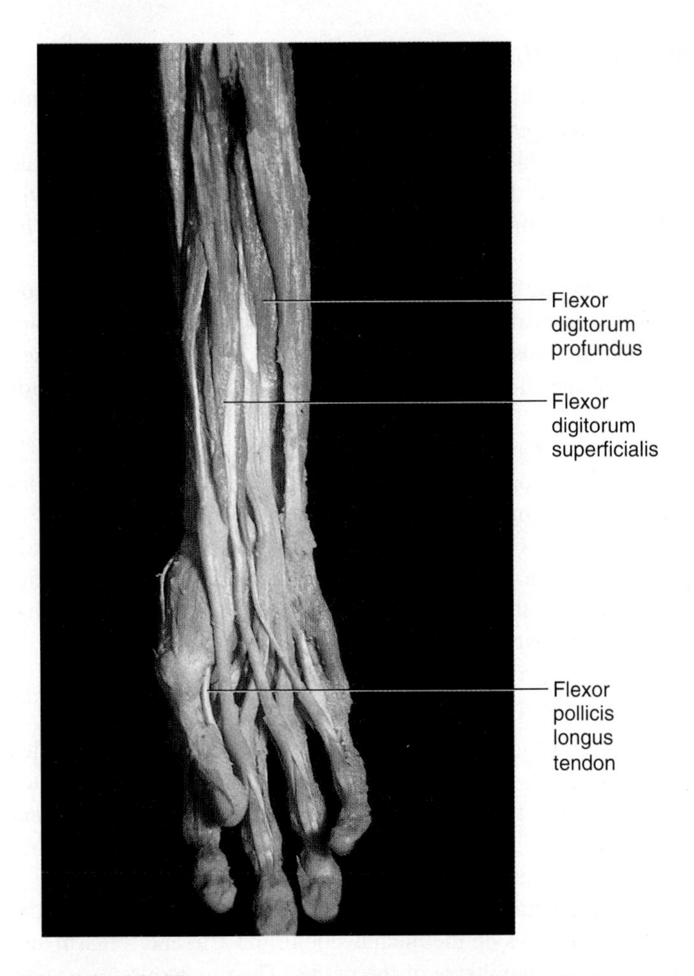

Figure 12.10 Deep Flexor Muscles of the Right Forearm and Hand

(up-pō´-nenz) **pollicis.** The pad of tissue at the base of digit 5 is known as the **hypothenar eminence,** and the muscles that arise there are the **abductor digiti minimi,** the **flexor digiti minimi brevis,** and the **opponens digiti minimi.** Find these muscles on models or cadavers in lab and compare them with figure 12.11.

Extensor Muscles

As a general rule, most of the extensors originate on the lateral epicondyle or lateral supracondylar ridge of the humerus. The extensor muscle tendons are held to the posterior surface of the wrist by a connective tissue band known as the **extensor retinaculum.** The **extensor carpi radialis longus** muscle originates on the humerus and inserts on the second metacarpal (on the radial side) of the hand. The **extensor carpi radialis brevis** is deep to the longus and inserts on the dorsum of the third metacarpal. Both of these muscles extend and abduct the hand. Examine these muscles in figure 12.12.

The **extensor carpi ulnaris** originates on the lateral epicondyle and inserts on the fifth metacarpal (on the ulnar side) of the hand. As it contracts, the hand is extended and adducted. The **extensor digitorum** is a singular muscle on the back of the hand (remember, there are two flexor digitorum muscles). As an extensor muscle, it originates on the lateral epicondyle of the humerus. It inserts on the middle and distal phalanges of the second through fifth digits. The tendons of this muscle can be seen on the dorsal surface of the hand and in figure 12.12.

The **abductor pollicis longus** muscle inserts on the metacarpal of the thumb. By pulling your thumb away from the index finger, in an anterior direction, you are abducting the thumb. There are two extensor pollicis muscles, the **extensor pollicis longus** and the **extensor pollicis brevis.** Extension of the thumb is done when you flip a coin. At the end of the flip, the thumb is extended. The tendons of the extensor pollicis muscles and the abductor pollicis muscles form a depression in the form of a triangle at the base of the thumb. This depression is known as the anatomical snuff box. These muscles can be seen in figures 12.12 and 12.13.

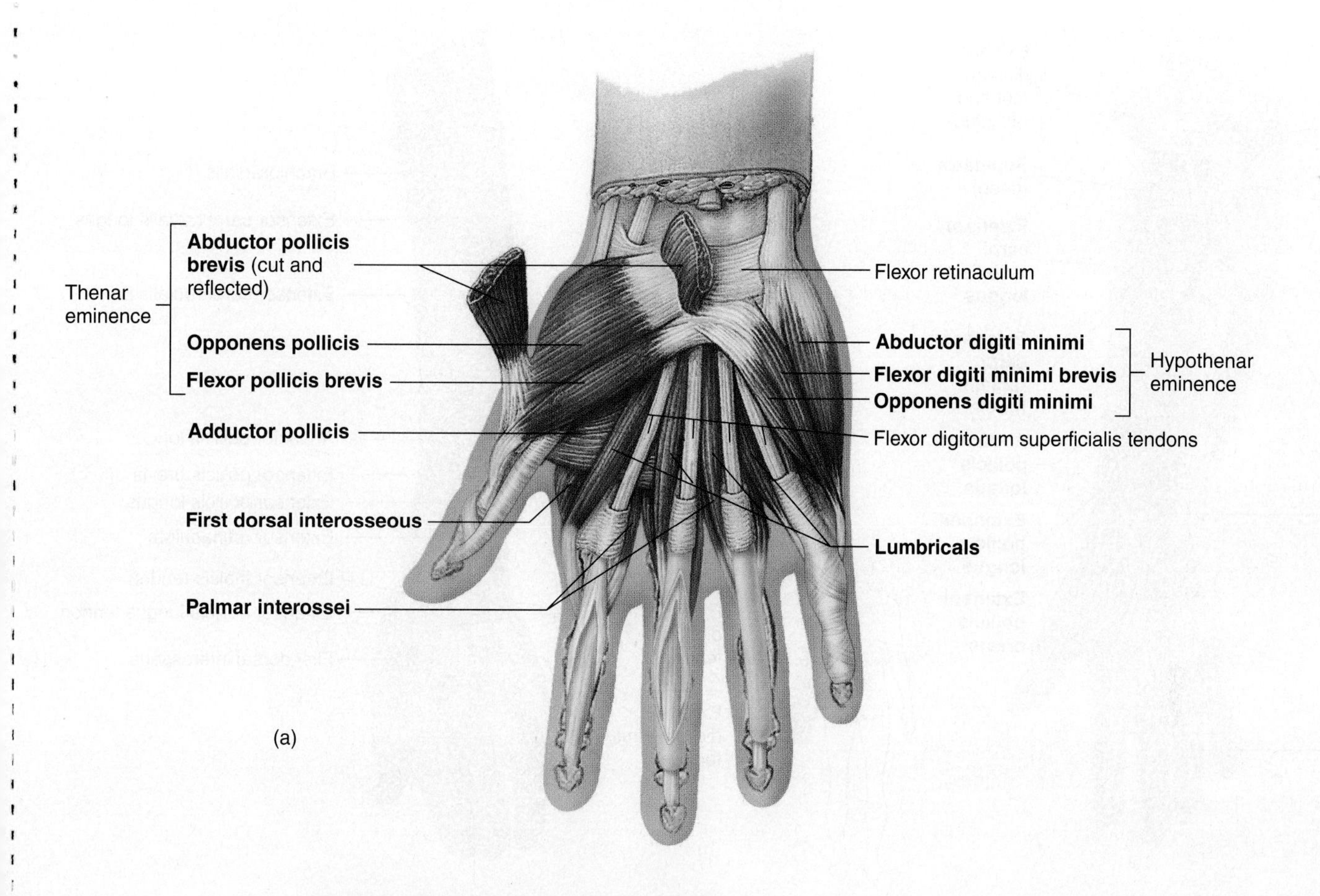

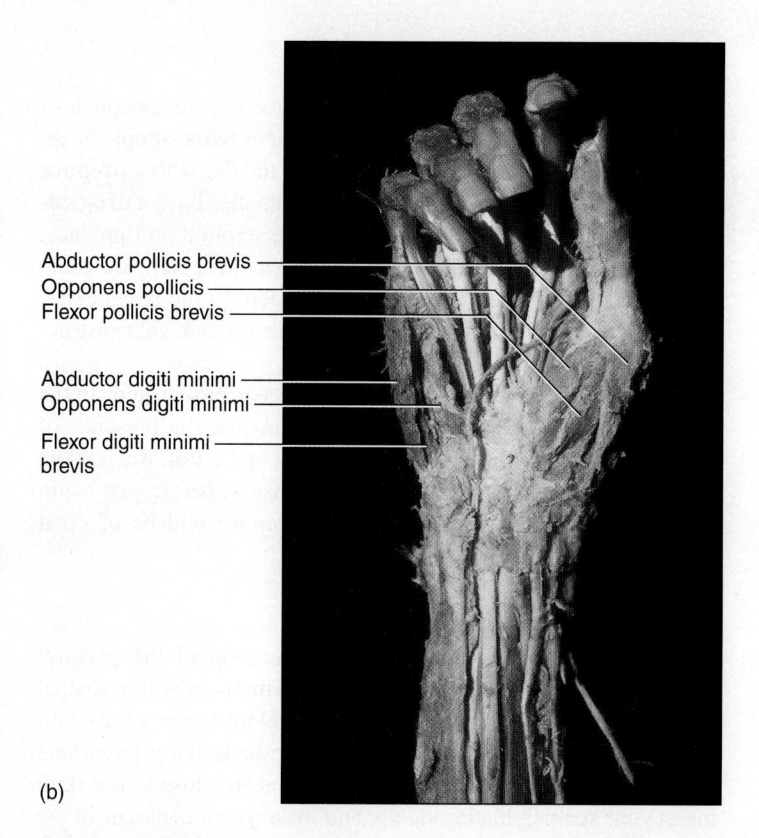

Figure 12.11 Intrinsic Muscles of the Right Hand
(a) Diagram; (b) photograph.

Cat Dissection

Ⓐ[3] The objective in using a cat for the exercise on musculature is to provide a study specimen for dissection and application to the human system. Differences occur between cat musculature and human musculature, but you should focus on similar structures in order to gain an appreciation of human musculature. Read all of the material in the exercise before beginning the dissection. Dissection is a skill in which you try to separate the overlying structures from those underneath while keeping intact as much material as you can.

Dissection Concerns

Wear an apron or an old overshirt as you dissect. Dissection instruments are sharp, and you must be careful when using scalpels. Do not cut *down* into the specimen but, rather, lift structures gently and try to make incisions so the scalpel blade cuts *laterally.* Cut *away* from yourself and your lab partners. *If you do cut yourself, notify the instructor immediately!* Wash the cut with antimicrobial soap and seek medical advice.

The laboratory should be well ventilated; if your eyes burn and you develop a headache, get some fresh air for a moment. Dissecting without having your face directly over the specimen may help. Occasionally, students have allergic reactions to formaldehyde. This usually consists of feeling that your

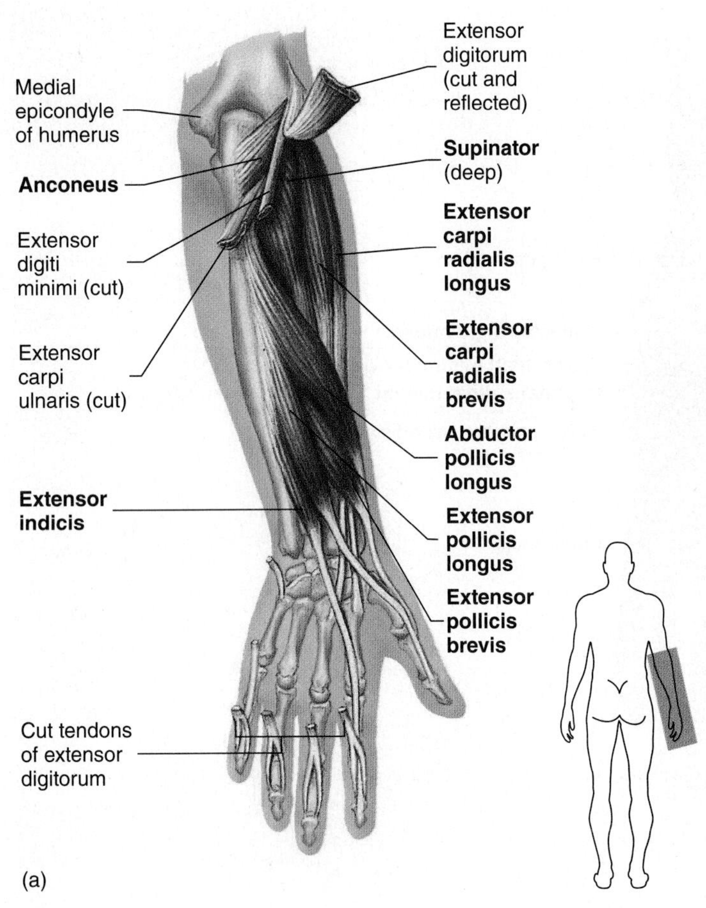

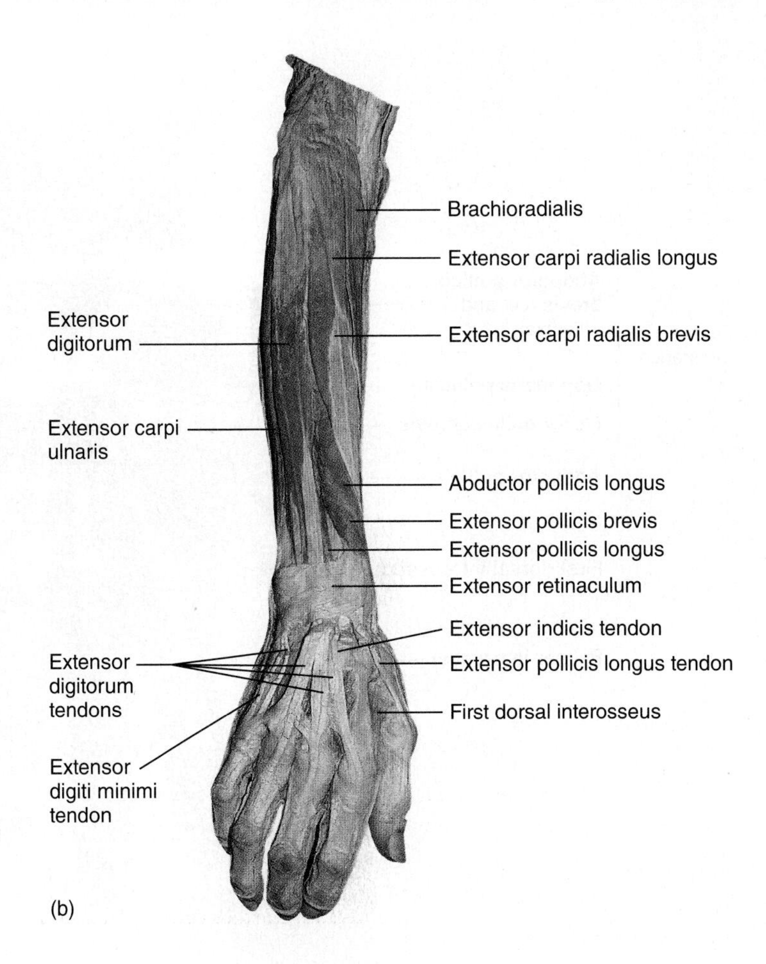

Figure 12.12 Extensor Muscles of the Right Hand
(a) Diagram; (b) photograph.

breathing is becoming restricted. If this happens, notify your instructor.

At the end of the exercise, remove the scapel blade and wash your dissection equipment with soap and water. Place all used blades and other sharp material in the **sharps container** in lab. Remove all of the excess animal material on the dissection trays and put it in the appropriate animal waste container. *Do not dump excess animal material in the lab sinks!* Wash the dissection trays and place them in the appropriate area to dry.

Cat Care

Keep your cat in a plastic bag. It is recommended that you retain the skin of the cat as a protective wrapping when you are finished with the day's dissection or that you take a small towel (or an old T-shirt) and wrap the specimen in the cloth, soaking it in a cat wetting solution before you place it back in the bag. Usually, one lab period is required to skin the cat and prepare the specimen for further study.

External Features

(A) 4 Take a dissection tray and a cat specimen to your table. Remove the cat from the plastic bag and place it on the tray. You may have excess fluid in the plastic bag, and this should be disposed of properly as directed by your instructor. Place the cat on its back and determine whether you have a male specimen or a female specimen. Both sexes have multiple **teats** (nipples), yet the males have a **scrotum** near the base of the tail with a **prepuce** (penile foreskin) anterior to the scrotum. Females have a **urogenital opening** anterior to the anus without a scrotum and prepuce. Once you have identified the sex of your specimen, compare yours with others in the class, so that you can identify the sexes externally. If you have difficulty determining the sex, ask your instructor for help.

Examine the cat and notice the **vibrissae,** or whiskers, in the facial region. Other variations from humans are the presence of **claws** and **friction pads** on the extremities and a **tail.** Review the planes of the body, as illustrated in Exercise 1, before you begin the dissection. The terms used in that exercise will be of great importance in the dissection procedures.

Removal of the Skin

(A) 5 Begin your dissection by lifting the skin in the pectoral region with a forceps and making a small cut with a scalpel or sharp scissors in the midline. Work a blunt probe gently into the cut, so that you free the skin from the underlying fascia and muscle somewhat. Be careful—the muscles are close to the skin. Insert your scalpel, blade side up, and make small incisions in the skin, cutting away from the underlying muscle, as illustrated in figure 12.14. In many cases you do not need to use a scalpel to

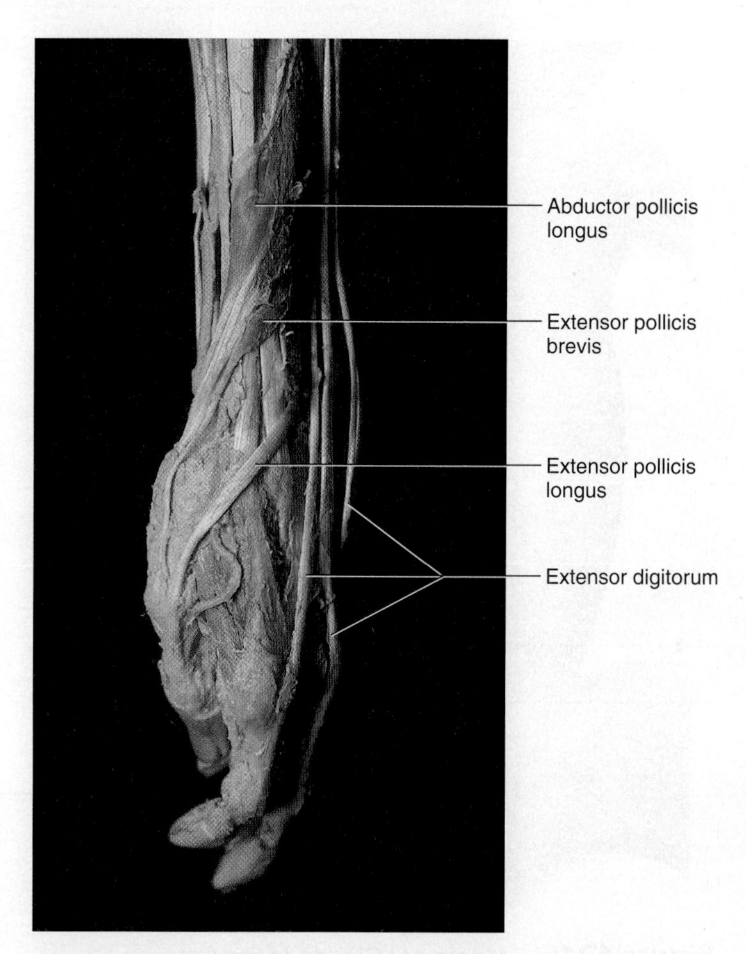

Figure 12.13 Posterior Muscles of the Thumb

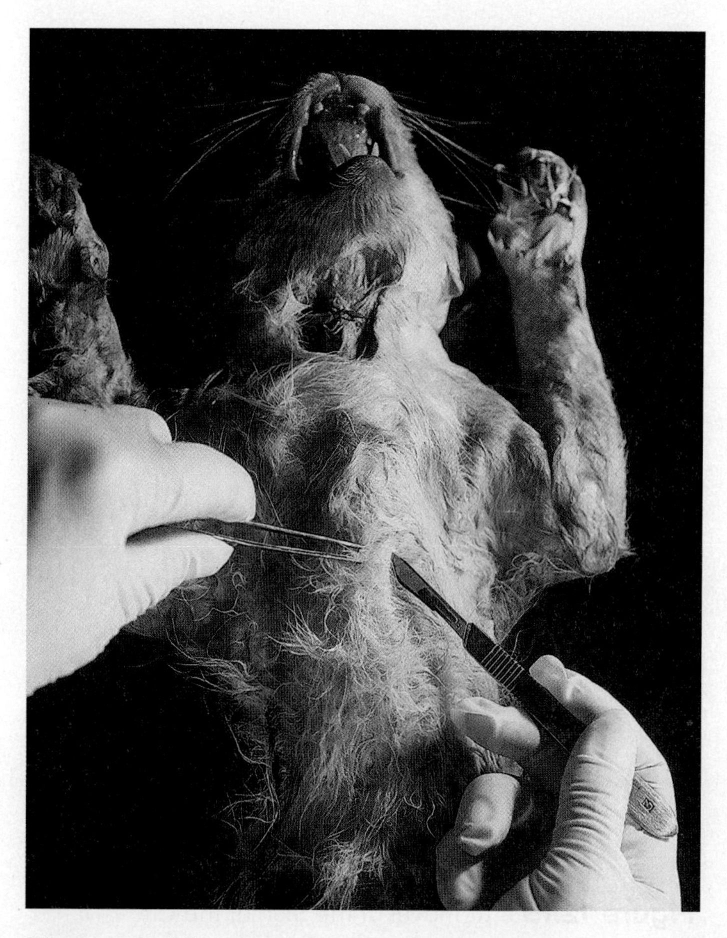

Figure 12.14 Lateral Cutting with a Scalpel

do dissection. Once an initial incision is made, you can use a pair of sharp, pointed dissection scissors and insert the closed scissors into the opening made by the scalpel. Open the scissors parallel to the surface of the muscle, separating the fascia from the skin. Once you have opened up an area that is a couple of inches wide, you will have to extend the incision with a scalpel or cut it with the scissors. This is known as a blunt dissection and, if done properly, tends to preserve the underlying structures better than scalpel dissection. You can then insert a gloved finger into the opening and tease away the skin from the underlying fascial plane.

Make a cut that runs up the midline of the sternal region until you reach the neck. Likewise, cut posteriorly until you reach an area craniad to the genital region. Leave the genitals intact and carefully make an incision that runs perpendicular transversely to your first cut. Likewise, make a transverse incision along the upper thoracic region (figure 12.15). Continue your caudal incision along the medial aspect of the thigh. Watch out for superficial blood vessels in this area (especially the great saphenous vein), and stop when you reach the knee. Cut the skin around the knee and begin removing it as a layer from the dorsal side of the cat. Cut the skin from the base of the tail and remove the skin from the back.

Once you reach the shoulders, turn the cat back to the ventral side and remove the skin on the medial side of the arm until you reach the elbow. Stop at this level and remove the skin from the lateral aspect of the arm, working back toward the shoulder. Be careful not to damage the superficial veins on the lateral side of the forearm. Continue your removal of the skin from the ventral body region. If your cat is a female, locate the mammary glands, which are elongated, beige, lobular masses on each side of the midline on the ventral side (figure 12.16).

Make an incision on the ventral side of the neck along the midline. Be careful—there are many blood vessels along the neck. Do not cut through these blood vessels. Continue up into the face and look for beige lumps of glandular tissue in the region of the mandible. These are the **salivary glands.** Cut the skin carefully from the face and remove it from the head by cutting anterior to the ears (figure 12.17).

You should be able to remove all the skin from the cat at this time. Examine the undersurface of the skin and note the superficial muscles that cause the skin to move. These are the **cutaneous maximus** and the **platysma.**

Return to the cat and remove as much fat as you can. **Subcutaneous fat** is variable from cat to cat, and your specimen may have little or it may have significant amounts. The muscle also is covered by fascia, a connective tissue wrapping. Remove the fascia from the muscle so that the fiber direction is apparent.

Once you have removed the skin from your cat and removed the superficial fascia, you should identify the major muscles of the cat. Look for the large latissimus dorsi muscle of the back and

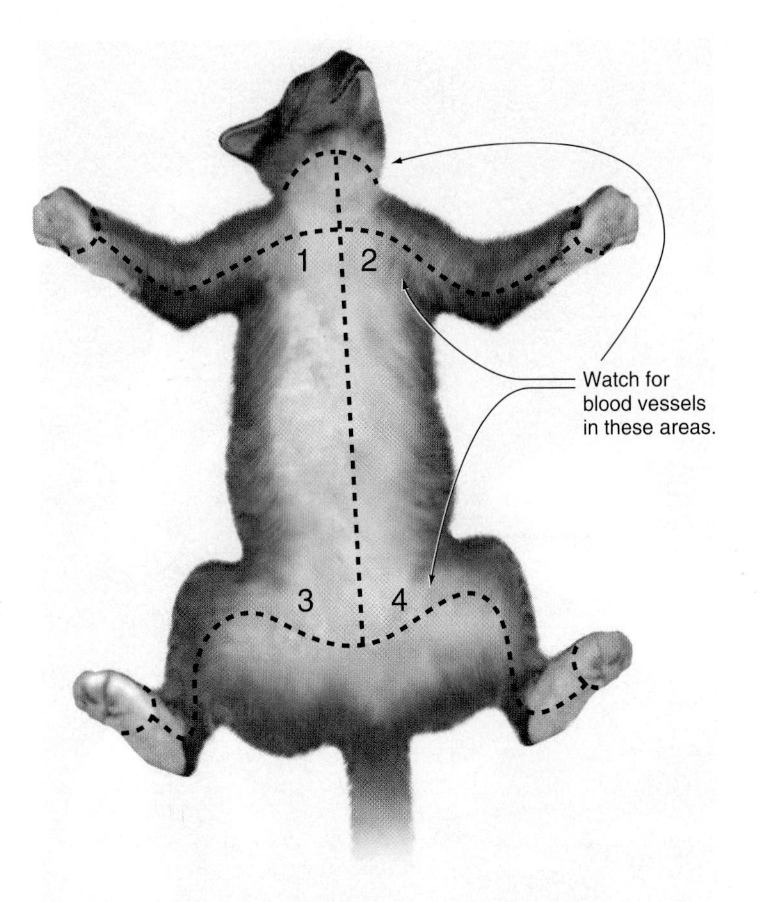

Figure 12.15 Removal of the Skin of the Cat

After cutting down the midline, incisions should be made into the limbs (follow the numbers) and the skin gently removed from underlying structures.

Figure 12.16 Mammary Glands of the Cat

the external abdominal oblique muscle. You should also find the deltoids and triceps brachii muscles of the shoulder region and the gluteus and biceps femoris muscles of the hip and thigh region. Compare your cat with figure 12.18.

Individual Muscles of the Cat

(A) 6 Begin your **dissection** of the muscles by understanding that the term *dissect* means to separate. When you isolate one muscle from another, locate the tendons of that muscle. The **tendon** is the attachment point of the muscle to a bone. Broad, flat tendons are known as **aponeuroses.**

If you have not already removed the skin from the forelimb of the cat, you should do so now. This can be accomplished by making a longitudinal incision along the length of the forelimb. Be very careful not to cut the tendons or the blood vessels or nerves. Pull the skin off as if you were removing a pair of knee socks.

As you get to the tips of the digits, cut the skin from the digits. Pay particular attention and keep the tendons intact. It is a good procedure to dissect only one side of the cat at a time. If you make an error on one side, you will have the other side to dissect.

As you make the dissection in the cat, you can leave many of the muscles of the forelimb intact. Deeper muscles can generally be seen by moving the more superficial muscles off to the side.

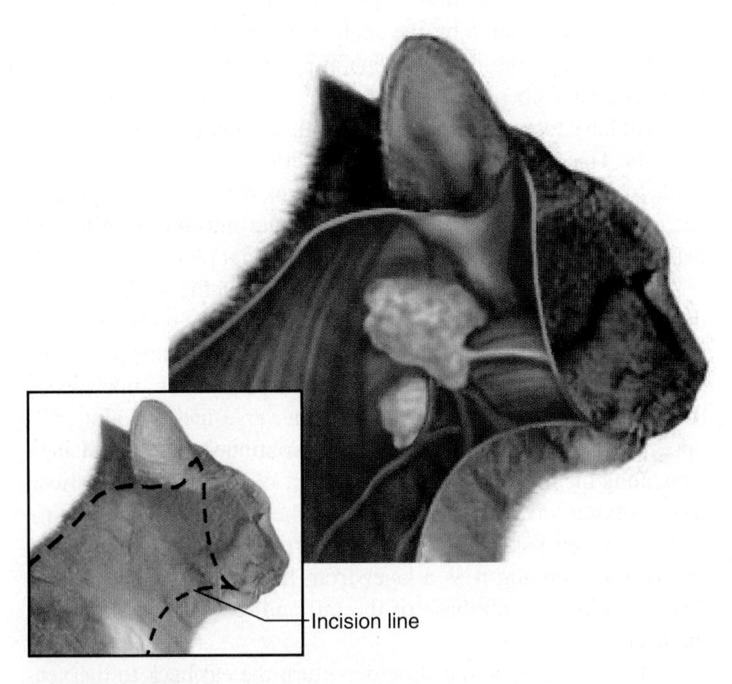

Figure 12.17 Removal of Skin from the Head of a Cat

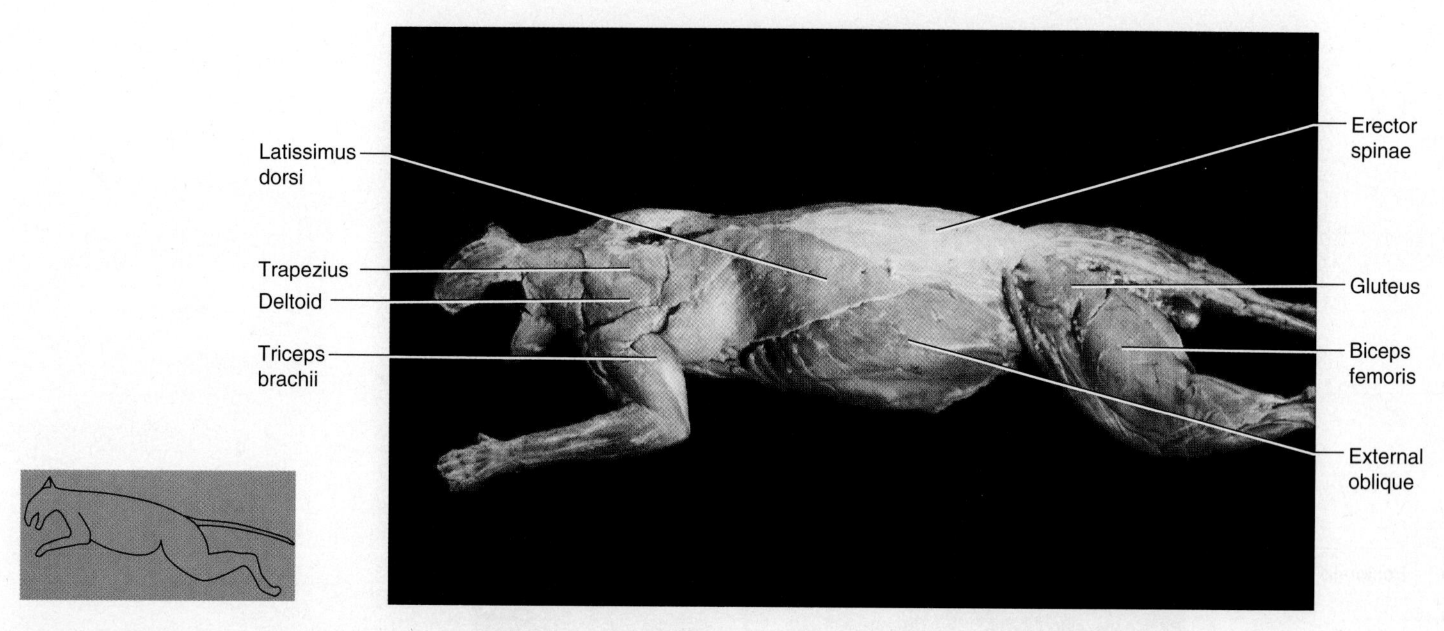

Figure 12.18 Major Muscles of the Cat

Once you locate the muscle, tug gently on it to locate its **origin** and **insertion.** If you pull too hard, you may rip the muscle. The outer wrapping of the muscle is the **fascia,** and it should be removed to find the fiber direction of the muscle. The main part of the muscle is the **belly.** You may need to cut a muscle to locate deeper muscles. This is done by **transecting** the muscle, which is to cut the muscle into two sections in the middle of the belly perpendicular to the fiber direction. After transecting the muscle, you will need to **reflect** it, or pull it toward its attachment site. When separating two muscles, you may find a cottonlike material between them. This is loose connective tissue that forms part of the fascia.

Pectoral Muscles of the Cat

(A) 7 Four major muscles are seen in a superficial view of the pectoral region. These are the pectoantebrachialis, pectoralis major, pectoralis minor, and xiphihumeralis. The **pectoantebrachialis** has no corresponding muscle in the human. It originates on the sternum and inserts on the forelimb. Transect and reflect the pectoantebrachialis to see the pectoralis major. The **pectoralis major** originates on the sternum and inserts on the proximal humerus. The **pectoralis minor** is a large muscle in cats and inserts on the upper humerus. The **xiphihumeralis** is another cat muscle that has no corresponding human muscle; it originates on the sternum and inserts on the proximal humerus along with the pectoralis major and minor. Locate these muscles on your cat and in figure 12.19.

Shoulder Muscles

(A) 8 In humans, there is a single trapezius on each side of the body. In cats, each trapezius consists of three muscles. Examine the cat from the dorsal side and locate the **clavotrapezius,** the **acromiotrapezius,** and the **spinotrapezius.** All of these muscles originate on the vertebral column, with the clavotrapezius also originating on the occipital bone. The clavotropezius inserts on the clavicle, the acromiotrapezius on the acromion process of the scapula, and the spinotrapezius on the spine of the scapula. Compare these muscles with figure 12.20.

Two other muscles of the region are the **latissimus dorsi** and the **levator scapulae ventralis.** The **levator scapulae** is deep to the trapezius. You will need to cut a part of the trapezius known as the clavotrapezius in order to see the levator scapulae. In cats, there is an additional muscle called the levator scapulae ventralis, which inserts on the scapular spine. Examine figure 12.20 for the levator scapulae muscles. These two muscles are similar to those in the human. Locate these muscles on the cat and compare them with figure 12.20.

In humans, there is a single deltoid muscle, but in cats the deltoid consists of three muscles, as does the trapezius. Like the trapezius, the deltoid muscles are named for their bony attachments. The **clavodeltoid** (clavobrachialis) originates on the clavicle, the **acromiodeltoid** on the acromion process, and the **spinodeltoid** on the spine of the scapula. Insertions of these muscles are on the forelimb. You will need to dissect the trapezius carefully to see the **rhomboideus** muscles. These muscles originate on the vertebral column and insert on the scapula. Locate these muscles on the cat and compare them with figures 12.20 and 12.21.

Scapular Muscles That Act on the Humerus

(A) 9 The deep muscles of the scapula can be seen by reflecting the overlying muscles. The **supraspinatus, infraspinatus, subscapularis, teres major,** and **teres minor** are roughly equivalent to the same muscles in the human. Locate the supraspinatus, infraspinatus, and teres major and find them in figure 12.21.

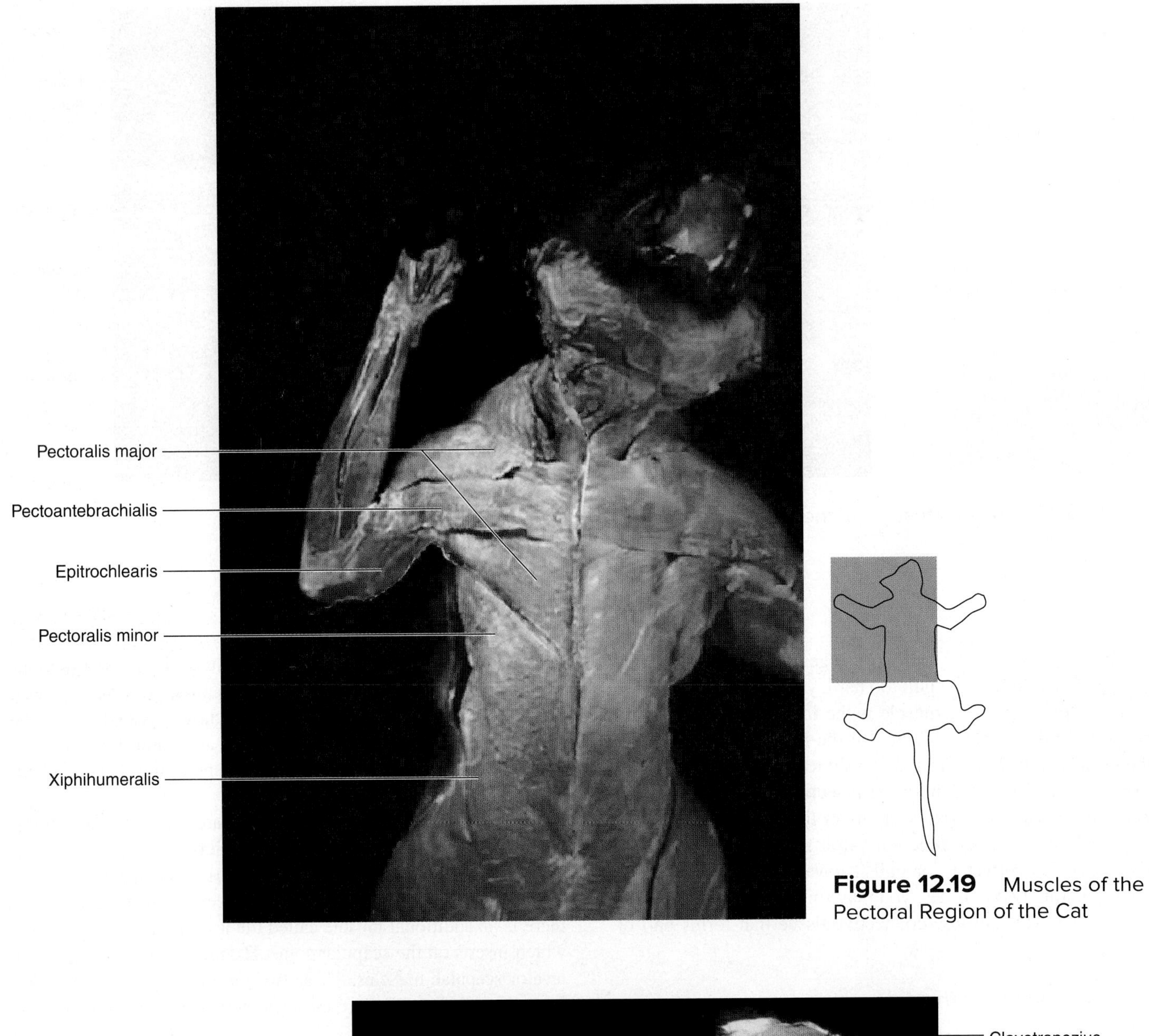

Figure 12.19 Muscles of the Pectoral Region of the Cat

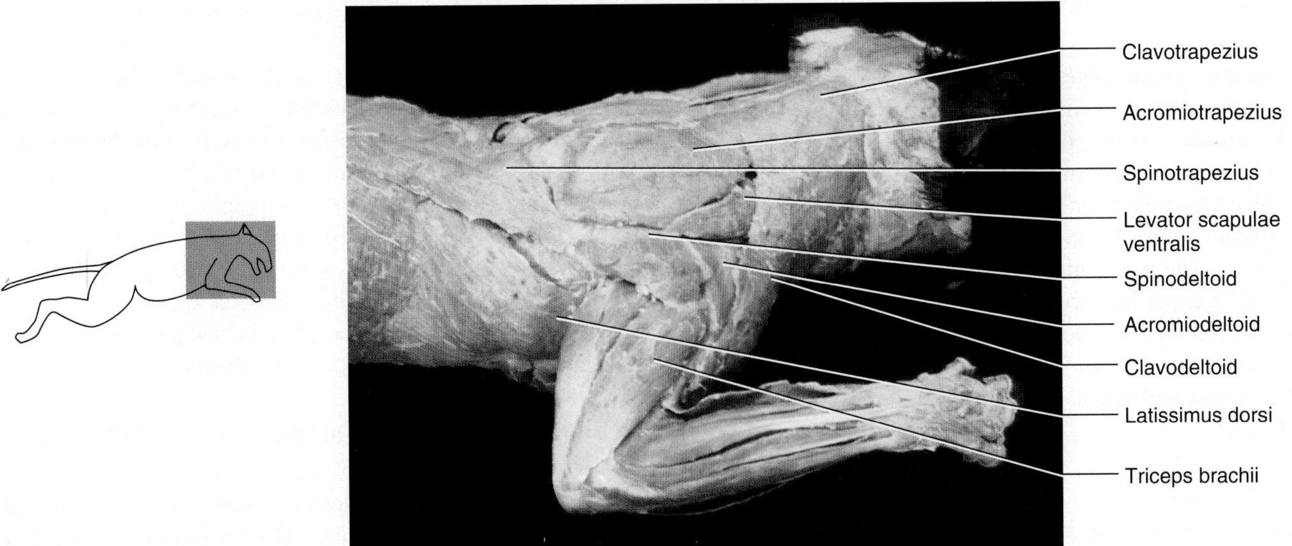

Figure 12.20 Superficial Muscles of the Right Shoulder of the Cat

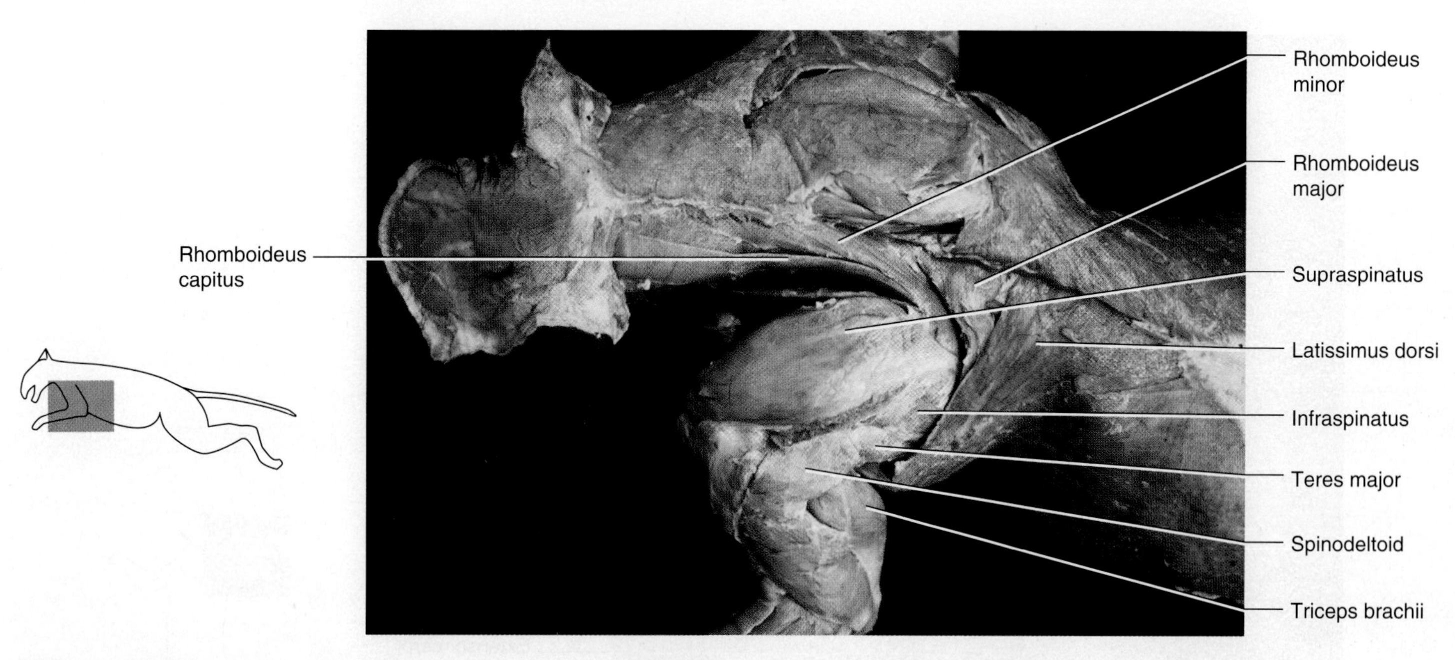

Figure 12.21 Deep Muscles of the Left Scapula of the Cat

Forelimb Muscles

A 10 The muscles that have an action on the forelimb of the cat typically either flex or extend the forelimb as their primary action. The **epitrochlearis** does not have a corresponding muscle in humans (figure 12.22). The epitrochlearis is on the medial side of the humerus, and it extends the forelimb. The **biceps brachii** is also a medial muscle, and it flexes the forelimb. The **triceps brachii, anconeus, brachioradialis,** and **brachialis** are lateral or posterior muscles. The triceps brachii and the anconeus extend the forelimb, whereas the brachialis flexes the forelimb. Find the lateral muscles, using figures 12.22 and 12.23 as a guide.

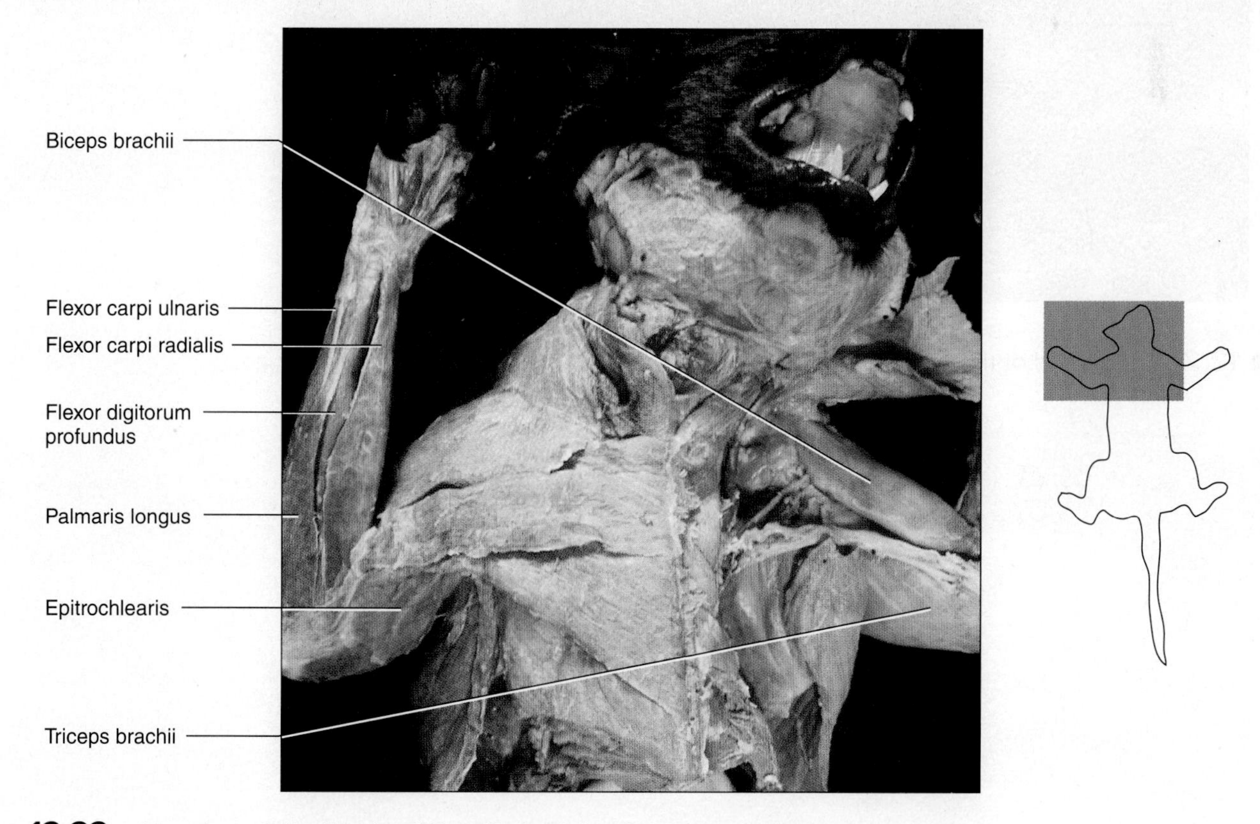

Figure 12.22 Muscles of the Proximal Forelimb of the Cat, Ventral View

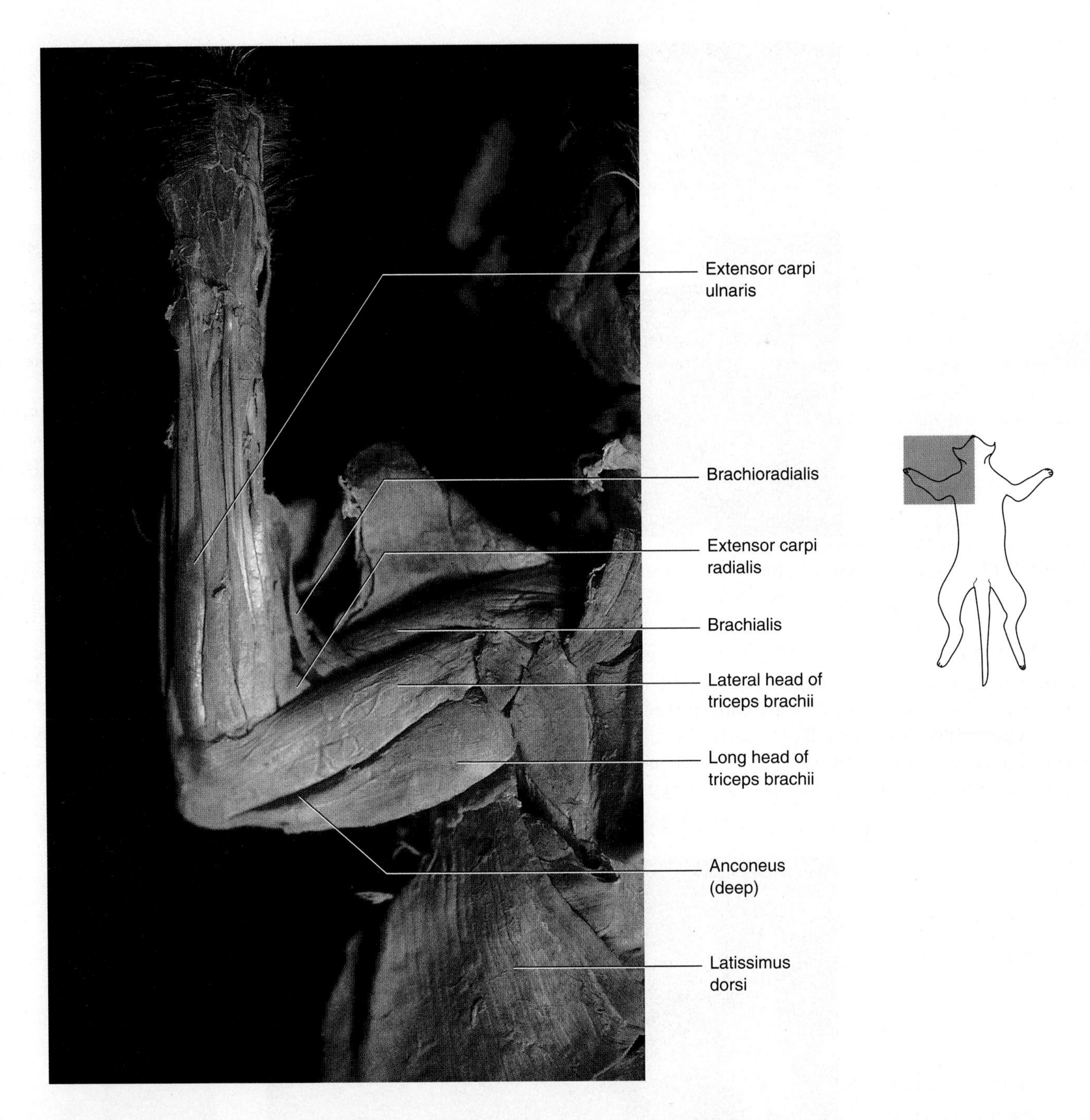

Figure 12.23 Muscles of the Left Proximal Forelimb of the Cat, Dorsal View

Superficial Muscles on the Medial Aspect of the Forelimb

(A) 11 Most of the muscles of the forelimb of the cat run parallel to the radius and ulna. An exception to this is the **pronator teres,** which is a small slip of muscle that runs obliquely down the forelimb. It pronates the wrist. The **palmaris longus** is a broad, flat muscle, superficially located on the forelimb with insertions into the digits. In humans, the palmaris longus terminates at the palmar fascia.

The **flexor carpi radialis** is a thin muscle inserting on the second and third metacarpals. It is named for its action, its insertion, and its location. The **flexor carpi ulnaris** originates on the humerus and ulna and inserts on the medial metacarpals and carpals. The **flexor digitorum superficialis (flexor digitorum sublimis)** is located in the forelimb as a middle-level muscle, underneath the palmaris longus. It originates on fascia of other forelimb muscles, as opposed to originating on bone. Examine the superficial muscles of the medial forelimb and compare them with figure 12.22.

Deep Muscles on the Medial Aspect of the Forelimb

(A) 12 Underneath the upper layer of muscles are many muscles with varied actions. The **supinator** is a deep muscle that runs diagonally from the lateral epicondyle of the humerus to the proximal radius. It is the deepest of the proximal forelimb muscles. The **flexor digitorum profundus** is an extensive muscle that inserts on the first through fifth digits. It replaces the flexor pollicis longus for the thumb flexion, as this muscle is absent in cats. The **pronator quadratus** is a square muscle located between the radius and ulna deep to the flexor digitorum profundus. Find the deep muscles in the cat and compare them with figure 12.24.

Muscles on the Lateral Aspect of the Forelimb

(A) 13 The lateral muscles of the forelimb are the extensor group; the **extensor carpi radialis longus** muscle is deep to the brachioradialis and inserts on the second metacarpal. The **extensor carpi radialis brevis** is deep to the extensor carpi radialis longus and inserts on the third metacarpal.

Locate the **extensor carpi ulnaris,** which is next to the **extensor digitorum lateralis,** inserting on the fifth metacarpal. The **extensor digitorum communis** can be located on the lateral aspect of the forelimb. It is a broad muscle, inserting by tendons on the second through fifth digits. Locate these muscles on the cat and in figure 12.25. The **extensor digitorum lateralis** is specific to the cat; it inserts with the tendons of the extensor digitorum communis to the digits. The **extensor pollicis brevis** is a well-developed muscle in the cat, whereas the abductor pollicis longus is absent in cats.

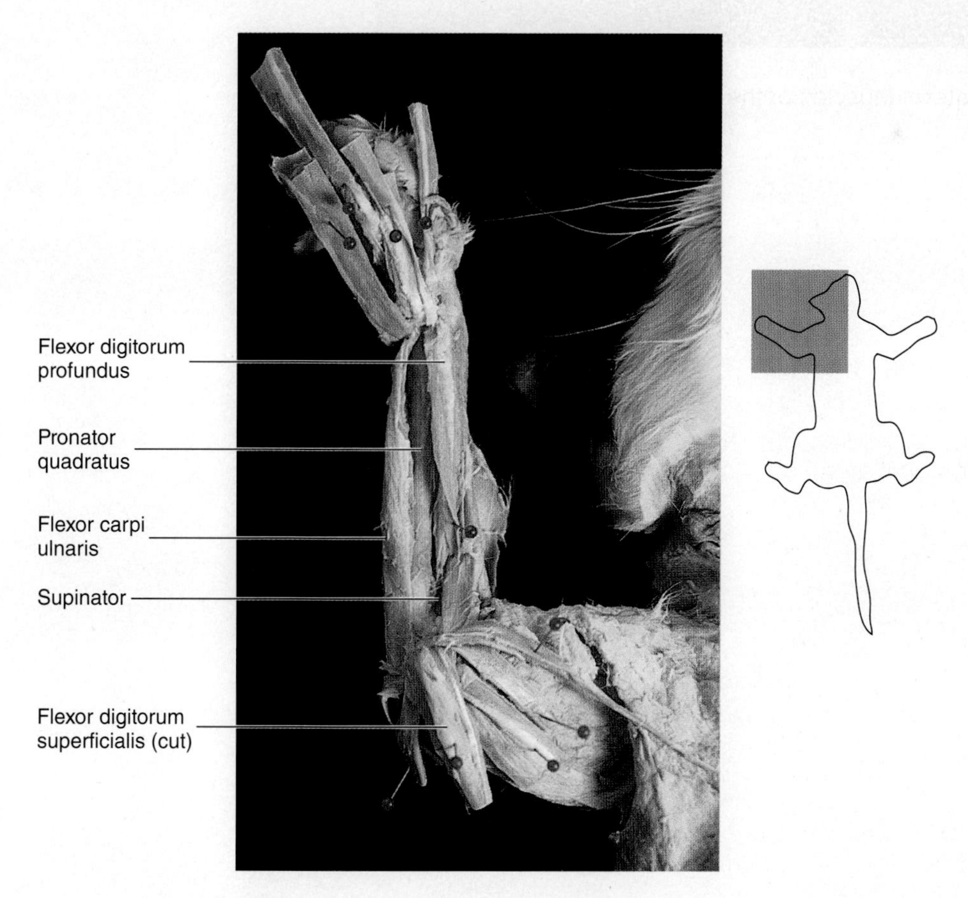

Figure 12.24 Deep Muscles of the Right Medial Forelimb of the Cat

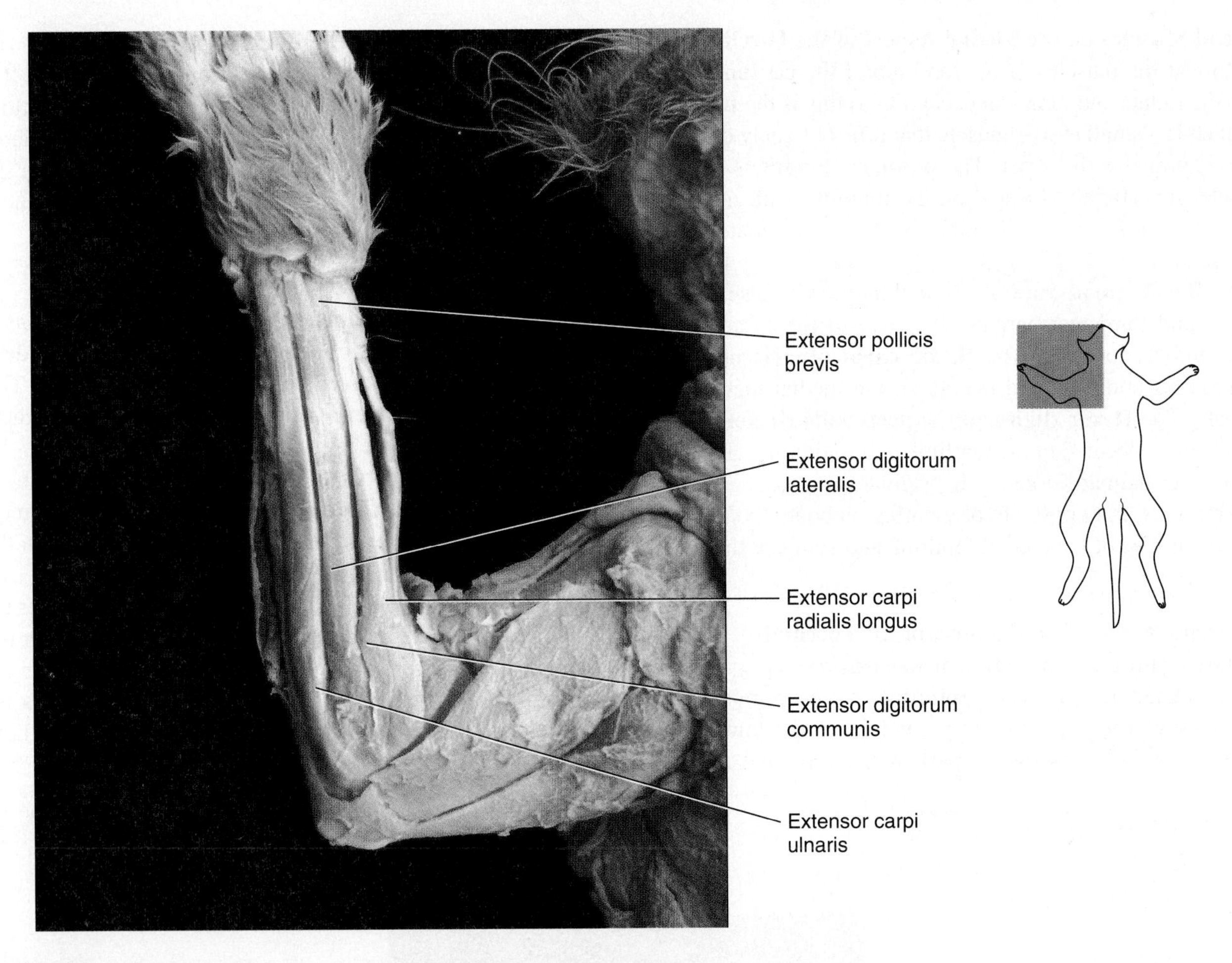

Figure 12.25 Lateral Muscles of the Left Forelimb of the Cat

Exercise 12 Review

Name: ____________________

Lab time/section: ____________________

Date: ____________________

Overview of Muscles and Muscles of the Shoulder and Upper Extremity

Answer the following questions in terms of human muscles.

1. What is the action of the deltoid muscle? ____________________

2. Name the origin of the supraspinatus muscle. ____________________

3. What is the insertion of the pectoralis minor? ____________________

4. What is the origin of the levator scapulae? ____________________

5. What is the insertion of the latissimus dorsi? ____________________

6. Does the biceps brachii muscle originate or insert on the humerus? ____________________

7. What is the origin of the trapezius? ____________________

8. The pectoralis major has what action? ____________________

9. What is the insertion of the subscapularis? ____________________

10. What is the action of the triceps brachii? ____________________

11. What is the origin of the brachialis? ____________________

12. Name all the muscles that flex the arm. ____________________

13. Which muscles are antagonists to the triceps brachii? ____

____

14. Label the muscles in the following illustration using the terms provided.

biceps brachii	brachialis	brachioradialis
deltoid	pectoralis major	triceps brachii

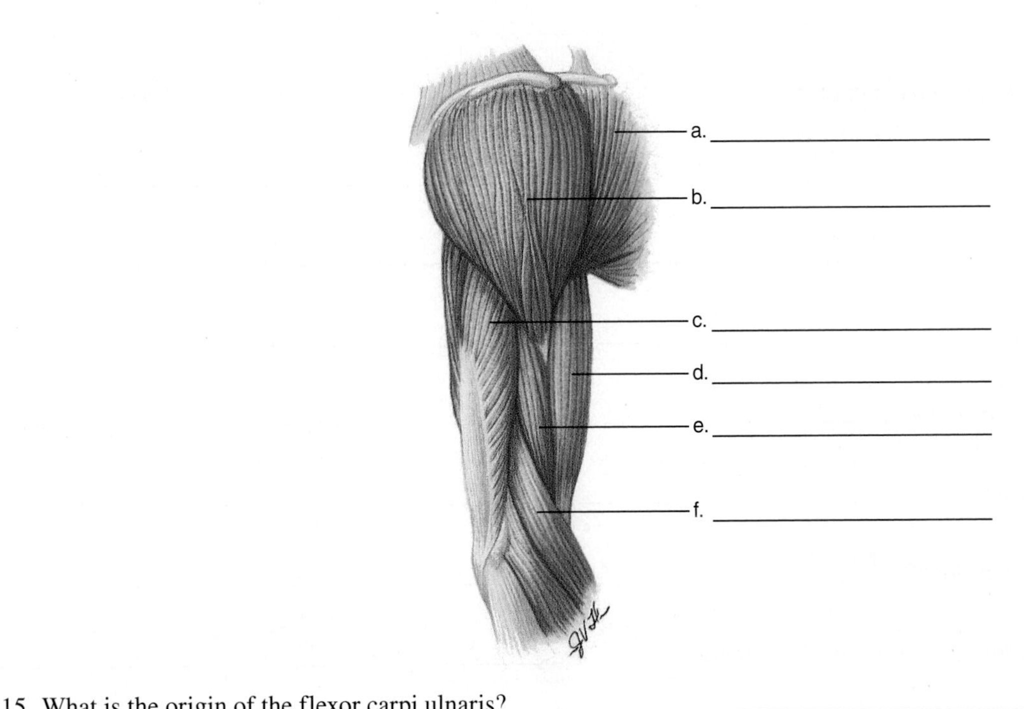

15. What is the origin of the flexor carpi ulnaris? ____

16. What muscles are antagonists to the supinator muscle? ____

17. Where does the flexor digitorum superficialis insert? ____

18. What is the insertion of the extensor carpi ulnaris muscle? ____

19. What is the action of the flexor carpi radialis muscles? ____

____

20. What muscle extends both the hand and the phalanges? ____

21. What muscles flex the hand? ____

____

22. What muscles extend the thumb? ____

____

23. Where are the extensor carpi muscles located, on the anterior or posterior side of the forearm? ______________________

24. Label the muscles or structures in the following illustration using the terms provided.

flexor carpi radialis   flexor carpi ulnaris   palmar aponeurosis
palmaris longus   pronator teres

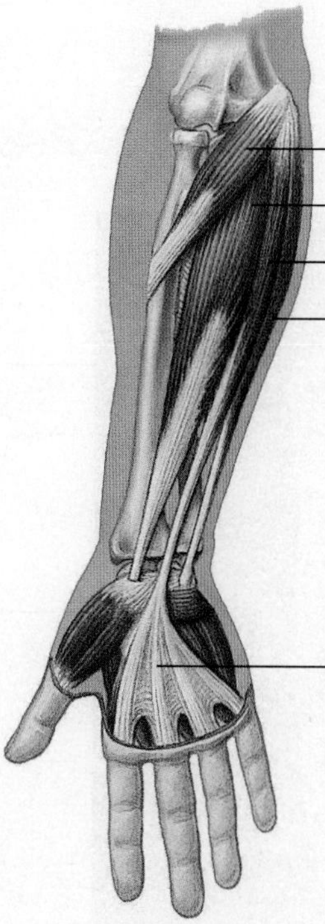

a. ______________________
b. ______________________
c. ______________________
d. ______________________
e. ______________________

Answer questions 25 to 29 in terms of cat muscles.

25. What are the three muscles that correspond to the single trapezius in humans? ______________________

____

26. What is the action of the epitrochlearis? ______________________

27. Which muscle does not correspond to a human muscle?

a. deltoid ______________________

b. pectoralis minor ______________________

c. xiphihumeralis ______________________

d. latissimus dorsi ______________________

28. How does the deltoid of the cat differ from the deltoid of the human? ______________________

29. Which muscle is more developed in cats, the flexor digitorum superficialis or the flexor digitorum profundus? ______________________

____

30. As you hit a nail with a hammer, what arm muscle are you using?

31. As you look up at the ceiling, what back muscle are you using?

32. Label the muscles in the following illustration using the terms provided.

    biceps brachii

    extensor carpi radialis longus

    flexor digitorum superficialis

    triceps brachii

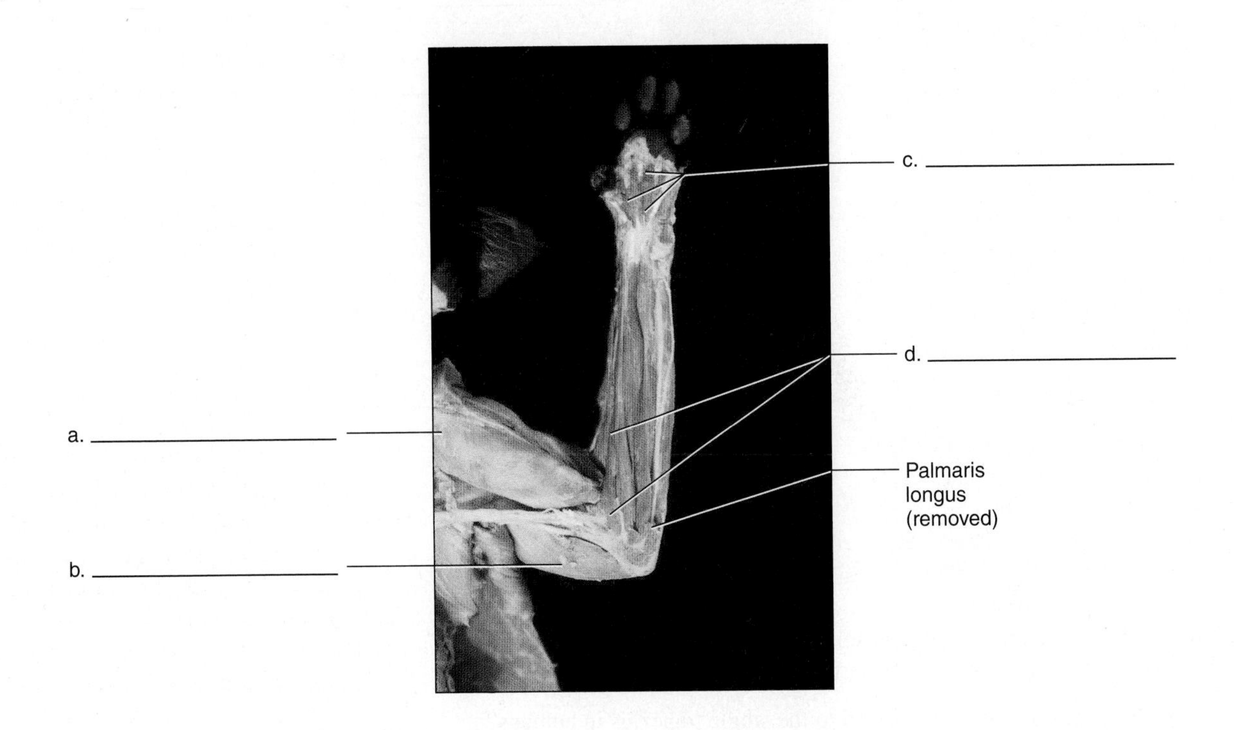

Exercise 13

Muscles of The Hip, Thigh, Leg, and Foot

Introduction

The muscles of the hip, thigh, leg, and foot are primarily muscles of locomotion. They can be grouped generally into muscles that have an action on the thigh, the leg, or the foot. These muscles are discussed in the Seeley text in chapter 10, "Muscular System: Gross Anatomy."

The size and shape of the muscles of the hip and thigh in humans are different from those of other mammals in that humans are bipedal. This two-legged walking habit is much different from a quadruped's locomotion in terms of balance and forward movement. When a quadruped walks, typically two legs remain on the ground at a time. When a biped walks, the inherent instability of standing on one leg brings into focus the importance of stabilizing the support limb and maintaining the center of balance. Think about what happens to the mechanics of movement and the center of balance when you lift a leg to walk or run.

The muscles of the leg originate typically from the femur, tibia, or fibula and insert on the bones of the foot. Anterior and posterior actions on the thigh and leg are flexion or extension. Those for the foot are described as **dorsiflexion** (decreasing the angle between the shin and dorsum or top of the foot) or **plantar flexion** (an action that is seen as you stand on your toes to reach something high). Review the other actions of the leg and foot in Exercise 10 in this lab manual. Some of the muscles of the feet are similar to those of the hand, such as the extensor digitorum muscles.

Objectives

At the end of this exercise, you should be able to

1. locate the muscles of the hip, thigh, leg, and foot on a model, chart, cadaver (if available), or cat;
2. list the origin, insertion, and action of each muscle presented;
3. describe what nerve controls each muscle;
4. list what muscles function as synergists or antagonists to the prime mover;
5. name all the muscles that have an action on a joint, such as all of the muscles that flex the thigh or those that dorsiflex the foot.

Materials

Human torso model
Human leg models
Human muscle charts
Articulated skeleton
Cadaver (if available)
Cat
Materials for cat dissection
- Dissection trays
- Scalpel with two or three extra blades
- Gloves (household latex gloves work well for repeated use)
- Blunt (Mall) probe
- String and tags
- Pins
- Forceps and sharp scissors
- First aid kit in lab or prep area
- Sharps container
- Animal waste disposal container

Procedure

(A) 1 Review the muscle nomenclature and the action of muscles as outlined in Exercises 10 and 12. Examine muscle models or charts in the lab and locate the muscles described in the following section and in table 13.1. Correlate the shape of the muscle with the name and study the models or charts. You may want to look at an articulated skeleton as you review the origins and insertions of the muscles, so that you can better see the muscle attachment points. Once you have learned the origin and the insertion, it should be easier to understand the action of the muscle. This is done, in part, by imagining how the bony attachments of the origin and insertion would come together if the muscle pulled them closer to one another. If you have a cadaver, pay particular attention to the origins and the insertions of the muscles of the thigh, leg, and foot.

Human Muscles

Muscles of the Hip and Thigh

(A) 2 The **iliopsoas** muscle is a major flexor of the thigh. It originates on the anterior surface of the ilium and crosses over the pelvic brim to insert on the posterior aspect of the femur. The iliopsoas is actually two muscles, the **iliacus** and **psoas major.** A companion muscle is the **psoas minor.** This exercise treats these muscles as one.

The **sartorius** (*sartorius* = tailor) is a slender muscle that originates in a superior, lateral position and inserts in an inferior, medial location. It allows you to sit cross-legged and is thus named the "tailor's muscle." The sartorius is the most superficial muscle on the anterior thigh. The most superficial muscle of the medial aspect of the thigh is the gracilis. The **gracilis** is a thin muscle that adducts the thigh. Locate these muscles in the lab and in figure 13.1.

TABLE 13.1 Muscles of the Hip, Thigh, Leg, and Foot

Name	Origin	Insertion	Action	Innervation
Muscles of the Medial Thigh				
Iliopsoas	T12, L1–5, iliac fossa	Lesser trochanter of femur	Flexes thigh and lumbar vertebrae	Femoral nerve, spinal nerves L1–3
Sartorius	Anterior superior iliac spine	Proximal, medial tibia	Flexes and laterally rotates thigh, flexes leg	Femoral nerve
Gracilis	Pubis, near symphysis	Proximal, medial tibia	Adducts thigh, flexes leg	Obturator nerve
Pectineus	Pubic crest	Pectineal line of femur	Adducts thigh	Femoral and obturator nerves
Adductor brevis	Pubis	Linea aspera at proximal portion of femur	Adducts, flexes, and laterally rotates thigh	Obturator nerve
Adductor longus	Pubis	Linea aspera at middle portion of femur	Adducts, flexes, and laterally rotates thigh	Obturator nerve
Adductor magnus	Pubis, ischium	Linea aspera, adductor tubercle of distal femur	Adducts, extends, and laterally rotates thigh	Obturator and tibial nerves
Muscles of the Anterior Thigh				
Quadriceps Femoris Muscles				
Rectus femoris	Anterior inferior iliac spine	Tibial tuberosity by the patellar ligament	Flexes thigh, extends leg	Femoral nerve
Vastus lateralis	Greater trochanter of femur, linea aspera of femur	Tibial tuberosity by the patellar ligament	Extends leg	Femoral nerve
Vastus intermedius	Proximal, anterior femur	Tibial tuberosity by the patellar ligament	Extends leg	Femoral nerve
Vastus medialis	Linea aspera of femur, on medial side	Tibial tuberosity by the patellar ligament	Extends leg	Femoral nerve
Muscles of the Posterior and Lateral Thigh				
Tensor fasciae latae	Anterior superior iliac spine	Iliotibial band of fasciae latae to lateral condyle of tibia	Flexes, abducts, and medially rotates thigh	Superior gluteal nerve
Gluteus maximus	Outer iliac blade, iliac crest, sacrum, coccyx	Gluteal tuberosity of femur, iliotibial band of fasciae latae	Extends, abducts, and laterally rotates thigh; braces knee	Inferior gluteal nerve
Gluteus medius	Outer iliac blade	Greater trochanter of femur	Abducts and medially rotates thigh	Superior gluteal nerve
Gluteus minimus	Outer iliac blade	Greater trochanter of femur	Abducts and medially rotates thigh	Superior gluteal nerve
Hamstrings				
Biceps femoris	Ischial tuberosity, linea aspera of femur	Head of fibula, lateral condyle of tibia	Extends thigh, flexes leg	Tibial and fibular nerves
Semitendinosus	Ischial tuberosity	Proximal, medial tibia	Extends thigh, flexes leg	Tibial nerve
Semimembranosus	Ischial tuberosity	Medial condyle of tibia, collateral ligament	Extends thigh, flexes leg	Tibial nerve
Muscles of the Anterior Leg				
Tibialis anterior	Tibia, interosseous membrane	Metatarsal 1 and cuneiform 1	Dorsiflexes and inverts foot	Deep fibular nerve
Extensor digitorum longus	Lateral condyle of tibia, shaft of fibula	Middle and distal phalanges of digits 2–5	Extends toes 2–5, dorsiflexes and everts foot	Deep fibular nerve
Extensor hallucis longus	Anterior shaft of fibula, interosseous membrane	Distal phalanx of hallux (big toe)	Extends hallux, dorsiflexes foot, inverts foot	Deep fibular nerve
Muscles of the Lateral Leg				
Fibularis longus	Head and shaft of fibula, lateral condyle of tibia	Metatarsal 1, cuneiform 1	Plantar flexes and everts foot	Superficial fibular nerve
Fibularis brevis	Shaft of fibula	Metatarsal 5	Plantar flexes and everts foot	Superficial fibular nerve
Fibularis tertius	Fibula, interosseous membrane	Dorsum of metatarsal 5	Dorsiflexes and everts foot	Deep fibular nerve

TABLE 13.1 Continued

Name	Origin	Insertion	Action	Innervation
Muscle of the Posterior Leg				
Gastrocnemius	Condyles of femur	Calcaneus by the calcaneal tendon	Flexes leg, plantar flexes foot	Tibial nerve
Soleus	Posterior, proximal tibia; fibula	Calcaneus by the calcaneal tendon	Plantar flexes foot	Tibial nerve
Popliteus	Lateral condyle of femur	Proximal tibia	Flexes leg	Tibial nerve
Tibialis posterior	Interosseous membrane, tibia, fibula	Metatarsals 2–5, navicular, cuneiforms, cuboid	Plantar flexes and inverts foot	Tibial nerve
Flexor digitorum longus	Posterior tibia	Distal phalanges of digits 2–5	Flexes toes 2–5, plantar flexes and inverts foot	Tibial nerve
Flexor hallucis longus	Midshaft of fibula	Distal phalanx of hallux (big toe)	Flexes hallux, plantar flexes and inverts foot	Tibial nerve

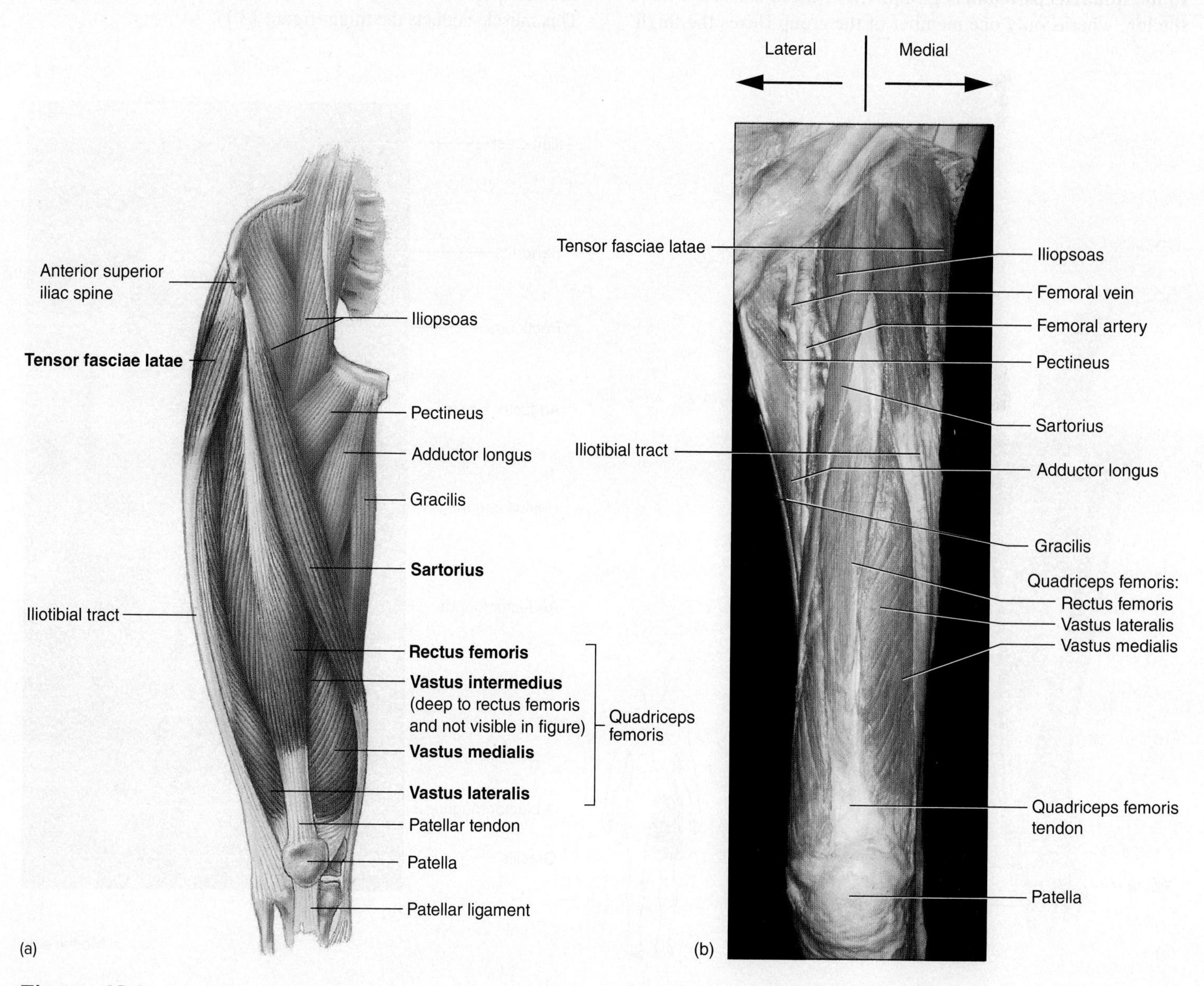

Figure 13.1 Muscles of the Thigh, Anterior View
(a) Diagram of right thigh; (b) photograph of left thigh.

Adductor muscles are located in the medial aspect of the thigh. The gracilis is an adductor muscle, as are the pectineus and the three adductor muscles proper. The **pectineus** is a small, triangular muscle that is the most proximal of the group. It is superficial to the adductor brevis muscle and proximal to the adductor longus muscle. Examine the material in the lab and compare the pectineus with figures 13.1 and 13.2.

The **adductor brevis** is the shortest adductor muscle; it is deep in the medial aspect of the thigh. It can be seen if the pectineus is reflected. The adductor brevis originates on the pubis and inserts on the proximal third of the femur. The **adductor longus** also has a pubic origin, but its insertion is in the middle third of the femur. The largest of the adductor muscles is the **adductor magnus,** and it originates along the length of the inferior hip bone (pubis and ischium) and inserts along the length of the femur. Examine these muscles in figure 13.2.

The muscles of the anterior aspect of the thigh belong to the **quadriceps femoris** group. All of these muscles extend the leg, wheras only one member of the group flexes the thigh. The quadriceps muscles all have a common ligament of insertion called the patellar ligament. This ligament inserts on the tibial tuberosity. The most superficial muscle of the group is the rectus femoris (figures 13.1, 13.2b, and 13.3). It originates directly inferior to the sartorius. The **rectus femoris** runs straight down the femur (*rectus* = straight). Because it crosses the hip joint, the rectus femoris is the only quadriceps muscle to flex the thigh.

The three other quadriceps muscles are the vastus muscles. The **vastus lateralis** is so-named due to its lateral position. The **vastus intermedius** is deep to the rectus femoris, and the **vastus medialis** is the most medial of the vastus muscles. These muscles originate on the femur, so they do not have an action on the hip. They insert on the tibia and extend the leg. Locate these muscles in the lab and compare them with figures 13.1 and 13.3.

The **tensor fasciae latae** is a lateral muscle of the thigh that attaches to a thick band of connective tissue that runs down the lateral aspect of the thigh like a stripe on a pair of tuxedo pants. This muscle abducts the thigh (figure 13.1).

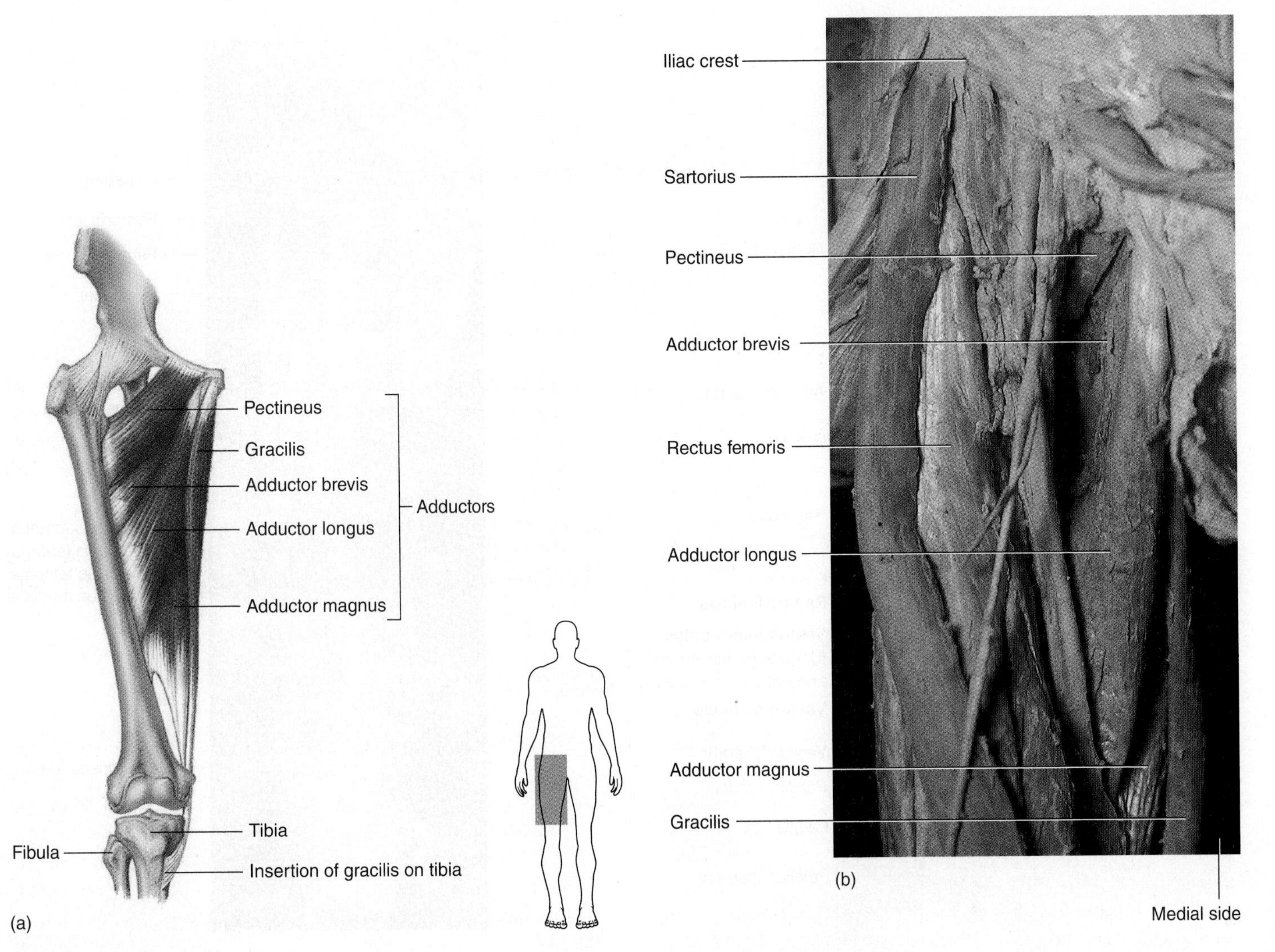

Figure 13.2 Adductor Muscles of the Right Thigh, Anterior View

(a) Diagram; (b) photograph.

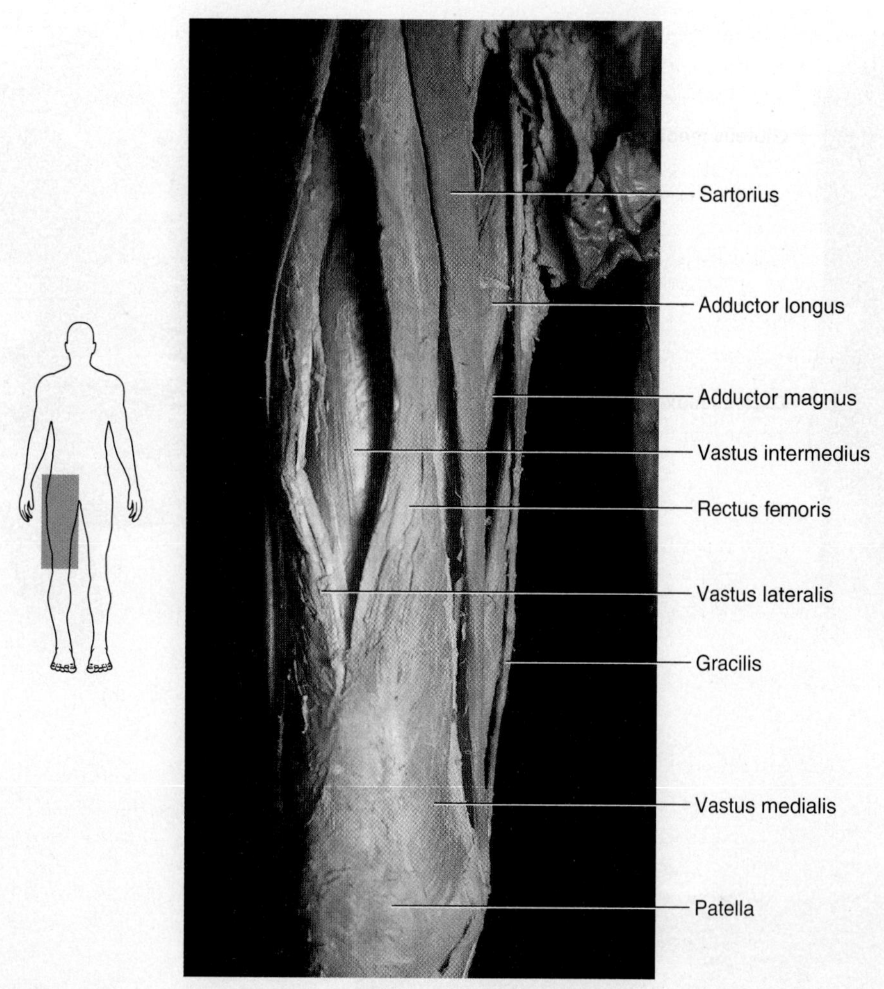

Figure 13.3 Quadriceps Muscles of the Right Thigh, Anterior View

The gluteus muscles in humans are different from those in other mammals, due to our walking upright. The largest of the gluteal muscles is the **gluteus maximus.** It extends the thigh when we stand up from a sitting position or when we climb stairs. The **gluteus medius** and the **gluteus minimus** both help keep our balance during walking in that they maintain the center of gravity. The gluteus medius is superficial to the gluteus minimus, but both of these muscles originate on the outer iliac blade. When we raise one leg to walk, gravity pulls the body in the direction of the lifted leg. The gluteus medius and minimus on the opposite side contract to maintain upright posture. Compare these muscles with figure 13.4.

The **hamstring** muscles are so-named because of an old practice of taking the ham muscles from a pig and tying them together by their tendons (the hamstrings) to hang in a smokehouse to cure. Three muscles make up the hamstrings: the biceps femoris, the semitendinosus, and the semimembranosus. All three of these muscles cross the hip joint and extend the thigh when walking and flex the leg.

The **biceps femoris** is the "two-headed muscle of the femur." One head originates, with the other hamstring muscles, on the ischial tuberosity, whereas the second head originates on the femur. The biceps femoris inserts laterally on the proximal leg. The **semitendinosus** also originates on the ischial tuberosity and can be distinguished as a superficial muscle with a long, pencil-like distal tendon. As opposed to the biceps femoris, the semitendinosus inserts medially. The **semimembranosus** originates on the ischial tuberosity as well and has a broad, flat, membranous tendon on the proximal part of the muscle. The semimembranosus inserts medially on the leg, deep to the semitendinosus. Locate these muscles on the material in lab and compare them with figure 13.5.

Muscles of the Leg and Foot

Anterior Muscles

(A) 3 The **tibialis anterior** is located just lateral to the crest of the tibia on the anterior side of the leg. The tendon of the tibialis anterior crosses to the medial side of the foot and inserts on the first metatarsal and first cuneiform. As the tibialis anterior contracts, it decreases the angle between the anterior tibial crest and the dorsum of the foot in an action known as dorsiflexion of the foot. Because the insertion of this muscle is medial, it also inverts the foot.

The **extensor digitorum longus** muscle is lateral to the tibialis anterior, and the tendons of this muscle splay out and insert

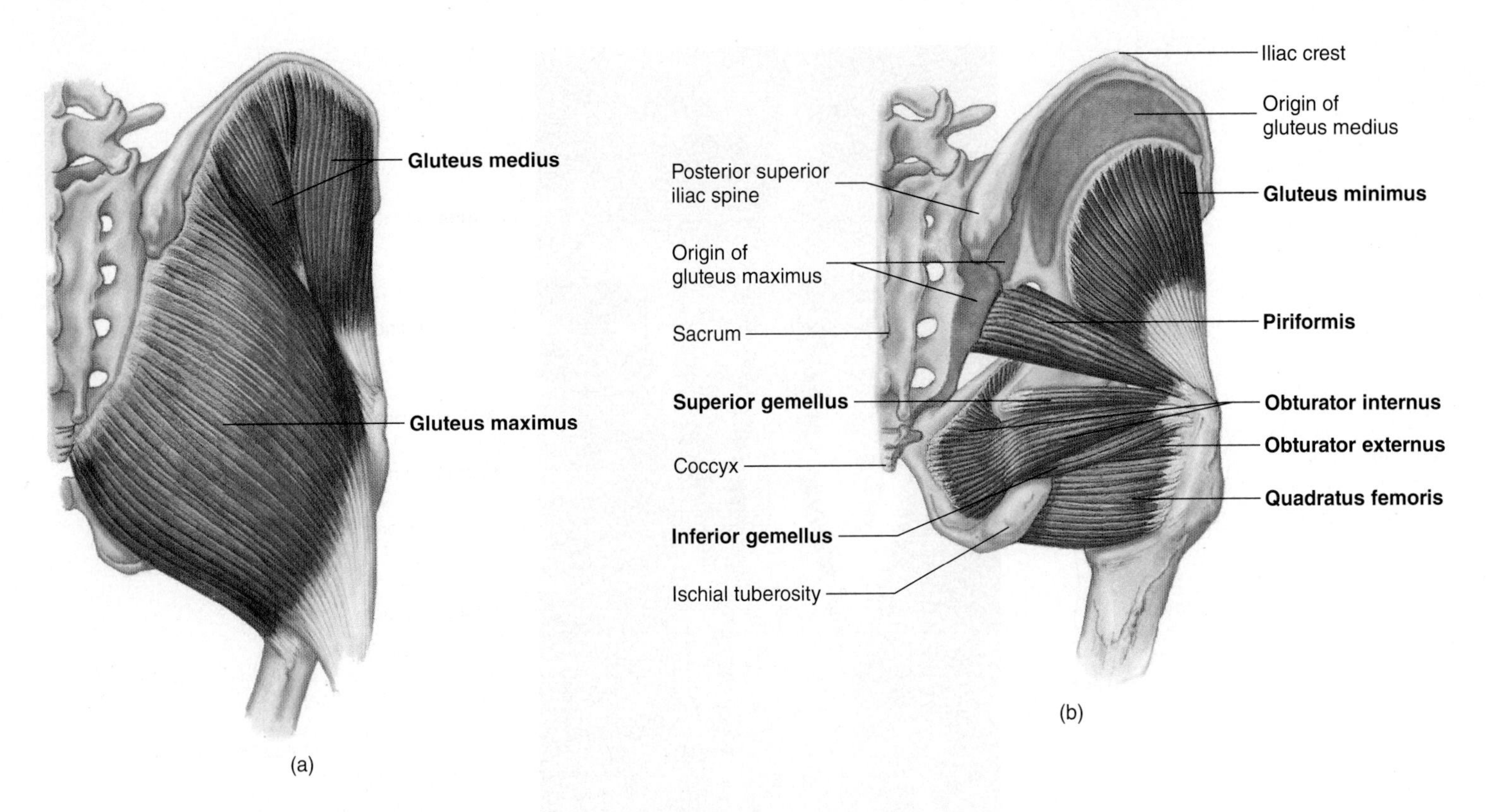

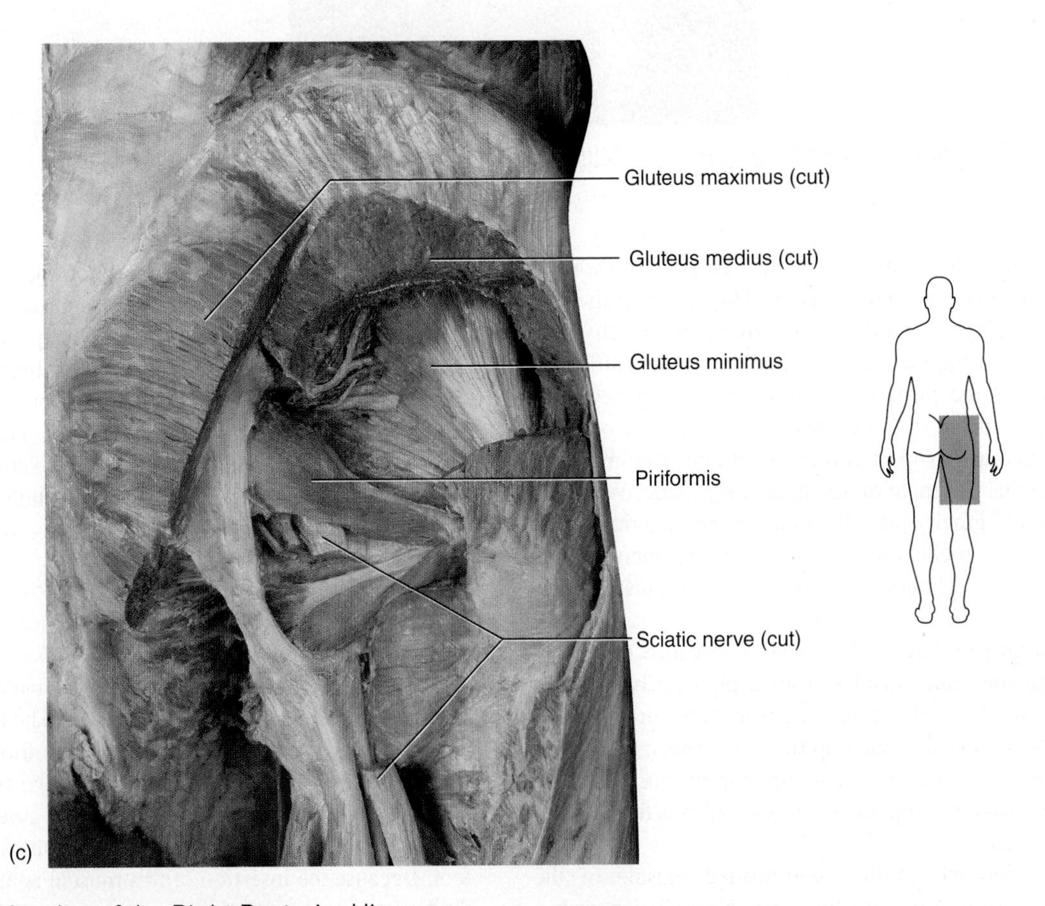

Figure 13.4 Muscles of the Right Posterior Hip
Diagram (a) superficial muscles; (b) deep muscles. Photograph (c) deep muscles.

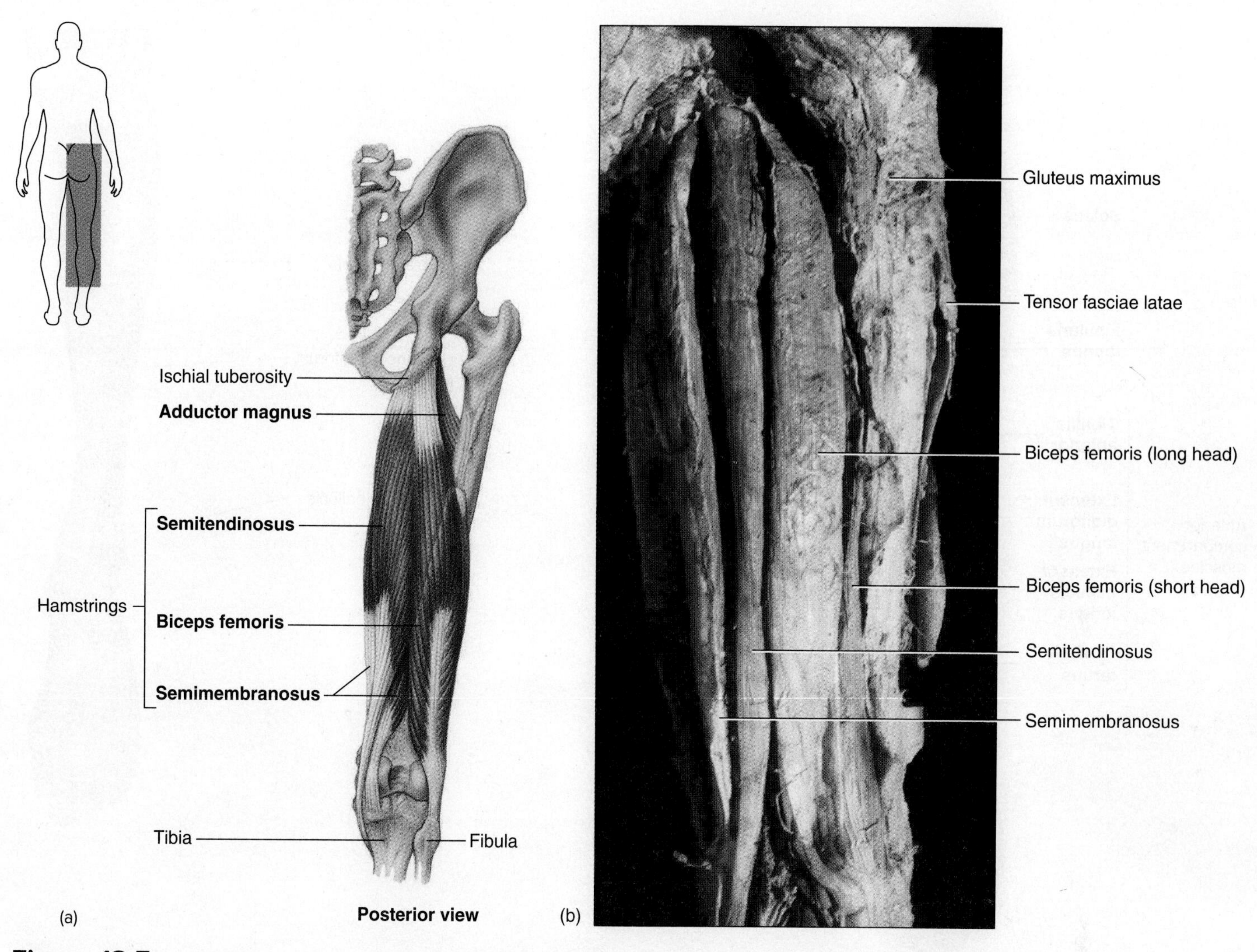

Figure 13.5 Muscles of the Right Thigh, Posterior View
(a) Diagram; (b) photograph.

on the middle and distal phalanges of all of the digits of the foot except the hallux. The **extensor hallucis longus** is the muscle that extends the hallux, as well as acting as a synergist to the extensor digitorum longus in dorsiflexion of the foot. Examine these muscles in the lab and compare them with figure 13.6.

The **fibularis** muscles are named for originating on and running the length of the fibula. The **fibularis longus** has a tendon that travels from the lateral side of the foot underneath to the medial side, crossing under the arch of the foot. The **fibularis brevis** parallels the fibularis longus, except that the tendon stops short and inserts on the lateral side of the foot at the fifth metatarsal. Both of these muscles have tendons that hook posterior to the lateral malleolus, and they both plantar flex and evert the foot. The third fibularis muscle, the **fibularis tertius,** does not arch behind the lateral malleolus, and it inserts on the dorsum of the fifth metatarsal. When it contracts, it dorsiflexes and everts the foot. Examine these muscles in figure 13.7.

Posterior Muscles

(A) **4** The **gastrocnemius** is a calf muscle that is the most superficial of the posterior leg group. The gastrocnemius crosses the knee joint, so it flexes the leg. It inserts on the calcaneus by way of the **calcaneal tendon,** plantar flexing the foot as well. The calcaneal tendon is also known as the Achilles tendon. Deep to the gastrocnemius is the **soleus** muscle. Unlike the gastrocnemius, the soleus does not originate on the femur; therefore, it does not cross the knee and has no action on the leg. It inserts on the calcaneus, sharing the calcaneal tendon with the gastrocnemius, and it plantar flexes the foot. The **popliteus** is a small muscle that crosses and unlocks the knee joint. Locate these muscles in figures 13.7 and 13.8. The **tibialis posterior** is a major muscle deep to the soleus that plantar flexes and inverts the foot. The tendon of the muscle runs along the medial aspect of the ankle. Near the tibialis posterior is the **flexor digitorum longus,** which is a muscle that inserts on the distal phalanges of all the digits of the foot except

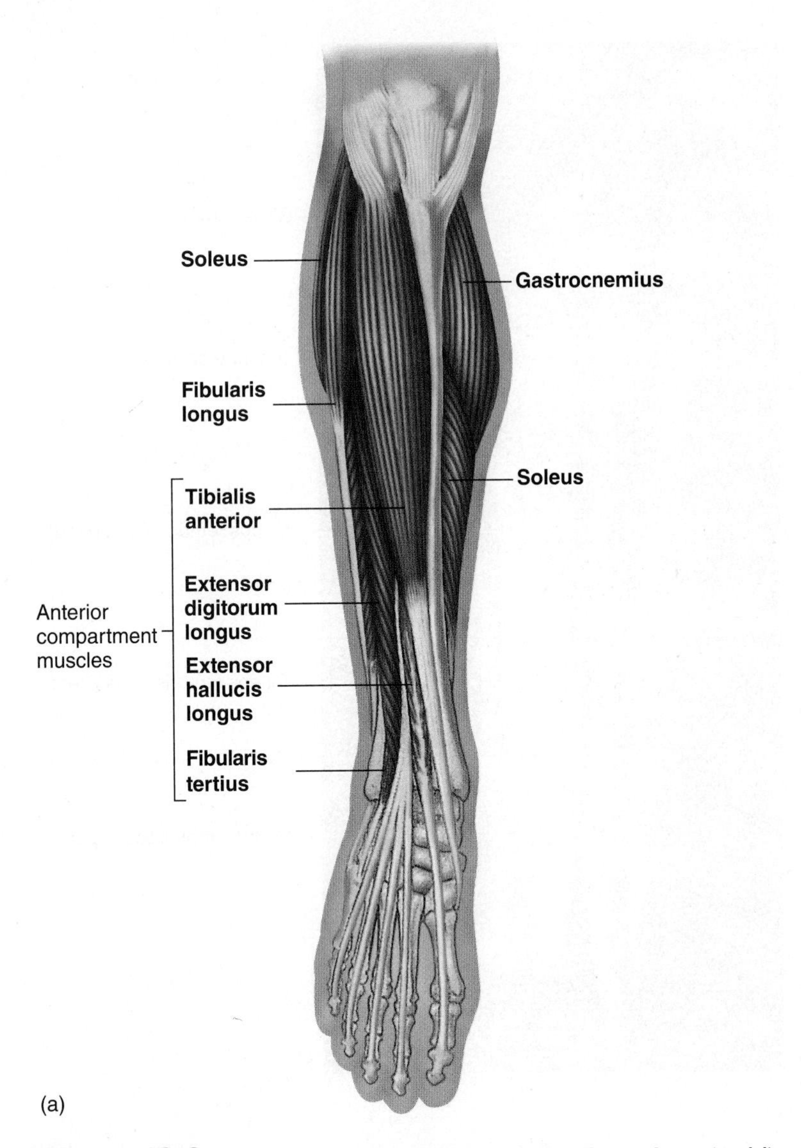

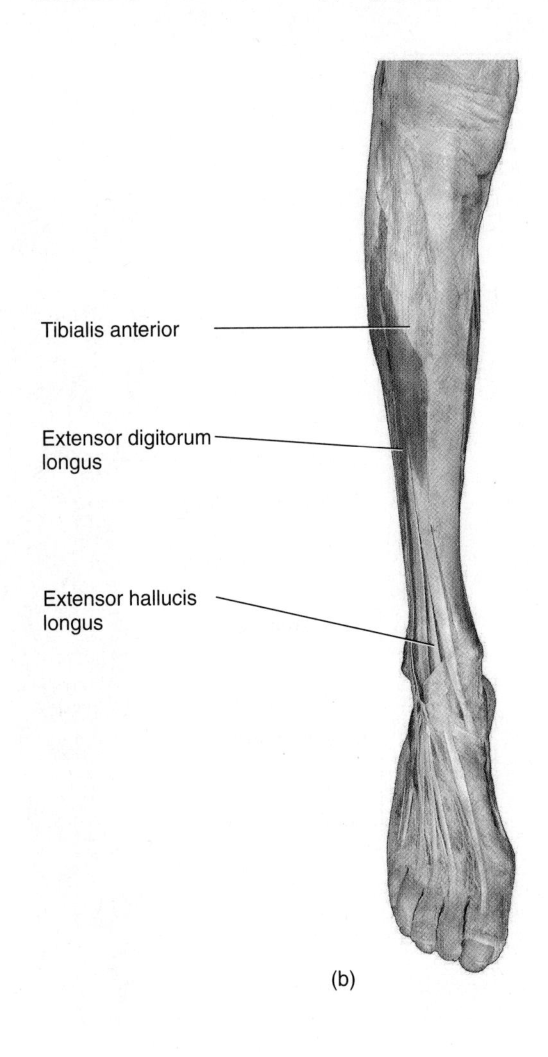

Figure 13.6 Muscles of the Right Leg and Foot, Anterior View
(a) Diagram; (b) photograph.

the hallux. The flexor digitorum longus has the action of flexing (curling) the toes in addition to plantar flexing and inverting the foot. The **flexor hallucis longus** flexes the hallux and aids the flexor digitorum longus in inverting the foot. Locate these muscles in the lab and compare them with figure 13.8.

Cat Dissection

If you have not done so already, remove the skin from the lower portion of the hind limb of the cat. Be careful cutting the skin from the distal portions of the leg, so that you do not cut through the tendons of the foot. Be careful in the dissection of the cat muscles that you *leave the blood vessels and nerves intact* as you dissect your specimen. You will be looking at these structures in later exercises. Pay particular attention to the **great saphenous vein,** which runs under the skin and should be preserved. Remove any excess fat and fascia from the muscles as you dissect the material. If you need to cut a muscle, bisect it perpendicular to the fiber direction, so that half of it is attached to the origin side and half of it is attached to the insertion side.

Cat Musculature

The cat has many muscles of the hip and thigh that are good models for studying human muscles. The quadrupedal nature of the cat lends itself to having a different size or placement of some of the thigh muscles, which are illustrated in this section.

Medial Muscles of the Thigh of the Cat

Ⓐ 5 Two major muscles on the medial aspect of the thigh in the cat are the **sartorius** and the **gracilis muscles.** Locate these muscles in figure 13.9 and note that they are much broader in the cat than in the human.

Cut through the sartorius and gracilis and reflect the ends of these muscles. You should see the deeper muscles of the thigh, including the **tensor fasciae latae,** the **vastus medialis,** the

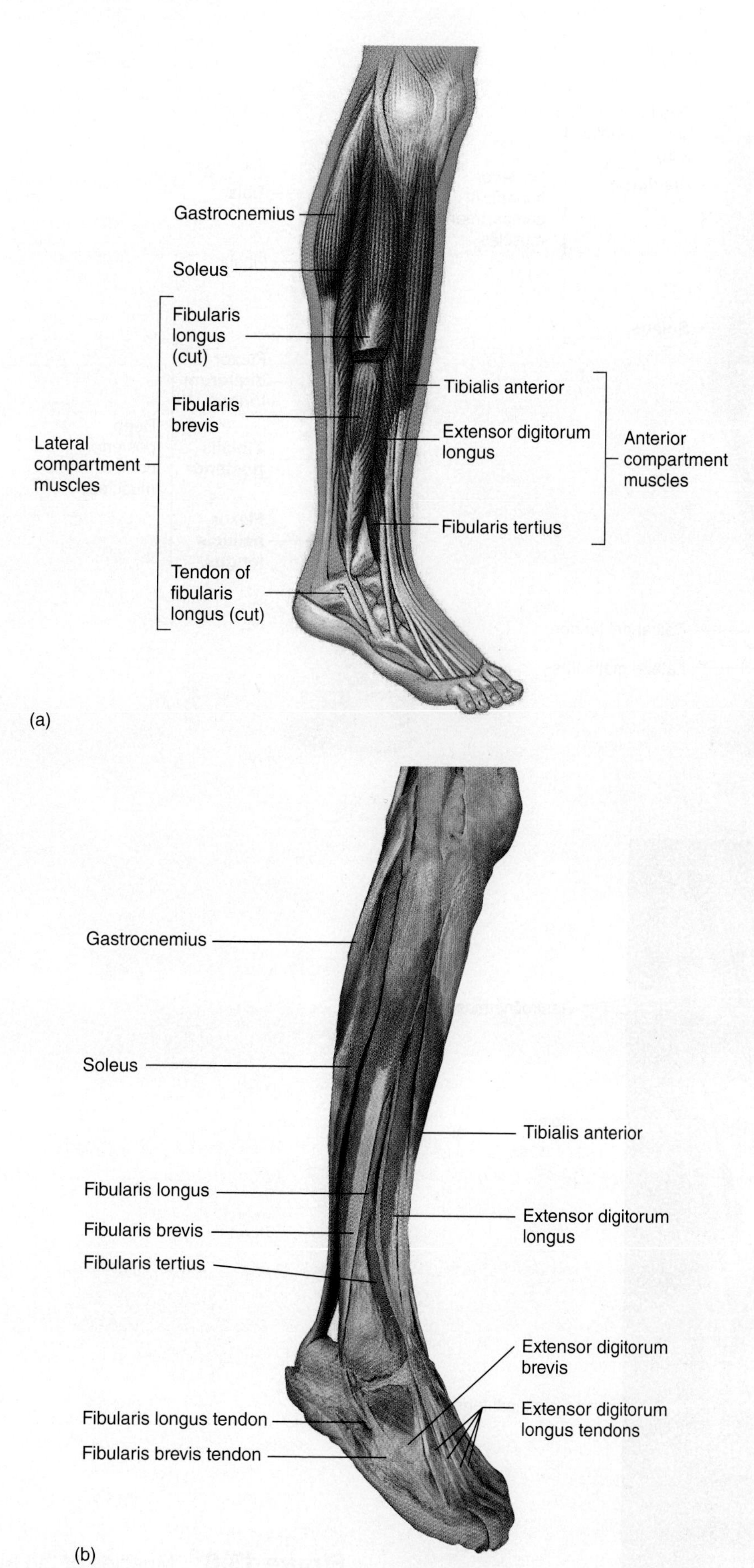

Figure 13.7 Muscles of the Right Leg and Foot, Lateral View
(a) Diagram; (b) photograph.

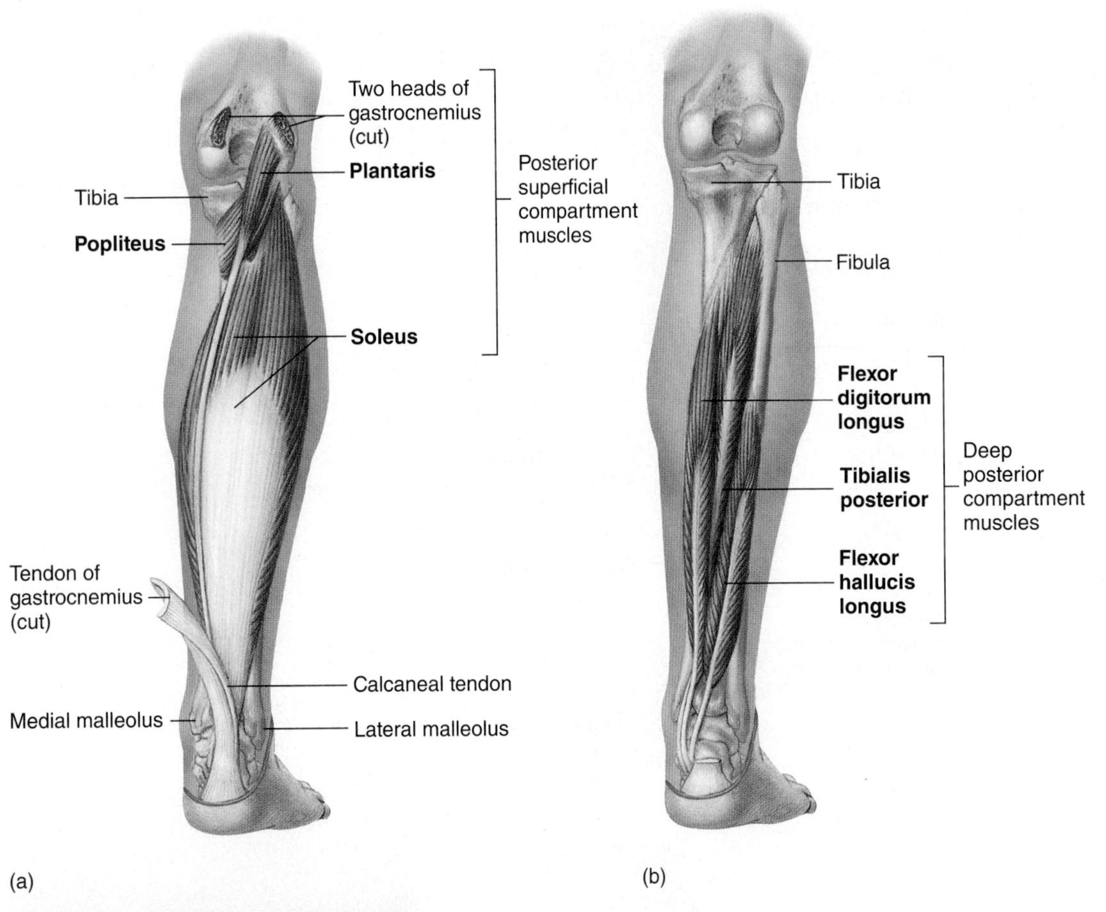

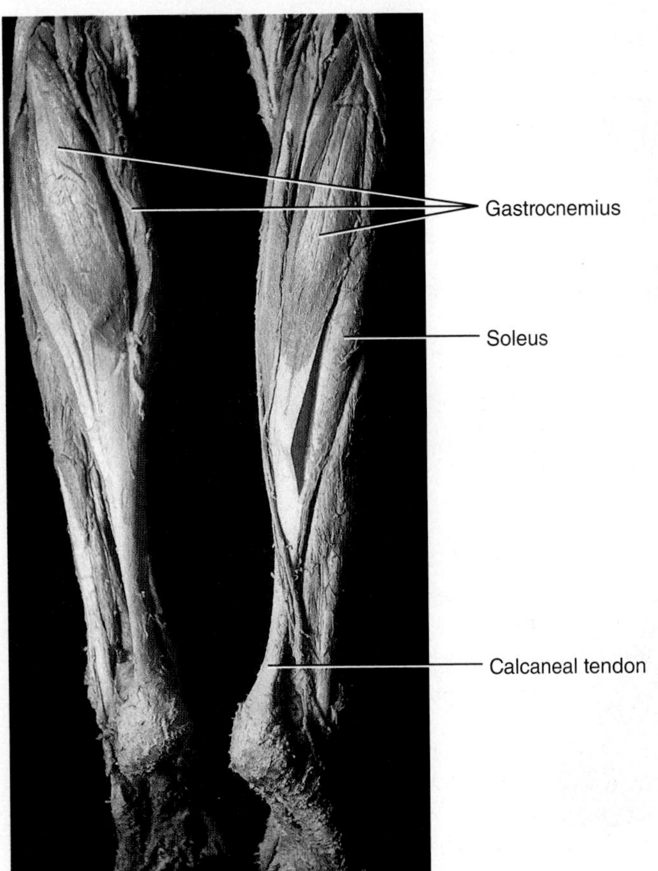

Figure 13.8 Muscles of the Right Leg and Foot, Posterior View

Diagram (a) superficial; (b) deep; (c) photograph.

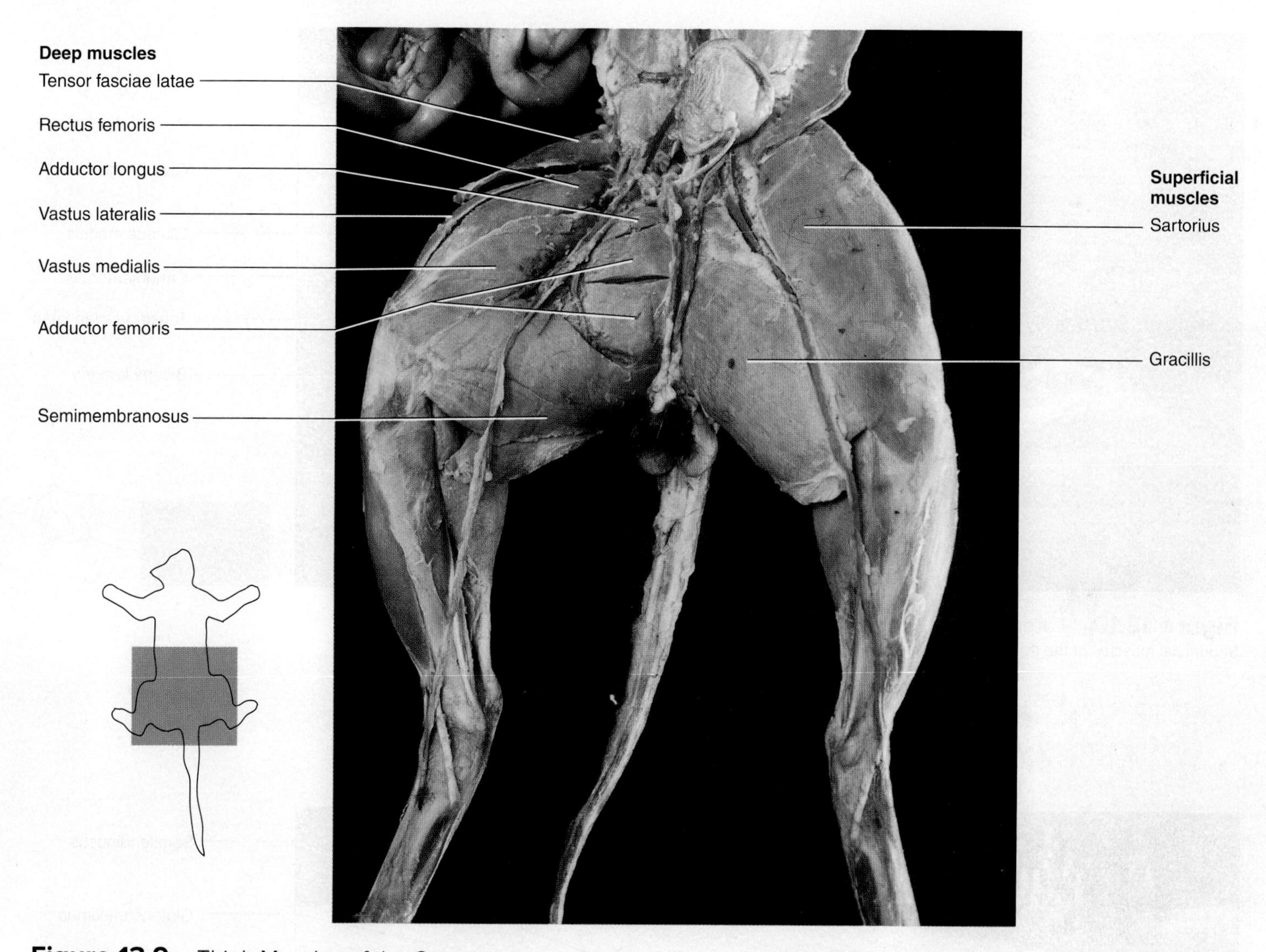

Figure 13.9 Thigh Muscles of the Cat
Superficial muscles (left side of cat); deep muscles (right side of cat).

adductor femoris (a large muscle that is specific to the cat), and the **semimembranosus.** The distal portion of the **semitendinosus** may also be seen from this view. Examine figure 13.9 for the deep muscles of the thigh and locate the **rectus femoris, vastus medialis,** and **vastus lateralis.** Bisect the rectus femoris in order to see the **vastus intermedius** muscle. The **adductor longus** is a thin muscle anterior to the **adductor femoris,** and it can be seen as a small muscle on the medial aspect of the thigh.

Lateral Muscles of the Thigh of the Cat

Ⓐ 6 On the lateral aspect of the thigh are numerous muscles, including the **biceps femoris;** the **tensor fasciae latae;** the **gluteus** muscles; and a muscle specific to the cat, the **caudofemoralis.** The biceps femoris is the largest and most lateral muscle of the thigh. The **gluteus medius** is larger than the **gluteus maximus** in cats due to the lengthening of the pelvic girdle. The **semimembranosus** is a large muscle in cats (much larger than the semitendinosus), and it can be seen from the lateral side of the cat. The adductor magnus and adductor brevis are not found in the cat. Examine these muscles in figures 13.10 and 13.11.

Posterior Muscles

Ⓐ 7 Examine the large **gastrocnemius** on the posterior aspect of the leg. The **soleus** is deep to the gastrocnemius and inserts with the gastrocnemius on the calcaneus. In cats the soleus has only one point of origin, the fibula; in humans the soleus originates on the tibia and fibula. Compare figure 13.12 with your dissection. Cut through the calcaneal tendon and lift the gastrocnemius and soleus, so that you can study the underlying muscles.

The **popliteus** is a small, triangular muscle that crosses the knee joint. Do not damage the nerves and blood vessels that pass over the popliteus because you will study them in the subsequent exercises.

The **flexor digitorum longus** runs along the medial side of the hind limb and flexes the digits of the cat. It joins with the **flexor hallucis longus,** which inserts on all the digits of the hind limb. The **tibialis posterior** is a narrow muscle that inserts on the tarsal bones of the foot. Locate these muscles in the cat and compare them with figure 13.13.

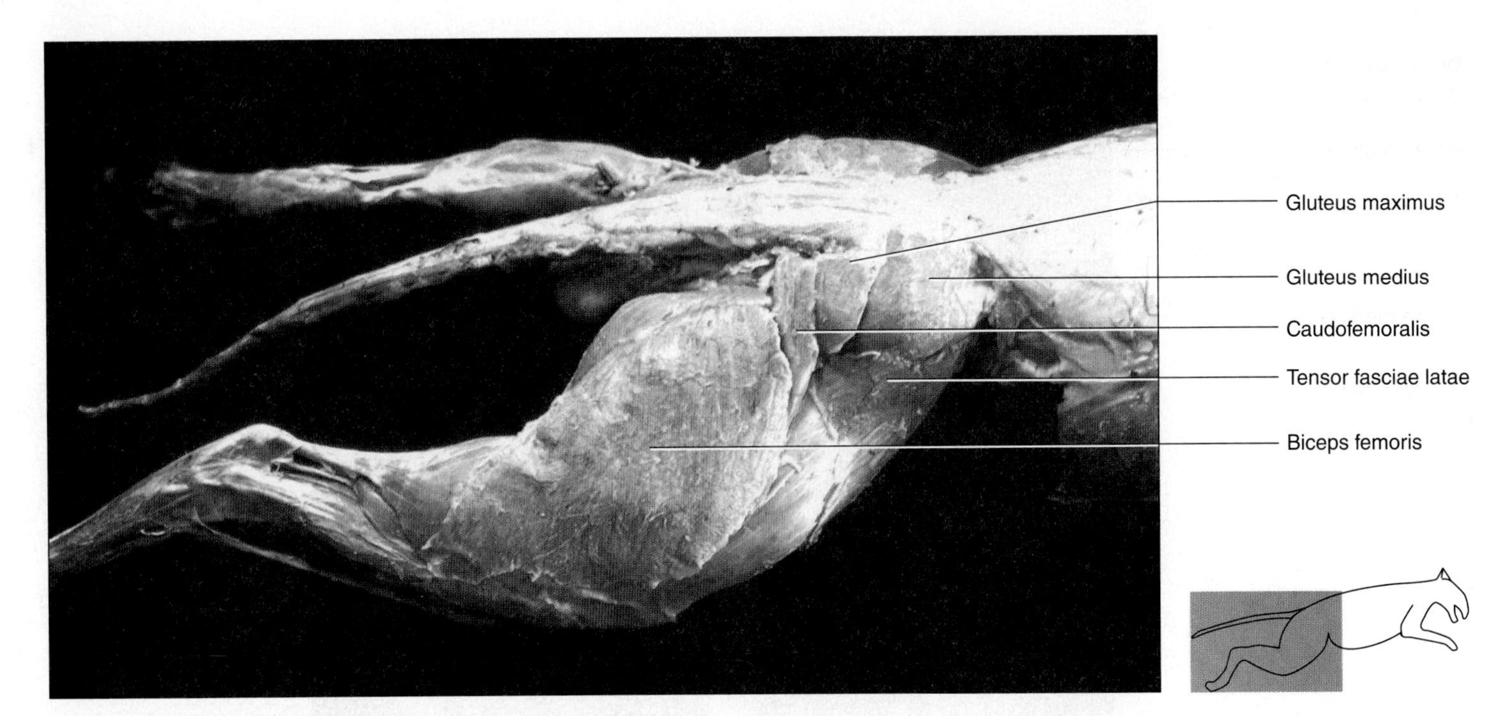

Figure 13.10 Lateral Thigh Muscles of the Cat
Superficial muscles of the right side.

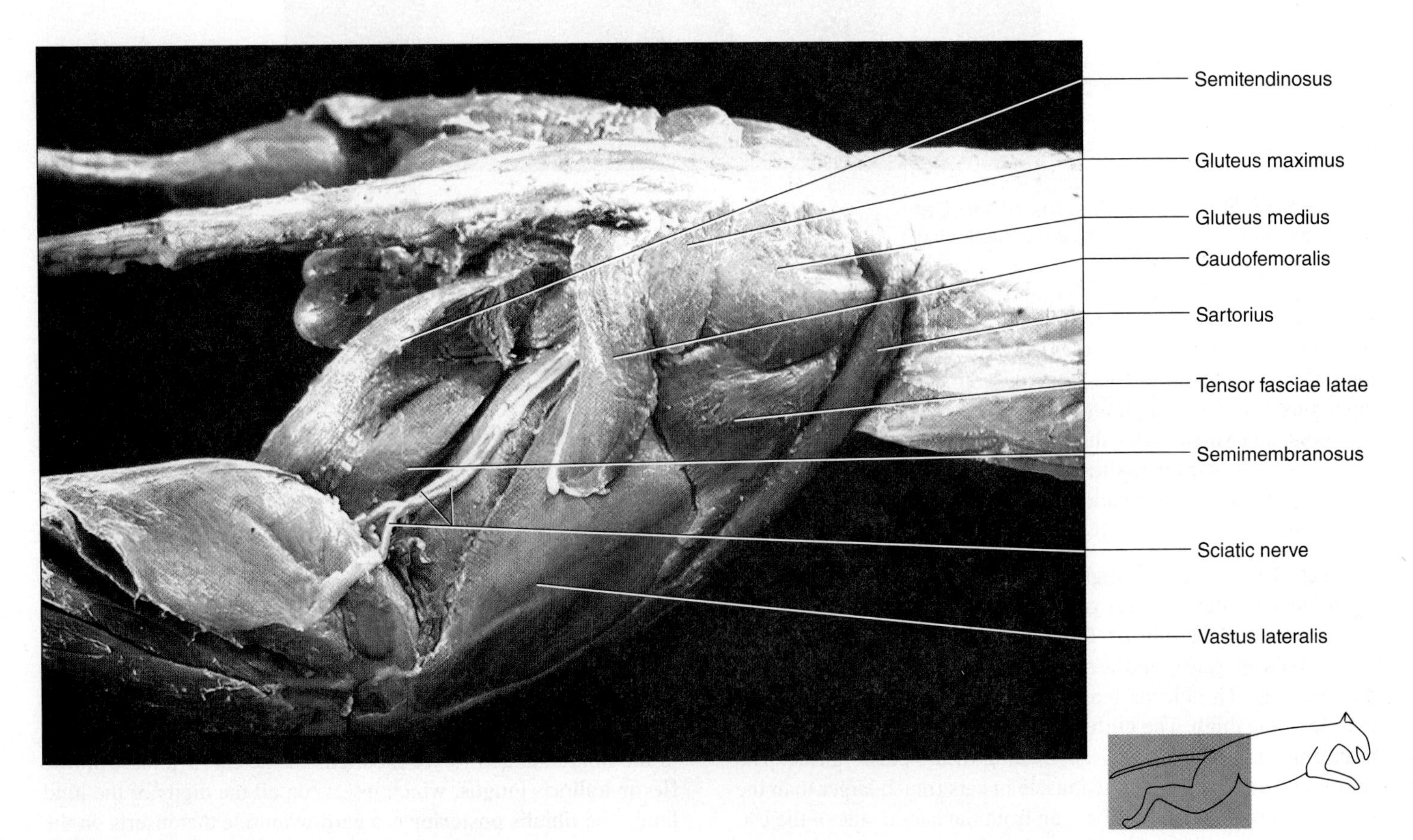

Figure 13.11 Lateral Thigh Muscles of the Cat
Deep muscles of the right side.

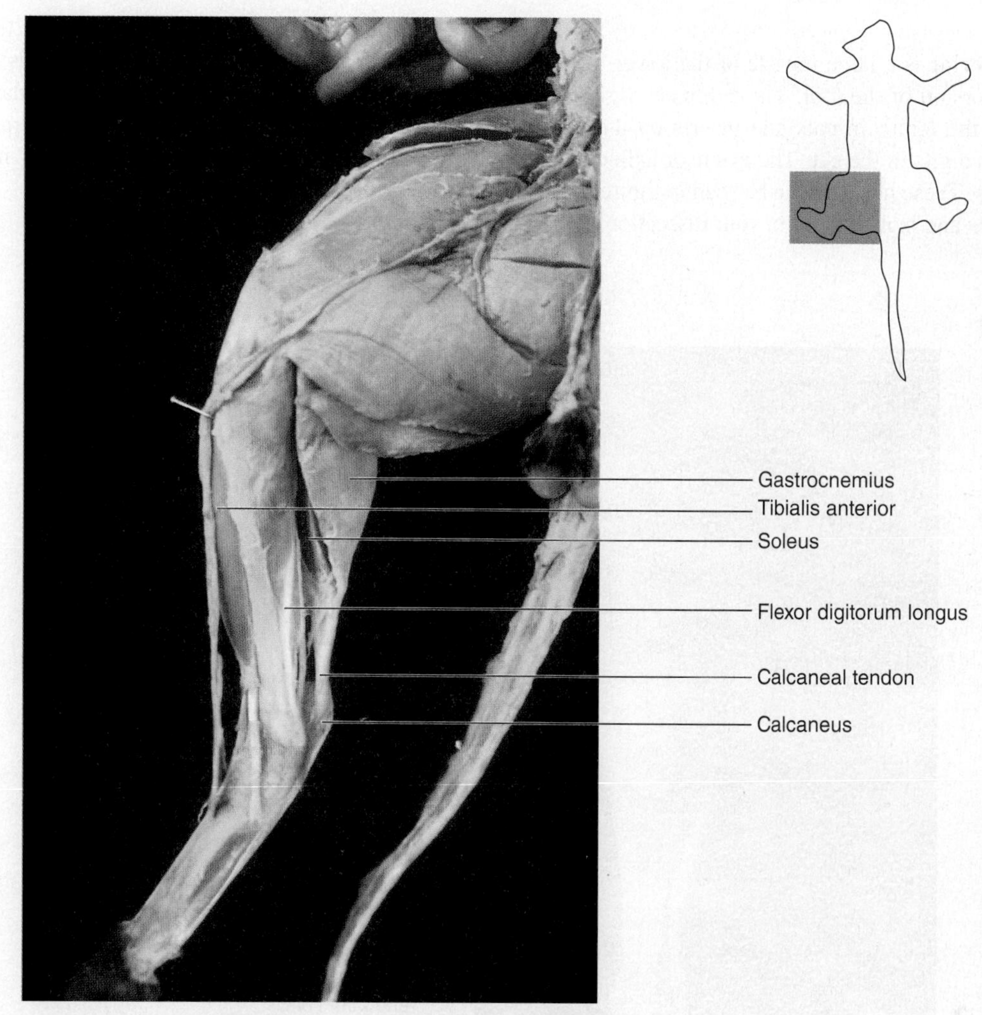

Figure 13.12 Superficial Muscles of the Right Leg of the Cat, Medial View

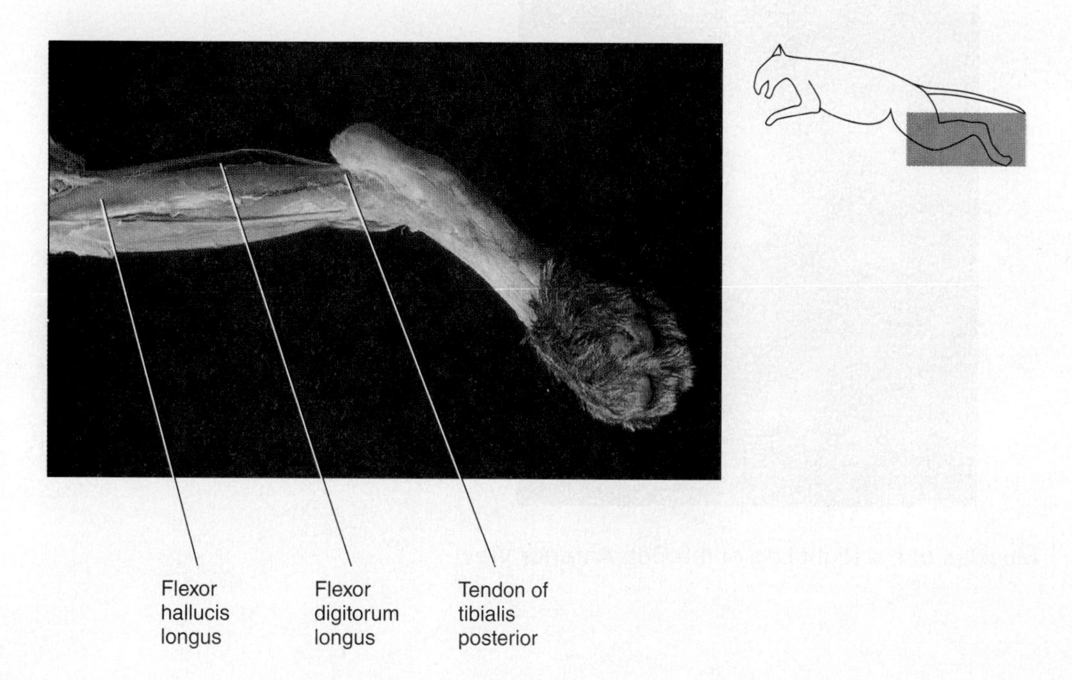

Figure 13.13 Deep Muscles of the Left Leg of the Cat, Lateral View

Anterior Muscles

(A) 8 The **tibialis anterior** is a large muscle of the lower limb; it inserts on the dorsum of the foot. The **extensor digitorum longus** originates on the femur in cats and inserts on the distal phalanges in all of the digits in the cat. The extensor hallucis longus is not found in cats. These muscles can be seen in figure 13.14. Examine these muscles and isolate them in your dissection.

The **fibularis longus** extends along the length of the fibula with the **fibularis brevis.** The fibularis longus inserts at the base of the metatarsals, whereas the fibularis brevis inserts on the fifth metatarsal. The **fibularis tertius** inserts into the tendon of the extensor digitorum muscle. These muscles can be seen in figure 13.15.

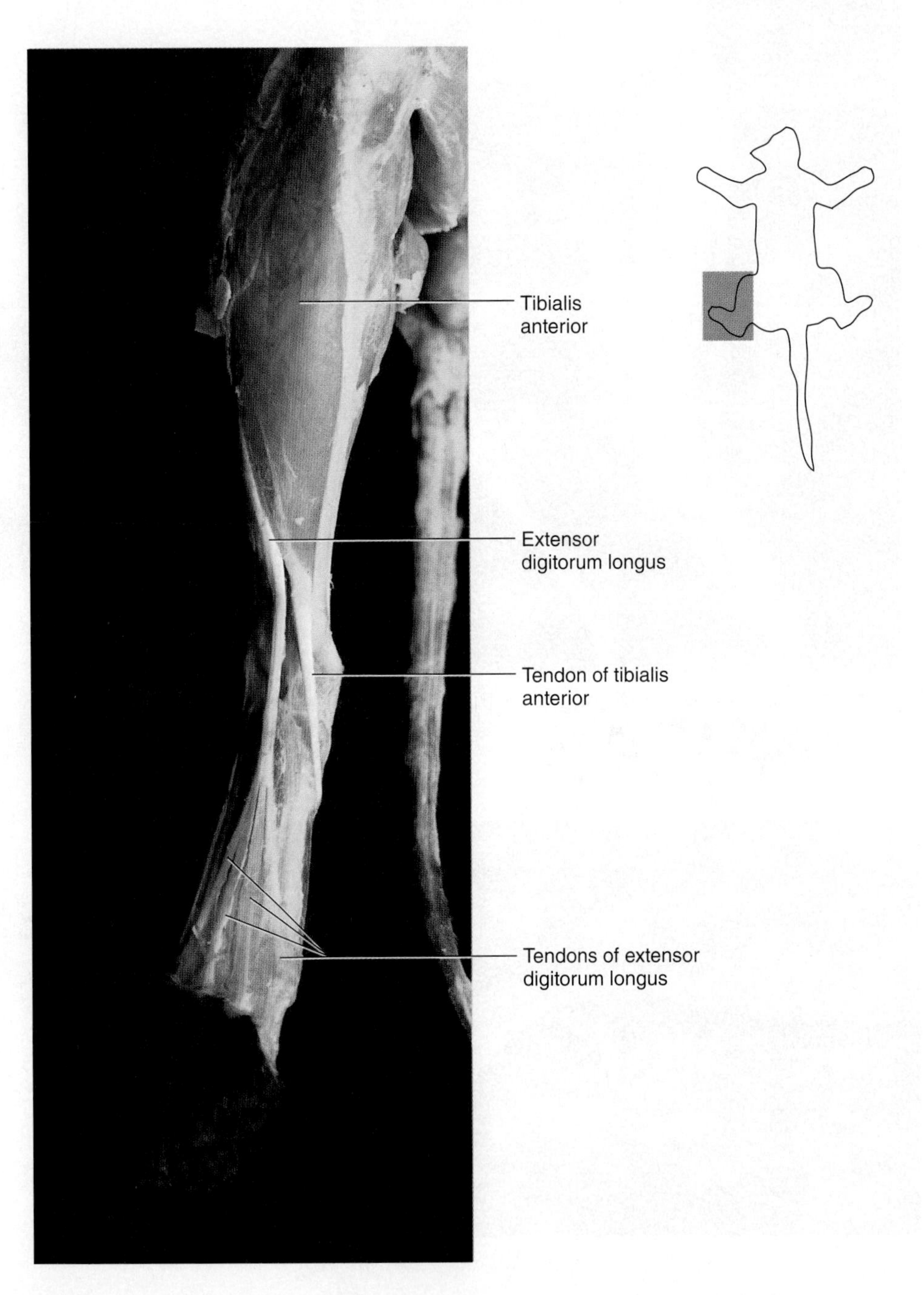

Figure 13.14 Muscles of the Right Leg of the Cat, Anterior View

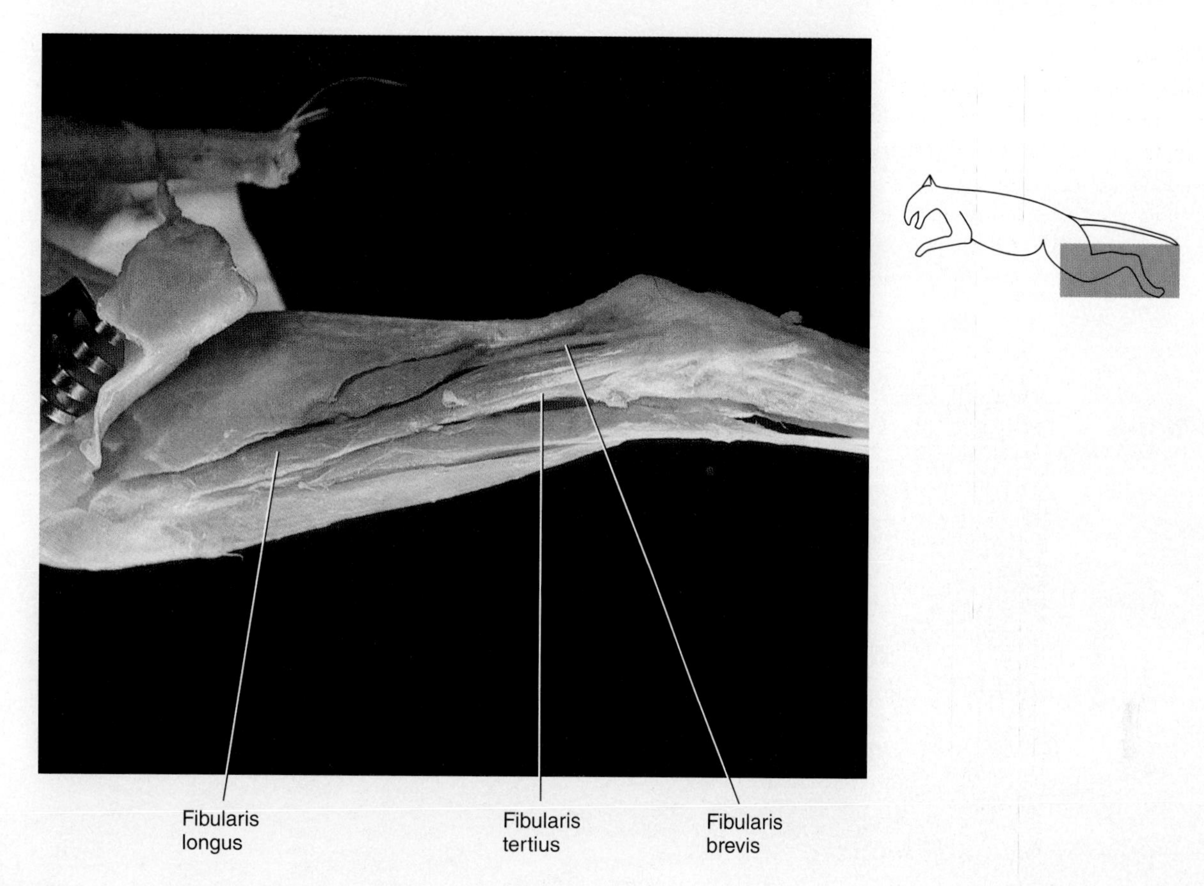

Figure 13.15 Muscles of the Left Leg of the Cat, Lateral View

NOTES

Exercise 13 Review

Muscles of the Hip, Thigh, Leg, and Foot

Name: ______________________

Lab time/section: ______________________

Date: ______________________

1. If you were to ride a horse, what muscles would you use to keep your seat out of the saddle as you rode? ______________________

2. How do the gluteus medius and gluteus minimus prevent you from toppling over as you walk? ______________________

3. What is a muscle that is an antagonist to the biceps femoris muscle? ______________________

4. What are two muscles that are synergists with the biceps femoris muscle? ______________________

5. What is the insertion of all the muscles of the quadriceps group? ______________________

6. How does the action of the rectus femoris differ from that of the other quadriceps muscles? ______________________

7. How many adductor muscles are there? ______________________

8. List two muscles in this exercise that are responsible for thigh flexion. ______________________

9. Where do the hamstring muscles originate as a group? ______________________

10. What is the action of the vastus lateralis? ______________________

11. Which muscle group is located on the anterior part of the thigh? ______________________

12. Is abduction of the thigh movement away from or toward the midline? ______________________

13. What muscle flexes the lumbar vertebrae as part of its action? ______________________________

____

14. Label the muscles in the following illustration using the terms provided.

gracilis	iliopsoas	pectineus
rectus femoris	sartorius	tensor fasciae latae
vastus lateralis	vastus medialis	

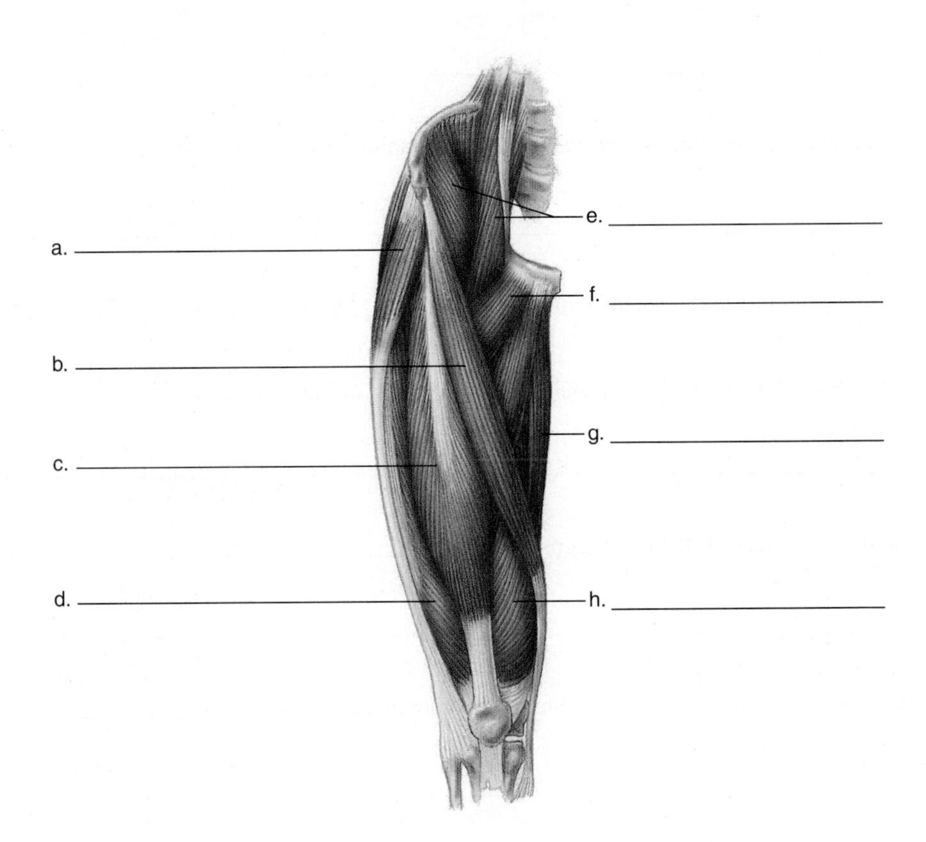

15. What is the origin of the gastrocnemius? ______________________________

16. What is the insertion of the tibialis anterior in humans? ______________________________

17. How does the action of the fibularis longus in humans differ from that of the fibularis tertius? ______________________________

____

18. What is the action of the extensor hallucis longus? ______________________________

19. Label the following illustration using the terms provided.

adductor femoris   gracilis   sartorius
semimembranosus   vastus medialis

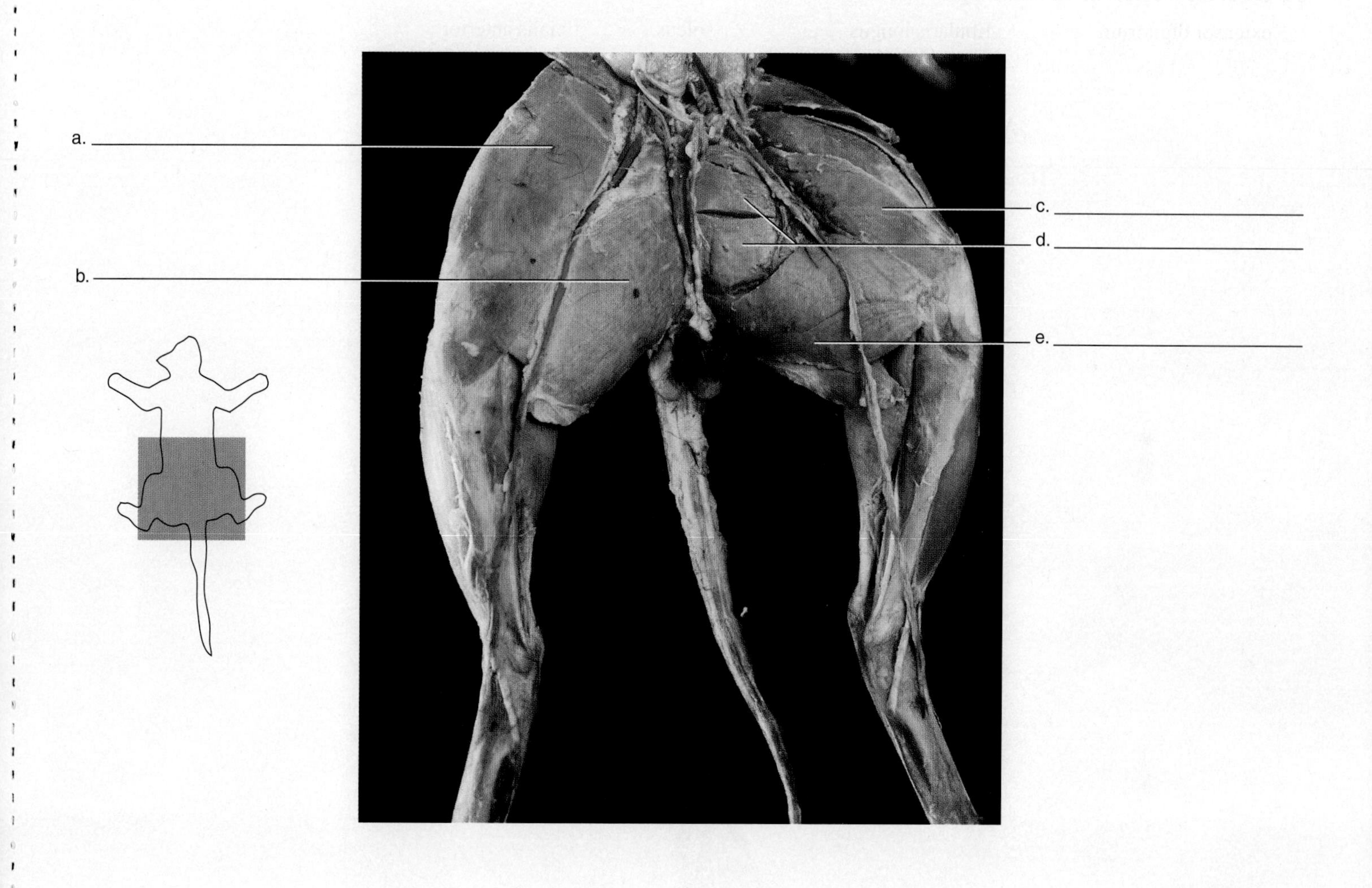

20. The calf is made up of what two major muscles? ________________________________

____

21. Plantar flexion occurs by what muscles? ________________________________

____

22. What muscle extends the toes? ________________________________

23. Name a muscle in this exercise that dorsiflexes the foot. ________________________________

24. Plantar flexion and eversion of the foot occur by the action of what muscle? ________________________________

25. What is the insertion of the soleus? ________________________________

26. What is the insertion of the fibularis tertius muscle? ______________________________

27. What is the insertion of the flexor digitorum longus? ______________________________

28. Label the muscles in the following illustration using the terms provided.

extensor digitorum | fibularis longus | soleus | tibialis anterior

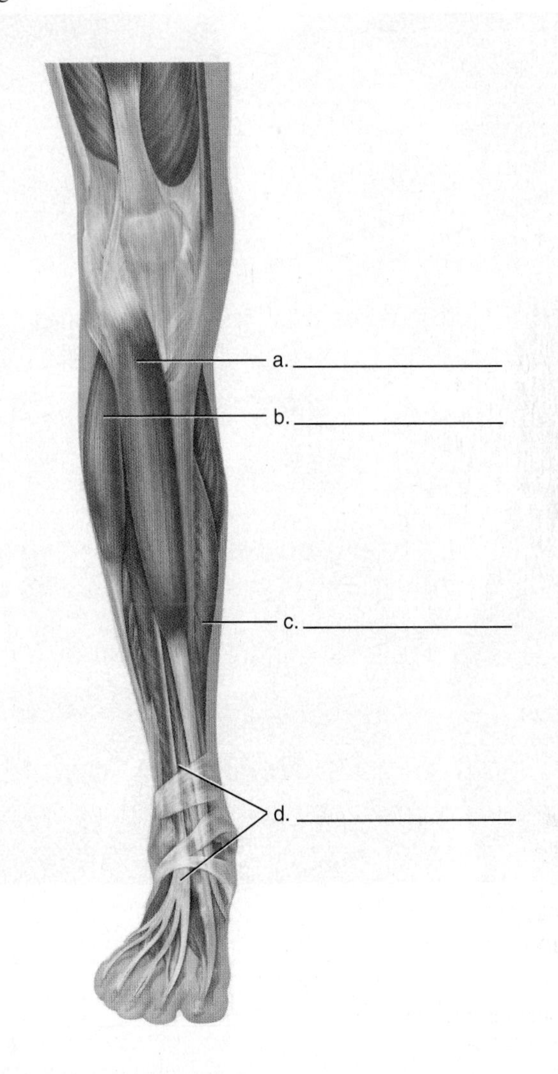

29. Label the muscles in the following illustration using the terms provided.

extensor digitorum longus | fibularis longus | gastrocnemius
soleus | tibialis anterior

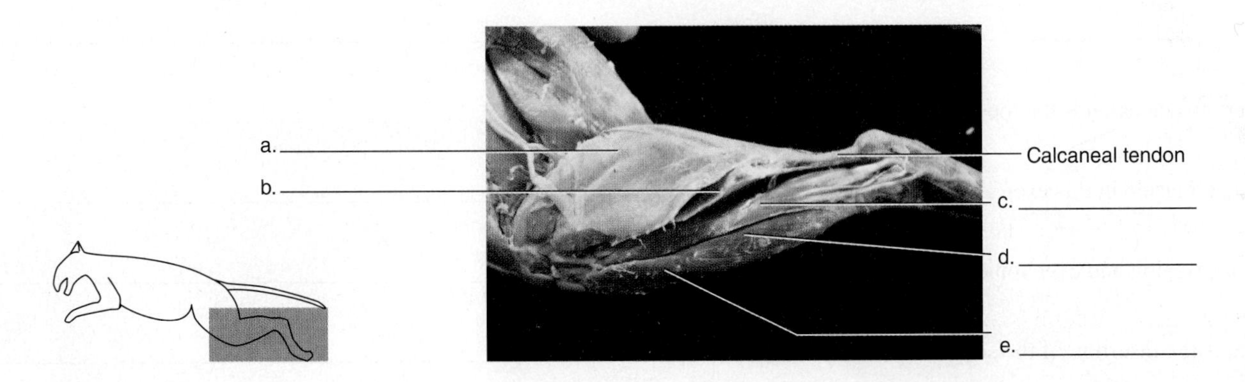

Exercise 14

Muscles of the Head and Neck

Introduction

In this exercise, you continue your study of muscles with the muscles of the head and neck. These muscles can be grouped into a few functional classifications. The head muscles can be subdivided into muscles of mastication (chewing), muscles of facial expression, and muscles that close the eyes and mouth. The neck muscles are those that move the neck or head or those that move the hyoid, larynx, or tongue during speech or swallowing. These muscles are discussed in the Seeley text in chapter 10, "Muscular System: Gross Anatomy."

Objectives

At the end of this exercise, you should be able to

1. locate the muscles of the head and neck on a torso model, chart, cadaver, or cat;
2. list the origin, insertion, and action of each muscle presented;
3. describe what nerve controls each muscle;
4. name all the muscles that have an action on a joint, such as all the muscles that flex the head;
5. reproduce the actions of selected muscles on your own head or neck.

Materials

Human torso model and head and neck model
Human muscle charts
Articulated skeleton
Cadaver
Cat
Cat wetting solution
Materials for cat dissection
- Dissection trays
- Scalpel and two or three extra blades
- Gloves (household latex gloves work well for repeated use)
- Blunt (Mall) probe
- String and tags
- Pins
- Forceps and sharp scissors
- First aid kit in lab or prep area
- Sharps container
- Animal waste disposal container

Procedure

Human Muscles

Muscles of the Neck

(A)[1] Examine a model or chart of the human musculature or cadaver (if available) as you read the following descriptions. The details of the muscles are listed in table 14.1. Examine a skull or an articulated skeleton and review the bony markings as you study the origins and insertions of the muscles in this exercise. Once you know the origins or insertions, the actions are easier to understand. The **trapezius** is a muscle that has an action on the neck. It was covered in exercise 12, as it also has an action on the shoulder.

The **scalene** muscles are found on the lateral side of the neck. They are bounded by the sternocleidomastoid in the front and the levator scapulae in the back. They rotate the neck or elevate the ribs. Locate these muscles in lab and examine figure 14.1. The scalene muscles are also illustrated in figure 15.3 in the following exercise.

The **sternocleidomastoid** muscle rotates the head in a unique way. Place your hand on your right sternocleidomastoid and turn your head to the right. Notice how the muscle does not contract. Now turn your head to the left, and you can feel the muscle contract. The right sternocleidomastoid turns the head to the left and the left sternocleidomastoid turns the head to the right.

The **sternohyoid, sternothyroid,** and **omohyoid** are all named for their origins and insertions. In the sternohyoid and sternothyroid, the sternum anchors the stable part of the muscle (the origin), and the hyoid and thyroid cartilage of the larynx moves when the respective muscles contract. In the omohyoid (*omo* = shoulder), the scapula is the origin. The hyoid is depressed when the omohyoid contracts. Examine the material in lab and compare it with figure 14.1 for the muscles of the anterior neck.

The **digastric** is so-named because it is a muscle with two bellies. Few muscles depress the mandible. The digastric is one that does. In addition to this, the digastric inserts on the hyoid, which is important in tongue movement for speech and swallowing. Deep to the digastric is the **mylohyoid,** a broad muscle of the floor of the mouth that pushes the tongue superiorly when swallowing. These muscles can be seen in figure 14.1.

The **platysma** is a broad, thin muscle that has a soft origin (on the fascia of the pectoral and deltoid muscles). It inserts on the mandible and skin of the lips and can be seen if you elevate your chin and subsequently pout. The thin wings that stick out on the side of your neck are the edges of the platysma muscle. Examine the platysma in figure 14.2 and in the models or on the cadaver in the lab.

TABLE 14.1 Muscles of the Head and Neck

Name	Origin	Insertion	Action	Innervation
Neck Muscles				
Scalenes (anterior, middle, and posterior)	Transverse processes of cervical vertebrae	Ribs 1 and 2	Flexes and rotates neck, elevates ribs 1 and 2	Spinal nerves C4–8
Sternocleidomastoid	Manubrium and medial clavicle	Mastoid process, superior nuchal line	One: rotates and extends neck; both: flex neck	Accessory nerve (XI)
Sternohyoid	Manubrium of sternum	Hyoid	Depresses hyoid	Spinal nerves C1–3
Sternothyroid	Manubrium of sternum	Thyroid cartilage of larynx	Depresses larynx	Spinal nerves C1–3
Omohyoid	Superior border of scapula	Hyoid	Depresses and fixes hyoid	Spinal nerves C1–3
Digastric	Mastoid process of temporal (posterior belly)	Mandible near midline (anterior belly)	Elevates, protracts, and retracts hyoid; depresses mandible	Trigeminal (V) and facial (VII) nerves
Mylohyoid	Body of mandible	Hyoid	Elevates floor of mouth and tongue	Trigeminal nerve (V)
Muscles of the Face				
Platysma	Fascia covering pectoralis major and deltoid	Skin over inferior border of mandible	Depresses lower lip	Facial nerve (VII)
Frontalis	Epicranial aponeurosis	Skin superior to orbit	Raises eyebrows, draws scalp anteriorly	Facial nerve (VII)
Occipitalis	Occipital bone	Epicranial aponeurosis	Draws scalp posteriorly	Facial nerve (VII)
Orbicularis oculi	Frontal bone, maxilla	Skin of eyelid	Closes eye	Facial nerve (VII)
Orbicularis oris	Fascia of facial muscles, maxilla, and mandible	Skin of lips	Closes lips	Facial nerve (VII)
Corrugator supercilii	Nasal bridge, orbicularis oculi	Skin of eyebrow	Depresses and pulls eyebrows together	Facial nerve (VII)
Zygomaticus (major and minor)	Zygomatic bone	Muscle and skin at angle of mouth	Elevates and abducts upper lip	Facial nerve (VII)
Risorius	Fascia of masseter, platysma	Skin at angle of orbicularis oris	Abducts corner of mouth	Facial nerve (VII)
Mentalis	Mandible	Skin of chin	Elevates lower lip	Facial nerve (VII)
Buccinator	Maxilla, mandible	Orbicularis oris at angle of mouth	Retracts angle of mouth, flattens cheek	Facial nerve (VII)
Depressor labii inferioris	Lower border of mandible	Skin of lower lip, orbicularis oris	Depresses lower lip	Facial nerve (VII)
Levator labii superioris	Maxilla	Skin and orbicularis oris of upper lip	Elevates upper lip	Facial nerve (VII)
Muscles of Mastication				
Temporalis	Temporal fossa	Coronoid process, mandibular ramus	Elevates and retracts mandible	Trigeminal nerve (V)
Masseter	Zygomatic arch	Lateral ramus of mandible	Elevates and protracts mandible	Trigeminal nerve (V)
Pterygoids (medial and lateral)	Pterygoid plate of sphenoid bone	Medial ramus and condylar process of mandible	Protracts, elevates, and depresses mandible (for chewing)	Trigeminal nerve (V)

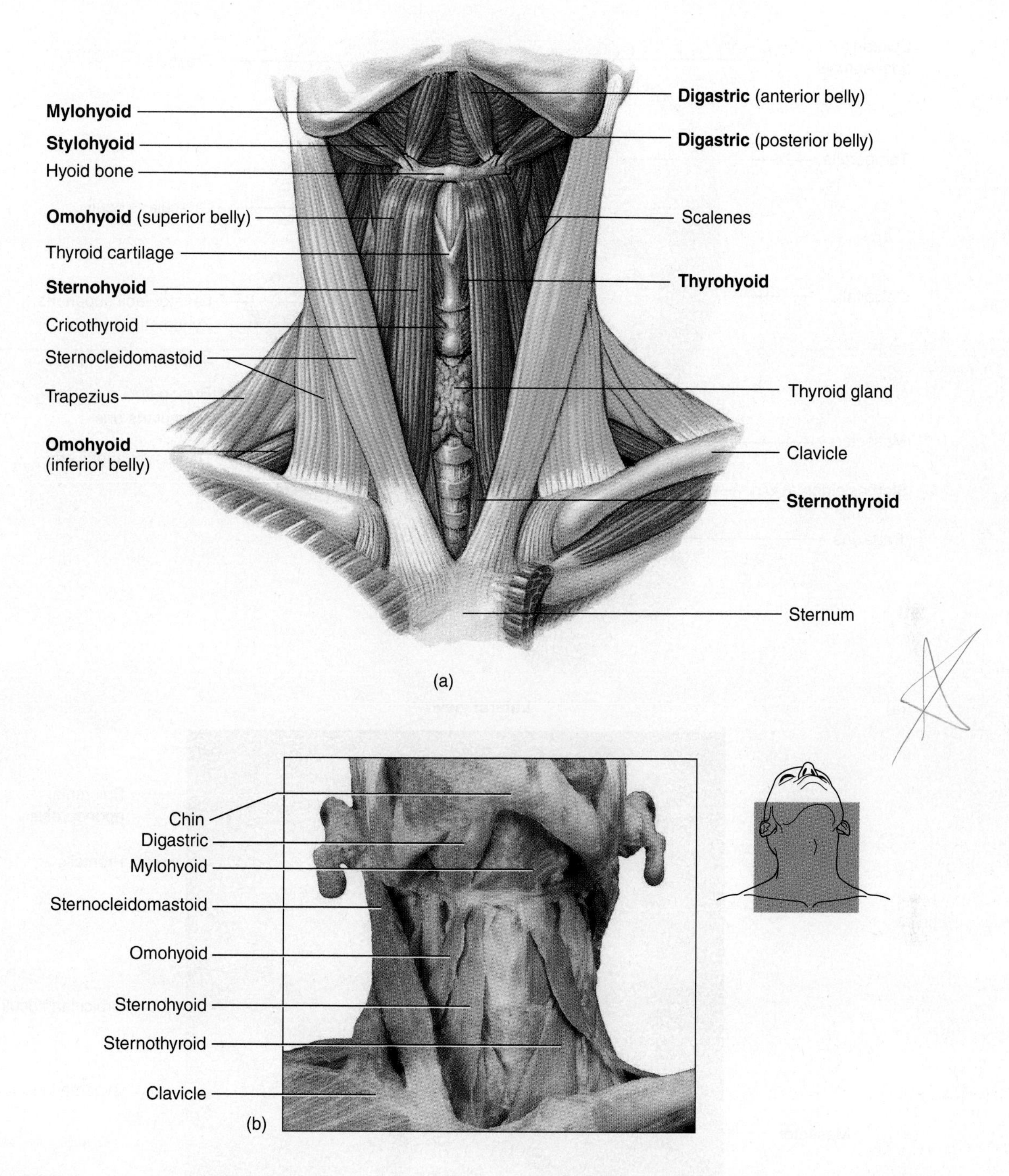

Figure 14.1 Muscles of the Anterior Neck
(a) Diagram; (b) photograph.

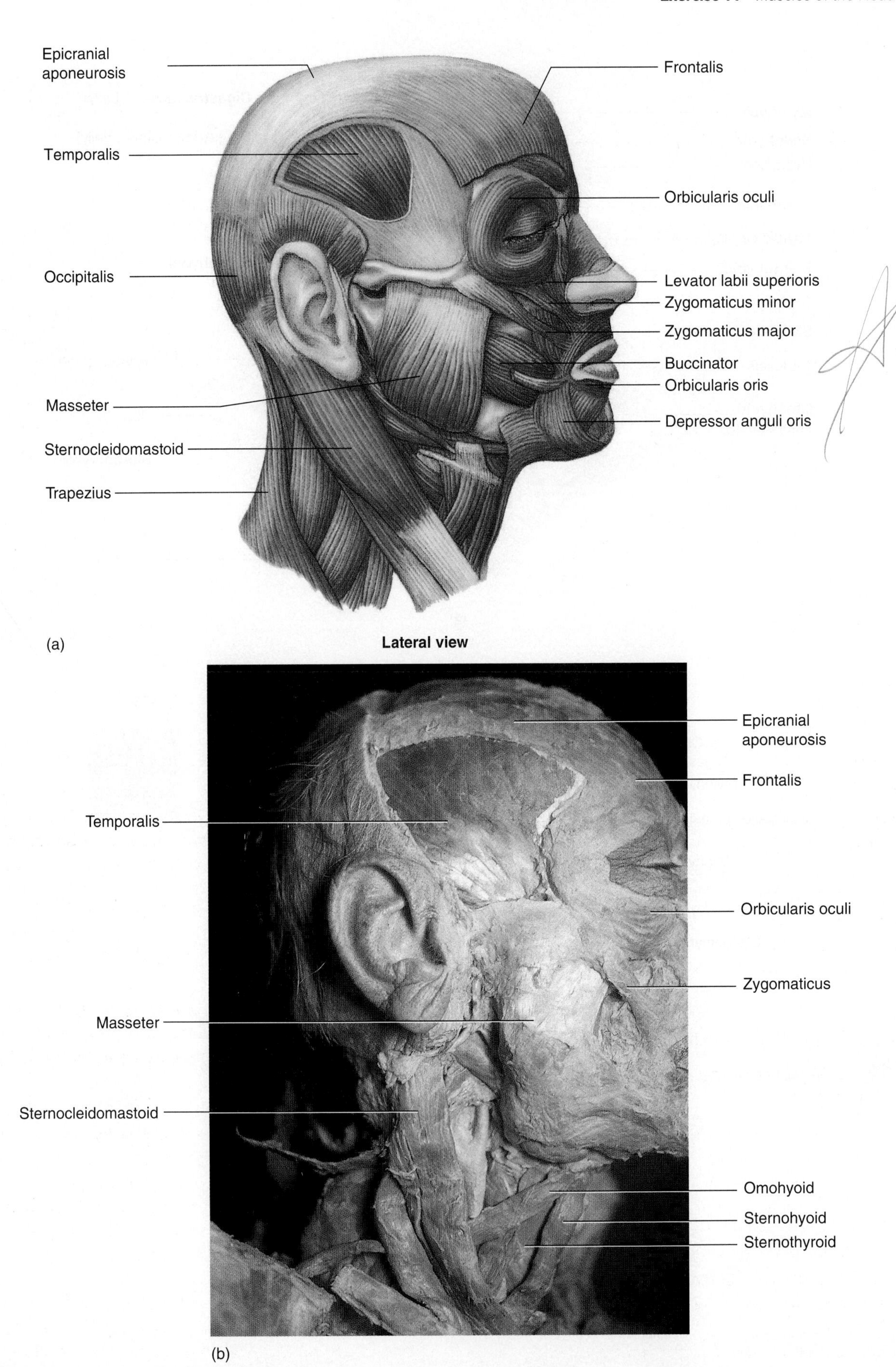

Figure 14.2 Muscles of the Head, Lateral View
(a) Diagram; (b) photograph.

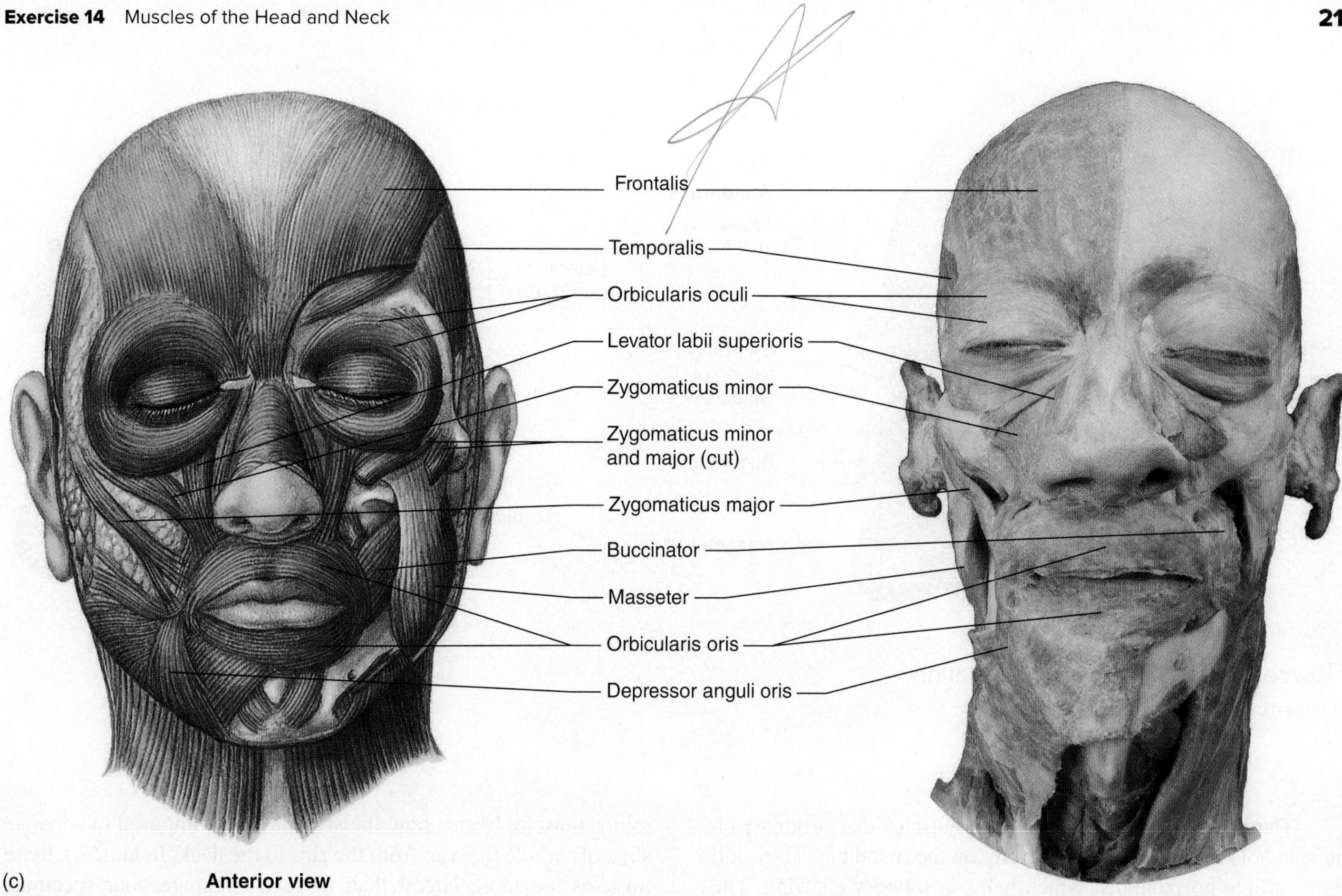

Figure 14.2 *Continued.*
(c) Diagram, anterior view.

Muscles of the Head

The **frontalis** muscle attaches to a broad, flat, tendinous sheet on the superior aspect of the skull known as the **epicranial aponeurosis.** The insertion of the frontalis is on the eyebrow region. If the occipitalis is contracted and the frontalis contracts, then the frontalis elevates the eyebrows. If the occipitalis is not contracting, then the scalp is brought forward, as in frowning. The **occipitalis** is a functional continuation of the frontalis muscle in that both muscles attach to the epicranial aponeurosis. The occipitalis pulls the scalp posteriorly. These muscles are sometimes discussed as one muscle, the **occipitofrontalis.** Examine these muscles in figure 14.2.

The **orbicularis oculi** and the **orbicularis oris** are sphincter muscles that close the eyes and the mouth, respectively. Sphincter muscles function similar to strings of a drawstring purse. The orbicularis oculi has a medial, bony origin and an insertion on the eyelid. The muscles encircle the eye and close the eyelids. The orbicularis oris originates on fascia and facial muscles near the mouth and inserts on the skin of the lips, closing the lips. Examine the facial muscles in lab and in figure 14.2.

The **corrugator supercilii** muscle has a medial point of origin between the eyebrows and inserts laterally on the eyebrow. It furrows the eyebrows.

The **zygomaticus major** and **zygomaticus minor** elevate the corners of the mouth by pulling them superiorly and laterally, as in smiling or laughing. They are named for their origin on the zygomatic bone. The **risorius** is known as the laughing muscle because it pulls the lips laterally. It does not have a bony point of origin but attaches to the fascia of the masseter and platysma.

The **mentalis** originates on the chin (anterior mandible) and, like the platysma, is a pouting muscle. The **buccinator** muscle of the cheek runs in a horizontal direction. It puckers the cheeks, as in trumpet playing, and pushes food toward the molars in chewing.

The **depressor labii inferioris** pulls the lower corners of the mouth inferiorly when pouting. It is named for its action (depressing the lower lip), whereas the **levator labii superioris** muscle raises the skin of the lips and expands the nostrils, as in the expression of extreme disgust. Look at these muscles in lab and in figure 14.2.

Muscles of Mastication

The **temporalis** is a powerful muscle that closes the jaw. It is a chewing muscle, or a muscle of *mastication.* The temporal fossa is so named because it is a depression medial to the zygomatic arch (although, when you examine the skull, the temporal fossa appears slightly domed). The temporalis muscle is deep to the zygomatic arch and inserts on the coronoid process and on the superior and medial ramus of the mandible.

The **masseter** is a large muscle of the head whose synergist is the temporalis. It powerfully closes the jaws. If you place your fingers on the ramus of the mandible and clench your teeth, you can feel the masseter tighten. These muscles are seen in figures 14.2 and 14.3.

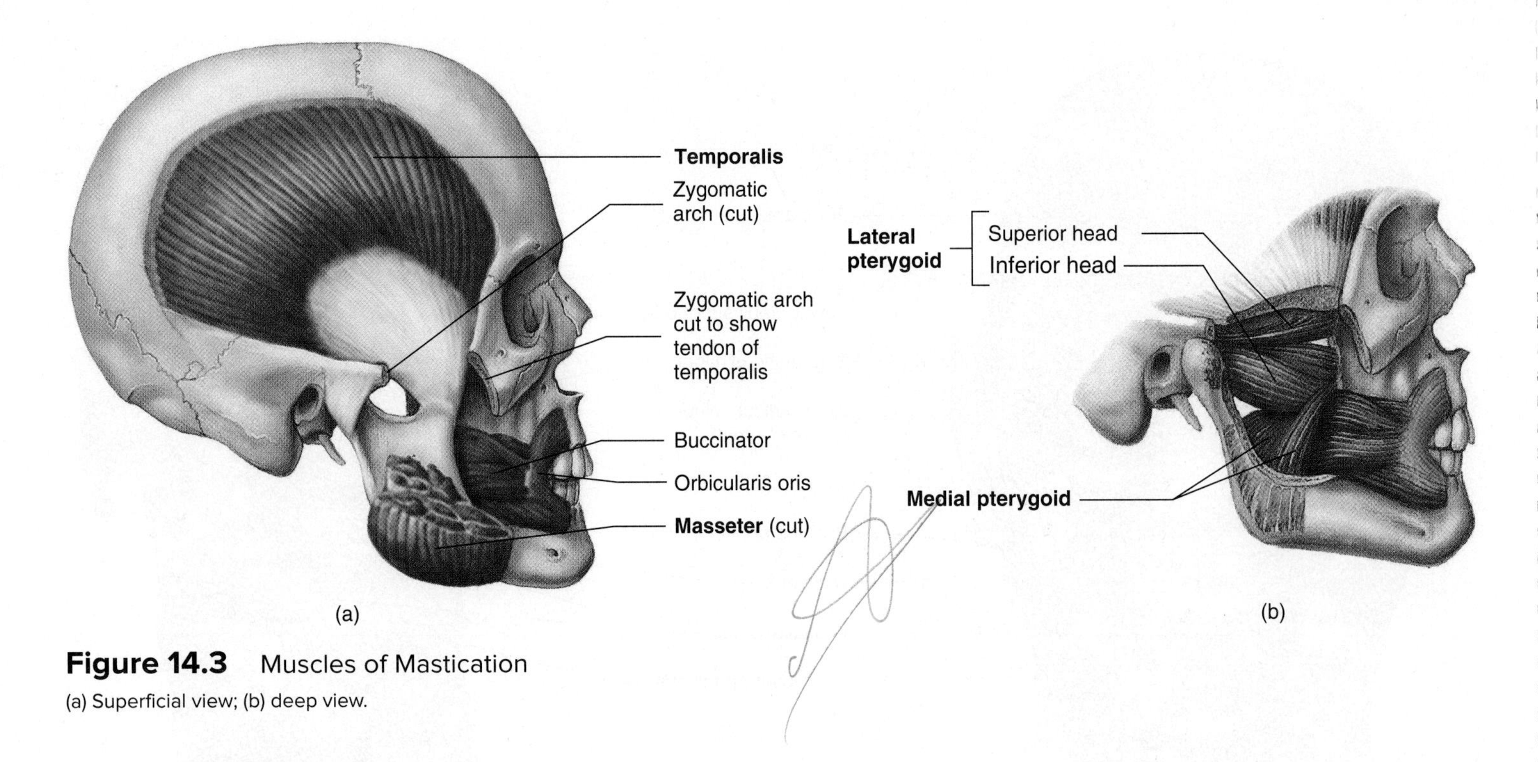

Figure 14.3 Muscles of Mastication

(a) Superficial view; (b) deep view.

The **pterygoid** muscles are deep muscles that originate on the sphenoid bone and insert laterally on the mandible. They pull the mandible horizontally, which helps in rotatory chewing. The temporalis and masseter open the jaw, whereas the pterygoids provide the sideways movement characteristic of a person chewing gum. These muscles can be seen in figure 14.3.

Ⓐ[2] *Activities*

1. When you turn your head to the left, which sternocleidomastoid muscle contracts?

   ____

2. Tilt your head so that your chin is elevated and pout your lower lip. What muscle forms a thin membrane along the anterolateral neck?

   ____

3. Purse your lips and feel the buccinator muscle as it contracts in the cheek.
4. Clench your teeth and palpate the temporalis muscle and the masseter.
5. Examine figure 14.4 for surface views of facial muscles. Fill in which muscles are represented in the photographs.

Muscles of the Head and Neck of the Cat

Ⓐ[3] In dissecting the muscles of the head and neck, take care not to cut through the digestive structures, such as the salivary glands, and the circulatory structures, such as the veins and arteries. You will study these structures in later exercises.

The **platysma** in the cat was removed during the skinning process, and it will not be seen unless you kept the skin with the cat. The **scalenes** can be dissected by reflecting the pectoralis minor muscle. Notice how the scalenes are composed of separate slips of muscle that run from the ribs to the neck. In humans, these muscles are more lateral than in cats. Compare your specimen with figure 14.5.

In the cat, the **sternocleidomastoid** consists of two muscles, the **sternomastoid** and the **cleidomastoid.** The sternomastoid extends from the sternum to the mastoid process of the skull, and the cleidomastoid runs from the clavicle to the mastoid process. Underneath the sternomastoid is the most medial muscle of the neck group, the **sternohyoid.** The **sternothyroid** is deeper and more lateral than the sternohyoid. These muscles can be seen in figure 14.6.

The **digastric** muscle runs parallel to the lower edge of the mandible and underneath the submandibular gland. Dissect only one side of the head, leaving the structures on the other intact for study of the digestive system. Deep to the digastric is the **mylohyoid,** which is a broad muscle. Notice how the muscle fibers run perpendicular to the direction of the digastric. Compare your dissection with figure 14.6. The **occipitalis** is a posterior head muscle in the cat that attaches to the **epicranial aponeurosis,** which connects to the anterior **frontalis** muscle. Lateral to these muscles are the temporalis and masseter muscles. The **temporalis** is located more dorsally than the masseter and can be dissected by removing the skin and fascia anterior to the ear. The **masseter** is a large, well-developed muscle in the cat that originates on the zygomatic arch and inserts on the lateral surface of the mandible. Examine these muscles in figure 14.7.

The **pterygoids** are usually not dissected because you have to cut through the ramus of the mandible to examine them. The muscles of facial expression are generally not studied in the cat. These muscles are small and are often removed with the skin. You should study these muscles on human models or a cadaver.

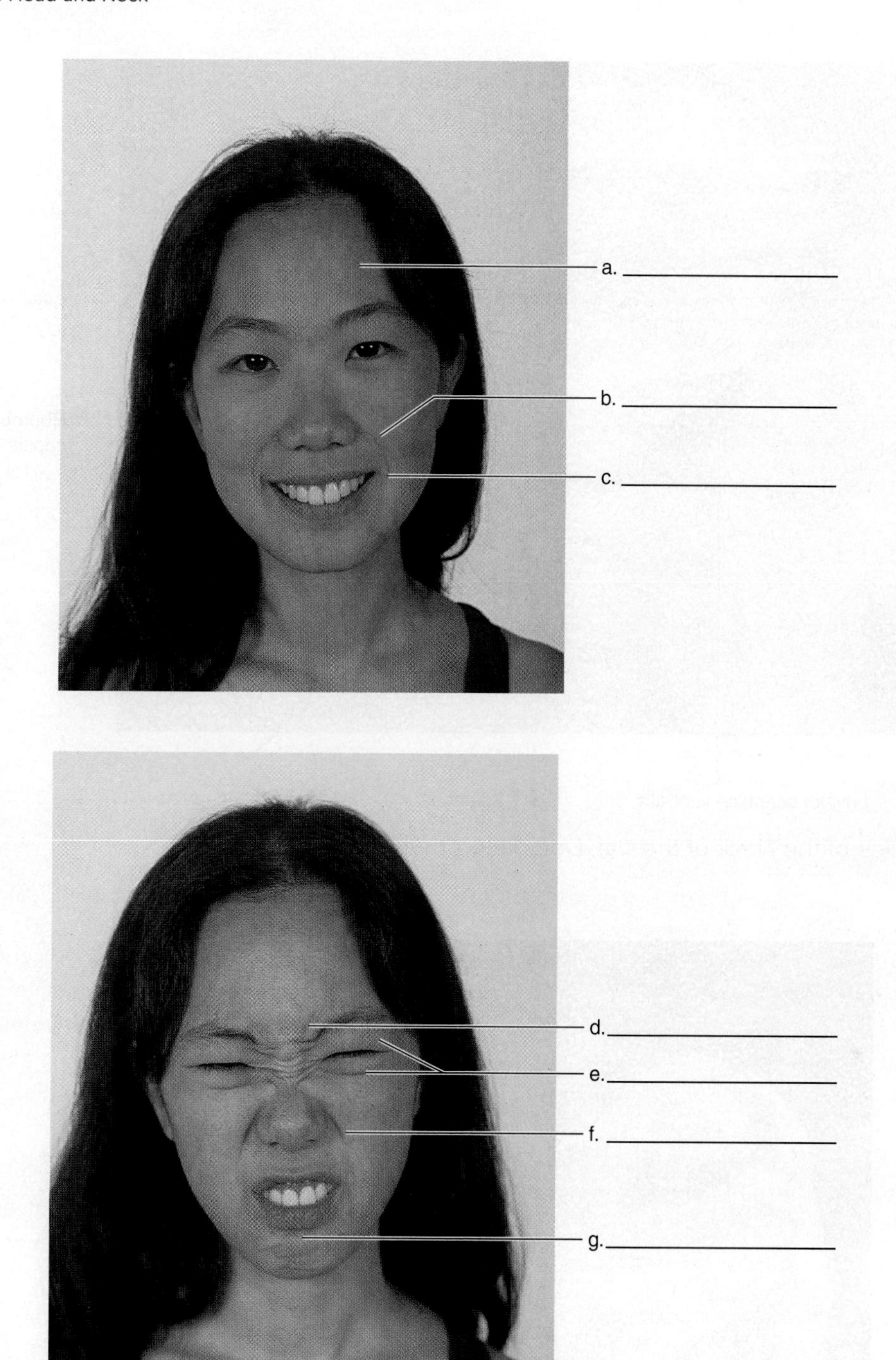

Figure 14.4 Muscles of Facial Expression

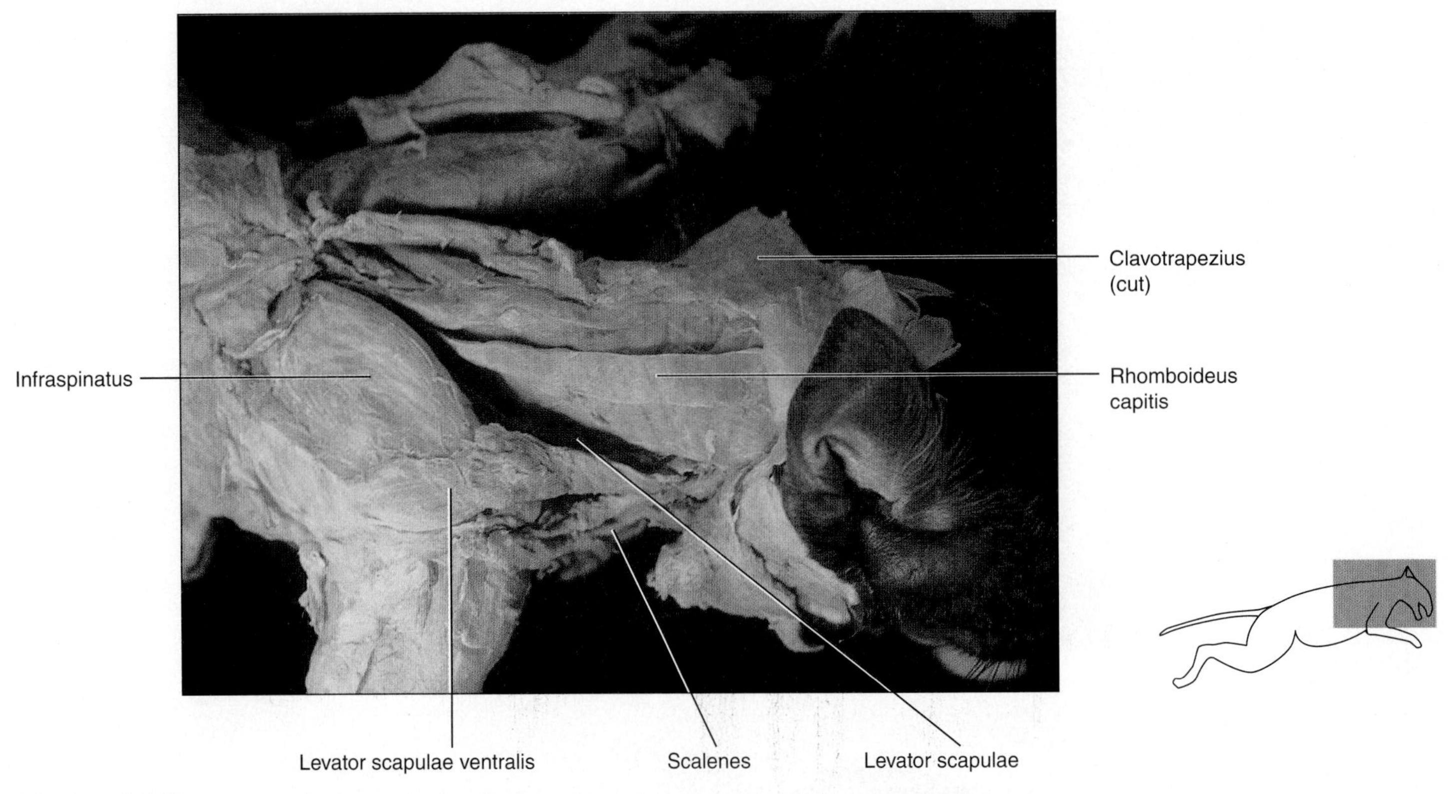

Figure 14.5 Muscles of the Neck of the Cat, Dorsolateral View

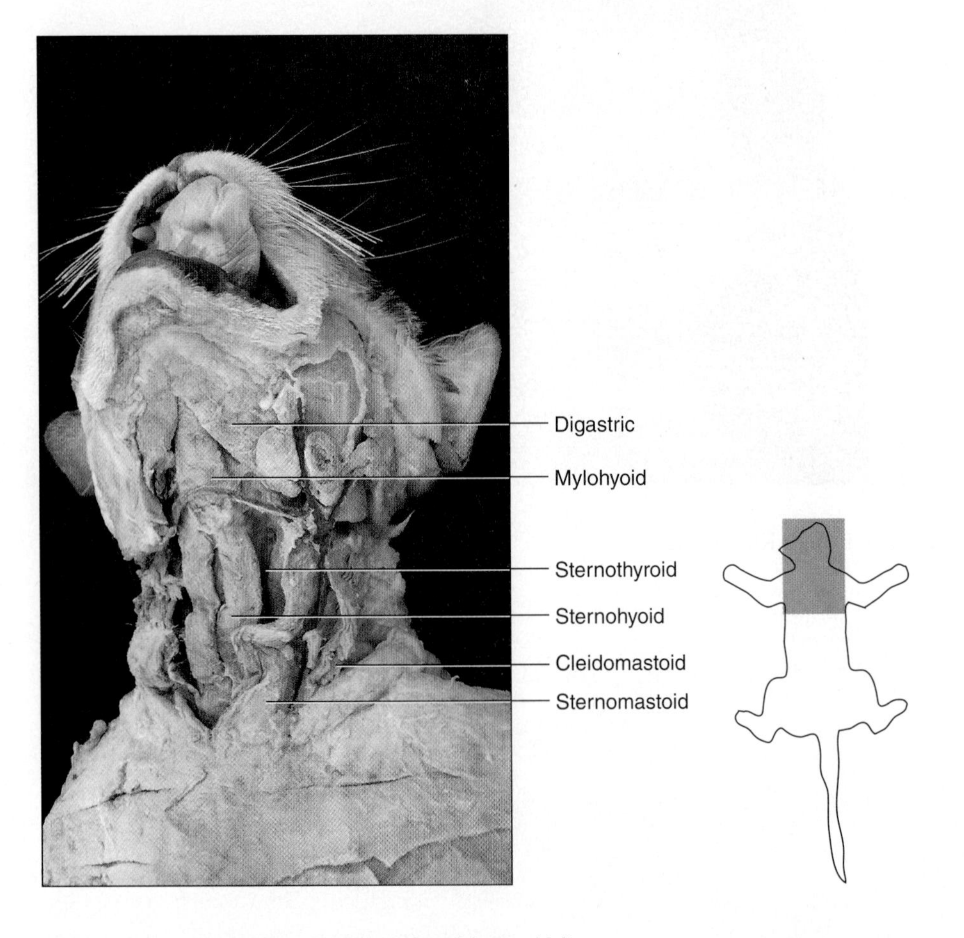

Figure 14.6 Muscles of the Head and Neck of the Cat, Ventral View

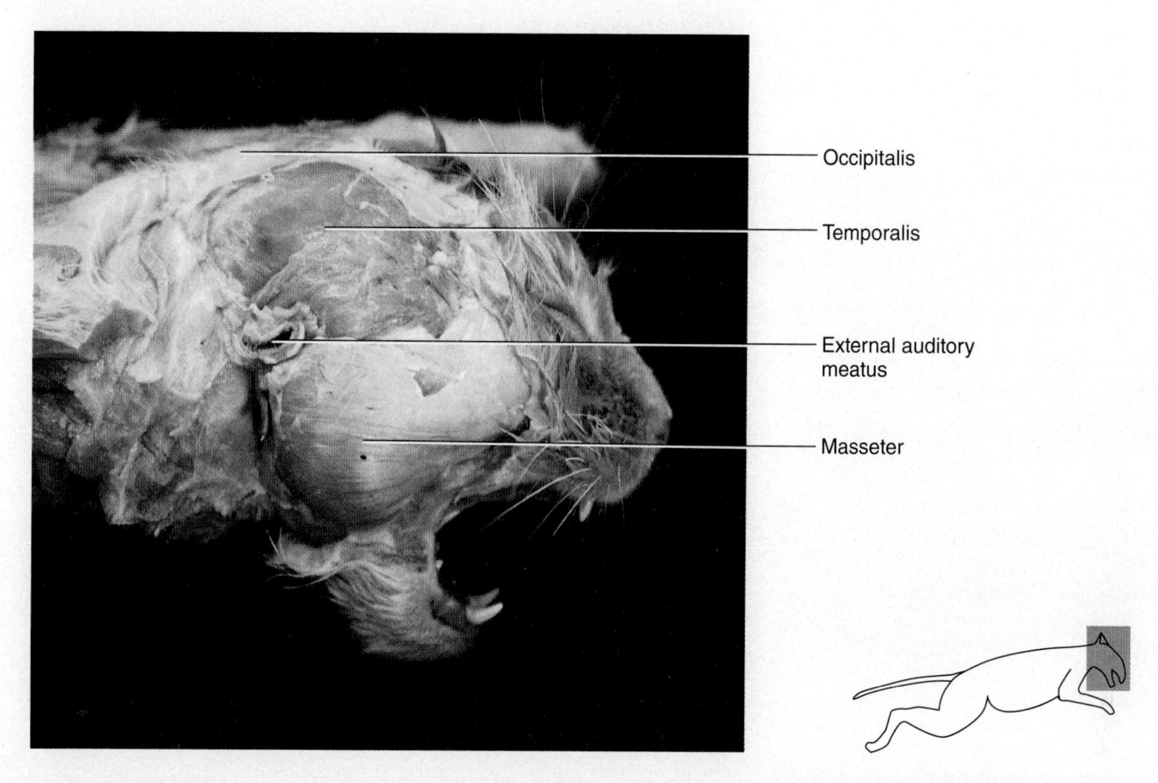

Figure 14.7 Muscles of the Head of the Cat, Lateral View

NOTES

Exercise 14 Review

Muscles of the Head and Neck

Name: ______________________

Lab time/section: ______________________

Date: ______________________

1. What is the origin of the masseter muscle? ______________________

2. What is the action of the risorius? ______________________

3. What kind of muscle is the orbicularis oculi or orbicularis oris muscle in terms of action? ______________________

4. Fill in the following illustration for the muscles of the neck using the terms provided.

   digastric mylohyoid scalenes sternocleidomastoid sternohyoid

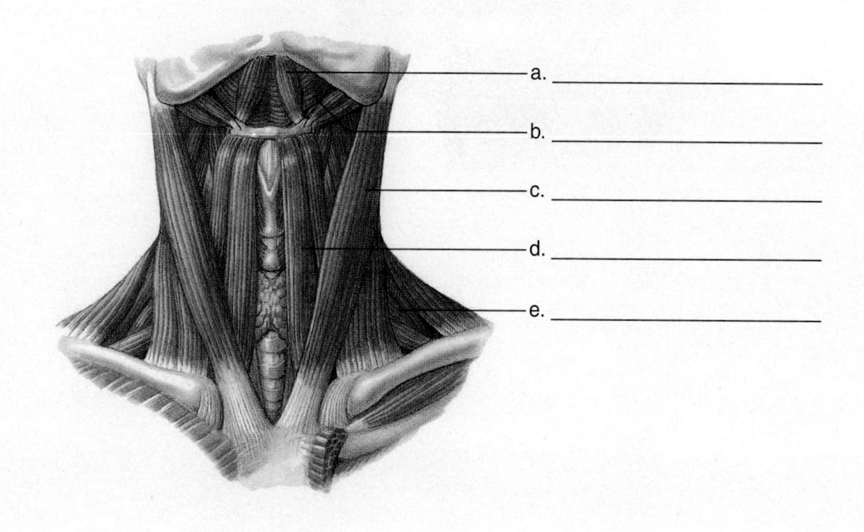

5. What muscle originates on the temporal fossa? ______________________

6. Name two muscles that close the jaw. ______________________

7. Where does the sternocleidomastoid muscle insert? ______________________

8. What muscle closes the lips? ______________________

9. Where does the orbicularis oculi insert? ______________________

10. What is the insertion of the temporalis? ______________________

11. Name a muscle that closes the eye. ______________________

12. What is the action of the sternocleidomastoid? ______________________

13. What muscle is a synergist with the masseter? ______________________________

14. Label the muscles of the head in the following illustration using the terms provided.

buccinator frontalis masseter orbicularis oculi orbicularis oris temporalis

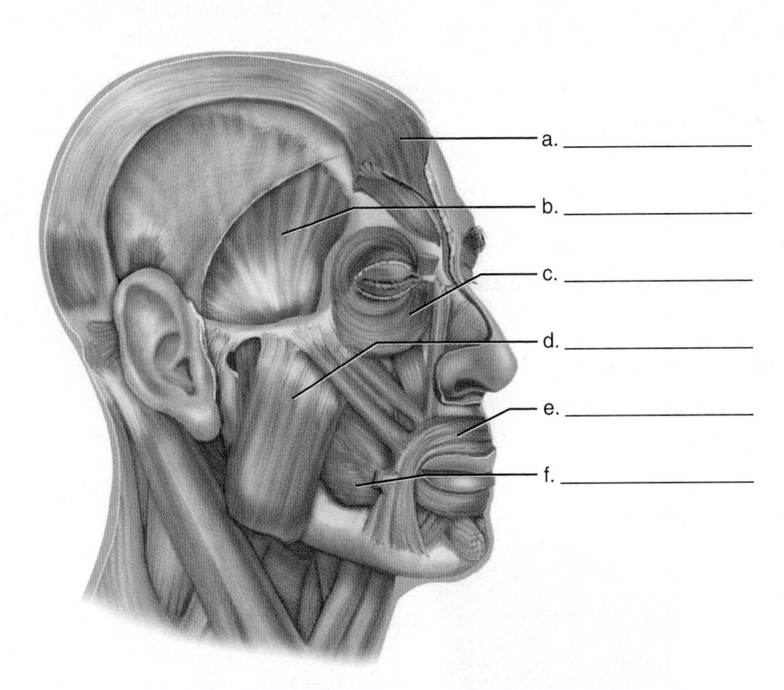

Exercise 15

Muscles of the Torso

Introduction

The muscles of the trunk can be grouped into a few functional groups. These are the abdominal muscles, which tighten the abdomen; the respiratory muscles, which assist in breathing; the postural muscles of the back; and the muscles that act on the scapula or on the head and neck. These muscles are discussed in the Seeley text in chapter 10, "Muscular System: Gross Anatomy."

Objectives

At the end of this exercise, you should be able to

1. locate the muscles of the torso on a model, chart, cadaver, or cat;
2. list the origin, insertion, and action of each muscle;
3. describe what nerve controls each muscle;
4. list what muscles function as synergists or antagonists to the prime mover;
5. name all the muscles that compress the abdomen.

Materials

Human torso model
Human muscle charts
Articulated skeleton
Cadaver
Cat
Materials for cat dissection
- Dissection trays
- Scalpel and two or three extra blades
- Gloves (household latex gloves work well for repeated use)
- Blunt (Mall) probe
- Forceps and sharp scissors
- First aid kit in lab or prep area
- Sharps container
- Animal waste disposal container

Procedure

(A)[1] Review the muscle nomenclature and the actions as outlined in Exercises 10 and 12. Examine a torso model or chart in the lab and locate the muscles described next and in table 15.1. Correlate the shape or fiber direction of the muscle with the name and visualize the muscles as you study the models or charts. You may want to look at an articulated skeleton as you study the origins and insertions of the muscles, so that you can better see the muscle attachment points. The descriptions in the text portion of this exercise can help you understand the nature of the muscle, whereas table 15.1 presents specific information about the muscle.

Once you have learned the origin and the insertion, you should be able to understand the action of the muscle. This is done, in part, by visualizing how the insertion of the muscle is pulled toward the origin when the muscle contracts.

Human Muscles

Abdominal Muscles

(A)[2] Locate the abdominal muscles in the lab and compare them with figure 15.1. Most of the muscles of the abdomen compress the viscera, which aids in food regurgitation and bowel movements. The **external abdominal oblique** is a broad, superficial muscle of the abdomen with fibers that run from a superior direction to an inferior, medial direction. The **internal abdominal oblique** is deep to the external abdominal oblique and has fiber directions that run perpendicular to the external abdominal oblique.

The deepest of the abdominal muscles is the **transversus abdominis,** which has fibers running in a horizontal direction. The **rectus abdominis** (*rectus* = straight) runs vertically up the abdomen. The rectus abdominis has small connective tissue bands, called **tendinous intersections,** located horizontally across the muscle, dividing it into small segments. If the abdominal fat is minimal and the muscles are well developed, the "washboard stomach," or "six-pack," is apparent due to the muscle fibers' increasing in girth while the tendinous intersections remain undeveloped. Another muscle in the area is the **quadratus lumborum,** which is a square muscle that runs from the iliac crest to the lower vertebrae and rib 12. These muscles can be seen in figure 15.2.

Postural Muscles

(A)[3] The **erector spinae** muscles make up most of the postural muscles of the back. The erector spinae are actually many muscles that are located between individual vertebrae or between the vertebrae and the ribs. These separate muscles are grouped into long strap muscles known collectively as the erector spinae. There are three major groups of erector spinae muscles: the **spinalis,** the **longissimus,** and the **iliocostalis.** Locate these muscles in lab and compare them with figure 15.2.

Thoracic Muscles

The **serratus anterior** is a broad, fan-shaped muscle that has slips of muscle originating on the upper ribs. These slips of muscle unite and insert on the medial border of the scapula. As the muscle contracts, it abducts the scapula by pulling it away from the spine. These muscles are shown in figure 15.1.

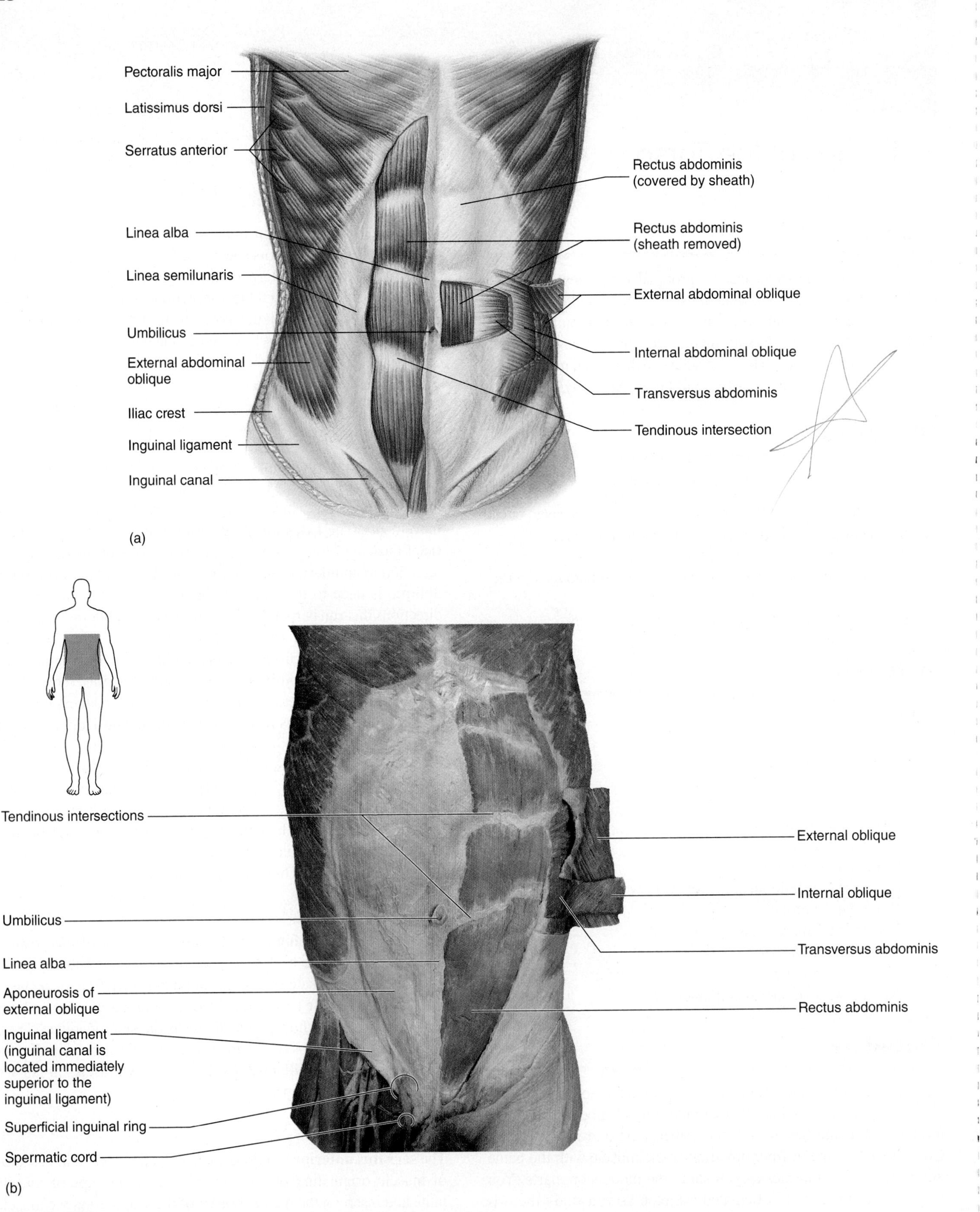

Figure 15.1 Muscles of the Abdomen, Anterior View

(a) Diagram; (b) photograph.

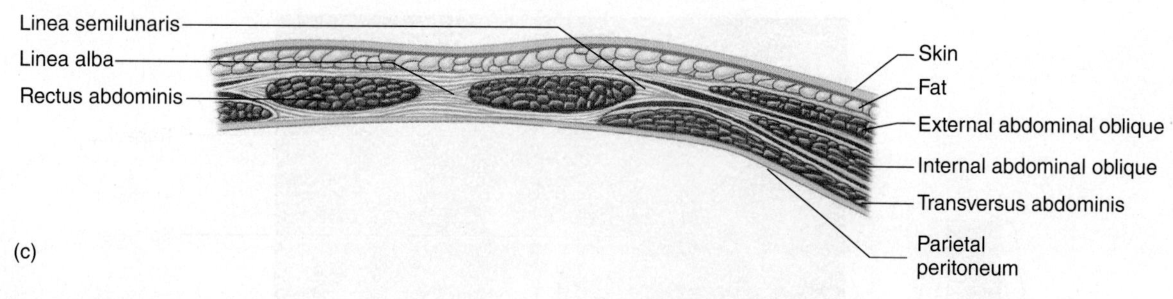

Figure 15.1 *Continued.*
(c) Diagram of cross section.

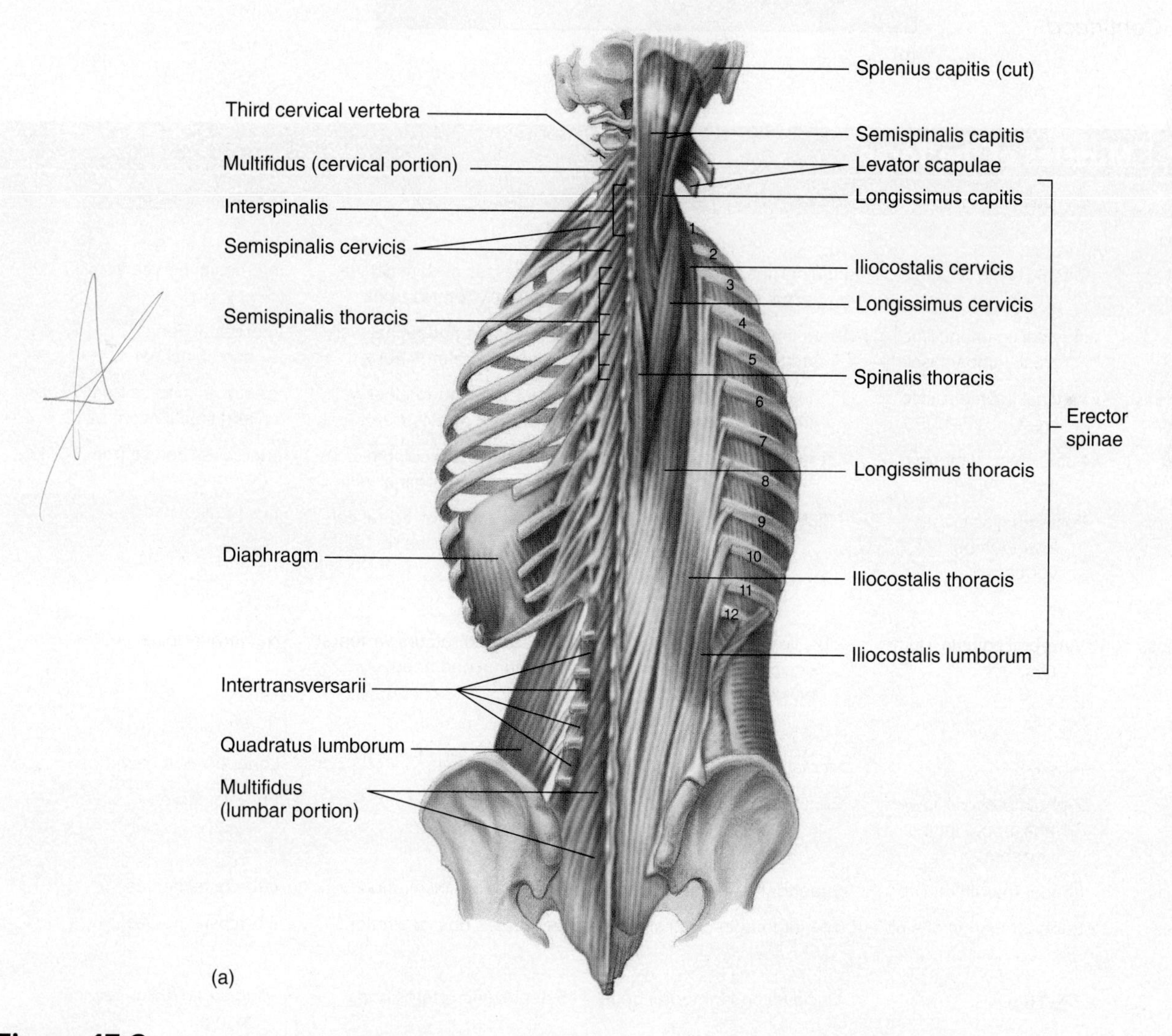

Figure 15.2 Deep Muscles of the Back
(a) Diagram.

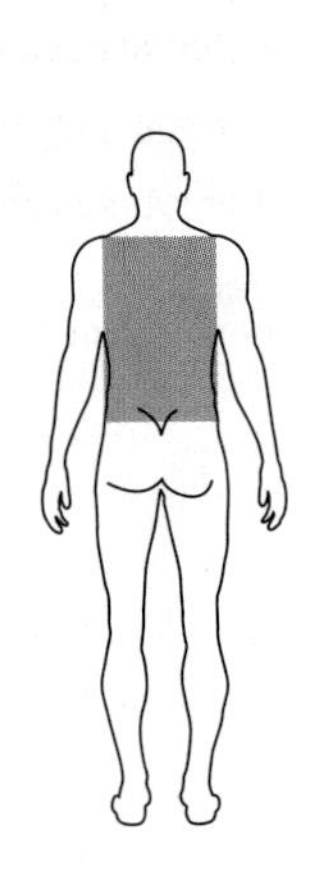

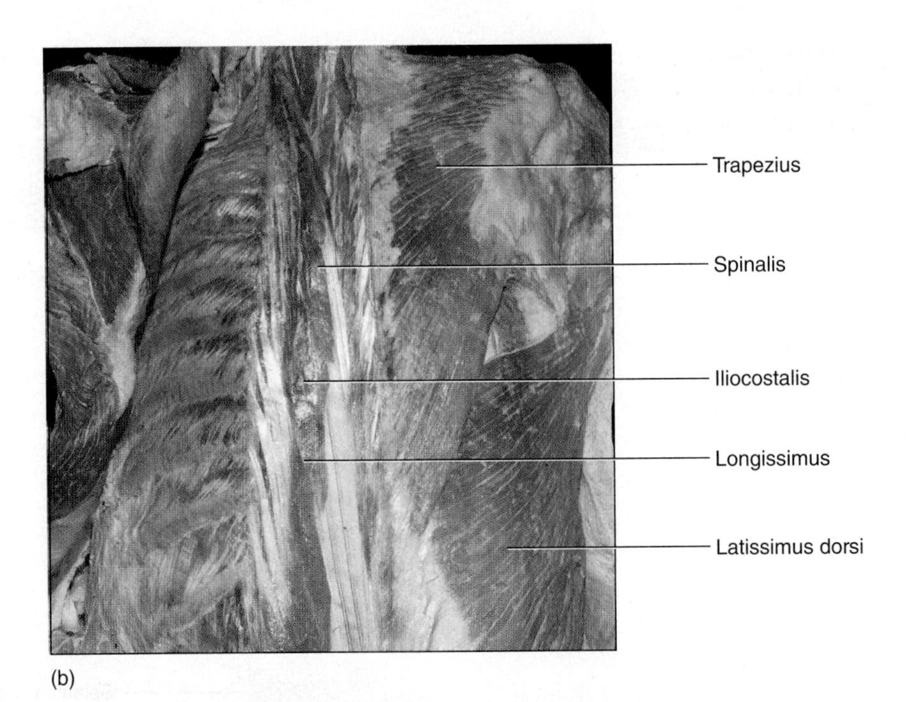

Figure 15.2 *Continued.*
(b) Photograph.

TABLE 15.1 Muscles of the Torso

Name	Origin	Insertion	Action	Innervation
Abdominal Muscles				
External abdominal oblique	Ribs 5–12	Inguinal ligament, iliac crest, rectus sheath	Compresses abdominal wall, laterally rotates trunk	Intercostal nerves from T7–12
Internal abdominal oblique	Inguinal ligament, iliac crest, lumbar fascia	Linea alba, ribs 10–12, rectus sheath	Compresses abdominal wall, laterally rotates trunk	Intercostal nerves from T7–12 and spinal nerve L1
Transversus abdominis	Inguinal ligament, iliac crest, ribs 7–12	Xiphoid process, linea alba, crest of pubis	Compresses abdominal wall, laterally rotates trunk	Intercostal nerves from T7–12 and spinal nerve L1
Rectus abdominis	Pubic crest, symphysis pubis	Inferior ribs, xiphoid process	Flexes vertebral column, compresses abdominal wall	Intercostal nerves from T6–12
Quadratus lumborum	Posterior iliac crest, lower lumbar vertebrae	T12, L1–4, rib 12	Laterally flexes vertebral column, depresses rib 12	T12, L1–4
Erector Spinae				
Spinalis, Longissimus, and Iliocostalis	Vertebral column, ilium, ribs	Ribs, vertebral column, occipital and temporal bones	Extends and rotates vertebral column and head	Numerous spinal nerves
Thoracic Muscles				
Serratus anterior	Ribs 1–9	Medial border of scapula	Abducts scapula	Long thoracic nerve
Diaphragm	Xiphoid process, lower ribs, upper lumbar vertebrae	Central tendon	Inspiration	Phrenic nerve
External intercostalis	Inferior margin of a rib	Superior margin of rib below	Elevates ribs, inspiration	Intercostal nerves
Internal intercostalis	Superior margin of a rib	Inferior margin of rib above	Depresses ribs, expiration	Intercostal nerves
Splenius Muscles				
Splenius	C4–T6	Occipital and temporal bone	Extends and rotates head	Middle and lower cervical nerves
Transversospinal Muscles				
Semispinalis	C4–T6	Occipital bone	Extends and rotates head	Cervical nerves
Multifidus	Vertebral column, ilium	Vertebral column	Extends and rotates vertebral column	Numerous spinal nerves

C = cervical vertebrae, T = thoracic vertebrae, L = lumbar vertebrae, S = sacral vertebrae.

(A)[4] The intercostal muscles and the diaphragm are respiratory muscles. Normally, the diaphragm is responsible for about 60% of the resting breath volume, whereas the external intercostals contribute to the remaining volume. The **diaphragm** is a domed muscle with a peripheral origin. The diaphragm inserts centrally at the base of the mediastinum. If you think of the diaphragm as a trampoline, the outer springs represent the origin, whereas the center (where you jump) represents the insertion. Examine the models in the lab and compare them with figure 15.3.

The intercostal muscles contribute to the breathing volume at rest, but they also contribute to a greater increase in the movement of the thorax during exercise. There is some debate as to the action of the intercostals. Some evidence suggests that both the intercostals are involved in inhalation. Other evidence suggests that the external intercostals are involved in inhalation, whereas the internal intercostals are involved in exhalation. Although there is some difference of opinion among anatomists, in this exercise we will treat the external intercostals as muscles that assist in inhalation and the internal intercostals as muscles that assist in exhalation. Locate the intercostal muscles and compare them with figure 15.3.

(A)[5] The deep muscles of the superior torso consist of the **splenius** and the **semispinalis** muscles. These extend and rotate the head and are illustrated in figures 15.2 and 15.4. The **multifidus** (figure 15.2) is a closely associated muscle with the erector spinae muscles (the spinalis, longissimus, and iliocostalis), and as a group they are known as the *s.l.i.m.* muscles as you move from medial to lateral and then inferior.

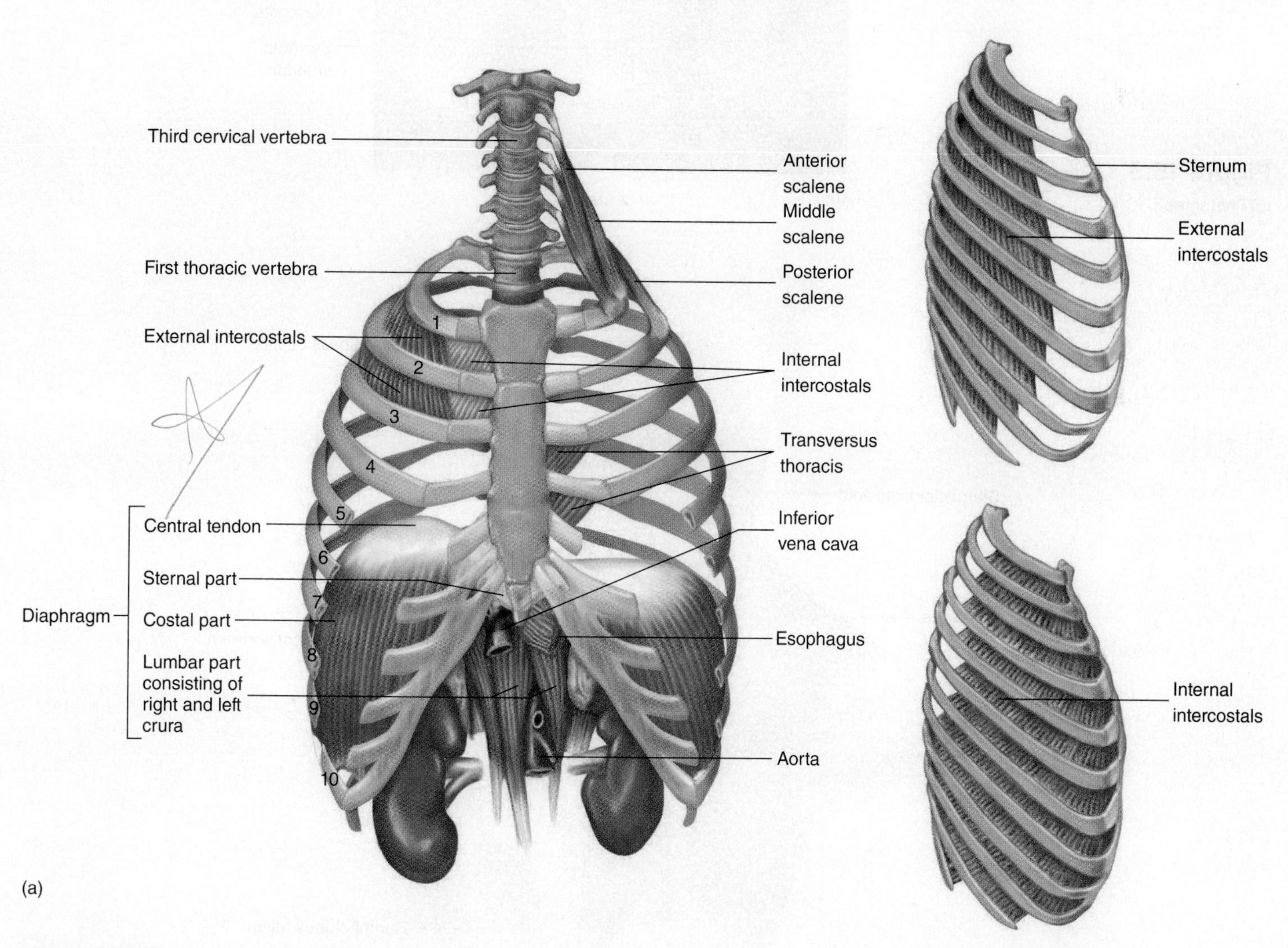

Figure 15.3 Muscles Involved in Respiration
(a) Diagram.

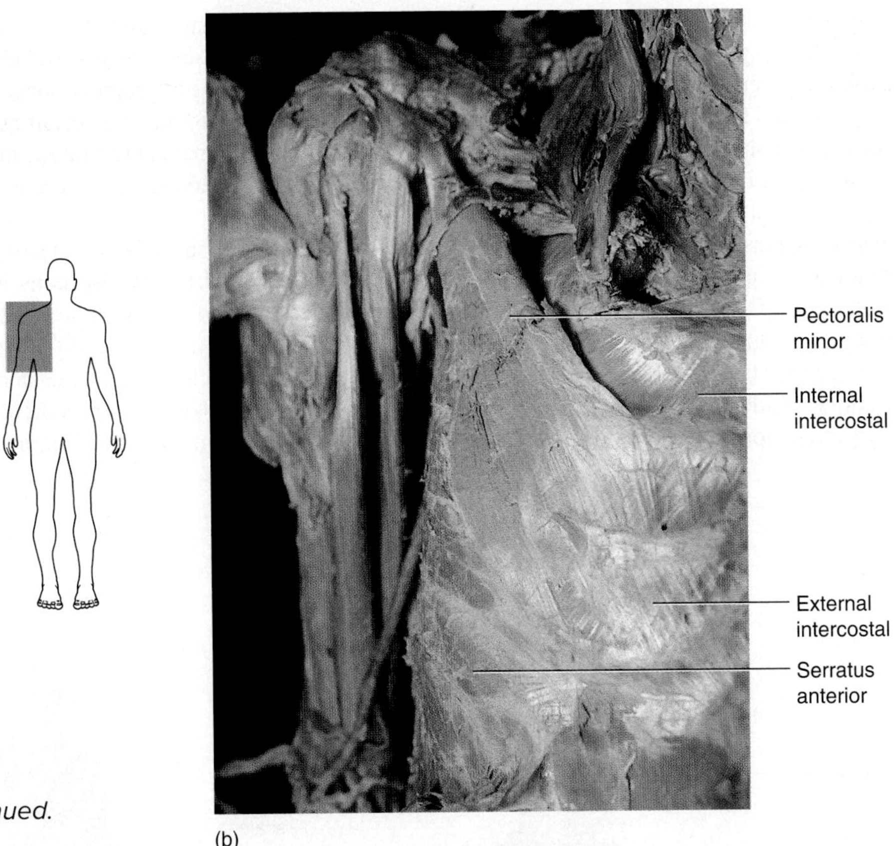

Figure 15.3 *Continued.*

(b) Photograph.

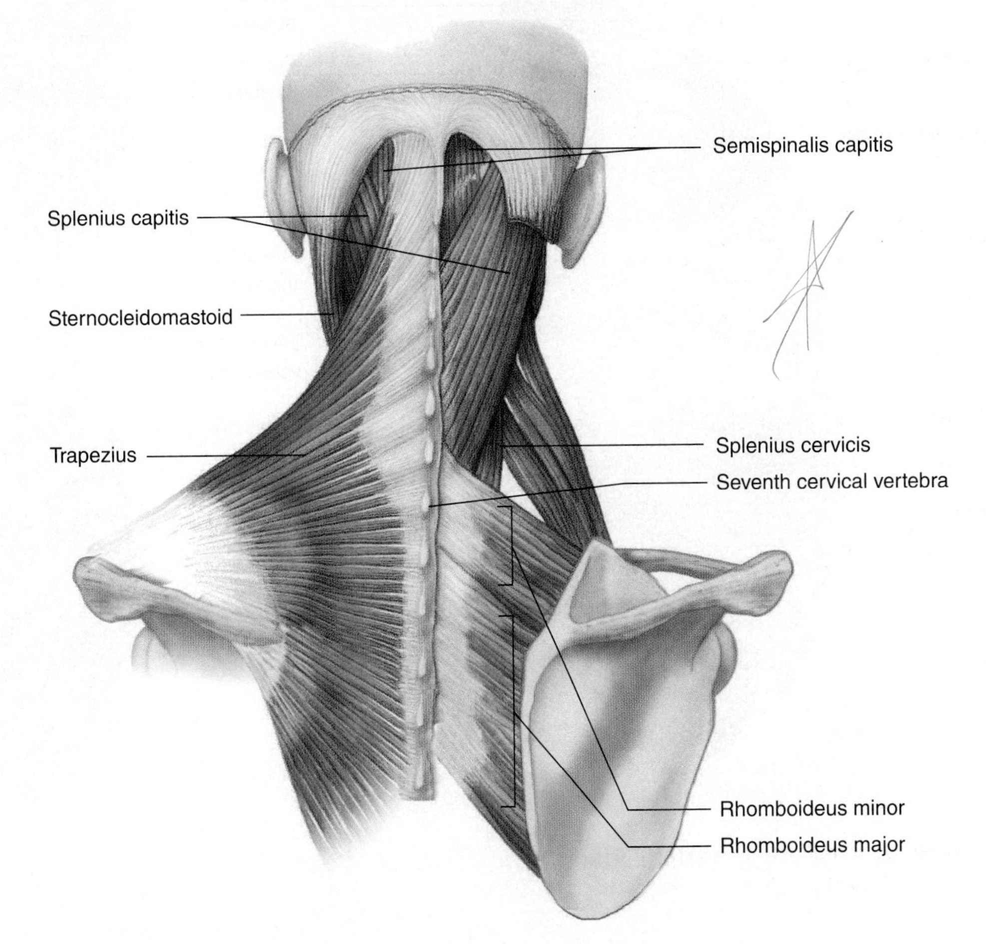

Figure 15.4 Muscles of the Neck, Posterior View

Cat Dissection

Abdominal Muscles

Ⓐ[6] Place the cat on its back and examine the abdominal muscles. The abdominal muscles in the cat are similar to those in the human in that the **external abdominal oblique** is a broad, superficial muscle on the ventral abdomen. Carefully cut through the external abdominal oblique to reveal the **internal abdominal oblique,** as illustrated in figure 15.5. Deep to this is the **transversus abdominis,** and it can be seen by carefully dissecting the internal abdominal oblique. If you cut too deeply, you will enter the abdominal cavity, so be careful in this part of the dissection. The **rectus abdominis** is the muscle that runs from the pubic region to the sternum.

Move to the thoracic region and examine the muscle of the lateral thorax dorsal to the xiphihumeralis. This is the **serratus anterior** (actually, the serratus ventralis in the cat), and you should see the scalloped edges of the muscle. Carefully separate this muscle from the others and trace its insertion to the scapula. To see the intercostal muscles, you will have to bisect the superficial chest muscles. If you have not done so already, cut through the middle of the belly of the pectoral muscles, exposing the ribs of the cat. Carefully remove the outer layer of fascia from the muscle between the ribs and locate the **external intercostal** muscle. You should be able to cut part of this muscle away and expose the **internal intercostal** muscle. Note how the fibers run perpendicular to one another. Do not look for the diaphragm at this time. You can see it in Exercise 35, as you examine the lungs. Examine these thoracic muscles in figure 15.6.

Deep Muscles of the Back

Ⓐ[7] Place the cat so that you can examine the dorsal surface. Locate the **splenius** muscle located in figure 15.7.

You will need to bisect the latissimus dorsi and the posterior portion of the external abdominal oblique to see the **erector spinae** muscles. The relative position of the cat erector spinae can be seen in figure 15.8. Compare this figure with your dissection. Locate the iliocostalis, longissimus, and spinalis of the erector spinae along with the multifidus in the cat.

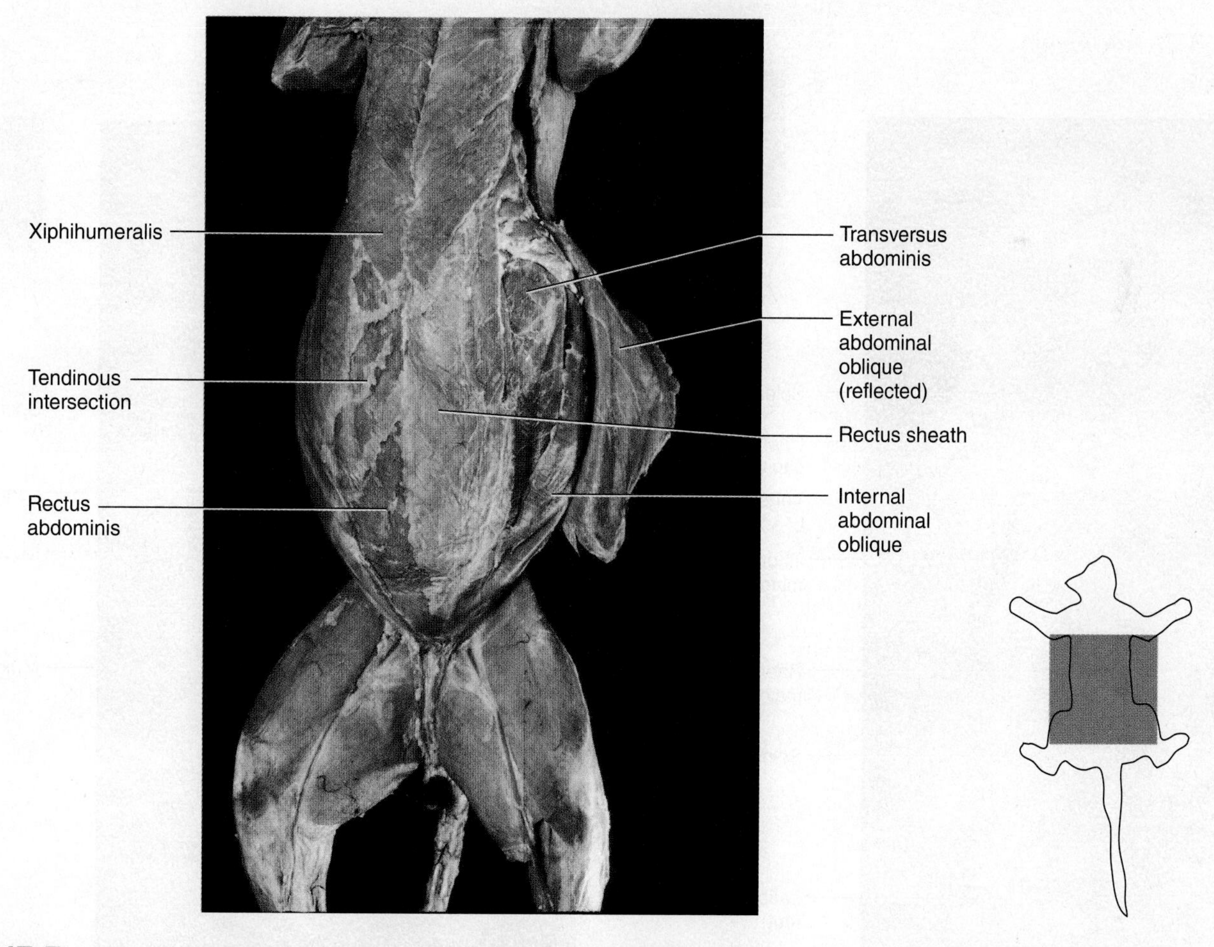

Figure 15.5 Muscles of the Abdomen of the Cat, Ventral View

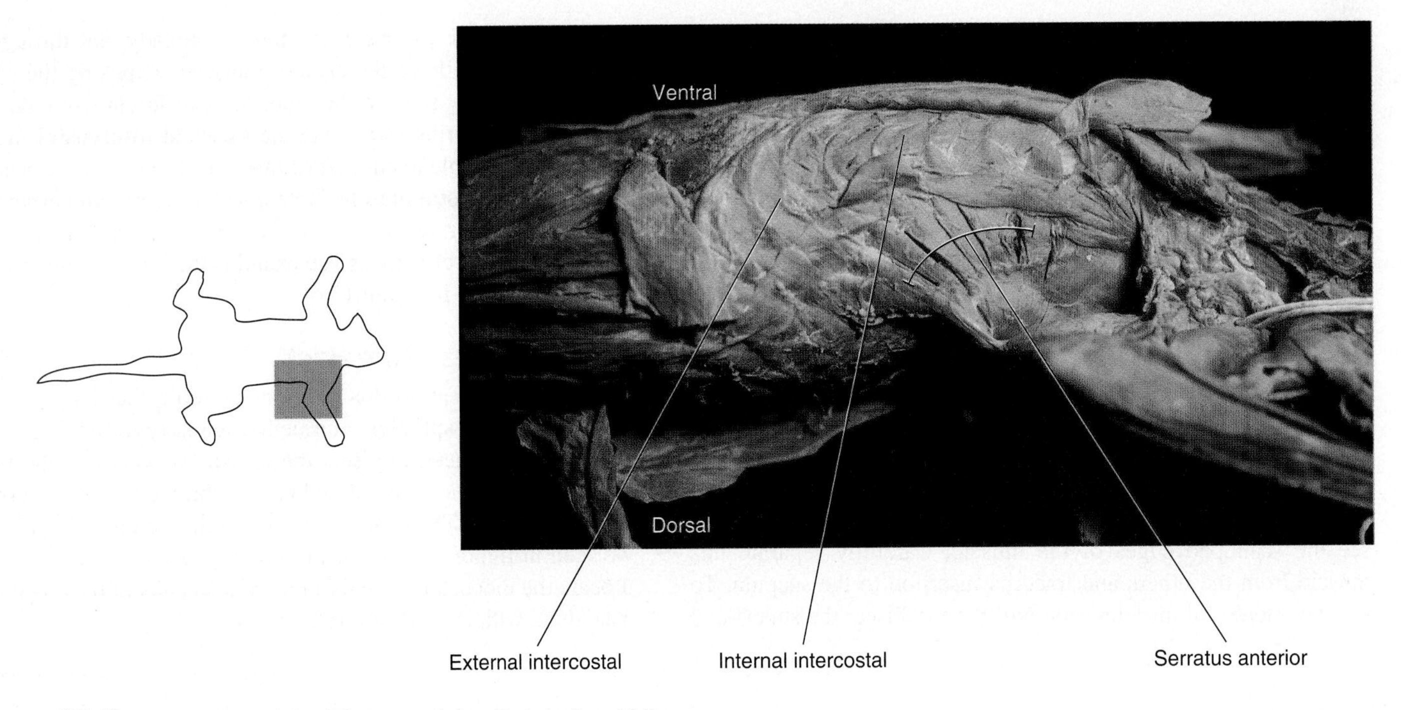

Figure 15.6 Muscles of the Thorax of the Cat, Lateral View

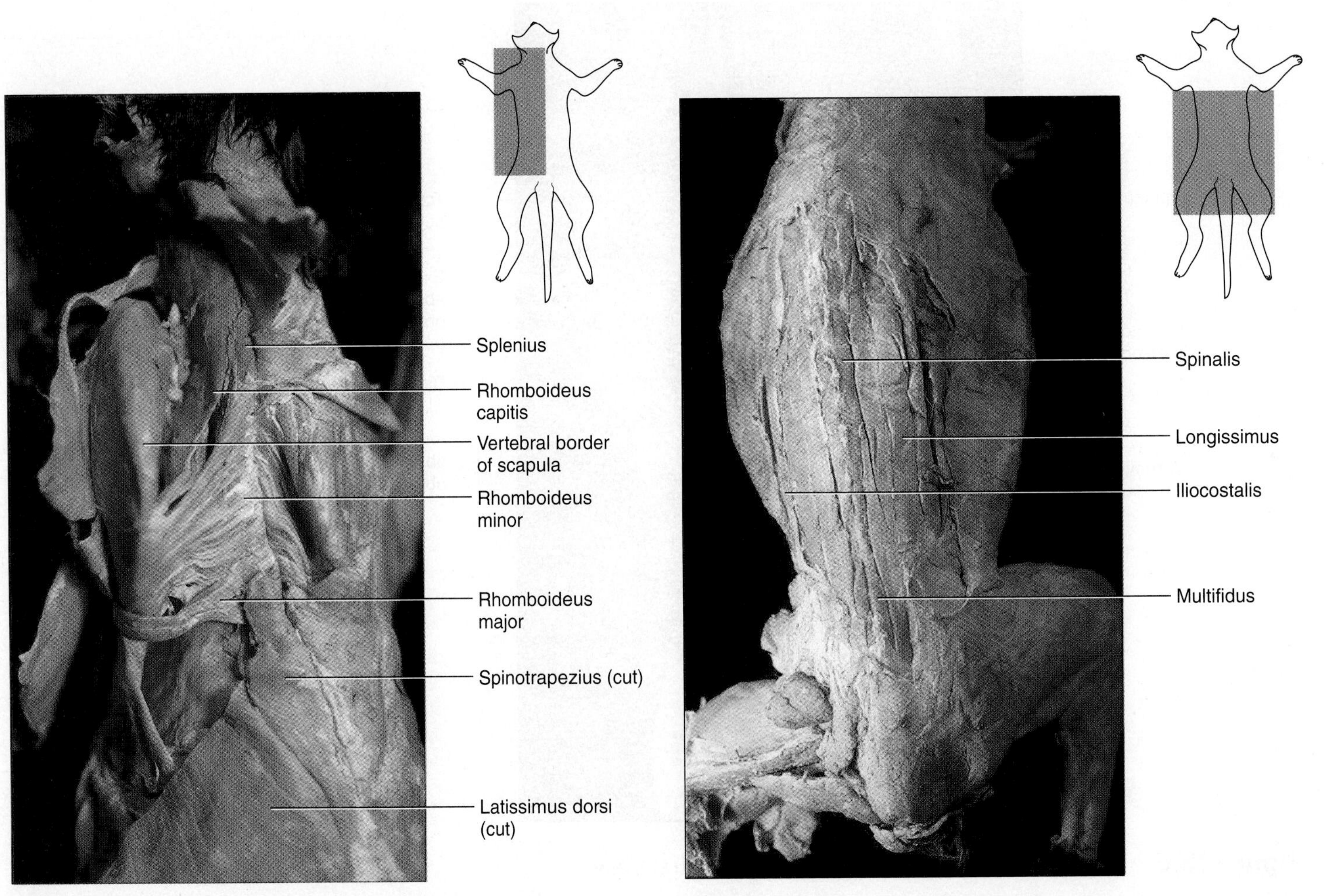

Figure 15.7 Anterior, Deep Muscles of the Back of the Cat, Dorsal View

Figure 15.8 Posterior, Deep Muscles of the Back of the Cat, Dorsal View

Exercise 15 Review

Muscles of the Torso

Name: ______________________

Lab time/section: ______________________

Date: ______________________

1. What is the action of the serratus anterior muscle? ______________________

____

2. Name four muscles that extend the vertebral column. ______________________

____

3. How does the serratus anterior function as an antagonist to the rhomboideus muscles? ______________________

____

4. How does the action of the rectus abdominis differ from that of the other abdominal muscles? ______________________

____

5. What is the physical relationship of the intercostal muscles to each other? ______________________

____

6. Extension and rotation of the vertebral column occur by what group of muscles? ______________________

7. What is the action of the intercostal muscles? ______________________

____

8. What muscle inserts on the central tendon? ______________________

9. The tendinous intersections are found in what muscle? ______________________

10. Which is the deepest abdominal muscle? ______________________

11. What is the action of the quadratus lumborum? ______________________

____

12. What two muscles originate on the neck and extend and rotate the head? ______________________

13. Label the muscles in the following illustration using the terms provided.

external abdominal oblique | internal abdominal oblique | rectus abdominis
serratus anterior | transversus abdominis

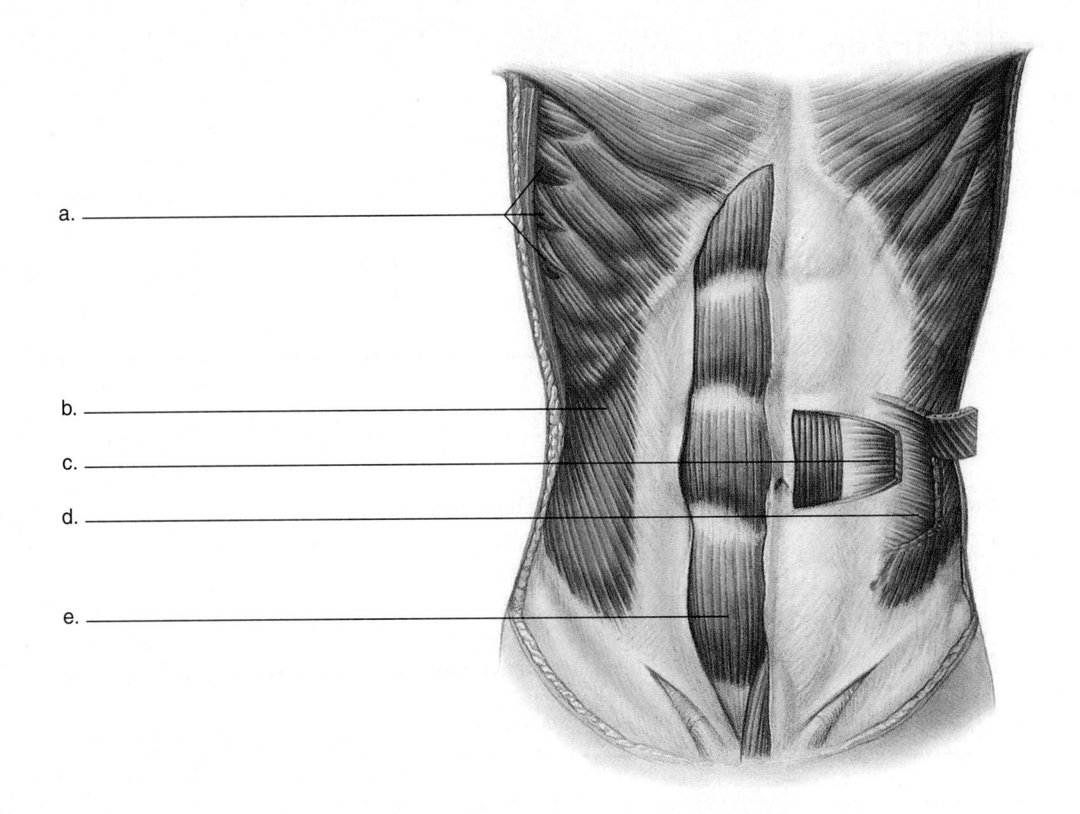

14. How do the abdominal muscles in the cat compare with those in the human in terms of relative position? ______________________

____

Exercise 16

Introduction to the Nervous System

Introduction

The functions of the nervous system, among other things, are **communication** between the various regions of the body, **coordination** of body functions (as in digestion or walking), **orientation** to the environment, and **assimilation** of information. These topics are further discussed in the Seeley text in chapter 11, "Functional Organization of Nervous Tissue."

In this exercise, you learn about the basic structure of the nervous system and the component cells of the nervous system. The functional unit of the nervous system is the neuron. It is the cell that carries out the activity of nervous tissue. Neurons are located in the nerves of the body, the spinal cord, and the brain. Glial cells are also important cells of the nervous tissue. They help increase the speed of neuron transmission, provide nutrients to the neurons, affect synapses, and protect the neurons, among other things.

Objectives

At the end of this exercise, you should be able to

1. describe the three parts of the neuron;
2. list the main divisions of the nervous system;
3. group the organs of the nervous system into the main divisions;
4. describe the functions of the various glial cells.

Materials

Charts or models of the nervous system
Charts or models of neurons
Microscopes
Prepared slides
    Spinal cord smear
    Longitudinal section of nerve
    Neuroglia (glial cells)

Procedure

Divisions

The nervous system can be divided into two general divisions based on location. The **central nervous system (CNS)** consists of the brain and the spinal cord. The **peripheral nervous system (PNS)** is composed of the spinal nerves, the dorsal root ganglia, the somatic nerves (those that radiate into the extremities and other regions of the body), and the cranial nerves.

The peripheral nervous system has a sensory, or afferent, division and a motor, or efferent, division. The **sensory division** conducts impulses *from* the regions of the body *to* the central nervous system. The **motor division** conducts impulses *from* the central nervous system *to* the various regions of the body. The motor division has two subunits—the **somatic motor nervous system,** which takes impulses from the CNS and innervates muscles, and the **autonomic nervous system (ANS),** which innervates glands, smooth muscle, cardiac muscle, and organs. The ANS has centers in the central nervous system (**midbrain, pons, medulla oblongata,** and **spinal cord**) and has peripheral branches. These function independently and provide automatic controls for activities normally under subconscious direction. When you walk up stairs, your heart rate increases automatically, along with your breathing rate, without your having to think about controlling these activities.

The activities of the ANS can be controlled consciously in some cases. The principle of biofeedback involves the conscious lowering of heart rate or blood pressure. Both of these activities are normally under the control of the ANS.

(A) 1 Examine the models or charts in the lab for the central and peripheral divisions of the nervous system. Compare the material in the lab with figure 16.1.

Neuron

The **neuron,** or **nerve cell,** is a remarkable cell, not only for its functional nature but also for the anatomical extremes it exhibits. The nerves in your thigh and leg are composed of neuron fibers, and the neurons that pick up sensation in your toes continue as single cells up the leg and thigh to synapse (join) with other neurons in the lower back. When you look at prepared slides of neurons in the microscope in this exercise, remember that these neurons are very long in some cases.

(A) 2 Neurons consist of three main parts: the **axon** or nerve fiber, the **dendrite,** and the **neuron cell body,** or *soma.* Examine figure 16.2 and models or charts in the lab to see the structure of neurons. The receptive end of the neuron is the dendrite or the neuron cell body. Dendrites are so named because they have branching structures that resemble a tree (*dendros* = tree). Neuron cell bodies consist of the **neuroplasm** (cytoplasm of the neuron), **Nissl bodies** (rough endoplasmic reticulum of the neuron), and the **nucleus.** The triangular region of the neuron cell body that is devoid of Nissl bodies is the **axon hillock,** and it leads to the axon that exits the neuron cell body. Axons consist of long strands of neurofibrils wrapped in myelin sheaths. These sheaths are discussed later in the exercise.

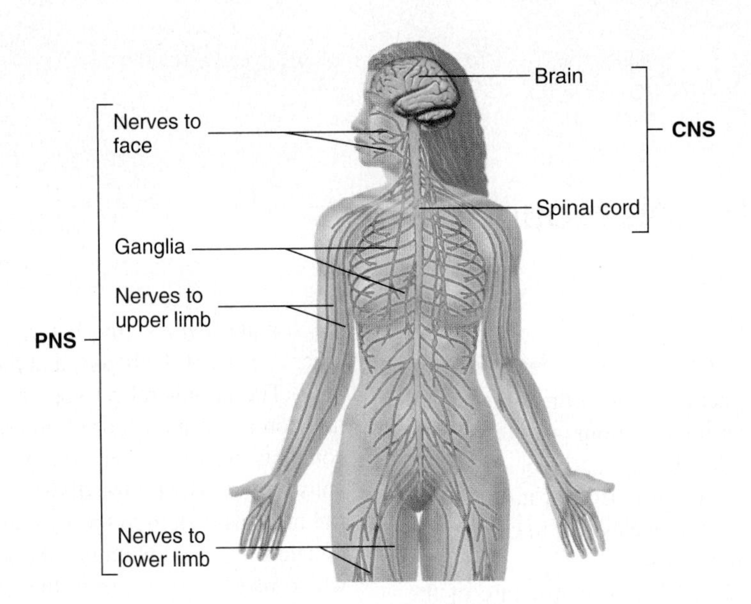

Figure 16.1 The nervous system is divided into two main divisions: the central nervous system and the peripheral nervous system.

Histology of the Neuron

Examine a prepared slide of a spinal cord smear under the microscope and locate the purple, star-shaped structures under low power. These are the neuron cell bodies of multipolar neurons. Switch to high power and locate the darker-staining Nissl bodies, the nucleus, and the axon hillock. The axon is attached to the axon hillock. All the other processes that are attached to the neuron cell body are dendrites. The small nuclei that are scattered throughout the smear belong to glial cells in the spinal cord. Compare your slide with figure 16.2*b*.

Functions of Neurons

There are three types of neurons based on function. **Sensory,** or **afferent, neurons** conduct impulses *to* the central nervous system. They convey information from the external or internal body environment to the spinal cord and/or the brain. **Motor,** or **efferent, neurons** conduct impulses *away from* the central nervous system. **Association neurons,** or **interneurons,** are located *between* sensory and motor neurons. They also transmit information to the brain for processing. These are illustrated in figure 16.3.

Neuron Shapes

Neurons can be classified according to shape. A **multipolar neuron** consists of several dendritic processes, a single neuron cell body, and a single axon. In most neurons, dendrites take information to the cell body and axons take information away from the cell body. This is the case in multipolar neurons but it is not universal for all neurons. The majority of the neurons of the body (such as those in the brain and spinal cord) are multipolar neurons. Compare material in the lab with figure 16.4. **Bipolar neurons** are so named because the neuron cell body has two poles. Dendrites receive information and conduct it to one pole of the neuron cell body. At the other pole, an axon leaves the neuron cell body and transmits the impulse away from the cell body. Bipolar neurons are located in nerves conducting the senses of smell and vision.

Pseudo-unipolar neurons have a cell body with a single process attached to it (figure 16.4). Dendrites receive impulses, which travel next to an axon. These axons transmit that information either to the cell body or via another axon directly to the spinal cord. Most of the common nerves of the body (somatic sensory nerves such as the ulnar nerve or femoral nerve) are pseudo-unipolar neurons.

Synapses

Neurons transmit information electrochemically along the length of the axon to the **presynaptic terminal.** Neurons are not physically attached to one another but communicate by signal molecules that flow across a short space between the neurons. The space is called the **synapse,** and the molecules such as acetylcholine, that move across the synapse are called **neurotransmitters.** Figure 16.5 illustrates a synapse.

Activity

(A) 3 Work with your lab partner and estimate how long it takes from the time you receive a verbal signal to move a finger or a toe to when you actually move the digit. Note that the pathway must travel from your ear to the interpretive regions of the brain to the limb and digit that is moved. Record your estimation in the following space.

Estimated time for reaction: ______________________

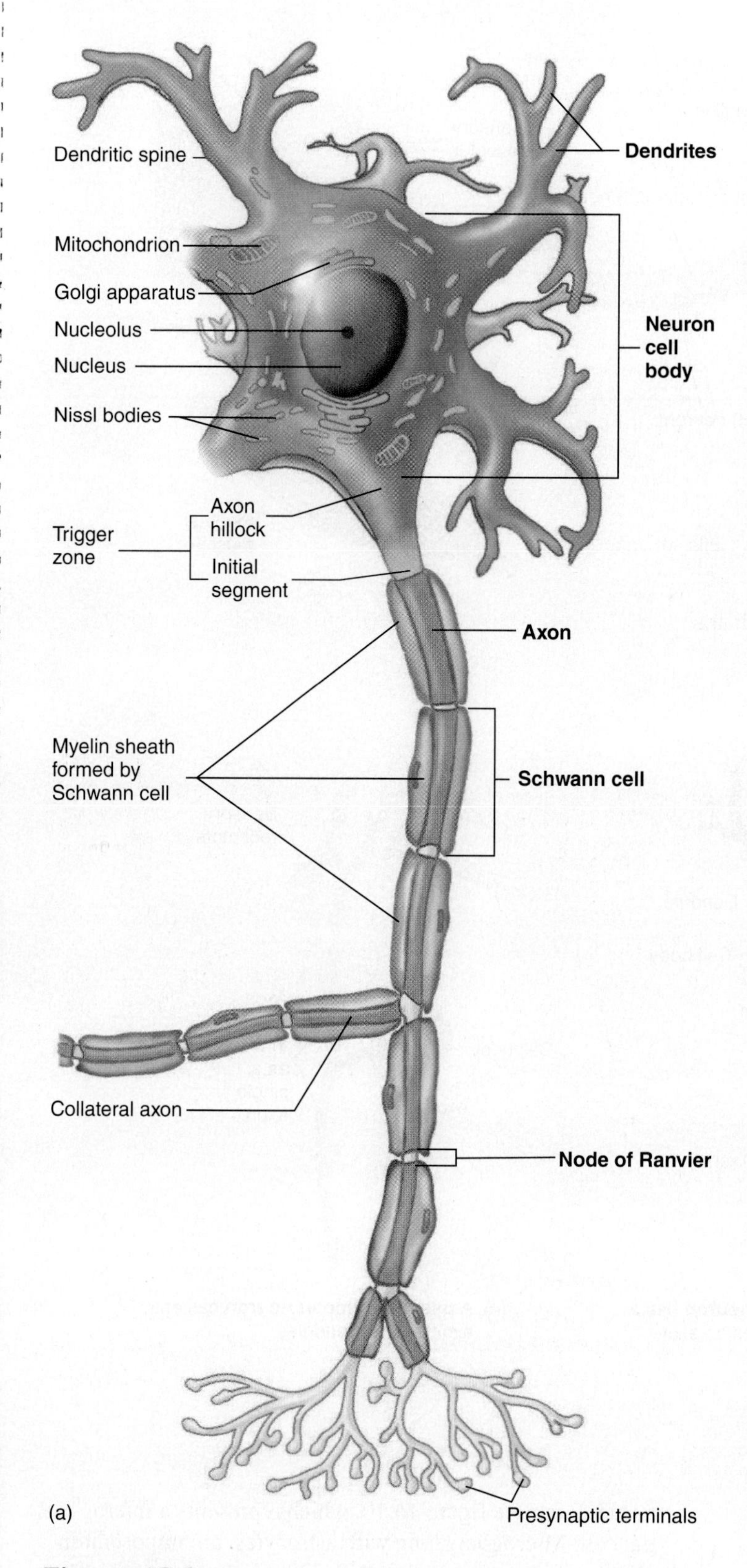

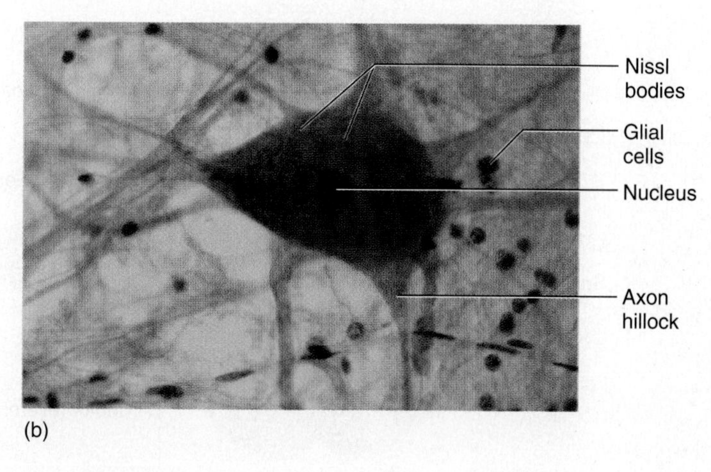

Figure 16.2 Parts of a Multipolar Neuron
(a) Diagram; (b) photograph.

Glial Cells

There are other types of cells that are important in the nervous system. These cells are called **glial cells.** About half of the weight of the brain is composed of glial cells. The common glial cell of the peripheral nervous system is the **Schwann cell,** or **neurolemmocyte** (figure 16.6). These glial cells wrap around the axon, leaving thin regions between successive cells called the **nodes of Ranvier,** as seen in figure 16.2. Neurolemmocytes are cells that wrap around the axon, much like a thin strip of paper wrapped around a pencil. The neurolemmocyte consists of a significant amount of a lipoprotein material called **myelin,** and the series of neurolemmocytes produces a **myelin sheath.**

Myelinated nerve fibers appear white, so this type of nervous tissue is called **white matter.** The nodes of Ranvier allow for the nerve transmission to jump from node to node, increasing the transmission speed of the neuron. This type of jumping transmission is called **saltatory conduction. Unmyelinated fibers** and neuron cell bodies form nervous tissue known as **gray matter.**

Histology of the Neurolemmocyte

(A) 4 Examine a prepared slide of a longitudinal section of nerve under high power. You should be able to see the axon fibers as long, dark threads in the microscope. What appear to be clear areas on each side of the axon fibers compose the myelin sheath. If you scan the slide closely, you should be able to see the junction of two neurolemmocytes. The gap between them is the node of Ranvier. Compare your slide with figure 16.7.

CNS Glial Cells

Neurolemmocytes are glial cells in the PNS. Other types of glial cells are located in the CNS, and these are described in the following section:

- **Oligodendrocytes** are cells that produce myelin in the CNS. Unlike the neurolemmocytes, oligodendrocytes often wrap around several neurons (figure 16.8). The white matter of the spinal cord and brain is due to the lipoprotein sheaths of the oligodendrocytes.

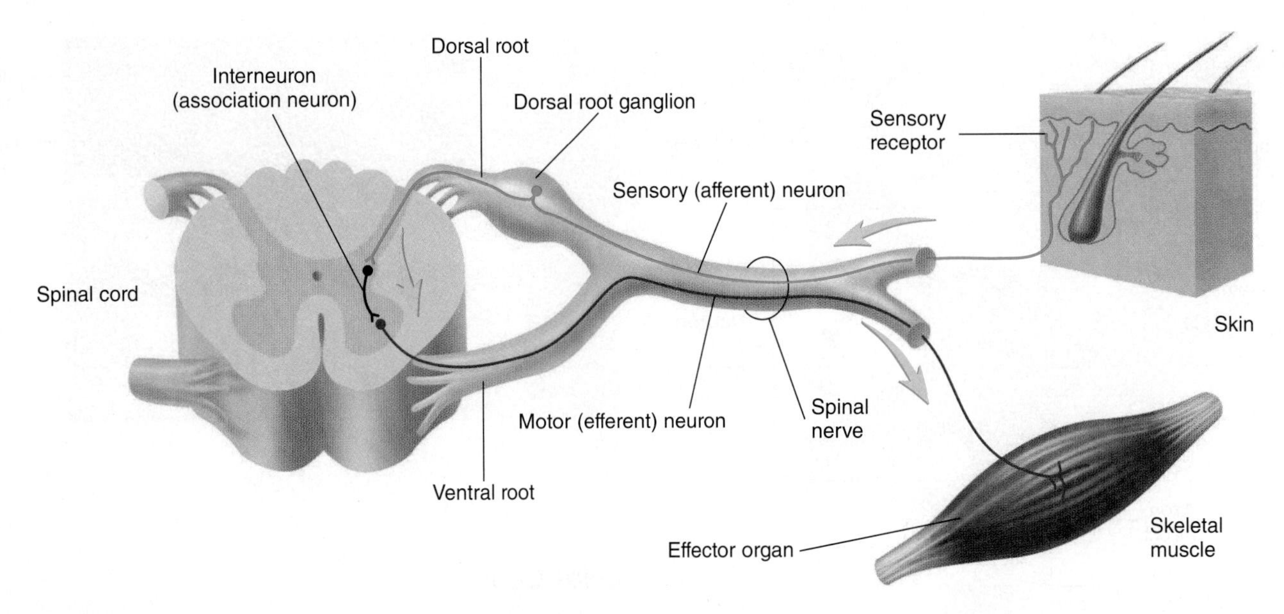

Figure 16.3 Sensory (Afferent) and Motor (Efferent) Neurons

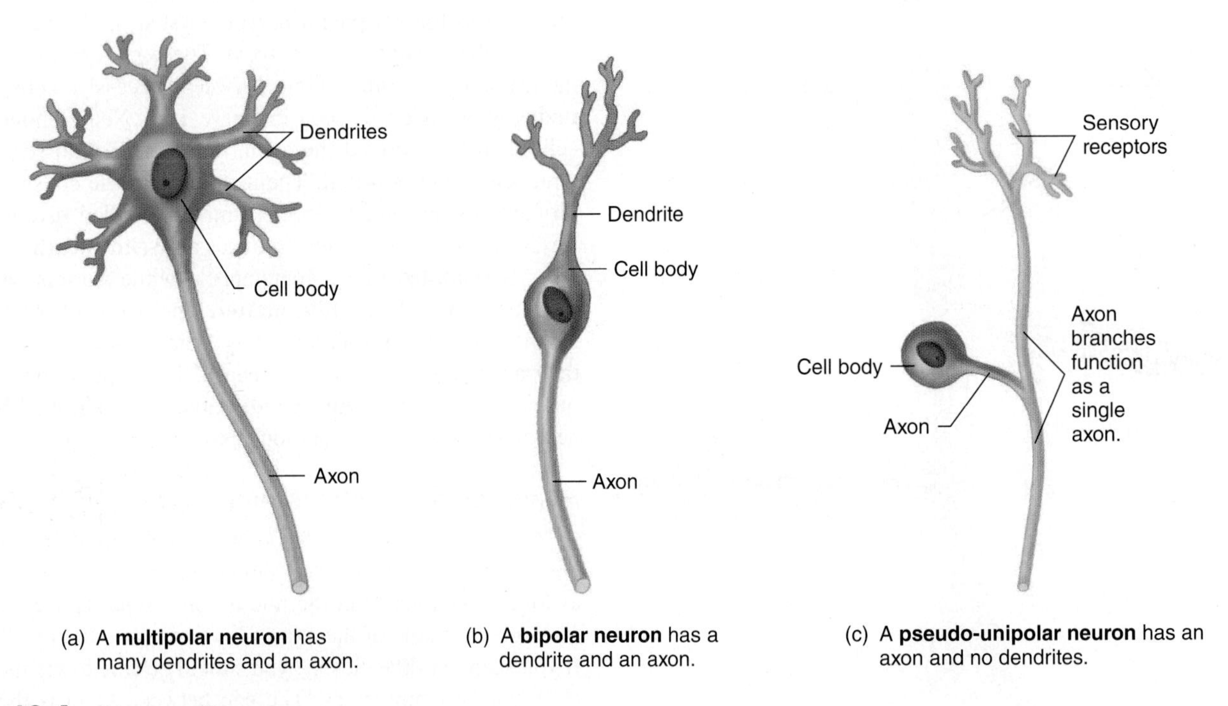

(a) A **multipolar neuron** has many dendrites and an axon.

(b) A **bipolar neuron** has a dendrite and an axon.

(c) A **pseudo-unipolar neuron** has an axon and no dendrites.

Figure 16.4 Neuron Shapes
(a) Multipolar; (b) bipolar; (c) pseudo-unipolar.

- **Astrocytes** are branched glial cells that nourish neurons and provide a barrier between the nervous tissue and the blood. Astrocytes along with capillary endothelial cells are responsible for the **blood-brain barrier** that protects the nervous tissue from some bloodborne infections. This barrier also inhibits some medications from reaching the brain. Astrocytes also seem to play a role in the development of synapses between neurons. Astrocytes are illustrated in figure 16.9.
- **Microglia** are small glial cells that are phagocytic. The microglia digest foreign particles that invade the nervous tissue. Examine figure 16.10, which represents a microglial cell. Microglia, along with astrocytes, are important in glial scar formation in the CNS. They form a type of scar tissue that isolates the brain from further damage caused by microorganisms and begins the healing process after trauma.
- **Ependymal cells** line the ventricles of the brain and serve as a barrier between the fluid in the area (the cerebrospinal fluid) and the nervous tissue. Ependymal cells are illustrated in figure 16.11.

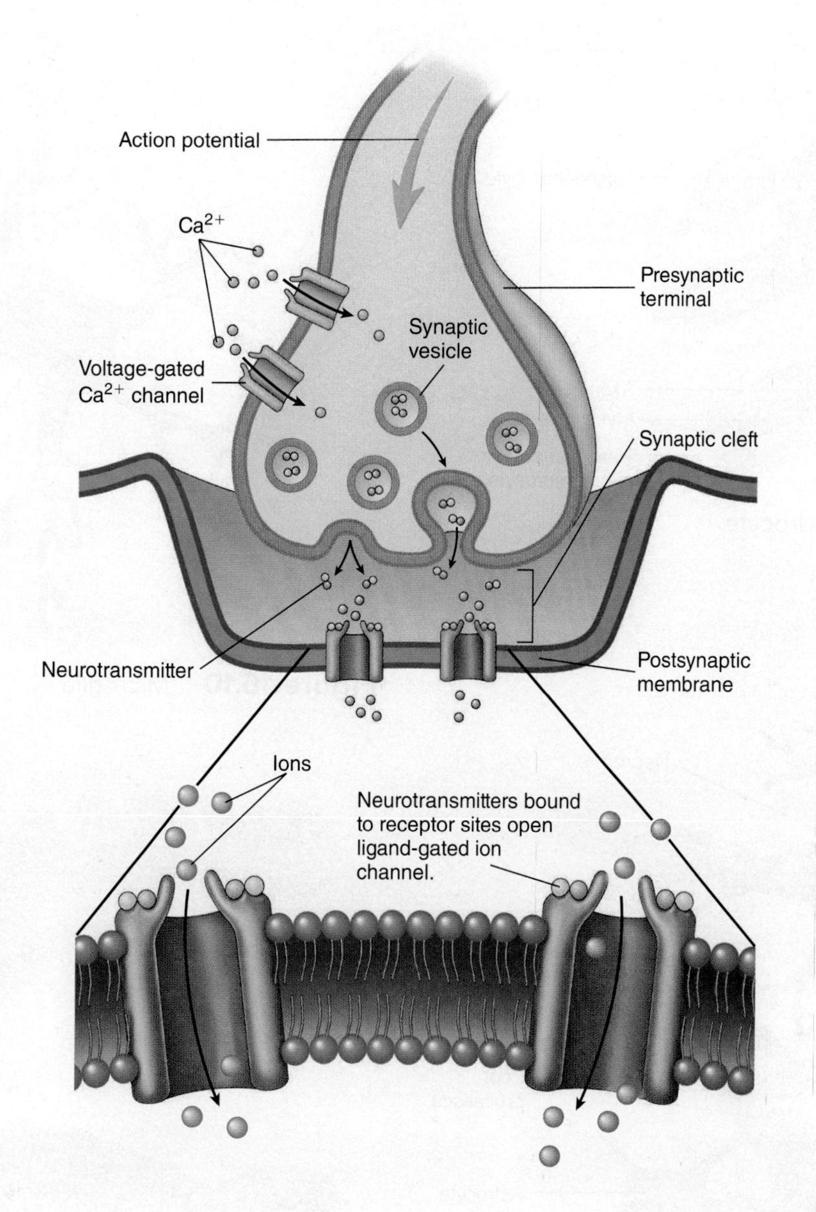

Figure 16.5 Synapse

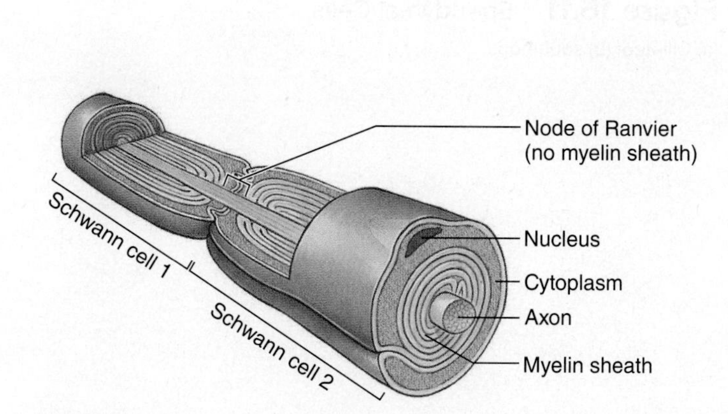

Figure 16.6 Schwann Cell

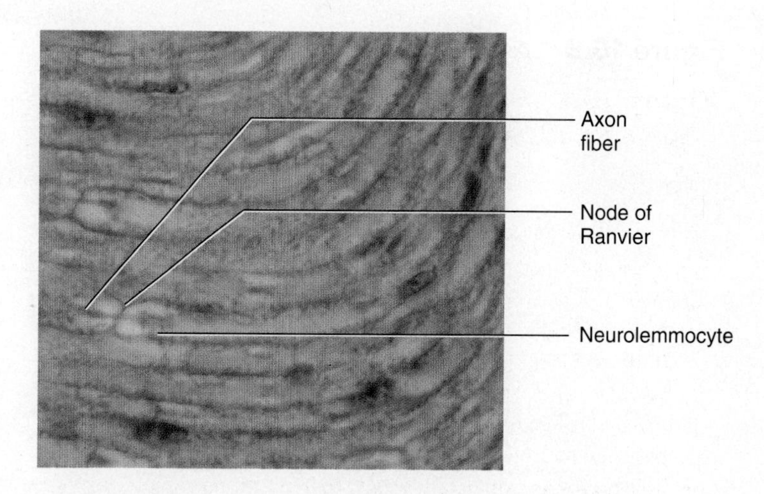

Figure 16.7 Nerve, Longitudinal Section (400X)

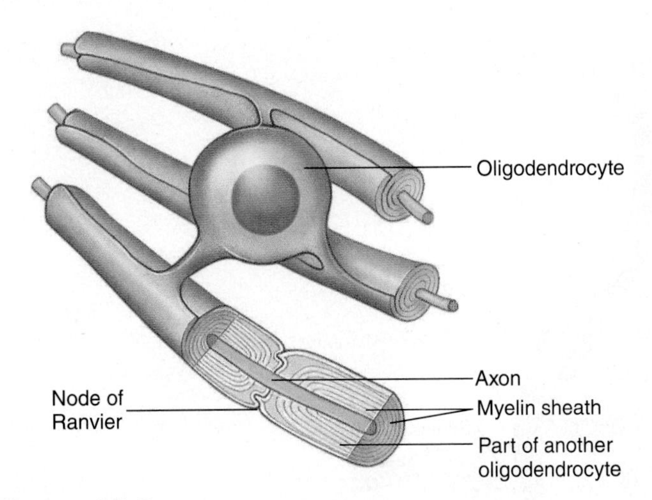

Figure 16.8 Oligodendrocyte

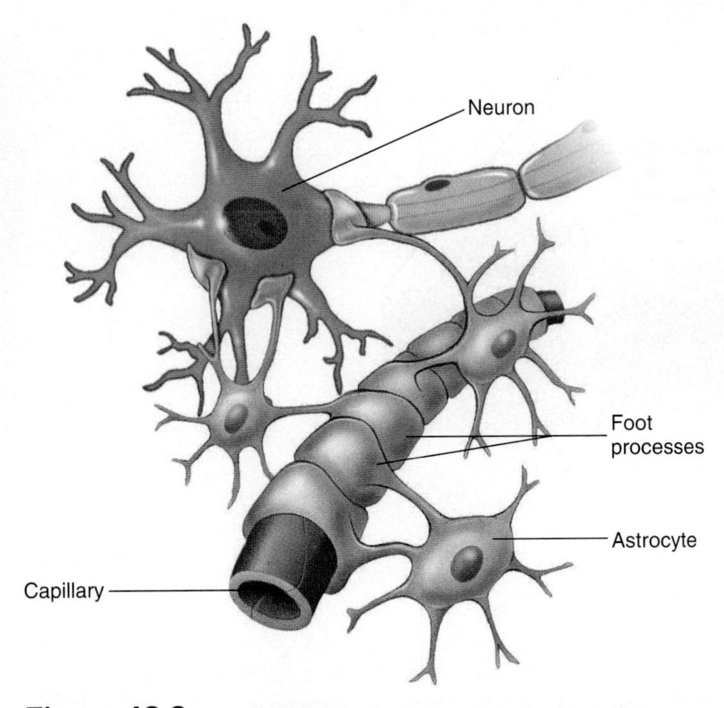

Figure 16.9 Astrocyte

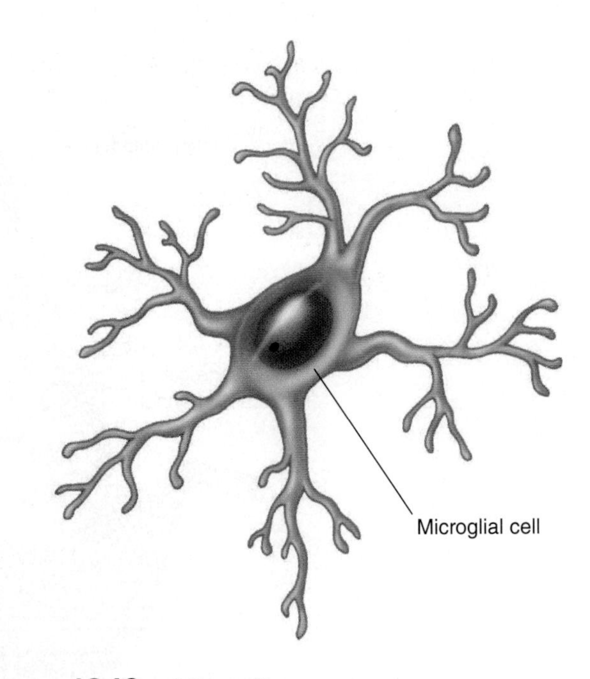

Figure 16.10 Microglia

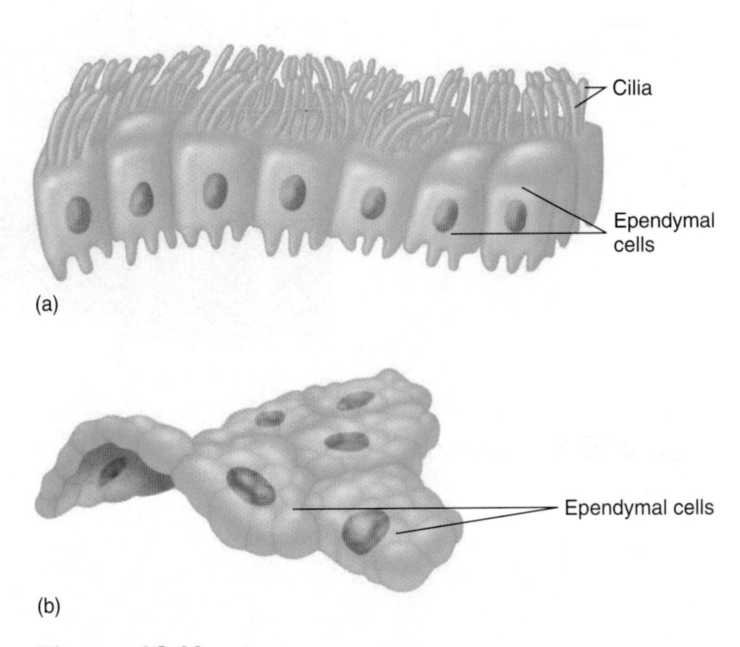

Figure 16.11 Ependymal Cells

(a) Ciliated; (b) squamous.

Exercise 16 Review

Introduction to the Nervous System

Name: ___________________________

Lab time/section: ___________________________

Date: ___________________________

1. What type of neuron (multipolar, bipolar, or pseudo-unipolar) is represented by the drawing? Label the parts with the terms provided.

axon dendrite neuron cell body Nissl body

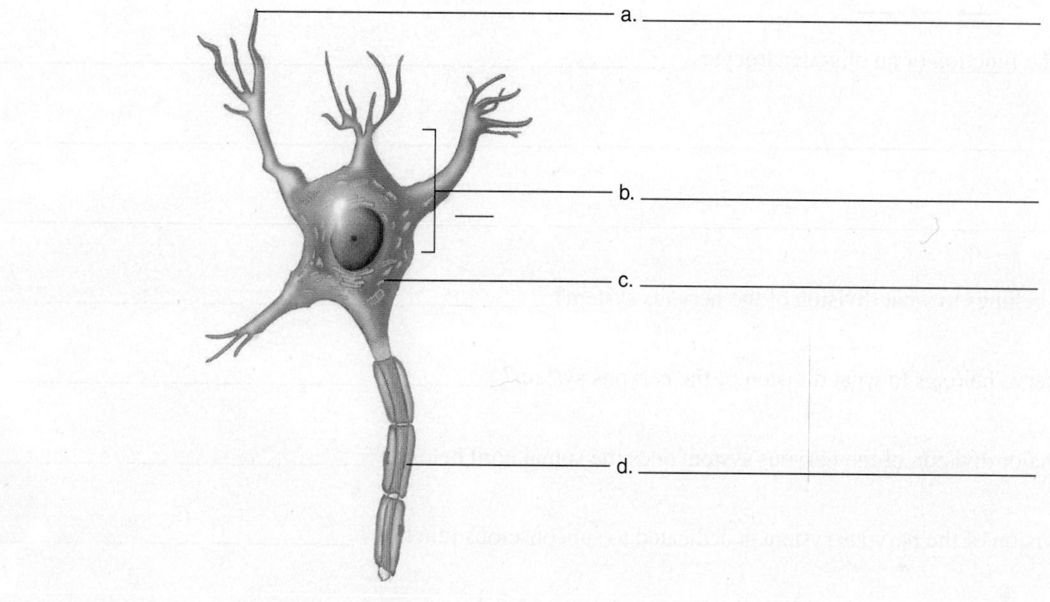

Neuron type ___________________________

2. What type of neuron (multipolar, bipolar, or pseudo-unipolar) is represented by the drawing? Label the parts with the terms provided.

axon neuron cell body nucleus

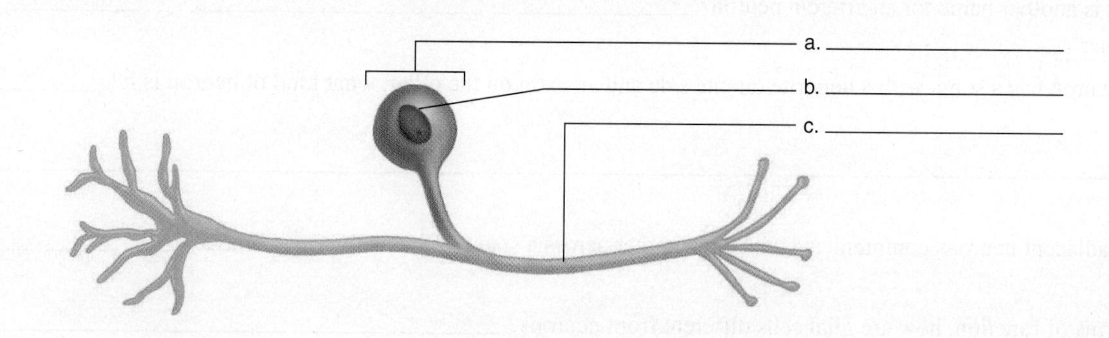

Neuron type ___________________________

3. Describe the function of an astrocyte. ______________________________

____

____

4. Describe the function of an ependymal cell. ______________________________

____

____

5. Describe the function of an oligodendrocyte. ______________________________

____

____

6. The brain belongs to what division of the nervous system? ______________________________

7. A spinal nerve belongs to what division of the nervous system? ______________________________

8. To what major division of the nervous system does the spinal cord belong? ______________________________

9. Which division of the nervous system is dedicated to subconscious function? ______________________________

10. What does *CNS* stand for? ______________________________

11. A neuron has three main parts. What are they? ______________________________

12. In what part of the neuron is the nucleus found? ______________________________

13. What is another name for an efferent neuron? ______________________________

14. If a neuron has a soma with a dendrite on one side and an axon on the other, what kind of neuron is it? ______________________________

____

15. Two adjacent neurons communicate with one another across a space. What is this space called? ______________________________

16. In terms of function, how are glial cells different from neurons? ______________________________

17. In which one of the three nervous system divisions are neurolemmocytes located? ______________________________

18. Myelin is made of what kind of material? ______________________________

Exercise 17

Brain and Cranial Nerves

Introduction

Two specific traits distinguish humans from other animals. One is our upright posture, and the other is the extensive development of the brain. In this exercise, you examine the anatomy of the human brain and cranial nerves. The brain and spinal cord compose the central nervous system and are covered in the Seeley text in chapter 13, "Brain and Cranial Nerves."

The brain is located in the cranial cavity of the skull and weighs approximately 1.4 kilograms (3 pounds). The brain is derived from three embryonic parts, each of which further develops into more specific areas. Cranial nerves are associated with the brain and should be studied after you know the anatomy of the brain.

You may also use a sheep brain as a model for understanding the human brain. There are many similarities between the organs in the two species. They have the same general parts, although sheep brains are smaller and some areas of the sheep brain are proportionally larger or smaller than the corresponding areas of human brains.

Objectives

At the end of this exercise, you should be able to

1. name the three meninges of the brain and describe their locations relative to one another;
2. locate the three major regions of the brain;
3. name the main structures in each of the three regions in intact brains, sectioned brains, or models of brains;
4. describe the function of the major structures in the brain;
5. trace the path of cerebrospinal fluid through the brain;
6. identify each of the 12 pairs of cranial nerves on an illustration, a model, or a real brain;
7. describe whether a cranial nerve is sensory, motor, or both sensory and motor;
8. identify the structure that a particular cranial nerve innervates.

Materials

Models and charts of the human brain
Preserved human brains (if available)
Cast of the ventricles of the brain
Chart, section, or illustration of the brain in coronal and transverse sections
Sheep brains
Dissection trays
Scalpel and two or three extra blades
Gloves (household latex gloves work well for repeated use)
Blunt (Mall) probe
First aid kit in lab or prep area
Sharps container
Animal waste disposal container

Procedure

Meninges

(A)[1] Locate the meninges in preserved brains in the lab (if available) and compare them with figure 17.1. Three layers, called meninges (mĕ-nin′-jēz), surround the brain. The outermost of these is the **dura mater** (doo′rămā′ter), a tough, dense connective tissue sheath that encircles the brain and has a series of shelves that extend into the brain. The dura mater is divided into an outer **periosteal dura** and an inner **meningeal dura.** The space deep to the dura is the **subdural space.** The next deeper layer is the **arachnoid** (ă-rak′noyd) **mater,** a thin membrane that resembles a spider's web. Deep to the arachnoid is the **subarachnoid space,** which contains **cerebrospinal fluid (CSF).** The deepest layer is the **pia mater,** which is a membrane directly on the outer surface of the brain.

Overview of the Brain

(A)[2] Examine a model of the brain and find the major parts. The brain can be subdivided into three major parts. The embryonic **prosencephalon** develops into the **forebrain,** which contains the cerebral hemispheres and diencephalon (figure 17.2 and table 17.1). The **mesencephalon** develops into the **midbrain,** which is small in the adult brain. The embryonic **rhombencephalon** becomes the **hindbrain,** and it consists of the pons, medulla oblongata, and cerebellum. The hindbrain is the most inferior portion of the brain and connects to the spinal cord at the foramen magnum. Table 17.1 lists the major structures of the brain covered in this exercise and the parts of the brain in which they are located.

Ventricles of the Brain

(A)[3] Locate the ventricles of the brain in figure 17.3. The hollow neural tube that develops in the first trimester of pregnancy becomes the ventricles of the brain in the adult. The two ventricles that occupy the center of each cerebral hemisphere are known as the **lateral ventricles.** Tufts of capillaries called **choroid plexuses** secrete cerebrospinal fluid into the ventricles. In preserved brains, the choroid plexuses are small, brown areas at the superior portions of the ventricles. Fluid from the lateral

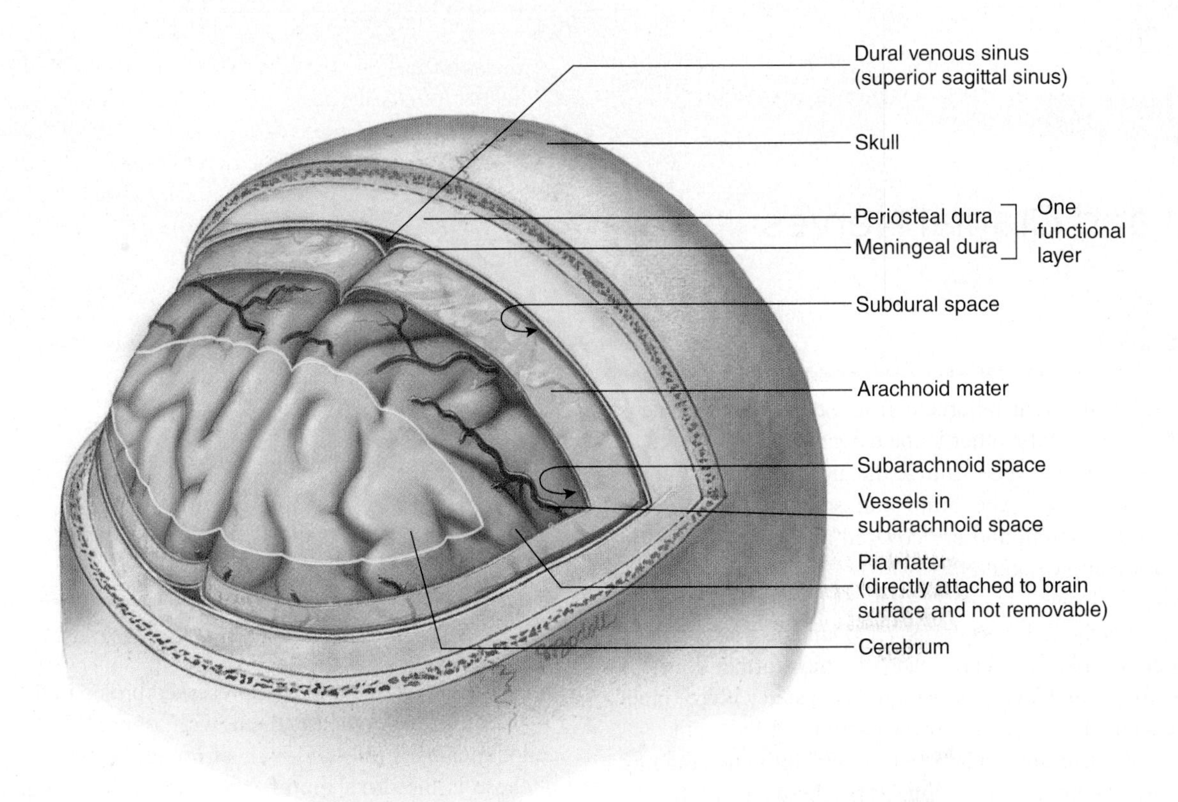

Figure 17.1 Meninges of the Brain

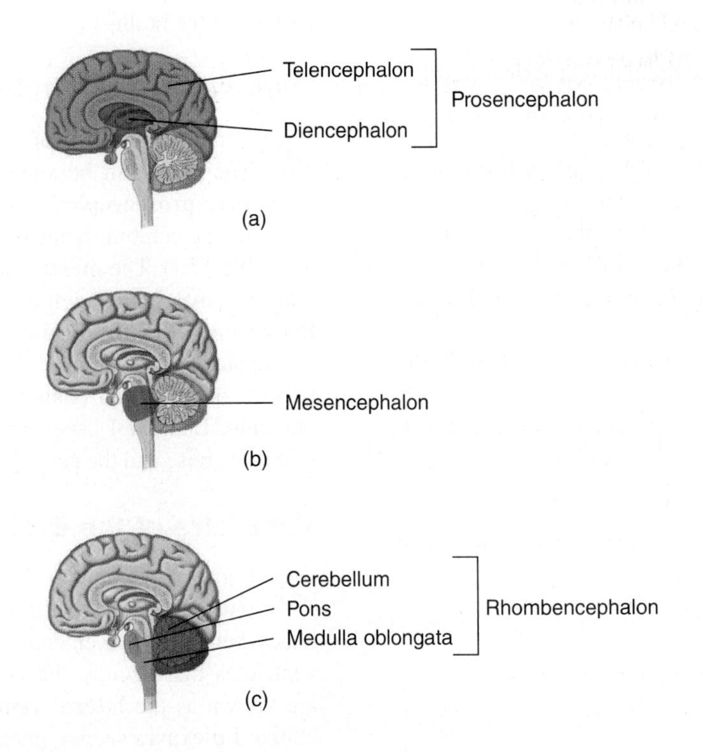

Figure 17.2 Embryonic Derivatives

(a) Forebrain, or prosencephalon; (b) midbrain, or mesencephalon; (c) hindbrain, or rhombencephalon.

TABLE 17.1 Parts of the Brain

Forebrain
Telencephalon
Cerebrum (cerebral hemispheres)
Cerebral cortex (gray matter)
Basal nuclei (gray matter)
Corpus callosum
Diencephalon
Pineal body
Thalamus
Hypothalamus
Pituitary gland
Mammillary bodies
Midbrain
Peduncles
Tectum
Corpora quadrigemina
Superior colliculus
Inferior colliculus
Hindbrain
Metencephalon
Pons
Cerebellum
Myelencephalon
Medulla oblongata

ventricles flows through the **interventricular foramina** into the **third ventricle.** The third ventricle is located in the thalamus and receives CSF from choroid plexuses in that area. The third ventricle drains into the **fourth ventricle** by way of the **cerebral aqueduct.** If this duct becomes occluded or if there is more fluid produced than can be absorbed, then CSF accumulates in the lateral and third ventricles. Commonly, this causes an increase in the size of the ventricles or puts pressure on the nervous tissue, causing a condition known as **internal hydrocephalus.** Normally, the cerebral aqueduct is open and CSF passes through the midbrain. Inferior to the cerebral aqueduct is the fourth ventricle, which occupies a space anterior to the cerebellum. The fourth ventricle also has a choroid plexus that secretes CSF. Cerebrospinal fluid flows from the fourth ventricle to the central canal of the spinal cord, as well as into the subarachnoid space of the spinal cord and brain. The CSF cushions and provides buoyancy to the brain. CSF flows through the arachnoid granulations under the dura mater of the skull to the venous sinuses, and the fluid returns to the rest of the cardiovascular system by the **internal jugular veins.** There are approximately 140 mL of CSF in the central nervous system, and it takes about 6 hours to circulate through the system.

Surface View of the Brain

(A) 4 Examine a model or chart of the brain and locate its major surface features. Of the major parts of the brain, you will be able to easily see the **forebrain** and the **hindbrain.** In the

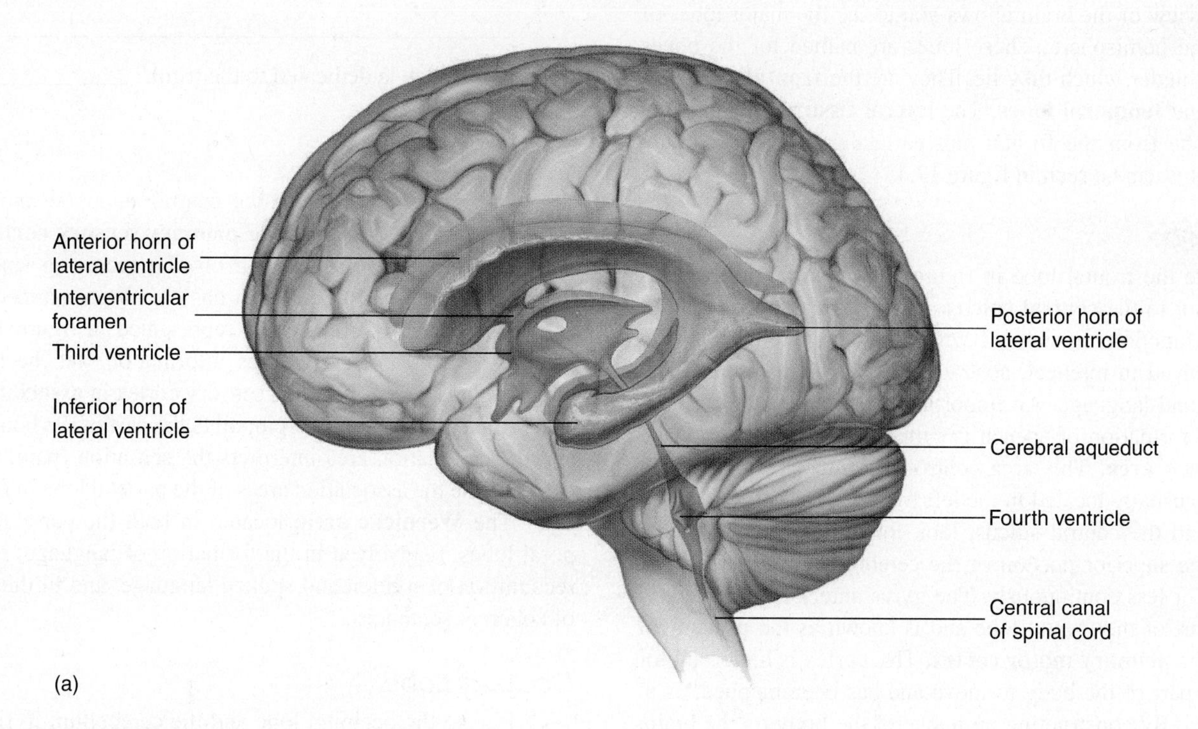

Figure 17.3 Ventricles of the Brain

(a) Lateral view.

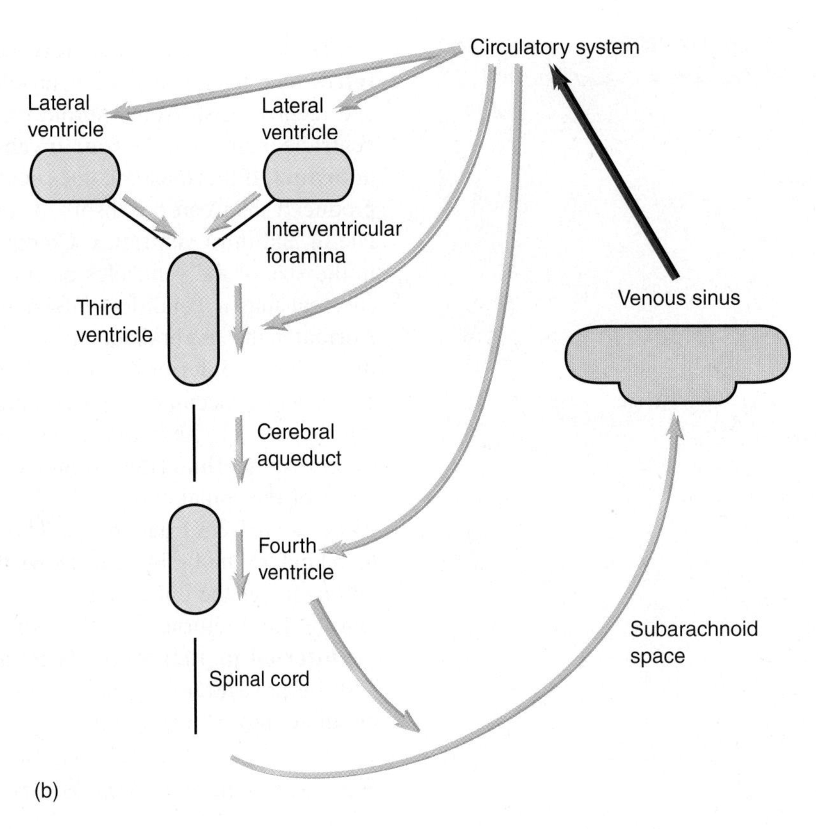

Figure 17.3 *Continued.*
(b) Schematic.

forebrain, you should examine the large **cerebrum,** with its folds and ridges known as **convolutions.** The ridges of the convolutions are **gyri** (sing. **gyrus**), and the depressions are either **sulci** (sing. **sulcus**) or **fissures.** Usually, fissures are deeper than sulci. The lateral view of the brain allows you to see the major lobes of each cerebral hemisphere. These lobes are named for the bones of the skull under which they lie. They are the **frontal, parietal, occipital,** and **temporal lobes.** The **lateral fissure** separates the temporal lobe from the frontal and parietal lobes of the brain. These features can be seen in figure 17.4.

Frontal Lobe

(A) **5** Locate the frontal lobe in figure 17.4. The **frontal lobe** is anterior to the **central sulcus.** It is responsible for many of the higher functions associated with being human. The frontal lobe is involved in intellect, abstract reasoning, creativity, social awareness, and language. An important area responsible for controlling the formation of speech is called the **Broca area,** or the **motor speech area.** This area controls the muscles involved in speech. It is usually located in the left frontal lobe.

To find the central sulcus, look for two convolutions that run from the superior portion of the cerebrum to the lateral fissure, more or less continuously. The gyrus anterior to the central sulcus is part of the frontal lobe and is known as the **precentral gyrus,** or the **primary motor cortex.** This cortex is important for directing a part of the body to move and has been mapped, as in figure 17.5*a*. By constructing an image of the body on the brain, this figure produces the image of a person known as a homunculus. A **motor homunculus** is seen in figure 17.5*a*.

How much of the precentral gyrus is dedicated to the face?

____

How much of the gyrus is dedicated to the hands?

____

How much is dedicated to the trunk? ________________

Parietal Lobe

(A) **6** The gyrus posterior to the central sulcus is known as the **postcentral gyrus,** or the **primary sensory cortex.** This is part of the **parietal lobe** and is involved in receiving sensory information from the body. This area has also been mapped, and you can see a **sensory homunculus** represented in figure 17.5*b*. The primary sensory cortex receives information, yet the material is integrated just posterior to the sensory cortex in **association areas.** The primary sensory cortex pinpoints the part of the body affected, and the association area interprets the sensation (pain, heat, cold, etc.). Locate the association areas of the parietal lobe in figure 17.4.

The **Wernicke area,** located in both the parietal and temporal lobes, is involved in the formation of language, such as the recognition of written and spoken language, and in the formation of coherent sentences.

Occipital Lobe

(A) **7** Locate the occipital lobe and the cerebellum in figure 17.4. Posterior to the parietal lobe is the **occipital lobe** of the brain. The occipital lobe is considered the **visual area** of the brain,

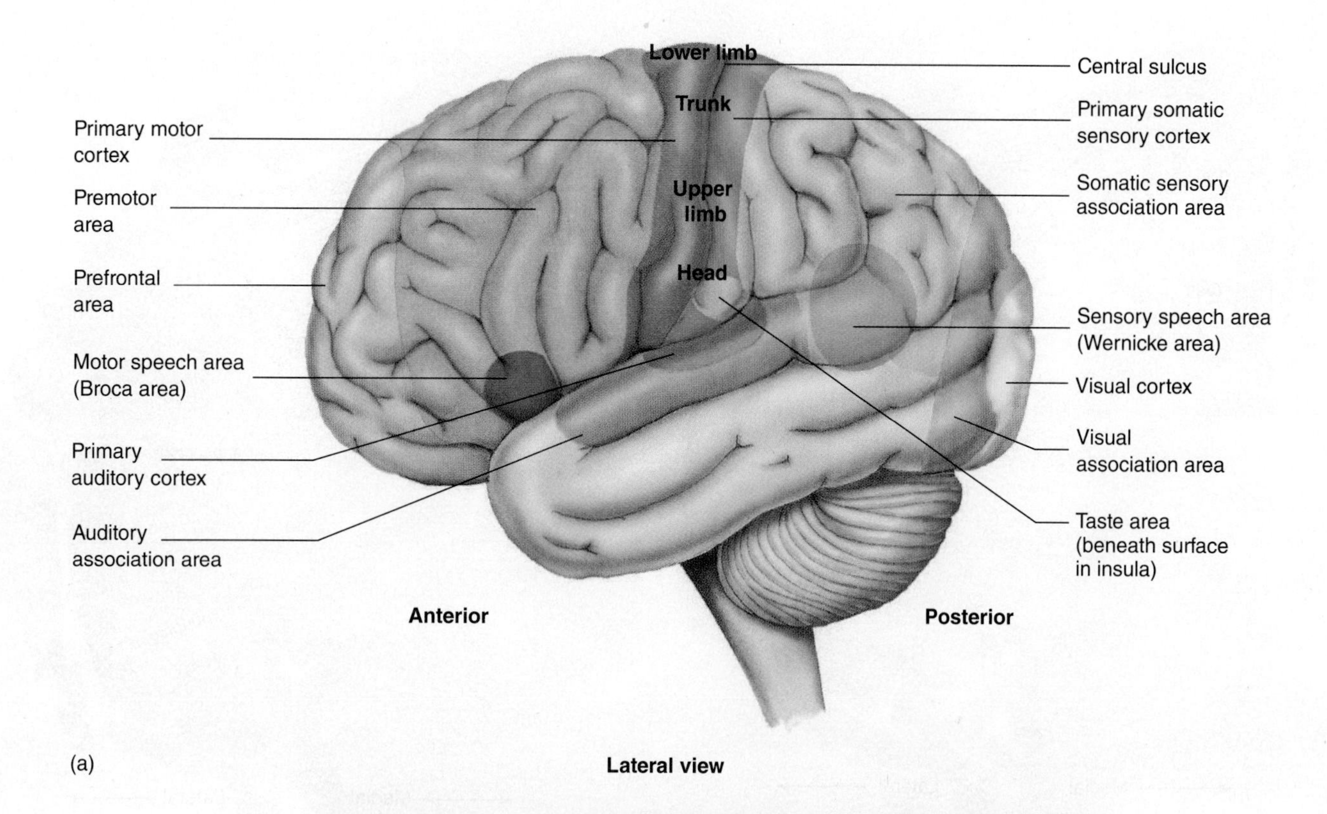

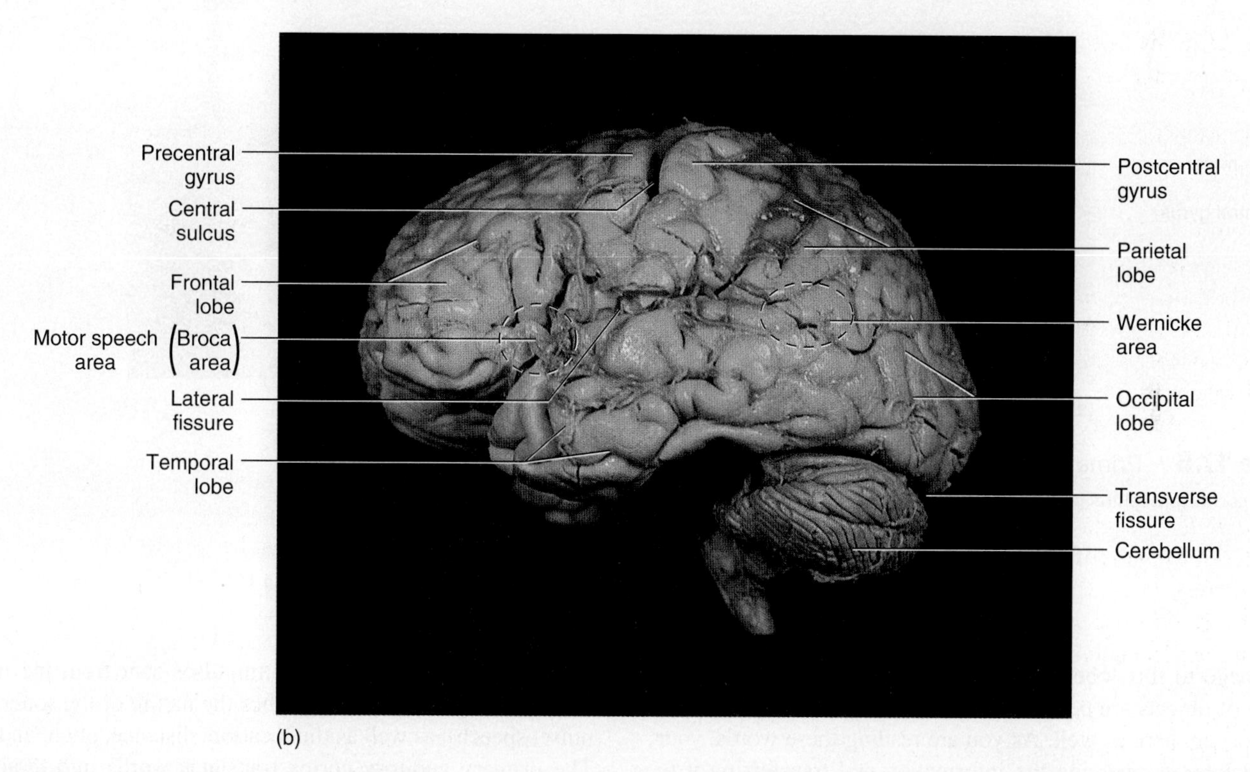

Figure 17.4 Brain, Lateral View

(a) Diagram; (b) photograph. Bold print terms on brain refer to areas of the cortex that receive information from or transmit information to those regions.

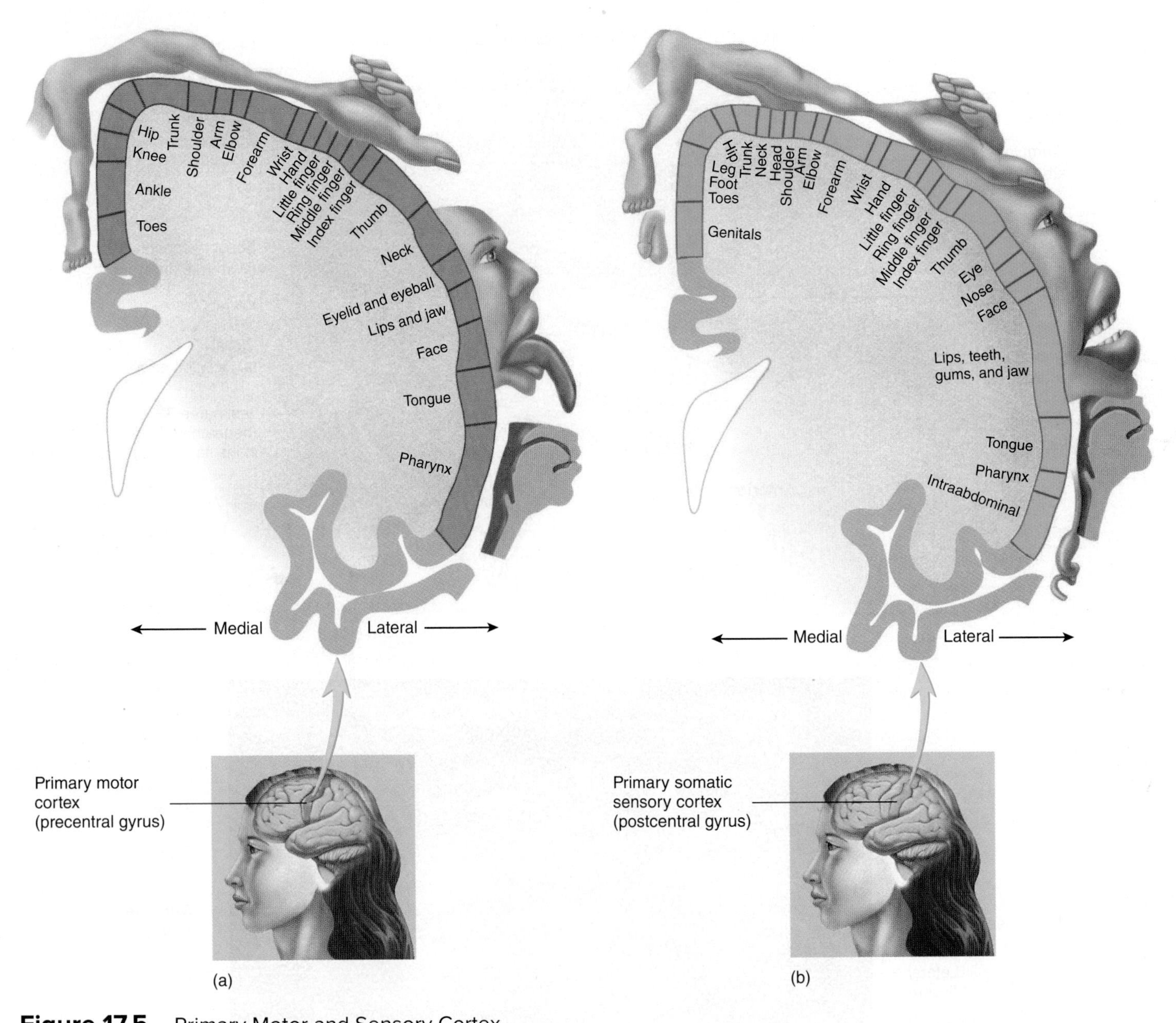

Figure 17.5 Primary Motor and Sensory Cortex

(a) Motor cortex (precentral gyrus); (b) sensory cortex (postcentral gyrus).

and damage to this lobe can cause blindness. Shape, color, and distance of objects are perceived here. Recollection of past visual images occurs here as well. As you are reading these words, your occipital lobe is receiving the information and transferring it to other regions, which convert the words to thought. Between the occipital lobe and the cerebellum is a **transverse fissure** that separates these two regions of the brain.

Temporal Lobe

(A) 8 The **temporal lobe** is separated from the frontal and parietal lobes by the **lateral fissure,** as seen in figure 17.4. The temporal lobe contains an area, known as the **primary auditory cortex,** that interprets hearing impulses sent from the inner ear. This auditory cortex distinguishes the nature of the sound (music, noise, speech), as well as the location, distance, pitch, and rhythm. The primary auditory cortex translates words into thought. The temporal lobe also has centers for the sense of smell **(olfactory centers)** and taste **(gustatory centers).** Deep to the temporal lobe is a small mass of cortical material called the **insula.**

Cerebral Hemispheres

In the surface regions of the cerebrum, associations are compared and contrasted as the brain integrates information about the environment. If you rotate the brain so that you are looking at it from a superior view, you should be able to see the **longitudinal fissure** that separates the cerebrum into the left and right cerebral hemispheres. The **left cerebral hemisphere** in most people is involved in language and reasoning. For example, the Broca area is typically on the left side of the brain.

(A) 9 Examine the surface features of the brain in figure 17.6. The **right cerebral hemisphere** of the brain in most people integrates space and pattern perceptions, artistic awareness, imagination, and music comprehension. The specialization of one hemisphere for one set of tasks is known as cerebral asymmetry, or hemispheric dominance.

Inferior Section of the Brain

Forebrain

(A) 10 You can see the frontal lobes of the cerebrum and the temporal lobes from the inferior view. You may be able to see the **pituitary gland** if it has not been removed. The **optic chiasm** (*chiasm* = cross) is anterior to the pituitary. It transmits visual impulses from the optic nerves to the brain. Two small processes posterior to the pituitary are the **mammillary bodies,** which function in olfactory reflexes and emotional responses to smells.

Hindbrain

(A) 11 Examine the **pons, medulla oblongata,** and **cerebellum** from the inferior aspect. The cerebellum has much finer folds of neural tissue called folia. The medulla oblongata is located inferior to the cerebellum and connects to the spinal cord at the level of the foramen magnum. The enlarged portion of the brain anterior to the medulla is the pons, which serves as a relay center for information. Examine these structures in figure 17.7.

Midsagittal Section of the Brain

Forebrain

(A) 12 Examine a midsagittal section of a brain, as illustrated in figure 17.8. This section is seen by cutting the brain through the longitudinal fissure. Locate the C-shaped **corpus callosum,** which connects the two cerebral hemispheres. The posterior portion of the corpus callosum is known as the **splenium,** and the anterior portion is the **genu.** Just inferior to the corpus callosum is the **septum pellucidum,** which separates the lateral ventricles from each other. If the septum pellucidum is missing, you will be able to look into the lateral ventricle without obstruction.

Examine the material in the lab and locate the **diencephalon** (dī-en-sef′-ă-lon), which consists of, in part, the thalamus and the hypothalamus. The **thalamus** (thal′ă-mus) forms the lateral wall around the third ventricle and is a relay center that receives almost all the sensory information from the body and sends it to the cerebral cortex.

Below the thalamus is the **hypothalamus,** which directs some of the ANS and is involved with the **pituitary gland** (in the hypothalamopituitary axis) in many endocrine functions. Centers for thirst, water balance, pleasure, rage, sexual desire, hunger, sleep patterns, temperature regulation, and aggression are located in the hypothalamus.

Locate the **mammillary bodies** on the inferior portion of the diencephalon and the **optic chiasm** just anterior to it. The **pineal gland** is located posterior to the thalamus and is an endocrine gland that secretes melatonin, a hormone that regulates daily rhythms. Both the pineal gland and the pituitary gland are covered in more depth in Exercise 24.

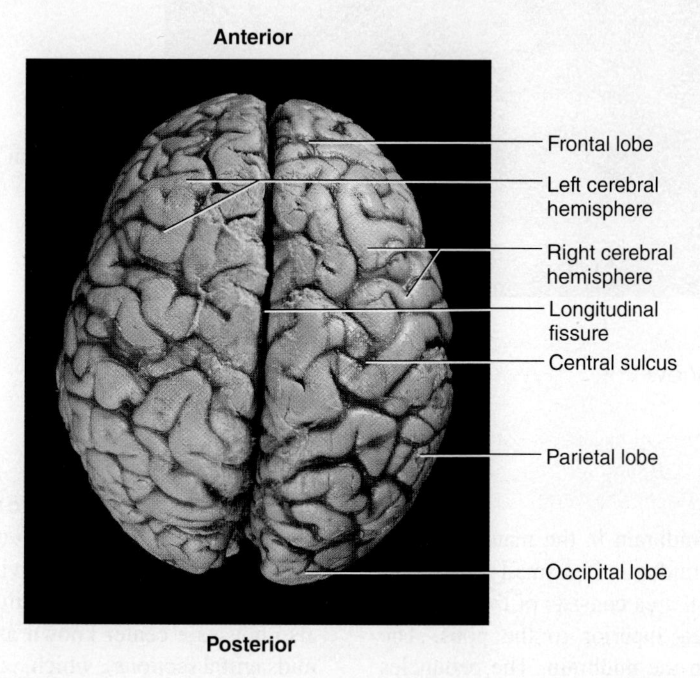

Figure 17.6 Brain, Superior View

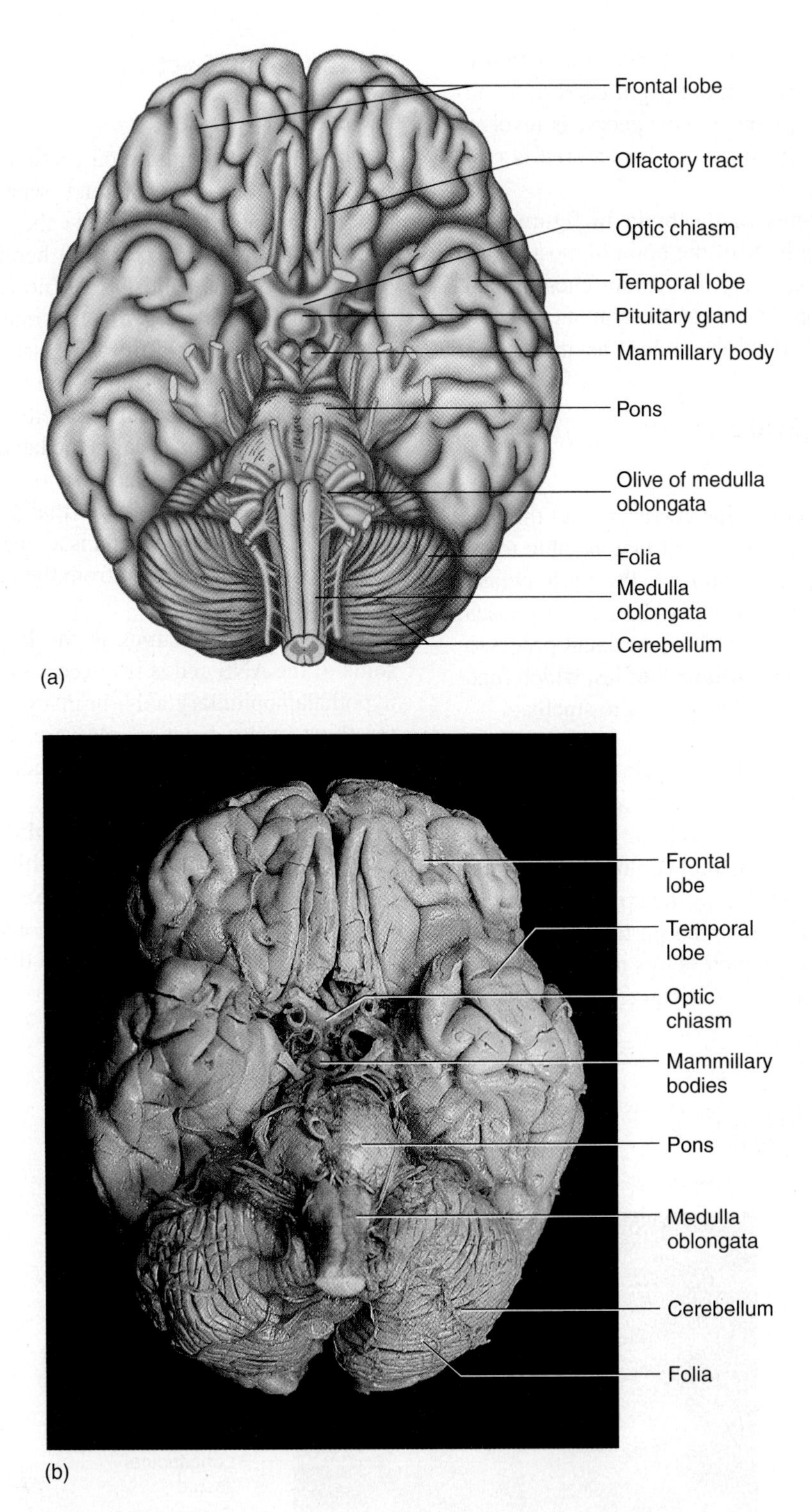

Figure 17.7 Brain, Inferior View

(a) Diagram; (b) photograph.

Midbrain

(A) 13 Locate the features of the midbrain in the material in the lab and in figure 17.8. The midbrain is a small area posterior to the diencephalon. This small area consists of the **cerebral peduncles,** which occupy an area superior to the pons. The **cerebral aqueduct** passes through the midbrain. The peduncles are anterior to the **tectum,** which is a membrane posterior to the cerebral aqueduct. Posterior to the tectum are four hemispheric processes known as the **corpora quadrigemina,** which consist of the **superior colliculi** (areas of visual and auditory reflexes) and the **inferior colliculi** (involved in auditory pathways). The midbrain also houses a center known as the **substantia nigra** (not seen in midsagittal sections), which, when not functioning properly, causes Parkinson disease.

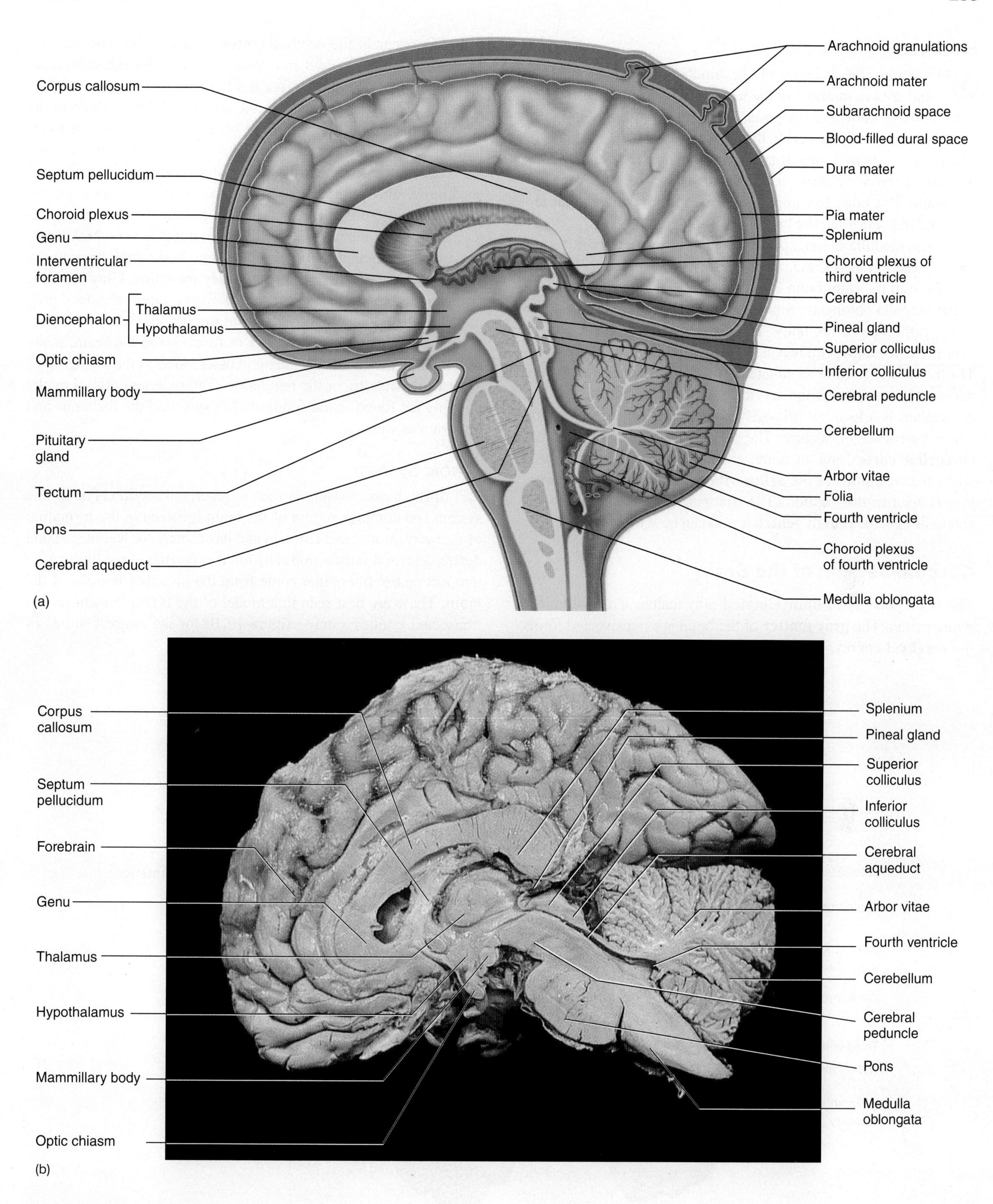

Figure 17.8 Brain, Midsagittal Section

(a) Diagram; (b) photograph.

Hindbrain

(A) 14 Examine the features of the hindbrain as described here and seen in material in the lab. The hindbrain consists of an anterior bulge known as the **pons,** a terminal **medulla oblongata,** and the highly folded **cerebellum** (ser-e-bel′ŭm) (figure 17.8). The pons is a relay center shunting information from the inferior regions of the body through the thalamus and to other areas of the brain. The pons has important **respiratory centers** that are involved in controlling the breathing rate.

The medulla oblongata has centers for respiratory rate control, centers for heart rate, and other vital centers. Some motor signals from the right brain cross over to the left side of the body in the medulla oblongata. Signals from the left brain crosses over to the right side of the body. The area where motor tracts cross over in the medulla is known as the **decussation of the pyramids.** The medulla oblongata terminates at the foramen magnum and the nerve tracts continue as the cervical region of the spinal cord. The cerebellum is a location primarily noted for muscle coordination and maintenance of posture. The cerebellum consists of an outer **cerebellar cortex** and an inner, extensively branched pattern of white matter known as the **arbor vitae.** The **folia** of the cerebellum is seen in this section. The triangular space anterior to the cerebellum is the **fourth ventricle** and can be seen in figure 17.8.

Coronal Section of the Brain

The brain consists of **unmyelinated** gray matter and **myelinated** white matter. The **gray matter** of the brain is extensive and forms the **cerebral cortex.** Most of the active, integrative processes of the brain occur in the cerebral cortex (figure 17.9). The cerebral cortex is approximately 4 mm thick and occupies the superficial regions of the brain. The cortex is where humans do most of their "thinking." It is the main metabolic area of the brain. Deep to the cortex is the **white matter** of the brain, which consists of tracts that take information from one region of the brain to the cerebral cortex for processing. Sensory information coming from the spinal cord moves through the inferior regions of the brain and through the white matter for integration in the cerebral cortex. White matter can also take information from one region of the cerebral cortex to another for integration or from the cerebral cortex back to the spinal cord and to other parts of the body for action. Gray matter is not restricted to the cerebral cortex, however. Deep islands of gray matter in the brain compose the **basal nuclei,** also seen in this section. Basal nuclei serve a number of functions in the brain, many of which involve subconscious processes, such as the swinging of arms while walking or the regulation of muscle tone. Basal nuclei not only are found in the cerebrum but surround the thalamus and midbrain as well.

Limbic System

Part of the limbic system is seen in a coronal section. The limbic system is a complex region of the brain involved in the formation of memory, mood, and emotion and has centers for feeding, sexual desire, fear, and satisfaction. The inferior portion of the limbic system has neural fibers that come from the olfactory regions of the brain. These are best seen in a model of the limbic system or in a transected brain. Examine figure 17.10 for the major features of the limbic system.

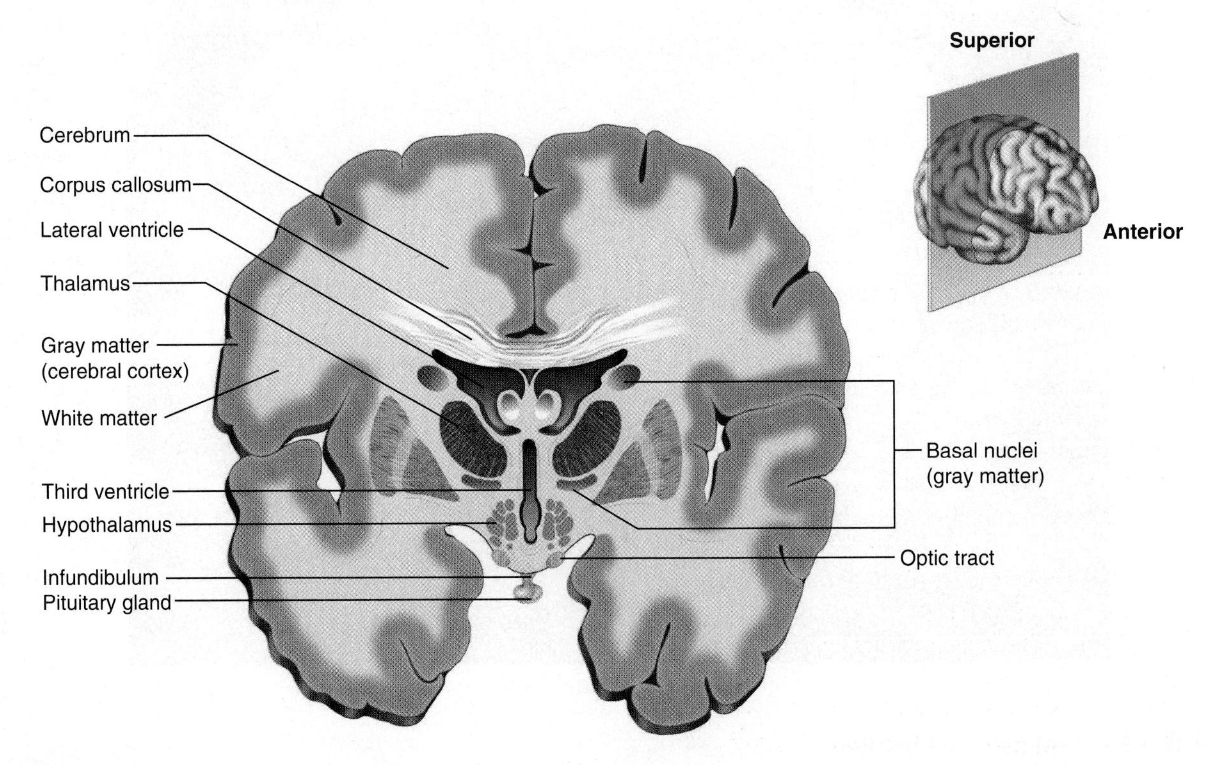

Figure 17.9 Coronal Section of the Brain

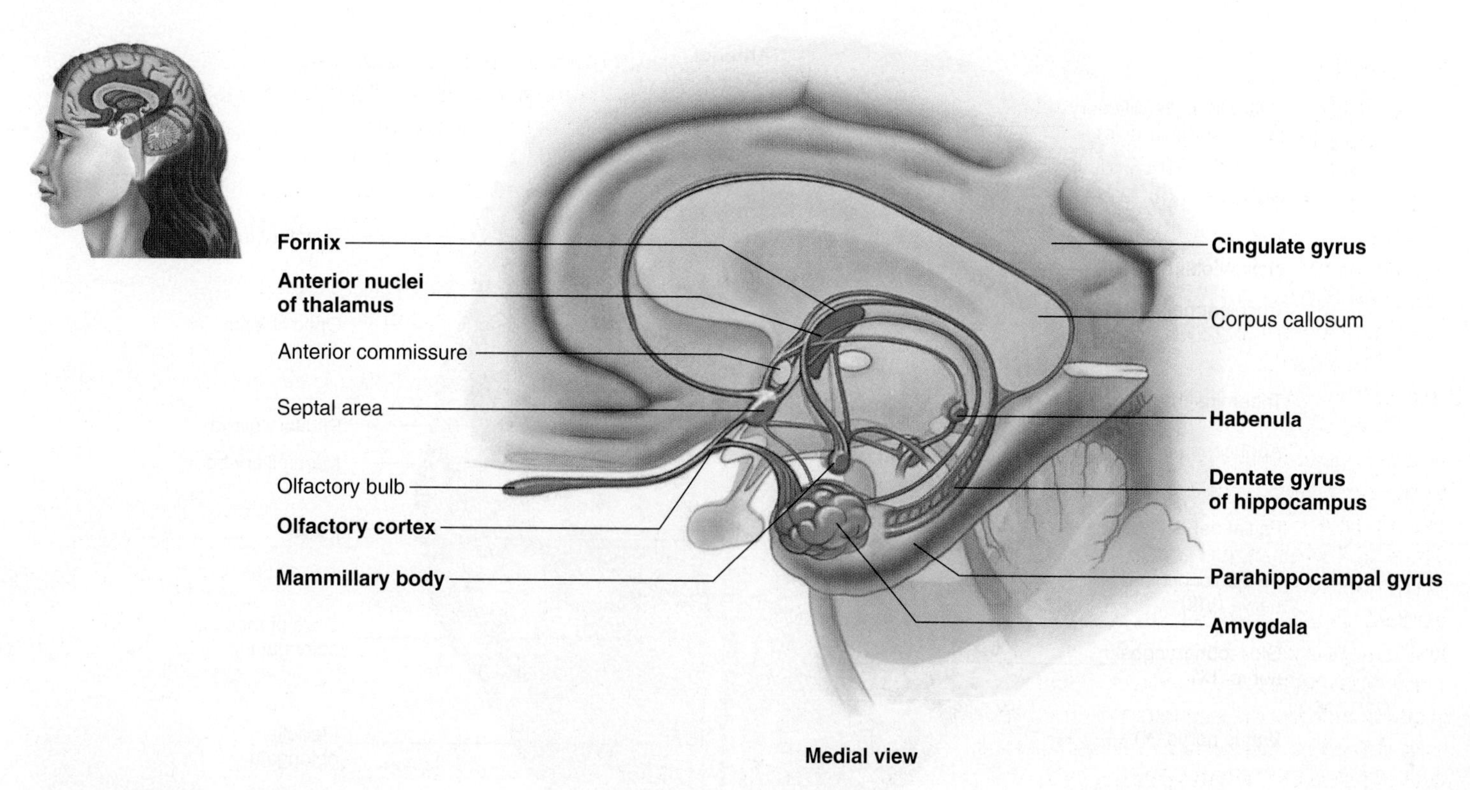

Figure 17.10 Limbic System

Red indicates the limbic system as it occurs in the whole brain.

Brainstem

The brainstem consists of the midbrain, the pons, and the medulla oblongata. Look at a model or section of brain that has had the cerebrum removed. Locate the corpora quadrigemina (figure 17.11), along with the medulla oblongata and the pons.

Cranial Nerves

(A) 15 Locate the cranial nerves in figure 17.12 and note their details in tables 17.2 and 17.3. There are 12 pairs of cranial nerves. The cranial nerves are part of the PNS, but they are often studied along with the brain. The cranial nerves are listed by Roman numeral, and you should know the nerve by name *and* by number. Examine a model of the brain along with figure 17.12 and note that all the cranial nerves except nerve XII are in sequence from anterior to posterior. Nerves may be sensory, motor, or mixed (both sensory and motor). Sensory nerves take information to the CNS, whereas motor nerves conduct information away from the CNS, resulting in some kind of action (usually skeletal muscle contraction). When you study nerves, you should examine the anatomy of the brain to see where the nerve emerges from the brain. If you know, for example, that the abducens is found between the pons and the medulla oblongata, then you can use that nerve as a way to locate other nerves.

(A) 16 The **olfactory nerves** (*olfaction* = smell) pass through the ethmoid bone and turn into the olfactory track, which runs along the anterior base of the brain at the inferior aspect of the frontal lobe and transmits the sense of smell to the brain. The **optic nerve** (*optic* = sight) is a sensory nerve from the eye that leads to the base of the brain and forms the optic chiasm. Some tracts from the optic nerve lead to one side of the brain,

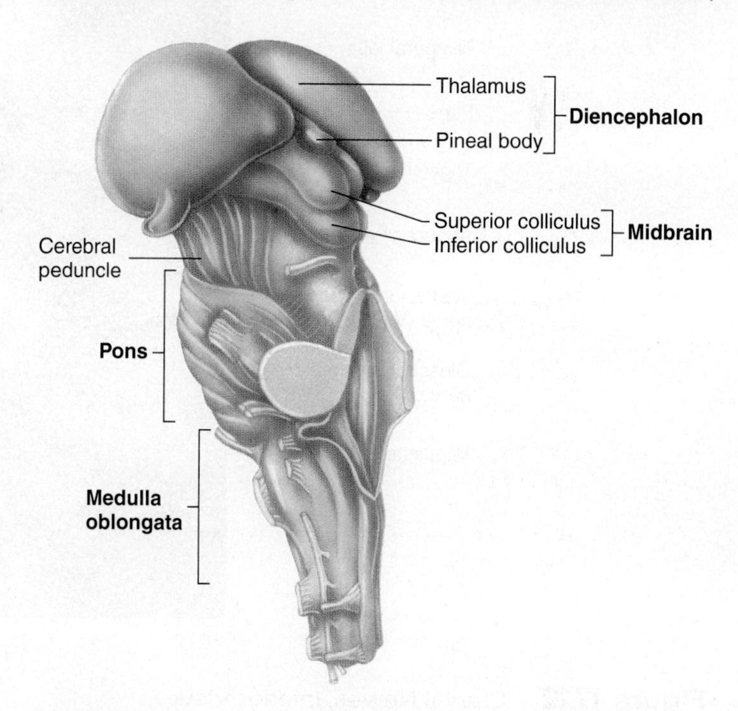

Figure 17.11 Brainstem, Posterolateral View

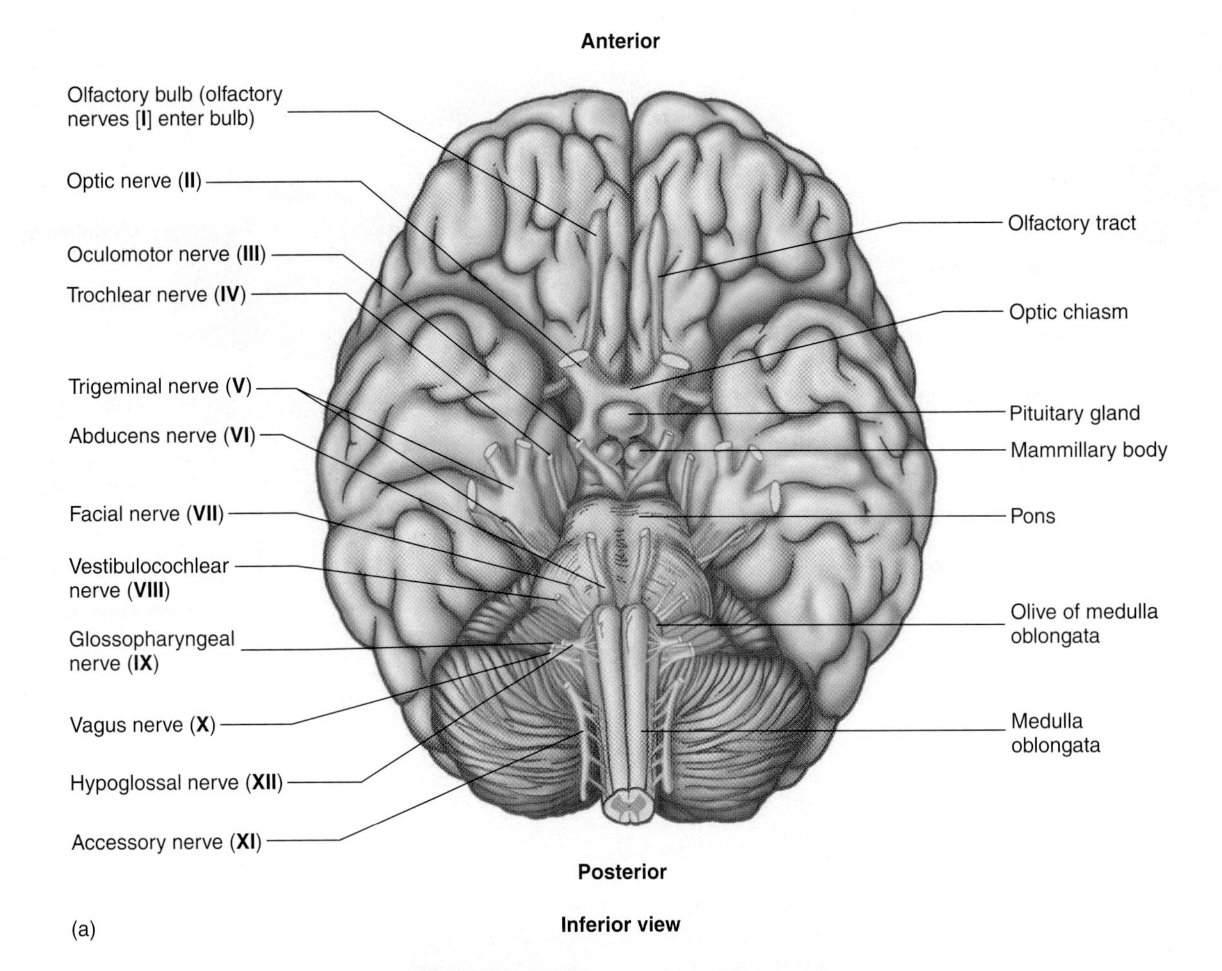

(a)

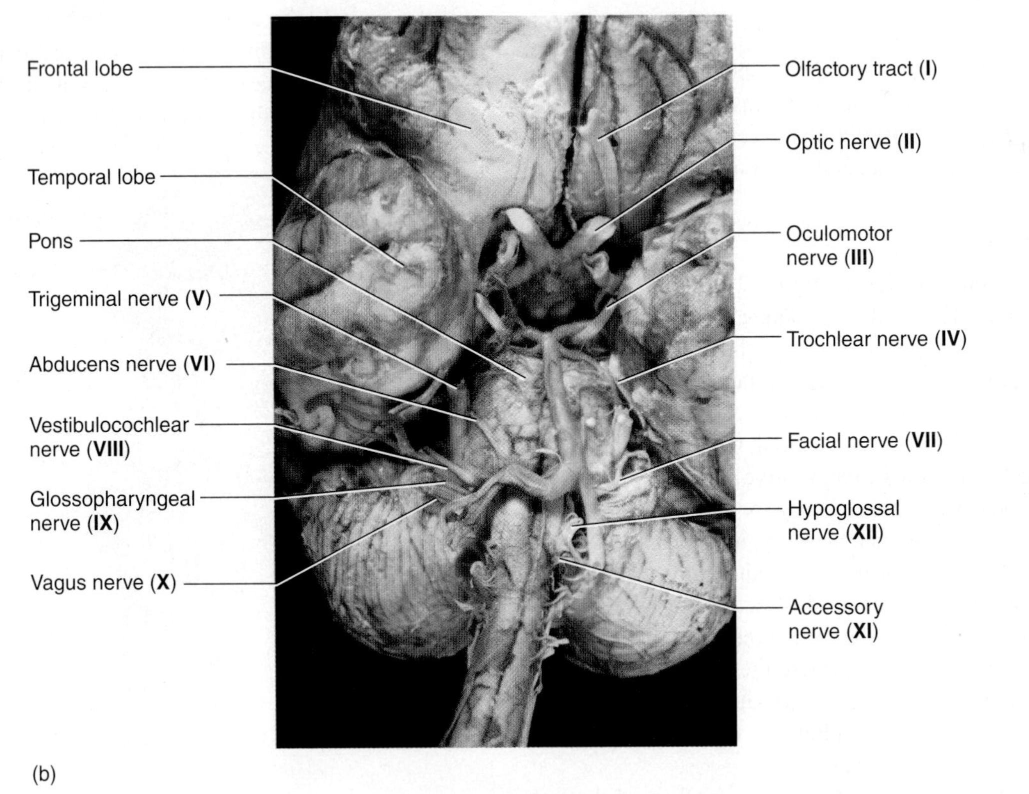

(b)

Figure 17.12 Cranial Nerves, Inferior View

(a) Diagram; (b) photograph.

TABLE 17.2 Cranial Nerves—Function

Number	Name	Function
I	Olfactory	Receives sensory information from the nose
II	Optic	Receives sensory information from the eye, transmitting the sense of vision to the brain
III	Oculomotor	Transmits motor information to move the eye muscles, particularly to the medial, superior, and inferior rectus muscles and to the inferior oblique muscle
IV	Trochlear	Transmits motor information to move the eye muscles, particularly the superior oblique muscle
V	Trigeminal	A three-branched nerve; transmits both sensory information from and motor information to the head
VI	Abducens	A motor nerve; moves the eye muscles, particularly the lateral rectus muscle
VII	Facial	Receives sensory information from the anterior tongue and takes motor information to the head muscles
VIII	Vestibulocochlear	Receives sensory information from the ear; the vestibular part transmits equilibrium information and the cochlear part transmits acoustic information
IX	Glossopharyngeal	A mixed nerve of the tongue and throat; receives information on taste
X	Vagus	Receives sensory information from abdomen, thorax, neck, and root of tongue; transmits motor information to pharynx and larynx; controls autonomic functions of heart, digestive organs, spleen, and kidneys
XI	Accessory	A motor nerve to the muscles of the neck; moves the head
XII	Hypoglossal	A motor nerve to the tongue

TABLE 17.3 Cranial Nerves—Location

Number	Name	Location
I	Olfactory	Begins in the upper nasal cavity and passes through the cribriform plate of the ethmoid bone. It synapses in the olfactory bulbs on each side of the longitudinal fissure of the brain. The fibers take information on the sense of smell and pass it through the olfactory tracts to be interpreted in the temporal lobe of the brain.
II	Optic	Takes sensory information from the retina at the back of the eye and transmits the impulses through the optic canal in the sphenoid bone. Some fibers cross at the optic chiasm and pass via the optic tracts to the occipital lobe, where vision is interpreted.
III	Oculomotor	Emerges from the surface of the brain near the midline and just anterior to the pons. It passes through the superior orbital fissure, innervates the inferior oblique muscle and the medial, superior, and inferior rectus muscles; it carries parasympathetic fibers to the lens and iris.
IV	Trochlear	Is at the sides of the pons at about a 45° angle from the midline of the brain. It passes through the superior orbital fissure to the superior oblique muscle.
V	Trigeminal	Is at a 90° angle from the midline at the lateral sides of the pons. The trigeminal has three branches: (1) The ophthalmic branch passes through the superior orbital fissure; (2) the maxillary branch passes through the foramen rotundum of the sphenoid bone; (3) the mandibular branch passes through the foramen ovale of the sphenoid bone and enters the mandible by the mandibular foramen and exits by the mental foramen.
VI	Abducens	Begins at the midline junction between the pons and the medulla oblongata and passes through the superior orbital fissure to carry motor information to the lateral rectus muscle of the eye.
VII	Facial	Begins as the first of a cluster of nerves on the anterolateral part of the medulla oblongata. It passes through the internal auditory meatus and through the inner ear to the stylomastoid foramen of the temporal bone to innervate facial muscles and glands. It receives sensory information from the anterior tongue. Sensory information of the tongue is interpreted in the temporal lobe of the brain.
VIII	Vestibulocochlear	Comes from the inner ear and passes through the internal auditory meatus. The conduction passes to the pons, and hearing and balance are interpreted in the temporal lobe.
IX	Glossopharyngeal	Passes through the jugular foramen to innervate muscles of the throat (pharyngeal branches) and the tongue (glossal branches). Motor portions of the nerve control some muscles of swallowing and salivary glands, whereas sensory nerves receive information from the posterior tongue and from baroreceptors of the carotid artery.
X	Vagus	Passes through the jugular foramen and along the neck to the larynx, heart, and abdominal region. The sensory impulses travel in this nerve from the viscera in the abdomen, the thorax, the neck, and the root of the tongue to the brain.
XI	Accessory	Multiple fibers arise from the lateral sides of the medulla oblongata and pass through the jugular foramen to numerous muscles of the neck.
XII	Hypoglossal	Begins at the anterior surface of the medulla and passes through the hypoglossal canal to innervate the muscles of the tongue.

whereas others cross to the other side. The **oculomotor nerve** (*oculo* = sight; *motor* = the function of the nerve) controls eye muscles and is located anterior to the pons, more or less in the midline of the brain, whereas the **trochlear nerve** (*trochlea* is in reference to the trochlea that the superior oblique muscle passes through) is found at about a 45° angle from midline on the lateral aspect of the pons. The large **trigeminal nerve** (*trigeminal* = triplets as the nerve divides into three parts) is at a 90° angle to the pons and is located on the lateral aspect of the pons. It is a mixed nerve that innervates much of the face. The **abducens nerve** (*abduce* = to pull away; the muscle controlled by this nerve pulls the eye laterally) is located at the midline junction of the pons and medulla oblongata, whereas the **facial nerve** (named for its innervation of the face) is more lateral. Posterior to the facial nerve is the **vestibulocochlear nerve** (the vestibule and cochlea are parts of the ear), and the nerve directly behind that is the **glossopharyngeal nerve** (*glosso* = tongue; *pharyngeal* = pharynx). The **vagus nerve** (*vagus* = wandering) is seen as a large nerve or large cluster of fibers on the lateral aspect of the medulla oblongata. The vagus travels to the thorax and abdomen. Posterior to the vagus nerve is the **accessory nerve,** which controls muscles of the neck. More toward the midline of the medulla oblongata is the **hypoglossal nerve** (*hypo* = below; *glossus* = tongue), which is the last of the cranial nerves. There is a mnemonic to help you remember the sequence of cranial nerves. If you use the first letter of the name of the cranial nerve, you can learn the sequence of these nerves: "Old Oliver Ogg Traveled To Africa For Very Good Vacations And Holidays."

The cranial nerves are listed in table 17.4 as sensory nerves, motor nerves, or both sensory and motor nerves. A mnemonic device is also listed in the table to help you remember the name and function of each nerve.

TABLE 17.4 Cranial Nerves—Types

Number	Name	Mnemonic	Type	Mnemonic
I	Olfactory	Old	Sensory	Sally
II	Optic	Oliver	Sensory	Sells
III	Oculomotor	Ogg	Motor*	Many
IV	Trochlear	Traveled	Motor	Monkeys
V	Trigeminal	To	Both	But
VI	Abducens	Africa	Motor	My
VII	Facial	For	Both	Brother
VIII	Vestibulocochlear	Very	Sensory	Sells
IX	Glossopharyngeal	Good	Both	Bigger
X	Vagus	Vacations	Both	Better
XI	Accessory	And	Motor	Mega
XII	Hypoglossal	Holidays	Motor	Monkeys

* Many of the motor nerves have sensory fibers that come from proprioreceptors in the muscles they innervate. Information about the tension of the muscle is sent back to the brain to make adjustments in contractile rate. Since the main function of these nerves is motor, they are listed as motor nerves, even though they have some sensory capabilities.

Dissection of the Sheep Brain

(A) 17 Work in pairs during the dissection of the sheep brain. Take a sheep brain to your table, along with a dissecting tray and appropriate dissection tools. If the brain still has the **dura mater,** examine this tough connective tissue coat located on the outside of the brain. Look for the pituitary gland before you remove the dura. Cut through this layer to examine the meninges underneath it. Deep to the dura mater is a filmy layer of tissue that contains blood vessels. This is the **arachnoid mater.** If you tease some of the membrane away from the brain, you will see that it has a cobweblike appearance in the **subarachnoid space.** The subarachnoid space may contain some fluid, which is the **CSF.** Underneath this layer and adhering directly to the brain convolutions is the **pia mater,** which is the surface lining of the **convolutions** of the brain.

(A) 18 Find the major lobes of the sheep brain: the cerebellum, pons, and medulla oblongata (figure 17.13). Since sheep are quadrupeds, the flexure (the 90-degree angle of the brain to the spinal cord) of the brain does not occur in them as it does in humans. Sheep have a horizontal spinal cord, whereas humans have a vertical one. Sheep also have a reduced cerebrum. Examine the inferior surface of the sheep brain. You should see the olfactory bulbs and tracts and the optic nerve and optic chiasm easily from this view. The pituitary gland will probably not be attached, but you should locate the infundibulum (the attachment region of the pituitary to the hypothalamus), caudad to the optic chiasm. Locate these structures in figure 17.14. If the sheep brain is intact, work with another student pair and decide which brain will be sectioned in the midsagittal plane (figure 17.15) and which will be sectioned in the coronal plane (figure 17.16). For the midsagittal section, divide the brain along the length of the longitudinal fissure. Your cut should reflect a section illustrated in figure 17.15. Note that the sheep brain has an enlarged corpora quadrigemina, compared with that of humans. Locate the corpus callosum, lateral ventricles, third ventricle, hypothalamus, pineal gland, superior and inferior colliculi, cerebellum, arbor vitae, pons, medulla oblongata, cerebral aqueduct, and fourth ventricle. After making a coronal section about midway through the cerebrum, you should locate the cerebral cortex, cerebral medulla, lateral ventricles, corpus callosum, third ventricle, thalamus, and hypothalamus.

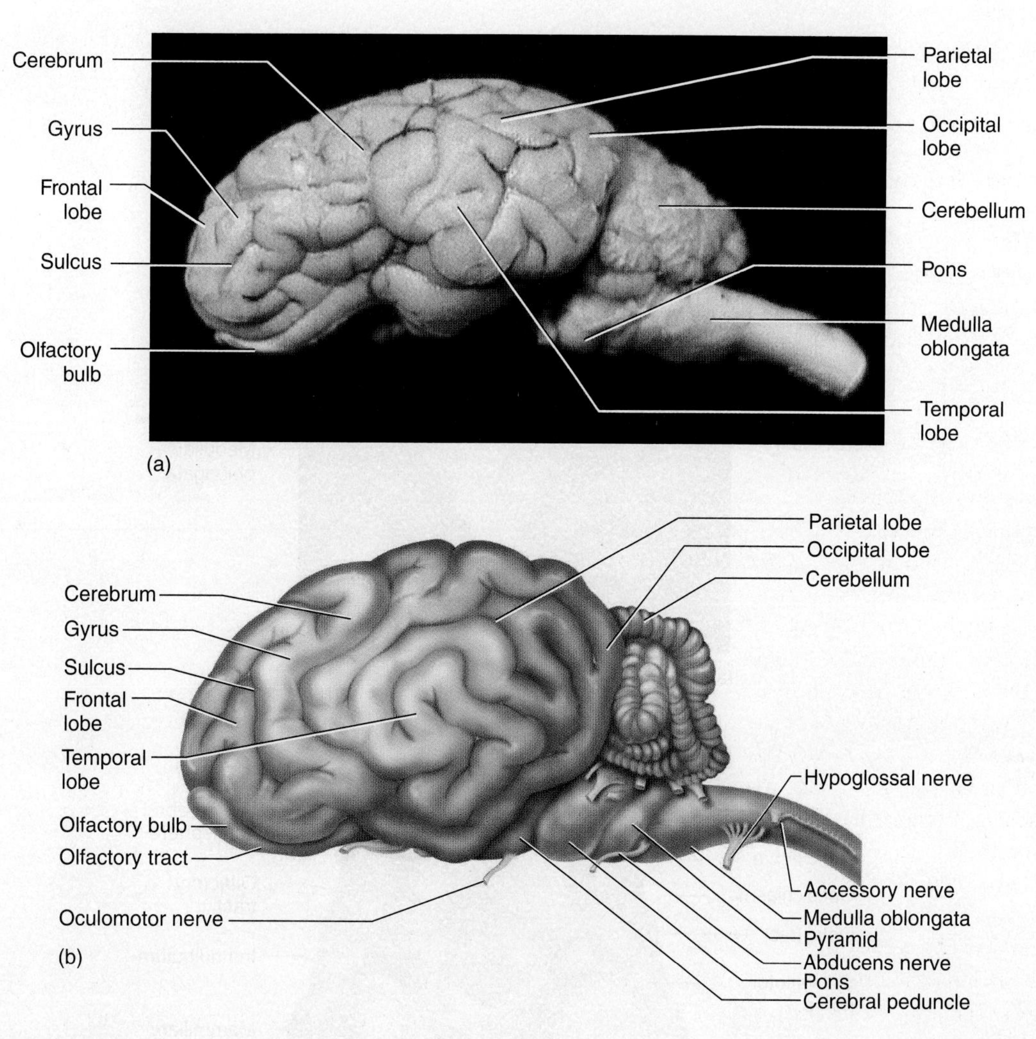

Figure 17.13 Brain of the Sheep, Lateral View
(a) Photograph; (b) diagram.

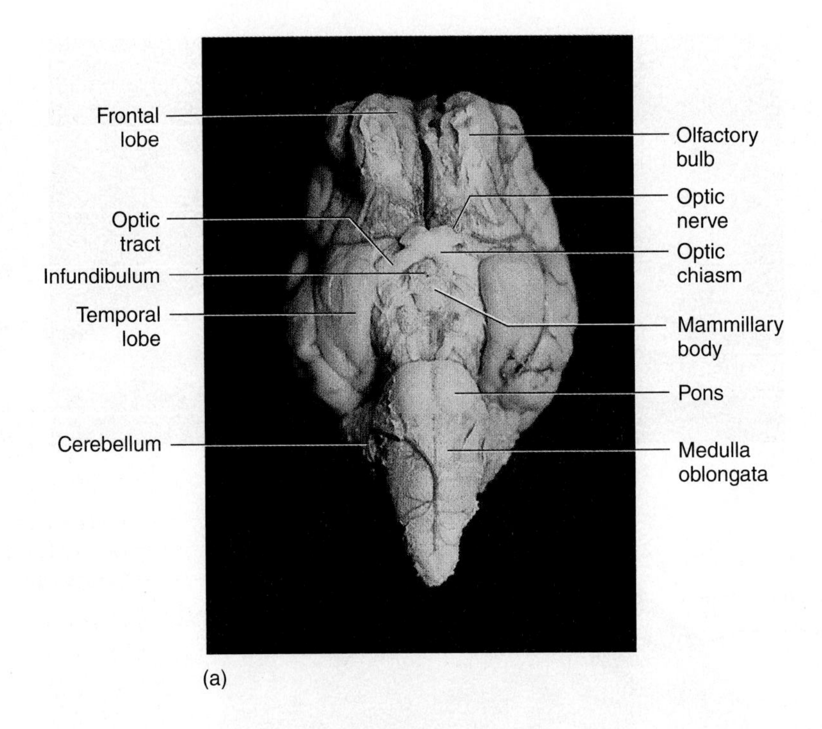

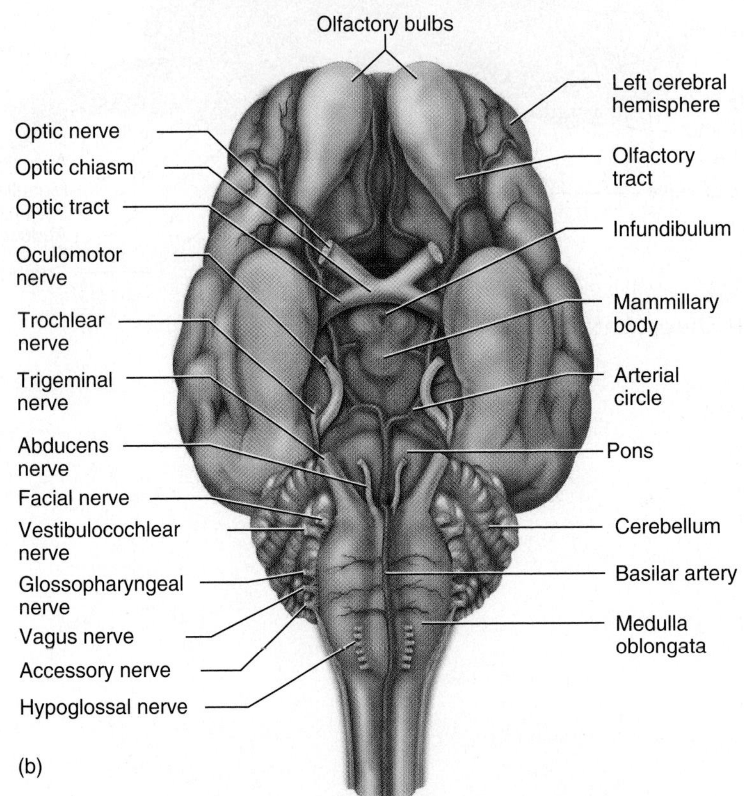

Figure 17.14 Brain of the Sheep, Inferior View

(a) Photograph; (b) diagram.

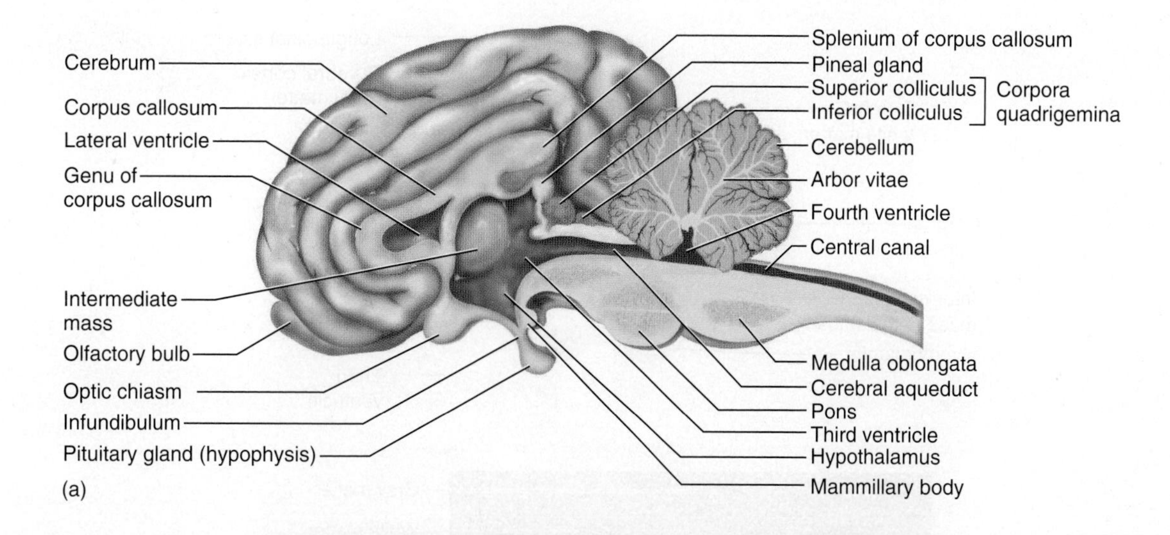

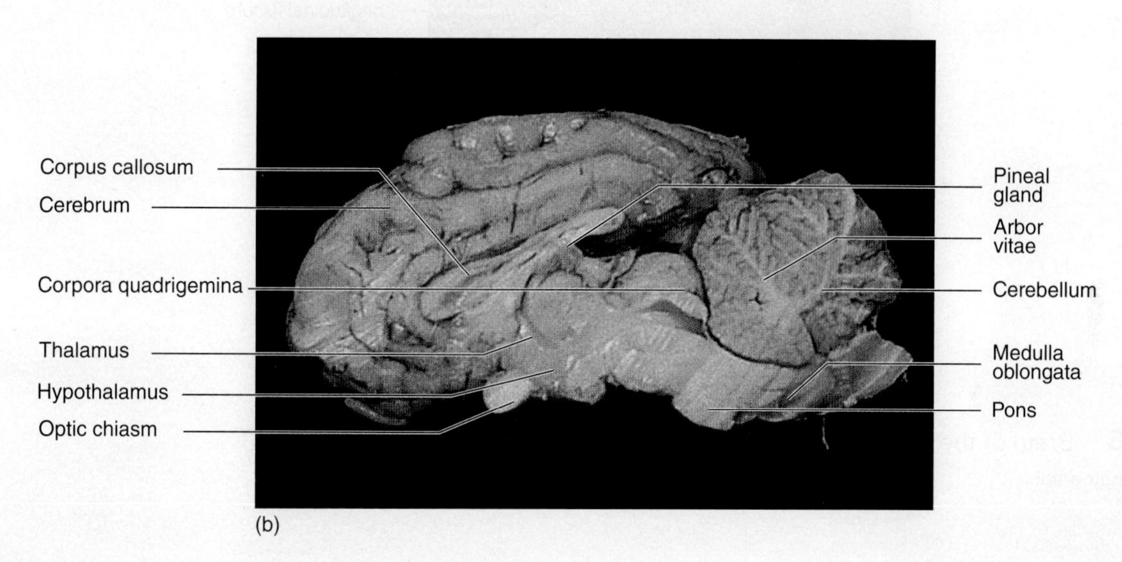

Figure 17.15 Brain of the Sheep, Longitudinal Section

(a) Diagram; (b) photograph.

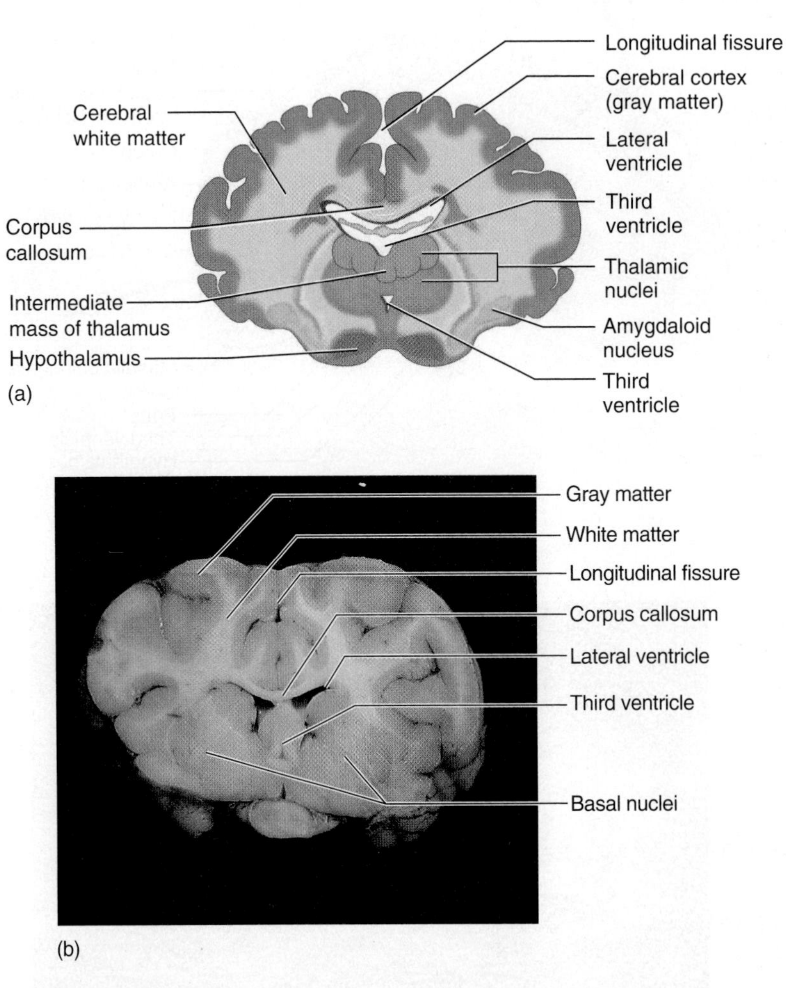

Figure 17.16 Brain of the Sheep, Coronal Section

(a) Diagram; (b) photograph.

Exercise 17 Review

Brain and Cranial Nerves

Name: ______________________

Lab time/section: ______________________

Date: ______________________

1. Which one of the meninges is just superficial to the surface of the brain? ______________________

2. What fluid is found in the ventricles of the brain? ______________________

3. Into what space does fluid flow from the cerebral aqueduct? ______________________

4. What is the difference between a gyrus and a sulcus? ______________________

5. Name all the lobes of the cerebrum. ______________________

6. What is the function of the precentral gyrus? ______________________

7. What sense does the temporal lobe alone interpret? ______________________

8. What physical depression separates the temporal lobe from the parietal lobe? ______________________

9. What structure connects the cerebral hemispheres? ______________________

10. Name the major regions of the midbrain. ______________________

11. What is the function of the cerebellum? ______________________

12. John "pulled a no-brainer" by hitting his forehead against a wall. What damage might he have done to the function of his brain, particularly the functions associated with the frontal lobe? ______________________________

____

____

13. If a stroke had affected all the sensations interpreted by the brain concerning only the face and the hands, how much of the post-central gyrus would be affected? ______________________________

____

14. Describe what effect the loss of an entire cerebral hemisphere would have on specific functions, such as spatial awareness or the ability to speak. ______________________________

____

____

____

15. Aphasia is loss of speech. Different types of aphasia can occur. If the Broca area were affected by a stroke, would the content of the spoken word be affected, or would the ability to pronounce the words be affected? ______________________________

____

16. Label the following illustration using the terms provided.

arbor vitae	cerebral aqueduct	corpora quadrigemina	corpus callosum
fourth ventricle	hypothalamus	medulla oblongata	pineal gland
pituitary gland	pons	thalamus	

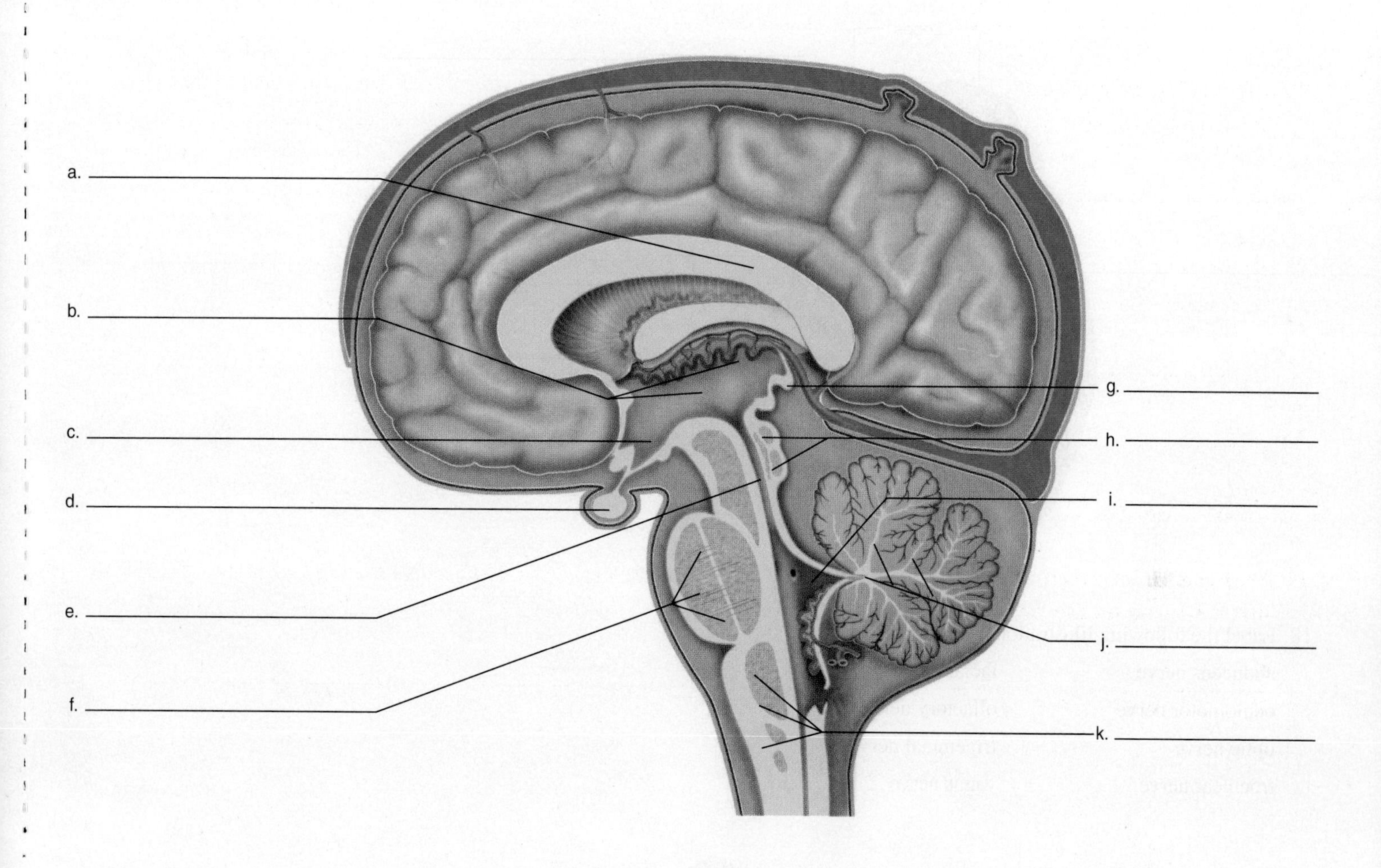

17. Label the brain using the terms provided.

frontal lobes	medulla oblongata
optic nerve	pituitary gland
pons	temporal lobe

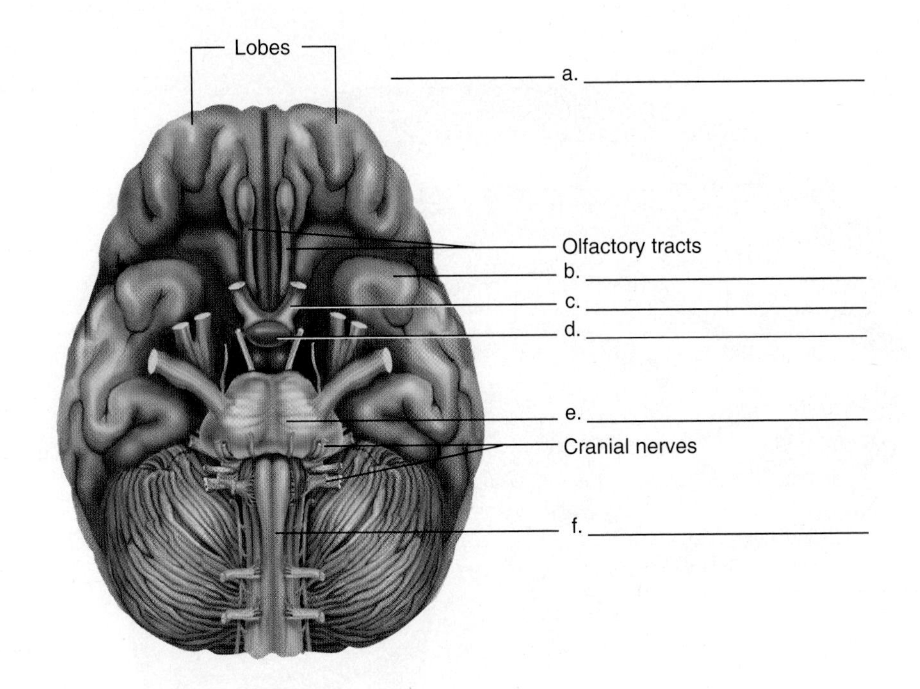

18. Label the following illustration using the terms provided.

abducens nerve	facial nerve
oculomotor nerve	olfactory nerve
optic nerve	trigeminal nerve
trochlear nerve	vagus nerve

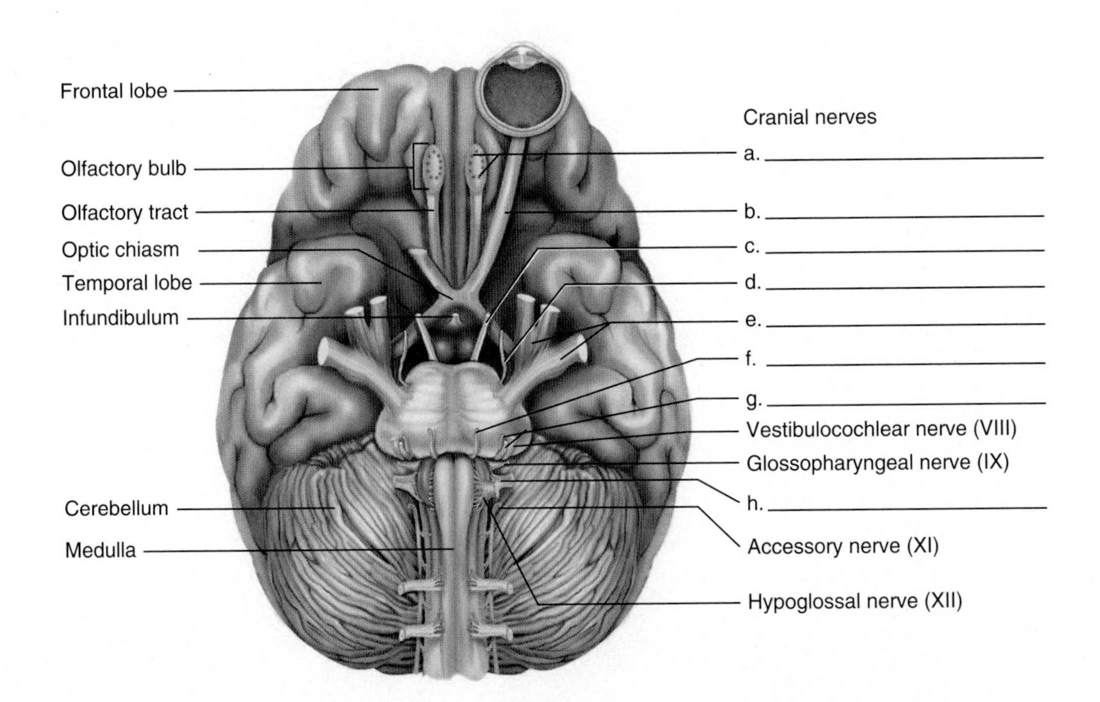

19. Describe the following nerves in terms of function (sensory, motor, or both).

Optic nerve ______________________________

Trochlear nerve ______________________________

Glossopharyngeal nerve ______________________________

Hypoglossal nerve ______________________________

Vagus nerve ______________________________

20. Name the cranial nerve that innervates each of the following areas.

Anterior tongue ______________________________

Ear ______________________________

Mandible ______________________________

Eye ______________________________

Stomach ______________________________

Lateral rectus muscle of the eye ______________________________

NOTES

Exercise 18

Spinal Cord and Somatic Nerves

Introduction

The spinal cord begins at the foramen magum of the skull and ends at approximately the second lumbar vertebra. This is because the vertebrae continue to grow after the spinal cord has reached its maximum length. The spinal cord receives sensory information from and transmits motor information to the spinal nerves, which radiate into the body as somatic nerves. In this exercise, you learn the major features of the spinal cord in longitudinal aspect and in cross section, as well as the nerves that are associated with the spinal cord.

The peripheral nervous system associated with the cord consists of the somatic nerves and their ganglia. Somatic nerves are those that take sensory information from the body to the spinal cord and motor information from the spinal cord to the body. Nerves such as the radial nerve and the femoral nerve are somatic nerves. These nerves run throughout the body, receiving sensory information and sending it to the CNS or taking motor information from the CNS to skeletal muscles. The material in this lab is covered in the Seeley text in chapter 12, "Spinal Cord and Spinal Nerves."

Objectives

At the end of this exercise, you should be able to

1. demonstrate the major regions in a cross section of spinal cord;
2. describe the anatomy seen in the longitudinal aspect of the spinal cord;
3. list the major nerves that arise from each plexus;
4. name all the major nerves of the upper and lower extremities.

Materials

Models or charts of the central and peripheral nervous systems
Cadaver (if available)
Prepared slide of a spinal cord in cross section
Model or chart of a spinal cord in cross section and longitudinal section

Procedure

Spinal Cord

Longitudinal Aspect of the Spinal Cord

(A) 1 Examine a model or chart in the lab of a longitudinal view of the spinal cord and locate the major features illustrated in figure 18.1. The spinal cord terminates inferiorly as the **conus medullaris** at approximately vertebra L2 and is attached to the coccyx by a thin thread of connective tissue known as the **filum terminale** (fī′lŭmter′mi-nal′-ĕ), which is an extension of the pia matter. The neural continuation of the spinal cord is an extension of parallel nerve fibers in the lumbar and sacral regions. These parallel fibers resemble a horse's tail and are called **cauda equina.**

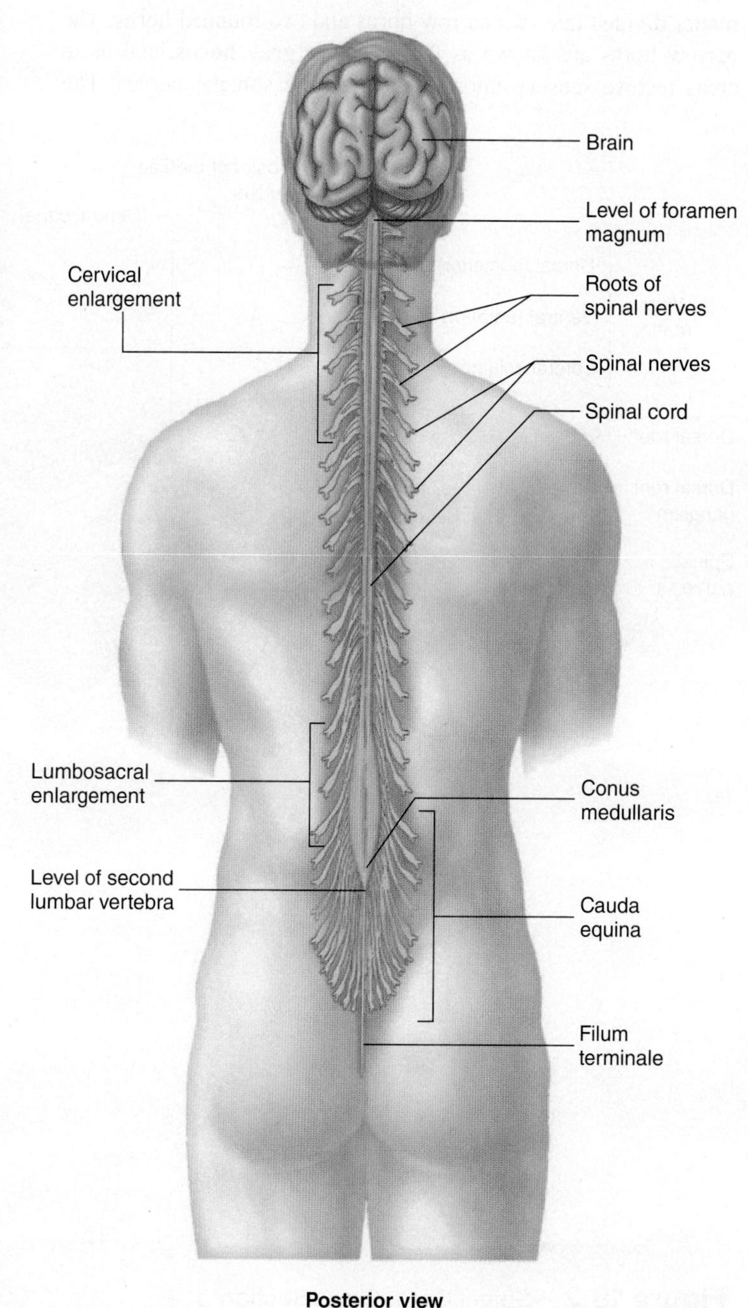

Figure 18.1 Spinal Cord, Longitudinal View

The spinal cord has a couple of expansions in two locations. The **cervical enlargement** is located at C3 through T2 because of nerves that supply the upper extremities. The **lumbosacral enlargement** is located at about T7 through T11, and this expanse is due to the nerves that supply the lower extremity. Compare the material in the lab with figure 18.1.

Cross Section of Spinal Cord

Ⓐ[2] Examine a model or chart in the lab of a cross section of the spinal cord and compare it to figure 18.2a. Locate the **gray matter,** which appears as the shape of an H or a butterfly in the middle of the spinal cord. The **white matter** is located on the periphery of the cord. Note how the distribution of gray and white matter in the spinal cord differs from that in the brain.

In a cross section of the spinal cord, you should see the gray matter divided into two narrow horns and two rounded horns. The narrow horns are known as the **posterior gray horns,** and these areas receive sensory information from the somatic nerves. The rounded horns are the **anterior gray horns,** and these send motor signals to the spinal nerves. The anterior gray horns are enlarged due to the motor neuron cell bodies. In some parts of the spinal cord, there are additional sections of gray matter known as the **lateral gray horns.** These contain autonomic motor neuron cell bodies.

Each side of the gray matter is connected to the other by a crossbar known as the **gray commissure.** In the middle of the gray commissure is the **central canal,** which runs the length of the spinal cord. The white matter of the cord is divided into **tracts,** or **fasciculi** (fă-sik′-ū-lī), which take sensory information to the brain or motor information from the brain. The tracts that take sensory information to the brain are called **ascending tracts,** and those that receive motor information are called **descending tracts.** Tracts can be bundled into larger units called **columns.** You should also see a depression in the posterior surface of the spinal cord. This is the **posterior median sulcus.** The deeper depression on the anterior side is known as the **anterior median fissure.**

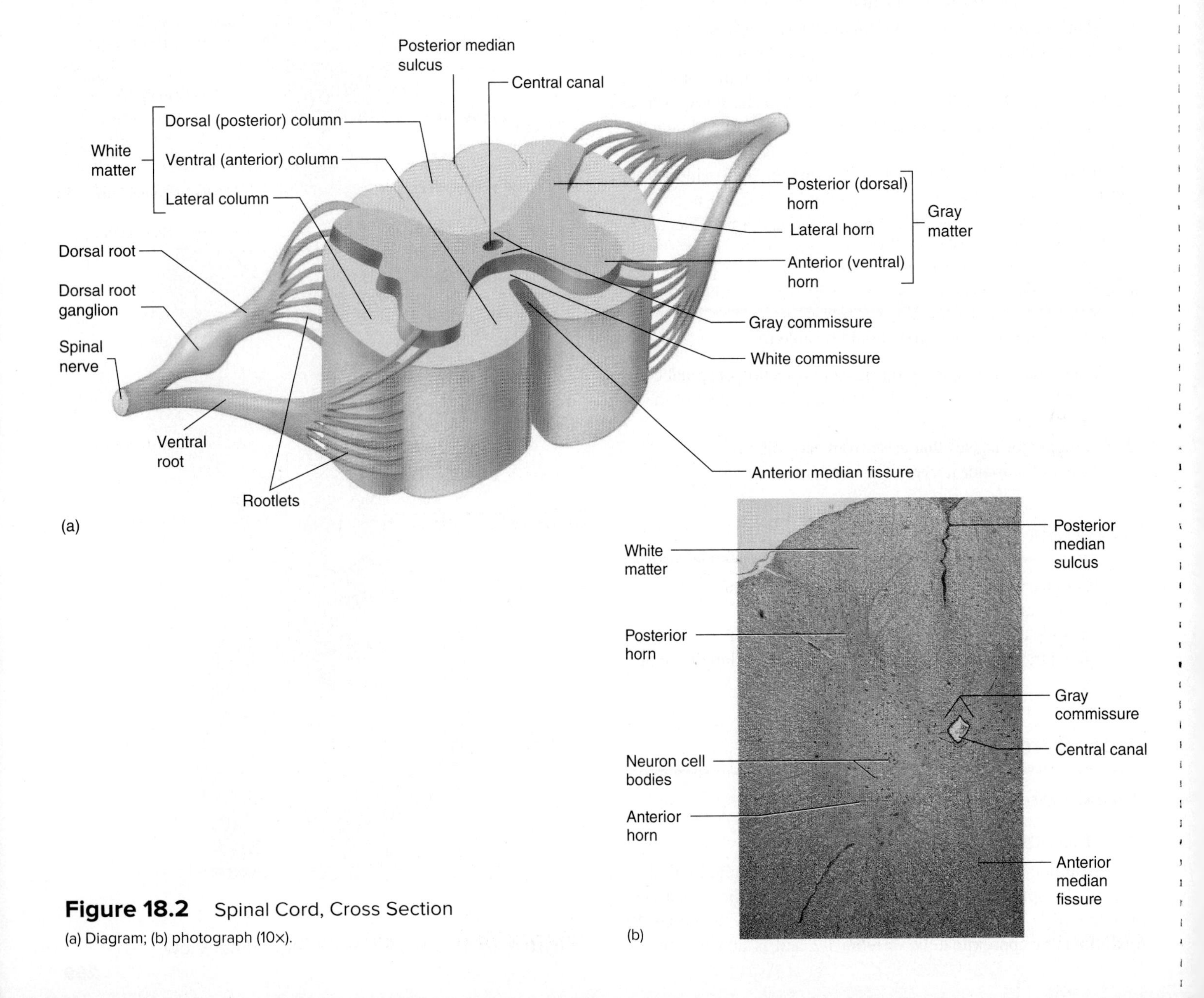

Figure 18.2 Spinal Cord, Cross Section

(a) Diagram; (b) photograph (10×).

Histology of the Spinal Cord

(A)[3] Examine a prepared slide of the spinal cord under low power and locate the **anterior gray horns** and **posterior gray horns,** the **gray commissure,** the **central canal,** the **posterior median sulcus,** and the **anterior median fissure.** Examine the anterior horn of the spinal cord and look for the nerve cell bodies of the **multipolar neurons** there. Compare your slide with figure 18.2*b*.

Meninges

The spinal cord is covered by **meninges,** similar to the brain. The outer covering of the cord consists of the **dura mater.** Between the dura mater and the bone of the spinal canal is the **epidural space,** a site used for the injection of anesthetics. The next layer deeper to the dura mater is the **arachnoid mater.** Deep to the arachnoid is the **pia mater,** which is the innermost of the meninges and a thin cover on the spinal cord proper. Between the pia mater and the arachnoid is the **subarachnoid space,** which contains the **cerebrospinal fluid.** Examine figure 18.3 for an illustration of the meninges of the spinal cord.

Nerves Associated with the Spinal Cord

Nerve Structure

Parallel neuron fibers that carry information from one area to another are called **nerves** in the peripheral nervous system or **tracts** if they occur in the central nervous system. Nerves have a number of connective tissue wrappings that envelop the individual nerve fibers, clusters of fibers, and the entire nerve. The sheath that wraps around single nerve fibers is the **endoneurium,** and the sheath that wraps around groups of nerve fibers is the **perineurium.** The wrapping that covers the entire nerve is called the **epineurium.** Examine figure 18.4 for these layers.

Spinal Nerves and Plexuses

From the periphery of the vertebral column, the **dorsal ramus,** a branch of sensory and motor nerve fibers, unites with the **ventral ramus** to form a **spinal nerve.** The spinal nerve is located in the intervertebral foramen and divides into a **dorsal root ganglion** and a **ventral root.** The dorsal root ganglion contains the neuron cell bodies of the **sensory neurons,** and the **dorsal root** carries sensory information to the dorsal horn of the spinal cord. The neuron cell bodies of the motor neurons are located in the ventral horn of the spinal cord and exit via the ventral root. Examine the material in the lab and compare it with figure 18.5.

There are 31 pairs of **spinal nerves** that exit from the spinal cord. All pairs of spinal nerves, except for the first cervical, exit the spinal cord via the intervertebral foramina. The first cervical spinal nerve exits superior to the first cervical vertebra. The spinal nerves are **mixed nerves** carrying both sensory and motor information. The spinal nerves are named according to their region of origin. There are **8** pairs of **cervical nerves, 12** pairs of **thoracic nerves, 5** pairs of **lumbar nerves, 5** pairs of **sacral nerves,** and **1** pair of **coccygeal nerves.** Some of these nerves exit the spinal cord and branch out to parts of the body individually, whereas others exit the spinal cord and form a branching network with other spinal nerves. These networks are called **plexuses,** and there are four

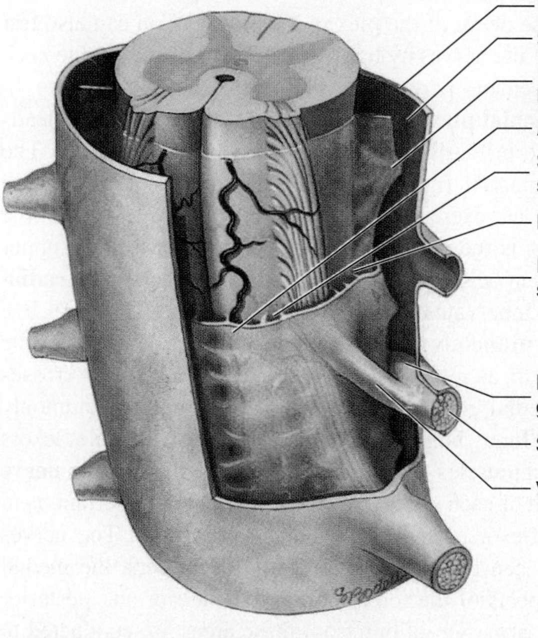

Dura mater
Subdural space
Denticulate ligament
Arachnoid mater
Subarachnoid space
Pia mater
Epineurium of spinal nerve
Dorsal root ganglion
Spinal nerve
Ventral root

Figure 18.3 Spinal Meninges

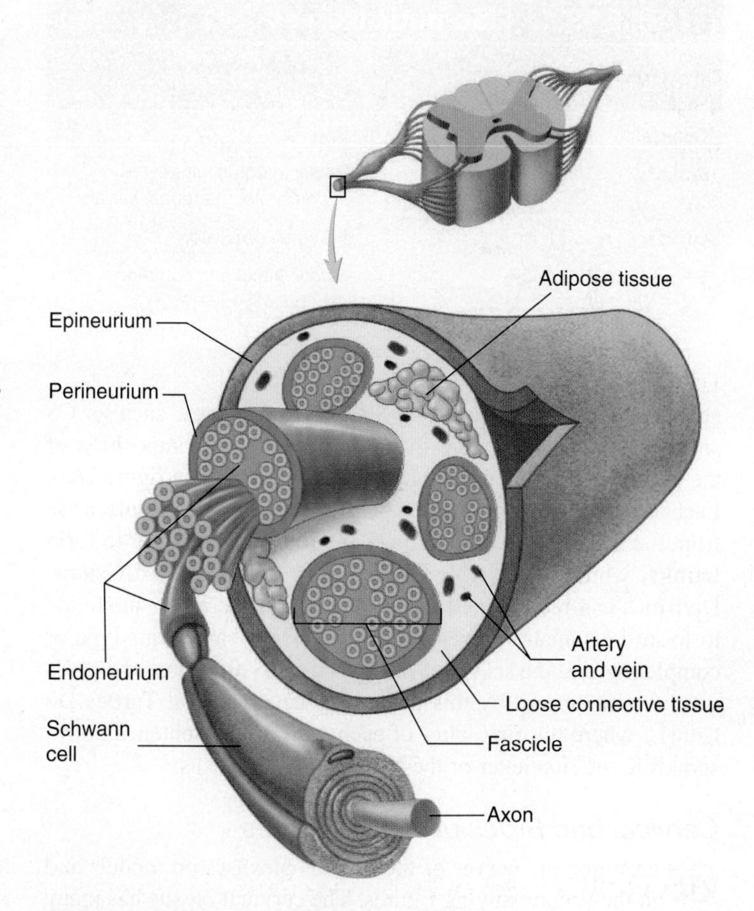

Figure 18.4 Coverings of the Nerve

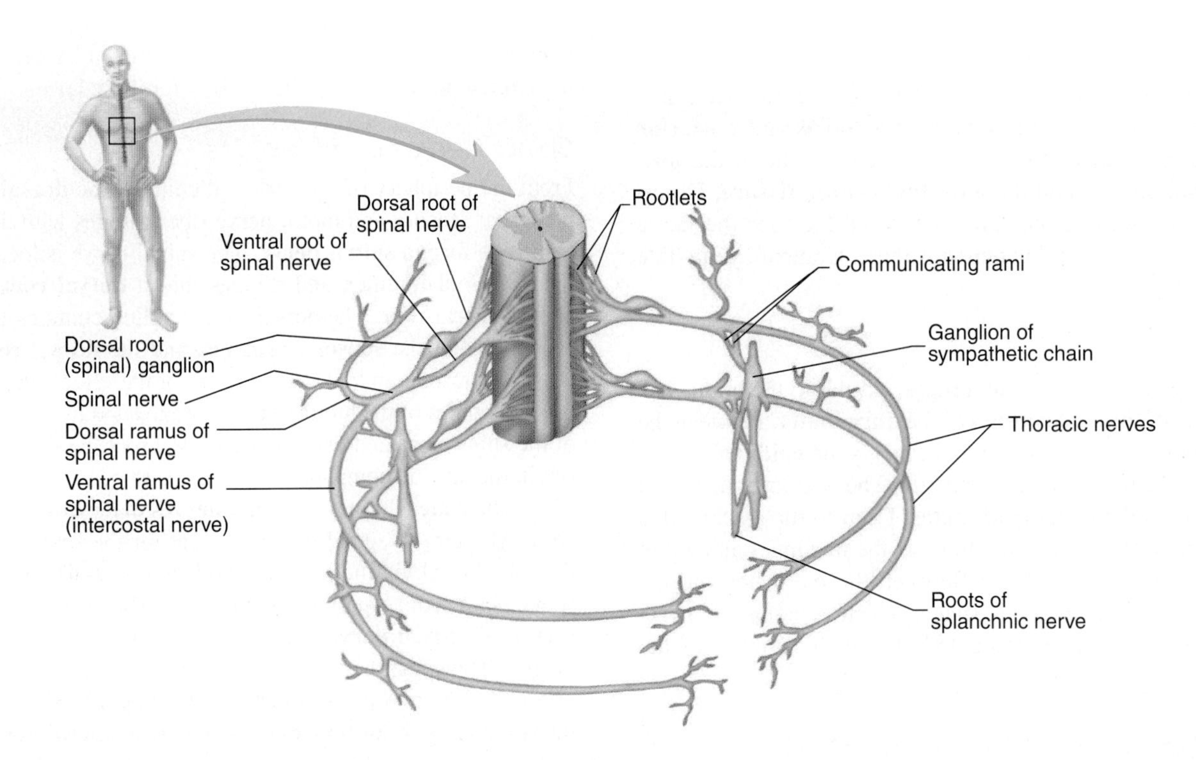

Figure 18.5 Thoracic Nerves Associated with the Spinal Cord

TABLE 18.1 Somatic Nerves

Name	Contributing to Plexus	Major Nerves of Plexus
Cervical	C1–4	Phrenic
Brachial	C5–T1	Radial, median, ulnar, musculocutaneous, axillary
Lumbar	L1–4	Femoral, obturator
Sacral	L4–S4	Sciatic (tibial and common fibular)

generally recognized plexuses. Some spinal nerves, such as C5 and L4, contribute to more than one plexus. The composition of the plexuses is outlined in table 18.1 and illustrated in figure 18.6. Each plexus in the body has a different structure. **Roots** arise from the spinal nerves. Sometimes the roots join together to form **trunks,** which may divide into anterior or posterior **divisions.** Divisions can recombine to form **cords,** and these can subdivide to form the somatic nerves. Not every plexus shares this type of complexity, and the sequence of roots, trunks, divisions, and cords can be remembered by this mnemonic device, "Real Turtles Do Crawl," where the first letter of each word in the sentence corresponds to the first letter of the structure in the plexus.

Cervical and Brachial Plexus Nerves

(A) 4 Examine the nerves of the various plexuses on models and in the accompanying figures. The cervical plexus has many nerves that innervate head and neck muscles. The cervical plexus exits from the upper spinal nerves of the neck. An unusual feature of the cervical plexus is the **ansa cervicalis,** which forms a loop in the plexus. An important nerve that comes from the cervical plexus is the **phrenic nerve.** This nerve runs to the diaphragm and is responsible for its contraction in breathing. Examine the nerves of the **cervical plexus** in figure 18.7. Many nerves from the cervical plexus innervate the muscles and skin of the neck and the skin of the ear. Since breathing is controlled by the phrenic nerve, the observation that a person is breathing is an obvious test that at least one nerve of the plexus is working! You can also test the function of the nerves by lightly pinching the sides of the neck or the ear for sensory perception.

The **brachial plexus** has branches from C5 to T1 and leads to nerves that primarily innervate the upper extremities. The plexus is illustrated in figure 18.8. The innervations of muscles are covered in the lab exercises on muscles. A major nerve from the brachial plexus is the **axillary nerve,** which innervates the upper shoulder and can be seen in figure 18.9. Another nerve, the **radial nerve,** mostly innervates the extensors of the hand (figure 18.10). The **musculocutaneous nerve** innervates the muscles that flex the arm and forearm, as seen in figure 18.11. The **ulnar nerve** crosses behind the medial epicondyle of the humerus and is commonly known as the "funny bone." The ulnar nerve involves some flexors and many hand muscles (figure 18.12), whereas the **median nerve** runs the length of each upper extremity and serves important hand and forearm flexors, as illustrated in figure 18.13. The nerves of this plexus can be tested by pinching the fingers, the medial and lateral aspects of the forearm, and the anterior and posterior aspects of the arm. Sensations from these areas are conducted to the brain via the brachial plexus.

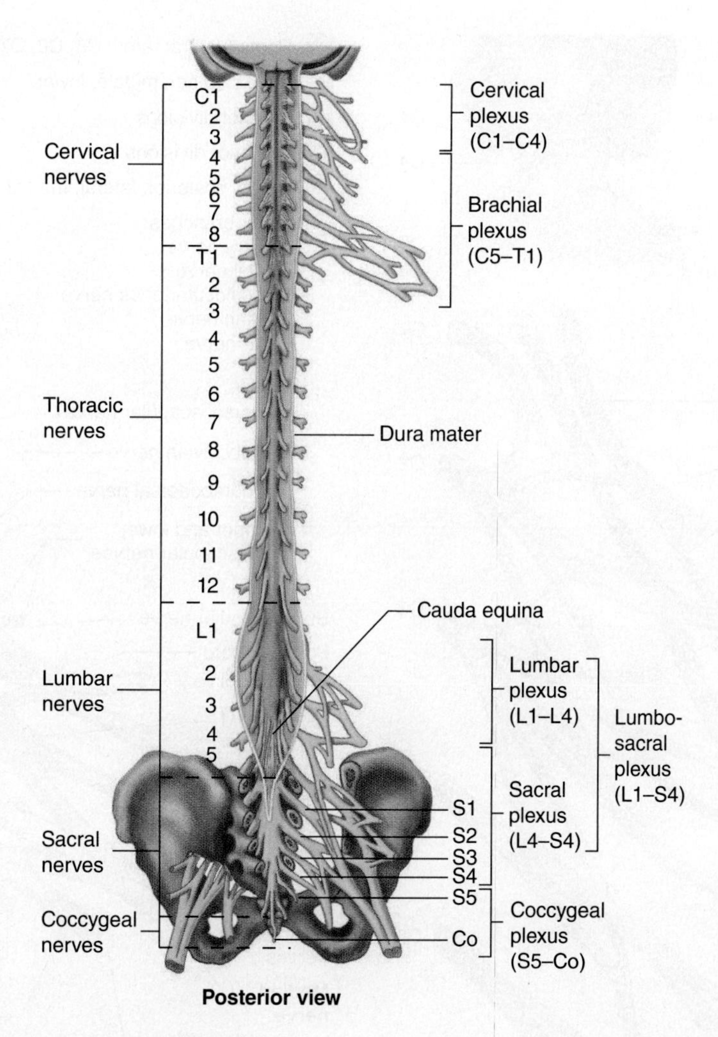

Figure 18.6 Spinal Cord and Nerve Plexuses

Thoracic Nerves

Many nerves are not associated with a plexus. The thoracic nerves are a good example of this. Many of the **thoracic nerves** exit through the intervertebral foramina of the vertebral column and innervate the ribs, muscles, and other structures of the thoracic wall. These nerves can be seen in figure 18.5.

Lumbar and Sacral Plexus Nerves

The lumbar and sacral plexuses take sensory information from and motor information to the lower limbs. Some authors combine the two plexuses into the **lumbosacral** plexus, as seen in figure 18.14. One of the nerves of the lumbar plexus is the **obturator nerve.** It innervates the adductor muscles of the thigh, as seen in figure 18.15. The **femoral nerve** is another nerve arising from the **lumbar plexus.** This large nerve passes anteriorly across the inguinal ligament and mostly innervates the muscles of the anterior thigh (figure 18.16). To test the nerves of this plexus, you can lightly pinch the anterior thigh for the femoral nerve and the medial thigh for the obturator nerve.

There are many nerves that come from the **sacral plexus.** Many innervate the pelvis and the muscles that move the hip, thigh, and leg. Two of the nerves from this plexus, the **tibial** and **common fibular (peroneal) nerves,** unite to form the **sciatic nerve** (figure 18.14). These nerves innervate the leg and foot, as indicated in figures 18.17 and 18.18.

Test for the sciatic nerve by lightly pinching the posterior aspect of the thigh. Examine the nerves on models, in charts, or in a cadaver in the lab and compare them with the illustrations.

Cat Dissection

Brachial Plexus

(A)[5] The dissection of the brachial plexus involves careful dissection in the axillary region. Your cat should be placed ventral side up as you begin the dissection. Remove any skin, if you have not done so already. Bisect the pectoralis major and pectoralis minor muscles and carefully fold them back to see the brachial plexus that is deep to these muscles. Remove adipose tissue and fascia to expose the blood vessels and the brachial plexus. Do not damage the blood vessels, as you will study these in future labs. Examine figure 18.19 as you study the cat.

The large and anterior nerve of the brachial plexus is the **musculocutaneous nerve.** It innervates the skin of the brachium and is the motor nerve of the biceps brachii muscle.

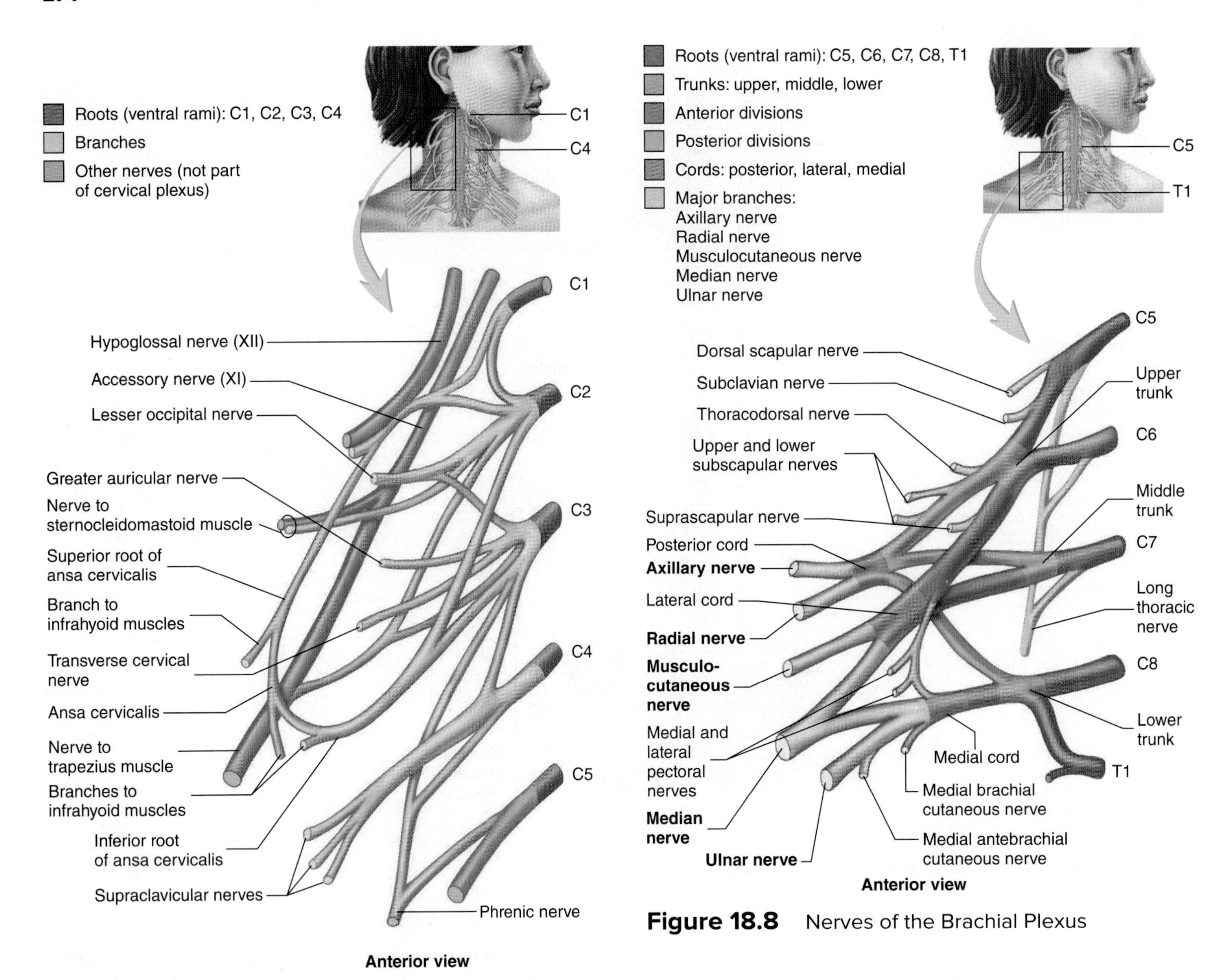

Figure 18.7 Nerves of the Cervical Plexus

Figure 18.8 Nerves of the Brachial Plexus

The **radial nerve** is the largest nerve of the plexus and is posterior to the musculocutaneous nerve. It is located on the posterior of the distal arm, where it innervates the muscles on the dorsal side of the forelimb, including many of the extensor muscles.

The **median nerve** is located alongside the brachial artery and is in the midline of the brachium and antebrachium. It innervates many of the flexor muscles.

The **ulnar nerve** is the most posterior of the brachial plexus nerves. It travels posterior to the medial epicondyle of the humerus to innervate the muscles on the ulnar side of the antebrachium.

Sacral Plexus

(A) 6 The sacral plexus is best seen from the dorsal view. Examine figure 18.20 and locate the large **sciatic nerve** by separating the biceps femoris on the dorsal thigh. The sciatic nerve is composed of two nerves, the medial **tibial nerve** and the lateral **common fibular nerve.** These can be seen at the distal part of the thigh.

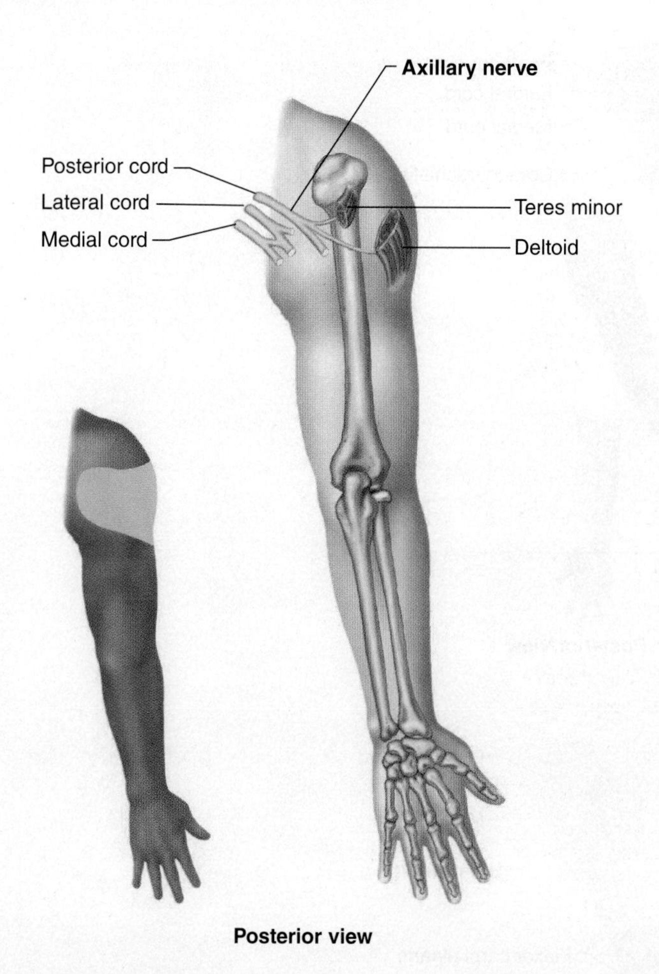

Figure 18.9 Axillary Nerve.
Muscles indicated in this figure are innervated by the axillary nerve.

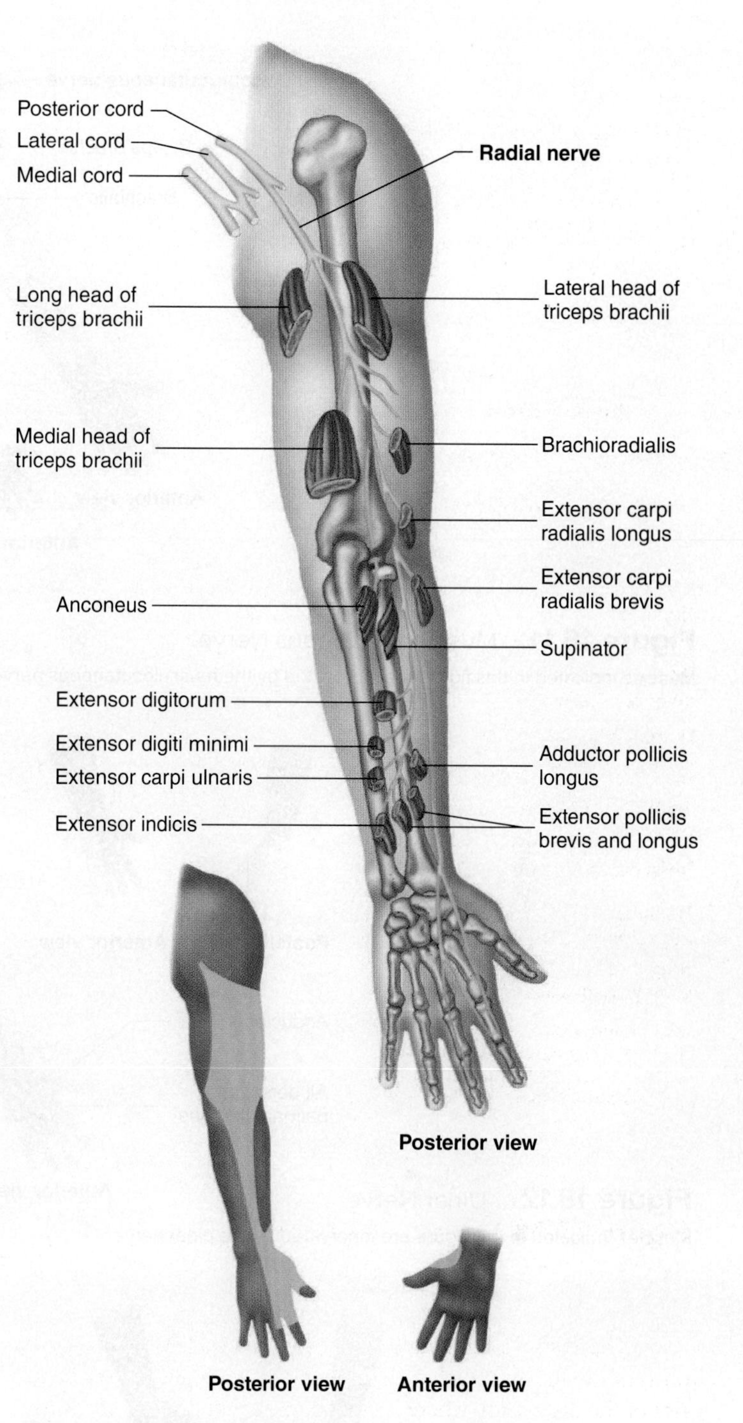

Figure 18.10 Radial Nerve.
Muscles indicated in this figure are innervated by the radial nerve.

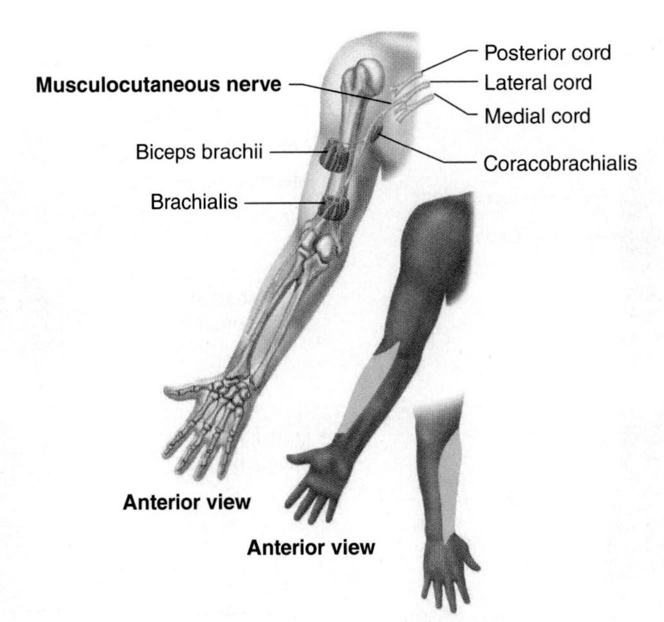

Figure 18.11 Musculocutaneous Nerve.

Muscles indicated in this figure are innervated by the musculocutaneous nerve.

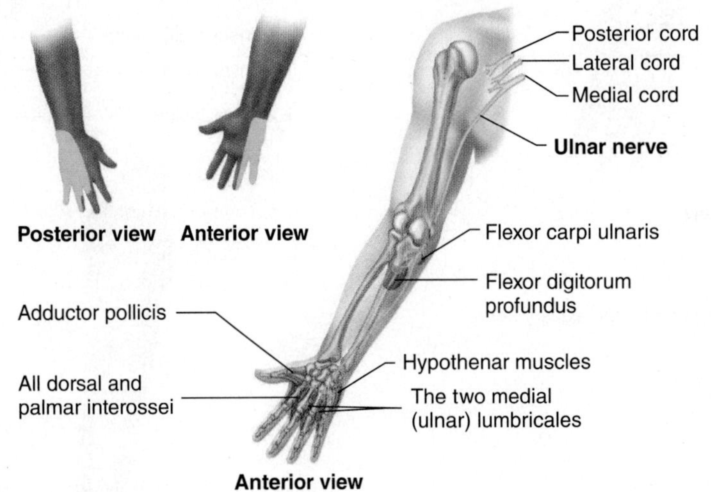

Figure 18.12 Ulnar Nerve.

Muscles indicated in this figure are innervated by the ulnar nerve.

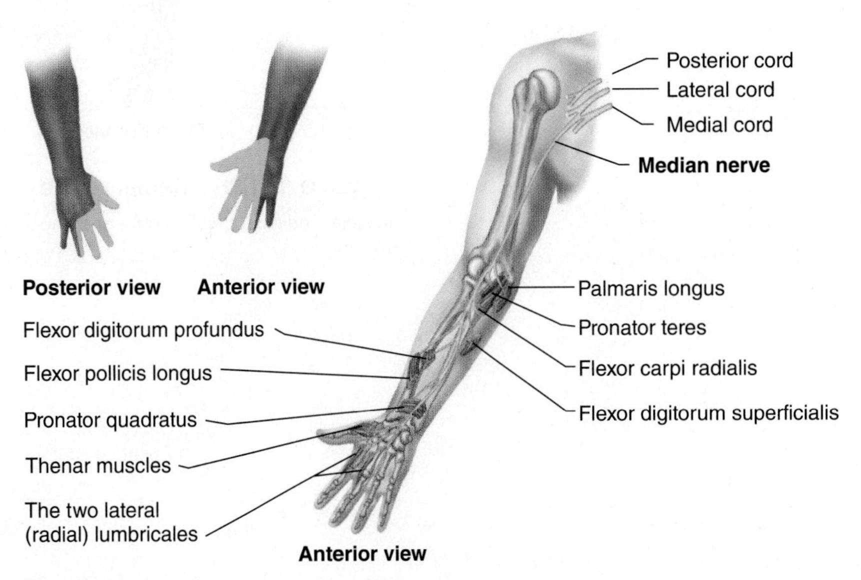

Figure 18.13 Median Nerve.

Muscles indicated in this figure are innervated by the median nerve.

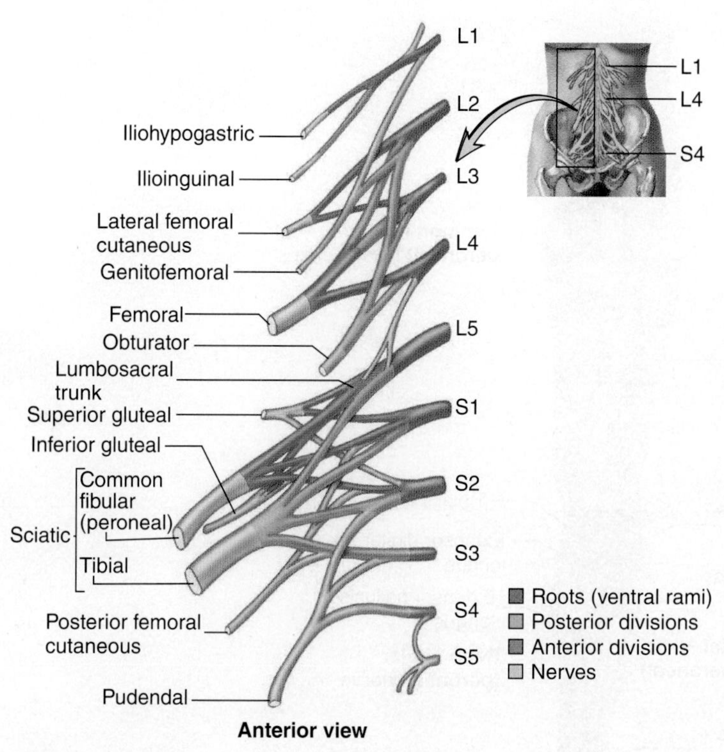

Figure 18.14 Lumbosacral Plexus

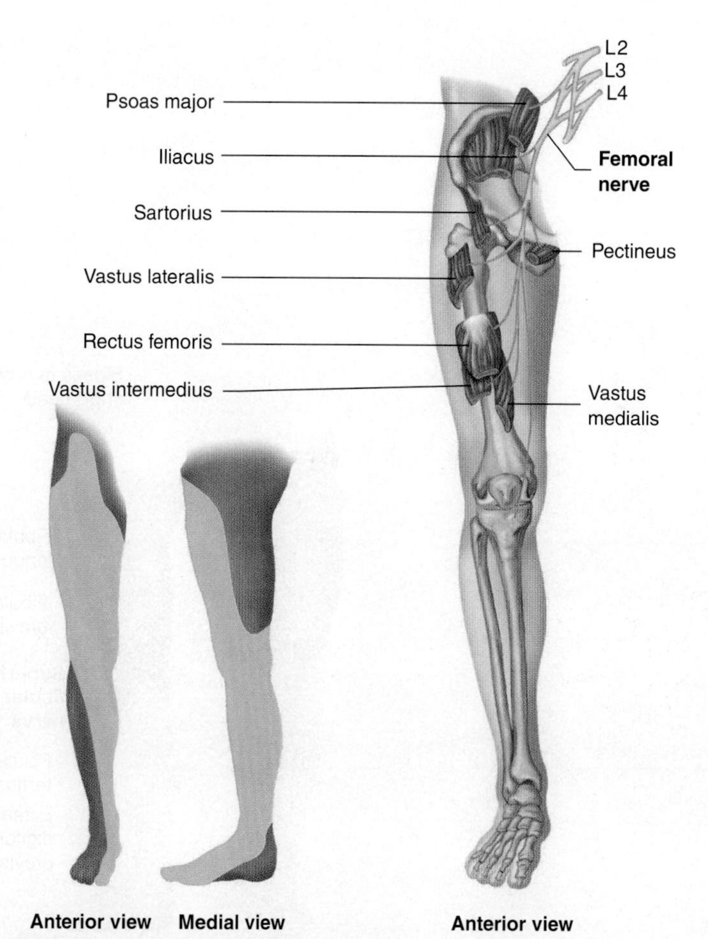

Figure 18.16 Femoral Nerve.

Muscles indicated in this figure are innervated by the femoral nerve.

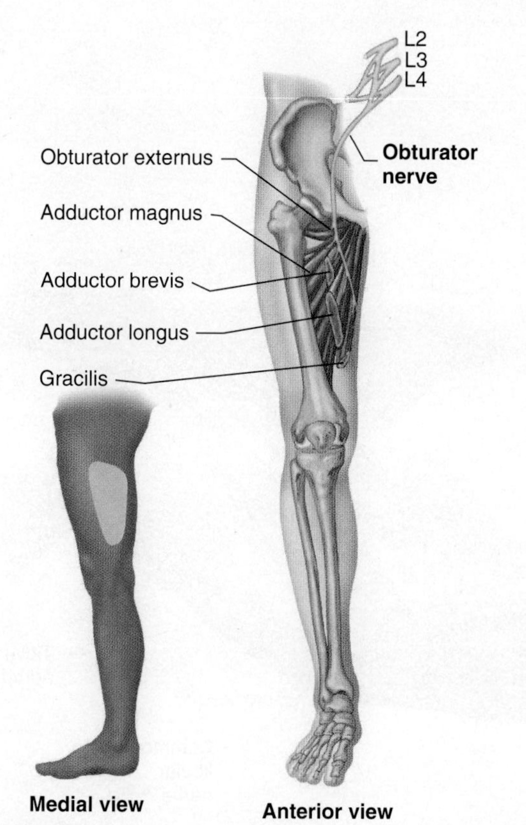

Figure 18.15 Obturator Nerve.

Muscles indicated in this figure are innervated by the obturator nerve.

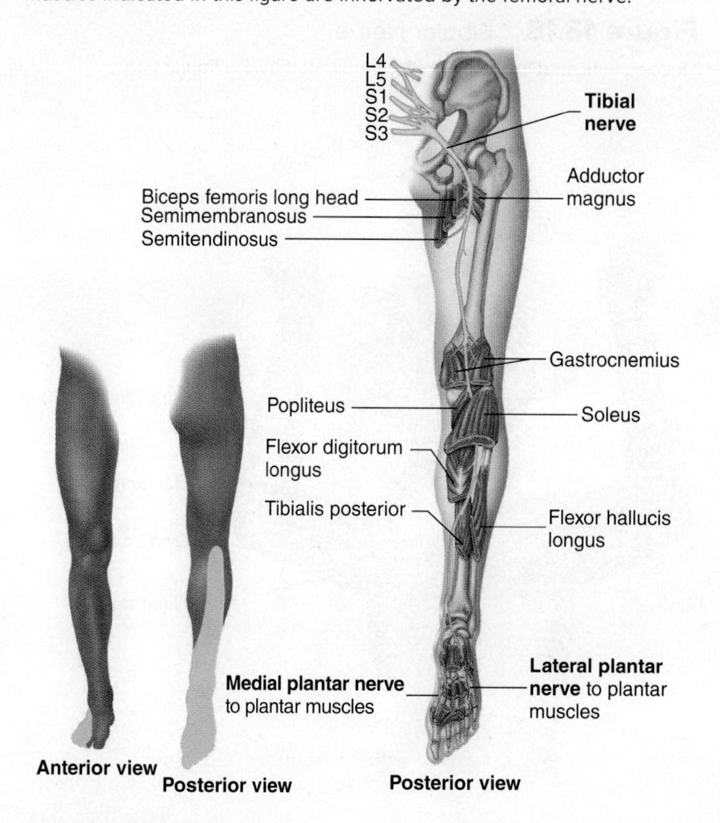

Figure 18.17 Tibial Nerve.

Muscles indicated in this figure are innervated by the tibial nerve.

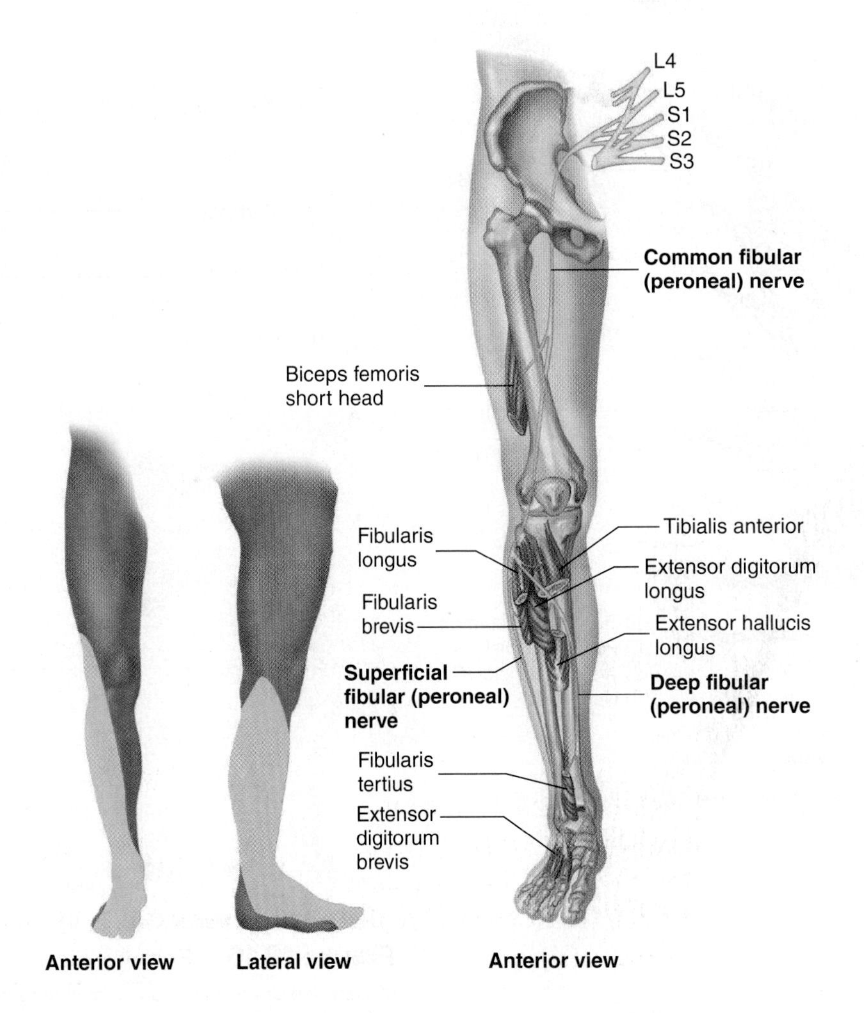

Figure 18.18 Fibular Nerve.

Muscles indicated in this figure are innervated by the fibular nerve.

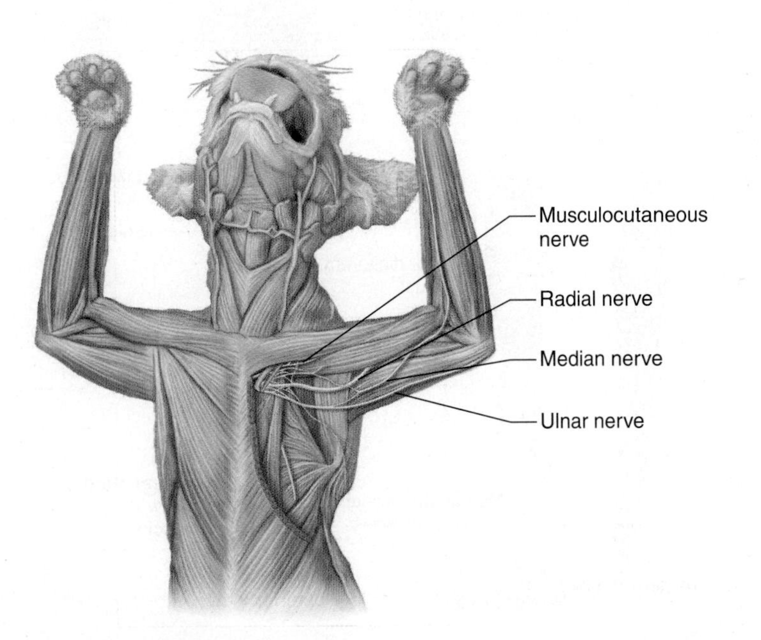

Figure 18.19 Brachial Plexus of the Cat

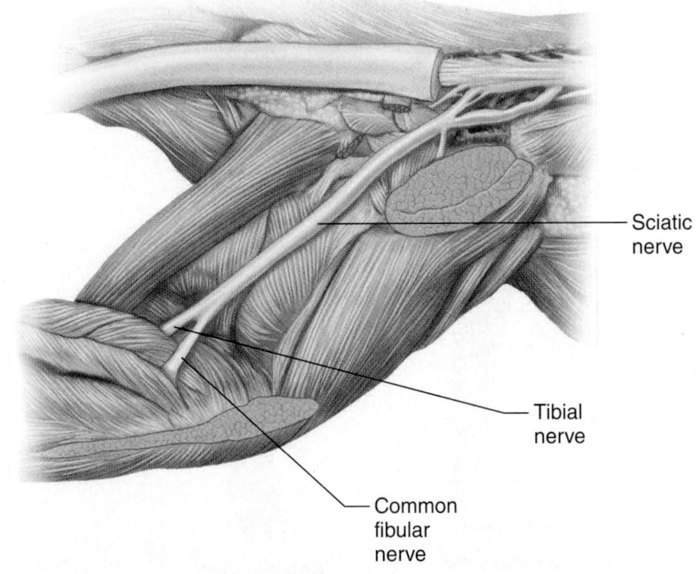

Figure 18.20 Sacral Plexus, Posterolateral View of Right Side

Exercise 18 Review

Spinal Cord and Somatic Nerves

Name: ______________________

Lab time/section: ______________________

Date: ______________________

1. In the spinal cord, what type of impulse (sensory/motor) travels through the

   a. anterior gray horn? ______________________

   b. posterior gray horn? ______________________

   c. ascending spinal tracts? ______________________

   d. descending spinal tracts? ______________________

2. What causes the cervical enlargement of the spinal cord? ______________________

   ____

3. Where is the filum terminale located? ______________________

   ____

4. What is the conus medullaris? ______________________

   ____

5. What is the cauda equina? ______________________

   ____

6. In the spinal cord, which is deep to the other, the white matter or the gray matter? ______________________

7. What is the area of gray matter found between the lateral halves of the spinal cord? ______________________

8. The subarachnoid space is filled with what fluid? ______________________

9. What major nerves arise from the following plexuses? ______________________

   a. Cervical ______________________

   b. Brachial ______________________

   c. Lumbar ______________________

   d. Sacral ______________________

10. In terms of function, how does the dorsal spinal root vary from the ventral spinal root? ______________________

    ____

11. What is the endoneurium? ______________________________

____

12. How do tracts differ from nerves? ______________________________

____

13. What is a mixed nerve? ______________________________

14. The diaphragm's contractions are regulated by what nerve? ______________________________

____

15. The muscles of the arm, such as the biceps brachii, have what innervation? ______________________________

____

16. The extensor muscles of the hand are controlled by what nerve? ______________________________

____

17. The sciatic nerve is composed of two nerves. What are they? ______________________________

18. A person has feeling from the deltoid and biceps brachii region but no feeling from the wrist extensors. Where on the spinal cord has injury occurred? ______________________________

____

19. Fill in the illustration using the terms provided.

anterior median fissure

anterior horn

central canal

gray commissure

posterior horn

posterior median sulcus

a. ______________

b. ______________

c. ______________

d. ______________

e. ______________

f. ______________

Exercise 19

Nervous System Physiology: Stimuli and Reflexes

Introduction

Nervous tissue shows two fundamental properties: irritability and conductivity. **Irritability** is the potential of a nerve to respond to some kind of stimulus (chemical, mechanical, electrical), and **conductivity** is the movement of a nerve impulse along the length of the neuron. Nerves receive information from a particular area (sense organ, regions in the CNS, etc.) and transmit impulses to either the brain for interpretation or an effector for some kind of action. The description of nerve impulses and reflexes is covered in the Seeley text in chapter 11, "Functional Organization of Nervous Tissue," and chapter 12, "Spinal Cord and Spinal Nerves."

In this exercise, you study the basic properties of neuronal conduction and sensitivity to various stimuli. In addition to studying the function of neurons, you experiment with nerves as they form reflex arcs in several parts of the body.

Objectives

At the end of this exercise, you should be able to

1. describe the threshold nature of nerve responses;
2. list three things that cause a nerve to be stimulated;
3. name one substance that stimulates nerves and one that inhibits them;
4. describe reflex arcs;
5. list all the parts of a monosynaptic and a polysynaptic reflex arc;
6. define *hyporeflexic* and *hyperreflexic*.

Materials

Nerve Physiology Section

Virtual Physiology Labs
Ph.I.L.S. Version 4.0
Compatible computer

Live Frog Physiology Labs
Frog
Latex or plastic gloves
Glass rod with hook at one end
Hot pad or mitt
Bunsen burner
Matches or flint lighter
Frog Ringer's solution in dropper bottles
Ice bath
Microscope slide or small glass plate
Filter paper or paper towel
Dissection equipment for live animals
Scalpel or scissors
Cotton sewing thread
Stimulator apparatus with probe
Myograph transducer
Physiograph or physiology computer
5% sodium chloride solution
0.1% hydrochloric acid solution (1 mL of concentrated HCl in 1 liter of water)
Procaine hydrochloride solution
Gauze squares
Small beakers (50 mL)
Sterile cotton applicator sticks

Reflex Section

Patellar reflex hammer
Rubber squeeze bulb
Models or charts of spinal cord and nerves

Procedure

Review the structure of the neuron for descriptions of the dendrites, neuron cell body, and axon and the anatomy of the nerve in Exercise 16.

Virtual Experiments in Neurophysiology

Ⓐ 1 Load the Physiology Interactive Lab Simulations (Ph.I.L.S.) Version 4.0. For all of the virtual muscle experiments you should follow the standard procedures that come up when you open the program.

1. Click on the number of the exercise that you want to complete. These are under the **Resting Potentials** or **Action Potentials** headings.

   After you have selected your choice you should

2. Read the objectives and introduction to the lab simulation. You can click on highlighted terms to see pictures or animations of the material in question.
3. Take the pre-lab quiz. If you get a question wrong, the program will let you know and provide you with the correct answer.
4. Click on the **Wet Lab** tab and read the material. The highlighted terms open up videos of an actual procedure.
5. Click on **Continue,** which opens up the **Laboratory Exercise.**

6. Perform the exercises as outlined below. Print out any information that your instructor directs after you have performed the experiment, or download the information to a portable data storage device.
7. Take the post-lab quiz to determine your understanding of the lab.

The information for the specific lab exercises follows.

(A)[3] *Resting Potentials 10. Resting Potential and External [K⁺]*

Open the Resting Potential and External [K^+] program. When the screen appears in the Ph.I.L.S. program, click on the **power** button on the virtual computer screen. Click on the **power** button on the data acquisition unit and the electrometer. You can then click and drag the blue cable to the Recording Inputs upper connection (#1). You should then click on the **Start** button, which is located in the upper right of the virtual computer screen. Follow the rest of the directions. In order to record the material in the journal, you must click on the journal and then click on the "X" to quit the journal. The micromanipulator control (up or down) is found on the circular button underneath the tube labeled "50" but at the level of where the electrometer is found.

At the bottom of each tube is a stopcock indicated by a black rectangle. Click and hold the 5 mM tube until it empties. Move the microelectrode by clicking on the gray adjustment knobs. These are located to the right of where the orange wire attaches to the microelectrode (above and to the right of the micromanipulator control knob). Record the data in the journal and continue to move the microelectrode and take readings for 5, 10, 20, 50, and 100 mM KCl solutions by sequentially opening the stopcock for the respective fluids and taking readings.

What impact does increasing the potassium ion concentration have on membrane depolarization? Record your answer in the following space.

Effect of increasing potassium ion concentration: ________

(A)[3] *Resting Potentials 11. Resting Potential and External [Na⁺]*

Follow the same procedure as in the Resting Potential and External [K^+] exercise for the **Resting Procedure and External Sodium Concentration** program. The setup is the same except that you will be starting with a more concentrated level of sodium and decreasing the sodium ion concentration rather than what you did when you increased the potassium ion concentration.

Comparing the two solutions, which has a greater impact on membrane dynamics, sodium or potassium? Record your answer in the following space.

Greater impact on membrane dynamics: ______________

Action Potentials 12. The Compound Action Potential

(A)[4] Open the program, read the introduction, and answer the pre-lab questions. Open the Wet Lab and click on the power switch to the virtual computer screen and the data acquisition unit. Place all of the color-coded cables on the appropriate attachment points of the chamber holding the nerve. The drain tap is the small circular wheel located at the top of the chamber on the left-hand side. You must click and hold the knob in order for it to drain.

If you examine the graph you will see the stimulus on the left side of the graph followed by a compound action potential on the right. As with the muscle physiology experiment, if you move the cursor to the graph you will see crosshairs as you move the cursor. Place the crosshairs on the top of the curve and click. Click again on the baseline of the curve. When you click on the **Journal** icon it will enter the data into the journal.

You must click the "X" on the journal in order to proceed with obtaining more data. The threshold voltage is the voltage below which you get no response from the nerve. What was your threshold voltage? Record your answer in the following space.

Threshold voltage: ______________________________

(A)[5] *Action Potentials 13. Conduction Velocity and Temperature*

In this exercise you examine the speed of nerve conduction by altering the temperature of the nerve. The experiment runs at room temperature (22°C) and cold temperature (10°C). Follow the instructions as presented. Begin by clicking on the power switch to the **Temperature Unit.** Make sure that you click on all power buttons to turn on the equipment. Insert all of the plugs into their color-coded receptacles. Place all of the color-coded cables on the attachment points on the chamber holding the nerve. The tap is the small circular wheel located at the top of the chamber on the left-hand side. You must click and hold the knob in order for it to drain. Follow the directions to decrease the length of the space between the electrodes. Move the cross hairs and note the conduction velocity. Record this velocity in the following space.

Conduction velocity: ______________________________

After you run the first experiment at room temperature, make sure that you enter your data into the journal by clicking on the **Journal** icon. Run the second experiment (make sure that you drain the tap and adjust the electrodes) and compare the results of nerve conduction at different temperatures.

In which sequence (room temperature or cold) did you see a slowing down of the nerve conduction? Remember that a lower conduction velocity refers to a slower speed. Record your answer in the following space.

Slower nerve response: ______________________________

How does this response correlate with decreasing enzyme function based on temperature? Record your answer in the following space.

Correlation: ______________________________

(A)[6] *Action Potential 14. The Refractory Period*

Click on the power switch to the virtual computer screen and the data acquisition unit. Connect the blue plug to the acquisition unit and attach the colored electrodes to their matching color posts. Attach the red and black plugs and electrodes and click and hold the drain plug until the saline is drained from the chamber. Increase the voltage until there is no greater peak produced with increasing voltage. This value is the maximum recruitment for the nerve, and it means that all of the axons of the nerve are firing. With less voltage only some of the axons are firing and there is less of a response.

What is the maximum recruitment voltage that you obtained? Record your answer in the following space.

Maximum recruitment voltage: ____________________

What happens to the CAP (compound action potential) as you decrease the time from 7 ms to 1 ms?

CAP with reduced time: ____________________

What happens to the CAP as you increase the time beyond 7 ms?

CAP with increased time: ____________________

Absolute refractory period: ____________________

Relative refractory period: ____________________

Frog Nerve Conduction

Make sure you read through all of this exercise before performing the experiments. Wear latex gloves as a general precaution when working with fresh specimens, such as frogs. In the first part of this lab exercise, you determine if nerves respond to only a specific stimulus or if they are more general and respond to many stimuli. You also determine if certain materials or environmental conditions inhibit nerve response. You observe the process of nerve impulse conduction by experimenting on frogs or watching a demonstration, depending on the wishes of your instructor. If you are to experiment on frogs, obtain a doubly pithed frog, dissection equipment, frog Ringer's solution, and various test solutions and take them to your table.

(A) 7 Keep the frog nerve preparation moist with Ringer's solution and cotton gauze during the entire experiment. Prepare the frog by cutting the skin away from the hip (figure 19.1). Do not cut, pinch, or otherwise damage the sciatic nerve on the posterior side of the thigh.

Gently remove the nerve from between the muscles with a glass rod (figure 19.2). Do not stretch the nerve; leave it intact alongside the muscles. You can attach the tendon of the gastrocnemius to a myograph transducer, as you did in Exercise 11, or just examine the muscle to see if there is a response.

Nerve Response to Physical Stimuli

Flush the nerve with Ringer's solution and make sure it remains moist. You measure the effects of the nerve stimulation by the contraction of the gastrocnemius muscle. If the nerve is stimulated, then the gastrocnemius muscle should contract. This is due to the neuromuscular junction where the nerve impulse crosses the junction and causes the muscle to contract.

Cut a small (10 cm) section of cotton thread and gently slip it under the sciatic nerve with a pair of fine forceps. Gently move the thread up toward the hip. When you reach the place where the nerve descends into the muscle, loop the thread and ligate (tie off) the nerve close to the sacrum (figure 19.2) while watching the gastrocnemius muscle.

Figure 19.1 Removal of the Skin and Nerve Preparation

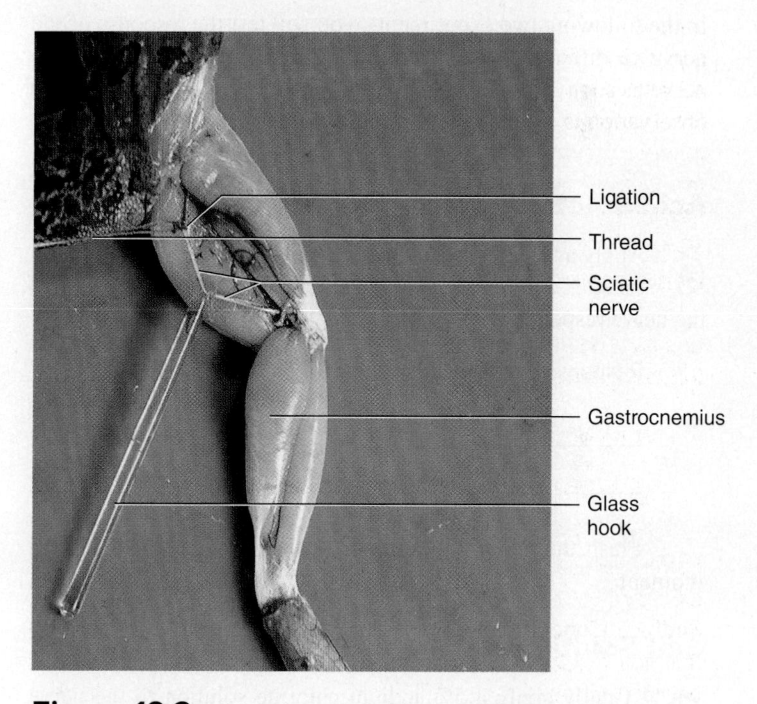

Figure 19.2 Ligation of the Sciatic Nerve of the Frog

As the thread begins to tighten on the nerve, record the response in the space provided.

Response of the nerve to physical stimulation: __________

____

____

After you have ligated the nerve, cut it from the anterior side, leaving the nerve attached to the gastrocnemius muscle. Moisten the nerve with Ringer's solution and prepare for the next experiment.

Nerve Response to Electrical Activity

(A) 8 Place a stimulator probe connected to a stimulator underneath the sciatic nerve, lifting the nerve away from the gastrocnemius muscle. Keep the nerve moist as you determine the minimum voltage **(threshold voltage)** required for nerve conduction. Set the stimulator to a frequency of two pulses per second and a duration of 10 milliseconds. The voltage should be set at zero (with the knob on 0.1 volt). Slowly increase the voltage until you see the gastrocnemius twitch. As soon as you see the gastrocnemius twitch at the lowest voltage, record this as threshold voltage in the space provided. Turn the voltage to zero, flush the nerve with Ringer's solution, and let it rest for a moment.

Threshold voltage: ______________________

Continue to increase the voltage until the muscle contracts maximally. Record this as the **maximum recruitment voltage.** This voltage is obtained when all the neurons of a particular nerve are stimulated.

Maximum recruitment voltage: ______________________

Nerve Response to Chemical Stimuli

In the following two experiments, you will test the response of the nerve to different chemical agents. Make sure you observe the nerve as soon as you apply the solution and rinse it as soon as the observation is made.

Acid Solution

(A) 9 Apply a 0.1% hydrochloric acid solution to a cotton applicator stick and touch the applicator gently to the nerve. Record the nerve response.

Response to hydrochloric acid: ______________________

____

____

Flush the nerve with Ringer's solution and let it rest for a moment.

Salt Solution

(A) 10 Gently apply a 5% sodium chloride solution to the nerve with a new cotton applicator stick. Record the response. Flush the nerve with Ringer's solution and let it rest for a moment.

Response to sodium chloride solution: ______________________

____

____

Nerve Response to Anesthetics

(A) 11 Apply a solution of **procaine hydrochloride (Novocain)** or other anesthetic, if provided by your instructor, to the nerve by soaking a small square of gauze or cotton with procaine solution and placing it on the nerve for a moment. As the gauze remains on the nerve, set up the stimulator apparatus and place the nerve over the stimulator probes. Remove the gauze and stimulate the nerve with a single pulse stimulus at the voltage that produced a maximum recruitment voltage in the previous experiment. If the nerve responds to the stimulus, leave the procaine hydrochloride on longer. When the nerve does not respond, remove the gauze and stimulate the nerve once every 30 seconds until it recovers from the local anesthetic. Keep the nerve moist at all times with frog Ringer's solution. Record the recovery time.

Recovery time: ______________________

Nerve Response to Changes in Temperature

(A) 12 Gently touch the nerve with a glass rod at room temperature. Record the response.

Response to gentle touch: ______________________

Place the glass rod in an ice bath. As it equilibrates in the ice water, place a small chip of ice on the nerve and let it stay there for a moment. Gently touch the nerve with the cold rod and record the response. Flush the nerve with room temperature Ringer's solution and let it rest for a moment.

Response to gentle touch with cold stimulation: ______________________

Take a new glass rod in a hot pad or mitts and heat one end of it in a Bunsen burner. Touch the nerve with the hot end of the glass rod. What is the response? Record your result.

Response to gentle touch with hot stimulation: ______________________

____

____

Cleanup

Make sure you clean your station before continuing. Place the specimen in the appropriate container, and use care when cleaning sharp instruments, such as scalpels or razor blades.

Reflexes

A reflex is defined as a motor response to a stimulus without conscious thought. Reflexes are involuntary, predictable responses to stimuli. Reflexes occur through **reflex arcs,** and these arcs have the following structure:

1. **Receptor** (the structure that receives the stimulus)
2. **Afferent (sensory) neuron** (the neuron taking the stimulus to the CNS)
3. **Integrating center** (the brain or spinal cord)
4. **Efferent (motor) neuron** (the neuron taking the response from the CNS)
5. **Effector** (the structure causing an effect)

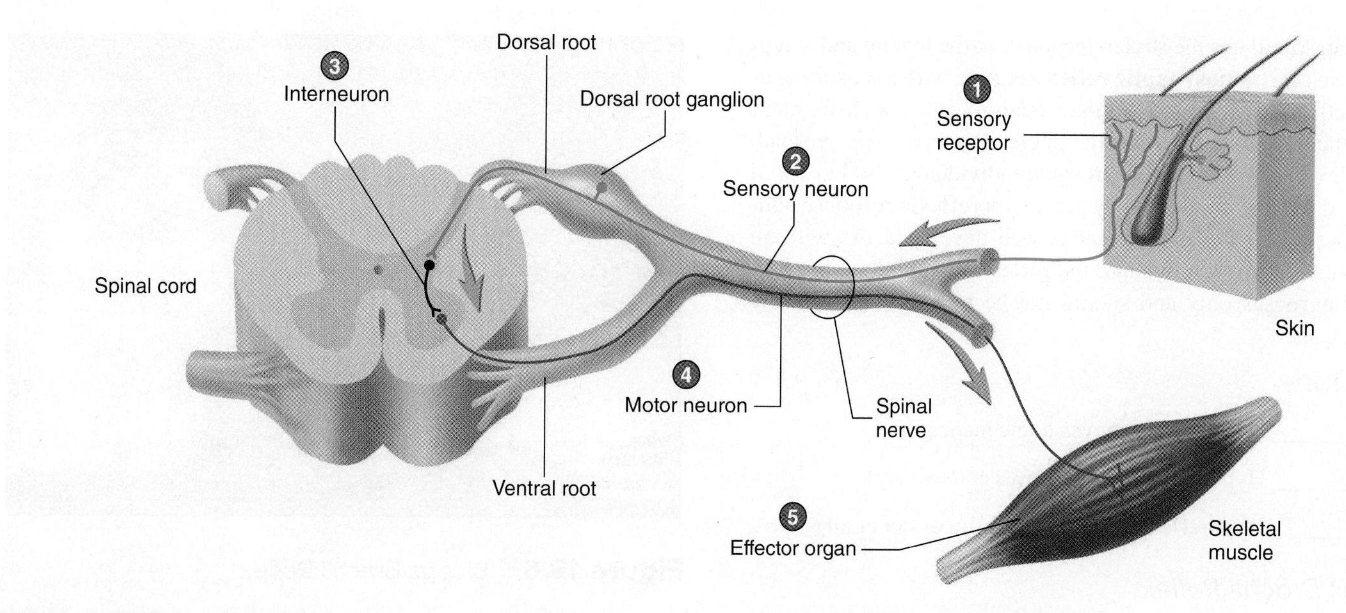

Figure 19.3 Reflex Arc

A stimulus is picked up by the sensory receptor (1), which transmits the stimulus to the sensory neuron (2). The stimulus travels either through the interneuron (3) or to the motor neuron (4), where it stimulates the effector (5) for a response.

If the effector is skeletal muscle, the reflex is called a **somatic reflex.** If the effector is a gland, smooth muscle, or cardiac muscle, the reflex is called a **visceral,** or **autonomic, reflex.**

Most reflexes involve many neurons with many synapses and are called **polysynaptic reflex arcs.** A few reflexes involve just two neurons, a sensory neuron and a motor neuron with one synapse between them, and these are called **monosynaptic reflex arcs.** A polysynaptic reflex arc is illustrated in figure 19.3.

Reflexes depend on a **stimulus,** or an environmental cue, a receptor that is sensitive to the stimulus, a sensory neuron, a motor neuron, and an effector. The polysynaptic reflex arc has these structures, as well as an **interneuron** located between the sensory and motor neurons. Interneurons can take information to the brain via ascending tracts in the spinal cord.

Testing for reflexes is very important for the clinical evaluation of the condition of the nervous system. Decreased response or even exaggerated response to a stimulus may indicate disease or damage to the nervous system.

In this experiment, you test several reflexes and determine if the response is **normal** (movement of an inch or two), **hyporeflexic** (showing less than average response), or **hyperreflexic** (showing an exaggerated response). In clinical settings, stretch reflexes are often used to determine one's neurological condition or to identify conditions such as hypothyroidism. In hypothyroidism, the reflexes are depressed. In some patients with CNS damage, the reflexes are exaggerated.

Stretch Reflexes

For stretch reflexes, receptors are in the muscle spindle. Stretching the muscle causes an increase in action potentials in the sensory neuron. The impulse travels to the motor neuron, causing contraction of the muscle that is stretched. A typical example of a stretch reflex is the patellar reflex, in which striking the patellar ligament stretches the quadriceps muscles. Sensory neurons transmit this information to the spinal column, where motor neurons stimulate the quadriceps muscle to contract, thus extending the leg.

Patellar Reflex

(A) 13 The **patellar reflex** tests the conduction of the femoral nerve. It is the most commonly performed reflex test in clinical settings. Sit down on the lab table and, with your leg hanging over the edge of the table, have your lab partner tap you on the patellar ligament with the blunt side of a patellar reflex hammer. The percussion should be placed 3 to 4 cm below the inferior edge of the patella, and it should be firm but not hard enough to hurt. Look for extension of the leg as a response to the patellar reflex (figure 19.4).

Figure 19.4 Patellar Reflex

The tap stimulates the stretch receptors in the tendon and is representative of a **monosynaptic reflex arc** (one with a sensory neuron directly synapsing with a motor neuron). Place a check mark next to the term that describes the degree (hyperreflexic, normal, hyporeflexic) of the response. In some individuals who have well developed musculature you may get a hyporeflexic response simply because the patellar ligament is well developed. As with all experiments in this lab manual, the reflex tests are done for educational purposes only and should not be regarded as clinically relevant tests.

Patellar Reflex

_______ Normal (foot moves a few inches)

_______ Hyperreflexic (foot moves extensively)

_______ Hyporeflexic (foot moves little or not at all)

Triceps Brachii Reflex

(A) 14 The **triceps brachii reflex** tests the radial nerve. Sit on a chair or lie down on your back on a clean lab table or cot and place your forearm on your abdomen. Have your lab partner tap the distal tendon of the triceps brachii muscle about 2 inches proximal to the olecranon process. Look for the triceps muscle to twitch (figure 19.5). Record your result by placing a check mark next to the term that describes the degree of the reflex response.

Triceps Brachii Reflex

_______ Normal

_______ Hyperreflexic

_______ Hyporeflexic

Biceps Brachii Reflex

(A) 15 The **biceps brachii reflex** tests the musculocutaneous nerve. Sit comfortably and have your lab partner place his or her fingers on the biceps tendon just proximal to the antecubital fossa (figure 19.6).

Your lab partner should tap his or her fingers with the reflex hammer, while they remain on the tendon, and look for the biceps brachii muscle contraction. Record your results by placing a check mark next to the term that describes the degree of the reflex response.

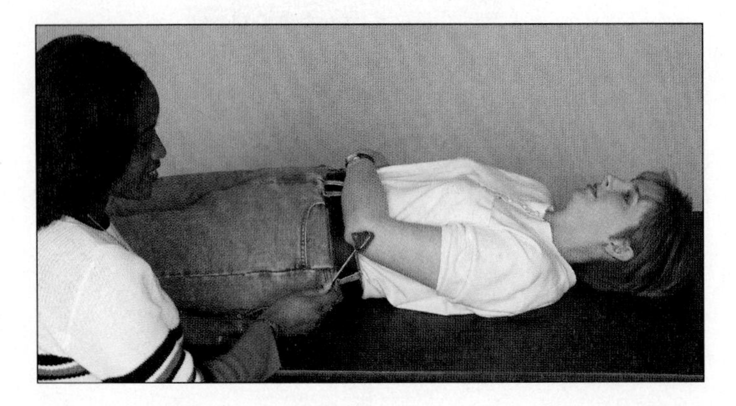

Figure 19.5 Triceps Reflex

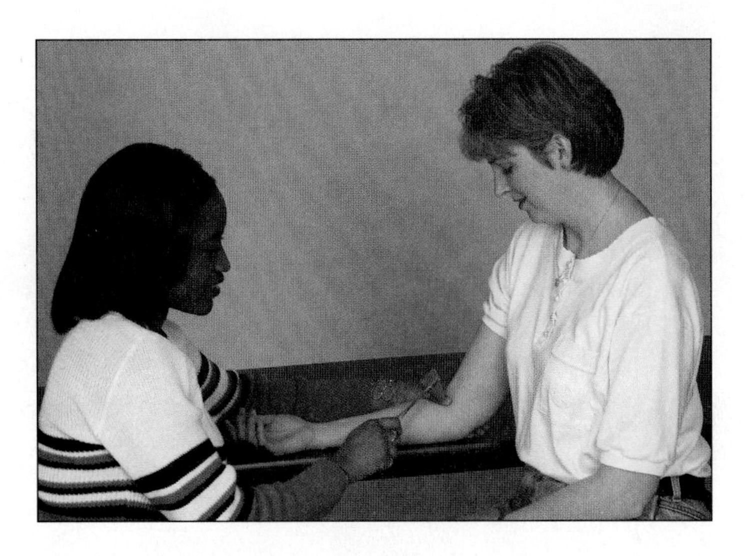

Figure 19.6 Biceps Brachii Reflex

Biceps Brachii Reflex

_______ Normal

_______ Hyperreflexic

_______ Hyporeflexic

Calcaneal (Achilles) Tendon Reflex

(A) 15 To test the **calcaneal tendon reflex,** kneel on a chair with your foot dangling over the edge (figure 19.7). Have your lab partner tap the calcaneal tendon in order to test the tibial nerve.

As your lab partner taps your calcaneal tendon, look for plantar flexion of the foot. You may see an initial movement of the foot due to the depression of the tendon by the reflex hammer, but there should be a slight pause and then another quick movement of the foot. Record your results.

Calcaneal Tendon Reflex

_______ Normal

_______ Hyperreflexic

_______ Hyporeflexic

Eye Reflexes

(A) 17 The automatic blinking of the eye is important to keep material, such as dust, away from the outer layer of the eye, known as the cornea. In the first part of this experiment, have your lab partner try to make you blink by carefully flicking his or her fingers near your eyes. They should not come close enough to touch your eyes! Can you prevent the blinking reflex? Record your answer.

Control of Blink Reflex

_______ Yes

_______ No

Figure 19.7 Calcaneal Reflex

For the next experiment, use either new rubber squeeze bulbs or ones that are free of debris to avoid damage to the eye.

(A) 18 Now have your lab partner take a *clean* rubber squeeze bulb (a large pipette bulb works well) and squirt a blast of air across the surface of the eye (figure 19.8). Can you inhibit the corneal reflex? Record your response.

Control of Corneal Reflex

_______ Yes

_______ No

Plantar Response

The extension of the hallux (big toe) in reponse to stroking the plantar surface of the foot is the **plantar response** or **Babinski reflex.** This response is perfectly normal in infants under the age of 1. The plantar response in infants is due to incomplete myelination of some nerve fibers. In adults this stimulation usually results in flexion of the toes. A plantar response in adults may occur due to damage to the pyramidal tracts and is important in determining spinal damage.

(A) 19 Using the metal end of the patellar hammer, stroke the foot from the heel along the lateral, inferior surface and then toward the ball of the foot (figure 19.9). The pressure should be firm but not uncomfortable.

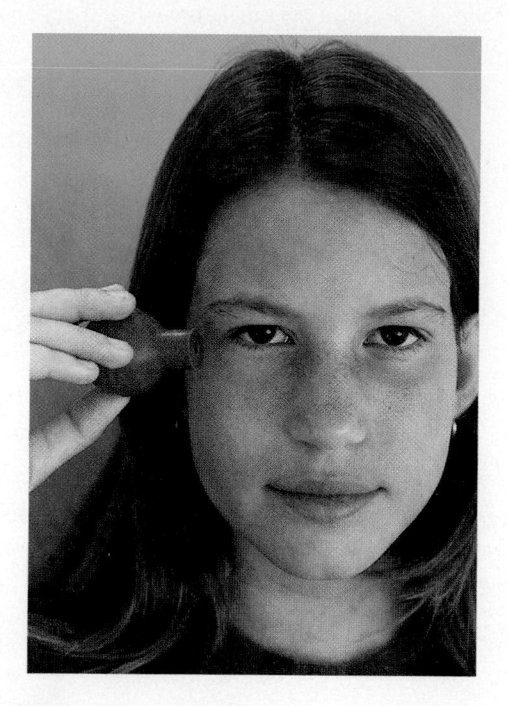

Figure 19.8 Corneal Reflex

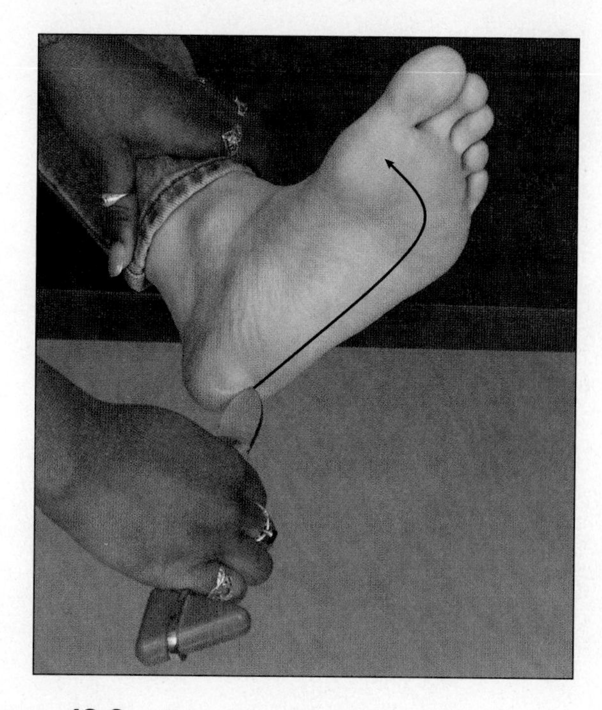

Figure 19.9 Plantar Response

NOTES

Exercise 19 Review

Name: ______________________

Lab time/section: ______________________

Date: ______________________

Nervous System Physiology: Stimuli and Reflexes

1. Define *threshold voltage* in nerve conduction. ______________________

____

2. Define *maximum recruitment voltage* in nerve conduction. ______________________

____

3. What structure receives a stimulus from the external environment and relays it to the sensory neuron? ______________________

4. What is another name for an efferent neuron? ______________________

5. Define *reflex.* ______________________

____

____

6. In what kind of reflex are there just two neurons? ______________________

7. Polysynaptic reflexes have a neuron specific to them. What is the name of that neuron? ______________________

8. In terms of numbers of synapses, what kind of reflex is a patellar reflex? ______________________

9. After surgery, patients leave the operating room and are transferred to an area called the "recovery room." Correlate the meaning of the word *recovery* in this context with what you have learned in this exercise about the recovery of nerves. ______________________

____

____

____

10. Draw a reflex arc in the space provided. Label your illustration with the terms provided.

effector interneuron motor neuron sensory neuron stimulus synapse

11. What action occurs with a hyperreflexic response? ______________________________

____

12. What action occurs with a hyporeflexic response? ______________________________

____

____

13. List the positive responses obtained in the frog experiment, and correlate this with the specificity of neuronal sensitivity. __________

____

____

____

____

14. What was the threshold voltage observed in the nerve response that you obtained by experimentation? __________

15. What was the maximum recruitment voltage in the nerve response that you obtained by experimentation? __________

Exercise 20

Introduction to Sensory Receptors

Introduction

The gateway to understanding our world comes from our ability to sense the environment around us. There are many types of sense receptors in the body, and each responds to a specific stimulus (for example, light, sound, touch). Sense receptors are not uniformly distributed throughout the body but are absent, or few in number, in some areas, while densely clustered in others. This pattern of uneven distribution is called **punctate distribution.**

For the sensory system to operate, several factors need to be present. There can be no perception without environmental input, or **stimulus,** which is listed by type, or **modality.** Examples of modalities are light, heat, sound, pressure, and specific chemicals. **Receptors** are the receiving units of the body that respond to stimuli. They transform the stimulus to neural signals that are transmitted by sensory nerves and neural tracts to the somatosensory cortex or other areas of the brain, which interpret the message. If any link in this sensory chain is broken, the perception of stimuli cannot occur.

Receptors are sensitive to specific modalities and can be classified according to them. **Photoreceptors** detect light (for example, the retina of the eye); **thermoreceptors,** located in the skin, detect changes in temperature; **proprioreceptors** detect changes in tension, such as those in joints; **nociceptors** that transmit the sensation of pain are present as naked nerve endings in areas such as the skin or stomach; **mechanoreceptors** perceive mechanical stimuli (for example, touch receptors or receptors that determine hearing or equilibrium in the ear); **baroreceptors** respond to changes in pressure, such as blood pressure; and **chemoreceptors** respond to changes in the chemical environment (for example, taste and smell). These receptors are discussed in the Seeley text in chapter 14, "Integration of Nervous System Functions."

The skin has several types of receptors and therefore makes a good starting point for understanding sense organs. There are receptors for pressure, pain, temperature, and light touch at your fingertips. The world is a complex environment, which is why it is important to distinguish something that is harmless, such as a bug walking on your arm, from something that is harmful, such as a hot pot handle burning your hand.

Objectives

At the end of this exercise, you should be able to

1. define the terms *modality* and *receptor;*
2. list the major receptor types in the body;
3. define the punctate distribution of sensory receptors;
4. distinguish between tonic and phasic receptors;
5. define *adaptation* in reference to a stimulus;
6. distinguish between relative and absolute determination of stimuli;
7. define *referred pain.*

Materials

Microscopes
Prepared slides of thick skin (with Pacinian and Meissner corpuscles)
Blunt metal probes
Three lab thermometers
Dishpan or large finger bowls of ice water (2 L)
Dishpan of room temperature water
Dishpan of warm water (45°C)
Towels
Small centimeter ruler
Tweezers
Hand lotion
Black, fine-tipped, washable felt markers
Red, fine-tipped, washable felt markers
Blue, fine-tipped, washable felt markers
Two-point discriminators (or a mechanical compass)
Von Frey hairs (horse hair glued onto a wooden stick)

Procedure

Touch Receptors

(A) 1 Examine a slide of thick skin for touch corpuscles. Two of these light touch corpuscles are the **Meissner corpuscles,** in the upper portion of the dermis, and **Merkel discs,** located in the upper dermis and lower epidermis. These two receptors allow for the perception of very light touch stimuli (such as a fly walking over your cheek). In addition, deep touch or pressure receptors are found in the dermis, farther away from the epidermis. **Pacinian (lamellated) corpuscles** sense pressure, as when you lean against a wall or feel a vibration (figure 20.1). Other receptors in the skin are warm receptors and cool receptors. When you are at a comfortable temperature, both of these receptors are firing. If you increase or decrease the skin temperature beyond the perception of these receptors, pain receptors are stimulated. Pain receptors are naked nerve endings in the dermis that respond to many environmental stimuli. Locate the Meissner corpuscles and Pacinian corpuscles in a prepared slide of skin.

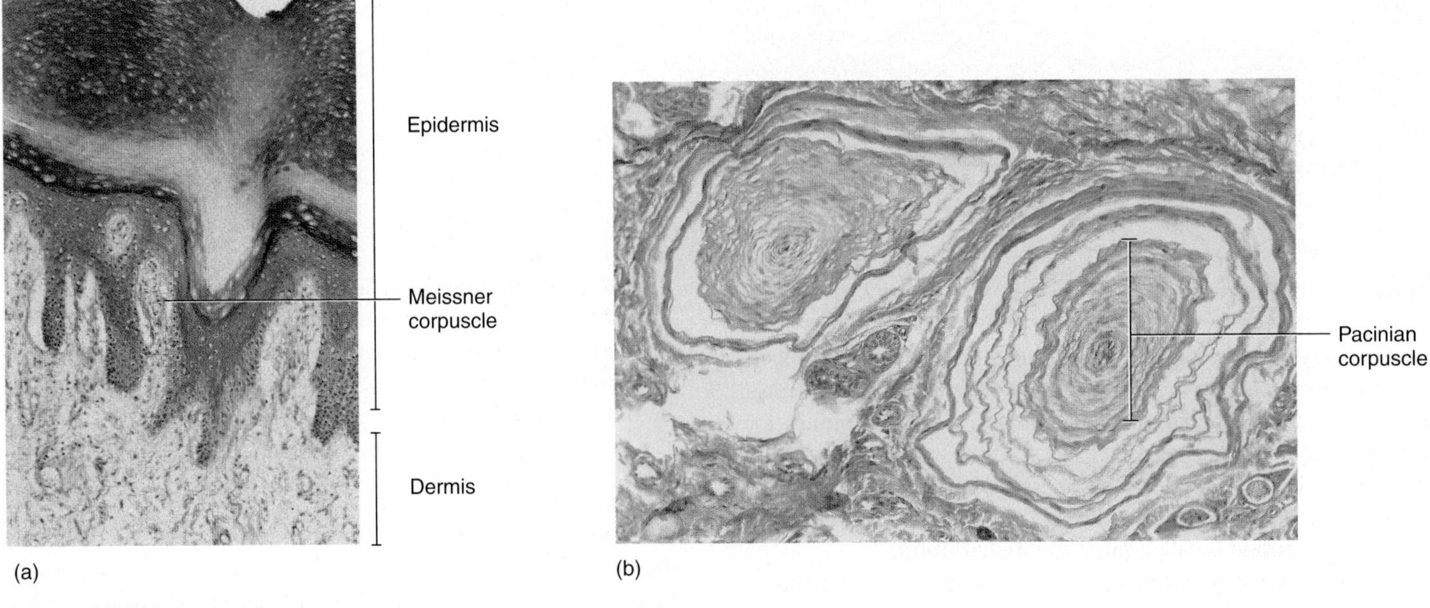

Figure 20.1 Skin Receptors (100×)
(a) Meissner corpuscle; (b) Pacinian corpuscle.

Draw them in the following space

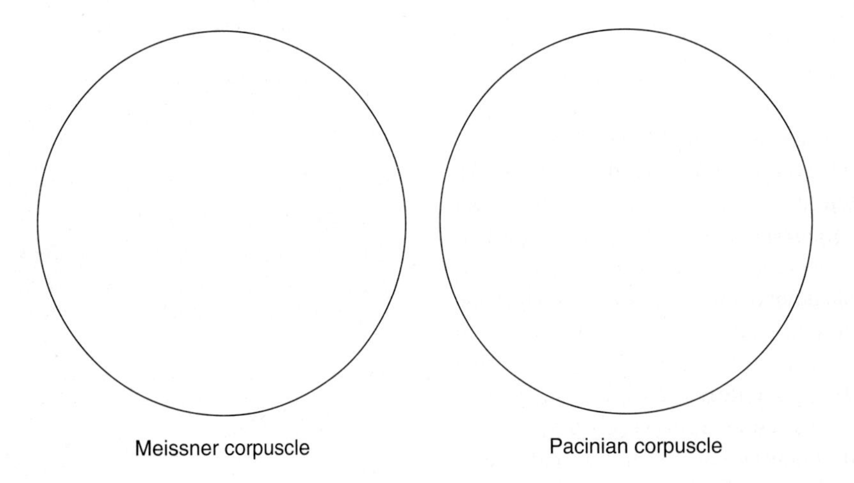

Two-Point Discrimination Test

Ⓐ 2 The ability to distinguish touch depends on the type and number of nerve endings in the skin. You can easily map the relative density of the receptors in the skin by performing a two-point discrimination test. Have your lab partner sit with eyes closed and his or her hand, palm up, resting on the lab counter. Using the two-point discriminator or a mechanical compass, touch your lab partner's fingertip with both points of the instrument and see if he or she can sense one or two points. This is illustrated in figure 20.2. Determine the *minimum* distance apart your lab partner senses as two points. Begin the procedure by determining the minimum distance on the finger. To establish an accurate reading, make sure you gently touch both points of the discriminator at the same time. If you sequentially touch the two-point discriminator on the finger, your lab partner may perceive two points in time and not in space. One good way to establish accuracy is to occasionally

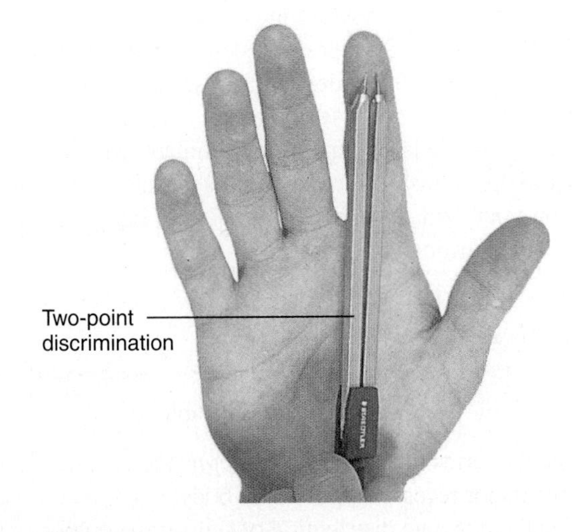

Figure 20.2 Two-Point Discrimination

touch just one of the points to your lab partner's finger. Another way is to vary the spread of the discriminator. You might begin with 2 cm and then adjust it to 0.5 cm followed by a 3 cm spread. Record your results.

Minimum distance perceived as two points on fingertip:

____

Now move to the posterior forearm. Establish the minimum distance that is perceived as two points by your lab partner. Record your results.

Distance on posterior forearm: ______________

Is there a difference in distance between the fingertip and the posterior forearm? ______________

If there is, what might be the explanation for the difference in terms of the number of nerve endings per unit area?

____

Now try the palm and then the back of the arm or neck. Record the results.

Distance on palm: ______________

Distance on posterior arm or neck: ______________

Warm and Cool Receptors

(A) 3 The skin has receptors that are sensitive to cool or warm temperatures. To determine the difference between the two, take two blunt metal probes and place the end of one in an ice water bath and one in a warm water bath (45°C). Let the probes reach the temperature of each respective bath, which should take a minute or two. Have your lab partner close his or her eyes and rest an arm on the lab counter. Remove one of the probes and quickly wipe it on a clean towel. Test the ability of your lab partner to distinguish between cool and warm by using the handle of the probe on your lab partner's forearm. Repeat the experiment for a total of five cool trials and five warm trials. Frequently return the metal probe handles to their respective water baths to maintain the appropriate temperature. Record your results.

Number of accurate cool recordings: ______________

Number of accurate warm recordings: ______________

Mapping Temperature Receptors

(A) 4 Mark off a square that is 2 cm on a side on the anterior forearm of your lab partner using a washable felt marker. If you apply a little hand lotion to the skin before doing this test, the ink comes off more easily after the experiment. Now take the *pointed end* of the blunt probe and place it in cold water. Let the probe sit in the cold water for a few minutes. Quickly wipe the probe and systematically test areas in the square. When your lab partner perceives cold (not just touch) in a location, mark it with the blue marker. Retest the area with the warm probe and place a red mark where warm is perceived.

Number of cool receptors in square: ______________

Number of warm receptors in square: ______________

What is the ratio of blue to red (cold/warm receptors) in the square? ______________

Mapping Light-Touch Receptors

(A) 5 You can map light-touch receptors by testing the ability of your lab partner to distinguish fine touch. Using washable markers, draw a square, 2 cm on a side, on the *anterior* surface of the *forearm.* Using a stiff bristle hair attached to a matchstick, or a Von Frey hair, map the number of areas in the square that your lab partner can perceive. Press only until the hair bends a little to stimulate the touch corpuscles. If you push too hard, you will stimulate other receptors. Use a washable black felt marker to record the location of each positive result. How many positive responses did you get in the square on the forearm? Record your result.

Number of anterior forearm responses: ______________

Repeat the experiment on the *lateral* surface of the *arm* using another square, 2 cm on a side. How does the number of receptors here compare with those on the forearm? Record your results.

Number of responses on lateral side of the arm: ______________

Adaptation to Touch

(A) 6 Adaptation is important because, if you were aware of all the stimuli that enter your brain, you would have too much information to sort out. Receptors can be classified according to the length of time that they can perceive stimuli. **Tonic receptors** constantly perceive stimuli, whereas **phasic receptors** adapt to a stimulus. In this experiment, you try to determine if the sense of light touch is tonic or phasic. Cut out a small piece of paper, about 2 cm on a side, and crumple it into a small ball the size of a pea. Have your lab partner place his or her hand, anterior side up, comfortably on the lab desk. With a pair of tweezers, place the ball of paper on your lab partner's palm. Is the paper ball perceived after a few seconds? Record your results in the following space and determine whether the sense of light touch is tonic or phasic.

Perception of paper ball: ______________

____

Locating Stimulus with Proprioception

(A) 7 In this exercise, you use washable markers of two colors. Have your lab partner close his or her eyes and rest a forearm on the lab counter. Touch your lab partner's forearm with a felt marker and have him or her try to locate the same spot with a felt marker of another color. Test at least five locations on various parts of the forearm, and repeat each location at least twice. Now try the fingertip and palm of the hand and record the result.

Maximum distance error on the forearm: ______________

On the palm: ______________

On the fingertip: ______________

Another method to test proprioception is to close your eyes and *gently* try to touch the lateral corner of your own eye with your fingertip. Have your lab partner watch you and determine the accuracy of your attempt. While your eyes are still closed, bring your hand far behind your head and then try to touch the bottom part of your earlobe or the exact tip to your chin. Record the error distance, if any, for each location.

Corner of eye: ______________________

Earlobe: ______________________

Tip of chin: ______________________

Temperature Judgment

(A) 8 In this exercise, you examine the adaptation of thermoreceptors to temperature and the ability to determine temperature by absolute value or by relative value.

On the lab counter, place three dishpans or large finger bowls full of water. One bowl, located on the right, should be marked "Cold" (it should be about 10°C); another bowl, located on the left, should be marked "Warm" (it should be about 45°C); and the middle bowl should be marked "Room Temperature." Place one hand in the cold dish and the other in the warm dish and let them adjust to the temperature for a few minutes. If your hand begins to ache in the cold water, you may remove it for a short time, but try to keep it in the cold water for as long as possible during the adjustment time. After your hands have equilibrated, place them both in the room temperature water and describe to your lab partner the temperature (cold, warm, hot) of the water as sensed by each hand. How does the hand that was in cold water feel in the room temperature water, and how does the hand that was in warm water feel in the room temperature water? Record your results.

Cold hand perception: ______________________

Warm hand perception: ______________________

If the determination of temperature is absolute, then the room temperature water should feel the same whether you are testing it with your warm hand or your cold hand. If it is relative, the room temperature water should feel warmer with your cold hand and colder with your warm hand.

Is the determination of the temperature of water absolute or relative? ______________________

How did your experiment prove this? ______________________

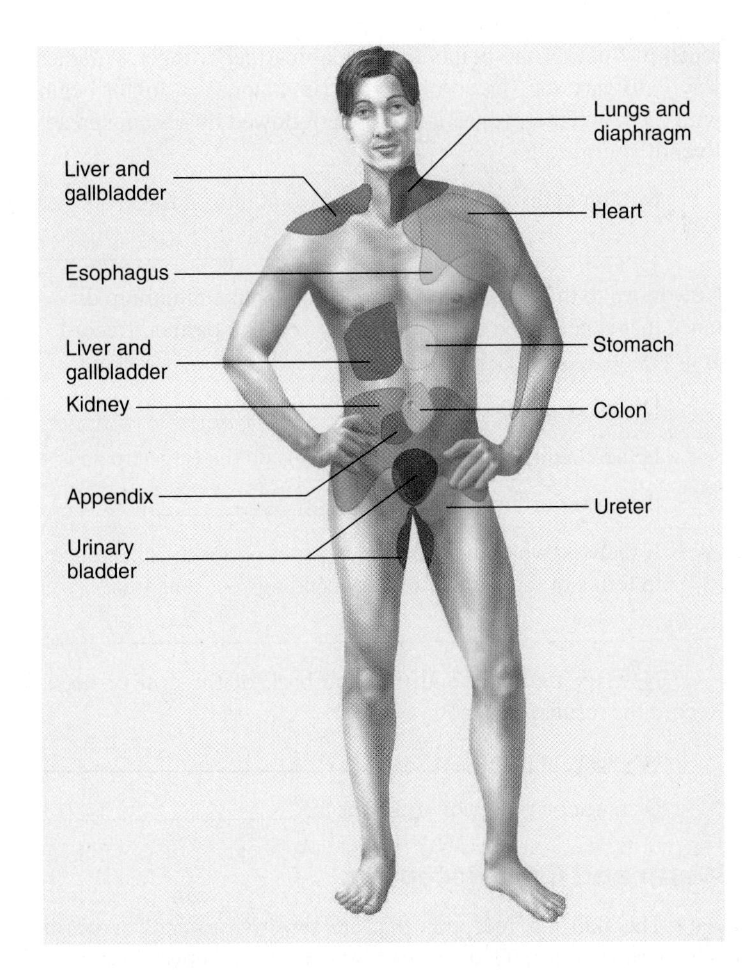

Figure 20.3 Regions of Referred Pain

Referred Pain

Referred pain is the perception of pain in one area of the body when the pain is actually somewhere else (see chapter 14 in the text). An example of referred pain is the pain felt in the left shoulder and arm when a person is suffering from a heart attack or chest pain (angina pectoris; figure 20.3). Referred pain may be caused by many pain impulses from different receptors traveling along the same neural tract to the brain. When one area sends a message of pain, the brain may interpret the sensation as coming from more than one location.

(A) 9 Place your elbow into a dish of ice water and leave it there for 2 painful minutes. Describe the sensation you feel and the location of the sensation. Record your results.

Description of sensation: ______________________

Initial location of sensation: ______________________

Sensation after 2-minute period: ______________________

Exercise 20 Review

Introduction to Sensory Receptors

Name: ______________________

Lab time/section: ______________________

Date: ______________________

1. An area with a great number of nerve endings is the upper lip. What can you predict about the ability of the upper lip to distinguish two points? ______________________

2. Distinguish among the functions of Pacinian corpuscles, Meissner corpuscles, and pain receptors in the skin. ______________________

3. When you extend the temperature beyond the level of cool and warm, the pain receptors in the skin are activated. Cool receptors are activated between 12°C and 35°C. Warm receptors are activated between 25°C and 47°C. You or your lab partner may have had an experience with very cold conditions, such as when cleaning out a freezer or holding dry ice. What perception is sensed?

4. Are there more cool receptors or warm receptors in the skin? What adaptive advantage might there be for an unequal number of receptors of one kind? ______________________

5. Adaptation occurs in some sensory stimulation. Why do you think this is important? ______________________

6. Why do pain receptors function as tonic receptors? ______

7. In terms of receptor density, describe why it is difficult to find the same location on the forearm when your eyes are closed

8. In reference to the sense organs, what is a modality? ______

9. What kinds of receptors are sensitive to the following modalities? ______

   a. light ______

   b. touch ______

   c. temperature ______

   d. sound ______

   e. smell ______

10. What kind of receptor is responsive to extremely hot sensations? ______

11. Meissner corpuscles respond to what kind of sensation? ______

12. What kind of receptor determines the weight of an object when you pick it up? ______

13. Which kind of receptor (phasic/tonic) adapts to low light in a darkened movie theater? ______

14. When you drink a liquid that is burning hot, the "chest pain" felt in the region of the sternum does not really occur there. What is this kind of pain called? ______

Exercise 21

Taste and Smell

Introduction

We take for granted our senses of taste and smell. They are not fully appreciated unless they are lost. People who have lost the sense of smell find food difficult to eat, since they derive little or no pleasure from the act of eating. In this exercise, you examine the structure and function of these two important senses.

Both taste and smell are examples of chemoreception in which specific chemical compounds are detected by the sense organs and interpreted by various regions of the brain. The sense of taste, or **gustation,** is received predominantly by taste buds in the tongue, although there are also receptors in the soft palate and pharynx. The sense of taste is transmitted by the facial, glossopharyngeal, and vagus nerves and interpreted in the postcentral gyrus of the parietal lobe of the brain. The sense of smell, or **olfaction,** originates when particles stimulate hair cells in the olfactory epithelium (a specialized neuroepithelium in the upper nasal cavities) and is transmitted by the olfactory nerves. This transmission occurs through the cribriform plate of the ethmoid bone to the olfactory bulb at the base of the frontal lobe of the brain. From there, the sense of smell is transmitted to two regions, one in the limbic system and another in the temporal lobe of the brain.

The sense of taste and the sense of olfaction are covered in the Seeley text in chapter 15, "The Special Senses."

Objectives

At the end of this exercise, you should be able to

1. list the two major chemoreceptors in the head;
2. draw a diagram of a taste bud;
3. trace the sense of smell from the nose to the integrative areas of the brain;
4. list the five tastes perceived by humans;
5. describe what happens in an olfactory reflex;
6. compare and contrast the senses of taste and smell.

Materials

Microscopes
Prepared slides of taste buds
Roll of household paper towels
Small bowl of salt crystals (household salt)
Small bowl of sugar crystals (household granulated sugar)
Flat toothpicks
Sterile cotton-tipped applicators
Five food-grade dropper bottles containing one of the following
- Solution of salt water (3%) labeled "salty"
- Quinine solution (tonic water) labeled "bitter"
- Vinegar solution (household vinegar or 5% acetic acid solution) labeled "acidic"
- Sugar solution (3% sucrose) labeled "sweet"
- Umami solution (15 g MSG in 500 mL water) labeled "Umami"

Biohazard bag
Small bottle (100 mL) of household ammonia
Four small vials colored red and labeled "wild cherry" filled with benzaldehyde solution
Four small vials of "almond" essence
Several small vials (10–20 mL), screw-cap bottles with cotton saturated with essential oil labeled "peppermint," "almond," "wintergreen," and "camphor" (keep vials in separate wide-mouthed jars to prevent cross-contamination of scents)
One vial of dilute perfume (one part perfume, five parts ethyl alcohol)
Selection of four or five fruit nectars (such as Kern's nectars), two cans each: apricot, coconut/pineapple, strawberry, mango, peach, apple
Small, 3 oz paper cups (89 mL), five per student (or student pair)
Marking pens
Noseclips and alcohol swabs
Napkins

Procedure

Examination of Taste Buds

Ⓐ[1] Examine the prepared slide of taste buds and compare them with figure 21.1. Note how they are located on the sides of the papilla on the tongue. The taste buds appear lighter than the surrounding tissue (like microscopic onions cut in long sections). Taste buds are composed of neural tissue and epithelial tissue. Draw the taste buds you see in the following space.

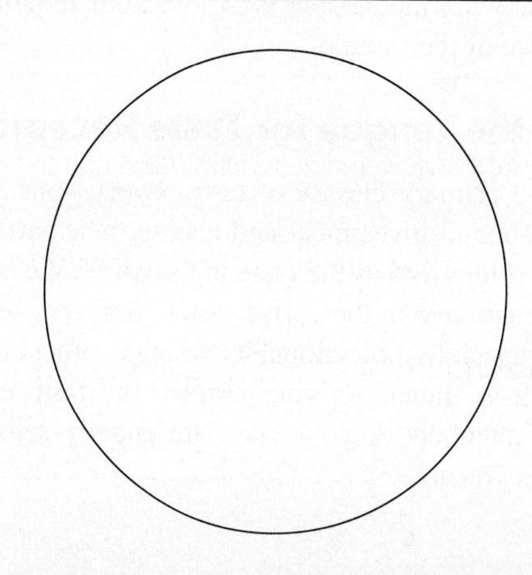

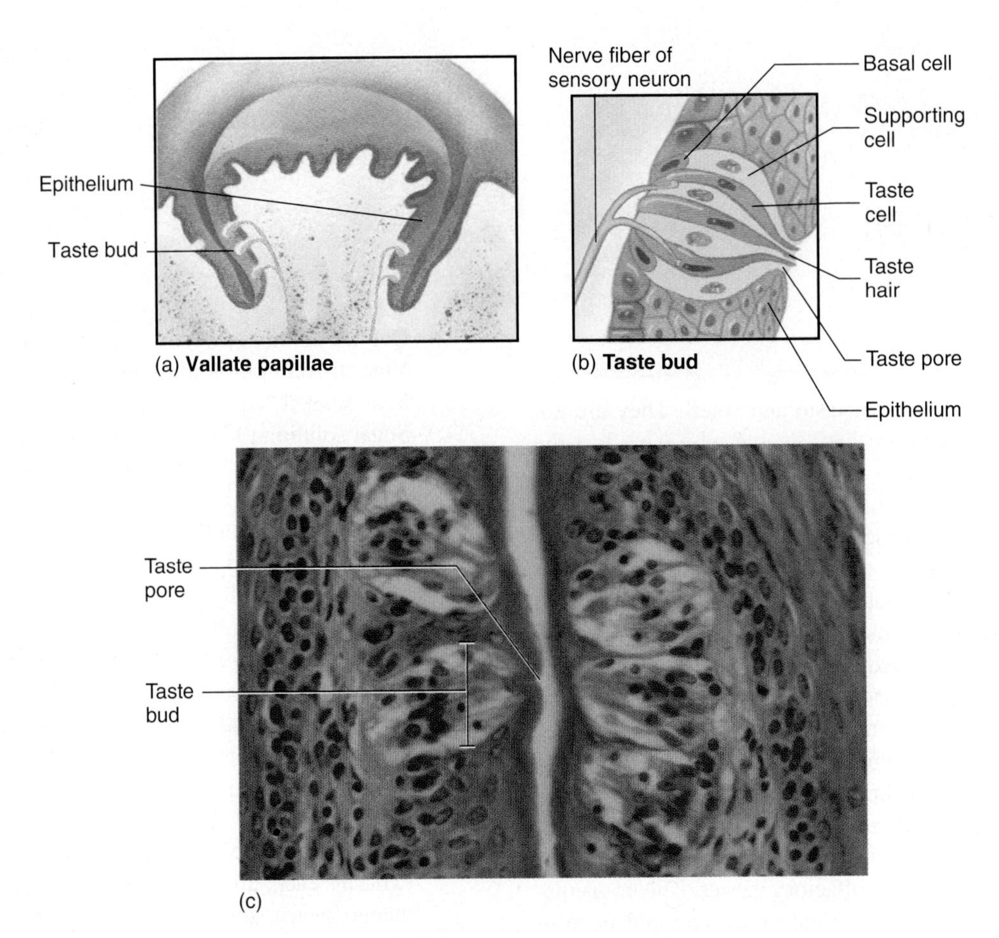

Figure 21.1 Taste Buds
(a) Taste buds on sides of a tongue papilla; (b) details of taste buds; (c) photomicrograph of taste buds (100×).

Taste Determination of Solid Materials

(A) 2 For something to be tasted, it must be in solution. This allows the fluid to run down the sides of the tongue papilla, where the taste buds are located. Blot your tongue thoroughly with a paper towel. Make sure the surface is relatively dry. Have your lab partner select either the sugar crystals or the salt crystals and place a small scoop (with the end of a flat toothpick) on your tongue. Keep your mouth open and do not swirl saliva around. Can you determine what the sample is? Close your mouth and determine the nature of the sample.

Mapping the Tongue for Taste Receptors

There are five primary classes of taste: sweet, sour, salty, bitter, and umami. Umami gives meat and cheese their particular tastes and it can be referred to as the taste of "savory." We have varying degrees of sensitivity to these five tastes. Some of us find bitter tastes to be especially objectionable, whereas others do not seem to mind them as much. As you perform the taste experiments, determine if members of your class are equally sensitive to the same tastes as you are.

Caution

If you have food allergies, you may wish to omit this part of the lab. People with allergies, migraines, or heart problems should avoid the tasting of umami, which is derived from monosodium glutamate (MSG).

(A) 3 Using a sterile cotton-tipped applicator stick and a dropper bottle, apply one of the tasting solutions (salty, bitter, acidic, umami, or sweet) to the cotton. Saturate the swab. Remember to keep track of the kind of substance that is on your applicator stick. Do *not* reuse the applicator sticks for more than one experiment. Dab the entire surface of your lab partner's tongue and determine by nod or hand signals when perception of the taste occurs. Do not have your lab partner tell you about the taste at first, since the movement of the tongue will wash the solution to other areas of the tongue and you may get a false reading. Use figure 21.2 and draw on the figure where you perceive the specific sensations of taste. Are there regions of the tongue that are more sensitive to some tastes? If so, note these areas on figure 21.2.

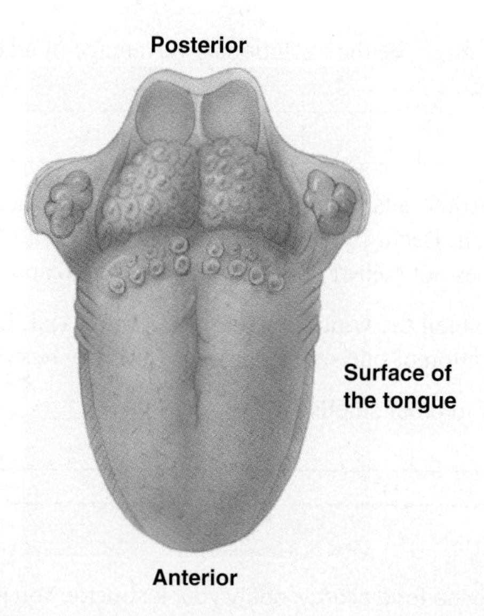

Figure 21.2 Map of Taste Receptors on the Tongue
Indicate the regions of your tongue that are sensitive to the specific tastes of sweet, sour, salty, bitter, and umami.

Throw the applicator stick in the biohazard bag. Repeat the test with a new applicator stick for each of the five solutions. Do not contaminate the solutions by reusing the applicator stick with the same or different solutions after it has been in your lab partner's mouth! Compare the taste map of members of the class with yours. Are they the same? If not, do the regions or relative area vary significantly?

Transmission of Sense of Olfaction to Brain

(A) 4 Examine a model, chart, or diagram of a midsagittal section of the head or a model of the brain and locate the **olfactory nerve fibers, cribriform plate, olfactory bulb,** and **olfactory tract.** Compare the lab charts or models with figure 21.3.

Olfactory Reflex

(A) 5 Place a small bottle of household ammonia under the nose of your lab partner. Have your lab partner take a brief sniff from the bottle. If there is a visible movement of the head in a posterior direction, then your lab partner demonstrates an olfactory

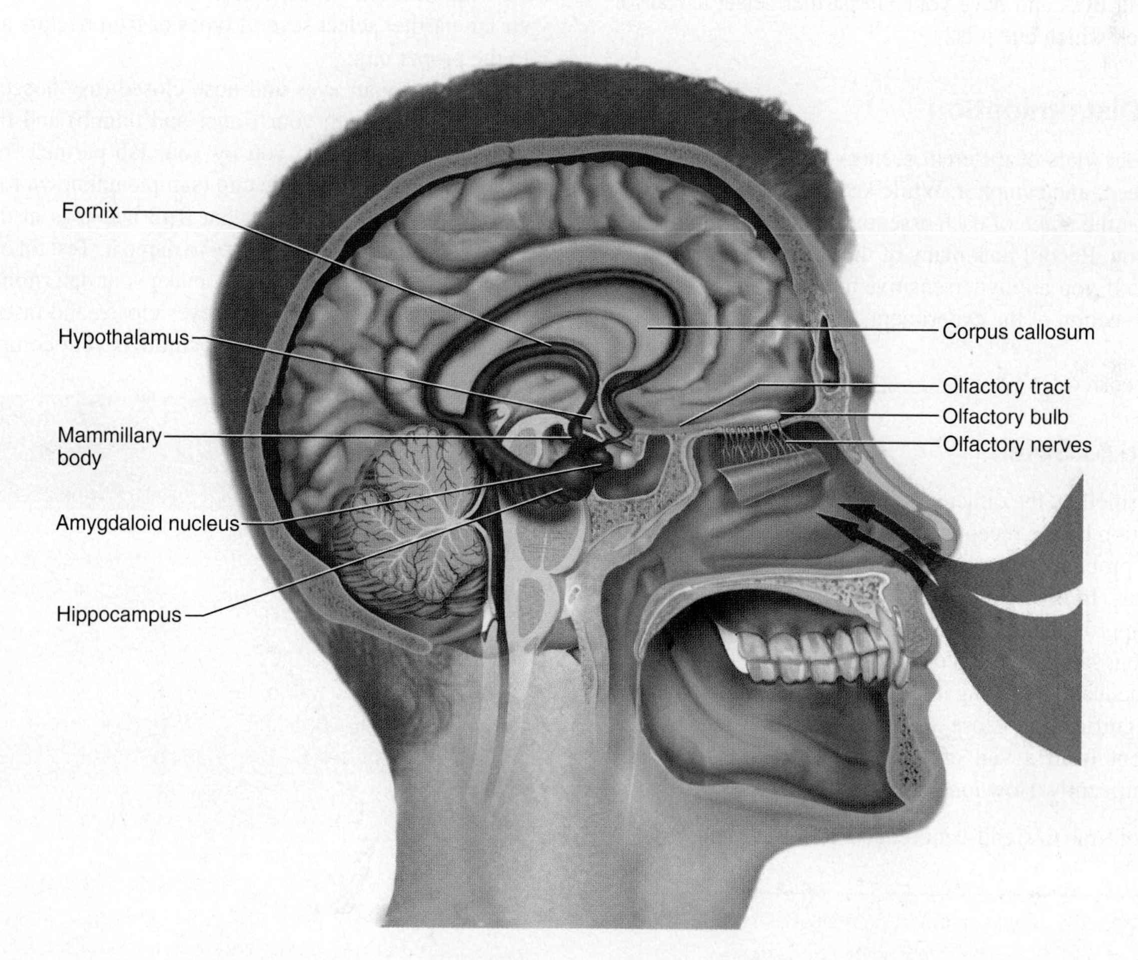

Figure 21.3 Olfactory Transmission. From the olfactory nerves the transmission goes to the olfactory bulb, nerve, and then is processed in parts of the limbic system, the temporal lobe, and the frontal lobe.

reflex to smell. Record the results of your experiment in the following space.

Olfactory Reflex

_____ yes

_____ no

What might be the adaptive benefit for people having an

olfactory reflex? ________________________________

____

____

Visual Cues in Smell Interpretation

(A) 6 In this section, you examine the influence of visual cues on the interpretation of smell. You seek to determine whether the color of a substance has any effect on what you perceive the smell to be.

Have your lab partner show you a small vial labeled "almond" and then smell it. Then examine and smell the small red vial labeled "wild cherry." Do you perceive these as two separate smells?

Close your eyes and have your lab partner select a vial for you. Can you tell which one it is?

Olfactory Discrimination

(A) 7 Obtain four vials of different scents—peppermint, almond, wintergreen, and camphor. While keeping your eyes closed, try to determine the name of each essential oil as your lab partner presents it to you. Record how many of the smells you get correct out of the four. If you are hypersensitive to smells, have your lab partner do this section of the experiment.

Number correct: ________________________________

Adaptation to Smell

Adaptation to smell by the olfactory receptors occurs very rapidly, but the adaptation by the receptors is incomplete. Complete adaptation to smell probably occurs by additional CNS inhibition of the olfactory signals. In this section of the experiment, you are trying to determine approximately how long olfactory adaptation takes.

(A) 8 Have your lab partner record the time when you begin the experiment and how long it takes for the perception of smell to decrease significantly. Close your eyes and plug one nostril. Inhale the scent from a vial of dilute perfume until the smell decreases significantly. How long does this take?

Length of time to significant reduction of the smell: ______

____

What might be the evolutionary advantage of adaptation to

smell? ________________________________

____

Predict whether adaptation to one smell causes adaptation to another smell. Record your prediction that the smell of one material does/does not (select one) cause adaptation to another smell.

Now smell the wintergreen or peppermint vial. Does the adaptation of one smell cause the olfactory receptors to

adapt to other smells? ________________________________

____

Caution

If you have a food allergy, notify your instructor. You may wish to omit part or all of this test.

Taste and Olfaction Tests

(A) 9 This experiment demonstrates the dependence of the sense of smell as a component of what we call taste. Obtain four or five small drinking cups (3 oz) and, using a marking pen, label each with the name of the fruit juice or nectar it will contain. Have your lab partner select several types of fruit nectars and pour each into the proper cup.

Sit with your eyes and nose closed (use noseclips or pinch off your nostrils with your finger and thumb) and try to identify the sample presented to you by your lab partner. Your lab partner should place the sample cup (sample unknown to you) in your hand and you should guess what fruit nectar is in the cup. After you have "tasted" the sample, try to name it. Test all of the samples with your nose closed. After you make your determination, release your nostrils but still keep your eyes closed and taste the samples again. Record the results. How accurate is your comparison?

Trial Number	Sample	Accuracy (Yes/No) Nose Closed	Nose Open
1	Apricot		
2	Mango		
3	Coconut/pineapple		
4	Strawberry		
5	Peach		
6	Apple		
7	Other		

Exercise 21 Review

Taste and Smell

Name: ______________________

Lab time/section: ______________________

Date: ______________________

1. Why does material have to be in solution for it to be sensed as taste? ______________________

2. What are the primary classes of taste? ______________________

3. Did everyone in your lab have the same reaction to tastes, such as sweet, sour, or bitter? ______________________

4. What nerves transmit the sense of smell to the brain? ______________________

5. What nerves transmit the sense of taste to the brain? ______________________

6. Where are the taste buds located? ______________________

7. What is the exact region of the nasal cavity that is sensitive to smell stimuli? ______________________

8. What is the adaptation for having taste buds that determine unpleasant bitter compounds in many plant species?

9. Some people with severe sinus infections can lose their sense of smell. How can an infection that spreads from the frontal or maxillary sinus impair the sense of smell? What structure or structures might be affected? ______________________

10. Material must be in solution for it to be tasted. What process would be used (olfaction or gustation) to perceive a lipid-based food? ____

____

11. How does a cold (rhinovirus) influence our perception of taste? ____

____

12. Does adaptation to one smell influence adaptation to another smell? ____

13. Some smells that we perceive as two separate smells are actually identical. What other cues do we use to distinguish between these two smells? ____

Exercise 22

Eye and Vision

Introduction

Eyesight accounts for much of our accumulated knowledge. In this exercise, you learn the features of the eye and some of the functions of the eye. The anatomy and physiology of the eye are discussed in the Seeley text in chapter 15, "The Special Senses."

The importance of the eye can be inferred in that 5 of the 12 pairs of cranial nerves are dedicated, at least in part, to either receiving visual stimuli or coordinating the eyes.

Anatomically, the eye consists of an anterior portion, visible as we look at a person's face, and a posterior portion, located in the orbit of the skull. Light travels through a number of transparent structures before it strikes the retina, which is the receptive layer of the eye that converts light energy to action potentials, which travel from the eyes through the optic nerves to the brain. The action potentials are interpreted as sight in the occipital lobes of the brain.

Objectives

At the end of this exercise, you should be able to

1. identify the major structures of the mammalian eye;
2. describe the six extrinsic muscles of the eye and their effect on the movement of the eye;
3. describe the position of the choroid in reference to the retina and the sclera;
4. distinguish between the pupil and the iris of the eye;
5. describe the function of the rods and the cones of the eye;
6. define the near point of the eye;
7. determine the visual field for both eyes;
8. demonstrate the Snellen vision tests and those for accommodation and astigmatism.

Materials

Models and charts of the eye
Microscopes
Prepared microscope slides of the eye in sagittal section
Paper with small print (8- or 9-point font)
Ruler (approximately 35 cm)
Paper card with a simple, colored image (red circle, blue triangle) printed on it
Vision Disk (Hubbard) or large protractor
Snellen charts
3- by 5-inch cards
Astigmatism charts
Ophthalmoscope and batteries
Penlight
Ishihara color book or colored yarn
Preserved sheep or cow eyes
Dissection trays
Protective gloves
Scalpel
Animal waste disposal container
Desk lamp with bright light bulb

Procedure

External Features of the Eye

(A)[1] The external anatomy of the eye and the accessory structures are illustrated in figure 22.1. Examine the eye of your lab partner and compare it with the figure. The **pupil** is an opening located in the center of the eye and is surrounded by the colored **iris.** The **sclera** is the white of the eye and it is covered by a membrane, known as the **conjunctiva,** that continues underneath the eyelids. The eyelids join together at the **lateral commissure** and the **medial commissure.** There is a small piece of tissue near the medial commissure known as the **lacrimal caruncle.** Examine the **upper eyelid** and **eyelashes** and the **lower eyelid** and eyelashes, which prevent material from entering the eyes and (in the case of the eyelids) reduce visual stimulation when we sleep. Look also at the **eyebrows** located superficial to the supraorbital ridge. The sclera is a protective portion of the eye that is an attachment point for the muscles of the eye and helps maintain the **intraocular pressure** (the pressure inside the eye). The pressure maintains the shape of the eye and keeps the retina attached to the posterior wall of the eye. Numerous blood vessels traverse the sclera, and if they become dilated anteriorly they give the eye the appearance of being "bloodshot." The sclera is continuous with the transparent

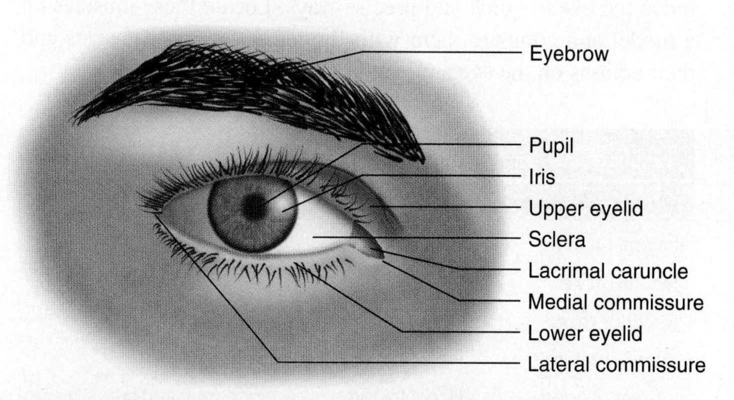

Figure 22.1 External Anatomy of the Eye

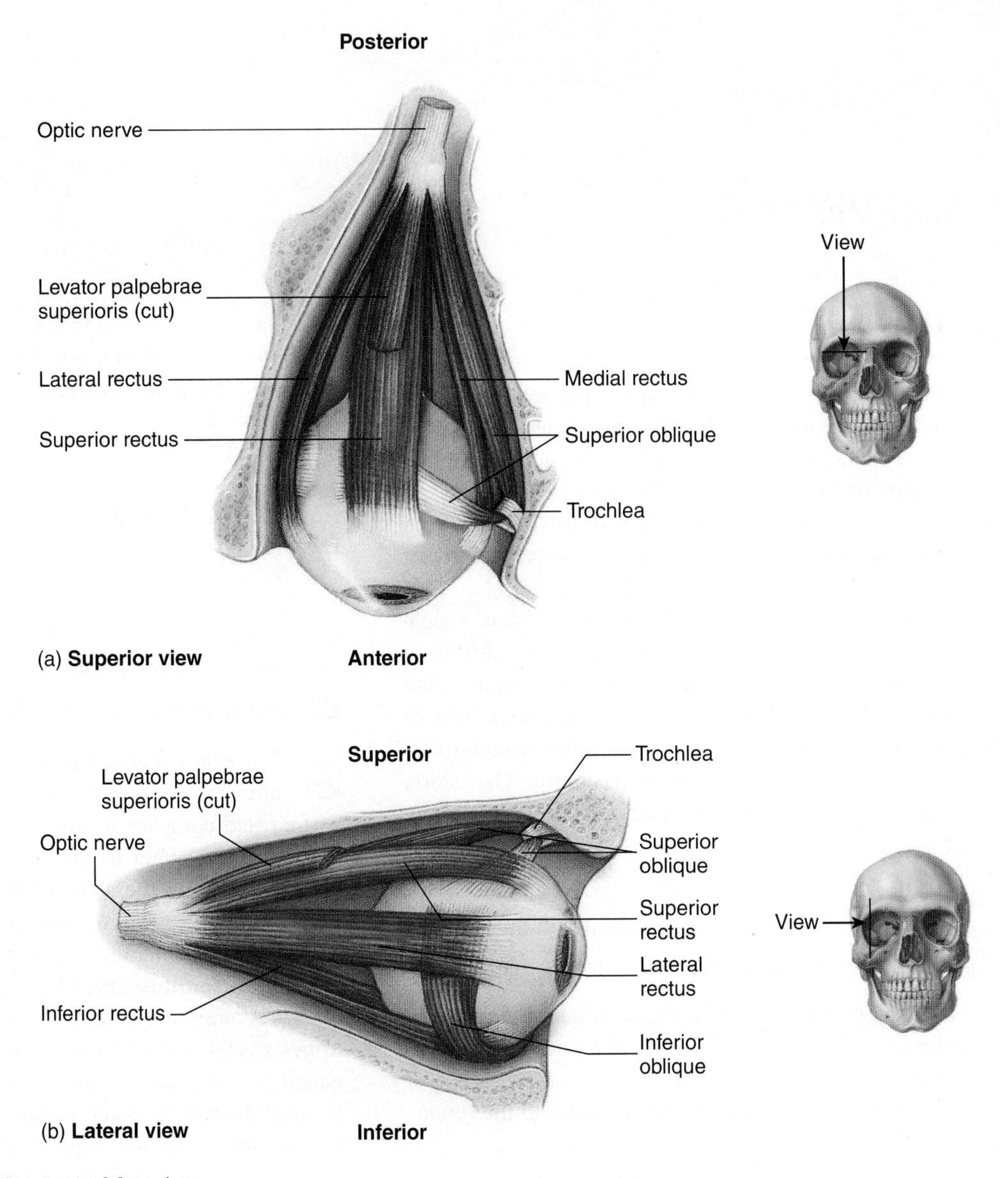

Figure 22.2 Right Eye, Extrinsic Muscles
(a) Superior view; (b) lateral view.

cornea of the anterior eye, and both of them together form the **fibrous tunic** of the eye.

(A)[2] Attached to the sclera are the **extrinsic muscles** of the eye. There are six extrinsic muscles that, in coordination, move the eye in quick and precise ways. Locate these muscles on a model and compare them with figure 22.2. These muscles and their actions on the eye are listed in table 22.1.

TABLE 22.1 **Extrinsic Muscles of the Eye**

Muscle Name	Innervation	Direction Eye Turns
Lateral rectus	VI (abducens)	Laterally
Medial rectus	III (oculomotor)	Medially
Superior rectus	III (oculomotor)	Superiorly
Inferior rectus	III (oculomotor)	Inferiorly
Inferior oblique	III (oculomotor)	Superiorly and laterally
Superior oblique	IV (trochlear)	Inferiorly and laterally

A structure important in the maintenance of the exterior of the eye is the **lacrimal apparatus.** This consists of the **lacrimal gland** located superior and lateral to the eye (figure 22.3). Lacrimal secretions (tears) bathe and protect the eye and clean dust from its surface. The fluid drains through the **nasolacrimal duct** into the nasal cavity.

Interior of the Eye

From the anterior of the eye, the first layer covering the sclera is the **conjunctiva.** The conjunctiva is composed of epithelial tissue and is an important indicator of a number of clinical conditions (for example, conjunctivitis). In the center of the eye is the transparent **cornea** (figure 22.4). The cornea is the structure of the eye important for bending light rays that strike the eye, allowing us to focus light. It is composed of dense connective tissue and is avascular. Why would the presence of blood vessels in the cornea be a visual liability?

Directly behind the cornea is the **anterior chamber,** between the cornea and the iris, and the **posterior chamber,** between the

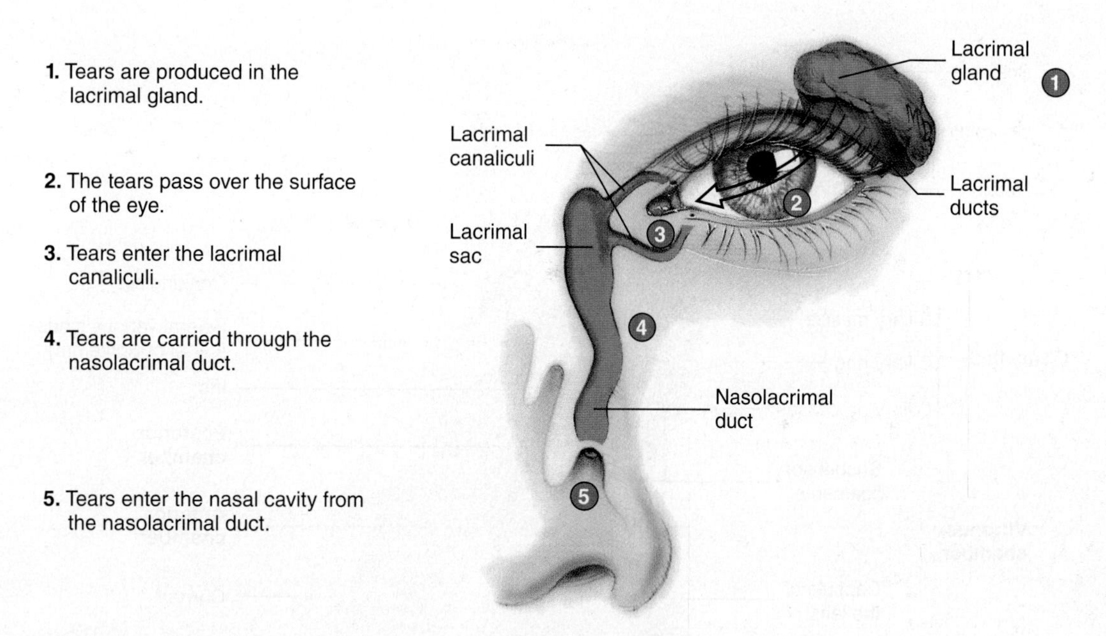

Figure 22.3 Lacrimal Apparatus

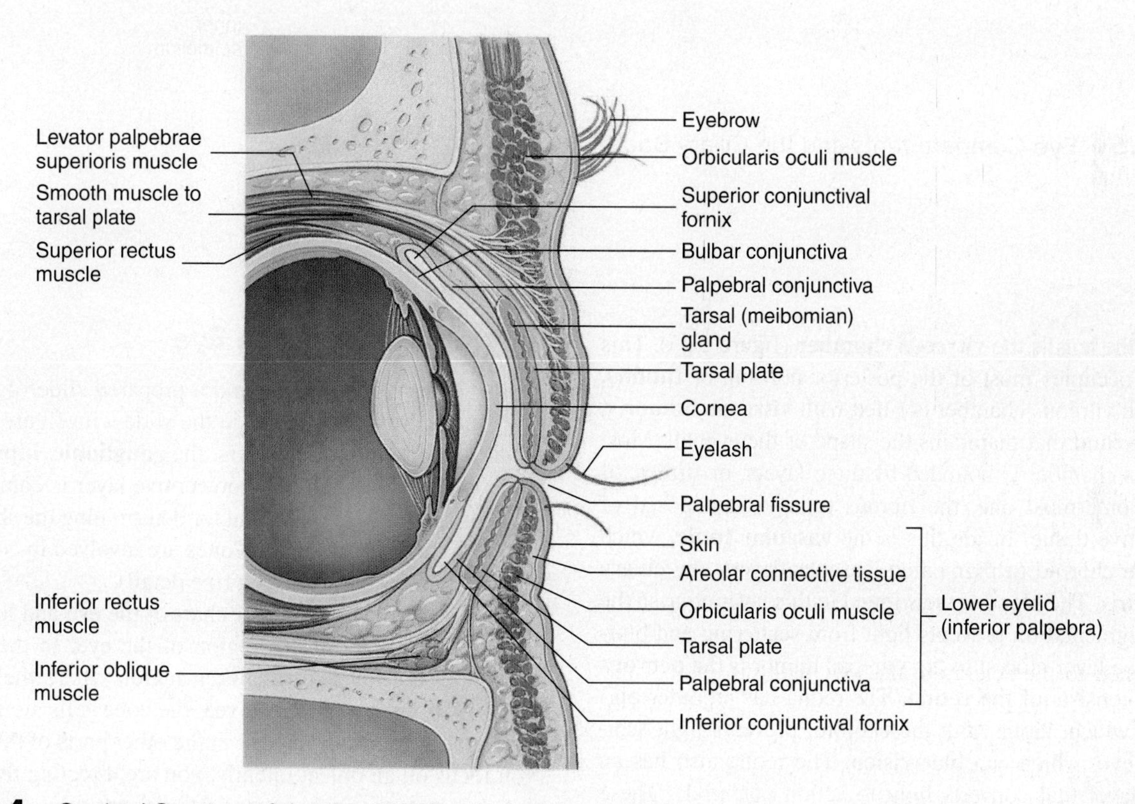

Figure 22.4 Sagittal Section of the Anterior Portion of the Eye

iris and the lens. These two chambers are filled with **aqueous humor,** which is produced by the **ciliary body.** Only a few milliliters of aqueous humor are produced each day, and this amount is absorbed by the **scleral venous sinus (canal of Schlemm),** as illustrated in figure 22.5.

The **iris** is what gives us a particular eye color. People with blue or gray eyes are more sensitive to ultraviolet light than those with brown eyes, due to more of the protective pigment **melanin** found in brown eyes. The **sphincter pupillae** (of circular muscles) of the iris constrict in bright light, reducing the diameter of the **pupil** (the space enclosed by the iris), and the **dilator pupillae** (of radial muscles) constrict in dim light, increasing the diameter of the pupil. Behind the pupil is the **lens,** which is made of a crystalline protein. The **ciliary muscle** in the **ciliary body** contracts and the **suspensory ligaments** that attach to the lens loosen, decreasing the pull on the lens. The lens becomes more round for focusing on closer images. The lens is more pliable in youth and becomes less elastic as a person ages, which accounts for the need for reading glasses later in life.

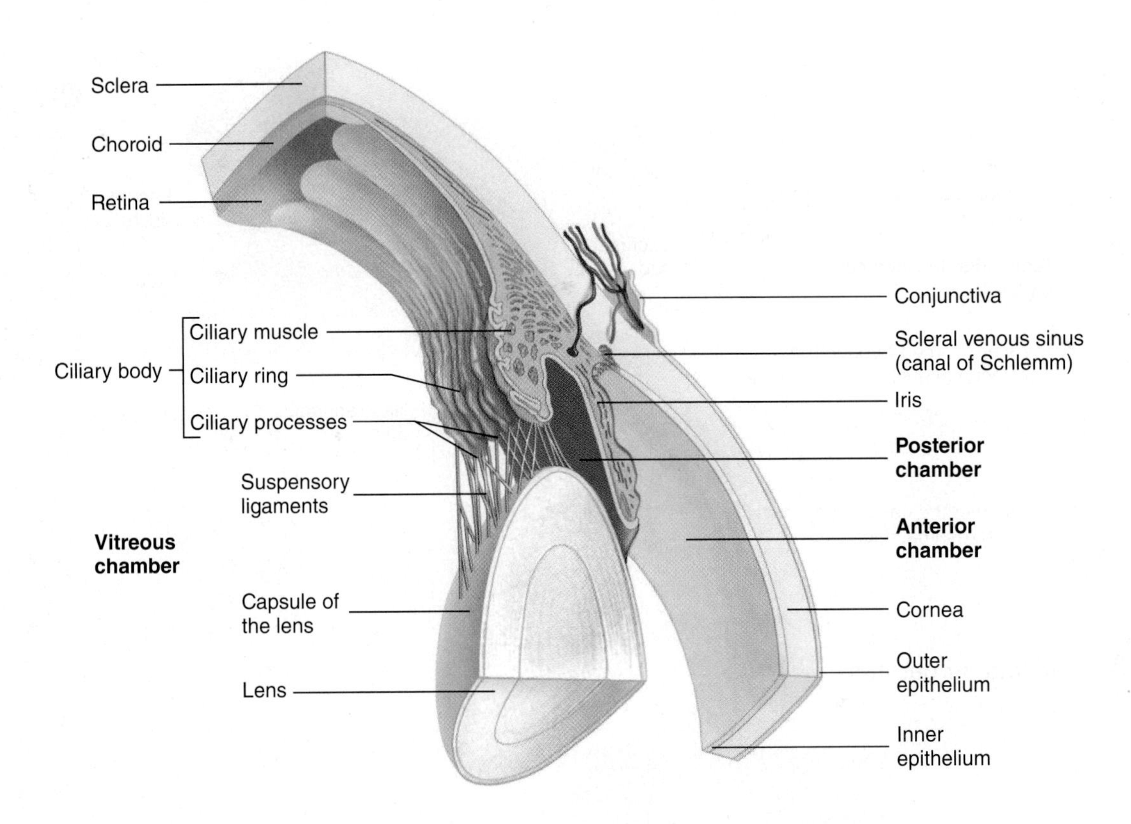

Figure 22.5 Eye Compartments and the Ciliary Body

Behind the lens is the **vitreous chamber** (figure 22.5). This compartment occupies most of the posterior portion, or **fundus,** of the eye. The vitreous chamber is filled with **vitreous humor,** a clear, jellylike fluid that maintains the shape of the eyeball. Most of the vitreous chamber is bounded by three layers, or **tunics,** of the eye. The outermost one, the fibrous tunic, is composed of dense connective tissue. Inside this is the **vascular tunic,** which consists of the choroid (a pigmented, vascular layer), the ciliary body, and the iris. The blood vessels found in this layer nourish the eye and the pigmentation prevents light from scattering and blurring vision. The layer closest to the vitreous humor is the **nervous tunic,** which consists of the **retina.** The retina has an outer pigmented layer which, along with the choroid, prevents light scattering in the eye, which can blur vision. The retina also has an inner neural layer that converts light to action potentials. These structures are seen in figure 22.6.

The retina is a neural tunic that converts light energy into nerve impulses. Light strikes the photoreceptive cells at the posterior portion of the retina, causing these cells to transmit signals to the **bipolar neurons.** The bipolar neurons synapse with the **ganglion cells;** thus, the visual stimulation that occurs in the posterior portion of the eye is transmitted anteriorly (toward the vitreous humor) to the ganglion cells. The axons of the ganglion cells exit the eye as the **optic nerve.** From there, the neural impulse is transmitted to the occipital region of the brain and integrated further in the temporal lobe (figure 22.7).

Posterior Wall of the Eye

(A) 3 Obtain a microscope and a prepared slide of the retina and compare what you see in the slide with figure 22.8. The retina is composed of three layers: the **ganglionic, bipolar,** and **photoreceptive** layers. The photoreceptive layer is composed of **rods** and **cones.** Rods are important for determining the shape of objects and for seeing in dim light. Cones are involved in color vision and in visual acuity (determining fine detail).

(A) 4 Examine a model or chart of the eye and locate the **macula** at the posterior region of the eye. In the center of the macula is the **fovea centralis,** a region where the concentration of cones is greatest. In the fovea, the cone cells are not covered by the nervous tunic, as they are in the other parts of the retina. When you focus on an object intently, you are directing the image to the fovea. Locate the fovea in models or charts available in the lab and compare them with figure 22.9.

Visual Tracking

(A) 5 Eye movements are very precisely controlled by the six extrinsic muscles of the eye. In order to determine their effectiveness, have your lab partner follow your finger as you move it in front of the eyes. Is the movement of the eye smooth (which indicates normal function), or do the eyes move in a jerky fashion (an abnormal condition)? Record your results in the following space.

Eye movement (select one): smooth/jerky

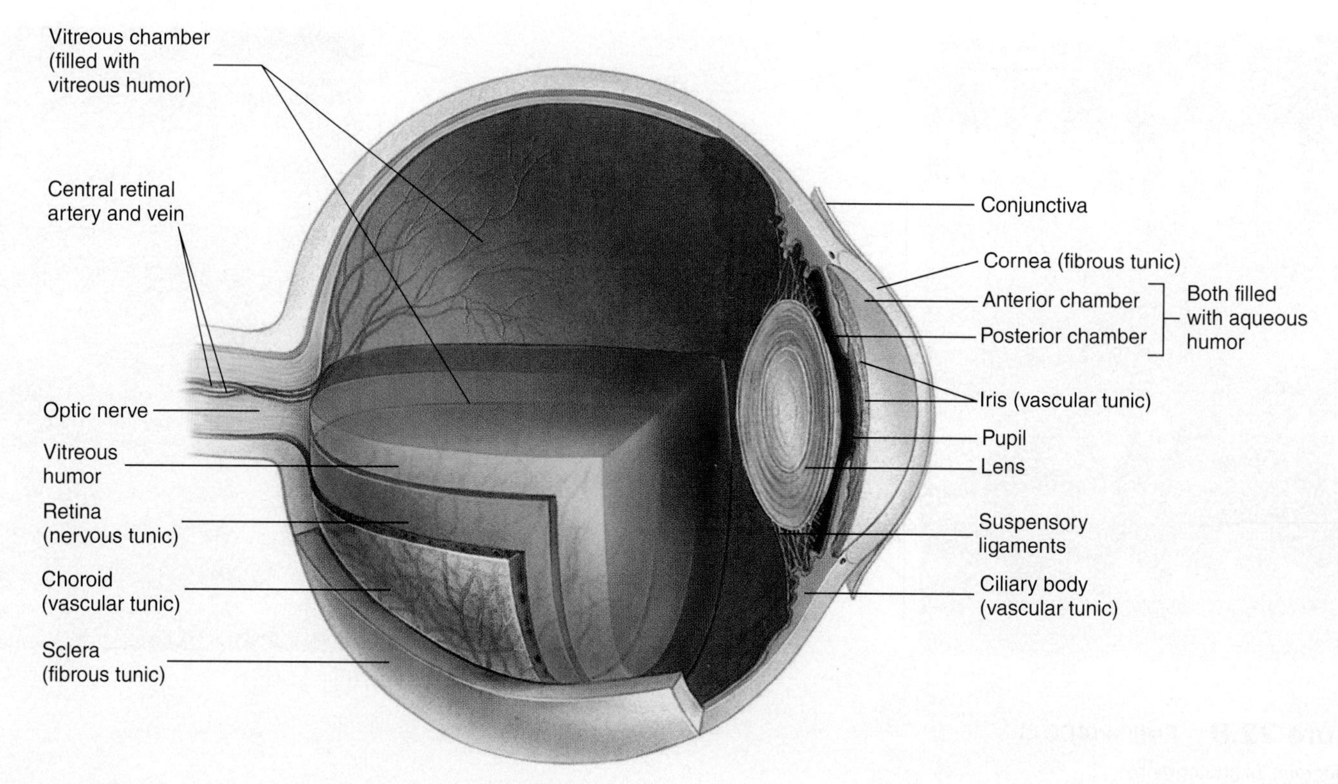

Figure 22.6 Sagittal Section of the Eye with Layers

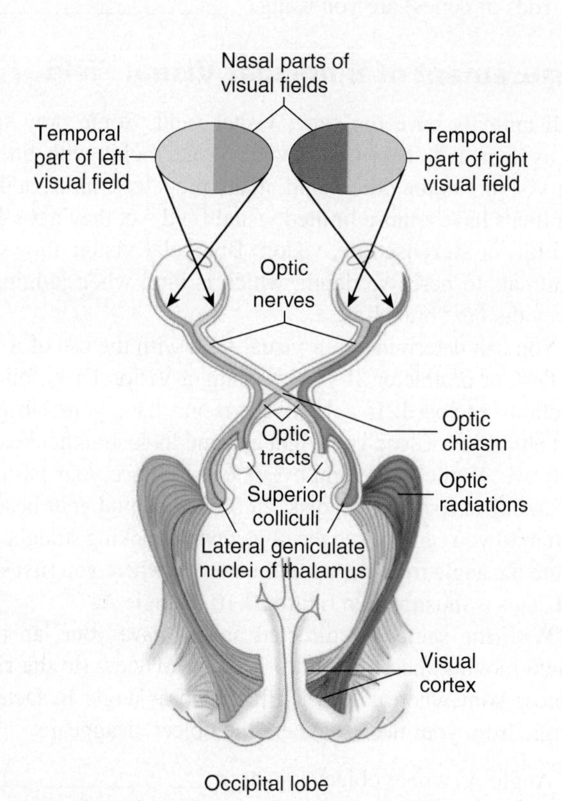

Figure 22.7 Visual Pathway to the Brain

Determination of the Near Point

The *minimum* distance an object can comfortably be held in focus is called the **near point.** The eye's ability to focus is due to the elasticity of the lens, which decreases with age. A 10 year old can focus 8 to 10 cm away from the eye, yet a 65 year old may not be able to focus closer than 80 to 100 cm. The stiffening of the lens becomes noticeable around 40 to 50 years of age, when many people find reading glasses necessary because their aging lenses do not accommodate as well as the lenses of a younger person.

(A)[6] You can measure your near point by holding a piece of paper with fine print (8- or 9-point font) at arm's length in front of you. Close one eye and slowly move the paper closer until the print becomes blurry. Have your lab partner measure the distance, in centimeters, from your eye to the paper. This is the near point distance. Measure the near point for both eyes, in centimeters, and record the data.

Near point of right eye: ____________________

Near point of left eye: ____________________

Measurement of the Distribution of Rods and Cones

(A)[7] To determine the distribution of rods and cones, have your lab partner look straight ahead. Slowly move a small, colored object (a pen, piece of chalk, comb, etc.), without letting your

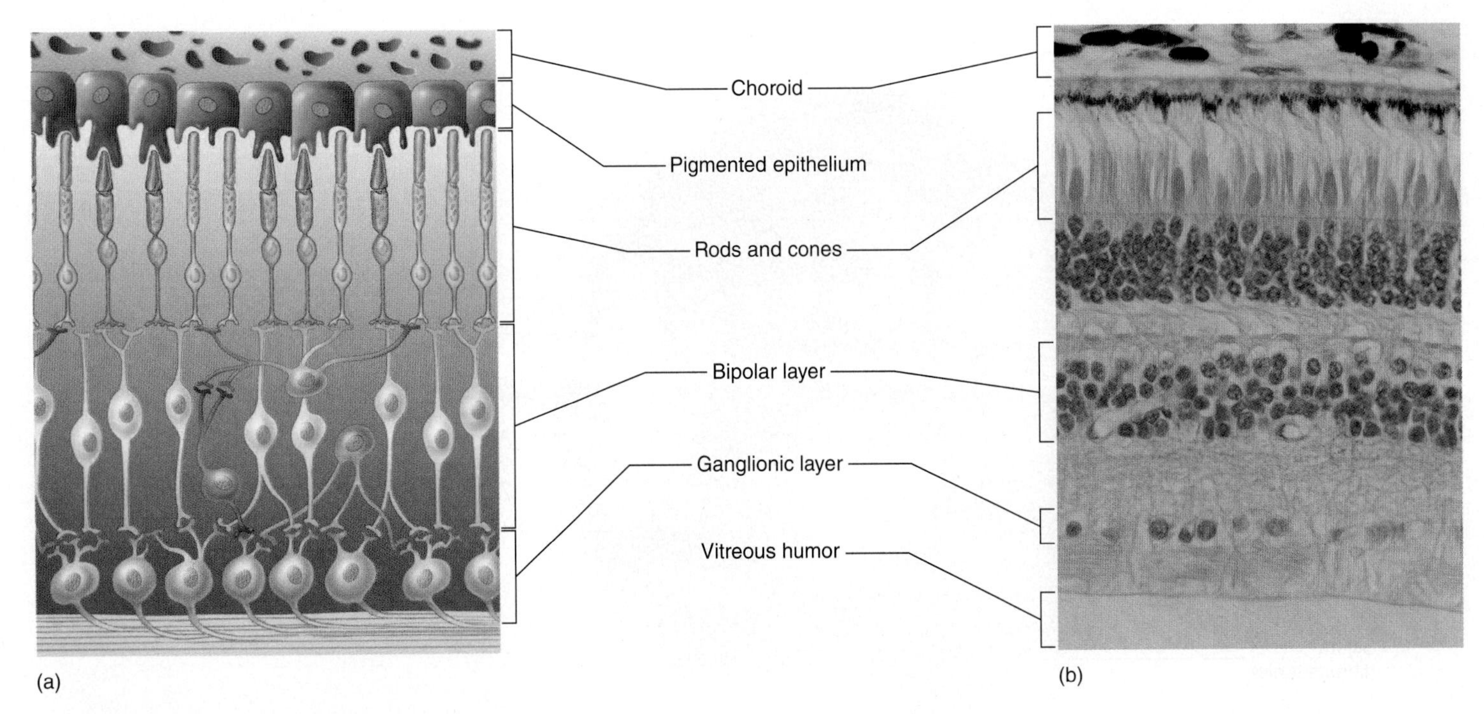

Figure 22.8 Retina (400×)
(a) Diagram; (b) photograph.

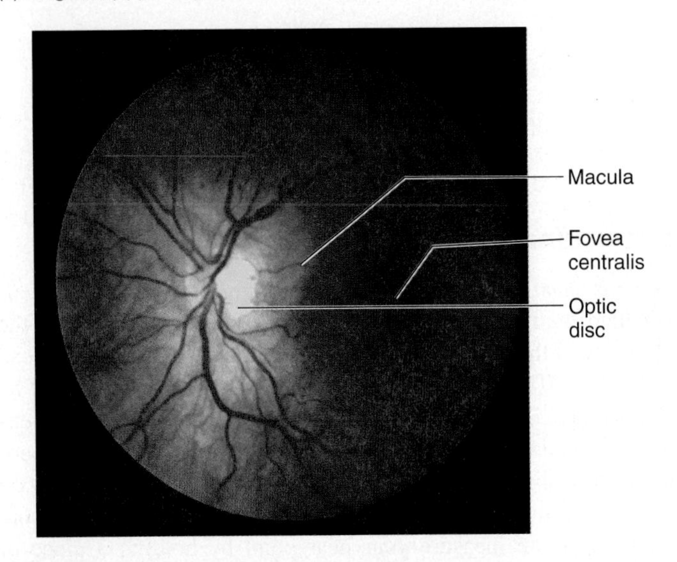

Figure 22.9 Eye, Posterior View

lab partner see it ahead of time, from the back of your lab partner's head, around the side, toward the front. With your lab partner still looking straight ahead, note the approximate angle when your lab partner is able to see the object (the use of rods). Continue to move the object forward slowly until your partner recognizes the color of the object. Note the approximate angle from the tip of your lab partner's nose. You can record the approximate angle by using an apparatus called a Vision Disk or by placing a protractor above your lab partner's head while you make your measurements.

Angle where object is perceived: ____________________

Angle where color is determined: ____________________

What is the difference in the distribution of rods and cones in the eye?

Rod distribution: ____________________

Cone distribution: ____________________

When you are intently focusing on a subject, what cell type (rods or cones) are you using? ____________________

Measurement of Binocular Visual Field

Not all animals have the same visual field. Some prey species (such as deer and sheep) have wide visual fields with little binocular vision. On the other hand, many predators, birds, and arboreal animals have a more limited visual field, yet they have greater **binocular,** or **stereoscopic, vision.** Binocular vision allows arboreal animals to perceive depth, which is vital when judging how far away the next branch is.

(A) 8 You can determine your visual field with the use of a Vision Disk or protractor. If you are using a Vision Disk, follow the instructions enclosed. If not, sit down and have your lab partner stand behind you. Close your right eye and look straight ahead with the left eye. While your right eye is closed, have your lab partner move an object (pen, paper, disk, etc.) from behind your head from the left until you can just see the object while looking straight ahead. Measure the angle from the tip of the nose to where you first saw the object. This is illustrated in figure 22.10 as angle A.

With the same eye directed ahead, have your lab partner continue moving the object until it is out of view (to the right of your nose somewhere). This is illustrated as angle B. Determine the angle from your nose to where the object disappears.

Angle A (where object appears): ____________________

Angle B (where object disappears): ____________________

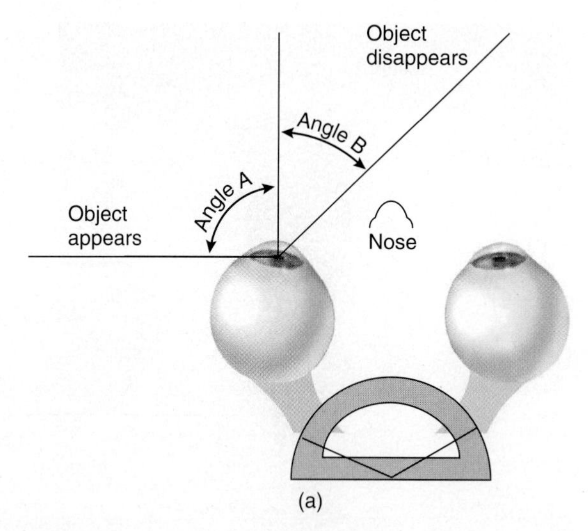

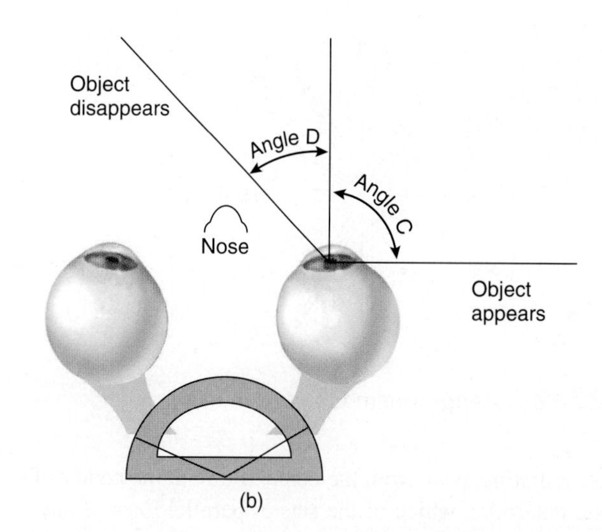

Figure 22.10 Measuring the Visual Field
(a) Left eye; (b) right eye.

Now close your left eye and repeat the exercise on the right side. This is illustrated as angles C and D in figure 22.10*b*.

Angle C: ______________________

Angle D: ______________________

Determine the total visual field for each eye (the sum of angles A and B for the left eye and the sum of angles C and D for the right eye) and then for both eyes (the sum of angles A and C) and the degree of overlap between the eyes (the sum of angles B and D; figure 22.11). Record these data.

Visual field for left eye (A and B): ______________

Visual field for right eye (C and D): ______________

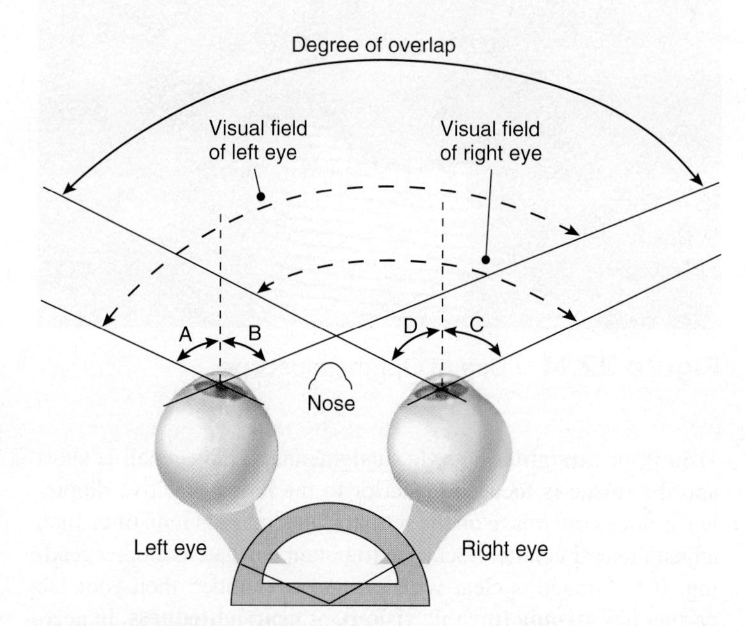

Figure 22.11 Visual Field

Complete visual field for both eyes (A and C): __________

Degree of overlap (B and D): ______________

Measurement of Visual Acuity (Snellen Test)

Ⓐ 9 Face the Snellen eye chart from 20 feet away and have your lab partner stand next to the chart. Cover one eye with a 3- by 5-inch card and have your lab partner point to the largest letter on the chart. Do *not* read the chart with both eyes open. Your lab partner should then progressively move down the chart and note the line that has the smallest print *in which you made no errors*. Your lab partner should record the number at the side of the line. This number refers to your **visual acuity.** Now switch the card to the other eye and repeat the test. Record your results.

Left eye: ______________________

Right eye: ______________________

A vision of 20/20 is considered normal. In 20/20 vision, you can see the same details at 20 feet that most other people can see at the same distance. If your vision is 20/15, then you can see at 20 feet what most people can see at 15 feet. If your vision is 20/100, then you see at 20 feet what most people see at 100 feet.

Astigmatism Tests

If the cornea, or lens, of the human eye were perfectly smooth, the incoming image would strike the retina evenly and there would be no blurry areas. In most people, the cornea, or lens, is not perfectly smooth, and this is known as astigmatism. Various degrees of **astigmatism** occur. In a normal optical exam, the distance correction is made first (to determine visual acuity) and then, with the corrective lenses in place, an astigmatism test is performed.

Ⓐ 10 If you do not normally wear corrective lenses, then cover one eye and examine the astigmatism chart (as in figure 22.12) from 20 feet away. The chart consists of a series of

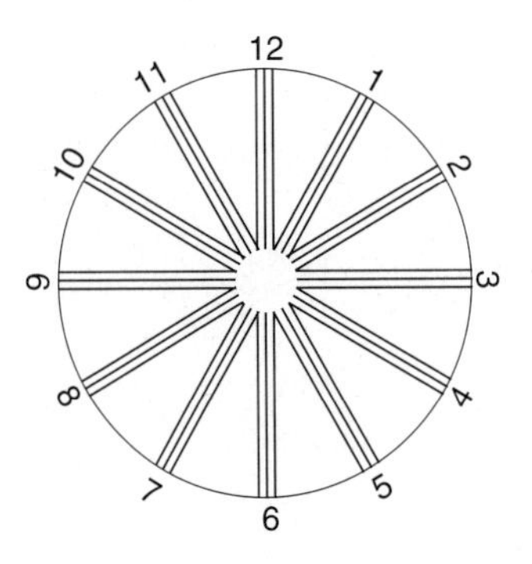

Figure 22.12 Astigmatism Chart

parallel lines radiating away from the center. Stare at the center of the chart and determine which of the sets of parallel lines, if any, appear light or blurry. Have your lab partner note the corresponding number on the chart and record the number.

Astigmatism numbers: ______________________

If you wear corrective lenses, then not only is the condition of nearsightedness or farsightedness corrected but astigmatism is corrected also. To test for astigmatism, move about 12 feet from the chart, hold your glasses slightly away from your face, and then rotate them 90°. If your glasses correct for astigmatism, some of the lines on the chart will be blurry as you rotate your glasses.

Ophthalmoscope

Clinical use of the ophthalmoscope is important not only for the diagnosis of variances in the eyeball but also as a potential indicator of diseases, such as **diabetes mellitus,** that affect the eye.

(A) **11** Familiarize yourself with the parts of the ophthalmoscope (figure 22.13). Locate the **ring** (rheostat control) at the top of the handle. Depress the button and turn the ring until the light comes on. The **head piece** of the ophthalmoscope consists of a **rotating disk** of lenses. You can change the **diopter** (strength) of the lenses by rotating the disk clockwise or counterclockwise. As the numbers get progressively larger, you are looking through more convex lenses. If you rotate the dial in the other direction, the lenses become more concave. At the zero reading, there are no lenses in place.

Observation with the Ophthalmoscope

Caution

Examine the posterior region of the eye only for a short period of time. Extensive use of the ophthalmoscope is irritating to the eye.

Sit facing your lab partner. Select the left eye to examine. Have the ophthalmoscope setting at zero and use your left hand to hold the ophthalmoscope. Use your left eye to look through the scope and move in close to examine your lab partner's eye (figure 22.14). Look into the pupil and examine the back of the eye. If the back appears fuzzy, rotate the dial clockwise (positive diopters) and see if it comes into focus. If a positive number provides a clear view of the eye, then your lab partner has **hyperopia (hypermetropic vision),** or **farsightedness.** In farsightedness, the eyeball is short and the image is focused posterior to the retina. Positive diopter lenses focus the image on the retina. If the image is indistinct, then adjust the dial counterclockwise to obtain a negative diopter reading. If the image is clear with a negative number, then your lab partner has **myopia (myopic vision),** or **nearsightedness.** In nearsightedness, the eyeball is long and lenses with negative diopters

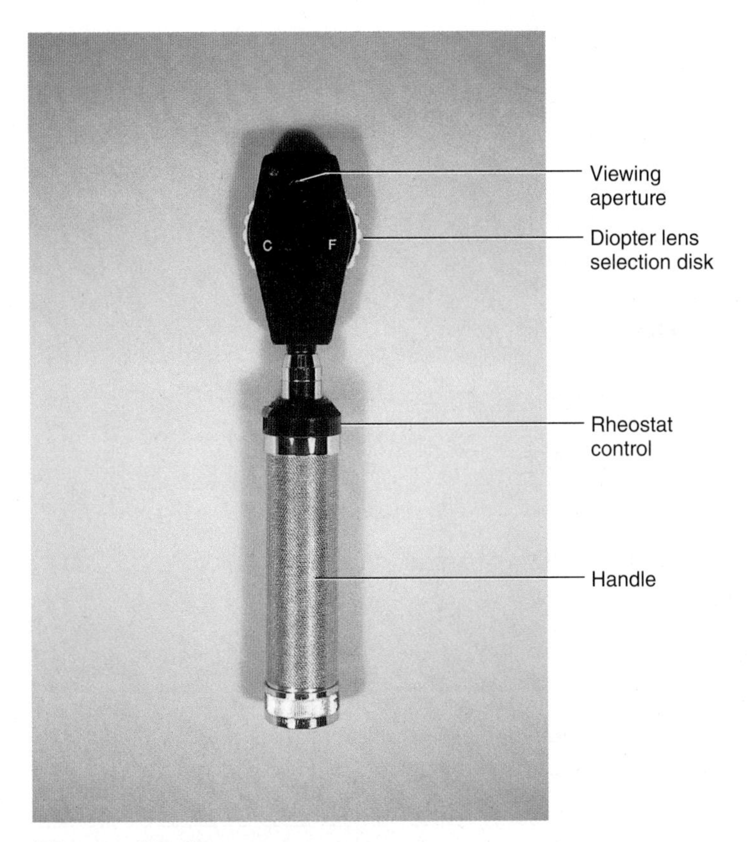

Figure 22.13 Ophthalmoscope

Figure 22.14 Use of Ophthalmoscope

focus the images farther back on the retina. This procedure is based on the assumption that your vision, as the examiner, is normal.

Pupillary Reactions

12 Have your lab partner sit in a dark room for a minute or two. Examine his or her eyes in dim light. Are the pupils dilated or constricted?

Pupil diameter in dim light: ______________________

Now shine a penlight briefly in the right eye. Record what happens to the pupil diameter.

Right pupil diameter: ______________________

As you shine the light into the right eye, what occurs in the left pupil? This is called a **consensual reflex.** Record the effect.

Left pupil diameter: ______________________

Color Blindness

Color vision is dependent on three separate cone cell sensitivities. Cones may be **red, green,** or **blue sensitive.** Changes in the genes on the X chromosome are the most common cause of color blindness. Some people may be unable to see a particular color at all, whereas others may have a reduced ability to see a particular color. **Color blindness** is most common in males and relatively rare in females. The male chromosomal makeup is XY. The Y chromosome does not carry the gene for color vision. If the X chromosome carries a gene for color blindness, then the male exhibits color blindness. On the other hand, if a female carries the gene for color blindness on the X chromosome, the chances are that she will have a normal gene on the other X chromosome. The normal gene is expressed, cone pigments are produced, and the female has normal color vision.

13 You can test for color blindness by matching various samples of colored yarn with a presented test sample or you can use Ishihara color charts. If you use the color charts, flip through the book with a lab partner and record which charts were accurately viewed and which plates, if any, were missed. You can calculate the type and degree of color blindness by following your instructor's directions or those that come with the color chart.

Your degree of color vision: ______________________

Dissection of a Sheep or Cow Eye

14 Rinse a sheep or cow eye in running water and place it on a dissection tray. Obtain a scalpel, forceps, and a blunt probe. Be careful with the sharp instruments and cut *away from* the hand holding the eye. Wear protective gloves while you perform the dissection. Remove the fat and muscles carefully from the eyeball. Examine the eye and find the cornea, extrinsic eye muscles, sclera, and optic nerve. Using a scalpel, make a coronal section of the eye behind the cornea (figure 22.15*b*). Do not squeeze the eye with force or thrust the blade sharply because you may squirt yourself with vitreous humor. Cut through the eye entirely and note the jellylike material in the posterior compartment (figure 22.15*c*). This is the **vitreous humor.** Look at the posterior portion of the eye. Note the beige **retina,** which may have pulled away from the darkened **choroid.** The choroid in humans is very dark, but you may see an iridescent color in your specimen. This is the **tapetum lucidum,** and it improves night vision in some animals. The tapetum lucidum produces the "eye shine" of nocturnal animals. Also examine the tough, white **sclera,** which envelops the choroid.

Now examine the anterior portion of the eye (figure 22.15*d*). Is the **lens** in place? The lens in your specimen probably will not be clear. Normally, the lens is transparent and allows for light penetration.

Note that the lens is held to the **ciliary body** by the **suspensory ligaments** (figure 22.5). These ligaments pull on the lens and alter its shape for distant vision. As suspensory ligaments relax, the lens becomes more round, which allows the eye to focus on objects that are close. Locate the ciliary body at the edge of the suspensory ligaments and the ciliary processes.

Now turn the eye over. Is there any **aqueous humor** left in the anterior and posterior chambers? Make an incision through the **conjunctiva** and **cornea** and find the **anterior chamber** and the **posterior chamber.** Locate the **iris** and the **pupil.** When you are finished, dispose of the specimen in a designated waste container and rinse your dissection tools.

Afterimages

The photosensitive pigment of the eye is called **rhodopsin,** which is composed of light-sensitive **retinal** (a metabolite of vitamin A) and the protein **opsin.** When light strikes the retina, the purple-colored rhodopsin splits into its two component parts and becomes pale (a process known as "bleaching").

15 You can test the time for separation and reassembly of the photopigments by staring at a colored image on a card under a moderately bright light for a few moments. Stare long enough to get an image (about 10 to 20 seconds) and then shut your eyes. Have your lab partner record the time. You should see a colored image against a dark background. This is known as a **positive afterimage,** which is due to the photoreceptors continuously firing. After a few moments, you should see the reverse of the original image (dark against a light background). This is known as a **negative afterimage.** A negative afterimage reflects the bleaching effects of rhodopsin.

Determination of the Blind Spot

The **optic disc** is a region where the retinal nerve fibers exit from the back of the eye and form the **optic nerve.** The mass exit of the nerve fibers leaves a small circle at the back of the eye devoid of photoreceptors. This region, the optic disc, is also known as the **blind spot.**

16 You can locate the blind spot by holding this lab manual at arm's length. Use your right eye (close your left eye) and stare at the following X. It should be in line with the middle of your nose. Slowly move the manual closer to you and stare only at the X. Notice that at a particular distance the dot disappears.

•

You can test for the blind spot in your left eye as well.

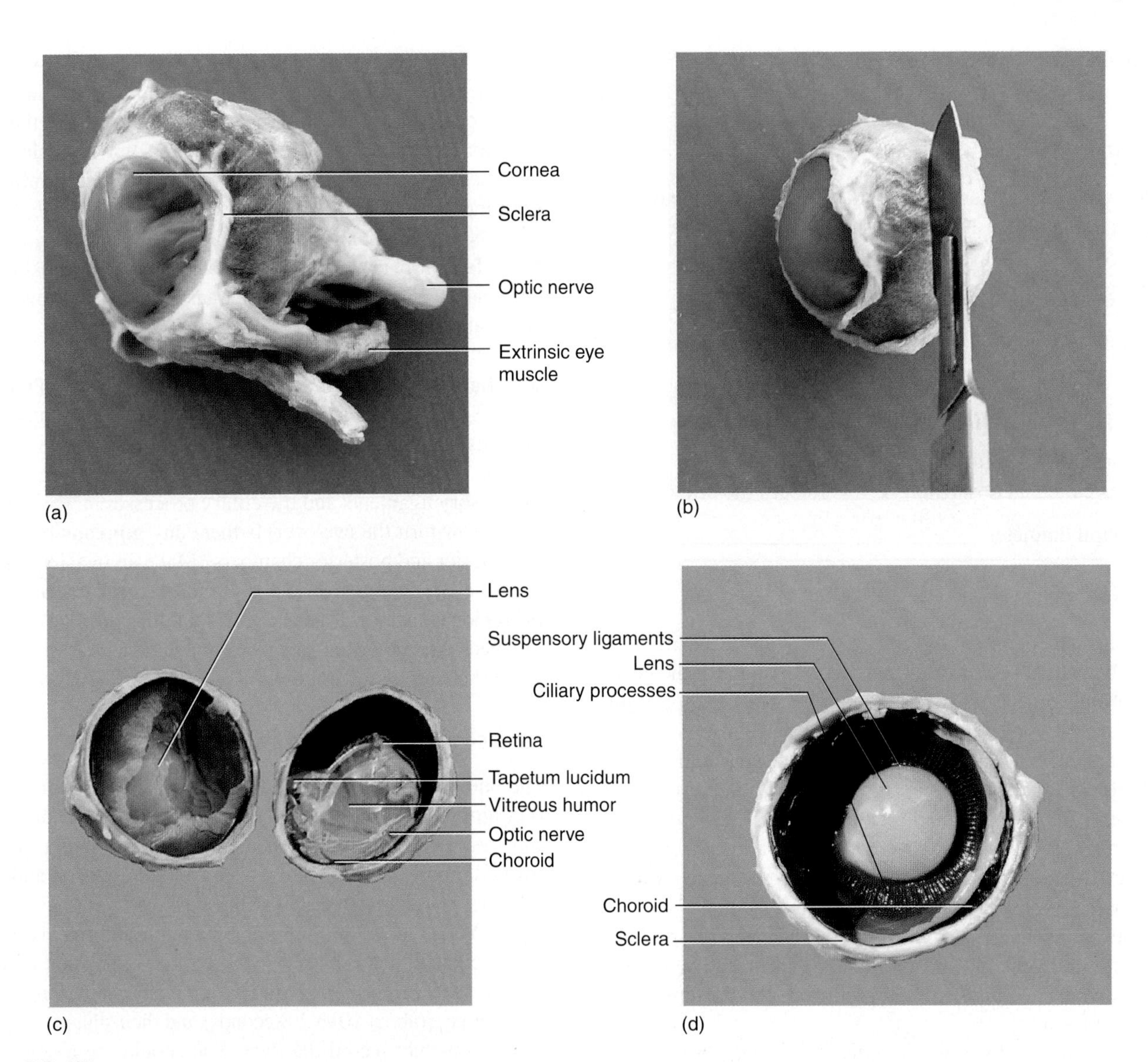

Figure 22.15 Dissection of a Sheep Eye

(a) External features; (b) coronal section; (c) eye with vitreous humor; (d) anterior eye without vitreous humor.

Exercise 22 Review

Eye and Vision

Name: ______________________

Lab time/section: ______________________

Date: ______________________

1. "Eye shine" in nocturnal mammals is different from the "red eye" seen in some flash photographs. Eye shine is the reflection of light off the tapetum lucidum. What visual mechanism might explain red eye? ______________________

2. Label the following illustration using the terms provided.

anterior chamber	choroid
lens	optic nerve
retina	sclera

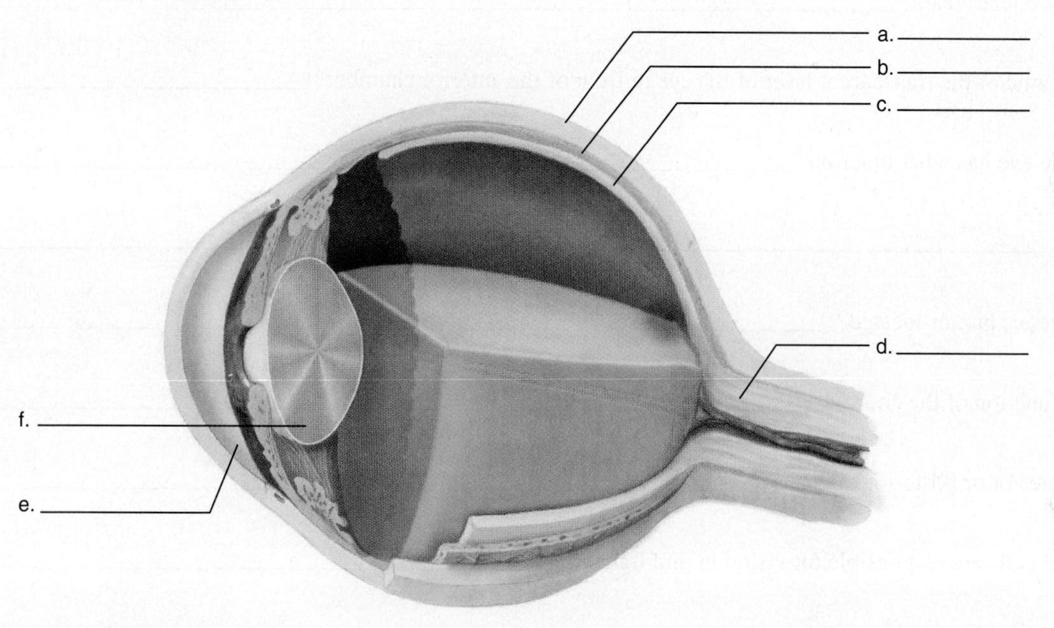

Sagittal section of the eye

3. Since the lens is made of protein, what effect might the preserving fluid used in lab have on the structure of the lens? How might this affect the clarity? ______________________

4. What is the consensual reflex of the pupil? ______________________

5. How does vitreous humor differ from aqueous humor in terms of location and viscosity? ______________________________

____

____

6. What tunic of the eye converts visible light into nerve impulses? ______________________________

7. What nerve takes the impulse of sight to the brain? ______________________________

8. What is another name for the sclera? ______________________________

9. How would you define an extrinsic muscle of the eye? ______________________________

____

10. What gland produces tears? ______________________________

11. What is the name of the transparent layer of the eye in front of the anterior chamber? ______________________________

12. The iris of the eye has what function? ______________________________

____

13. Where is vitreous humor located? ______________________________

14. What is the function of the choroid? ______________________________

15. Is the lens anterior or posterior to the iris? ______________________________

16. Which retinal cells are responsible for vision in dim light? ______________________________

17. How would you define the near point of the eye? ______________________________

____

18. What do the numbers 20/100 mean for visual acuity? ______________________________

____

____

19. What is astigmatism? ______________________________

____

20. In what area of the eye is the blind spot located? ______________________________

Exercise 23

Ear, Hearing, and Balance

Introduction

The ear is a complex sense organ that performs two major functions, hearing and balance. It consists of three regions: an external ear, a middle ear, and an inner ear. The structure and function of the ear are covered in the Seeley text in chapter 15, "The Special Senses."

Hearing is considered **mechanoreception,** because the ear receives mechanical vibrations (sound waves) and translates them into nerve impulses. This process begins with the vibrations reaching the external ear and ends up being interpreted as sound in the temporal lobe of the brain. Sound waves can be close together; thus, many of them reach the ear in a particular time period. These are sound waves of high frequency and they produce sound of high pitch. The height of the wave determines the loudness, which is measured in units called **decibels.** Sound waves that are longer have lower frequency and produce low pitch. Balance, on the other hand, involves receptors in the inner ear, visual cues, and **proprioception** (the perception of gravity or of forces applied to a structure). Two types of balance are sensed by the inner ear: **static balance** and **dynamic,** or **kinetic, balance.** In static balance, a person can determine his or her nonmoving position (such as standing upright or lying down). In dynamic balance, motion is detected. Sudden acceleration, abrupt turning, and spinning are examples of dynamic balance.

Objectives

At the end of this exercise, you should be able to

1. explain how mechanical sound vibrations are translated into nerve impulses;
2. list the structures of the external, middle, and inner ear;
3. describe the structure of the cochlea;
4. perform conduction deafness tests, such as the Rinne and Weber tests;
5. compare dynamic and static balance and the structures involved in their perception.

Materials

Models and charts of the ear
Microscope
Prepared slides of the cochlea
Tuning fork (256 Hz)
Rubber reflex hammer
Audiometer
Model of ear ossicles
Meter stick
Ticking stopwatch
Bright desk lamp

Procedure

Anatomy of the Ear

(A) **1** Examine a model in lab for the structures of the ear. The ear can be divided into three regions: the external, middle, and inner ear. The **external ear** consists of auditory structures superficial to the **tympanic membrane** (eardrum). The **middle ear** contains the tympanic cavity, ear ossicles, and auditory tube, and the **inner ear** consists of the cochlea, vestibule, and semicircular canals (figure 23.1).

Structure of the External Ear

The external ear consists of the **auricle,** or **pinna,** which can further be subdivided into the **helix** and the **lobule,** or earlobe. The helix is composed of stratified squamous epithelium overlying elastic cartilage. This cartilage allows the ears to bend significantly. Deep to the pinna, the external ear forms the **external auditory canal,** which penetrates into the temporal bone. The **tympanic membrane** is the border between the external ear and the middle ear. It is composed of connective tissue covered by epithelial tissue. The membrane is sensitive to sound and vibrates as sound is funneled down the external auditory canal.

Structure of the Middle Ear

The middle ear consists of a main cavity known as the **tympanic cavity;** three small bones, or **ossicles;** and the **auditory, eustachian,** or **pharyngotympanic** (fă-ring′-go-timpăn′-ĭk) **tube** (figure 23.1).

The ossicle that is attached to the tympanic membrane is the **malleus** (*malleus* = hammer). As the membrane vibrates, the malleus rocks back and forth, carrying the sound waves. The malleus is attached to the **incus** (*incus* = anvil), which is attached to the **stapes** (*stapes* = stirrup; figure 23.2).

The ear ossicles magnify sound about 20 times. Sound consists of pressure waves. As these waves strike the tympanic membrane, it vibrates. This vibration is conducted by the ossicles to the oval window. The process of moving from a large-diameter structure (tympanic membrane) to a smaller-diameter structure (oval window) magnifies the sound.

Examine the ossicles in figure 23.2 and on models in the lab. In addition to the ossicles, the auditory tube is in the middle ear. This tube connects the middle ear to the nasopharynx and provides for the equalization of pressure between the middle ear and the

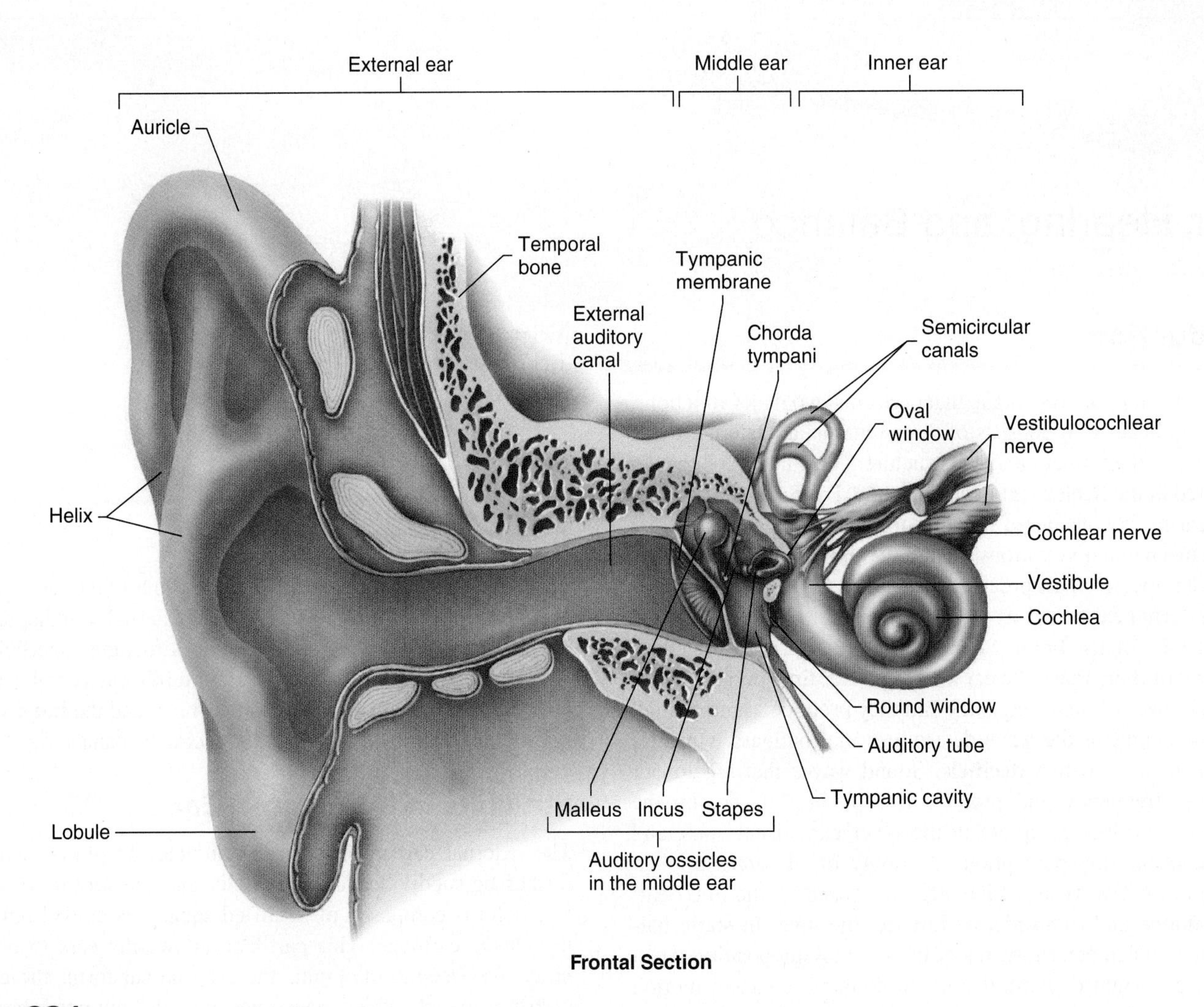

Figure 23.1 Anatomy of the Ear

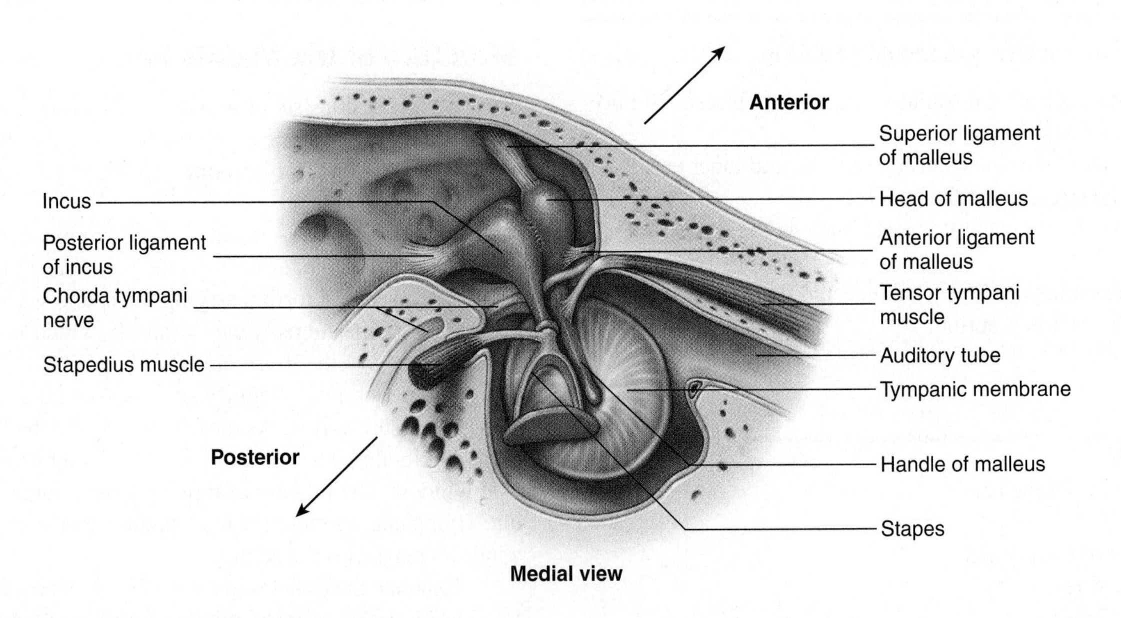

Figure 23.2 Middle Ear, Medial View

This view is as if you were looking from the inside of the head (deep) toward the outside (superficial).

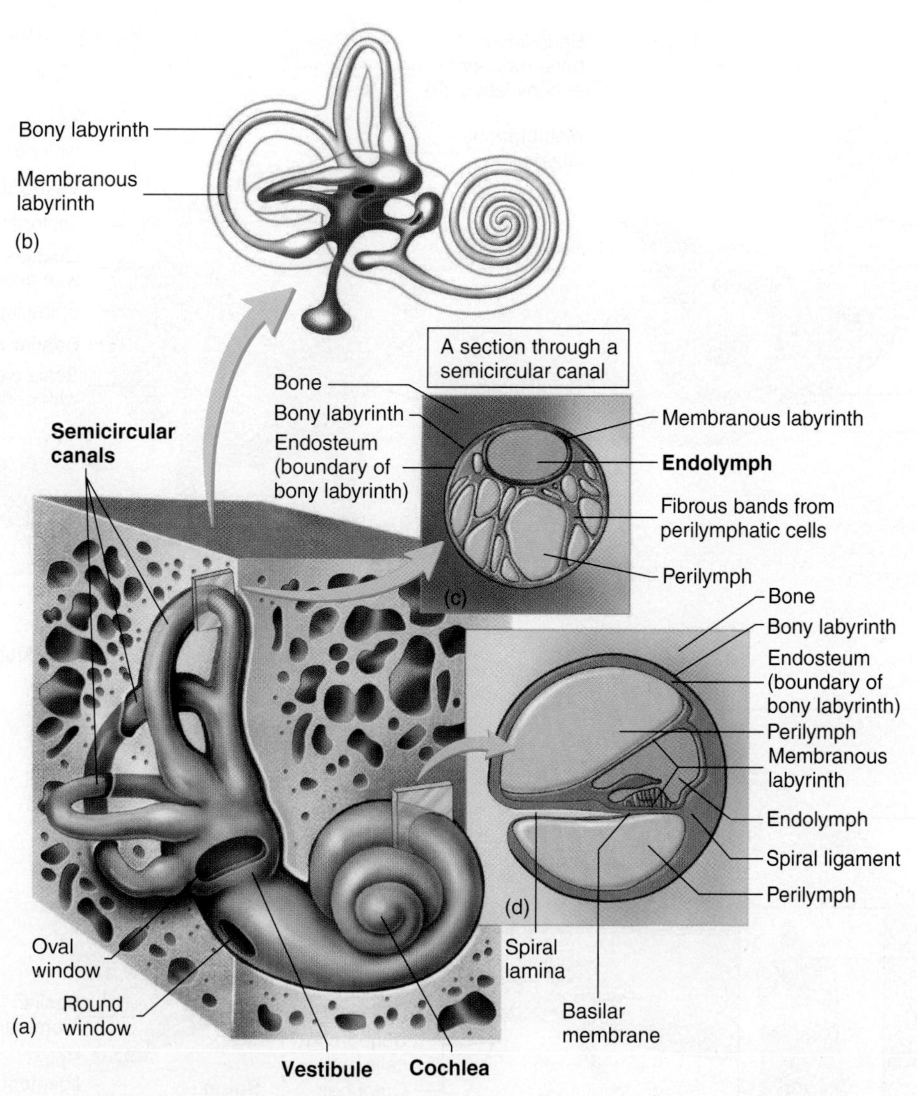

Figure 23.3 Anatomy of the Inner Ear, Labyrinths

(a) Inner ear situated in bone, (b) labyrinths as if isolated from bone, (c) section through semicircular canal, (d) section of cochlea.

external environment when pressure changes occur (such as during changes of elevation). The auditory tube can be a conduit for microorganisms that travel from the nasopharynx to the middle ear and lead to middle-ear infections, particularly in young children.

Structure of the Inner Ear

Ⓐ 2 Examine models of the inner ear in lab. The inner ear is encased in two labyrinthine structures and filled with two separate fluids. The outermost structure is the **bony labyrinth.** Inside the bony labyrinth is **perilymph,** a clear fluid that is external to the **membranous labyrinth.** The fluid enclosed by the membranous labyrinth is the **endolymph,** which is important in both hearing and balance.

The inner ear is a complex structure composed of three separate regions: the cochlea, the vestibule, and the semicircular canals (figures 23.1 and 23.3).

Cochlea

The **cochlea** (figures 23.1 and 23.4) is involved in hearing. As the sound waves travel down the external auditory meatus, they cause the tympanic membrane to vibrate. This vibration rocks the ear ossicles, which are connected to the inner ear. As the stapes vibrates, it moves back and forth in the **oval window,** and this causes fluid to move back and forth in the cochlea. The cochlea also has a **round window,** which allows the vibration from the ossicles to move fluid back and forth. Without the round window, the fluid that transmits the sound would not move as easily and hearing would be greatly reduced. Sound waves are measured by their wavelengths. These wavelengths are measured in cycles per second, also known as hertz (Hz).

High-pitched sounds with vibrations of short wavelength, up to 20,000 Hz, stimulate the region of the cochlea closest to the middle ear. Low-pitched sounds with longer wavelengths, down to 20 Hz, stimulate the region of the cochlea farther from the middle ear. In this way, the cochlea can perceive sounds of varying wavelengths at the same time.

Microscopic Section of Cochlea

Ⓐ 3 If you examine a microscopic section of the cochlea in cross section, you will see a number of chambers. The chambers are clustered in threes. Find the **scala vestibuli, scala media (cochlear duct),** and **scala tympani** on the microscope slide.

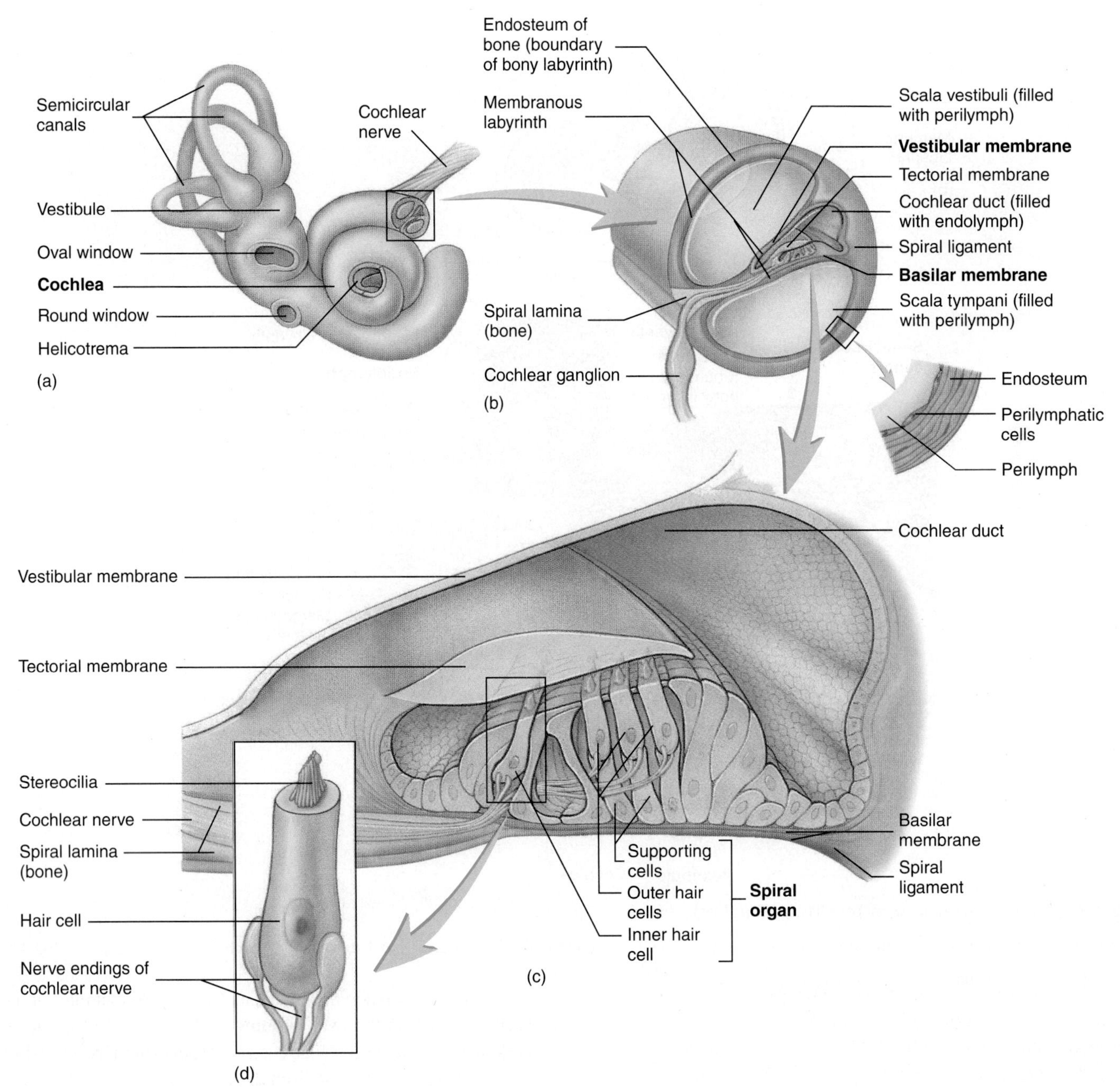

Figure 23.4 Anatomy of the Inner Ear

(a) Overview; (b) details of cochlear chambers; (c) spiral organ; (d) enlarged hair cell.

Compare them with figures 23.4 and 23.5. Note the **spiral organ,** or the **organ of Corti,** which is in the area between the **vestibular membrane** and the **basilar membrane.** The spiral organ is seen here in cross section, but remember that it runs the length of the cochlea. The spiral organ is sensitive to sound waves. As a particular region of the spiral organ is stimulated, the basilar membrane and **tectorial membrane** vibrate independently of one another. The movement of the basilar membrane tugs on the cilia that attach the hair cells to the tectorial membrane. The hair cells send impulses to the cochlear branch of the **vestibulocochlear nerve.** These impulses travel to the **auditory cortex** of the **temporal lobe,** where they are interpreted as sound (figure 23.6).

Vestibule

Another part of the inner ear is the vestibule. The **vestibule** consists of the **static labyrinth** which includes the **utricle** and the **saccule** (figure 23.7). These two chambers are involved in the interpretation of static balance and acceleration. The utricle and saccule have regions known as **maculae,** which consist of **cilia** grouped together with an overlying gelatinous mass and calcium carbonate stones, called **otoliths.** These can be seen in figure 23.7. As the head is accelerated or tipped by gravity, the otoliths cause the cilia to bend, indicating that the position of the head has changed. Static balance is perceived not only from the vestibule but from visual cues as well. When the visual cues and the vestibular cues are not synchronized, then a sense of imbalance or nausea can occur.

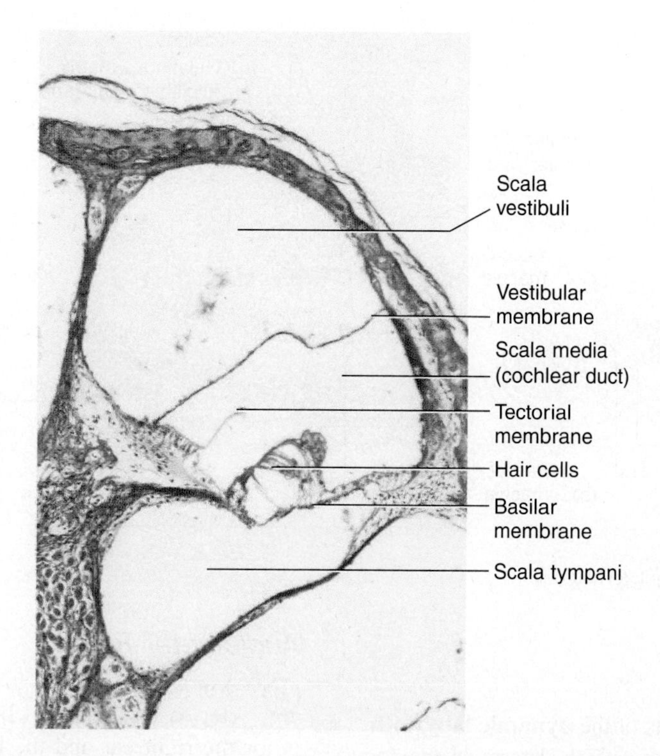

Figure 23.5 Photomicrograph of Cross Section of the Cochlea (100×)

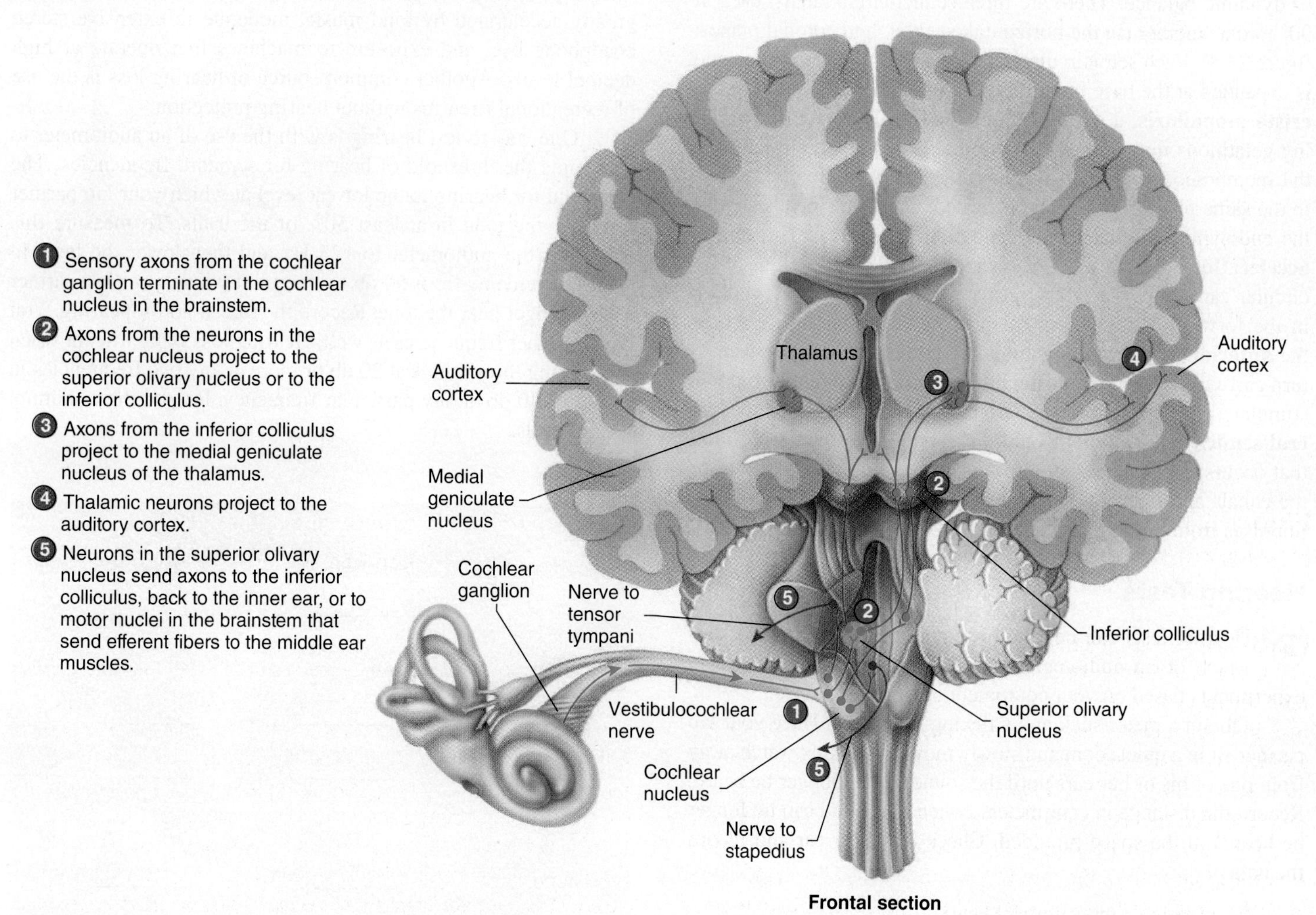

Figure 23.6 Interpretive Pathway of Hearing

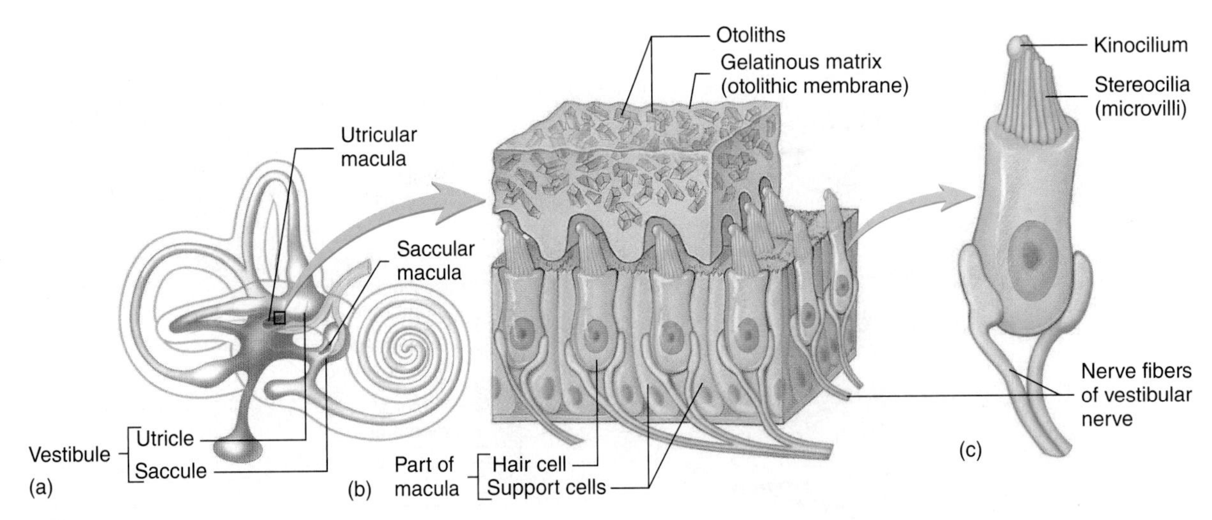

Figure 23.7 Vestibule

(a) Inner ear; (b) macula when head is upright; (c) hair cell.

Semicircular Canals

The third part of the inner ear consists of the **dynamic labyrinth,** consisting of the **semicircular canals,** which determine moving or dynamic balance. There are three semicircular canals, each at 90° to one another (in the horizontal, sagittal, and coronal planes; figure 23.8). Each semicircular canal is filled with endolymph and is expanded at the base into an **ampulla.** Inside each ampulla is a **crista ampullaris,** a cluster of **hair cells** (cilia) with an overlying gelatinous mass called the **cupula.** The endolymph that is in the membranous labyrinth has inertia; that is, it tends to remain in the same place. As the head is turned, the cupula bends against the endolymph, the hair cells bend, and **angular,** or **rotational, acceleration** is perceived, as shown in figure 23.9. The three semicircular canals are placed at right angles to each other. If motion in the forward plane occurs (such as by doing back flips), then the **anterior semicircular canals** are stimulated. If you were to turn cartwheels, then the **posterior semicircular canals** would be stimulated. If you were to spin around on your heels, then the **lateral semicircular canals** would pick up the information. Motion that occurs in between these areas is picked up by two or more of the canals and interpreted as movement due to a combination of impulses from the semicircular canals.

Hearing Tests

(A) 4 Hearing tests can be administered by either a ticking stopwatch or an audiometer. Select one of the two following experiments based on your equipment in lab.

Obtain a meter stick and a ticking stopwatch. Have your lab partner sit in a quiet room and slowly move the ticking watch away from one of his or her ears until the sound can no longer be heard. Record the distance in centimeters (when the sound can no longer be heard) in the space provided. Check the other ear and record the data.

Maximum distance for right ear: ____________________

Maximum distance for left ear: ____________________

Audiometer Test

(A) 5 Ask your instructor to demonstrate how to use an audiometer to test hearing. In a typical audiometer, the red side is for the right ear and the blue side is for the left ear. Hearing normally decreases somewhat with age, although hearing loss is greatly accelerated by loud music, moderate to extensive stereo headphone use, and exposure to machines that operate at high decibel levels. Another common source of hearing loss is the use of recreational firearms without hearing protection.

One way to test hearing is with the use of an audiometer to determine the threshold of hearing for standard frequencies. The threshold for hearing is the lowest level at which your lab partner can hear the tone in at least 50% of the trials. To measure this value, set the audiometer to 125 Hz and then lower the level in 10 db increments from 50 db to the point where your lab partner can no longer hear the tone. Record the threshold for hearing. You can test other frequencies as well. Hearing loss is significant when the hearing threshold is at 20 db or more at any two frequencies in an ear or 30 db at any particular frequency. Record the minimum sound levels.

Threshold (in Decibels) for Each Ear		
	Left Ear	Right Ear
125:	________	________
250:	________	________
500:	________	________
1000:	________	________
2000:	________	________
4000:	________	________
8000:	________	________

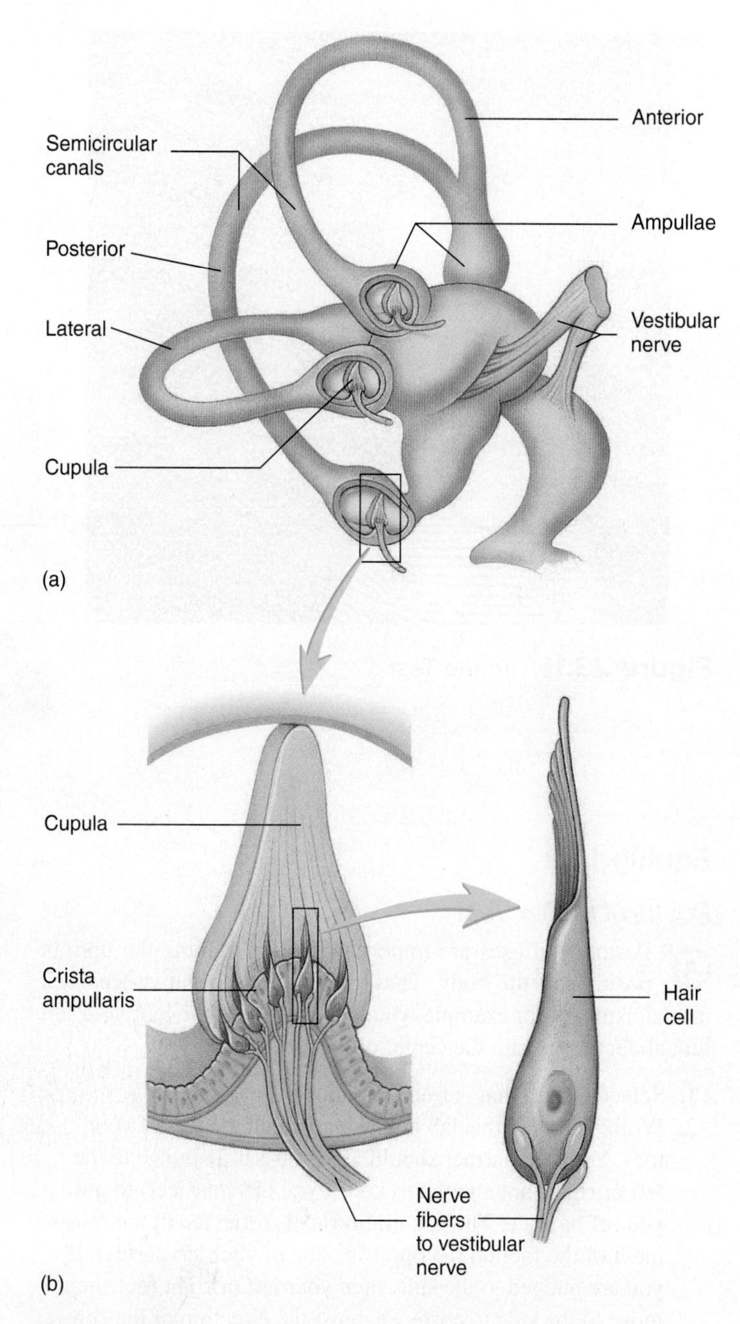

Figure 23.8 Semicircular Canals

(a) Overview; (b) details of canals, crista ampullaris, and cupula.

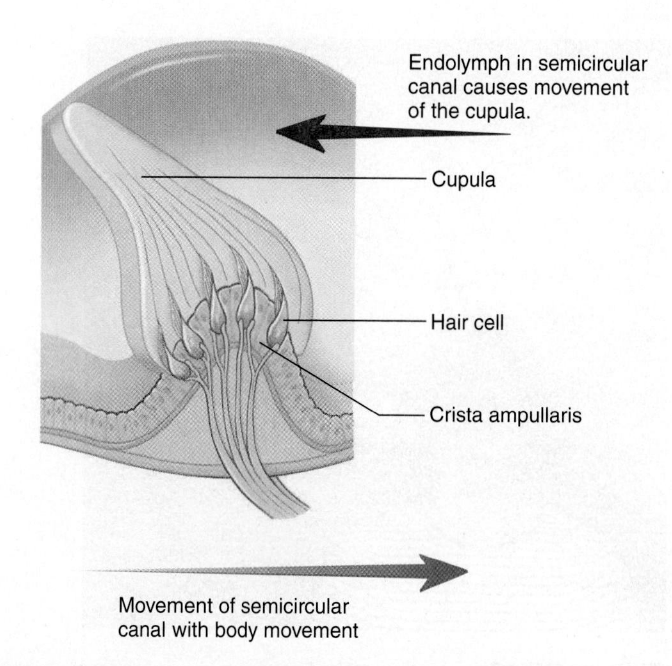

Figure 23.9 Movement of Semicircular Canals

There are two kinds of hearing loss, **conduction deafness** and **sensorineural deafness.** Conduction deafness involves the outer or middle ear where the mechanical vibrations in the external environment do not reach the cochlea. This may be due to a number of causes, including a perforated tympanic membrane, a blocked external auditory meatus, or a middle-ear infection. Sensorineural hearing loss takes place in the cochlea, in the vestibulocochlear nerve, or in the auditory centers of the temporal lobe of the brain. There are a couple of tuning fork tests that can indicate either conduction or senorineural hearing loss. These are the Weber test and Rinne test.

Weber Test

(A)[6] Strike a tuning fork (at 256 Hz, or middle C) with a patellar hammer or strike the fork on your knee or elbow. Hold the handle of the tuning fork to the forehead of your lab partner (figure 23.10) while the tines (the forked portion) vibrate. If there is asymmetrical hearing loss, the sound will be louder in one ear. If your lab partner has conductive hearing loss, the sound will be louder in the ear with the hearing loss. You can demonstrate this by having your lab partner plug one ear and do the test. The sound will be louder in the ear that is plugged. Record the results of your test in the following space.

Sound perception (right/left or equal in both ears): ____________

____

You can interpret your test in the following way.

1. Normal—sound is the same in both ears
2. Conductive hearing loss—sound is louder in the ear with conductive loss
3. Sensorineural hearing loss—sound is louder in the ear without hearing loss

Rinne Test

(A)[7] Strike the tuning fork as you did in the Weber test and place the handle of the fork on your lab partner's mastoid process for a few seconds (figure 23.11). Once the sound diminishes significantly, hold the fork a few centimeters from the external auditory canal. Can your lab partner hear the sound? If hearing is normal, your lab partner should hear the sound through the external auditory canal (air conduction) louder than when the tuning fork was held to the mastoid process (bone conduction). If bone conduction is better than air conduction, then there is conductive

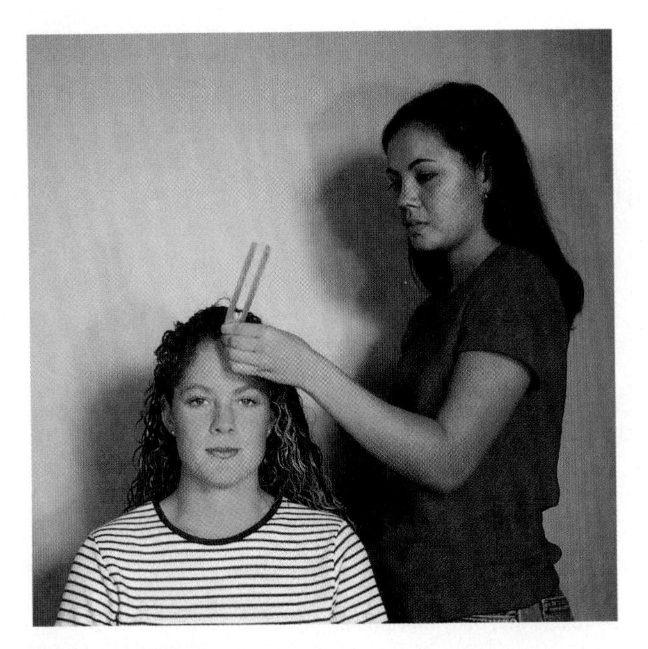

Figure 23.10 Weber Test

Figure 23.11 Rinne Test

hearing loss. This is a negative Rinne test. Do the test for both ears and record your results.

Right ear: ______________________

Left ear: ______________________

Sound Location

(A) 8 Have your lab partner sit with eyes closed. Strike the tuning fork with a rubber reflex hammer above his or her head. Have your lab partner describe to you where the sound is located. Strike the tuning fork behind the head, to each side, in front, and below the chin of your lab partner. Strike the tuning fork at the location to be tested. Do not strike the tuning fork and move it to the spot for location of sound. Record the results.

Where Sound Was Struck	Where Sound Was Perceived
Above head	______________
Behind head	______________
Right side	______________
Left side	______________
In front of head	______________
Below chin	______________

Equilibrium

Postural Reflex Test

(A) 9 Postural reflexes are important for maintaining the upright position of the body. These reflexes are negative-feedback mechanisms. If, for example, you lean slightly to the left, your left foot abducts to regain the center of balance.

1. Select an area that is free from any obstacles.
2. While you read this lab manual, stand on the tips of your toes. Your lab partner should give you a little nudge to the left or right (not enough to knock you off your feet) to push you off balance. The postural reflex is reflected in the movement of the foot on the opposite side of your lab partner. If you are nudged to the side, then your left or right foot should move to the side to correct against the direction of the contact. Did your postural reflex work?
3. Record your result in the following space.

Postural reflex: ______________________

Barany's Test

(A) 10 Barany's test examines visual responses to changes in dynamic balance. As the head turns, one of the reflexes that occurs is movement of the eyes in the opposite direction of the rotation. Nerve impulses from the semicircular canals innervate the eye muscles and cause the eye movement. When the head rotates in one direction, the eyes move in the opposite direction, so that there is enough time for a visual image to be fixed. The volunteer for this test should not be subject to dizziness or nausea.

1. Place the subject in a swivel chair with four or five students close by and equally spaced around the chair in case the

subject loses balance and begins to fall. The student chosen for this exercise should grab onto the chair firmly so as not to fall off. The subject should tilt his or her head forward about 30°, which will place the lateral semicircular duct horizontally for maximum stimulation.

2. One member of the group should spin the chair around about 10 revolutions while the subject keeps his or her eyes open.
3. Stop the chair and have the subject look forward. The twitching of the eyes is called **nystagmus** and is due to the stimulation of the endolymph flowing in the semicircular ducts. When the chair is stopped, the fluid in the endolymph will have overcome the inertia and will continue to flow in the ducts. Did the eyes move in the direction of the rotation or in the opposite direction? Record your results.

Direction of chair rotation relative to subject: ___________

Direction of eye rotation: ___________________________

Romberg Test

(A) 11 The Romberg test involves testing the static balance function of the body.

1. Place the subject in front of a chalkboard or any surface on which you can see the shadow of the subject. Do not lean against the chalkboard or against the wall.
2. Have the subject stand in that direction for 1 minute and determine if there are any exaggerated movements to the left or right. The feet should be close together.
3. Record your results in the following space.

Amount of lateral sway: ___________________________

You should then have your lab partner close his or her eyes and repeat the test. Is the swaying motion greater or less than before? Record your results in the following space.

Amount of lateral sway with eyes closed: ______________

A positive Romberg's test (indicated by significant swaying) may indicate inner ear problems or nerve, brain, or proprioreceptor dysfunctions in muscles or joints.

NOTES

Exercise 23 Review

Ear, Hearing, and Balance

Name: ______________________

Lab time/section: ______________________

Date: ______________________

1. What are the three general regions of the ear? ______________________

2. The pinna of the ear consists of what two main parts? ______________________

____

3. The ear is what kind of receptor? ______________________

4. The ear performs two major sensory functions. What are they? ______________________

____

5. What structure separates the external ear from the middle ear? ______________________

6. Name the three ear ossicles. ______________________

____

7. From the following choices, select the function of the cochlea.

   a. static balance b. taste

   c. hearing d. dynamic balance

8. What area is found between the scala vestibuli and the scala tympani? ______________________

9. What is the name of the nerve that takes information about balance and hearing to the brain? ______________________

10. What units are used to measure sound energy? ______________________

11. What part of the inner ear is involved in perceiving static balance? ______________________

12. Name the parts of the ear that might be impaired if a person demonstrates conduction deafness. ______________________

____

13. What two diagnostic tests are used to determine conduction deafness? ______________________

____

14. What is the name of the tube that runs from the auricle to the tympanic membrane? ______________________

15. The auditory tube connects what two cavities? ____

____

16. What tube is responsible for the equalization of pressure when you change elevation? ____________________

17. What is the name of the space that encloses the ear ossicles? ____________________

18. Place the ear ossicles in sequence from the tympanic membrane to the oval window. ____________________

____

19. Name all the parts of the inner ear. ____

____

20. Background noise affects hearing tests. In the ticking watch test or audiometer test, what kind of result, in terms of auditory sensitivity, would you have recorded if moderate background noise were present? ____________________

____

21. In the Weber test, the ear that perceives the sound as being louder is the deaf ear. Why is this the case? ____________________

____

____

22. Label the following illustration of the ear using the terms provided.

auditory tube	cochlea	ear ossicles
external auditory canal	semicircular canals	tympanic membrane

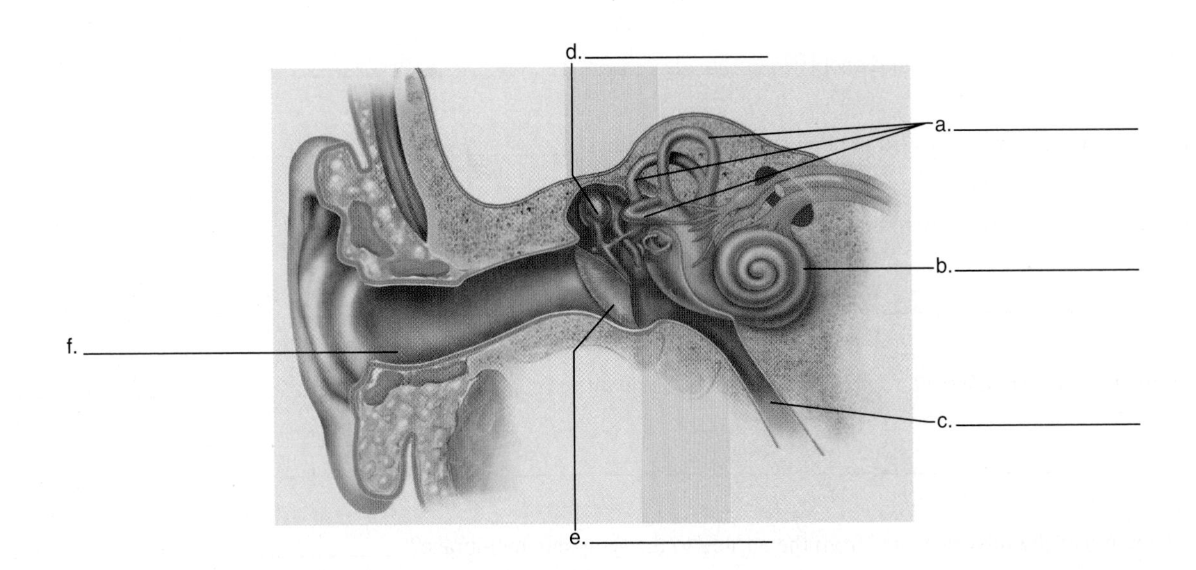

23. Label the following illustration of the cross section of the cochlea using the terms provided.

hair cells	scala media (cochlear duct)	scala tympani
scala vestibuli	tectorial membrane	vestibular membrane

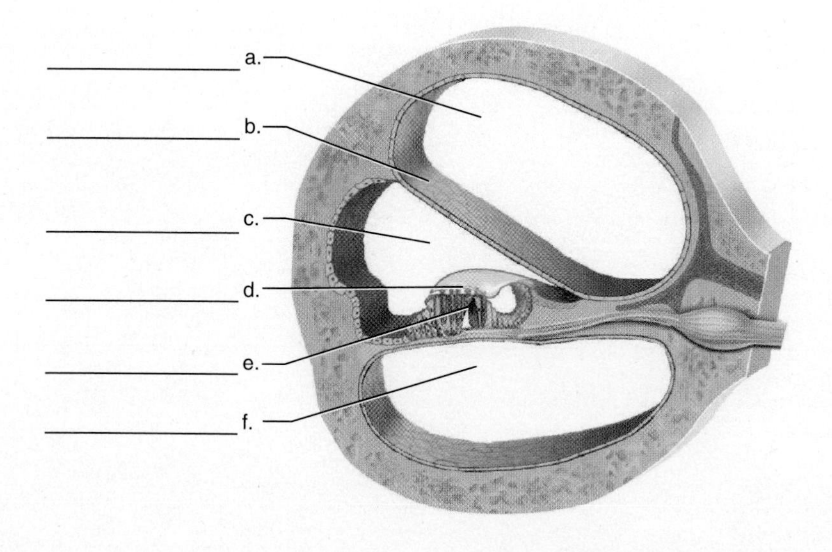

NOTES

Exercise 24

Endocrine System

Introduction

The **endocrine system** consists of glands that produce chemical messengers called **hormones,** which are picked up by blood capillaries. This type of release is called a "ductless secretion," and it is characteristic of the endocrine system. Glands that secrete material in tubules or ducts (such as sweat and salivary glands) are known as **exocrine glands.** The secretions of the endocrine glands enter the interstitial fluid and then travel by blood vessels, which act as highways to carry the hormones throughout the body. Areas that are receptive to hormones are called **target areas;** these may be organs or tissues. Hormones have many effects. Some of these are growth, changes and development, maturation, metabolism, sexual development, regulation of the sexual cycle, and homeostasis. Many organs of the body, such as the stomach, heart, and kidney, produce hormones and thus have endocrine functions. This exercise focuses on the glands that have a major endocrine component. Hormones and the endocrine glands that produce them are covered in the Seeley text in chapter 17, "Functional Organization of the Endocrine System," and chapter 18, "Endocrine Glands."

Objectives

At the end of this exercise, you should be able to

1. discuss how the secretions of the endocrine glands differ from those of the exocrine glands;
2. list the major endocrine organs of the human body;
3. identify endocrine organs in histological slides;
4. name the hormones produced by the endocrine organs;
5. describe the detection of hormones by monoclonal antibodies.

Materials

Models and charts of endocrine glands
Model or chart of midsagittal section of head
Microscopes
Microscope slides
- Thyroid
- Pituitary
- Adrenal gland
- Pancreas
- Testis
- Ovary

Urine samples
Ovulation test kit
Disposable gloves (latex or plastic) and biohazard container

Procedure

Anatomy of the Major Endocrine Organs

Ⓐ1 Locate the major endocrine glands in charts or models in lab and compare them with figure 24.1, including the hypothalamus, pineal gland, pituitary gland, thyroid, parathyroids, thymus, pancreas, adrenals, and gonads (testes or ovaries). Once you have noted their location, you can proceed with a more detailed study.

Hypothalamus

Ⓐ2 Locate the hypothalamus on models or charts in lab and in figure 24.2. The **hypothalamus** has a major role in controlling endocrine functions. The hypothalamus secretes both stimulating and inhibitory hormones that either cause the release of hormones from their target areas or prevent the release of hormones from these areas.

Pineal Gland

Ⓐ3 Locate the pineal gland in a model or chart of a midsagittal section of the head and compare it with figure 24.2. The **pineal gland (pineal body)** develops from the diencephalon of the

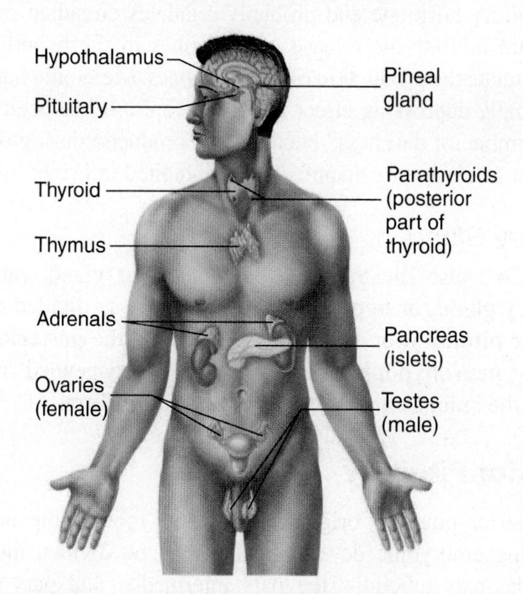

Figure 24.1 Major Endocrine Glands

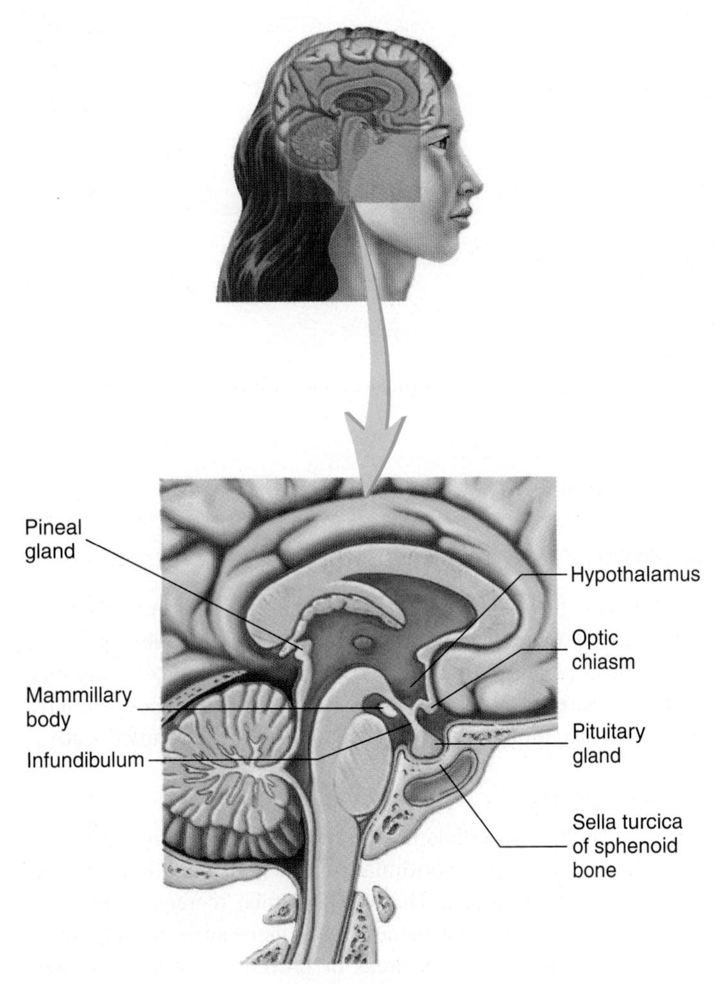

Figure 24.2 Endocrine Glands of the Head (Midsagittal Section)

brain. The gland is so named because it resembles a pine nut. One of the hormones the pineal gland secretes is **melatonin,** which is an inhibitory hormone and probably regulates circadian rhythms. Melatonin inhibits the release of hormones in the hypothalamus that stimulate the production of sex hormones. Melatonin has a psychologically depressing effect in some people. It has been named the "hormone of darkness" because it is produced during times of low light levels and its manufacture is inhibited in bright light.

Pituitary Gland

Figure 24.2 also illustrates another endocrine gland, called the **pituitary gland,** or **hypophysis.** The pituitary is divided into the **anterior pituitary,** or **adenohypophysis,** and the **posterior pituitary,** or **neurohypophysis.** The pituitary is suspended from the base of the brain by a stalk called the **infundibulum.**

Anterior Pituitary

The anterior pituitary originates from the roof of the oral cavity during embryonic development. It can be divided into three parts: the pars tuberalis, the pars intermedia, and pars distalis. The cells of the anterior pituitary have the same embryonic origin, yet they have differentiated into several types of specialized cells.

(A) **4** The result of this development can be seen if you examine a prepared slide of the pituitary gland. In the histological section, note that the adenohypophysis is composed of **epithelial cells** (figure 24.3). These cells produce a number of hormones that have broad effects throughout the body. Draw a section of the anterior pituitary in the following space.

Some of the hormones produced in the anterior pituitary and their functions are as follows:

- **Thyroid-stimulating hormone (TSH),** or **thyrotropin,** stimulates the thyroid gland to produce thyroid hormones.
- **Growth hormone (GH),** or **somatotropin,** promotes the growth of most of the cells and tissues of the body.
- **Prolactin,** or **luteotropin (LTH),** stimulates mammary glands to begin the production of milk.
- **Gonadotropins** are hormones that stimulate the growth and control the function of the gonads (ovaries and testes). There are two gonadotropins, **follicle-stimulating hormone (FSH)** and **luteinizing hormone (LH),** and these stimulate the production of sex cells in the ovaries and testes and regulate the production of hormones, estrogen and progesterone in the ovaries and testosterone in the testes.
- **Adrenocorticotropin (ACTH)** controls hormone production in the adrenal cortex.

Posterior Pituitary

(A) **5** Compare the tissue of the neurohypophysis in a prepared slide with figure 24.3. The tissue of the posterior pituitary is very different from that of the adenohypophysis. The neurohypophysis is composed of **nervous tissue** that originated from the base of the brain. Draw a representative image of the tissue in the following space.

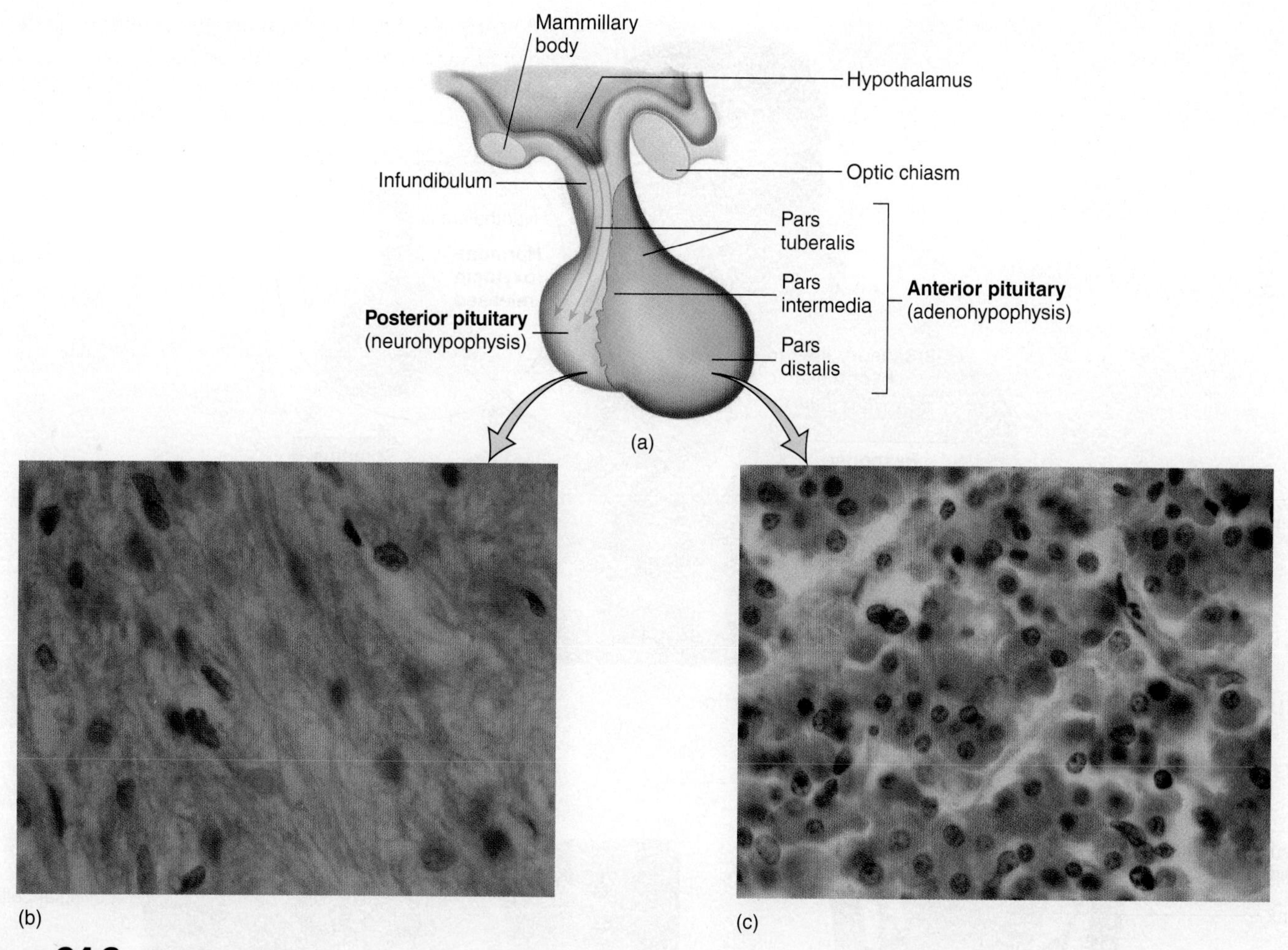

Figure 24.3 Pituitary Gland

(a) Regions of the entire gland; (b) photomicrograph of the posterior pituitary (400×); (c) photomicrograph of the anterior pituitary (400×).

Hormones released from the posterior pituitary are secreted by the **hypothalamus** and flow through axons to be stored in the posterior pituitary. These hormones and their actions include the following:

- **Antidiuretic hormone (ADH)** stimulates the reabsorption and retention of water by the kidneys. It is also known as vasopressin because it causes arterioles to constrict, which elevates blood pressure.
- **Oxytocin** stimulates the contraction of the cells of the mammary glands, resulting in the release of milk, and causes uterine contractions. External influences on hormone actions can be seen in the release of oxytocin. Look at figure 24.4 and note that the sucking action of an infant on the mother's breast sends impulses to the hypothalamus. The hypothalamus stimulates the posterior pituitary to release oxytocin, which causes milk ejection.

Thyroid

(A) 6 Examine models, charts, and prepared slides of the thyroid gland. Figure 24.5 illustrates the **thyroid gland,** which is named for its location near the thyroid cartilage of the larynx. The thyroid gland has two lateral **lobes** and a medial **isthmus** that connects them.

The thyroid is easy to recognize histologically. Examine a slide of thyroid gland and compare it with figure 24.5. Locate the **follicle cells** that surround the **colloid.** Colloid holds large amounts of thyroglobulin, which is a storage molecule consisting of many thyroid molecules. Thyroid hormones are manufactured in the thyroid and are bound to the larger thyroglobulin molecule. They are cleaved from it when needed by the follicular cells. Also locate the **parafollicular cells (C cells),** which are located in the spaces between the follicles.

The thyroid gland secretes three specific hormones. Two of these are **T3,** or **triiodothyronine** (a molecule that includes three iodine atoms), and **T4,** or **thyroxine** (containing four iodine atoms per molecule). These hormones increase **basal metabolic rates** and protein synthesis and are stored in the colloid. Another hormone of the thyroid gland is **calcitonin.** Calcitonin causes the deposition of calcium in bone by decreasing osteoclast activity. Calcitonin is produced by the parafollicular cells. Calcitonin is antagonistic to parathyroid hormone (discussed next).

Parathyroid Glands

In the posterior portion of the thyroid are typically two pairs of organs called the **parathyroid glands. Chief cells** in the

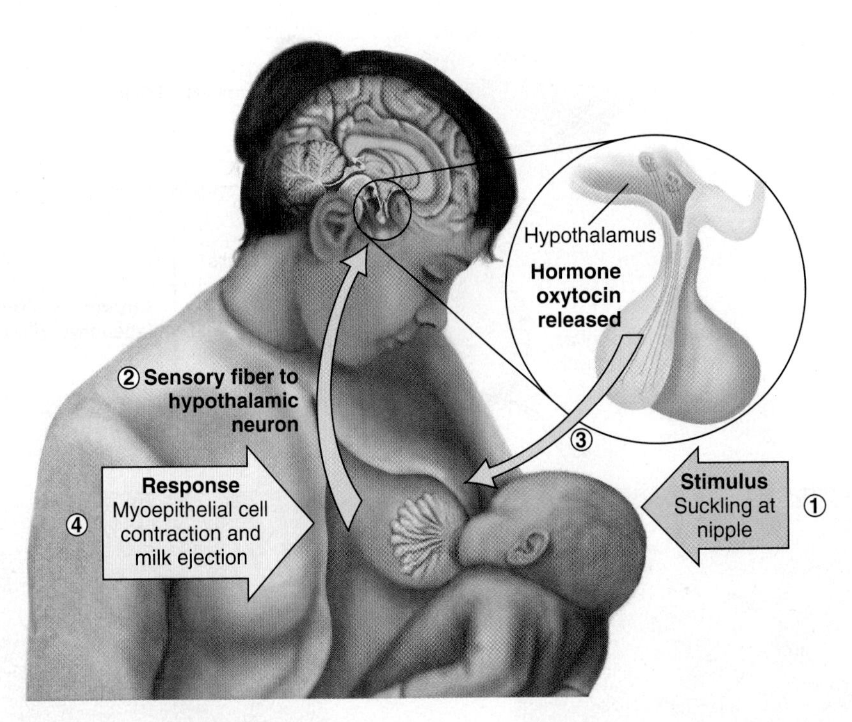

Figure 24.4 External Influences on Hormonal Action

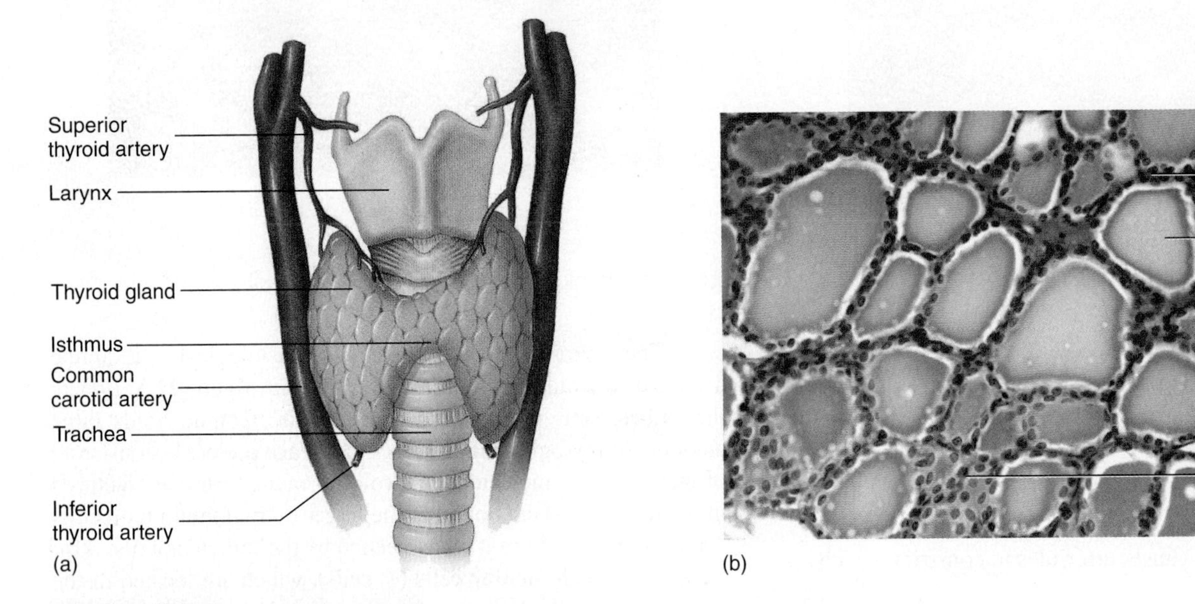

Figure 24.5 Thyroid Gland

(a) Diagram; (b) photomicrograph (100×).

parathyroid gland secrete **parathyroid hormone (PTH),** or **parathormone,** which is responsible for increasing calcium levels in the blood. It accomplishes this by increasing calcium uptake in the intestines, increasing the kidneys' reabsorption of calcium, and releasing calcium from bone by stimulating osteoclasts. Examine the location of the parathyroid glands embedded in the posterior surface of the thyroid gland (figure 24.6).

Thymus

(A) **7** Locate the thymus on models or charts anterior and superior to the heart (figure 24.1). The **thymus** is active in young people and plays an important part in immunocompetency. You will not find the thymus on models of adults. It secretes a hormone, **thymosin,** that causes the maturation of T cells. These cells originate as stem cells in the bone marrow and migrate to the thymus. Under the influence of thymosin, the cells mature to provide **cell-mediated immunity** against **antigens** (foreign material, such as viruses and tumors, that cause immune reactions). The T cells migrate from the thymus predominantly to the lymph nodes and the spleen to carry out their functions.

Pancreas

(A) **8** Locate the pancreas in figures 24.1 and 24.7. The **pancreas** is a mixed gland in that it has an exocrine function and an

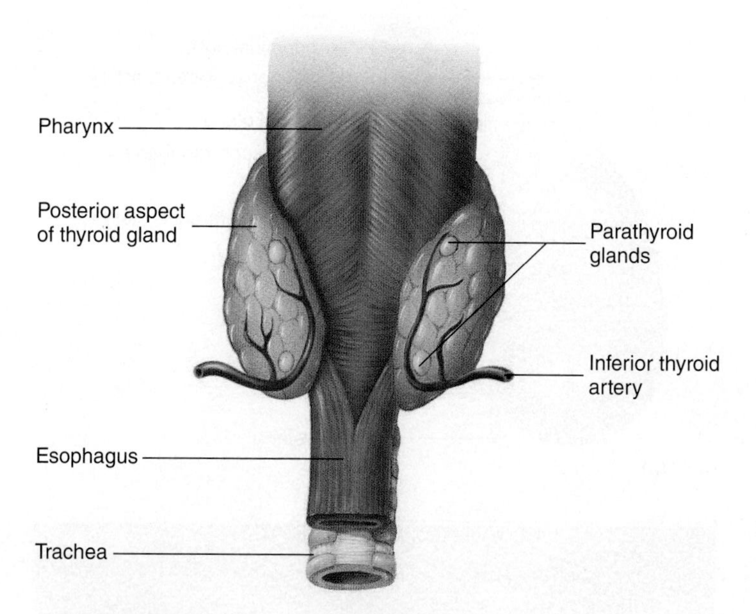

Figure 24.6 Parathyroid Glands

endocrine function. The exocrine function is digestive, because pancreatic juice contains both buffers and digestive enzymes. These are secreted by the acinar cells. The endocrine function of the pancreas consists of the secretion of the hormones **insulin** and **glucagon,** which regulate blood glucose levels. When blood glucose levels drop, glucagon converts glycogen (a starch storage product) to glucose. Glucagon is produced in specialized cells **(alpha cells)** on the periphery of clusters called **pancreatic islets (islets of Langerhans;** figure 24.7*b*). Pancreatic islets also produce insulin, which lowers the blood glucose level. Insulin, which is produced in **beta cells,** stimulates the conversion of glucose to glycogen. The pancreatic islets also contains **delta cells,** which secrete **somatostatin,** a hormone that inhibits insulin and glucagon secretion. Locate the pancreatic islets on a microscope slide and compare them with those in figure 24.7. Draw an islet in the space provided.

Islet:

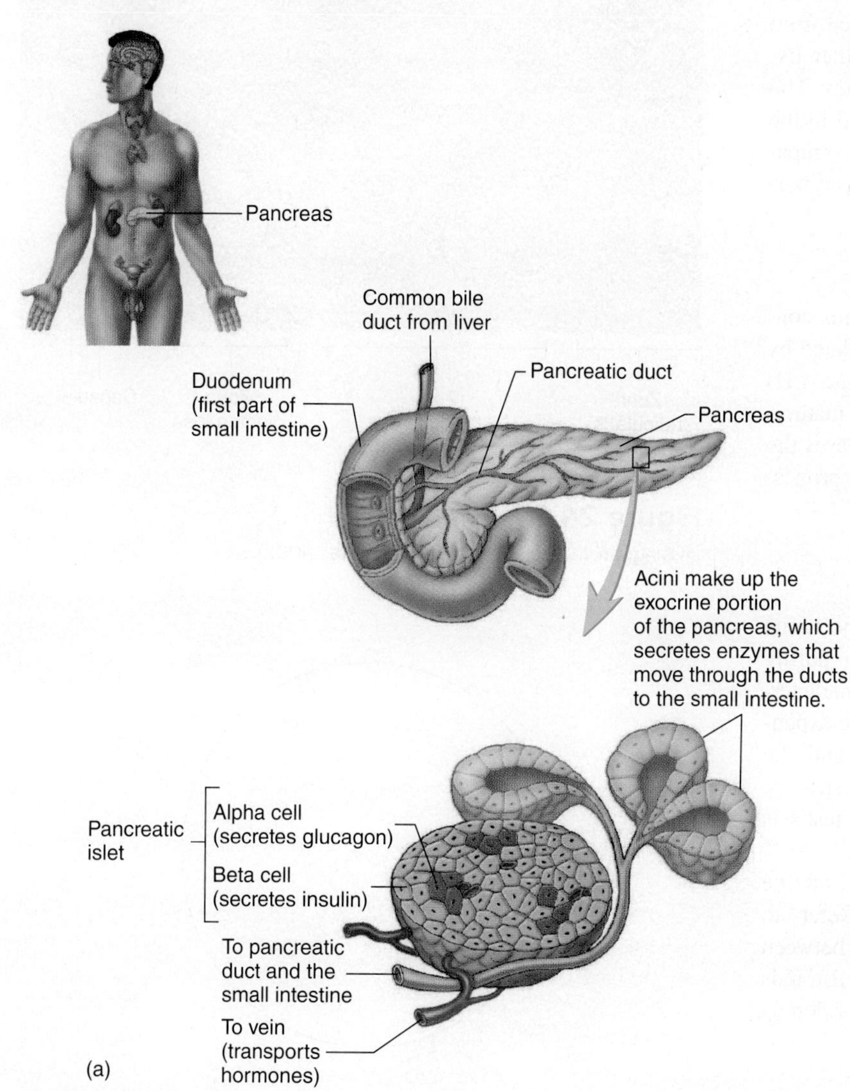

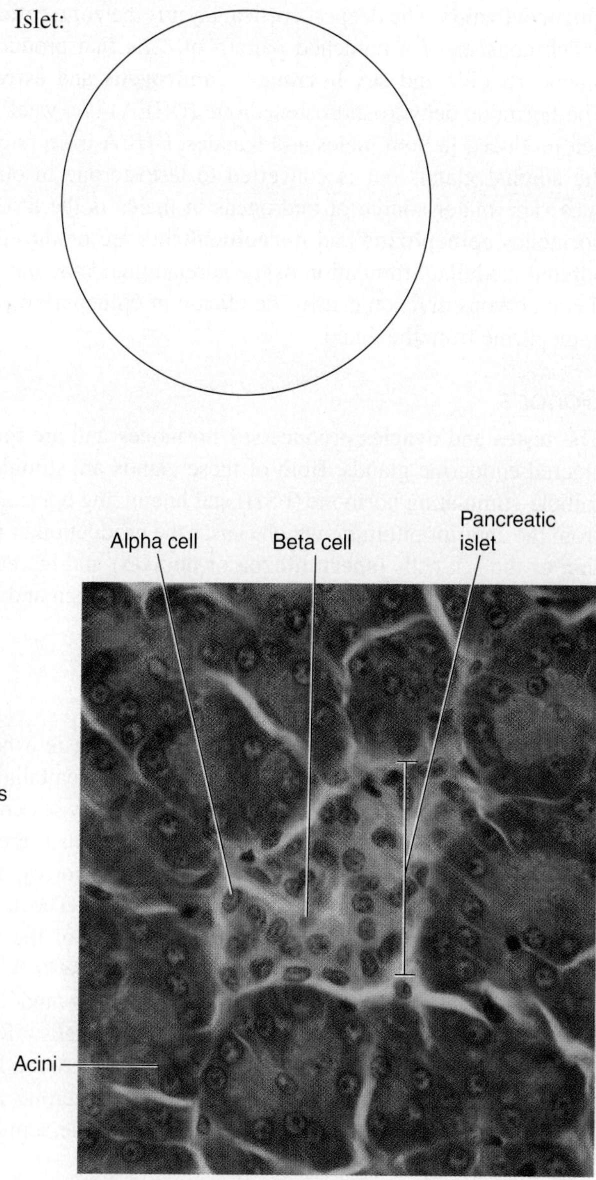

Figure 24.7 Pancreas

(a) Diagram; (b) photomicrograph (400×).

Adrenal Glands

9 Locate the adrenal glands on models or charts in lab. The **adrenal glands** (*ad* = next to, *renal* = kidney) are located superior to the kidneys. Each adrenal gland is composed of an outer **cortex** and an inner **medulla.** Examine a microscope slide of an adrenal gland and locate the cortex and the medulla. Compare this with figures 24.1 and 24.8.

The hormones secreted from the adrenal cortex are called **corticosteroid hormones.** They are important in water and electrolyte balance (Na^1, K^1) in the body. They are also important for carbohydrate, protein, and fat metabolism, as well as stress management. The cortex can be divided into three regions. The outermost is the **zona glomerulosa** (figure 24.8), which consists of clusters of cells that secrete **mineralocorticoids** (especially **aldosterone**). Inside this layer (closer to the medulla) is the **zona fasciculata,** which consists of parallel bundles of cells that secrete **glucocorticoids.** The deepest cortical layer is the **zona reticularis,** which consists of a branched pattern of cells that produce both glucocorticoids and **sex hormones** (**androgens** and **estrogens**). The hormone dehydroandrostenedione (DHEA) is a weak androgen produced in both males and females. DHEA is secreted from the adrenal glands but is converted to testosterone in other tissues. The major source of androgens in males is the testes. The hormones **epinephrine** and **norepinephrine** are produced in the adrenal medulla. Stimulation of the adrenal glands by the sympathetic nervous division causes the release of epinephrine and norepinephrine from the gland.

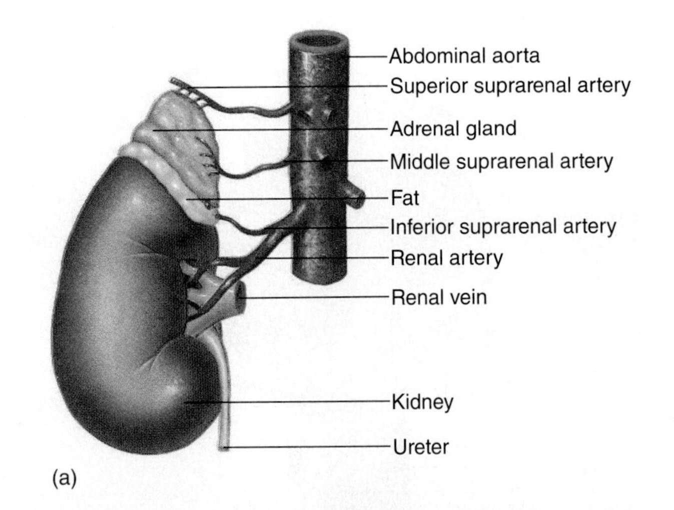

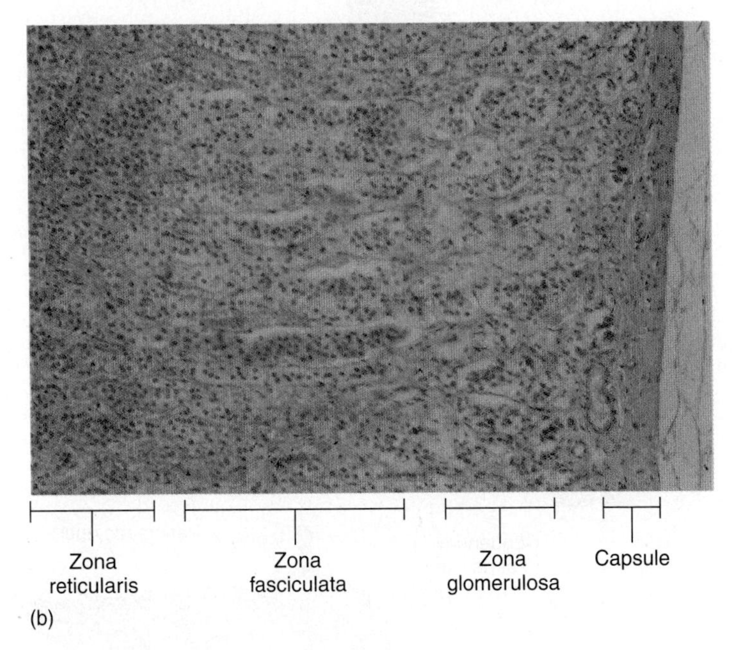

Figure 24.8 Adrenal Gland

(a) Diagram; (b) photomicrograph of cortex (100×).

Gonads

The **testes** and **ovaries** produce sex hormones and are thus considered endocrine glands. Both of these glands are stimulated by follicle-stimulating hormone (FSH) and luteinizing hormone (LH) from the anterior pituitary, which causes the production or maturation of the sex cells (**spermatozoa** or **oocytes**) and increases the level of sex hormone (testosterone in males, estrogen and progesterone in females) production.

Testes

In males, the **testes** produce **testosterone,** which is a hormone responsible for the development of the male genitalia during embryonic and fetal development and for secondary sex characteristics, such as the development of facial and body hair, the expansion of the larynx (which produces a deeper voice), and the increased muscle and bone mass seen in males. The testes are illustrated in figure 24.1. The exocrine function of the testis is explored in Exercise 42, on the male reproductive system.

10 Examine a microscope slide of a testis and find the **seminiferous tubules** and **interstitial cells.** Refer to figure 24.9 for assistance. The interstitial cells are found between the tubules, and they produce testosterone. Testosterone also aids FSH in the production of **spermatozoa.** Draw the seminiferous tubule and interstitial cells in the space provided.

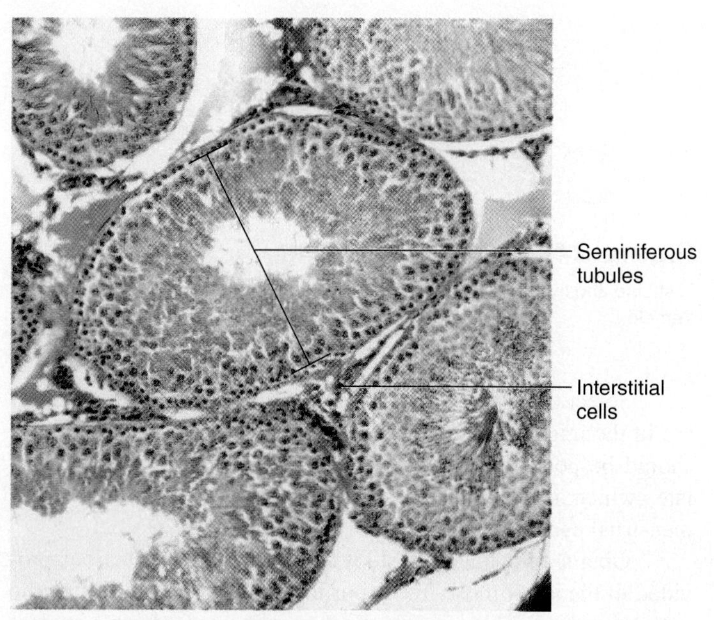

Figure 24.9 Histology of the Testis (100×)

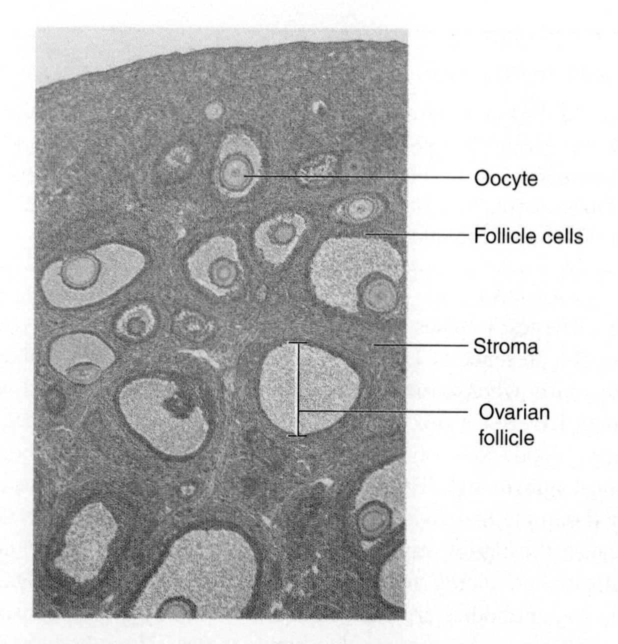

Figure 24.10 Histology of the Ovary (40×)

Ovaries

(A) 11 Examine a prepared slide of an ovary and draw an ovarian follicle in the following space. In females, the **ovaries** produce oocytes (eggs) and have an endocrine function in that they produce **estrogen** and **progesterone** (figure 24.1). The ovaries are illustrated in figure 24.10. Estrogen is produced by the **follicle cells** and progesterone is produced in the corpus luteum. Female hormones are also responsible for secondary sex characteristics in women, such as the development of breasts, an additional subcutaneous adipose layer, and a higher voice. Estrogen and progesterone also influence the development of the endometrium, cause the maturation of the oocytes, and regulate the menstrual cycle. *Estrogen* is a generic term for several hormones produced by the female, including **estradiol.** During pregnancy, the placenta functions as an endocrine gland in the secretion of estrogen and progesterone.

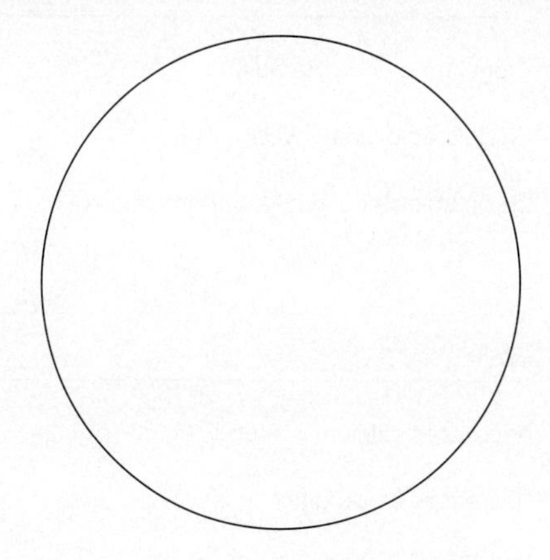

Endocrine Physiology—Detection of Hormones

(A) 12 Hormones are secreted in small but detectable amounts in the bloodstream. Some of these hormones may be filtered in the urine and detected from a urine sample. Luteinizing hormone (LH) is one of these hormones. LH stimulates the final maturation of the oocyte and causes ovulation. Normally, small amounts of LH occur in a woman's body but an increase, or a spike, in the levels of LH occurs about 24–36 hours prior to ovulation (see figure 24.11). This increase lasts typically for 10–30 hours. In this part of the exercise you will use a test kit to determine the presence of LH. Your instructor may provide an unknown sample of urine, or female students in the middle of their ovarian cycle may be asked to volunteer to test for the presence of LH.

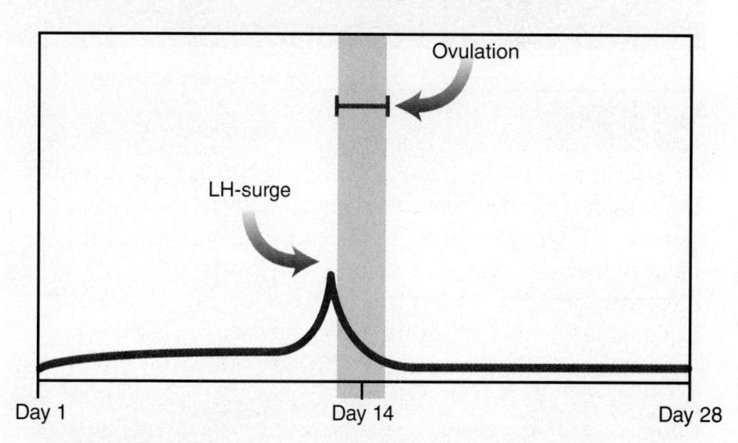

Figure 24.11 Surge of Luteinizing Hormone (LH) Prior to Ovulation

Day 1 of the menstrual cycle begins with menstrual flow.

Caution!

Remember to treat bodily fluids as if they carried pathogens. Wear protective gloves and eyewear when handling fluids! Dispose of all contaminated disposable materials in the biohazard container. The reusable materials should be placed in a 10% bleach solution.

The test kit uses **monoclonal antibody** techniques to determine the presence of LH. The development of a monoclonal antibody starts when a mouse (or another mammal) is injected with human LH. Since this hormone is foreign to the mouse, specific white blood cells, called beta lymphocytes, make antibodies against human LH. These can be removed from the mouse in a blood sample and fused with tumor cells. Tumor cells are used because they grow rapidly. The hybrid tumor-lymphocyte cells replicate, producing more cells that have the antibody against LH. The antibodies are isolated and absorbed by a solid substrate (the test strip). An enzyme that bonds to the LH antibody is also attached to the substrates. This enzyme is attached to a dye marker and it changes color when attached to the antibody.

In summary, the test strip contains antibodies to LH, an enzyme that will bind to the LH-antibody complex, and a dye that changes color when the enzyme binds to the LH-antibody complex. If a woman has significant amounts of LH, then the test strip produces a color (usually blue). The generic name for tests that use antibodies and enzymes to determine the presence of biological materials is *enzyme-linked immunosorbent assay (ELISA)*. In some cases, pregnancy, endometriosis, hyperthyroidism, and ingestion of some prescription drug may produce false test results.

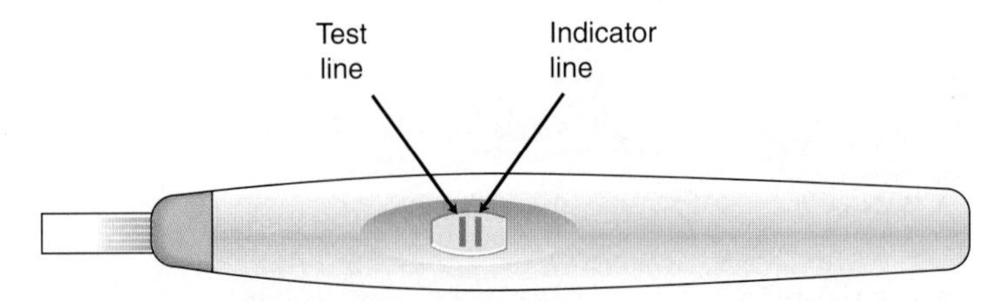

Figure 24.12 Ovulation Test Kit

Test Line and Indicator Line should be the same from sample of Ovulating Female.

Select the unknown urine sample or have three women who are in the middle of their ovarian cycle perform the test. The test should be positive for women who are just about ready to ovulate, which is normally 2 weeks prior to the next anticipated menstrual cycle.

Obtain an ovulation test kit and follow the instructions provided in the kit or those from your instructor. Record whether the urine sample provided is an unknown or is from a female student in the class and whether it has significant levels of LH. Typically, the test involves collecting a urine sample and dipping the absorbent sampler in the urine for a specific amount of time. The sample is then compared to a reference line to indicate ovulation, as seen in figure 24.12.

Sample: ________________________ (unknown/student)

Presence of LH: ________________________

Once you have studied the endocrine glands and materials, complete table 24.1. You should fill in the blank spaces where needed.

TABLE 24.1

Organ	Hormones Produced	Effect of Hormones
______________	Antidiuretic hormone	______________
Thyroid	Thyroxine	______________
______________	Corticosteroid hormones	Regulate electrolyte balance
Ovary	______________	Regulates ovarian cycle
______________	Melatonin	Regulates sleep cycles, inhibits release of reproductive hormones
______________	Luteinizing hormone	______________
Neurohypophysis	Oxytocin	______________
Parathyroid	______________	Increases calcium in blood
______________	Glucagon	Increases blood glucose levels
Pancreas	______________	Decreases blood glucose levels
______________	Calcitonin	______________

Exercise 24 Review

Endocrine System

Name: ______________________

Lab time/section: ______________________

Date: ______________________

1. What is the general name for organs that produce hormones? ______________________

2. What name is given to regions that are receptive to hormones? ______________________

3. Melatonin is secreted by what gland? ______________________

4. In what specific part of what gland is ADH stored? ______________________

5. What is the effect of TSH, and where is it produced? ______________________

6. What does glucagon do as a hormone, and where is it produced? ______________________

7. Which hormones in the adrenal gland control water and electrolyte balance? ______________________

8. What is the primary gland that secretes epinephrine? ______________________

9. Where is growth hormone produced? ______________________

10. What is another name for T3? ______________________

11. What connects the two lobes of the thyroid gland? ______________________

12. Does parathormone increase or decrease calcium levels in the blood? ______________________

13. Label the endocrine glands in the following illustration using the terms provided.

adrenal glands	ovaries	pancreas	parathyroid glands
testes	thymus	thyroid gland	

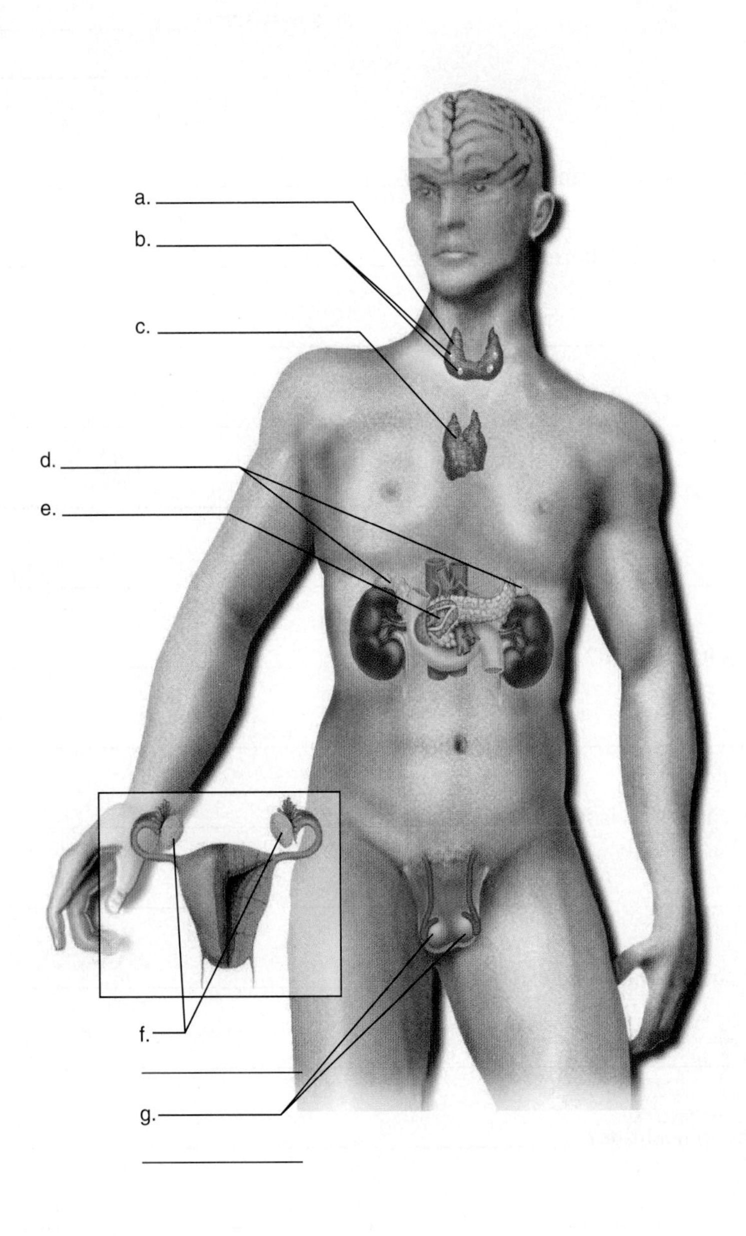

14. Identify the parts of the pituitary, as seen in the following illustration, and label two hormones located in each.

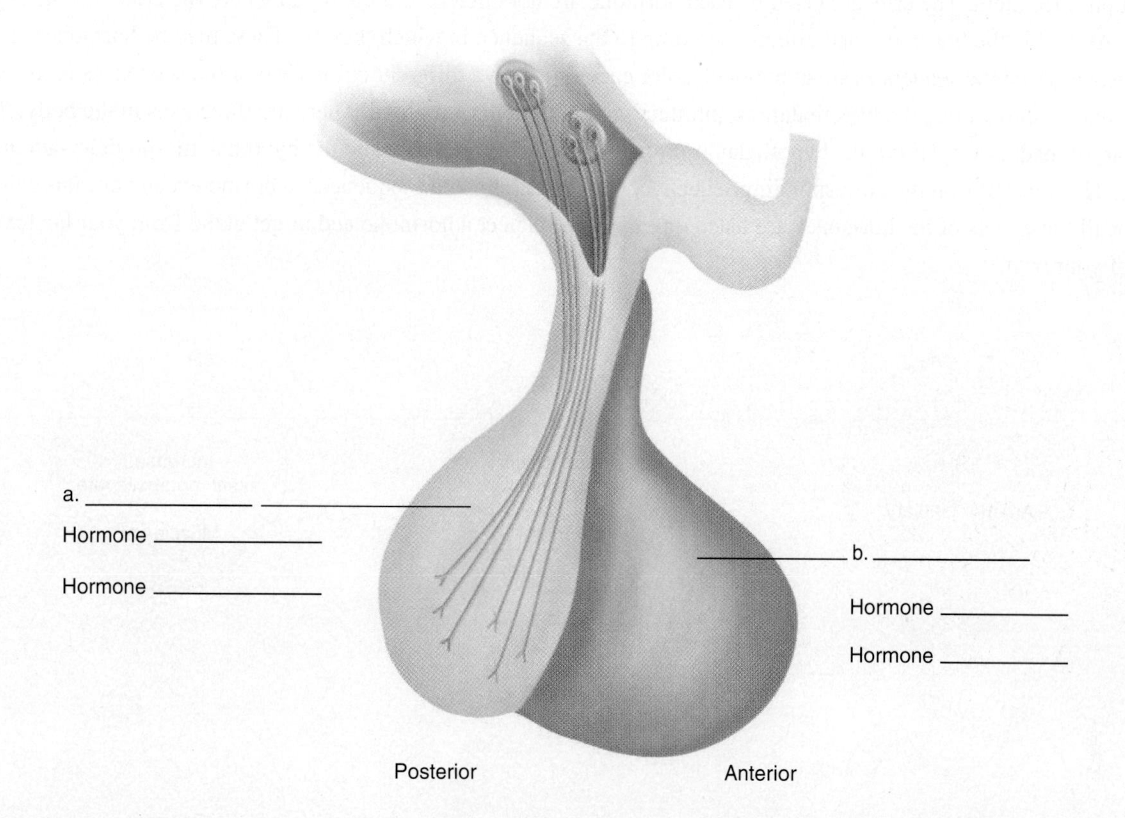

15. Identify the three layers of the adrenal cortex, as illustrated, and list the hormones produced by each layer.

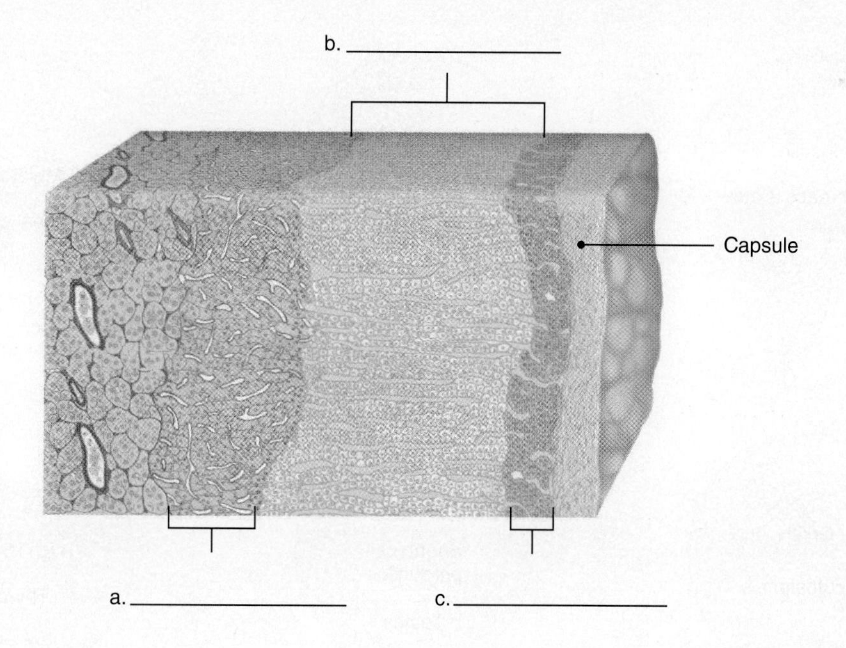

16. Interstitial cells of the testis produce which hormone? ______________________________

17. Target cells must have the correct receptor site in order for hormones to influence those cells. Cut out the following 12 outlines, which represent either hormones or an effect caused by a hormone. In the structures, hormones are circled above the endocrine glands that produce them. The effects caused by each hormone are not circled, and the organ where the effect occurs is listed below the effects. Assemble the hormones and effects according to the sequence in which they fit. There may be hormones or cells that do fit the sequence. Once the sequences are assembled, color each one with a different color. Two of the sequences represent an axis, which is a relationship among the hypothalamus, pituitary, and a more distant gland. There are three axes in the body: the hypothalamo-pituitary-gonadal axis (HPG), the hypothalamo-pituitary-thyroid axis (HPT), and the hypothalamo-pituitary-adrenal axis (HPA). Write HPG, HPT, or HPA on the corresponding sequence. Notice that, in some sequences, a hormone can have more than one effect (though not all the effects of the hormones are listed on this page). Select a hormone and target gland from your lab text and make a sequence of your own.

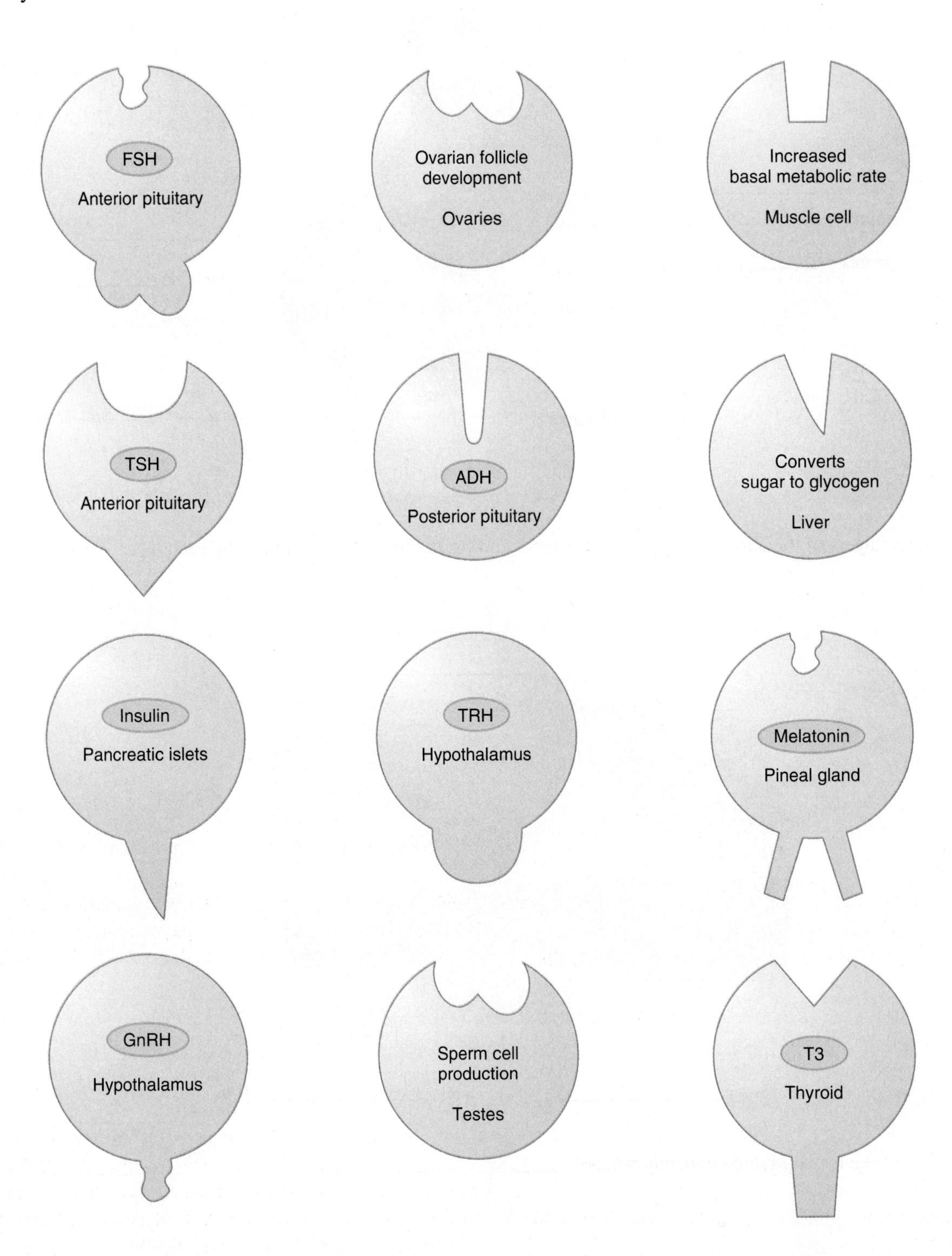

Exercise 25

Blood

Introduction

Blood is made of two major components, formed elements and plasma. Formed elements make up about 45% of the blood volume. They can be subdivided into red blood cells (erythrocytes), white blood cells (leukocytes), and platelets (thrombocytes). Plasma, which constitutes about 55% of the blood volume, contains water, lipids, dissolved substances, colloidal proteins, and clotting factors.

The study of blood is significant both because it is the fluid medium of the cardiovascular system and because it has significant clinical importance. Changes in the numbers and types of blood cells can be used as indicators of disease. In this exercise, you examine the nature of blood cells, which are discussed in the Seeley text in chapter 19, "Cardiovascular System: Blood."

Caution!

The risk of bloodborne diseases is significantly minimized by the use of sterilized human blood and nonhuman mammal blood in this exercise. However, use the same precautions as if you were handling fresh and potentially contaminated human blood. Your instructor will determine whether to use animal blood (from nonhuman mammals), sterilized blood, or synthetic blood. In all these cases, follow strict procedures for handling potentially pathogenic material.

1. Wear protective gloves during the procedures.
2. Do not eat or drink in lab.
3. If you have an open wound, do not participate in this exercise, or make sure the wound is securely covered.
4. Your instructor may elect to have you use your own blood. Due to the potential for disease transmission, such as HIV or hepatitis B and C, in fresh blood samples, **make sure you keep away from other students' blood and keep them away from your blood.**
5. Talk to your instructor about this exercise if you have recently skipped a meal or if you are not well.
6. Place all disposable material in the biohazard bag.
7. Place all used lancets in the sharps container and all glassware that is to be reused in a 10% bleach solution.
8. After you have finished the exercise, clean and disinfect the countertops with a 10% bleach solution.

Objectives

At the end of this exercise, you should be able to

1. discuss the composition of blood plasma;
2. distinguish among the various formed elements of blood;
3. describe hematopoiesis;
4. determine the percentage of each type of leukocyte in a differential white cell count;
5. describe what an elevated level of a particular white blood cell may indicate about a disease state or an allergic reaction.

Materials

Prepared slides of human blood with Wright's or Giemsa stain
Compound microscopes
Lab charts or illustrations showing the various blood cell types
Protective gloves
Roll of paper towels
Vial of mammal blood or
- Sterile cotton balls
- Alcohol swabs
- Sterile, disposable lancets
- Adhesive bandages

Dropper bottle of Wright's stain
Squeeze bottle of distilled water or phosphate buffer solution
Large finger bowl or staining tray
Toothpicks
Clean microscope slides
Coverslips
Pasteur pipette and bulbs
Hand counter
Biohazard bag or container
10% bleach container
Sharps container

Procedure

Plasma

Plasma is the fluid portion of blood and is about 91% water. The remainder consists mostly of proteins, such as albumins, globulins, and fibrinogen. **Albumins** are produced by the liver and make up the majority of the plasma proteins. Some **globulins** are made by the plasma cells, and these make up the next largest share of proteins. **Fibrinogen** is a clotting protein; this and other clotting factors are produced by the liver. Plasma also contains electrolytes (Na^+, K^+, and Cl^+), nutrients, hormones, and wastes.

Examination of Blood Cells

(A)[1] In this part of the exercise, you need to distinguish among red blood cells, white blood cells, and platelets. If you are using a prepared slide of blood, you can move to the section entitled "Microscopic Examination of Blood." If you are to make a smear, then read the following directions.

Preparing a Fresh Blood Smear

Your lab instructor will direct you to use either fresh, nonhuman mammal blood or your own blood. If you are using provided mammal blood, wear protective gloves and withdraw a small amount of blood from the vial with a clean Pasteur pipette. Place a drop of the blood on a clean microscope slide. If you are using your own blood, make sure you follow the safety precautions as discussed at the beginning of the exercise.

Withdrawal of Your Own Blood

Obtain the following materials: a clean, sterile lancet; an alcohol swab; an adhesive bandage; protective gloves; sterile cotton balls; two clean microscope slides; and a paper towel.

1. Arrange the materials on the paper towel in front of you on the countertop.
2. Clean the end of the donor finger with the alcohol swab and let the hand from which you will withdraw blood hang by your side for a few moments to collect blood in the fingertips. You will be puncturing the pad of your fingertip (where the fingerprints are located).
3. Peel back the covering of the lancet and hold on to the blunt end as you withdraw it from the package. Do not touch the sharp end or lay the lancet on the table before puncturing your finger.
4. Jab your finger quickly and wipe away the first drop of blood that forms with a sterile cotton ball. Never reuse the lancet or set it down on the table. Place the lancet in the sharps container.
5. Place the second drop of blood that forms on the microscope slide about 2 cm away from one of the ends of the slide (figure 25.1) and place another cotton ball on your finger.
6. You can now put an adhesive bandage on your finger.
7. Whether you are using prepared blood or your own blood, use another clean slide to spread the blood by touching the drop with the edge of the slide and push the blood across the slide (figure 25.1). This should produce a smooth, thin smear of blood. Let the blood smear dry completely.
8. After the slide is dry, place it in a large finger bowl elevated on toothpicks or on a staining tray.
9. Cover the blood smear with several drops of Wright's stain from a dropper bottle. Let the stain remain on the slide for 1 to 2 minutes.
10. After this time, add water or a prepared phosphate buffer solution to the slide. You can rock the slide gently with gloved hands or gently blow on it to stir the stain and water. A metallic green material should come to the surface of the slide. Let the slide remain covered with stain and water for 3 to 8 minutes.
11. Wash the slide gently with distilled water until the material is light pink, and then stand it on edge to dry. You can also stain blood using an alternate stain (such as Giemsa stain), following your lab instructor's procedure.

Place all blood-contaminated disposable material in the biohazard container. Place the slide used to spread the blood in the 10% bleach solution and make sure the lancets are in the sharps container. Once the slide is completely dry, it can be examined under the high-power or oil immersion lens of your microscope.

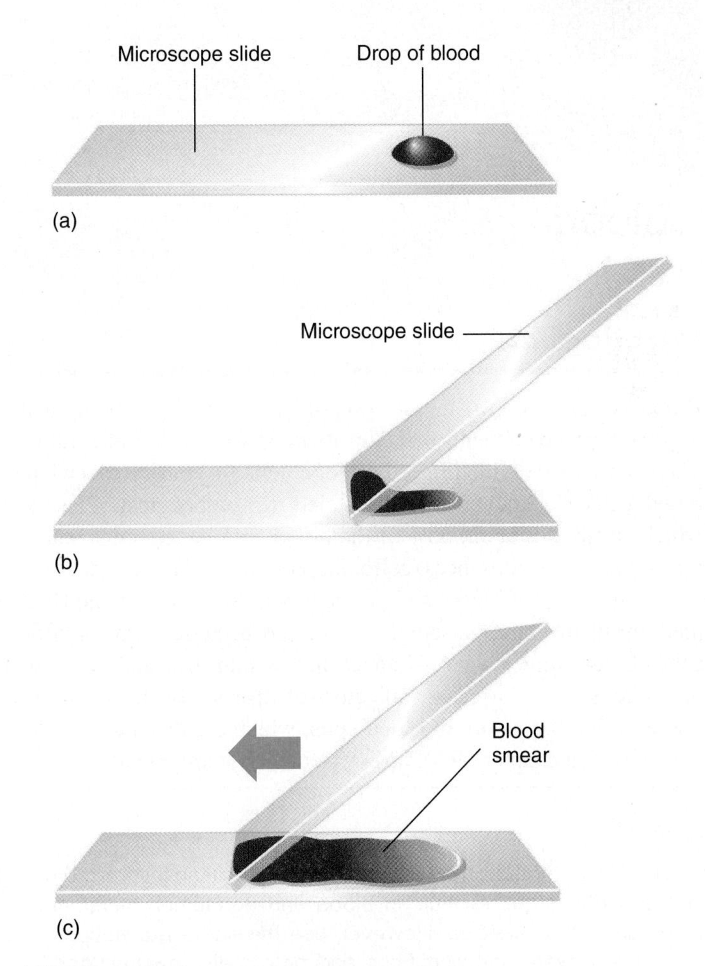

Figure 25.1 Making a Blood Smear
(a) Placing blood on a slide; (b) touching glass slide to front of blood drop; (c) spreading blood across slide.

Microscopic Examination of Blood

Overview

Ⓐ2 As you examine the blood slide, refer to figure 25.2 and to the photomicrographs of the blood cells. You will have to look at many blood cells in order to see all of the cells that are included in this exercise.

Red Blood Cells

Examine a slide of blood stained with either Wright's or Giemsa stain. Red blood cells are the most common cells you will find on the slide. There are about 5 million red blood cells per cubic millimeter. They do not have a nucleus but appear as pink, biconcave disks (as if you had placed two dinner plates back to back). This shape increases surface area and provides for greater oxygen-carrying capabilities. Blood formation is called **hematopoiesis,** or **hemopoiesis.** The specific production of red blood cells is called **erythropoiesis.**

Red blood cells are about 7.5 μm in diameter, on average, and have a life span of about 120 days, after which time they are broken down by the spleen or liver. Compare what you see under the microscope with figure 25.3.

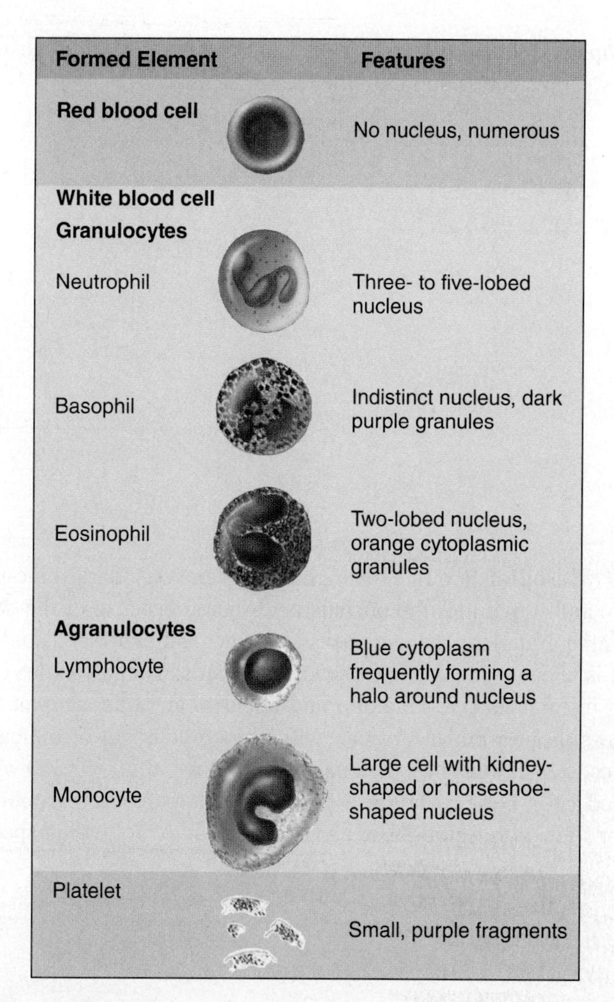

Formed Element	Features
Red blood cell	No nucleus, numerous
White blood cell	
Granulocytes	
Neutrophil	Three- to five-lobed nucleus
Basophil	Indistinct nucleus, dark purple granules
Eosinophil	Two-lobed nucleus, orange cytoplasmic granules
Agranulocytes	
Lymphocyte	Blue cytoplasm frequently forming a halo around nucleus
Monocyte	Large cell with kidney-shaped or horseshoe-shaped nucleus
Platelet	Small, purple fragments

Figure 25.2 Formed Elements of the Blood

Platelets

There are about 250,000 to 400,000 platelets per cubic millimeter of blood. Platelets, or thrombocytes, are involved in clotting and consist of small fragments of megakaryocytes. Examine the slide for small, purple fragments that may be single or clustered and compare them with the platelets in figure 25.3.

White Blood Cells

There are far fewer white blood cells, or leukocytes, in the blood than red blood cells. The number of white blood cells in a healthy adult is about 7,000 cells per cubic millimeter, with a range of 5,000 to 9,000 cells per cubic millimeter of blood. White blood cells are formed in bone marrow tissue. The lifespan of a white blood cell varies from a few hours to several months. Many are capable of ameboid movement as they squeeze between cells (a process termed *diapedesis*), engulfing foreign particles and cellular debris.

White blood cells can be divided into two groups based on the presence or absence of granules in their cytoplasm. These two groups are the **granular** and **agranular leukocytes.** Examine the slide under high power or oil immersion and identify the different white blood cells as presented in the following text.

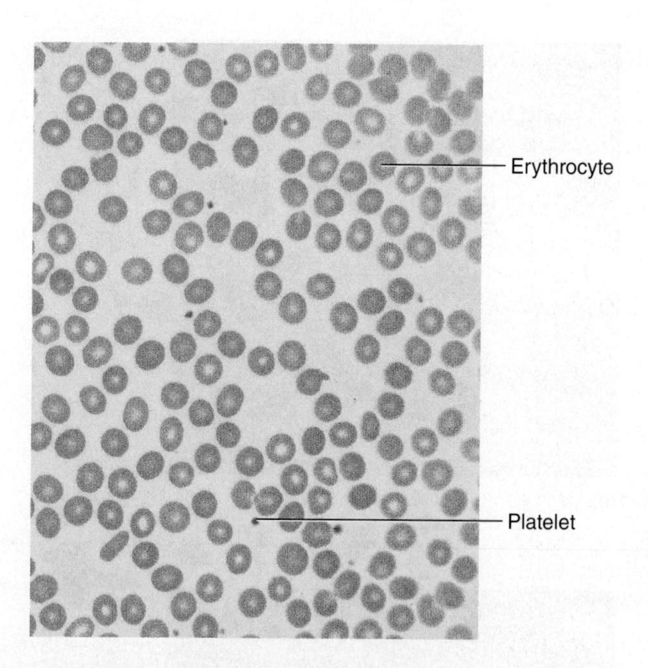

Figure 25.3 Blood Smear (400×)

Granular Leukocytes

Granular leukocytes (granulocytes) are so named because they have granules in their cytoplasm. They are also known as **polymorphonuclear (PMN) leukocytes** due to the variable shapes of their nuclei (which are lobed, not round).

Neutrophils are the most common of all white blood cells. Neutrophils typically live for about 1 to 2 days and have ameboid capabilities. They move by diapedesis between blood capillary cells and in the interstitial areas between the cells of the body. Neutrophils move toward infection sites and destroy foreign material. They are the major phagocytic white blood cells. The granules of neutrophils absorb very little stain, but neutrophils can be distinguished from other white blood cells by their two- to five-lobed nuclei. They are about one and a half times the size of red blood cells. Examine your slide for neutrophils, compare them with figure 25.4*a,* and draw one in the following space.

Eosinophils typically have a two-lobed nucleus with red-orange granules in the cytoplasm. The term *eosinophil* actually means eosin-loving (eosin is a red-orange stain and is picked up by the granules). They are about twice the size of red blood cells.

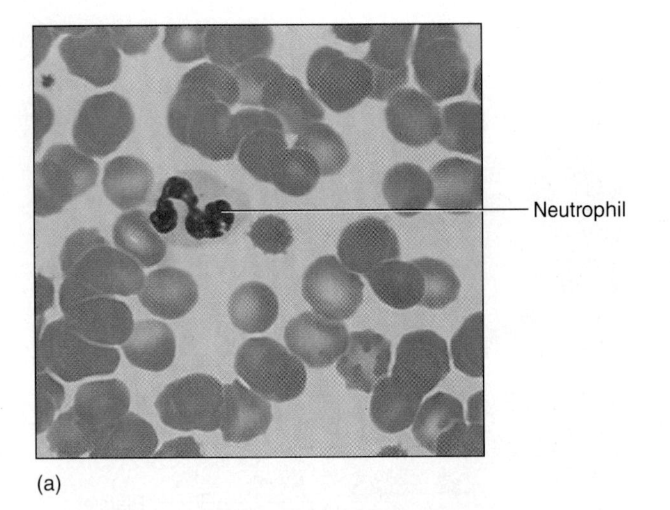

(a)

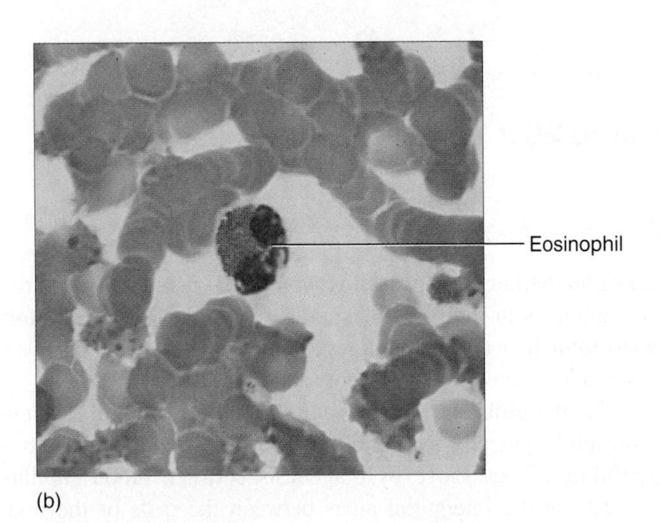

(b)

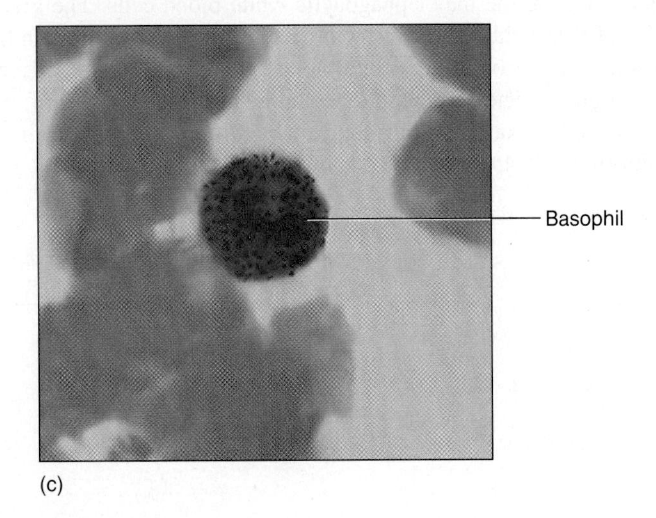

(c)

Figure 25.4 Granular Leukocytes

(a) Neutrophil (1,000×); (b) eosinophil (1,000×); (c) basophil (1,500×).

Compare figure 25.4*b* with your slide as you locate the eosinophils. Draw an eosinophil in the following space.

Basophils are rare. The granules stain very dark (blue-purple), and sometimes the nucleus is obscured because of the dark-staining granules. The nucleus is indistinct and two-lobed, and the cell is about twice the size of red blood cells. Their granules contain histamines (vasodilators) and heparin, an anticoagulant that allows the movement of other white blood cells out of the capillaries to infection sites. You may have to look at 200 to 300 white blood cells before finding a basophil. Compare the basophil in your slide with figure 25.4*c* and draw one in the following space.

Agranular Leukocytes

Agranular leukocytes (agranulocytes) are so named because they lack cytoplasmic granules. The nuclei are not lobed but may be dented or kidney bean–shaped. There are two types of agranular white blood cells, lymphocytes and monocytes.

Lymphocytes have a large, unlobed nucleus that usually has a flattened or dented area. The cytoplasm is clear and may appear as a blue halo around the purple nucleus. In terms of function, lymphocytes do not need prior exposure to recognize antigens, and they are remarkable in that separate cells have specific antibodies for specific antigens. There are two groups of lymphocytes, the B cells and the T cells. These cells cannot normally be distinguished from one another in standard histological preparations (for example, Wright's stain), and they are simply considered lymphocytes in this exercise. Both B cells and T cells arise from fetal bone marrow. **B cells** probably mature in the fetal liver and spleen, and **T cells**

mature in the thymus gland. B-cell lymphocytes mature into **plasma cells,** which make **antibodies,** also called **immunoglobulins.** Plasma cells provide **antibody-mediated immunity** (that is, the plasma cells secrete antibodies that travel in the fluid portion of the blood).

T cells, on the other hand, provide **cell-mediated immunity.** In cell-mediated immunity, the cells themselves (not antibodies in the blood plasma) move close to and destroy some types of bacteria or virus-infected cells. T cells also attack tumors and transplanted tissues. Most of the T cells are found in the lymph nodes, thymus, and spleen. They enter the bloodstream via the lymphatics. Compare your slide with figure 25.5*a* for lymphocytes and draw one in the following space.

Monocytes are very large and have a kidney–, or horseshoe-shaped nucleus. They are about three times the size of red blood cells and are activated by T cells. Locate the large cells on your slide, compare them with figure 25.5*b,* and draw one in the following space.

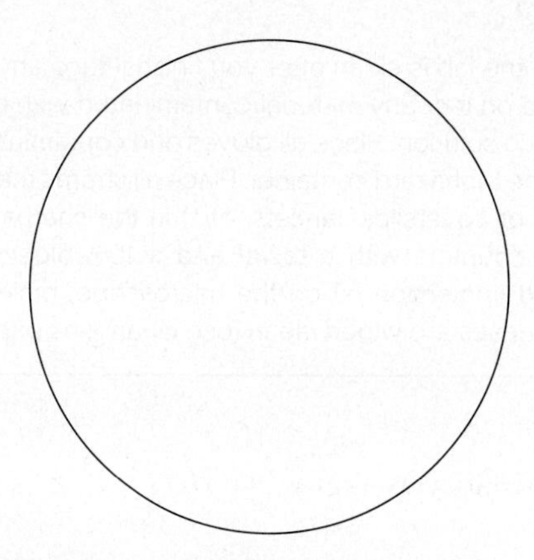

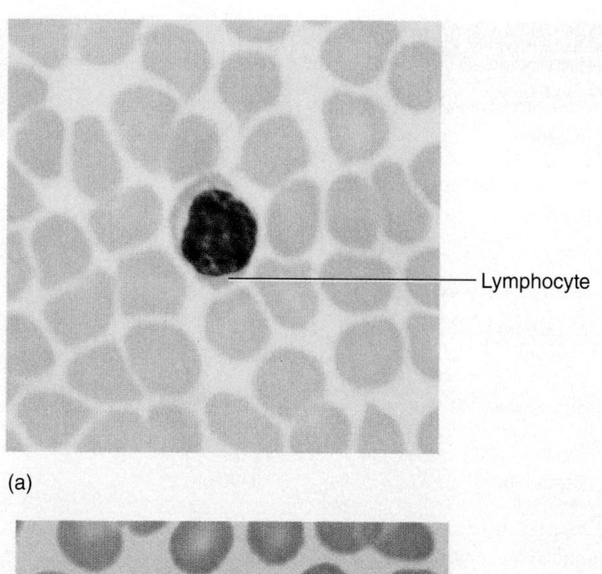

(a)

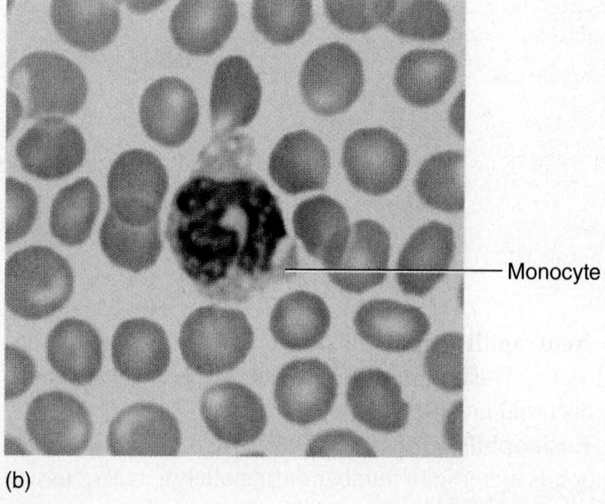

(b)

Figure 25.5 Agranular Leukocytes (1,000×)
(a) Lymphocyte; (b) monocyte.

Differential White Blood Cell Count

(A)3 Review the formed elements in table 25.1. Once you have identified the various white blood cells on the blood slide, conduct a differential white blood cell count. This is a very important count. Changes in the relative percentages of white blood cells may indicate the presence of a disease. Remember that diagnostic evaluations of differential white blood cell counts should be done by qualified professionals.

1. Use a hand counter and count 100 leukocytes. One lab partner should look into the microscope and methodically call out the names of the different types of leukocytes seen.
2. The other lab partner should record how many of each type are found and keep track of the overall number with the use of a hand counter until 100 cells are counted.
3. Scan the slide in a systematic way, so that you do not count any cell twice. One method is illustrated in figure 25.6. Tally your results.

Neutrophils: ____________________

Eosinophils: ____________________

Basophils: ____________________

Lymphocytes: ____________________

Monocytes: ____________________

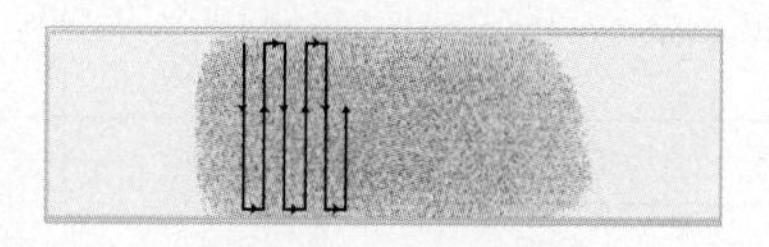

Figure 25.6 Counting Leukocytes

TABLE 25.1 Summary of Formed Elements in Blood			
Formed Element	**Size**	**Number**	**Characteristics**
Erythrocytes	7.7 μm	5 million/mm^3	Live 120 days on average
			No nucleus
Platelets	2–4 μm	250,000–400,000/mm^3	Cell fragments
Leukocytes		7,000/mm^3	Nucleus present
Granular leukocytes			
Neutrophils	9–12 μm	60–70% of leukocytes	Two- to five-lobed nucleus
			Granules indistinct
Eosinophils	10–14 μm	2–4% of leukocytes	Two-lobed nucleus
			Red-orange granules
Basophils	8–10 μm	0.5–1% of leukocytes	Indistinct nucleus
			Large, dark granules
Agranular leukocytes			
Lymphocytes	5–17 μm	20–25% of leukocytes	Nucleus appearing dented
			Thin rim of cytoplasm in some
Monocytes	12–15 μm	3–8% of leukocytes	Large, kidney-shaped nucleus

Neutrophils represent about 60–70% of all of the white blood cells. Their numbers increase in cases of appendicitis or acute bacterial infections.

Eosinophils represent about 2–4% of all white blood cells. Eosinophils increase in number during allergic reactions and parasitic infections (for example, trichinosis).

Basophils represent about 0.5–1% of all white blood cells. They increase in number during allergic and inflammatory reactions.

Lymphocytes make up 20–25% of the white blood cells. Their numbers increase in cases of bacterial infection and antibody-antigen reactions.

Monocytes make up about 3–8% of the white blood cells. Their numbers increase during chronic infections, such as tuberculosis.

Fill in the following table with what you know about the formed elements. You should read all of the information given before filling in the table.

Cleanup

Make sure the lab is clean after you finish. Place any slide with fresh blood on it or any material contaminated with bodily fluid in the bleach solution. Place all gloves and contaminated paper towels in the biohazard container. Place all sharps material (broken slides or coverslips, lancets, etc.) in the sharps container. Clean the counters with a towel and a 10% bleach solution. If you used immersion oil on the microscope, make sure the objective lenses are wiped clean (use clean lens paper only).

Characteristics of Formed Elements			
Formed Element	**Granules (If Present)**	**Shape of Nucleus (If Present)**	**Cause for Increase**
Erythrocyte	________	________	________
________	Not obvious	________	Mononucleosis
________	Orange-staining	________	Parasitic infections
________	________	Two- to five-lobed	________
________	Not obvious	Kidney bean	________
Basophil	________	________	________

Exercise 25 Review

Blood

Name: ______________________

Lab time/section: ______________________

Date: ______________________

1. Formed elements consist of three main components. What are they? ______________________

____

2. What is the most common plasma protein? ______________________

3. What is another name for a thrombocyte? ______________________

4. Which is the most common blood cell? ______________________

5. What is another name for a white blood cell? ______________________

6. What white blood cell is most numerous in a normal blood smear? ______________________

7. How many red blood cells are normally found per cubic millimeter of blood? ______________________

8. What is an average number of white blood cells found per cubic millimeter of blood? ______________________

9. B cells and T cells belong to what class of agranular leukocytes? ______________________

10. What value is there to a change in the percentage of white blood cells to diagnostic medicine? ______________________

11. In counting 100 white blood cells, you are accurately able to distinguish 15 basophils. Is this a normal number for the differential white blood cell count, and what possible health implications can you draw from this? ______________________

____

____

____

12. What is the function of the platelets? ______________________

____

____

____

13. Formed elements constitute what percentage of the total blood volume? ______________________________

14. Label the formed elements in the following illustration using the terms provided.

basophil	eosinophil	lymphocyte	monocyte
neutrophil	platelets	red blood cell	

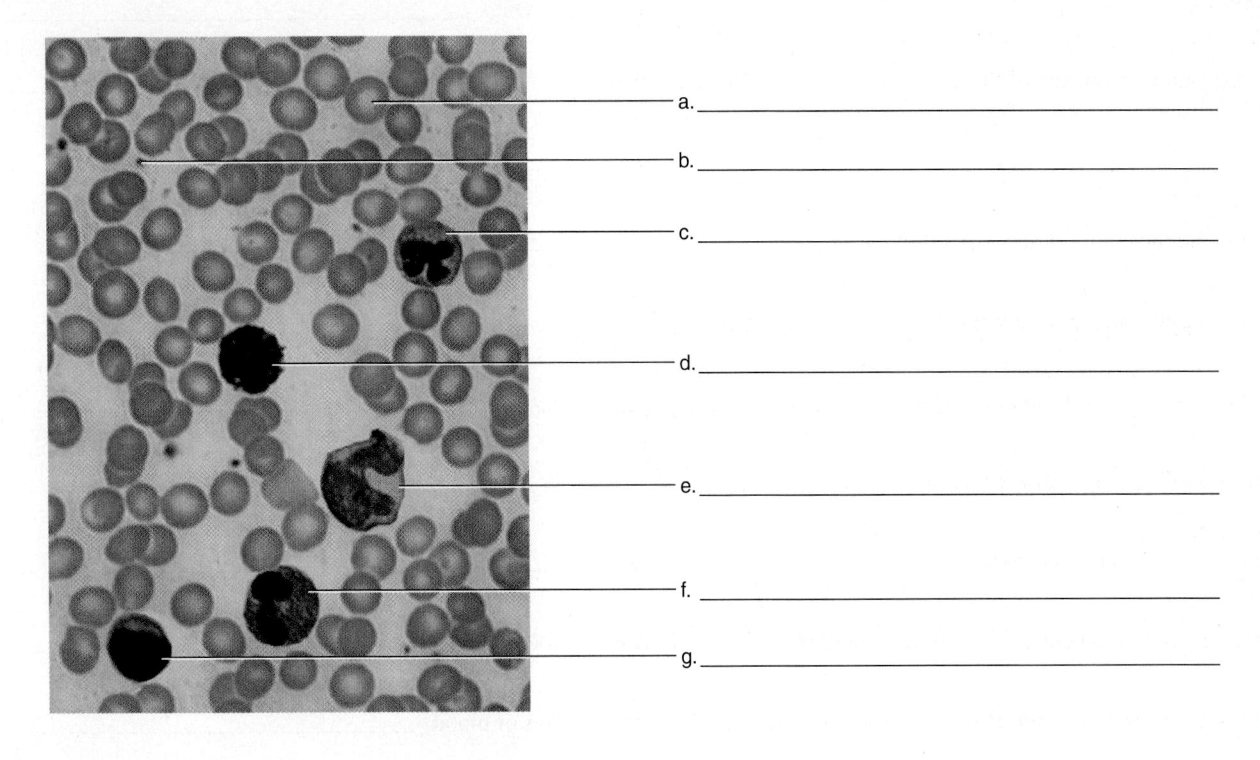

15. In terms of volume, does the blood normally contain more plasma or more formed elements? ______________________________

Exercise 26

Blood Tests and Typing

Introduction

Blood is a complex medium that has multiple functions in the body. In Exercise 25, the visual characteristics of blood were explored. In this exercise, you perform selected tests that allow you to evaluate different samples of blood. This is discussed in the Seeley text in chapter 19, "Cardiovascular System: Blood."

Many clinical tests are performed on blood to determine the number of red or white blood cells in a known volume of blood, a person's blood type, hemoglobin concentration, or other information. As in the previous exercise, handle the samples with care. The caution statement from the previous exercise is repeated here because of the importance of handling bodily fluids safely.

Caution!

The risk of bloodborne diseases has been significantly minimized by the use of sterilized human blood and nonhuman mammal blood in this exercise. However, use the same precautions as if you were handling fresh and potentially contaminated human blood. Your instructor will determine whether to use animal blood (from nonhuman mammals), sterilized blood, or synthetic blood. In any of these cases, follow strict procedures for handling potentially pathogenic material.

1. Wear protective gloves during the procedures.
2. Do not eat or drink in lab.
3. If you have an open wound, do not participate in this exercise, or make sure the wound is *securely* covered.
4. Your instructor may elect to have you use your own blood. Due to the potential for disease transmission, such as HIV or hepatitis B and C, in fresh blood samples, **make sure you keep away from other students' blood and keep them away from your blood.**
5. Talk to your instructor about this exercise if you have recently skipped a meal or if you are not well.
6. Place all disposable material in the biohazard bag.
7. Place all used lancets in the sharps container and all glassware that is to be reused in a 10% bleach solution.
8. After you have finished the exercise, clean and disinfect the countertops with a 10% bleach solution.

Objectives

At the end of this exercise, you should be able to

1. determine the antigens (agglutinogens) present in a particular ABO blood type;
2. list the antibodies (agglutinins) present in a particular ABO blood type;
3. relate Rh-positive or Rh-negative blood to antigens present;
4. perform specific diagnostic tests, such as hematocrit and hemacytometer tests;
5. demonstrate the procedure for typing blood;
6. correlate hematocrit with the red blood cell counts.

Materials

5 mL of fresh, nonhuman mammal blood (dog, sheep, or cow)
Sterilized human blood (Carolina #k3-70-0120 or other supply company), labeled 1 to 4
Blood typing antisera (anti-A, anti-B, anti-D), test cards, and toothpicks
Rh warming tray
Heparinized blood microcapillary tubes
Capillary tube centrifuge
Hematocrit reader (Criticorp Micro-hematocrit tube reader, Damon Micro-capillary reader, or mm ruler)
Seal-ease®
300 mL Erlenmeyer flask
1000 µL micropipette or 1mL pipette with pipette pump
0.9% saline solution
Hemoglobinometer
Hemacytometer, coverglass
Microscope
Protective gloves
Goggles
Tallquist paper
Tallquist chart
Cotton balls
Contamination bucket with autoclave bag
Container with 10% household bleach solution
Sharps container

Procedure

Blood Typing

An understanding of blood type is critical in clinical work. Proper matching of blood between donor and recipient is a vital process, and you learn the essentials of this process in this exercise. There are many different typing systems that match blood, such as the Kell, Lewis, MNS, and Duffy blood groups, but the ABO and Rh systems are the two most common in clinical settings. These are antigen/antibody groups, and they are often used in forensic studies. Blood types are genetically determined.

Blood cells have surface membrane molecules **(glycoproteins)** that are **antigens (agglutinogens).** If this blood is injected into a

TABLE 26.1 ABO Blood System		
Blood Type	**Agglutinogens (Antigens)**	**Agglutinins (Antibodies)**
A	A	Anti-B
B	B	Anti-A
AB	A and B	None
O	None	Anti-A and anti-B

person with **antibodies (agglutinins)** against that blood, then the injected blood clumps, or **agglutinates.** In the ABO system, no prior exposure to the antigen is needed for granulation, or agglutination, to occur. For example, if a person has **type A blood,** that person has antibodies against **type B blood.** If the person with type A blood receives a transfusion of type B blood, then the **anti-B antibodies** attack the antigens in the blood introduced into the system. This causes a **transfusion reaction,** which is the agglutination and **hemolysis** of the transfused red blood cells. If the reaction is severe enough, death may occur. Table 26.1 gives details of the ABO blood system.

In a normal blood transfusion, donated blood is matched to the exact blood type of the recipient (for example, type AB matched with type AB). In emergencies, **type O blood,** with no antigens, can be used for individuals with A, B, or AB blood. Therefore, a person with O blood is considered a **universal donor** (however, there are other factors to consider such as the reaction of other blood groups between donor and recipient or reactions between the antibodies in the donor blood). A **universal recipient** has which blood type?

Universal recipient: ______________________

Procedure for Blood Typing

(A)[1] 1. Read the entire blood typing procedure before you begin this part of the exercise.
2. Obtain a sample of sterilized blood as directed by your instructor. The vial should be labeled 1, 2, 3, or 4.
3. Record the sample number you are using in the space provided.
4. Place two separate drops of the sample blood on a blood test card or on a very clean glass microscope slide (figure 26.1).
5. Place **antiserum A** on one drop and **antiserum B** on the other drop.
6. Use separate toothpicks to stir each sample of blood and antiserum.
7. Keep the antisera separate from one another. Examine the sample for clumping, or granulation, within 2 minutes. Make sure that you dispose of your toothpicks in the biohazard bag. If both blood samples coagulate, you have a sample of type AB blood. If neither sample agglutinates, you have a sample of type O blood. If the blood with antiserum A agglutinates, you have a sample of type A blood, and if blood with antiserum B agglutinates, you have a sample of type B blood.
8. Compare your sample with figure 26.2.

Blood sample number: ______________________

Blood type: ______________________

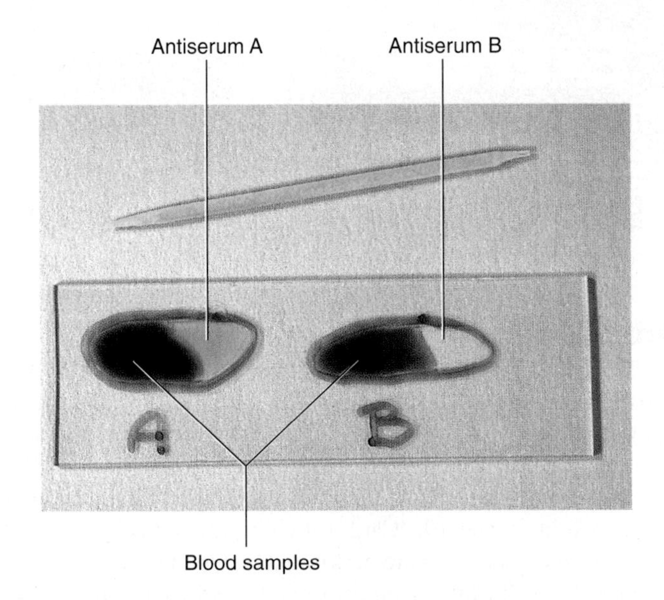

Figure 26.1 Blood Testing Procedure

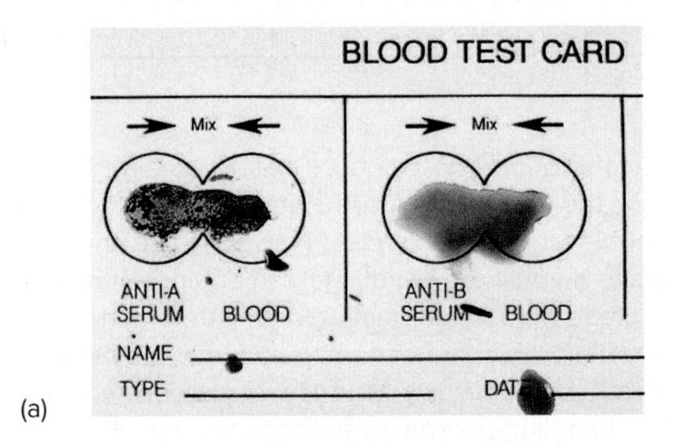

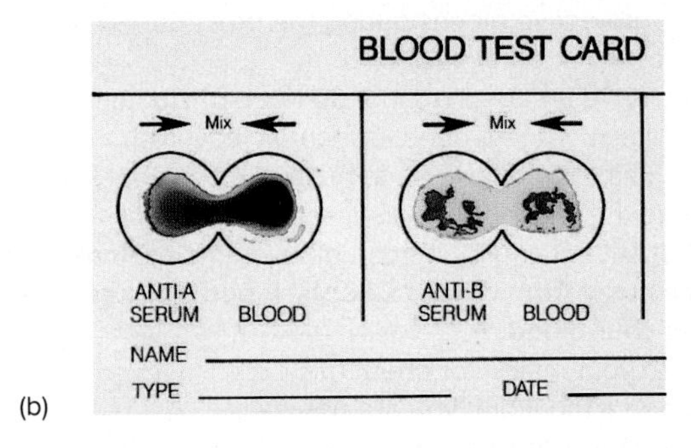

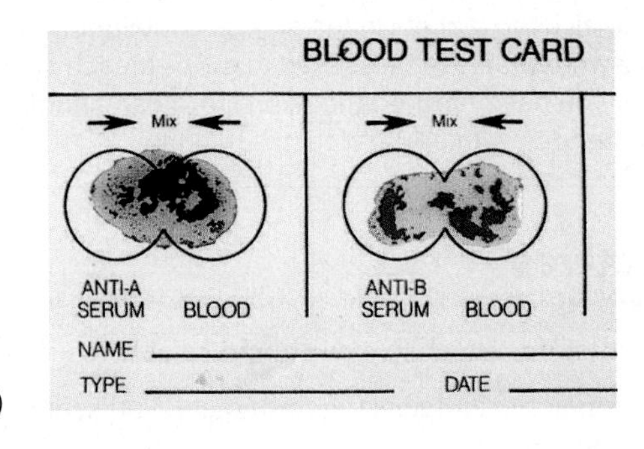

Figure 26.2 Blood Types
(a) Type A; (b) type B; (c) type AB.

TABLE 26.2 Percent U.S. Population by Blood Type

Blood Type	Caucasians	African Americans	Asian Americans	Native Americans
A	41	27	28	16
B	9	20	23	4
AB	3	7	13	1
O	47	46	36	79
Total	100	100	100	100

Common ABO blood types for various groups in the United States are listed by percentages in table 26.2.

Determination of Rh Factor

(A)[2] The other blood system of clinical significance is the Rh system. *Rh* stands for *rhesus monkey* (the animal in which the Rh system was discovered). About 85% of the U.S. population is Rh-positive; 15% is Rh-negative. An Rh-positive individual carries the **Rh antigen** (surface membrane marker). The Rh-negative individual does not carry the Rh antigen but develops **antibodies** to the antigen after exposure to the Rh antigen.

Determination of the Rh type is a much more subtle test than ABO typing. The **Rh antiserum (anti-D)** is more fragile in shipping and storage and should not be considered clinically relevant in this exercise. Use the same precautions as outlined at the beginning of the exercise regarding bodily fluids.

1. Place one drop of blood on a slide and place a drop of antiserum-D (Rh antiserum) on the slide.
2. Use a new toothpick and stir the two drops together.
3. Place this mixture on a warming tray and gently rock the slide.
4. Examine the slide for clumping after a minute or two. If clumping occurs, then the sample is Rh-positive (the Rh antiserum reacted with the Rh antigen in the blood). If no clumping occurs, then the sample is Rh-negative. This test is more difficult to determine, so look for slight granulation. Rh antiserum should be used fresh.
5. Record your results.

   Rh factor determination: ______________________________

Rh determination is very important if a pregnant woman is Rh-negative. If her unborn child is Rh-positive, then she may develop antibodies against the Rh antigen in the fetal blood during pregnancy or delivery (a time when fetal and maternal blood often mix). If her second child is Rh-positive, then the antibodies that she produced during her first pregnancy may cross the placenta and cause severe reactions in the fetus. This is called **hemolytic disease of the newborn (HDN),** or *erythroblastosis fetalis.* Rho-GAM, an Rh immune globulin, prevents antibody formation in the mother against the Rh antigen and is often given to the mother during pregnancy and again after delivery. A woman who is Rh-positive does not have the antibodies for the Rh factor and therefore will not produce antibodies against the developing fetus, regardless of the Rh factor of the fetus.

Blood Typing Problems

Extensive blood transfusions can lead to problems for the recipient of the transfusions. Blood that is donated from one person carries antibodies, and these may react with the recipient's blood. There are other blood types in addition to the ABO and Rh systems, and sensitivities may develop due to antibodies present in the recipient's blood.

Hematocrit

Hematocrit, or **packed cell volume (PCV),** is the percentage of red blood cells (or blood cells in general) in the total blood volume. The percentage of red blood cells, or total cell volume (red blood cells and white blood cells), can be calculated after centrifuging a sample of blood. The cells end up as a large sediment in the bottom of the microcentrifuge tube, leaving the less dense plasma on top. Normal hematocrit values for females are 38–47%, and values of 40–54% are normal for males. An increase in the hematocrit above normal is known as **polycythemia** and can exceed 65%. When blood is lost faster than it is replaced or when the production of red blood cells is low, **anemia** occurs. In anemia, the hematocrit may drop to 15% or less. Anemia may also be due to low levels of hemoglobin in the blood. **Hemoglobin** is a complex molecule composed of an iron-containing **heme** group and the protein **globin.** You will next compare the hematocrit of sterilized human blood with that of another mammal.

Procedure for Determining Hematocrit

(A)[3] Refer to the caution statement at the beginning of the exercise concerning working with blood.

1. While wearing protective gloves, fill two capillary tubes with blood. One should have commercially prepared, sterilized human blood and the other should have nonhuman mammal blood.
2. Fill each tube by touching the red end of the capillary tube to the blood sample (figure 26.3). Let the blood flow up the tube by capillary action.

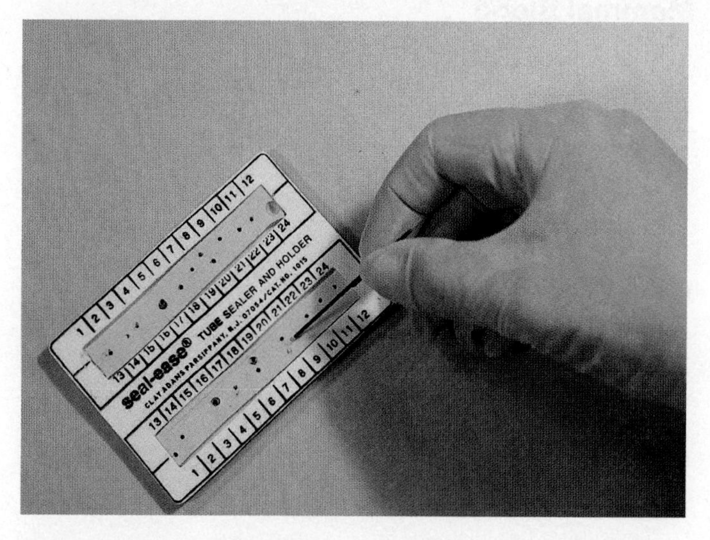

Figure 26.3 Preparation of a Hematocrit Tube

3. Once the tube is filled, place your finger on the other end of the tube (the nonred end) to prevent blood from flowing back out. Fill each tube to about three-quarters of the tube length. Both tubes should have about the same volume of blood in them.
4. Seal the red end of the tube with Seal-ease® or modeling clay. Be careful—the capillary tubes are both fragile and sharp. They can break and puncture the skin.
5. Place the capillary tube in the centrifuge with the clay plug against the outer rubber ring.
6. Place your tubes opposite each other and record the number of each tube. Make sure you keep track of the sample from each tube and do not mix them up.

Mammal blood tube #: ______________________

Sterilized blood tube #: ______________________

After many tubes have been placed into the centrifuge, make sure that each has the clay seal facing the rubber gasket. If the seals are facing the center of the centrifuge, the blood will spray out of the tube when the centrifuge begins turning. The tubes should be opposite one another, so that the centrifuge is balanced. **Make sure the metal top is on the centrifuge and screwed in place.** Close the cover and spin the capillary tubes for 3 to 5 minutes.

Once the centrifuge has come to a complete stop, remove each tube and calculate the percentage of red blood cells in the tube (figure 26.4). This can be done with hematocrit readers or a ruler. Your instructor can help you use the hematocrit readers. If you use a ruler, you need to measure the total height of the blood column in millimeters and subsequently the total height of the red blood cells in the column. If you use a ruler to measure the hematocrit, record your results here.

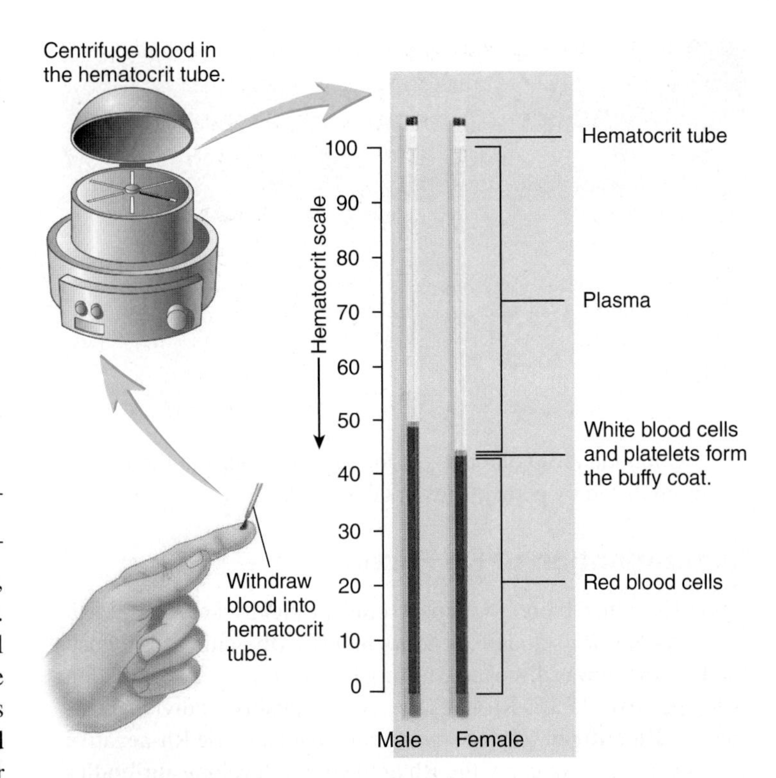

Figure 26.4 Hematocrit

What commercial procedure might lead to a difference in the hematocrit between these two samples? ______________

Plasma with red coloration indicates hemolysis of the red blood cells. Normally, the plasma should be straw-colored or light yellow. Does either of your samples show hemolysis?

____

Note the buff-colored layer between the red blood cell layer and the plasma layer. This is the white blood cell layer.

Sterilized Blood

Red blood cell height (mm): ______________________

Total column height (mm): ______________________

Mammal Blood

Red blood cell height (mm): ______________________

Total column height (mm): ______________________

The hematocrit can be figured by the following equation:

$$\frac{\text{Millimeters of red blood cell}}{\text{Millimeters of total blood}} \times 100 = \text{Hematocrit}$$

Record the hematocrit.

Sterilized blood: ______________________

Mammal blood: ______________________

How do these values compare? ______________________

____

____

Hemoglobin Determination

Anemia is a condition whereby the oxygen-carrying capacity of the blood is diminished, which may be due to a reduced number of erythrocytes in the blood and show up as a low hematocrit. In some cases the hematocrit is normal but there is a decrease in hemoglobin levels in the individual red blood cells.

Hemoglobinometer Method

(A) 4 Levels of hemoglobin can be examined with the use of a hemoglobinometer. This instrument uses differences in the absorption of green light by hemoglobin to measure its concentration in the blood. Different hemoglobinometers are available. The following test is for the Leica (AO) hemoglobinometer. You can determine the level of hemoglobin in the blood with the following procedure.

1. Make sure that the hemoglobinometer has fresh batteries installed and that the light turns on when the light switch is depressed.

2. Place two glass plates together, so that there is a chamber on the inside, and insert them into the metal clip.
3. Fill the space between the glass plates with blood.
4. Insert the metal clip with the glass plates into the hemoglobinometer.
5. Move the lever on the side of the hemoglobinometer until the green colors are the same shade.
6. Read the scale to determine the hemoglobin concentration.
7. Carefully remove the glass plates and clean them with a cotton ball soaked in alcohol. Dispose of the cotton ball in a biohazard container and place the glass plates in a secure area where they can be used by other students or placed in a 10% bleach solution.
8. You may want to compare two different samples of blood, such as a sample of sterilized blood versus nonhuman mammal blood. Record your results.

Hemoglobin (Hb) levels are expressed as grams Hb/100 mL of blood. The average value for humans is 12 to 16 grams/100 mL of blood. In males the normal range is 14 to 18 g/100 mL and for females it is 12 to 16 g/100 mL.

Hemoglobin concentration:

Sample 1______________

Sample 2 ______________

Tallquist Method

(A) 5 Another method to determine the level of hemoglobin in the blood is by comparing the color of the oxyhemoglobin in a drop of blood on Tallquist paper to a comparative chart.

1. Place a piece of Tallquist paper on a paper towel on your desk.
2. Place a drop or two of blood on the paper and wait 15 seconds.
3. Compare the color on your paper to the Tallquist chart.

Hemoglobin from sterile blood sample: ______________

Hemoglobin from mammal blood sample: ______________

Blood Cell Counts

Another way to determine the number of blood cells is to count the number of cells in a known volume and subsequently determine the total number of cells per cubic millimeter (mm^3). Modern evaluation of erythrocyte and leukocyte counts is performed by injecting blood samples into an optical computer system, which automatically calculates the cell counts. Most hospitals use computer-driven optical systems to count cells.

The average number of erythrocytes in blood is approximately 5 million per cubic millimeter. This is too large a number to count during the lab period, so another technique will be used. To determine the number of cells in blood, you will dilute the blood and count the cells that occur in a small volume. The process used in lab will involve the use of a **hemacytometer,** a glass slide that has very fine lines etched on its surface. When a coverslip is placed on the hemacytometer and a sample of blood is introduced, the blood fills a precise volume. This volume is $1/50^{th}$ of a cubic millimeter.

Even with this small volume there are too many erythrocytes in the blood in a given volume to be counted accurately. The solution to this is to dilute the blood so that the cells can be counted individually. In the procedure that you will perform, you will take 1 part of blood and dilute it to 200 parts diluent (dilution solution).

Procedure for Red Blood Cell Counts

For males, the average count is about 5.4 million red blood cells, or erythrocytes, per cubic millimeter of blood, with a range of 4.6 to 6.2 million cells per cubic millimeter. For females, the average count is about 4.8 million erythrocytes per cubic millimeter of blood, with a range of 4.2 to 5.4 million cells per cubic millimeter.

(A) 6 You will perform this experiment twice: once using commercially prepared, sterilized human blood and once using fresh, nonhuman mammal blood. Obtain the following:

Micropipette or pipette and pipette pump
300 mL Erlenmeyer flask
0.9% saline solution
Sample of nonhuman mammal blood
Sample of sterilized blood
Hemacytometer and coverslip
Protective gloves and eyewear
Microscope
Cotton balls

Because of the volume of solution made, you may want to use one dilution of blood for the entire class, depending on the wishes of your instructor. It is particularly important to accurately measure the blood withdrawn and the amount of diluent in order to make an accurate erythrocyte count. For withdrawing blood from your samples you may use either a micropipette or a volumetric pipette. The following instructions provide directions for both. Make sure that the blood sample you use is well mixed, so that blood cells are evenly distributed in the plasma. Wear protective gloves and eyewear during the entire procedure.

1. Add exactly 200 mL of 0.9% saline solution to an Erlenmeyer flask.
2. Use either a 1,000-microliter micropipette or a 1 mL pipette with a pipette pump. If you use a micropipette, depress the plunger until it stops and insert the tip into the blood sample. Slowly release the plunger to pull blood into the micropipette. If using the 1 mL pipette, use a pipette pump attached to the blunt end of the pipette and draw up exactly 1 mL of blood to the calibration line of the pipette. Never pipette anything by mouth! Use a pipette bulb or pump.
3. Transfer the blood to the Erlenmeyer flask. If you use a micropipette, depress the plunger until it stops and then push it a little farther. It you are using a volumetric pipette (one that has an expanded middle section), the blood should drain out of the pipette and you should NOT forcibly expel the remainder of the fluid in the pipette. Dispose of the pipette in a tray for that purpose or in a beaker of 10% bleach.

4. Gently swirl the flask to distribute the blood cells. Withdraw the diluted sample with a Pasteur pipette.
5. Place the hemacytometer coverslip (not a regular coverslip) over the hemacytometer chamber and fill the chamber with the Pasteur pipette. Do not overfill the chamber.
6. Place the hemacytometer under the microscope and examine the cells under high power (400–450×).
7. Count all of the erythrocytes in each of the five areas marked with an "R" (see figure 26.5). Some cells will be on the border of the counting grid. Count only those cells touching the left line and the upper line as part of the sample. Do not count the cells touching the right line or bottom line.

The blood was diluted 1:200, and the sum of the volume in the five areas represents 1/50th of a cubic millimeter. If you multiply the dilution factor (200) and the volume that would occur to fill 1 mm^3 (50), then you would multiply the number of erythrocytes counted by 10,000 to obtain the number of erythrocytes per cubic millimeter.

8. Record your results.

Number of erythrocytes counted: ____________________

Number of erythrocytes per cubic millimeter: __________

Use the sample of nonhuman mammal blood. Use the chamber on the opposite side of the hemacytometer and follow the same procedure you used for human blood. Record the results.

Number of erythrocytes counted: ____________________

Number of erythrocytes per cubic millimeter: __________

How do the numbers of cells in sterilized blood and nonhuman mammal blood compare? ____________________

Is there a correlation between the numbers of erythrocytes counted from these samples and the hematocrit taken from the same samples? ____________________

After you finish your counting, place the hemacytometer and coverslip *gently* into a 10% bleach solution.

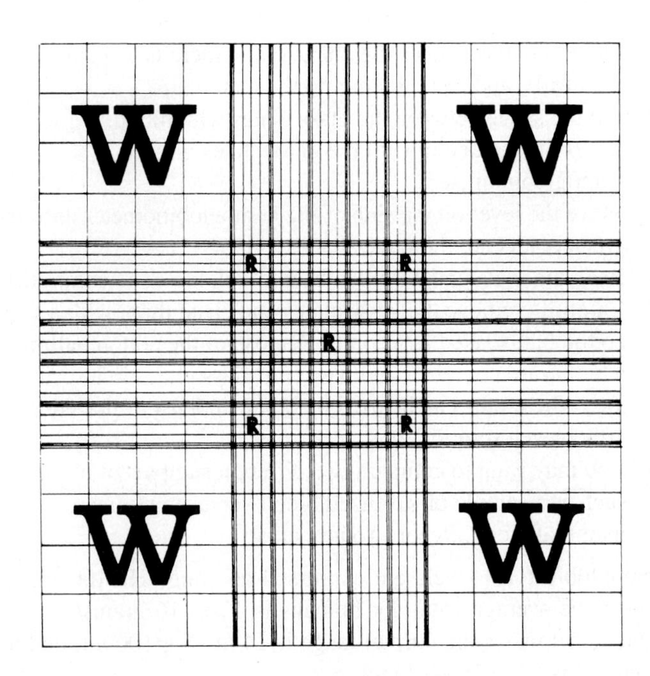

Figure 26.5 Erythrocyte Counts Using a Neubauer Hemacytometer

Regions indicated by *R* are for erythrocyte counts.

Leukocyte Counts (Optional)

Ⓐ7 Leukocytes are counted in a similar fashion except that the four large corner regions of the hemacytometer are used and the sum of the cell count from these four areas is multiplied by 50. Your instructor may provide you with additional instructions for a leukocyte count, using a specific solution for leukocytes.

Number of leukocytes counted: ____________________

Number of leukocytes per cubic millimeter: __________

Normal blood levels for leukocytes are about 7,000 cells per cubic millimeter, with a range of 5,000 to 10,000 cells per cubic millimeter. Leukocytosis is an elevated white blood cell count typically above 11,000 cells per cubic millimeter. Leukocytosis indicates inflammation. Leukopenia, on the other hand, is a decreased white blood cell count, with less than 4,000 cells per cubic millimeter. This may be due to chemotherapy, radiation therapy, HIV infection, infectious hepatitis, cirrhosis of the liver, typhoid fever, or chronic infections, such as tuberculosis.

Cleanup

Make sure the lab is clean before you leave. Wipe any blood from the microscope objective lenses with lens cleaner and lens paper. Make sure you do not leave glass or blood on the lab tables. Use a 10% bleach solution and a towel to clean off your lab table before you leave.

Exercise 26 Review

Blood Tests and Typing

Name: ______________________

Lab time/section: ______________________

Date: ______________________

1. What is the name of a surface membrane molecule on a blood cell that causes an immune reaction? ______________________

2. What ABO blood type is found in a person who is a universal donor? ______________________

3. What is the average range of hematocrit for a normal female? ______________________

4. What is the average range of hematocrit for a normal male? ______________________

5. What percentage of the blood volume consists of formed elements? ______________________

6. A person with blood type B has what kind of antibodies? ______________________

7. A person has antibody A and antibody B in his or her blood with no Rh antibody. What blood type does this person have? ______________________

8. A total of 240 red blood cells are counted in the hemacytometer chamber. What is the red blood cell count of this person in cells per cubic millimeter? ______________________

9. A person with blood type B negative is injected with type A positive blood. From an immunological (antigen/antibody) standpoint, what will happen after the injection? ______________________

10. How might changes in the pipette technique alter the final determined value of red blood cells? What kinds of errors might you expect? ______________________

11. Using the following illustration, calculate the hematocrit of the individual. Determine if it falls within normal limits.

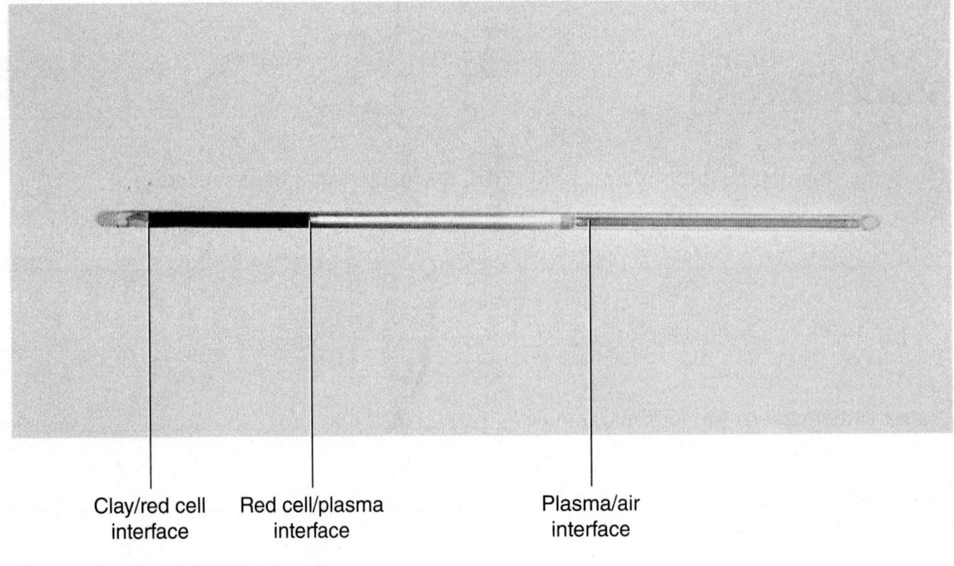

Determination of Hematocrit

12. Define *anemia.* ______________________________

____

____

____

____

13. Explain the possible erroneous results that you might get if you used just one toothpick to stir the various blood types in the ABO blood test. ______________________________

Exercise 27

Structure of the Heart

Introduction

In this exercise, you study the structure of the heart, which is covered in the Seeley text in chapter 20, "Cardiovascular System: The Heart." In the general circulatory pattern, vessels that carry blood from the heart are called **arteries,** and those that return blood to the heart are called **veins.** The **cornary arteries** carry blood from the heart (thus, they are arteries), but they carry the blood to the heart muscle. The **cardiac veins** take blood from the heart muscle back to the heart for recirculation.

In this exercise, you examine models of the heart, preserved sheep hearts, and human hearts, if available. As you look at the preserved material, determine how the structure of the heart relates to its function.

Objectives

At the end of this exercise, you should be able to

1. list the three layers of the heart wall;
2. describe the position of the heart in the thoracic cavity;
3. describe the significant surface features of the heart;
4. describe the internal anatomy of the heart;
5. find and name the anatomical features on models of the human heart and in the sheep heart;
6. describe the blood flow through the heart and the function of the internal parts of the heart;
7. discuss the functioning of the atrioventricular valves and the semilunar valves and their role in circulating blood through the heart.

Materials

Models and charts of the heart
Preserved sheep hearts
Preserved human hearts (if available)
Blunt probes (Mall probes)
Dissection pans
Scalpels
Sharps container
Protective gloves
Waste container
Microscopes

Procedure

Heart Wall

The heart is located deep in the thorax between the lungs in a region known as the mediastinum. The **mediastinum** contains the heart, the coverings of the heart (the pericardia), and other structures, such as the esophagus and descending aorta. The mediastinum is located between the sternum, the lungs, and the thoracic vertebrae and is illustrated in figure 27.1.

If you open the chest cavity, the first structure you see is the **fibrous pericardium.** This tough, outer connective tissue sheath encloses the heart. The lining deep to the fibrous pericardium is the **parietal pericardium.** Deep to the parietal pericardium is the **pericardial cavity,** which contains a small amount of **serous pericardial fluid.** This fluid reduces the friction between the outer surface of the heart and the parietal pericardium. The layer closest to the heart is the **serous pericardium.** Locate these pericardial layers in figures 27.1 and 27.2.

The heart wall is composed of three major layers. The outermost layer is the **epicardium,** or **serous pericardium,** which is composed of epithelial and connective tissue. The middle layer, the **myocardium,** is the thickest of the three layers. It is mostly made of cardiac muscle. You may wish to review the slides of involuntary cardiac muscle and note the intercalated disks, branching fibers, and fine striations of cardiac tissue (as described in Exercise 4). The cardiac muscle is arranged spirally around the heart, providing a more efficient wringing motion to the heart. The inner layer of the heart wall is the **endocardium,** which is a serous membrane and consists of **endothelium** (simple squamous epithelium) and connective tissue. These layers are seen in figure 27.2.

It is best to examine heart models before dissecting a sheep heart, unless your instructor directs you to do otherwise. Heart models are color-coordinated and labeled to make the structures easier to locate.

Examination of the Heart Model

Overview

The heart is commonly described as an organ about the size of your fist, located in the chest cavity, and tipped slightly to the left. It is a four-chambered pump with two superior atria and two inferior ventricles. Blood enters the heart in the **right atrium** (figure 27.3) and flows into the **right ventricle.** From the right

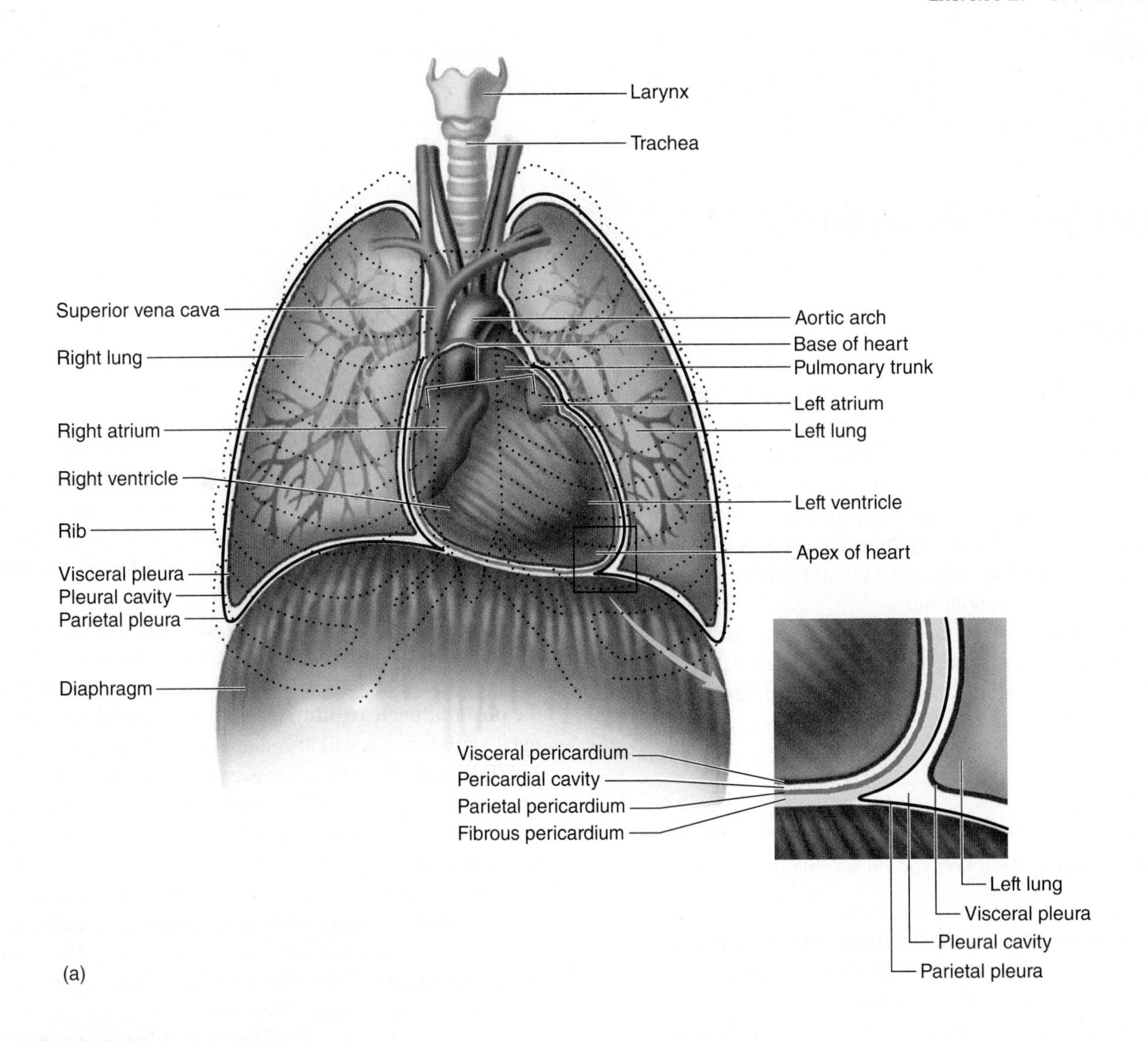

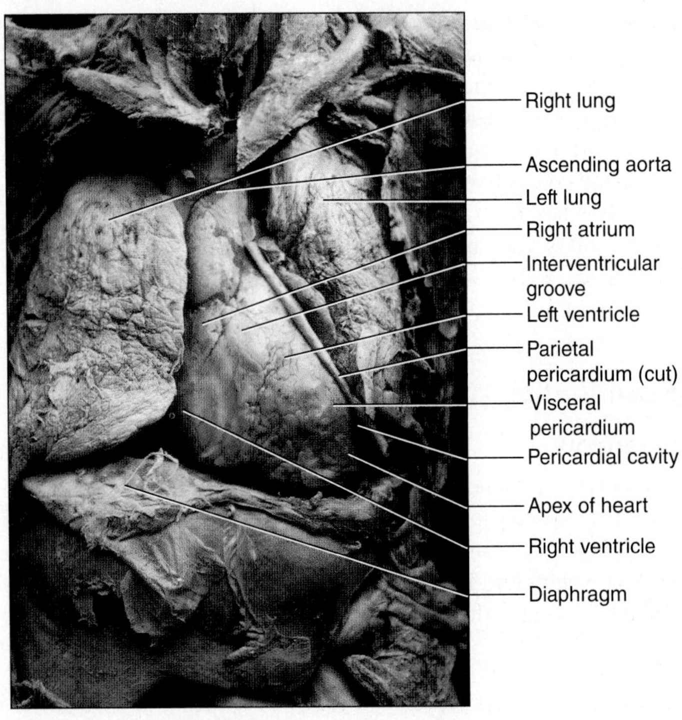

Figure 27.1 Heart in Thoracic Cavity

(a) Diagram; (b) photograph of heart with parietal and visceral pericardium.

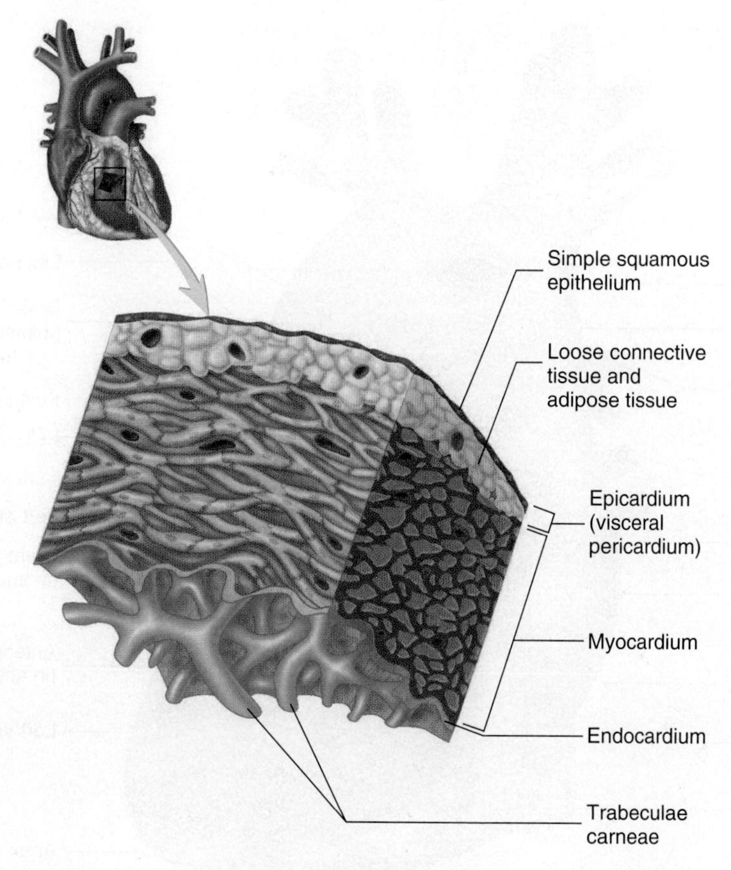

Figure 27.2 Layers of the Heart Wall

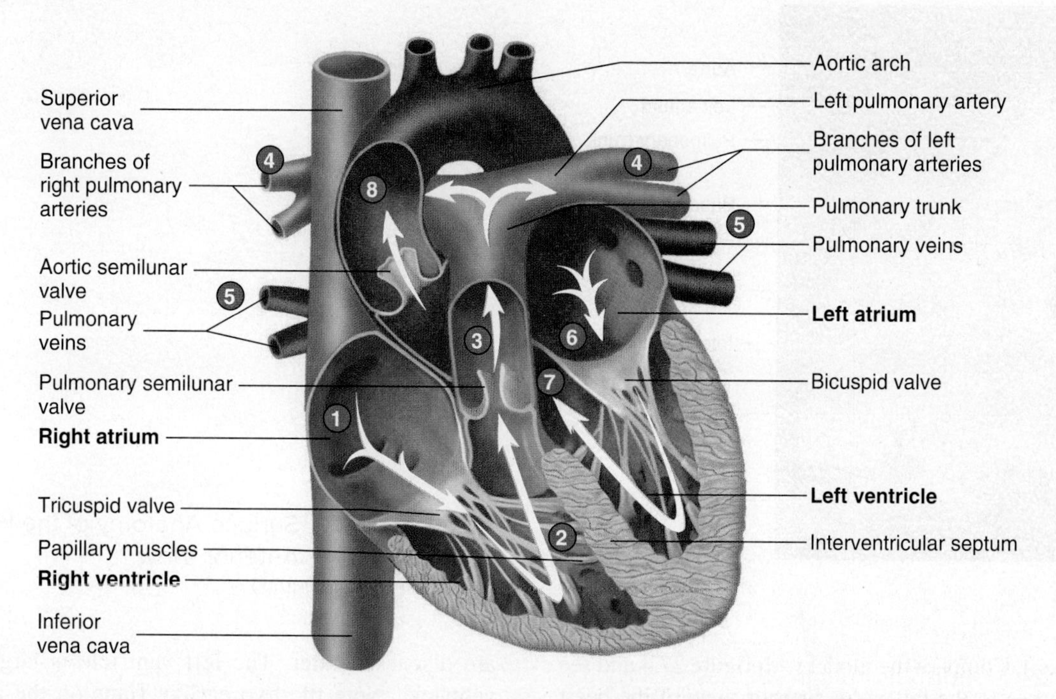

Figure 27.3 Chambers of the Heart. Numbers indicated blood flow through the heart.

ventricle, the contraction of the ventricular wall sends blood to the lungs. The blood is oxygenated in the lungs and returns to the heart by entering the **left atrium.** Blood moves from the left atrium to the **left ventricle** and is then pumped from there to the rest of the body.

Exterior of the Heart

A[1] Examine the heart model and notice that the heart has a pointed end, or **apex,** and a blunt end, or **base.** The apex of the heart is inferior, and the great vessels leaving the heart are located at the base (therefore, in the case of the heart, the base is

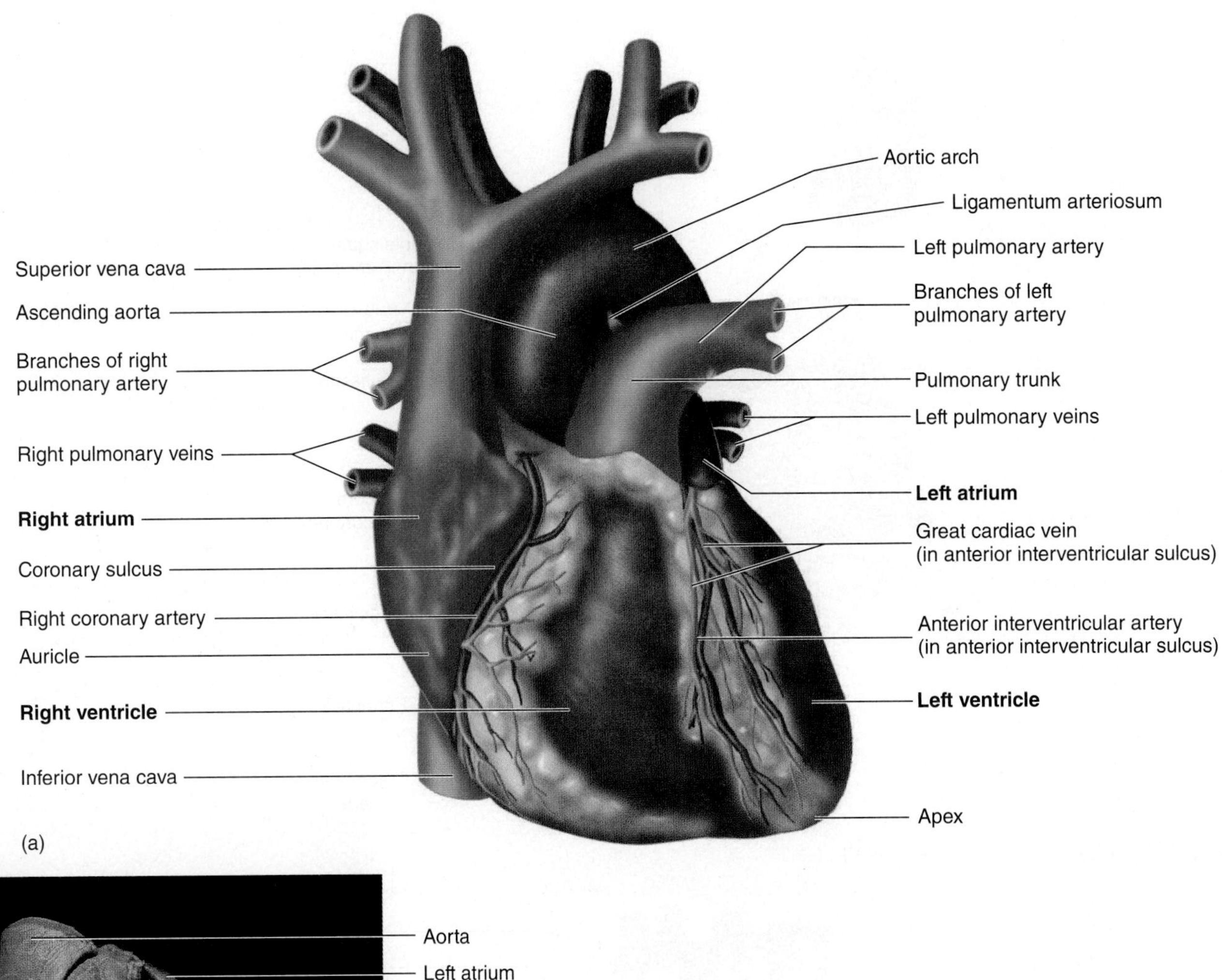

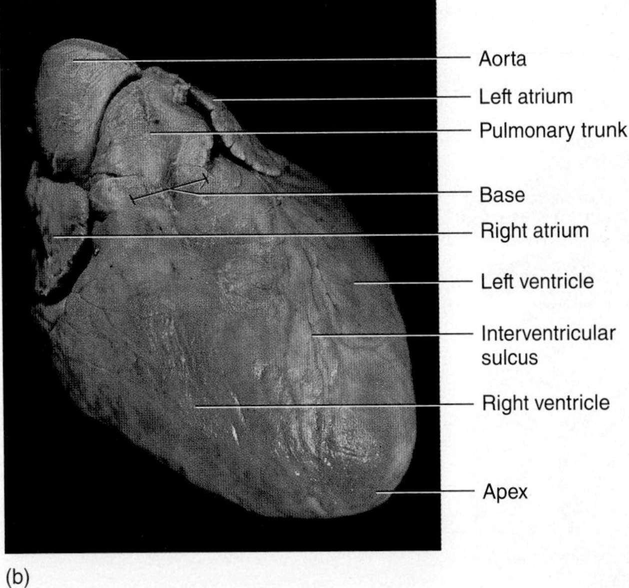

Figure 27.4 Surface Anatomy of the Heart, Anterior View

(a) Diagram; (b) photograph.

superior to the apex). Compare the model with figure 27.4 and see how the **aorta** curves to the left in an anterior view of the heart and is posterior to the **pulmonary trunk.**

Locate the anterior features of the heart. The heart is composed of two large inferior **ventricles** and two smaller and superior **atria.** The left ventricle extends to the apex of the heart and is delineated from the right ventricle by the **interventricular sulcus,** or **groove.** In this interventricular sulcus are some of the **coronary arteries** and **cardiac veins,** which are discussed later. The left ventricle is larger than the right ventricle. Note the two earlike flaps on the anterior, superior region of the heart. These structures are the **auricles,** which are part of the atria.

If you examine the heart from the posterior side, you will see the atria more clearly. At the junction of the right atrium and the right ventricle is the **atrioventricular sulcus,** or **groove.** The **coronary sinus,** a large venous chamber that carries blood from the cardiac veins to the right atrium, is in this sulcus. Locate the

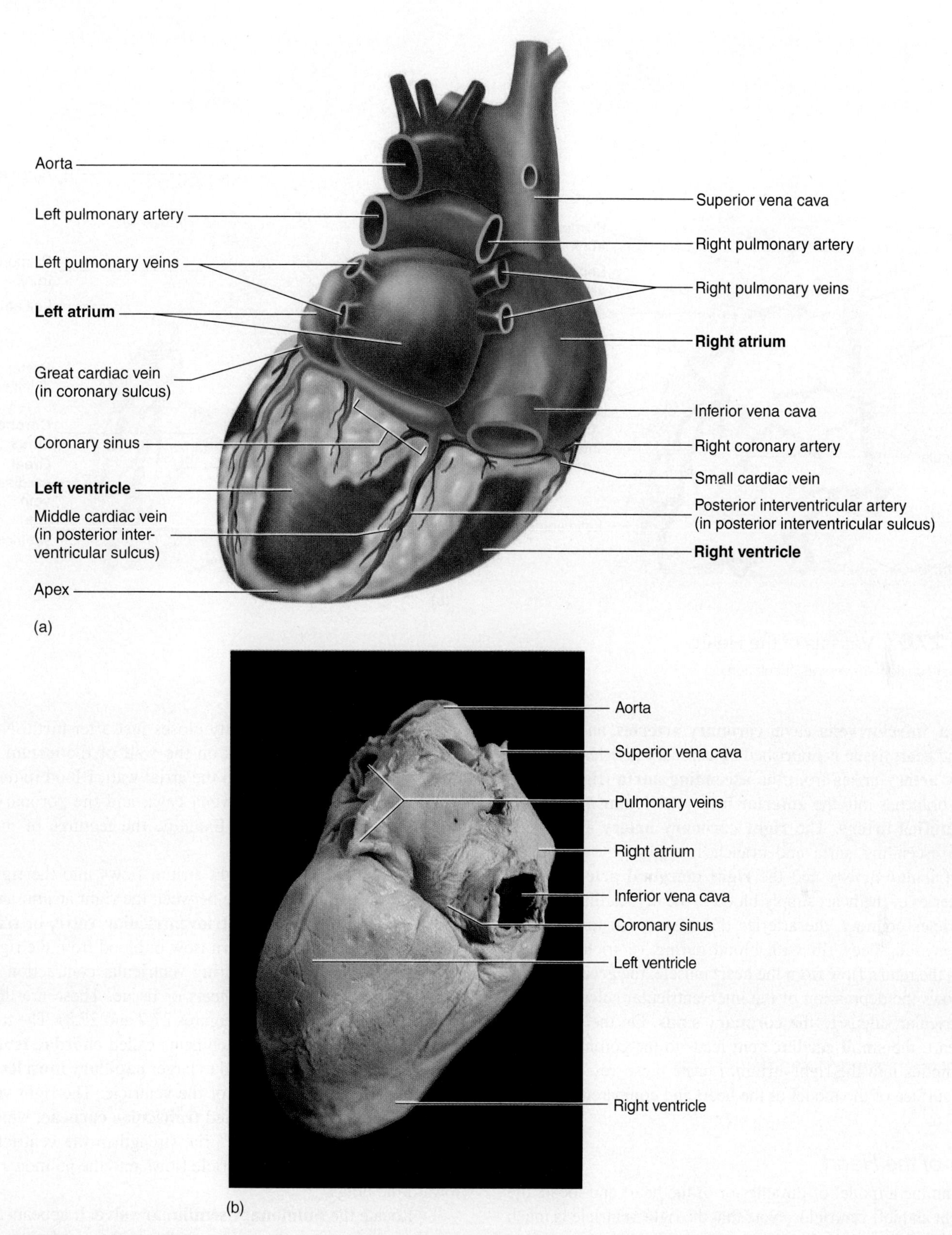

Figure 27.5 Surface Anatomy of the Heart, Posterior View
(a) Diagram; (b) photograph of cadaver heart.

superior vena cava and the **inferior vena cava,** two vessels that also return blood to the right atrium. Locate the **pulmonary veins,** which carry blood from the lungs to the left atrium. Compare the heart model with figure 27.5.

The major vessels of the heart are illustrated in figures 27.4 and 27.5. Locate the **pulmonary trunk, pulmonary arteries, ligamentum arteriosum** (between the pulmonary trunk and aortic arch), **ascending aorta, pulmonary veins, superior**

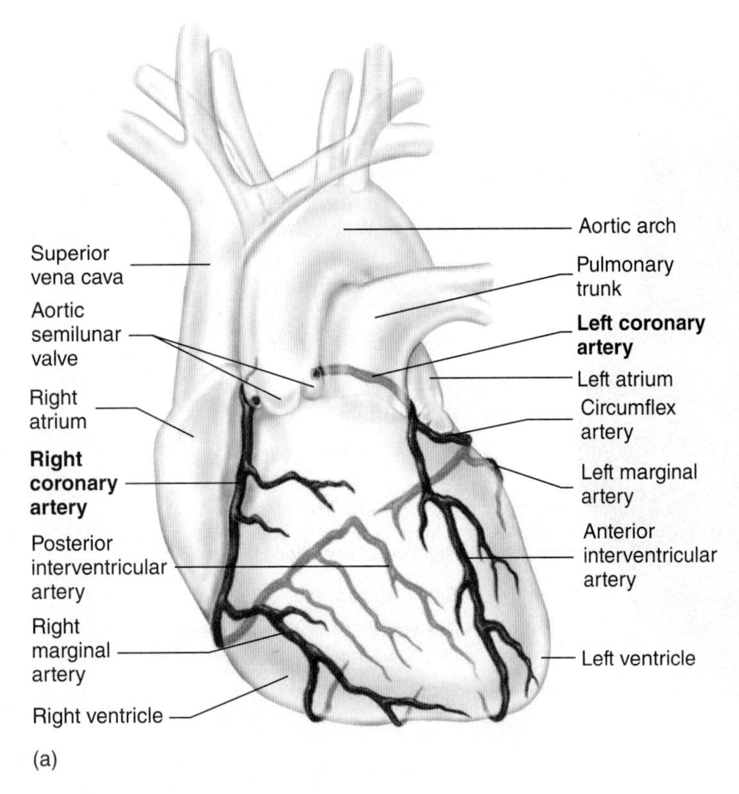

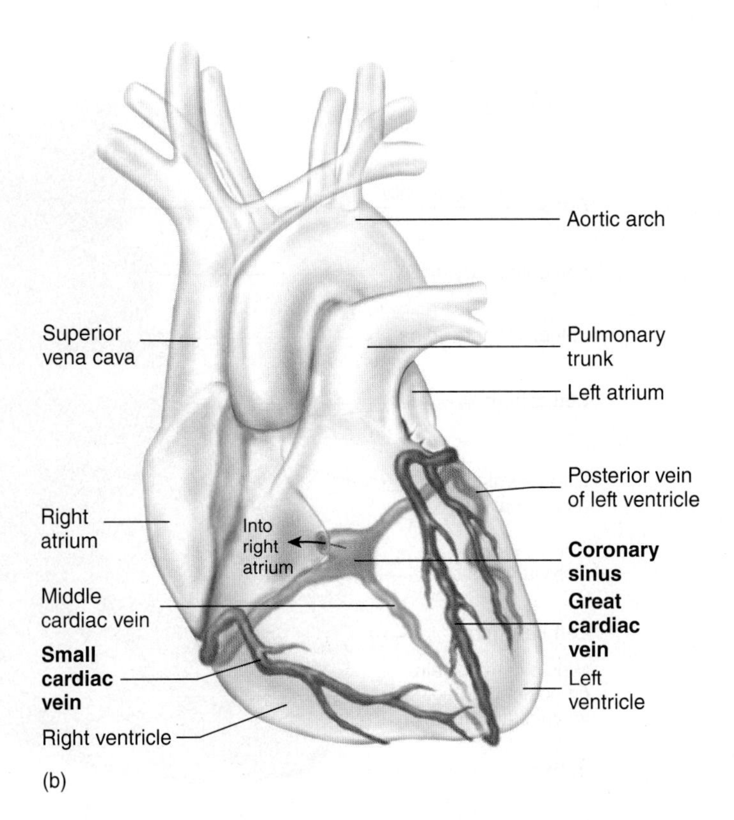

Figure 27.6 Vessels of the Heart
(a) Arterial circulation; (b) venous circulation.

vena cava, inferior vena cava, coronary arteries, and **cardiac veins.** The heart tissue is nourished by coronary arteries. The **left coronary artery** arises from the **ascending aorta** (figure 27.6) and then branches into the **anterior interventricular artery** and the **circumflex artery.** The **right coronary artery** also arises from the ascending aorta and branches to form the **posterior interventricular artery** and the **right marginal artery.** These major arteries of the heart supply blood to the myocardium. When the ventricles contract, the arteries that take blood to the heart are compressed. They fill with blood during ventricular relaxation. On the return flow from the heart muscle, the **great cardiac vein** follows the depression of the interventricular sulcus and the atrioventricular sulcus to the **coronary sinus.** On the right side of the heart, the **small cardiac vein** leads to the coronary sinus, which empties into the right atrium. Locate these vessels on the external surface of the model of the heart and compare them with figure 27.6.

Interior of the Heart

(A)[2] Examine a model of the interior of the heart and locate the right and left ventricles. Note that the right ventricle is much thinner-walled than the left ventricle. The ventricles are separated by the **interventricular septum,** which forms a wall between the two ventricular chambers. Compare the model with figure 27.7.

Examine the right atrium and note how thin the wall is, compared with the ventricles. Examine the medial wall of the atrium, known as the **interatrial septum,** and locate a thin, oval depression in the atrial wall. This depression is the **fossa ovalis** (figure 27.8). In fetal hearts, this is the site of the **foramen ovale,** but the foramen usually closes just after birth. Note the extensive **pectinate muscles** on the wall of the atrium. These provide additional strength to the atrial wall. Blood in the superior vena cava, the inferior vena cava, and the coronary sinus returns to the right atrium. Examine the features of the right atrium in figure 27.8.

The blood from the right atrium flows into the right ventricle. Now examine the valve between the right atrium and right ventricle. This is the **right atrioventricular valve,** or **tricuspid valve,** and it prevents the return flow of blood from the right ventricle into the right atrium during ventricular contraction. Examine the valve for three flat sheets of tissue. These are the three **cusps** of the tricuspid valve (figures 27.7 and 27.8). The tricuspid valve has thin, threadlike attachments called **chordae tendineae.** These tough cords are attached to larger **papillary muscles,** which are extensions from the wall of the ventricle. The right ventricle wall has small extensions called **trabeculae carneae,** which, like the pectinate muscles of the atria, strengthen the ventricle wall. The blood from the right ventricle flows into the pulmonary trunk toward the lungs.

Locate the **pulmonary semilunar valve.** It appears as three small cusps between the right ventricle and the pulmonary trunk and keeps blood from flowing in reverse from the pulmonary trunk into the right ventricle during ventricular relaxation. Examine the details of the right ventricle in models in the lab and in figures 27.7 and 27.8.

Blood from the pulmonary trunk flows through the pulmonary arteries before entering the lungs. Blood in the lungs releases carbon dioxide and picks up oxygen. The **pulmonary veins** carry

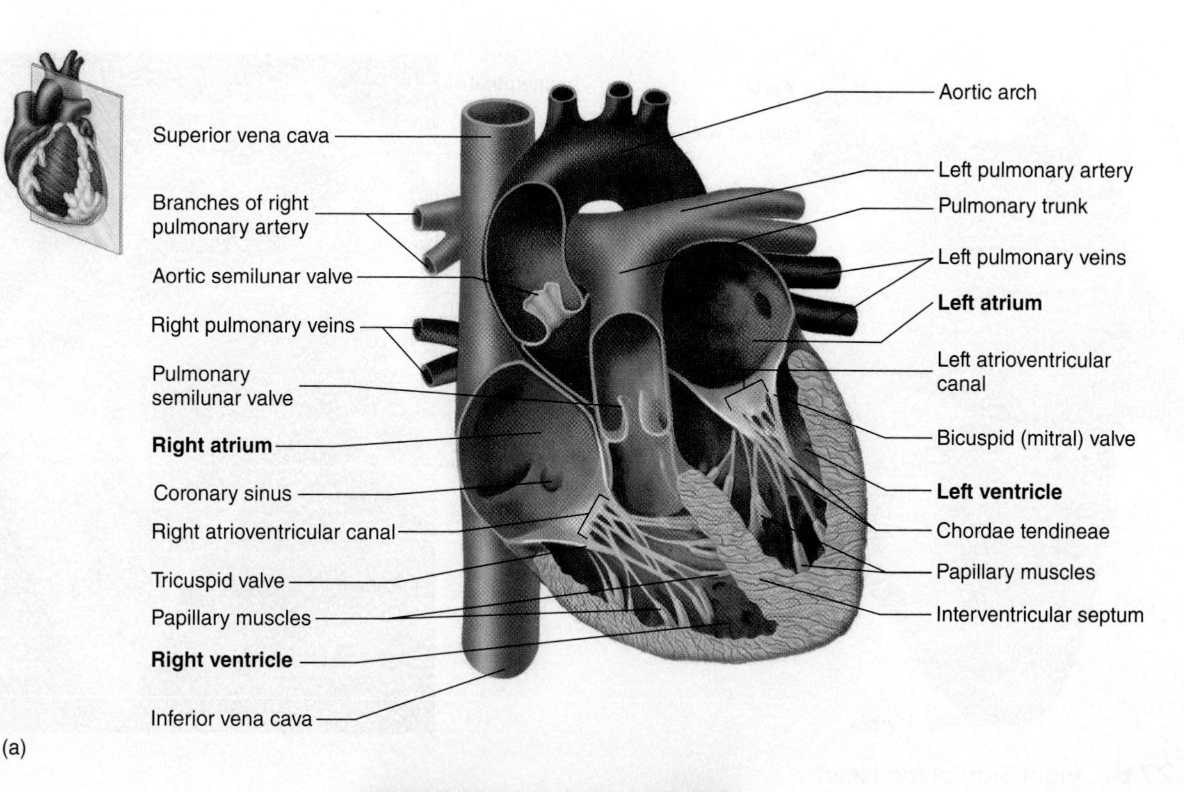

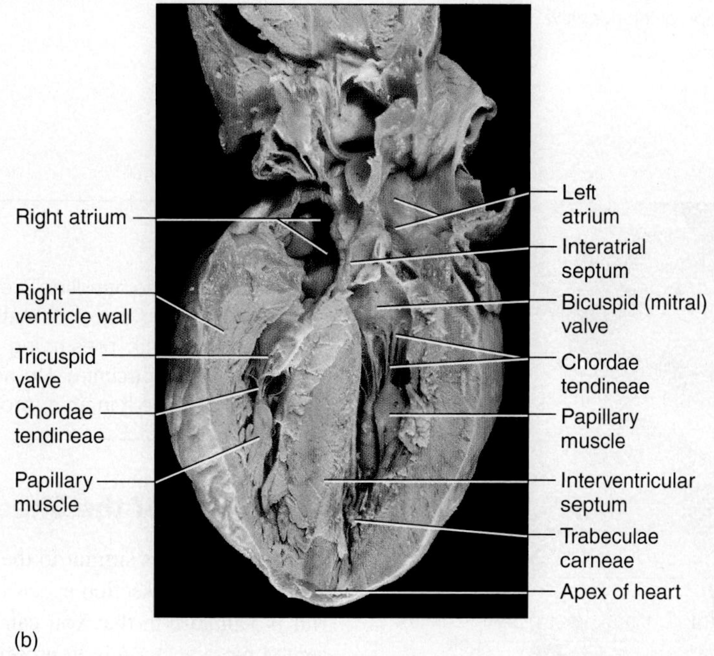

Figure 27.7 Coronal Section of the Heart
(a) Diagram; (b) photograph of cadaver heart.

oxygenated blood from the lungs into the **left atrium.** These vessels are located in the superior, posterior portion of the left atrium. Blood from the left atrium flows into the left ventricle. Locate the two large cusps of the **bicuspid valve** between the left atrium and left ventricle. The bicuspid valve is also known as the **mitral valve,** or **left atrioventricular valve.** It also has attached chordae tendineae and papillary muscles, which you should locate in the models in the lab. Note the thickness of the left ventricle wall, compared with the wall of the right ventricle. Compare the left side of the heart with figures 27.7 and 27.9.

The **aortic semilunar valve** is located at the junction of the left ventricle and the ascending aorta. It has the same basic structure and general function as the pulmonary semilunar valve in that it prevents the flow of blood from the aorta into the left ventricle. Blood from the left ventricle moves into the aorta and subsequently to the rest of the body.

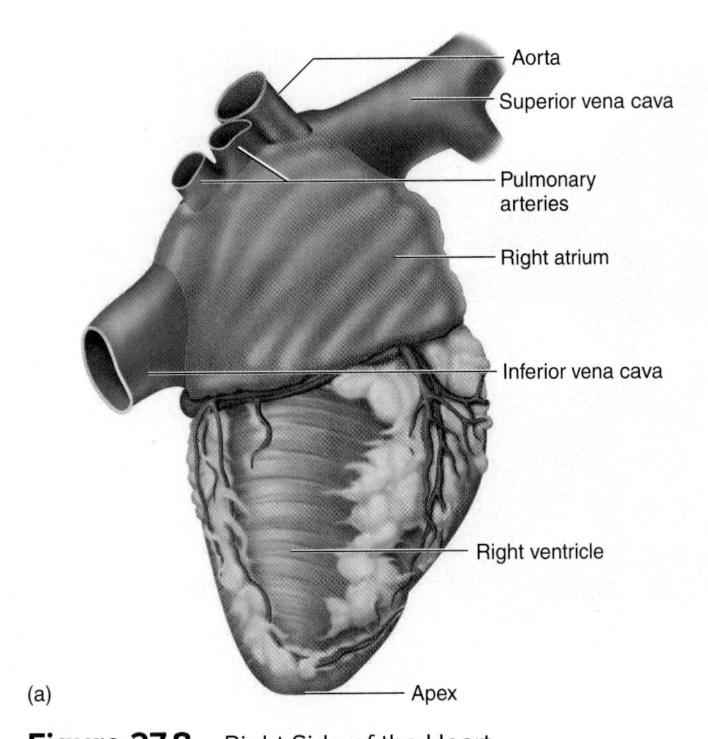

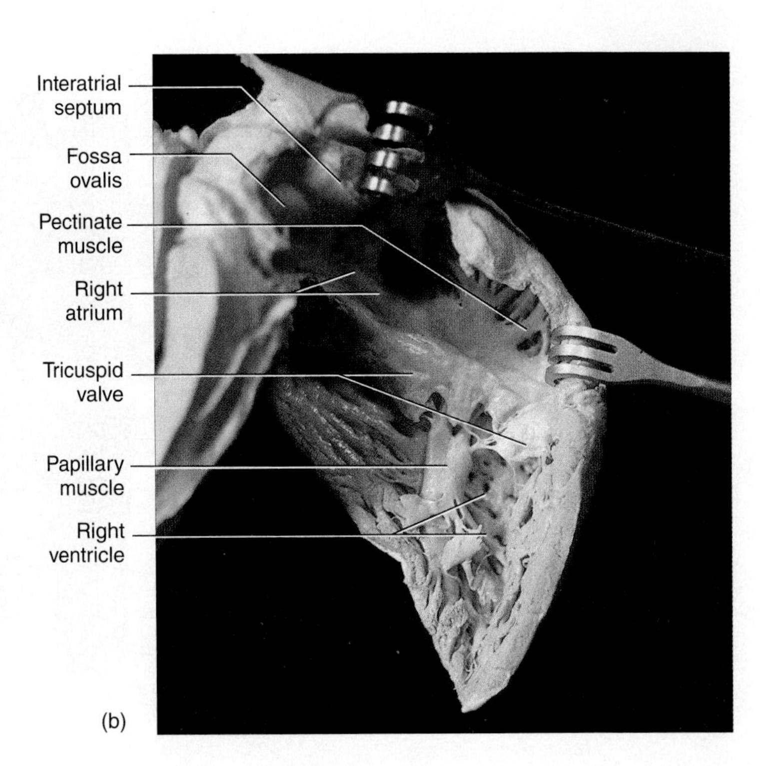

Figure 27.8 Right Side of the Heart

(a) Diagram of surface view; (b) photograph of interior view.

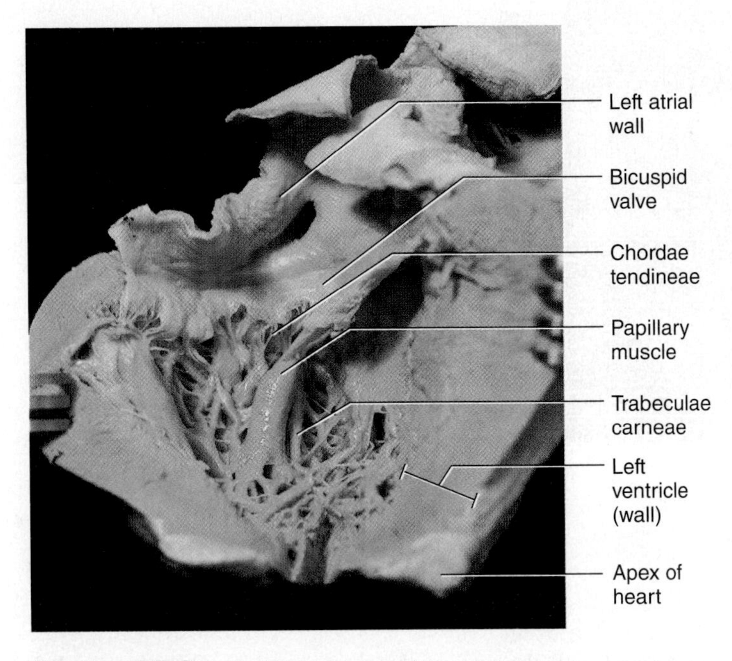

Figure 27.9 Left Atrium and Ventricle

Caution!

Be careful when handling preserved materials. Put on protective gloves and ask your instructor for the proper procedure for working with preserving fluid and for handling and disposal of the specimen. Do not dispose of animal material in the sinks. Place it in an appropriate waste container.

Dissection of the Sheep Heart

The sheep heart is similar to the human heart and usually is readily available as a dissection specimen. Dissection of anatomical material is valuable in that you can examine structures that are represented more accurately in preserved material than in models. Also, the preserved material has greater flexibility and is more easily manipulated. There are differences between sheep hearts and human hearts, especially in the position of the **superior** and **inferior venae cavae.** In sheep, these are called the anterior and posterior venae cavae, but we refer to them using the human terminology.

(A)[3] If your sheep heart has not been dissected, then you will need to open the heart. If your sheep or other mammalian heart has been previously dissected, then you can skip the next three paragraphs.

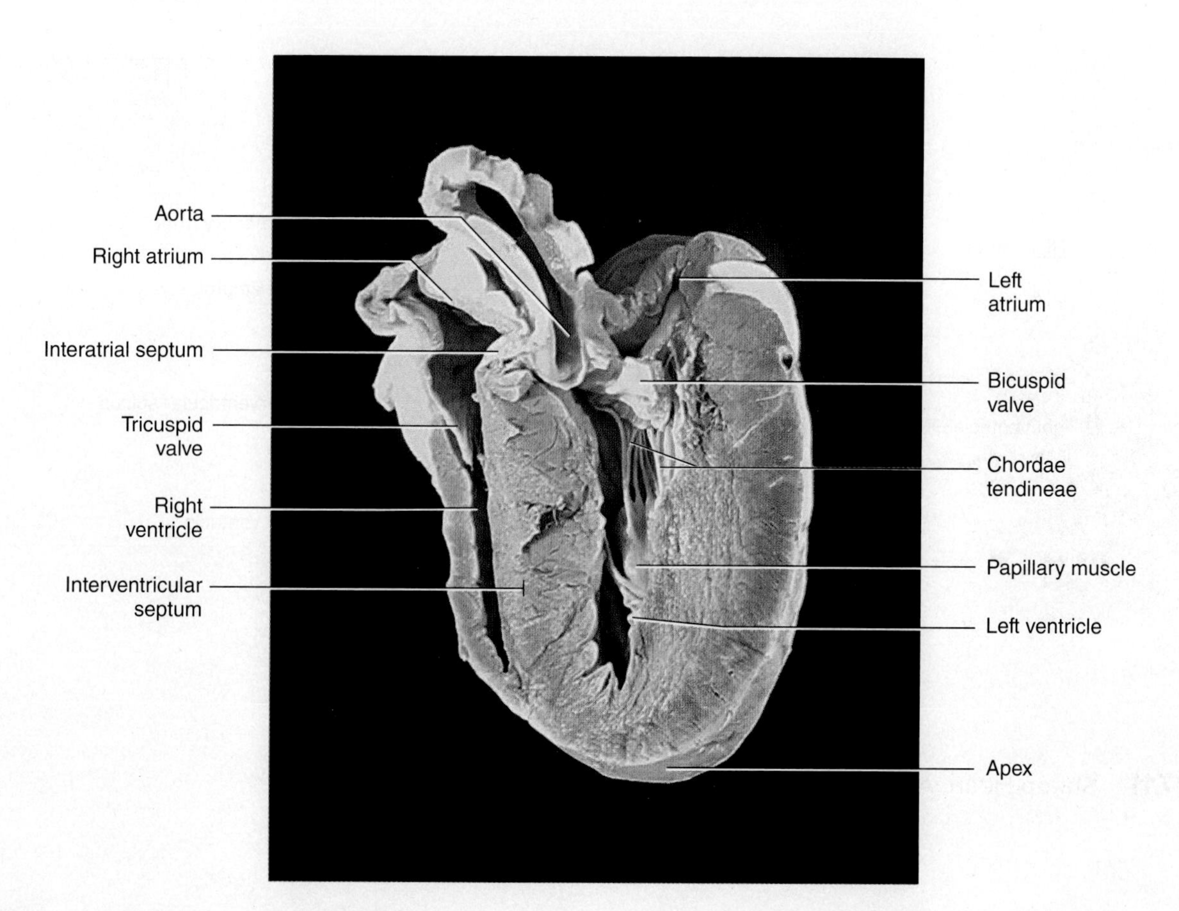

Figure 27.10 Sheep Heart, Coronal Section

Place the heart under running water for a few moments to rinse off the preserving fluid. Examine the external features of the heart. Note the fat layer on the heart. The amount of fat on the human or sheep heart varies.

Locate the **left ventricle,** the **right ventricle,** the **interventricular sulcus,** the **right atrium,** and the **left atrium.** Note the **auricles** that extend on the anterior surface of the atria.

Using a sharp scalpel, make an incision along the right side of the heart (lateral side) from the apex of the heart to the lateral side of the right atrium. If you are unsure about how to proceed during any part of the dissection, ask your instructor for directions. Make another long cut from the lateral side of the left atrium through the lateral side of the left ventricle. You will have made a coronal section of the heart if you have cut through the **interventricular septum.** Once you have opened the heart, compare the structures of the sheep heart with figure 27.10.

You can locate the vessels of the heart by inserting a blunt metal probe into the vessels and determining which chamber the vessel goes to or comes from. Place the heart in anatomical position and insert the probe into the large, anterior vessel that exits toward the specimen's left side. The blunt end of the probe should enter the **right ventricle.** This vessel is the **pulmonary trunk.** The pulmonary trunk may still have the **pulmonary arteries** attached. Locate the large vessel directly behind the pulmonary trunk (figure 27.11). This is the **ascending aorta.** If the vessels are cut farther away from the heart, you can see the **aortic arch.** Insert the probe into this vessel and into the left ventricle.

(A)[4] Turn the heart to the posterior surface and locate the **superior vena cava** and **inferior vena cava.** Insert the probe into the superior and inferior venae cavae, pushing the probe into the **right atrium.** If you find only one large opening in the atrium, you may have cut through either the superior or inferior vena cava during your initial dissection. The probe can be felt through the wall more easily here than in a ventricle because the atrial walls are thinner than those of the ventricles. On the left side, the **pulmonary veins** may appear either as four separate veins or as a large hole on each side of the left atrium if the vessels were cut close to the atrial wall. Locate the same structures in the sheep heart that you found on the model and compare them with figure 27.12.

Cut into the right atrium and use your blunt probe to locate the opening of the **coronary sinus** in the posterior, inferior portion of the atrium. It is small and somewhat difficult to find.

Examine the opening between the right atrium and the right ventricle to locate the **tricuspid valve.** You may have dissected through one of the cusps as you opened the heart. Locate the major features of the right ventricle. Find the **chordae tendineae** and the **papillary muscles.** You can find the **pulmonary trunk** by inserting a blunt probe into the superior portion of the right ventricle. Make an incision in the pulmonary trunk near the right ventricle to expose the three thin cusps of the **pulmonary semilunar valve.** Note how the cusps press against the wall of the pulmonary trunk when the probe is pushed against them in a superior direction. These cusps close when blood begins to flow back into the right ventricle as the ventricle relaxes.

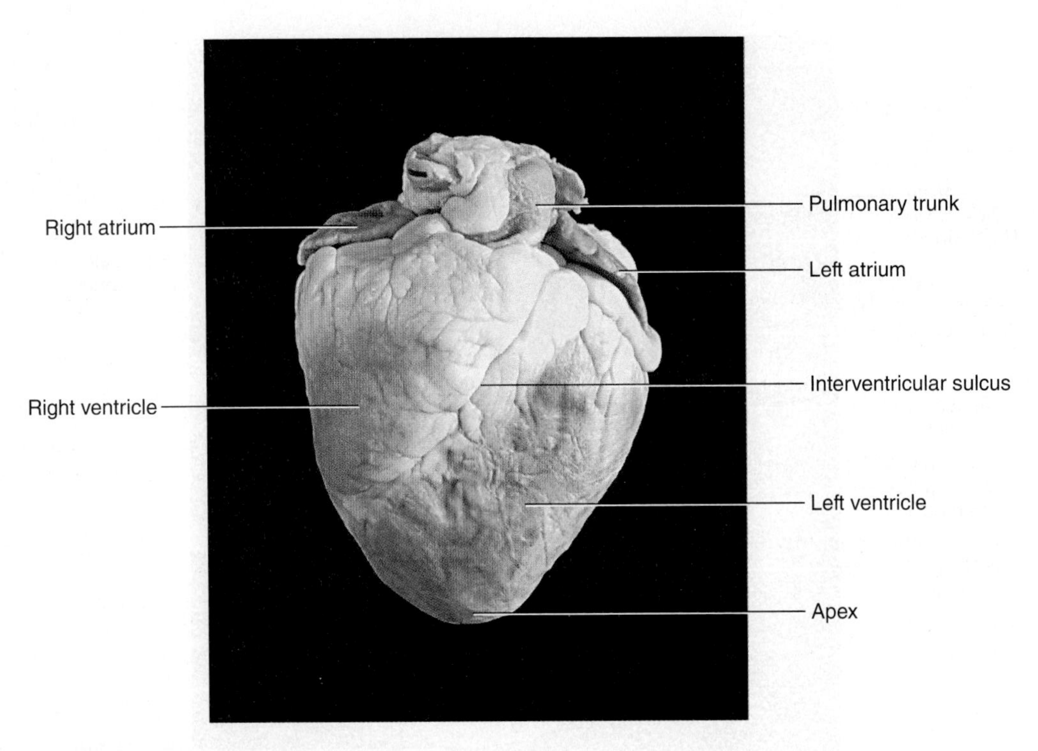

Figure 27.11 Sheep Heart, Anterior View

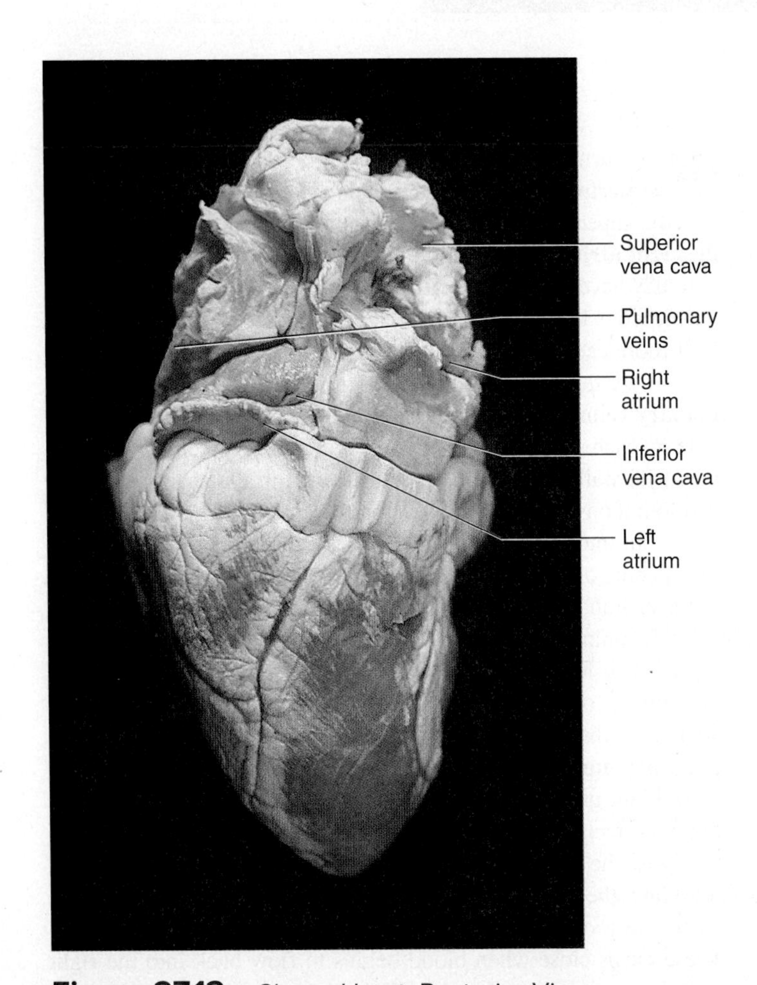

Figure 27.12 Sheep Heart, Posterior View

Locate the **left atrium, bicuspid valve,** and **left ventricle.** In the left ventricle, you should find the papillary muscles, chordae tendineae, and **trabeculae carneae.** You can find the **aorta** and **aortic semilunar valve** by inserting a blunt probe toward the superior end of the left ventricle toward the middle of the heart.

Cleanup

When you have finished your study of the sheep heart, make sure you clean your dissection equipment with soap and water. Be careful with sharp blades. Place the sheep heart either back in the preserving fluid or in the appropriate waste container, as directed by your instructor.

Exercise 27 Review

Name: ____________________

Lab time/section: ____________________

Date: ____________________

Structure of the Heart

1. The heart is located between the lungs in an area known as the ____________________.

2. What is the name of the layer that is superficial to the pericardial cavity? ____________________

3. What is the innermost layer of the heart wall called? ____________________

4. Is the apex of the heart superior or inferior to the rest of the heart? ____________________

5. What is the name of the depression between the two ventricles on the anterior surface of the heart? ____________________

6. Are auricles extensions of the atria or the ventricles? ____________________

7. What three vessels take blood to the right atrium? ____________________

____

8. Where do the great cardiac vein and the small cardiac vein take blood? ____________________

____

9. What blood vessels nourish the heart tissue? ____________________

____

10. What structure separates the left atrium from the right atrium? ____________________

____

11. What is the name of the thin spot between the atria? ____________________

____

12. The bicuspid valve is located between what two chambers of the heart? ____________________

____

13. Name the structure between the atrioventricular valve and the papillary muscle. ____________________

14. What is the function of the aortic semilunar valve? ______________________________

____

15. What is another name for the tricuspid valve? ______________________________

____

16. What cell type makes up most of the myocardium? ______________________________

____

17. What adaptation do you see in the walls of the left ventricle being thicker than those of the right ventricle? ______________________________

____

____

____

18. How does cardiac muscle resemble skeletal muscle? ______________________________

____

____

19. In terms of function, how is cardiac muscle different from skeletal muscle? ______________________________

____

____

20. Label the following illustration using the terms provided.

apex	chordae tendineae	interatrial septum	interventricular septum
left atrium	left ventricle (wall)	right atrium	right ventricle (wall)

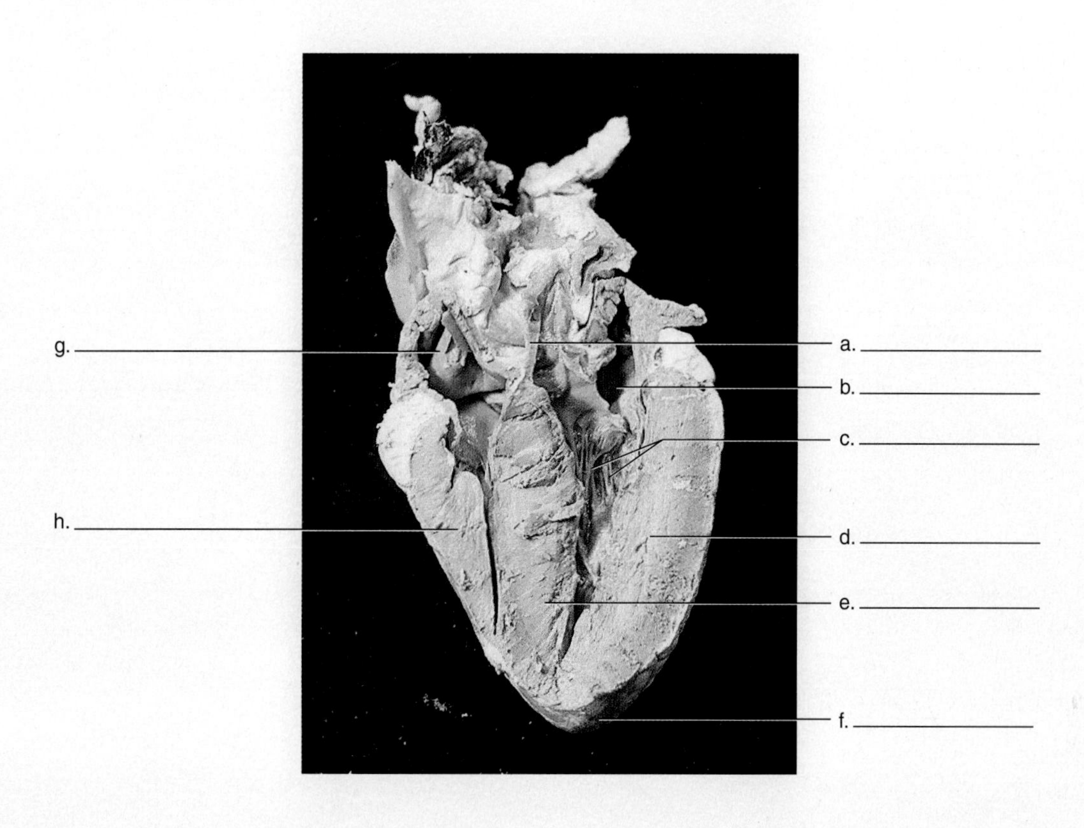

NOTES

Exercise 28

Electrical Conductivity of the Heart

Introduction

Before you begin the study of the function of the heart, review the anatomy of the heart in Laboratory Exercise 27. The topics for this exercise are covered in the Seeley text in chapter 20, "Cardiovascular System: The Heart."

The living heart is a phenomenal organ. It contracts an average of 72 times per minute for the lifetime of an individual. The heart has been described as a muscular pump or, more accurately, two pumps acting in unison. As a pump, the heart has a contraction mode and a relaxation mode. **Systole** is contraction of the heart muscle and can be described more specifically as **atrial systole** and **ventricular systole** (contraction of the atria and ventricles, respectively). **Diastole** is the relaxation of the heart muscle, and there are **atrial diastole** and **ventricular diastole.**

Electrical activity of the heart stimulates the heart muscle to contract. In this exercise you observe the electrical activity of the heart and correlate it to the mechanical functioning of the heart. You learn the conductive structures of the heart, make a recording of the electrical activity of the heart, and correlate the recording with the activity of the heart.

The initiation of the electrical impulse in the heart begins at the **sinoatrial (SA) node,** which is commonly known as the **pacemaker.** The sinoatrial node is located in the superior portion of the right atrium. Cells of the sinoatrial node use the sodium-potassium pump to generate a difference in electrical voltage across the cells' membranes. The cells are said to be **polarized** because the inside of the cell membrane has more negative ions than the outside of the membrane. This separation of charged particles is reflected in the voltage difference across the membrane. The sinoatrial node spontaneously **depolarizes,** which causes a change in the total amount of charge across the cell membrane. This change in membrane potential occurs approximately 72 times per minute (the average heart rate). Conduction from the sinoatrial node travels across the atria, causing the muscles of the atria to contract. The impulse that spreads out across the atria reaches the **atrioventricular node (AV) node.** The impulse has a slight delay (about 0.1 second) in the node before being conducted further. This delay allows the atrial cardiac muscle to contract prior to ventricular firing (see figure 28.1).

The electrical impulse then travels from the AV node to the **atrioventricular bundle (bundle of His),** to the **right** and **left bundle branches,** and finally to the **Purkinje (conduction) fibers.** The Purkinje fibers stimulate the cardiac muscle of the ventricles to contract. The ventricles are thus stimulated from the apex toward the base, and the contraction proceeds from the inferior end of the ventricles toward the atria. Locate the structures of the heart's conduction system in figure 28.1. Contraction of the heart muscle occurs just after depolarization of the heart muscle cells. Relaxation of the heart occurs just after repolarization of the heart muscle cells.

Two concepts are important in understanding the electrical activity of the heart. One is that the heart produces low-voltage **electrochemical impulses** in a similar way to the production of impulses in the nervous system. The "average" potential difference of **– 90 millivolts** is a little less than one-tenth of a volt; therefore, the heart is operating electrically at slightly less than 0.1 volt. The other important concept in understanding the electrical activity of the heart is that these impulses can travel through the saline medium of the body and be picked up by **sensors (electrode plates)** attached to the skin. The cells of the body are bathed in a saline solution of a little less than 1% salt, which is an excellent conducting medium for electrical impulses. The electrical impulse generated by the atria and ventricles depolarizing and repolarizing can be recorded using an electrocardiograph machine attached to the electrode plates.

Early work done by Willem Einthoven established the modern techniques of recording the electrical activity of the heart by an **electrocardiograph (ECG)** machine. It is an instrument that measures slight changes in the voltage related to cardiac activity. The electrocardiograph produces a chart paper recording or a data file called an **electrocardiogram (ECG, or EKG).** Time is measured along the horizontal axis (or *x* axis, with each millimeter equal to 0.04 second), and voltage difference is measured along the vertical axis (or *y* axis, with 1 millimeter equal to 1 millivolt). The ECG has a **baseline** known as the **isoelectic line,** and deflections from that line record the electrical activity of the heart. Einthoven established three standard leads that record the heart's electrical events with the heart sitting in the middle of a theoretical shape known as Einthoven's triangle (figure 28.2).

Lead I connects the right arm and the left arm (RA-LA). It measures the potential voltage across the horizontal axis of the heart.
Lead II connects the right arm and the left leg (RA-LL). This is the lead that records the potential voltage from the base to the apex of the heart.
Lead III connects the left arm and the left leg (LA-LL). This is the lead that records the potential voltage along the left side of the heart.

In many college physiology labs, the ECG electrode plates are attached to four areas. These are the medial side of the **left** and **right ankles** and the anterior surface of the **left** and **right wrists.** The attachment of the electrode plate to the right ankle serves as

1. Action potentials originate in the sinoatrial (SA) node and travel across the wall of the atrium (*arrows*) from the SA node to the atrioventricular (AV) node.

2. Action potentials pass through the AV node and along the atrioventricular (AV) bundle, which extends from the AV node, through the fibrous skeleton, into the interventricular septum.

3. The AV bundle divides into right and left bundle branches, and action potentials descend to the apex of each ventricle along the bundle branches.

4. Action potentials are carried by the Purkinje fibers from the bundle branches to the ventricular walls and papillary muscles.

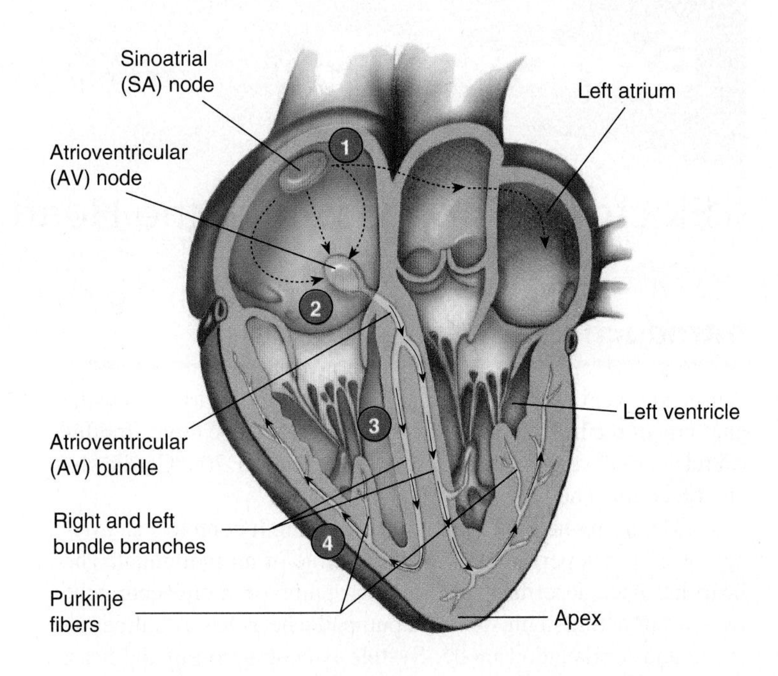

Figure 28.1 Conduction System of the Heart
Conduction starts at 1 and finishes at 4.

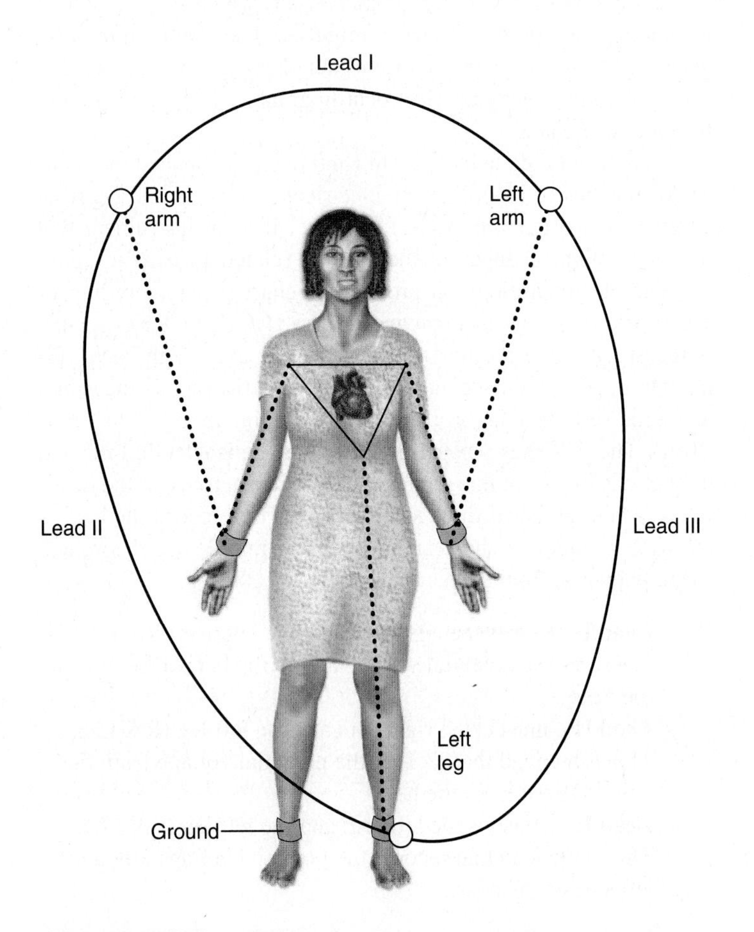

Figure 28.2 Leads of a Standard ECG

an **electrical ground** and is not used for measurement. The ECG is measured as the potential voltage difference between selected electrode plates. There are numerous "V" electrodes, which are chest electrodes. You will not use the V electrodes in this exercise unless directed to do so by your instructor.

Before you record an ECG, examine the ECG in figure 28.3 and read the accompanying description. The ECG typically records three major events. The first event is the **P wave,** which is a small bump called a **deflection wave.** The P wave represents **atrial depolarization.** The **QRS complex** represents the **ventricular depolarization.** Because the ventricles are more massive than the atria, the electrical events produced as the ventricles depolarize are much larger than the P wave generated by the atria. The **T wave** represents the **ventricular repolarization.** The **atrial repolarization** occurs during ventricular depolarization and is masked by the larger QRS complex.

Objectives

At the end of this exercise you should be able to

1. distinguish between systole and diastole;
2. list the structures in the conductive system of the heart;
3. describe how the electrocardiogram can be used to monitor electrical events in the heart;
4. correlate the electrical potentials of the heart to mechanical activity in a cardiac cycle;
5. describe the sequence of electrical conductivity of the heart;
6. associate the P wave, QRS complex, and T wave of an ECG with electrical events that occur in the heart;
7. relate how the electrical activity of the heart is transferred to the ECG;

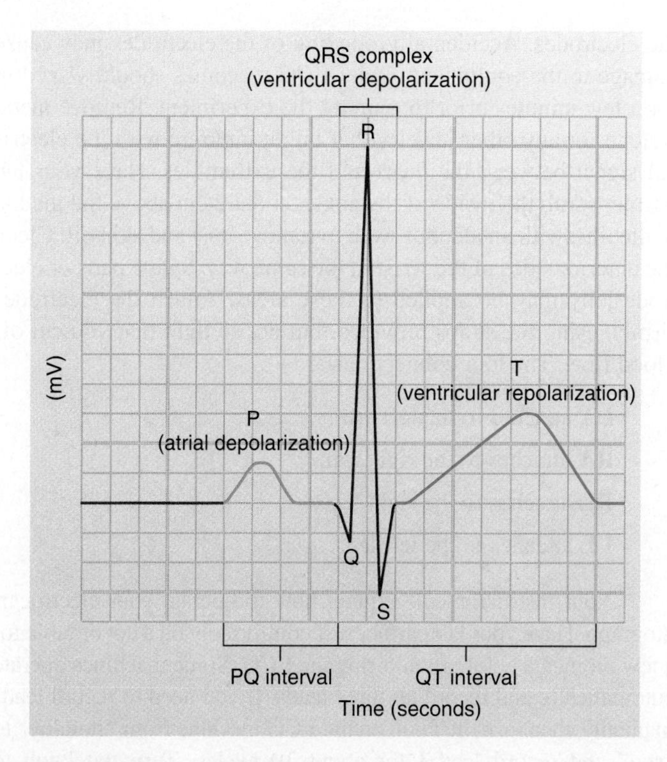

Figure 28.3 Electrocardiogram

8. calculate heart rate, PR interval, QRS interval, and QT interval from an ECG recording;
9. describe the variance in an ECG from normal if a person has a heart block;
10. describe the changes in an ECG immediately after exercise;
11. determine the mean electrical axis of the heart.

Materials

Cot or covered lab table
Alcohol swabs
Electrode jelly, paste, or saline pads

BIOPAC Section
EL 503—general purpose disposable electrode pads
SS2L—lead cables
CBLSERA—CBL serial for MP30
MP30 or MP150—data acquisition unit and power supply (AC100A)
Compatible computer

ECG Machine Section
Electrocardiograph machine

Procedure

Recording ECGs

There are a few ways to record ECGs. One is with a conventional ECG machine that puts out a paper chart strip, another is with a physiograph, and a third is with a computer program and hardware, such as BIOPAC. This exercise will describe the BIOPAC procedure and the ECG machine procedure.

BIOPAC Procedure

You will collect data for the electrical activity of your heart by attaching receiving electrodes to your skin. In this part of the exercise you will be using lead II. The resulting electrocardiogram is displayed on the computer monitor and saved as a file in your BIOPAC folder. If you are going to determine the mean electrical axis of the heart later in this exercise you will need to use BIOPAC Lesson 6.—L06 Electrocardiography 2. The following instructions are for determining a simple ECG and are found in BIOPAC Lesson 5. These are the steps you will take:

(A)[1] ***1. Setup***

a. Make sure the computer is ON.
b. Connect the data acquisition unit to an electrical outlet with the power supply.
c. With the data acquisition unit OFF, plug in the SS2L lead cables, into the CBLSERA cable, which should be connected to CH2 on the data acquisition unit.
d. Turn the data acquisition unit ON.
e. Have your lab partner remove all jewelry and lie comfortably. Make sure that you have three of the disposable electrode pads.
f. Clean the inner side of each ankle of your lab partner with an alcohol swab and attach one of the disposable EL 503 electrode pads to the inner left leg just proximal to the medial malleolus, as seen in figure 28.2. You may want to place a small amount of electrode gel in the center of the pad in order to get proper electrical conduction. Attach another electrode pad to the right leg at the same height. Clean the distal, anterior, right forearm just proximal to the wrist with an alcohol swab and attach the third electrode pad there. You should have three electrode cables, a white, black, and red one.
g. Place the **white lead** on the right forearm. The clip has a squeeze lever and attaches to the metal button of the electrode pad. The clip connects on only one surface. If you cannot get it to attach one way, try flipping the connector over.
h. Place the **black lead** on the electrode pad of the right leg. This is a ground.
i. Place the **red lead** on the left leg.
j. Open the BIOPAC Student Lab (BSL) Software on the computer desktop and select BIOPAC Lesson 5.—L05 Electrocardiography 1.
k. Type in your (folder) name. When done, click OK. If you have a folder on this computer station a window should appear with the message "A folder with this name already exists. Would you like to use it or create a new folder?" Choose USE IT.

2. *Calibration* Make sure that your test subject is comfortable and relaxed. It is often beneficial to talk to the subject calmly as you prepare the test. Any extraneous movement may cause electrical interference. People who are nervous or jittery do not produce good ECGs.

Click on CALIBRATE in the upper left corner of the computer screen. The calibration will stop automatically after approximately 8 seconds. You should see a small ECG tracing on the computer screen with a flat baseline. If not, check your connections and click REDO CALIBRATION.

3. ***Record Data*** Have your lab partner relax, click on the RECORD button, and let several cycles occur. After about 15 seconds, click on SUSPEND. If the recording does not have a flat baseline, then click on the REDO button. If your recording looks similar to figure 28.3 or 28.4*a,* then you have made a successful recording.
4. ***Analyze Data*** Find the **Review Saved Data** folder from the lessons menu and look for your file.

You can determine the length of various sections of your ECG by selecting the **Delta T** mode. This mode determines the time from one part of the ECG tracing that you select to another part of the tracing that you select. You can use the **I-beam** tool to select the appropriate area of your ECG. Choose the magnifying tool in the lower right side of the computer screen to give you a close-up view of the area you are going to choose.

Beats per minute—bpm Select the area from one R peak to another R peak using the I-beam tool. The bpm will show up in the channel measurement box. Record your bpm in table 28.1.

PR interval Use the I-beam and highlight a section of the ECG from the beginning of the P wave to the beginning of the QRS complex. Record this time (use the Delta T mode) in table 28.1.

QRS complex Use the same procedure as you did with the PR interval to determine the time for the QRS complex. The length of the QRS complex begins at the first deflection of the Q wave and ends when the S wave returns to the baseline. Record this time in table 28.1.

QT interval This interval is determined with the beginning of the Q wave to the end of the T wave. Record this time in table 28.1.

Procedure for Conventional ECG Machine

(A)2 The ECG machine should be plugged in and turned on. Make sure that the ECG machine is on "standby" when you attach the electrodes. Accidental grounding of the electrodes may cause damage to the equipment. Older ECG machines should warm up for a few minutes prior to running the experiment. Remove metal watches or any other jewelry that might interfere with the electrical signal between the heart and the extremities. Have your lab partner scrub the inside of the ankle, about 2 cm above the medial malleolus, with an alcohol swab to remove dust and skin oil. Clean the anterior sides of the wrist in the same way. Saline pads or electrode jelly may be applied to these areas. Attach the electrodes firmly, using the straps provided, but not so tight that you cut off blood flow. The four connections are

LA attaches to the left arm.

RA attaches to the right arm.

RL attaches to the right leg.

LL attaches to the left leg.

Your instructor will explain how to operate your electrocardiograph. Have your lab partner rest comfortably on a cot or table for a few moments before monitoring the ECG. Some machines operate automatically and record all three leads. If you need to record leads manually, then turn the knob on the ECG machine from "standby" to "run" and record lead I for about 10 cycles. Turn the knob to "standby" after recording each lead. Then record lead II and lead III for about 10 cycles each. Tear the ECG from the machine and make sure you write your lab partner's name on the paper and write the appropriate lead on the paper. It is important that you provide a relaxed atmosphere for your lab partner to produce a good ECG recording. People who are fidgeting or nervous generate extraneous interference and do not produce a level baseline recording. You can demonstrate this electrical "noise" by the following activity.

(A)3 ***Interference in an ECG*** While your lab partner is still attached to the ECG machine, set the dial to lead II. Run three to four cardiac cycles and then have your lab partner clench his or her fists. Note the electrical interference that occurs as the skeletal muscles depolarize and repolarize. Emotional state also plays a part in electrical activity. If a patient is nervous or excitable, this will show in the ECG. Background electrical noise in the ECG varies from person to person, but you should be able to get a good ECG in the lab.

TABLE 28.1 ECG Data

Record your value for the various ECG data either from the BIOPAC ECG or the ECG machine and determine if your data fall within the normal values.

bpm	Yours ____________	Above 100 Tachycardia	Below 60 Bradycardia	Within norm? ____________
PR interval	Yours ____________	Normal 0.16–0.18 sec	Longer than 0.2 Partial AV heart block	Within norm? ____________
QRS complex	Yours ____________	Normal 0.08 sec	Longer than 0.12 sec Rt/lft bundle branch block	Within norm? ____________
QT interval	Yours ____________	Normal 0.3–0.4 sec	Shortens with increased heart rate	Within norm? ____________

Ⓐ[4] ***ECG and Exercise*** Once you have compared your ECG to normal, you should exercise vigorously for a few minutes. You can do this by running outside of the lab room or doing jumping jacks for a minute or two. Once you feel that your heart is pumping faster and harder, quickly lie down on the cot and re-measure your ECG.

Prerecorded ECG Strip

If, for some reason, you do not have access to ECG recordings or you did not obtain a good recording from your equipment, you can use the recordings in figure 28.4 to fill in the data in table 28.1. Record the data for ECG and Exercise and analyze the ECG.

How does your ECG compare to normal in terms of distance between the P waves?

How does the height of the QRS complex compare to the resting ECG?

The time between the T wave and the next P wave is the resting interval of the heart. As you exercise, this interval shortens.

ECG Evaluation Select one of the leads (lead II usually works best) for the following evaluations.

Beats per Minute The standard rate of paper travel in an ECG machine is 25 millimeters/second. Therefore, each millimeter is equal to 0.04 second. The small squares are millimeters and the darker lines indicate 5 mm or 0.5 cm. You can calculate the resting heart rate by counting how many millimeters ("Y" mm) occur between two peaks (such as R to R). This value is going to be multiplied by 0.04 second in the following equation:

______ mm/beat × 0.04 second/mm = "Y" seconds/beat

Y = ______

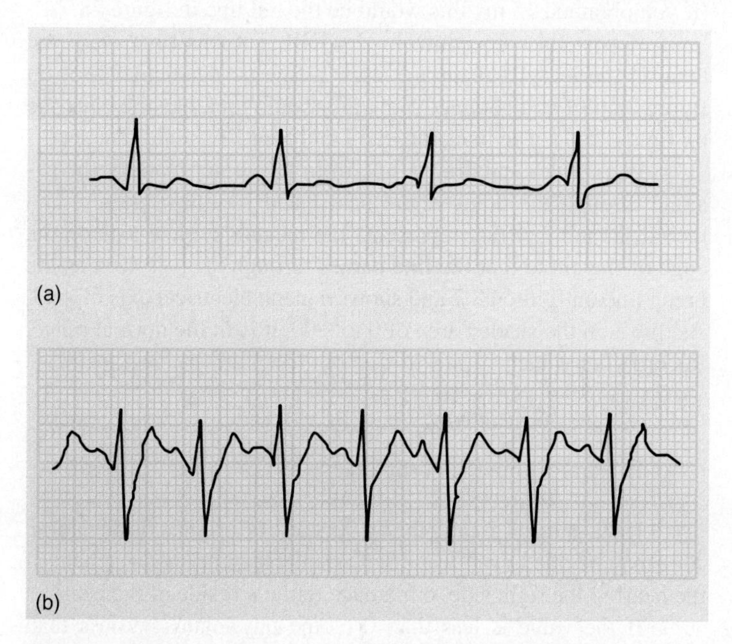

Figure 28.4 Representative ECG Recordings

(a) Normal ECG of person at rest; (b) normal ECG of person after exercise. Chart rate is 25 mm/second.

There are 60 seconds/minute and you want to calculate the beats/minute, so you can use the following equation to find out the beats per minute.

$$\frac{60 \text{ seconds/minute}}{\text{"Y" seconds/beat}} = \text{your number of beats/minute}$$

Calculated beats/minute: ______________________.

Record your bpm in table 28.1.

PR Interval Select a section of the ECG from the beginning of the P wave to the beginning of the QRS complex. Record this interval in table 28.1.

QRS Complex The length of the QRS complex begins at the first deflection of the Q wave and ends when the S wave returns to the baseline. Record this time in table 28.1.

QT Interval This interval is determined with the beginning of the Q wave to the end of the T wave. Record this time in table 28.1.

Analysis of the ECG ECG recordings are important in assessing the health of the heart. **Conduction problems, myocardial infarcts (heart attacks),** and **heart blocks** are a few of the problems that may show up as variances on an ECG. The interpretation of ECGs in this exercise is for educational purposes only. Clinical evaluations of ECGs should be done only by trained health-care workers. Locate the P wave, the QRS complex, and the T wave. The whole cardiac cycle should take, on average, 0.7 to 0.8 second.

Irregularities in Heart Rate An excessively high heart rate is termed **tachycardia.** In young adults a heart rate above 100 beats per minute is considered **tachycardia.** However, 100 beats per minute in small children may be perfectly normal. An excessively low heart rate is termed bradycardia. In young adults a rate below 60 beats per minute is considered bradycardia unless they are highly trained aerobic athletes.

PR Intervals PR intervals (also known as PQ intervals) are normally about 0.16 second. The PR interval is the time between the beginning of atrial depolarization and the beginning of ventricular depolarization. If the PR interval is longer than 0.2 second (5 mm on the chart paper), this might indicate a **heart block.** Heart blocks result from reduced conduction between the atria and the ventricles. This may be caused by damage to the AV node or decreased transmission in the AV bundle. In a **complete heart block** the atria do not stimulate the ventricular depolarization at all; therefore, the atria fire independently from the ventricles. In a complete heart block the P waves may be spaced at 0.8 second apart, the SA node depolarization rate, but the ventricles fire at a much lower rate (1.5 to 2.0 seconds apart). Are the times measured on your recording within normal limits?

QRS Complex The **QRS complex** is 0.08 to 0.10 second, on average. If the QRS complex spans longer than 0.12 second this may indicate a **right** or **left bundle branch block.** In this condition the two ventricles contract at slightly different times, increasing the length of the QRS complex.

QT Interval The QT interval is 0.3 second, on average. The interval is shorter as the heart rate increases, and the interval becomes longer as the heart rate slows down.

Cardiac Arrhythmias One of the major diagnostic uses of the ECG is to detect arrhythmias, abnormal rhythms, of the heart. Arrhythmias range from mild variations due to emotions or stimulants to severe abnormalities causing life-threatening conditions. Compare your ECG to figure 28.5, which shows only a few examples of irregular ECGs. The normal ECG is represented in figure 28.3 and 28.4a.

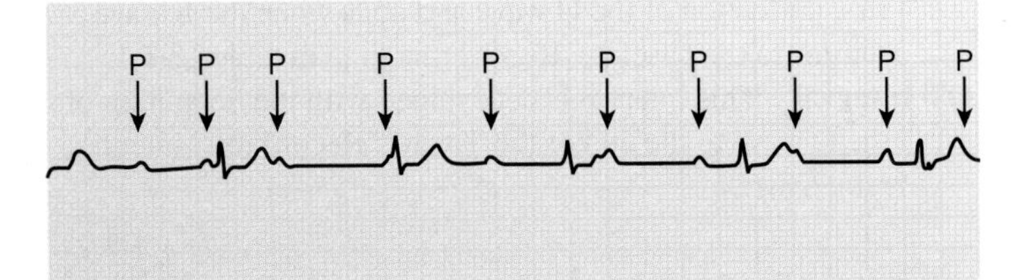

Complete heart block (P waves and QRS complexes are not coordinated)

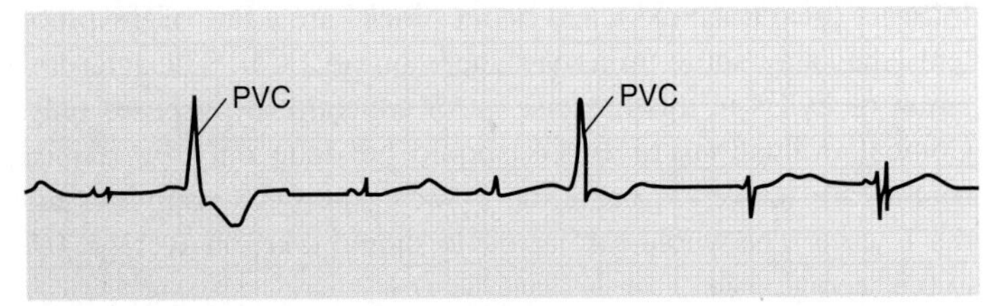

Premature ventricular contraction (PVC) (no P waves precede PVCs)

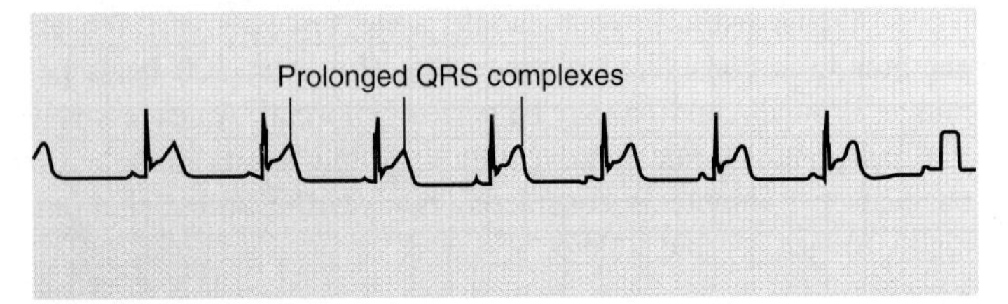

Bundle branch block

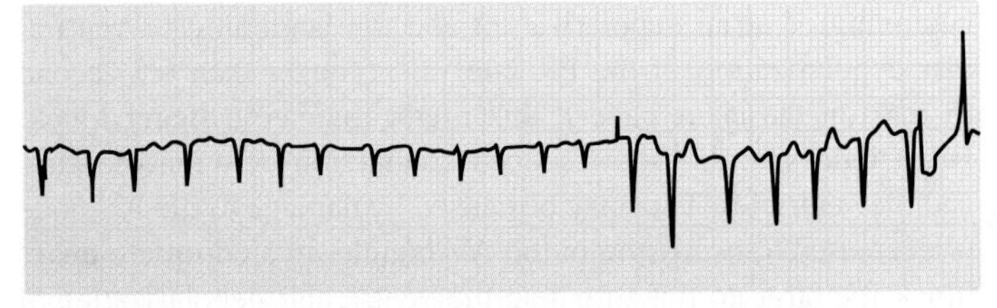

Atrial fibrillation (no clear P waves and rapid QRS complexes)

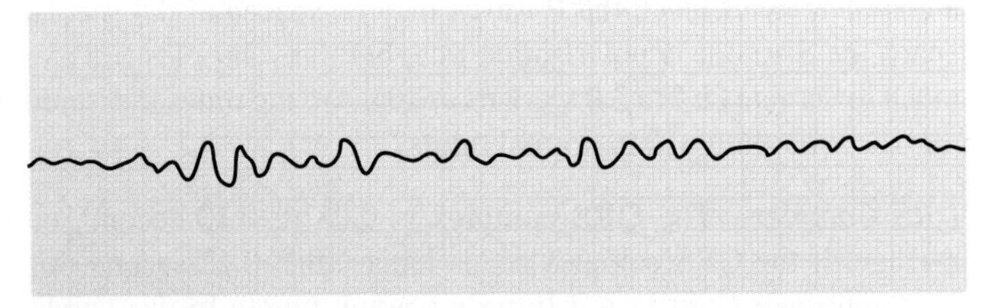

Ventricular fibrillation (no P, QRS, or T waves)

Figure 28.5 Alterations in an Electrocardiogram

Mean Electrical Axis of the Heart

(A) **5** If you have recorded an ECG from lead I and lead III, you can determine the electrical axis of the heart. If you are using BIOPAC you must use BIOPAC Lesson 6.—L06 Electrocardiography 2. If you are using a standard strip chart ECG you will need to have lead I and lead III in hand. It is possible to calculate the relative position of the heart in the thoracic cavity by using the electrical data as was first analyzed by Einthoven. The mean electrical axis of the heart is essentially parallel to the interventricular septum. If you examine the three standard leads, they form a triangle known as Einthoven's triangle (see figure 28.2). Normally, the electrical axis of the heart runs from the superior right side to the inferior left side, meaning that the heart is in the middle of the chest but the apex points to the inferior and left side. There are many reasons for deviations from the normal pattern, including loss of electrical activity in a portion of the heart (a bundle branch block), death of heart tissue, or an enlarged ventricle. If you are using a standard ECG you can calculate the mean electrical axis using the following information. You may be using a program that automatically calculates the mean electrical axis; however, the interpretation of the axis will be more meaningful if you also read the following description.

To determine the mean electrical axis of the heart, you must acquire data from leads I and III. Using your ECG, find the QRS complex in lead I. Count the number of millivolts (this is equal to the number of millimeters on chart paper) that the QRS complex projects above the isoelectric line (the baseline of the recording). If you are using BIOPAC L06, you can determine the amplitude of the wave by selecting the **delta mode.** Select the region of the graph using the I-beam tool. Subtract from that the sum of the number of millivolts of both the Q and the S waves that project below the isoelectric line. Mark this number on the scale for lead I and draw a vertical line perpendicular to the lead I axis on figure 28.6 (if you obtained 15 mv this would be the red line in figure 28.7).

Repeat this procedure for lead III, but this time mark the area on the lead III scale and draw a line perpendicular to it on figure 28.6 (if you obtained 5 mv this would look like the blue line on figure 28.7). The lines for lead I and lead III should intersect.

Use a ruler and draw a line from the center mark of the triangle through the point where lead I and III lines intersect and record the number that represents the mean electrical axis on the edge of the circle that the line passes through. This is seen as the green line on figure 28.7 and shows a mean electrical axis of +49. As this is in the shaded area of 0 to +90, it is in the normal range. Record your value.

Mean electrical axis: ____________________

Is yours within the normal range? ________________

If the value is greater than +90, there is a right axis deviation. This may be due to right ventricular enlargement, displacement of the heart to the right side, or damage to the left side of the heart.

If the value is less than 0 (some physicians say less than −30), there is a left axis deviation. This may be due to left ventricular enlargement, left bundle branch block, displacement of the heart to the left side, or damage to the right side of the heart.

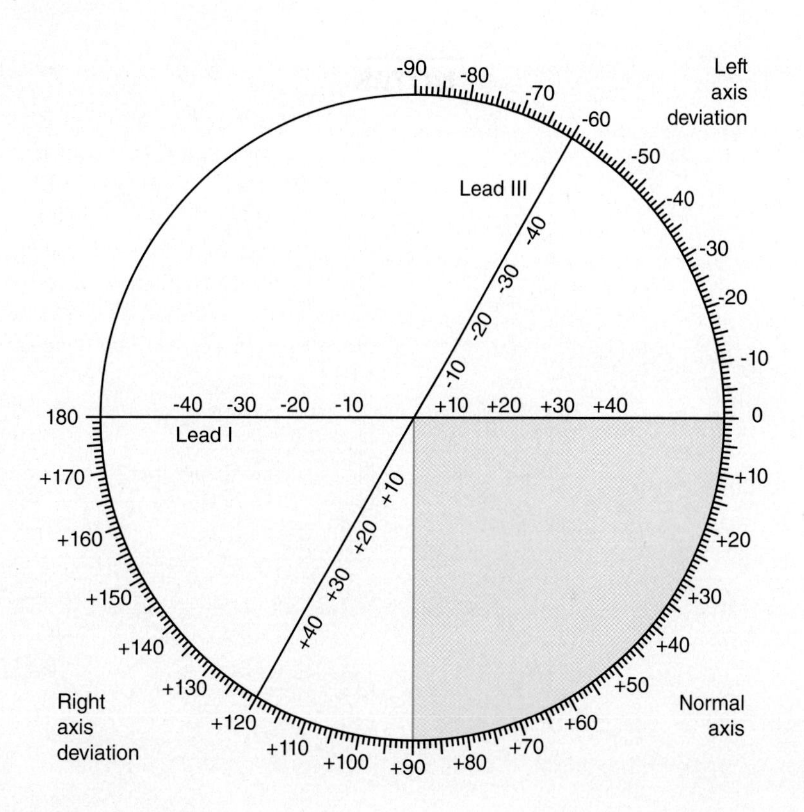

Figure 28.6 Electrical Axis of the Heart

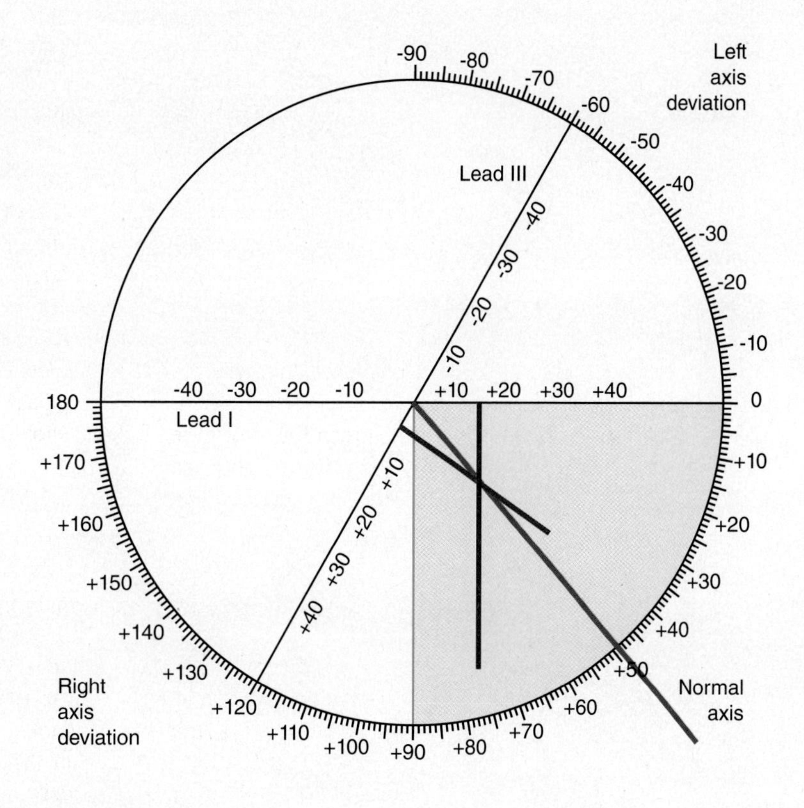

Figure 28.7 Electrical Axis of the Heart

Sample shown to determine mean electrical axis.

NOTES

Exercise 28 Review

Name: ____________________

Lab time/section: ____________________

Date: ____________________

Electrical Conductivity of the Heart

Review Questions

1. The sinoatrial node has a common name. What is it? ____________________

2. Which two chambers of the heart (atria or ventricles) contract last in a normal cardiac cycle? ____________________

3. What two chambers are stimulated immediately after the SA node depolarizes? ____________________

4. After the AV node depolarizes, what structures conduct the impulse to the myocardium of the ventricles? ____________________

____

5. What are the main events recorded by an ECG? ____________________

____

6. What electrical event in the heart does the QRS complex represent? ____________________

____

7. Ventricular repolarization is represented by what part of an ECG? ____________________

____

8. What ECG wave is represented by atrial depolarization? ____________________

____

9. Why is the ECG event indicating atrial repolarization not seen in an ECG? ____________________

____

10. What does a heart block do to impulse transmission in the heart? ____________________

____

11. Fibrillation is uncoordinated cardiac muscle contraction. Predict what an ECG would look like if there were no uniform conduction of electrical activity in the heart. Draw what it might look like.

12. What consequence does fibrillation have for cardiac muscle contraction and for the pumping efficiency of the heart? Which is more serious—atrial or ventricular fibrillation?

____

____

____

13. If a myocardial infarct (heart attack) destroyed a portion of the right or left bundle branches, what potential change might you see in an ECG? ____

____

____

____

14. Tape or paste your ECG in the following space. Label the P wave, the QRS complex, and the T wave. You can also mark up figure 28.4*a* and label the parts.

15. What was your value for the mean electrical axis of the heart? ____________________

Was this within normal limits? ______________________________

Exercise 29

Functions of the Heart

Introduction

One of the fundamental physiological activities of the body is to maintain adequate blood pressure. If blood pressure is too low, cells may not function correctly or they may die. If blood pressure is too high, damage may occur to the organs of the body or excess fluid may be expressed from the capillaries. One way to control blood pressure is through changes in **cardiac output.** When more oxygen is required, cardiac output increases with an elevated heart rate or an increase in the volume ejected per contraction, or both. The heart has an **intrinsic control,** which is a pacemaker located in the heart wall that determines heart rate, and it has **extrinsic controls,** from organs other than the heart such as hormones or input from the nervous system that controls the heart rate or contraction strength.

In this exercise, you explore some of the functions of the human heart, with particular emphasis on changes in heart rate and contraction strength. These topics are covered in the Seeley text in chapter 20, "Cardiovascular System: The Heart." You examine the resting rate of human hearts, the natural rate of heart contraction in a frog, and changes that occur when various solutions are added to the frog heart muscle. The heart of a frog is physically somewhat different from a human heart. The frog has two atria and only a single ventricle; however, the physiological response of the frog to various cardiac-influencing substances is similar to that of humans.

Due to the decline in some frog populations (the "leopard frogs" in North America), bullfrogs may be preferable as study specimens. Bullfrogs are not only plentiful but frequently a pest species in many areas and may be used without significant impact on their population levels.

Cardiac Muscle Characteristics

Cardiocytes (cardiac muscle cells), like neurons, undergo a normal **polarization** process. This produces a **resting membrane potential.** The mechanism by which electrochemical impulses occur in heart muscle depends on three ion channels, **fast sodium channels, slow calcium channels,** and **potassium channels.** In cardiocytes polarization occurs by activation of the **sodium-potassium pump.** The result is an increase in sodium and potassium ions outside the cell membrane. This produces a resting membrane potential.

When the cardiocyte is stimulated, the fast sodium channels open, causing depolarization. This leads to a twitch of the muscle fiber. In skeletal muscle the twitch is quick (2 msec). In cardiac muscle the twitch is much longer (250 msec), caused by a **plateau** in the repolarization of the cell. This may be due to the closing of slow calcium channels or the slow movement of calcium to the cytosol. The plateau can be seen in figure 29.1. Depolarization travels across the cardiocyte. **Repolarization** occurs relatively quickly by reestablishing the resting membrane potential, and the heart muscle is ready for its next contraction.

Cardiac Pacemaker

Pacemaker cells in the **sinoatrial (SA) node** of the heart wall are specialized muscle cells that spontaneously and periodically depolarize, sending action potentials across the heart wall, causing the heart to contract. In neurons, by contrast, the resting membrane potential is stable. In the sinoatrial node the depolarization is thought to occur due to the leaking of sodium ions into the pacemaker cells without the corresponding outflow of potassium ions. When threshold is reached, fast calcium-sodium channels open and calcium and sodium ions flow into the cell. When 0 mV is reached, the **potassium channels** open and potassium ions leave the cell.

Once this happens, a **field effect** occurs, depolarizing cardiocytes adjacent to the pacemaker. This depolarization spreads throughout the atria and eventually to the ventricles, stimulating the cardiac muscle to contract. Action potentials from the sinoatrial node open **voltage-gated slow calcium channels** in the cardiocytes. **Ligand-gated calcium channels** in the sarcoplasmic reticulum open and calcium moves into the cell, binding to troponin, as it does in skeletal muscle.

Because the heart muscle cells are linked together by intercalated disks with gap junctions between cells, they act as one electrical unit. An impulse generated in the nodal tissue spreads to

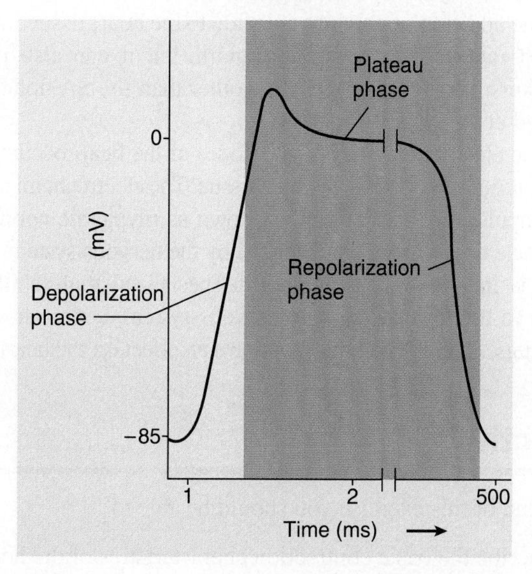

Figure 29.1 Cardiac Muscle Action Potential

the surrounding muscle, causing it first to depolarize (an electrical event) and subsequently to contract (a mechanical event). Review the conduction pathway of the heart in Laboratory Exercise 28.

Normally, the resting heart rate of mammals is lower than the rate of excised hearts. In human hearts that have been removed for heart transplant surgery, the cardiac rate is about 100 beats per minute, which is the native sinoatrial depolarization rate.

This slowing of the heart rate from 100 beats per minute to an average of 72 beats per minute is due to the action of the **parasympathetic nervous system.** The **vagus nerve** conveys parasympathetic fibers that innervate the SA node. **Acetylcholine (ACh)** is a neurotransmitter released by this nerve, and it promotes potassium (K^+) leakage from the nodal cells of the heart to the interstitial fluid, increasing the flow of potassium ions to the outside of the membrane, thus **hyperpolarizing** the cells. By increasing the difference in voltage between the inside and the outside of the cells, it takes longer for the cells to reach threshold and depolarize. In this way the heart rate is slowed. Acetylcholine is a **parasympathomimetic substance,** one that mimics the effects of innervation by the parasympathetic nervous system.

The heart rate can be elevated by increasing the firing rate of the SA node. An increase in the heart rate from the normal resting rate to above 100 bpm is due to the progressive *inhibition* of the **parasympathetic nervous system** and subsequent *stimulation* of the **sympathetic nervous system.** As the vagus nerve secretes less acetylcholine at the node, the firing rate increases. Above 100 beats per minute, the heart rate is controlled by the sympathetic nervous system. Nerves from the cervical region of the spinal cord release **norepinephrine,** which binds to **beta-adrenergic receptors.** This causes calcium channels to open, *decreasing* the threshold and causing more rapid firing of the SA node. Like norepinephrine, **epinephrine** increases the heart rate and strength of contraction by making the calcium channels more permeable. Epinephrine is a **sympathomimetic drug,** one that mimics the effect of sympathetic nervous innervation. Other chemicals function as stimulants. Caffeine is a central nervous system stimulant, and it has an effect on heart muscle. It increases heart rate and the strength of contraction. High concentrations of caffeine cause **arrhythmias,** or irregular heart rates.

The addition of calcium chloride to the heart tissue increases the heart rate and contraction strength, but it can also produce pacemakers in regions of the heart other than the SA node. These are called **ectopic foci.**

The electrical activity of the nodes of the heart occurs in specialized *muscle* cells, not in nervous tissue. The electrochemical transfer of impulses by cardiocytes is known as **myogenic conduction.** The muscle tissue may be influenced by the nervous system, but the activity is initiated and generated in specialized muscle fibers. In addition to the influence of the nervous system, certain hormones, some drugs, and ion concentration have an effect on the heart rate.

Objectives

At the end of this exercise you should be able to

1. describe the basic contraction characteristics of the heart;
2. list the effects of various substances, such as epinephrine, calcium chloride, and acetylcholine, on the heart rate;
3. outline the mechanisms by which these substances change heart rate;
4. measure the resting pulse rate of the heart;
5. correlate the sounds the heart makes with the action of the heart;

Materials

Pulse Rate and Heart Sounds

Clock or watch with second hand
Stethoscope
Alcohol wipes

Virtual Experiment Lab

Ph.I.L.S. Physiology Interactive Lab Simulations Version 4.0 (Available through McGraw-Hill)
Compatible computer

Frog Heart Rate and Contraction Strength

Frog Preparation

Live frog
Clean dissection instruments
- Scissors
- Scalpels
- Dissection tray
- Forceps
- Dissection pins

Small heart hook (fishhook, copper wire, or Z wire)
Nylon thread
Frog anchoring board
Ring stand
Goggles
Animal waste container

BIOPAC Equipment

Force transducer assembly—SS12LA
HDW100A tension adjuster
Data acquisition unit MP30 or MP150
BIOPAC *BSL PRO* A04 Frog Heart (FrogHeart.gtl)
50-gram weight with hook

Physiograph Equipment

Myograph transducer
Physiograph or duograph
Cable
Channel amplifier

Solutions

Frog Ringer's solution—room temperature, 37°C, and iced
2% calcium chloride solution
0.1% acetylcholine chloride solution
0.1% epinephrine solution
Saturated caffeine solution

Procedure

Pulse Rate

(A)[1] Measure the pulse rate (heart rate) and listen to the heart sounds of your lab partner.

1. Place your fingers on the radial artery or the carotid artery of your lab partner and count the number of beats that occur in 1 minute and record the resting pulse rate.

   Resting pulse rate: ____________________ beats/minute

2. You can calculate the average pulse rate for your class by having all the members of your class record their pulse rate in beats per minute on the chalkboard or whiteboard. Add all the pulse rates together and divide by the total number to determine the average. Compare the value you calculate from your class with the average heart rate.

   Average pulse rate for class: ____________________ bpm

   There are differences in pulse rate, depending on whether you are sitting, lying, or standing. The normal resting pulse rate is measured when you are sitting. One of the functions of changes in pulse rate is to alter blood pressure. The carotid sinus is one of the areas of the body that senses changes in blood pressure.

3. After you measure the pulse rate while sitting, lie down on a table or cot and measure the pulse rate. Record any change in the pulse rate in the following space.

   Pulse rate while lying down: ____________________ bpm

4. Stand up and record the change in the pulse rate in the following space.

   Pulse rate while standing: ____________________ bpm

   Can you explain any changes in the rate from the one you recorded when you were sitting?

Heart Sounds

The **lubb/dupp** sounds of the heart reflect the closure of the heart valves. The first heart sound is the lubb sound, and it occurs due to the closing of the atrioventricular valves. The second heart sound is the dupp sound, and it is due to the closing of the semilunar valves. You may hear an additional whooshing sound as you listen through the stethoscope. This is attributed to the imperfect closure of the valves, known as a **heart murmur.**

(A)[2] 1. Clean the earpieces of a stethoscope with alcohol swabs and examine the stethoscope. The earpieces should point in an anterior direction as you place them into your ears.

2. Listen for heart sounds generated by your lab partner by placing the diaphragm of the stethoscope inferior to the second rib on the left side of the body. The pulmonary semilunar valves are best heard in this location.
3. Place the diaphragm of the stethoscope on the right side of the body, in the intercostal space between the second and third ribs, for the location of the aortic semilunar valve.
4. Use figure 29.2 as a guide and listen to heart sounds in these locations.

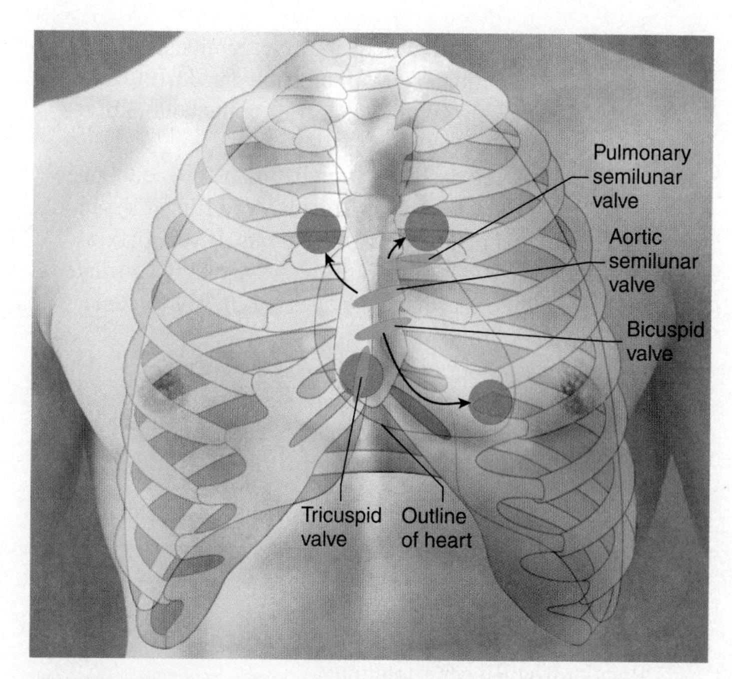

Figure 29.2 Auscultation Areas
Circles are locations where specific valvular sounds can best be heard.

Virtual Experiment Ph.I.L.S.

(A)[3] 1. Open the program for Frog Heart Function and find number 21—Thermal and Chemical Effects. Click on and read the Objectives and Introduction tab.

2. Take the pre-lab quiz to make sure that you understand the concepts presented in the virtual experiment. The **Wet Lab** portion describes the process as it is done on a live frog and the video clips demonstrate the technique.
3. Follow the instructions carefully in the **Laboratory Exercise.**
4. Click the power button to turn on the computer screen and also the power button on the data acquisition unit. Connect the blue plug to the Recording Input 1. Touch the tip of the plug to the unit and it should insert. Click on the **Start** button on the computer screen. You should see the heart contractions.
5. Click the pipette once and drag it to the location where the simulated heart is attached to the myograph transducer.
6. Click the bulb of the pipette again to get it to dispense the regular frog Ringer's solution.
7. Let the heart beat for three or four cycles and then add the cold frog Ringer's solution. Once you get three or four more cycles, click on the **Stop** button. In order to record the data you will have to click on the left arrow at the bottom of the chart to get to the first recording.
8. Move the crosshair cursor to the top of one of the curves and click it there.
9. Move the crosshair cursor to the bottom of one of the curves and click it there. This will set the data points for the amplitude of the contraction.
10. Click on the **Journal** tab to get a record of the amplitude written in the journal.
11. Use the right cursor and move it to the peak of the next wave in order to determine the frequency of the contraction. Click on the **Journal** tab to enter the data.

12. Repeat the amplitude and frequency measurements for the cold Ringer's solution. You will probably need to move the chart by clicking on the right arrow at the bottom of the computer screen.
13. Make sure that you push the **Start** button before adding new solutions. You will need to add regular frog Ringer's solution again before you record the new solutions. If you make a mistake you have to re-record the entire regular frog Ringer's solution and the experimental values before you can enter them into the journal.
14. Wait for three or four cycles; then add the adrenaline solution via a dropper bottle. Record the amplitude of the contraction and the interval between contractions, as you did in the first recording.

    Do the same with the acetylcholine solution as you did with the adrenaline.

    Record the amplitude of the various solutions below and at the end of the exercise.

    Regular frog Ringer's solution ______________________

    Cold frog Ringer's solution ______________________

    Frog Ringer's solution with adrenaline ______________

    Frog Ringer's solution with acetylcholine ____________

    Record the interval between contractions below and at the end of the exercise.

    Regular frog Ringer's solution ______________________

    Cold frog Ringer's solution ______________________

    Frog Ringer's solution with adrenaline ______________

    Frog Ringer's solution with acetylcholine ____________

Examination of Frog Heart Rate and Contraction Strength in Frogs

Either you have a frog prepared for you or your instructor may wish for you to double pith a frog for this exercise. If you need to pith a frog, follow the instructions outlined in Laboratory Exercise 11 and then follow the procedure described next. You may be using a BIOPAC setup or a physiograph setup. In either case, the frog will be prepared the same way.

Frog Preparation

Ⓐ[4] **Procedure**

1. Once the frog is pithed, pin the frog down to a board.
2. Cut through the pectoral skin with a scissors and through the sternum. The heart should be beating inside the pericardial sac. Moisten the heart with room temperature frog Ringer's solution.
3. Carefully cut the pericardial membrane of the heart away from the dorsal surface of the frog.
4. Cover the heart with a moist paper towel and prepare either the BIOPAC setup or the physiograph setup.

Biopac Section

Biopac Setup for MP30 or MP150 Data Acquistion Unit You can access detailed instructions or customize the experiment at www.biopac.com.

Ⓐ[5] 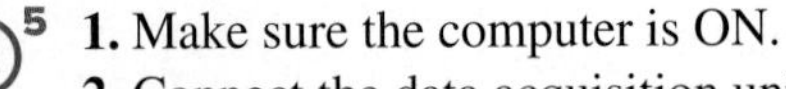

1. Make sure the computer is ON.
2. Connect the data acquisition unit to an electrical outlet with the power supply.
3. With the data acquisition unit OFF, connect it to the computer.
4. Plug SS12LA into Channel 1.
5. Turn the data acquisition unit ON.

Computer Setup

1. Start the BSL *PRO* software. An untitled window should appear.
2. Go to the File menu and choose "Graph Template (*GTL) > File Name: FrogHeart.gtl."
3. Physically attach the HDW100A tension adjuster to a ring stand.
4. Attach the force transducer assembly to the tension adjuster. Make sure that the holes where the S-hook attaches are facing down. The adjuster should be set so that, when the frog heart is hooked up, the string will be vertical.
5. On the computer, select 0–50 grams force range.
6. Select the Setup Channels from the menu, choose the wrench icon from the Channel 1 menu, and locate the wrench icon and scaling window.
7. With only the S-hook on the tension adjuster, click Cal 1.
8. Attach a 50-gram weight to the adjuster and click Cal 2.
9. Click OK and exit the scaling window.

Running the Experiment

1. Place the frog on the board underneath the force transducer assembly and remove the wet paper towel from the frog heart.
2. Attach a metal hook or wire to the membrane attached to the dorsal portion of the heart. You can pierce the apex of the heart with a wire or hook, but you must be very careful not to puncture the ventricle.
3. Attach the hook to the force transducer assembly (see figure 29.3).
4. Adjust the tension, so that the nylon thread is taut but not so much that you tear the hook from the heart.
5. Click the "Start" button and record 10 cycles of the heart.
6. Save your data by selecting the **File menu.** Choose **Save As** and select the file type: **BSL Pro files (*.ACQ) File name: (your name).** Choose **Save.**
7. Alter the temperature or add chemicals as described after "Physiograph Section."

Physiograph Section

Ⓐ[6] **Procedure**

1. Attach a metal hook or wire to the membrane at the dorsal portion of the heart. You can pierce the apex of the heart with a wire or hook, but you must be very careful not to puncture the ventricle.
2. Attach the hook to a thread and tie the thread to the myograph transducer, adjusting the tension until you get

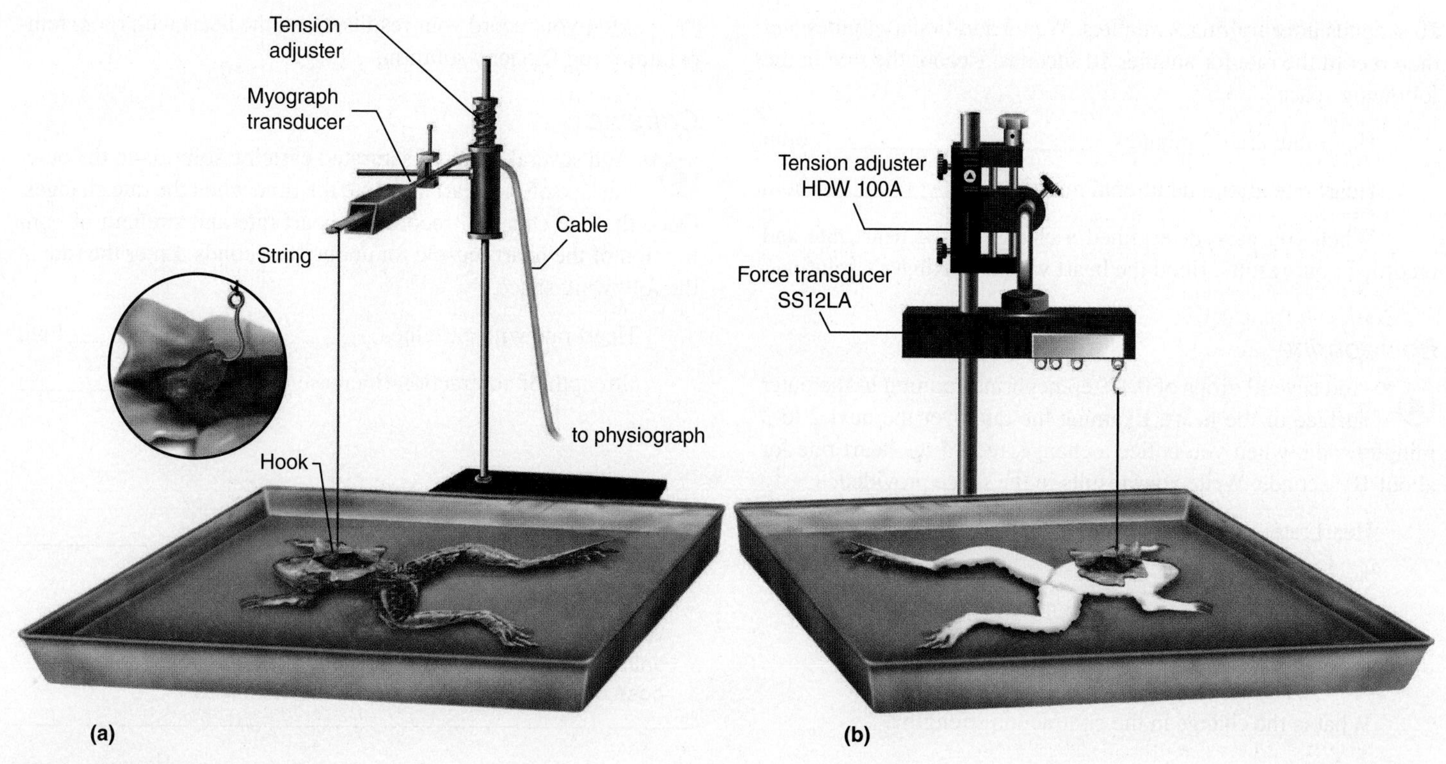

Figure 29.3 Frog Setup
(a) For physiograph; (b) for BIOPAC.

a deflection in the myograph leaf. Your setup should look similar to figure 29.3.

3. Set the chart speed to 0.5 cm/second and record the contraction rate and strength of contraction of the heart muscle. The force that is generated by the heart is measured by the height of the stylus on the chart paper. The rate can be determined by the distance between the peaks on the chart paper.

   Record these values in the following spaces.

   Heart contraction strength: ____________________

   Heart contraction rate: ____________________

4. Alter the temperature or add chemicals as described in the following section.

Data Acquisition

Keep the heart moistened with frog Ringer's solution. Unless directed to do otherwise, use the Ringer's solution that is at room temperature.

Changes in Heart Rate with Temperature You have already recorded the heart rate at room temperature.

(A)[7] 1. Run tests by flooding the heart with iced Ringer's solution. Note the change in heart rate.

   Heart rate with iced Ringer's solution: __________ bpm

2. You can further study the effect of temperature on enzymes by flooding the frog heart with frog Ringer's solution at 37°C. Record the contraction rate.

   Heart rate at 37°C: ____________________ bpm

The **Q10** is an enzymatic relationship that states that, for every 10°C rise in temperature (from 0°C to 50°C), there is a doubling of enzyme activity. Does this correlate with your findings on the temperature relationships with the heart? ____________________

Calcium

(A)[8] ***Effect of Various Chemicals on Heart Rate and Contraction Strength***

Add several drops of 2% calcium chloride solution to the heart muscle. Wait until you get a change in the heart rate before you record the data. Turn on the recording device (BIOPAC or physiograph) and produce a tracing. Calculate the heart rate and record the number in the following space.

Heart rate: ____________________ bpm

What happens to the contraction strength?

Rinse the heart with room temperature frog Ringer's solution.

Acetylcholine

(A)[9] Add several drops of 0.1% acetylcholine solution to the outer surface of the heart. You may have to wait a few minutes to observe any changes in heart rate. Record the heart rate for about

10 seconds after waiting 2 minutes. Wait 3 additional minutes and then record the rate for another 10 seconds. Record the rate in the following space.

Heart rate after 2 minutes: ______________________________ bpm

Heart rate after 3 additional minutes: ______________ bpm

When you have determined a change in the heart rate and recorded your results, flood the heart with frog Ringer's solution.

Epinephrine

(A) **10** Add several drops of 0.1% epinephrine solution to the outer surface of the heart. Examine the rate over the next 2 to 3 minutes, and, when you notice a change, record the heart rate for about 10 seconds. Write your results in the space provided.

Heart rate: ______________________________________ bpm

What is the change in the contraction strength?

After you record your results, flush the heart with room temperature frog Ringer's solution.

Caffeine

(A) **11** Add several drops of saturated caffeine solution to the outer surface of the heart and note the time when the rate changes. Once the rate changes, record the heart rate and strength of contraction of the heart muscle for about 10 seconds. Enter the rate in the following space.

Heart rate with caffeine: ______________________________ bpm

Strength of contraction (increase/decrease): ____________

Cleanup

When you are finished, dispose of the frog in the appropriate animal waste container. Clean your desk and equipment. Dispose of scalpel blades in the sharps container.

Exercise 29 Review

Name: ______________________

Lab time/section: ______________________

Date: ______________________

Functions of the Heart

1. Decreasing heart rate is under the control of what nervous division? ______________________

2. What is the resting heart rate of the average person? ______________________

3. Are there more sodium ions inside or outside a cardiac muscle cell during the resting membrane potential? ______________________

4. What happens to sodium ions when a membrane depolarizes? ______________________

5. What region in the heart depolarizes spontaneously? ______________________

6. What happens to the heart when an action potential is generated in the SA node? ______________________

7. The movement of electrochemical impulses in the myocardium is called ______________________ conduction.

8. What effect do calcium slow channels have on shortening or lengthening the contraction time of the heart muscle? ______________________

9. Beta-adrenergic blockers bind to norepinephrine sites, preventing these neurotransmitters from having an effect. What effect would the use of "beta blockers" have on heart rate? ______________________

10. How much of a change in the heart rate of the frog did you see after the addition of calcium chloride? ______________________

    What was the change in the contraction strength? ______________________

    What process might account for this in terms of cardiac muscle interactions with calcium? ______________________

11. What heart sound is produced by the closure of the atrioventricular valves in the heart? ______

12. A heart murmur is normally caused by what event? ______

______

13. When would a murmur occur in the lubb/dupp cycle if the AV valves were not closing properly? ______

______

______

14. Pilocarpine stimulates the release of acetylcholine from the vagus nerve, thereby increasing parasympathetic stimulation.

    What impact would this drug have on heart rate? ______

    ______

15. In the Ph.I.L.S. virtual heart experiment, what effect did the addition of the adrenaline solution have on heart rate and strength?

    ______

    ______

16. Did the addition of the adrenaline solution increase or decrease the threshold in the heart muscle tissue?

    ______

    ______

17. In the Ph.I.L.S. virtual heart experiment, what effect did the addition of the acetylcholine solution have on heart rate and strength?

    ______

    ______

18. Did the addition of acetylcholine solution hyperpolarize or depolarize the cardiac muscle plasma membrane?

    ______

    ______

Exercise 30

Introduction to Blood Vessels and Arteries of the Upper Body

Introduction

Blood vessels are the conduits that carry oxygen, nutrients, and other materials to the cells and remove wastes from the extracellular fluid. These vascular conduits consist of numerous vessels, such as the large **arteries,** smaller **arterioles,** and **capillaries,** which are the sites of exchange between the blood and cells of the body. **Venules** return blood to the **veins,** which carry it back to the heart.

Arteries are blood vessels that carry blood *away from* the heart. Arteries have thicker walls than veins, and the thickness of arterial walls reflects the higher blood pressure found in them.

Arteries are often named for the region of the body they pass through. The **brachial** artery is found in the arm, whereas the **femoral** artery is in the thigh. In this exercise, you learn the basic structure of blood vessels and locate the arteries of the upper body. The nature of vessels and the descriptions of specific vessels are found in the Seeley text in chapter 21, "Cardiovascular System: Blood Vessels and Circulation."

Objectives

At the end of this exercise, you should be able to

1. draw a cross section of the wall of a generalized blood vessel and label the three layers;
2. distinguish between arteries and veins;
3. distinguish between elastic arteries and muscular arteries;
4. describe the sequence of major arteries that branch from the aortic arch;
5. list the major blood vessels that take blood to the brain;
6. list the arteries of the upper extremity;
7. trace the arteries that supply blood to the head;
8. describe the major organs that receive blood from the upper arteries.

Materials

Microscopes
Prepared slides of arteries and veins
Models of the blood vessels of the body
Charts and illustrations of the arterial system
Cadaver
Microscope slides of arteries and veins in cross section
Cats
Dissection equipment
- Dissection tray or pan
- Scalpels
- Pins
- Blunt (Mall) probes
- Protective gloves
- Scissors
- Forceps
- Animal waste container
- First aid kit

Procedure

Overview of Blood Vessels

(A) 1 Obtain a prepared microscope slide of a cross section of artery and vein. Both arteries and veins have walls that consist of three layers. The outer layer is known as the **tunica adventitia** consisting of a connective tissue sheath. The middle layer, or **tunica media,** is composed of smooth muscle in both arteries and veins. The tunica media is thicker in arteries, and there are pronounced elastic fibers in the wall of the tunica media in arteries. The innermost layer (near the blood) is the **tunica intima**, which consists of a thin layer of connective tissue and a thin layer of simple squamous epithelium, known as **endothelium.** Endothelium is the layer closest to the blood. In arteries, there is an **inner elastic membrane,** which is a thin layer of elastic fibers. These layers are seen in figure 30.1. Examine a prepared slide of an artery and a vein and locate these layers.

Arteries do not have valves, but some veins do, and they are described in Exercise 32. You can easily distinguish veins from arteries in a prepared slide where the vessels are cut in cross section. Veins have a large lumen (although the vein is often collapsed in prepared sections) and a thin tunica media relative to their overall size. You may also find nerves in the prepared slide. These are also circular structures, but they are solid, not hollow. Compare your slide with figure 30.2.

Types of Arteries

Elastic (conducting) arteries are larger arteries close to the heart. Their appearance is made distinctive by the presence of significant amounts of elastic fibers in the tunica media. **Muscular (distributing) arteries** are farther away from the heart and have more smooth muscle in the tunica media. There is a gradual transition from elastic arteries to muscular arteries.

Specific Arteries

(A) 2 Examine the models, charts, and illustrations of the cardiovascular system in the lab and locate the major arteries of the upper body. Read the following descriptions of the arteries. Name the vessels that take blood to an artery and those that receive blood from the artery in question. An overview of the major arteries of

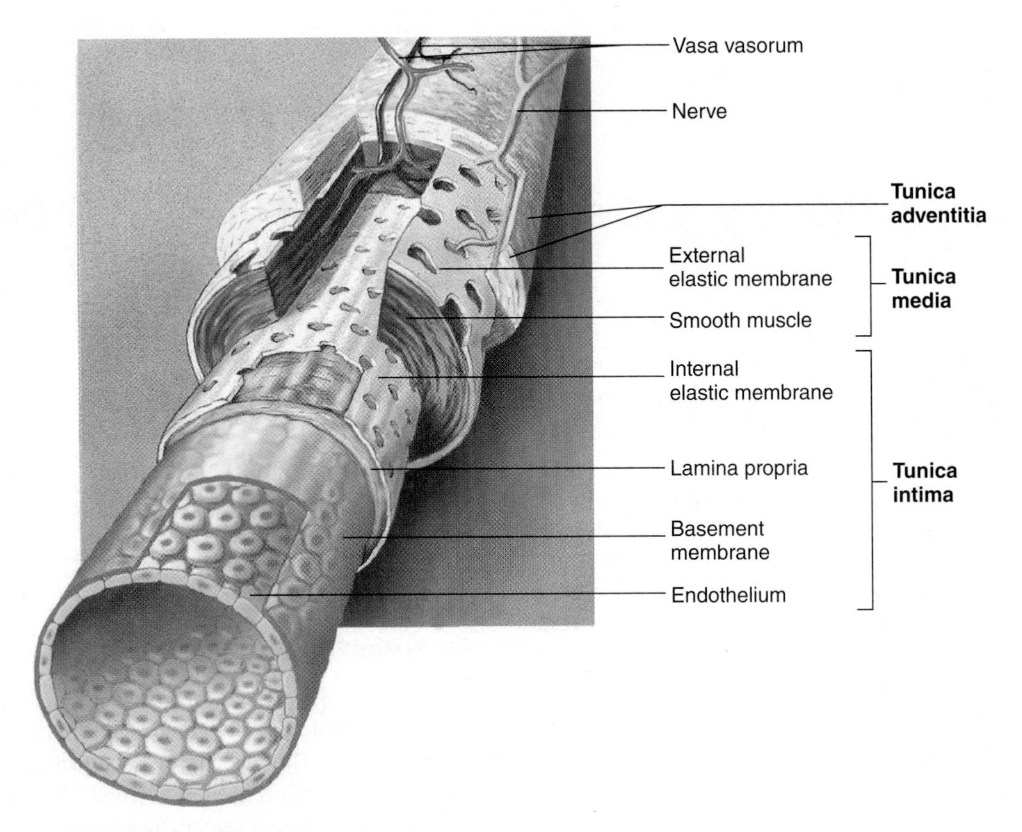

Figure 30.1 Layers of a Generalized Blood Vessel

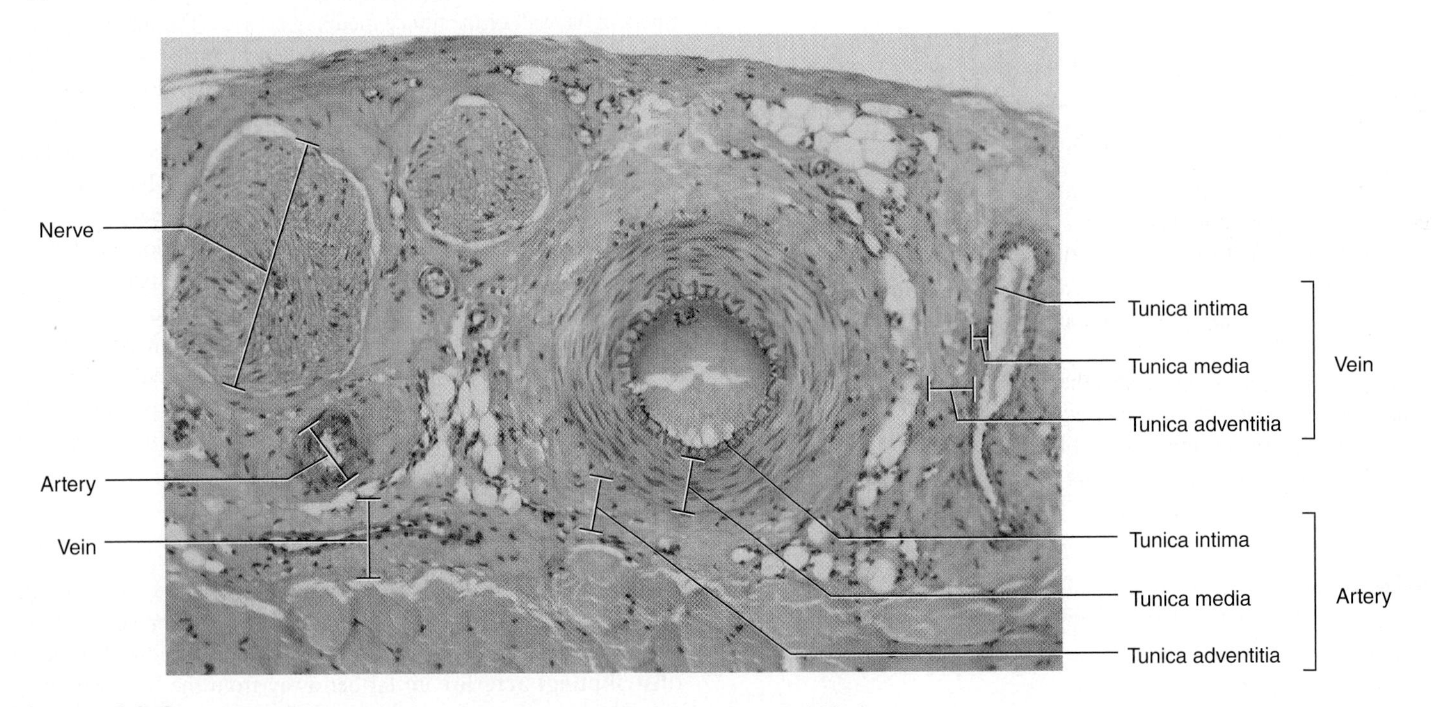

Figure 30.2 Photomicrograph of an Artery, a Vein, and a Nerve (40×)

the body is shown in figure 30.3. You should refer to this illustration for this exercise and for the vessels in Exercises 31 and 32.

Aortic Arch Arteries

(A) 3 Locate the heart on models, charts, or a cadaver and find the large **aorta** that exits from the left ventricle. This is the **ascending aorta,** and it is a large vessel about the size of a garden hose. The ascending aorta is relatively thick-walled due to the high pressure of the blood coming from the heart. The first two arteries that arise from the ascending aorta are the **coronary arteries,** which were covered in detail in Exercise 27. The ascending aorta curves to the left side of the body and forms the **aortic arch.** In humans, there

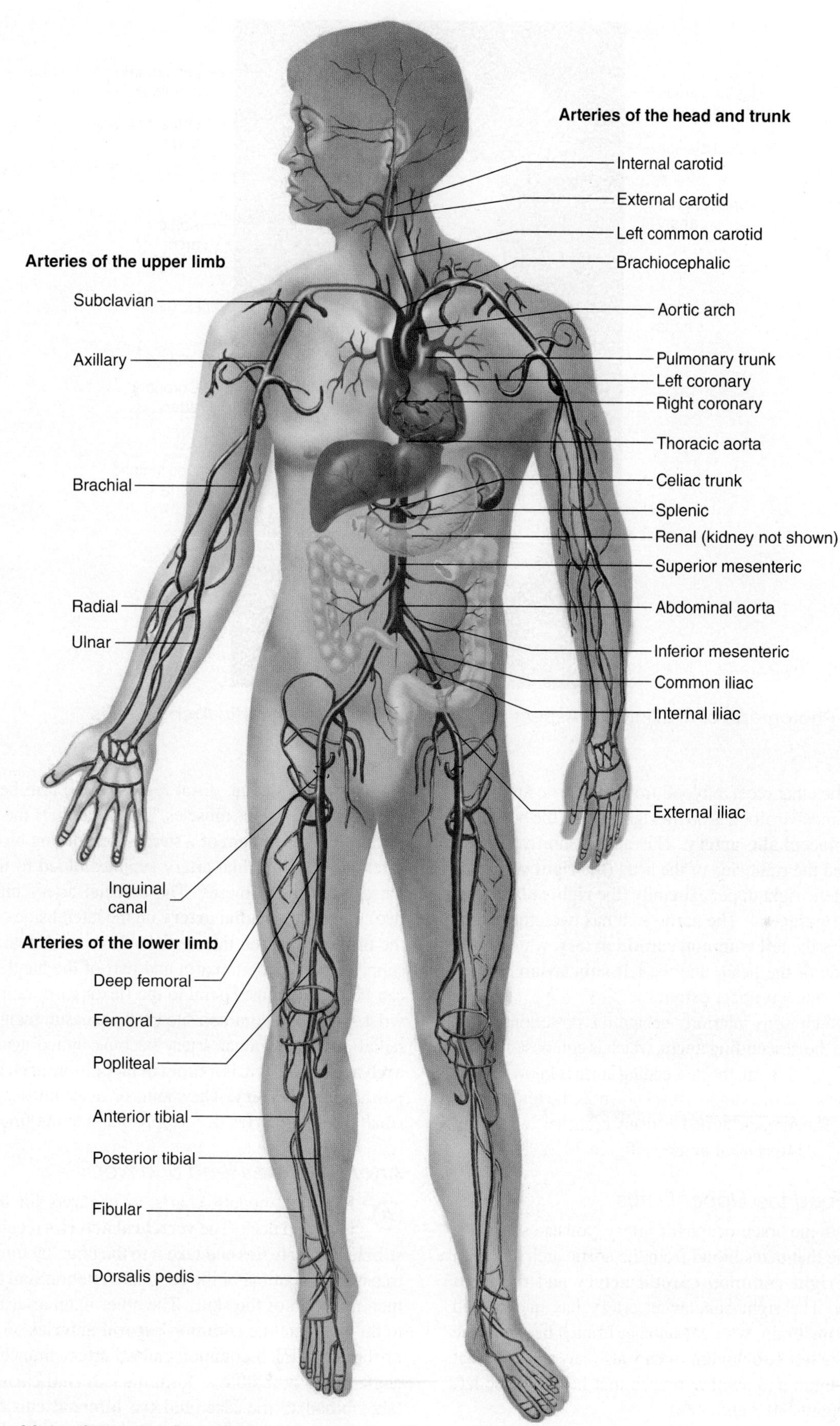

Figure 30.3 Major Arteries of the Body

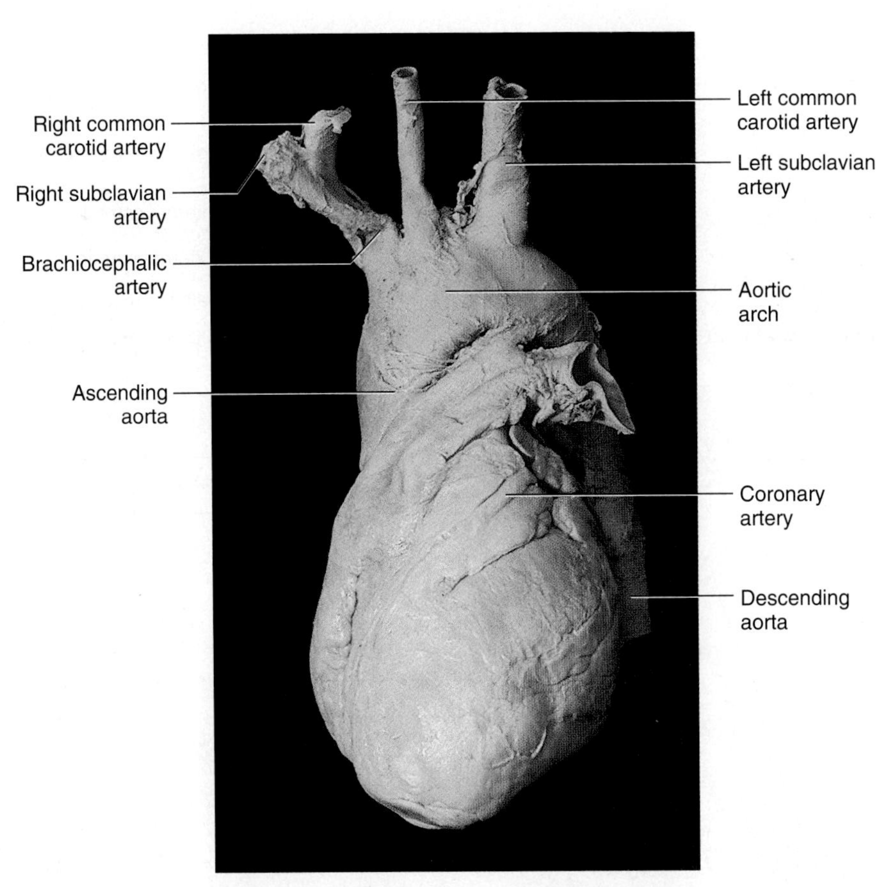

Figure 30.4 Photograph of the Anterior Aspect of the Human Heart and Aortic Arch Arteries

are three main arteries that receive blood from the aortic arch. The first major artery to receive blood is on the right side of the body and is called the **brachiocephalic artery.** This artery shortly divides into arteries that feed the right side of the head (the **right common carotid artery**) and the right upper extremity (the **right subclavian artery**), as seen in figure 30.4. The aortic arch has two other arteries. On the left side is the **left common carotid artery,** which takes blood to the left side of the head, and the **left subclavian artery,** which takes blood to the left upper extremity.

As the aortic arch turns inferiorly behind the posterior part of the heart, it becomes the **descending aorta,** which is composed of two segments. Above the diaphragm, the descending aorta is known as the **thoracic aorta;** below the diaphragm, it is known as the **abdominal aorta** (figure 30.3). The thoracic aorta has many branches, which run between the ribs, called **intercostal arteries** (figure 30.5).

Arteries That Feed the Upper Limbs

Ⓐ 4 If you return to the brachiocephalic artery, you can see that it is a short tube that takes blood from the aortic arch and then branches into the **right common carotid artery** and the **right subclavian artery.** The right subclavian artery has one branch that takes blood to the brain, whereas another branch becomes the **axillary artery.** The **left subclavian artery** also has a branch that takes blood to the brain and another branch that becomes the **left axillary artery** (figures 30.3 and 30.6).

The axillary artery on each side of the body turns into the **brachial artery,** which has a pulse that can be palpated by finding a groove on the medial, distal region of the arm between the biceps brachii and brachialis muscles. This location is the site for the placement of the diaphragm of a stethoscope during blood pressure measurement. The brachial artery supplies blood to the triceps brachii muscle and the humerus. The brachial artery bifurcates (splits in two) to form the **radial artery** on the lateral side of the forearm and the **ulnar artery** on the medial side of the forearm. These arteries supply blood to the forearm and part of the hand. The radial artery can be palpated just lateral to the flexor carpi radialis muscle at the wrist, which is a common site for the measurement of the pulse. The radial artery and ulnar artery become united again as the **palmar arch arteries.** There is a **superficial palmar arch artery** and a **deep palmar arch artery.** They join, or anastomose, and send out the small **digital arteries** that supply blood to the fingers (figure 30.7).

Arteries of the Head and Neck

Ⓐ 5 Examine models, charts, or cadavers for the arteries of the head and neck. The **vertebral arteries** receive blood from the **subclavian arteries** and take it to the brain by traveling through the transverse foramina of the cervical vertebrae and then into the foramen magnum of the skull. The other main arteries that take blood to the brain are the **common carotid arteries** on the anterior sides of the neck. Each common carotid artery branches just below the angle of the mandible to form the **external carotid artery,** which takes blood to the face, and the **internal carotid artery,** which passes through the carotid canal and takes blood to the brain. The external carotid artery on each side has several branches, including

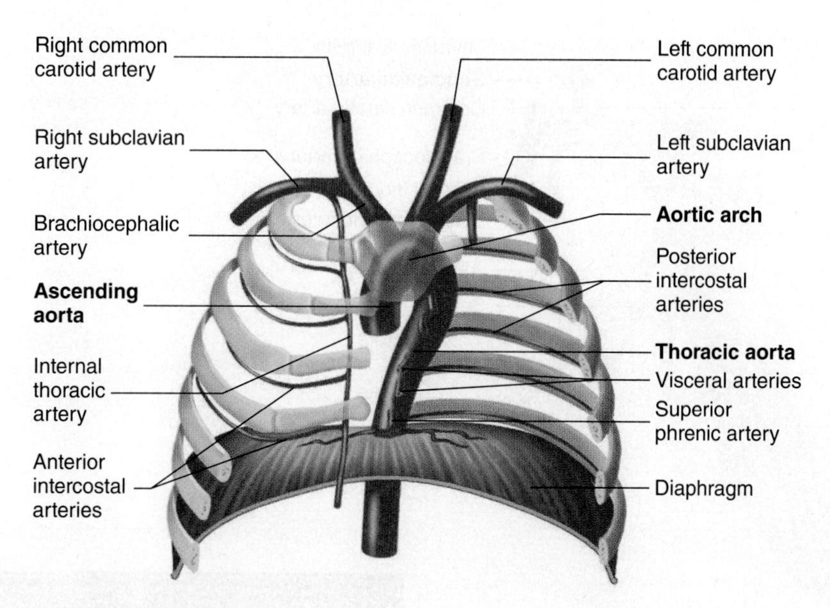

Figure 30.5 Intercostal Arteries

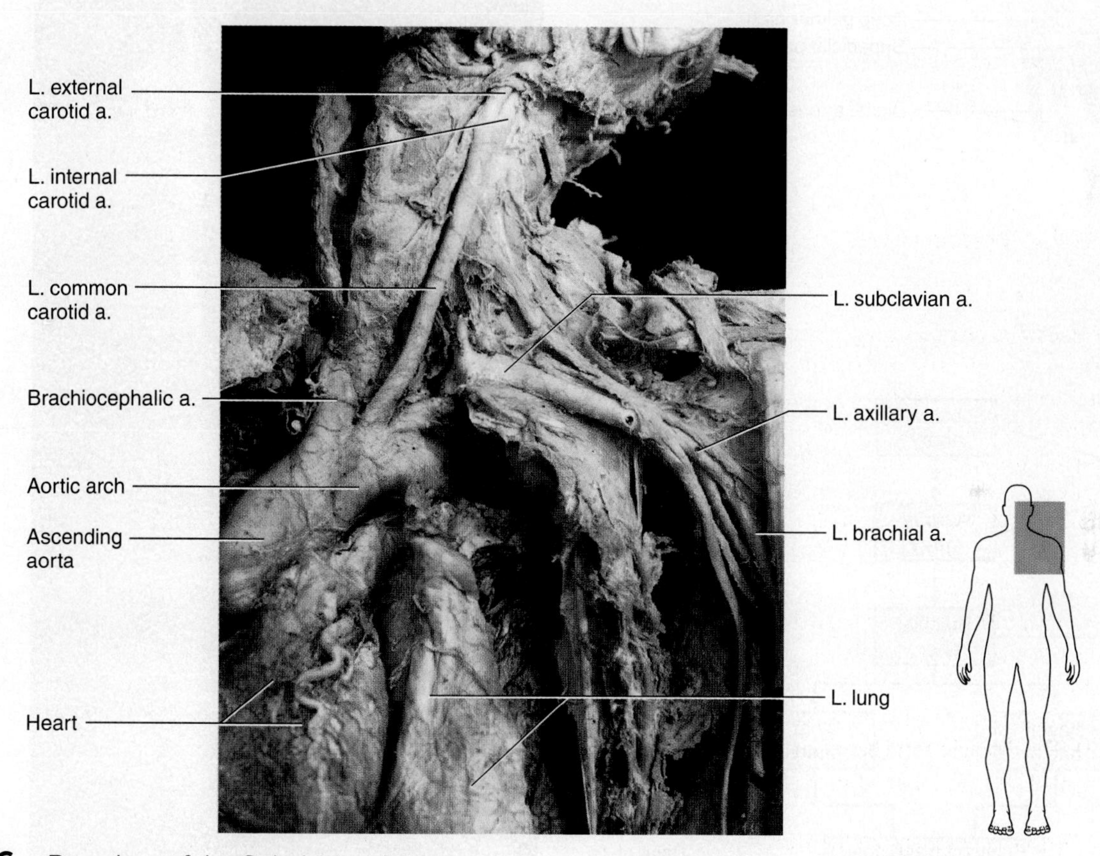

Figure 30.6 Branches of the Subclavian Artery

The abbreviation for artery is a.

the **facial artery,** the **superficial temporal artery,** the **maxillary artery,** and the **occipital artery** (figure 30.8).

Blood Supply to the Brain

(A)6 The blood vessels that supply nutrients and oxygen to the brain do not actually penetrate the brain tissue itself but, rather, form a meshwork around the brain and into the open spaces in the interior. The brain receives blood from four major arteries. These are the **left** and **right vertebral arteries** and the **left** and **right internal carotid arteries.**

The vertebral arteries pass through the foramen magnum and unite to form the **basilar artery** at the base of the brain (figure 30.9). The two internal carotid arteries pass through the

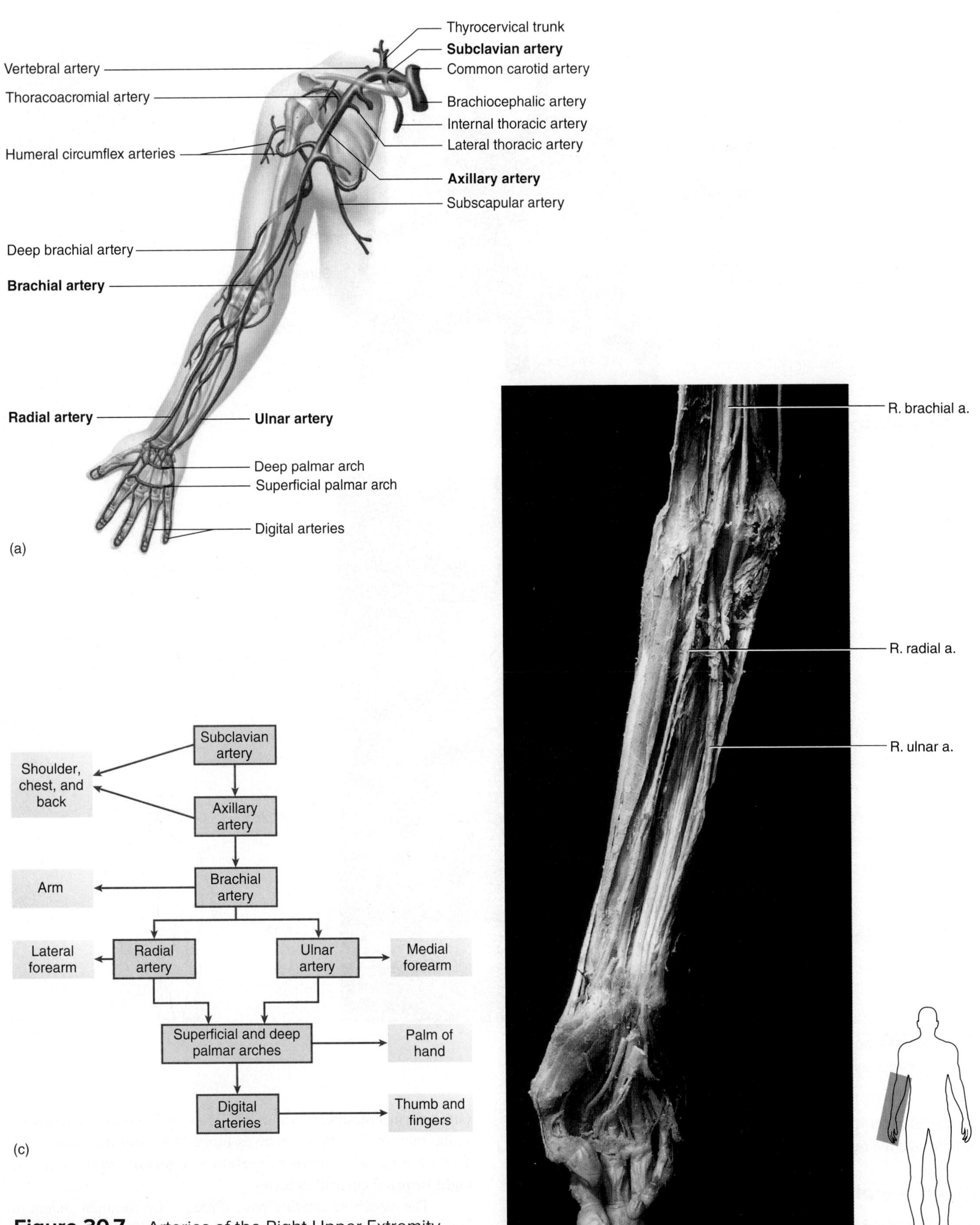

Figure 30.7 Arteries of the Right Upper Extremity
(a) Diagram; (b) photograph; (c) schematic.

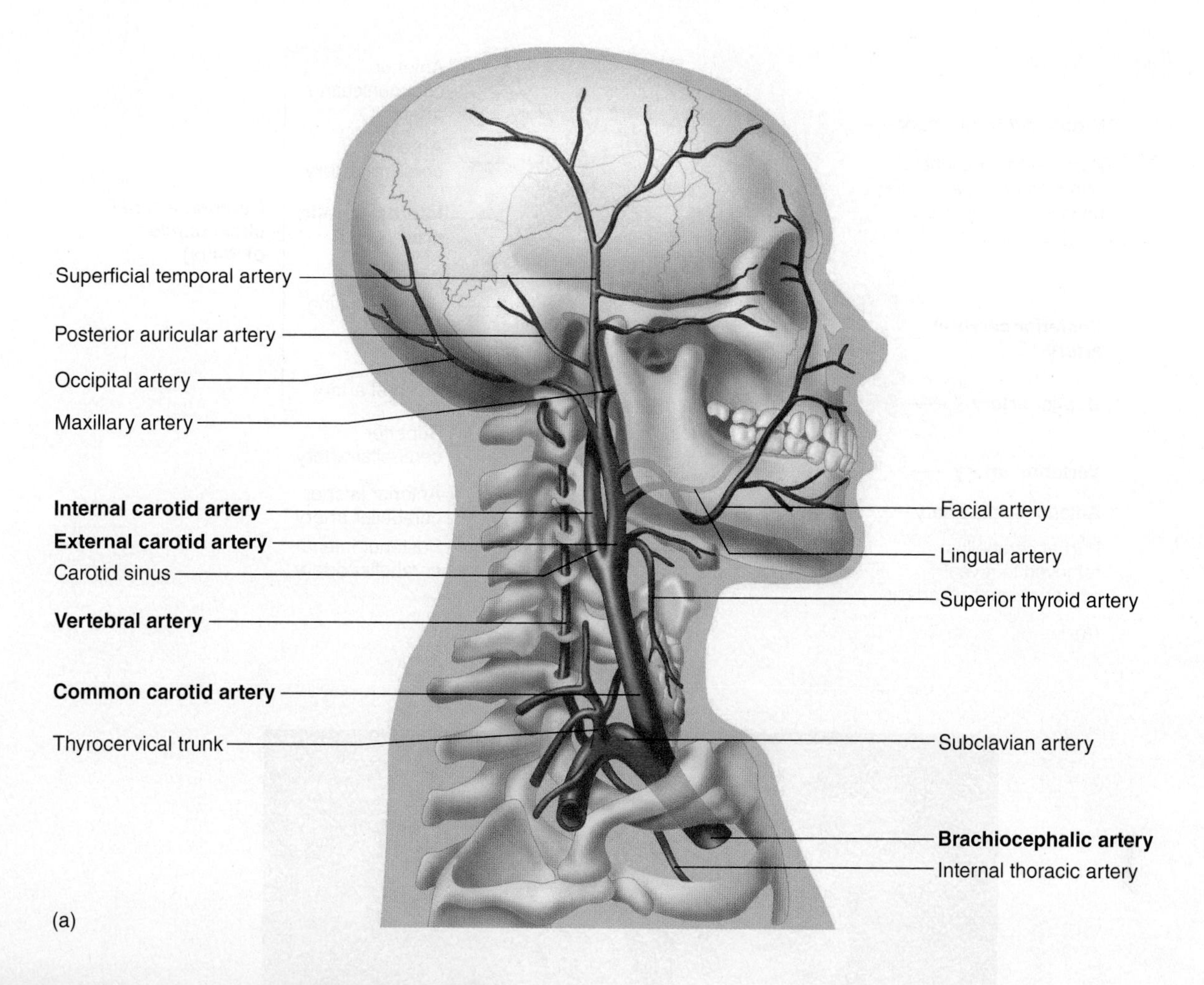

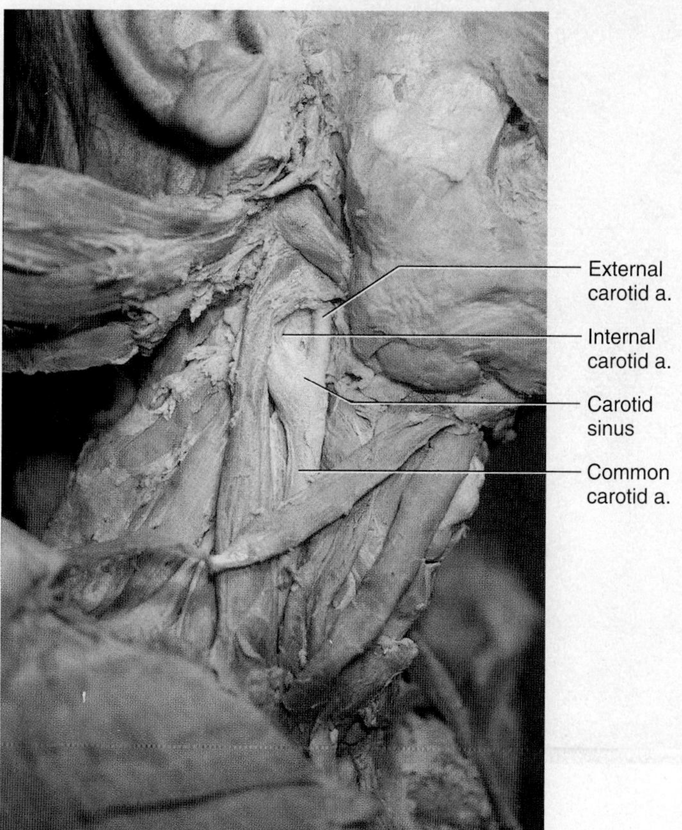

Figure 30.8 Arteries of the Head and Neck
(a) Diagram; (b) photograph.

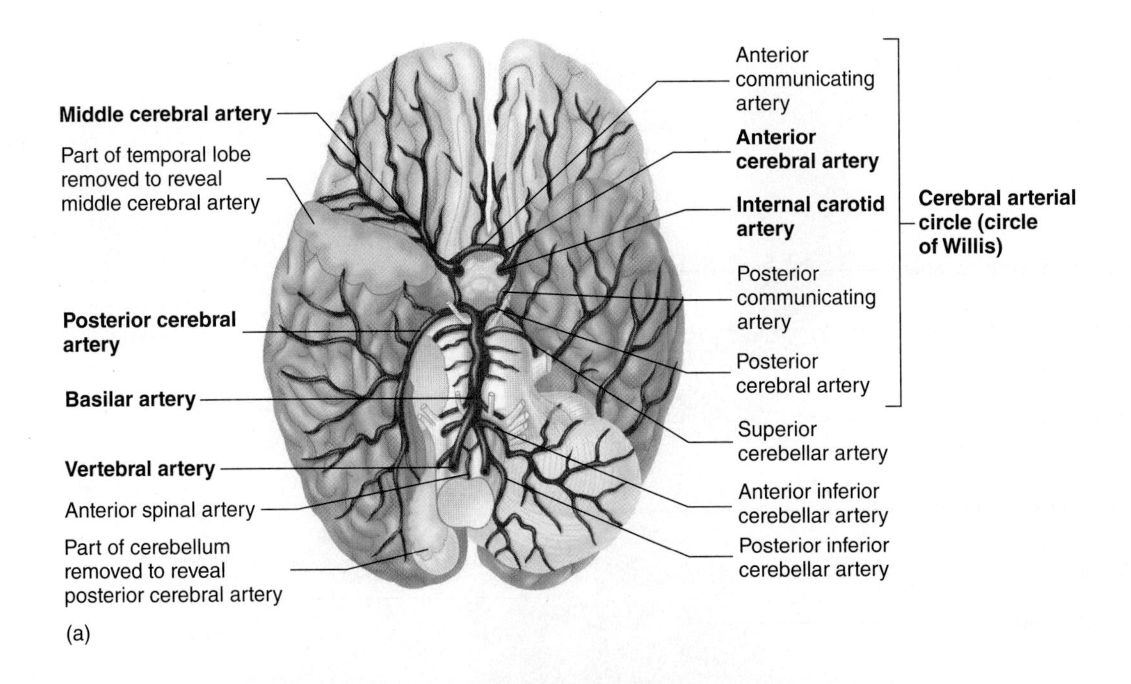

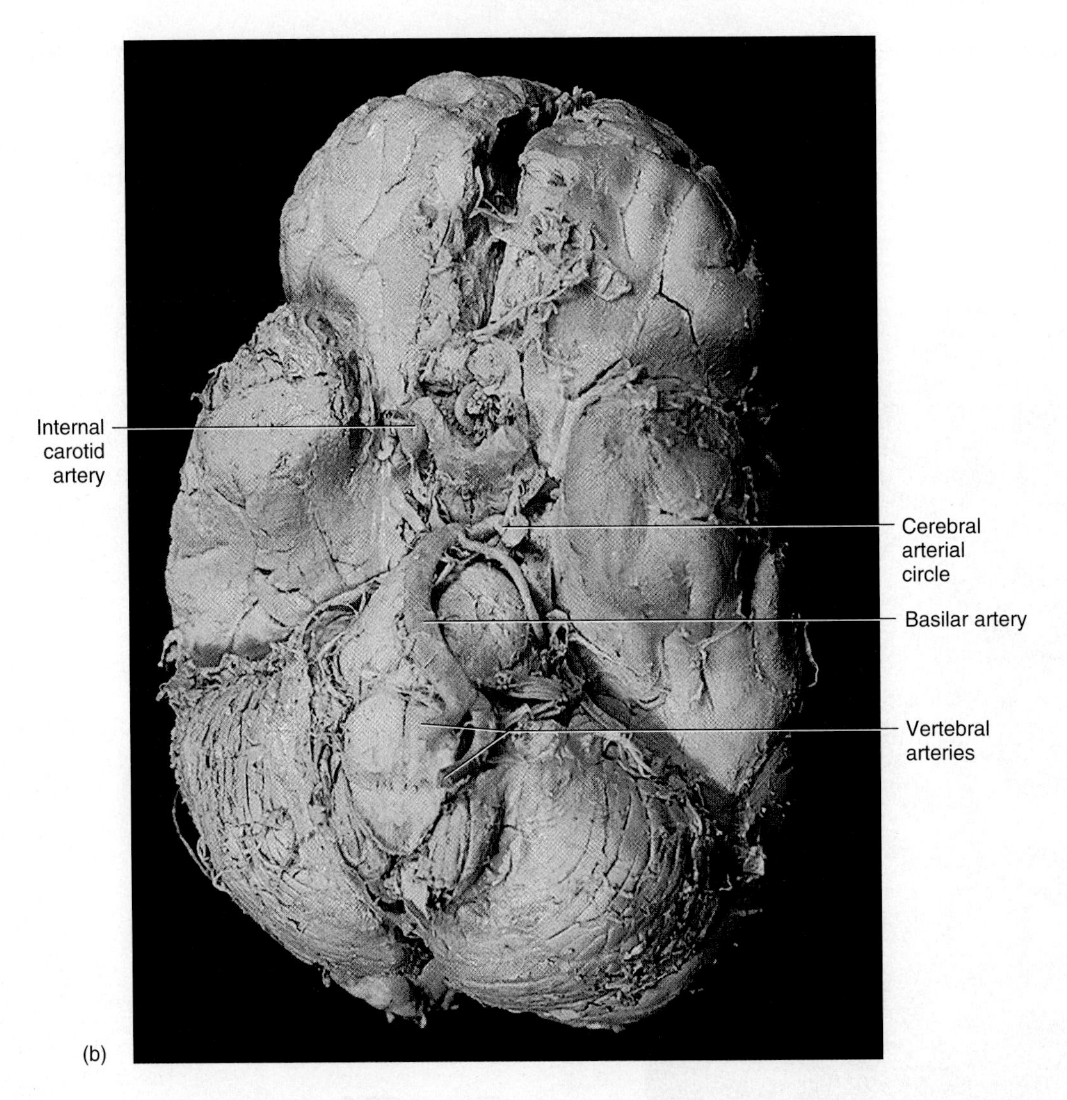

Figure 30.9 Arteries of the Base of the Brain

(a) Diagram; (b) photograph.

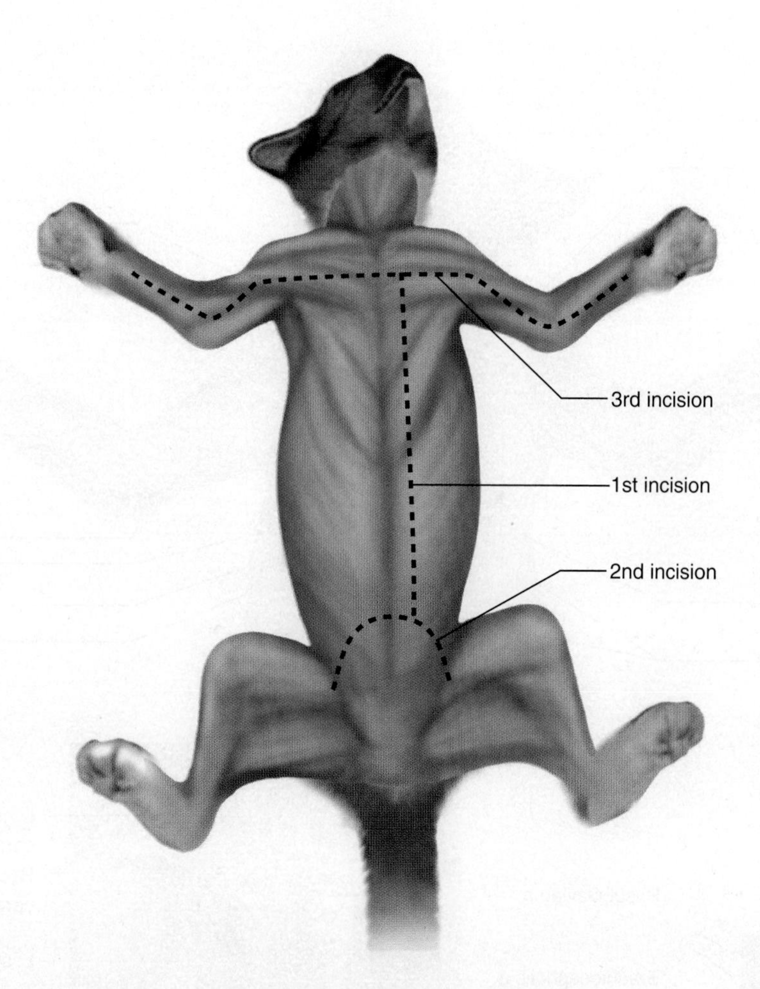

Figure 30.10 Incisions to Open the Thoracic and Abdominal Cavities of the Cat

carotid canal and take blood to the base of the brain. The blood from the basilar and the two internal carotid arteries combines again in a vessel called the **cerebral arterial circle (circle of Willis).** The cerebral arterial circle forms a loop around the pituitary and is composed of the **anterior** and **posterior communicating arteries.** This reuniting of arteries is called **collateral circulation,** which is important in that, if one artery becomes congested, the region of the brain fed by that artery can be supplied by blood from other collateral arteries. The major arteries that branch off the cerebral arterial circle and take blood directly to the brain are the **anterior, middle,** and **posterior cerebral arteries.** The basilar artery gives rise to the **cerebellar arteries,** which take blood to the cerebellum.

Cat Dissection

(A) 7 Prepare for the cat dissection by taking a dissection tray back to your table, along with a scalpel, scissors or a bone cutter, string, forceps, and, of course, a cat. Remember to place all excess tissue in the appropriate waste container and not in a standard wastebasket or down the sink!

Be careful as you make an incision into the body cavity of the cat. Cut with the scalpel blade facing to one side or, even better, use a lifting motion and cut upward through the tissue. If you use a downward sawing motion, you will damage the internal organs of the cat.

Open the thoracic and abdominal regions of the cat if you have not done so already. Make an incision into the body wall of the cat in the belly, slightly lateral to midline. Cut into the muscle; then grasp the tissue with a forceps and lift the body wall away from the viscera. Continue cutting anteriorly through the belly. Stop at the level of the diaphragm, and then continue cutting a little off center as you cut through the costal cartilages. Use a scalpel or scissors to cut through the thoracic region.

You can now make a transverse incision in the lower region of the belly, so that you have an inverted T–shaped incision (figure 30.10). Lift the tissue near the ribs and expose the diaphragm. Gently and carefully snip or cut the diaphragm away from the ventral region by cutting as close to the ribs as possible.

Next make an incision by cutting transversely across the upper part of the chest. Be careful not to cut the blood vessels of the neck that were exposed during the original skinning process. Open the body cavity by pulling the two flaps laterally, exposing the thoracic and abdominal cavities. You may want to cut through the ribs at their most lateral point to facilitate the opening of the thoracic cavity.

Carefully expose the thoracic region of the cat and locate the two lungs and the heart. Note that the heart is enclosed in the pericardial sac. Look for the fatty material, called the **greater omentum,** that covers the **intestines** in the abdominal region. As you dissect the arteries of the cat, pay careful attention to the veins

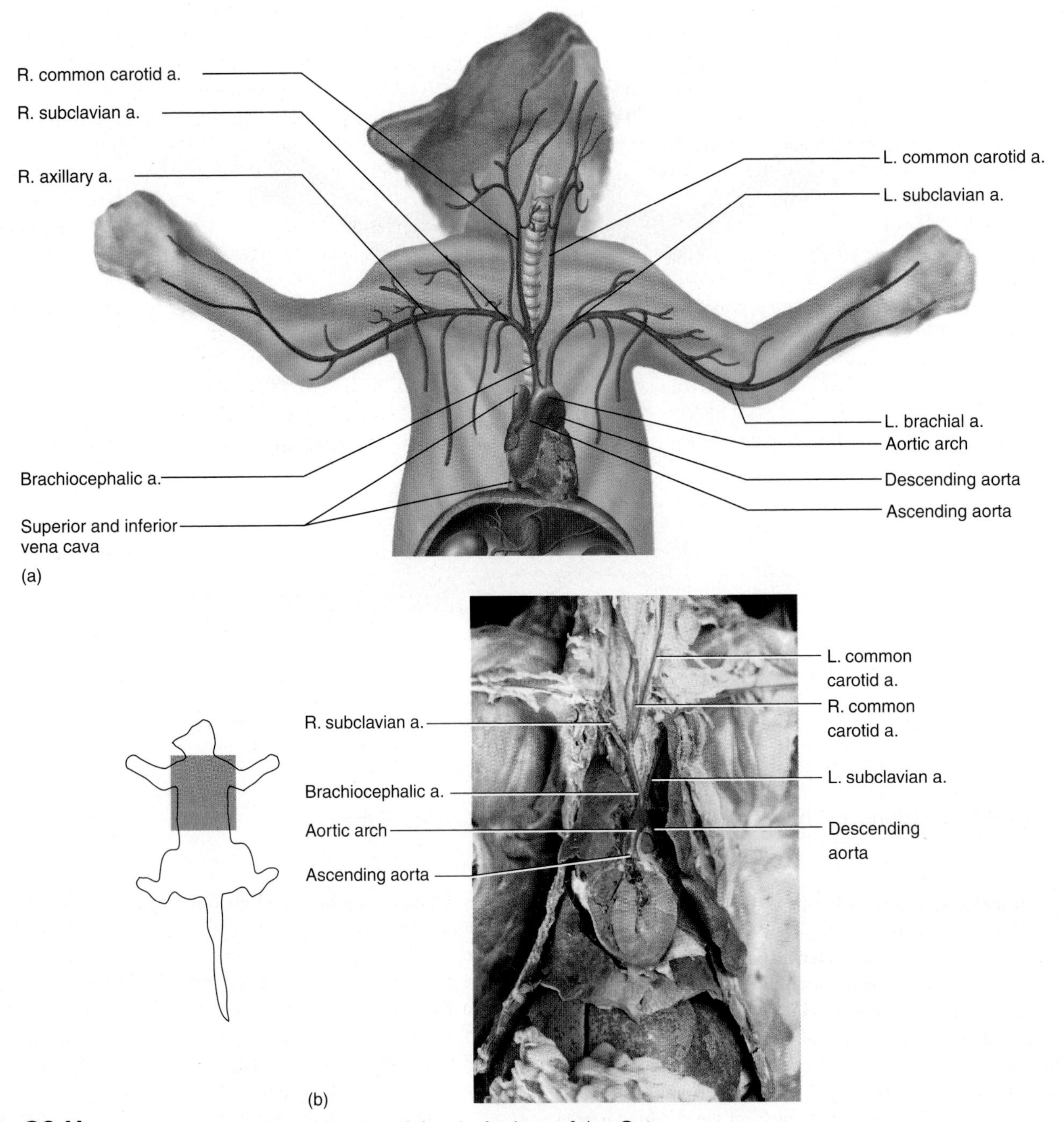

Figure 30.11 Anterior View of the Heart and Aortic Arches of the Cat
(a) Diagram; (b) photograph.

and nerves that travel along with the arteries. *Do not dissect other organs.* You will study these systems later (for example, respiratory, digestive, and urogenital systems).

As you open the abdominal cavity, you will see the space that contains the internal organs, known as the **coelom,** or **body cavity.** The lining on the inside of the body wall is the **parietal peritoneum,** and the lining that wraps around the outside of the intestines is the **visceral peritoneum.**

Arteries of the Cat

The blood vessels in the cat have been injected with colored latex. If the cat is doubly injected, the **arteries are red** and the **veins are blue.** If the cat has been triply injected, the hepatic portal vein is typically injected with yellow latex. Be careful locating the arteries in the cat. You can tease away some of the connective tissue with a dissection needle to see the vessel more clearly. Note the difference in the aortic arch arteries in the cat compared with those of the human. Look for the **coronary arteries,** which take blood from the ascending aorta to the heart tissue.

(A) 8 You can begin your study of the arteries of the cat by examining the ascending aorta. Note that the **aorta** is composed of the **ascending aorta,** which bends and leads to the **aortic arch** and then moves posteriorly to form the **descending aorta.** The descending aorta is composed of the **thoracic aorta** above the diaphragm and the **abdominal aorta** below the diaphragm. The pattern of the arteries that leave the aortic arch is somewhat different in cats than in humans. In the cat, typically only two large vessels leave the aortic arch, the large **brachiocephalic artery** and the **left subclavian artery.** The brachiocephalic artery further divides into the **left common carotid artery,** the **right common carotid artery,** and the **right subclavian artery.** Examine figure 30.11 for the pattern in the cat. How is the pattern in the cat different from that in the human?

Several arteries arise from the **subclavian artery.** The **vertebral artery** takes blood to the head, and the **subscapular artery** takes blood to the pectoral girdle muscles, along with the **ventral thoracic artery** and **long thoracic artery.** The ventral thoracic and long thoracic arteries supply the latissimus dorsi and pectoralis muscles with blood. The **left** and **right subclavian arteries** become the **left** and **right axillary arteries,** which in turn take blood to the brachial arteries. The **brachial artery** takes blood to the **radial** and **ulnar arteries.** Additional arteries that branch from the subclavian artery are the **internal mammary artery,** which takes blood to the ventral body wall, and the **thyrocervical artery,** which takes blood to the region of the shoulder and neck. The **costocervical artery** reaches the deep muscles of the back and neck (figure 30.12).

The **common carotid artery** takes blood along the ventral surface of the neck to feed the small **internal carotid artery,** which supplies the brain. The common carotid artery also empties into the **external carotid artery,** which takes blood to the external portions of the head. The **occipital artery** receives blood from the common carotid artery and supplies the muscles of the neck.

The thoracic aorta not only takes blood to the abdominal aorta but also supplies the rib cage with blood from the **intercostal arteries. Esophageal arteries** also branch from the thoracic aorta, supplying the esophagus with blood. The thoracic aorta continues as the **abdominal aorta** as it passes through the diaphragm.

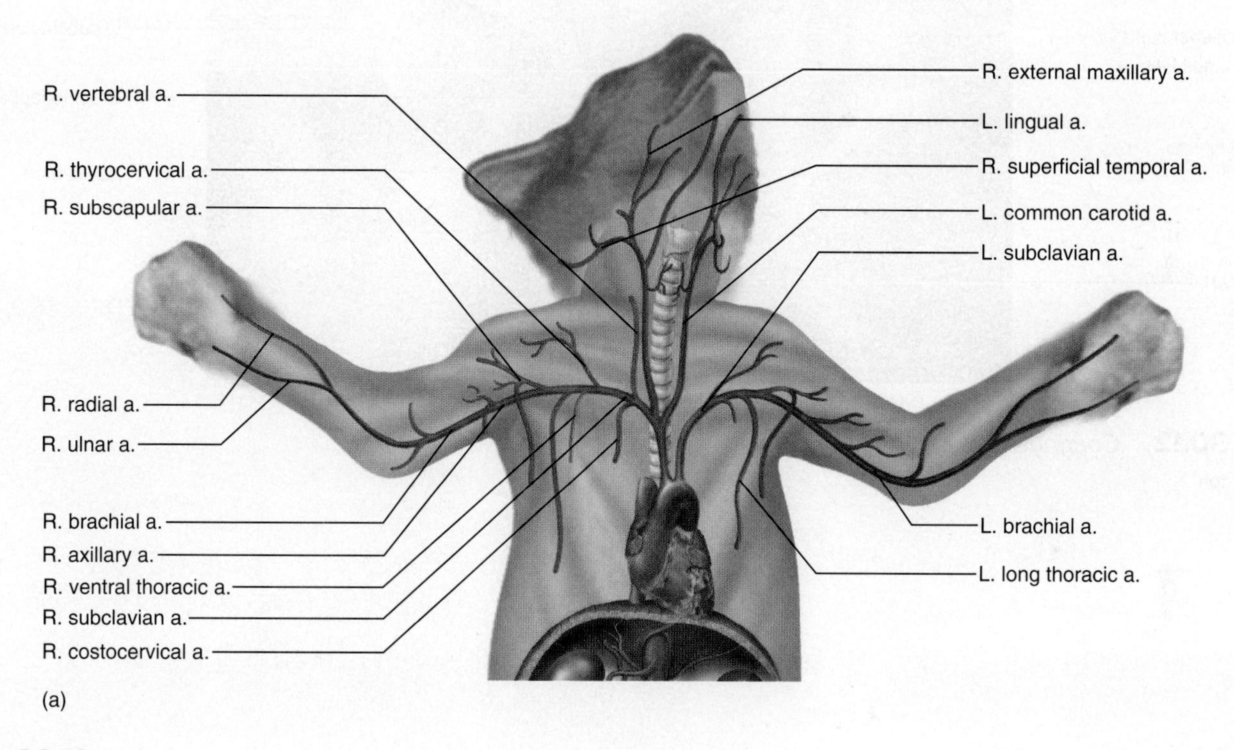

Figure 30.12 Arteries of the Shoulder and Upper Extremity of the Cat
(a) Diagram.

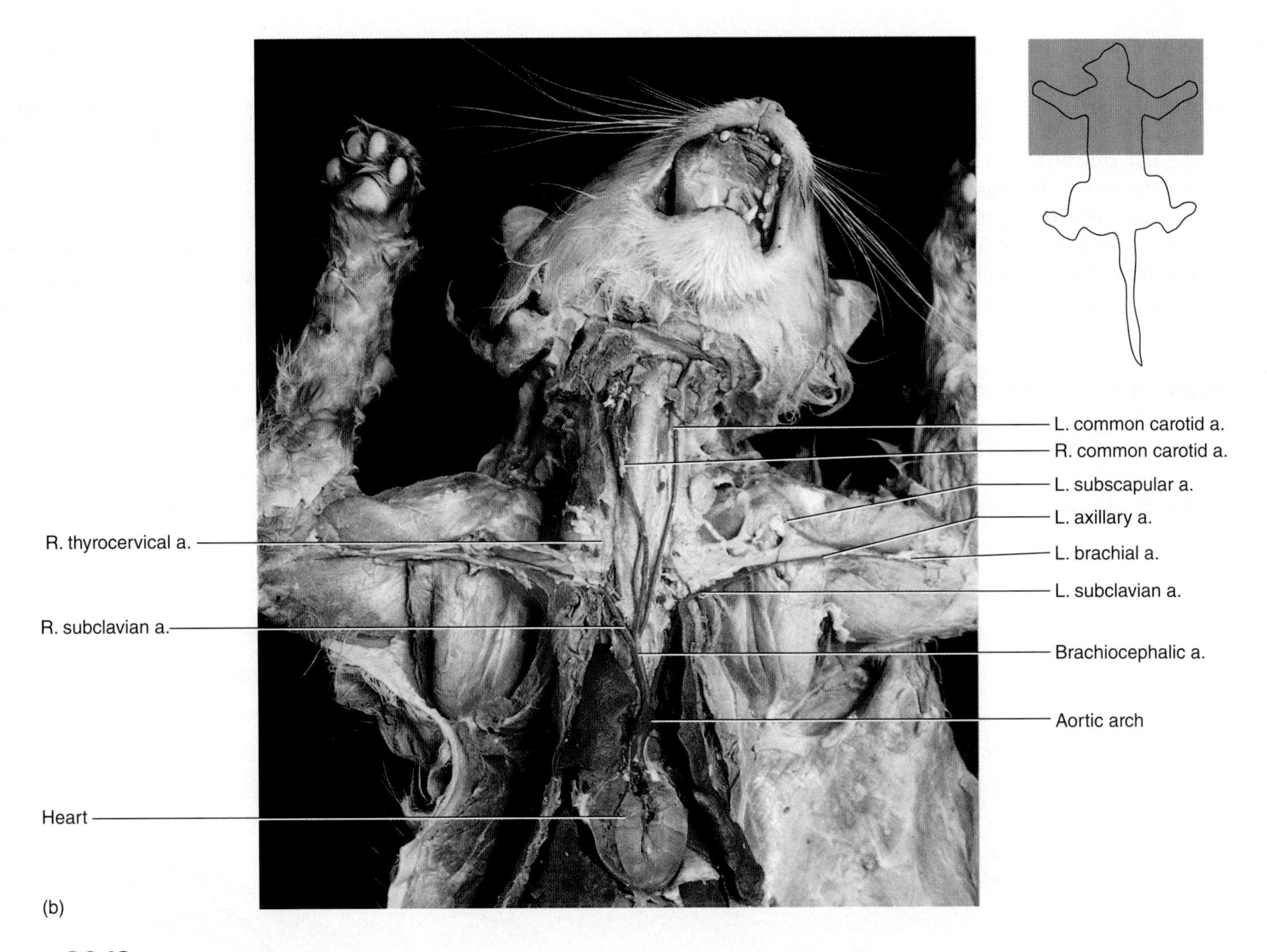

Figure 30.12 *Continued.*
(b) Photograph.

Exercise 30 Review

Introduction to Blood Vessels and Arteries of the Upper Body

Name: ______________________

Lab time/section: ______________________

Date: ______________________

1. Label the following illustration with the major arteries of the body using the terms provided. Try to complete the illustration first and then review the material in this exercise to determine your accuracy.

abdominal aorta
aortic arch
brachiocephalic trunk
common carotid artery
external carotid artery
internal carotid artery
subclavian artery
thoracic aorta

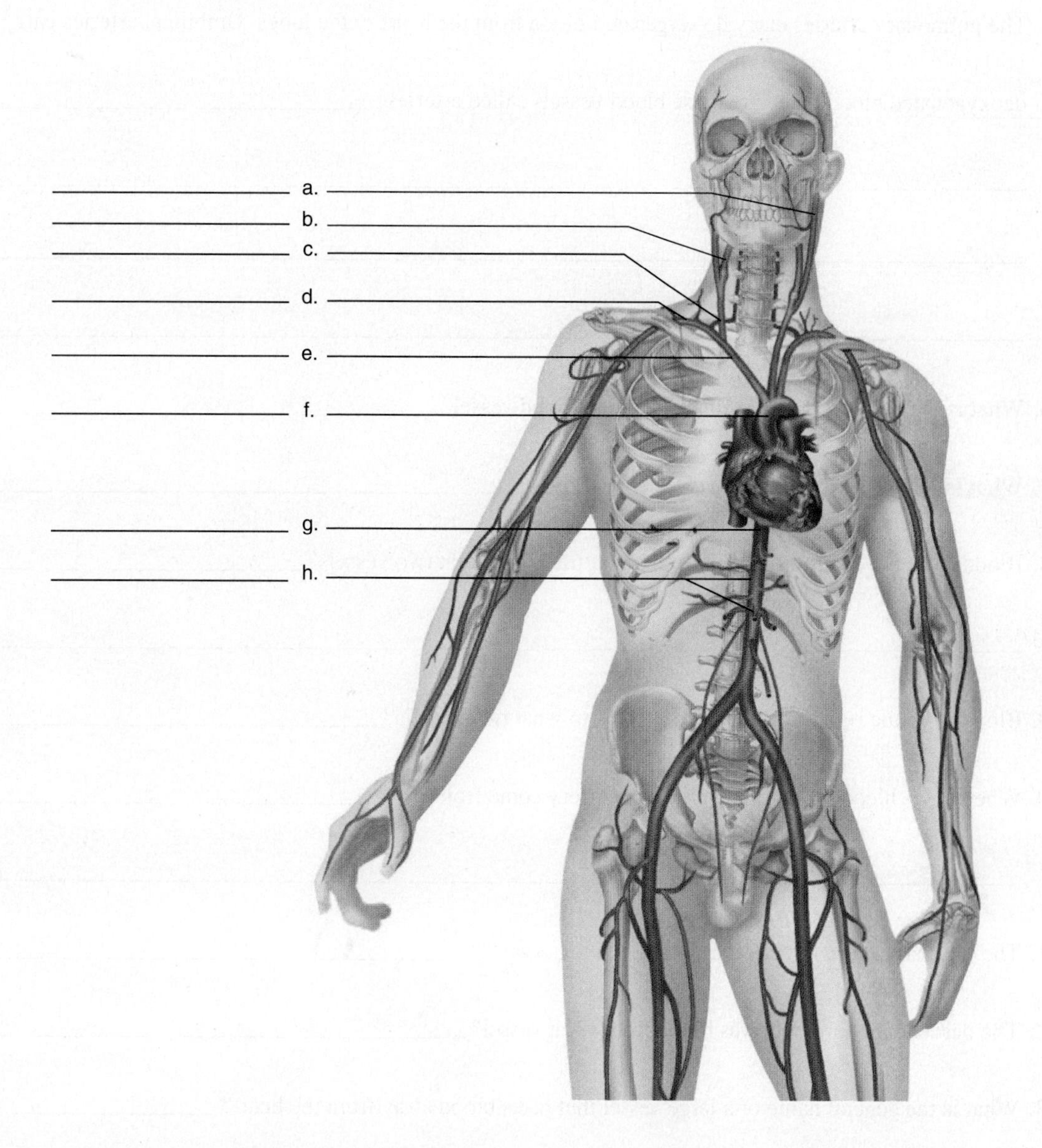

2. Blood from the left subclavian artery flows into what vessels as it moves toward the left arm? ______________________

____

____

3. Blood in the radial artery comes from what blood vessel? ______________________

4. An aneurysm is a weakened, expanded portion of an artery. Ruptured aneurysms can lead to rapid blood loss. Describe the significance of an aortic aneurysm versus a digital artery aneurysm. ______________________

5. The pulmonary arteries carry deoxygenated blood from the heart to the lungs. Umbilical arteries carry a mixture of oxygenated and deoxygenated blood. Why are these blood vessels called arteries? ______________________

6. What is the name of the outermost layer of a blood vessel? ______________________

7. What kind of blood vessel has valves? ______________________

8. Blood from the common carotid artery next travels to what two vessels? ______________________

9. Blood from the right brachial artery travels to what two vessels? ______________________

10. Where does blood in the right subclavian artery come from? ______________________

11. The internal carotid artery takes blood to what organ? ______________________

12. The descending aorta receives blood from what vessel? ______________________

13. What is the general name of a large vessel that takes blood away from the heart? ______________________

14. Blood in the left common carotid artery receives blood from what vessel? ______________________

15. Name three blood vessels that exit from the aortic arch. ______________________________

____

16. How do the aortic arch arteries of a cat differ from those of a human? ______________________________

____

____

17. Working in pairs, have your lab partner select an artery for you to name. Quiz each other on the material learned in this exercise.

18. Name the vessels or the layers in the illustration using the terms provided.

artery
tunica adventitia
tunica intima
tunica media
vein

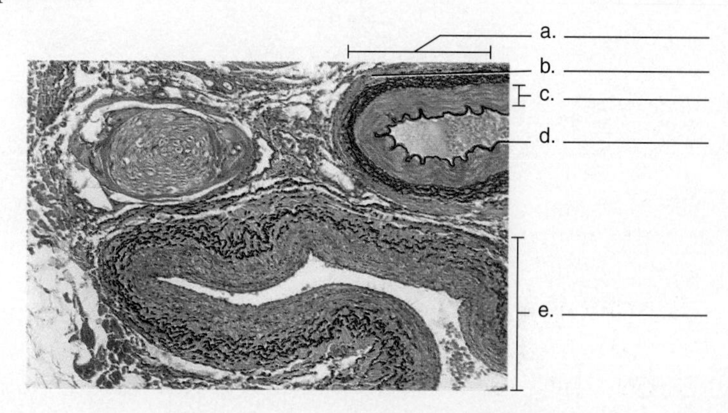

19. Name the specific vessels using the terms provided.

axillary artery
brachial artery
digital artery
radial artery
subclavian artery
superficial palmar arch artery
ulnar artery

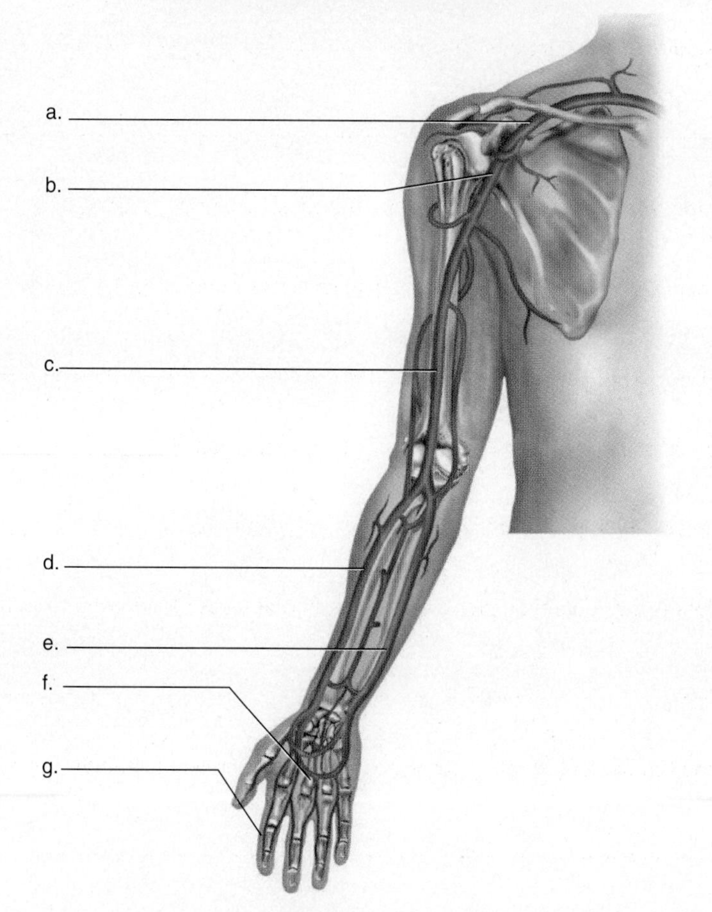

NOTES

Exercise 31

Arteries of the Lower Body

Introduction

The arteries of the lower body receive blood directly or indirectly from the descending aorta. As with the arteries of the upper extremity, these vessels are often named for their locations, such as the *iliac* artery, the *femoral* artery, and the *mesenteric* arteries, or they are named for the organs they serve, such as the common *hepatic* (*hepatic* = liver) artery and the *splenic* artery. These arteries are presented in the Seeley text in chapter 21, "Cardiovascular System: Blood Vessels and Circulation."

It is very important that you are careful when dissecting the arteries of the abdominal region of the cat. Do as little damage as possible to the other structures as you examine the arteries in this region.

Objectives

At the end of this exercise, you should be able to

1. describe the sequence of major arteries that originate from the abdominal aorta or its derivatives;
2. list the arteries of the lower extremities;
3. describe the major organs that receive blood from the lower arteries;
4. name the vessels that take blood to a particular artery and the vessels that receive blood from a particular artery.

Materials

Microscope
Prepared slide of arteriosclerosis
Models or charts of the blood vessels of the body
Cadaver
Cats
Materials for cat dissection

- Dissection trays
- Scalpel
- Protective gloves
- Blunt (Mall) probe
- Pins
- Forceps and sharp scissors
- First aid kit in lab or prep area
- Sharps container
- Animal waste disposal container

Procedure

Examine the models and charts in the lab and the accompanying illustrations to locate the major arteries in the abdomen and lower body. Read the following descriptions of the vessels and locate the individual arteries. You should refer to figure 30.3 for an overview of the major arteries of the body.

Abdominal Arteries

(A)[1] Examine models, charts, or a cadaver in lab and look for the following arteries as you read along in this lab manual. The **abdominal aorta** is the portion of the **descending aorta** inferior to the diaphragm. The first major branch of the abdominal aorta is the **celiac trunk (celiac artery).** The celiac trunk splits into three separate arteries: the **splenic artery,** which takes blood to the spleen, pancreas, and part of the stomach; the **left gastric artery,** which takes blood to the stomach and esophagus; and the **common hepatic artery,** which takes blood to the liver. Locate the celiac trunk and the branches of the celiac in figure 31.1.

Inferior to the celiac trunk is the **superior mesenteric artery.** This vessel takes blood from the abdominal aorta and continues through the mesentery until it reaches the small intestine and proximal portions of the large intestine, including the cecum, ascending colon, and part of the transverse colon. The major branches of the superior mesenteric artery are the intestinal arteries, the ileocolic arteries, and the right and middle colic arteries.

The next vessels to branch from the aorta are the paired **suprarenal arteries,** which take blood to the adrenal glands. Inferior to the suprarenal arteries are the **left** and **right renal arteries.** These arteries take blood to the kidneys. Locate these vessels in the lab and in figures 31.1 and 31.2.

The **gonadal arteries** branch inferior to the renal arteries and descend to either the testes or the ovaries. The **inferior mesenteric artery** is the next vessel to leave the aorta, and it takes blood to the lower portion of the large intestine, including part of the transverse colon, the descending colon, and the rectum. These can be located in figures 31.1 and 31.2.

The abdominal aorta terminates by dividing into the two **common iliac arteries.** The common iliac arteries take blood to the internal and **external iliac arteries.** The **internal iliac artery** takes blood to the pelvic region, including branches that take blood to the rectum, pelvic floor, external genitalia, groin muscles, and hip muscles, as well as the uterus, ovary, and vagina in females. These arteries are illustrated in figure 31.2, and a schematic of the abdominal and pelvic arteries is presented in figure 31.3.

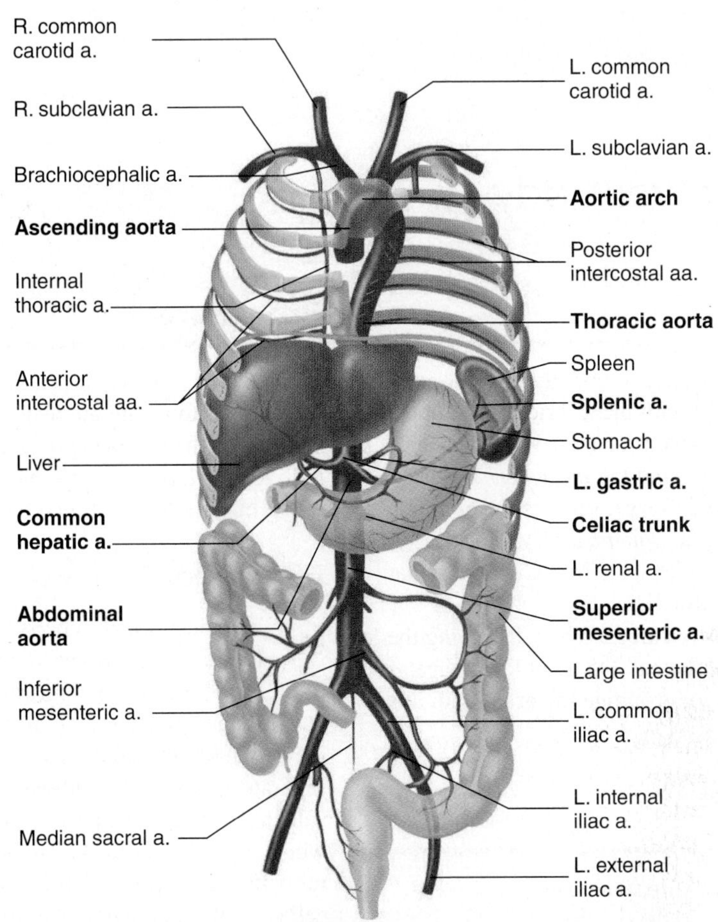

Figure 31.1 Arteries of the Thoracic and Abdominal Regions

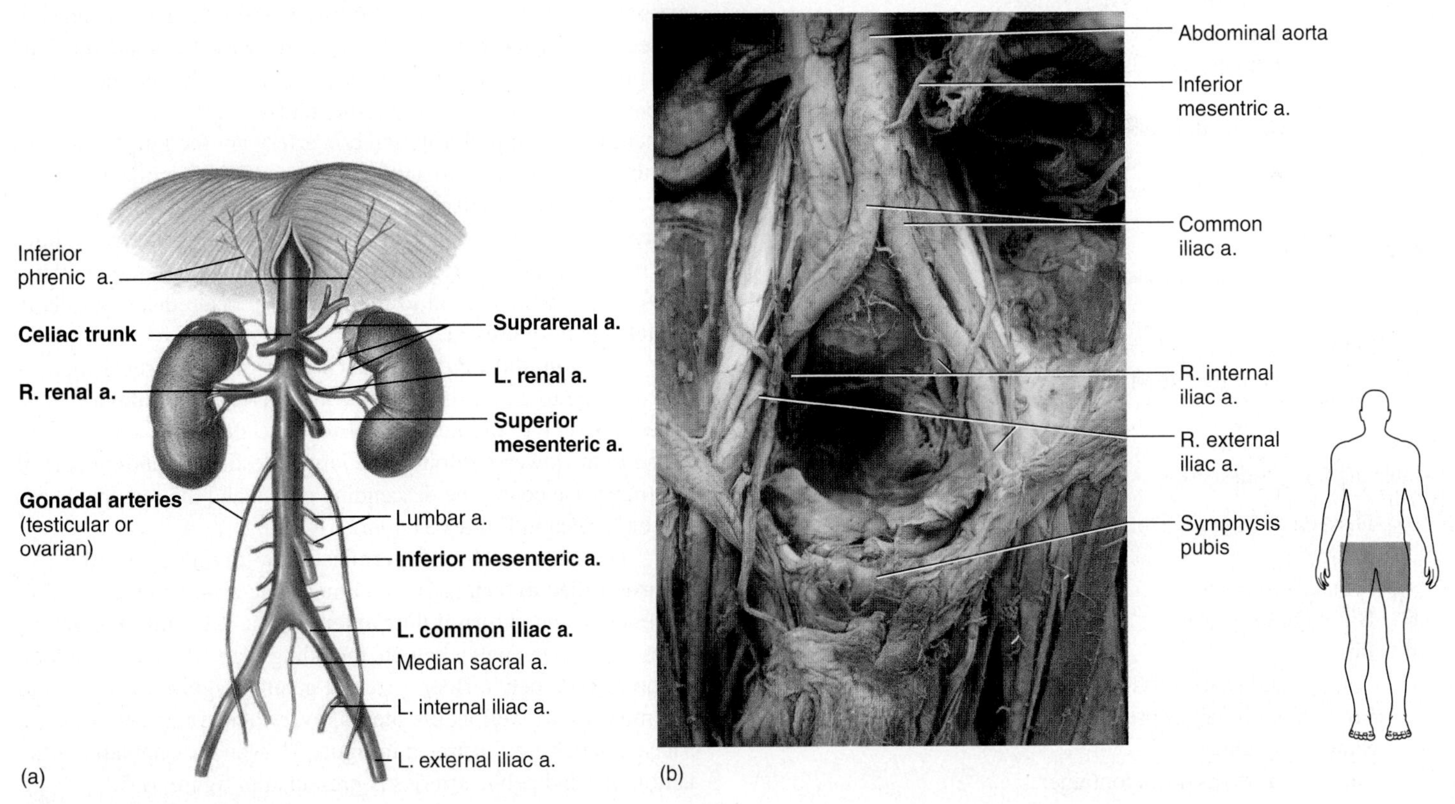

Figure 31.2 Deep Abdominal and Pelvic Arteries

(a) Diagram; (b) photograph.

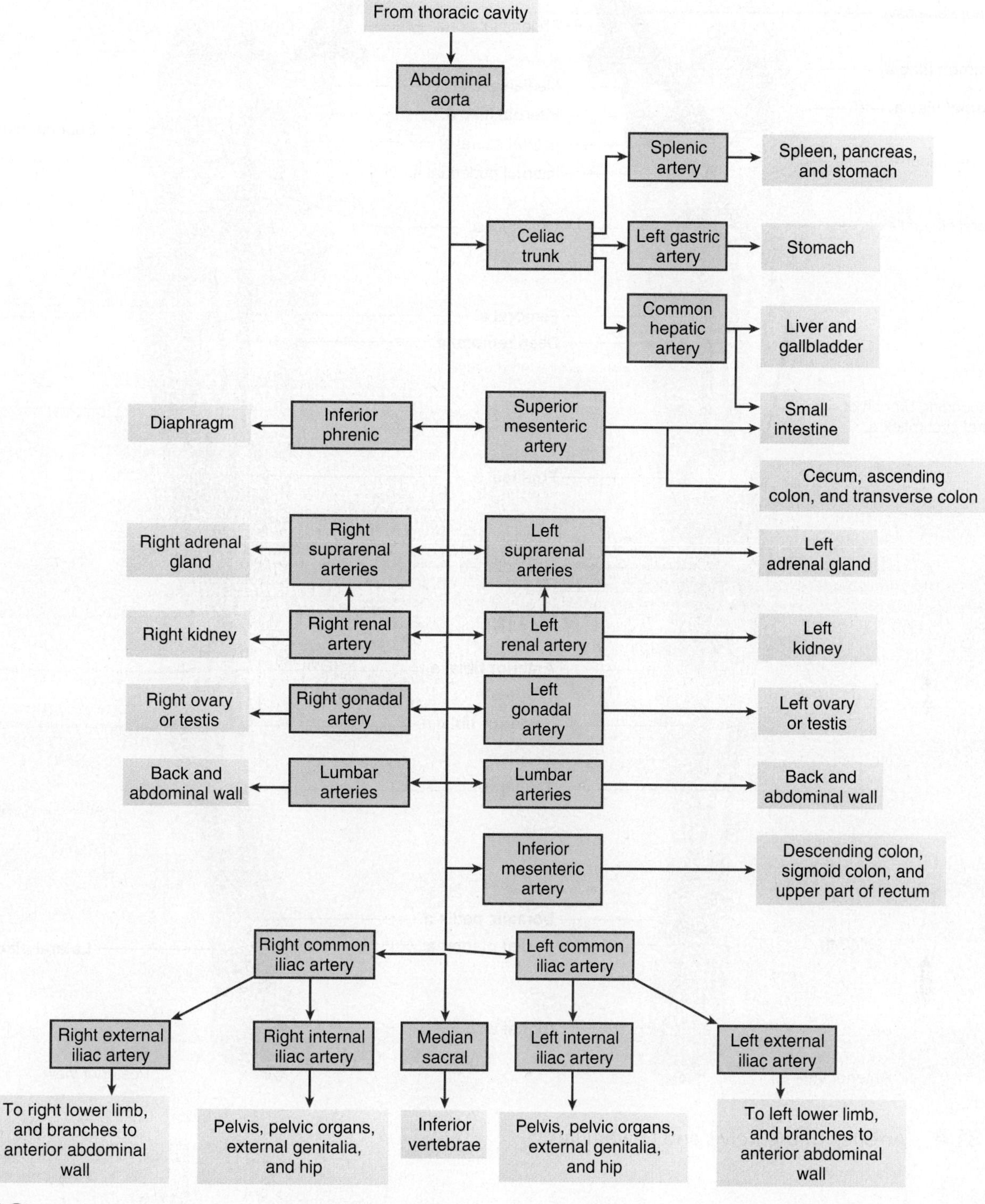

Figure 31.3 Major Arteries of the Abdomen and Pelvis

Arteries of the Lower Limb

(A) 2 Each external iliac artery branches from a common iliac artery and exits the body wall near the **inguinal canal,** continuing into the thigh as the **femoral artery.** The femoral artery has a superficial branch that feeds the thigh and a deeper branch, called the **deep femoral artery,** that takes blood to the muscles of the thigh, knee, and femur. The femoral artery continues posterior to the knee as the **popliteal artery** and then divides into the **posterior tibial artery** and **anterior tibial**

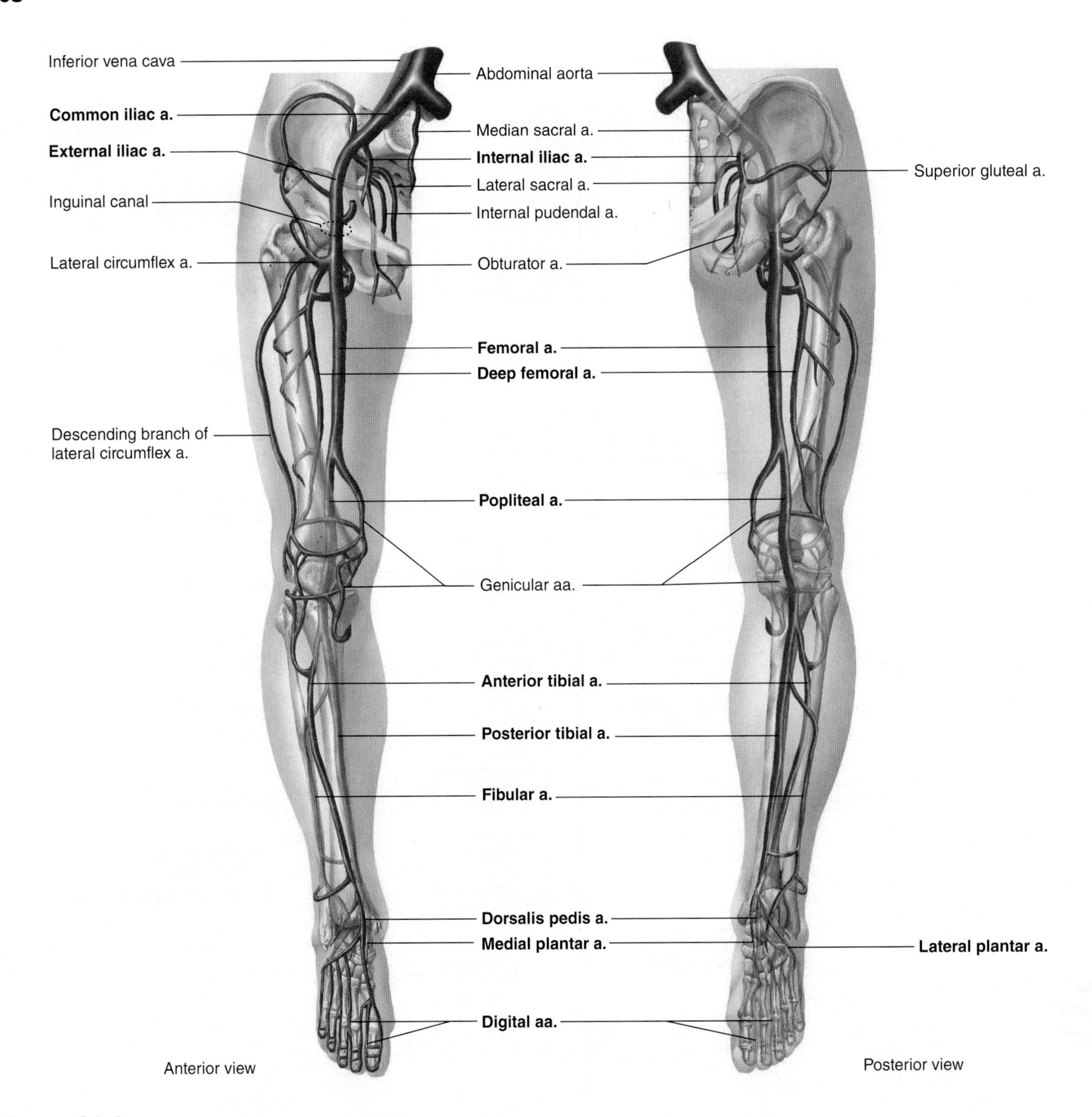

Figure 31.4 Arteries of the Pelvis and Lower Limb

artery, which take blood to the knee and leg. Another branch of the popliteal artery is the **fibular artery,** which supplies the muscles on the lateral side of the leg. The tibial and fibular arteries anastomose and supply blood to the **plantar arteries,** the **dorsalis pedis artery,** and the **digital arteries.** Locate these arteries in figure 31.4 and examine the schematic of the arteries in figure 31.5.

Study Hint

Draw a map of the inferior arteries from memory after you have studied the flowcharts in the lab manual. This gives you a good understanding of where the blood flows in the inferior arteries.

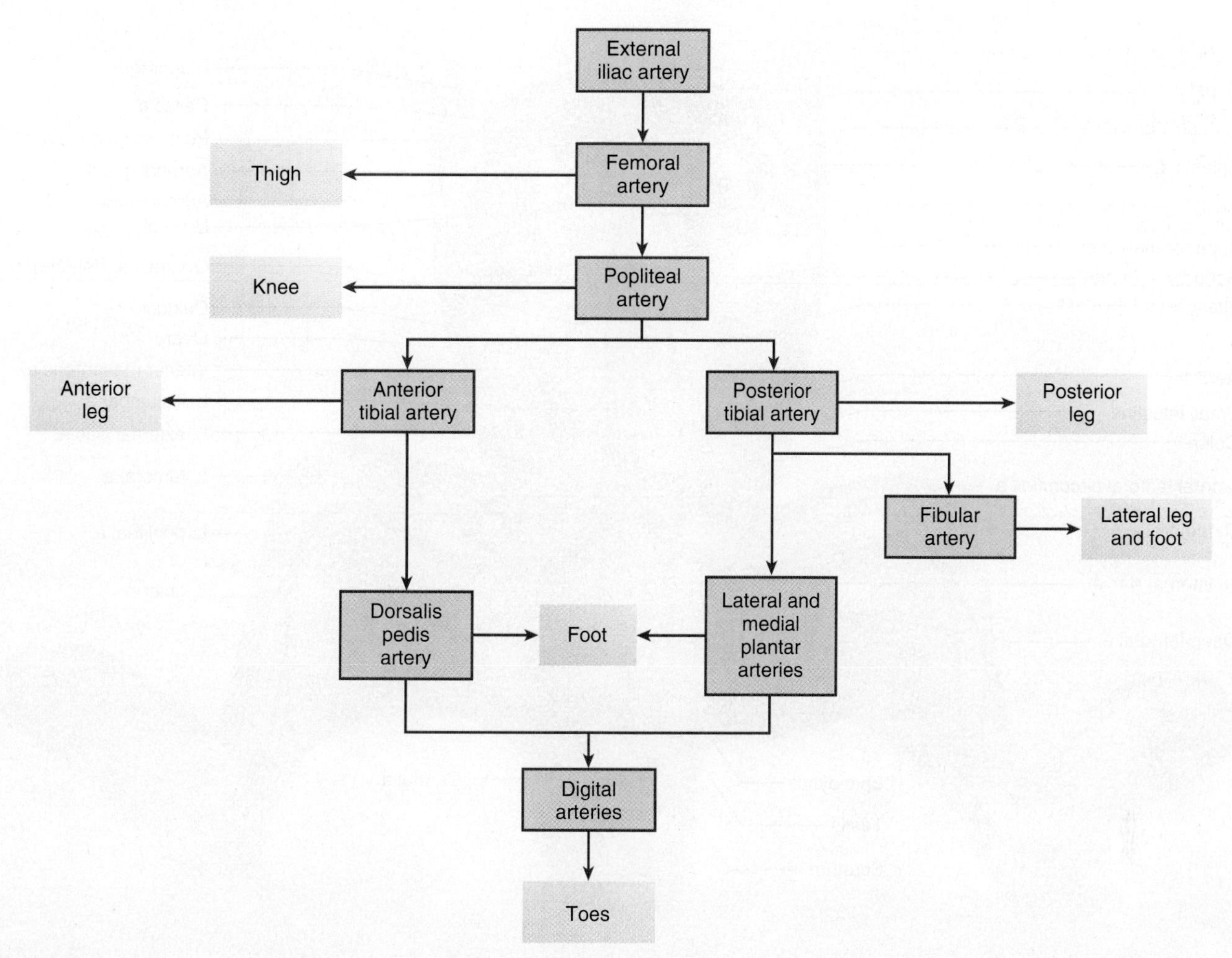

Figure 31.5 Major Arteries of the Lower Limb

Arteriosclerosis

(A) 3 Examine a microscope slide or an illustration of arteriosclerosis and note the development of **cholesterol plaque** under the endothelial layer. **Arteriosclerosis** is a condition known commonly as hardening of the arteries and is caused by the development of cholesterol plaque. Draw what you see in the space provided and compare it with the illustration in figure 31.6.

Drawing of an artery with arteriosclerosis:

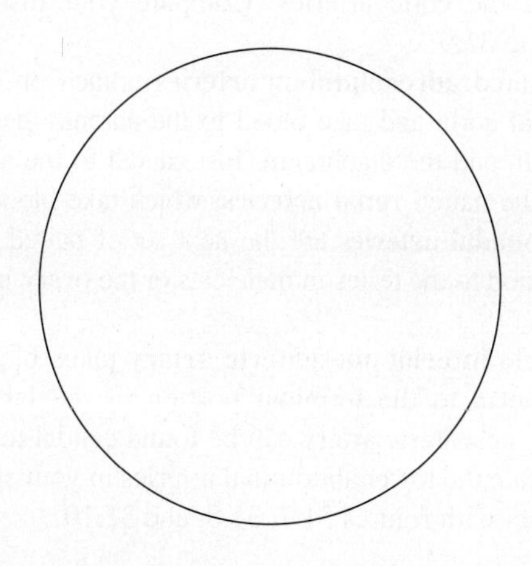

Cat Dissection

(A) 4 Take a cat and your dissection material to your lab table. Follow the dissection procedures discussed in the previous exercise. Make sure you do not cut the blood vessels away from the organs that will be studied later. Note the major arteries of the caudal region of the cat in figure 31.7. Refer to this figure and the photographs for the remainder of this exercise. We will use human terms for most of the arteries of the cat.

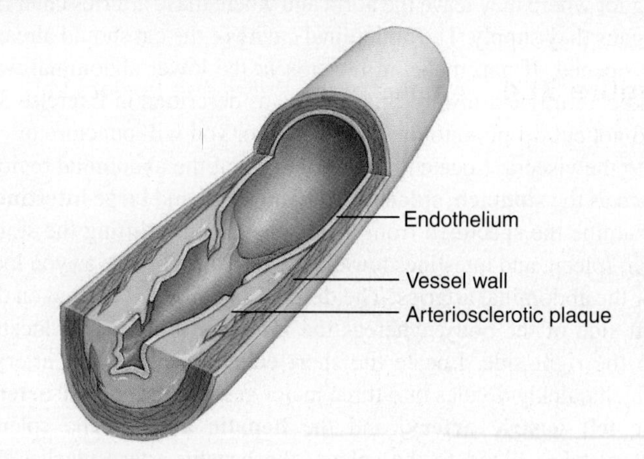

Figure 31.6 Arteriosclerotic Plaque

Note how the plaque develops deep to the endothelial layer.

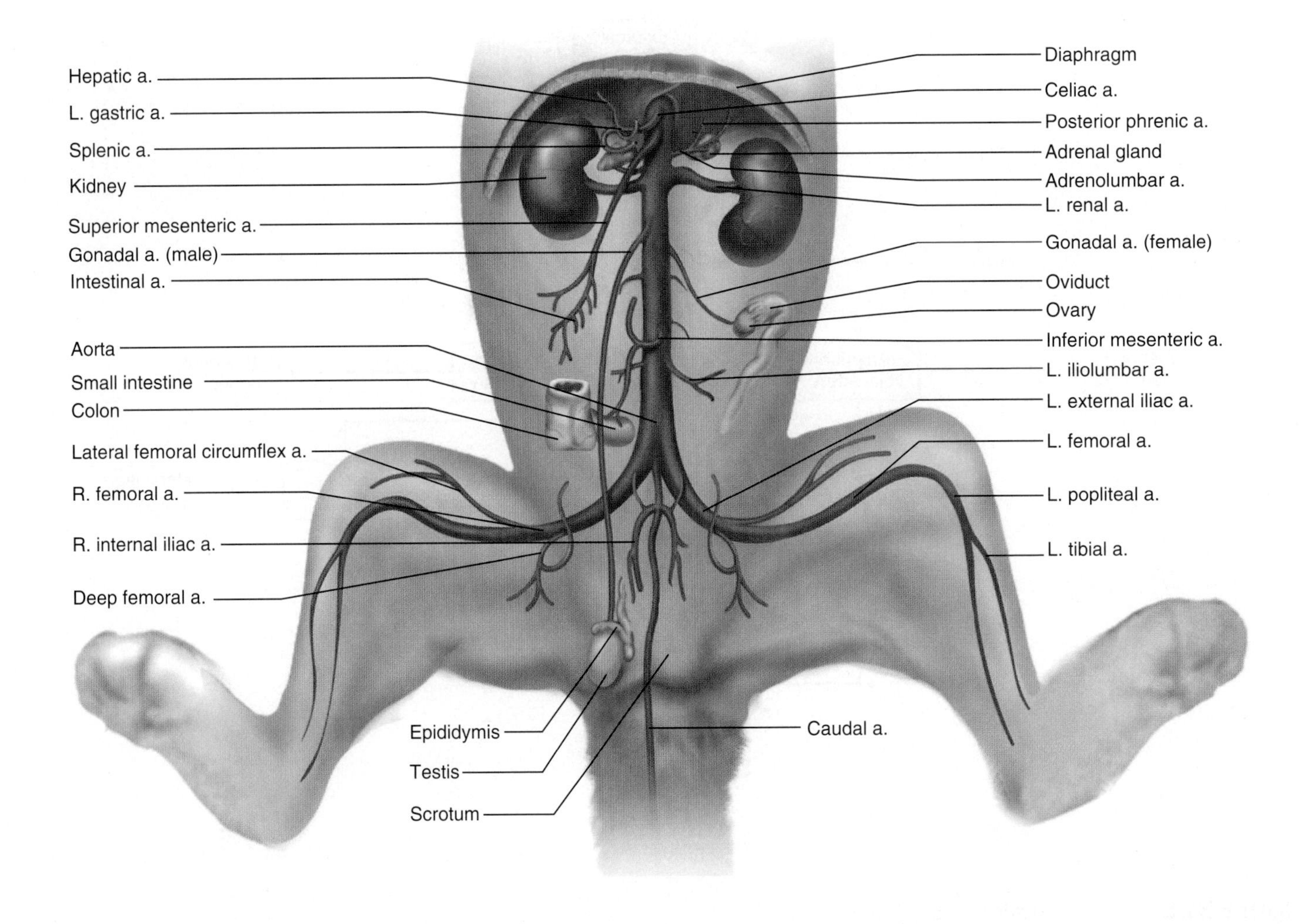

Figure 31.7 Diagram of the Caudal Arteries of the Cat

Abdominal Arteries

The determination of the abdominal arteries is best done by looking for where they leave the aorta and where these arteries enter the organs they supply. The abdominal cavity of the cat should already be opened. If not, make an incision in the lower abdominal wall and carefully cut toward the sternum as described in Exercise 30. Do not cut deeply into the body cavity or you will puncture or cut into the viscera. Locate the major organs of the abdominal region, such as the **stomach, spleen, liver,** and **small** and **large intestines.** Examine the specimen from the left side, gently lifting the stomach, spleen, and intestines toward the ventral right side as you look for the abdominal arteries. The **descending aorta** is located on the left side of the body, whereas the **inferior vena cava** is located on the right side. Locate the short **celiac trunk** (celiac artery), which quickly divides into three major vessels, the **splenic artery,** the **left gastric artery,** and the **hepatic artery.** The splenic artery takes blood to the spleen, the hepatic artery reaches the liver, and the left gastric artery supplies the stomach, as seen in figure 31.8.

The **superior mesenteric artery** is the next major vessel to take blood from the abdominal aorta. The superior mesenteric artery travels to the small intestine and the proximal portion of the large intestine. It branches into the jejunal arteries, the ileal arteries, and the colic arteries. Compare your dissection with figures 31.7 to 31.9.

The paired **adrenolumbar arteries** branch on each side of the abdominal aorta and take blood to the adrenal glands, parts of the body wall, and the diaphragm. Just caudal to the adrenolumbar arteries are the paired **renal arteries,** which take blood to the kidneys. The **gonadal arteries** are the next set of paired arteries, and these take blood to the testes in male cats or the ovary in female cats (figure 31.7).

A single **inferior mesenteric artery** takes blood from the abdominal aorta to the terminal portion of the large intestine. The inferior mesenteric artery can be found caudal to the gonadal arteries. Locate the lower abdominal arteries in your specimen and compare them with figures 31.7, 31.8, and 31.10.

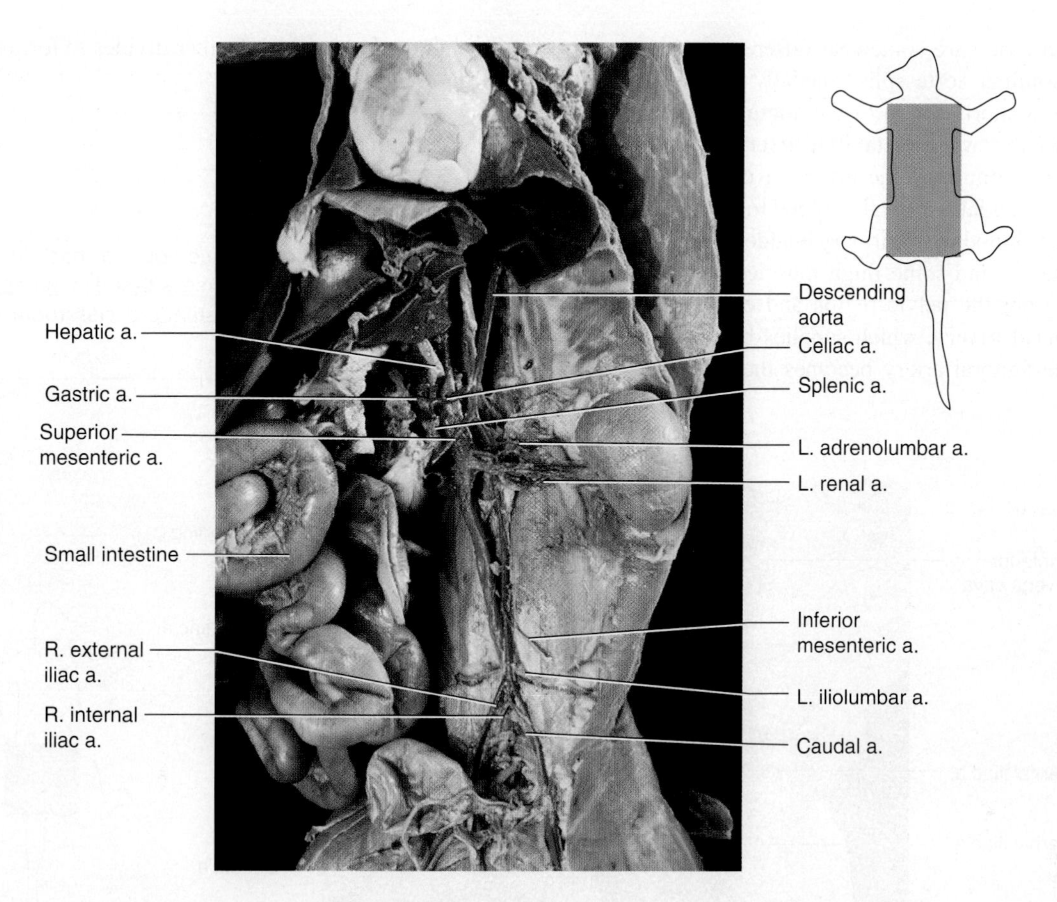

Figure 31.8 Photograph of the Caudal Arteries of the Cat

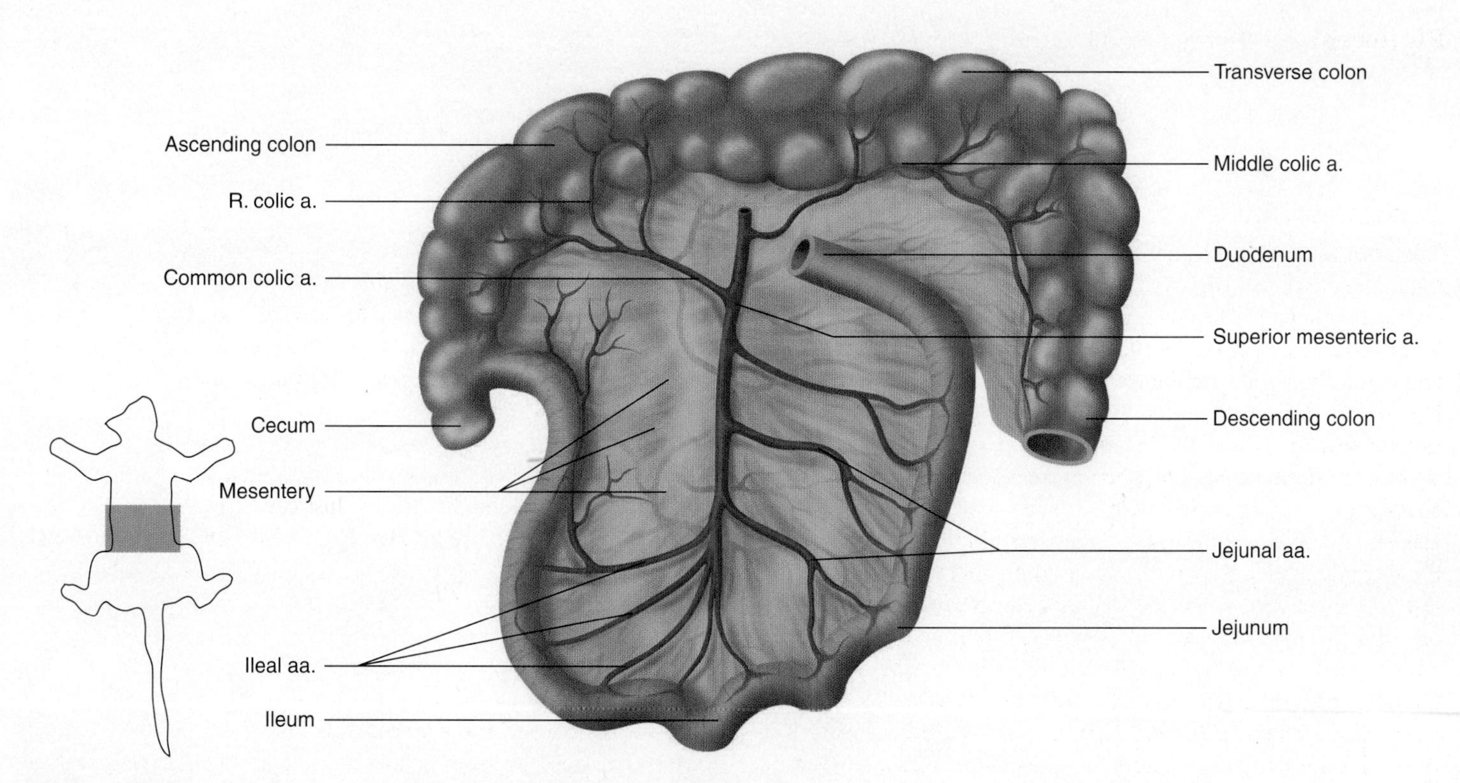

Figure 31.9 Branches of the Superior Mesenteric Artery of the Cat

The pelvic arteries are somewhat different in cats than in humans. The abdominal aorta splits caudally into the **external iliac arteries,** and a short section of the **aorta** continues on and then divides to form the **two internal iliac arteries** and the **caudal artery.** There is no common iliac artery in cats as there is in humans. In cats, the caudal artery takes blood to the tail. The internal iliac artery takes blood to the urinary bladder, rectum, external genital organs, uterus, and some thigh muscles. As the external iliac artery exits from the pelvic region and enters the thigh, it becomes the **femoral artery,** which supplies blood to the thigh, leg, and foot. The femoral artery becomes the **popliteal artery** behind the knee, which further divides to form the **tibial arteries** (figures 31.7 and 31.10).

Cleanup

When you are done, place your cat back in the plastic bag. Remember to place all excess tissue in the appropriate waste container and not in a standard classroom wastebasket or down the sink!

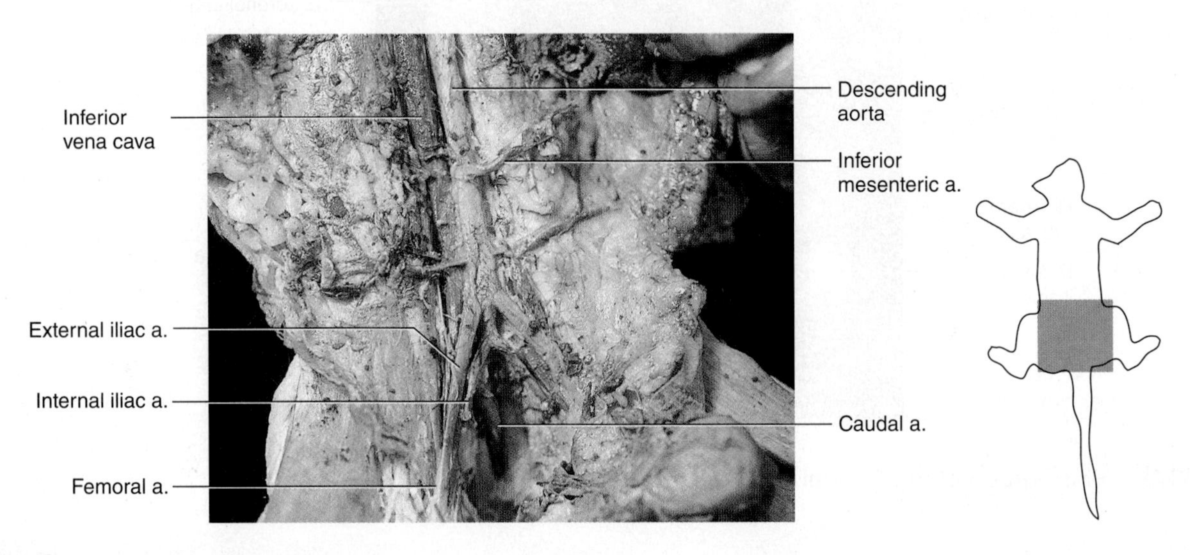

Figure 31.10 Pelvic Arteries of the Cat

Exercise 31 Review

Arteries of the Lower Body

Name: ______________________

Lab time/section: ______________________

Date: ______________________

1. Name the section of the descending aorta inferior to the diaphragm. ______________________

2. Blood from the celiac artery flows into three different blood vessels. What are these vessels? ______________________

3. Blood from the superior mesenteric artery goes to what major abdominal organs? ______________________

4. What vessels take blood to the kidneys? ______________________

5. The ovaries or testes receive blood from which arteries? ______________________

6. Blood in the inferior mesenteric artery travels to what organs? ______________________

7. In humans, where does blood in the external iliac artery come from? ______________________

8. What artery takes blood directly to the femoral artery? ______________________

9. Blood from the popliteal artery comes directly from what artery? ______________________

10. What is arteriosclerosis? ______________________

11. In what part of the arterial wall does cholesterol plaque develop? ______________________

12. What is the vessel that takes blood to the adrenal glands in the cat called? ______________________________

____

13. How do the lower pelvic arteries in humans differ from those in cats? ______________________________

____

14. Label the following illustration using the terms provided.

celiac trunk
common iliac artery
femoral artery
gonadal artery
inferior mesenteric artery
internal iliac artery
popliteal artery
renal artery
tibial artery

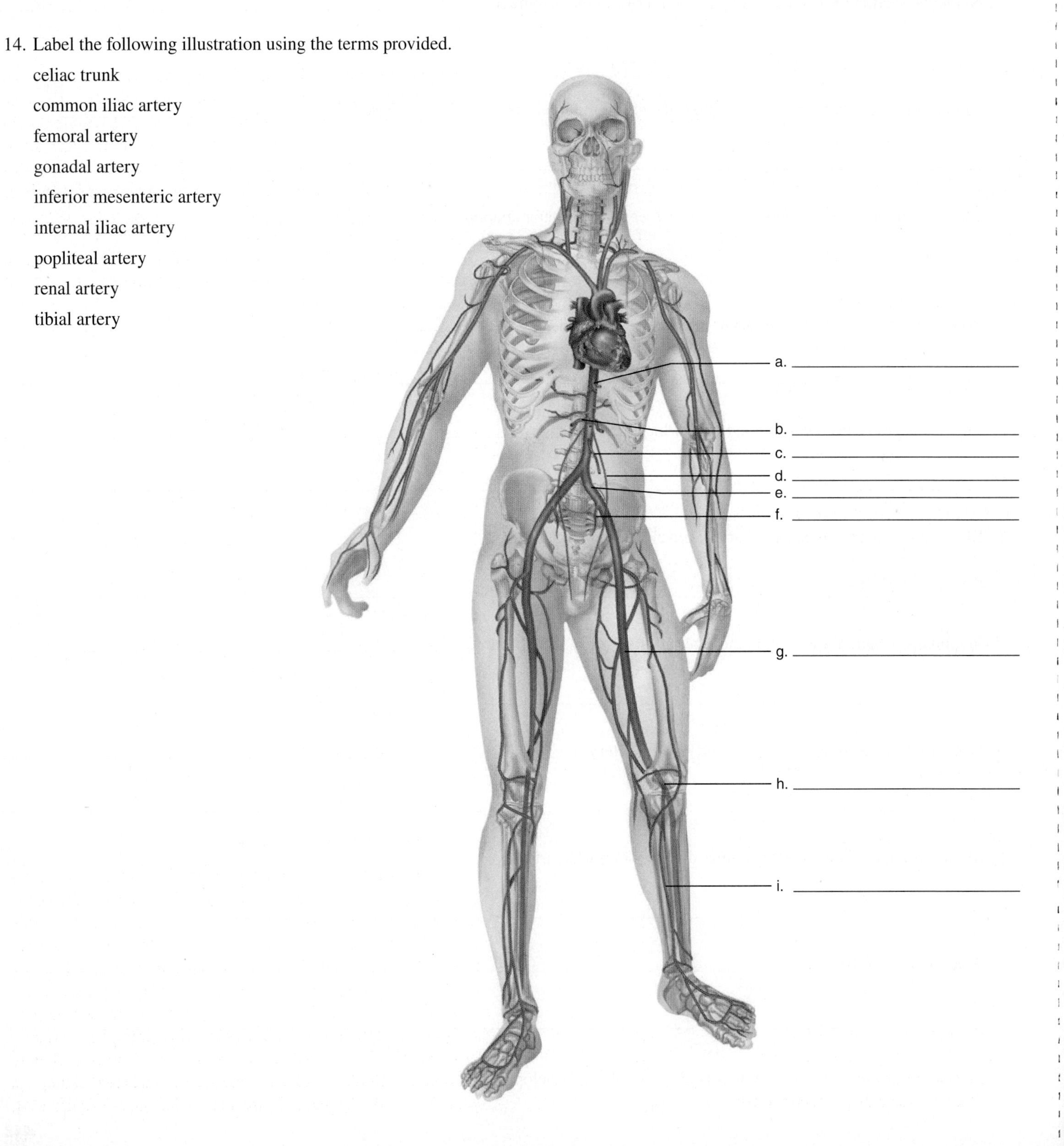

Exercise 32

Veins and Special Circulations

Introduction

Veins are blood vessels that carry blood toward the heart. They can be superficial (underneath the skin) or deep. The deep veins of the body often travel alongside the major arteries and take on the arterial names, such as the femoral vein, brachial vein, and subclavian vein. The superficial veins have names of their own, such as the cephalic vein, basilic vein, and great saphenous vein. Veins have thinner walls than arteries, and superficial veins contain valves, which maintain a one-way flow of blood to the heart. They are covered in the Seeley text in chapter 21, "Cardiovascular System: Blood Vessels and Circulation."

A description of the differences between artery walls and vein walls was presented in Exercise 30. In this exercise, you begin the study of veins by examining charts or models of the entire venous system and then study the veins of specific regions of the human body and the cat.

Objectives

At the end of this exercise, you should be able to

1. describe the sequence of major veins that drain the upper limbs;
2. list the veins of the lower limbs;
3. trace the blood flow from the brain to the heart;
4. distinguish a portal system from normal venous return flow;
5. describe the major digestive organs that supply blood to the hepatic portal system;
6. identify major veins in models, a cadaver, or a cat.

Materials

Models of the blood vessels of the body
Charts and illustrations of venous system
Cadaver (if available)
Cats
Materials for cat dissection

- Dissection trays
- Scalpel and new blades
- Protective gloves
- Blunt (Mall) probe
- String and tags
- Pins
- Forceps and sharp scissors
- First aid kit in lab or prep area
- Sharps container
- Animal waste disposal container

Procedure

(A) 1 Examine the models, charts, and illustrations in the lab and locate the major veins of the body. Read the following descriptions of the veins and find them as they are represented in lab. As you locate a specific vein, name the vessel that takes blood to the vein and those that receive blood from the vein. Veins in this exercise are studied in the direction of their flow from the cells of the body to the heart. In terms of their flow, veins resemble tributaries of rivers, as smaller veins flow into larger veins.

An overview of the major veins of the body is shown in figure 32.1. Compare that figure with the following list of some of the major veins of the body.

_____ Internal jugular vein

_____ External jugular vein

_____ Brachiocephalic vein

_____ Superior vena cava

_____ Axillary vein

_____ Cephalic vein

_____ Basilic vein

_____ Inferior vena cava

_____ Common iliac vein

_____ External iliac vein

_____ Internal iliac vein

_____ Femoral vein

_____ Great saphenous vein

_____ Tibial veins

Veins of the Head and Neck

(A) 2 Examine the models and charts in the lab and locate the various veins that drain blood from the head. The drainage of the brain occurs as veins take blood from the brain and pass through the subarachnoid membrane to the venous sinuses in the subdural spaces. The blood from the brain flows into the **internal jugular veins.**

The vertebral veins take blood from the bones and muscles of the neck and, like the vertebral arteries, travels through the transverse foramina of the cervical vertebrae. The vertebral veins take blood to the **subclavian veins,** which in turn flow to the **brachiocephalic veins.** The veins of the head are illustrated in figures 32.2 and 32.3 and presented in a schematic in figure 32.4.

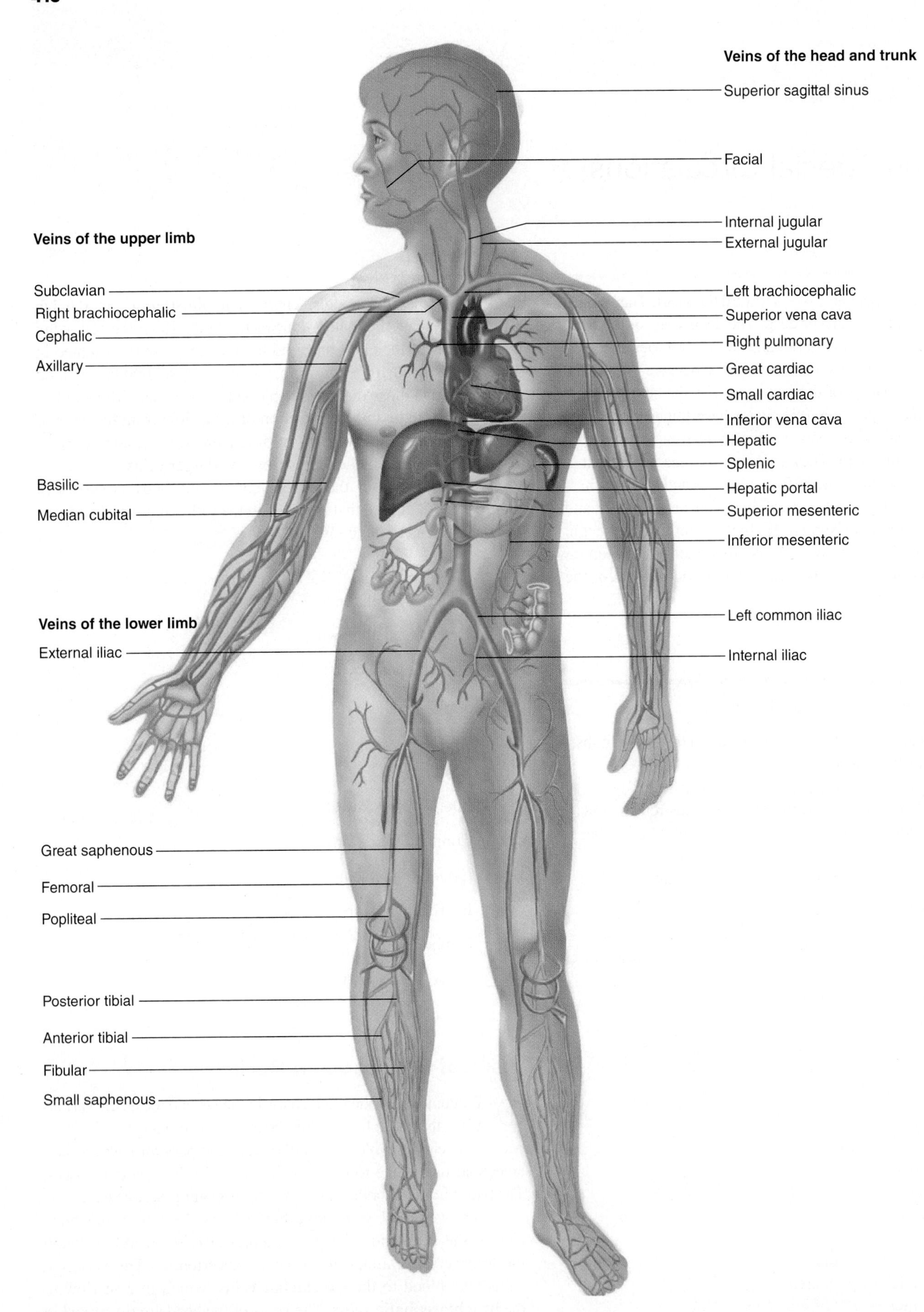

Figure 32.1 Major Systemic Veins

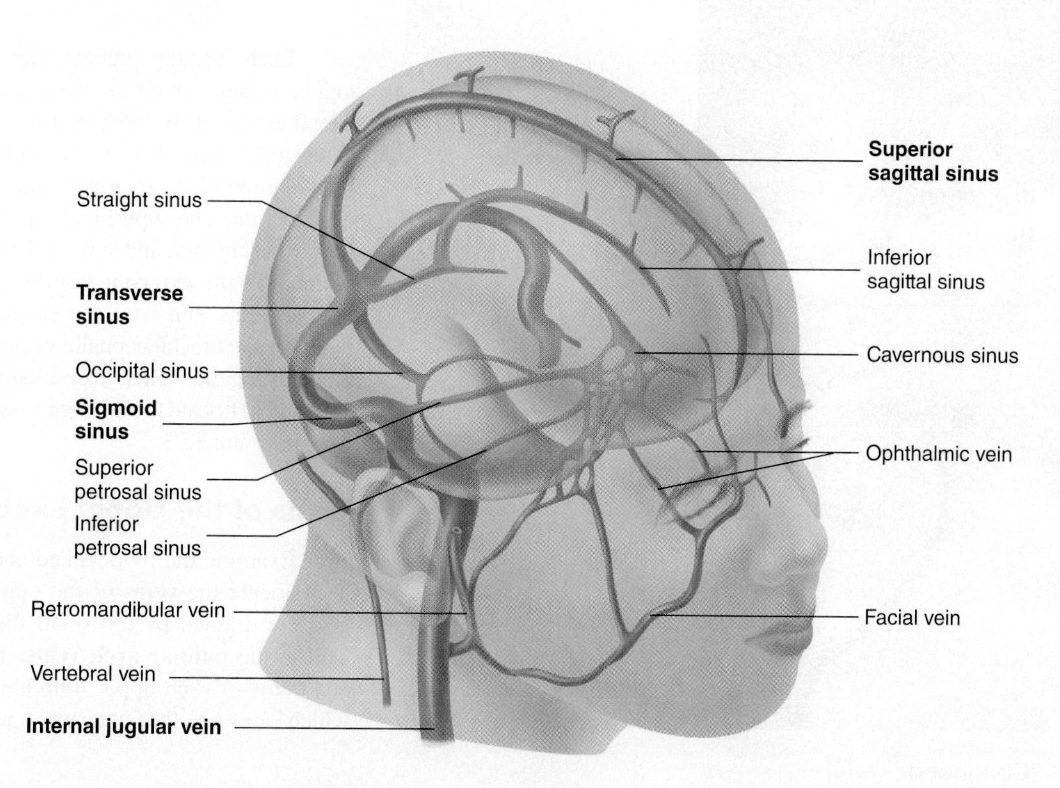

Figure 32.2 Venous Drainage of the Brain

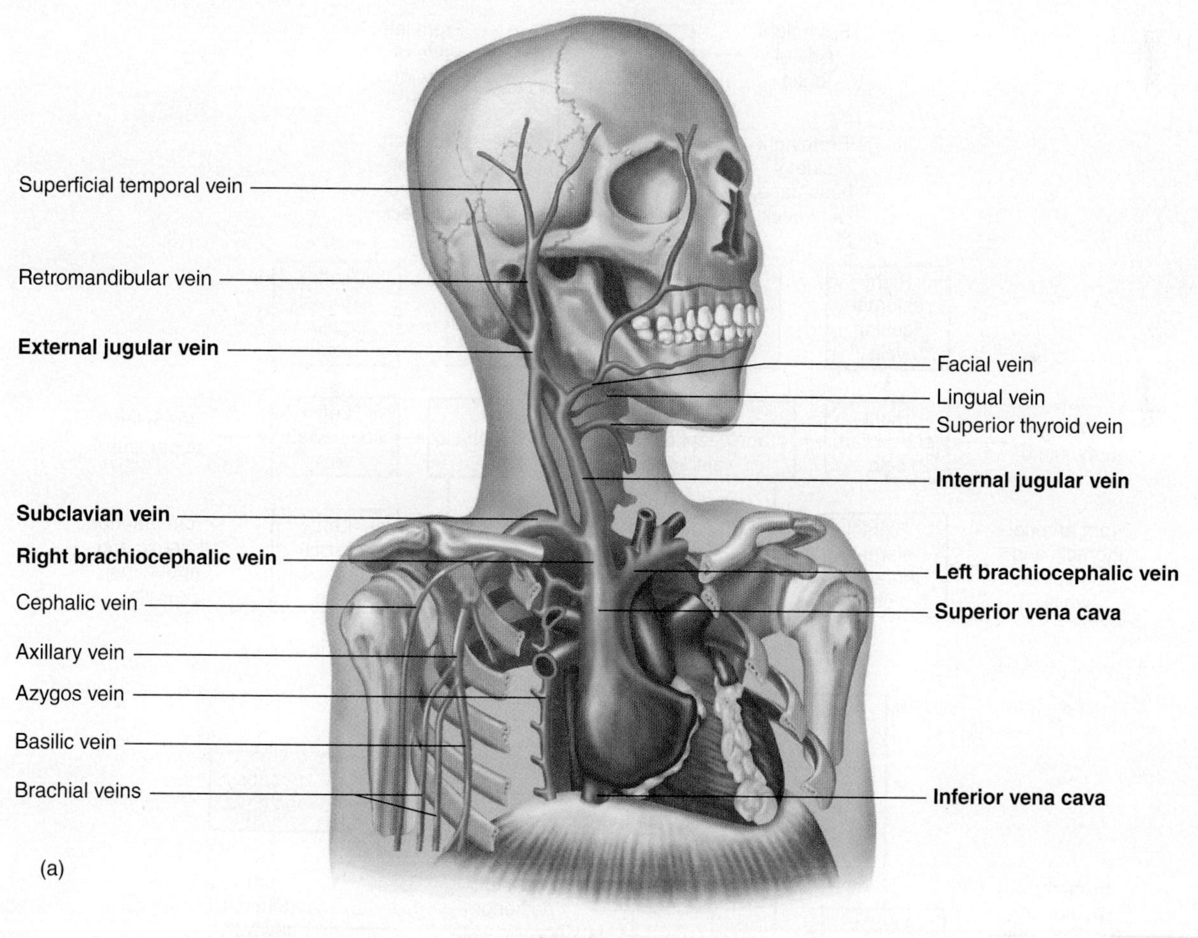

Figure 32.3 Veins of the Head and Neck

(a) Diagram;

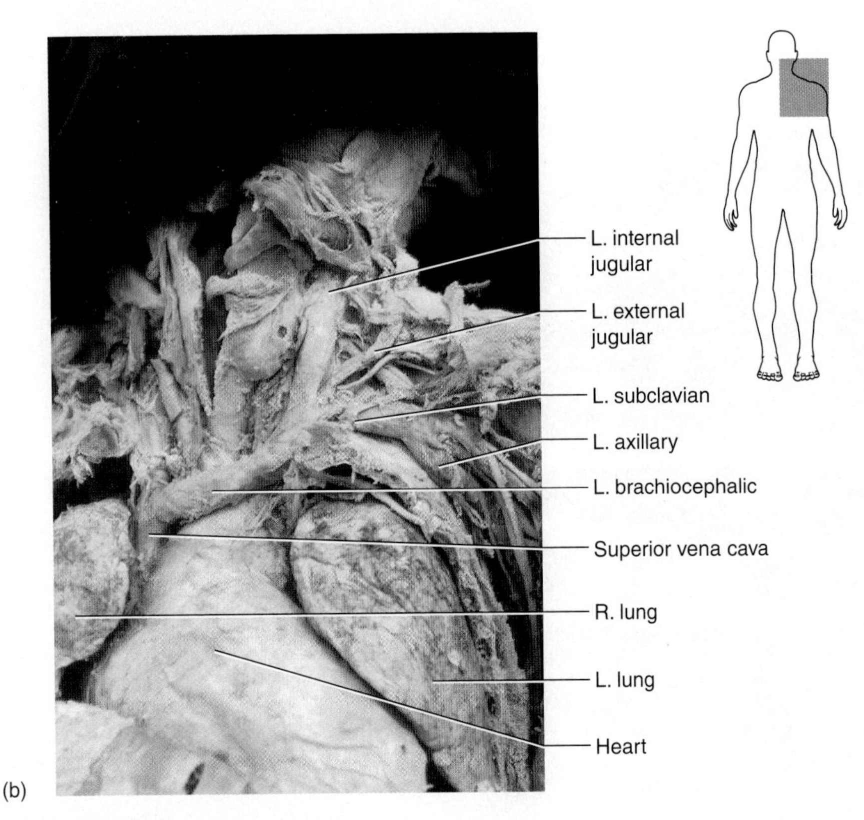

Figure 32.3 Continued

(b) photograph of cadaver.

Each internal jugular vein passes through a jugular foramen of the skull and then passes along the lateral aspect of the neck as it moves toward the brachiocephalic vein. The brachiocephalic vein is formed by the union of the internal jugular vein and the subclavian vein. The superficial regions of the posterior head (musculature and skin of the scalp and face) are drained by the **external jugular vein.** The external jugular veins join with the subclavian veins before reaching the brachiocephalic veins. The left and right brachiocephalic veins take blood to the **superior vena cava.** Locate these major vessels of the head and neck in figure 32.3.

Veins of the Upper Limbs

(A) 3 Examine the models and charts in the lab and locate the veins of the upper limbs. The fingers are drained by the small **digital veins,** which lead to the **palmar arch veins.** The major superficial veins of each upper limb are the **basilic veins,** which are on the anterior, medial side of the

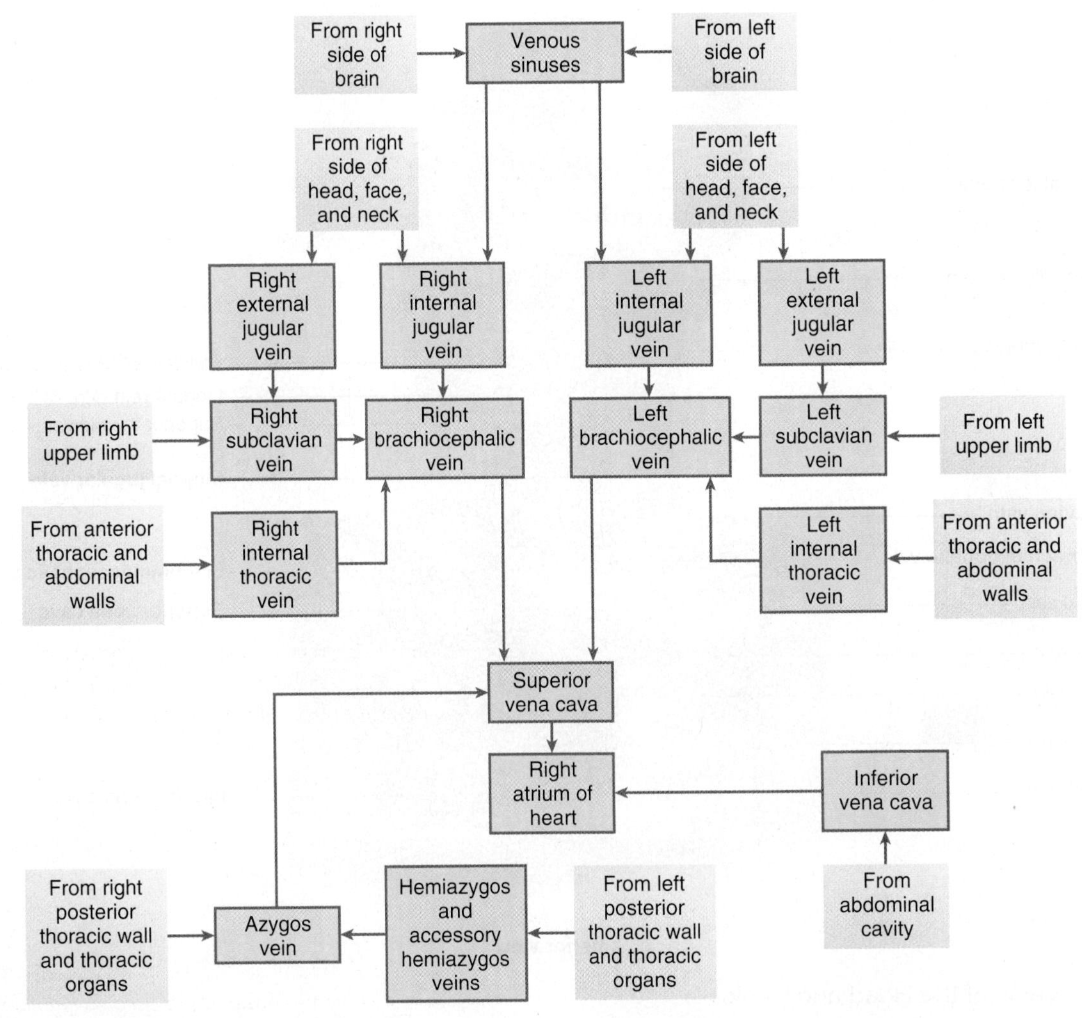

Figure 32.4 Major Veins of Head and Thorax

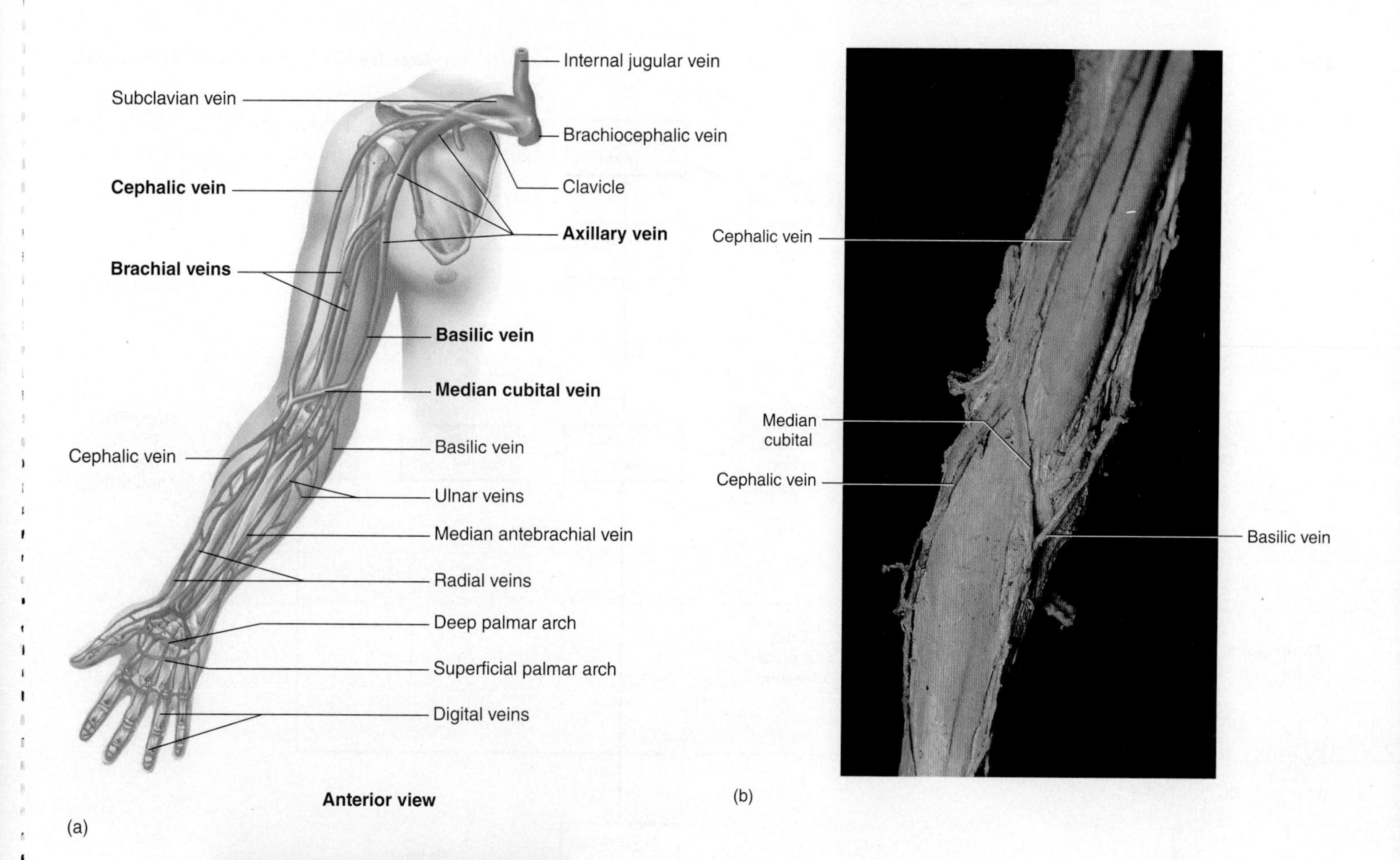

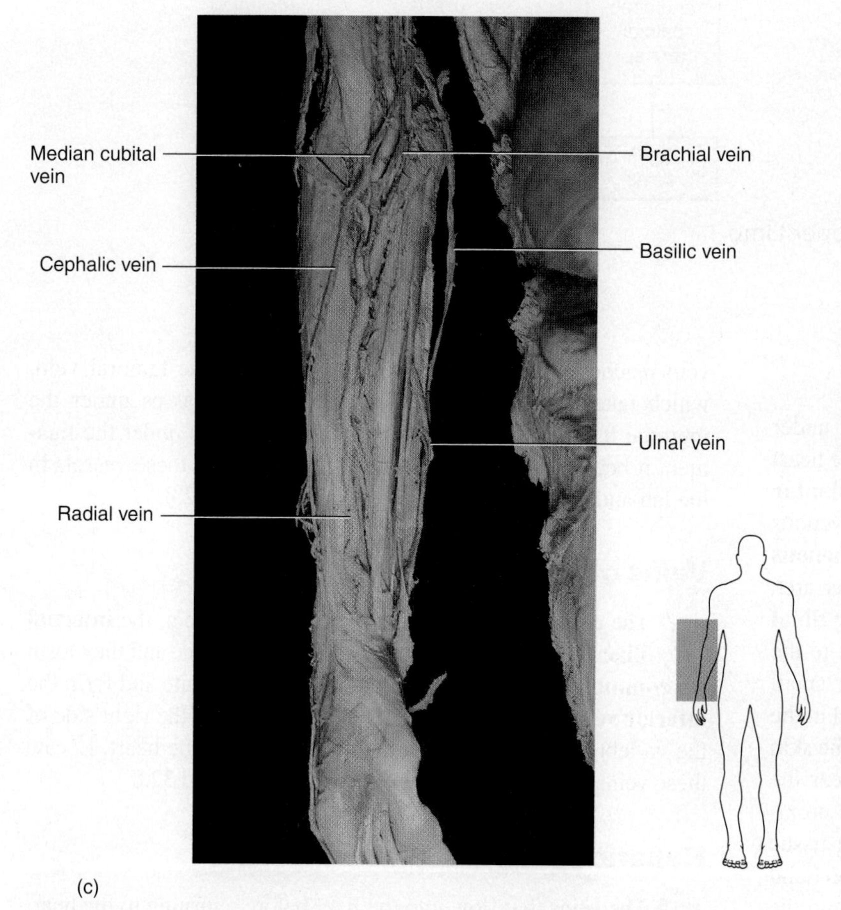

Figure 32.5 Veins of the Upper Limb
(a) Diagram; (b) photograph of superficial veins; (c) photograph of deep veins.

forearms and arms, and the **cephalic veins,** which are on the anterior, lateral side of the forearms and arms. The two vessels have many anastomosing branches (cross-connections) between them. One of the significant anastomosing veins is the **median cubital vein,** which crosses the anterior cubital fossa and is a common site for the withdrawal of blood. Locate these superficial veins in figure 32.5.

The deep veins of the forearm are the **radial vein** and the **ulnar vein,** each of which can be found traveling near the artery of the same name. The deep veins of the arm are the **brachial veins,** which are next to the brachial artery. The brachial veins are formed by the union of the radial and ulnar veins and merge superiorly with the **basilic vein** to form the short **axillary vein.** The axillary vein connects with the cephalic vein to form the subclavian vein. The drainage of the upper limbs is carried to the heart by the left and right subclavian veins, which take the blood via the brachiocephalic veins to the superior vena cava and finally to the right atrium of the heart. Locate the deep veins of the upper limb in figure 32.5 and see the schematic in figure 32.6.

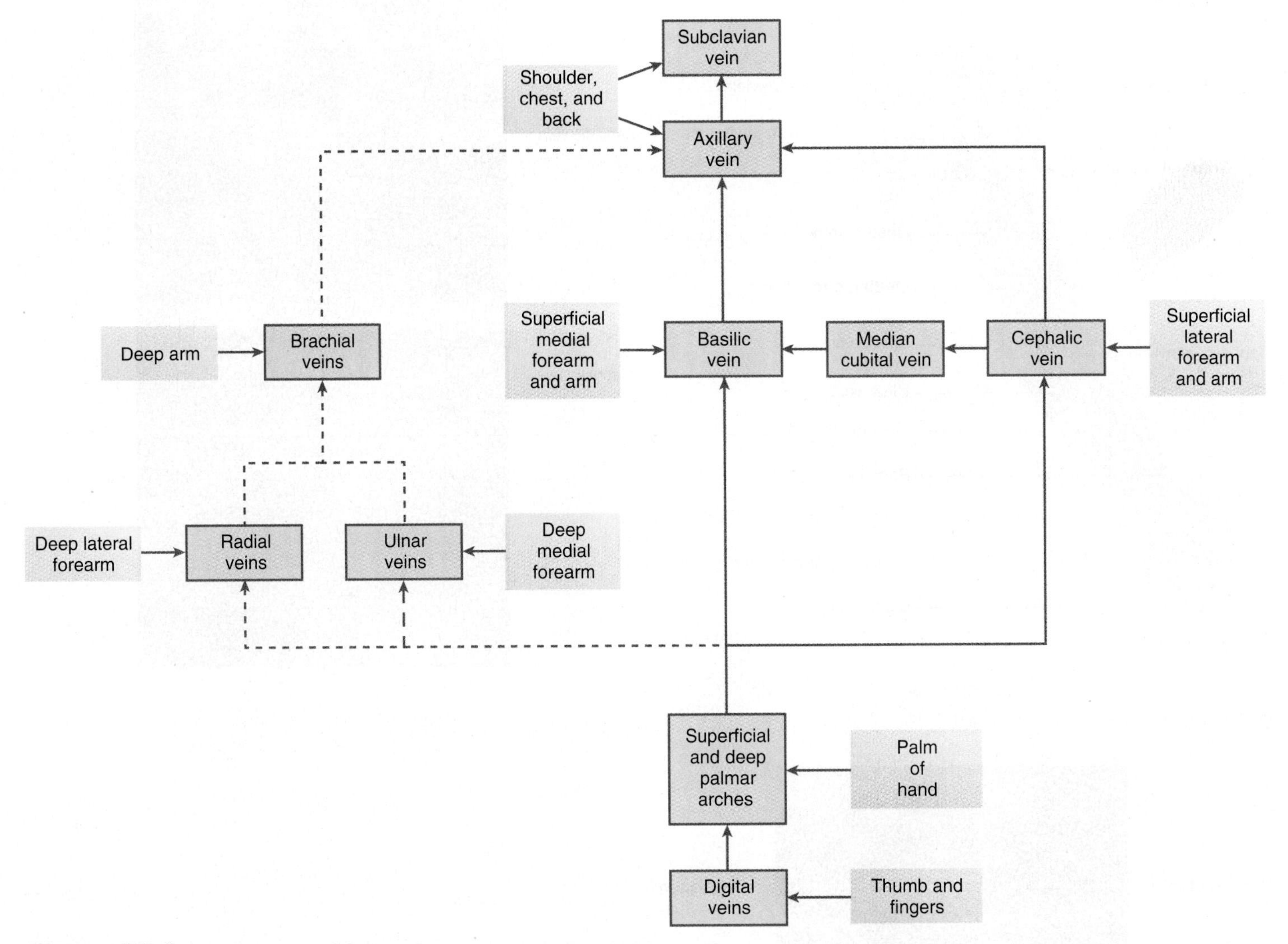

Figure 32.6 Major Veins of the Shoulder and Upper Limb
Deep veins are indicated by connections with dotted lines.

Veins of the Lower Limb

(A) 4 The blood vessels that drain the lower limbs operate under relatively low pressure and must take blood back to the heart against gravity. **Digital veins** in the feet take blood to the **plantar venous arch** and **dorsal venous arch** veins. The dorsal venous arch takes blood to the **anterior tibial vein,** the **great saphenous vein,** and the **small saphenous vein.** The plantar venous arch takes blood to the small saphenous vein and the **posterior tibial vein.** The anterior and posterior tibial veins unite posterior to the knee and form the **popliteal vein,** which joins with the small saphenous vein to form the **femoral vein.** The longest vessel in the human body is the **great saphenous vein,** just underneath the skin on the medial aspect of the lower extremity beginning near the medial malleolus and traversing the lower extremity to the proximal thigh. This vessel frequently is embedded in adipose tissue below the skin in humans, yet it is still considered a superficial vein. Two other vessels in the thigh are the **femoral vein** and the **deep femoral vein,** which take blood from the thigh. The femoral vein travels alongside the femoral artery. As the great saphenous vein reaches the inguinal region, it joins with the femoral vein, which takes blood from the thigh region and passes under the inguinal ligament. Once the **femoral vein** crosses under the ligament it becomes the **external iliac vein.** Examine these vessels in the lab and compare them with figures 32.7 and 32.8.

Veins of the Pelvis

(A) 5 The **external iliac vein** joins with another vein, the **internal iliac vein,** which drains the region of the pelvis, and they form the **common iliac vein.** The common iliac veins unite and form the **inferior vena cava,** which travels superiorly along the right side of the vertebrae, taking blood to the right atrium of the heart. Locate these veins and compare them with figures 32.7 and 32.8.

Hepatic Portal Circulation

(A) 6 The veins that flow into the liver before returning to the heart are part of the **hepatic portal system.** Most veins take blood from capillaries and venules and return the blood to the heart. In

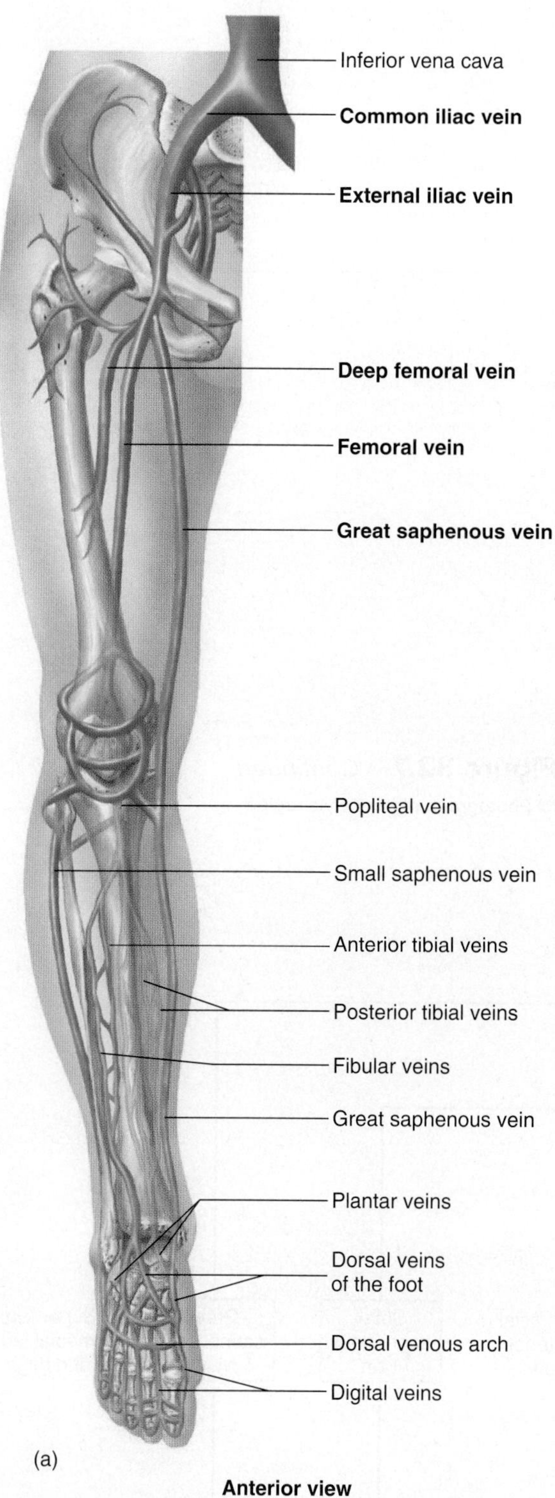

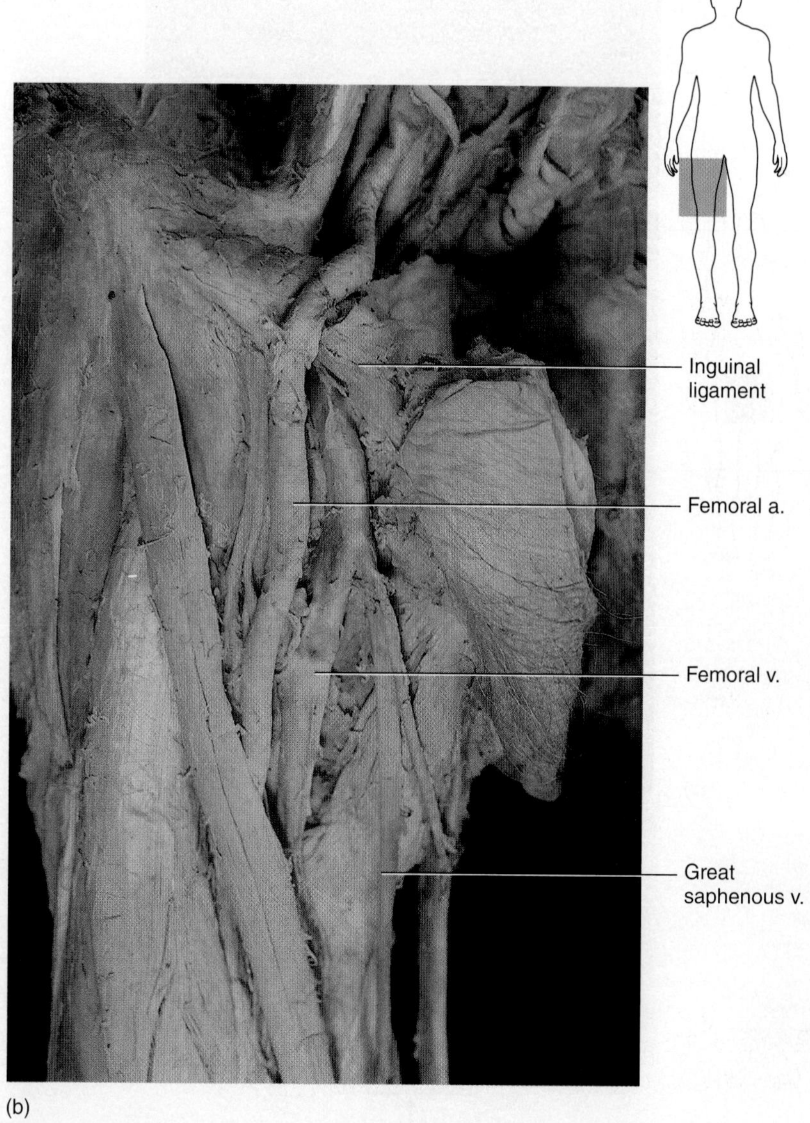

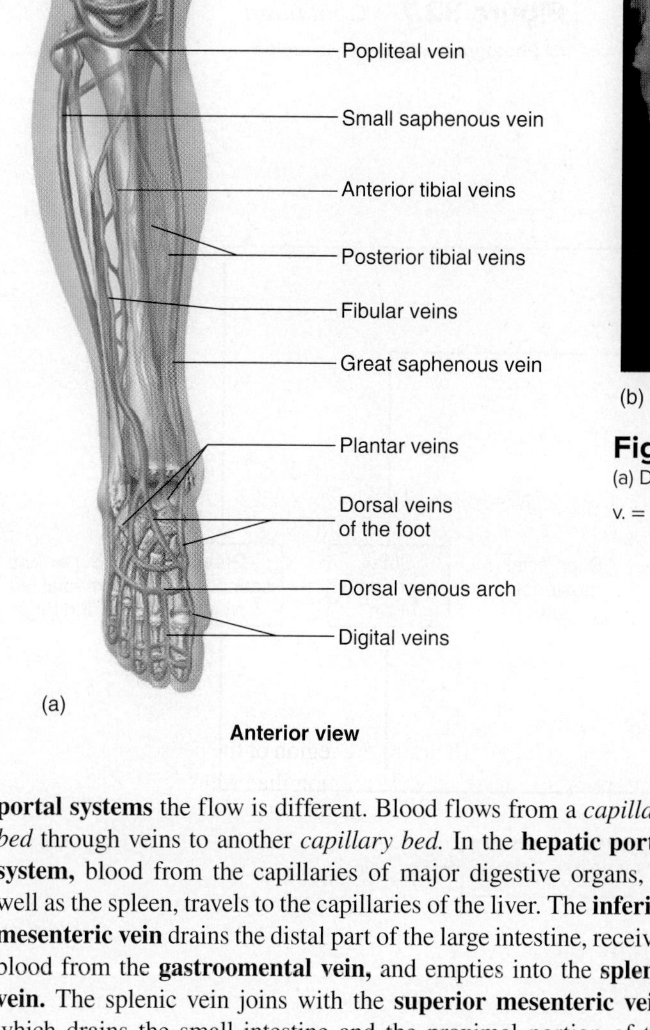

Figure 32.7 Veins of the Pelvis and Lower Limb
(a) Diagram; (b) photograph of the femoral region.
v. = vein; a. = artery

portal systems the flow is different. Blood flows from a *capillary bed* through veins to another *capillary bed.* In the **hepatic portal system,** blood from the capillaries of major digestive organs, as well as the spleen, travels to the capillaries of the liver. The **inferior mesenteric vein** drains the distal part of the large intestine, receives blood from the **gastroomental vein,** and empties into the **splenic vein.** The splenic vein joins with the **superior mesenteric vein,** which drains the small intestine and the proximal portion of the large intestine. The splenic vein and the superior mesenteric vein unite to form the **hepatic portal vein,** which takes blood to the capillaries of the liver, where the blood is cleaned by macrophages and nutrients are either stored in the liver or passed into the bloodstream. The liver receives the blood from the digestive organs and processes it before sending it through the **hepatic veins** to the inferior vena cava and toward the heart. Examine models or charts in lab and compare them to figure 32.10 for the main vessels of the hepatic portal system and identify the following veins:

_____ Inferior mesenteric vein

_____ Superior mesenteric vein

_____ Splenic vein

_____ Gastroomental vein

_____ Hepatic portal vein

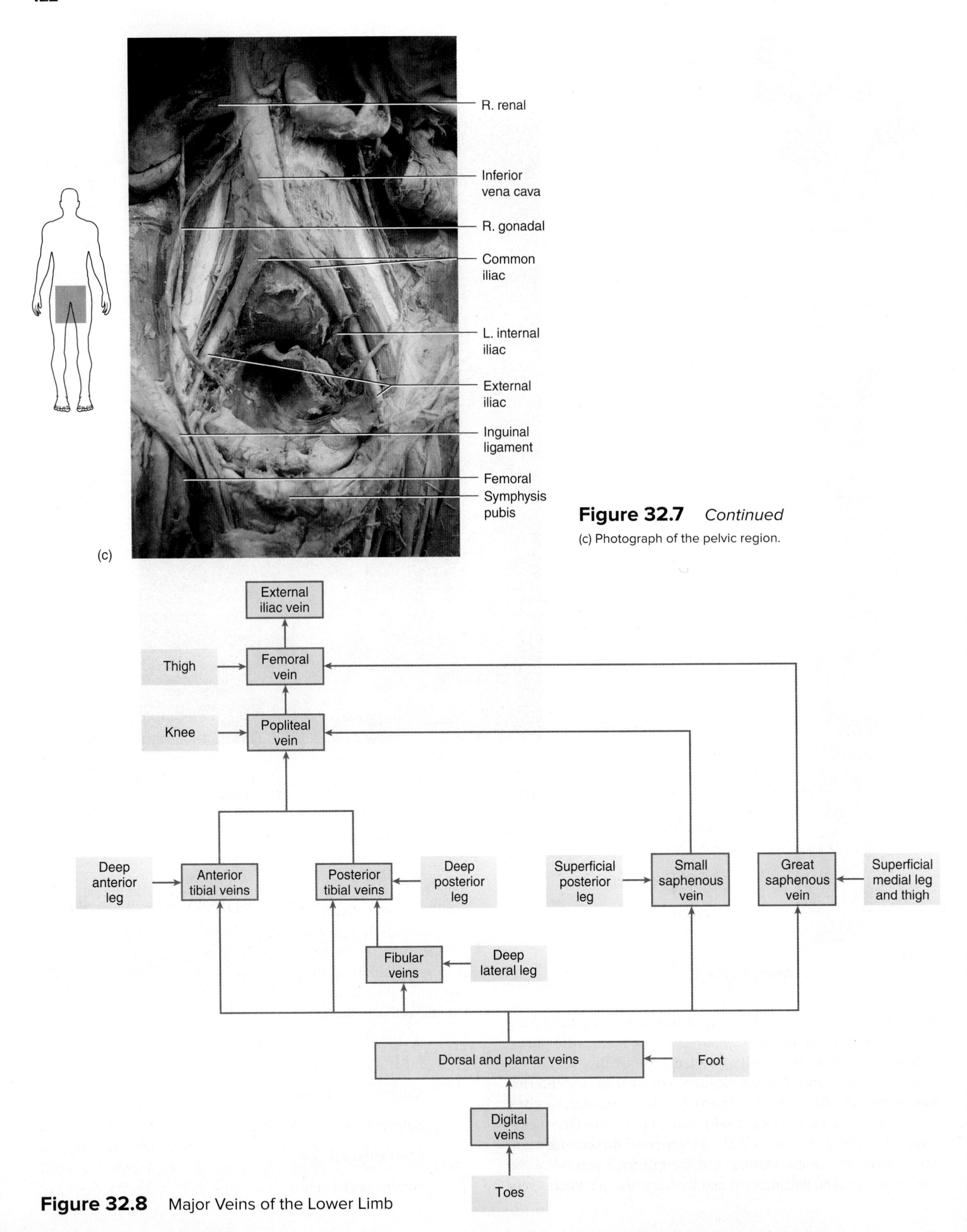

Figure 32.7 *Continued*

(c) Photograph of the pelvic region.

Figure 32.8 Major Veins of the Lower Limb

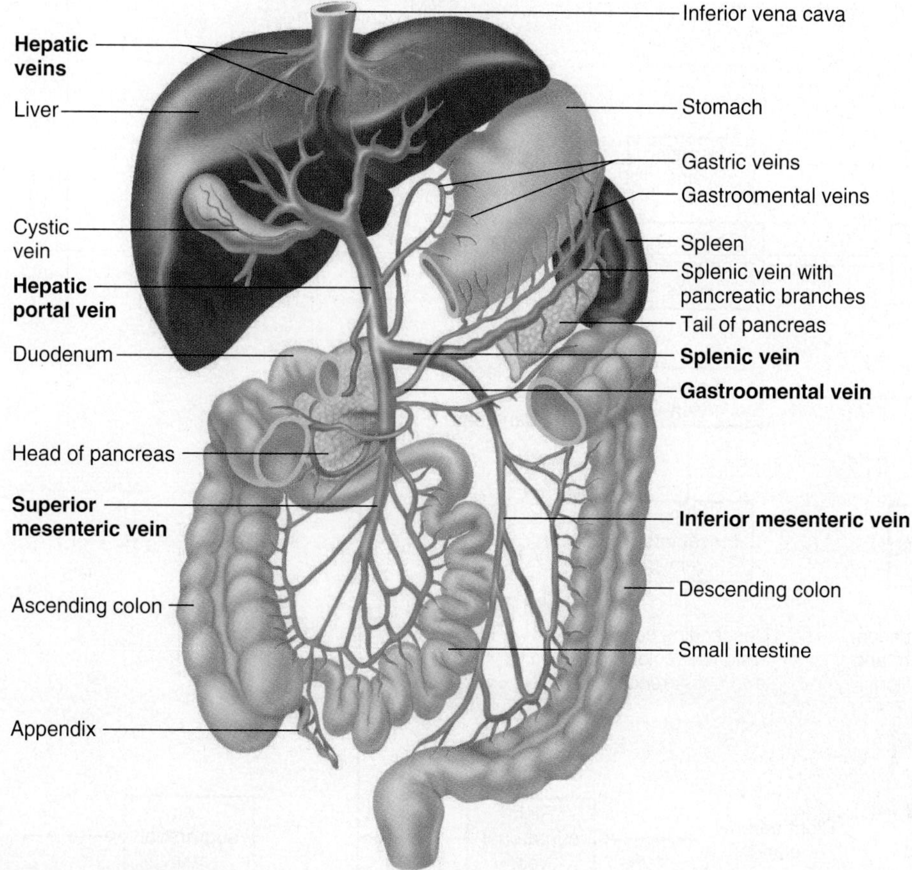

Figure 32.9 Hepatic Portal System

Veins of the Abdomen and Pelvis

(A)[7] Some abdominal veins take blood directly to the inferior vena cava, and some pass through the liver before reaching the inferior vena cava. Those that take blood directly to the inferior vena cava are the **renal veins,** the **suprarenal veins,** and the **lumbar veins.** The **right gonadal vein** takes blood directly to the inferior vena cava, but the **left gonadal vein** takes blood to the renal vein before flowing into the inferior vena cava. These vessels can be seen in figures 32.11 and 32.12.

Thoracic Veins

(A)[8] The major veins of the thoracic region consist of the **intercostal veins,** which drain the intercostal muscles, and the **azygos vein** and **hemiazygos vein,** which drain blood from the thoracic region. Locate these vessels in the lab and compare them to figure 32.12.

Fetal Circulation

The pathway of human fetal blood is somewhat different from that of the adult in that the lungs are nonfunctional in the fetus. Oxygen and nutrients move from the maternal side of the **placenta** to the fetal bloodstream, whereas carbon dioxide and metabolic wastes move from the fetal bloodstream to the placenta. From the placenta, the blood flows through the **umbilical vein,** which is located in the **umbilical cord.** The blood from the umbilical vein travels through the **ductus venosus,** which is a shunt to the **inferior vena cava** of the fetus. The maternal blood, which is relatively high in oxygen and nutrients, mixes with the deoxygenated, nutrient-poor fetal blood, and thus the fetus receives a mixture of blood. Examine figure 32.13 for an overview of fetal circulation.

The blood from the inferior vena cava travels to the right atrium of the heart. While in the right atrium, the blood can travel either to the right ventricle (which pumps blood to the lungs) or through a hole in the right atrium called the **foramen ovale.** Since the lungs do not oxygenate blood in the fetus, the foramen ovale is a bypass route away from the lungs and to the chambers of the heart that will pump blood to the body. Blood in the right ventricle is pumped to the pulmonary trunk, where another shunt vessel, the **ductus arteriosus,** carries blood to the aortic arch, bypassing the lungs. The lungs do receive some blood, but it is for the nourishment of the lung tissue, as opposed to gas exchange. Locate the structures of the fetal heart in figure 32.13.

Blood from the heart exits the left ventricle and passes through the aorta to the systemic arteries. Blood travels down the internal iliac arteries to the lower limbs, and some moves into the **umbilical arteries,** which carry blood to the placenta. At birth, the pressure changes in the newborn's heart cause the closing of

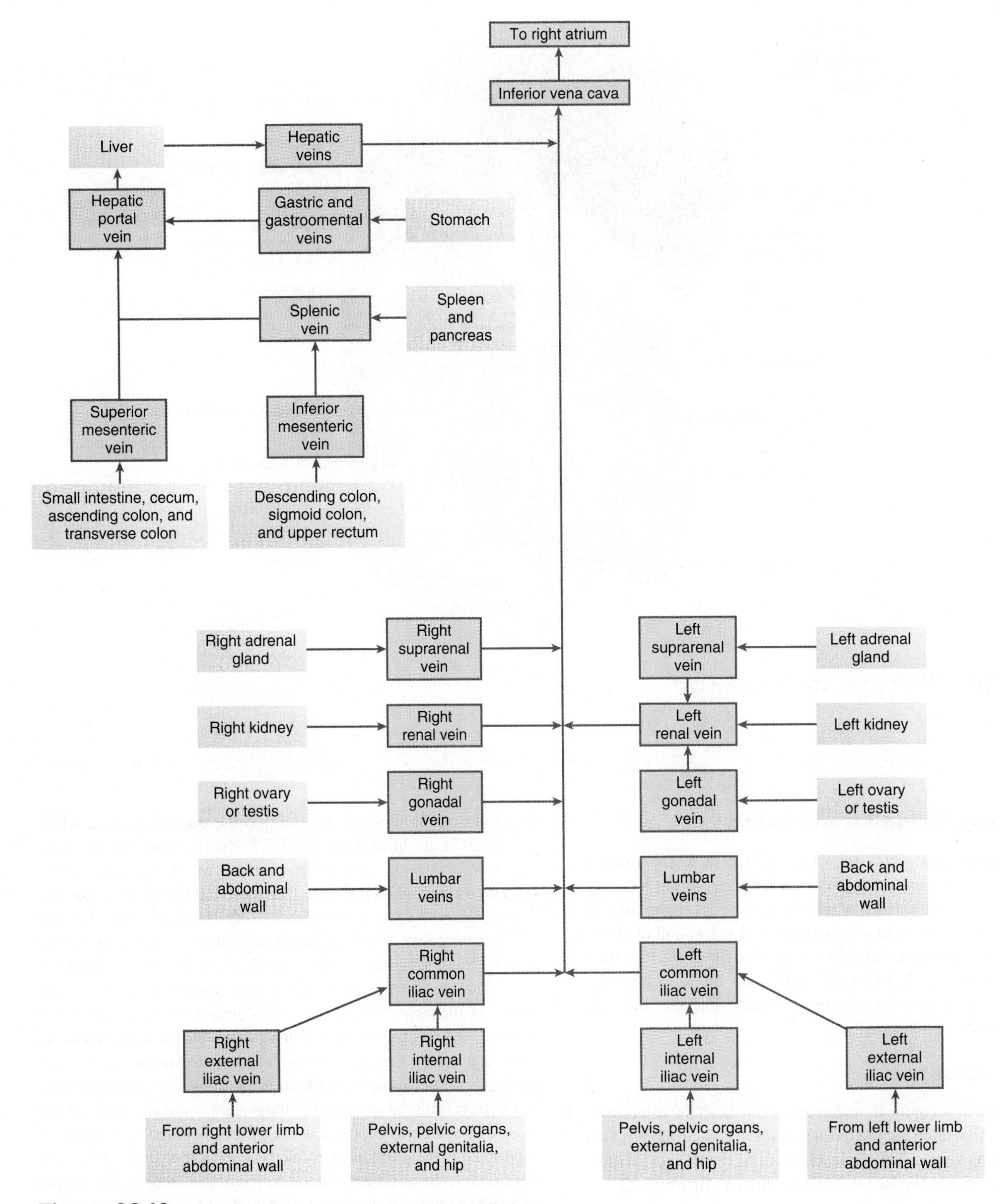

Figure 32.10 Major Veins of the Abdomen and Pelvis

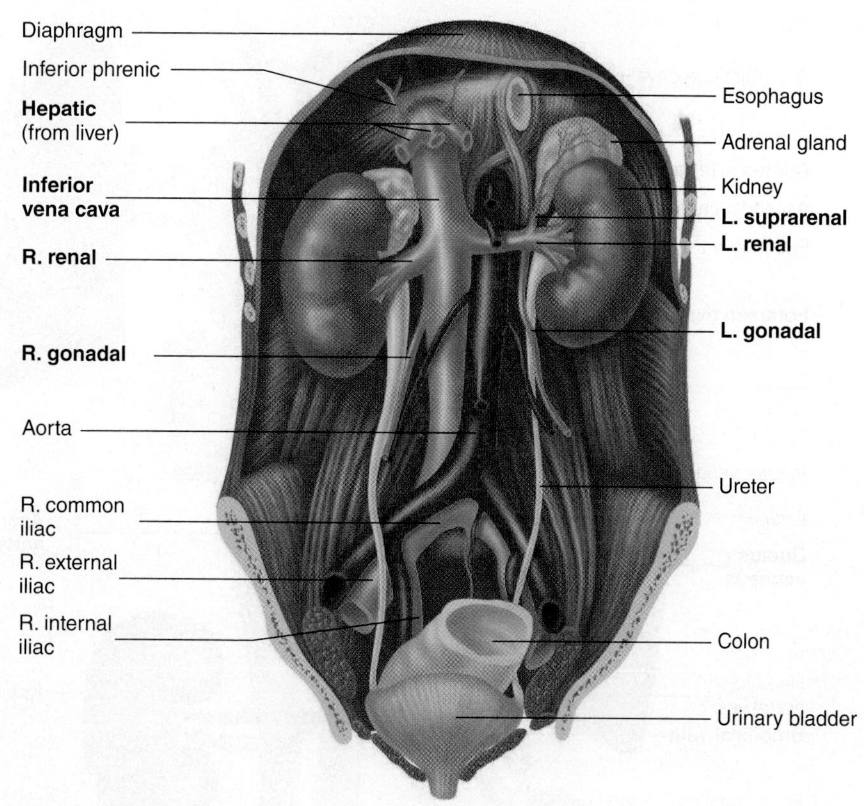

Figure 32.11 Inferior Veins of the Abdomen

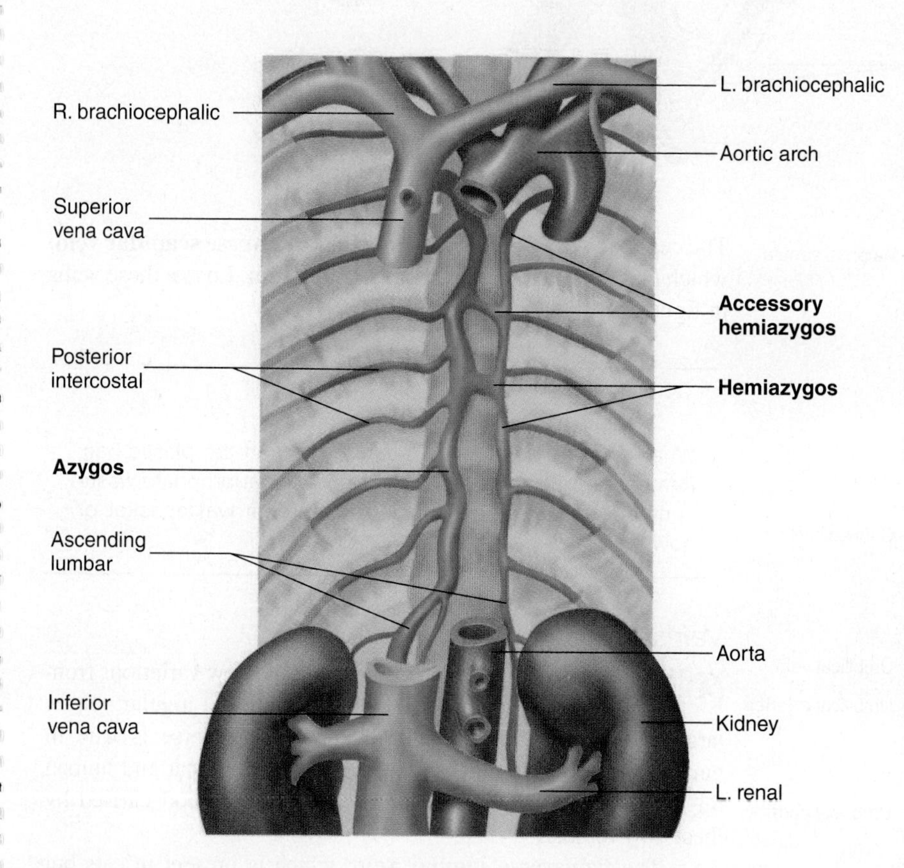

Figure 32.12 Veins of the Thorax

a flap of tissue over the foramen ovale, leaving a thin spot in the interatrial septum known as the **fossa ovalis.** Lack of closure of this foramen can lead to a condition known as "blue baby."

Cat Dissection

Ⓐ 9 Take a cat and dissection material back to your lab table. Begin your dissection by reviewing the arterial system previously studied in Exercises 30 and 31. As stated in previous exercises, the blood vessels in the cat have been injected with colored latex. If the cat has been doubly injected, the arteries are red and the veins are blue. If the cat has been triply injected, the hepatic portal vein is also injected, typically with yellow latex.

Veins of the Upper Limb

Ⓐ 10 Begin your dissection by examining the veins of the upper limb. The basilic vein is not found in cats, but you should be able to find the **ulnar vein** and note that it joins the **radial vein** to form the **brachial vein.** The brachial vein is next to the brachial artery. Locate the **median cubital vein** and the **cephalic vein.** The **axillary vein** and **subscapular vein** (from the shoulder) join to form the **subclavian vein,** which takes blood to the **brachiocephalic vein.**

1. Blood bypasses the lungs by flowing from the pulmonary trunk through the ductus arteriosus to the aorta.

2. Blood also bypasses the lungs by flowing from the right to the left atrium through the foramen ovale.

3. Oxygen-rich blood is returned to the fetus from the placenta by the umbilical vein.

4. Blood bypasses the liver sinusoids by flowing through the ductus venosus.

5. Oxygen-poor blood is carried from the fetus to the placenta through the umbilical arteries.

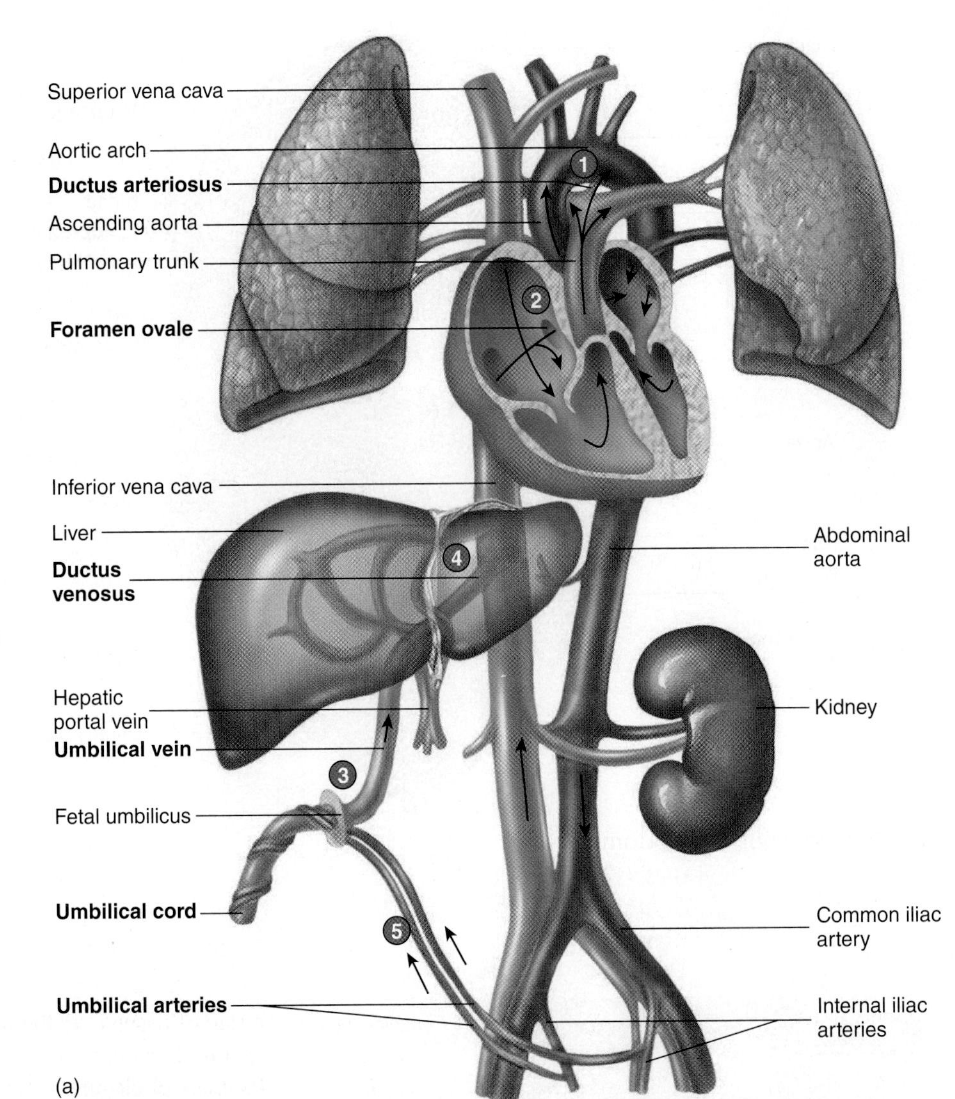

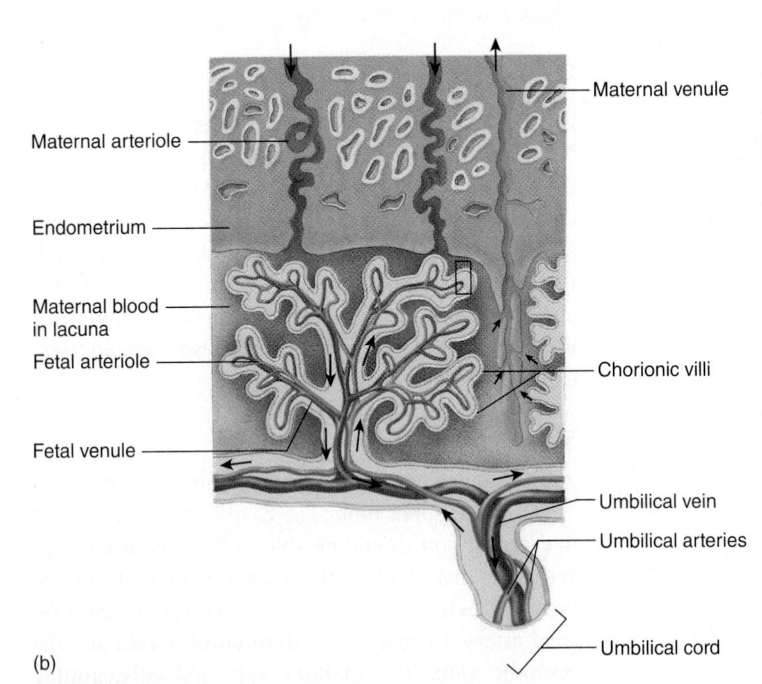

Figure 32.13 Fetal Circulation in the Human

(a) Overview of circulation; (b) details of the placenta.

The cephalic vein continues on into the **transverse scapular vein,** which takes blood to the **external jugular vein.** Locate these veins in figure 32.14.

Cleanup

When you are done, place your cat back in the plastic bag. Remember to place all excess tissue in the appropriate waste container and not in a standard classroom wastebasket or down the sink!

Veins of the Head and Neck

(A) **11** The veins in this region of the cat have a few variations from those in the human. In the cat, the external jugular vein is larger than the **internal jugular vein.** The reverse is true in humans. What anatomical difference between the cat and human might explain the difference in the volume of blood carried by these two vessels?

The **transverse jugular vein,** which is present in cats but absent in humans, connects the two external jugular veins. The

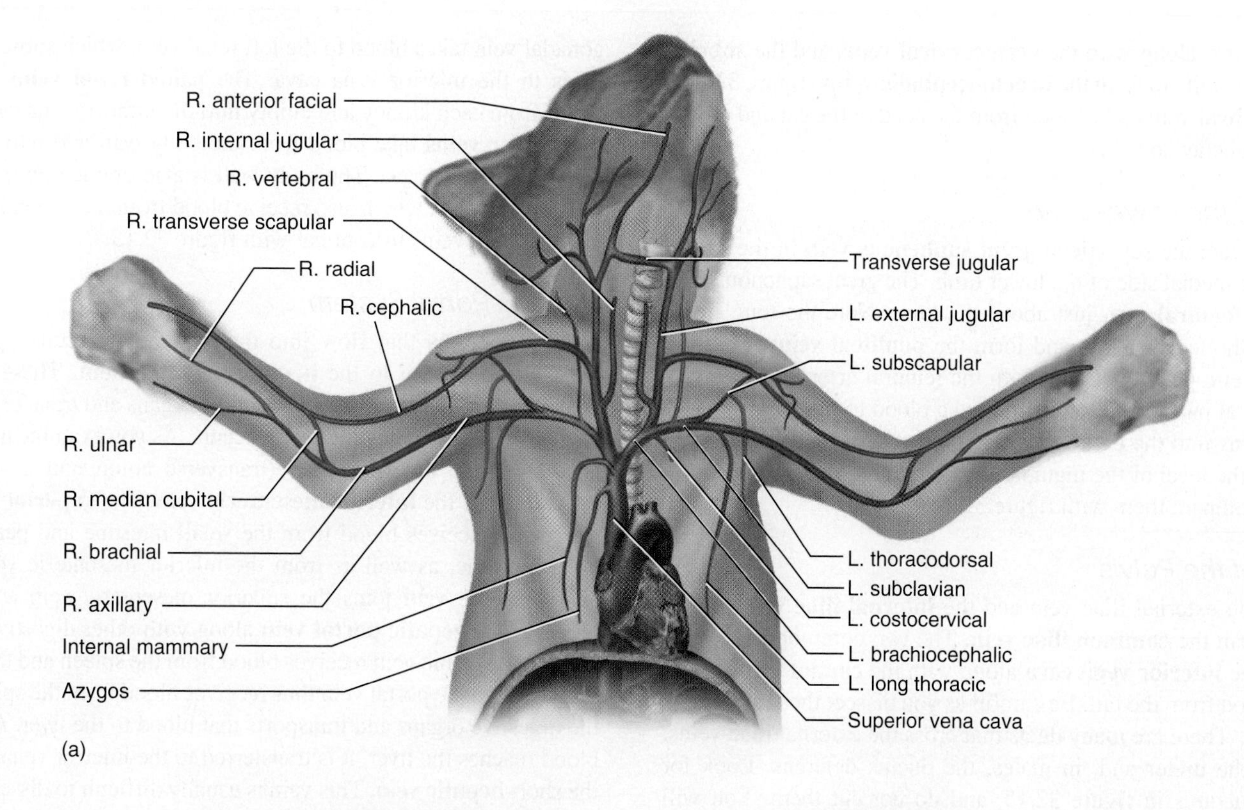

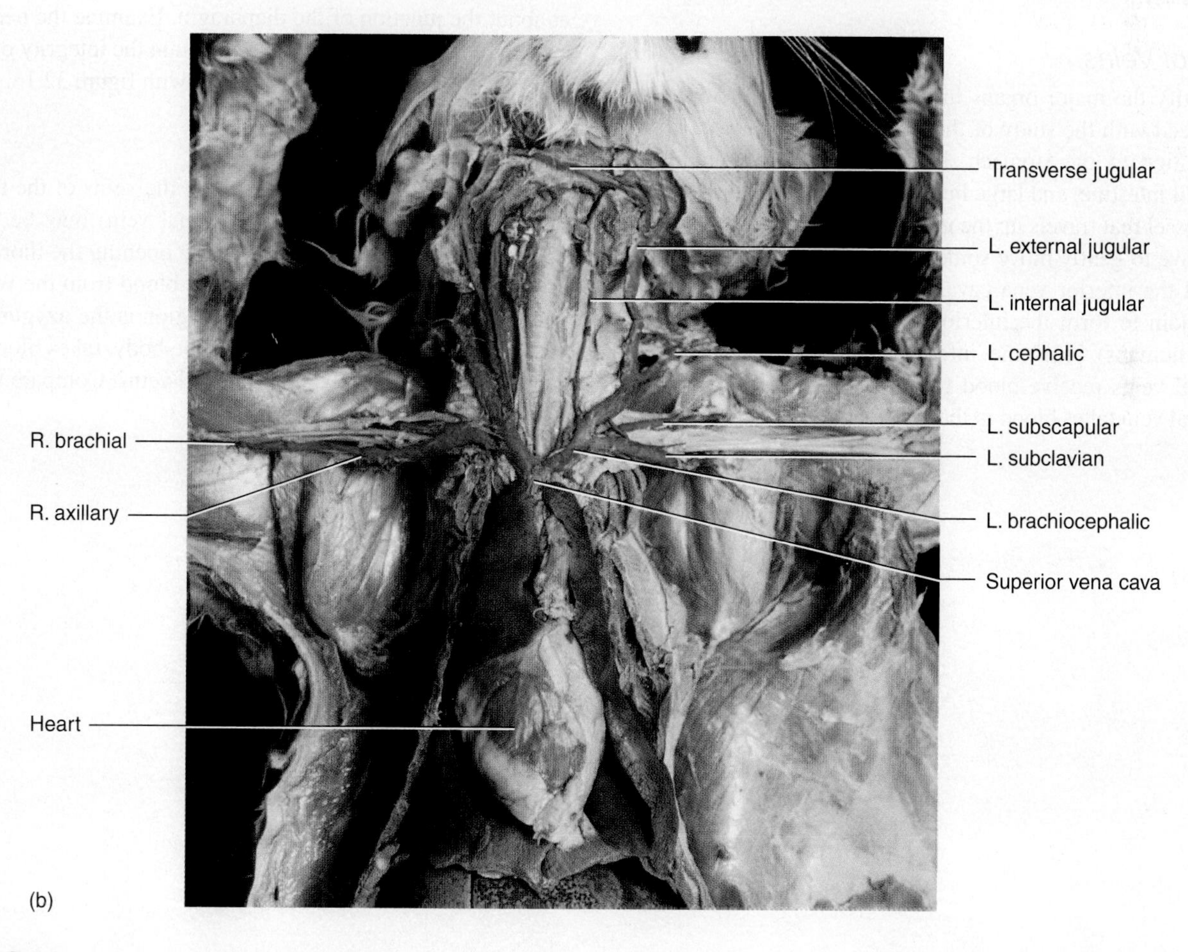

Figure 32.14 Anterior Veins of the Cat

(a) Diagram; (b) photograph.

jugular veins, along with the **costocervical veins** and the **subclavian veins,** unite to form the **brachiocephalic veins** (figure 32.14). The **vertebral veins** take blood from the head of the cat and empty into the subclavian veins.

Veins of the Lower Limb

(A) 12 Locate the superficial **great saphenous vein** in the cat on the medial side of the lower limb. The great saphenous vein joins the **femoral vein** just above the knee. Note the long **tibial veins** of the leg that join and form the **popliteal veins.** Find the femoral vein, which is found with the femoral artery. The vessels of the distal part of the lower limb take blood to the femoral vein, which turns into the **external iliac vein** as it passes into the body cavity at the level of the inguinal ligament. Find these veins in the cat and compare them with figure 32.15.

Veins of the Pelvis

(A) 13 The external iliac vein and the **internal iliac vein** join to form the **common iliac vein.** The two common iliac veins lead to the **inferior vena cava** along with the **caudal vein,** which takes blood from the tail. Be careful as you dissect these structures in the cat. There are many ducts that cross the external iliac veins, such as the ureter and, in males, the ductus deferens. Look for these structures in figure 32.15, and do not cut them. You will study them later.

Abdominal Veins

(A) 14 Identify the major organs in the digestive tract before you proceed with the study of the veins of this region. Pay particular attention to the stomach, liver, spleen, kidneys, adrenal glands, small intestine, and large intestine. The **inferior vena cava** is a large vessel that travels up the right side of the body in the cat. You may have to gently move some of the digestive organs to the side to find the inferior vena cava. Examine where the common iliac veins join to form the inferior vena cava. The **caudal vein** (absent in humans) joins the inferior vena cava at this point. The gonadal veins receive blood from the testes or ovaries. The right gonadal vein takes blood to the inferior vena cava, and the left gonadal vein takes blood to the left renal vein, which subsequently leads to the inferior vena cava. The paired **renal veins** receive blood from each kidney and empty into the inferior vena cava. The **iliolumbar veins** take blood from the body wall and return it via the inferior vena cava. The body wall is also drained by the **adrenolumbar veins,** which also receive blood from the adrenal glands. Compare the veins in your cat with figure 32.15.

Hepatic Portal System

(A) 15 The veins that flow into the liver before returning to the heart belong to the **hepatic portal system.** These vessels take blood from a number of abdominal organs and transfer it to the liver, where metabolic processing occurs. As you examine the lower portion of the digestive tract (transverse colon and descending colon), locate the **inferior mesenteric vein.** The **superior mesenteric vein** receives blood from the small intestine and part of the large intestine, as well as from the inferior mesenteric vein. The **gastrosplenic vein** joins the superior mesenteric vein and takes blood to the **hepatic portal vein** along with other digestive veins. The gastrosplenic vein receives blood from the spleen and the stomach. The hepatic portal vein thus receives blood from the spleen and the digestive organs and transports that blood to the liver. After the blood reaches the liver, it is transferred to the inferior vena cava by the short **hepatic vein.** This vein is usually difficult to dissect, since it is at the dorsal aspect of the liver and joins the inferior vena cava at about the junction of the diaphragm. Examine the hepatic portal system of the cat while trying to maintain the integrity of the digestive organs. Compare your dissection with figure 32.16.

Thoracic Veins

(A) 16 Anterior to the diaphragm are the veins of the thorax. The internal mammary vein (sternal vein) may be difficult to locate, since it is frequently cut while opening the thoracic cavity. The internal mammary vein receives blood from the ventral body wall. Another vein of the thoracic region is the **azygos vein.** This vessel, located on the right side of the body, takes blood from the esophageal, intercostal, and bronchial veins. Compare the veins in your cat with figure 32.14.

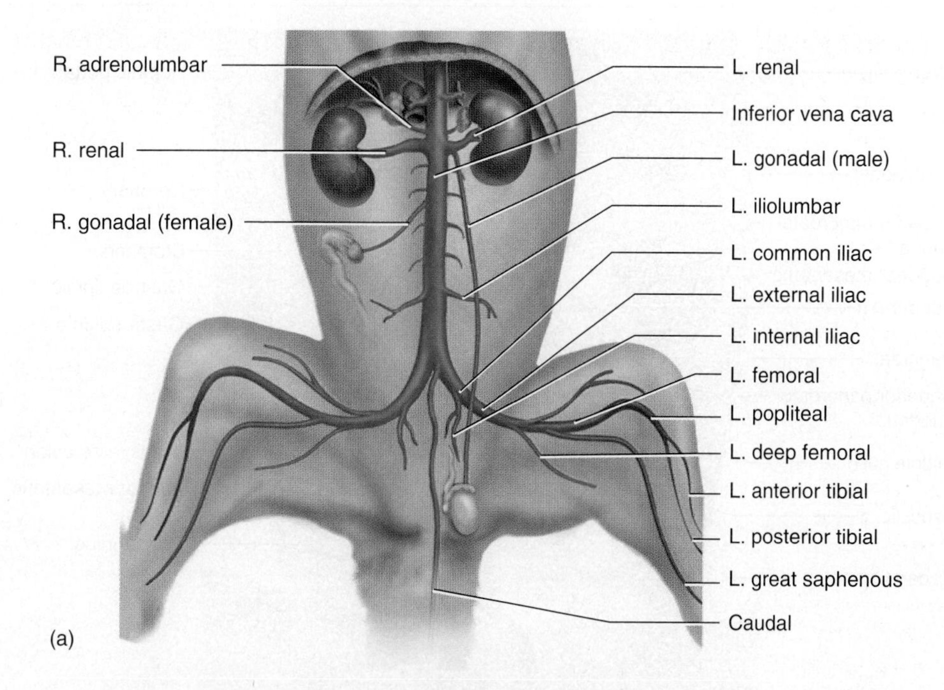

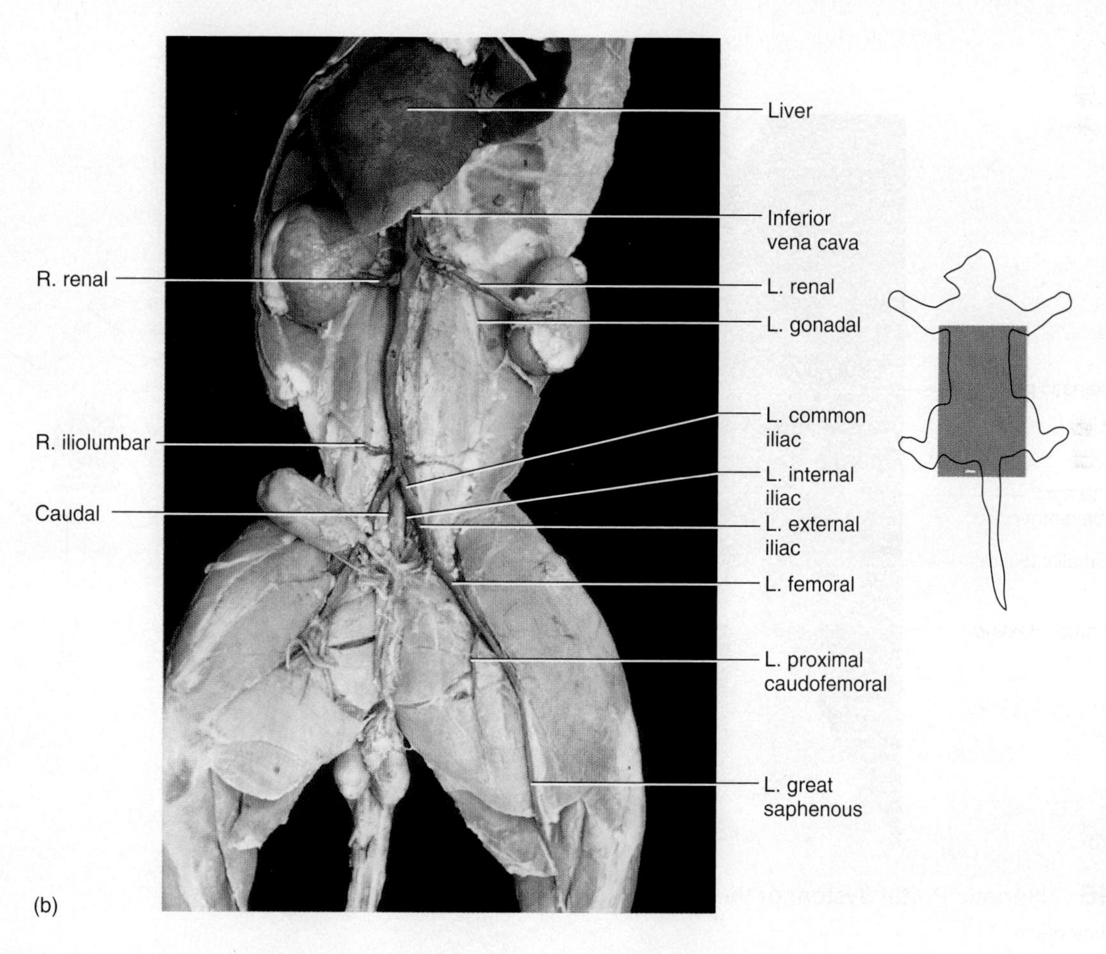

Figure 32.15 Posterior Veins of the Cat
(a) Diagram; (b) photograph.

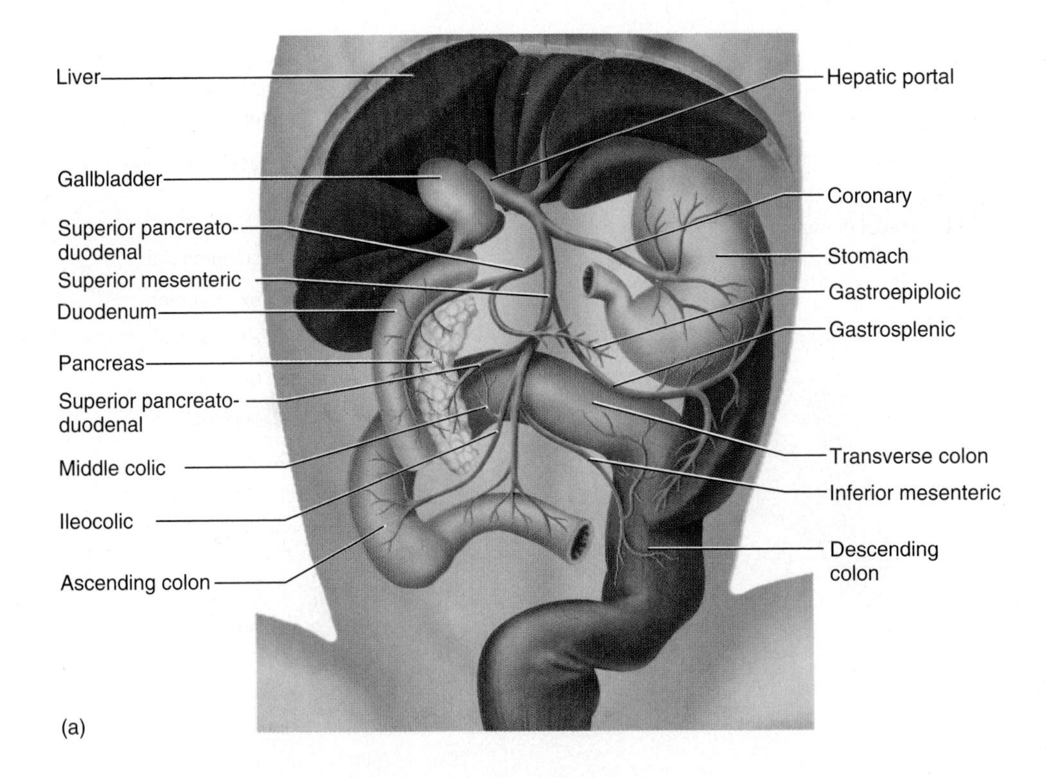

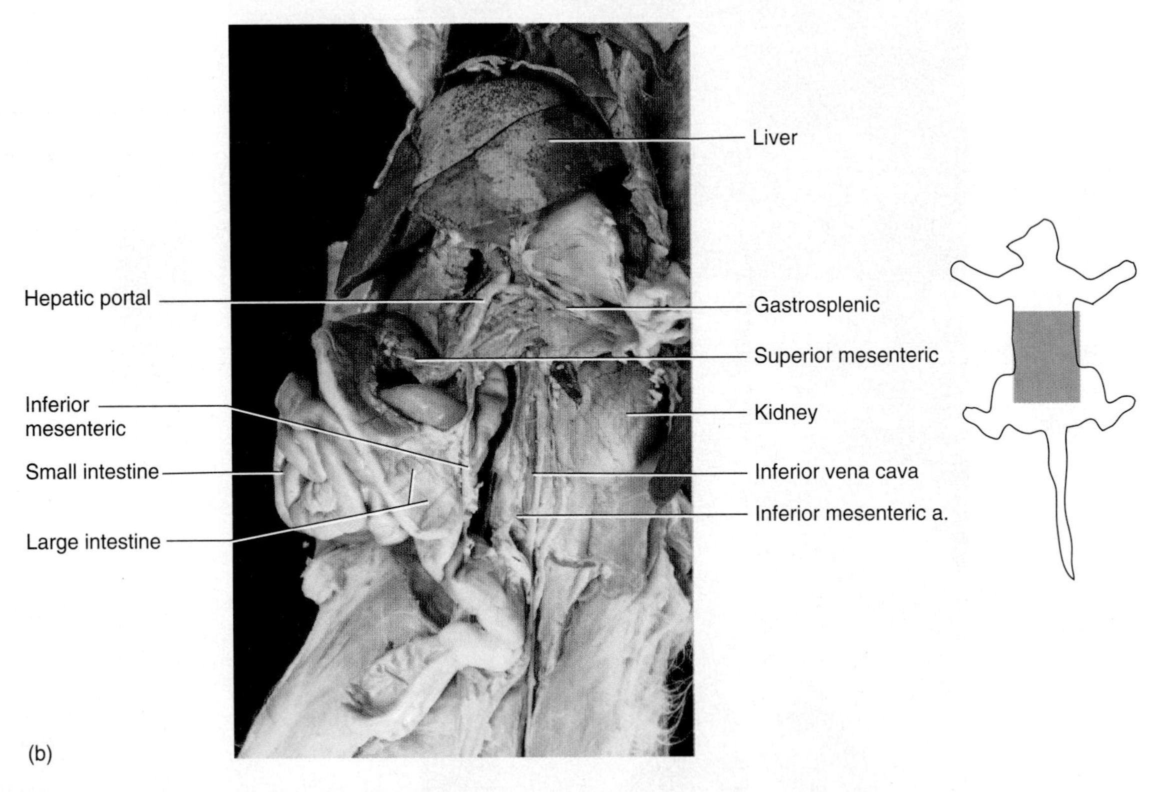

Figure 32.16 Hepatic Portal System of the Cat
(a) Diagram; (b) photograph.

Exercise 32 Review

Veins and Special Circulations

Name: ____________________

Lab time/section: ____________________

Date: ____________________

1. Label the following illustration using the terms provided.

basilic vein	great saphenous vein
brachial vein	inferior vena cava
brachiocephalic vein	internal iliac vein
cephalic vein	internal jugular vein
common iliac vein	superior vena cava
femoral vein	

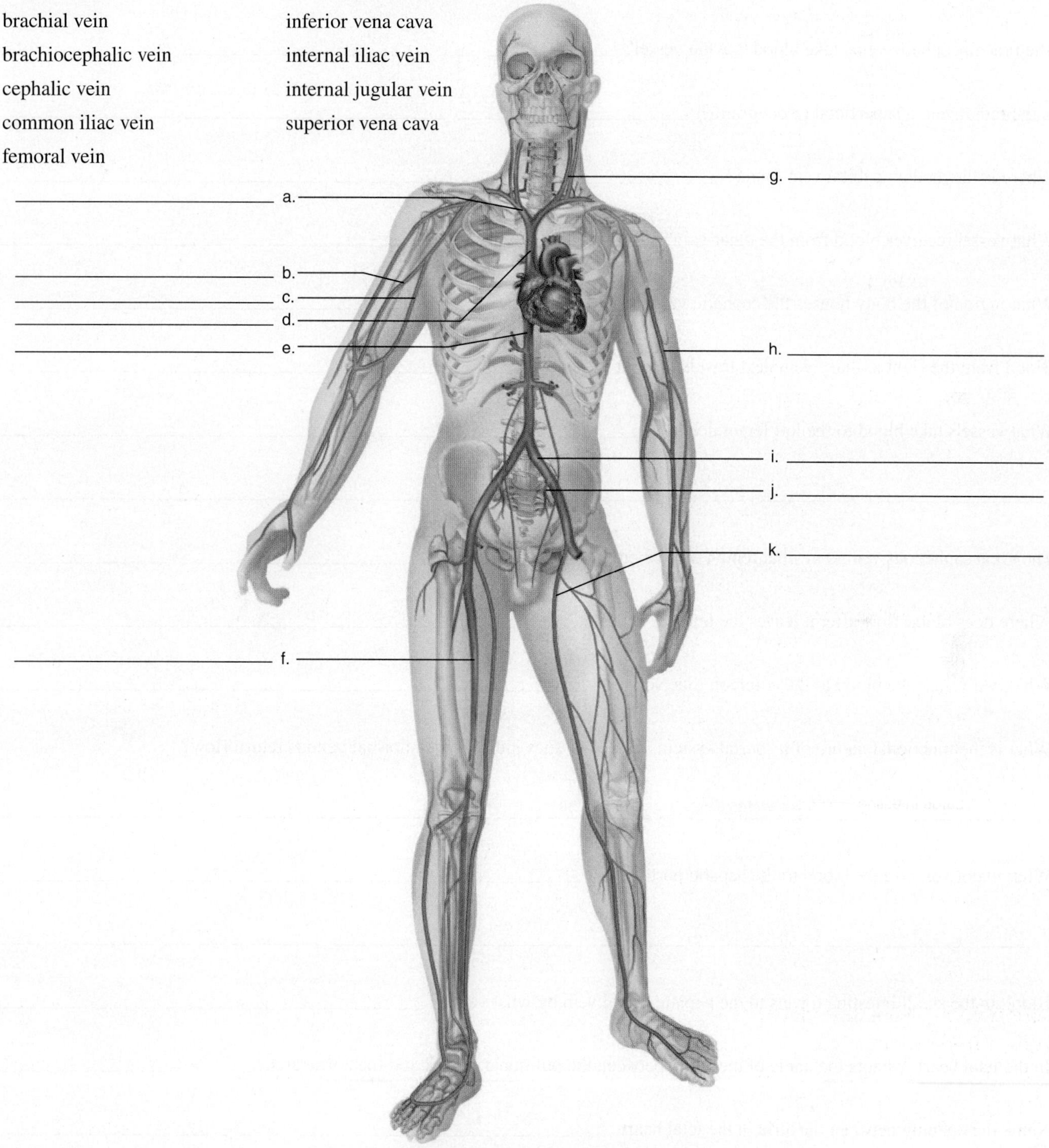

Veins of the Human

2. Which veins (superficial/deep) have names that do not correlate to arteries (circle one)?

____

3. The internal jugular vein takes blood from what area? _______________

4. What veins pass through the transverse vertebral foramina? _______________

5. What area do the right and left external jugular veins drain? _______________

6. The brachiocephalic veins take blood to what vessel? _______________

7. Is the radial vein a superficial or deep vein? _______________

8. Where is the median cubital vein found? _______________

9. What vessel receives blood from the ulnar vein? _______________

10. What region of the body houses the cephalic vein? _______________

11. Blood from the right axillary vein next travels to what vessel? _______________

12. What vessels take blood to the left femoral vein? _______________

____

13. The great saphenous vein is in what region of the body? _______________

14. Where does blood flow after it leaves the femoral vein? _______________

15. What two veins take blood to the common iliac vein? _______________

16. What is the functional nature of a "portal system," and how does it differ from normal venous return flow?

____

17. What major vessels take blood to the hepatic portal vein? _______________

____

18. Blood in the small intestine travels to the hepatic portal vein by what vessel? _______________

19. In the fetal heart, what is the name of the shunt between the pulmonary trunk and the aortic arch? _______________

20. Name the opening between the atria in the fetal heart. _______________

Exercise 33

Functions of Vessels and the Lymphatic System

Introduction

Blood vessels and the lymphatic system are intimately associated in the circulatory pattern. Fluid in the circulatory route may travel by several different pathways. As described in Exercise 30, blood is pumped from the **heart** through **arteries** and distributed through the **arterioles** to the **capillaries,** where nutrients, water, and oxygen are exchanged with the cells of the body. The blood returns via **venules** to **veins** and back to the heart under relatively low pressure. The valves in veins keep blood flowing in the direction of the heart. Their role will be examined in this exercise.

The arteries are thicker than veins, which reflects the greater amount of pressure to which they are subjected. Arteries and veins transport blood cells and plasma. Blood cells and plasma protein typically stay in the vessels, yet a significant amount of fluid from the plasma leaks from the capillaries and bathes the cells of the body. This fluid flows between the cells of the body and is known as **interstitial fluid.** It provides nutrients to the cells and receives dissolved wastes from the cells, along with cellular debris. Most of the interstitial fluid returns to the capillaries, but some is picked up by **lymphatic capillaries,** which return the fluid, now known as **lymph,** to the **lymphatic vessels.** Lymphatic vessels take the lymph back to the venous system, returning it to the cardiovascular system. This process is illustrated in figure 33.1

In addition to the functions just described, the lymphatic system is also instrumental in absorbing lipids from the digestive system. Lipids in the cells that line the small intestine are converted to **chylomicrons** (phospholipids and other molecules). These chylomicrons are conducted into the lymphatic system, which then transports the material through the thoracic duct and into the subclavian vein, where it enters the cardiovascular system.

As the fluid flows through the lymphatic vessels, macrophages in the lymph nodes clean the lymph of cellular debris and foreign material (such as bacteria and viruses) that may have entered the lymphatic system. These topics are covered in the Seeley text in chapter 21, "Cardiovascular System: Blood Vessels and Circulation," and in chapter 22, "Lymphatic System and Immunity." In this exercise, you examine the nature of the lymphatic system.

Objectives

At the end of this exercise, you should be able to

1. describe the function of valves in veins;
2. identify the valves in lymphatic vessels;
3. locate the major histological features of a lymph node;
4. demonstrate on a model the location of the tonsils and other lymphatic organs;
5. describe the functions of the spleen and thymus.

Materials

Live frog
Dissection scopes
Paper towels
Small squeeze bottle of water
Microscopes
Microscope slides of lymphatic vessels with valves, lymph nodes
Charts, diagrams, and models of blood vessels
Torso models

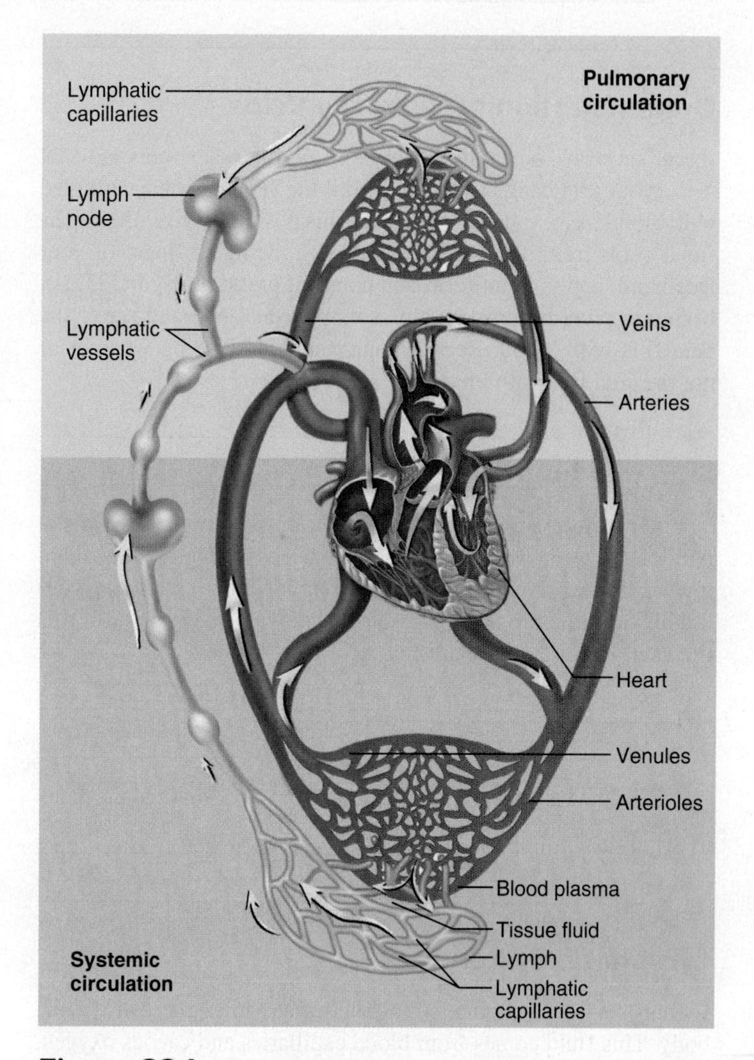

Figure 33.1 Lymph Flow and Its Relationship to the Cardiovascular System

Interstitial fluid becomes lymph (yellow arrows) and is returned to the vascular system by lymphatic vessels and valves.

Procedure

Movement Through Capillaries

1 Blood capillaries have a relatively simple structure in that they are composed of endothelium. The **endothelium** consists of a single layer of **simple squamous epithelium** that forms a tube that the erythrocytes pass through. You can observe capillaries in living tissue by examining the blood flow through the webbed foot of a live frog. Obtain a living frog and wrap its body in a wet paper towel. Leave the head and the foot exposed. *Do not let the frog dry out while you are examining it.* Hold the foot of the frog under a dissecting microscope and fan out the webbing of the foot. Make sure to hold the frog securely. Observe the flow of the blood through the capillaries, keeping the foot moist with water. Describe the flow of the blood through the capillaries of the foot in terms of speed and in terms of the size of the capillary relative to the diameter of an erythrocyte.

Your description: ______________________________

____

____

Demonstration of Valves in Veins

2 One way to examine the nature of **valves** in veins is to let your arm hang at your side until the veins become engorged with blood. As you hold your arm in this position, stroke the superficial veins from distal to proximal with the index finger of your free hand, applying uniform and constant pressure (figure 33.2a). Maintain pressure on the vein at its proximal end and see if the vein fills with blood. Record your results, indicating whether or not the vein fills with blood.

Results: ______________________________

____

Try the experiment again, but keep pressure on the distal part of the vein with your index finger as you push the blood toward the heart with your thumb (figure 33.2b). Release your thumb from the proximal area of the vein and see if blood refills the vein. Record your results.

Results: ______________________________

____

Do the valves prevent the blood from flowing in a proximal or distal direction? ______________________________

____

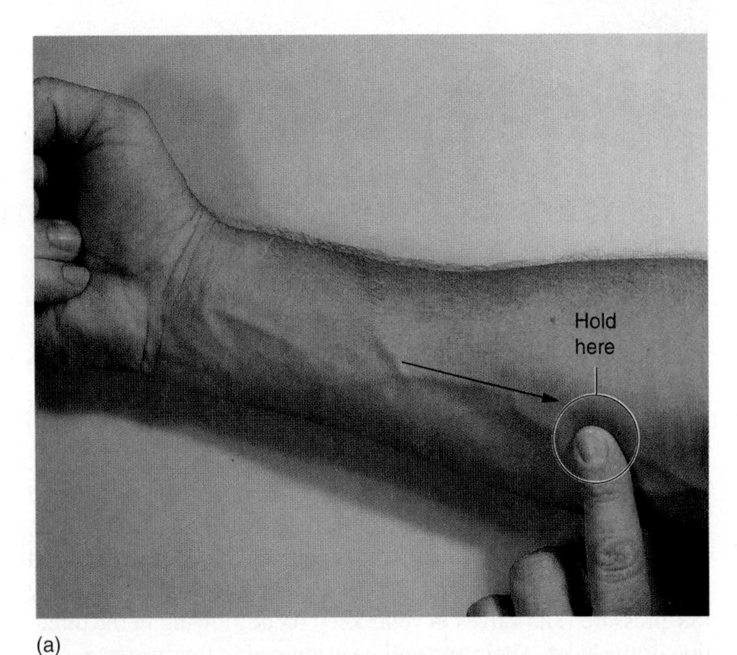

(a)

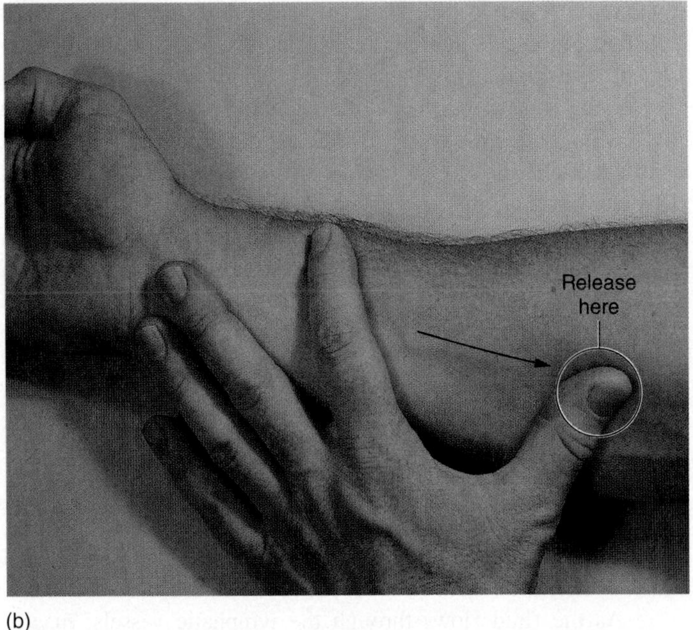

(b)

Figure 33.2 Testing for Valves in Veins

(a) Stroke the vein and keep pressure on the vein; (b) press on the distal part of the vein, stroke proximally, and then release.

Lymphatic System

Lymph originates as interstitial fluid that bathes the cells of the body. This fluid comes from blood capillaries and carries oxygen, nutrients, and other dissolved materials to the cells. The fluid enters the lymphatic capillaries by way of small, valvelike slits in the lymphatic capillary wall. These slits act as one-way valves, preventing lymph from returning to the interstices (spaces) between the cells. This process is illustrated in figure 33.3

The lymphatic system is difficult to study in preserved specimens because it collapses at death. The lymphatic system is composed of **lymphatic capillaries, lymphatic vessels, lymph nodes, lymphatic organs,** and **lymphatic tissue.** An overview of the system is illustrated in figure 33.4

3 Examine charts and models in lab and compare them with this figure.

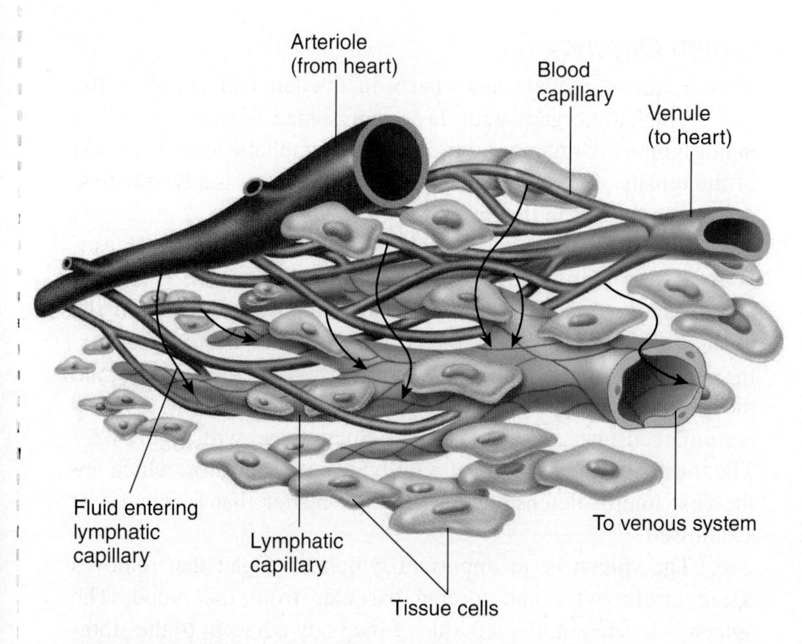

Figure 33.3 Fluid Movement from Blood Capillaries to Interstitial Fluid to Lymphatic Capillaries

Lymphatic Vessels

Ⓐ **4** Once the lymph is in the lymphatic capillaries, it travels through the **lymphatic vessels (lymphatics).** As with the lymphatic capillaries, a one-way flow occurs in the lymphatics due to the presence of valves. Examine a prepared slide of a lymphatic and note the presence of the **valve.** Compare your slide with figure 33.5

Use figure 33.4 to examine the drainage pattern of lymph in the body. The **thoracic duct** drains most of the body, taking lymph to the left subclavian vein, where the fluid is returned to the cardiovascular system. Identify the **cisterna chyli,** which is an enlarged portion of the thoracic duct in the abdominal region. The **right lymphatic duct** drains the right side of the head and neck, the right thoracic region, and the right upper extremity. The right lymphatic duct returns lymph to the right subclavian vein.

Note that the lymphatic vessels lead to regions of the body where lymph nodes are found. Nodes are clustered in the groin (inguinal region), axilla, antecubital fossa, popliteal region, neck, thorax, and abdomen.

Lymph Nodes

Ⓐ **5** Look at a slide of a **lymph node.** The lymph node is enclosed by a sheath of tissue called the **capsule.** Locate the dark purple **lymphatic nodules** in the lymph node. The outer portion of the lymph node is called the **cortex,** and the inner part is

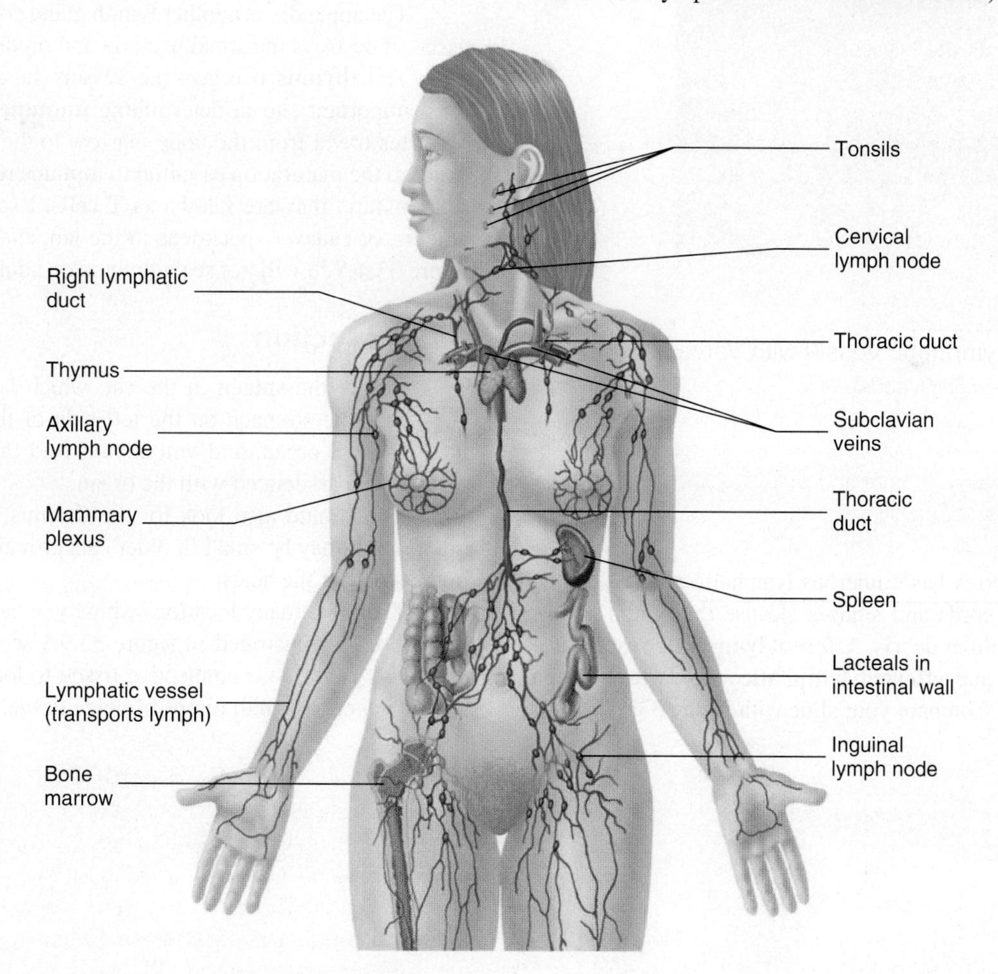

Figure 33.4 Overview of the Lymphatic System

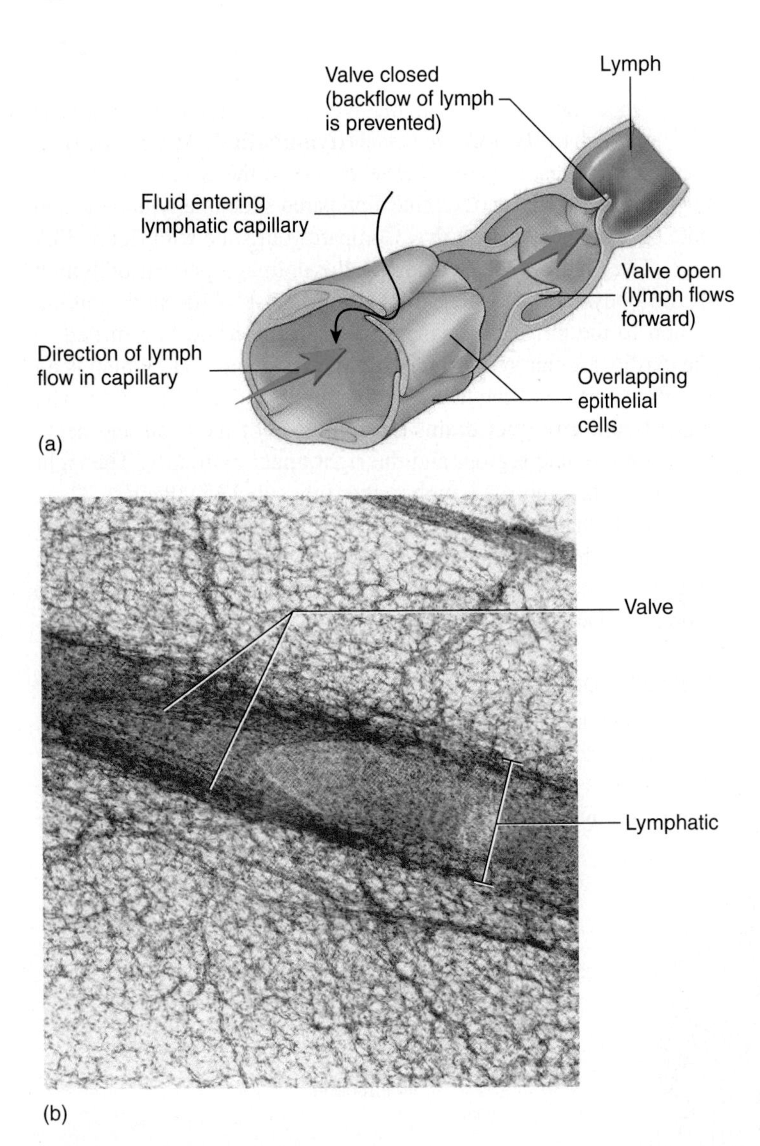

Figure 33.5 Lymphatic Vessel with Valve
(a) Diagram; (b) photomicrograph (100×).

the **medulla.** The cortex has numerous lymphatic nodules. In the medulla, medullary cords and sinuses cleanse the lymph of foreign particles and cellular debris. **Afferent lymphatic vessels** take lymph to the node, and **efferent lymphatic vessels** take lymph away from the node. Compare your slide with figure 33.6

Lymph Organs

(A) 6 Examine models and charts in the lab and compare the lymphatic organs with those represented in figure 33.4 The major lymphatic organs and tissue of the lymphatic system consist of the tonsils, thymus, spleen, lymph nodes along the lymphatics, and Peyer patches in the intestine.

Tonsils are lymphatic organs in the oral cavity and nasopharynx. Look at illustrations in your text and models in the lab and locate the three pair of tonsils. These are the **palatine tonsils,** located along the sides of the oral cavity near the oropharynx, the **lingual tonsils** at the posterior portion of the tongue, and the **pharyngeal tonsils,** located in the nasopharynx. **Adenoids** are enlarged pharyngeal tonsils. Compare these with figure 33.7 The tonsils have **crypts** lined with **lymphatic nodules,** which are the first line of defense against foreign matter that is inhaled or swallowed.

The **spleen** is an important lymphatic organ that removes aging erythrocytes and foreign particles from the blood. The spleen is located on the left side of the body adjacent to the stomach. It is a highly vascular organ that filters blood and produces lymphocytes. The spleen contains **red pulp,** which filters blood, and **white pulp,** which produces lymphocytes. Locate the spleen in models and charts in the lab, or in cadaver specimens, and compare it with figures 33.4 and 33.8.

The appendix is another lymph gland consisting of lymphoid tissue. It destroys intestinal bacteria and produces lymphocytes.

The **thymus** overlays the vessels superior to the heart. It is an important site in determining **immune competence.** Lymphocytes travel from the bone marrow to the thymus, where they undergo the maturation essential to immune responses. Once these cells mature, they are known as **T cells.** Examine the charts and models, or cadaver specimens in the lab, and compare them with figure 33.4 You will not see a thymus on adult models.

Cat Dissection

(A) 7 Locate the **spleen** in the cat, which is an elongated organ near the stomach on the left side of the abdominal cavity. It is a brown organ, and you should find the splenic artery and splenic vein associated with the organ.

You should also look for the **thymus,** which is anterior to the heart. It may be small in older cats or may have been removed in the study of the heart.

There are many locations where you can find **lymph nodes,** and these are illustrated in figure 33.9 Use a forceps or a probe and tease away loose connective tissue to locate the nodes. They should appear as small lumps of beige tissue.

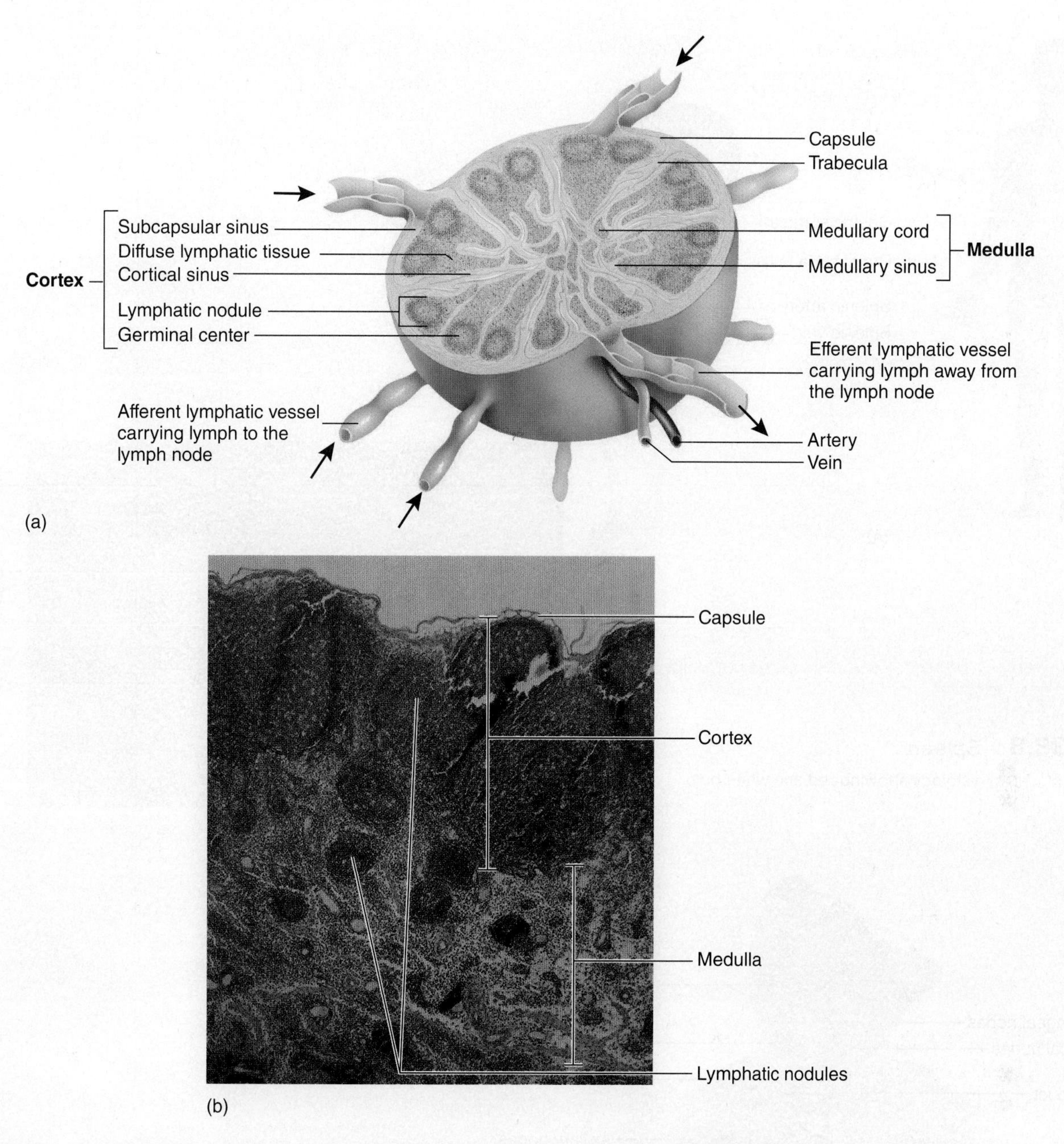

Figure 33.6 Lymph Node

(a) Diagram showing lymph flow by arrows; (b) photomicrograph (40×).

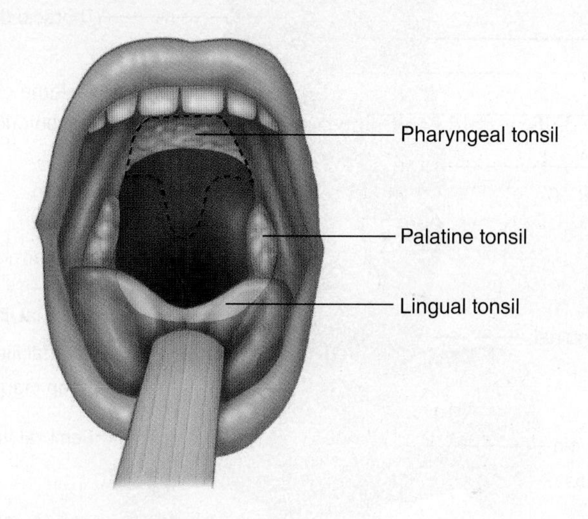

Figure 33.7 Tonsils

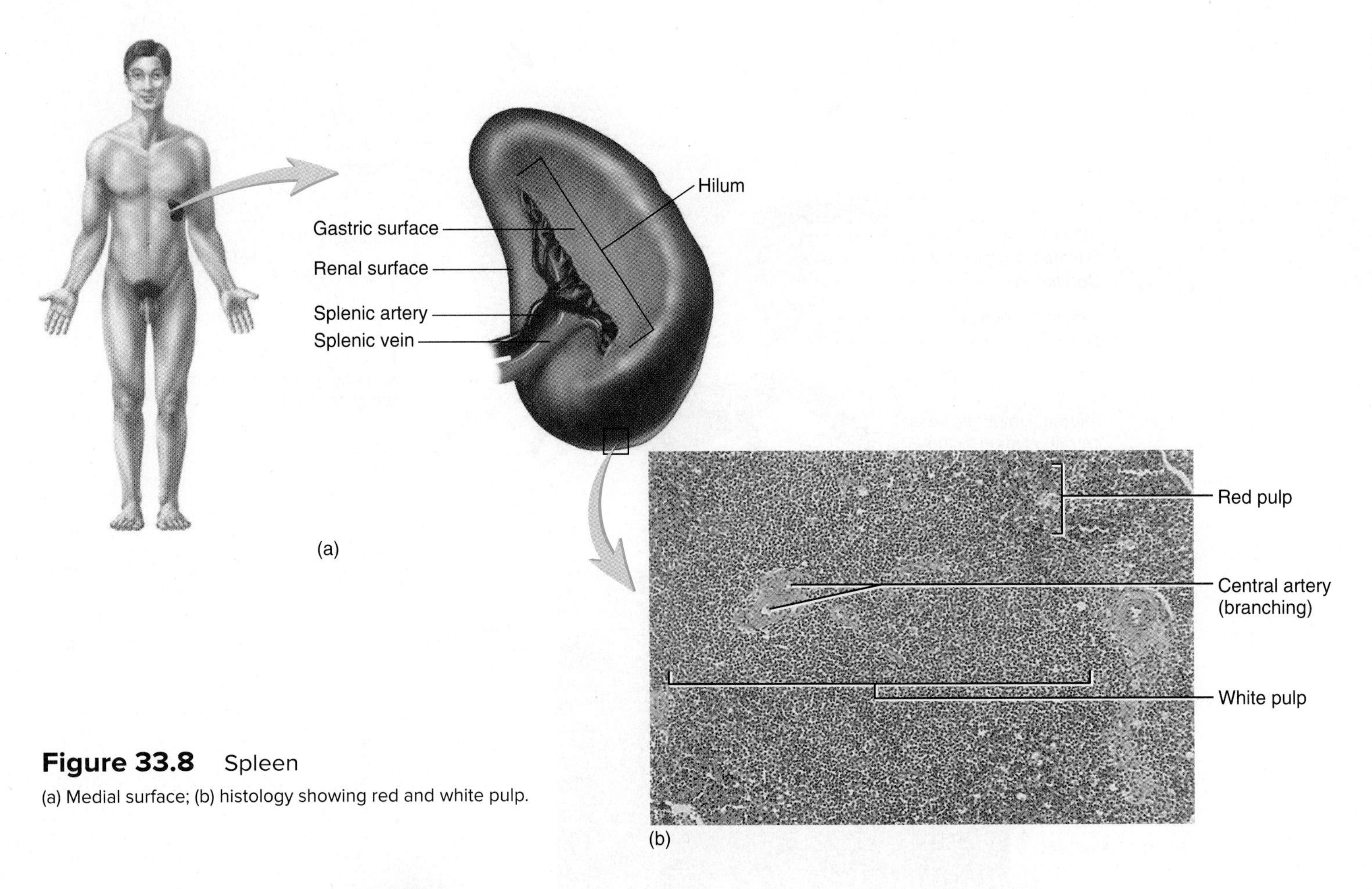

Figure 33.8 Spleen
(a) Medial surface; (b) histology showing red and white pulp.

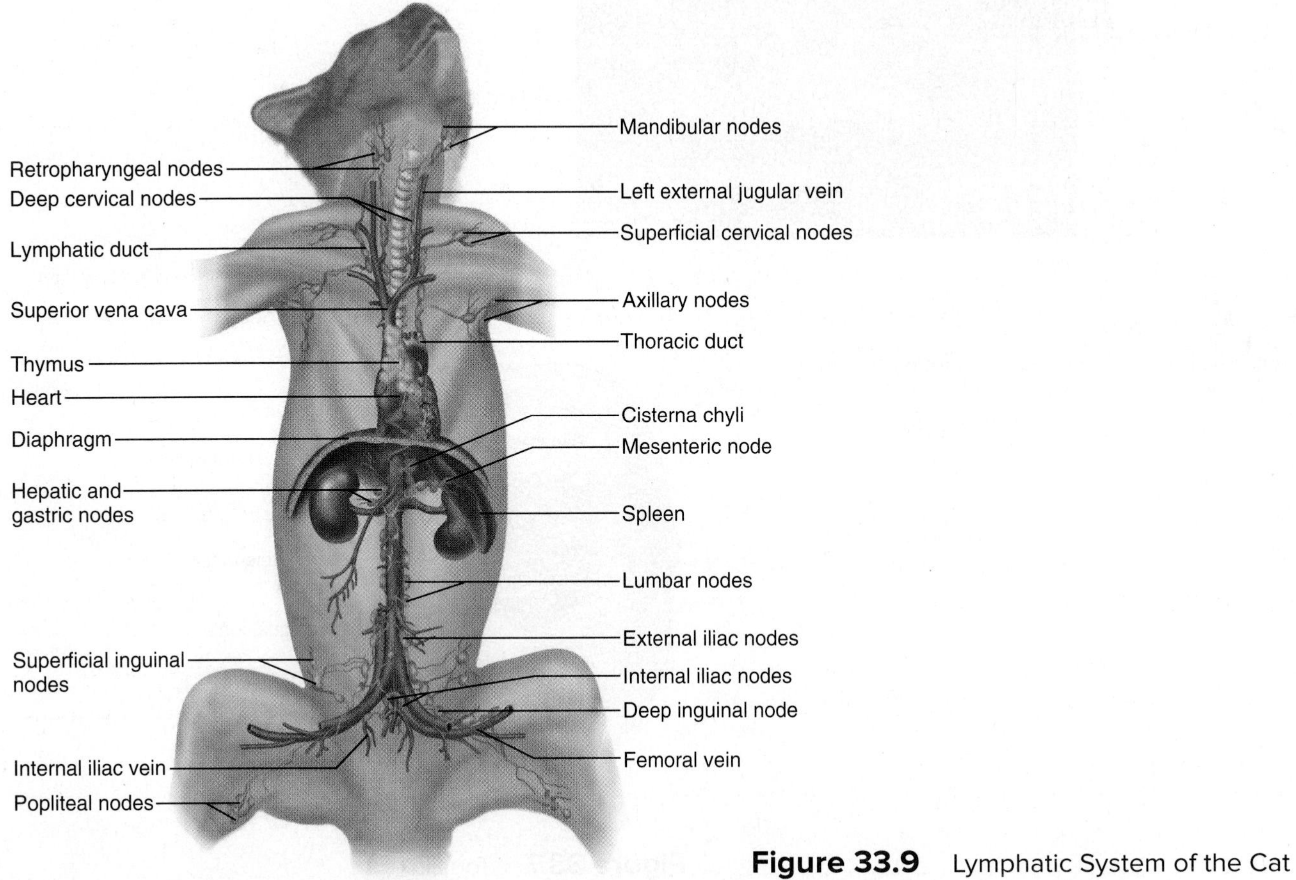

Figure 33.9 Lymphatic System of the Cat

Exercise 33 Review

Name: ______________________

Lab time/section: ______________________

Date: ______________________

Functions of Vessels and the Lymphatic System

1. What cell type makes up the endothelium of capillaries? ______________________

2. What is the name of the vessels that carry lymph from the lymphatic capillaries to the veins? ______________________

3. Once interstitial fluid enters the lymphatic vessels, what is it called? ______________________

4. What is the name of the inner region of a lymph node? ______________________

5. What kind of vessel takes lymph away from a lymph node? ______________________

6. The adenoids are enlarged ______________________ tonsils.

7. Which tonsils are found on the sides of the oral cavity? ______________________

8. Which tonsils are located at the back of the tongue? ______________________

9. Blood is filtered by which lymphatic organ in the adult? ______________________

10. What part of the spleen is involved in producing lymphocytes? ______________________

11. Where do T cells mature? ______________________

12. Lymphatic vessels have a one-way flow from the extremities to the heart. Damage to the lymphatic system can lead to edema, or an increase in tissue fluid. From the standpoint of reducing edema, how does the use of medical leeches (segmented worms that drain tissue fluid) work for a region that has suffered trauma? ______________________

    ____

    ____

    ____

13. From what you know of the functions of lymph nodes, predict the difference between lymph entering a node and lymph leaving a node. What materials may be missing from the lymph leaving the node? ______________________

    ____

    ____

14. Elephantiasis is a disease that is caused, in some cases, by a parasitic worm blocking the lymphatic vessels. Examine the following illustration and predict where the lymphatic vessel blockage occurs. ______________________________

____

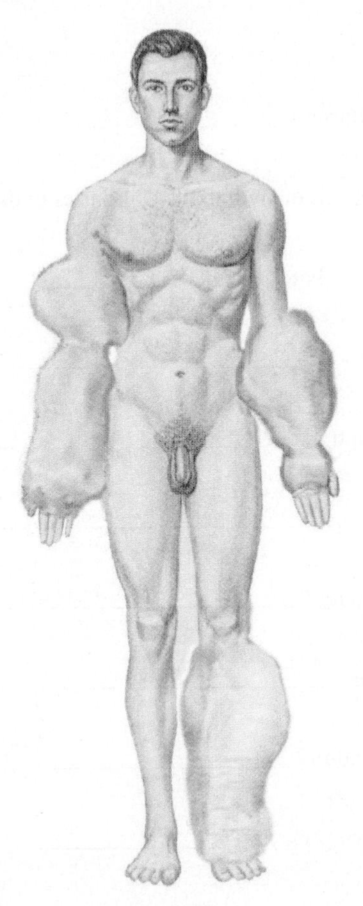

Damage to the Lymphatic System by Elephantiasis

15. In an analysis of breast cancer, lymph nodes of the axillary region are removed and a biopsy is performed. The removal of the nodes is done to determine if cancer has spread from the breast to other regions of the body. What effect would the removal of lymph nodes have on the drainage of the pectoral region? ______________________________

____

____

16. Superficial veins contain valves, yet deep veins do not. The deep veins are surrounded by muscles. People who are inactive may have problems with their veins. Can you propose a mechanism by which blood from the deep veins can be returned to the heart (other than by standing on your head)? ______________________________

____

____

Exercise 34

Blood Vessels and Blood Pressure

Introduction

Maintenance of blood pressure is important for the health of the heart and for proper functioning of various organs. **Hypertension** is elevated blood pressure. Hypertension increases the workload of the heart by increasing the force with which the heart must pump to provide blood to the body. Normal blood pressure is approximately 120/80 mmHg. Hypertension may be due to a number of factors; however, the majority of hypertensive cases occur for unknown reasons. Hypertension is typically a blood pressure in excess of 140/90 mmHg for young adults, though the greatest concern is in the diastolic reading. The diastolic pressure should be less than 90 mmHg. Hypertension can harm the arteries, causing fatty deposits in the walls of the arteries, a condition called **atherosclerosis** (ath'er-ō-skler-ō'sis), which narrows the arteries and reduces blood flow to organs, such as the brain, heart, and kidneys. These deposits can harden, causing **arteriosclerosis,** or hardening of the arteries.

As people age, their arteries typically become less elastic and blood pressure increases. Kidney disease and other illnesses can cause hypertension. Poor diets, such as those rich in fats or salt; lack of exercise; and elevated caffeine, nicotine, or alcohol intake can also increase blood pressure, as can obesity. Men generally have higher blood pressure than women until about the age of 55; women show an increase in hypertension after that age.

Hypotension can also be dangerous. Low blood pressure may occur when you stand up quickly. This is known as orthostatic hypotension. Severe dehydration, genetics, and the time of early pregnancy can all lead to low blood pressure. A decrease in pressure may lead to fainting or dizziness.

Historically, blood pressure has been indirectly measured with a **sphygmomanometer** (blood pressure cuff), which measures pressure in millimeters of mercury rising in a glass column. The greater the pressure, the higher the rise of mercury. Mercury-filled sphygmomanometers are still used, but aneroid sphygmomanometers (using air) and differential pressure transducers are more common. Electronic blood pressure measurement is commonly used in clinics and homes. The advantage of the electronic measurement for home use is that it is simple to use for the average person, producing fewer reading errors. It allows for daily monitoring of the blood pressure, so that a baseline is established.

Blood pressure is not uniform throughout the body but is influenced by gravity; therefore, the pressure in the arteries of the head and neck is less than the blood pressure from the heart. The pressure in the arteries of the leg is greater than the blood pressure in the heart. For this reason, blood pressure is measured in the **brachial artery,** which is at the level of the heart and has the approximate pressure of that leaving the heart. Auscultatory measurement of blood pressure involves listening to sounds as blood passes through the brachial artery. Normally, no sound is heard through a stethoscope as blood passes through the brachial artery. The blood passes smoothly through the vessel, like water through a garden hose. When pressure is applied by pinching off a hose, the turbulence creates sound. When pressure is applied to the arm due to the constriction from the blood pressure cuff, the turbulence of the blood passing through the vessel creates sound. The significant difference between the flow of blood in the body and water in a hose is that the blood pulses through the arteries as the heart contracts during ventricular systole. The sounds made by the flow of blood in a partially constricted artery are known as **Korotkoff sounds.** Completely open arteries and completely closed arteries do not produce Korotkoff sounds.

These topics are covered in the Seeley text in chapter 21, "Cardiovascular System: Blood Vessels and Circulation." In this exercise you learn how to determine blood pressure and some factors affecting blood pressure.

Objectives

At the end of this exercise, you should be able to

1. determine the pulse rate of an individual;
2. define hypertension in terms of millimeters of mercury pressure;
3. distinguish between systolic pressure and diastolic pressure;
4. properly take and record blood pressure;
5. list three factors that affect blood pressure.

Materials

Blood Pressure Experiment

Stethoscope
Alcohol wipes or isopropyl alcohol and sterile cotton swabs
Washable felt pen
Sphygmomanometer
Watch or clock with accuracy in seconds

Flow Experiment

Corn syrup
1 yard of 1/4-inch plastic flexible tubing
1 yard of 1/2-inch plastic flexible tubing
Small funnel to fit small tubing
Large funnel to fit large tubing
Stopwatch
Two ring stands

Beakers
Clamps
Wash basin
Roll of paper towels

Procedure

Measurement of Pulse Rate in Beats per Minute (bpm)

(A)[1] Initially record a baseline pulse rate. The pulse rate is measured in beats per minute (bpm) by locating regions of the body where the pulse can be palpated—the **radial artery** or **carotid artery.** If you are recording your lab partner's pulse, make sure you use the tips of your fingers to measure the pulse rate, not your thumb. If you use your thumb you may feel the pulse in your thumb and not the pulse of the person you wish to record. Determine beats per minute by counting the pulse for 1 minute.

Pulse rate of lab partner: ______________________ bpm

Measurement of Arterial Blood Pressure

1. To determine blood pressure, locate the pulse of the brachial artery on your lab partner. This can be done by placing two fingers on the medial side of the biceps brachii muscle near the antecubital fossa. You can place a small "X" with a felt pen on this location on the arm (figure 34.1).
2. Place the blood pressure cuff around the arm of your lab partner at the level of the heart. Make sure the inflatable portion of the cuff is on the anterior medial side. Some cuffs have a metal bar that provides a loop through which a part of the cuff goes through. This bar should *not* be located on the medial side of the arm because if the bar is located there, it may not constrict the brachial artery effectively as the cuff is inflated (figure 34.2).
3. Clean the ear pieces of the stethoscope with alcohol wipes, place the diaphragm of the stethoscope on the "X" where you located the brachial pulse, and have your lab partner rest his or her forearm on the lab counter. You should *not* hear any sound at this time.
4. Hold on to the rubber squeeze bulb with the attached rubber tubing leading away from you. Turn the metal dial clockwise until it is closed. You can now begin to inflate the cuff (figure 34.3).
5. Pump the cuff up to about 80 mmHg. Look at the mercury in the glass tube or the dial on an aneroid instrument. If it seems to bounce up and down a little, then listen closely for the sounds in the stethoscope. (The ear pieces are inserted into the ears, facing forward.) If you do not hear any sound, then inflate the mercury or the dial on an aneroid instrument to 100 or 120 mmHg. If you see the mercury or needle pulsing, listen again for the sound. Make sure the diaphragm of the stethoscope is in the right place.
6. Once you are sure you have heard the sound of the heart, remove the cuff and place it on the other arm. Excessive constriction of the arm may elevate your lab partner's blood pressure.
7. Inflate the cuff on the other arm, but make sure you exceed the level where you see motion in the needle or pulsing in the mercury (usually around 150 mmHg). Do not leave the cuff inflated for a long period.

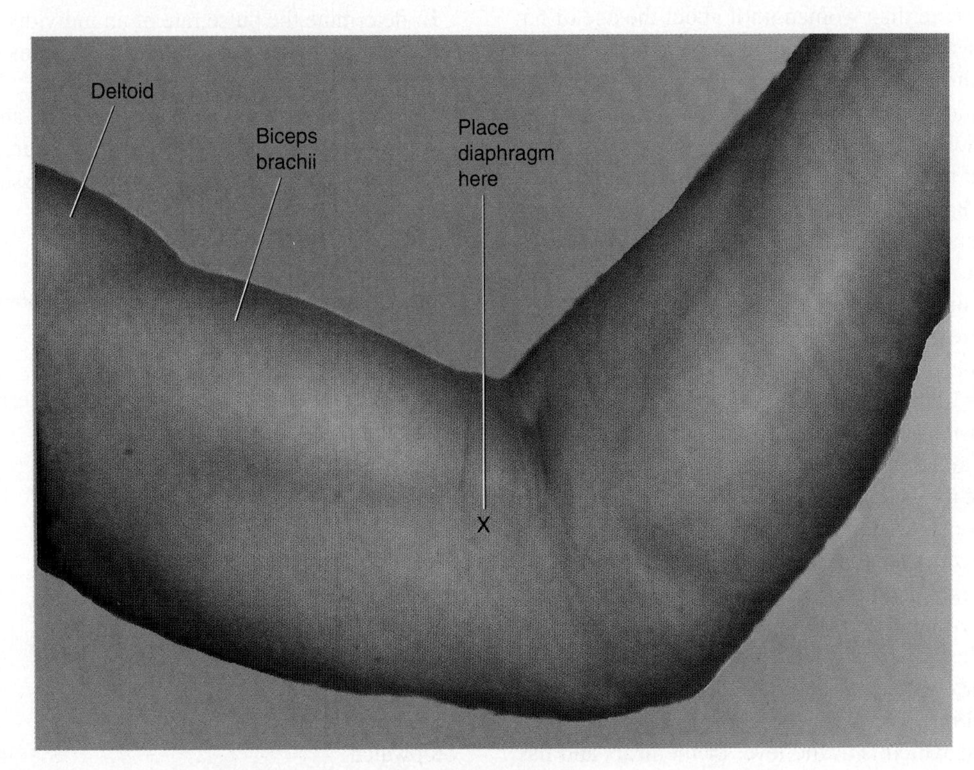

Figure 34.1 Location of Stethoscope Diaphragm, Left Medial Arm

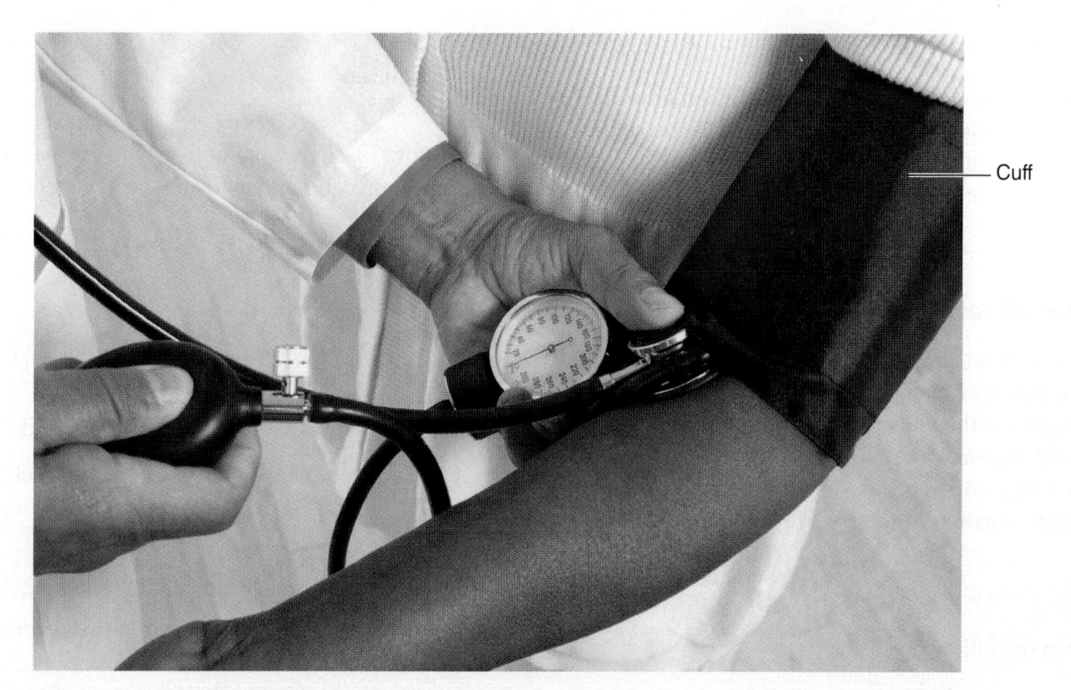

Figure 34.2 Proper Placement of Cuff

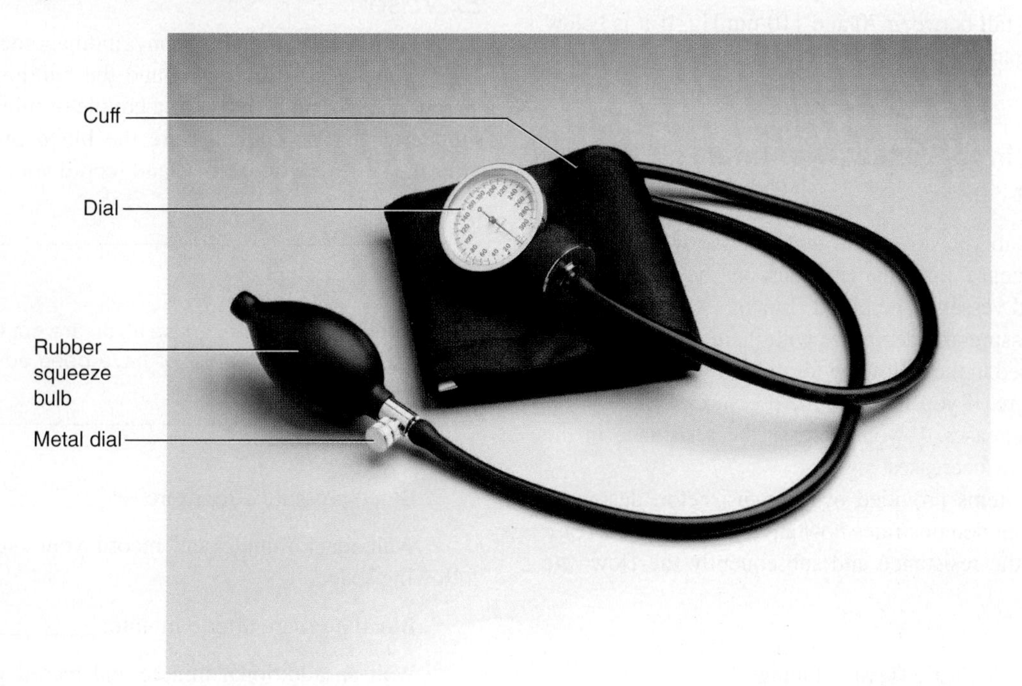

Figure 34.3 Parts of Sphygmomanometer

8. Release the knob slowly, so that the mercury or needle slowly begins to descend.
9. Record the number when the first sound is heard. This sound represents the **systolic pressure** of the ventricles, which is the pressure the heart generates that exceeds the pressure of the cuff. The sound may muffle a bit and then come in strong.
10. Continue to let air out of the cuff slowly until the sound starts to muffle. Listen very carefully until the sound completely disappears. The exact level at which the sound disappears is the **diastolic pressure.**
11. Record the level of blood pressure as the systolic pressure over the diastolic pressure:

$$\frac{\text{Systolic pressure}}{\text{Diastolic pressure}} = \rule{6cm}{0.4pt}$$

If you are unsure of exactly where the pressure is, you can fine-tune your readings by increasing the cuff pressure at each approximate end to obtain an accurate reading. Do not subject your lab partner to more than two or three attempts. Ask your instructor for help if you have difficulty determining blood pressure.

Pulse Pressure

(A) 2 After you determine the blood pressure, you can now determine the pulse pressure of your lab partner. The pulse pressure is determined by subtracting the diastolic pressure from the systolic pressure. It is normally around 50 mmHg.

Pulse pressure: ______________________________

Mean Arterial Blood Pressure

(A) 3 Mean arterial blood pressure (MAP) is used clinically to measure the average blood pressure of a person. If the MAP is too low, then organs of the body are not receiving enough blood pressure and could become ischemic. Once you measure your systolic blood pressure (SP) and diastolic blood pressure (DP), you can get an approximation of the MAP by using the following formula.

Approximated MAP = DP + 1/3 (SP – DP)

Enter your MAP in the following space:

MAP ______________________________

Normally, it should fall between 70 and 110 mmHg. If it is below 60 mmHg, the organs of the body are not receiving adequate blood flow.

Influence of Blood Vessel Diameter on Resistance

In this part of the lab you do a little investigative work to try to determine the influence that diameter has on the resistance to blood flow in blood vessels. The blood flow in a vessel increases with increasing pressure and decreases with increasing resistance. This can be expressed in the following formula: **Flow = pressure/resistance.** Therefore, if you increase the pressure of the system, the rate of flow increases. If you increase the resistance in the blood vessel, the flow decreases.

(A) 4 Using the items provided by your instructor, design an experiment that demonstrates how an increase or a decrease in diameter alters the resistance and subsequently the flow rate in blood vessels.

1. corn syrup
2. 1 yard of 1/4-inch plastic flexible tubing
3. 1 yard of 1/2-inch plastic flexible tubing
4. small funnel to fit small tubing
5. large funnel to fit large tubing
6. stopwatch
7. ring stand
8. clamps
9. wash basin
10. beakers

Flow is altered in humans by cardiac output, peripheral resistance, precapillary sphincters, and general vasoconstriction.

Factors Affecting Blood Pressure

Body Position

(A) 5 Measure the blood pressure of your lab partner while he or she lies down. Record this value.

Measurement lying down: ______________________________

Have your lab partner stand suddenly, and quickly take a new measurement. *Be careful not to drop the sphygmomanometer!* Record the results.

Blood pressure on immediately standing: ______________________________

How can you account for these two readings? ______________________________

____

Exercise

(A) 6 Without having the sphygmomanometer attached, have your lab partner run around the building for a while or do strenuous physical activity for a couple of minutes. Put the sphygmomanometer on and measure the blood pressure immediately after the cessation of exercise and record your results.

Caution!

Do not do strenuous exercise if you are not feeling well, have a history of heart trouble, or have been advised by a physician to avoid exercise.

Blood pressure after exercise: ______________________________

Wait for 1 minute and record your blood pressure in the following space.

Blood pressure after 1 minute: ______________________________

Wait an additional minute and record your blood pressure 2 minutes after exercise.

Blood pressure after 2 minutes: ______________________________

Is there a difference between individuals in class who are in shape and those who are out of shape?

Exercise 34 Review

Blood Vessels and Blood Pressure

Name: ______________________

Lab time/section: ______________________

Date: ______________________

1. The letters bpm stand for what phrase in cardiac measurement? ______________________

2. What does a sphygmomanometer measure? ______________________

3. To measure blood pressure, what artery would you most commonly use? ______________________

4. If you have a blood pressure of 140/80, what does the 80 represent? ______________________

5. What is the clinical threshold for high blood pressure in young adults? ______________________

6. When the first sound is heard during measurement with a blood pressure cuff, what is measured—systolic or diastolic pressure?

   ____

7. Emotions have an effect on blood pressure. Predict the blood pressure of an individual who recently had a heated argument with a roommate about rent money. ______________________

   ____

8. Illness can also affect blood pressure. Illness tends to increase stress responses. Predict the blood pressure of an individual with a sinus headache and postnasal drip. ______________________

9. Nicotine and caffeine both elevate blood pressure. Explain how an increase in blood pressure could have a negative effect on the cardiac output. ______________________

   ____

   ____

10. Record your blood pressure. ______________________

    ____

11. How did the change in blood pressure after exercise vary from those people who were in shape versus those who were out of shape? ______________________________

____

12. According to the potential risk factors for hypertension, which, if any, do you have? ______________________________

____

Exercise 35

Structure of the Respiratory System

Introduction

The respiratory system provides oxygen to the cells of the body and removes carbon dioxide. The body exchanges oxygen and carbon dioxide with the atmosphere. This vital exchange occurs due to the respiratory system, which is discussed in the Seeley text in chapter 23, "Respiratory System." Atmospheric oxygen moves into the lungs and diffuses into the circulatory system. It subsequently reaches the individual cells of the body while the metabolic waste product, carbon dioxide, is released from the intercellular environment and travels via the blood to the lungs, where it is released by exhalation. In this exercise, you examine the anatomy of the respiratory system. As you study the anatomy of the system, be aware of the role that other systems, such as the cardiovascular system, play in respiration.

Objectives

At the end of this exercise, you should be able to

1. list the organs and significant structures of the respiratory system;
2. define the role of the respiratory system in terms of the overall function of the body;
3. explain the physical reason for the tremendous surface area of the lungs;
4. identify the cartilages of the larynx;
5. distinguish among a bronchus, a bronchiole, and a respiratory bronchiole;
6. recognize a bronchiole and an alveolus in a prepared slide of lung.

Materials

Lung models or detailed torso model, including midsagittal section of head
Model of larynx
Microscopes
Prepared microscope slides of lung tissue
Prepared microscope slides of "smoker's lung"
Charts and illustrations of the respiratory system
Cats
Materials for cat dissection

Dissection trays
Scalpel and new blades
Protective gloves
Blunt (Mall) probe
Forceps and sharp scissors
First aid kit in lab or prep area
Sharps container
Animal waste disposal container

Procedure

Overview

Look at the charts and models of the respiratory system and compare them with figure 35.1. Locate the following structures:

Nose
Nasal cavity
Pharynx
Larynx
Trachea
Bronchi
Lungs

Nose and Nasal Cartilages

(A)1 Examine a midsagittal section of a model or chart of the head and look for the **nose, nasal cartilages, external nares (nostrils),** and **nasal septum.** The nasal septum separates the nasal cavities and is composed of the **perpendicular plate of the ethmoid bone,** the **vomer,** and the **septal cartilage.** Examine these features in figure 35.2.

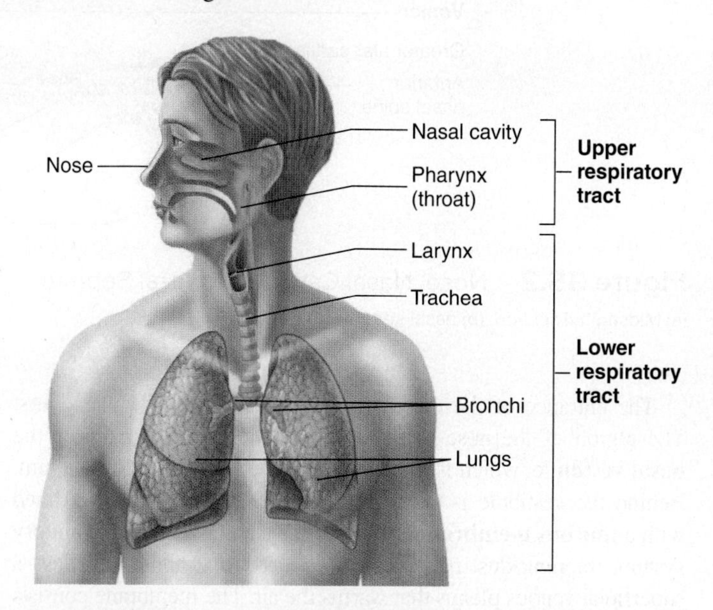

Figure 35.1 Structures of the Respiratory System

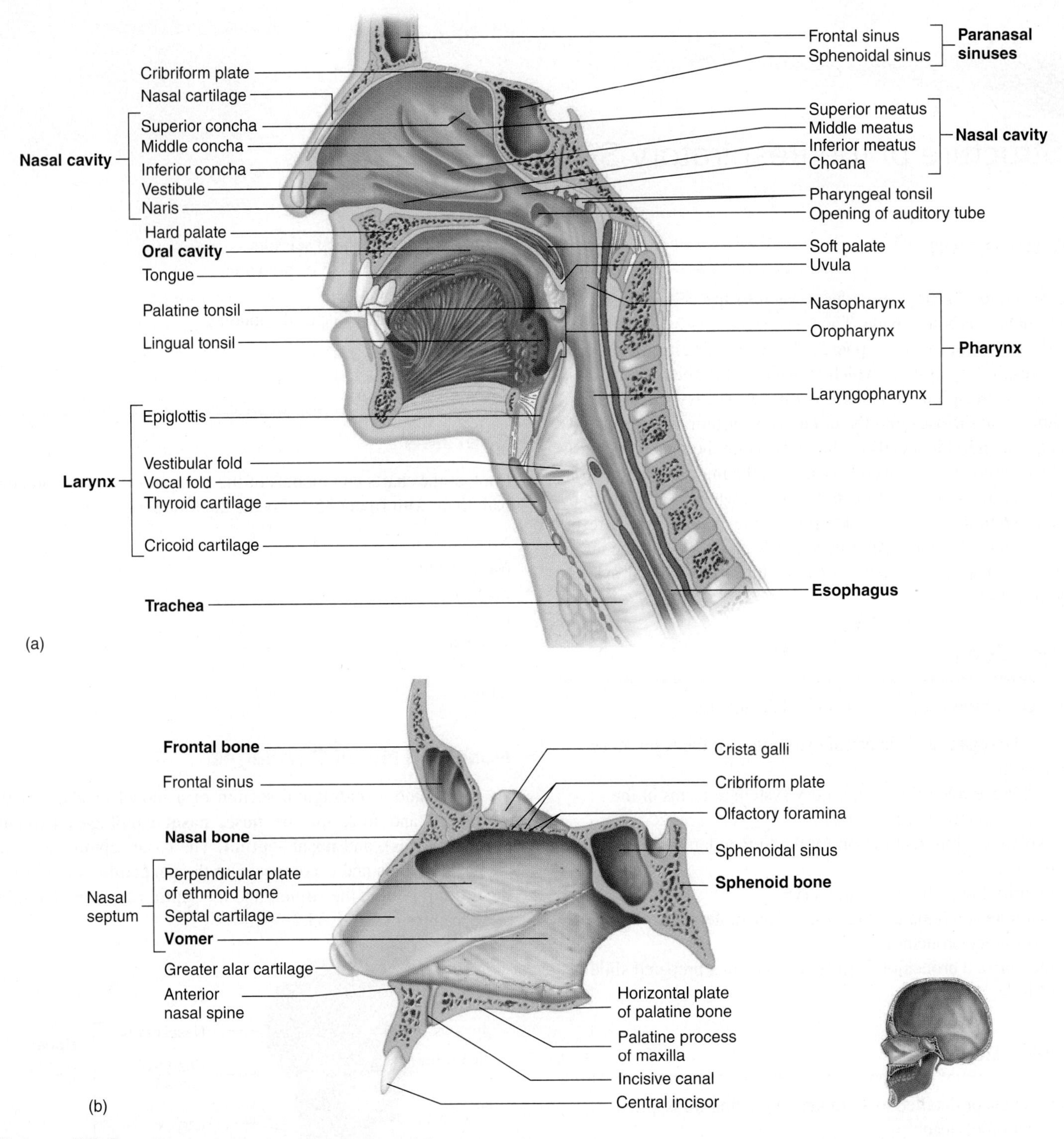

Figure 35.2 Nose, Nasal Cavity, and Nasal Septum
(a) Midsagittal section; (b) nasal septum.

The entrance of the external nares is protected by guard hairs. The region of the nose just posterior to the external nares is the **nasal vestibule,** which is lined with stratified squamous epithelium. Behind the vestibule is the **nasal cavity.** The nasal cavity is lined with a **mucous membrane** that moistens air entering the respiratory system, trapping dust particles. This mucous membrane overlays a superficial venous plexus that warms the air. The membrane consists of connective tissue and **respiratory epithelium,** which is composed of **pseudostratified ciliated columnar epithelium** with **goblet cells.** The lateral walls of the cavity have three protrusions that push into the nasal cavity. These are the **nasal conchae.** They cause the air to swirl in the nasal cavity and come into contact with the mucous membrane.

(A) 2 Locate the **superior, middle,** and **inferior conchae** in the nasal cavity on a model in lab. The nasal cavity terminates where two openings, the **internal nares** (sing. **naris**), or **choanae,** lead to the **pharynx.**

Pharynx

Ⓐ[3] You should examine a model of a midsagittal section of the head and neck to find the pharynx. The pharynx can be divided into three regions based on location. The uppermost area is the **nasopharynx,** which is directly posterior to the nasal cavity. The nasopharynx has two openings on the lateral walls, which are the openings of the **auditory,** or **eustachian, tubes.** As the pharynx descends behind the oral cavity, it becomes the **oropharynx.** The oropharynx is a common passageway for food, liquid, and air. The **uvula** is a small, pendulous structure that partially separates the **oral cavity** from the oropharynx. The uvula flips upward during swallowing, helping prevent fluids from entering the nasopharynx. The most inferior portion of the pharynx is the **laryngopharynx,** which is located superior to the larynx. Find the regions of the pharynx in figure 35.2.

Larynx

Ⓐ[4] Examine models and charts in the lab and compare the structures of the larynx with figure 35.3. The **larynx** is commonly known as the "voice box" because it is an important organ for sound production in humans. It controls the pitch of the voice, while the shape of the oral cavity and the placement and size of the paranasal sinuses are responsible for the sonority (sound quality) of the voice. Look at a model or chart of the larynx in lab. The larynx is located at about the level of the fourth through sixth cervical vertebrae and consists of a number of cartilages. The most prominent cartilage in the larynx is the **thyroid cartilage,** which is a shield-shaped structure made of hyaline cartilage.

The thyroid cartilage is more prominent in males because of the influence of testosterone. Inferior to the thyroid cartilage is the **cricoid cartilage.** The cricoid cartilage is also composed of hyaline cartilage, and it is relatively narrow when seen from the anterior but increases in size at its posterior surface. Superior to the cricoid cartilage in the posterior wall of the larynx are the paired **arytenoid cartilages.** These cartilages attach to the posterior end of the **vocal folds (true vocal cords).** Muscles pull the arytenoid cartilages, stretching the true vocal folds, increasing the pitch of the voice. This occurs by the contraction of **intrinsic muscles** attached to the arytenoid cartilages from the back, while the vocal cords are held stationary by the thyroid cartilage in the front. Above the true vocal folds are the **vestibular folds (false vocal cords;** figure 35.3).

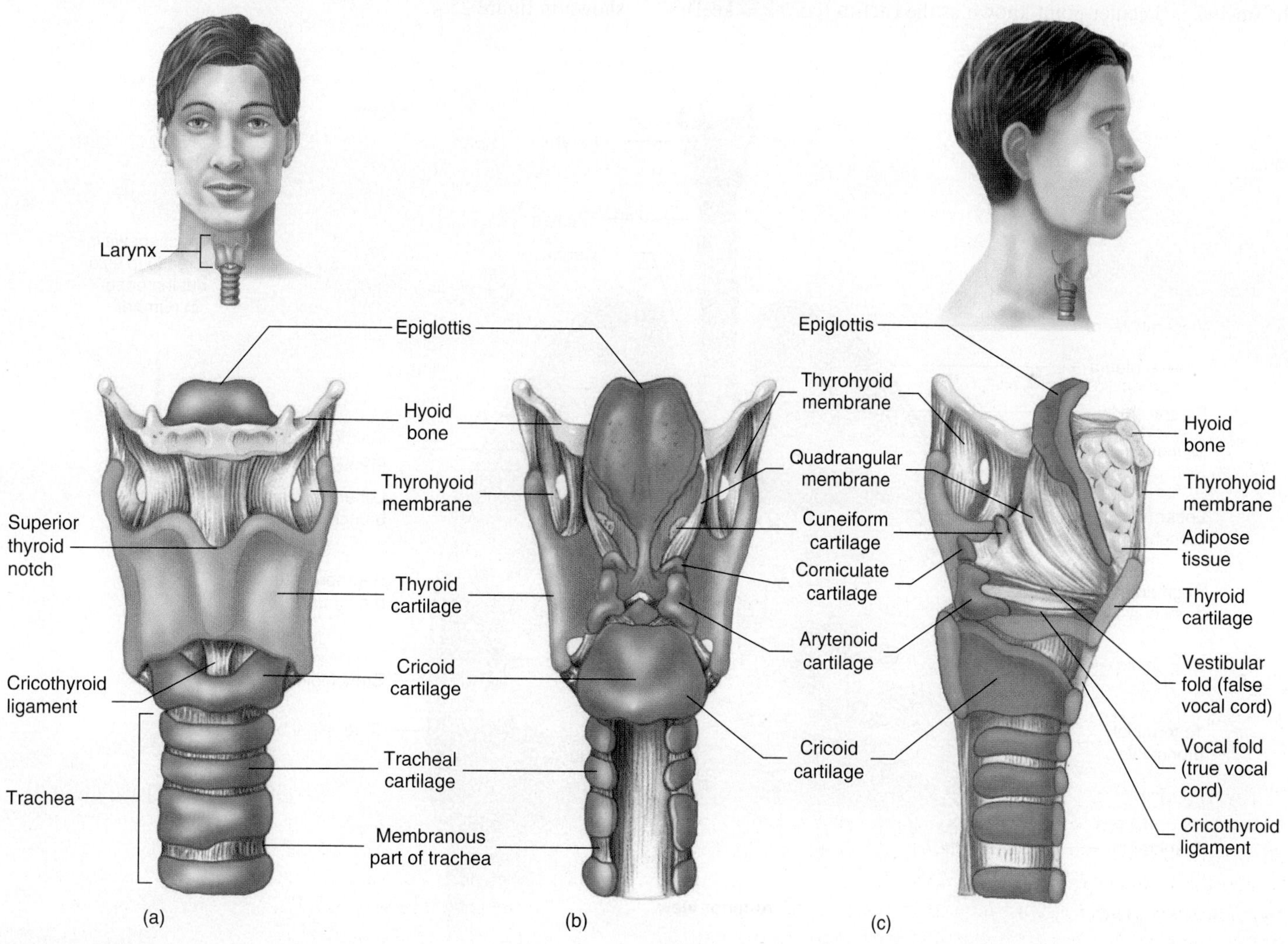

Figure 35.3 Larynx

(a) Anterior view; (b) posterior view; (c) midsagittal section.

At the very posterior, superior edge of the larynx are the **corniculate** and **cuneiform cartilages.** These are also made of hyaline cartilage. The most superior cartilage of the larynx is the **epiglottis,** which is composed of elastic cartilage. During swallowing, the epiglottis covers the opening of the larynx, which is known as the **glottis.** This is not a perfect system, as anyone knows who has started laughing at a joke while swallowing a liquid. Inhalation at the beginning of the laugh causes fluid to move into the larynx and trachea, irritating the respiratory lining. This causes another reflex, called the **cough reflex,** which propels the liquid out of the respiratory system (frequently, through the nose). Locate the structures of the larynx on models or charts in the lab and in figure 35.3.

Trachea and Bronchi

(A) 5 The trachea is commonly known as the "windpipe" because it conducts air from the larynx to the lungs. The **trachea** is a straight tube whose lumen is kept open by **tracheal cartilages.** These are C-shaped cartilages. Examine these cartilages by running your fingers gently down the outside of your throat. Palpate the cartilage rings below the larynx. The tracheal cartilages are composed of hyaline cartilage. At the most inferior portion of the trachea is a center point known as the **carina** (*carina* = keel). Locate the features of the trachea in figure 35.4 and 35.5 and in a model or chart in lab.

The trachea is also lined with respiratory epithelium. Obtain a prepared slide of the trachea and find the tracheal cartilage, respiratory epithelium, and **posterior tracheal membrane** (trachealis muscle; figure 35.5).

The trachea splits into two tubes, which enter the lungs. These tubes are the **main,** or **primary, bronchi.** Each lung receives air from a primary bronchus. These primary bronchi contain hyaline cartilage and are lined with respiratory epithelium. The primary bronchi of the lung divide into the **lobar,** or **secondary, bronchi,** and these further divide to form **segmental,** or **tertiary, bronchi.** The extensive branching of the bronchi produces a structure called the tracheobronchial tree (figure 35.4 and 35.6).

Lungs

(A) 6 There are two **lungs** in humans; the right lung has **three lobes** and the left lung has **two lobes.** The **right lung** consists of **superior, middle,** and **inferior lobes,** and the **left lung** has **superior** and **inferior lobes,** as well as an indentation occupied by the heart. This indentation is known as the **cardiac notch.** Look at models or charts in the lab and identify the major features, as shown in figure 35.4.

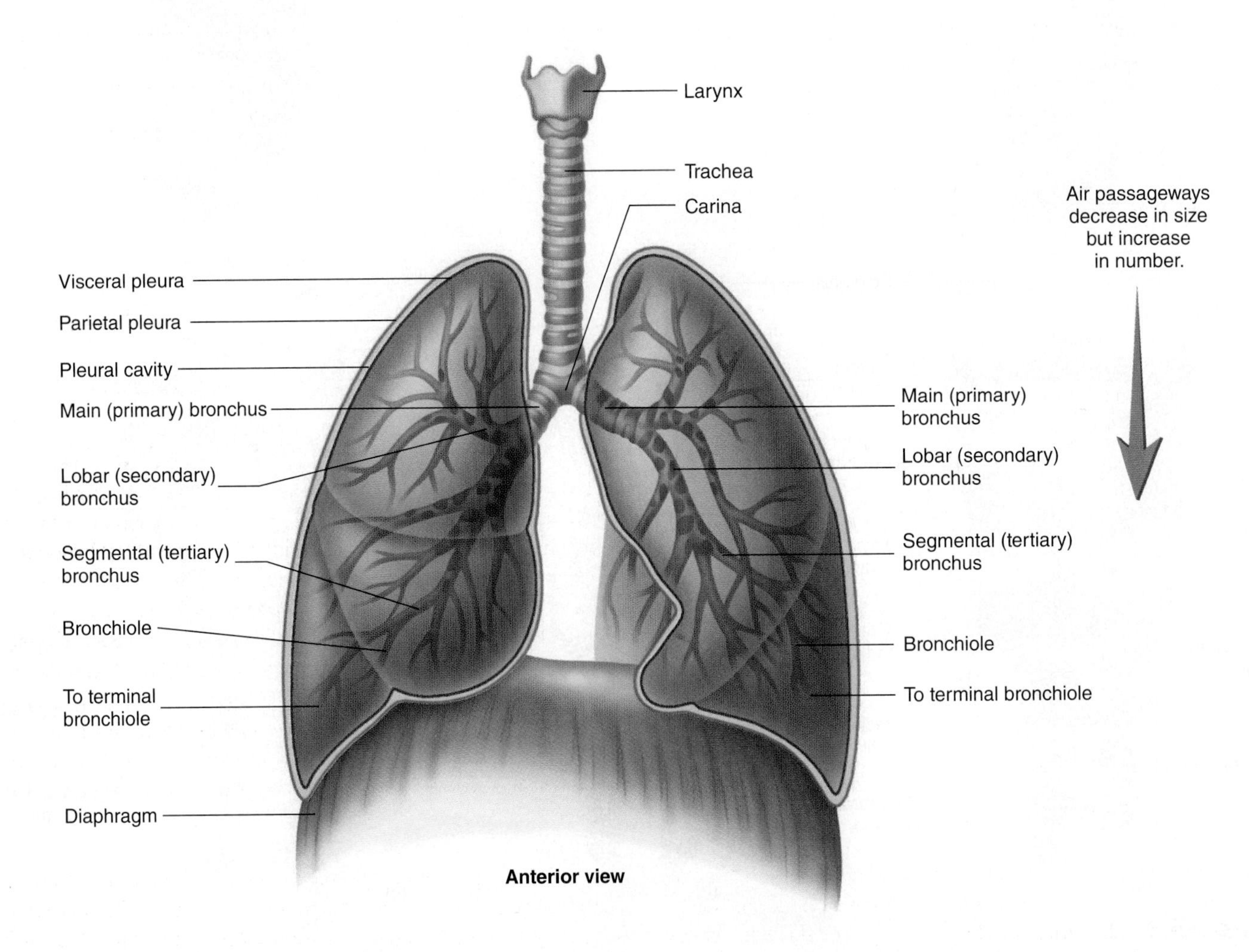

Figure 35.4 Trachea and Bronchial Tree, Anterior View

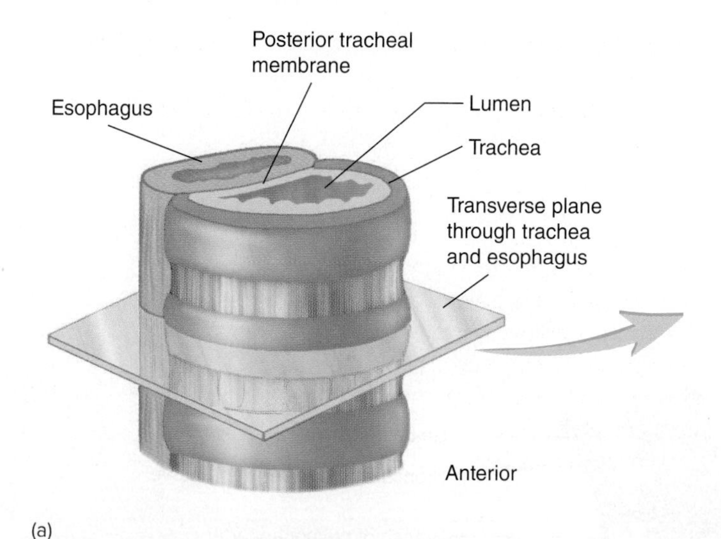

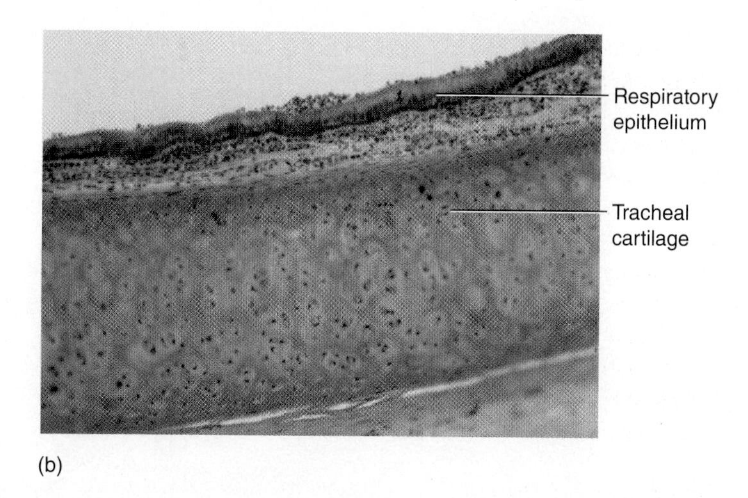

Figure 35.5 Trachea, Cross Section
(a) Diagram; (b) photomicrograph (100x).

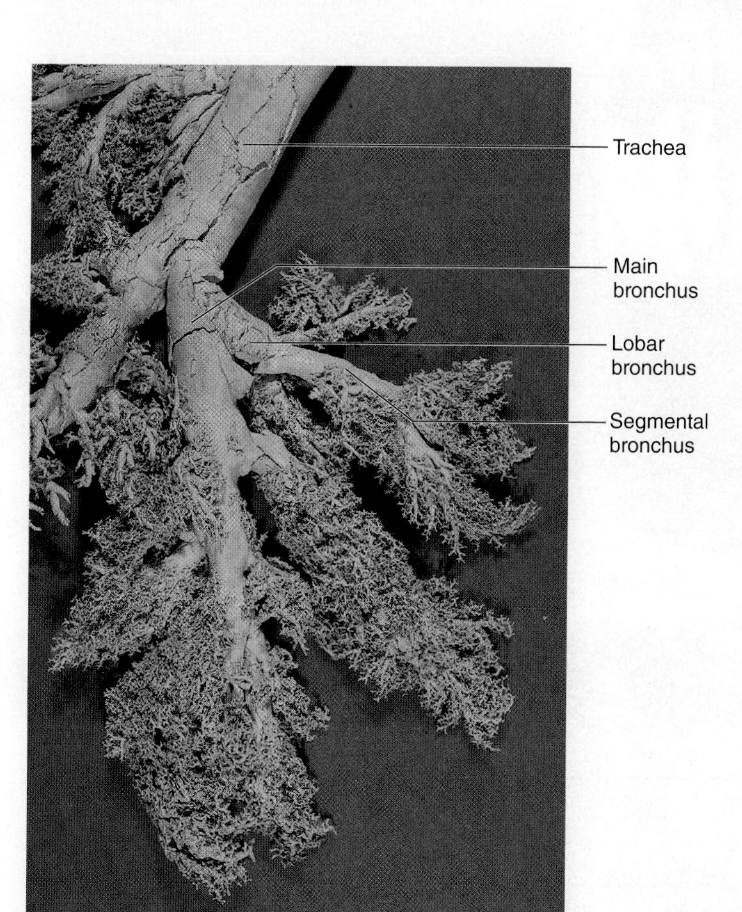

Figure 35.6 Cast of Human Bronchial Tree

The lungs are located in the **pleural cavities,** one on each side of the mediastinum. The **parietal pleura** is the outer membrane on the chest cavity wall, and the membrane on the surface of the lungs is the **visceral pleura.** The space between the membranes is known as the pleural cavity. These are shown in figure 35.7.

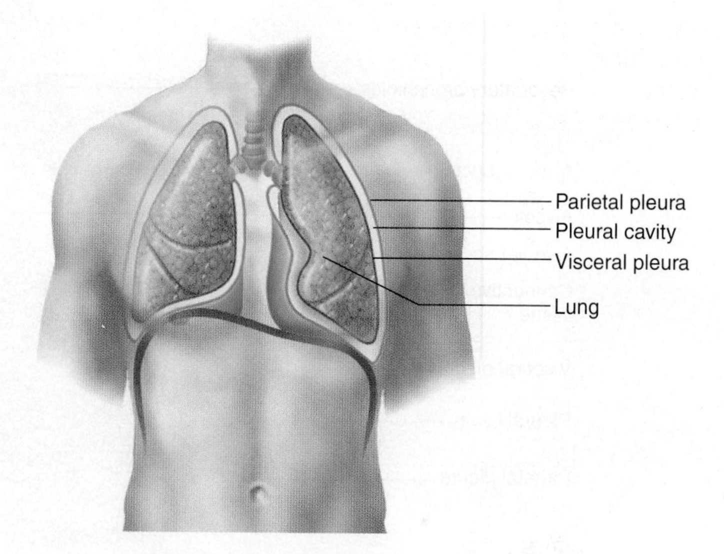

Figure 35.7 Pleural Membranes and Cavity

Histology of the Lung

The bronchi continue to divide until they become **bronchioles,** which are small respiratory tubules with smooth muscle in their walls. The bronchioles in the lung further divide into **respiratory bronchioles,** which are so named because of the small structures, called alveoli, attached to their walls. The respiratory bronchioles lead to passageways known as **alveolar ducts,** which branch into alveoli. **Alveoli** are air sacs in the lung that exchange oxygen and carbon dioxide with the blood capillaries that are adjacent to them. Obtain a prepared slide of lung and scan first under low power and then under higher powers.

(A)[7] Examine your slide and locate the bronchi, bronchioles, respiratory bronchioles, alveolar ducts, and alveoli. Alveoli that are clustered around an alveolar duct are collectively known as **alveolar sacs** (figure 35.8 and 35.9).

Type I pneumocytes make up about 90% of the alveoli and are composed of simple squamous epithelium. The division of the lung into many small sacs tremendously increases the surface area of the lung. This increase is vital for the rapid and extensive

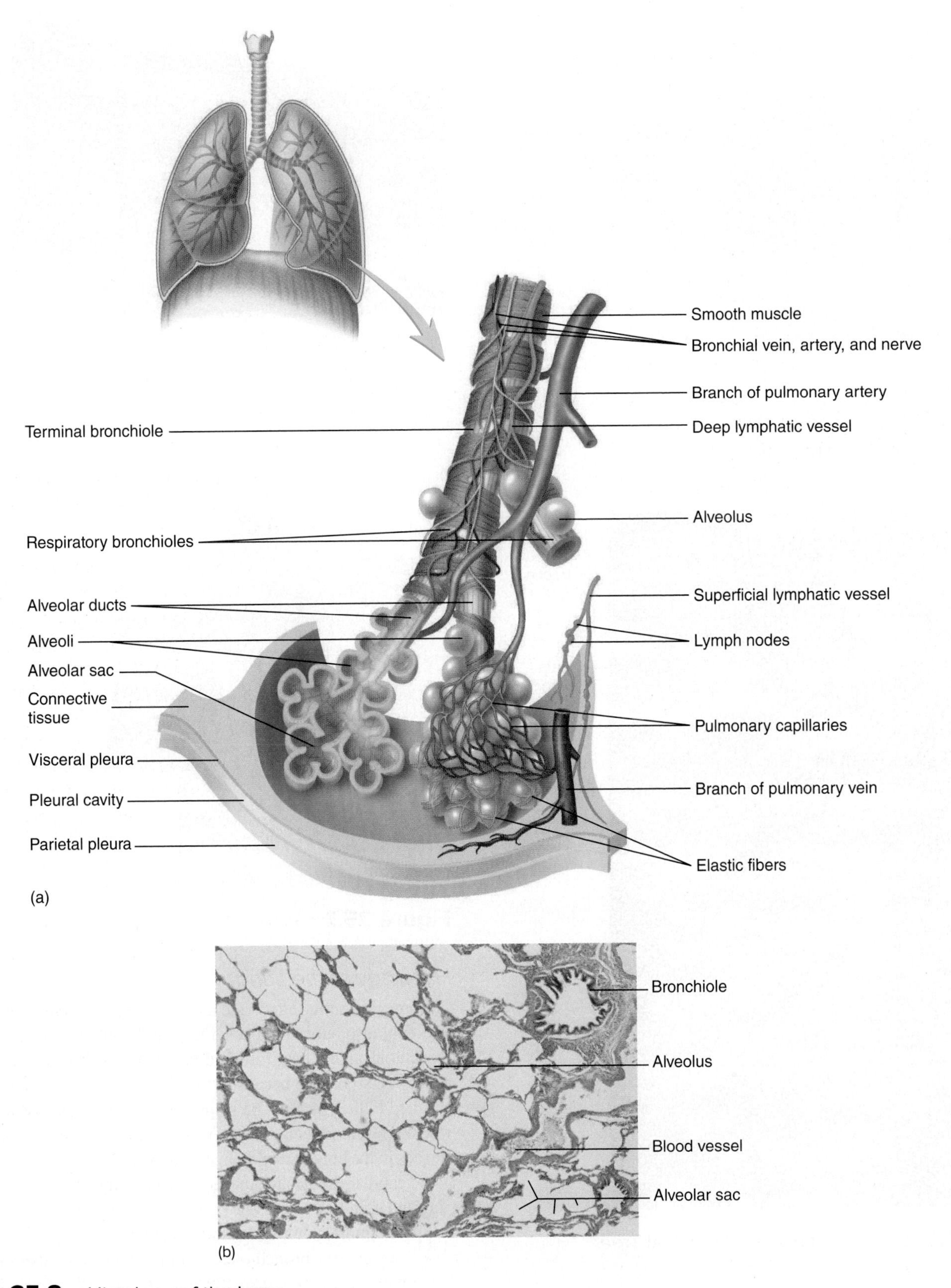

Figure 35.8 Histology of the Lung

(a) Diagram; (b) photomicrograph of lung (40×).

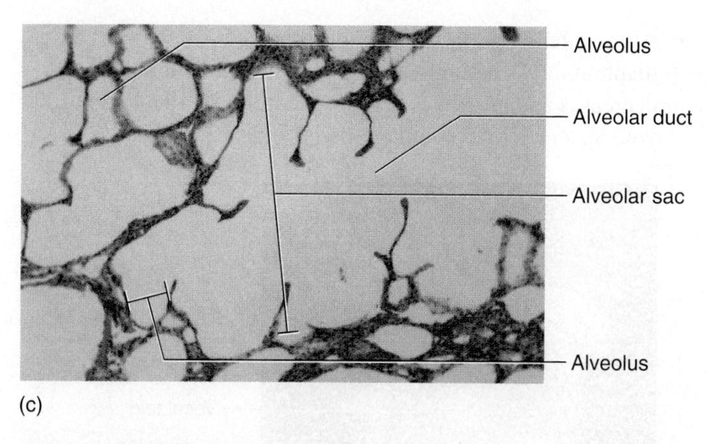

(c)

Alveolus

(d)

Figure 35.8 *Continued.*

(c) Photomicrograph of alveolar duct (100×); (d) photomicrograph of alveolus (400×).

diffusion of oxygen across the respiratory membranes. If the lungs were just two open spaces (like shoe boxes), you would not have enough surface area to breathe. By adding many small boxes to the model of the shoe box lungs and connecting them with tubes, you increase the surface area dramatically, increasing the diffusion of oxygen across the lung surface. Oxygen moves across the **respiratory membrane,** which consists of the alveolus, the capillary epithelium, and the basement membrane between the two.

You may see other cell shapes in the prepared lung sections. Some of these cells are **type II pneumocytes (septal cells);** they decrease the surface tension of the lung by secreting **surfactant.** These are seen in figure 35.9.

(A) 8 Examine a prepared slide of smoker's lung. Note the dark material in the lung tissue and the general destruction of the alveoli. Breakdown of the alveoli leads to a disease known as **emphysema,** which is a severe reduction in the surface area of the lung. Oxygen diffusion is reduced as a result.

Cat Dissection

(A) 9 Prepare your cat for dissection and remember to place all excess tissue in the appropriate waste container and not in a standard wastebasket or down the sink!

1. If you have not opened the chest cavity in your study of the cat, then you should do so now. Removal of the skin is discussed in Exercise 12, and the procedure for opening the thoracic cavity is described in Exercise 30.
2. Locate the **larynx** of the cat above the **trachea.** Notice the broad, wedge-shaped structure in the front. This is the **thyroid cartilage.**
3. Make a midsagittal incision through the thyroid cartilage and continue carefully cutting until you have cut completely through the larynx. To examine the larynx more completely, you may wish to continue your midsagittal cut partway through the trachea.

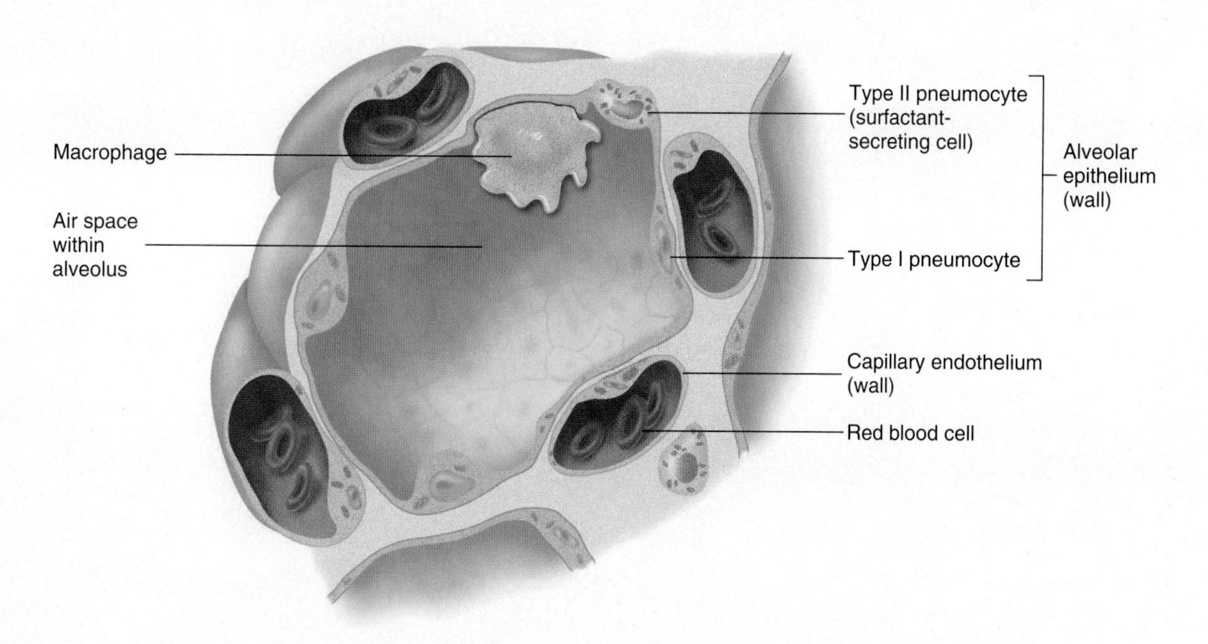

Figure 35.9 Details of a Single Alveolus

4. Open the larynx and find the **epiglottis** of the cat. Notice how the epiglottis is lighter in color than the thyroid cartilage. This is because the epiglottis is made of elastic cartilage, while the thyroid cartilage is composed of hyaline cartilage.
5. Find the **cricoid cartilage,** the **arytenoid cartilages,** and the **vestibular fold** and **vocal folds** (see figure 35.3 and 35.10).
6. Examine the trachea as it passes from the larynx and into the thoracic cavity. Ask your instructor for permission before you cut the trachea in cross section. If you do, you should see the **tracheal cartilages** and the **posterior tracheal membrane.** The posterior portion of the trachea is located ventrally to the esophagus. Notice how the trachea splits into the two **bronchi,** which then enter the lungs. The lobes of the lungs are different in the cat than in the human. The right lung in the cat has four lobes, while the left lung has three lobes. How does this compare with the pattern in humans?
7. Examine the structure of the lungs in your specimen and compare it with figure 35.11. The lungs are covered with a thin serous membrane called the **visceral pleura,** while the outer covering (on the deep surface of the ribs and intercostal muscles) is called the **parietal pleura.** The space between these two membranes is the **pleural cavity.**
8. Cut into one of the lungs of the cat and examine the lung tissue. Note how the lungs appear as a very fine mesh sponge. The alveoli of the lungs are microscopic.
9. At the inferior portion of the thoracic cavity is the diaphragm. As it contracts, the pressure in the thoracic cavity decreases and air fills the lungs. Examine the diaphragm in your specimen and compare it with figure 35.11.

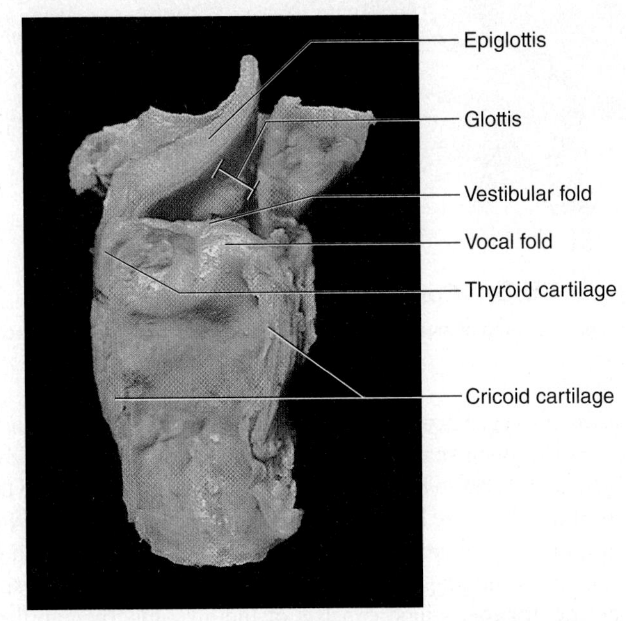

Figure 35.10 Larynx of the Cat, Midsagittal Section

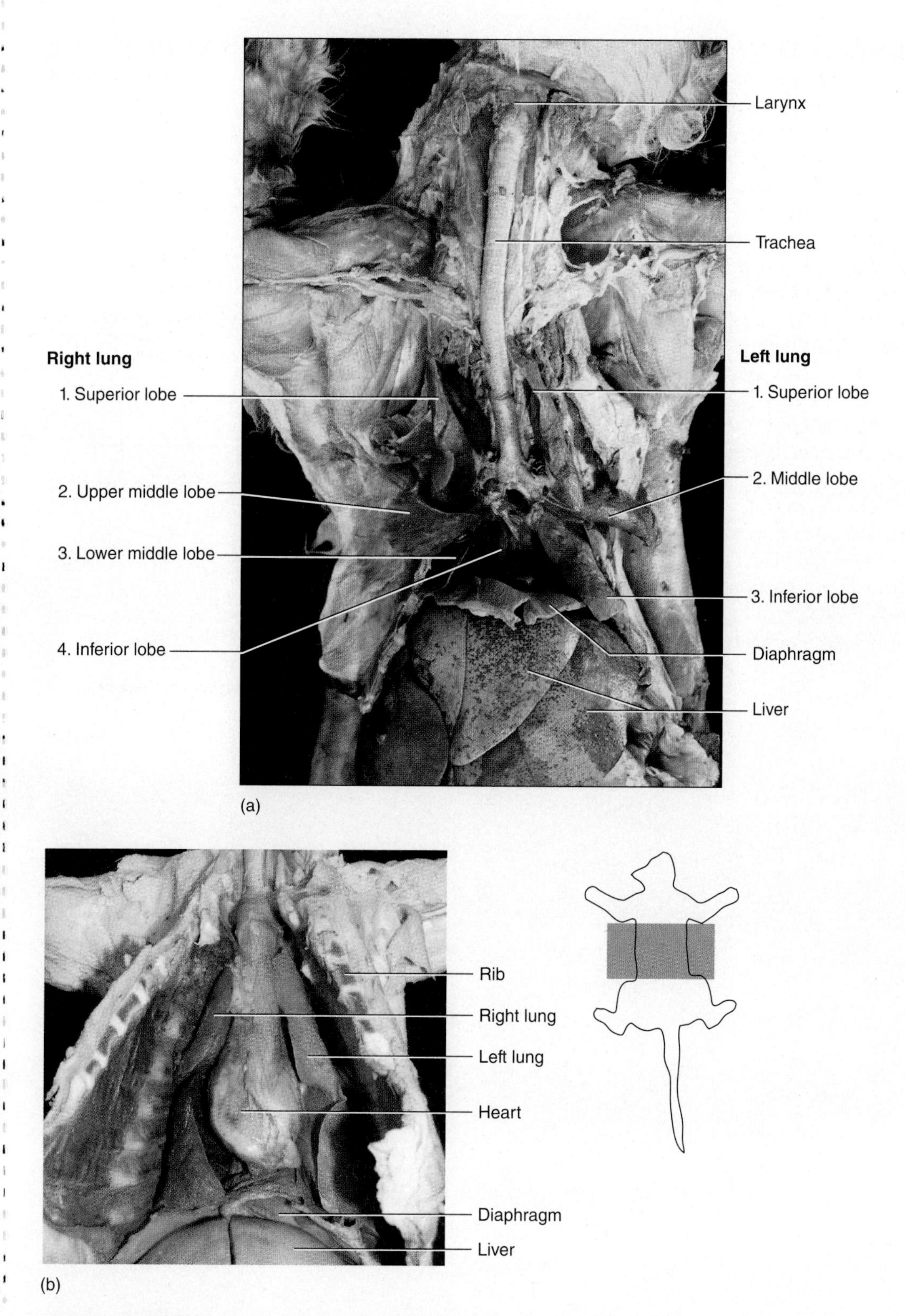

Figure 35.11 Respiratory System of the Cat

(a) Overview; (b) lungs in thoracic cavity.

NOTES

Exercise 35 Review

Structure of the Respiratory System

Name: ______________________

Lab time/section: ______________________

Date: ______________________

1. What is the common name for the external nares? ______________________

2. The nasal cartilages are made of hyaline cartilage. What functional adaptation does cartilage have over bone in making up the external framework of the nose? ______________________

3. The nasal cavities are separated from each other by what structure? ______________________

4. Three structures make up the nasal septum. What are they? ______________________

5. What is the function of respiratory epithelium and the superficial blood vessels in the nasal cavity? ______________________

6. Name the openings between the nasal cavity and the pharynx. ______________________

7. What is the name of the space behind the oral cavity and above the laryngopharynx? ______________________

8. What is the name of the structure that prevents fluid from entering the nasopharynx during swallowing? ______________________

9. What is the name of the large cartilage of the anterior larynx? ______________________

10. What is the structure that protects the glottis from fluid entering the larynx? ______________________

11. Which lung has just two lobes in the human? ______________________

12. What membrane attaches directly to the lungs? ______________________________

13. The trachea branches into two tubes that go to the lungs. What are these tubes called? ______________________________

14. Where is the tracheobronchial tree located? ______________________________

15. What small structure in the lung is the site of oxygen exchange with the blood capillaries? ______________________________

16. The surface area of the lungs in humans is about 70 square meters. How can this be so if the lungs are located in the small space of the thoracic cavity? What role do alveoli play in the nature of surface area? ______________________________

____

17. Emphysema is a destruction of the alveoli of the lungs. What effect does this have on the surface area of the lungs? ______________________________

____

____

18. Label the following illustration using the terms provided.

cricoid cartilage	epiglottis	inferior lobe	internal nares	main bronchus
middle lobe	nasal cavity	superior lobe	trachea	

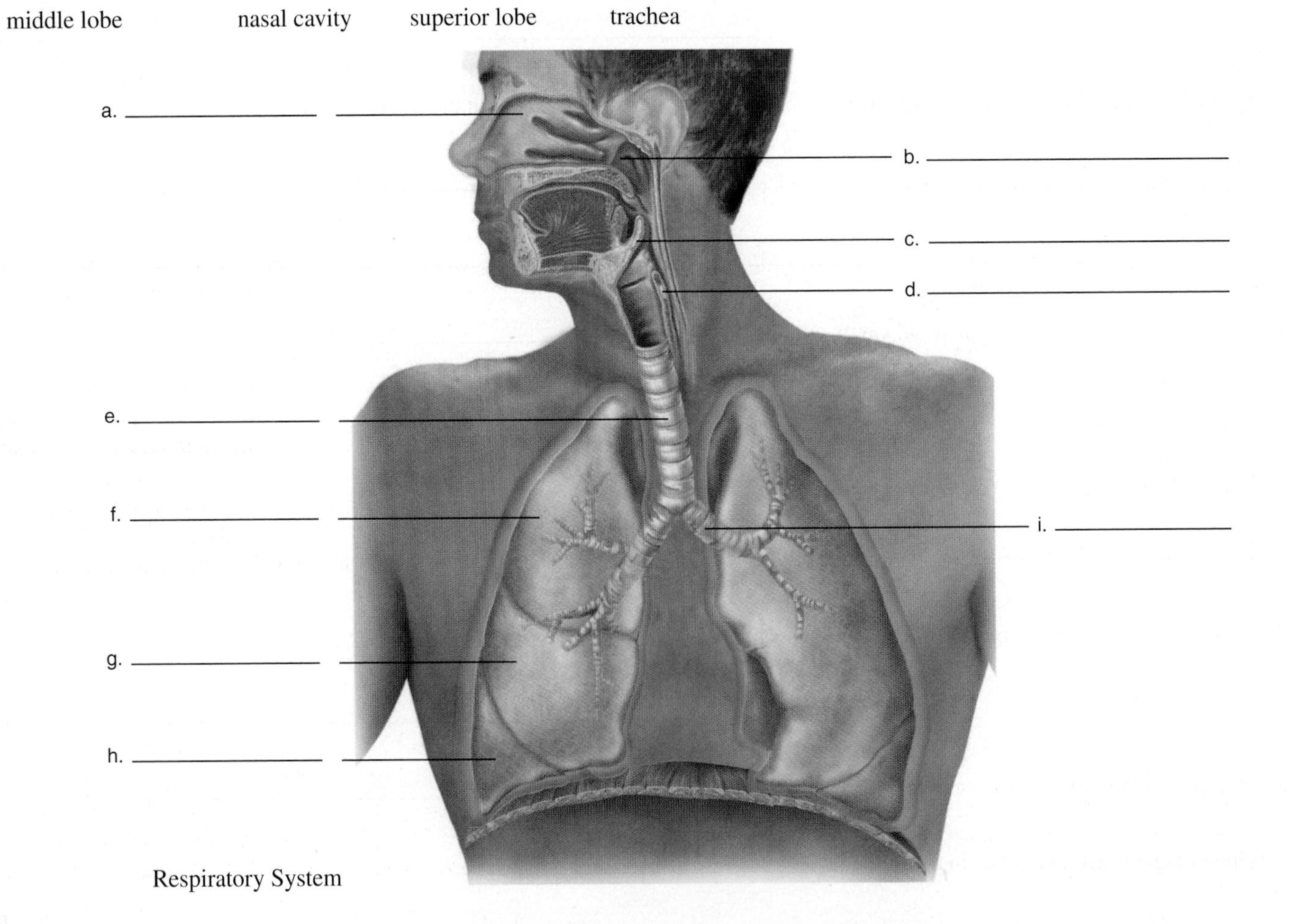

Respiratory System

Exercise 36

Respiratory Function, Breathing, and Respiration

Introduction

The function of the respiratory system is to exchange gases between the external air and the blood. Oxygen diffuses from the air into the blood of the lungs, and carbon dioxide diffuses from the blood into the air. The process can be divided into **pulmonary ventilation,** the mechanical process of moving air into and out of the lungs, and the **exchange of gases** across the respiratory membrane. Oxygen requirements vary with the body's needs. Variations in the loading of oxygen by the lungs can be accomplished by increasing/decreasing the volume of air with each inhalation, the breathing rate, or both. Removal of carbon dioxide can also be accomplished by an increase in volume or breathing rate. This is important in maintaining an appropriate acid-base balance in the blood. These topics are discussed in the Seeley text in chapter 23, "Respiratory System." In this exercise, you measure breathing rates, lung volumes and capacities, and the effects of carbon dioxide on the acid-base balance of a solution.

Objectives

At the end of this exercise, you should be able to

1. measure and understand the relationship of pulmonary volumes and capacities, including vital capacity, tidal volume, inspiratory reserve volume, and expiratory reserve volume;
2. describe the relationship between oxygen and carbon dioxide levels in the blood and breathing rates;
3. describe the mechanical process of breathing;
4. describe the use of the spirometer or airflow transducer available in your lab;
5. demonstrate the use of the stethoscope in obtaining respiratory sounds;
6. predict the flow of air in an individual, with changes in air pressure between the lungs and the external air or with changes in the resistance in the respiratory passages;
7. identify the tidal volume, expiratory reserve volume, and vital capacity on a spirogram.

Materials

Respiration Model
Bell jar respiration model

Pulmonary Volume Setup
BIOPAC data acquisition unit and air transducer, wet spirometer, or handheld spirometers
Disposable mouthpieces to fit respirometer or spirometers
Noseclips
Watch or clock with accuracy in seconds
Biohazard bag
AFT6—600 mL calibration syringe
AFT1—disposable bacteriological filter
AFT2 disposable mouthpiece
SS11LA airflow transducer

Breathing Sounds and Breathing Rate Setup

Stethoscope
Alcohol wipes

Acid-Base Setup (One Setup per Table)

Litmus solution (2 g litmus powder in 600 mL water)
NaOH solution (1 N) in dropper bottles
Straws
100 mL Erlenmeyer flasks
Tape or parafilm®
Safety glasses

Procedure

Mechanics of Breathing

Air moves from regions of higher pressure to regions of lower pressure. The lungs fill with air or deflate due to changes in air pressure. Normal atmospheric air pressure (barometric pressure) is measured in millimeters of mercury (mmHg), and standard air pressure at sea level is 760 mmHg. When there is no movement of air into or out of the lungs, the pressure in the lungs, (alveolar air pressure) is equal to that of the surrounding air, as illustrated in figure 36.1a. During inspiration, the diaphragm contracts, increasing the volume in the thoracic cavity. This leads to a decrease in the pressure in the lungs, and air moves from the atmosphere into the lungs. This is illustrated in figure 36.1b. If the diaphragm relaxes, then the abdominal pressure forces the diaphragm upwards and increases the pressure in the thoracic cavity beyond that of the atmospheric pressure, and the air moves out of the lungs, as illustrated in figure 36.1c. You can use a bell jar model to demonstrate this in lab.

(A)[1] This model illustrates the movement of air into or out of the lungs, depending on air pressure differentials between the lungs and the external environment. The glass housing represents the thoracic cage; the balloons represent the lungs; the latex sheeting represents the diaphragm; and the rubber stopper represents the nose. Pull *gently* on the latex diaphragm.

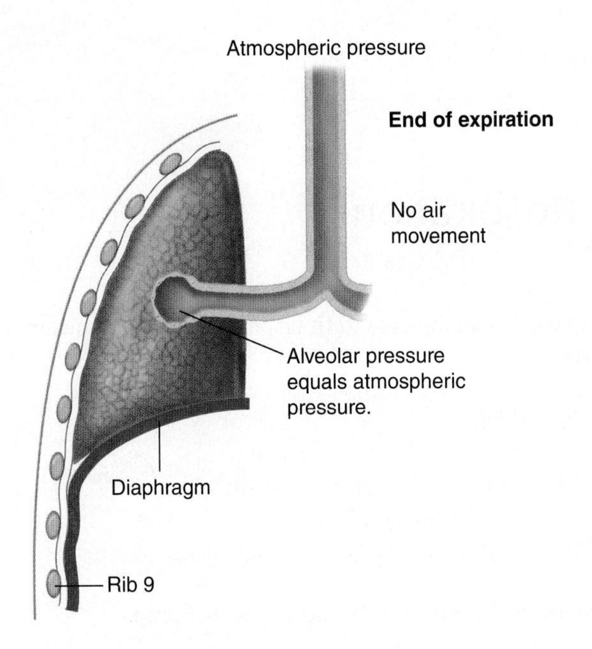

(a) At the end of expiration, alveolar pressure is equal to atmospheric pressure, and there is no air movement.

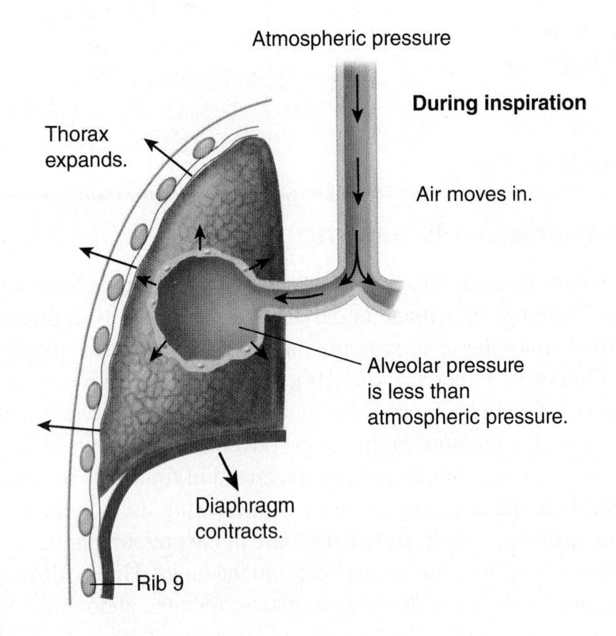

(b) During inspiration, increased thoracic volume results in increased alveolar volume and decreased alveolar pressure. Atmospheric pressure is greater than alveolar pressure, and air moves into the lungs.

Figure 36.1 Mechanics of Breathing

(a) No movement of air; (b) inspiration; (c) expiration.

As you pull down gently on the latex diaphragm, does the thoracic volume increase or decrease?

What effect does this have on the pressure in the balloon "lungs"?

Where does the air move because of this effect?

Why does the air stop moving?

As you allow the diaphragm to relax by gently moving it back to its original position, what happens to the space between the balloon "lungs" and the glass "thoracic cage"?

What effect does this have on the pressure in the balloon "lungs"?

Where does the air move because of this effect?

What causes the lungs to stop deflating?

Although the mechanics of human breathing differ somewhat from this model, the bell jar model is valuable to demonstrate how changes in pressure cause changes in airflow.

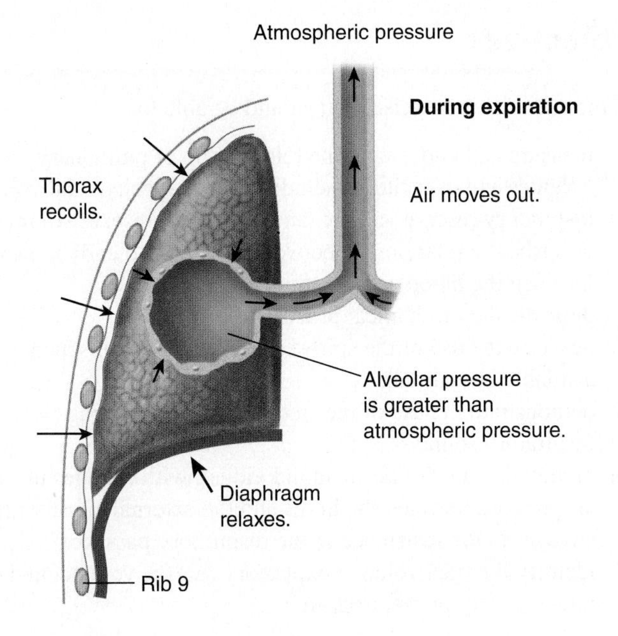

(c) During expiration, decreased thoracic volume results in decreased alveolar volume and increased alveolar pressure. Alveolar pressure is greater than atmospheric pressure, and air moves out of the lungs.

Measurement of Relaxed Breathing Rate

(A)[2] It is important that your lab partner be distracted from thinking about breathing.

1. Therefore, your lab partner should read the remainder of the laboratory exercise while you count the number of breaths he or she takes for a total of 2 minutes.
2. Divide the number by 2 to calculate the average number of breaths per minute.
3. Record that number for your lab partner in the space provided.

Breaths per minute: ____________________

Breathing rate varies with oxygen demand and carbon dioxide levels in the blood. Before doing any exercise, *estimate* the breaths per minute your lab partner might take after completing 2 minutes of strenuous exercise. Record your estimation.

Estimation: ____________________

Caution!

If you have a heart condition, a family history of heart failure, or another medical condition that prevents you from doing strenuous exercise, then do not do the exercise sections of this lab.

Have your lab partner do strenuous exercise for 2 minutes (jumping jacks, running in place or outside of lab, running up and down stairs, etc.). As soon as your lab partner is finished, record the number of breaths per minute for the *first minute* after exercise and write this number in the following space.

Number of breaths per minute after 2 minutes of exercise:

____

How does this compare with your estimation?

____

Measurement of Pulmonary Volumes and Capacities

Pulmonary volumes are the amount of air that flows into or out of the lungs during a particular event. **Capacities** are the summation of volumes. There are many different ways to record pulmonary volumes. An affordable way to measure pulmonary volume is to use a handheld spirometer in which the exhaled air spins the vanes of the spirometer, estimating the volume of air exhaled. This is like the anemometer used by weather stations to measure wind speed. These are relatively inexpensive pieces of equipment and are good for measuring vital capacity, but they are not as accurate as other equipment. The wet spirometer measures the volume of air exhaled into a chamber, so it measures the actual amount of air exhaled. Some digital recorders, such as those made by BIOPAC, use a flow meter to estimate volume.

Caution!

You should never inhale using a spirometer.

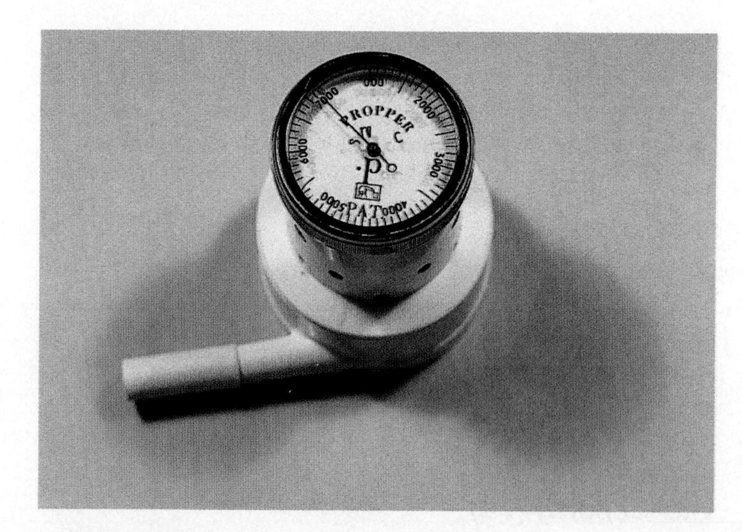

Figure 36.2 Handheld Spirometer

Tidal Volume with Handheld Spirometer

(A)[3]

1. Place a disposable mouthpiece on the spirometer tube.
2. Make sure you set the indicator dial to zero by twisting the knurled ring on the top of the spirometer (zero may be the same as the maximum volume, 7,000 in some spirometers; figure 36.2).
3. As you exhale, estimate what a normal breath volume will be and exhale this amount rather forcefully. Do not exhale more than what you would for a normal breath. If you breathe gently into a handheld spirometer, you may not cause the vanes to spin enough to get a significant recording.
4. Exhale five total breaths (without resetting the spirometer to zero between breaths) and record your results.

Total of five breaths recorded by spirometer: ____________

Divide the total number by 5 and enter the average tidal volume.

Average tidal volume: ____________________

You may find that the tidal volume is variable among members of your class. This is due to the difficulty of measuring tidal volumes with a standard lab apparatus. The average tidal volume is about 500 mL.

Tidial Volume with a Wet Spirometer

(A)[4] The tidal volume can be measured if your wet spirometer is accurate. Place a disposable mouthpiece into the flexible hose and breathe a normal breath into the mouthpiece. Record your data in the following space. Throw the disposable mouthpiece into the biohazard container when you are finished.

Tidal volume: ____________________

Expiratory Reserve Volume

(A)[5] You can determine your **expiratory reserve volume (ERV)** with the wet spirometer or the handheld spirometer. The expiratory reserve volume is the maximal amount of air you can exhale after a *normal exhalation.* This is typically around 1,100 mL.

1. Make sure to close off your nostrils and, after a normal exhalation, forcibly expel the remainder of your breath through the mouthpiece into the spirometer.
2. Repeat this two more times for a total of three exhalations, and calculate the average expiratory reserve by dividing the sum of the volumes by 3.
3. Record your results.

Trial 1: ______________________

Trial 2: ______________________

Trial 3: ______________________

Average expiratory reserve volume: ______________

Vital Capacity (VC)

(A) 6 The vital capacity (VC) is the total volume of air that can be forcefully expelled from the lungs after a maximum inhalation. Measuring vital capacity is like participating in the national championship of exhalation.

1. Measure the vital capacity with the use of a wet spirometer or handheld spirometer by first breathing in as deeply as you possibly can.
2. Close your nostrils, and then exhale through the mouthpiece completely until you cannot exhale anymore.
3. Force as much air from your lungs as you can.
4. Record your vital capacity.

Vital capacity: ______________________

Lung volumes are illustrated in figure 36.3.

Percent of Normal Vital Capacity

(A) 7 You can compare your vital capacity with those of other individuals of your sex, height, and age by using one of the two charts (one for females, one for males) in table 36.1. As you examine the chart for a person of your height, notice that the vital capacity decreases with age.

Calculate your percent of normal vital capacity by dividing your data by the normal data for a person of your age, sex, and height and multiplying the result by 100.

$$\frac{\text{Your vital capacity}}{\text{Normal vital capacity}} \times 100 = \text{Percent of normal vital capacity}$$

Your percent of normal vital capacity: ______________

If you come out at 100% of normal, then you have an average vital capacity of individuals in your bracket. If your value is less than 100%, then you have a smaller vital capacity than normal for your bracket. A percent vital capacity of 80% of average, or better, is in the normal range. If your value is larger than 100%, then you have a larger vital capacity than average.

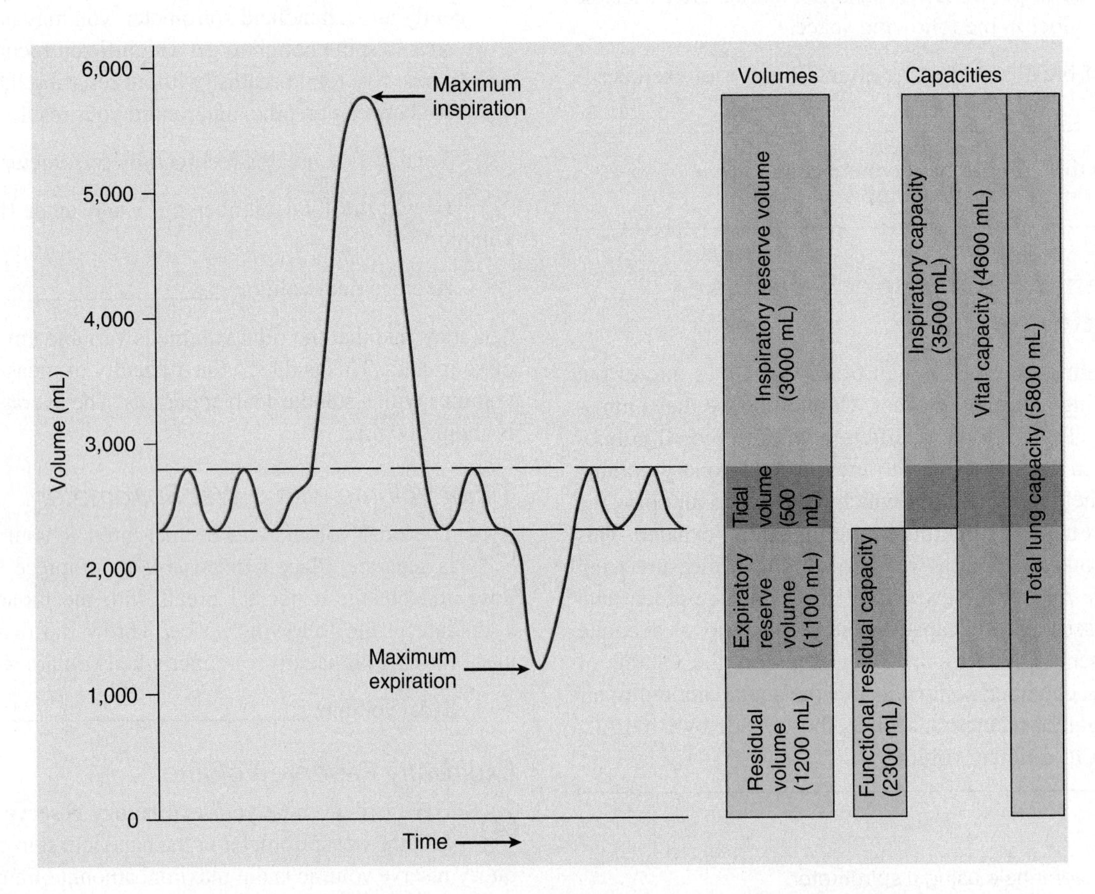

Figure 36.3 Pulmonary Volumes and Capacities

TABLE 36.1 Predicted Vital Capacities for Females and Males

Females

Height in Centimeters and Inches

	cm	152	154	156	158	160	162	164	166	168	170	172	174	176	178	180	182	184	186	188
Age	in.	**59.8**	**60.6**	**61.4**	**62.2**	**63.0**	**63.7**	**64.6**	**65.4**	**66.1**	**66.9**	**67.7**	**68.5**	**69.3**	**70.1**	**70.9**	**71.7**	**72.4**	**73.2**	**74.0**
16		3070	3110	3150	3190	3230	3270	3310	3350	3390	3430	3470	3510	3550	3590	3630	3670	3715	3755	3800
17		3055	3095	3135	3175	3215	3255	3295	3335	3375	3415	3455	3495	3535	3575	3615	3655	3695	3740	3780
18		3040	3080	3120	3160	3200	3240	3280	3320	3360	3400	3440	3480	3520	3560	3600	3640	3680	3720	3760
20		3010	3050	3090	3130	3170	3210	3250	3290	3330	3370	3410	3450	3490	3525	3565	3605	3645	3695	3720
22		2980	3020	3060	3095	3135	3175	3215	3255	3290	3335	3370	3410	3450	3490	3530	3570	3610	3650	3685
24		2950	2985	3025	3065	3100	3140	3180	3220	3260	3300	3335	3375	3415	3455	3490	3530	3570	3610	3650
26		2920	2960	3000	3035	3070	3110	3150	3190	3230	3265	3300	3340	3380	3420	3455	3495	3530	3570	3610
28		2890	2930	2965	3000	3040	3070	3115	3155	3190	3230	3270	3305	3345	3380	3420	3460	3495	3535	3570
30		2860	2895	2935	2970	3010	3045	3085	3120	3160	3195	3235	3270	3310	3345	3385	3420	3460	3495	3535
32		2825	2865	2900	2940	2975	3015	3050	3090	3125	3160	3200	3235	3275	3310	3350	3385	3425	3460	3495
34		2795	2835	2870	2910	2945	2980	3020	3055	3090	3130	3165	3200	3240	3275	3310	3350	3385	3425	3460
36		2765	2805	2840	2875	2910	2950	2985	3020	3060	3095	3130	3165	3205	3240	3275	3310	3350	3385	3420
38		2735	2770	2810	2845	2880	2915	2950	2990	3025	3060	3095	3130	3170	3205	3240	3275	3310	3350	3385
40		2705	2740	2775	2810	2850	2885	2920	2955	2990	3025	3060	3095	3135	3170	3205	3240	3275	3310	3345
42		2675	2710	2745	2780	2815	2850	2885	2920	2955	2990	3025	3060	3100	3135	3170	3205	3240	3275	3310
44		2645	2680	2715	2750	2785	2820	2855	2890	2925	2960	2995	3030	3060	3095	3130	3165	3200	3235	3270
46		2615	2650	2685	2715	2750	2785	2820	2855	2890	2925	2960	2995	3030	3060	3095	3130	3165	3200	3235
48		2585	2620	2650	2685	2715	2750	2785	2820	2855	2890	2925	2960	2995	3030	3060	3095	3130	3160	3195
50		2555	2590	2625	2655	2690	2720	2755	2785	2820	2855	2890	2925	2955	2990	3025	3060	3090	3125	3155
52		2525	2555	2590	2625	2655	2690	2720	2755	2790	2820	2855	2890	2925	2955	2990	3020	3055	3090	3125
54		2495	2530	2560	2590	2625	2655	2690	2720	2755	2790	2820	2855	2885	2920	2950	2985	3020	3050	3085
56		2460	2495	2525	2560	2590	2625	2655	2690	2720	2755	2790	2820	2855	2885	2920	2950	2980	3015	3045
58		2430	2460	2495	2525	2560	2590	2625	2655	2690	2720	2750	2785	2815	2850	2880	2920	2945	2975	3010
60		2400	2430	2460	2495	2525	2560	2590	2625	2655	2685	2720	2750	2780	2810	2845	2875	2915	2940	2970
62		2370	2405	2435	2465	2495	2525	2560	2590	2620	2655	2685	2715	2745	2775	2810	2840	2870	2900	2935
64		2340	2370	2400	2430	2465	2495	2525	2555	2585	2620	2650	2680	2710	2740	2770	2805	2835	2865	2895
66		2310	2340	2370	2400	2430	2460	2495	2525	2555	2585	2615	2645	2675	2705	2735	2765	2800	2825	2860
68		2280	2310	2340	2370	2400	2430	2460	2490	2520	2550	2580	2610	2640	2670	2700	2730	2760	2795	2820
70		2250	2280	2310	2340	2370	2400	2425	2455	2485	2515	2545	2575	2605	2635	2665	2695	2725	2755	2780
72		2220	2250	2280	2310	2335	2365	2395	2425	2455	2480	2510	2540	2570	2600	2630	2660	2685	2715	2745
74		2190	2220	2245	2275	2305	2335	2360	2390	2420	2450	2475	2505	2535	2565	2590	2620	2650	2680	2710

source: "Predicted Vital Capacities for Females and Males (tables)" by E. A. Gaensler, M.D. and G. W. Wright, M.D., AEH, Vol. 12, pp. 146–189, February 1966.

TABLE 36.1 Continued

	Males Height in Centimeters and Inches																			
	cm	152	154	156	158	160	162	164	166	168	170	172	174	176	178	180	182	184	186	188
Age	in.	59.8	60.6	61.4	62.2	63.0	63.7	64.6	65.4	66.1	66.9	67.7	68.5	69.3	70.1	70.9	71.7	72.4	73.2	74.0
16		3920	3975	4025	4075	4130	4180	4230	4285	4335	4385	4440	4490	4540	4590	4645	4695	4745	4800	4850
18		3890	3940	3995	4045	4095	4145	4200	4250	4300	4350	4405	4455	4505	4555	4610	4660	4710	4760	4815
20		3860	3910	3960	4015	4065	4115	4165	4215	4265	4320	4370	4420	4470	4520	4570	4625	4675	4725	4775
22		3830	3880	3930	3980	4030	4080	4135	4185	4235	4285	4335	4385	4435	4485	4535	4585	4635	4685	4735
24		3785	3835	3885	3935	3985	4035	4085	4135	4185	4235	4285	4330	4380	4430	4480	4530	4580	4630	4680
26		3755	3805	3855	3905	3955	4000	4050	4100	4150	4200	4250	4300	4350	4395	4445	4495	4545	4595	4645
28		3725	3775	3820	3870	3920	3970	4020	4070	4115	4165	4215	4265	4310	4360	4410	4460	4510	4555	4605
30		3695	3740	3790	3840	3890	3935	3985	4035	4080	4130	4180	4230	4275	4325	4375	4425	4470	4520	4570
32		3665	3710	3760	3810	3855	3905	3950	4000	4050	4095	4145	4195	4240	4290	4340	4385	4435	4485	4530
34		3620	3665	3715	3760	3810	3855	3905	3950	4000	4045	4095	4140	4190	4245	4285	4330	4380	4425	4475
36		3585	3635	3680	3730	3775	3825	3870	3920	3965	4010	4060	4105	4155	4200	4250	4295	4340	4390	4435
38		3555	3605	3650	3695	3745	3790	3840	3885	3930	3980	4025	4070	4120	4165	4210	4260	4305	4350	4400
40		3525	3575	3620	3665	3710	3760	3805	3850	3900	3945	3990	4035	4085	4130	4175	4220	4270	4315	4360
42		3495	3540	3590	3635	3680	3725	3770	3820	3865	3910	3955	4000	4050	4095	4140	4185	4230	4280	4325
44		3450	3495	3540	3585	3630	3675	3725	3770	3815	3860	3905	3950	3995	4040	4085	4130	4175	4220	4270
46		3420	3465	3510	3555	3600	3645	3690	3735	3780	3825	3870	3915	3960	4005	4050	4095	4140	4185	4230
48		3390	3435	3480	3525	3570	3615	3655	3700	3745	3790	3835	3880	3925	3970	4005	4060	4105	4150	4190
50		3345	3390	3430	3475	3520	3565	3610	3650	3695	3740	3785	3830	3870	3915	3960	4005	4050	4090	4135
52		3315	3355	3400	3445	3490	3530	3575	3620	3660	3705	3750	3795	3835	3880	3925	3970	4010	4055	4100
54		3285	3325	3370	3415	3455	3500	3540	3585	3630	3670	3715	3760	3800	3845	3890	3930	3975	4020	4060
56		3255	3295	3340	3380	3425	3465	3510	3550	3595	3640	3680	3725	3765	3810	3850	3895	3940	3980	4025
58		3210	3250	3290	3335	3375	3420	3460	3500	3545	3585	3630	3670	3715	3755	3800	3840	3880	3925	3965
60		3175	3220	3260	3300	3345	3385	3430	3470	3500	3555	3595	3635	3680	3720	3760	3805	3845	3885	3930
62		3150	3190	3230	3270	3310	3350	3390	3440	3480	3520	3560	3600	3640	3680	3730	3770	3810	3850	3890
64		3120	3160	3200	3240	3280	3320	3360	3400	3440	3490	3530	3570	3610	3650	3690	3730	3770	3810	3850
66		3070	3110	3150	3190	3230	3270	3310	3350	3390	3430	3470	3510	3550	3600	3640	3680	3720	3760	3800
68		3040	3080	3120	3160	3200	3240	3280	3320	3360	3400	3440	3480	3520	3560	3600	3640	3680	3720	3760
70		3010	3050	3090	3130	3170	3210	3250	3290	3330	3370	3410	3450	3480	3520	3560	3600	3640	3680	3720
72		2980	3020	3060	3100	3140	3180	3210	3250	3290	3330	3370	3410	3450	3490	3530	3570	3610	3650	3680
74		2930	2970	3010	3050	3090	3130	3170	3200	3240	3280	3320	3360	3400	3440	3470	3510	3550	3590	3630

Calculation of the Inspiratory Reserve Volume (IRV)

(A) 8 The vital capacity consists of the expiratory reserve volume, the tidal volume, and the inspiratory reserve volume. You can *indirectly* determine the **inspiratory reserve volume (IRV)** by subtracting the expiratory reserve volume (ERV) and the tidal volume (TV) from the vital capacity (VC). This is indicated in figure 36.3. You cannot measure the inspiratory reserve volume directly with the use of a handheld spirometer (it records exhalations only). An average IRV is around 3,000 mL. Calculate your inspiratory reserve.

$$IRV = VC - (ERV + TV)$$

Your IRV: ____________________

A decrease in lung capacity may be due to a decrease in lung elasticity (which reduces the pulmonary compliance) caused by disorders such as tuberculosis or pulmonary fibrosis. The obstructive disorders, such as asthma and emphysema, decrease the volume of the lung capacity as well.

BIOPAC Respiration Lab

(A) 9 Pulmonary volumes and capacities can be measured directly with the use of a respirometer or spirometer, or they can be measured indirectly with the use of a flow meter. One such device is an airflow transducer made by BIOPAC. You should have all the necessary material for measuring pulmonary flow, including an MP30 or MP150 unit, transducer hardware, mouthpieces, bacteriological filters, a compatible computer, and software (BIOPAC Pro or BIOPAC Student Lessons). Every student in lab should record a spirogram. Make sure that everyone is able to have access to the equipment prior to doing any data analysis.

Caution!

When you change subjects always use a new disposable bacteriological filter.

Calibrating the Machine

1. Make sure that the computer is turned on, that the data acquisition unit is connected to the computer and turned on, and that the transducer is plugged into the data acquisition unit.
2. Select **Lesson 12: Pulmonary 1—Volumes & Capacities (L012-Lung-1)** and type your name where appropriate. If you have a folder with your name already on the computer, then select "Use it" if prompted.
3. Place a new bacteriological filter into the end with the calibration syringe. Insert the syringe into the side of the transducer labeled "Inlet."
4. Pull the plunger of the calibration syringe all the way out by holding the syringe by the body. Do not hold the unit by the transducer.
5. Click "Calibrate," read all the directions, and click "OK."
6. Push the plunger in and out for a total of five cycles (five in and five out). This should take about 30 seconds.
7. Click on "End Calibration."
8. The peaks should be even. If the calibration looks abnormal on the screen, then click "Redo." If it looks OK, then begin recording the data.

Recording Data

1. Place a new mouthpiece into the transducer where the calibration syringe was.
2. Pinch off your nose with a noseclip.
3. Click on "Record" and begin breathing easily into the transducer.
4. Take three normal breaths **(tidal volume)** and then inhale maximally and exhale normally **(inspiratory reserve volume),** as seen in figure 36.3.
5. Take a few normal breaths again and then exhale maximally **(expiratory reserve volume).**
6. Take a few normal breaths; then inhale maximally and exhale maximally (**vital capacity).**
7. Click on "Stop."

If you are not satisfied with your recording (compare what you see with figure 36.3), click on "Redo." If you are happy with your recording, then click on "Done" and remove the bacteriological filter from the apparatus.

You can select "Review Saved Data" from the menu.

Data Analysis

(A) 10 Select **CH2** (which is the volume) and select **p–p** (peak to peak) which selects the range of minimum and maximum values. Select CH2 and click on "max"; then select CH2 and click on "min." Select CH2 again and click on "delta." Compare your data with figure 36.3.

Select the I-beam and highlight one of the tidal volume measurements from the peak to the depth of a breath. Save these data as "Tidal Volume."

Select the area from the peak of a normal inhalation to the maximum air inhaled and save the data as "Inspiratory Reserve Volume."

Find the point of a normal exhalation (after a tidal volume) to the complete exhalation and save the data as "Expiratory Reserve Volume."

Find the point of the maximum inhalation to the maximum exhalation and save the data as "Vital Capacity."

Enter your data from the BIOPAC Respiration Lab in the following section.

Tidal volume: ____________________

Inspiratory reserve volume: ____________________

Expiratory reserve volume: ____________________

Vital capacity: ____________________

Your percent normal vital capacity can be calculated as it was using wet spirometry or handheld spirometers. Record your percent normal vital capacity in the following space.

Percent normal vital capacity: ____________________

Residual Volume

There is still air left in the lungs after a maximal exhalation. This is known as the residual volume (RV), and it is approximately 1,000 mL. The total lung capacity (TLC) is the sum of the vital capacity and the residual volume.

Minute Volume

11 The total amount of air inspired and expired in 1 minute of normal (tidal volume) breathing is known as the minute volume. This is calculated by multiplying the number of breaths per minute by the tidal volume. Record your minute volume in the following space.

Minute volume: ______________________

Construct a bar graph using your data in chart 36.1. The vital capacity should be composed of the ERV, TV, and IRV.

Flow and Resistance

12 The flow of air is proportional to the pressure of the air and inversely proportional to the resistance in the airways. As pressure difference between the outside air and the air inside the lungs increases, the flow of air increases. If the resistance in the air passageways increases, then the flow rate decreases. You can show the effects of changing the resistance of the airways in a simple demonstration. Do not do this experiment if you have a respiratory condition, such as asthma or severe sinus allergies.

1. Using a stopwatch or a watch with a second hand, record how long it takes to forcibly inhale the maximum amount of air that your lungs can hold. Do this by breathing through both your mouth and your nose. Record the time in the following space.

   Time for maximum inhalation: ____________ seconds

2. Close your mouth and one nostril and try the experiment again. Breathe only through one nostril and record the time that it takes to maximally inhale. Record this in the following space.

   Time for inhalation through one nostril: ____________

   ______________________________ seconds

Closing off the respiratory passages (mouth and one nostril) increases the resistance in the respiratory system. How does this change the rate of airflow? Asthma occurs due to a restriction of the bronchioles in the lung. If this occurs, is there a change in the pressure between the alveoli and the external air or an increase in the resistance to airflow into the lungs?

Respiratory Sounds

13 In this section, you listen to breathing sounds, which can be heard with a stethoscope.

1. Before you begin with the experiment, clean the earpieces of the stethoscope with an alcohol wipe and let them dry. The stethoscope earpieces should point toward the anterior as you insert them into your ears.
2. Locate the larynx of your lab partner and place the diaphragm of the stethoscope just inferior to it.
3. Listen for the sound as your lab partner inhales and exhales. These are the tracheal and bronchial sounds.

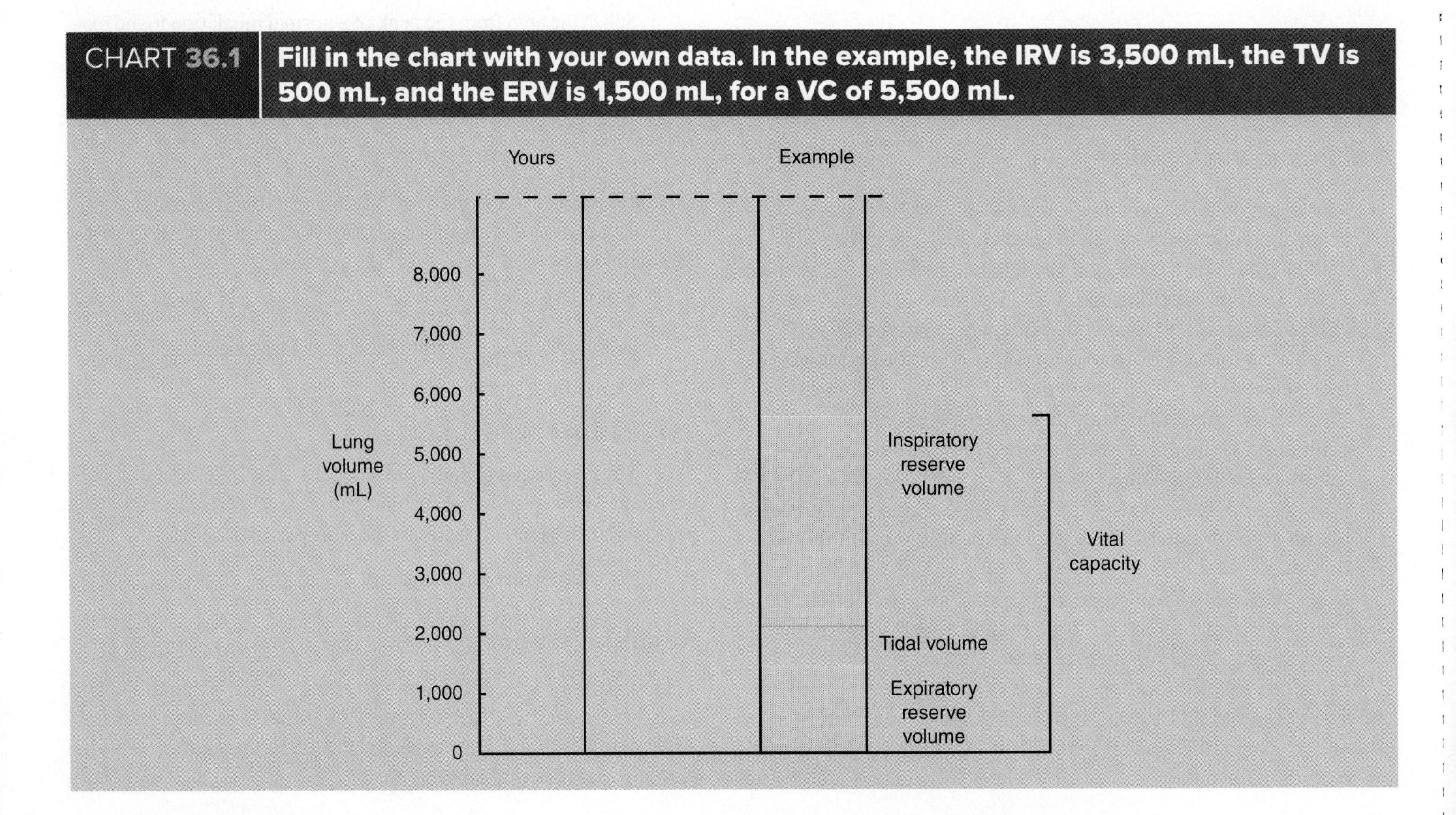

4. Locate the triangle of auscultation (figure 36.4), an area just medial to the inferior angle of the scapula. This is an ideal area for listening to sounds because the thoracic cage is not covered by muscles in this location.
5. Have your lab partner inhale and exhale deeply several times.
6. Listen for a smooth flow of air into and out of the lungs. Wheezing and rattling are indicators of congestion in the lungs.
7. Record the sounds you hear in the following space and indicate the condition of your lab partner. ______________

____

____

Acid-Base Effects of the Respiratory Gases

Carbon dioxide combines with water to form carbonic acid, which subsequently dissociates into bicarbonate and hydrogen ions, as represented in the following equation:

CO_2 +	H_2O →	H_2CO_3 →	HCO_3^- +	H^+
Carbon dioxide	Water	Carbonic acid	Bicarbonate ion	Hydrogen ion

(A) 14 You can see the effects of increasing carbon dioxide levels in the blood in the following experiment.

1. Add about 50 mL of litmus solution to a small (100 mL) Erlenmeyer flask. If the solution is red, add the sodium hydroxide solution (NaOH) drop by drop until the color just begins to turn blue.
2. Cover the flask with tape or parafilm.

Caution!

Wear safety goggles if you add NaOH to the flask.

3. Insert a drinking straw into the flask and gently blow air into the flask.
4. Bubble your exhaled breath into the flask and look for a color change. A blue color indicates an alkaline condition, and a red color indicates an acid condition.
5. Using the preceding formula, determine how the exhalation into the flask alters the acid-base conditions.
6. If your lab is equipped with a pH meter, you can measure the increase in hydrogen ion concentration. A change from a pH of 7 to 6 represents a 10-fold increase in the hydrogen ion concentration. If the pH drops to 5, it has 100 times the hydrogen ion concentration that it would at pH 7.

Cardiopulmonary Resuscitation

Cardiopulmonary resuscitation, or **CPR,** is typically used for people suffering from myocardial infarcts (heart attacks), drug overdoses, drowning or trauma, and obstruction of the airways, among other things. This is a technique that combines the use of chest compression of about 100 times per minute on the body of the sternum with mouth-to-mouth ventilation. The ratio is about 30 chest compressions followed by 2 ventilations. It is important to check to make sure that the airway is clear prior to ventilating the lungs. When the lungs are temporarily nonfunctional, CPR may keep a person alive until medical help arrives.

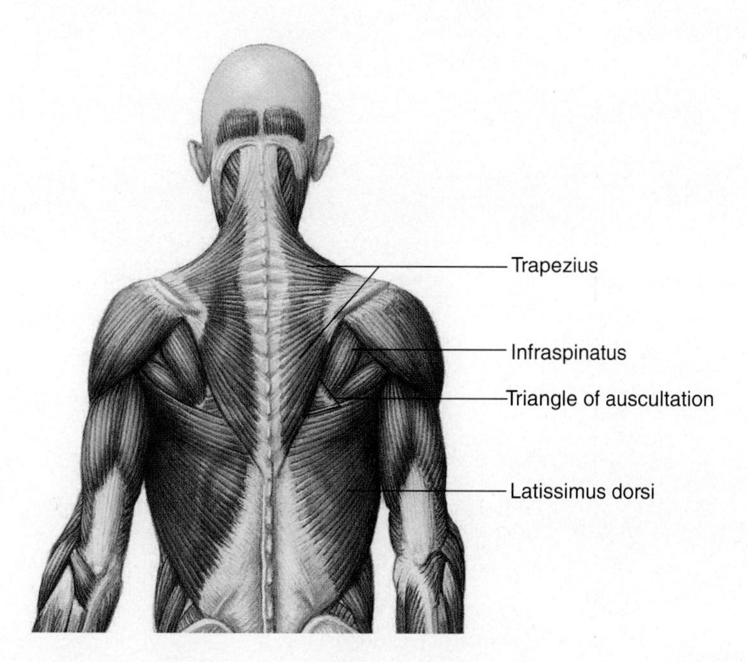

Figure 36.4 Triangle of Auscultation

Find the area between the trapezius, the medial border of the scapula, and the latissimus dorsi to listen to respiratory sounds.

NOTES

Exercise 36 Review

Respiratory Function, Breathing, and Respiration

Name: ______________________

Lab time/section: ______________________

Date: ______________________

1. What is pulmonary ventilation? ______________________

2. What was your measured breathing rate in breaths per minute? ______________________

3. What is the tidal volume in liters for an average adult? ______________________

4. Define *tidal volume.* ______________________

5. What instrument do you use to measure breathing volumes? ______________________

6. If you inhale maximally, what is the name of the volume of air that you completely exhale? ______________________

7. Examine the predicted vital capacity chart. What is the approximate percent decrease of vital capacity in the same person from age 25 to age 75? ______________________

8. How does the decrease in vital capacity potentially influence a person's athletic performance or aerobic condition as he or she ages? ______________________

9. What is the pressure difference between the external air and the pleural cavity when inhalation just begins? ______________________

10. Calculate the IRV of a person with a vital capacity of 4,360 mL, an expiratory reserve volume of 1,300 mL, and a tidal volume of 500 mL. ______

11. How does carbon dioxide change the acid-base condition of a solution when present in excess? ______

12. Why would carbon monoxide (it binds to hemoglobin) and carbon dioxide in cigarette smoke cause smokers to generally have a higher-than-average vital capacity? ______

13. For people of the same height in your lab, did they all have the same vital capacity? ______

14. If they varied in vital capacity, what might explain the difference? ______

Exercise 37

Physiology of Exercise and Pulmonary Health

Introduction

An understanding of exercise physiology is important not only for athletic training but also for wellness and general health. Significant changes occur during exercise that affect many organ systems of the body. The changes vary, depending on the type of exercise. Exercise is broadly classified as either **aerobic exercise** or **anaerobic exercise.** Aerobic exercises increase heart rate and breathing rates at moderate levels for extended periods of time. Aerobic exercises do not deplete the level of oxygen consumption by the tissues of the body. Anaerobic exercises result in the consumption of available oxygen faster than it can be supplied to the body tissue. In anaerobic respiration, muscle tissue uses glucose anaerobically and produces lactic acid. Exercise has a profound effect on the muscular, skeletal, cardiovascular, and respiratory systems. Consider the increased metabolic demands placed on skeletal muscles during repeated contractions. Skeletal muscle is more metabolically active when contracting than when at rest; it uses additional oxygen and nutrients. Oxygen diffuses from the blood to the muscle tissue due to the concentration gradient of oxygen between these two areas. In addition to the diffusion of oxygen, the precapillary sphincters of the circulatory system open and the arterioles dilate, providing greater blood flow to the muscles. This additional blood flow requires an increase in the volume of blood passing through the heart with each contraction. Heart size increases and pulse rate decreases with long-term rigorous exercise. The respiratory system responds to this greater oxygen demand with increased volume per breath and a greater number of breaths per minute. This increases the minute volume. The lung capillaries expand as well, and a greater diffusion of oxygen occurs between the alveoli and the blood capillaries. This respiratory response is covered in the Seeley text in chapter 23, "Respiratory System." In this exercise, you examine respiratory health, the effects of exercise on the body, and the body's comparative responses to aerobic exercise.

Objectives

At the end of this exercise, you should be able to

1. list the major organ systems directly involved in fitness;
2. describe basic physiological differences between a person who is physically active and one who is physically inactive;
3. determine the forced expiratory volume exhaled in 1 second;
4. calculate the personal fitness index of a subject who performs the Harvard step test or Cooper's 12-minute run test.

Materials

BIOPAC Setup

AFT6—6000 mL calibration syringe
AFT1—disposable bacteriological filter
AFT2—disposable mouthpiece
SS11L or SS11LA—airflow transducer
MP30 or MP150—data acquisition unit
Compatible computer
Noseclips
16-inch step
20-inch step
Metronome or clock with second hand
Treadmill

Procedure

Pulmonary Health

Forced Expiratory Vital Capacity (FEV)

Indications of health can be roughly correlated with the amount of air expelled from the lungs in 1 second. This is usually expressed as a percent when compared to the person's vital capacity (VC) as FEV_1/VC%. This should be approximately 75% of the VC in healthy adults. In this exercise you will use the BIOPAC setup, which is similar to the one used in Exercise 36. Decreases in forced expiratory vital capacity may be caused by asthma, emphysema, or other pulmonary conditions.

BIOPAC Lesson 13—Pulmonary Function II

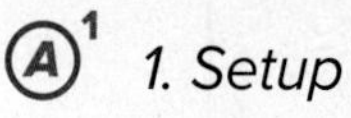

1. Setup

Make sure that the computer is on but the MP30 or MP150 data acquisition unit is off. Plug in the airflow transducer to the data acquisition unit and turn the unit on. Open the BIOPAC Student Lab (BSL) Software and select BIOPAC Lesson 13—Pulmonary Function II. You will collect data for your respiratory volumes by breathing into a flow meter that measures the force of air passing through it, which the BIOPAC program will then convert to a volume. The resulting "spirogram" is displayed on the computer monitor and saved as a file in your BIOPAC folder. The following are the steps you will take.

Click on the BIOPAC icon on the computer desktop to start BIOPAC. A menu of lessons will appear. Select **BIOPAC Lesson 13—Pulmonary Function II.** Type in your (folder) **name.** If you have a folder on this computer station, a window should appear with the message "A folder with this name already exists. Would you like to use it or create a new folder?" Choose **Use it.**

Place the bacterial filter onto the end of the calibration syringe. Insert the calibration syringe/filter assembly into the side of the airflow transducer labeled **"Inlet."**

2. Calibration

Pull the calibration syringe plunger all the way out and hold the calibration syringe horizontal, so that the airflow transducer is upright. It must remain vertical for the calibration and experimentation, as the membrane is sensitive to changes in deflection due to gravity. Hold the calibration syringe by the barrel (body), not by the airflow transducer (connecting duct). Click on **Calibrate.** After the first stage of the calibration is recorded, a second dialog box will appear, prompting you to read all the directions. Do so and click on **Yes.** Cycle the syringe plunger in and out 5 times (10 strokes). Use a rhythm of about 1 second per stroke, with 2 seconds between each stroke. Click on **End Calibration.** If the calibration looks good, then detach the calibration syringe and proceed to Step 3 **(Record).** (You should be able to see 5 downward and 5 upward deflections on the screen.) If the calibration peaks are uneven or significantly vary from one another, click **Redo.**

3. Record Data

Insert a clean mouthpiece (where the calibration syringe had been attached). Place a clean noseclip on your nose. Make sure that you hold the transducer upright at all times. Click on **Record** and breathe as follows (be sure to start the recording before taking the first normal breath):

1. Take three normal breaths (three inhales and three exhales).
2. Inhale as deeply as you can (take in as much air as you can). Hold it for an instant.
3. Exhale as fast and as much as you can.
4. Click on **Stop.**

If the recording is not good, click on **Redo** to repeat it. If the recording looks good, as seen in figure 37.1, click on **Done.** Select **Record from Another Subject** or, if you have recorded everyone's data, select **Analyze Current Data File.**

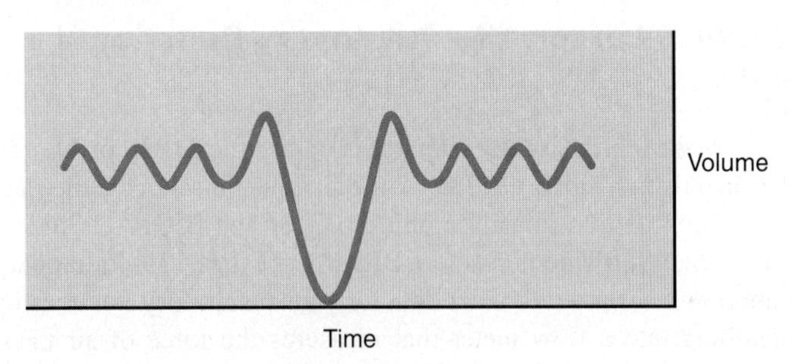

Figure 37.1 Spirogram

4. Data Analysis for FEV

Open your folder in the **Data Files** folder (probably listed as *"your name FEV-L13"*) and then open the data file. Select **Display Preferences** from the **File** menu, choose **Grids,** and click on **Show Grids** and then **OK.** Use the **I-beam** to highlight your graph. The **p–p** (peak to peak) will represent your vital capacity, as represented in figure 37.2.

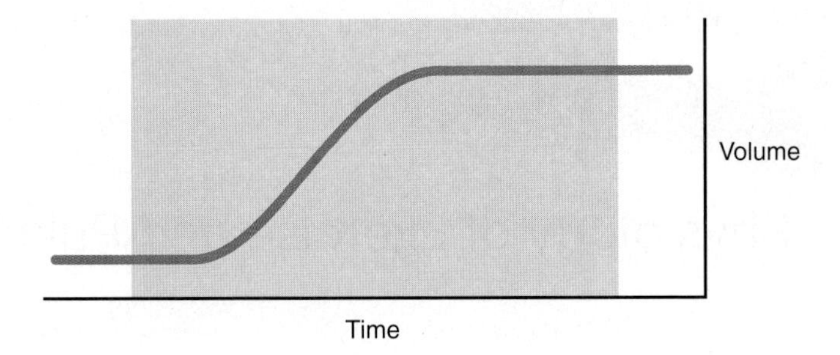

Figure 37.2 Peak to Peak Determination of Vital Capacity

Use the **I-beam** to select the volume of air you exhaled for the first second. This will be the FEV_1, as seen in figure 37.3.

Calculate your $FEV_1/VC\%$ and enter the value here:

____

Save and print the entire spirogram you record to include in your lab report.

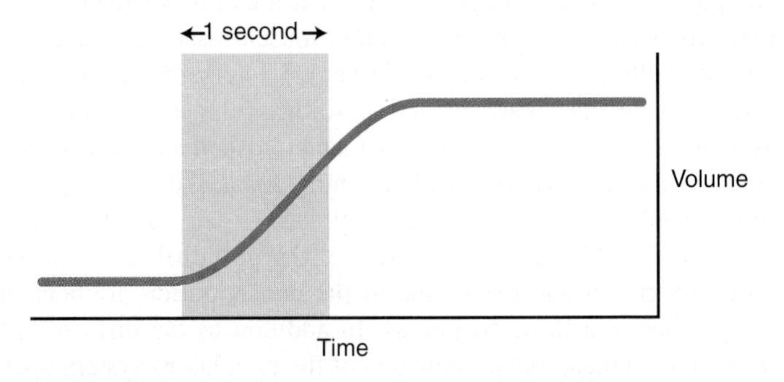

Figure 37.3 FEV_1 Highlight 1 Second of Time from Beginning of Expiration

5. Interpretation

Spirometry is an important measurement tool to assist in diagnosing asthma, chronic obstructive pulmonary disease (COPD), and cystic fibrosis. Most cases of COPD in North America are caused by smoking.

Pulmonary obstruction occurs when the respiratory passages are narrowed. This can be due to conditions such as asthma, excess mucus, and inflammation, such as bronchitis. In these cases, the vital capacity of the individual is normal but the $FEV_1/VC\%$ is low. It takes the subject longer to exhale completely because of increased resistance in the airways.

Pulmonary restriction occurs when the lungs cannot fully inspire or expire the full volume of air. In these cases, the vital capacity of the individual is reduced. This can be due to fibrosis of the lungs (cystic fibrosis or fibrosis due to asbestos or silica); scarring of the lung tissue, as when a subject has had chronic lung infections; adhesions of the lung to the chest wall due to extreme emphysema; or removal of a section of lung. It may also be due to damage to the phrenic nerve, which stimulates the diaphragm. In these cases, the vital capacity of the individual is low but the $FEV_1/VC\%$ is normal. Clinical values for FEV_1 compared to vital capacity are listed in table 37.1.

TABLE 37.1

FEV_1 /VC%	Status
100–75	Normal
74–60	Mild COPD
59–50	Moderate COPD
<50	Severe COPD

Measurement of Heart Rate

The measurement of fitness in this part of the exercise is on a voluntary basis. It is best if the class can obtain data from people who regularly participate in aerobic exercises (three to six times per week) and from a people who do not exercise. The Harvard step test and the Cooper's 12-minute run test are two reliable measurements of fitness.

Caution!

Do not do these exercises if you are at risk for heart disease or have a family history of heart disease or another condition, such as asthma, for which exercise is harmful. Stop if you feel exhausted or faint. If you develop chest pain or pain radiating down the left arm, seek medical attention immediately.

A correlation exists between heart rate and fitness level. The resting heart rate of people involved in regular, active aerobic exercise is lower than the rate of sedentary people. In addition, the heart recovers faster in people who have a regular exercise program than in those who do not exercise. In this experiment record the measurements of at least two volunteers from the class. The greater the number of students who participate, the better the data. Read the entire exercise before beginning.

Harvard Step Test

(A) [2] The Harvard step test was developed during World War II at Harvard University to determine a person's physical fitness. In this exercise students should select either the 20-inch step for people 5′8″ or taller or the 16-inch step for people under 5′8″. With one person acting as an observer, the subject should step up on the step in 1 second and down on the floor in another second, thus completing 30 complete cycles in 1 minute. The subject should keep the body upright and keep pace with the observer's count or with a metronome set at 60 beats per minute. The subject should exercise for at least 3 minutes but no more than 5 minutes. If the subject stops due to exhaustion, then the observer should note the time of exercise. After the period of exercise, the subject should sit and rest for 1 minute. The pulse should be taken from 1 minute to 1 minute 30 seconds and recorded in the following space. (See "A" on figure 37.4.)

Pulse from 1 minute to 1 minute 30 seconds: __________

The subject should rest for another 30 seconds before the pulse is taken from 2 minutes to 2 minutes 30 seconds after exercise and recorded here. (See "B" on figure 37.4.)

Pulse from 2 minutes to 2 minutes 30 seconds: __________

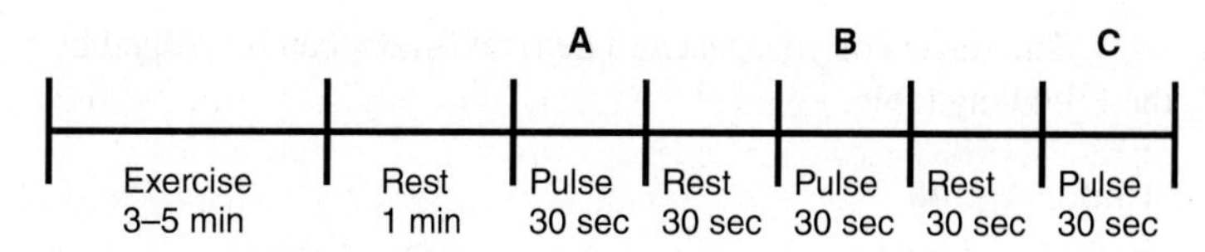

Figure 37.4 Harvard Step Test Periods

The subject should remain seated for another 30 seconds before the pulse is taken from 3 minutes to 3 minutes 30 seconds and recorded here. (See "C" on figure 37.4.)

Pulse from 3 minutes to 3 minutes 30 seconds: __________

Add the three pulse counts recorded.

Sum of three pulse counts: __________

To determine the personal fitness index (PFI), use the following formula:

$$PFI = \frac{\text{Number of seconds of exercise}}{2\ (\text{sum of three pulse counts})} \times 100$$

Therefore, if you exercised for 4 minutes and your pulse counts were 50, 48, and 45, then the PFI was

$$PFI = \frac{(4 \times 60) \times 100}{2(50 + 48 + 45)} = \frac{240 \times 100}{2(143)} = \frac{24{,}000}{286} = 84$$

Determine the fitness evaluation of the subject using the following data:

Below 55: poor physical condition
55–64: low average physical condition
65–79: high average physical condition
80–89: good physical condition
90 and above: excellent physical condition

Record the PFI of the subject: __________

What is the fitness evaluation for this person? __________

Cooper's 12-Minute Run Test

Another fitness test involves determining how far a person can run and/or walk in 12 minutes.

Caution!

As with the Harvard step test, do not do these exercises if you are at risk for heart disease or have a family history of heart disease or another condition, such as asthma, for which exercise is harmful.

(A) [3] *Procedure* Warm up for 15 minutes prior to doing this test. Find a suitable running track or use a treadmill that is set to no incline and run or walk as fast and as far as you can in 12 minutes. You will not be able to sprint for the entire time, so pace yourself. If you cannot run anymore during the test, you should finish the test by walking or jogging or a combination of running and walking.

Record the distance you traveled in 12 minutes: __________

The results of your test and general fitness can be judged by the following table.

Under Age 40

Distance in 12 Minutes	Physical Condition
Over 2,700 m	Excellent
2,300 m to 2,700 m	Good
1,900 m to 2,299 m	Average
1,500 m to 1,899 m	Below average
Less than 1,500 m	Poor

Over Age 40

Distance in 12 Minutes	Physical Condition
Over 2,500 m	Excellent
2,100 m to 2,500 m	Good
1,700 m to 2,099 m	Average
1,500 m to 1,699 m	Below average
Less than 1,400 m	Poor

Target Heart Rate Zone for Exercise

The target heart rate zone for exercise is 60–80% of the maximum heart rate (MHR) for healthy adults. The maximum heart rate for an individual is his or her age subtracted from 220. You can calculate a basic target heart rate zone with the following procedure:

1. To calculate your maximum heart rate, subtract your age from 220. Enter the MHR value here: ________________.
2. For 60% of the MHR, multiply the MHR by .6 and enter it in the space: ________________.
3. For 80% of the MHR, multiply the MHR by .8 and enter it in the space: ________________.
4. Your target heart rate zone for exercise should be between these two values.

Guidelines for Exercise

The American Heart Association recommends that healthy adults from the ages of 18 to 65 need moderate exercise (such as brisk walking) 30 minutes a day, 5 days a week, or vigorous exercise (such as jogging or running) for 20 minutes 3 days a week. Physical inactivity is a major public health issue. Any person of normal mobility who is inactive can increase his or her fitness with a program of exercise that may start simply with walking. Gradual increase in the intensity and duration of exercise will improve the general health of the individual.

Body Mass Index

(A) 4 The body mass index (BMI) is a general guide to fitness that makes a couple of assumptions. One is that the person is of average build. Fitness level, gender, muscle mass, bone structure, and ethnicity can all influence the BMI. One way to get a general idea of the BMI is to use this calculation:

$$\text{BMI} = \frac{\text{Weight in pounds}}{(\text{Height in inches})^2} \times 703$$

Record your BMI here: ________________

According to the National Heart, Lung, and Blood Institute, the BMI for average adults can be interpreted this way:

- Underweight = <18.5
- Normal weight = 18.5–24.9
- Overweight = 25–29.9
- Obesity = BMI of 30 or greater

If a person has a high muscle mass, the BMI values may be overestimated, as muscle will show a person with higher weight than predicted. In older individuals who have less muscle mass, the BMI may be underestimated.

Waist/Hip Ratio

(A) 5 Another way to calculate fitness is to use the waist/hip ratio (WHR). According to the American Heart Association, people who carry more weight in their waist region (with "apple-shaped bodies") are more at risk for health problems than people with more weight in their hips (with "pear-shaped bodies"). Of course, increased weight of any kind is a health risk. You can use a flexible tape measure (U.S. standard or metric) and do the following:

1. Measure the circumference of your hips at their widest part. Record the value.

   Circumference of hips: ________________

2. Measure your waist just superior to the umbilicus (belly button). Record the value.

   Circumference of waist: ________________

3. Use the formula WHR = Circumference of waist/Circumference of hips

   Calculate the WHR and record the value.

   WHR: ________________

4. Compare it to table 37.2 and determine your health risk.

   Health risk: ________________

TABLE 37.2 Waist/Hip Ratio

Females	Males	Health Risk Based on WHR
0.80 or below	0.95 or below	Low risk
0.81 to 0.85	0.96 to 1.00	Moderate risk
0.85+	1.0+	High risk

Personal Goals

If your personal fitness index or body mass index indicates that you are out of shape or overweight, list three things that you can do to increase your fitness.

Exercise 37 Review

Physiology of Exercise and Pulmonary Health

Name: ______________________

Lab time/section: ______________________

Date: ______________________

1. Define $FEV_1/VC\%$. ______________________

____

____

2. Record your $FEV_1/VC\%$ or the one you measured in lab. ______________________

3. Does your $FEV_1/VC\%$ value fall within normal limits? ______________________

4. How does the $FEV_1/VC\%$ compare in a person with a pulmonary obstructive condition, such as asthma? Why? ______________________

____

____

5. How does the $FEV_1/VC\%$ compare in a person with a pulmonary restrictive condition, such as asbestosis? Why? ______________________

____

____

6. If a person had a smaller body and therefore a smaller vital capacity, would the $FEV_1/VC\%$ necessarily change? ______________________

____

7. What is your personal fitness index or the one measured in lab? ______________________

8. If the heart rate after 5 minutes of exercise was 70, 68, and 66 beats in the consecutive 30-second trials, what was the personal fitness index and what condition does that represent? ______________________

____

9. Record the results of the Harvard step test or the Cooper's 12-minute run test for the members in the class who performed them. Next to the Personal Fitness Index (Harvard) or Physical Condition (Cooper's) for your own results, write an *S* if you smoke cigarettes or an *N* if you are a nonsmoker.

PFI (Personal Fitness Index)
Physical Condition

1. ______________________
2. ______________________
3. ______________________
4. ______________________
5. ______________________
6. ______________________
7. ______________________
8. ______________________
9. ______________________
10. ______________________
11. ______________________
12. ______________________
13. ______________________
14. ______________________
15. ______________________
16. ______________________
17. ______________________
18. ______________________
19. ______________________
20. ______________________
21. ______________________
22. ______________________
23. ______________________
24. ______________________

Exercise 38

Anatomy of the Digestive System

Introduction

The digestive system can be divided into two major parts, the **digestive tract** and the **accessory organs.** The digestive tract is a long tube that runs from the mouth to the anus and comes into contact with food and the breakdown products of digestion. The digestive tract is also known as the **alimentary canal** and is commonly called the "food tube." Some of the organs of the digestive tract are the esophagus, stomach, intestines, colon, rectum, and anus. The accessory organs are important in that they secrete many substances necessary for digestion, yet these organs do not come into direct contact with food. Examples of accessory organs are the salivary glands, liver, gallbladder, and pancreas.

The functions of the digestive system are many. They include the ingestion of food, the physical breakdown of food, the chemical breakdown of food, food storage, food and water absorption, vitamin synthesis, and the elimination of indigestible material. The anatomy and physiology of the digestive system are covered in the Seeley text in chapter 24, "Digestive System."

In this exercise, you examine the anatomy of the digestive system in both human and cat, correlating the structure of the digestive organs with their functions.

Objectives

At the end of this exercise, you should be able to

1. list, in sequence, the major organs of the digestive tract;
2. describe the basic functions of the accessory digestive organs;
3. note the specific anatomical features of each major digestive organ;
4. describe the layers of the wall of the digestive tract;
5. describe the major functions of the stomach and the small and large intestines.

Materials

Models, charts, or illustrations of the digestive system
Hand mirror
Cats
Dissection trays
Pins
Scalpels
Protective gloves
Animal waste container
Skull, human teeth, or cast of teeth
Cadaver (if available)
Microscopes
Microscope slides

- Esophagus
- Stomach
- Small intestine
- Large intestine
- Liver

Procedure

Overview of the Digestive System

(A) 1 Begin this exercise by examining a torso model or charts in the lab, and compare them with figure 38.1. Locate the major digestive organs and place a check mark in the appropriate space.

_______Oral cavity
_______Pharynx
_______Esophagus
_______Stomach
_______Small intestine
_______Large intestine
_______Liver
_______Pancreas
_______Anus
_______Gallbladder
_______Salivary glands

Organs of the Upper Digestive Tract

(A) 2 Examine models and charts in lab. The structures of the upper digestive tract have been numbered here for ease of reference.

1. Examine a midsagittal section of the head, as represented in figure 38.2, and locate the **oral cavity (mouth),** which starts as the opening surrounded by lips, or **labia.**
2. Find the **labial frenulum,** a membranous structure that keeps the lip adhered to the gums, or **gingivae.**
3. The hard and soft palates form the roof of the oral cavity, and the floor of the chin is the inferior border. The **hard palate** is composed of the palatine processes of the maxillae and palatine bones. The **soft palate** is composed of connective tissue and a mucous membrane. The posterior border of the oral cavity is the oropharynx. At the posterior portion of the oral cavity is the **uvula,** a small, grapelike structure suspended from the posterior edge of the soft palate. The uvula helps prevent food and liquid from moving into the nasal cavity during swallowing. The oral cavity is

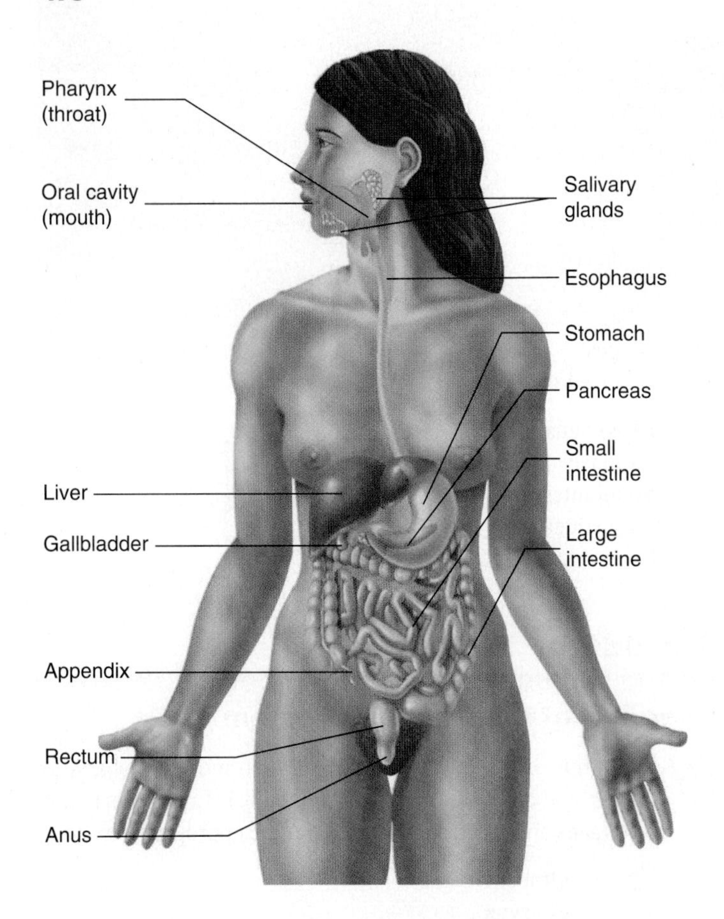

Figure 38.1 Overview of the Digestive System

lined with **nonkeratinized stratified squamous epithelium,** which protects the underlying tissue from abrasion.

4. Using a mirror, examine your tongue, which is made of skeletal muscle. One of the major muscles of the tongue is the **genioglossus.** The tongue is important in speech, taste, the movement of food toward the teeth for chewing, and swallowing. The tongue acts as a piston to propel food to the **oropharynx,** the space behind the oral cavity. The tongue is held down to the floor of the mouth by a thin mucous membrane called the **lingual frenulum.**
5. Find the four types of **papillae,** or raised areas, on the tongue—the **foliate, fungiform, filiform,** and **circumvallate** papillae—in figure 38.3. Foliate papillae are leaf-shaped, fungiform are mushroom-shaped, filiform are threadlike, and circumvallate are large papillae. Papillae increase the frictional surface of the tongue. The tongue also has taste receptors in **taste buds.** Taste buds are located on the tongue along the sides of some of the papillae. The sense of taste is covered in Exercise 21.

   The oral cavity is important for the physical breakdown of food. This process is driven by powerful muscles called the **muscles of mastication.**
6. Examine a model of the muscles of the head and find the masseter, temporalis, pterygoids, digastric, and platysma muscles. The **masseter** and the **temporalis muscles** are involved in the closing of the jaws, and the **pterygoid muscles** are important in the sideways grinding action of the molar and premolar teeth. The **digastric** and **platysma** muscles open the mandible.

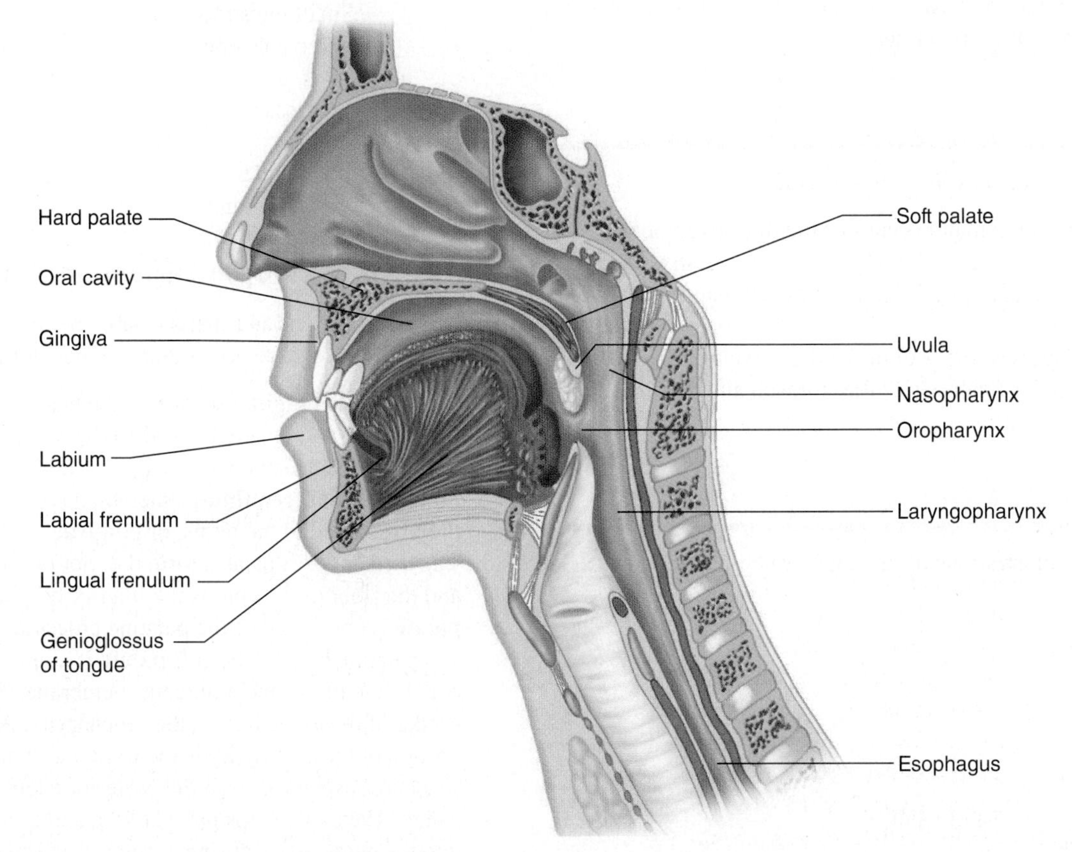

Figure 38.2 Oral Cavity, Midsagittal Section

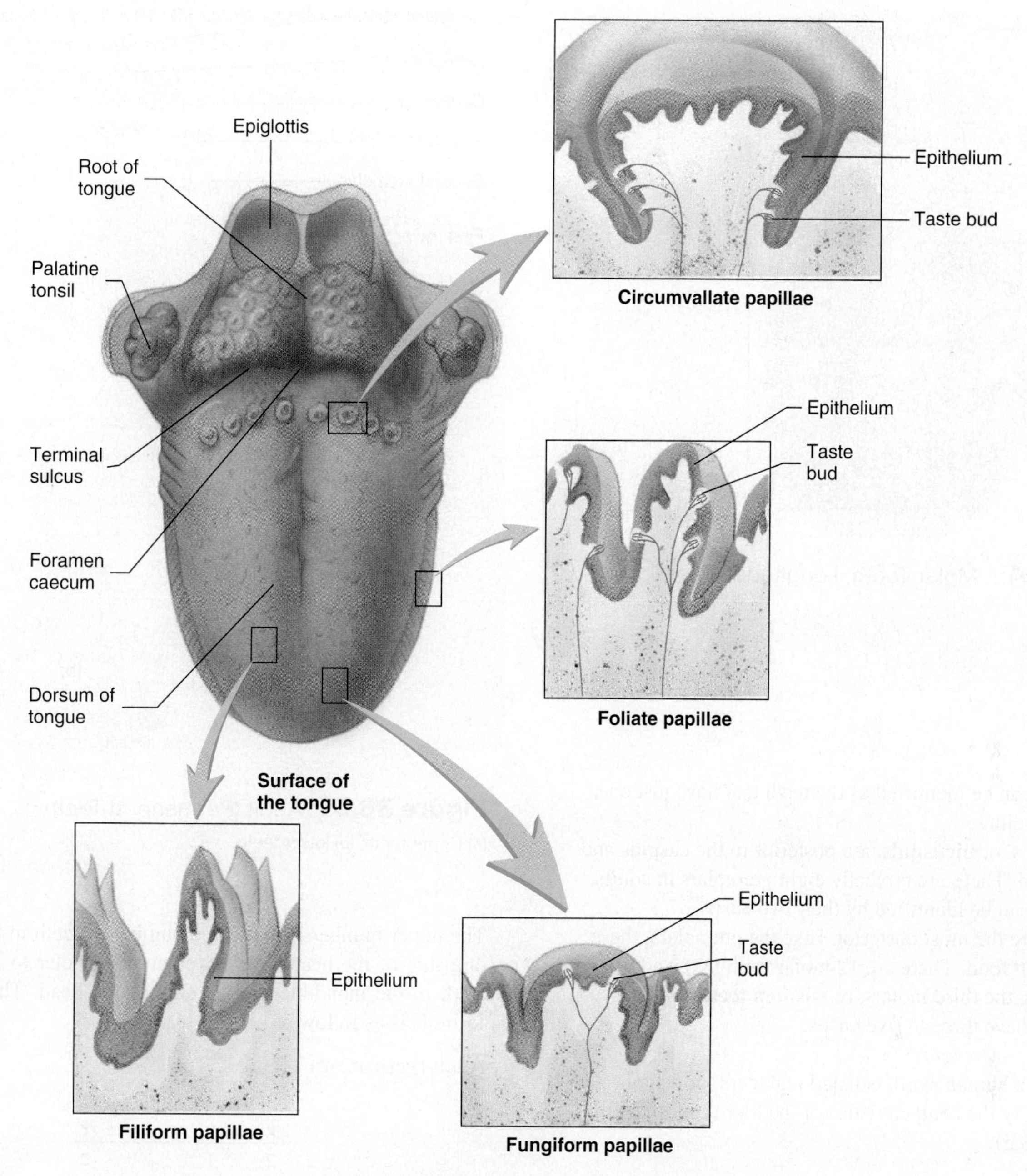

Figure 38.3 Tongue, Location of the Papillae on the Tongue, and Types of Papillae

Structure of Teeth

7. Examine models of teeth, dental casts, or real teeth on display in the lab and compare them with figure 38.4. A tooth consists of a **crown, neck,** and **root.** The crown is the exposed part of the tooth, the neck is a constricted portion of the tooth normally located at the surface of the gingivae, and the root is embedded in the jaw.

8. Examine a model or an illustration of a longitudinal section of a tooth and find the outer **enamel,** an extremely hard material. Inside this layer is the **dentin,** which is made of bonelike material. The innermost portion of the tooth consists of the **pulp cavity,** which leads to the **root canal,** a passageway for nerves and blood vessels into the tooth. The nerves and blood vessels enter the tooth through the **apical foramen** at the tip of the root of the tooth. The teeth are in depressions in the mandible or maxilla called **alveolar sockets,** and they are anchored to the bone by periodontal ligaments.

There are four types of teeth in the adult mouth:

- **Incisors** are flat, bladelike front teeth that nip food. There are eight incisors in the adult mouth.
- **Canines,** or **cuspids,** are the pointed teeth just lateral to the incisors that shear food. There are four cuspids in the adult,

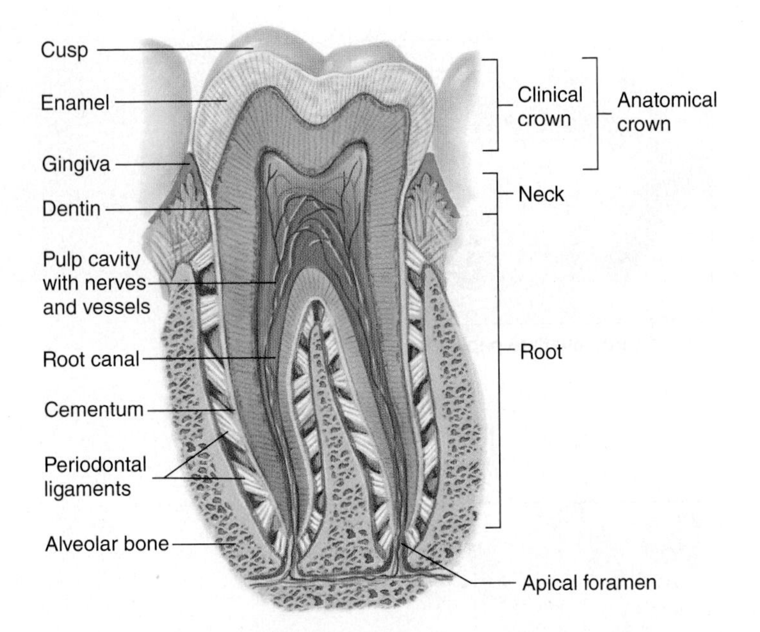

Figure 38.4 Molar Tooth, Longitudinal Section

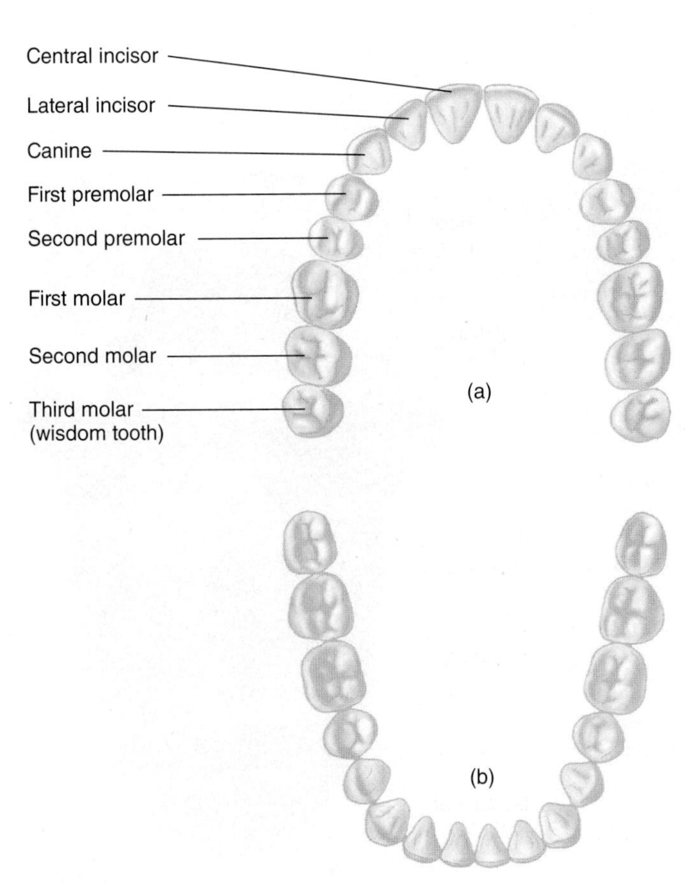

Figure 38.5 Adult (Permanent) Teeth

(a) Upper teeth; (b) lower teeth.

and they can be identified as the teeth that have just one cusp, or point.

- **Premolars,** or **bicuspids,** are posterior to the cuspids and grind food. There are typically eight premolars in adults, and they can be identified by their two cusps.
- **Molars** are the most posterior. Like the premolars, these teeth grind food. There are 12 molar teeth in the adult mouth (including the third molars, or **wisdom teeth**). Molars typically have three to five cusps.

9. Examine a human skull, isolated real teeth, or dental cast and identify the characteristics of the four types of teeth (figure 38.5).

Humans have two sets of teeth. The **primary,** or **deciduous, teeth** (milk teeth) appear first, and these are replaced by the **secondary,** or **permanent, teeth.** There are 20 deciduous teeth. There are no deciduous premolar teeth, and there are only 8 deciduous molar teeth. In adults, there are 8 premolar teeth and 12 molar teeth.

10. Compare figure 38.5, the adult pattern, with figure 38.6, the deciduous teeth.

Often, the pattern of tooth structure is represented by a dental formula, which describes the teeth by quadrants. The dental formula for the primary teeth is illustrated here: I = incisor, C = cuspid, P = premolar, and M = molar.

Deciduous Teeth (20 Total) Dental Formula

I	C	P	M	One side (quadrant)
2	1	0	2	Top
2	1	0	2	Bottom

The upper numbers refer to the number of teeth in the maxilla on one side of the head. The lower numbers refer to the number of teeth in the mandible on one side of the head. The adult dental formula is as follows:

Adult (Permanent) Teeth (32 Total)

I	C	P	M
2	1	2	3
2	1	2	3

Pharynx

11. Locate the oropharynx and the esophagus in figure 38.2. The space behind the oral cavity is the **oropharynx.** Superior to the oropharynx is the **nasopharynx,** which leads to the nasal cavity, and inferior to the oropharynx is the **laryngopharynx,** which leads to the larynx and the esophagus. The nasopharynx begins at the internal nares and ends at the soft palate. The oropharynx begins there and continues to the larynx. Posterior to the larynx is the laryngopharynx. The oropharynx is composed of nonkeratinized stratified squamous epithelium and is a common passageway for food, liquids, and air. Muscles around the wall of the oropharynx are the **pharyngeal constrictor muscles** and are involved in swallowing. Food is moved by the tongue to the region of the pharynx, where it is propelled into the **esophagus.**

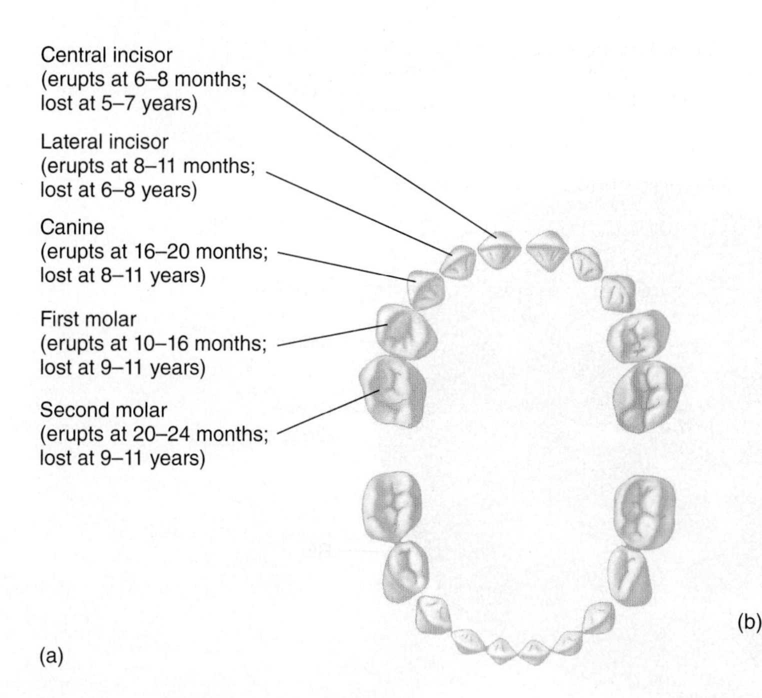

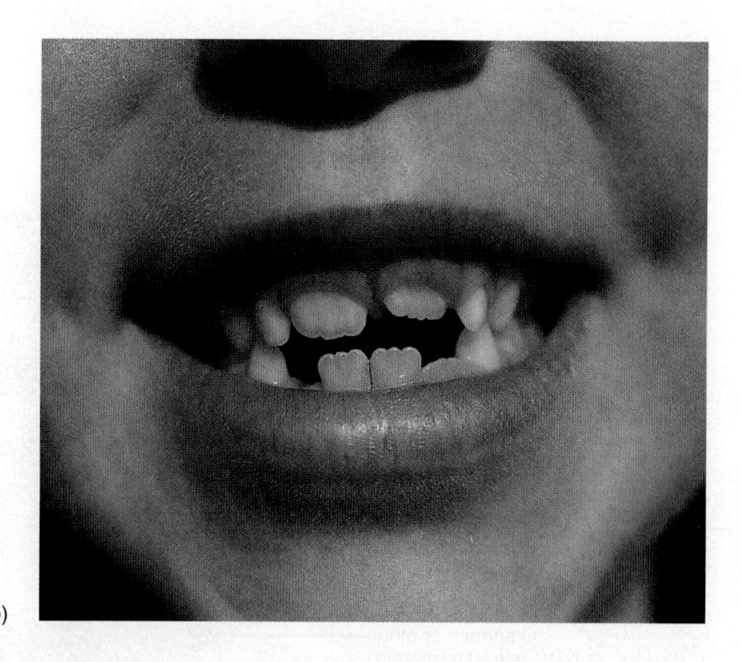

Figure 38.6 Deciduous Teeth
(a) Diagram; (b) photograph of replacement of teeth in a child.

Esophagus

12. Locate the esophagus in charts and models in lab and compare it with figure 38.1. The esophagus conducts food and liquid from the oropharynx, through the diaphragm, and into the stomach. Normally, the esophagus is a closed tube that begins at about the level of the sixth cervical vertebra. As a lump of food, or **bolus,** enters the esophagus, **skeletal muscle** begins to move it toward the stomach. The middle portion of the esophagus is composed of both skeletal and smooth muscle, whereas the inferior portion of the esophagus is made of **smooth muscle.** In the esophagus, muscle contracts, moving the bolus by a process known as **peristalsis.**

The esophagus has an inner epithelial lining of stratified squamous epithelium and an outer connective tissue layer called the **adventitia.** The space in the esophagus that the food passes through is called the **lumen,** which continues through the digestive tract. The inferior portion of the esophagus has an **esophageal sphincter,** which prevents the backflow of stomach acids. Heartburn, or gastroesophageal reflux disease (GERD), occurs if the stomach contents pass through the esophageal sphincter and irritate the esophageal lining.

Abdominal Portions of the Digestive Tract

The inner structure of the body has been referred to as a "tube within a tube." The body wall forms the outer tube, and the digestive tract, including the stomach, small intestine, and large intestine, forms the inner tube. Specialized serous membranes cover the various organs and line the inner wall of the **coelom** (the **body cavity**). The membrane lining the outer surface of the digestive tract is called the **visceral peritoneum (serosa).** It continues as a double-folded membrane called the **mesentery,** which attaches the tract to the back of the body wall. The mesentery is continuous with the membrane on the inner side of the body wall, where it is called the **parietal peritoneum.**

Ⓐ[3] The following structures of the abdominal portions of the digestive tract have been numbered here for ease of reference. Locate these structures on materials in lab as you read the material and examine the illustrations in the lab manual.

1. Locate these three membranes in figure 38.7.
2. Examine a microscope slide of the digestive tract (small intestine) under the microscope and locate the layers, as represented in figure 38.8. In general, the stomach, small intestine, and large intestine have the same layers from the lumen to the coelom. The innermost layer is the **mucosa,** which consists of **epithelium** closest to the lumen, a connective tissue layer called the **lamina propria,** and an outer, muscular layer called the **muscularis mucosae.** These three layers are called the **mucous membrane.**

The next layer is called the **submucosa,** which is mostly made of **connective tissue** and contains many blood vessels. The next layer is the **muscularis (muscularis externa),** which is typically made of two or three layers of smooth muscle. The muscularis propels material through the digestive tract and mixes ingested material with digestive juices. The outermost layer is called the **serosa,** or **visceral peritoneum,** and this layer is closest to the coelom.

Stomach

The **stomach** is located on the left side of the body; it receives its contents from the esophagus. The food that enters the stomach is stored and mixed with enzymes and hydrochloric acid to form a soupy material called **chyme.** The stomach can have a pH as low as 1 or 2. Chyme remains in the stomach as the acids denature proteins and enzymes reduce proteins to shorter fragments.

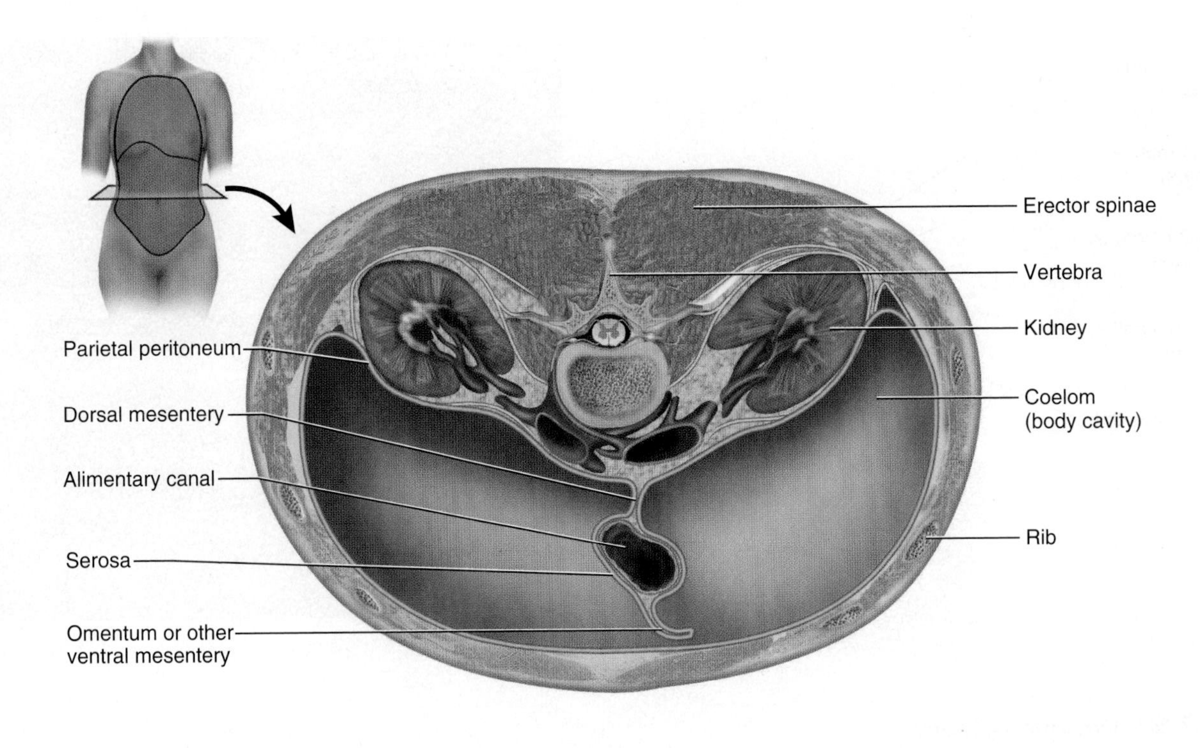

Figure 38.7 Membranes of the Digestive Tract

3. Examine a model of the stomach or charts in the lab and compare them with figure 38.9. Locate the upper portion of the stomach, called the **cardiac part,** or **cardia.** A portion of the cardiac part extends superiorly as a domed section called the **fundus.** The main part of the stomach is called the **body,** and the terminal portion of the stomach, closest to the small intestine, is called the **pyloric part.**

The pyloric part has an expanded area called the antrum and a narrowed region called the pyloric canal. The pylorus leads to the duodenum, and this opening is controlled by the pyloric sphincter.

The left side of the stomach is arched and forms the **greater curvature,** whereas the right side of the stomach is a smaller arch, forming the **lesser curvature.** The inner surface of the stomach has a series of folds called **rugae,** which allow for significant expansion of the stomach. The contents are held in the stomach by two sphincters. The **esophageal sphincter** prevents stomach contents from moving into the esophagus, and the **pyloric sphincter** prevents the premature release of stomach contents into the small intestine.

4. Locate the pyloric sphincter at the terminal portion of the stomach.

Stomach Histology

5. Examine a prepared slide of stomach tissue. Identify the four primary layers: **mucosa, submucosa, muscularis,** and **serosa.** Notice how the mucosa in the prepared slide has a series of indentations. These depressions are **gastric glands,** which are in the inner lining of the stomach.

6. Find the three layers of the mucosa, including the **epithelial layer,** the **lamina propria,** and the **muscularis mucosa.** The epithelial layer consists of **simple columnar epithelium,** as well as specialized cells, such as **surface mucous cells, chief cells,** and **parietal cells.** Surface mucous cells secrete mucus, which protects the stomach lining from erosion by stomach acid and proteolytic (protein-digesting) enzymes. Chief cells secrete **pepsinogen** (the inactive state of a proteolytic enzyme). In some prepared slides, chief cells can be recognized by their blue-staining granules. Parietal cells secrete **HCl.** They typically contain orange-staining granules. When pepsinogen comes into contract with HCl, it is activated as **pepsin.** Deep to the epithelial layer is the lamina propria, a connective tissue layer. The deepest layer of the mucosa is the muscularis mucosa, a smooth muscle layer that moves the mucosa. The submucosa is the next layer and is typically lighter in color in prepared slides.

The next layer of the stomach is the **muscularis.** In some parts of the stomach, there are three layers of the muscularis—an inner **oblique layer,** a middle **circular layer,** and an outer **longitudinal layer.**

7. Locate these layers in your slide. The muscularis moves chyme from the stomach through the pyloric sphincter and into the small intestine.

The outermost layer is the **serosa,** and it is composed of a thin layer of connective tissue and **simple squamous epithelium.**

8. Locate the serosa and compare it with figure 38.9.

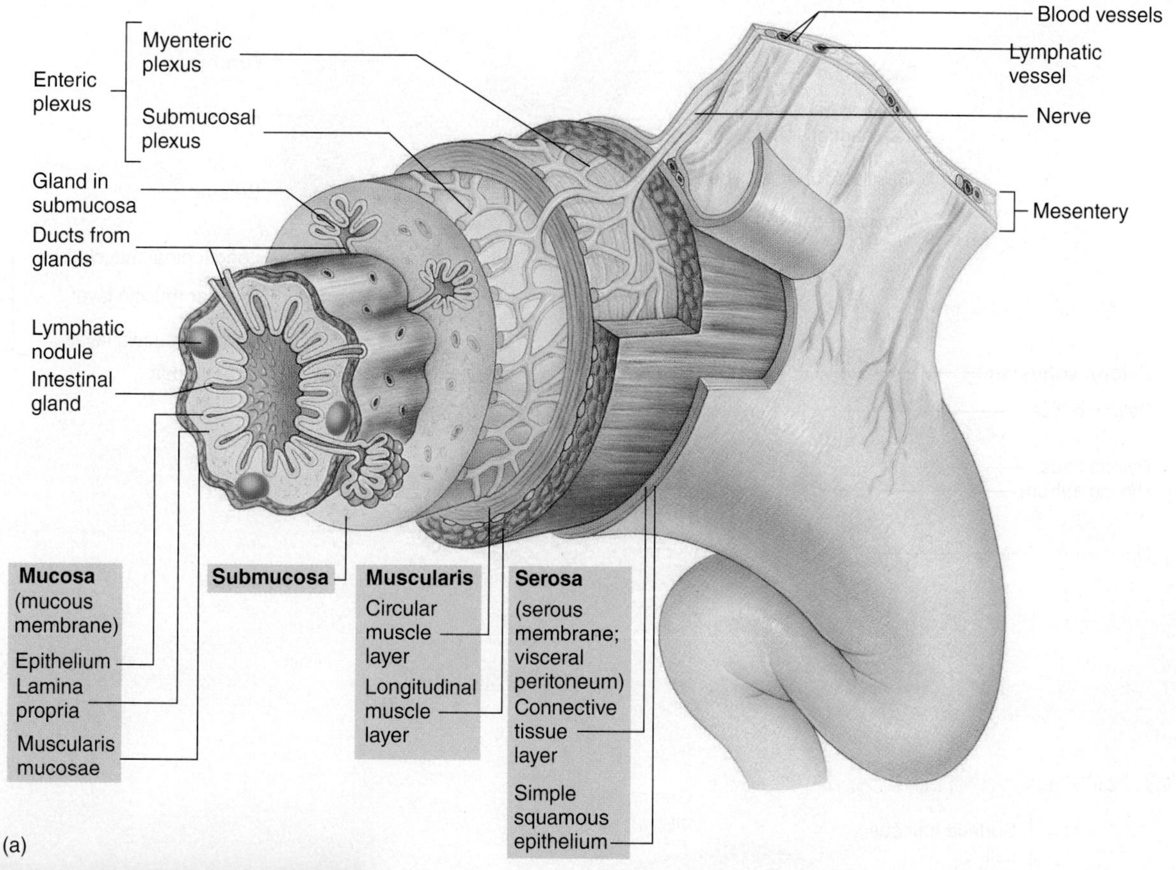

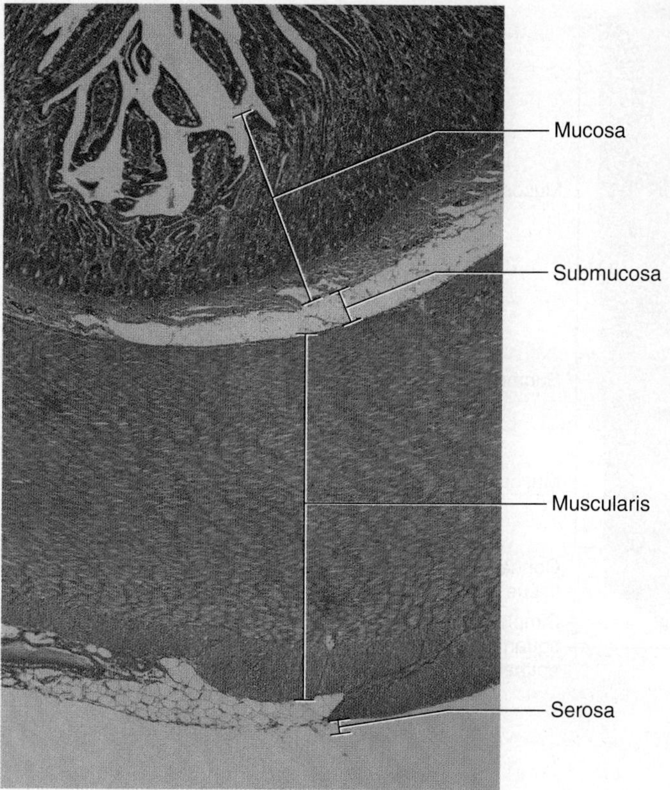

Figure 38.8 Gastrointestinal Tract (Small Intestine), Cross Section

(a) Diagram; (b) photomicrograph (100×).

Small Intestine

The **small intestine,** so named because it is small in diameter, is approximately 5 m (17 feet) long in living humans, yet it can be longer in cadaveric specimens due to the relaxing of the smooth muscle. It is typically 3 to 4 cm (1.5 inches) in diameter when empty. Movement through the small intestine occurs by peristalsis, which is smooth muscle contraction. The primary function of the small intestine is nutrient absorption.

9. Locate the three major regions of the small intestine on a model or chart and compare them with figure 38.10. The first part of the small intestine is the **duodenum,** a C-shaped structure attached to the pyloric region of the stomach. The duodenum begins in the body cavity and then moves behind the parietal peritoneum and returns to the body cavity. The duodenum is approximately 25 cm (10 inches) long. It receives fluid from both the **pancreas** and the **gallbladder.** The junction between the duodenum and the jejunum is known as the duodenojejunal flexure.

The second portion of the small intestine is the **jejunum,** which is approximately 2 m (6.5 feet) long. The terminal portion of the small intestine is the **ileum,** which is approximately 3 m (10 feet) long. A closure between the small intestine and large intestine is called the **ileocecal valve.** This valve keeps material in the large intestine from reentering the small intestine.

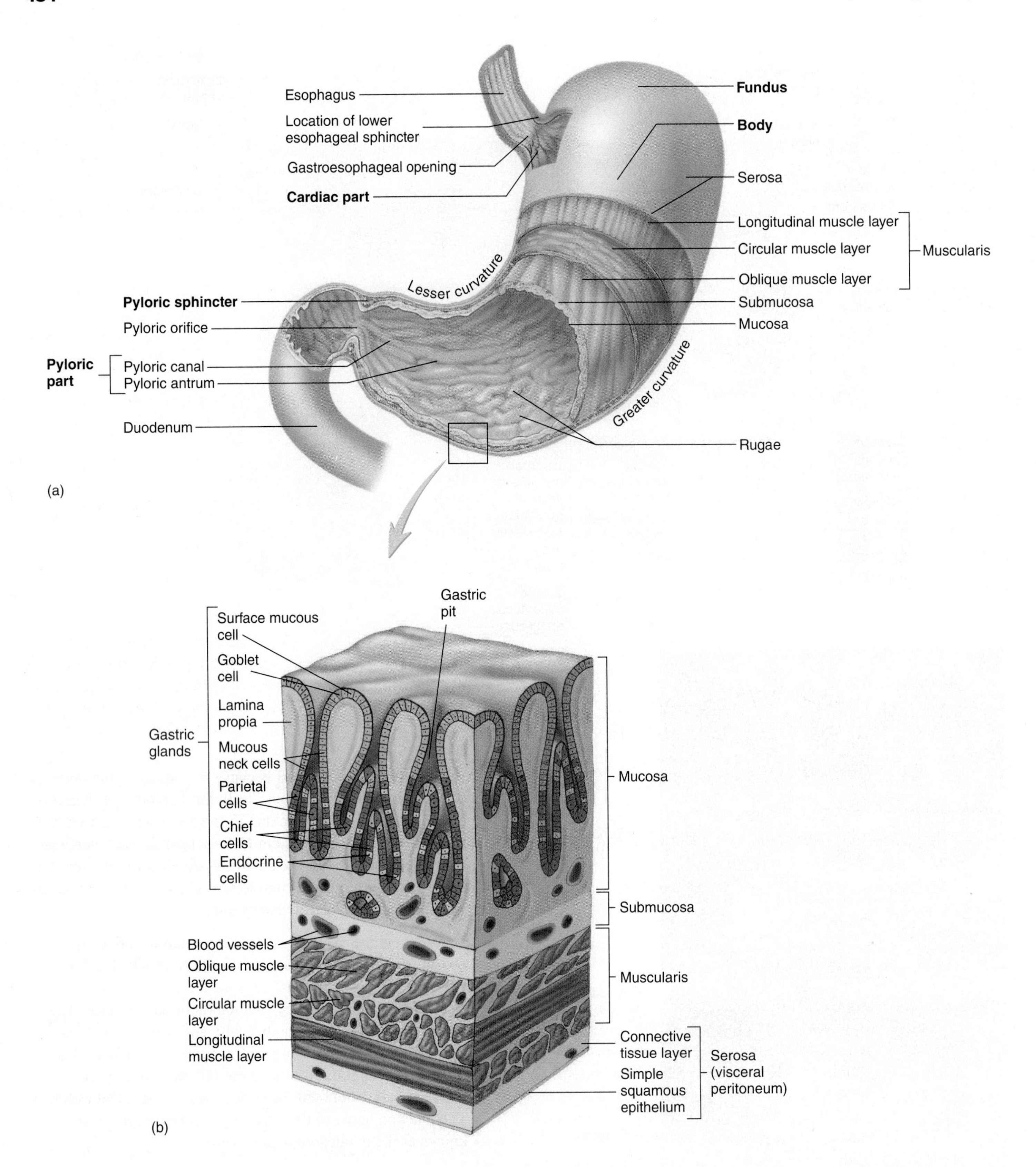

Figure 38.9 Anatomy of the Stomach

(a) Gross anatomy; (b) diagram of histology. (c) Photomicrograph (40×); (d) mucosa (100×).

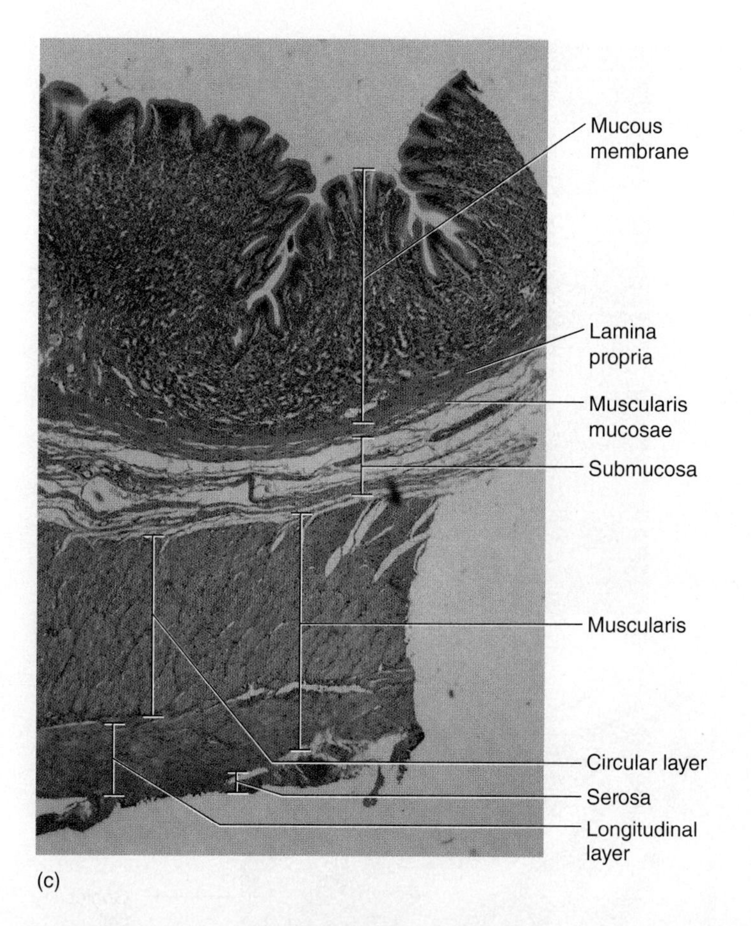

(c)

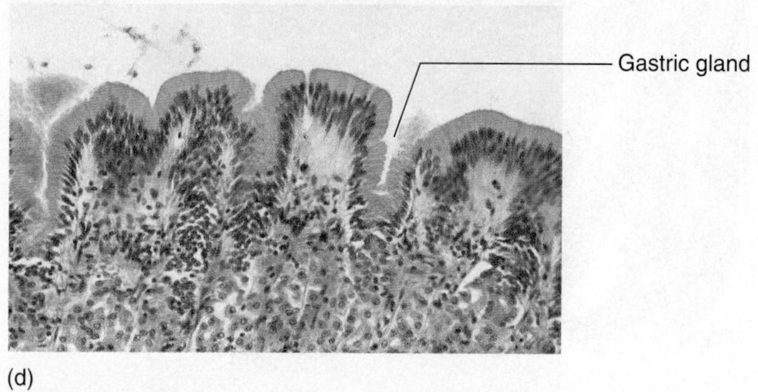

(d)

Figure 38.9 *Continued.*

Histology of the Small Intestine

Villi distinguish the small intestine from both the stomach and the large intestine. Villi are fingerlike projections that increase the surface area of the mucosa. Each villus contains **blood vessels** that transport sugars and amino acids from the intestine to the liver. In addition to this, the villi also contain **lacteals,** which transport fatty acids via lymphatics to the venous system. The villi give the lining of the small intestine a velvety appearance to the naked eye. As with the stomach, the inner lining of the small intestine consists of **simple columnar epithelium** with **goblet cells.**

10. Examine the prepared slides of the small intestine and compare it with figure 38.11. You may distinguish the three sections of the small intestine by noting that the duodenum has **duodenal glands** in the portion of the mucosa away from the lumen. The jejunum and the ileum lack these glands. The ileum is distinguished by the presence of **aggregated lymph nodules,** or **Peyer patches,** in the mucosa and submucosa. These lymphatic nodules produce lymphocytes, which protect the body from the bacterial flora in the lumen of the small intestine.

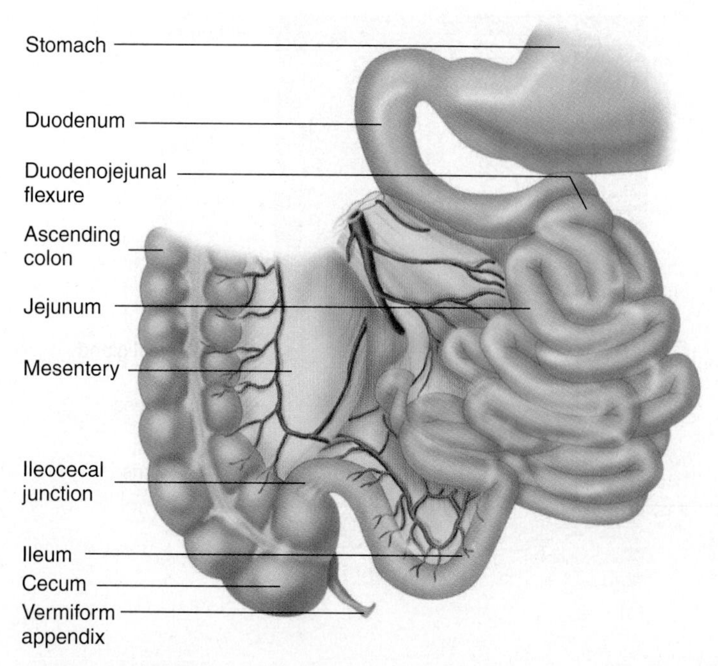

Figure 38.10 Gross Anatomy of the Small Intestine

Large Intestine

The **large intestine** is so named because it is large in diameter. The large intestine is approximately 7 cm (3 inches) in diameter and 1.4 m (4.5 feet) in length. The functions of the large intestine are the absorption of water and the formation of feces. The mucosa of the large intestine is made of **simple columnar epithelium** with a large number of **goblet cells.**

11. Look at a model or chart of the large intestine and note the major regions. Compare them with figure 38.12.

- **Cecum:** first part of the large intestine. The cecum is a pouchlike area that articulates with the small intestine at the ileocecal valve.
- **Ascending colon:** on the right side of the body. It becomes the transverse colon at the **hepatic (right colic) flexure.**
- **Transverse colon:** traverses the body from right to left. It leads to the descending colon at the **splenic (left colic) flexure.**
- **Descending colon:** passes inferiorly on the left side of the body and joins with the sigmoid colon.
- **Sigmoid colon:** an S-shaped segment of the large intestine in the left inguinal region.

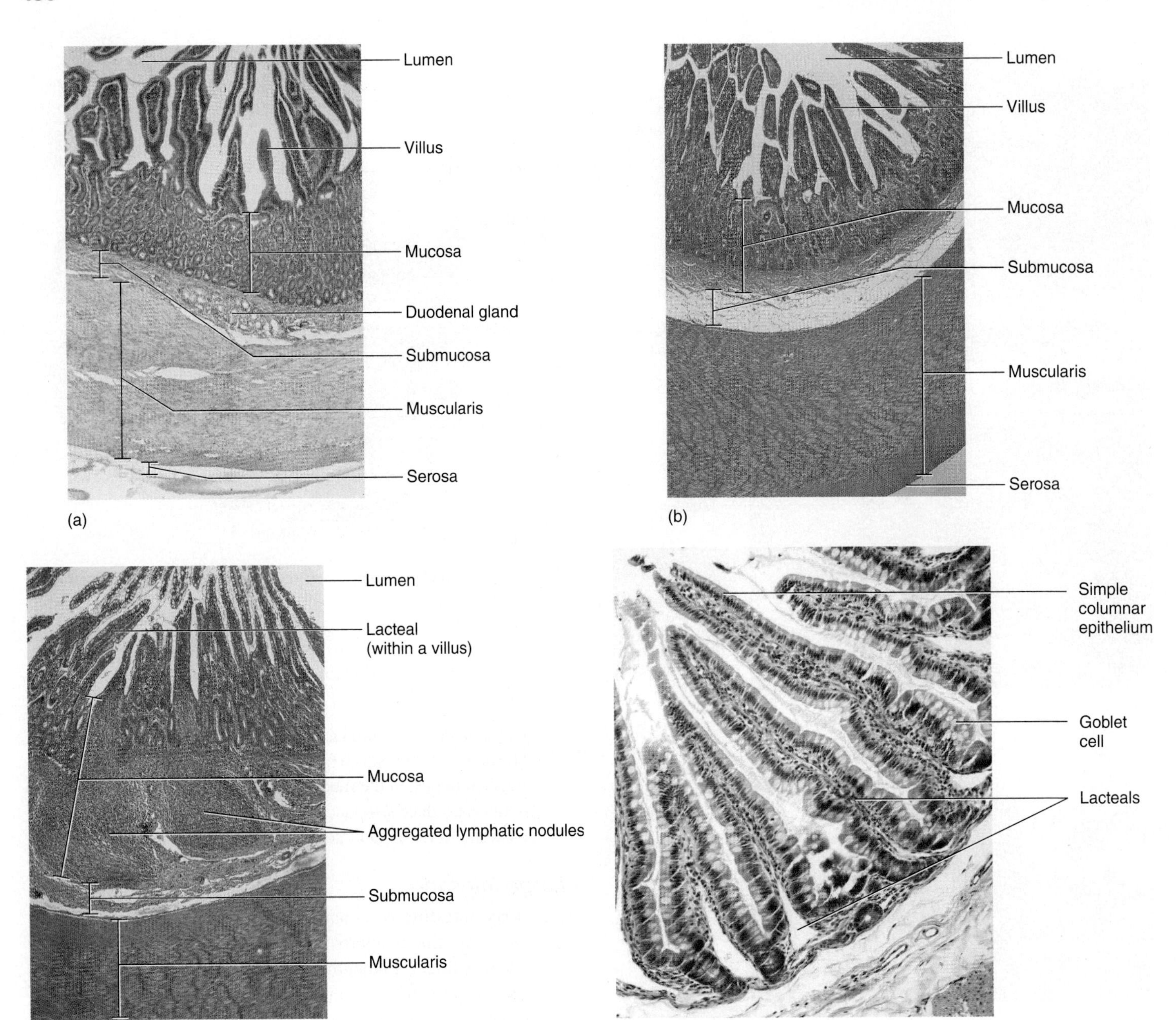

Figure 38.11 Sections of Small Intestine (40×)

(a) Duodenum; (b) jejunum; (c) ileum; (d) close-up of villus (100×).

- **Rectum:** a straight section of colon in the pelvic cavity. The rectum has superficial veins in its wall called **hemorrhoidal veins.** They may enlarge and cause the uncomfortable condition known as **hemorrhoids.**

The large intestine has some unique structures. The longitudinal muscle of the large intestine is not a continuous sheet but is located along the length of the large intestine as three bands called **teniae coli.** These muscles contract and form pouches or puckers in the intestinal tract called **haustra** (sing. **haustrum**). Another unique feature of the outer wall of the large intestine is the fat lobules called **omental appendages.** Locate these structures in figure 38.12.

Fecal material passes through the large intestine by peristalsis and is stored in the rectum and sigmoid colon. Defecation occurs as mass peristalsis causes a bowel movement.

Histology of the Large Intestine

12. Examine a prepared slide of the large intestine and compare it with figure 38.13. The large intestine is distinguished from the small intestine by the absence of villi and from the stomach by the presence of large numbers of goblet cells. There are no villi or aggregated lymphatic nodules present, yet the wall of the large intestine has solitary lymphatic nodules. Examine your slide for these characteristics.

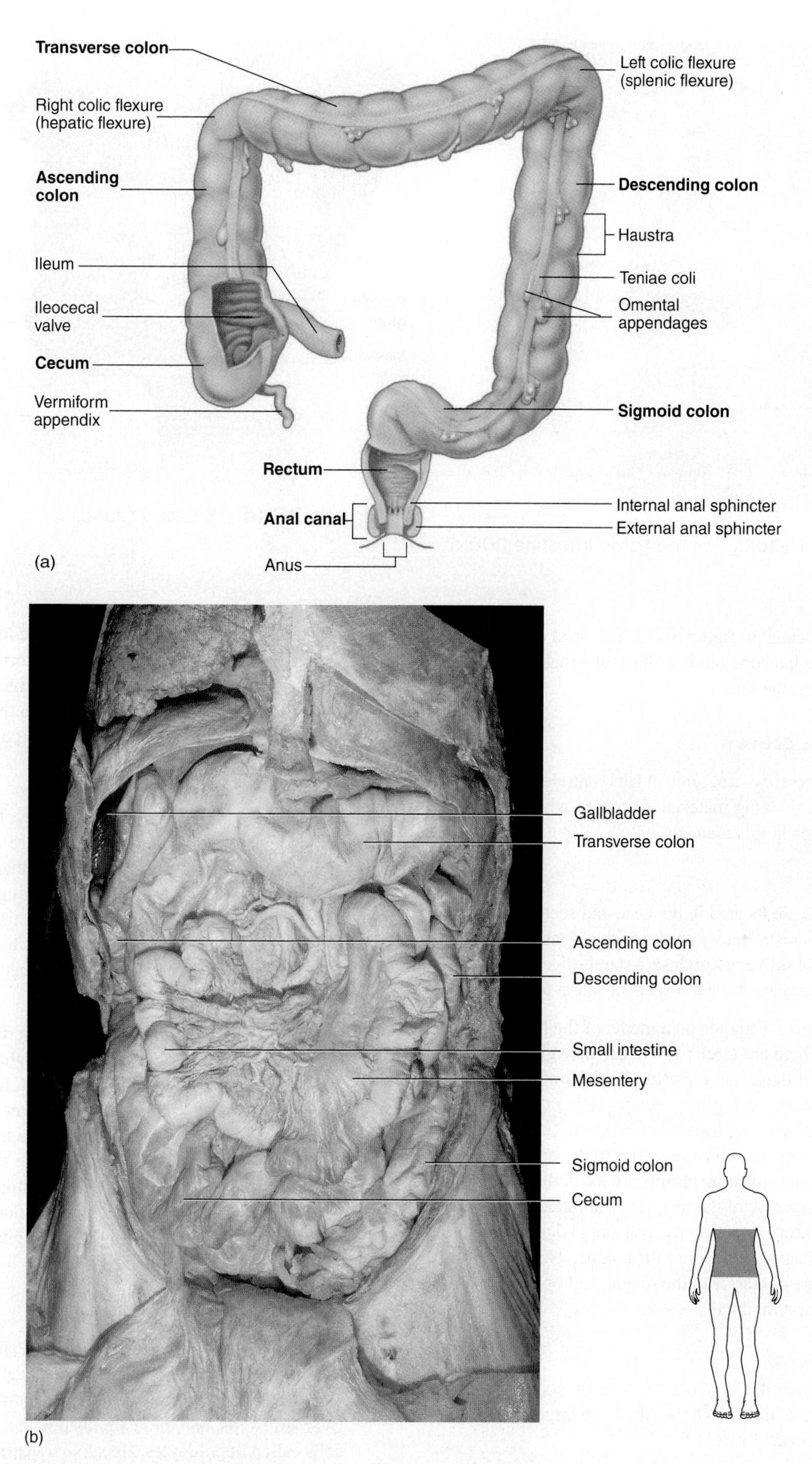

Figure 38.12 Gross Anatomy of the Large Intestine and Anal Canal
(a) Diagram; (b) photograph of cadaver.

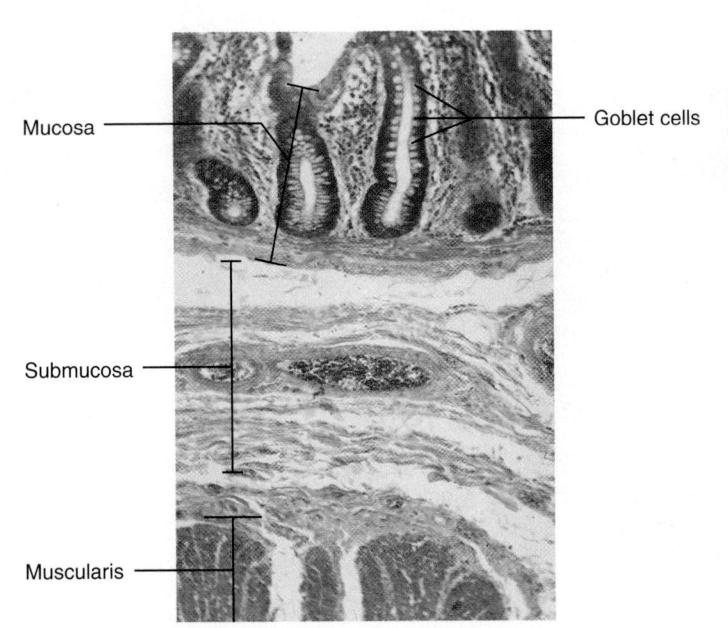

Figure 38.13 Histology of the Large Intestine (100×)

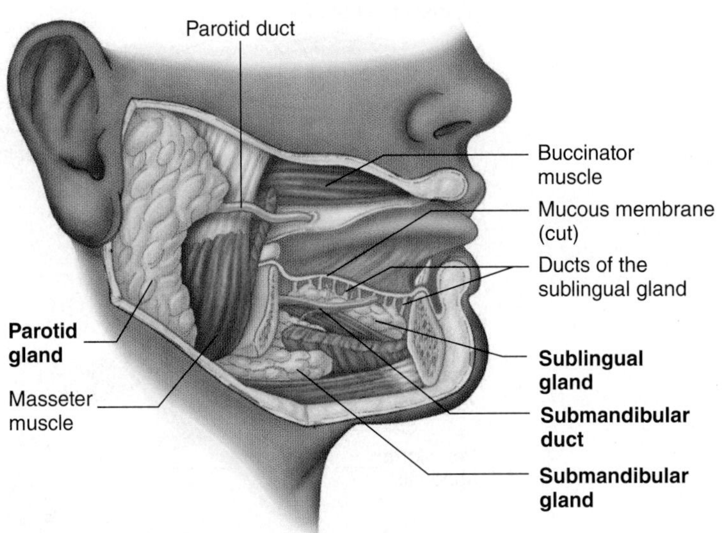

Figure 38.14 Salivary Glands

Anal Canal

13. Locate the anal canal in figure 38.12. The **anal canal** is not part of the large intestine but is a short tube that leads to an external opening, the **anus.**

Accessory Structures

(A) 3 Locate the accessory structures in lab on models or charts as you read the following material. Compare what you find in lab to the illustrations in this manual.

Salivary Glands

The **salivary glands** are located in the head and secrete saliva into the oral cavity. Saliva is a watery secretion that contains **mucus,** a protein lubricant, and **salivary amylase,** a starch-digesting enzyme. The average adult secretes about 1.5 liters of saliva per day.

1. Locate these salivary glands on a model of the head and in figure 38.14. There are three pairs of salivary glands. The most superior of these, the **parotid glands,** are located just anterior to the ears. Each gland secretes saliva through a **parotid duct,** a tube that traverses the buccal (cheek) region and enters the oral cavity just posterior to the upper second molar. The **submandibular glands** are located medial to the mandible on each side of the face. The submandibular glands secrete saliva into the oral cavity by a single duct on each side of the oral cavity inferior to the tongue. The **sublingual glands** are located inferior to the tongue and open into the oral cavity by several ducts.

Vermiform Appendix

The **vermiform appendix** is about the size of your little finger and is located near the junction of the small and large intestines (at the region of the ileocecal valve).

2. Locate the appendix on a torso model or chart in the lab and compare it with figures 38.12 and 38.15.

Omenta

3. Locate the omentum in figure 38.7. The **lesser omentum** is an extension of the peritoneum that forms a double fold of tissue between the stomach and the liver. The **greater omentum** is a section of peritoneum on the transverse colon and drapes over the intestines as a fatty curtain.

Liver

The **liver** is a complex organ with many functions, some of which are digestive but many of which are not. The liver processes digestive material from the vessels that return blood from the intestines and has a role in both moving nutrients into the bloodstream and storing them in the liver tissue. The liver also produces blood plasma proteins, detoxifies harmful material that has been produced by the body or introduced into the body, and produces bile.

4. Examine a model or chart of the liver and note its relatively large size. The liver is located on the right side of the body and is divided into four lobes: the **right, left, quadrate,** and **caudate lobes.** The right lobe of the liver is the largest lobe; the left is also fairly large. The quadrate lobe is located in the middle portion of the liver ventral to the caudate lobe. Traversing through the liver is the **falciform ligament,** which attaches the liver to the inferior side of the diaphragm. Locate the liver, and compare it with figure 38.16.

Liver Histology

5. Examine a prepared slide of liver tissue. Note the hexagonal structures in the specimen. These are the **liver lobules.** Each lobule has a blood vessel in the middle called the **central vein.** Vessels that carry blood to the central vein are the **liver sinusoids,** and these are lined with a double row of cells called **hepatocytes.** Hepatocytes carry out the various functions of the liver, as just described. The **hepatic arteries, hepatic portal veins,** and **hepatic ducts** frequently are

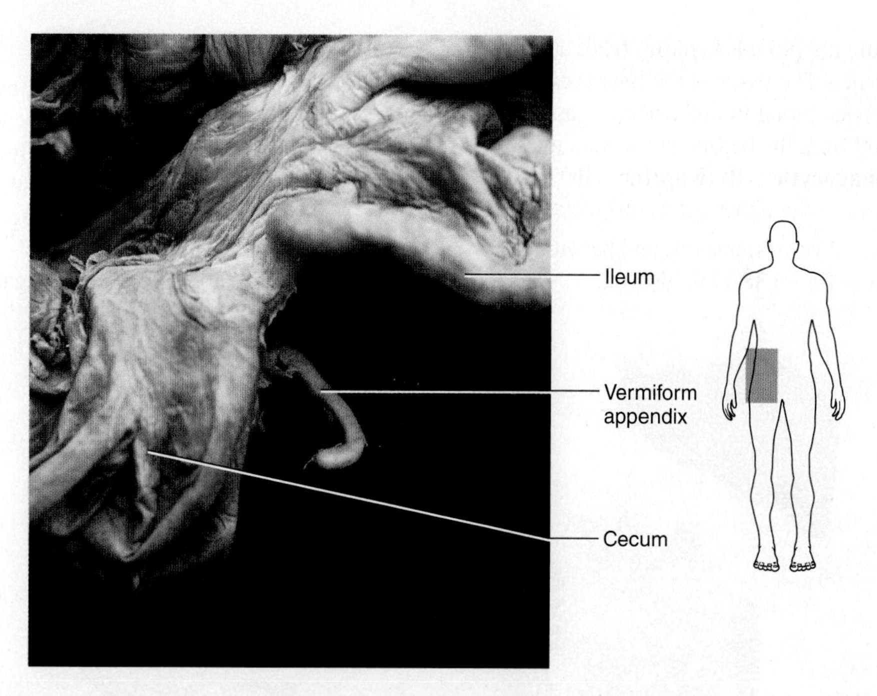

Figure 38.15 Vermiform Appendix

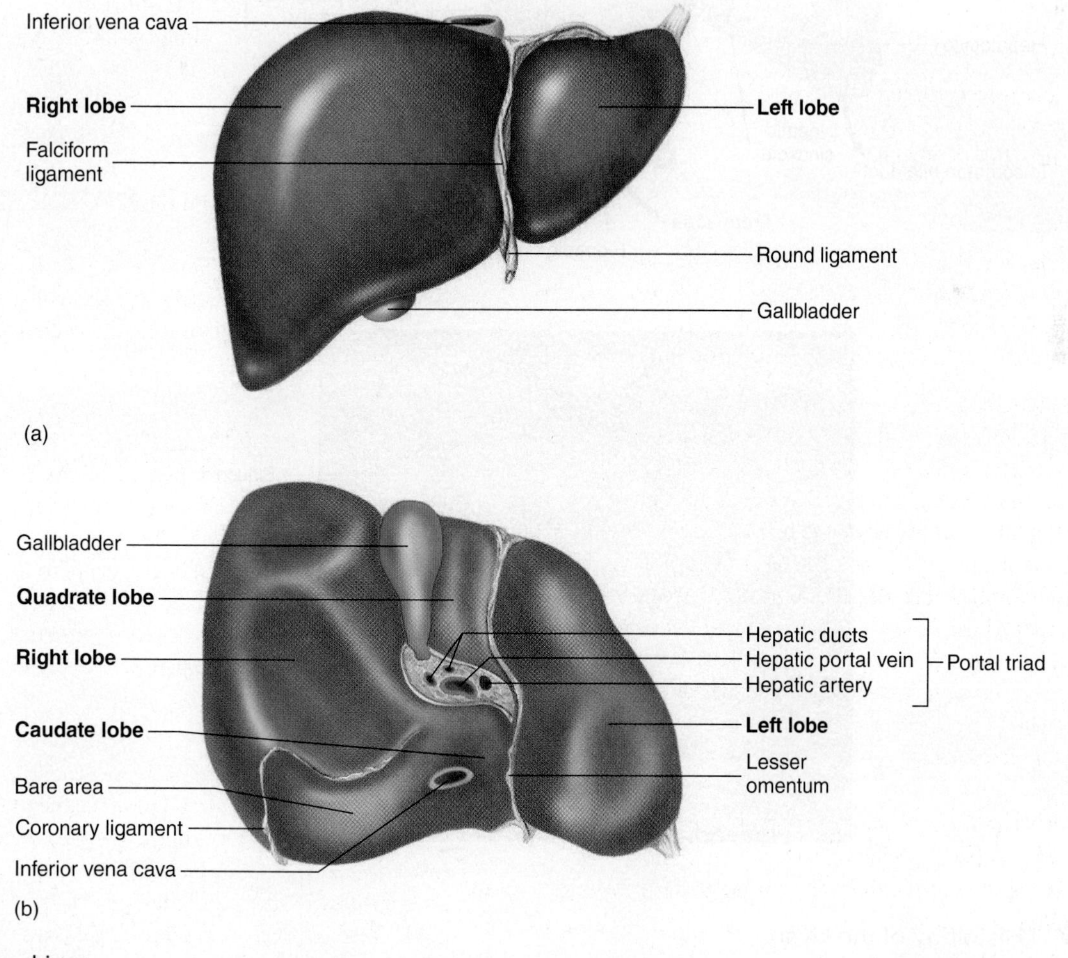

Figure 38.16 Liver

(a) Anterior view; (b) inferior view.

located together, forming the **portal (hepatic) triad** at the corners of the liver lobules. The tissue of the liver is extremely vascular. Fresh, oxygenated blood from the hepatic artery and deoxygenated blood from the hepatic portal vein mix in the liver. **Hepatic phagocytic cells (Kupffer cells)** exist throughout the liver tissue, where they act as phagocytic cells.

6. Locate the lobules, central vein, sinusoids, and hepatocytes in a prepared slide, using figure 38.17 to aid you.

Pancreas

7. Locate the pancreatic structures, as represented in figure 38.18. The **pancreas** is located inferior to the stomach and on the left side of the body. It has both endocrine and exocrine functions. The hormonal function of the pancreas is covered in Exercise 24. The exocrine function of the pancreas is studied in Exercise 39. The pancreas consists of a **tail** near the spleen; an elongated **body;** and a rounded

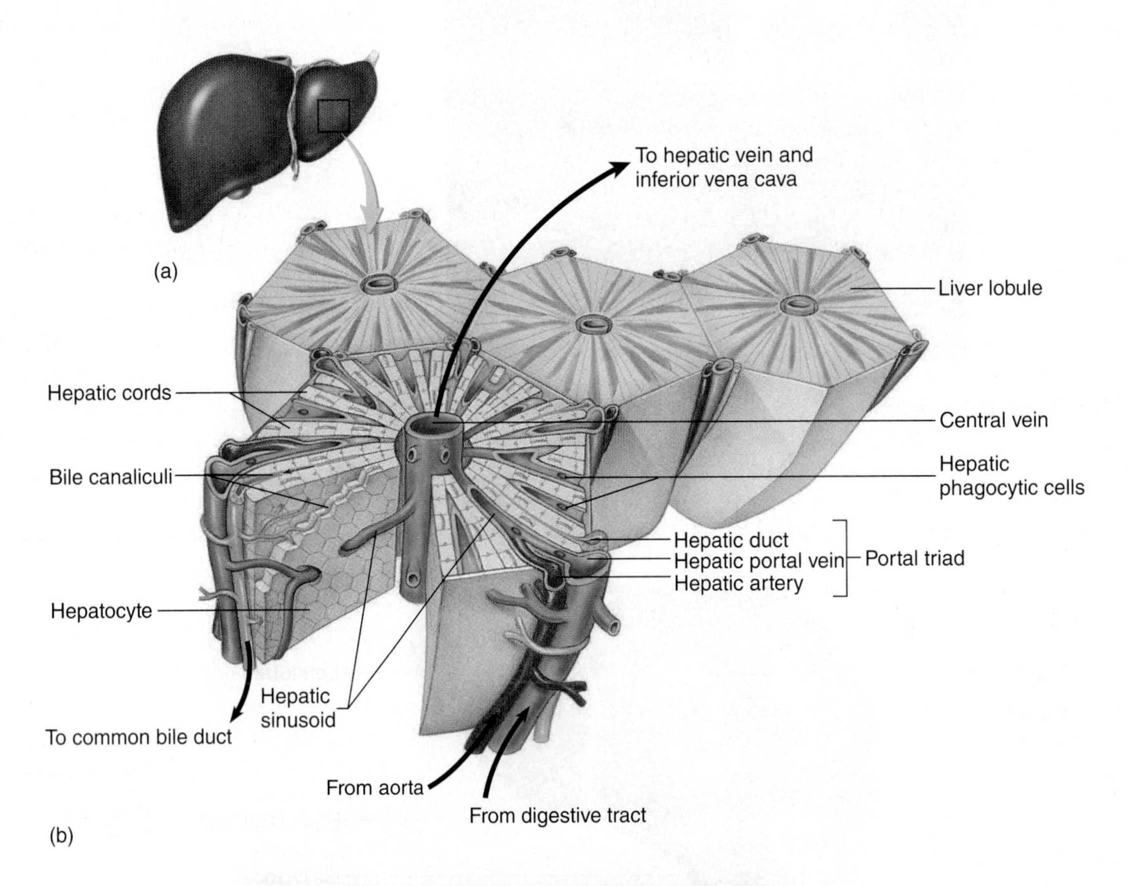

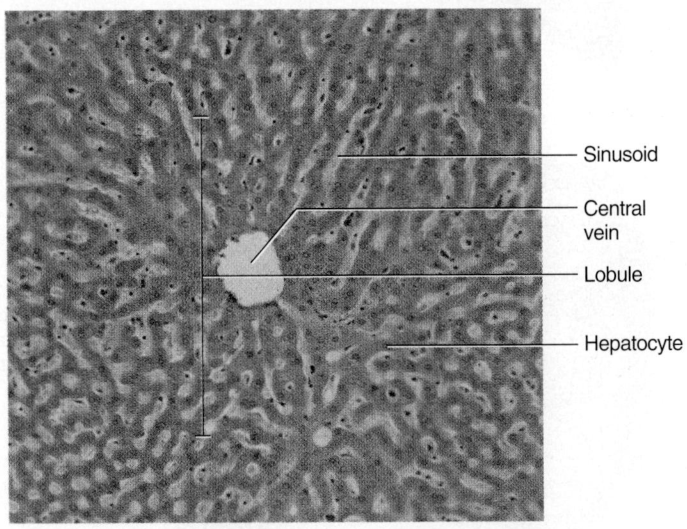

Figure 38.17 Histology of the Liver

(a) Anterior view of liver; (b) diagram of liver lobule; (c) photomicrograph of liver lobule (100×).

head near the duodenum. Enzymes and buffers pass from the tissue of the pancreas into the **pancreatic duct** and then into the duodenum.

Gallbladder

The **gallbladder** releases bile, which emulsifies lipids, into the duodenum. The lipids break into smaller droplets, which increase the surface area for digestion. The gallbladder is located just inferior to the liver. The liver is the site of bile production, and the bile flows from the liver through **left** and **right hepatic ducts** (figure 38.18, step 1) to enter the **common hepatic duct.** Once in the common hepatic duct, the sphincter (4) closes. Bile fills the ducts and flows into the **cystic duct** (2) and then is stored in the gallbladder. As the stomach begins to empty its contents into the duodenum, the gallbladder constricts, and bile flows from the gallbladder back into the cystic duct and into the **common bile duct,** which empties into the duodenum (4).

8. Locate these structures on a model and compare them with figure 38.18.

The bile may be transported to the duodenum by the common bile duct. The pancreas secretes many digestive enzymes and buffers, which neutralize the stomach acids. These are secreted into the duodenum by the **pancreatic duct.** In some cases, the pancreatic duct joins with the **common bile duct** to form the **hepatopancreatic ampulla** (ampulla of Vater).

Cat Dissection

(A)[4] 1. Prepare for the cat dissection by obtaining a dissection tray, a scalpel, scissors or bone cutter, string, forceps, and a plastic bag with a label. Remember to place all excess tissue in the appropriate waste container and not in a standard wastebasket or down the sink!
2. Begin your study of the digestive system of the cat by locating the large **salivary glands** around the face. You must first remove the skin from in front of the ear. Use figure 38.19 to locate these glands. If you have dissected the face for the musculature, you may have already removed the **parotid gland,** which is a spongy, cream-colored pad anterior to the ear.
3. Locate the **submandibular gland** as a pad of tissue slightly anterior to the angle of the mandible and lateral to the digastric muscle. The **sublingual gland** is just anterior to the submandibular gland and is an elongated gland that parallels the mandible.
4. Examine the **tongue** of the cat and note the numerous papillae on the dorsal surface. These are **filiform papillae,** and they have both a digestive and a grooming function. You can lift up the tongue and examine the **lingual frenulum** on the ventral surface. Dissection of the head of the cat should be done only if your instructor gives you permission to do so.
5. Use a scalpel and make a cut in the midsagittal plane from the forehead of the cat to the region of the occipital bone.

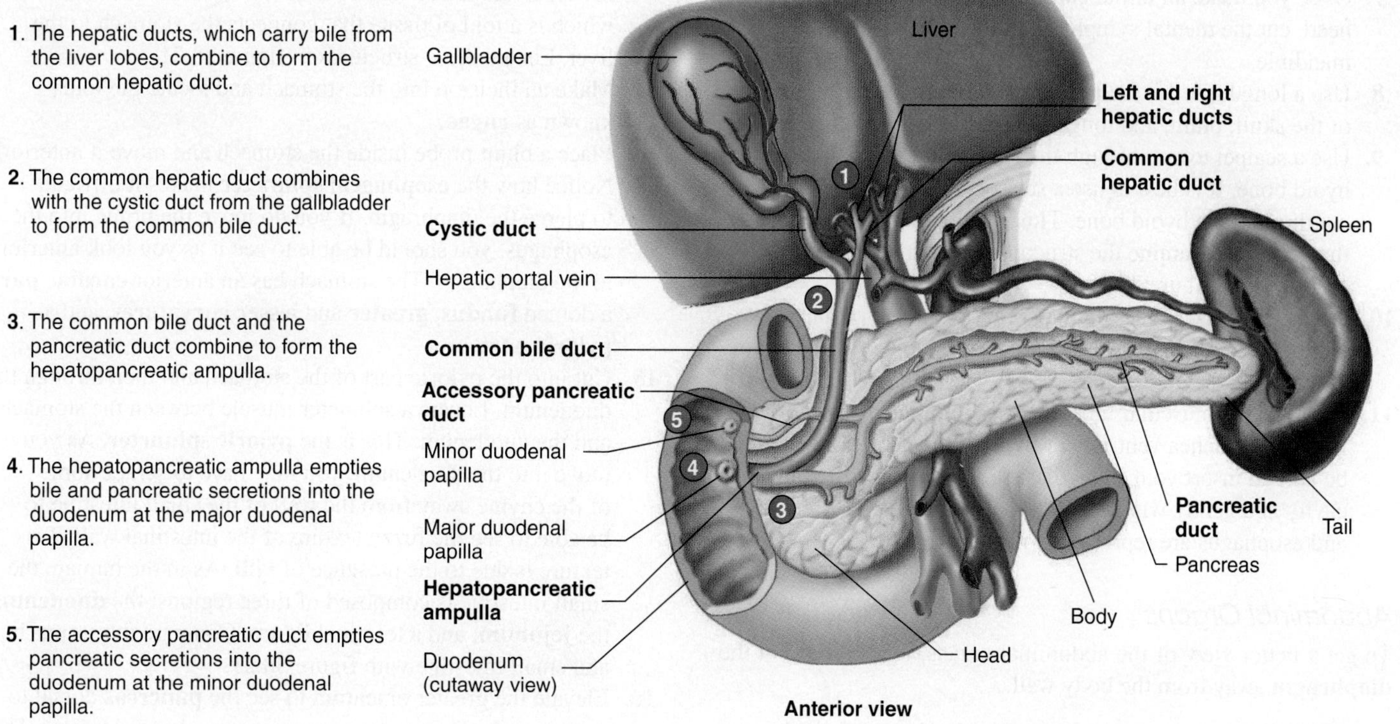

Figure 38.18 Liver, Gallbladder, and Pancreas

Bile flow moves from the liver through the hepatic ducts (1), to the cystic duct (2). Bile is stored in the gallbladder, then flows back out the cystic duct to the common bile duct (3) and then into the duodenum (4). The pancreas secretes enzymes through to pancreatic duct, to the hepatopancreatic ampulla (4) and through the accessory pancreatic duct (5).

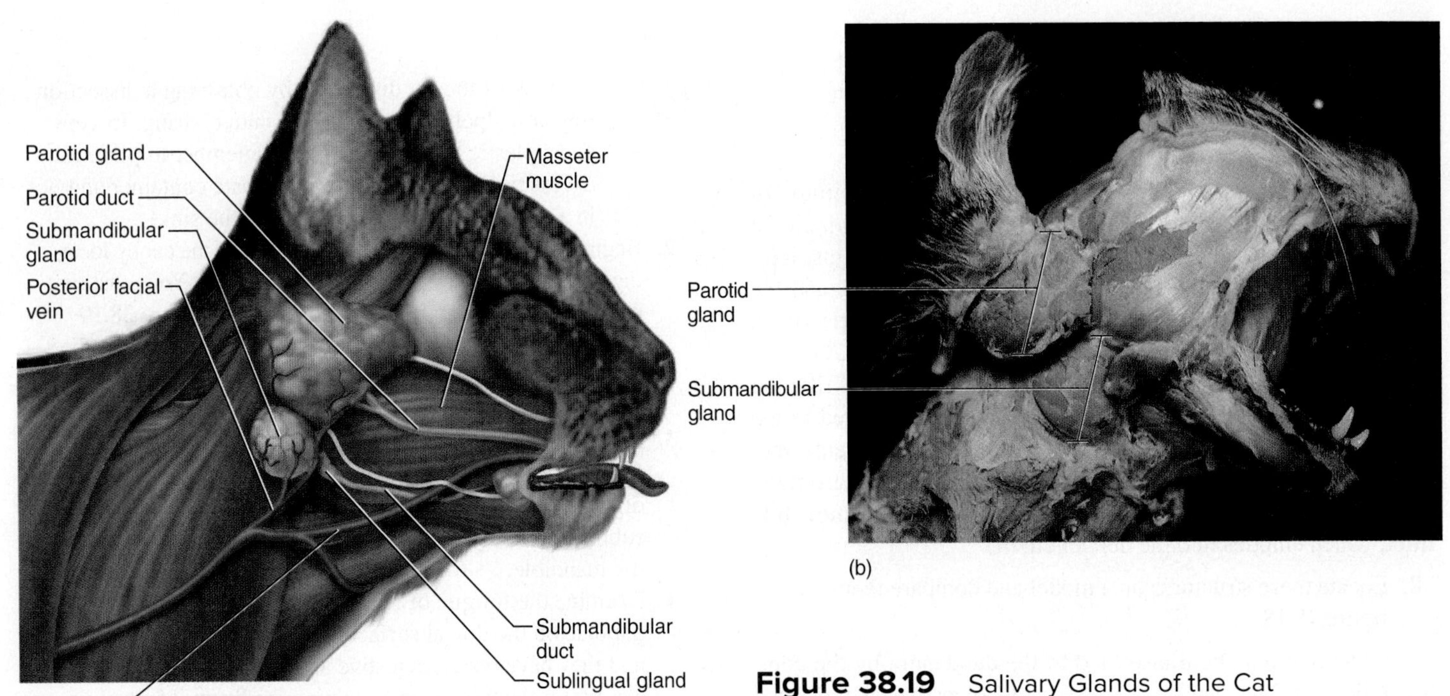

Figure 38.19 Salivary Glands of the Cat
(a) Diagram; (b) photograph.

6. Cut through any overlying muscle in the cat. With a small saw, gently cut through the cranial region of the skull, being careful to stay in the midsagittal plane.
7. Once you make an initial cut through the dorsal side of the head, cut the mental symphysis of the cat, separating the mandible.
8. Use a long knife (or scalpel) to cut through the softer regions of the skull, brain, and tongue.
9. Use a scalpel to cut through the floor of the oral cavity to the hyoid bone. It is best to use a scissors or small bone cutter to cut through the hyoid bone. This should allow you to open the head and examine the structures seen in a midsagittal section, as in figure 38.20.
10. Locate the hard palate, tongue, oral cavity, and oropharynx. Note how the larynx in the cat is closer to the tongue than in humans.
11. Look for the muscular tube of the **esophagus** by carefully lifting the trachea ventrally away from the neck. You should be able to insert your blunt dissection probe into the oral cavity and gently wiggle it into the esophagus. The tongue and esophagus are represented in figure 38.20.

Abdominal Organs

To get a better view of the abdominal organs, it is best to cut the **diaphragm** away from the body wall.

12. Carefully cut the lateral edges of the diaphragm from the ribs. You should see a fatty curtain of material covering the intestines. This is the **greater omentum.** Just caudal to the diaphragm, note the dark brown, multilobed **liver.** In the middle of the liver is the green **gallbladder.** To the left of the liver (in reference to the cat) is the J-shaped **stomach.** If you lift the liver, you should be able to see the **lesser omentum,** which is a fold of tissue that connects the stomach to the liver. Locate these structures in figure 38.21.
13. Make an incision into the stomach and locate the folds known as **rugae.**
14. Place a blunt probe inside the stomach and move it anteriorly. Notice how the **esophageal sphincter** makes it difficult to pierce the diaphragm. If you do move the probe into the esophagus, you should be able to see it as you look anterior to the diaphragm. The stomach has an anterior **cardiac part,** a domed **fundus, greater** and **lesser curvatures,** and a **pyloric part.**
15. Cut into the pyloric part of the stomach and then through the duodenum. Locate a sphincter muscle between the stomach and the duodenum. This is the **pyloric sphincter.** As you move into the duodenum, you may have to scrape some of the chyme away from the wall of the small intestine to be able to see the fuzzy texture of the intestinal wall. This texture is due to the presence of **villi.** As in the human, the small intestine is composed of three regions: the **duodenum,** the **jejunum,** and a terminal **ileum.** Compare the stomach and small intestine with figure 38.21.
16. Elevate the greater omentum to see the **pancreas** dorsal to the stomach. The pancreas appears granular and brown. The **tail** of the pancreas is near the **spleen,** which is a brown, elongated organ on the left side of the body.

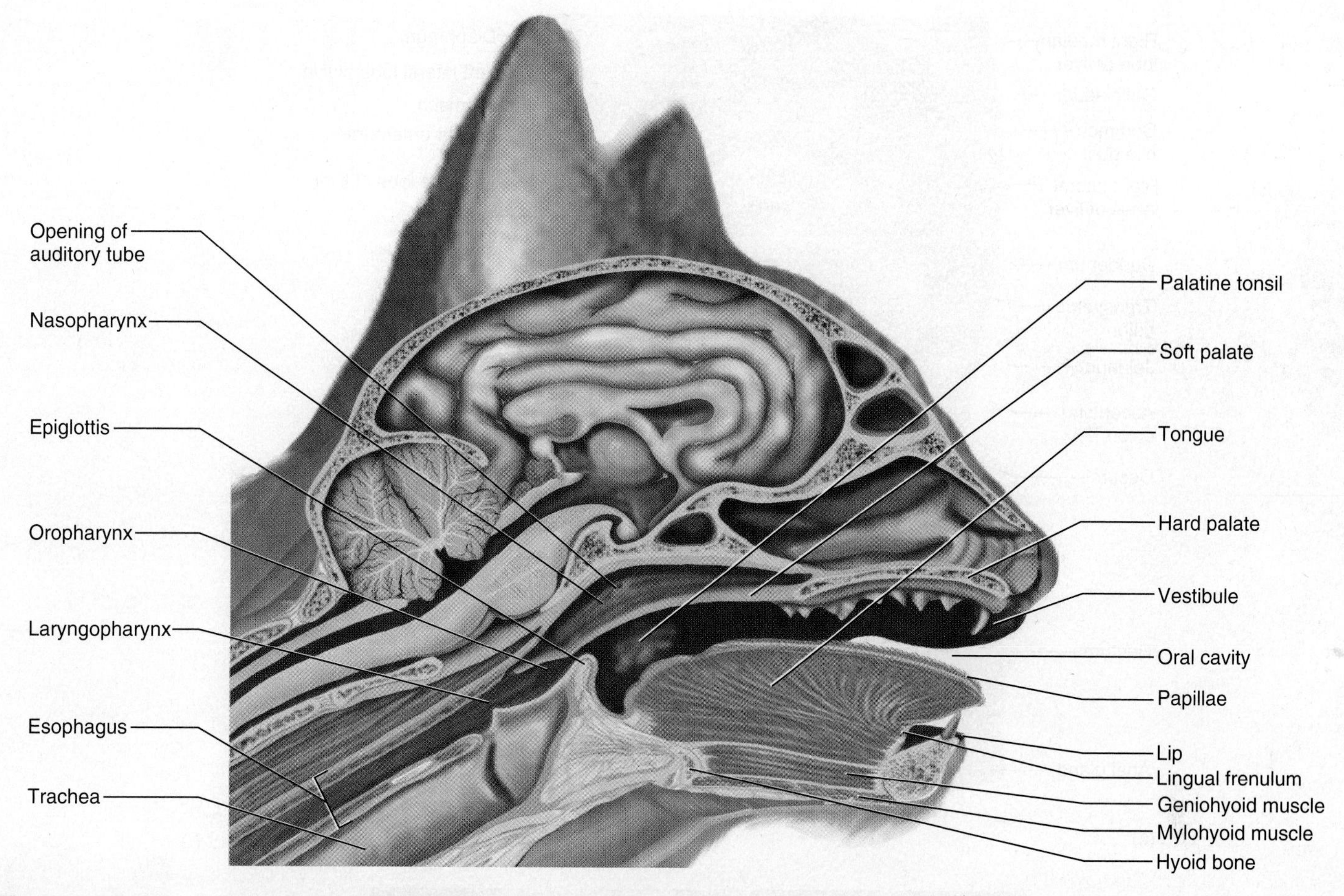

Figure 38.20 Midsagittal Section of the Head and Neck of the Cat

The small intestine is an elongated, coiled tube about the diameter of a wooden pencil. Note the **mesentery,** which holds the small intestine to the posterior body wall near the vertebrae. The small intestine is extensive in the cat and rapidly expands into the large intestine. There is no vermiform appendix in the cat. In humans, it is found at the ileocecal junction.

The **large intestine** in the cat is a fairly short tube with a diameter slightly larger than your thumb. The first part of the large intestine is a pouch called the **cecum.** The remainder of the large intestine can be further divided into the **ascending, transverse,** and **descending colon** and the **rectum.**

17. Compare these with figure 38.21. Examine the **parietal peritoneum** along the inner surface of the body wall and the **visceral peritoneum** that envelopes the intestines.

Cleanup

When you are done with your dissection, carefully place your cat back in the plastic bag. Place all waste material in the appropriate container.

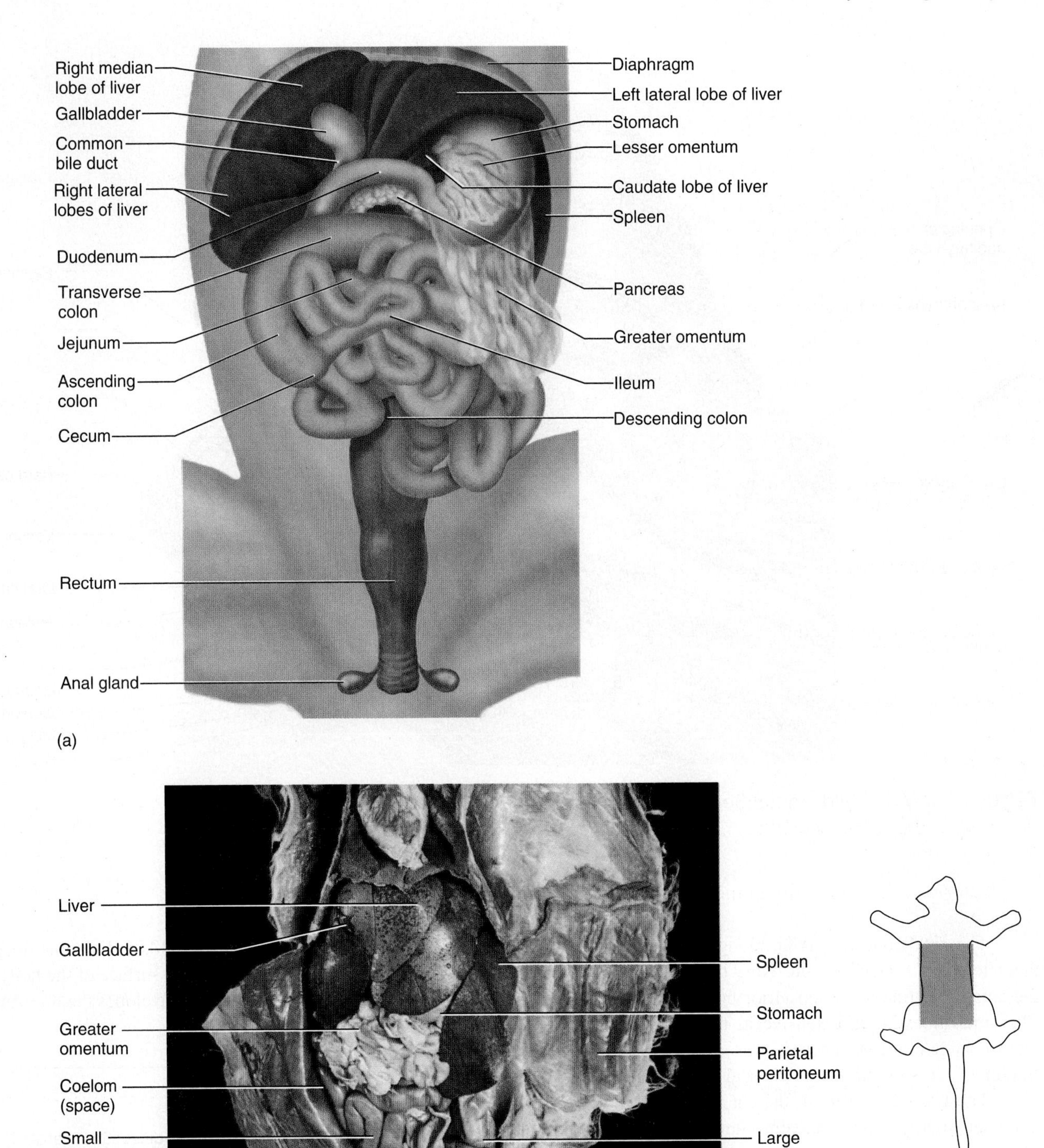

Figure 38.21 Abdominal Organs of the Cat

(a) Diagram; (b) photograph.

Exercise 38 Review

Name: ______________________

Lab time/section: ______________________

Date: ______________________

Anatomy of the Digestive System

Matching

Match the terms on the left with those on the right. A term may be used more than once or not at all.

Set One

1. pancreas	a. jejunum
2. descending colon	b. pyloric region
3. part of the stomach closest to the small intestine	c. ileum
4. middle portion of the small intestine	d. alimentary canal
5. distal portion of the small intestine	e. accessory organ

Set Two

6. outer surface of the stomach	a. submucosa
7. layer adjacent to the lumen of intestine	b. serosa
8. cell type in the muscularis	c. mucosa
9. location of the villi	d. smooth muscle

10. In the stomach, what is partially digested food called? ______________________

11. Where are lacteals located in the digestive tract? ______________________

12. What membrane holds the tongue to the floor of the oral cavity? ______________________

13. What part of the tooth is found above the neck? ______________________

14. What is the layer of a tooth superficial to the dentin? ______________________

15. What are the adult teeth that are directly posterior to the canine teeth called? ______________________

16. The segments, or pouches, of the large intestine have what name? ______________________

17. What are the names of the salivary glands located anterior to the ear? ______

18. Where is the lesser omentum found? ______

19. Where does the cystic duct take bile for storage? ______

20. Trace the flow of bile from the liver to the duodenum, listing all the structures that come into contact with the bile on its journey. ______

21. How does the large intestine differ from the small intestine in terms of length? ______

22. How does the large intestine differ from the small intestine in terms of diameter? ______

23. Name two functions of the pancreas. ______

24. Label the following illustration using the terms provided.

ascending colon	cecum	descending colon
duodenum	esophagus	liver
oral cavity	parotid gland	rectum
sigmoid colon	small intestine	stomach
submandibular gland	tongue	transverse colon
vermiform appendix		

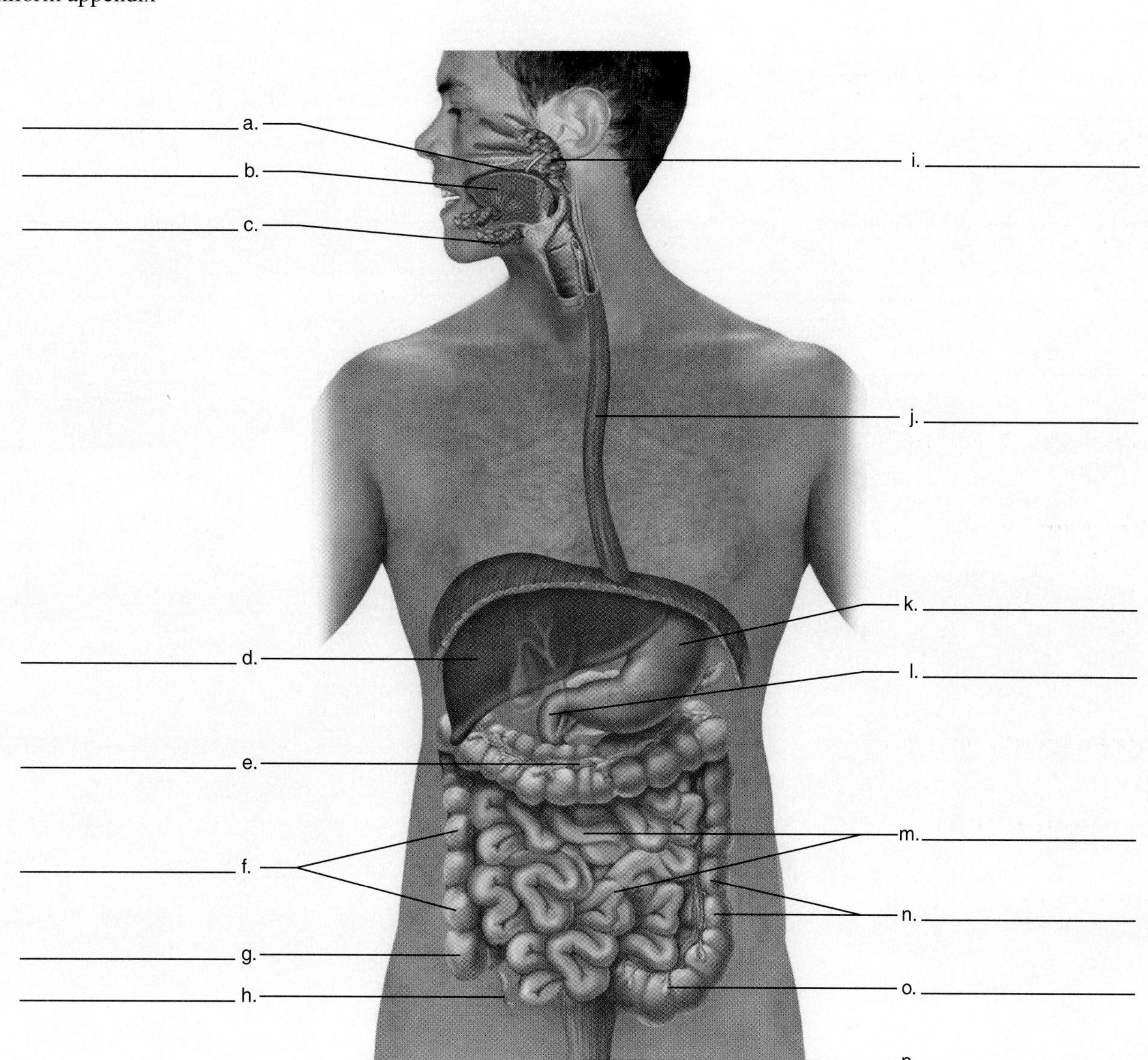

Digestive Organs

NOTES

Exercise 39

Digestive Physiology

Introduction

An understanding of the digestive process is fundamental to the study of human physiology. Digestive physiology is covered in the Seeley text in chapter 24, "Digestive System," and chapter 25, "Nutrition, Metabolism, and Temperature Regulation."

We obtain almost all the raw materials the body uses for growth, development, life-sustaining energy, and other metabolic functions from digestion. Digestion and nutrition have significant health implications. Poor diet has been linked to inadequate body functions and to conditions such as heart disease and colon cancer. Digestive physiology also involves the physical and chemical breakdown of material, the absorption of nutrients, coordination between sections of the digestive tract and accessory organs (liver secretions and processing of absorbed foods), and the elimination of fecal material. In this exercise, you look at the physical (mechanical) and chemical digestion of food.

To develop some understanding of the digestive process, you examine the digestive functions of the enzymes. In this experiment, you analyze **pancreatin** (an extract of the pancreas containing several enzymes important in digestion) and amylase. You determine the effectiveness of these enzymes in digesting four materials commonly found in plant and animal tissue: starch, lipid, cellulose, and protein. The goal of the experiment is to determine if these enzymes can break down each of the four substances into smaller products. If the enzymes are effective, then starch should be broken down into sugars, lipids broken down into glycerol and fatty acids, cellulose reduced to sugars, and proteins split into amino acids. Important biomolecules are illustrated in your text (in Seeley, they are in chapter 2, "The Chemical Basis of Life" in the "Organic Chemistry" section).

Furthermore, you examine the mechanical breakdown of food that precedes most of the chemical processes. By taking solid food and increasing its surface area, you determine if the rate of digestion increases. This is an example of the role of mechanical breakdown in the digestion of food.

Objectives

At the end of this exercise, you should be able to

1. discuss the importance of catabolism in the digestive process;
2. describe how an enzyme is important in chemical digestion;
3. list possible human digestive enzymes as determined by experiment;
4. demonstrate the effectiveness of enzymes in food digestion;
5. describe the mechanisms involved in the mechanical digestion of food;
6. state one of the limitations of human digestive enzymes.

Materials

Microscopes

General Supplies

80 test tubes (15 mL each)
12 test tube holders
6 test tube racks
12 test tube brushes and soap
Warm water bath, set at 37°C, with test tube racks and thermometer
Hot water bath or hot plates and 400 mL beakers for hot bath (100°C) and thermometer
6 permanent markers
18 pipettes of 5 mL each
6 pipette pumps
7 100 mL beakers to pour from stock bottle for distribution
1 250 mL bottle of tap water
6 small spatulas
100 mL 1% pancreatin solution, fresh
300 mL 1% alpha amylase solution, fresh
100 mL 0.1% alpha amylase solution, fresh
300 mL Benedict's reagent (6 50 mL bottles)
Parafilm squares (10 per table)
Biohazard container or container with 10% bleach solution

Starch Digestion

6 dropper bottles of iodine solution
250 mL 0.5% potato starch solution
6 boxes of microscope slides (1 per table)
6 boxes of coverslips (1 per table)
6 eyedroppers

Sugar Test

100 mL 1% maltose solution
6 10 mL graduated cylinders

Cellulose Digestion

Cellulose—2 cotton balls cut into small pieces
6 small dropper bottles of water

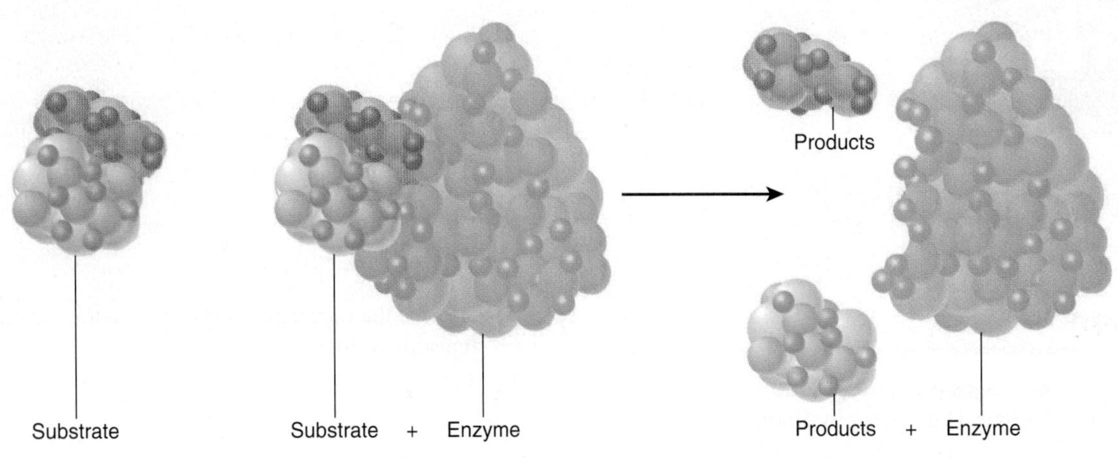

Figure 39.1 Catabolic Reactions with Enzymes

Lipid Digestion

200 mL litmus cream
6 dropper bottles of "acid solution" (lemon juice or vinegar)
6 dropper bottles of 1% sodium hydroxide solution (NaOH)
12 pair of goggles (during use of NaOH)
12 pair of protective gloves

Protein Digestion

25 mL 1% BAPNA solution; BAPNA (benzoyl-DL-arginine-p-nitroanilide), available from Sigma B4875 or Aldrich 85,711-4

Surface Area and Digestion

2 large potatoes per class
2 breadboards
2 number 4 cork borers
2 small rods (applicator sticks) to push through the borers
12 rulers (15 cm)
6 razor blades
2 mortars and pestles

Procedure

General Procedures

Read all of each experiment before beginning. Your instructor may want you to do these experiments in student pairs or as members of a larger group. If you do these experiments as part of a larger group, make sure that you *witness* and *record* the results of your group. **Do not discard any material until everyone in your group has seen the result** or until your instructor directs you to do so. While you are waiting for one set of materials to incubate, you can begin the next set of experiments.

Label all test tubes with your group name and the test tube number. This is very important, because removing the wrong tube will give you erroneous results, and removing a tube belonging to another group will produce the wrong result not only for your group but for the other group as well.

Tests in this exercise are divided into **reagent tests** and **experimental tests.** The reagent tests allow you to see the results of an experiment when using known samples. The experimental tests determine if a digestive reaction has taken place.

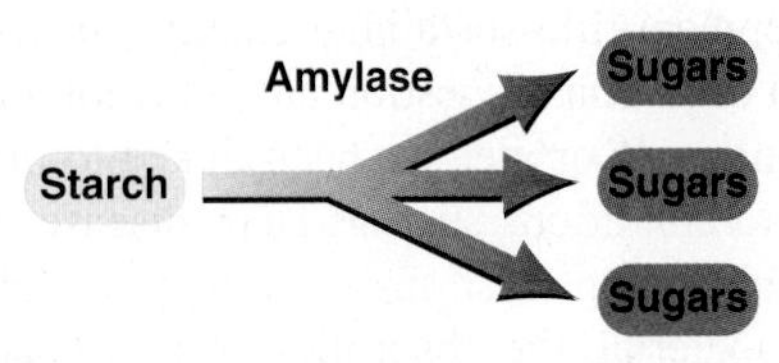

Figure 39.2 Starch Digestion

The Nature of Enzymes

The function of a digestive enzyme is to break down large **macromolecules** in the stomach and small intestine into smaller biomolecules in a process known as **catabolism.** This enzymatic process is shown in figure 39.1.

Starch Digestion

Starch digestion occurs by the breakdown of a large molecule of starch into smaller molecules of sugar by enzymes called **amylases.** Amylases catabolize starches as represented by figure 39.2. One way to analyze the effectiveness of amylases is to see if the enzyme removes starch from solution. The decrease or absence of starch in solution indicates that amylases are present.

(A)[1] *Experiment 1: Reagent Test—Iodine*

Use **iodine** as a reagent test to determine the presence of starch. If starch is present, iodine produces a blue-black color, which is a **positive result.** If starch is absent, then the solution remains yellow, and this is a **negative result.**

1. Label a test tube with your group name and the number 1.
2. Pour a small amount of starch solution from the stock bottle into a small, labeled beaker. *Never return excess solution to the stock bottle!*
3. Take a pipette pump and pipette and place 1 to 2 mL of starch solution from the small beaker into the test tube labeled number 1.

Caution!

Never pipette material by mouth; use a pipette pump or bulb.

---

4. Add four or five drops of iodine solution to the starch solution in test tube 1. Cover with parafilm. The solution should turn blue or black if starch is present. Save this test result for future comparison and record the outcome of your test.

   Color of reaction (blue/black or yellow): ________________

   Test result (positive/negative for starch): ________________

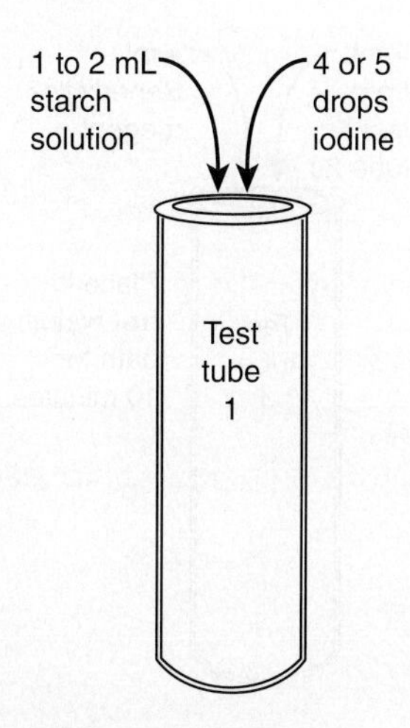

Experiment 2: Determination of Starch Digestion by Amylase

1. Using a pipette, remove 5 mL of starch solution from the beaker that received the stock solution and place it in a small test tube that you have labeled with your group name and the number 2a.
2. Pour a small amount of 1% amylase from the stock bottle into a small, labeled beaker. With a clean pipette, withdraw 2 mL of amylase solution from the beaker and add it to the starch solution. Discard any remaining solution in your pipette. Do not return it to the beaker. You can substitute 2 mL of saliva for the amylase if directed by your instructor. Do not use your saliva if you recently had sugar in your mouth. Remember to be careful with bodily fluids. Stir the solutions in the test tube by flicking the bottom of the tube gently with your finger.
3. Incubate the mixture in test tube 2a in a **warm** water bath (37°C) for 60 minutes.

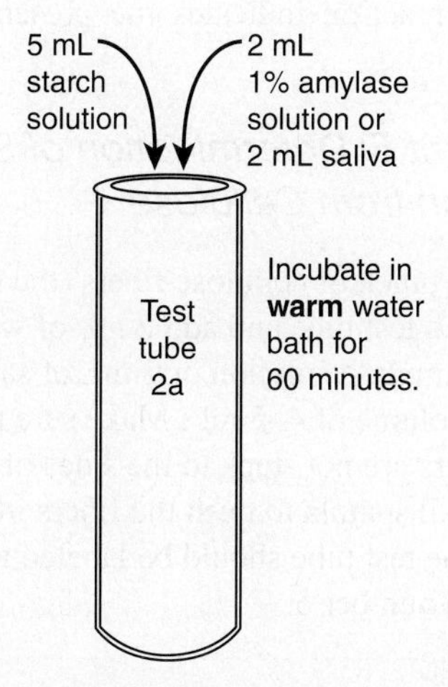

4. After 60 minutes, take 1 to 2 mL of the solution from test tube 2a and place it in a new test tube labeled 2b.
5. Test the solution in test tube 2b with four or five drops of iodine solution and record your results.
6. Compare the results of your experiment with test tube 1.

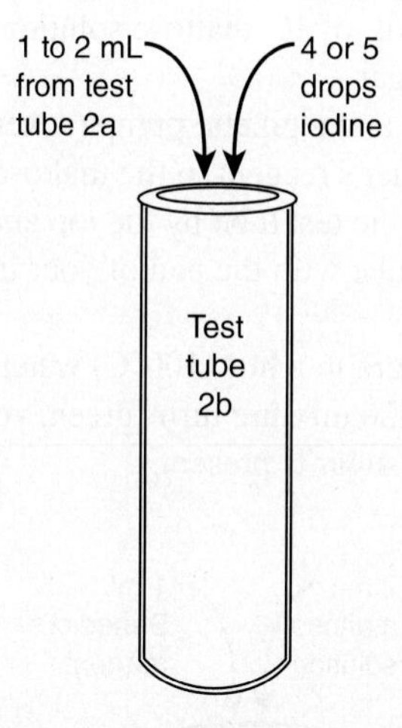

   Color of the solution with iodine: ________________

   Test results (positive/negative): ________________

7. From test tube 2b, withdraw a small sample with an eyedropper and place a drop on a microscope slide.
8. Examine the slide under low power and look for starch grains. Compare this slide with one that you made from a drop of starch from the *undigested* starch solution (from test tube 1). Is there a difference? Record your observations in the following space. Save this reaction mixture.

   Observations: ________________

   ____

   ____

   ____

   ____

Sugar Production

The preceding experiment tested for the *decrease* in starch as a way to determine if a catabolic reaction has taken place. Another way to look at the effectiveness of amylase is to determine the *presence* of sugar after the exposure of starch to the enzyme. In this experiment, instead of looking for the absence of the **substrate** (starch) to determine enzyme effectiveness, you can test for the formation of the **product** (sugar). If amylase is effective, then the presence of sugar in solution, after exposure to the enzyme, indicates the conversion of starch to sugar.

Experiment 3: Reagent Test—Benedict's Test

The first test in this section involves the detection of sugar in a solution. This is done with the Benedict's test.

Caution!

Benedict's reagent is poisonous. Use caution when mixing the solutions in the test tube.

1. Label a test tube with the number 3 and your group name and then pipette 2 mL of 1% maltose solution into the test tube. Maltose is a sugar.
2. Using a pipette and a pipette pump (never by mouth) add 1 mL of Benedict's reagent to the maltose solution in the test tube. Hold the test tube by the top and gently flick the bottom of the tube with the pad of your index finger to mix the contents.
3. Place this mixture in a **hot** (100°C) water bath for about 10 minutes. If the mixture turns green, yellow, orange, or brick red, then sugar is present.

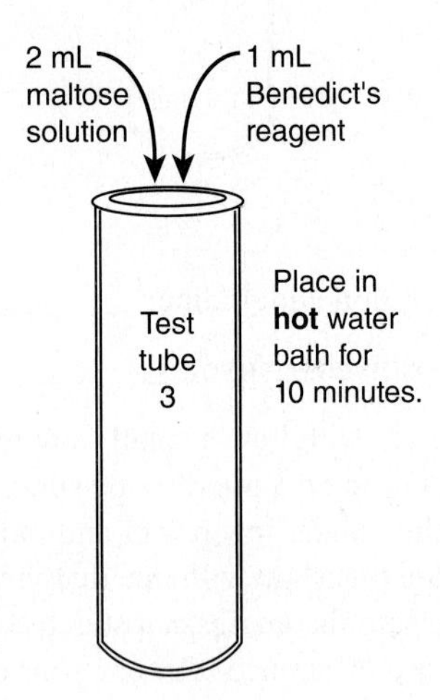

The sequence of colors indicates the presence of sugar in increasing amounts.

Presence of sugar:

Blue = 0

Green = trace

Yellow = moderate

Orange, red = large amounts

If there is no color change (if the solution stays blue) after 10 minutes, then remove the tube from the bath and note the absence of sugar (a negative result). Record your results.

Color of test solution: ______________________________

Test result (positive/negative for sugar): ______________

Ⓐ4 Experiment 4: Determination of Sugar (Maltose) Production by Amylase

1. Take 4 mL of the solution from test tube 2a (the starch and amylase reaction) and put it in a test tube labeled with your group name and the number 4.
2. To the solution in test tube 4, add 4 mL of Benedict's reagent.
3. Place the tube in a **hot** water bath (100°C) for 10 minutes. Record your results.

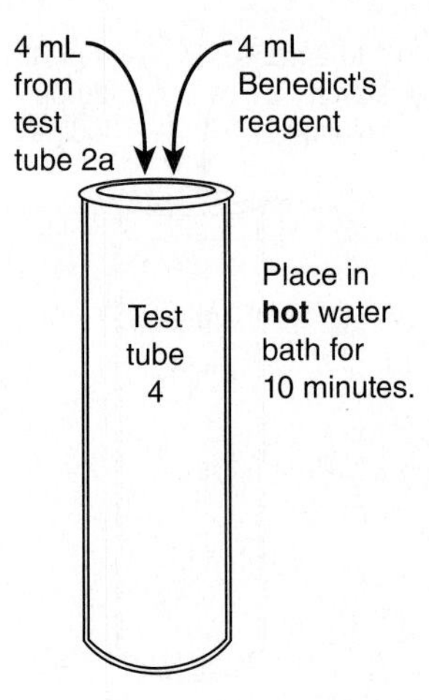

Color of solution with Benedict's reagent: ______________

Compare these results with the sequence of colors in experiment 3.

Presence of sugar: ______________________________

Effectiveness of digestion: ______________________

Cellulose Digestion

Cellulose is composed of repeating **glucose** units. Starch is also composed of repeating glucose units. If amylase converts cellulose to sugar, then the presence of sugar in solution after incubation indicates that the enzyme amylase is effective in breaking down cellulose into glucose. This is the same process as in experiment 4, except that the substrate is changed from starch to cellulose. To test for the effectiveness of amylase in digesting cellulose, you incubate a cellulose mixture with amylase in a warm water bath. A positive reaction indicates the presence of sugar after incubation.

Ⓐ5 Experiment 5: Determination of Sugar Production from Cellulose

1. Place a small pinch of cellulose fibers (the size of a green pea) in a test tube and add 2 mL of water and either 3 mL of 1% amylase solution or 2 mL of saliva. You should have a total volume of 4–5 mL. Make sure the majority of cellulose fibers are not stuck to the sides of the test tube. You can use a small spatula to push the fibers into the liquid, if necessary. The test tube should be labeled with your group name and the number 5.

2. Incubate for 60 minutes in a **warm** (37°C) water bath.

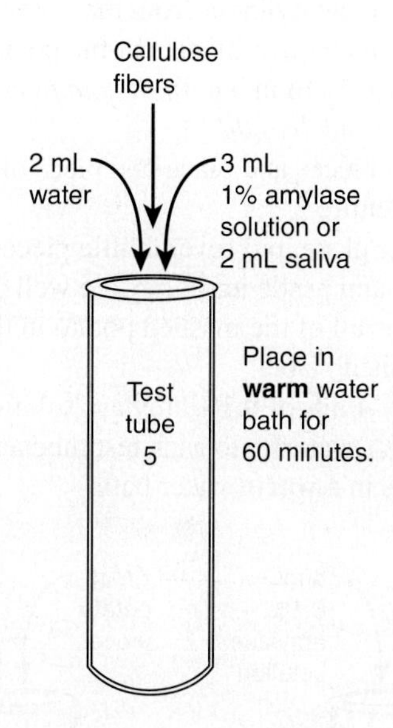

3. After 60 minutes, remove the solution from the bath and add 4 mL of Benedict's reagent.
4. Place the tube in a **hot** water bath (100°C) for 10 minutes. Record the presence or absence of sugar. A positive reaction to the Benedict's test indicates the presence of sugar after incubation.

Color of Benedict's test: ____________________

Presence of sugar: ____________________

Effectiveness of digestion: ____________________

Lipid Digestion

In the digestive process, some **lipids (triglycerides)** are broken down into **monoglycerides** and **free fatty acids.** The fatty acids make the solution more **acidic,** thus lowering the **pH.** The decrease in pH can be used as a measure of digestion. The greater the acidity of the solution, the greater the digestion.

(A)[6] *Experiment 6: Reagent Test—Litmus Test*

1. To determine the effectiveness of the litmus reagent, pour 3 mL of **litmus cream** into a test tube labeled with your group name and the number 6. Litmus cream consists of dairy cream (containing fat) and litmus powder (a pH indicator).
2. Add an acid solution (lemon juice or vinegar) drop by drop while flicking the bottom of the test tube until the color changes from blue to pink. Do not add too much acid but just enough to change the color. A pink color indicates an acidic condition.

Caution!

Sodium hydroxide is a caustic solution that can dissolve your skin, so be careful! Wear protective goggles and gloves as you add the sodium hydroxide solution.

---

3. Now add 1% sodium hydroxide solution (NaOH) a drop at a time until you reverse the color. This indicates that the solution is alkaline.

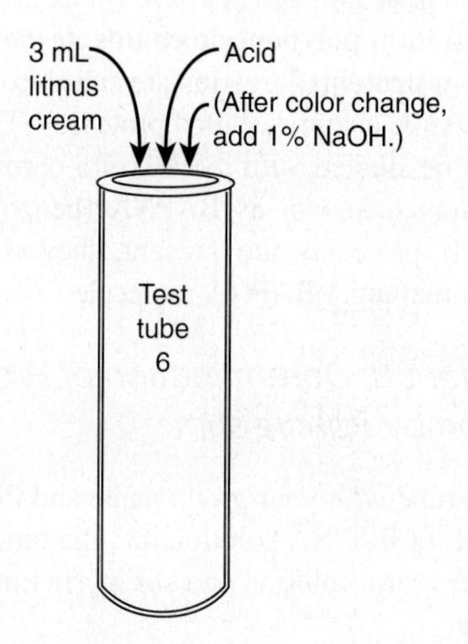

(A)[7] *Experiment 7: Determination of Lipid Digestion by Pancreatin*

1. Pipette 3 mL of litmus cream into a small test tube labeled with your group name and the number 7.
2. Add 1 mL of pancreatin solution (pancreatin contains many digestive enzymes) to the litmus cream and stir well with a spatula.
3. Incubate for 60 minutes in a **warm** (37°C) water bath.

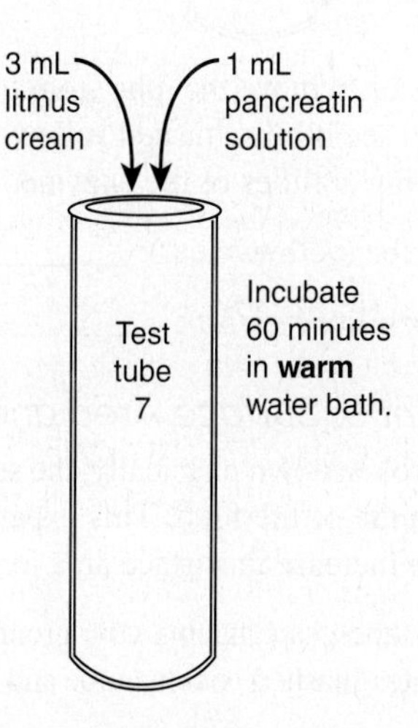

4. After 60 minutes, look for a color change. If the cream turns pink, then digestion has occurred. If the color remains blue, then no digestion has occurred. Record your results.

   Color of solution after incubation (pink/blue): ___________

   Condition of the litmus cream (acidic/alkaline): ___________

   Effectiveness of digestion: ___________

   Change in smell from normal cream: ___________

Protein Digestion

Proteins are made of **amino acids.** These amino acids are linked by **peptide bonds** to form **polypeptide chains.** If the chains are long enough, they form **proteins.** Proteins are split into amino acids by a number of digestive enzymes called proteases. The effectiveness of proteases can be studied with the use of a chromogenic (color-producing) substance known as BAPNA (benzoyl-DL-arginine-p-nitroanilide). If proteases are present, they release a yellow aniline dye from the larger BAPNA molecule.

(A)[8] *Experiment 8: Determination of Protein Digestion by Pancreatin*

1. Label a test tube with your group name and the number 8.
2. Pipette 1 mL of BAPNA solution into the tube. To this, add 1 mL of pancreatin solution and stir by flicking the bottom of the test tube.
3. Place the tube in a **warm** (37°C) water bath for 15 minutes.

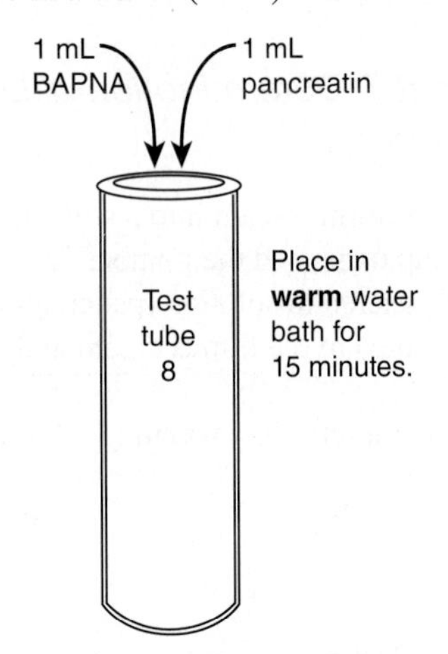

4. After 15 minutes, remove the tube from the bath and examine the test tube to see if it has turned yellow. Yellow indicates protein-digesting abilities of the enzyme. Record your results.

   Color of the tube (yellow/clear): ___________

   Digestion capability (yes/no): ___________

(A)[9] *Experiment 9: Surface Area and Digestion*

As food is broken down physically, the surface area increases relative to the volume of the food. This experiment examines the effectiveness of an increase in surface area for digestion.

1. Label two test tubes, each having your group name and the number 9. Write "mashed" on one tube and "entire" on the other.
2. Carefully plunge a number 4 cork borer into the center of a raw potato. *Make sure your hand is not on the receiving end of the cork borer! Use a breadboard.*
3. Remove the cork borer from the potato and, using a small rod, push the potato cylinder from the borer.
4. Using a ruler and a razor blade, cut the potato cylinder into two pieces, each 1 cm in length. *It is important that the two pieces are the same length!*
5. Rinse off both pieces and place one piece of potato in the test tube labeled "entire."
6. Chop the other piece into several little pieces and mash these with a mortar and pestle until they are well pulverized.
7. Carefully place all of the mashed potato in the other test tube with the "mashed" label.
8. Pipette exactly 4 mL of **0.1%** amylase solution* (or 0.5 mL saliva and 4 mL water) into each test tube and incubate them for **20** minutes in a **warm** water bath.

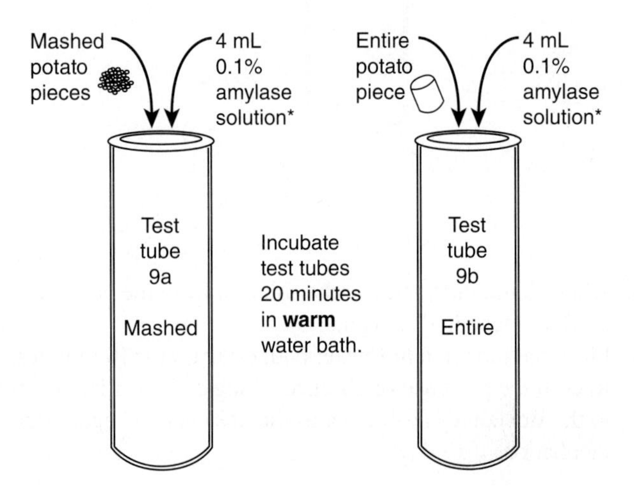

9. After incubation, place exactly 2 mL of Benedict's reagent into each tube and place the two tubes in the hot water bath for 10 minutes.

   The amount of digestion can be estimated by the color of the test. As described in experiment 4, blue indicates no sugar, with green, yellow, orange, and brick red indicating increasing amounts of sugar. In which tube did the greatest amount of digestion take place? In which tube did the least amount of digestion take place? Record your results.

   Color of solution with intact potato piece: ___________

   Color of solution with mashed potato piece: ___________

   Relative amount of digestion of intact piece: ___________

   Relative amount of digestion of mashed piece: ___________

Cleanup

Place all test tubes, pipettes, and other materials in either a biohazard container or a 10% bleach solution.

*You can use a solution of 0.1% alpha amylase or dilute the 1% stock solution by adding 1 mL of 1% amylase solution to 9 mL of water.

Exercise 39 Review

Name: ___

Lab time/section: ___

Date: ___

Digestive Physiology

1. Why is pancreatin used in a digestion experiment? ___

2. Enzymes convert substrates into what substances? ___

3. Name the group of enzymes that digest starch. ___

4. Starch is reduced by enzymes into smaller molecules, known as ___.

5. What is catabolism? ___

6. What are the bonds that hold amino acids together? ___

7. Explain why a negative iodine test for starch indicates a positive result for the enzymatic degradation of starch by amylase.

8. What would a positive iodine test indicate in the reaction in question 7? ___

9. Record all your negative results. Determine if these negative results indicate digestion or no digestion. ______________________

____

____

____

10. Explain why cellulose could or could not be digested by amylase. (Hint: Look in your text in the organic chemistry section.)

____

____

____

11. What effect does chewing food have on digestion? What experiment did you perform to evaluate the effectiveness of chewing for digestion? ____

____

____

____

12. Lipase is an enzyme that converts some lipids to monoglycerides and free fatty acids. Which substance used in lab probably contained lipase? ____

____

13. What does a negative Benedict's test (a blue color) indicate? ____

____

14. What does a pink color in the lipid digestion experiment indicate? ____

____

15. Pancreatin (which has amylase) and amylase are often prepared with lactose or other sugars as extenders. Determine which experiments would give erroneous results if pancreatin or amylase were contaminated with sugar. ______________________

____

____

Exercise 40

Anatomy of the Urinary System

Introduction

The urinary system filters dissolved material from the blood, regulates electrolytes and fluid volume, concentrates and stores waste products, and reabsorbs metabolically important substances, returning them to the circulatory system. **Filtration** occurs when one or more substances pass through a selectively permeable membrane, whereas others do not. Filtration in the kidney involves both metabolic waste products (urea) and material beneficial to the body. It is not desirable to have all filtered material removed from the body. Glucose and other materials, such as sodium and potassium ions, are **reabsorbed** from the kidney back into the circulatory system. The kidney **secretes** urea; some drugs; hydrogen and hydroxide ions; and hormones, such as renin and erythropoietin. Finally, the kidneys excrete metabolic wastes, hydrogen ions, toxins, water, and salts. The system consists of two kidneys, two ureters, a single urinary bladder, and a single urethra. These topics are covered in the Seeley text in chapter 26, "Urinary System."

In this exercise, you examine the gross and microscopic anatomy of the urinary system, study the major organs as represented in humans, and dissect a mammal kidney.

Objectives

At the end of this exercise, you should be able to

1. list the major organs of the urinary system;
2. describe the blood flow through the kidney;
3. describe the flow of filtrate through the kidney;
4. name the major parts of the nephron;
5. trace the flow of urine from the kidney to the exterior of the body;
6. distinguish among the parts of the nephron in histological sections.

Materials

Models and charts of urinary system
Models and illustrations of kidney and nephron system
Microscopes
Microscope slides of kidney and bladder
Samples of renal calculi (if available)
Preserved specimens of sheep or other mammal kidney
Dissection trays and materials
Scalpels
Forceps
Blunt (Mall) probes
Protective gloves
Waste container

Procedure

(A) 1 You can begin the study of the urinary system by locating its principal organs. Compare figure 40.1 with charts and models available in the lab. Find the **kidneys, ureters, urinary bladder,** and **urethra.**

Kidneys

The kidneys are **retroperitoneal** (posterior to the parietal peritoneum) and are embedded in **renal fat pads (perirenal fat).** These adipose pads cushion the kidneys, which are located mostly below the protection of the rib cage (figure 40.1). The kidneys are located adjacent to the vertebral column about at the level of T12 to L3. The right kidney is slightly more inferior than the left.

(A) 2 Examine a model of the kidney and compare it with figure 40.2. Locate the outer **renal capsule,** a tough connective tissue layer, the outer **cortex,** and the inner **medulla** of the kidney. The kidney has a depression on the medial side where the **renal artery** enters the kidney and the **renal vein** and the **ureter** exit the kidney. This depression is called the **hilum.**

Examine a coronal section of the kidney, and locate the **renal pyramids,** which are separated by the **renal columns** in the renal medulla. Each renal pyramid ends in a blunt point called the **renal papilla** (figure 40.2). Urine drips from many papillae toward the middle of the kidney.

The urine drips into the **minor calyces** (sing. **calyx**), which enclose the renal papillae. Minor calyces are somewhat like funnels that collect fluid. Minor calyces lead to the **major calyces** and these, in turn, conduct urine into the large **renal pelvis.** The renal pelvis is located in a space known as the **renal sinus.** The renal pelvis is like a glove in a coat pocket. The pocket is the renal sinus, and the membranous glove that occupies the space is the renal pelvis. Locate these structures in figure 40.2. The renal pelvis is connected to the ureter at the medial side of the kidney.

Blood Flow Through the Kidney

(A) 3 There are two fluid flows in the kidney. One is the flow of the filtrate and the other is the flow of blood. The blood enters the kidney by way of the **renal artery** and then passes through the **segmental arteries** to the **interlobar arteries** and into the **arcuate arteries.** From the arcuate arteries, the blood moves to the **interlobular arteries.** These blood vessels can be seen in figure 40.3. From one of the interlobular arteries, blood flows into an **afferent arteriole,** through the **glomerulus** (a tuft of capillaries), and to the **efferent arteriole.** The blood then enters the **peritubular capillaries** and returns via the **interlobular veins,** the **arcuate veins,** and the **interlobar veins** to the **renal vein.** The separation between the cortex and the medulla occurs at the level of the arcuate arteries and veins (figure 40.3).

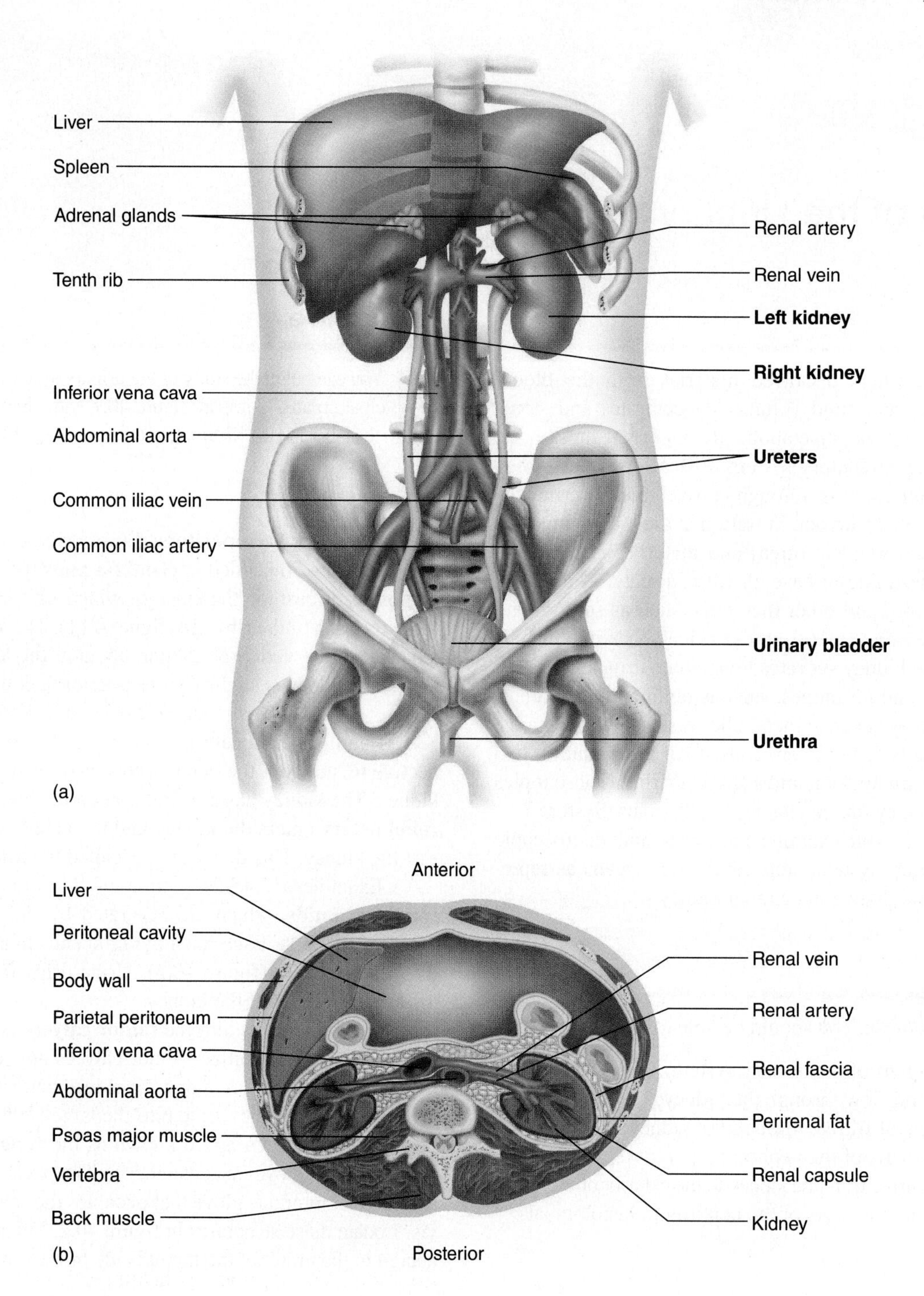

Figure 40.1 Major Organs of the Urinary System

(a) Anterior view; (b) cross section of the system, showing kidneys in retroperitoneal position.

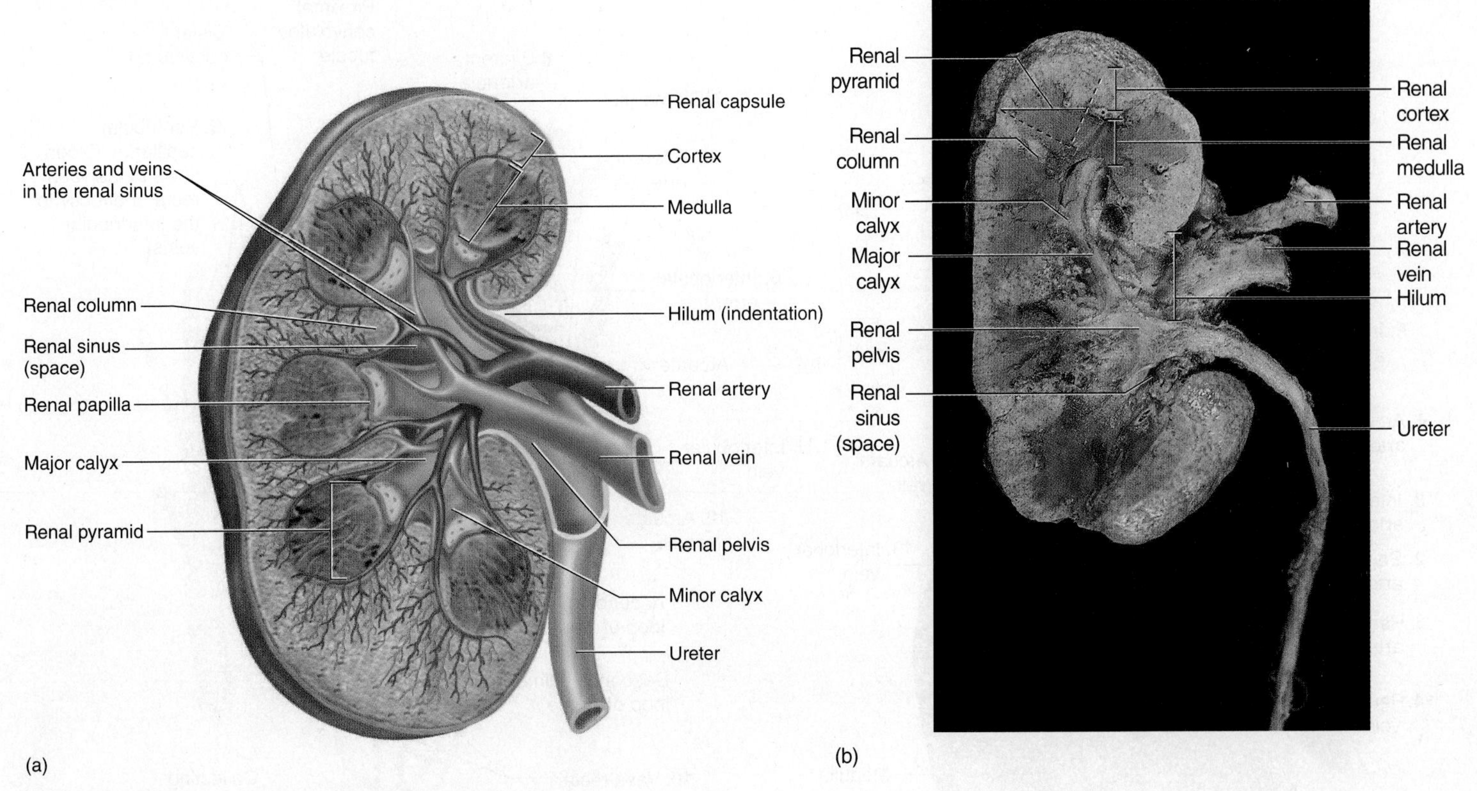

Figure 40.2 Coronal Section of the Kidney
(a) Diagram; (b) photograph.

There are vessels that branch off of the efferent arterioles and form a network of blood vessels known as the **vasa recta.** The vasa recta vessels are found primarily in the medullary area, and though they represent only a small fraction of the capillary bed of the kidney, they are important in producing a concentrated urine. The vasa recta are found in association with **juxtamedullary nephrons.**

The blood flow in the kidney forms a portal system, in which blood flows from one capillary bed (the glomerulus) to another capillary bed (the peritubular capillaries) prior to returning to the heart. The capillaries of the glomerulus are the site of filtration, whereas the peritubular capillaries are the primary site of reabsorption.

Ureters

Ⓐ 4 The **ureters** are long, thin tubes that conduct urine from the kidneys to the urinary bladder. The ureters have **transitional epithelium** as an inner lining and smooth muscle in their outer wall. Urine is expressed by **peristalsis** from the kidney to the urinary bladder. Examine the models in the lab and compare them with figure 40.4.

Urinary Bladder

The **urinary bladder** is located anterior to the parietal peritoneum and is thus described as being **anteperitoneal.** Locate the urinary bladder in the torso model and note that it is found just posterior to the symphysis pubis (figure 40.1). **Transitional epithelium** lines the inner surface of the bladder, whereas layers of smooth muscle known as **detrusor muscles** are located in the wall of the bladder. On the posterior wall of the urinary bladder is a triangular region known as the trigone. The **trigone** is defined by the superior entrances of the ureters and the inferior exit of the urethra, as illustrated in figure 40.4.

Histology of the Urinary Bladder

Ⓐ 5 Transitional epithelium has a special role in the urinary bladder. This epithelium can withstand a significant amount of stretching (distension) when the bladder fills with urine. Examine a slide of transitional epithelium under the microscope and compare it with figure 40.5. Look at the inner surface of the prepared section for the epithelial layer. Note that the cells are shaped somewhat like teardrops. Transitional epithelium can be distinguished from stratified squamous epithelium in that the cells of transitional epithelium from an empty bladder do not flatten at the surface of the tissue.

Urethra

The terminal organ of the urinary system is the **urethra.** The urethra is approximately 3–4 cm long in females. It passes from the urinary bladder to the **external urethral orifice,** located anterior to the vagina and posterior to the clitoris (figure 40.6). The urethra is about 20 cm long in males. It begins at the urinary bladder, passes through the body wall, and then exits through the penis to the external urethral orifice at the tip of the glans penis. Urinary bladder infections are more common in females than in males. How can you explain this in terms of microorganism movement

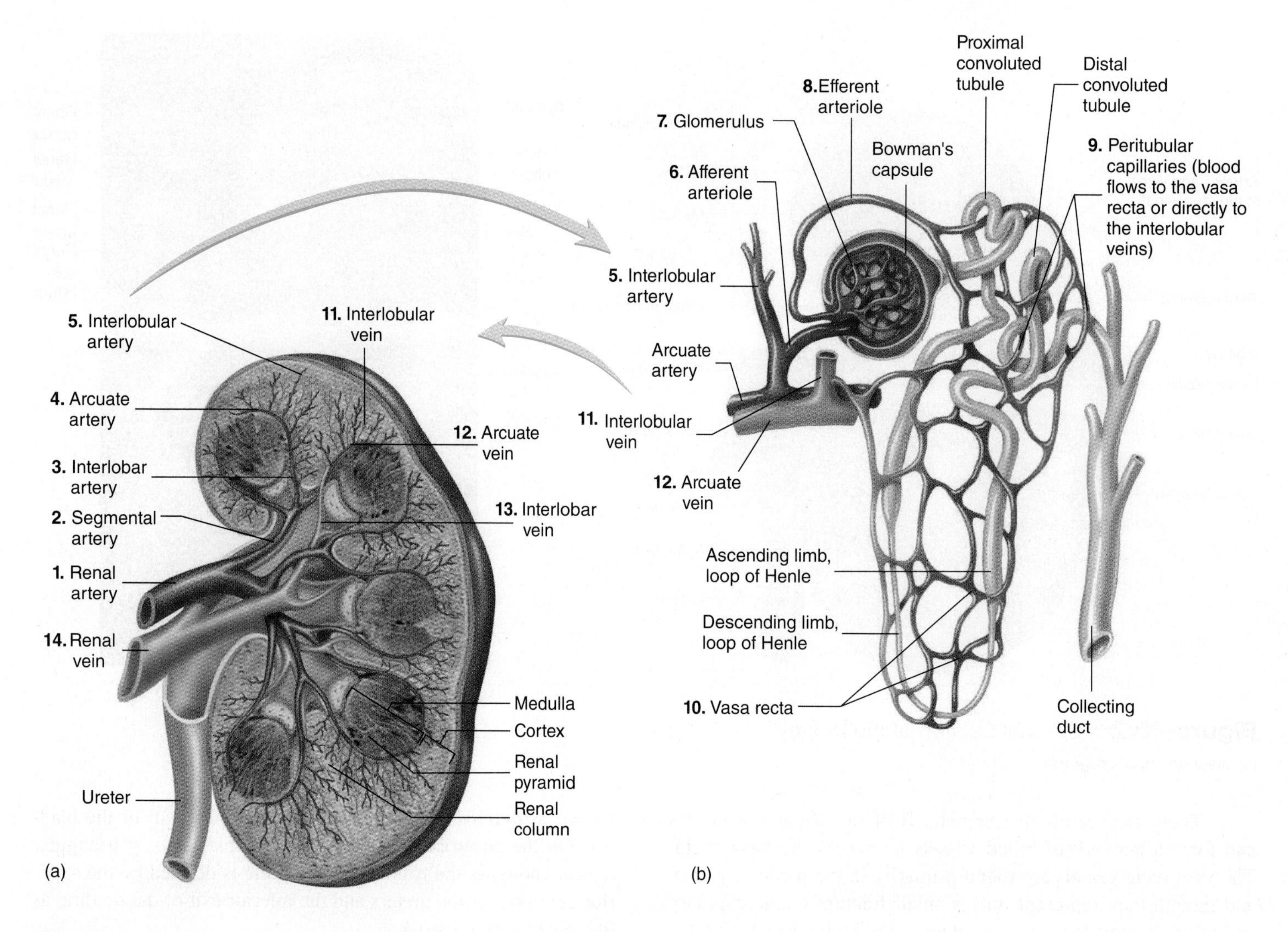

Figure 40.3 Blood Vessels and Ultrastructure of the Kidney
(a) Major vessels; (b) nephron and associated vessels. The blood flow can be followed by the numbers on the illustration.

into the bladder through the urethra? ______________________

____

____

____

Microscopic Examination of the Kidney

(A) 6 Before you examine the sections of kidney under the microscope, first become familiar with the structure of the **nephron.** Look at figures 40.3 and 40.7 and locate the **renal corpuscle (Bowman capsule** and **glomerulus), proximal convoluted tubule, loop of Henle,** and **distal convoluted tubule.** These structures make up the nephron.

Blood travels to the nephron via the **afferent arteriole.** When the blood reaches the **glomerulus,** a cluster of capillaries in the renal cortex, the plasma is filtered by blood pressure forcing fluid across the capillary membranes. This fluid, which is present in the nephron, is called filtrate.

As the filtrate flows through the nephron, water, glucose, and many electrolytes are returned to the blood. The urea is concentrated as it passes through the entire nephron and **collecting duct,** a tube that receives the end product of the nephrons. Locate the glomerulus, Bowman capsule, proximal convoluted tubule, loop of Henle, distal convoluted tubule, and collecting duct in figure 40.3 and on models or charts in the lab.

Examine a kidney slide under low power. You should see the **cortex** of the kidney, which has a number of round structures scattered throughout. These are the **glomeruli,** and they are composed of capillary tufts. The other part of the kidney slide should have fairly open, parallel spaces. These spaces are the **collecting ducts,** and they are located in the **medulla.** Compare the slide with figure 40.7a.

Examine the slide under higher magnification and locate the glomerulus and **Bowman capsule.** The capsule is composed of simple squamous epithelium and specialized cells called podocytes. Examine the outer edge of the capsule around the glomerulus. If you move the slide around in the cortex, you should

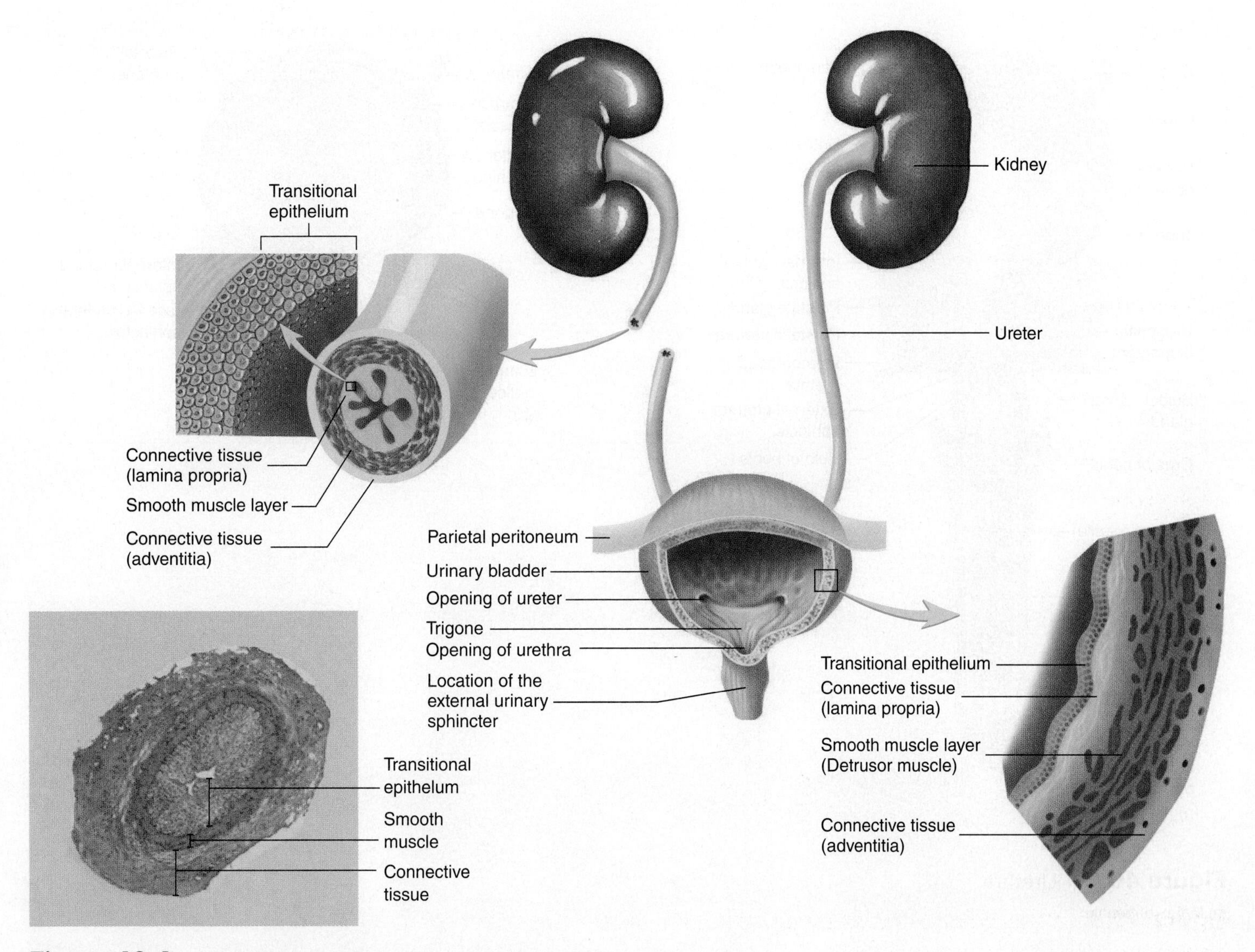

Figure 40.4 Ureter and Bladder, Cross and Coronal Sections

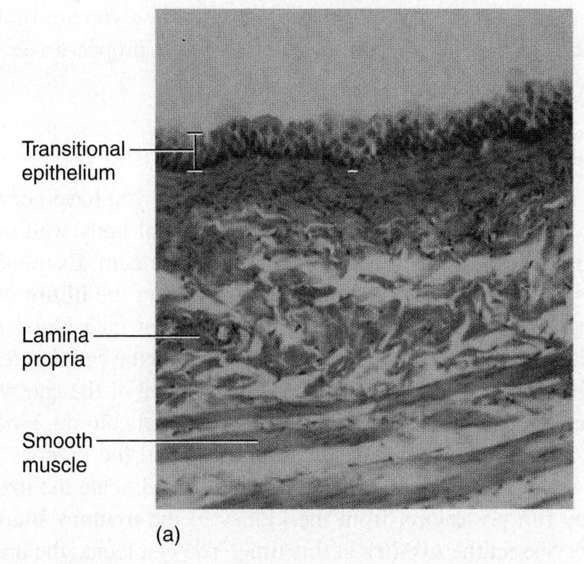

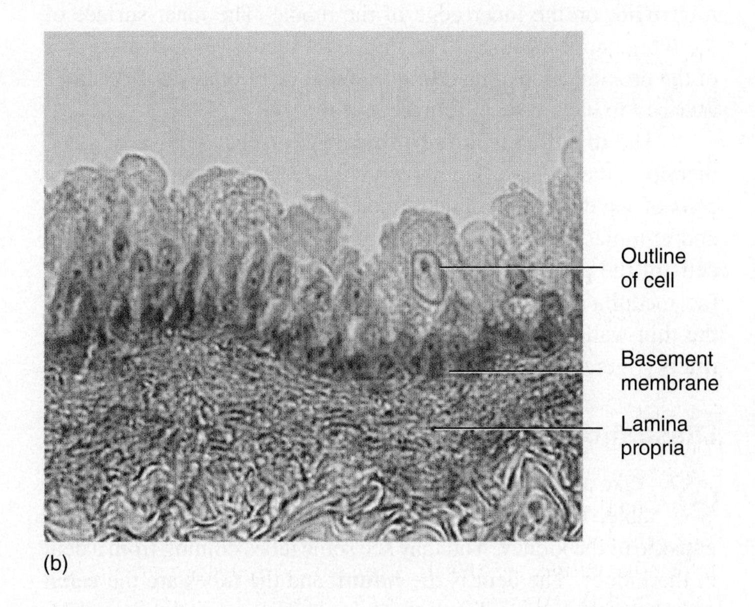

Figure 40.5 Histology of the Bladder (100×)

(a) Overview (100×); (b) detail of epithelium (100×).

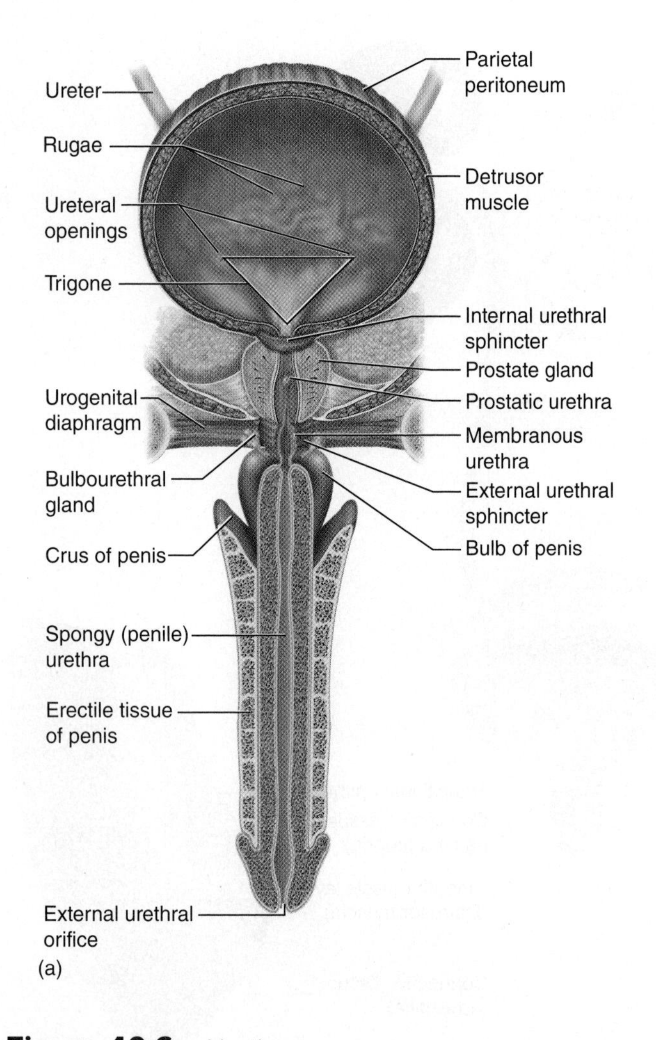

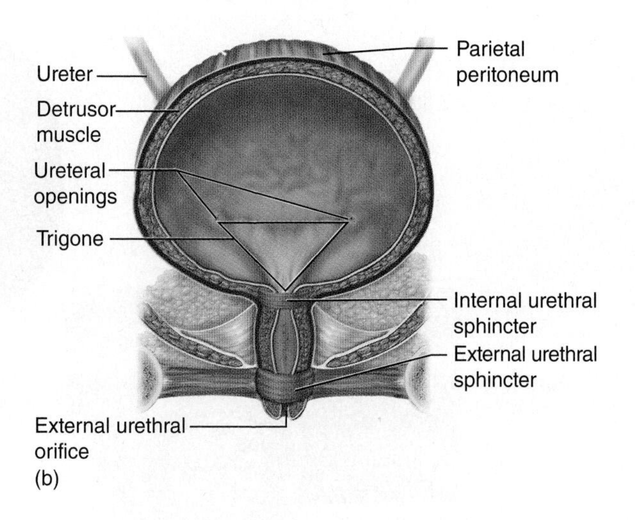

Figure 40.6 Urethra
(a) Male; (b) female.

find the **proximal convoluted tubules** with the **brush border,** or **microvilli,** on the inner edge of the tubule. The inner surface of the tubule appears fuzzy. The microvilli increase the surface area of the proximal convoluted tubule. What functional value could be ascribed to having such a surface area?

The **distal convoluted tubules** do not have brush borders; therefore, the inner surface of a tubule does not appear fuzzy. The cells of the distal convoluted tubules generally have darker nuclei and cytoplasm that is relatively clear when compared with the cells of the proximal convoluted tubules (figure 40.7c). Examine the medulla of the kidney under high magnification and locate the thin-walled **loop of Henle** and the larger-diameter **collecting ducts** (figure 40.7d).

Dissection of the Sheep Kidney

(A) 7 Take a sheep kidney and dissection equipment back to your table. Place the kidney before you and examine the outer **capsule** of the kidney. You may see some tubes coming from a dent in the kidney. The dent is the **hilum,** and the tubes are the **renal artery, renal vein,** and **ureter.** Make an incision in the sheep kidney a little off center in the coronal plane (figure 40.8). Locate the **renal cortex, renal medulla, renal pyramids, papillae, minor** and **major calyces,** and **renal pelvis.** Lift the renal pelvis somewhat to pull it away from the **renal sinus.** When you are finished with the dissection, place the material in the proper waste container provided by your instructor.

Cat Dissection

(A) 8 Open the abdominal region of the cat, if you have not done so already. The **kidneys** are on the dorsal body wall of the cat and are located dorsal to the parietal peritoneum. Examine the kidneys and the structures that lead to and from the **hilum** of the kidney. You should locate the **renal veins** that take blood from the kidney. The veins are larger in diameter than the **renal arteries,** and they are attached to the posterior vena cava of the cat, which runs along the ventral, right side of the vertebral column. Find the renal arteries that take blood from the aorta to the kidneys. The aorta lies to the left of the posterior vena cava. Locate the **ureters** as they run posteriorly from the kidney to the **urinary bladder.** Do not dissect the **urethra** at this time. You can locate the urethra during the dissection of the reproductive structures in Exercises 42 and 43. Compare your dissection with figure 40.9.

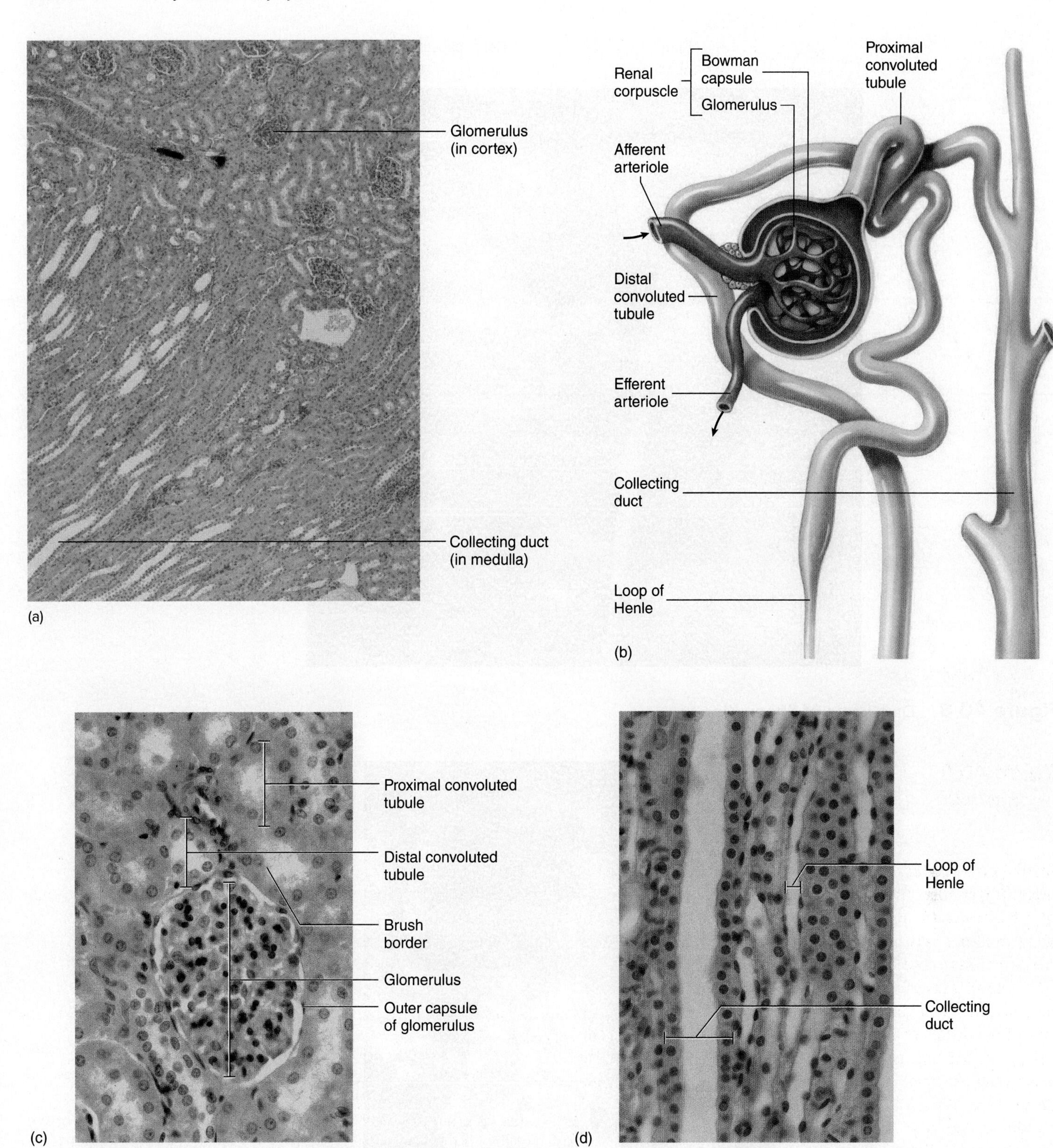

Figure 40.7 Histology of the Kidney

(a) Overview (40×); (b) diagram of nephron; (c) cortex (400×); (d) medulla (400×).

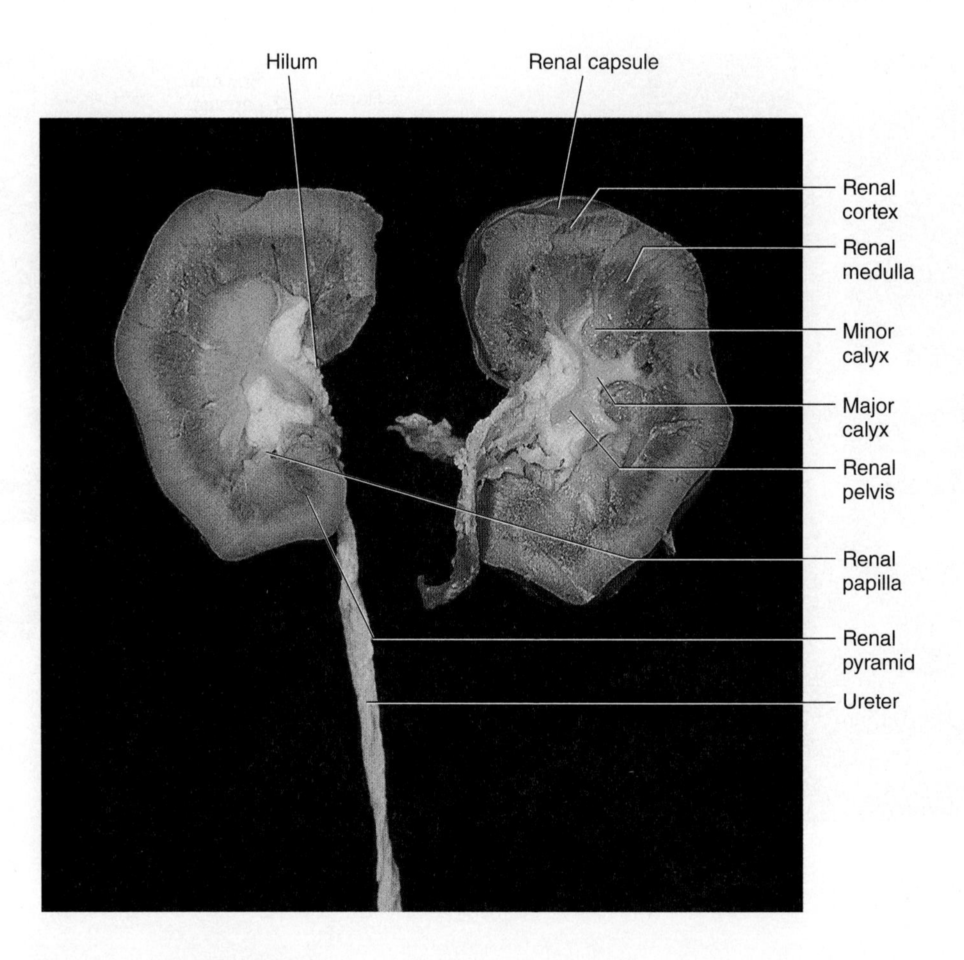

Figure 40.8 Dissection of Sheep Kidney

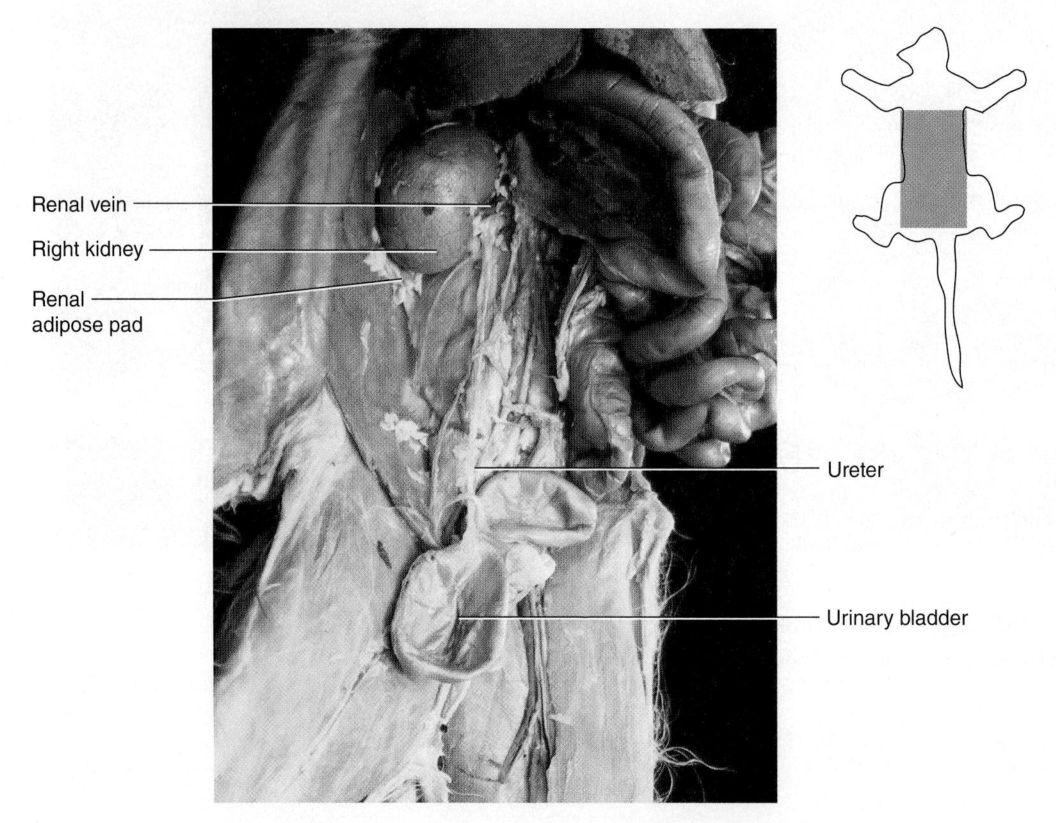

Figure 40.9 Urinary System in the Cat

Exercise 40 Review

Name: ______________________

Lab time/section: ______________________

Date: ______________________

Anatomy of the Urinary System

Matching

Match the descriptions in the left column with the terms in the right column.

1. outermost part of the kidney	a. renal vein
2. storage organ of the urinary system	b. major calyx
3. takes blood from the kidney	c. minor calyx
4. separates the renal cortex from the medulla	d. urinary bladder
5. receives urine from the renal papilla	e. renal capsule
6. leads directly to the renal pelvis	f. arcuate arteries

Fill-in

7. The ______________________ is found between the kidney and the urinary bladder.

8. The ______________________ is the terminal part of the nephron.

9. Filtration occurs at the ______________________ part of the nephron.

10. What is the outer region of the kidney called that contains glomeruli? ______________________

11. Describe the kidneys' position with regard to the parietal peritoneum. ______________________

____

12. What takes urine from the bladder to the exterior of the body? ______________________

13. What blood vessel takes blood to the kidney? ______________________

14. On the posterior bladder is a triangular region. What is it called? ______________________

15. Distal convoluted tubules flow directly into what structures? ______________________

____

16. What is a renal papilla? ______________________________

____

17. A renal corpuscle consists of what structures? ______________________________

18. Blood in the glomerulus next flows to what arteriole? ______________________________

19. Which shows greater anatomical difference between the sexes: ureters, urinary bladder, or urethra? ______________________________

____

20. Blood in an arcuate vein would next flow into what structure? ______________________________

21. What histological feature distinguishes a proximal convoluted tubule from a distal convoluted tubule? ______________________________

____

____

22. Name the parts of the nephron. ______________________________

____

23. What type of cell lines the bladder? ______________________________

24. Label the following illustration using the terms provided.

major calyx	minor calyx	renal artery	renal capsule
renal cortex	renal medulla	renal papilla	renal pelvis
renal pyramid	renal vein	ureter	

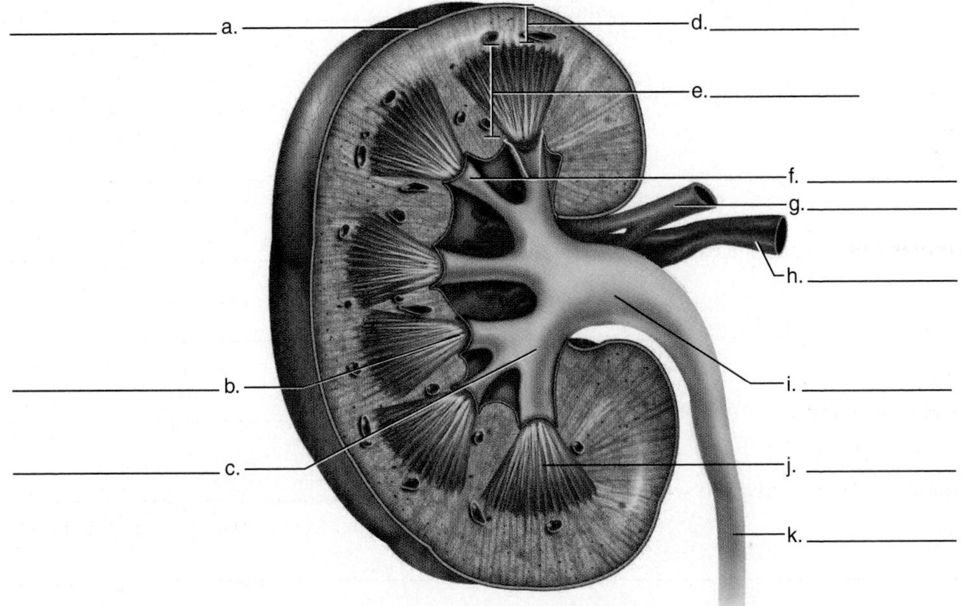

Exercise 41

Urinalysis

Introduction

Urinalysis is important in the clinical assessment of an individual's physical condition. Traces of blood in the urine can indicate the presence of kidney stones or renal damage; elevated levels of glucose can indicate diabetes mellitus; high bilirubin levels can indicate liver problems; and high bacterial counts can indicate bladder or kidney infections. The analysis of urine provides a very broad diagnostic indicator of general health. The function of the urinary system is covered in the Seeley text in chapter 26, "Urinary System." In this exercise, you analyze a sample of urine or synthetic urine for materials dissolved in the urine or suspended in it. The suspended material may include cells.

Normal values of urine vary, based on water intake and diet. A person who drinks 3 quarts of water per day has a very different urine composition than one who drinks 3 cups of water per day. Not only does the volume of water in the urine vary, but the concentration of other urinary solutes varies as well. Some cations, such as potassium and sodium, are also diet-dependent. The process of collecting urine and examining it under the microscope may seem unappealing at first, yet the exercise is usually appreciated after it is completed.

Objectives

At the end of this exercise, you should be able to

1. determine the specific gravity of a urine sample and compare it with normal values;
2. list the sediments commonly found in urine;
3. discuss the importance of urinalysis as a general diagnostic tool;
4. distinguish among casts, crystals, and microbes in a urine sample;
5. prepare a stained sediment slide and identify the major components of the sediment.

Materials

Sterile urine collection containers
Permanent marker or wax pencil
Microscope slides
Coverslips
Urine sediment stain (Sedistain, Volusol, etc.)
Chemstrip or Multistix 10SG urine test strips
Tapered centrifuge tubes
Test tube racks
Pasteur pipettes and bulbs
Centrifuge
Protective gloves
Protective eyewear
Biohazard bag
Microscopes
Urine specimen (yours or synthetic/sterilized urine)
10% bleach or designated disinfecting solution

Caution!

Urine is potentially contaminated with pathogens. Wear barrier gloves and protective eyewear during the entire exercise. Place all disposable material that comes into contact with urine in the biohazard bag. Work only with your own urine and avoid any contact between urine and an open wound or cut. If you spill your sample, notify your instructor, wipe up the spill, and swab the countertop with a 10% bleach or designated disinfecting solution. When you are finished with the exercise, place reusable glassware in a 10% bleach or designated disinfecting solution. Read all procedures before beginning this exercise.

Procedure

Ⓐ 1 You can use either your own urine for urinalysis or synthetic urine. If you use synthetic urine, you can test for color, pH, specific gravity, glucose, and protein. Some tests are provided in kit form from biological supply houses. If you use your own urine, follow these directions: Using a marker or wax pencil, write your name on a sterile collection container. Proceed to the restroom and obtain a urine sample for testing. If you collect your urine at home, do so in the morning and store the sample in a collection cup in a sealed plastic bag in the refrigerator. The sample should be a **midstream** collection, because the urethra has microbes and the first volume of urine will contain abnormally elevated levels of microorganisms. Take the specimen back to the lab and note the color of the sample.

Urine color: ______________________________

The color of urine should normally be a very pale yellow. This is due to the pigment **urochrome,** which is a metabolic product of hemoglobin breakdown. Higher levels of B vitamins may artificially color urine a brighter yellow, and a low fluid intake may also cause the urine to be a darker yellow. Urine is normally clear, but the sample may be cloudy due to bacterial infection or other contaminants. Record the turbidity of your urine sample.

Turbidity: ______________________________

Smell the urine sample and record the odor.

Odor: ______________________________

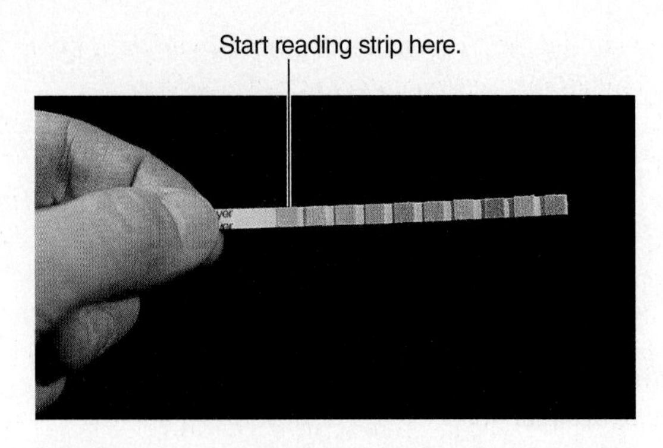

Figure 41.1 Reading a Urine Test Strip

Read the strip near the end where you hold on to the strip with your fingers and progress to the other end of the strip.

Urine should normally have a faint but characteristic odor. The consumption of certain foods, such as asparagus, may produce sulfur compounds in urine, which produce stronger odors.

Urine Characteristics

(A) 2 *Gently* swirl the urine before testing and pour a 10 mL sample into a clean test tube. Follow your instructor's directions for determining standard values in urine. You can test for specific compounds in urine individually, or you can use a Chemstrip or Multistix to run a battery of tests in a few minutes. The Multistix procedure is outlined next. You should have the urine sample, the test strip, a pencil or pen, and a container (so you can read the values) before you proceed with the test. The test covers 10 specific exams in 2 minutes, so you have to be organized to record the results accurately. Consider any result read after 2 minutes invalid. Multistix test strips are composed of sections of paper with test reagents embedded into the fibers. They react with urine components if present. Examine figure 41.1 and note the way the stick is read. Read the test strips first from the region near where you hold the strip and continue away from that area to the end of the strip.

Multistix/Chemstrip Procedure

If you are using a Multistix or Chemstrip, be sure to pay attention to the time limits provided on the container for reading the results. Wear gloves to dip the test strip into the urine sample and leave the strip in the urine for no longer than 1 second. Follow the directions on the test container precisely. As you read the test strip, record the results in the following spaces. When you finish the test, compare your results with the normal values in each paragraph following a specific test.

Glucose: ______________________________

Although glucose is present in blood plasma, there is normally no glucose in urine. **Glycosuria** is the condition of having glucose in the urine. Levels of glucose should be minimal (less than 40 mg/dL) or absent after fasting. Trace amounts of glucose can be found in the urine after ingesting food that is high in carbohydrates (especially sugar), and significant levels can be found in people with diabetes mellitus.

Bilirubin: ______________________________

The bilirubin test should be negative, since bilirubin is not normally present in urine. Bilirubin is a metabolic waste product from the destruction of erythrocytes. In the condition of **bilirubinuria,** bile pigments and bilirubin are present in urine. This can be due to erythrocyte destruction (hemolytic anemia), blockage of the bile duct, or liver damage, such as hepatitis or cirrhosis.

Ketones: ______________________________

There should be only trace amounts of (5 mg/mL) or no ketones (acetone) in the urine. Ketone bodies in urine **(ketonuria)** represent the general body condition known as **ketosis.** This is due to the mobilization and use of fat stores in people during periods of starvation, in people with diabetes mellitus, and in people with an abnormally high-fat diet.

Specific gravity: ______________________________

The specific gravity of pure water is 1.000. The specific gravity of urine varies from a dilute 1.001 to 1.035 in concentrated urine. When urine has a high specific gravity, there are more dissolved solutes in the urine. A high specific gravity may be due to lower water intake or diseases such as diabetes mellitus.

Hematuria: ______________________________

There should be no blood present in urine. The presence of erythrocytes or hemoglobin in urine may be due to some forms of anemia, erythrocyte destruction after incompatible blood transfusions, renal disease or infection, kidney stones, or urinary contamination during the menses (bleeding period) of a female's menstrual cycle. If the indicator strip shows a spotted pattern, then the erythrocytes are intact (nonhemolyzed). If the red blood cells have ruptured, then the presence of **hematuria** shows up as uniform color changes on the indicator strip.

pH: ______________________________

The pH of urine can vary from around 4.5 to 8.2. This represents more than a 1000-fold change in hydrogen ion concentration. The dramatic fluctuation in pH is seen as the kidneys regulate blood pH by removing ions from the blood. Urine is normally slightly acidic. In a diet high in protein, the urine is more acidic, whereas a diet high in vegetable material yields a urine that is more alkaline.

Protein: ______________________________

There should be no protein in urine. The most common blood protein is albumin, and its presence in urine is known as **albuminuria.** This may be due to diabetes mellitus, renal damage (kidney disease), extreme physical activity, or hypertension.

Urobilinogen: ______________________________

Normal ranges of urobilinogen are 0.2 to 1 mg/dL. Increases in the secretion of urobilinogen indicate significant **hemolysis** of erythrocytes to the point that the liver cannot process the bilirubin. The bilirubin increases in the plasma, and the formation of urobilinogen in the intestines increases as well. The urobilinogen diffuses into the blood, where it is filtered by the kidneys.

Nitrites: ______________________________

The normal urine value of nitrites should be negative. A positive value for nitrites indicates the presence of large numbers of bacteria in the urinary tract.

Leukocytes: ______________________________

The normal value for leukocytes is negative or present in trace amounts. An elevated level of leukocytes in urine is known as **pyuria.** This indicates a potential urinary tract infection—typically, bladder or kidney infection.

Note that some drugs and vitamins can give false positive results in urinalysis tests. Any test that indicates renal disease should be further validated by professionals before you take any course of action.

After you have finished collecting your results, complete table 41.1.

Sediment Study

(A)[3] 1. Obtain a centrifuge tube from the supply area and write your name on the tube for identification.
2. Pour 10 mL of urine into the tube and place it in the centrifuge opposite another tube containing an equal volume of urine. If there is an uneven number of tubes, make sure to fill a tube with water to the same level as the unpaired tube and place it opposite that tube in the centrifuge. This balances the centrifuge while it is operating.
3. Spin the tubes for about 4 minutes at a slow speed (1500 rpm) and let the rotor of the centrifuge slow down over time. If you brake the spinning, you may resuspend the sediments.
4. Carefully remove your tube and place it in a test tube rack on your desk.
5. Examine the tube. You should have an upper fluid layer and a small amount of urine sediment at the bottom of the tube.
6. Place one drop of urine sediment stain (Sedistain or other urine stain) on a clean microscope slide.
7. Withdraw a small sample from the bottom of the centrifuge tube with a Pasteur pipette and place a drop of the sediment on the slide. *Do not put the urine on the slide before you put on the stain, because you may contaminate the stain if the stain bottle comes in contact with the sediment.*
8. Place a coverslip on the sample and examine the slide under the microscope. Examine the slide under low power first, and look at the free edge of the coverslip. Red-orange, elongated crystals may appear here later as the stain evaporates. Do not confuse the stain crystals with sediment crystals.
9. Now examine the slide under high power and compare your sample with the common urine sediments in figures 41.2 through 41.4.
10. Record the material you find in urine sediment in the review section at the end of this laboratory exercise.
11. Some of the common items found in urine are as follows:

TABLE 41.1 Adult Urine Values

Characteristic	Normal Value or Range	Your Results
Appearance	Almost clear to deep amber	
Odor	Faint odor	
Glucose	0–trace (less than 40 mg/dL)	
Bilirubin	None	
Ketones	None–trace (5 mg/dL)	
Specific gravity	1.001–1.035	
Free hemoglobin	None	
pH	4.5–8.2	
Albumin	None–trace	
Urobilinogen	Trace (0.2–1.0 mg/dL)	
Nitrites	None	
Leukocytes	None–trace	

Organisms (Figure 41.2)

Yeasts (other than *Candida*)—many yeasts are a normal constituent.

Trichomonas vaginalis—common protozoan infection

Candida albicans—causes vaginal yeast infections

Cells (Figure 41.2)

Epithelial cells

Cells of urethra (squamous cells)—normal constituent

Cells of bladder or ureter (transitional cells)—normal constituent

Bacterial cells in urine—indicate bacteriuria

Erythrocytes (RBCs)—menstrual blood in females or blood from the passage of kidney stones

Leukocytes (WBCs)—indicate urinary tract (bladder/kidney) infections

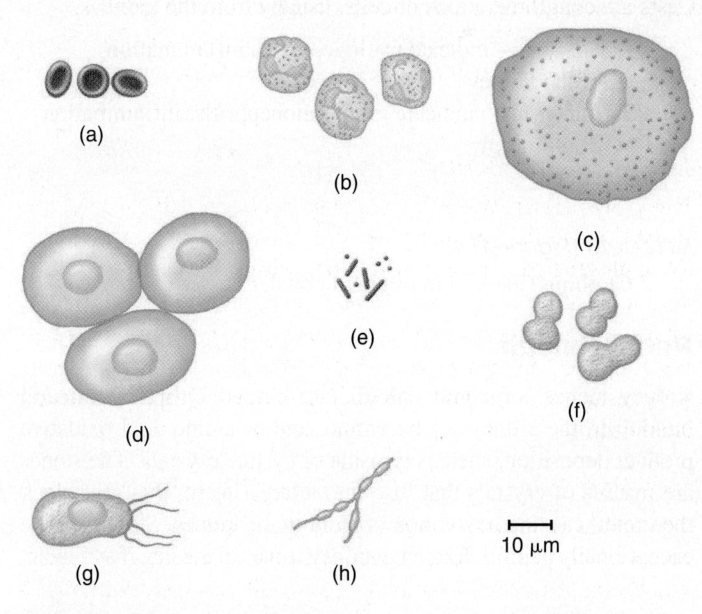

Figure 41.2 Urine Sediments—Cells and Organisms

Not to scale. (a) Red blood cells; (b) white blood cells; (c) squamous epithelial cells; (d) transitional epithelial cells; (e) bacteria; (f) yeast (g) *Trichomona;* (h) *Candida.*

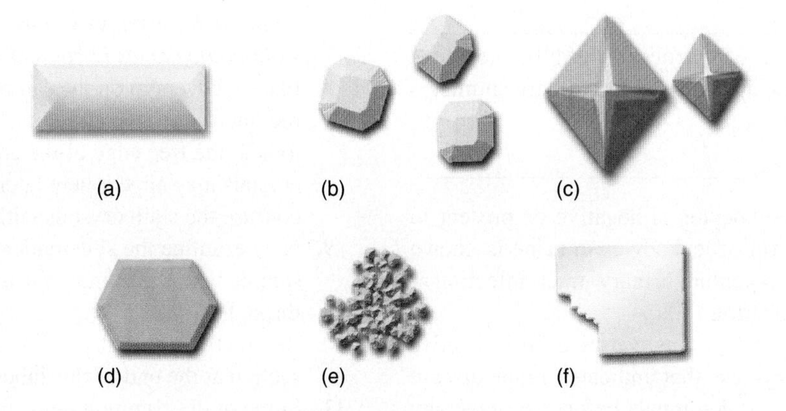

Figure 41.3 Urine Sediments—Crystals

Not to scale. Sizes are variable: (a) struvite; (b) calcium carbonate; (c) calcium oxalate; (d) cystine; (e) ureate; (f) cholesterol.

Crystals (*Figure 41.3*)

Crystals usually indicate reduced water intake.

Struvite
Calcium oxalate—small, green crystals
Calcium carbonate
Calcium phosphate
Uric acid
Ammonium ureates
Cholesterol
Tyrosine—amino acid
Cystine—oxidation product of an amino acid

Casts (*Figure 41.4*)

Casts are conglomerations of cells, usually from the kidneys.

Leukocytes—indicate pyelonephritis, inflammation of the kidney
Erythrocytes—indicate glomerulonephritis, inflammation of glomeruli
Hyaline

Artifacts (*Figure 41.4*)

Clothing fibers, skin oil, bath powder

Renal Calculi

Kidney stones, or **renal calculi,** are caused either by mineral buildup in the kidneys or by amino acid or amino acid oxidative product deposition, such as tyrosine or cystine crystals. The stones are masses of crystals that fuse into a larger form; they can block the ureter, causing a retention of fluid in the kidney. Stones can be exceptionally painful. Examine kidney stones in the lab, if available.

Cleanup

Place urine-contaminated material to be disposed of in the biohazard container and all the urine-contaminated material to be reused in the 10% bleach container. Swab the countertops with 10% bleach (or designated disinfecting solution), which destroys infectious agents.

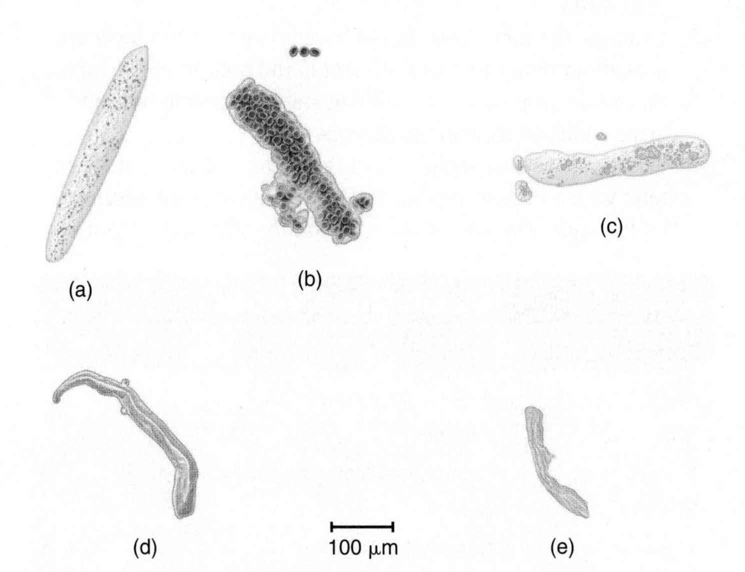

Figure 41.4 Urine Sediments—Casts and Fibers

(a) Hyaline cast; (b) erythrocyte cast; (c) leukocyte cast; (d) vegetable fiber; (e) mucus thread.

Exercise 41 Review

Urinalysis

Name: ____________________

Lab time/section: ____________________

Date: ____________________

1. Assume that a person did not collect a midstream sample of urine but collected a sample from the beginning of urination.

   What additional materials might be in greater numbers in this sample? ____________________

2. What metabolic by-product from hemoglobin colors the urine yellow? ____________________

3. How can adequate water intake be judged by the color of urine? ____________________

4. What is the name of the condition of having measurable amounts of sugar in urine? ____________________

5. What is the normal value for sugar in urine? ____________________

6. What substance in urine creates ketonuria? ____________________

7. What is the range of the specific gravity in normal urine values? ____________________

8. How much water do you normally drink each day? ____________________

9. Does the amount of water you drink each day correspond to the specific gravity of your urine? ____________________

10. What is hematuria? ____________________

11. The kidneys are very efficient at balancing blood pH. It is critical that blood acidity remain constant. If excess hydrogen ions are present in the blood and increase blood acidity, the kidneys secrete the hydrogen ions. What effect does this have on the pH of urine? ______________________

____

12. There are about 200 grams of protein in the blood plasma and normally none in urine. What mechanism keeps protein out of the urine, and what structure might be damaged (review Exercise 40, if necessary) if protein is found in significant amounts in urine? ______________________

13. Elevated levels of white blood cells produce what condition in urine? ______________________

14. If you process 180 liters of water through the kidney each day yet produce only 1.8 liters of urine, approximately how efficient are your kidneys at reabsorbing the water passing through them? ______________________

____

15. List the materials you found in the urine sample. ______________________

____

____

16. Which one of the following urine sediments is probably due to reduced water intake?

a. vegetable fibers
b. crystals
c. epithelial cells
d. bacterial cells

17. What cells found in the urine originally come from the walls of the urethra? ______________________

18. What cells in the urine come from the wall of the urinary bladder? ______________________

____

19. If your urine contained large numbers of calcium crystals, what might this tell you about the amount of water you drink?

____

____

Exercise 42

Male Reproductive System

Introduction

The male reproductive system produces **male gametes (sperm cells),** transports the gametes to the female reproductive tract, and secretes the male reproductive hormone, **testosterone.** The gonad, or gamete-producing structure, of the male reproductive system is the testis (plural *testes*). The testes produce sperm cells in the seminiferous tubules. The endocrine function of the testes is secretion of testosterone. In this exercise, you examine the gross anatomy of the male reproductive system, the histology of the system, and the male reproductive system of the cat. The structure and function of the male reproductive system are covered in the Seeley text in chapter 28, "Reproductive System."

Objectives

At the end of this exercise, you should be able to

1. describe the gamete-producing organ of the male reproductive system;
2. describe the anatomy of the major structures of the male reproductive system;
3. describe the formation of sperm cells in the testis;
4. list the pathway that sperm cells follow from production to expulsion;
5. describe the anatomy of the spermatic cord;
6. list the four components of semen;
7. name the three cylinders of erectile tissue in the penis.

Materials

Charts, models, and illustrations of the male reproductive system
Microscopes
Prepared slides of a cross section of testis and sperm smear
Cats
Materials for cat dissection

- Dissection trays
- Scalpel and new blades
- Protective gloves
- Blunt (Mall) probe
- Forceps and sharp scissors
- Sharps container
- Animal waste disposal container
- First aid kit

Procedure

Overview of the Gross Anatomy of the Male Reproductive System

(A) 1 Examine the models and charts of the male reproductive system available in the lab, and locate the following structures in figure 42.1:

Testis
Epididymis
Scrotum
Ductus (vas) deferens
Seminal vesicle
Prostate gland
Bulbourethral (Cowper's) gland
Penis

Testes

(A) 2 Examine charts and models of the testes and compare them with figures 42.1 and 42.2. The **testes** are paired organs wrapped in a tough connective tissue sheath called the **tunica albuginea** (figure 42.2). They lie outside of the body cavity, where the temperature is somewhat cooler, and are surrounded by the **scrotum,** which envelops the testes. The testes are the site of **spermatozoa (sperm cells)** production, and this process must occur at about 35°C (a few degrees cooler than the core body temperature). The scrotum is lined with a layer of muscle called the **dartos muscle** (figure 42.3). These muscle fibers contract when the testes are cold, bringing the testes closer to the body. When the environment around the testes is warm, the dartos muscles relax and the testes descend from the body, becoming cooler. Examine charts and models of the testes and compare them with figures 42.1 through 42.3.

Histology of the Testis

(A) 3 The testis, as the gamete-producing organ of the male reproductive system, should be viewed in a prepared slide. Many tubules are seen in cross section. These are the **seminiferous tubules.** The gametes, or sperm cells, are produced in seminiferous tubules in the testis (figures 42.2 and 42.4).

Find the clusters of cells that often appear as triangles between the tubules. These are called **interstitial cells.** They produce the male sex hormone, **testosterone.** Examine the seminiferous tubules under high magnification. You should be able to see the outer row of cells, called the **spermatogonia.** These cells reproduce by mitosis to produce **primary spermatocytes.**

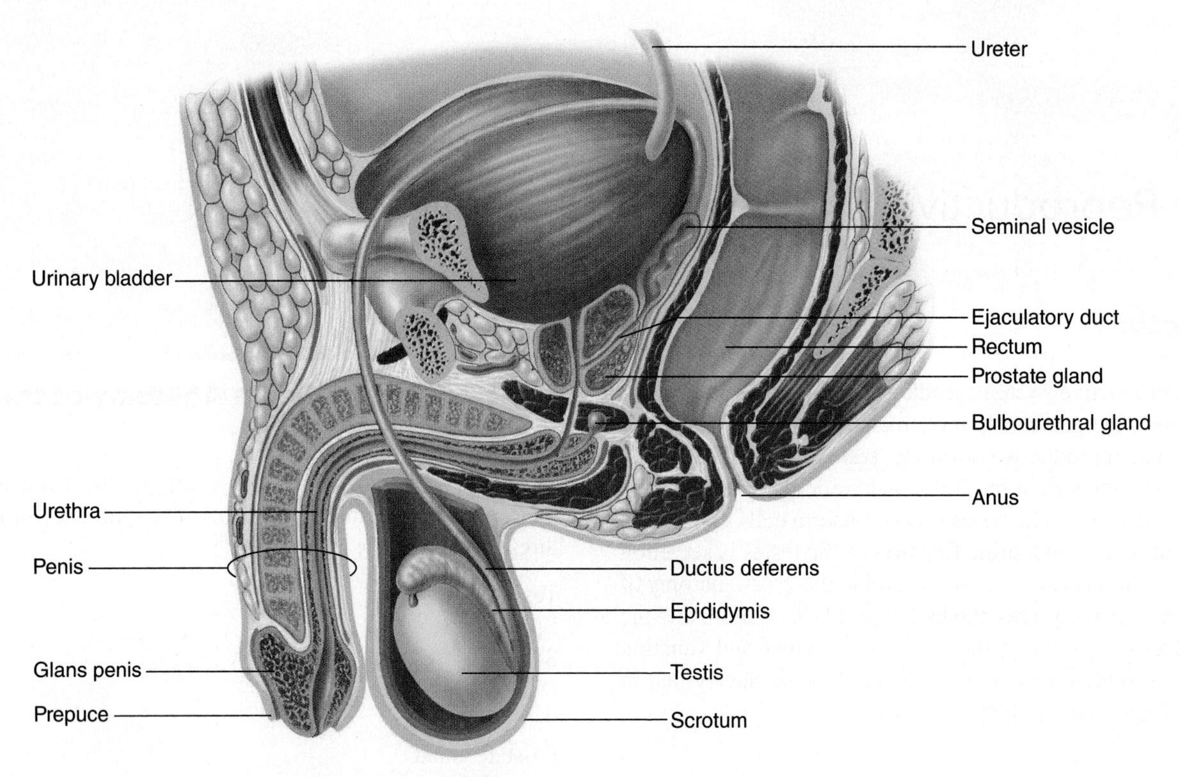

Figure 42.1 Male Reproductive System, Midsagittal View

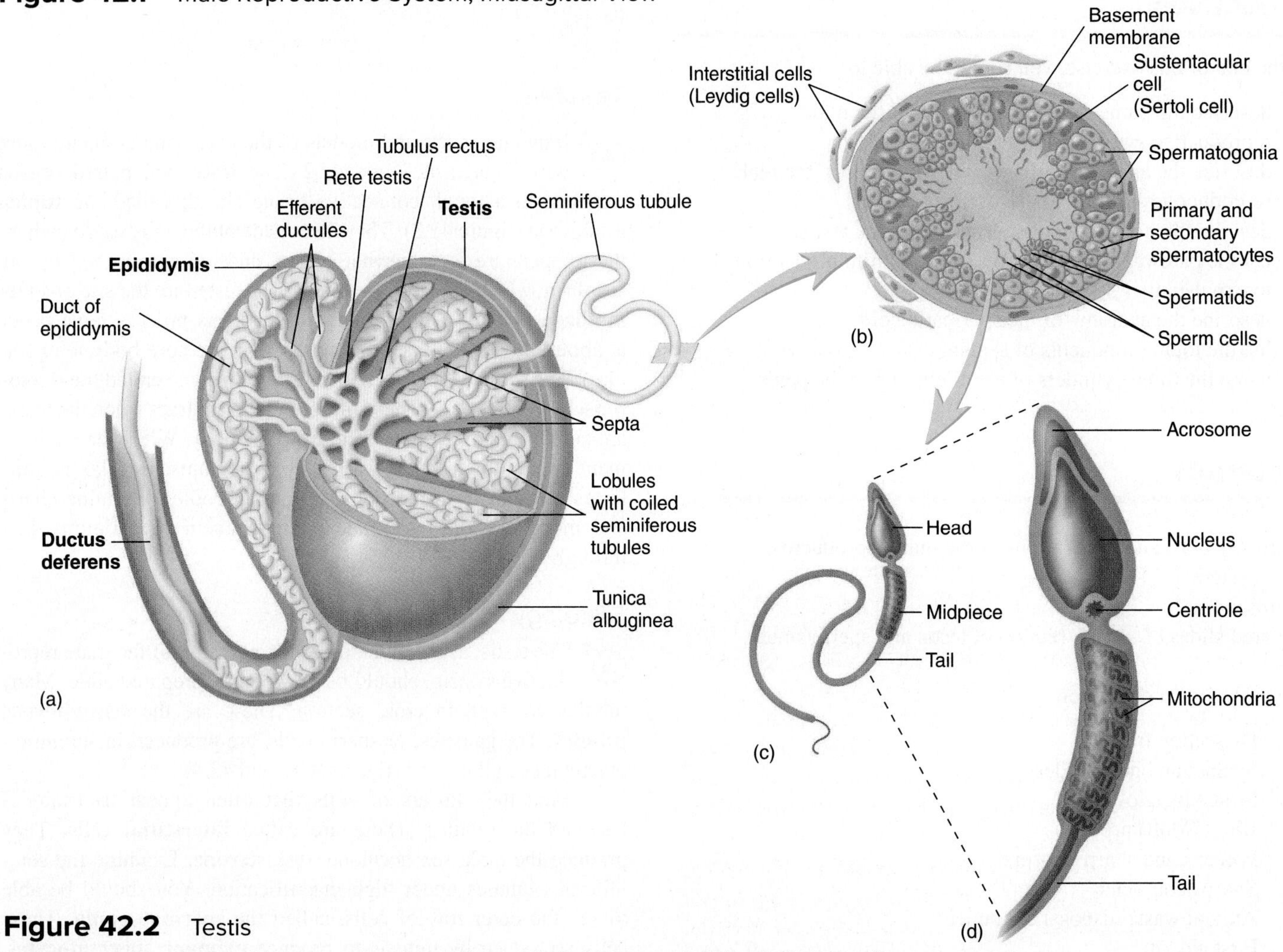

Figure 42.2 Testis

(a) Overview with epididymis; (b) cross section of seminiferous tubule; (c) sperm cell; (d) head of sperm cell.

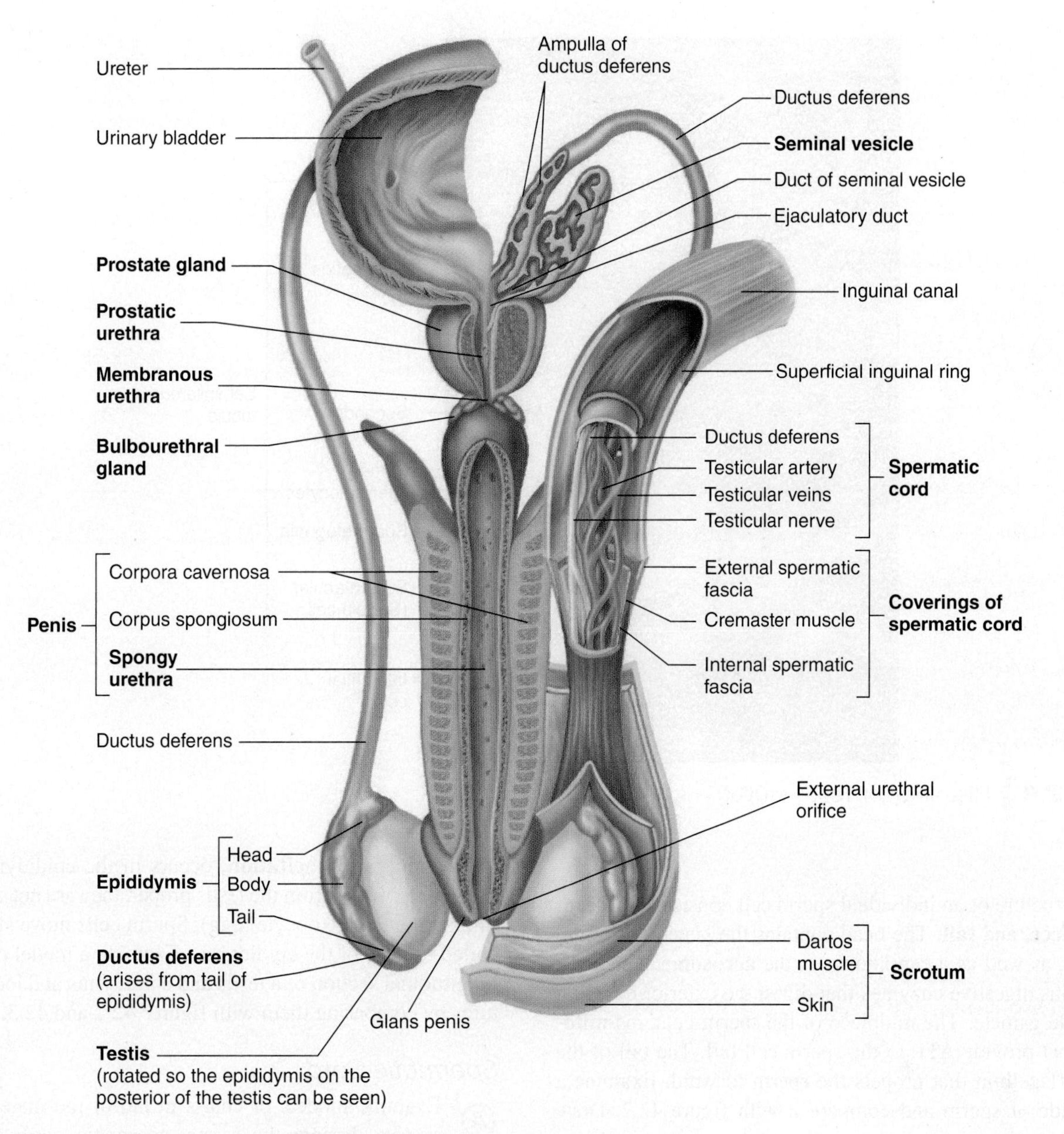

Figure 42.3 Scrotum, Testis, Spermatic Cord, and Penis, Anterior View

Sustentacular (Sertoli) cells assist in the movement of the spermatocytes and isolate the sperm cells with the blood testis barrier (BTB).

The primary spermatocytes undergo meiosis, or reduction division, to eventually produce the sex cells (sperm cells). The primary spermatocytes divide to form **secondary spermatocytes,** which are located closer to the lumen. The secondary spermatocytes become **spermatids.** Spermatids lose their remaining cytoplasm and mature into **sperm cells.** The process of sperm formation from spermatogonia to sperm cells is called **spermatogenesis.** Examine a prepared slide of testis and compare it with figure 42.4. Locate the spermatogonia, primary and secondary spermatocytes, spermatids, and sperm cells.

After you have examined your prepared slide of a section of testis, you should draw what you see in the following space. You should include the interstitial cells, seminiferous tubules, spermatogonia, spermatocytes, spermatids, and spermatozoa. You may need to look at several sections to see all these items. Your drawing may be a composite of many tubules.

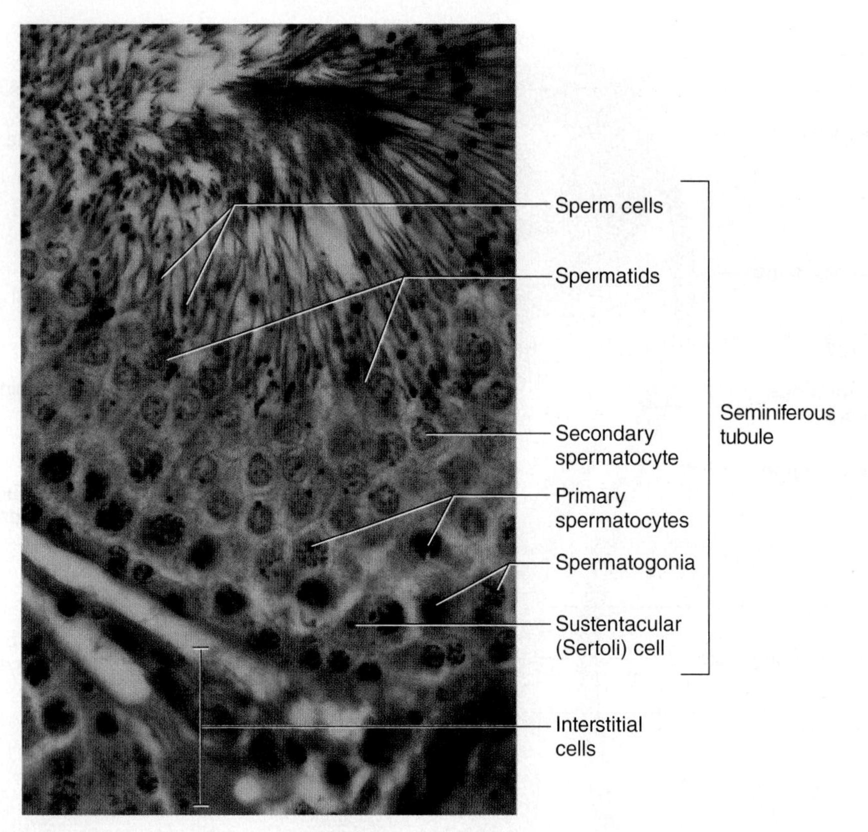

Figure 42.4 Histology of Testis (400×)

Sperm

(A) 4 The structure of an individual sperm cell consists of a **head, midpiece,** and **tail.** The head contains the genetic information (DNA), as well as a cap known as the **acrosome.** The acrosome contains digestive enzymes that digest the exterior covering of the female gamete. The midpiece of the sperm contains **mitochondria** that provide ATP to the sperm cell tail. The tail of the sperm is a flagellum that propels the sperm forward. Examine a prepared slide of sperm and compare it with figure 42.2. Draw what you see in the space below.

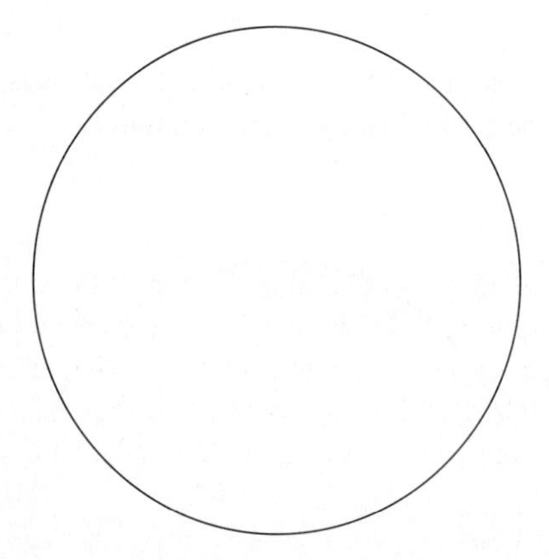

Epididymis

Sperm cells from each testis travel from tubules in the testis to the **rete testis** and into the **epididymis,** where they are stored and mature. Each epididymis has a blunt, rounded **head;** an elongated **body;** and a tapering **tail** that leads to the ductus deferens. Sperm maturation, or **capacitation,** occurs in the epididymis. If sperm cells are removed from the testis proper, they are not capable of fertilizing the female oocyte (egg). Sperm cells move slowly through coiled tubules of the epididymis. Examine a model or chart of the longitudinal section of a testis and epididymis and locate the structures by comparing them with figures 42.2 and 42.3.

Spermatic Cord

(A) 5 Examine models or charts in lab of the male reproductive system. Locate the testes, spermatic cord, seminal vesicle, prostate gland, and bulbourethral glands on the model or chart. Sperm cells travel from the epididymis into the **ductus deferens.** The ductus deferens is enclosed in the **spermatic cord,** which is a complex cable consisting of two layers of fascia, the ductus deferens, the **testicular artery** and **vein,** the **testicular nerves,** and the **cremaster muscle.** The cremaster muscle is a cluster of skeletal muscle fibers. The spermatic cord is longer on the left side than on the right; therefore, the left testis is lower than the right. Locate the structures of the spermatic cord in figure 42.3.

As the spermatic cord reaches the **inguinal canal,** each ductus deferens travels around the posterior of the urinary bladder. You can trace the course of the ductus deferens until it reaches the inferior portion of the bladder. The ductus deferens enlarges somewhat here to form the **ampulla.** Each ductus deferens joins with a **seminal vesicle,** which is a gland that adds fluid to the sperm cells. The union of the ductus deferens and the seminal vesicles produces the **ejaculatory duct.** Locate the seminal vesicle and the ejaculatory duct in figure 42.3. The fluid from the seminal vesicles adds about 60% to the final volume of semen.

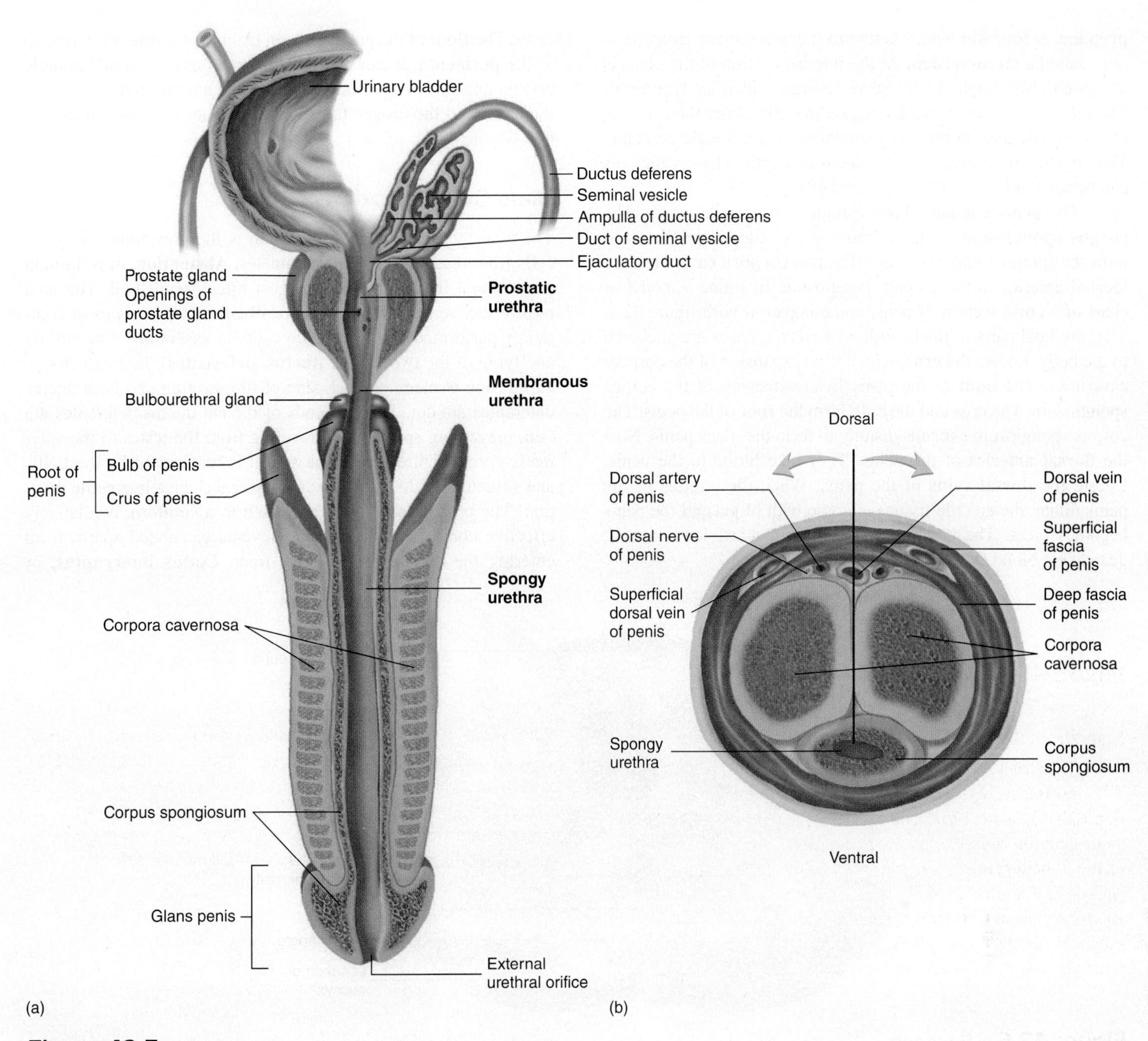

Figure 42.5 Penis
(a) Anterior view; (b) cross section.

From this location, the ejaculatory ducts lead to the inferior portion of the bladder and join with the urethra, which passes through the prostate gland. The **prostate gland** is located just inferior to the urinary bladder, and the urethra that passes through the gland is known as the **prostatic urethra.** The prostate gland adds a buffering fluid to the secretions of the testes and seminal vesicles. The prostate fluid makes up about 30% of the final semen volume. As the prostatic urethra exits the prostate gland, it becomes the **membranous urethra** and passes through the body wall. Here the paired **bulbourethral (Cowper's) glands** are located. They add a lubricant to the seminal fluid. **Seminal fluid** consists of secretions from the seminal vesicles, prostate gland, and bulbourethral glands. **Semen** consists of seminal fluid plus the sperm cells from the testes. The testes add about 5% to the total volume of semen. The urethra passes out of the body cavity and becomes the **spongy,** or **penile, urethra** of the penis. Locate the portions of the urethra and the accessory glands in figures 42.3 and 42.5.

External Genitalia

Penis

(A) 6 Examine a model of the penis and find the features in figure 42.5. The **penis** consists of an elongated shaft and a distally expanded **glans penis.** The glans is covered with the

prepuce, or **foreskin,** which is removed in some males by a procedure called a **circumcision.** At the inferior portion of the glans is a region richly supplied with nerve endings, called the **frenulum.** The glans penis is an expanded region that stimulates the genitalia of the female and, in turn, is stimulated by the female genitalia. The penis is, on average, about 16 cm in length. These structures can be seen in figures 42.1, 42.3, and 42.5.

The penis contains three cylinders of erectile tissue. The **corpus spongiosum** is the cylinder of erectile tissue that contains the spongy (penile) urethra. The two **corpora cavernosa** are located anterior to the corpus spongiosum. Examine a model or chart of a cross section of penis and compare it with figure 42.5. The proximal parts of the cylinders of erectile tissue are anchored to the body. Locate the **crus,** which is an expansion of the corpora cavernosa. The **bulb** of the penis is an extension of the corpus spongiosum. The crus and the bulb form the **root** of the penis. The corpus spongiosum expands distally to form the glans penis. Note the **dorsal arteries** of the penis. They take blood to the penis. Locate the **dorsal veins** of the penis. When the arteries of the penis dilate, the erectile tissues engorge with blood and the penis becomes erect. The erection subsides as the arteries constrict, decreasing blood flow into the penis.

The floor of the pelvis as seen from the outside is referred to as the **perineum.** It can be divided into a posterior **anal triangle** and an anterior **urogenital triangle.** The anal triangle surrounds the anus, and the urogenital triangle encloses the penis and scrotum (figure 42.6).

Male Contraception

The main goal in male contraception is the prevention of sperm cells from reaching the female gametes. **Abstention,** or refraining from sexual intercourse, is the most effective method. The next most effective method is **male sterilization,** which is most commonly performed by a procedure called a **vasectomy** (the cutting and tying of the two **vas,** or **ductus, deferentes**). In a vasectomy, an incision is made on each side of the scrotum and both ductus deferentes are cut. The free ends of the cut ductus deferentes are tied, preventing sperm from traveling from the testes to the spermatic cords (figure 42.7). The sperm is reabsorbed by the body, and vasectomies do not cause any physical change in erectile function. The proper use of a barrier, such as a **condom,** is relatively effective contraception, since it prevents ejaculated sperm from entering the female reproductive tract. **Coitus interruptus,** or

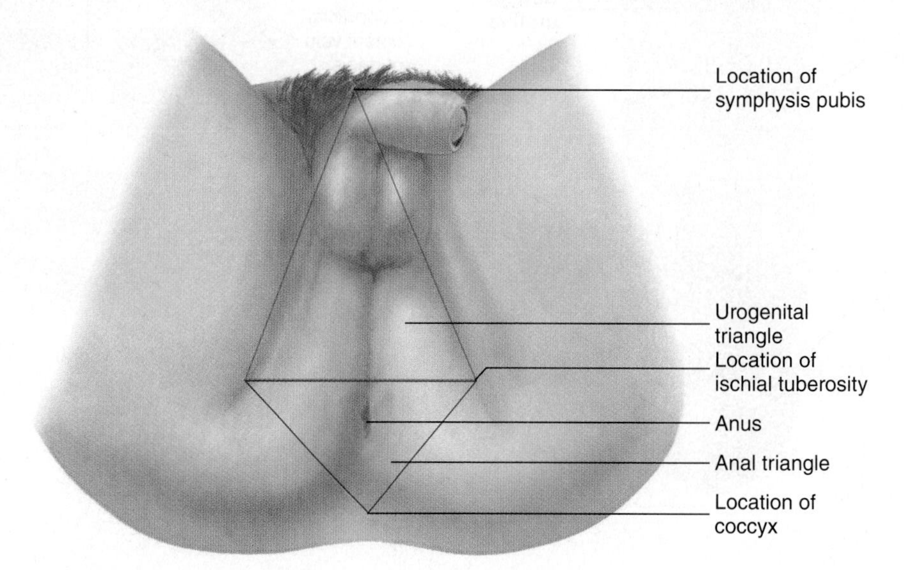

Figure 42.6 Perineum

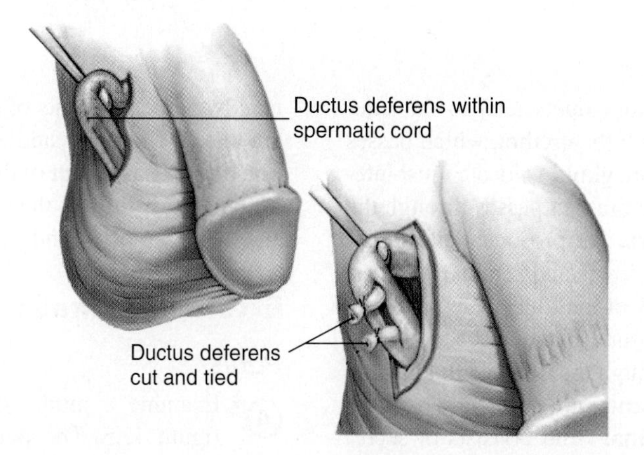

Figure 42.7 Vasectomy

preejaculatory withdrawal, is not a very effective method, since sperm may be present in seminal fluid prior to ejaculation, and pregnancy can result.

Cat Dissection

(A)7 Prepare for the cat dissection by obtaining a dissection tray, a scalpel, scissors or bone cutter, string, forceps, and a plastic bag with a label. Remember to place all excess tissue in the appropriate waste container and not in a standard wastebasket or down the sink!

Check to see whether the cat is male or female. The **penis** may be retracted in your specimen, so look for the opening of the penile urethra and the scrotum. The penis in male cats has numerous, backward-pointing barbs made of keratinized epithelium. These function to stimulate the female to ovulate. Team up with a lab partner or group that has a cat of a different sex than your specimen, so that you can learn both male and female reproductive systems. Once you are sure you have a male cat, locate the **scrotum** and paired **testes.** Make an incision on the lateral side of the scrotum and locate the testis inside the scrotal sac. If the cat was neutered, you will not be able to locate the testis. If the testes are present, cut through the connective tissue of the scrotum **(tunica vaginalis)** and observe both the testis and the **epididymis.** The testis is covered by a tough connective tissue membrane called the **tunica albuginea.** Use figure 42.8 as a guide. Sperm cells move from the testis and into the epididymis, where the sperm cells mature.

Once you have located the epididymis, proceed in an anterior direction and trace the thin **ductus deferens** from the epididymis into the **spermatic cord.** The spermatic cord traverses the body wall on the exterior and enters the body of the cat at the **inguinal canal** (an opening that pierces the inguinal ligament). You may want to gently insert a blunt probe into the inguinal canal, so that you can locate the ductus deferens as it passes into the **coelom.** Locate the ductus deferens as it enters the body cavity and notice how it arches around the ureter on the dorsal side of the urinary bladder. You may have cut the ductus deferens in an earlier exercise, so, if you cannot find it on one side, look for it on the other side.

(A)8 You may want to look for the accessory organs of the male reproductive system, but this takes some significant dissection. *Check with your instructor before cutting through the pelvis of the cat.* If your instructor directs you to do so, then begin by cutting through the musculature of the cat at the level of the **symphysis pubis.** Make a midsagittal incision through the groin muscles, and carefully cut through the cartilage of the symphysis pubis. You should now be able to open the pelvic cavity and locate the single **prostate gland** and the paired **bulbourethral glands** (figure 42.8). Much of the anatomy of the cat is similar to that of the human, except that there are no seminal vesicles in the cat. Trace the ductus deferens from the posterior surface of the bladder through the prostate gland and the penis. You can make either a longitudinal section through the penis to trace the urethra in the erectile tissue or a cross section of the penis to see the three cylinders of erectile tissue—the **corpus spongiosum** and the **two corpora cavernosa.**

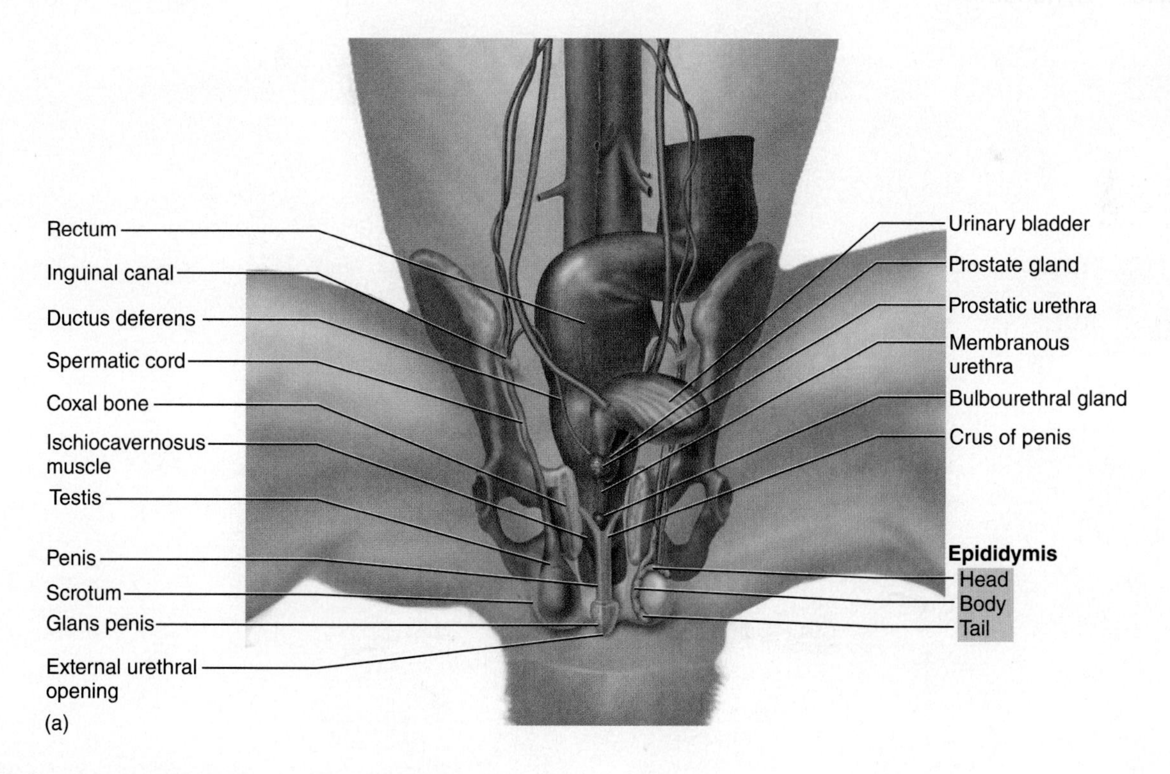

Figure 42.8 Male Reproductive Organs of the Cat

(a) Diagram. (b) Photograph.

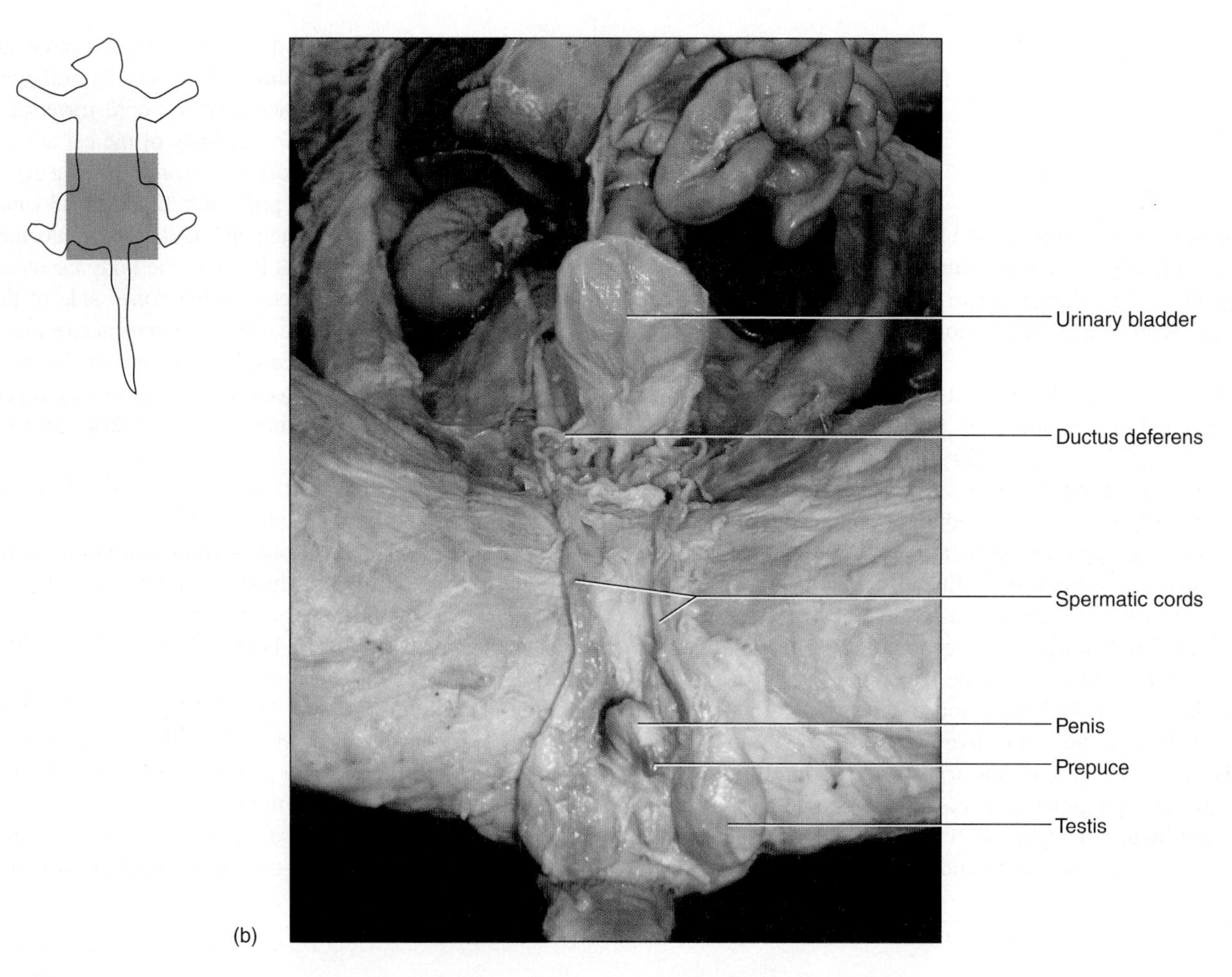

Figure 42.8 *Continued.*

Exercise 42 Review

Male Reproductive System

Name: ______________________

Lab time/section: ______________________

Date: ______________________

1. What is the gonad in the male reproductive system? ______________________

2. What cells and what hormone come from the testes? ______________________

3. Proper sperm production must occur at what temperature? ______________________

4. The condition cryptorchidism (undescended testicles) means that the testes are kept at body temperature (37°C). What effect do you think this has on male fertility? ______________________

5. Name the lining of the scrotal sac that consists of a smooth muscle layer. ______________________

6. What structure of the testes produces sperm cells? ______________________

7. What are the cells called that initiate sperm cell production? ______________________

8. Where do sperm cells move after leaving the epididymis? ______________________

9. Where is the cremaster muscle found? ______________________

10. List all the structures involved in producing semen. ______________________

11. How do sperm cells differ from seminal fluid? ______________________________

____

12. A vasectomy is the cutting and tying of the two ductus deferentes at the level of the spermatic cords. Review the percentage of sperm cells that compose the total volume of semen and determine the effect a vasectomy has on semen volume. ______________________________

____

13. Which one of the seminal fluid glands is not a paired gland? ______________________________

____

14. Name the three sections of the urethra and where they occur. ______________________________

____

15. Where is the glans penis located? ______________________________

____

16. What is the cylinder of erectile tissue below the corpora cavernosa? ______________________________

____

17. Label the following illustration using the terms provided.

bulb of penis
bulbourethral gland
corpus cavernosum
ductus (vas) deferens
epididymis
glans penis
prepuce
prostate gland
scrotum
seminal vesicle
testis
urinary bladder

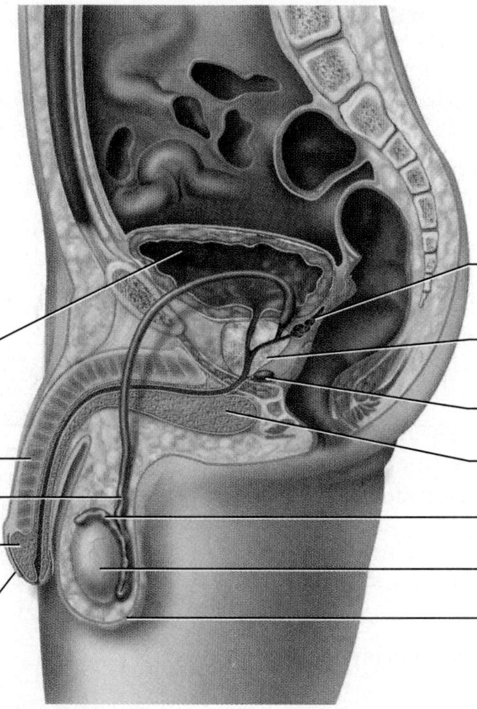

a. ______________
b. ______________
c. ______________
d. ______________
e. ______________
f. ______________
g. ______________
h. ______________
i. ______________
j. ______________
k. ______________
l. ______________

Male Reproductive System

18. What male reproductive gland is missing in the cat but present in the human? ______________________________

Exercise 43

Female Reproductive System

Introduction

The female reproductive system is functionally more complex than the male reproductive system. In the male, the reproductive system produces gametes and delivers them to the female reproductive system. The female reproductive system not only produces gametes and receives them from the male but also, under the hormonal influence of hCG from the early cell mass and later from the placenta, provides space and maternal nutrients for the developing **conceptus.** Finally, the female reproductive system delivers the child into the outer environment. The female reproductive system is covered in the Seeley text in chapter 28, "Reproductive System."

The ovaries are the gamete-producing organs of the female reproductive system. They produce oocytes and the female sex hormones, estrogen (estradiol) and progesterone. In this exercise, you learn about the structure and function of the female reproductive system.

Objectives

At the end of this exercise, you should be able to

1. identify the gamete-producing organ of the female reproductive system;
2. trace the pathway of a gamete from the ovary to the usual site of implantation;
3. list the structures of the vulva;
4. describe the function of each structure in the female reproductive system;
5. name the layers of the uterus from superficial to deep;
6. describe the internal anatomy of the breast;
7. describe the early development in humans from the zygote to the blastocyst; and
8. compare and contrast the anatomy of the cat reproductive system with that of the human.

Materials

Charts, models, and illustrations of the female reproductive system
Microscopes
Prepared slides of ovary and uterus
Cats
Materials for cat dissection

Dissection trays
Scalpel and new blades
Protective gloves
Blunt (Mall) probe
First aid kit
Forceps and sharp scissors
Sharps container
Animal waste disposal container

Procedure

Overview of the Gross Anatomy of the Female Reproductive System

Ⓐ 1 Examine a model or chart of the female reproductive system and locate the following major reproductive organs there and in figure 43.1.

Ovary
Uterine tube
Uterus
Vagina
Clitoris
Labia minora (sing. *labium minus*)
Labia majora (sing. *labium majus*)

Ovaries

Each **ovary** is approximately 3–4 cm long and oblong in form. The ovaries produce **oocytes** (eggs), which are shed from the outer surface of the ovary during **ovulation.** From there, the oocytes (which mature into ova if fertilized) move into the **uterine (fallopian) tube.** The ovaries are not directly attached to the uterine tube, and the oocytes must move from the surface of the ovary into the uterine tube.

Histology of the Ovary

Ⓐ 2 Examine a prepared slide of the ovary under the microscope on low power. Locate the background substance of the ovary, which is known as the **stroma.** Look for circular structures in the ovary. These are the **ovarian follicles.** The smallest of the follicles are the **primordial follicles.** Locate the primordial follicles in your slide and compare them with the follicles in figure 43.2. The process of oogenesis involves the production of the gametes.

Also locate the **primary** and **secondary follicles.** Some of the follicles may contain **oocytes.** Primordial and primary follicles contain **primary oocytes;** secondary follicles contain **secondary oocytes.** The largest follicles in the ovary are the **graafian follicles,** or **mature ovarian follicles,** and you may be able to see one if it is present in your slide. During ovulation in humans, usually one secondary oocyte is shed from the ovary. In cats, many secondary oocytes may be shed.

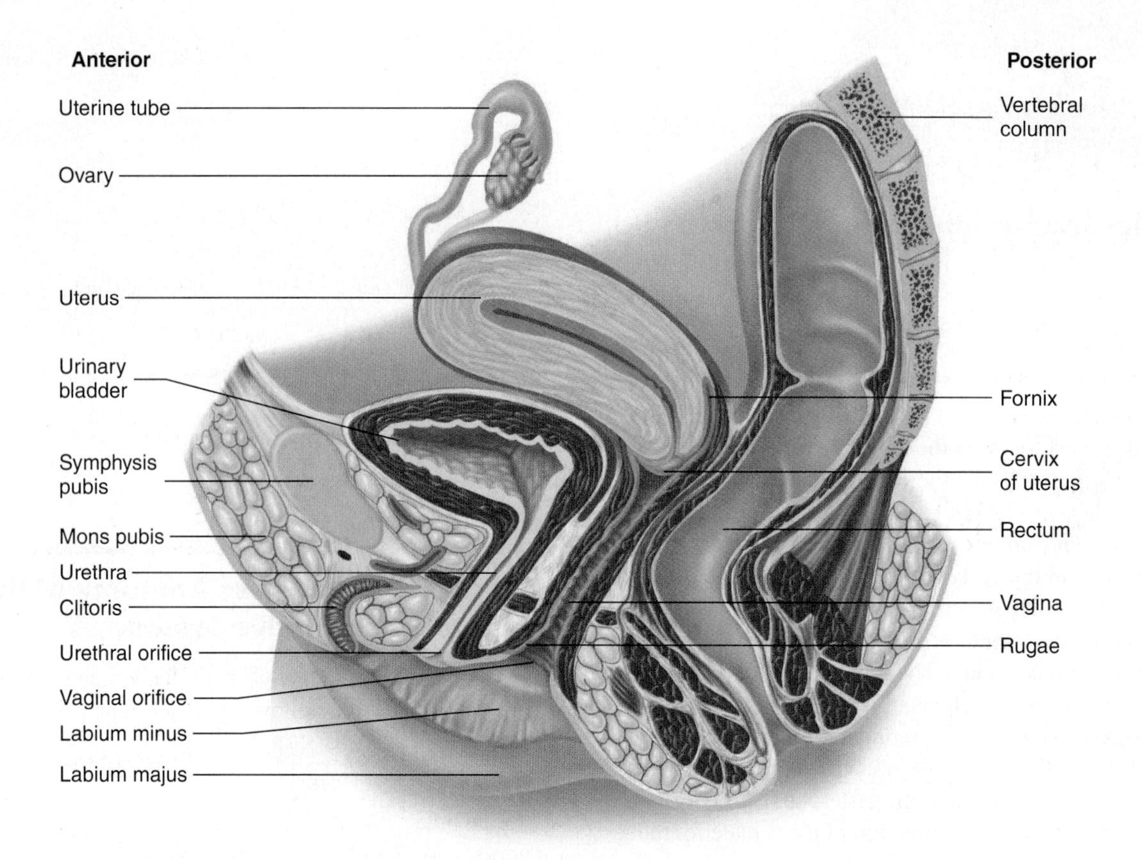

Figure 43.1 Female Reproductive System, Midsagittal View

Examine your slide and compare it with figures 43.2 and 43.3. Draw what you see in your slide in the following space.

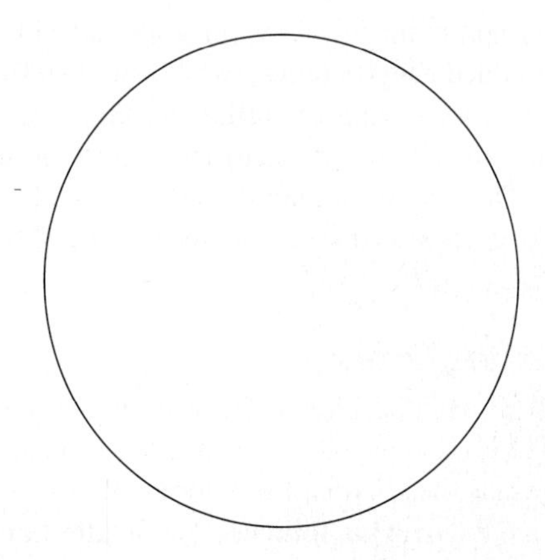

After ovulation, the remains of a mature ovarian follicle become a **corpus luteum,** which primarily secretes progesterone. If pregnancy does not occur, the corpus luteum decreases in size and becomes the **corpus albicans.** This progression is seen in figure 43.4. Examine a prepared slide of the ovary with a corpus luteum or corpus albicans and compare it with figures 43.4 and 43.5.

Uterine Tubes

(A)[3] Examine a model or chart of the female reproductive system and compare it with figure 43.6. The uterine tube has a small fringe on the distal region comprising the **fimbriae.** These are small, fingerlike projections attached to an expanded region known as the **infundibulum.** The uterine tube also has an enlarged region known as the **ampulla** and a narrower portion toward the uterus.

Uterus

(A)[4] The **uterus** is a pear-shaped organ with a domed **fundus,** a **body,** and a circular, inferior end called the **cervix.** The **uterine tubes** enter the uterus at about the junction of the fundus with the uterine body. A constricted portion of the inferior uterus is called the **isthmus.** The uterine wall is composed of three layers. The outer (superficial) surface of the uterus is called the **perimetrium.** This is located near the body cavity. The majority of the uterine wall consists of the **myometrium,** a thick layer of smooth muscle, and the innermost (deepest) layer of the uterus is the **endometrium.** Examine the models or charts in the lab and locate the structures in figure 43.6.

Histology of the Uterus

(A)[5] Examine a prepared slide of the uterus and locate its three layers. Locate the outer **perimetrium,** the smooth muscle of the **myometrium,** and the inner **endometrium.** Compare them with figures 43.6 and 43.7.

Now examine the two layers of the endometrium of the uterus under higher magnification. The **functional layer** is the one that is shed during menstruation. It is composed of **spiral arterioles** and **uterine glands** that appear as wavy lines toward the

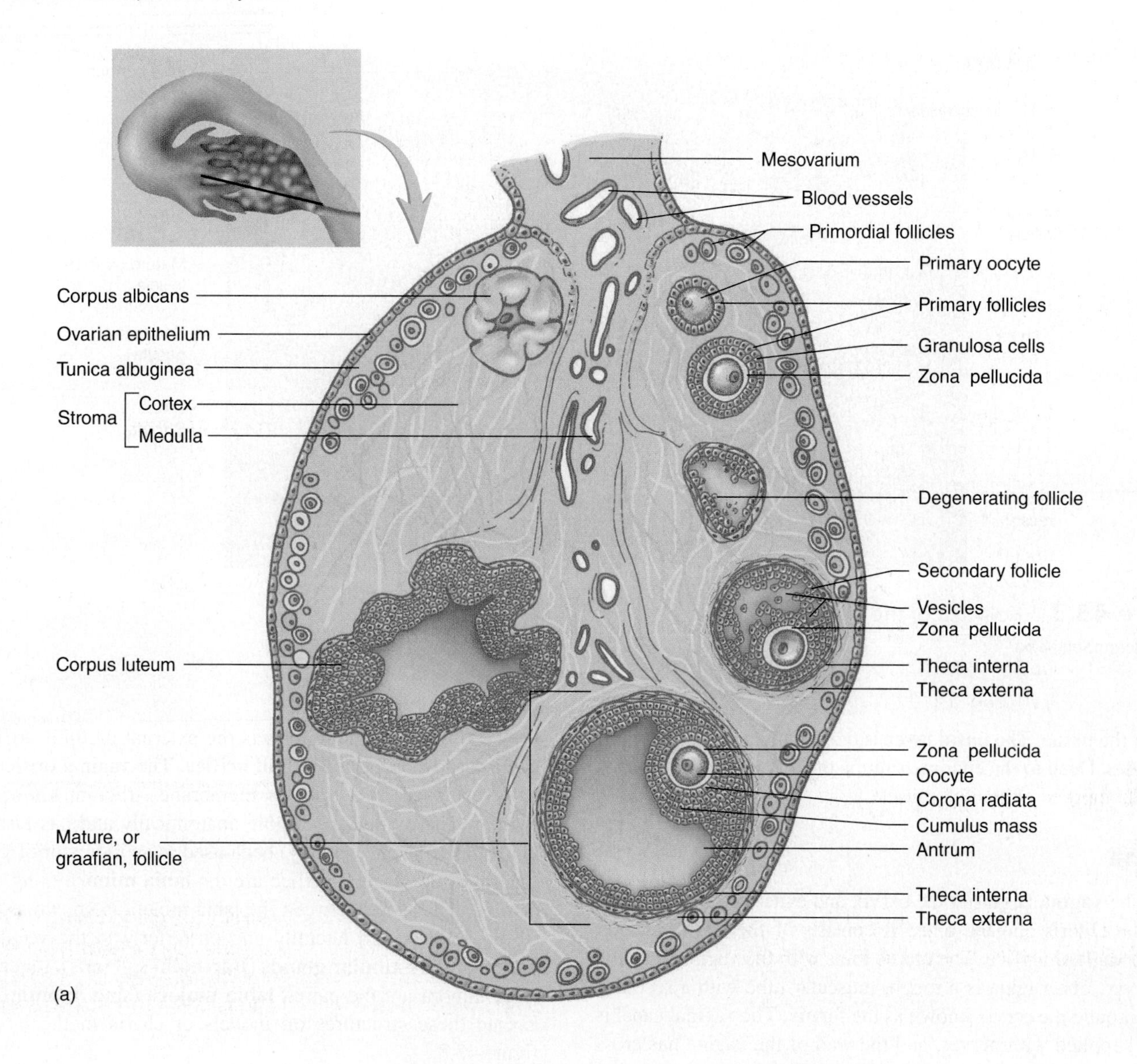

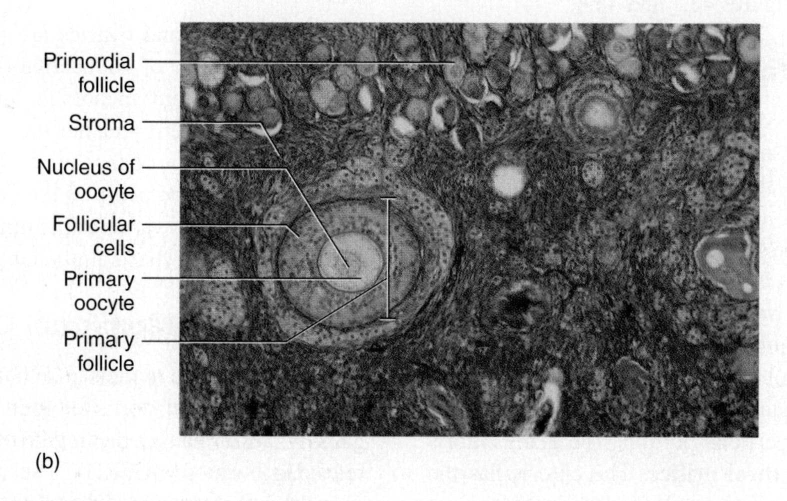

Figure 43.2 Histology of the Ovary

(a) Diagram; (b) photomicrograph (100×).

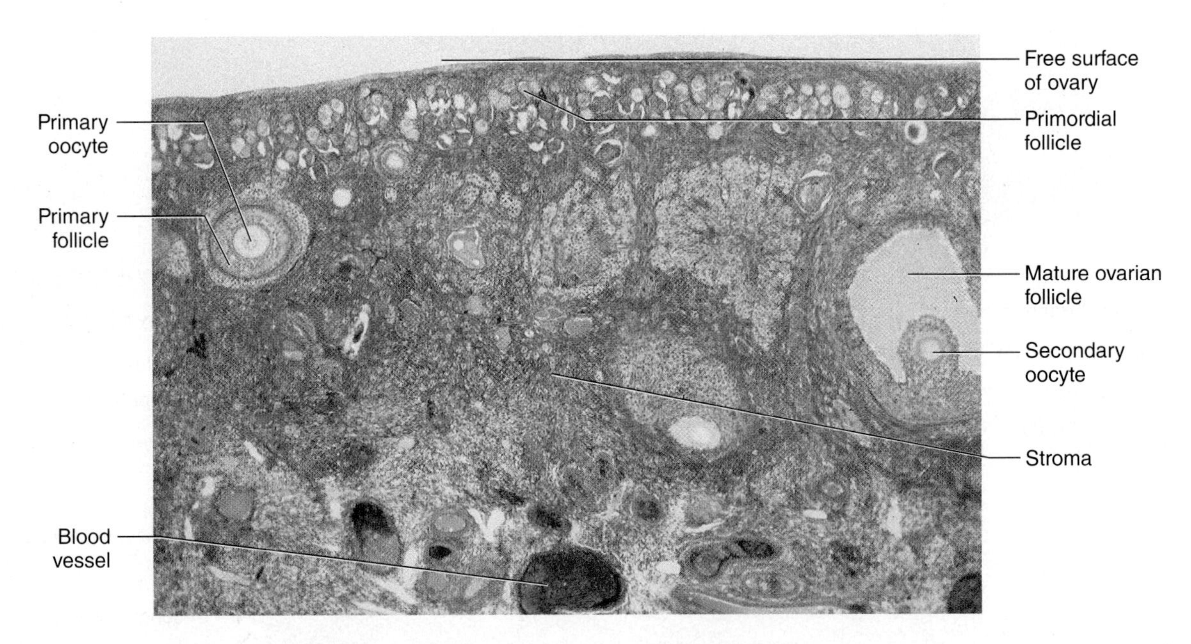

Figure 43.3 Follicles of the Ovary
Photomicrograph (40×).

edge of the tissue. The **basal layer** is deeper and contains **straight arterioles.** Deep to the endometrium is the myometrium, which is mostly composed of smooth muscle.

Vagina

(A) 6 The **vagina** begins at the cervix and exits the body between the clitoris and the anus. It consists of the **vaginal canal** and the **vaginal orifice.** The uterus joins with the vaginal canal at the cervix. The vagina is a tough, muscular tube with a recessed region around the cervix known as the **fornix.** The vaginal canal is poorly supplied with nerves, and the wall of the vagina has cross ridges called **rugae.** The vagina is lined with stratified squamous epithelium. The vagina expands greatly during the delivery of a child. Locate these features in figures 43.1 and 43.6.

External Female Genitalia

(A) 7 As in the male, the floor of the pelvis can be divided into the **anal triangle** and the **urogenital triangle.** The anal triangle consists of the **anus,** and the urogenital triangle consists of the **external female reproductive structures,** or **female genitalia.** The external female reproductive structures are collectively referred to as the **vulva.** The **mons pubis** is the anteriormost structure of the vulva. It is an adipose pad that overlies the symphysis pubis. Posterior to the mons pubis is the **clitoris,** which is a cylinder of erectile tissue embedded in the body wall that terminates as the **glans clitoris.** The body of the clitoris is curved, as illustrated in figure 43.1, and the glans clitoris is the superficial portion. The glans clitoris is anterior to the **external urethral orifice.** The clitoris has the same embryonic origin as the penis in males, and like the penis it is richly supplied with nerve endings. It is a major center for sexual pleasure in females. The anterior edge of the clitoris is enclosed by the **prepuce,** which is an extension of the labia minora.

Posterior to the clitoris is the external urethral orifice, and posterior to this is the **vaginal orifice.** The vaginal orifice is partially enclosed by a mucous membrane structure known as the **hymen.** The hymen is variable anatomically and has historically (and sometimes incorrectly) been used as an indicator of virginity. Lateral to the vaginal orifice are the **labia minora** (sing. **labium minus**). The space between the labia minora is known as the **vestibule,** and located laterally and posteriorly to the vestibule are the **greater vestibular glands (Bartholin's glands).** Lateral to the labia minora are the paired **labia majora** (sing. **labium majus**). Locate these structures on models or charts in the lab and in figure 43.8.

Ligaments

(A) 8 The uterus and ovaries are suspended in the pelvic cavity by a number of connective tissue sheaths called ligaments. The **broad ligament** anchors the uterus to the anterior body wall. The **round ligament** attaches the uterus to the anterior body wall at about the region of the inguinal canal. The **ovarian ligament** directly attaches the ovary to the uterus, and the **suspensory ligament** attaches the ovaries to the lumbar region. Locate these structures in models or charts in the lab and in figure 43.6.

Ovarian and Menstrual Cycles

(A) 9 Compare the models and charts in lab of the cross section of the ovary and uterus to Figure 43.9. The endometrium undergoes dynamic changes during the **menstrual cycle.** Gonadotropin-releasing hormone (GnRH) is secreted by the hypothalamus and stimulates the anterior pituitary to release hormones. **Luteinizing hormone (LH)** and **follicle-stimulating hormone (FSH)** from the anterior pituitary, and subsequently **estrogen** and **progesterone** from the ovary, have significant effects on the endometrium.

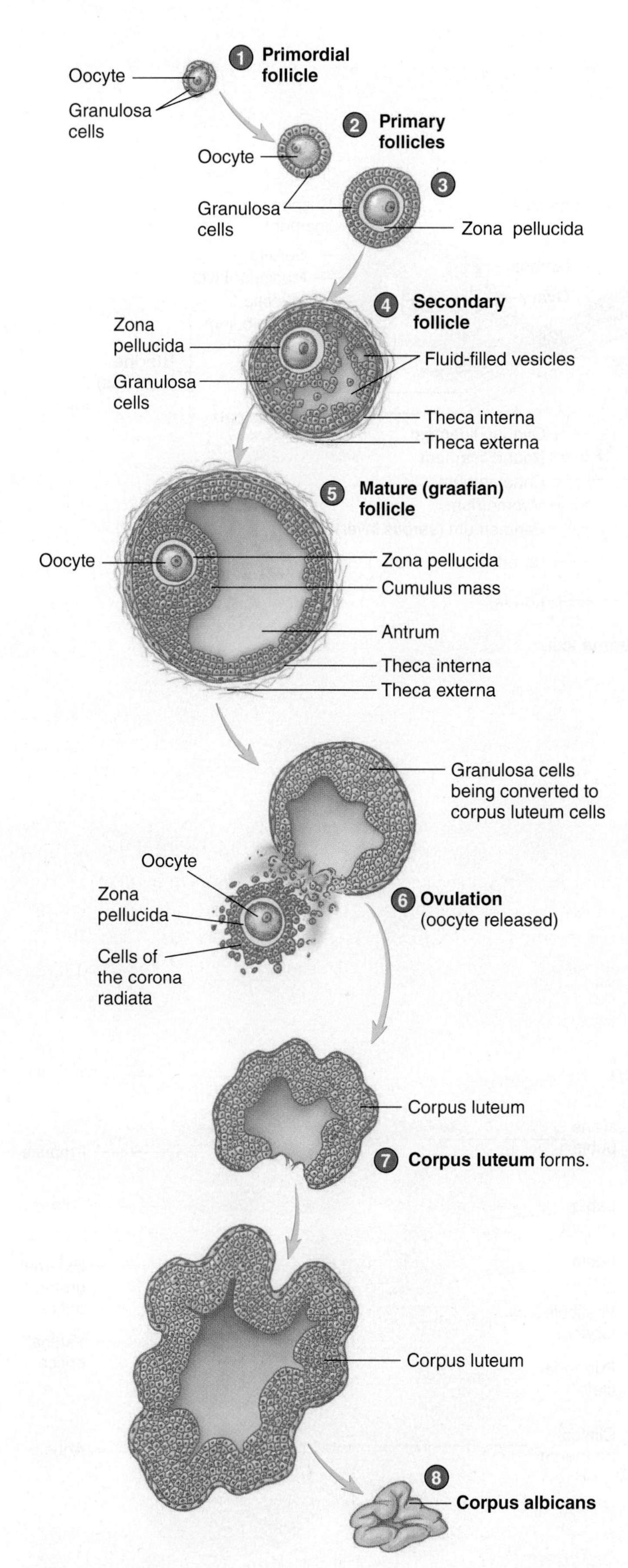

Figure 43.4 Maturation of the Oocyte and the Follicle

Arrows indicate a time line of development from primordial follicles to the corpus albicans.

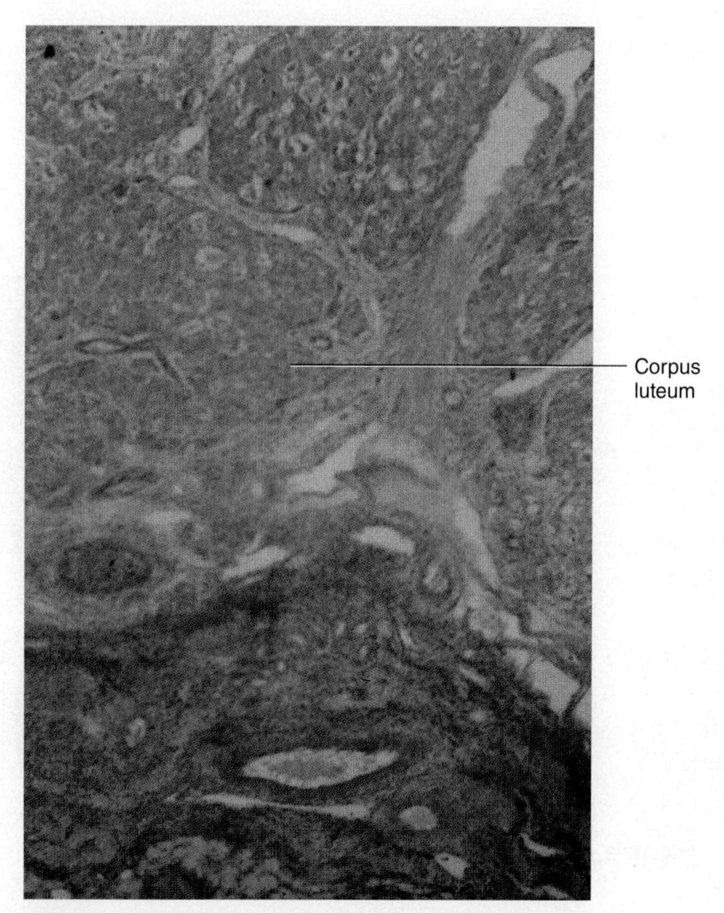

Figure 43.5 Postovulatory Ovary (40×)

FSH stimulates the ovarian follicles to secrete estrogens and some progesterone. Elevated levels of these hormones promote the thickening of the endometrium, as indicated in figure 43.9.

After ovulation, progesterone levels increase and the endometrial cells remain intact. Toward the end of a woman's menstrual cycle, estrogen and progesterone levels drop, causing the functional layer of the endometrium to slough off. The loss of the endometrial layer is the beginning of a woman's period, or **menstruation.** These events are briefly outlined in figure 43.9. Examine the figure and note how the endothelial layer begins to decrease when estrogen and progesterone levels fall (about day 25 in figure 43.9).

Female Contraception

As with male contraception, preventing male and female gametes from uniting is the goal in female contraception. Abstention, or refraining from intercourse, is the most effective method of birth control. Other methods involve oral contraceptives or hormonal implants; tubal ligation (cutting and tying the uterine tubes—see figure 43.10); the use of barrier methods, such as the condom or diaphragm; and the use of spermicidal foams. The use of oral contraceptives, which are synthetic estrogens and progesterones, decreases the levels of LH and FSH in the pituitary (by negative-feedback mechanisms), thus preventing ovulation. Progesterone implants elevate hormone levels, which also prevent ovulation. Examine the levels of progesterone in figure 43.9. When these

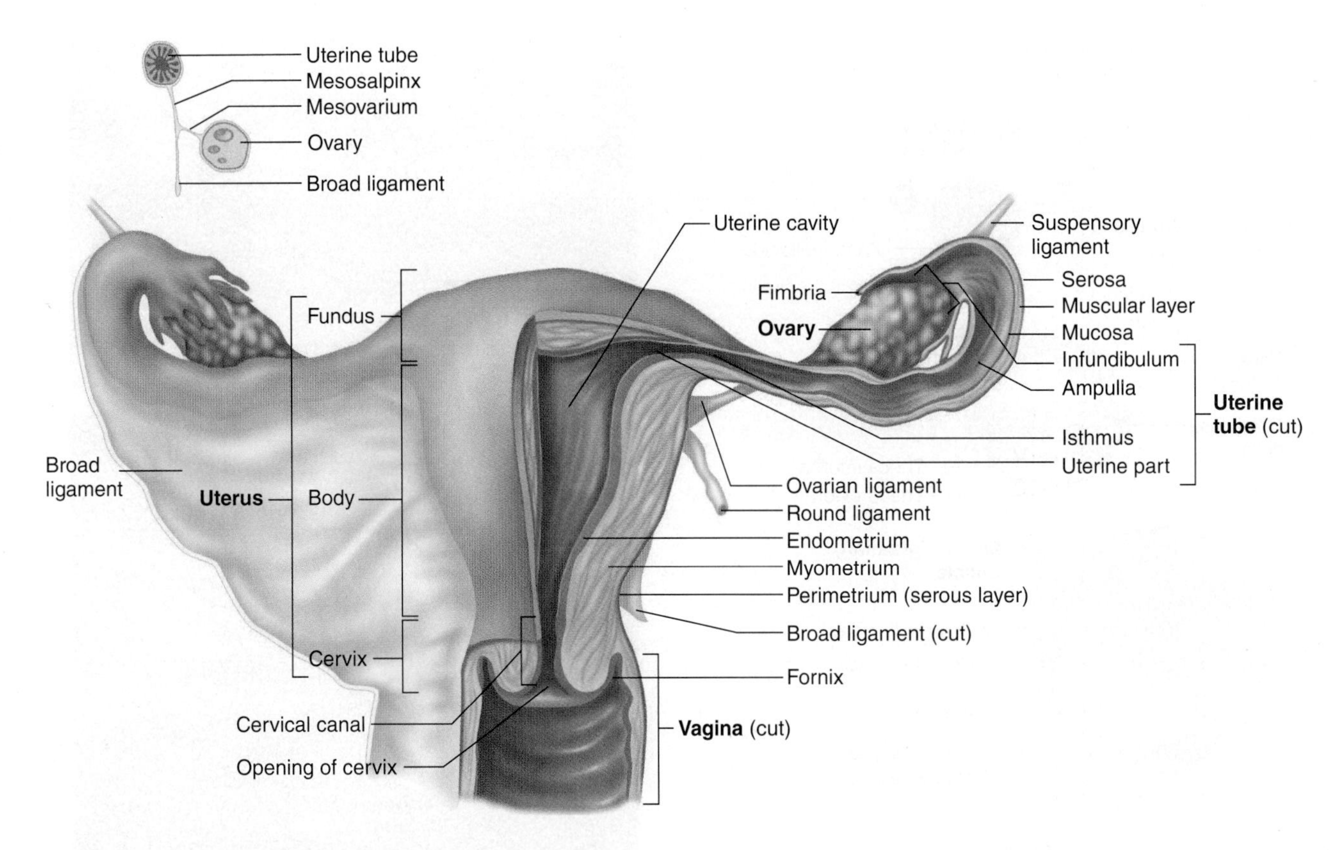

Figure 43.6 Female Reproductive System, Anterior View

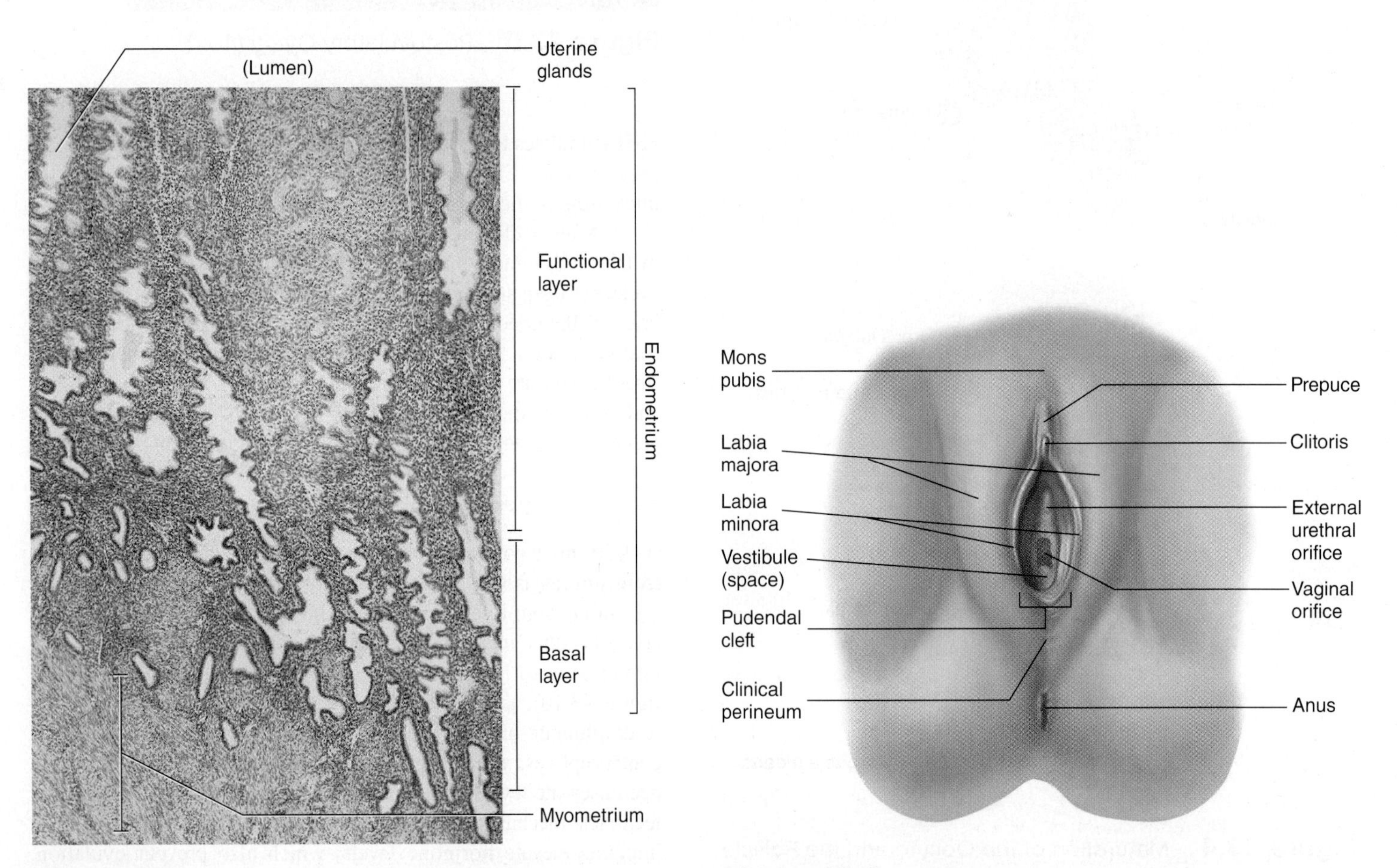

Figure 43.7 Histology of the Uterus (40×)

Figure 43.8 External Female Genitalia

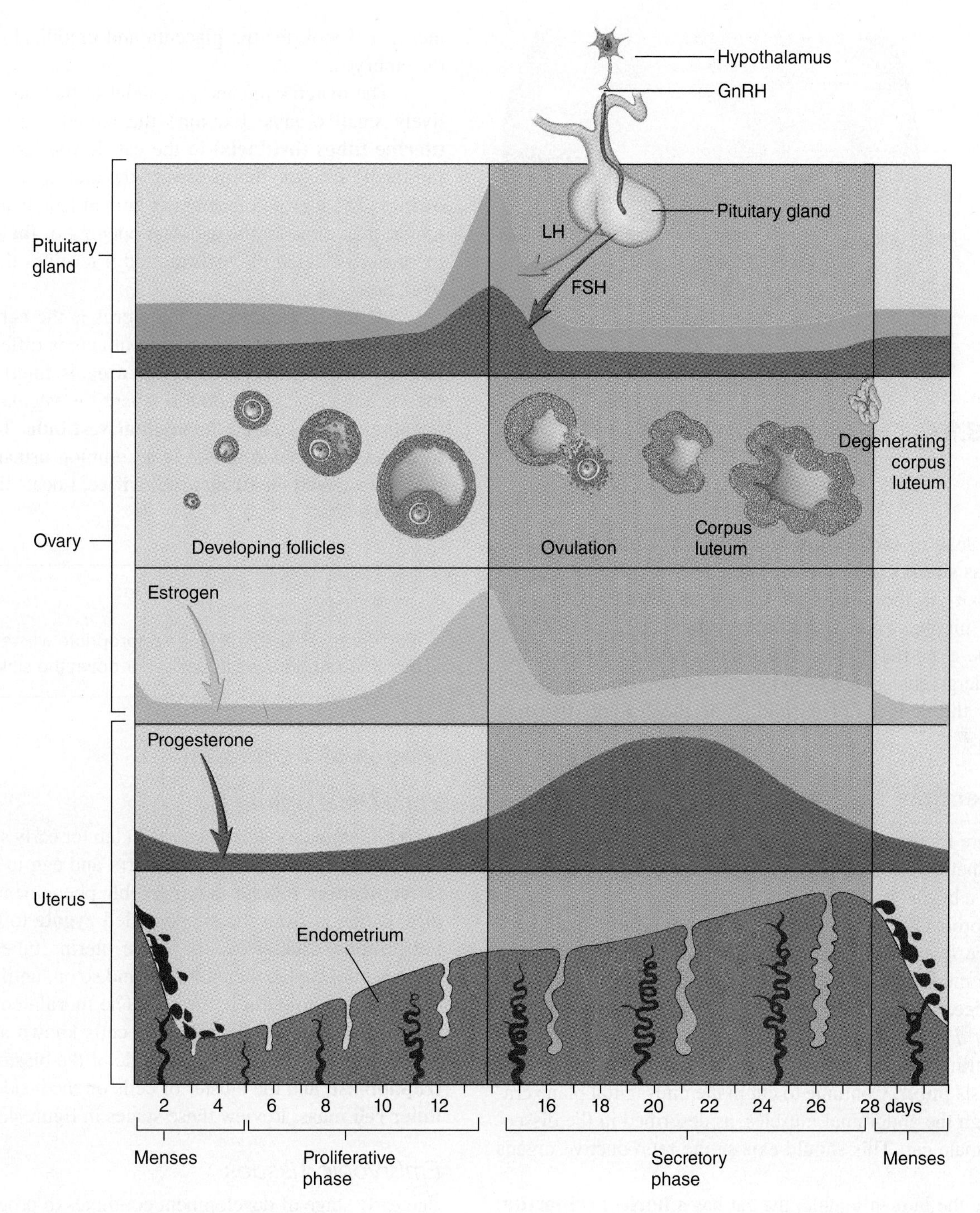

Figure 43.9 Ovarian and Menstrual Cycles

levels are elevated in the later part of the ovarian cycle, note how the levels of FSH and LH are low. An antiprogesterone compound, mifepristone, terminates pregnancy and is known as the "morning after" pill.

Anatomy of the Breast

Ⓐ 10 Examine the surface features of the breast in figure 43.11 and locate the structures listed. The structure of the breast derives from the integumentary system, yet the female breast is discussed as a reproductive structure due to its importance as a source of nourishment for offspring. The major structures of the external breast are the pigmented **areola,** the protruding **nipple,** the **body** of the breast, and the **axillary tail (tail of Spence).** The axillary tail is of clinical importance in that breast tumors often occur there.

Compare models or charts in the lab and locate the internal structures of the breast. Note that the breast is anchored to the pectoralis major muscle with **suspensory ligaments.** Much of the breast is composed of **adipose tissue,** and embedded in the adipose tissue are the **mammary glands.** The mammary glands are responsible for the production of milk in lactating females. The glands are clustered in **lobes,** and there are about 15 to 20 lobes in each breast. The mammary glands increase in size in nursing

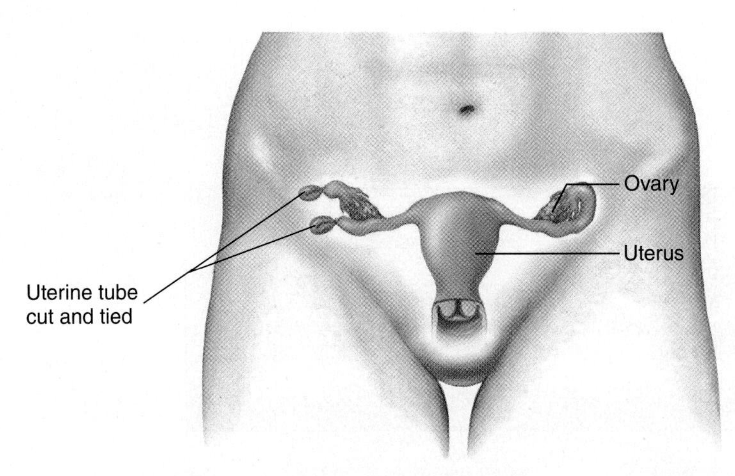

Figure 43.10 Female Contraception

women and lead to **lactiferous ducts,** which subsequently lead to **lactiferous sinuses (ampullae).** These lead to other lactiferous ducts that exit via the nipple. Humans have several ducts leading to each nipple, whereas in some mammals, such as cows, just one tube exits the nipple. The mammary glands in females begin to undergo changes prior to puberty and become functional glands after the delivery of a child. Note the features listed in figure 43.11.

Cat Dissection

(A) 11 Prepare for the cat dissection by obtaining a dissection tray, a scalpel, scissors or bone cutter, string, forceps, and a plastic bag with a label.

You probably already removed the multiple mammary glands of the female cat during the removal of the skin and the study of the muscles. Examine the external genitalia for the **urogenital orifice.** *Check with your instructor before cutting through the pelvis of the cat.* If your instructor directs you to do so, then begin by cutting through the musculature of the cat at the level of the **symphysis pubis.** Continue to cut in the midsagittal plane cranially through the abdominal muscles, as described in the dissection of the male cat. This should expose the reproductive organs (figure 43.12).

Unlike the human female, the cat has a **horned (bipartite) uterus.** The uterus of the cat has the appearance of a Y, with the upper two branches being the **uterine horns** and the stem of the Y the **body** of the uterus. Humans normally have single births from a pregnancy, whereas the expanded uterus in cats facilitates multiple births. If your cat is pregnant, the uterus will be greatly enlarged, and you may find many fetuses inside. Examine the inside of the uterus and look for the placenta and umbilical cords attached to the embryos.

The **ovaries** in cats are caudal to the kidneys and are relatively small organs. Examine the paired ovaries and the short **uterine tubes (oviducts)** in the cat. If you have difficulty locating them, trace the uterus toward the uterine horns and locate the ovaries. The uterine tubes in the human female are proportionally longer than those in the cat. The opening of the uterine tube near the ovary is called the **ostium,** and it receives the oocytes during ovulation.

At the termination of the uterus is the **cervix,** which leads to the **vaginal canal.** The vagina in cats is different from that in humans in that the **urethral opening** is internally enclosed in the vaginal canal. This region where the vagina and the urethral opening occur is called the **vaginal vestibule.** Thus, the opening to the external environment is a common urinary and reproductive outlet called the **urogenital orifice.** Locate these structures in figure 43.12.

Cleanup

Place all excess tissue in the appropriate waste container and not in a standard wastebasket or down the sink!

Stages of Development

Early Development

(A) 12 Examine models or charts in lab for early stages of development. The union of the sperm and egg in a process known as **fertilization** initiates a remarkable phenomenon of growth and differentiation from the single-celled **zygote** to the adult human. Fertilization usually occurs in the uterine tube, and the zygote divides into 2 cells, then 4, 8, 16, and so on, until a solid cluster of cells called a **morula** is formed. The morula continues to divide until it becomes a hollow ball of cells known as the **blastocyst.** The covering of cells on the outside of the blastocyst is called the **trophoblast,** and the cluster of cells on the inside is known as the **inner cell mass.** Review these stages in figure 43.13.

Embryonic Tissues

The early stage of development continues to progress, and the formation of three embryonic tissues occurs. These are the **ectoderm,** the **mesoderm,** and the **endoderm.** The ectoderm gives rise to the outer layer of skin and the nervous tissue, the mesoderm gives rise to bones and muscles, and the endoderm gives rise to many internal organs, such as digestive and respiratory organs. The early development of these layers is illustrated in figure 43.14.

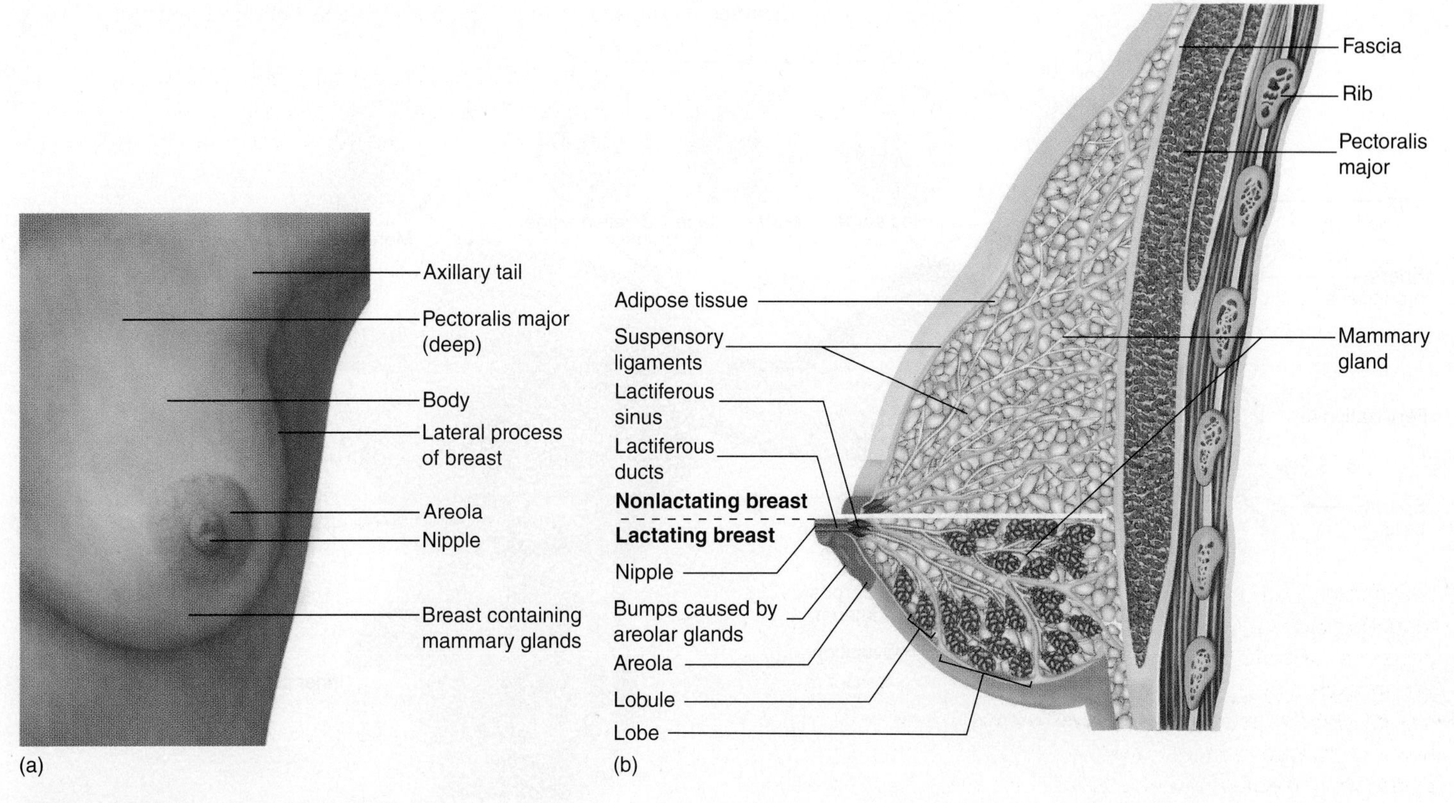

Figure 43.11 Anatomy of the Female Breast

(a) Surface features; (b) interior of the female breast.

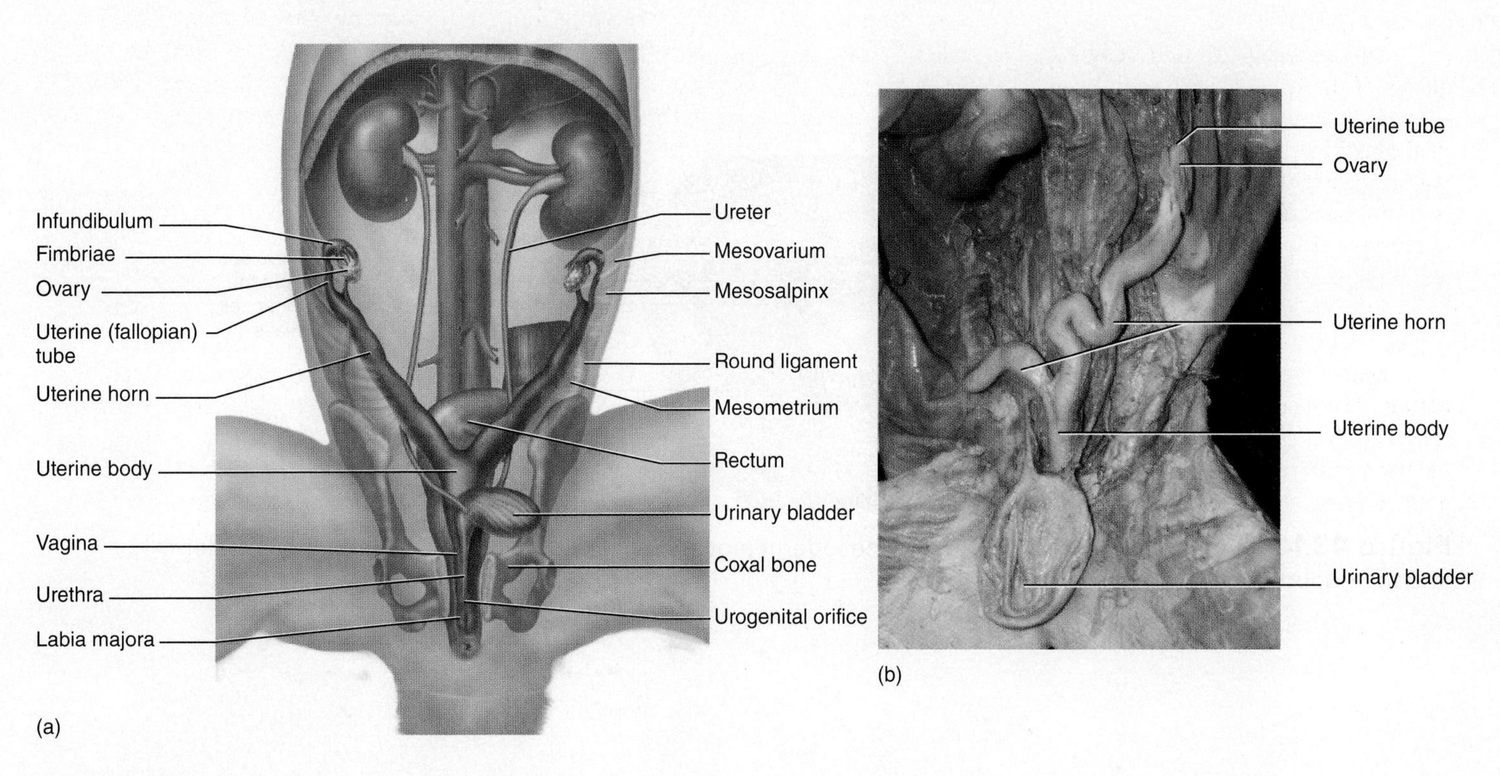

Figure 43.12 Female Reproductive Organs of the Cat

(a) Diagram; (b) photograph.

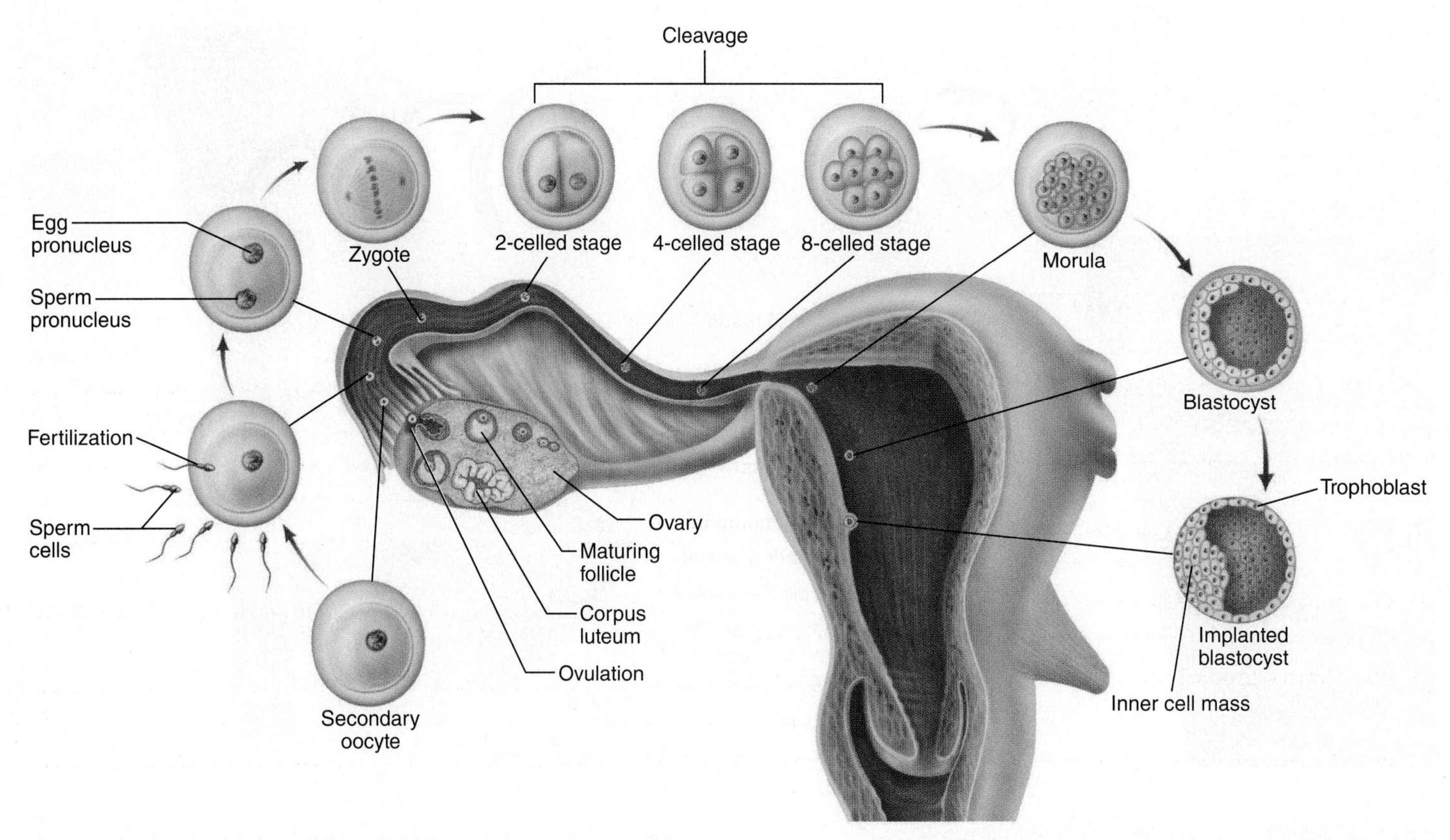

Figure 43.13 Early Stages of Development

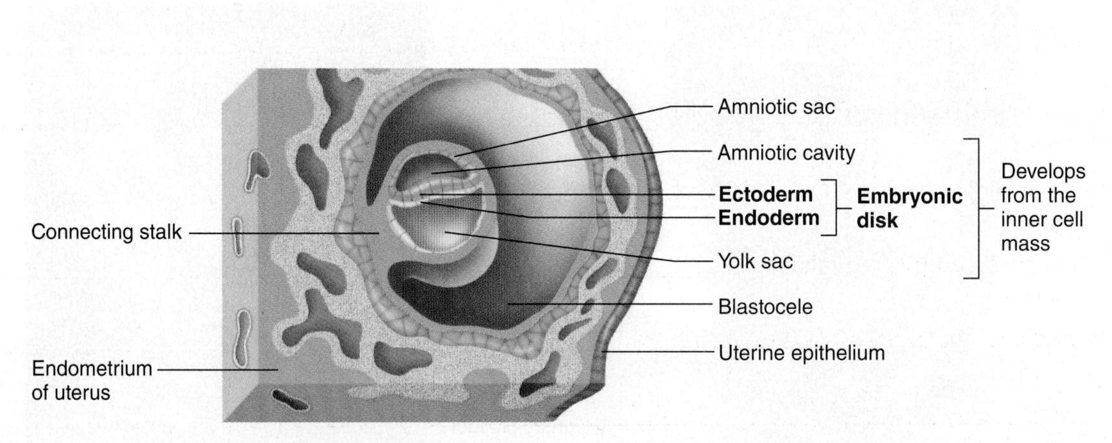

Figure 43.14 Embryonic Development. The mesoderm (not shown here) develops between the ectoderm and the endoderm.

Exercise 43 Review

Name: ______________________

Lab time/section: ______________________

Date: ______________________

Female Reproductive System

1. What is the gonad in the female reproductive system? ______________________

2. What is the inner layer of the uterus called? ______________________

3. What happens to estrogen and progesterone levels just prior to menstruation? ______________________

4. The ovaries attach to the uterus by what structure? ______________________

5. What three hormones are at elevated levels just prior to ovulation? ______________________

6. Where is the fornix in the female reproductive system? ______________________

7. Which is more anterior, the urethral opening or the clitoris? ______________________

8. What is the background substance of the ovary called? ______________________

9. What is the name for the expulsion of the secondary oocyte from the ovary? ______________________

10. What is the layer of the endometrium closest to the myometrium called? ______________________

11. What is the name of the part of the breast that is nearest to the shoulder? ______________________

12. What are the milk-producing glands of the breast called? ______________________

13. A zygote is formed from the fusion of what two cells? ______________________

14. Embryonic tissue consists of three layers. What are these called? ______________________

15. What is a morula? ______________________

16. Trace the pathway of milk from the mammary glands to expulsion. ______________________________

____

____

17. Ectopic pregnancies are those that occur outside the endometrial layer of the uterus. Explain how pregnancies may occur in the uterine tube (thus, a tubal pregnancy) or in the abdominopelvic cavity. ______________________________

____

____

____

18. How does the uterus of the human differ from that of the cat? ______________________________

____

____

19. How does the structure of the uterus in the cat correlate to multiple births from each pregnancy? ______________________________

____

____

20. Label the following illustration using the terms provided. ______________________________

cervix	clitoris	fornix
fundus	labium minus	rugae
urinary bladder	vaginal canal	vaginal orifice

a. ______________________

b. ______________________

c. ______________________

d. ______________________

e. ______________________

f. ______________________

g. ______________________

h. ______________________

i. ______________________ (ridges)

Appendix A

Measurement Conversions

The **metric system** is universally used in science to measure certain **values** or **quantities.** These values are **length, volume, mass** (weight), **time,** and **temperature.** The metric system is based in units of 10, and conversion to higher or lower values is relatively easy when compared with using the U.S. customary system. In the United States, length is typically measured in inches, feet, yards, and miles. The **base unit** for length in the metric system is the meter. As shown in table A.1, 1 meter is equivalent to 1.09 yards or 39.4 inches. When the measurement of great lengths makes the use of meters impractical, then the kilometer (1000 meters) is used. When the measurement of extremely small lengths is needed, then the centimeter (1/100 of a meter), millimeter, micrometer, nanometer, or angstrom is used. The common multiples or fractions for the values are outlined in table A.1.

TABLE A.1 Conversions of Standard and Metric Systems

Value	Base Unit	×1,000	1/100	1/1,000	1/1,000,000
Length	Meter (m) 1 m = 39.4 in. 1 m = 1.09 yd	Kilometer (km) 1 km = 5/8 mi	Centimeter (cm) 2.54 cm = 1 in.	Millimeter	Micrometer 1 nanometer = 10^{-9}m (1 angstrom = 10^{-10}m)
Mass	Gram (g) 454 g = 1 lb 1 g = 0.036 oz	Kilogram (kg) 1 kg = 2.2 lb	Centigram	Milligram	Microgram
Volume	Liter (L) 1 liter = 1.06 qt 1 gal = 3.78 L	Kiloliter	Centiliter	Milliliter (mL) 1 mL = 1 cc	Microliter
Time	Second (s)	Kilosecond	Centisecond	Millisecond	Microsecond
Temperature	Degrees 0°C = freezing 100°C = boiling °F = 9/5 °C + 32	Celsius			

Appendix B

Preparation of Materials

The following preparations are designed for a lab of 24 students. It is best to estimate how many milliliters of stock solutions are required for all the students in a lab and then double that amount. The preparations are listed alphabetically. The number of the lab exercise follows the solution description for cross-referencing.

Acetylcholine chloride solution (0.1%)
Add 0.1 g of acetylcholine chloride to 100 mL frog Ringer's solution. Pour into a small, labeled dropper bottle. (Exercise 29)

Acid solution
Pour 200 mL lemon juice or vinegar into six dropper bottles labeled "Acid Solution." (Exercise 39)

Agar plates
Add 22 g agar to enough water to make 1 L of solution. Boil and stir the agar until it all dissolves. Pour into petri dishes for three dishes per table. (Exercise 3)

Alpha amylase solution (1%)
Add 3 g of alpha amylase to a graduated cylinder and add water to the 300 mL mark. Pour into a small, labeled bottle or beaker. (Exercise 39)

Alpha amylase solution (0.1%)
Add 0.1 g of alpha amylase to a graduated cylinder and fill to 100 mL with water or make a 0.1 serial dilution by taking 10 mL of the above solution and adding 90 mL of water. Pour into a small, labeled bottle or beaker. (Exercise 39)

Ammonia
Place 100 mL of household ammonia in a small bottle. Label "Ammonia." (Exercise 21)

BAPNA solution (1%)
N-alpha-Benzoyl-Dl-arginine-4-nitroanilide hydrochloride 1% BAPNA solution (50 mL) in two bottles of 25 mL each. BAPNA is an expensive material, but you do not need much of it for the lab. Available from Sigma B4875 or Aldrich 85,711–4. (Exercise 39)

Benedict's reagent
A copper sulfate solution that turns color if reducing sugars are present and remains blue in the absence of reducing sugars. Add 35 g sodium citrate and 20 g sodium carbonate (Na_2CO_3) to 160 mL of water. Filter through paper into a glass beaker. Dissolve 3.5 g copper sulfate ($CuSO_4$) in 40 mL of water. Pour the copper sulfate solution into the 160 mL, stirring constantly. (Exercise 39)

Bleach solution (10%)
Mix 100 mL of household bleach (sodium hypochlorite) with 900 mL tap water. (Exercises 25, 26, 41)

Caffeine solution, saturated
Add small amounts of caffeine to 50 mL of water until no more will dissolve. Decant the solution into small dropper bottles. (Exercise 29)

Calcium chloride solution (2%)
Weigh 5 g of calcium chloride and place in a graduated cylinder. Add frog Ringer's solution to make 250 mL. Pour into dropper bottles. (Exercise 29)

Cat wetting solution
Numerous formulations are available for keeping preserved specimens moist. Some commercial preparations are available that reduce the exposure of students to formalin or phenol. You may not need any wetting solution at all if the cats are kept in a plastic bag that is securely tied closed. You can make a wetting solution by putting 75 mL of formalin, 100 mL glycerol, and 825 mL distilled water in a 1 L squeeze bottle. Another mixture consists of equal parts Lysol and water. (Exercises 12–15)

Cellulose
Cut several (3–4) grams of pure cotton (cotton wool, cotton balls) into fine pieces (0.5 cm or less). Label "Cellulose." (Exercise 39)

Dark corn syrup or concentrated sucrose solution (20%)
Use undiluted molasses or a 20% sugar solution. To make the sugar solution, add 100 g of table sugar (sucrose) to water to make 500 mL of solution. Add food coloring to the sucrose solution. Make sure that the sucrose is completely dissolved. (Exercise 3)

Epinephrine solution (0.1%)
Add 0.1 g of adrenalin chloride in 100 mL of frog Ringer's solution. Label and pour the solution into small dropper bottles. (Exercise 29)

Essential oils preparation
Fill several small, screw-top vials with peppermint, almond, wintergreen, and camphor oils (available from local drug stores). Label "Peppermint," "Almond," "Wintergreen," and "Camphor," respectively, and keep vials in separate wide-mouthed jars to prevent cross-contamination of scent. (Exercise 21)

Fill four small vials halfway to the top with cotton and color them red with food coloring. Label the vials "Wild Cherry" and add benzaldehyde solution until the cotton is moist. Fill four small vials halfway to the top with cotton but do NOT color them red. Label the vials "Almond." You can use the almond vials from the previous preparation. (Exercise 21)

Filtration solution (1% starch, charcoal, and copper sulfate solution)
Place 5 g of starch, 5 g of powdered charcoal, and 5 g of copper sulfate ($CUSO_4$) in a 1 L beaker. Add enough water to make 500 mL. Stir well and pour into a 500 mL bottle. Label "Filtration Solution. "(Exercise 3)

Frog Ringer's solution
Weigh and place the following materials in a 1 L graduated cylinder.
6.5 g NaCl (sodium chloride)
0.2 g $NaHCO_3$ (sodium bicarbonate)
0.1 g $CaCl_2$ (calcium chloride)
0.1 g KCl (potassium chloride)
To these, add enough water to make 1000 mL. This solution should be prepared fresh and used within a few weeks. Place in large dropper bottles. (Exercises 11, 19, and 29: In Exercise 29, there should be three solutions prepared—one at room temperature, one at 37°C, and one in an ice bath.)

Hydrochloric acid solution (0.1%)
Add 1 mL of concentrated HCl to 1 L of water. Remember "AAA"—Always Add Acid to water. Pour into a small, labeled dropper bottle. (Exercise 19)

India ink solution
Place 10 mL of India ink in a 100 mL dropper bottle. Fill with water and label "India ink solution."

Iodine solution
See Lugol's iodine solution.

Litmus cream
Use approximately 250 mL of heavy cream. To this, add powdered litmus until the cream is a light blue. Pour into two separate bottles and label "Litmus Cream." (Exercise 39)

Litmus solution
Weigh 2 g litmus powder and dissolve in 600 mL water. Titrate HCl into the solution while stirring until the color begins to turn from blue to red. Add NaOH drop by drop until the solution just turns back to blue. Pour into two bottles labeled "Litmus Solution." (Exercise 36)

Lugol's iodine solution
Prepare by adding 10.0 g I_2 (iodine) and 20.0 g KI (potassium iodide) to 1 L of distilled water. Store in small, dark dropper bottles. Label "Lugol's Iodine." (Exercises 3 and 39)

Maltose solution (1%)
Add 2 g of maltose in enough water to make 200 mL. Stir until dissolved and pour into two clean bottles. Label "1% Maltose Solution." (Exercise 39)

Methylene blue (1%)
Add 5 g methylene blue powder in 500 mL distilled water. Pour into labeled dropper bottles. (Exercise 2)

Methylene blue solution (0.01 M)
Add 3.2 g methylene blue (MW 320) to distilled water to make 1 L of solution. Place in dropper bottles. (Exercise 3)

Nitric acid (1 N) for decalcifying bones
Add 64 mL of concentrated nitric acid (70%) slowly to water to make 1 L of solution. Remember "AAA"—Always Add Acid to water. (Exercise 6)

Pancreatin solution (1%)
Place 2 g of pancreatin in a graduated cylinder and add water to make 200 mL. Stir well. Adjust the pH with 0.05 M sodium bicarbonate until neutral (pH 7). Pour into two different stock bottles and label "1% Pancreatin Solution." Preparation note: Use fresh pancreatin and not that which has been on the shelf for years. Pancreatin may be stored frozen (not in a frost-free freezer that regularly cycles between freezing and defrosting). Also note that commercially prepared pancreatin has an optimum pH. If the pH is too low, then the reaction will be slowed or stopped. (Exercise 39)

Perfume, dilute solution
Add 10 mL inexpensive perfume to 100 mL of isopropyl alcohol. (Exercise 21)

Phosphate buffer solution
Add 3.3 g of potassium phosphate (monobasic) and 1.3 g of sodium phosphate (dibasic) to 500 mL of water. Label and pour into squeeze bottles. (Exercise 25)

Potassium permanganate solution (0.01 M)
Add 1.58 g of potassium permanganate (MW 158) crystals to water to make 1 L of solution. Label and pour into dark brown dropper bottles. (Exercise 3)

Procaine hydrochloride
Place 1 g of procaine hydrochloride solution in 1 mL water. Add to this 30 mL of pure ethanol. Place in small, screw-capped bottles labeled "Procaine Hydrochloride Solution." (Exercise 19)

Quinine solution
Pour commercial tonic water into clean, food grade dropper bottles labeled "Bitter" for lab. (Exercise 21)

Saltwater solution (3%)
Using a clean, food-grade container, add 15 g of NaCl crystals (table salt is acceptable) to water to make 500 mL of solution. Pour into clean, food grade dropper bottles labeled "Salty" for lab. (Exercise 21)

Sodium bicarbonate (Exercise 36)

Sodium chloride solution (5%)
Add 25 g of NaCl crystals to water to make 500 mL of solution. Label the solution "Sodium Chloride Solution 5%" and place in small dropper bottles. (Exercise 3)

Sodium chloride solution (5%)
Add 2.5 g of NaCl crystals to water to make 50 mL of solution. Label the solution and place in a small beaker. (Exercise 19)

Sodium chloride solution 0.9% (physiologic saline)
Put 9.0 g NaCl in 1000 mL water and pour into small dropper bottles. Label the solution "Sodium Chloride Solution 0.9%." (Exercise 3, 26)

Sodium hydroxide solution (1%)
Add water to 2 g NaOH to make 200 mL of solution. Stir the mixture carefully. The reaction is exothermic and will generate quite a bit of heat. (Exercise 39)

Caution

NaOH is caustic. Wear gloves and goggles. Pour into dropper bottles and label "1% NaOH Solution—Caustic."

Caution

The reaction is exothermic and will generate heat. Wear protective gloves and eyewear. If you spill this on your skin, make sure that you flush your skin immediately with cold water.

Starch solution (1%)
Boil 500 mL of distilled water. Remove the water from the heat and add 5 g of cornstarch (or 5 g of potato starch). Filter the mixture through cheesecloth into bottles. (Exercise 3)

Starch solution (0.5%)
A potato starch solution is made by first boiling 500 mL of water. Remove the water from the hot plate and add 2.5 g of potato starch powder. Stir and cool the mixture. Do not boil the starch and water mixture, as this will lead to some hydrolysis of starch to sugar. Test for the presence of sugar by using Benedict's reagent. There should be no sugar present. Place into 250 mL bottles and label "0.5% Starch Solution." (Exercise 39)

Sugar solution (3%)
Using a clean food-grade container, dissolve 15 g of table sugar in enough water to make 500 mL of solution. Pour into clean, food grade dropper bottles labeled "Sweet" for lab. (Exercise 21)

Sugar solutions
Four table sugar solutions of 2 L each. (Exercise 3)
0%—2 L of water
5%—Dissolve 100 g sugar in enough water to make 2 L of solution.
15%—Dissolve 450 g of sugar in enough water to make 3 L of solution; pour into a 2 L bottle and a 1 L bottle.
30%—Dissolve 600 g of sugar in enough water to make 2 L of solution.

Umami solution
Using a clean food-grade container, dissolve 15 g of monosodium glutamate in 500 mL water. Pour into clean, food grade dropper bottles labeled "Umami." (Exercise 21)

Vinegar solution
Select a clean food-grade container. Use household vinegar or make a 5% acetic acid solution by adding 5 mL of concentrated food-grade acetic acid to about 50 mL of water and then adding water to make 100 mL. Pour into clean, food grade dropper bottles labeled "Acidic" for lab. (Exercise 21)

Wright's stain
Wright's stain is available as a commercially prepared solution from a number of biological supply houses. Place in labeled dropper bottles. (Exercise 25)

Appendix C

Lab Reports

Part of working in science involves writing lab reports. You should write lab reports in a certain style and follow the basic guidelines that are generally accepted in the field. The general format for lab reports includes four sections: **introduction, materials and methods, results,** and **conclusion.** Each of these parts is important in the write-up, and all these sections must be included in the lab report. Your report should have a title, your name, your instructor's name, the course name and semester, and your lab section.

The purpose of a scientific report is to explain an investigation. Your instructor is your primary audience for your report in this class, so you should communicate that you understand the basic information. Part of a grade in a lab report depends on doing the experiment correctly and demonstrating that you are aware of the outcome of the experiment and its relationship to theory or applications you have learned in lecture.

If your experiment did not come out as you thought it would, you still can do well in your lab report. Results from your experiment may have come out satisfactorily and be perfectly fine even if you think that the data should have been different from what you obtained.

Introduction

The introduction section consists of a description of the problem and the subject of your study. In professional journals, the introduction often includes a history of experimentation or current knowledge in the field, but you will probably not have this in your lab report. You should pose the question, or hypothesis, that your experiment is trying to resolve in the introduction. You may be conducting an experiment to determine whether an enzyme functions on a substrate or whether a particular effect occurs when you perform an action.

Materials and Methods

The materials and methods section is a clear description of what equipment, animals, chemicals, and so on were used and the experimental procedure followed. You must write this section in very clear and precise terms. From this section, you should expect that someone could repeat your experiment and produce the same results.

In one way, this is the "recipe" for your experiment. As in baking a cake, it is not good enough simply to list the materials. You must include quantities, in what sequence the materials were added, and how long things were stirred. You must provide specific details about your procedure. Make sure that you state how much material was added to a sample. For example, you should write "We added 5 mL." This is a known quantity that people can duplicate, whereas "We added a little" is vague. You must make sure that you include all steps in your procedure. If you leave something out in your description, then a person following your directions might get very different results.

Results

In the results section, you state the outcome of your experiment in a clear and defined way. Your data must be clearly presented. This may consist of the tabulation of data that you acquired from your experiment in the form of a line graph, bar graph, or table. Remember that any graph should have a complete description. If you are studying the effect of exercise on heart rate, then exercise is the **independent variable** and is listed on the horizontal axis, whereas heart rate is the **dependent variable** and is listed on the vertical axis, as illustrated in the following graph.

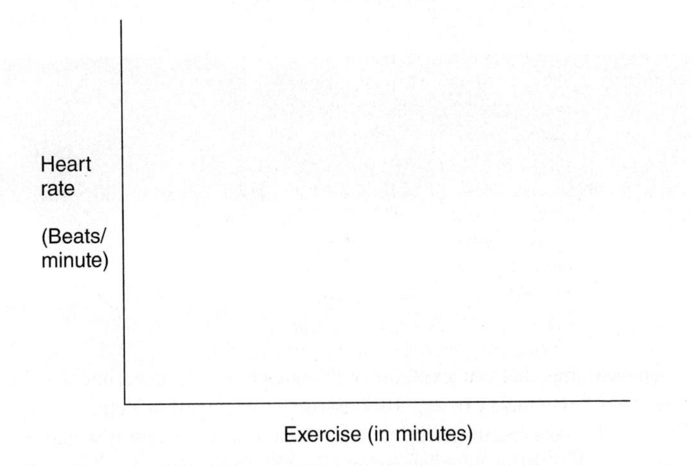

The data for your lab results section may be in raw form (direct measurements from your experiment) or organized by calculating the mean (average) data and range (high and low), as determined from your experiment. You should not try to explain your results at this point but simply state what the outcome was. Save the interpretation of the data for the conclusion.

Conclusion

In the conclusion, you can analyze and interpret your data and determine whether the results of your experiment proved your initial question (make sure you remember to answer your original question). This is the most important part of your lab report in terms of finding out if you understood the lab. Did you prove your hypothesis, did you disprove your hypothesis, or are the results inconclusive? Where should you go from here?

Sometimes experiments go awry and the results do not support the hypothesis. This can be due to many things, such as experimental error, which is a big factor in lab experiments, especially at the undergraduate level. Someone may not have followed the experimental procedure correctly and added too much, too little, or none of the materials that should have been part of the procedure. Sometimes it is a matter of timing—the process did not go on as long as it should or went on too long. In some cases, the data are recorded incorrectly or read wrong. A 3 in the data might be mistaken for an 8, or the data are entered in the wrong location. There are hundreds of possibilities as to why error plays a part in experiments, and your instructor is probably very familiar with many of them.

Another factor is faulty equipment, supplies, or setup. Other problems may be due to the variation in living organisms. Humans and other animals have their own idiosyncratic physiologic responses. Not every person and not all experimental animals respond in the same way. For example, almost all product information sheets that come with pharmaceutical drugs list adverse reactions seen in some people. If everyone responded the same way, there would be no need for warnings on drug products.

If you obtain variances from what you expect to get, then you should write this in your report. Sometimes experimental procedures are included that produce results other than what you expect to see. You should be honest in your recording of data and resist the temptation to "fudge" the data so that you can get a better result and therefore a better grade on your report.

Sometimes people in industry change their data to produce more desirable results. If this is discovered by independent investigations, the corporation that allowed this to happen usually pays significant fines. Students who change data and are discovered are also subject to penalty, which is usually far more severe than what they would have received due to the "bad" data they think they have gotten. In addition to being honest in the reporting of data, you also want to make sure that your words are your own. Even though you may have performed an experiment word for word from this lab manual or another source, do not plagiarize the material. You should paraphrase material if you want to have the same meaning but not the same words as someone else.

Your lab report should follow the format described here unless your instructor decides to make significant changes in the write-up. Lab reports tend to take a long time to write, but the analysis in the lab report is where you work toward understanding the experiment.

Appendix D

Aids to Understanding Words

acetabul-, vinegar cup: *acetabul*um
adip-, fat: *adip*ose tissue
agglutin-, to glue together: *agglutin*ation
aliment-, food: *aliment*ary canal
allant-, sausage-shaped: *allant*ois
alveol-, small cavity: *alveol*us
an-, without: *an*aerobic respiration
ana-, up: *ana*bolic
andr-, man: *andr*ogens
append-, to hang something: *append*icular
ax-, axis: *ax*ial skeleton
bil-, bile: *bil*irubin
-blast, budding: osteo*blast*
brady-, slow: *brady*cardia
bronch-, windpipe: *bronch*us
calat-, something inserted: inter*calat*ed disc
calyc-, small cup: *calyc*es
cardi-, heart: peri*cardi*um
carp-, wrist: *carp*als
cata-, down: *cata*bolic
chondr-, cartilage: *chondr*ocyte
chorio-, skin: *chorio*n
choroid, skinlike: *choroid* plexus
chym-, juice: *chym*e
-clast, broken: osteo*clast*
cleav-, to divide: *cleav*age
cochlea, snail: *cochlea*
condyl-, knob: *condyl*e
corac-, beaklike: *corac*oid process
cort-, covering: *cort*ex
cribr-, sievelike: *cribr*iform plate
cric-, ring: *cric*oid cartilage
crin-, to secrete: endo*crin*e
crist-, ridge: *crist*a galli
cut-, skin: sub*cut*aneous
cyt-, cell: *cyt*oplasm
de-, undoing: *de*amination
decidu-, falling off: *decidu*ous
dendr-, tree: *dendr*ite
derm-, skin: *derm*is
detrus-, to force away: *detrus*or muscle
di-, two: *di*saccharide
diastol-, dilation: *diastol*e
diuret-, to pass urine: *diuret*ic
dors-, back: *dors*al
ejacul-, to shoot forth: *ejacul*ation
embol-, stopper: *embol*us
endo-, within: *endo*plasmic reticulum
epi-, upon: *epi*thelial tissue
erg-, work: syn*erg*ist
erythr-, red: *erythr*ocyte
exo-, outside: *exo*crine gland
extra-, outside: *extra*cellular
fimb-, fringe: *fimb*riae
follic-, small bag: hair *follic*le
fov-, pit: *fov*ea
funi-, small cord or fiber: *funi*culus
gangli-, a swelling: *gangli*on
gastr-, stomach: *gastr*ic gland
-gen, to be produced: aller*gen*
-genesis, origin: spermato*genesis*
germ-, to bud or sprout: *germ*inal
glen-, joint socket: *glen*oid cavity
-glia, glue: neuro*glia*
glom-, little ball: *glom*erulus
glyc-, sweet: *glyc*ogen
-gram, something written: electrocardio*gram*
hema-, blood: *hema*toma
hemo-, blood: *hemo*globin
hepat-, liver: *hepat*ic duct
homeo-, same: *homeo*stasis
humor-, fluid: *humor*al
hyper-, above: *hyper*tonic
hypo-, below: *hypo*tonic
im-, (or in-), not: *im*balance
immun-, free: *immun*ity
inflamm-, to set on fire: *inflamm*ation
inter-, between: *inter*phase
intra-, inside: *intra*membranous
iris, rainbow: *iris*
iso-, equal: *iso*tonic
kerat-, horn: *kerat*in
labi-, lip: *labi*a
labyrinth, maze: *labyrinth*
lacri-, tears: *lacri*mal gland
lacun-, pool: *lacun*a
laten-, hidden: *laten*t
-lemm, rind or peel: neuri*lemm*a
leuko-, white: *leuko*cyte
lingu-, tongue: *lingu*al tonsil
lip-, fat: *lip*ids
-logy, study of: physio*logy*
-lyte, dissolvable: electro*lyte*
macro-, large: *macro*phage
macula, spot: *macula* lutea
meat-, passage: auditory *meat*us
melan-, black: *melan*in
mening-, membrane: *mening*es
mens-, month: *mens*trual cycle
meta-, change: *meta*bolism
mict-, to pass urine: *mict*urition
mit-, thread: *mit*osis
mono-, one: *mono*saccharide
mons-, mountain: *mons* pubis
morul-, mulberry: *morul*a
moto-, moving: *mot*or
mut-, change: *mut*ation
myo-, muscle: *myo*fibril
nat-, to be born: pre*nat*al
nephr-, kidney: *nephr*on
neutr-, neither one nor the other: *neutr*al
nod-, knot: *nod*ule
nutri-, nourish: *nutri*ent
odont-, tooth: *odont*oid process
olfact-, to smell: *olfact*ory
-osis, abnormal increase in production: leukocyt*osis*
oss-, bone: *oss*eous tissue
papill-, nipple: *papill*ary muscle
para-, beside: *para*thyroid glands
pariet-, wall: *pariet*al membrane
patho-, disease: *patho*gen
pelv-, basin: *pelv*ic cavity
peri-, around: *peri*cardial membrane
phag-, to eat: *phag*ocytosis
pino-, to drink: *pino*cytosis
pleur-, rib: *pleur*al membrane
plex-, interweaving: choroid *plex*us
poie-, to make: hemoto*poie*sis
poly-, many: *poly*unsaturated
pseudo-, false: *pseudo*stratified epithelium
puber-, adult: *puber*ty
pylor-, gatekeeper: *pylor*ic sphincter
sacchar-, sugar: mono*sacchar*ide
sarco-, flesh: *sarco*plasm
scler-, hard: *scler*a
seb-, grease: *seb*aceous gland
sens-, feeling: *sens*ory neuron
-som, body: ribo*som*e
squam-, scale: *squam*ous epithelium
stasis-, standing still: homeo*stasis*
strat-, layer: *strat*ified
syn-, together: *syn*thesis
systol-, contraction: *systol*e
tachy-, rapid: *tachy*cardia
tetan-, stiff: *tetan*ic
thromb-, clot: *thromb*ocyte
toc-, birth: oxy*toc*in
-tomy, cutting: ana*tomy*
trigon-, triangle: *trigon*e
troph-, well fed: muscular hyper*trophy*
-tropic, influencing: adrenocortico*tropic*
tympan-, drum: *tympan*ic membrane
umbil-, navel: *umbil*ical cord
ventr-, belly or stomach: *ventr*icle
vill-, hair: *vill*i
vitre-, glass: *vitre*ous humor
zym-, ferment: en*zym*e

Credits

Photographs

UNLESS OTHERWISE CREDITED: Courtesy of Eric Wise

Chapter 01

Figure 1.6: © Mediscan Agency/Alamy.

Chapter 04

4.14: © Dennis Strete/McGraw-Hill Education; 4.15: © McGraw-Hill Education/Al Telser, photographer; 4.16a: © Educational Images LTD/ Custom Medical Stock Photo/Newscom; 4.18: © Eric Grave/Science Source; 4.20: © Victor Eroschenko.

Chapter 12

Figure 12.2b: © McGraw-Hill Education/Photo and dissection by Christine Eckel; 12.6c: © McGraw-Hill Education/Rebecca Gray, photographer/Don Kincaid, dissections; 12.12b: © McGraw-Hill Education/Rebecca Gray, photographer/Don Kincaid, dissections.

Chapter 13

Figure 13.1b: © McGraw-Hill Education/Rebecca Gray, photographer/ Don Kincaid, dissections; 13.4c, 13.6b, 13.7B: © McGraw-Hill Education/Photo and dissection by Christine Eckel.

Chapter 14

Figure 14.1b: © McGraw-Hill Education/Photo and dissection by Christine Eckel; 14.6: © McGraw-Hill Education/Smith.

Chapter 15

Figure 15.1b: © McGraw-Hill Education/Photo and Dissection by Christine Eckel; 15.2b: © McGraw-Hill Education/Rebecca Gray, photographer/Don Kincaid, dissections.

Chapter 22

Figure 22.9: © Steve Allen/Getty Images, RF.
Figure 24.5b: © McGraw-Hill Education/Al Telser, photographer.

Chapter 30

Figure 30.2: © Dennis Strete/McGraw-Hill Education.

Chapter 33

Figure 33.8b: © McGraw-Hill Education/Dennis Strete, photographer.

Chapter 34

Figure 34.2: © Comstock Images/Thinkstock/Jupiter Images; 34.3: © Creatas/PunchStock.

Chapter 35

Figure 35.11b: © McGraw-Hill Education/Smith.

Chapter 38

Figure 38.11d: © McGraw-Hill Education/Al Telser, photographer; 38.12b: McGraw-Hill Education/Rebecca Gray, photographer/Don Kincaid, dissections.

Chapter 42

Figure 42.8b: © McGraw-Hill Education/Smith.

Chapter 43

Figure 43.12b: © McGraw-Hill Education/Smith.

Index

A

B

D

J

K

L

M

N

O

P

Q

R

T